College Algebra & Trigonometry

Seventh Edition

Michael L. **Levitan**
Villanova University

Bernard **Kolman**
Emeritus, Drexel University

Arnold **Shapiro**
Emeritus, Temple University

Publisher and Managing Director: Richard Schofield

Project Development Manager: Christine Davies

Production and Fulfillment Manager: Janai Escobedo

Designer and Typesetter: Suzanne Schmidt

Proofreader: Celeste Hernandez

All rights reserved. Printed in the United States of America. No part of this book may be used or reproduced in any manner whatsoever without written permission, except in the case of brief quotations embodied in critical articles and reviews. For information, address BVT Publishing, P.O. Box 492831, Redding, CA 96049-2831.

Some ancillaries, including electronic and print components, may not be available to customers outside the United States.

Cover image: Shutterstock

LAB BOOK^{Plus} ISBN: 978-1-5178-1118-1
TEXTBOOK^{Plus} (Loose-Leaf Bundle) ISBN: 978-1-5178-0158-8
eBook^{Plus} ISBN: 978-1-5178-0159-5
Loose-Leaf ISBN: 978-1-5178-0157-1
Soft Cover ISBN: 978-1-5178-0160-1

Copyright © 2022 by BVT Publishing

Dedications

To my parents, Joseph and Ruth; to my wife Mary Davis, marvelous, astonishing and iridescent; to my daughter Cheryl, intelligent and determined; to my exceptional, affable and accomplished son Eric (handsome and humble like his father); to the memory of our son Benjamin, our "beloved munchkin"; and to our daughter Laura, luminous, melodious, effervescent and scintillating (my one and only "Sugarplummy").

—MLL

To the memory of my mother, Eva.

—BK

To the memory of my parents.

—AS

Brief Contents

Contents

Preface

This seventh edition of *College Algebra & Trigonometry* maintains our objective of providing a textbook designed for use *by the student*. The book is written in a supportive style and includes abundant pedagogic tools to encourage the student to read the text with care, follow the numerous examples, and tackle the exercises. The skills and confidence the student acquires will enable him or her to succeed in more advanced mathematics courses (calculus in particular) that are required in the study of engineering, the natural sciences, business, and management.

New to This Edition

This edition introduces new and updated exercises, as well as attention-grabbing chapter openings and projects, which will help instructors and students make connections between algebra and the real world. This edition has gone through extensive copy editing and proofing in an effort to further ensure the accuracy of this textbook. This edition supports the use of graphing calculators as powerful problem-solving and visualization tools, without sacrificing the emphasis of the authors on traditional approaches.

Pedagogic Tools

- Concepts are introduced gradually and supported by fully worked examples, figures, and realistic applications to help students build problem-solving skills.
- Many algebraic procedures are described with the aid of a "split screen" that displays, side by side, both the steps of an algorithm and a worked-out example.
- A *Progress Check* following many worked-out examples provides an exercise with its answer, enabling the student to test his or her understanding of the presented material.
- *Warnings* reinforce good mathematical habits by pointing out incorrect practices most commonly found in homework and exams.
- Numerous vignettes have been included throughout the book. These are independent of the text, yet related to the mathematical concepts discussed nearby. They are intended to provide additional interesting material for the student and instructor.
- To help students check their understanding of the concepts, each chapter concludes with a list of terms and symbols and a list of key ideas for review.
- In addition, the *Chapter Review Exercises* and *Review Tests* give students a chance to practice what they have learned. The *Cumulative Review Exercises* at the end of every third chapter provide additional review practice.

Exercises

Abundant exercises provide practice in the mechanical and conceptual aspects of algebra. Exercises requiring the use of a graphing calculator are indicated by the calculator icon shown to the left. Although some exercises require student skills in graphing as well as graphing calculators, the latter may be disregarded, if desired. Answers to selected odd-numbered exercises, *Review Exercises*, *Review Tests*, and *Cumulative Exercises* appear in an appendix at the back of the book, and fully worked solutions are available in the *Student Solutions Manual*. The solved *Review Exercises* reassure students that they have mastered the concepts in preparation for the *Review Tests*.

Supplements and Resources

Instructor Supplements

A complete teaching package is available for instructors who adopt this book. This package includes an **online lab, instructor's resource manual with test bank, instructor's solutions manual, LMS Integration**, and **LMS exam bank files.**

Online Lab	BVT's online lab is available for this textbook on two different platforms—BVT*Lab* (at www.BVTLab.com), and LAB BOOK™ (at www.BVTLabBook.com). These are described in more detail in the corresponding sections below. Both platforms allow instructors to set up graded homework, quizzes and exams.
Instructor's Resource Manual with Test Bank	This resource helps first-time instructors develop the course, while also offering seasoned instructors a new perspective on the materials. Each section of the Instructor's Manual coincides with a chapter in the textbook. The user-friendly format begins with lecture outlines and includes section-by-section test questions. The test questions consist of multiple choice questions, open-ended questions, and word problems. Each chapter has additional practice exercises and practice tests. Answers to all the problem sets, test questions, practice exercises, and practice tests are provided at the back of the manual.
Instructor's Solutions Manual	This resource contains complete solutions to every exercise in the textbook, including *Progress Checks, Review Tests,* and *Cumulative Review Exercises.*
LMS Integration	BVT offers basic integration with Learning Management Systems (LMSs), providing single-sign-on links (often called LTI links) from Blackboard, Canvas, Moodle (or any other LMS) directly into BVT*Lab*, eBook^Plus or the LAB BOOK platform. Gradebooks from BVT*Lab* and the LAB BOOK can be imported into most LMSs.
LMS Exam Bank Files	Exam banks are available as Blackboard files, QTI files (for Canvas) and Respondus files (for other LMSs) so they can easily be imported into a wide variety of course management systems.

Student Resources

Student resources are available for this textbook on the BVT LAB BOOK and BVT*Lab* platforms, as described below. These resources are geared toward students needing additional assistance, as well as those seeking complete mastery of the content. The following resources are available:

Practice Questions	Students can work through hundreds of practice questions online. Questions are multiple choice or true/false in format, and are graded instantly for immediate feedback.
Student Solutions Manual	The SSM includes complete solutions to the odd-numbered exercises, and to every exercise in the *Chapter Reviews, Progress Checks,* and *Cumulative Review Exercises.* In addition, each chapter of this manual ends with review exercises and a review test.
Instructional Videos	These videos offer students expanded instruction and narrated examples of key concepts discussed in the text. These tutorials demonstrate how to interpret the questions and provide step-by-step instructions to help students master the concepts discussed.
Graphing Calculator Supplement	This resource provides basic and advanced instructions demonstrating how to use a graphing calculator effectively. Each *Power User's Corner* vignette illustrates the special features of the graphing calculator as applied to more advanced calculations. Many examples are included to reinforce concepts and allow for practice.
Additional LAB BOOK Resources	On the LAB BOOK platform, comprehension questions are sprinkled throughout each chapter of the eBook, in addition to the step-by-step solution tutorials. Study tools such as text highlighting and margin notes are also available. These resources are not available in BVT*Lab*.

LABBOOK

LAB BOOK is a web-based eBook platform with an integrated lab providing comprehension tools and interactive student resources. Instructors can build homework and quizzes right into the eBook. LAB BOOK is either included with Textbook^{Plus} (looseleaf bundle) or offered as a stand-alone product.

Course Setup	LAB BOOK allows instructors to set up their courses and grade books and replicate them from section to section and semester to semester.
Grade Book	Using an assigned passcode, students register into their section's grade book, which automatically grades and records all homework, quizzes, and tests. The gradebook can be exported to any LMS at any time.
Advanced eBook	LAB BOOK is a mobile-friendly, web-based eBook platform designed for PCs, MACs, tablets and smartphones. LAB BOOK allows highlighting, margin notes, and a host of other study tools.
Student Resources	All student resources for this textbook are available in the LAB BOOK, as described in the **Student Resources** section above.
Online Classes	LAB BOOK provides a host of instructor resources and tools to support online and digital learning environments.

Customization

BVT's Custom Publishing Division can help you modify this book's content to satisfy your specific instructional needs. The following are examples of customization:

- Rearrangement of chapters to follow the order of your syllabus
- Deletion of chapters not covered in your course
- Addition of paragraphs, sections, or chapters you or your colleagues have written for this course
- Editing of the existing content, down to the word level
- Customization of the accompanying student resources and online lab
- Addition of handouts, lecture notes, syllabus, and so forth
- Incorporation of student worksheets into the textbook

All of these customizations will be professionally typeset to produce a seamless textbook of the highest quality, with an updated table of contents and index to reflect the customized content.

List of Graphing Calculator Topics

The graphing calculator material in this textbook has been designed to provide a flexible approach to using calculator technology. The *Graphing Calculator Alert* sections instruct students on using a graphing calculator effectively. The *Power User's Corner* sections illustrate the special features of the graphing calculator in more advanced calculations. The lists below show the locations of these two types of boxes and the topics they cover. The *Instructor's Manual* lists the exercises requiring the use of a graphing calculator and those for which it is optional.

Graphing Calculator Alert Topics

Graphing Calculator Power User's Corner Topics

1 The Foundations of Algebra

Suppose you asked a friend of yours, who is a physics major, "How long does it take for a rock to reach the ground after being thrown into the air?" She will tell you that an object thrown straight up with a velocity of 20 meters per second would reach the ground in a little more than 4 seconds, if air resistance was not a factor. This is true, however, only on the Earth. What if we were on another planet, or even a large moon like Ganymede? An object thrown straight up from the surface of Ganymede, with the same initial velocity of 20 m/s, would take almost 20 seconds to reach the ground. (Check out the Chapter Project.)

If you asked your friend how she arrived at these conclusions, she could use words like *algebraic expression, factoring,* and *polynomial.* Before you read this chapter, explore one of these words at http://mathworld.wolfram.com/Polynomial.html. This site can help you discover the meanings of many other terms as well.

Many problems that each of us encounters in the real world require the use and understanding of mathematics. Often, the methods used to solve these problems share certain characteristics, and it is both helpful and important to focus on these similarities. Algebra is one branch of mathematics that enables us to learn basic problem-solving techniques applicable to a wide variety of circumstances.

For example, if one starts with 2 apples and gets 3 more apples, how many apples does one have? If the travel time between Philadelphia and New York was 2 hours in the morning and 3 hours in the afternoon, how much time was spent traveling? The solution to the first problem is

$$2 \text{ apples} + 3 \text{ apples} = 5 \text{ apples}$$

The solution to the second problem is

$$2 \text{ hours} + 3 \text{ hours} = 5 \text{ hours}$$

Algebra focuses on the fact that

$$2x + 3x = 5x$$

It does not matter what meaning one gives to the symbol x.

Although this level of abstraction can create some difficulty, it is the nature of algebra that permits us to distill the essentials of problem solving into such rudimentary formulas. In the examples noted above, we used the *counting* or *natural numbers* as the number system needed to describe the problems. This number system is generally the first that one learns as a child. One can create other formulas for more general number systems.

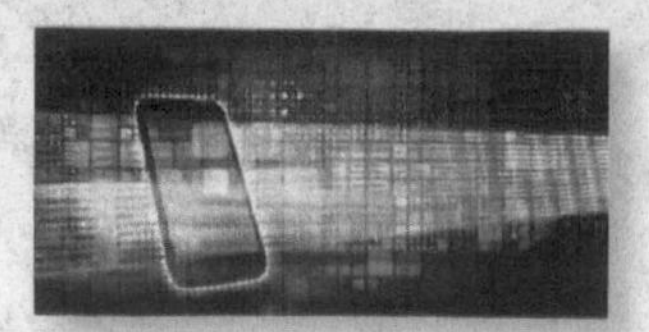

http://mathworld.wolfram.com/Polynomial.html

We shall begin our presentation with a discussion of the *real number system* and its associated properties. We note a correspondence between the real numbers and the points on a real number line, and give a graphical presentation of this correspondence. The remainder of this chapter is devoted to a review of some fundamentals of algebra: the meaning and use of variables, algebraic expressions and polynomial forms, scientific notation, factoring, operations with algebraic fractions, and an introduction to the *complex number system*.

1.1 The Real Number System

1.1a *Sets*

We will need to use the **notation** and terminology of sets from time to time. A **set** is simply a collection of objects or numbers that are called the **elements** or **members** of the set. The elements of a set are written within braces so that the notation

$$A = \{4, 5, 6\}$$

tells us that the set A consists of the numbers 4, 5, and 6. The set

$$B = \{\text{Exxon, Ford, Sony}\}$$

consists of the names of these three corporations. We also write $4 \in A$, which we read as "4 is a member of the set A." Similarly, Ford $\in B$ is read as "Ford is a member of the set B," and Chrysler $\notin B$ is read as "Chrysler is not a member of the set B."

If every element of a set A is also a member of a set B, then A is a **subset** of B. For example, the set of all robins is a subset of the set of all birds.

EXAMPLE 1 ***Set Notation and Properties***

The set C consists of the names of all coins whose denominations are less than 50 cents. We may write C in set notation as follows:

$$C = \{\text{penny, nickel, dime, quarter}\}$$

We see that dime $\in C$, but half dollar $\notin C$. Further, the set $H = \{\text{nickel, dime}\}$ is a subset of C. ■

The set V consists of the vowels in this particular sentence.

a. Write V in set notation.

b. Is the letter k a member of V?

c. Is the letter u a member of V?

d. List the subsets of V having four elements.

Answers

a. $V = \{a, e, i, o, u\}$ b. No c. Yes

d. $\{a, e, i, o\}$, $\{e, i, o, u\}$, $\{a, i, o, u\}$, $\{a, e, o, u\}$, $\{a, e, i, u\}$

1.1b *The Set of Real Numbers*

Since much of our work in algebra deals with the real numbers, we begin with a review of the composition of these numbers.

The numbers 1, 2, 3, ... , used for counting, form the set of **natural numbers.** If we had only these numbers to use to show the profit earned by a company, we would have no way to indicate that the company had no profit or had a loss. To indicate no profit we introduce 0, and for losses we need to introduce negative numbers. The numbers

$$\ldots, -2, -1, 0, 1, 2, \ldots$$

form the set of **integers.** Thus, every natural number is an integer, and the set of natural numbers is seen to be a subset of the set of integers.

When we try to divide two apples equally among four people, we find no number in the set of integers that expresses how many apples each person should get. We need to introduce the set of **rational numbers**, which are numbers that can be written as a ratio of two integers.

$$\frac{p}{q} \text{ with } q \text{ not equal to zero}$$

Examples of rational numbers are

$$0 \qquad \frac{2}{3} \qquad -4 \qquad \frac{7}{5} \qquad \frac{-3}{4}$$

By writing an integer n in the form $\frac{n}{1}$, we see that every integer is a rational number. The decimal number 1.3 is also a rational number since $1.3 = \frac{13}{10}$.

We have now seen three fundamental sets of numbers: the set of natural numbers, the set of integers, and the set of rational numbers. Each successive set includes the previous set or sets, and each is more complicated than the one before. However, the set of rational numbers is still inadequate for sophisticated uses of mathematics, since there exist numbers that are not rational, that is, numbers that cannot be written as the ratio of two integers. These are called **irrational numbers.** It can be shown that the number a that satisfies $a \cdot a = 2$ is such a number. The number π, which is the ratio of the circumference of a circle to its diameter, is also such a number.

The decimal form of a rational number always forms a repeating pattern, such as

$$\frac{1}{2} = 0.500\ldots$$

$$\frac{13}{10} = 1.3000\ldots$$

$$\frac{1}{3} = 0.333\ldots$$

$$\frac{-2}{11} = -0.\underbrace{18}\underbrace{18}\underbrace{18}\ldots$$

(*Note:* The three dots, known as an ellipsis, following the numbers in each of the examples above means that the pattern continues in the same manner forever.)

The decimal form of an irrational number *never* forms a repeating pattern. The rational and irrational numbers together form the set of **real numbers.** (See Figure 1-1.)

FIGURE 1-1
The Set of Real Numbers

Calculator ALERT

(1) Rational Numbers

A calculator display shows only a finite number of digits, which is often an approximation of the exact answer. Use your calculator to convert the rational numbers $\frac{1}{2}$, $\frac{13}{10}$, $\frac{1}{3}$, and $\frac{-2}{11}$ to decimal form, and note how these representations differ from those shown above.

$$\begin{aligned} 1 \div 2 &= 0.5 \\ 13 \div 10 &= 1.3 \\ 1 \div 3 &= 0.3333333333 \\ -2 \div 11 &= -0.1818181818 \end{aligned}$$

Which representations are exact and which are approximate?

(2) Irrational Numbers

Most calculators provide a rational decimal *approximation* to irrational numbers. For example, ten-digit approximations to $\sqrt{2}$ and π are

$$\sqrt{2} = 1.414213562$$
$$\pi = 3.141592654$$

1.1c *The System of Real Numbers*

The system of real numbers consists of the set of real numbers together with the operations of addition and multiplication; in addition, this system satisfies the properties listed in Table 1-1, where *a*, *b*, and *c* denote real numbers.

EXAMPLE 2 *Properties of Real Numbers*

Specify the property in Table 1-1 illustrated by each of the following statements.

a. $2 + 3 = 3 + 2$

b. $(2 \cdot 3) \cdot 4 = 2 \cdot (3 \cdot 4)$

c. $2 \cdot \frac{1}{2} = 1$

d. $2(3 + 5) = (2 \cdot 3) + (2 \cdot 5)$

SOLUTION

a. commutative under addition

b. associative under multiplication

c. multiplicative inverse

d. distributive law ■

TABLE 1-1 Properties of Real Numbers

Example	Algebraic Expression	Property
$3 + 4$ is a real number.	$a + b$ is a real number.	**Closure under addition** The sum of two real numbers is a real number.
$2 \cdot 5$ is a real number.	$a \cdot b$ is a real number.	**Closure under multiplication** The product of two real numbers is a real number.
$4 + 8 = 8 + 4$	$a + b = b + a$	**Commutative under addition** We may add real numbers in any order.
$3(5) = 5(3)$	$a(b) = b(a)$	**Commutative under multiplication** We may multiply real numbers in any order.
$(2 + 5) + 3 = 2 + (5 + 3)$	$(a + b) + c = a + (b + c)$	**Associative under addition** We may group the addition of real numbers in any order.
$(2 \cdot 5)3 = 2(5 \cdot 3)$	$(ab)c = a(bc)$	**Associative under multiplication** We may group the multiplication of real numbers in any order.
$4 + 0 = 4$	$a + 0 = a$	**Additive identity** The sum of the unique real number 0 and any real number leaves that number unchanged.
$3(1) = 3$	$a(1) = a$	**Multiplicative identity** The product of the unique real number 1 and any real number leaves that number unchanged.
$5 + (-5) = 0$	$a + (-a) = 0$	**Additive inverse** The number $-a$ is called the **negative, opposite,** or **additive inverse** of a. If $-a$ is added to a, the result is the additive identity 0.
$7\left(\frac{1}{7}\right) = 1$	If $a \neq 0$, $a\left(\frac{1}{a}\right) = 1$	**Multiplicative inverse** The number $\frac{1}{a}$ is called the **reciprocal,** or **multiplicative inverse,** of a. If $\frac{1}{a}$ is multiplied by a, the result is the multiplicative identity 1.
$2(5 + 3) = (2 \cdot 5) + (2 \cdot 3)$ $(4 + 7)2 = (4 \cdot 2) + (7 \cdot 2)$	$a(b + c) = ab + ac$ $(a + b)c = ac + bc$	**Distributive laws** If one number multiplies the sum of two numbers, we may add the two numbers first and then perform the multiplication; or we may multiply each pair and then add the two products.

1.1d *Equality*

When we say that two numbers are **equal**, we mean that they represent the same value. Thus, when we write

$$a = b$$

(read "a equals b"), we mean that a and b represent the same number. For example, $2 + 5$ and $4 + 3$ are different ways of writing the number 7, so we can write

$$2 + 5 = 4 + 3$$

Equality satisfies four basic properties shown in Table 1-2, where a, b, and c are any real numbers.

EXAMPLE 3 ***Properties of Equality***

Specify the property in Table 1-2 illustrated by each of the following statements.

a. If $5a - 2 = b$, then $b = 5a - 2$.

b. If $a = b$ and $b = 5$, then $a = 5$.

c. If $a = b$, then $3a + 6 = 3b + 6$.

SOLUTION

a. symmetric property b. transitive property c. substitution property ■

TABLE 1-2 Properties of Equality

Example	Algebraic Expression	Property
$3 = 3$	$a = a$	Reflexive property
If $\frac{6}{3} = 2$ then $2 = \frac{6}{3}$.	If $a = b$ then $b = a$.	Symmetric property
If $\frac{6}{3} = 2$ and $2 = \frac{8}{4}$, then $\frac{6}{3} = \frac{8}{4}$.	If $a = b$ and $b = c$, then $a = c$.	Transitive property
If $\frac{6}{3} = 2$, then we may replace $\frac{6}{3}$ by 2 or we may replace 2 by $\frac{6}{3}$.	If $a = b$, then we may replace a by b or we may replace b by a.	Substitution property

1.1e *Additional Properties*

Using the properties of real numbers, the properties of equalities, and rules of logic, we can derive many other properties of the real numbers, as shown in Table 1-3, where a, b, and c are any real numbers.

TABLE 1-3 Additional Properties of Real Numbers

Example	Algebraic Expression	Property
If $\frac{6}{3} = 2$ then $\frac{6}{3} + 4 = 2 + 4$.	If $a = b$, then $a + c = b + c$.	The same number may be added to both sides of an equation.
$\frac{6}{3}(5) = 2(5)$	$ac = bc$	Both sides of an equation may be multiplied by the same number.
If $\frac{6}{3} + 4 = 2 + 4$ then $\frac{6}{3} = 2$.	If $a + c = b + c$ then $a = b$.	Cancellation law of addition
If $\frac{6}{3}(5) = 2(5)$ then $\frac{6}{3} = 2$.	If $ac = bc$ with $c \neq 0$ then $a = b$.	Cancellation law of multiplication
$2(0) = 0(2) = 0$	$a(0) = 0(a) = 0$	The product of two real numbers can be zero only if one of them is zero.
$2(3) = 0$ is impossible.	If $ab = 0$ then $a = 0$ or $b = 0$.	The real numbers a and b are said to be **factors** of the product ab.
$-(-3) = 3$ $(-2)(3) = (2)(-3) = -6$ $(-1)(3) = -3$ $(-2)(-3) = 6$ $(-2) + (-3) = -(2 + 3) = -5$	$-(-a) = a$ $(-a)(b) = (a)(-b) = -(ab)$ $(-1)(a) = -a$ $(-a)(-b) = ab$ $(-a) + (-b) = -(a + b)$	Rules of signs

We next introduce the operations of subtraction and division. If a and b are real numbers, the *difference* between a and b, denoted by $a - b$, is defined by

$$a - b = a + (-b)$$

and the operation is called *subtraction*. Thus,

$$6 - 2 = 6 + (-2) = 4 \qquad 2 - 2 = 0 \qquad 0 - 8 = -8$$

We can show that the distributive laws hold for subtraction, that is,

$$a(b - c) = ab - ac$$
$$(a - b)c = ac - bc$$

If a and b are real numbers and $b \neq 0$, then the *quotient* of a and b, denoted $\frac{a}{b}$ or a/b, is defined by

$$\frac{a}{b} = a \cdot \frac{1}{b}$$

and the operation is called *division*. We also write $\frac{a}{b}$ as $a \div b$ and speak of the *fraction a* over b. The numbers a and b are called the *numerator* and *denominator* of the fraction $\frac{a}{b}$, respectively. Observe that we have not defined division by zero, since 0 has no reciprocal.

In Table 1-4, a, b, c, and d are real numbers with $b \neq 0$, $c \neq 0$, and $d \neq 0$.

Progress Check

Perform the indicated operations.

a. $\frac{3}{5} + \frac{1}{4}$ b. $\frac{5}{2} \cdot \frac{4}{15}$ c. $\frac{2}{3} + \frac{3}{7}$

Answers

a. $\frac{17}{20}$ b. $\frac{2}{3}$ c. $\frac{23}{21}$

TABLE 1-4 Additional Properties of Real Numbers

Example	Algebraic Expression	Property
$\frac{6}{10} = \frac{2 \cdot 3}{2 \cdot 5} = \frac{3}{5}$	$\frac{ac}{bc} = \frac{a}{b}$	Rules of fractions
$\frac{2}{3} \cdot \frac{5}{7} = \frac{10}{21}$	$\frac{a}{b} \cdot \frac{c}{d} = \frac{ac}{bd}$	
$\frac{4}{6} = \frac{2}{3}$ since $4 \cdot 3 = 6 \cdot 2$	$\frac{a}{b} = \frac{c}{d}$ if $ad = bc$	
$\frac{2}{9} + \frac{5}{9} = \frac{7}{9}$	$\frac{a}{d} + \frac{c}{d} = \frac{a+c}{d}$	
$\frac{\frac{2}{9}}{\frac{5}{9}} = \frac{\frac{2}{9}}{\frac{5}{9}} \cdot \frac{(9)}{(9)} = \frac{2}{5}$	$\frac{\frac{a}{d}}{\frac{b}{d}} = \frac{\frac{a}{d}}{\frac{b}{d}} \cdot \frac{(d)}{(d)} = \frac{a}{b}$	
$\frac{2}{3} + \frac{5}{4} = \frac{2}{3} \cdot \frac{(4)}{(4)} + \frac{5}{4} \cdot \frac{(3)}{(3)} = \frac{8}{12} + \frac{15}{12} = \frac{23}{12}$	$\frac{a}{b} + \frac{c}{d} = \frac{a}{b} \cdot \frac{(d)}{(d)} + \frac{c}{d} \cdot \frac{(b)}{(b)} = \frac{ad + cb}{bd}$	
$\frac{\frac{2}{3}}{\frac{5}{7}} = \frac{\frac{2}{3}}{\frac{5}{7}} \cdot \frac{(3 \cdot 7)}{(3 \cdot 7)} = \frac{2(7)}{5(3)} = \frac{14}{15}$	$\frac{\frac{a}{b}}{\frac{c}{d}} = \frac{\frac{a}{b}}{\frac{c}{d}} \cdot \frac{(bd)}{(bd)} = \frac{ad}{cb}$	

Exercise Set 1.1

In Exercises 1–8, write each set by listing its elements within braces.

1. The set of natural numbers from 3 to 7, inclusive
2. The set of integers between -4 and 2
3. The set of integers between -10 and -8
4. The set of natural numbers from -9 to 3, inclusive
5. The subset of the set $S = \{-3, -2, -1, 0, 1, 2\}$ consisting of the positive integers in S
6. The subset of the set $S = \{-\frac{2}{3}, -1.1, 3.7, 4.8\}$ consisting of the negative rational numbers in S
7. The subset of all $x \in S$, $S = \{1, 3, 6, 7, 10\}$, such that x is an odd integer
8. The subset of all $x \in S$, $S = \{2, 5, 8, 9, 10\}$, such that x is an even integer

In Exercises 9–22, determine whether the given statement is true (T) or false (F).

9. -14 is a natural number.
10. $-\frac{4}{5}$ is a rational number.
11. $\frac{\pi}{3}$ is a rational number.
12. $\frac{1.75}{18.6}$ is an irrational number.
13. -1207 is an integer.
14. 0.75 is an irrational number.
15. $\frac{4}{5}$ is a real number.
16. 3 is a rational number.
17. 2π is a real number.
18. The sum of two rational numbers is always a rational number.
19. The sum of two irrational numbers is always an irrational number.
20. The product of two rational numbers is always a rational number.
21. The product of two irrational numbers is always an irrational number.
22. The difference of two irrational numbers is always an irrational number.

In Exercises 23–36, the letters represent real numbers. Identify the property or properties of real numbers that justify each statement.

23. $a + x = x + a$
24. $(xy)z = x(yz)$
25. $xyz + xy = xy(z + 1)$
26. $x + y$ is a real number
27. $(a + b) + 3 = a + (b + 3)$
28. $5 + (x + y) = (x + y) + 5$
29. cx is a real number.
30. $(a + 5) + b = (a + b) + 5$
31. $uv = vu$
32. $x + 0 = x$
33. $a(bc) = c(ab)$
34. $xy - xy = 0$
35. $5 \cdot \frac{1}{5} = 1$
36. $xy \cdot 1 = xy$

In Exercises 37–40, find a counterexample; that is, find real values for which the statement is false.

37. $a - b = b - a$
38. $\frac{a}{b} = \frac{b}{a}$
39. $a(b + c) = ab + c$
40. $(a + b)(c + d) = ac + bd$

In Exercises 41–44, indicate the property or properties of equality that justify the statement.

41. If $3x = 5$, then $5 = 3x$.
42. If $x + y = 7$ and $y = 5$, then $x + 5 = 7$.
43. If $2y = z$ and $z = x + 2$, then $2y = x + 2$.
44. If $x + 2y + 3z = r + s$ and $r = x + 1$, then $x + 2y + 3z = x + 1 + s$.

In Exercises 45–49, a, b, and c are real numbers. Use the properties of real numbers and the properties of equality to prove each statement.

45. If $-a = -b$, then $ac = bc$.
46. If $a = b$ and $c \neq 0$, then $\frac{a}{c} = \frac{b}{c}$.
47. If $a - c = b - c$, then $a = b$.
48. $a(b - c) = ab - ac$
49. Prove that the real number 0 does not have a reciprocal. (*Hint:* Assume $b = \frac{1}{0}$ is the reciprocal of 0.) Supply a reason for each of the following steps.

$$1 = 0 \cdot \frac{1}{0}$$
$$= 0 \cdot b$$
$$= 0$$

Since this conclusion is impossible, the original assumption must be false.

50. Give three examples for each of the following:
    a. a real number that is not a rational number
    b. a rational number that is not an integer
    c. an integer that is not a natural number

51. Give three examples for each of the following:
    a. two rational numbers that are not integers whose sum is an integer
    b. two irrational numbers whose sum is a rational number

52. Find a subset of the real numbers that is closed with respect to addition and multiplication but not with respect to subtraction and division.

53. Perform the indicated operations.
    a. $(-8) + 13$ b. $(-8) + (-13)$
    c. $8 - (-13)$ d. $(-5)(3) - (-12)$
    e. $\left(\frac{8}{9}+3\right)+\left(\frac{-5}{9}\right)$ f. $\frac{-5}{\frac{3}{2}}$
    g. $\frac{\frac{5}{8}}{\frac{1}{2}}$ h. $\frac{\frac{-2}{3}}{\frac{-4}{3}}$
    i. $\left(\frac{3}{4}\right)\left(\frac{21}{37}\right)+\left(\frac{3}{4}\right)\left(\frac{16}{37}\right)$
    j. $\frac{\frac{1}{3}-\left(\frac{-1}{4}\right)}{\frac{7}{8}-\frac{3}{16}}$ k. $\frac{\left(\frac{3}{5}\right)+\left(\frac{1}{7}\right)}{\frac{1}{2}+\frac{1}{3}}$
    l. $\frac{2}{5}\left(\frac{3}{2}\cdot\frac{4}{7}\right)$

54. What is the meaning attached to each of the following?
    a. $\frac{6}{0}$ b. $\frac{0}{6}$
    c. $\frac{6}{6}$ d. $\frac{0}{\frac{1}{2}}$
    e. $\frac{0}{0}$

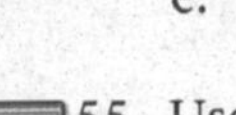

55. Use your calculator to convert the following fractions to (repeating) decimals. Look for a pattern that repeats.
    a. $\frac{1}{4}$ b. $-\frac{3}{5}$ c. $\frac{10}{13}$ d. $\frac{2}{7}$
    e. Does your calculator round off the final digit of an approximation, or does your calculator "drop off" the extra digits? To answer this question, evaluate $2 \div 3$ to see if your calculator displays 0.6666666666 or 0.6666666667.

56. A proportion is a statement of equality between two ratios. Solve the following proportions for x.
    a. $\frac{7}{8} = \frac{x}{12}$ b. $\frac{7}{x} = \frac{11}{3}$

57. On a map of Pennsylvania, 1 inch represents 10 miles. Find the distance represented by 3.5 inches.

58. A car travels 135 miles on 6 gallons of gasoline. How far can it travel on 10 gallons of gasoline?

59. A board 10 feet long is cut into two pieces, the lengths of which are in the ratio of 2:3. Find the lengths of the pieces.

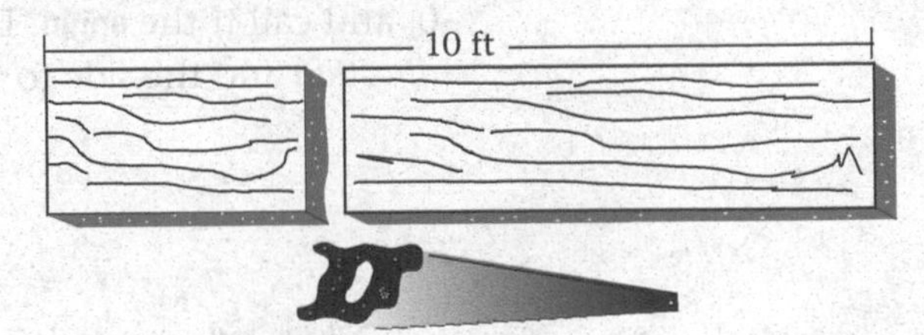

60. An alloy is $\frac{3}{8}$ copper, $\frac{5}{12}$ zinc, and the balance lead. How much lead is there in 282 pounds of alloy?

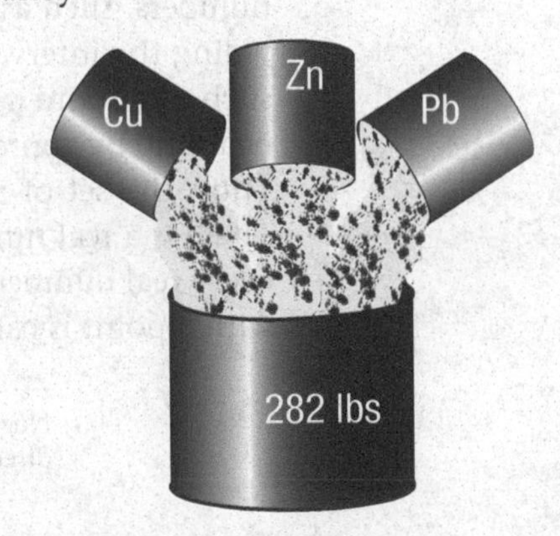

61. Which is the better value: 1 pound 3 ounces of beans for 85 cents or 13 ounces for 56 cents? (*Hint*: 1 pound = 16 ounces.)

62. A piece of property is valued at \$28,500. What is the real estate tax at \$75.30 per \$1000.00 evaluation?

63. A student's take-home pay is \$210.00 after deducting 18% withholding tax. What is her pay before the deduction?

64. List the set of possible ways of getting a total of 7 when tossing two standard dice.

65. A college student sent a postcard home with the following message:

    SEND
    MORE
    MONEY

    If each letter represents a different digit, and the calculation represents a sum, how much money did the student request?

66. Eric starts at a certain time driving his car from New York to Philadelphia going 50 mph. Sixty minutes later, Steve leaves in his car en route from Philadelphia to New York going 40 mph. When the two cars meet, which one is nearer to New York?

1.2 The Real Number Line

There is a simple and very useful geometric interpretation of the real number system. Draw a horizontal line. Pick a point on this line, label it with the number 0, and call it the **origin**. Designate the side to the right of the origin as the *positive direction* and the side to the left as the *negative direction.*

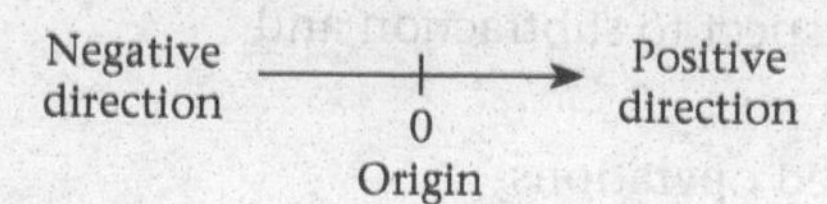

Next, select a unit for measuring distance. With each positive integer n, we associate the point that is n units to the right of the origin. With each negative number $-n$, we associate the point that is n units to the left of the origin. Rational numbers, such as $\frac{3}{4}$ and $-\frac{5}{2}$, are associated with the corresponding points by dividing the intervals between integers into equal subintervals. Irrational numbers, such as $\sqrt{2}$ and π, can be written in decimal form. The corresponding points can be found by approximating these decimal forms to any desired degree of accuracy. Thus, the set of real numbers is identified with all possible points on this line. There is a real number for every point on the line; there is a point on the line for every real number. The line is called the **real number line**, and the number associated with a point is called its *coordinate*. We can now show some points on this line.

Negative direction $-\frac{5}{2}$ $\frac{3}{4}$ $\sqrt{2}$ π Positive direction
-3 -2 -1 0 1 2 3

The numbers to the right of zero are called *positive*; the numbers to the left of zero are called *negative*. The positive numbers and zero together are called the **nonnegative numbers.**

We will frequently use the real number line to help picture the results of algebraic computations. For this purpose, we are only concerned with relative locations on the line. For example, it is adequate to show π slightly to the right of 3 since π is approximately 3.14.

EXAMPLE 1 *Real Number Line*

Draw a real number line and plot the following points: $-\frac{3}{2}$, 2, $\frac{13}{4}$.

SOLUTION

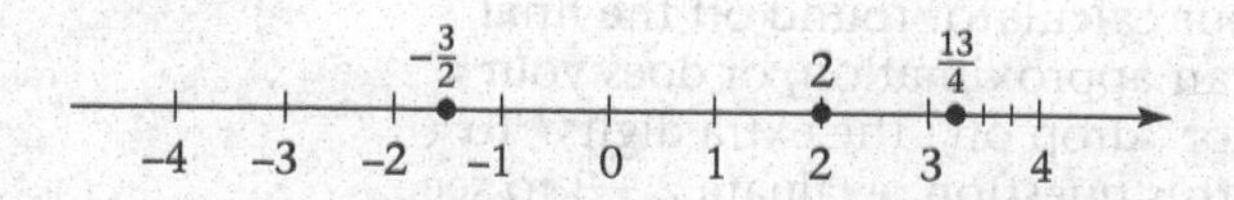

■

1.2a *Inequalities*

If a and b are real numbers, we can compare their positions on the real number line by using the relations *less than, greater than, less than or equal to,* and *greater than or equal to,* as shown in Table 1-5.

TABLE 1-5 Inequalities

Symbol	Meaning
$<$	Less than
$>$	Greater than
$\leq$	Less than or equal to
$\geq$	Greater than or equal to

Table 1-6 describes both algebraic and geometric interpretations of the **inequality symbols**, where a and b are real numbers.

Expressions involving inequality symbols, such as $a < b$ and $a \geq b$, are called **inequalities**. We often combine these expressions so that $a \leq b < c$ means both $a \leq b$ and $b < c$. (*Note:* $a < c$ is also true.) For example, $-5 \leq x < 2$ is equivalent to $-5 \leq x$ and $x < 2$. Equivalently, x is between -5 and 2, including -5 and excluding 2.

Progress Check

Verify that the following inequalities are true by using either the "Equivalent Statement" or the "Geometric Statement" of Table 1-6.

a. $-1 < 3$ b. $2 \leq 2$ c. $-2.7 < -1.2$

d. $-4 < -2 < 0$ e. $-\frac{7}{2} < \frac{7}{2} < 7$

TABLE 1-6 Inequalities

Algebraic Expression	Meaning	Equivalent Statement	Geometric Statement
$a > 0$	a is greater than 0.	a is positive.	a lies to the right of the origin.
$a < 0$	a is less than 0.	a is negative	a lies to the left of the origin.
$a > b$	a is greater than b.	$a - b$ is positive.	a lies to the right of b.
$a < b$	a is less than b.	$a - b$ is negative.	a lies to the left of b.
$a \geq b$	a is greater than or equal to b.	$a - b$ is positive or zero.	a lies to the right of b or coincides with b.
$a \leq b$	a is less than or equal to b.	$a - b$ is negative or zero.	a lies to the left of b or coincides with b.

TABLE 1-7 **Properties of Inequalities**

Example	Algebraic Expression	Property
Either $2 < 3$, $2 > 3$, or $2 = 3$.	Either $a < b$, $a > b$, or $a = b$.	Trichotomy property
Since $2 < 3$ and $3 < 5$, then $2 < 5$.	If $a < b$ and $b < c$ then $a < c$.	Transitive property
Since $2 < 5$, then $2 + 4 < 5 + 4$ or $6 < 9$.	If $a < b$ then $a + c < b + c$.	The sense of an inequality is preserved if any constant is added to both sides.
Since $2 < 3$ and $4 > 0$, then $2(4) < 3(4)$ or $8 < 12$.	If $a < b$ and $c > 0$, then $ac < bc$.	The sense of an inequality is preserved if it is multiplied by a positive constant.
Since $2 < 3$ and $-4 < 0$, then $2(-4) > 3(-4)$ or $-8 > -12$.	If $a < b$ and $c < 0$, then $ac > bc$.	The sense of an inequality is reversed if it is multiplied by a negative constant.

The real numbers satisfy the properties of inequalities shown in Table 1-7, where a, b, and c are real numbers.

EXAMPLE 2 ***Properties of Inequalities***

a. Since $-2 < 4$ and $4 < 5$, then $-2 < 5$.

b. Since $-2 < 5$, $-2 + 3 < 5 + 3$, or $1 < 8$.

c. Since $3 < 4$, $3 + (-5) < 4 + (-5)$, or $-2 < -1$.

d. Since $2 < 5$, $2(3) < 5(3)$, or $6 < 15$.

e. Since $-3 < 2$, $(-3)(-2) > 2(-2)$, or $6 > -4$. ■

1.2b *Absolute Value*

Suppose we are interested in the *distances* between the origin and the points labeled 4 and -4 on the real number line. Each of these points is four units from the origin; that is, the *distance is independent of the direction* and is nonnegative. (See Figure 1-2.) Furthermore, the distance between 4 and -4 is 8 units.

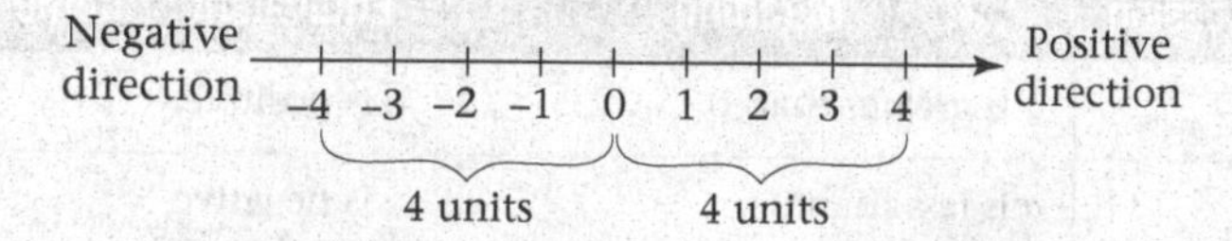

FIGURE 1-2
Distance on the Real Number Line

When we are interested in the magnitude of a number a, and do not care about the direction or sign, we use the concept of **absolute value**, which we write as $|a|$. The formal definition of absolute value is stated as follows:

$$|a| = \begin{cases} a \text{ if } a \geq 0 \\ -a \text{ if } a < 0 \end{cases}$$

Since distance is independent of direction and is always nonnegative, we can view $|a|$ as the distance from the origin to either point a or point $-a$ on the real number line.

EXAMPLE 3 ***Absolute Value and Distance***

a. $|4| = 4$ $\quad |-4| = 4$ $\quad |0| = 0$

b. The distance on the real number line between the point labeled 3.4 and the origin is $|3.4| = 3.4$. Similarly, the distance between point -2.3 and the origin is $|-2.3| = 2.3$. ■

In working with the notation of absolute value, it is important to perform the operations within the bars first. Here are some examples.

EXAMPLE 4 ***Absolute Value***

a. $|5 - 2| = |3| = 3$

b. $|2 - 5| = |-3| = 3$

c. $|3 - 5| - |8 - 6| = |-2| - |2| = 2 - 2 = 0$

d. $\dfrac{|4-7|}{-6} = \dfrac{|-3|}{-6} = \dfrac{3}{-6} = -\dfrac{1}{2}$ ■

Table 1-8 describes the properties of absolute value where a and b are real numbers.

We began by showing a use for absolute value in denoting distance from the origin without regard to direction. We conclude by demonstrating the use of absolute value to denote the distance between *any* two points a and b on the real number line. In Figure 1-3, the distance between the points labeled 2 and 5 is 3 units and can be obtained by evaluating either $|5 - 2|$ or $|2 - 5|$. Similarly, the distance between the points labeled -1 and 4 is given by either $|4 - (-1)| = 5$ or $|-1 - 4| = 5$.

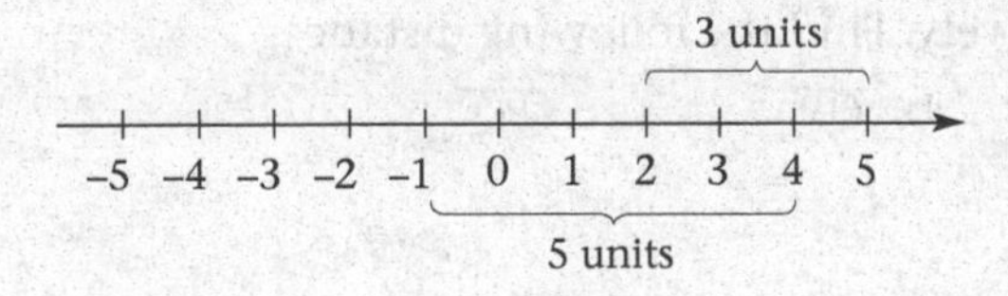

FIGURE 1-3 Distance on the Real Number Line

TABLE 1-8 Basic Properties of Absolute Value

Example	Algebraic Expression	Property
$\|-2\| \geq 0$	$\|a\| \geq 0$	Absolute value is always nonnegative.
$\|3\| = \|-3\| = 3$	$\|a\| = \|-a\|$	The absolute value of a number and its negative are the same.
$\|2 - 5\| = \|-3\| = 3$ $\|5 - 2\| = \|3\| = 3$	$\|a - b\| = \|b - a\|$	The absolute value of the difference of two numbers is always the same, irrespective of the order of subtraction.
$\|(-2)(3)\| = \|-2\|\|3\| = 6$	$\|ab\| = \|a\|\|b\|$	The absolute value of a product is the product of the absolute values.

Using the notation $\overline{AB}$ to denote the distance between the points A and B, we provide the following definition:

Distance on the Real Number Line
The **distance** $\overline{AB}$ between points A and B on the real number line, whose coordinates are a and b, respectively, is given by

$$\overline{AB} = |b - a|.$$

The third property of absolute value from Table 1-8 tells us that $\overline{AB} = |b - a| = |a - b|$. Viewed another way, this property states that the distance between any two points on the real number line is independent of the direction.

EXAMPLE 5 *Distance on the Real Number Line*

Let points A, B, and C have coordinates -4, -1, and 3, respectively, on the real number line. Find the following distances.

a. $\overline{AB}$ b. $\overline{CB}$ c. $\overline{OB}$, where O is the origin

SOLUTION

Using the definition, we have

a. $\overline{AB} = |-1 - (-4)| = |-1 + 4| = |3| = 3$

b. $\overline{CB} = |-1 - 3| = |-4| = 4$ c. $\overline{OB} = |-1 - 0| = |-1| = 1$ ■

Progress Check

The points P, Q, and R on the real number line have coordinates -6, 4, and 6, respectively. Find the following distances.

a. $\overline{PR}$ b. $\overline{QP}$ c. $\overline{PQ}$

Answers

a. 12 b. 10 c. 10

Exercise Set 1.2

1. Draw a real number line and plot the following points.
   a. 4 b. -2
   c. $\frac{5}{2}$ d. -3.5
   e. 0

2. Draw a real number line and plot the following points.
   a. -5 b. 4
   c. -3.5 d. $\frac{7}{2}$
   e. -4

3. State the real numbers associated with the points A, B, C, D, O, and E on the real number line below.

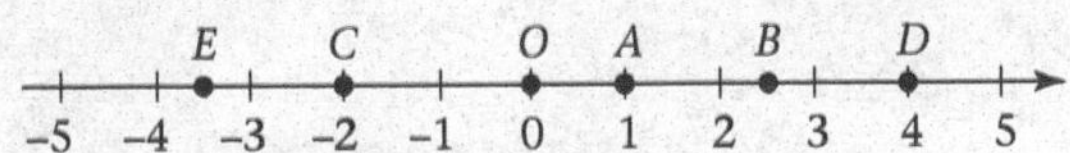

4. Represent the following by real numbers.
   a. a profit of \$10
   b. a loss of \$20
   c. a temperature of 20°F above zero
   d. a temperature of 5°F below zero

In Exercises 5–10, indicate which of the two given numbers appears first, viewed from left to right, on the real number line.

5. 4, 6
6. $\frac{1}{2}, 0$
7. $-2, \frac{3}{4}$
8. 0, -4
9. $-5, -\frac{2}{3}$
10. 4, -5

In Exercises 11–14, indicate the set of numbers on a real number line.

11. The natural numbers less than 8
12. The natural numbers greater than 4 and less than 10
13. The integers that are greater than 2 and less than 7
14. The integers that are greater than -5 and less than or equal to 1

In Exercises 15–24, express the statement as an inequality.

15. 10 is greater than 9.99.
16. -6 is less than -2.
17. a is nonnegative.
18. b is negative.
19. x is positive.
20. a is strictly between 3 and 7.
21. a is strictly between $\frac{1}{2}$ and $\frac{1}{4}$.
22. b is less than or equal to -4.
23. b is greater than or equal to 5.
24. x is negative.

In Exercises 25–30, give a property of inequalities that justifies the statement.

25. Since $-3 < 1$, then $-1 < 3$.
26. Since $-5 < -1$ and $-1 < 4$, then $-5 < 4$.
27. Since $14 > 9$, then $-14 < -9$.
28. Since $5 > 3$, then $5 \neq 3$.
29. Since $-1 < 6$, then $-3 < 18$.
30. Since $6 > -1$, then 7 is a positive number.

In Exercises 31–44, find the value of the expression.

31. $|2|$
32. $\left|-\frac{2}{3}\right|$
33. $|1.5|$
34. $|-0.8|$
35. $-|2|$
36. $-\left|-\frac{2}{5}\right|$
37. $|2-3|$
38. $|2-2|$
39. $|2-(2)|$
40. $|2|+|-3|$
41. $\frac{|14-8|}{|-3|}$
42. $\frac{|2-12|}{|1-6|}$
43. $\frac{|3|-|2|}{|3|+|2|}$
44. $\frac{|3-2|}{|3+2|}$

In Exercises 45–50, the coordinates of points A and B are given. Find $\overline{AB}$.

45. 2, 5
46. -3, 6
47. -3, -1
48. $-4, \frac{11}{2}$
49. $-\frac{4}{5}, \frac{4}{5}$
50. 2, 2
51. For what values of x and y is $|x+y| = |x| + |y|$?
52. For what values of x and y is $|x+y| < |x| + |y|$?
53. Find the set of integers whose distance from 3 is less than or equal to 5.
54. List the set of integers x such that
    a. $-2 < x < 3$ b. $0 < x < 5$
    c. $-1 < 2x < 10$

55. For the inequality $-1 < 5$, state the resulting inequality when the following operations are performed on both sides.
   a. add 2 b. subtract 5
   c. multiply by 2 d. multiply by -5
   e. divide by -1 f. divide by 2
   g. square

56. A computer sales representative receives \$400 monthly plus a 10% commission on sales. How much must she sell in a month for her income to be at least \$600 for that month?

In Exercises 57–62, use the coordinates given in Exercises 45–50 to find the midpoint of the interval.

63. For what values of x does each of the following hold?
   a. $|3 - x| = 3 - x$ b. $|5x - 2| = -(5x - 2)$

64. Evaluate $\frac{|x-3|}{x-3}$ for $x = -2, -1, 0, 1, 2$. Make a conjecture about the value of this expression for all values of x.

1.3 Algebraic Expressions and Polynomials

A **variable** is a symbol to which we can assign values. For example, in Section 1.1 we defined a rational number as one that can be written as $\frac{p}{q}$, where p and q are integers and q is not zero. The symbols p and q are variables since we can assign values to them. A variable can be restricted to a particular number system (for example, p and q must be integers) or to a subset of a number system.

If we invest P dollars at an annual interest rate of 6%, then we will earn $0.06P$ dollars interest per year; and we will have $P + 0.06P$ dollars at the end of the year. We call $P + 0.06P$ an **algebraic expression**. Note that an algebraic expression involves **variables** (in this case P), **constants** (such as 0.06), and **algebraic operations** (such as $+, -, \times, \div$). Virtually everything we do in algebra involves algebraic expressions.

An algebraic expression takes on a **value** when we assign a specific number to each variable in the expression. Thus, the expression

$$\frac{3m+4n}{m+n}$$

is **evaluated** when $m = 3$ and $n = 2$ by substitution of these values for m and n:

$$\frac{3(3)+4(2)}{3+2} = \frac{9+8}{5} = \frac{17}{5}$$

We often need to write algebraic expressions in which a variable multiplies itself repeatedly. We use the notation of exponents to indicate such repeated multiplication. Thus,

$$a^1 = a \qquad a^2 = a \cdot a \qquad a^n = \underbrace{a \cdot a \cdot \cdots \cdot a}_{n \text{ factors}}$$

where n is a natural number and a is a real number. We call a the **base** and n the **exponent** and say that a^n is the nth **power** of a. When $n = 1$, we simply write a rather than a^1.

It is convenient to define a^0 for all real numbers $a \neq 0$ as $a^0 = 1$. We will provide motivation for this seemingly arbitrary definition in Section 1.7.

EXAMPLE 1 *Multiplication with Natural Number Exponents*

Write the following without using exponents.

a. $\left(\frac{1}{2}\right)^3$

b. $2x^3$

c. $(2x)^3$

d. $-3x^2y^3$

SOLUTION

a. $\left(\frac{1}{2}\right)^3 = \frac{1}{2} \cdot \frac{1}{2} \cdot \frac{1}{2} = \frac{1}{8}$

b. $2x^3 = 2 \cdot x \cdot x \cdot x$

c. $(2x)^3 = 2x \cdot 2x \cdot 2x = 8 \cdot x \cdot x \cdot x$

d. $-3x^2y^3 = -3 \cdot x \cdot x \cdot y \cdot y \cdot y$ ■

Note the difference between

$$(-3)^2 = (-3)(-3) = 9$$

and

$$-3^2 = -(3 \cdot 3) = -9$$

Later in this chapter we will need an important rule of exponents. Observe that if m and n are natural numbers and a is any real number, then

$$a^m \cdot a^n = \underbrace{a \cdot a \cdot \dots \cdot a}_{m \text{ factors}} \cdot \underbrace{a \cdot a \cdot \dots \cdot a}_{n \text{ factors}}$$

Since there are a total of $m + n$ factors on the right side, we conclude that

$$a^m a^n = a^{m+n}$$

EXAMPLE 2 ***Multiplication with Natural Number Exponents***

Multiply.

a. $x^2 \cdot x^3$ b. $(3x)(4x^4)$

SOLUTION

a. $x^2 \cdot x^3 = x^{2+3} = x^5$ b. $(3x)(4x^4) = 3 \cdot 4 \cdot x \cdot x^4 = 12x^{1+4} = 12x^5$ ■

Progress Check

Multiply.

a. $x^5 \cdot x^2$ b. $(2x^6)(-2x^4)$

Answers

a. x^7 b. $-4x^{10}$

1.3a *Polynomials*

A polynomial is an algebraic expression of a certain form. Polynomials play an important role in the study of algebra since many word problems translate into equations or inequalities that involve polynomials. We first study the manipulative and mechanical aspects of polynomials. This knowledge will serve as background for dealing with their applications in later chapters.

Let x denote a variable and let n be a constant, nonnegative integer. The expression ax^n, where a is a constant real number, is called a **monomial in *x*.** A **polynomial in *x*** is an expression that is a sum of monomials and has the general form

$$P = a_n x^n + a_{n-1} x^{n-1} + \dots + a_1 x + a_0, \quad a_n \neq 0 \tag{1}$$

Each of the monomials in Equation (1) is called a **term** of P, and $a_0, a_1, \ldots, a_n$ are constant real numbers that are called the **coefficients** of the terms of P. Note that a polynomial may consist of just one term; that is, a monomial is considered to be a polynomial.

EXAMPLE 3 *Polynomial Expressions*

a. The following expressions are polynomials in x:

$$3x^4 + 2x + 5 \qquad 2x^3 + 5x^2 - 2x + 1 \qquad \frac{3}{2}x^3$$

Notice that we write $2x^3 + 5x^2 + (-2)x + 1$ as $2x^3 + 5x^2 - 2x + 1$.

b. The following expressions are not polynomials in x:

$$2x^{1/2} + 5 \qquad 3 - \frac{4}{x} \qquad \frac{2x-1}{x-2}$$

Remember that each term of a polynomial in x must be of the form ax^n, where a is a real number and n is a nonnegative integer. ■

The **degree of a monomial in *x*** is the exponent of x. Thus, the degree of $5x^3$ is 3. A monomial in which the exponent of x is 0 is called a **constant term** and is said to be of *degree zero*. The nonzero coefficient a_n of the term in P with highest degree is called the **leading coefficient** of P, and we say that P is a **polynomial of degree *n***. The polynomial whose coefficients are all zero is called the **zero polynomial.** It is denoted by 0 and is said to have no degree.

EXAMPLE 4 *Vocabulary of Polynomials*

Given the polynomial

$$P = 2x^4 - 3x^2 + \frac{4}{3}x - 1$$

The terms of P are

$$2x^4, \quad 0x^3, \quad -3x^2, \quad \frac{4}{3}x, \quad -1$$

The coefficients of the terms are

$$2, \quad 0, \quad -3, \quad \frac{4}{3}, \quad -1$$

The degree of P is 4 and the leading coefficient is 2. ■

A *monomial in the variables x and y* is an expression of the form ax^my^n, where a is a constant and m and n are constant, nonnegative integers. The number a is called the **coefficient** of the monomial. The *degree of a monomial in x and y* is the sum of the exponents of x and y. Thus, the degree of $2x^3y^2$ is $3 + 2 = 5$. A *polynomial in x and y* is an expression that is a sum of monomials. The **degree of a polynomial in *x* and *y*** is the degree of the highest degree monomial with nonzero coefficient.

EXAMPLE 5 ***Degree of Polynomials***

The following are polynomials in x and y:

$2x^2y + y^2 - 3xy + 1$	Degree is 3.
xy	Degree is 2.
$3x^4 + xy - y^2$	Degree is 4. ■

1.3b Operations with Polynomials

If P and Q are polynomials in x, then the terms ax^r in P and bx^r in Q are said to be **like terms**; that is, like terms have the same exponent in x. For example, given

$$P = 4x^2 + 4x - 1$$

and

$$Q = 3x^3 - 2x^2 + 4$$

then the like terms are $0x^3$ and $3x^3$, $4x^2$ and $-2x^2$, $4x$ and $0x$, -1 and 4.

We define equality of polynomials in the following way:

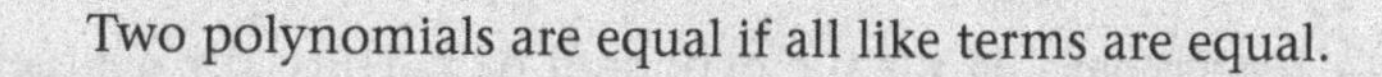

Two polynomials are equal if all like terms are equal.

EXAMPLE 6 ***Equality of Polynomials***

Find A, B, C, and D if

$$Ax^3 + (A + B)x^2 + Cx + (C - D) = -2x^3 + x + 3$$

SOLUTION

Equating the coefficients of the terms, we have

$A = -2$	$A + B = 0$	$C = 1$	$C - D = 3$
	$B = 2$		$D = -2$ ■

If P and Q are polynomials in x, the *sum* $P + Q$ is obtained by forming the sums of all pairs of like terms. The sum of ax^r in P and bx^r in Q is $(a + b)x^r$. Similarly, the *difference* $P - Q$ is obtained by forming the differences, $(a - b)x^r$, of like terms.

EXAMPLE 7 ***Addition and Subtraction of Polynomials***

a. Add $2x^3 + 2x^2 - 3$ and $x^3 - x^2 + x + 2$.
b. Subtract $2x^3 + x^2 - x + 1$ from $3x^3 - 2x^2 + 2x$.

SOLUTION

a. Adding the coefficients of like terms,

$$(2x^3 + 2x^2 - 3) + (x^3 - x^2 + x + 2) = 3x^3 + x^2 + x - 1$$

b. Subtracting the coefficients of like terms,

$$(3x^3 - 2x^2 + 2x) - (2x^3 + x^2 - x + 1) = x^3 - 3x^2 + 3x - 1$$ ■

$$(x + 5) - (x + 2) \neq (x + 5) - x + 2$$

The coefficient -1 must multiply each term in the parentheses. Thus,

$$-(x + 2) = -x - 2$$

Therefore,

$$(x + 5) - (x + 2) = x + 5 - x - 2 = 3$$

while

$$(x + 5) - x + 2 = x + 5 - x + 2 = 7$$

Multiplication of polynomials is based on the rule for exponents developed earlier in this section,

$$a^m a^n = a^{m+n}$$

and on the distributive laws

$$a(b + c) = ab + ac$$
$$(a + b)c = ac + bc$$

EXAMPLE 8 ***Multiplication of Polynomials***

Multiply $3x^3(2x^3 - 6x^2 + 5)$.

SOLUTION

$3x^3(2x^3 - 6x^2 + 5)$

$= (3x^3)(2x^3) + (3x^3)(-6x^2) + (3x^3)(5)$ — Distributive law

$= (3)(2)x^{3+3} + (3)(-6)x^{3+2} + (3)(5)x^3$ — $a^m a^n = a^{m+n}$

$= 6x^6 - 18x^5 + 15x^3$ ■

EXAMPLE 9 ***Multiplication of Polynomials***

Multiply $(x + 2)(3x^2 - x + 5)$.

SOLUTION

$(x + 2)(3x^2 - x + 5)$

$= x(3x^2 - x + 5) + 2(3x^2 - x + 5)$ — Distributive law

$= 3x^3 - x^2 + 5x + 6x^2 - 2x + 10$ — Distributive law and $a^m a^n = a^{m+n}$

$= 3x^3 + 5x^2 + 3x + 10$ — Adding like terms ■

Multiply.

a. $(x^2+2)(x^2-3x+1)$ b. $(x^2-2xy+y)(2x+y)$

Answers

a. $x^4-3x^3+3x^2-6x+2$ b. $2x^3-3x^2y+2xy-2xy^2+y^2$

The multiplication in Example 9 can be carried out in "long form" as follows:

$$\begin{array}{rl}
3x^2-x+5 & \\
x+2 & \\
\hline
3x^3-x^2+5x & = x(3x^2-x+5) \\
6x^2-2x+10 & = 2(3x^2-x+5) \\
\hline
3x^3+5x^2+3x+10 & = \text{sum of above lines}
\end{array}$$

In Example 9, the product of polynomials of degrees 1 and 2 is seen to be a polynomial of degree 3. From the multiplication process, we can derive the following useful rule:

The degree of the product of two nonzero polynomials is the sum of the degrees of the polynomials.

Products of the form $(2x+3)(5x-2)$ or $(2x+y)(3x-y)$ occur often, and we can handle them by the method sometimes referred to as FOIL: F = first, O = outer, I = inner, L = last.

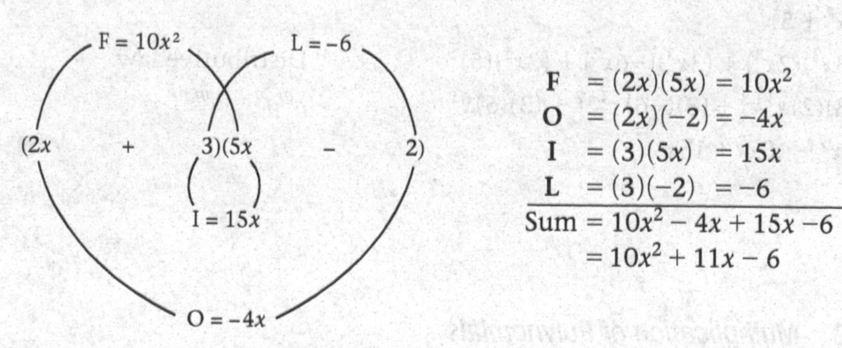

$$\begin{aligned}
F &= (2x)(5x) = 10x^2 \\
O &= (2x)(-2) = -4x \\
I &= (3)(5x) = 15x \\
L &= (3)(-2) = -6 \\
\hline
\text{Sum} &= 10x^2-4x+15x-6 \\
&= 10x^2+11x-6
\end{aligned}$$

A number of special products occur frequently, and it is worthwhile knowing them.

Special Products

$(a+b)(a-b)=a^2-b^2$

$(a+b)^2=(a+b)(a+b)=a^2+2ab+b^2$

$(a-b)^2=(a-b)(a-b)=a^2-2ab+b^2$

$(a+b)^3=a^3+3a^2b+3ab^2+b^3$

$(a-b)^3=a^3-3a^2b+3ab^2-b^3$

EXAMPLE 10 ***Multiplication of Polynomials***

Multiply.

a. $(x+2)^2$ b. $(x-3)^2$ c. $(x+4)(x-4)$

SOLUTION

a. $(x+2)^2 = (x+2)(x+2) = x^2 + 4x + 4$

b. $(x-3)^2 = (x-3)(x-3) = x^2 - 6x + 9$ c. $(x+4)(x-4) = x^2 - 16$ ■

Progress Check

a. Multiply $(2x^2 - xy + y^2)(3x + y)$ b. Multiply $(2x-3)(3x-2)$

Answers

a. $6x^3 - x^2y + 2xy^2 + y^3$ b. $6x^2 - 13x + 6$

Graphing Calculator Power User's CORNER

Values may be assigned to variables in your graphing calculator by using the [STO>] button. For example, to set $A = 3$, press the buttons 3 [STO>] [ALPHA] [A]. Now, whenever A is in an expression that you want to evaluate, it will have the value of 3.

Exercise Set 1.3

In Exercises 1–6, evaluate the given expression when $r = 2$, $s = -3$, and $t = 4$.

1. $r + 2s + t$
2. rst
3. $\dfrac{rst}{r+s+t}$
4. $(r + s)t$
5. $\dfrac{r+s}{rt}$
6. $\dfrac{r+s+t}{t}$
7. Evaluate $\frac{2}{3}r + 5$ when $r = 12$.
8. Evaluate $\frac{9}{5}C + 32$ when $C = 37$.
9. If P dollars are invested at a simple interest rate of r percent per year for t years, the amount on hand at the end of t years is $P + Prt$. Suppose you invest \$2000 at 8% per year ($r = 0.08$). Find the amount you will have on hand after
   a. 1 year b. $\frac{1}{2}$ year c. 8 months.
10. The perimeter of a rectangle is given by the formula $P = 2(L + W)$, where L is the length and W is the width of the rectangle. Find the perimeter if
    a. $L = 2$ feet, $W = 3$ feet
    b. $L = \frac{1}{2}$ meter, $W = \frac{1}{4}$ meter
11. Evaluate $0.02r + 0.314st + 2.25t$ when $r = 2.5$, $s = 3.4$, and $t = 2.81$.
12. Evaluate $10.421x + 0.821y + 2.34xyz$ when $x = 3.21$, $y = 2.42$, and $z = 1.23$.

Evaluate the given expression in Exercises 13–18.

13. $|x| - |x| \cdot |y|$ when $x = -3$, $y = 4$
14. $|x + y| + |x - y|$ when $x = -3$, $y = 2$
15. $\dfrac{|a - 2b|}{2a}$ when $a = 1$, $b = 2$
16. $\dfrac{|x|+|y|}{|x|-|y|}$ when $x = -3$, $y = 4$
17. $\dfrac{-|a - 2b|}{|a+b|}$ when $a = -2$, $b = -1$
18. $\dfrac{|a-b|-2|c-a|}{|a-b+c|}$ when $a = -2$, $b = 3$, $c = -5$

Carry out the indicated operations in Exercises 19–24.

19. $b^5 \cdot b^2$
20. $x^3 \cdot x^5$
21. $(4y^3)(-5y^6)$
22. $(-6x^4)(-4x^7)$
23. $\left(\frac{3}{2}x^3\right)(-2x)$
24. $\left(-\frac{5}{3}x^6\right)\left(-\frac{3}{10}x^3\right)$
25. Evaluate the given expressions.
    a. 1^3 b. 10^8
    c. 2^5 d. 7^1
26. Evaluate the given expressions using your calculator.
    a. 9^{10} b. 0.8^6
27. Which of the following expressions are *not* polynomials?
    a. $-3x^2 + 2x + 5$ b. $-3x^2y$
    c. $-3x^{2/3} + 2xy + 5$ d. $-2x^{-4} + 2xy^3 + 5$
28. Which of the following expressions are *not* polynomials?
    a. $4x^5 - x^{1/2} + 6$ b. $\frac{2}{5}x^3 + \frac{4}{3}x - 2$
    c. $4x^5y$ d. $x^{4/3}y + 2x - 3$

In Exercises 29–32, indicate the leading coefficient and the degree of the given polynomial.

29. $2x^3 + 3x^2 - 5$
30. $-4x^5 - 8x^2 + x + 3$
31. $\frac{3}{5}x^4 + 2x^2 - x - 1$
32. $-1.5 + 7x^3 + 0.75x^7$

In Exercises 33–36, find the degree of the given polynomial.

33. $3x^2y - 4x^2 - 2y + 4$
34. $4xy^3 + xy^2 - y^2 + y$
35. $2xy^3 - y^3 + 3x^2 - 2$
36. $\frac{1}{2}x^3y^3 - 2$
37. Find the value of the polynomial $3x^2y^2 + 2xy - x + 2y + 7$ when $x = 2$ and $y = -1$.
38. Find the value of the polynomial $0.02x^2 + 0.3x - 0.5$ when $x = 0.3$.
39. Find the value of the polynomial $2.1x^3 + 3.3x^2 - 4.1x - 7.2$ when $x = 4.1$.
40. Write a polynomial giving the area of a circle of radius r.
41. Write a polynomial giving the area of a triangle of base b and height h.

42. A field consists of a rectangle and a square arranged as shown below. What does each of the following polynomials represent?

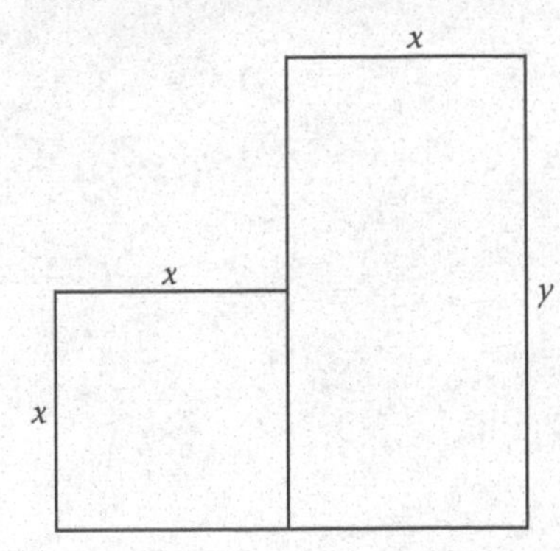

a. $x^2 + xy$ b. $2x + 2y$

c. $4x$ d. $4x + 2y$

43. An investor buys x shares of G.E. stock at \$35.5 per share, y shares of Exxon stock at \$91 per share, and z shares of AT&T stock at \$38 per share. What does the polynomial $35.5x + 91y + 38z$ represent?

Perform the indicated operations in Exercises 44–62.

44. $(4x^2 + 3x + 2) + (3x^2 - 2x - 5)$

45. $(2x^2 + 3x + 8) - (5 - 2x + 2x^2)$

46. $4xy^2 + 2xy + 2x + 3 - (-2xy^2 + xy - y + 2)$

47. $(2s^2t^3 - st^2 + st - s + t) - (3s^2t^2 - 2s^2t - 4st^2 - t + 3)$

48. $3xy^2z - 4x^2yz + xy + 3 - (2xy^2z + x^2yz - yz + x - 2)$

49. $a^2bc + ab^2c + 2ab^3 - 3a^2bc - 4ab^3 + 3$

50. $(x + 1)(x^2 + 2x - 3)$

51. $(2 - x)(2x^3 + x - 2)$

52. $(2s - 3)(s^3 - s + 2)$

53. $(-3s + 2)(-2s^2 - s + 3)$

54. $(x^2 + 3)(2x^2 - x + 2)$

55. $(2y^2 + y)(-2y^3 + y - 3)$

56. $(x^2 + 2x - 1)(2x^2 - 3x + 2)$

57. $(a^2 - 4a + 3)(4a^3 + 2a + 5)$

58. $(2u^7 + ub + b^7)(3u - b^7 + 1)$

59. $(-3a + ab + b^2)(3b^2 + 2b + 2)$

60. $5(2x - 3)^2$

61. $2(3x - 2)(3 - x)$

62. $(x - 1)(x + 2)(x + 3)$

63. An investor buys x shares of IBM stock at \$98 per share at Thursday's opening of the stock market. Later in the day, the investor sells y shares of AT&T stock at \$38 per share and z shares of TRW stock at \$20 per share. Write a polynomial that expresses the amount of money the buyer has invested at the end of the day.

64. An artist takes a rectangular piece of cardboard whose sides are x and y and cuts out a square of side $\frac{x}{2}$ to obtain a mat for a painting, as shown below. Write a polynomial giving the area of the mat.

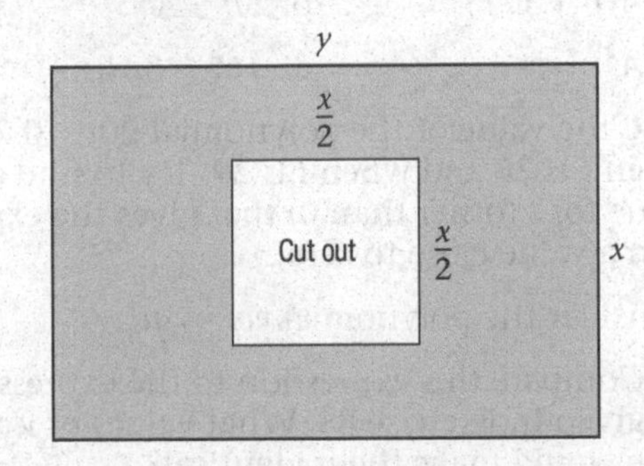

In Exercises 65–78, perform the multiplication mentally.

65. $(x - 1)(x + 3)$ 66. $(x + 2)(x + 3)$

67. $(2x + 1)(2x + 3)$ 68. $(3x - 1)(x + 5)$

69. $(3x - 2)(x - 1)$ 70. $(x + 4)(2x - 1)$

71. $(x + y)^2$ 72. $(x - 4)^2$

73. $(3x - 1)^2$ 74. $(x + 2)(x - 2)$

75. $(2x + 1)(2x - 1)$ 76. $(3a + 2b)^2$

77. $(x^2 + y^2)^2$ 78. $(x - y)^2$

79. Simplify the following.

a. $3^{10} + 3^{10} + 3^{10}$ b. $2^n + 2^n + 2^n + 2^n$

80. A student conjectured that the expression $N = m^2 - m + 41$ yields a prime number, for integer values of m. Prove or disprove this statement.

81. Perform the indicated operations.

a. $\left(\frac{2}{x} - 1\right)\left(\frac{2}{x} + 1\right)$ b. $\left(\frac{wx}{y} - z\right)^2$

c. $(x + y + z)(x + y - z)$

82. Find the surface area and volume of the open-top box below.

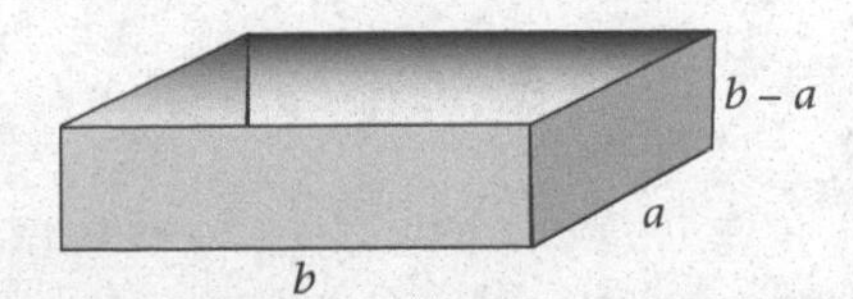

83. Eric can run a mile in 4.23 minutes, and Benjamin can run 4.23 miles in an hour. Who is the faster runner?

84. a. Let $P = \$1000$; that is, store 1000 in memory location P. Evaluate $P + 0.06P$ by entering the expression $P + 0.06P$ into your calculator.

b. Repeat part (a) for $P = \$28{,}525$.

85. Let $A = 8$ and $B = 32$; that is, store 8 in memory location A and 32 in memory location B. Evaluate the following expressions by entering them into your calculator as they appear below. (Use a multiplication sign if your calculator requires you to do so.)

a. $A(B + 17)$ b. $5B - A^2$

c. A^A d. $16^A - 3AB$

86. Find the value of the polynomial $20t - 0.7t^2$ when t is 28 and when t is 29. Try to find a value for t (other than 0) that gives the expression a value close to zero.

87. Consider the polynomial $vt - \frac{1}{2}at^2$.

a. Compare this expression to the expression given in Exercise 86. What values of v and a would make them identical?

b. Using your calculator, experiment with different values of v, a, and t. Try to put your data in an organized chart. In physics, this expression represents position of a body in free fall: v is the initial velocity, and a is the acceleration due to gravity.

1.4 Factoring

Now that we can find the product of two polynomials, let us consider the reverse problem: given a polynomial, can we find **factors** whose product yields the given polynomial? This process, known as **factoring**, is one of the basic tools of algebra. In this chapter, a polynomial with *integer* coefficients is to be factored as a product of polynomials of lower degree with *integer* coefficients; a polynomial with *rational* coefficients is to be factored as a product of polynomials of lower degree with *rational* coefficients. We will approach factoring by learning to recognize the situations in which factoring is possible.

1.4a *Common Factors*

Consider the polynomial

$$x^2 + x$$

Since the factor x is common to both terms, we can write

$$x^2 + x = x(x + 1)$$

EXAMPLE 1 ***Factoring with Common Factors***

Factor.

a. $15x^3 - 10x^2$ b. $4x^2y - 8xy^2 + 6xy$ c. $2x(x + y) - 5y(x + y)$

SOLUTION

a. 5 and x^2 are common to both terms. Therefore,

$$15x^3 - 10x^2 = 5x^2(3x - 2)$$

b. Here we see that 2, x, and y are common to all terms. Therefore,

$$4x^2y - 8xy^2 + 6xy = 2xy(2x - 4y + 3)$$

c. The expression $(x + y)$ is found in both terms. Factoring, we have

$$2x(x + y) - 5y(x + y) = (x + y)(2x - 5y)$$ ■

Progress Check

Factor.

a. $4x^2 - x$ b. $3x^4 - 9x^2$ c. $3m(2x - 3y) - n(2x - 3y)$

Answers

a. $x(4x - 1)$ b. $3x^2(x^2 - 3)$ c. $(2x - 3y)(3m - n)$

1.4b *Factoring by Grouping*

It is sometimes possible to discover common factors by first grouping terms. Consider the following examples:

EXAMPLE 2 *Factoring by Grouping*

Factor.

a. $2ab + b + 2ac + c$ b. $2x - 4x^2y - 3y + 6xy^2$

SOLUTION

a. Group those terms containing b and those terms containing c.

$$\begin{aligned} 2ab + b + 2ac + c &= (2ab + b) + (2ac + c) && \text{Grouping} \\ &= b(2a + 1) + c(2a + 1) && \text{Common factors } b, c \\ &= (2a + 1)(b + c) && \text{Common factor } 2a + 1 \end{aligned}$$

Alternatively, suppose we group terms containing a.

$$\begin{aligned} 2ab + b + 2ac + c &= (2ab + 2ac) + (b + c) && \text{Grouping} \\ &= 2a(b + c) + (b + c) && \text{Common factors } 2, a \\ &= (b + c)(2a + 1) && \text{Common factor } (b + c) \end{aligned}$$

b. $2x - 4x^2y - 3y + 6xy^2$

$$\begin{aligned} &= (2x - 4x^2y) - (3y - 6xy^2) && \text{Grouping with sign change} \\ &= 2x(1 - 2xy) - 3y(1 - 2xy) && \text{Common factors } 2x, 3y \\ &= (1 - 2xy)(2x - 3y) && \text{Common factor } 1 - 2xy \end{aligned}$$

■

Progress Check

Factor.

a. $2m^3n + m^2 + 2mn^2 + n$ b. $2a^2 - 4ab^2 - ab + 2b^3$

Answers

a. $(2mn + 1)(m^2 + n)$ b. $(a - 2b^2)(2a - b)$

1.4c *Factoring Second-Degree Polynomials*

To factor a second-degree polynomial, such as

$$x^2 + 5x + 6$$

we first note that the term x^2 can have come only from $x \cdot x$, so we write two incomplete factors:

$$x^2 + 5x + 6 = (x \quad)(x \quad)$$

The constant term +6 can be the product of either two positive numbers or two negative numbers. Since the middle term $+5x$ is the sum of two other products, both signs must be positive. Thus,

$$x^2 + 5x + 6 = (x + \quad)(x + \quad)$$

Finally, the number 6 can be written as the product of two integers in only two ways: $1 \cdot 6$ and $2 \cdot 3$. The first pair gives a middle term of $7x$. The second pair gives the actual middle term, $5x$. So

$$x^2 + 5x + 6 = (x + 2)(x + 3)$$

EXAMPLE 3 *Factoring Second-Degree Polynomials*

Factor.

a. $x^2 - 7x + 10$ b. $x^2 - 3x - 4$

SOLUTION

a. Since the constant term is positive and the middle term is negative, we must have two negative signs. Integer pairs whose product is 10 are 1 and 10, and 2 and 5. We find that

$$x^2 - 7x + 10 = (x - 2)(x - 5)$$

b. Since the constant term is negative, we must have opposite signs. Integer pairs whose product is 4 are 1 and 4, and 2 and 2. Since the coefficient of $-3x$ is negative, we assign the larger integer of a given pair to be negative. We find that

$$x^2 - 3x - 4 = (x + 1)(x - 4)$$ ■

When the leading coefficient of a second-degree polynomial is an integer other than 1, the factoring process becomes more complex, as shown in the following example.

EXAMPLE 4 *Factoring Second-Degree Polynomials*

Factor $2x^2 - x - 6$.

SOLUTION

The term $2x^2$ can result only from the factors $2x$ and x, so the factors must be of the form

$$2x^2 - x - 6 = (2x \quad)(x \quad)$$

The constant term, -6, must be the product of factors of opposite signs, so we may write

$$2x^2 - x - 6 = \begin{cases} (2x + \quad)(x - \quad) \\ \text{or} \\ (2x - \quad)(x + \quad) \end{cases}$$

The integer factors of 6 are

$$1 \cdot 6 \qquad 6 \cdot 1 \qquad 2 \cdot 3 \qquad 3 \cdot 2$$

By trying these we find that

$$2x^2 - x - 6 = (2x + 3)(x - 2)$$ ■

Factor.

a. $3x^2 - 16x + 21$ b. $2x^2 + 3x - 9$

Answers

a. $(3x - 7)(x - 3)$ b. $(2x - 3)(x + 3)$

The polynomial $x^2 - 6x$ can be written as

$$x^2 - 6x = x(x - 6)$$

and is then a product of two polynomials of positive degree. Students often fail to consider x to be a "true" factor.

1.4d *Special Factors*

There is a special case of the second-degree polynomial that occurs frequently and factors easily. Given the polynomial $x^2 - 9$, we see that each term is a perfect square, and we can verify that

$$x^2 - 9 = (x + 3)(x - 3)$$

The general rule, which holds whenever we are dealing with a difference of two squares, may be stated as follows:

Difference of Two Squares

$$a^2 - b^2 = (a + b)(a - b)$$

EXAMPLE 5 *Special Factors*

Factor.

a. $4x^2 - 25$ b. $9r^2 - 16t^2$

SOLUTION

a. Since

$$4x^2 - 25 = (2x)^2 - (5)^2$$

we may use the formula for the difference of two squares with $a = 2x$ and $b = 5$. Thus,

$$4x^2 - 25 = (2x + 5)(2x - 5)$$

b. Since

$$9r^2 - 16t^2 = (3r)^2 - (4t)^2$$

we have $a = 3r$ and $b = 4t$, resulting in

$$9r^2 - 16t^2 = (3r + 4t)(3r - 4t)$$

■

Factor.

a. $x^2 - 49$ b. $16x^2 - 9$ c. $25x^2 - y^2$

Answers

a. $(x + 7)(x - 7)$ b. $(4x + 3)(4x - 3)$ c. $(5x + y)(5x - y)$

The formulas for a sum of two cubes and a difference of two cubes can be verified by multiplying the factors on the right-hand sides of the following equations:

Sum and Difference of Two Cubes

$$a^3 + b^3 = (a + b)(a^2 - ab + b^2)$$
$$a^3 - b^3 = (a - b)(a^2 + ab + b^2)$$

These formulas provide a direct means of factoring the sum or difference of two cubes and are used in the same way as the formula for a difference of two squares.

EXAMPLE 6 *Special Factors*

Factor.

a. $x^3 + 1$ b. $27m^3 - 64n^3$ c. $\frac{1}{27}u^3 + 8v^3$

SOLUTION

a. With $a = x$ and $b = 1$, the formula for the sum of two cubes yields the following result:

$$x^3 + 1 = (x + 1)(x^2 - x + 1)$$

b. Since

$$27m^3 - 64n^3 = (3m)^3 - (4n)^3$$

we can use the formula for the difference of two cubes with $a = 3m$ and $b = 4n$:

$$27m^3 - 64n^3 = (3m - 4n)(9m^2 + 12mn + 16n^2)$$

c. Note that

$$\frac{1}{27}u^3 + 8v^3 = \left(\frac{1}{3}u\right)^3 + (2v)^3$$

and then use the formula for the sum of two cubes:

$$\frac{1}{27}u^3 + 8v^3 = \left(\frac{u}{3} + 2v\right)\left(\frac{u^2}{9} - \frac{2}{3}uv + 4v^2\right)$$ ■

1.4e Combining Methods

We conclude with problems that combine the various methods of factoring that we have studied. As the factoring becomes more complicated, it may be helpful to consider the following strategy:

Remove common factors before attempting any other factoring techniques.

EXAMPLE 7 ***Common Factors, Grouping, and Special Factors***

Factor.

a. $2x^3 - 8x$ b. $3y(y+3) + 2(y+3)(y^2-1)$

SOLUTION

a. Observing the common factor $2x$, we find that

$$\begin{aligned} 2x^3 - 8x &= 2x(x^2-4) \\ &= 2x(x+2)(x-2) \end{aligned}$$

b. Observing the common factor $y + 3$, we see that

$$\begin{aligned} 3y(y+3) + 2(y+3)(y^2-1) &= (y+3)[3y + 2(y^2-1)] \\ &= (y+3)(3y+2y^2-2) \\ &= (y+3)(2y^2+3y-2) \\ &= (y+3)(2y-1)(y+2) \end{aligned}$$

■

FOCUS on "Magical" Factoring for Second-Degree Polynomials

Factoring involves a certain amount of trial and error that can become frustrating, especially when the lead coefficient is not 1. You might want to try a scheme that "magically" reduces the number of candidates. We demonstrate the method for the polynomial

$$4x^2 + 11x + 6 \tag{1}$$

Using the lead coefficient of 4, write the pair of incomplete factors

$$(4x \quad)(4x \quad) \tag{2}$$

Next, multiply the coefficient of x^2 and the constant term in Equation (1) to produce $4 \cdot 6 = 24$. Now find two integers whose product is 24 and whose sum is 11, the coefficient of the middle term of (1). Since 8 and 3 work, we write

$$(4x + 8)(4x + 3) \tag{3}$$

Finally, within each parenthesis in Equation (3) discard any common numerical factor. (Discarding a factor may only be performed in this "magical" type of factoring.) Thus $(4x + 8)$ reduces to $(x + 2)$ and we write

$$(x + 2)(4x + 3) \tag{4}$$

which is the factorization of $4x^2 + 11x + 6$.

Will the method always work? Yes—if you first remove all common factors in the original polynomial. That is, you must first write

$$6x^2 + 15x + 6 = 3(2x^2 + 5x + 2)$$

and apply the method to the polynomial $2x^2 + 5x + 2$.

(For a proof that the method works, see M. A. Autrie and J. D. Austin, "A Novel Way to Factor Quadratic Polynomials," *The Mathematics Teacher* 72, no. 2 [1979].) We use the polynomial $2x^2 - x - 6$ of Example 4 to demonstrate the method when some of the coefficients are negative.

Try the method on these second-degree polynomials:

$3x^2 + 10x - 8$

$6x^2 - 13x + 6$

$4x^2 - 15x - 4$

$10x^2 + 11x - 6$

Factoring $ax^2 + bx + c$	Example: $2x^2 - x - 6$
Step 1. Use the lead coefficient a to write the incomplete factors $(ax \quad)(ax \quad)$	*Step 1.* The lead coefficient is 2, so we write $(2x \quad)(2x \quad)$
Step 2. Multiply a and c, the coefficient of x^2, and the constant term.	*Step 2.* $a \cdot c = (2)(-6) = -12$
Step 3. Find integers whose product is $a \cdot c$ and whose sum equals b. Write these integers in the incomplete factors of *Step 1*.	*Step 3.* Two integers whose product is -12 and whose sum is -1 are 3 and -4. We then write $(2x + 3)(2x - 4)$
Step 4. Discard any common factor *within each parenthesis* in *Step 3*. The result is the desired factorization.	*Step 4.* Reducing $(2x - 4)$ to $(x - 2)$ by discarding the common factor 2, we have $2x^2 - x - 6 = (2x + 3)(x - 2)$

Progress Check

Factor.

a. $x^3 + 5x^2 - 6x$ b. $2x^3 - 2x^2y - 4xy^2$ c. $-3x(x+1) + (x+1)(2x^2+1)$

Answers

a. $x(x+6)(x-1)$ b. $2x(x+y)(x-2y)$ c. $(x+1)(2x-1)(x-1)$

1.4f *Irreducible Polynomials*

Are there polynomials that cannot be written as a product of polynomials of lower degree with integer coefficients? The answer is yes. Examples are the polynomials $x^2 + 1$ and $x^2 + x + 1$. A polynomial is said to be **prime**, or **irreducible**, if it cannot be written as a product of two polynomials, each of positive degree. Thus, $x^2 + 1$ is irreducible over the integers.

Exercise Set 1.4

Factor completely.

1. $5x - 15$
2. $\frac{1}{4}x + \frac{3}{4}y$
3. $-2x - 8y$
4. $3x - 6y + 15$
5. $5bc + 25b$
6. $2x^4 + x^2$
7. $-3y^2 - 4y^5$
8. $3abc + 12bc$
9. $3x^2 + 6x^2y - 9x^2z$
10. $9a^3b^3 + 12a^2b - 15ab^2$
11. $x^2 + 4x + 3$
12. $x^2 + 2x - 8$
13. $y^2 - 8y + 15$
14. $y^2 + 7y - 8$
15. $a^2 - 7ab + 12b^2$
16. $x^2 - 49$
17. $y^2 - \frac{1}{9}$
18. $a^2 - 7a + 10$
19. $9 - x^2$
20. $4b^2 - a^2$
21. $x^2 - 5x - 14$
22. $x^2y^2 - 9$
23. $\frac{1}{16} - y^2$
24. $4a^2 - b^2$
25. $x^2 - 6x + 9$
26. $a^2b^2 - \frac{1}{9}$
27. $x^2 - 12x + 20$
28. $x^2 - 8x - 20$
29. $x^2 + 11x + 24$
30. $y^2 - \frac{9}{16}$
31. $2x^2 - 3x - 2$
32. $2x^2 + 7x + 6$
33. $3a^2 - 11a + 6$
34. $4x^2 - 9x + 2$
35. $6x^2 + 13x + 6$
36. $4y^2 - 9$
37. $8m^2 - 6m - 9$
38. $9x^2 + 24x + 16$
39. $10x^2 - 13x - 3$
40. $9y^2 - 16x^2$
41. $6a^2 - 5ab - 6b^2$
42. $4x^2 + 20x + 25$
43. $10r^2s^2 + 9rst + 2t^2$
44. $x^{12} - 1$
45. $16 - 9x^2y^2$
46. $6 + 5x - 4x^2$
47. $8n^2 - 18n - 5$
48. $15 + 4x - 4x^2$
49. $2x^2 - 2x - 12$
50. $3y^2 + 6y - 45$
51. $30x^2 - 35x + 10$
52. $x^4y^4 - x^2y^2$
53. $18x^2m + 33xm + 9m$
54. $25m^2n^3 - 5m^2n$
55. $12x^2 \quad 22x^3 \quad 20x^4$
56. $10r^2 - 5rs - 15s^2$
57. $x^4 - y^4$
58. $a^4 - 16$
59. $b^4 + 2b^2 - 8$
60. $4b^4 + 20b^2 + 25$
61. $x^3 + 27y^3$
62. $8x^3 + 125y^3$
63. $27x^3 - y^3$
64. $64x^3 - 27y^3$
65. $a^3 + 8$
66. $8r^3 - 27$
67. $\frac{1}{8}m^3 - 8n^3$
68. $8a^3 - \frac{1}{64}b^3$
69. $(x + y)^3 - 8$
70. $27 + (x + y)^3$
71. $8x^6 - 125y^6$
72. $a^6 + 27b^6$
73. $4(x + 1)(y + 2) - 8(y + 2)$
74. $2(x + 1)(x - 1) + 5(x - 1)$
75. $3(x + 2)^2(x - 1) - 4(x + 2)^2(2x + 7)$
76. $4(2x - 1)^2(x + 2)^3(x + 1) - 3(2x - 1)^5(x + 2)^2(x + 3)$
77. Show that the difference of the squares of two positive, consecutive odd integers must be divisible by 8.
78. A perfect square is a natural number of the form n^2. For example, 9 is a perfect square since $9 = 3^2$. Show that the sum of the squares of two odd numbers cannot be a perfect square.
79. Find a natural number n, if possible, such that $1 + n(n + 1)(n + 2)(n + 3)$ is a perfect square.
80. Determine whether the expression is a perfect square:
$$1 + n(n + 1)(n + 2)(n + 3)$$
(*Hint:* Consider $[1 + n(n + 3)^2]$.)
81. Factor completely.
    a. $(x + h)^3 - x^3$
    b. $2^n + 2^{n+1} + 2^{n+2}$
    c. $16 - 81x^{12}$
    d. $z^2 - x^2 + 2xy - y^2$
82. Factor completely.
    a. $\left[\frac{n(n+1)}{2}\right]^2 + (n + 1)^3$
    b. $\frac{n(n+1)(2n+1)}{6} + (n + 1)^2$
    c. $\frac{1}{b}(a+bx)^2 - \frac{a}{b}(a+bx)$
83. Factor the following expressions that arise in different branches of science.
    a. biology (blood flow): $C[(R + 1)^2 - r^2]$
    b. physics (nuclear): $pa^2 + (1 - p)b^2 - [pa + (1 - p)b]^2$
    c. mechanics (bending beams): $X^2 - 3LX + 2L^2$
    d. electricity (resistance): $(R_1 + R_2)^2 - 2r(R_1 + R_2)$
    e. physics (motion): $-16t^2 + 64t + 336$
84. Factor this expression, used to find the answers given in the chapter opening.
$$20t - 0.7t^2$$
85. Factor the general expression $vt - \frac{1}{2}at^2$.

86. Suppose we alter the expression from Exercise 85 by adding a constant:

$$s + vt - \frac{1}{2}at^2$$

Experiment with different values of s, v, and t. Which ones give you an expression that is easy to factor? (Reread Exercise 87 in Section 1.3. The s we have added could represent the original position of the object.)

87. *Mathematics in Writing:* Write a short paragraph explaining the differences in the techniques you used to factor the scientific expressions in Exercise 83 parts a, b, and c. Find at least one other problem in this problem set that uses a technique similar to each of the three you have described.

1.5 Rational Expressions

Much of the terminology and many of the techniques for the arithmetic of fractions carry over to **algebraic fractions**, which are the quotients of algebraic expressions. In particular, we refer to a quotient of two polynomials as a **rational expression**.

1.5a *Notation*

> Any symbol used as a divisor in this text is always assumed to be different from zero.

Therefore, we will not always identify a divisor as being different from zero unless it disappears through some type of mathematical manipulation.

Our objective in this section is to review the procedures for adding, subtracting, multiplying, and dividing rational expressions. We are then able to convert a complicated fraction, such as

$$\frac{1-\frac{1}{x}}{\frac{1}{x^2}+\frac{1}{x}}$$

into a form that simplifies evaluation of the fraction and facilitates other operations with it.

1.5b *Multiplication and Division of Rational Expressions*

The symbols appearing in rational expressions represent real numbers. We may, therefore, apply the rules of arithmetic to rational expressions. Let a, b, c, and d represent any algebraic expressions.

$\frac{a}{b}\cdot\frac{c}{d}=\frac{ac}{bd}$	Multiplication of rational expressions
$\frac{\frac{a}{b}}{\frac{c}{d}}=\frac{\frac{a}{b}}{\frac{c}{d}}\cdot\frac{(bd)}{(bd)}=\frac{ad}{cb}$	Division of rational expressions

EXAMPLE 1 ***Multiplication and Division of Rational Expressions***

Divide $\frac{2x}{y}$ by $\frac{3y^3}{x-3}$.

SOLUTION

$$\frac{\frac{2x}{y}}{\frac{3y^3}{x-3}}=\frac{\frac{2x}{y}}{\frac{3y^3}{x-3}}\cdot\frac{(y)(x-3)}{(y)(x-3)}=\frac{2x(x-3)}{3y^4},\ x\neq 3$$ ■

The basic rule that allows us to simplify rational expressions is the cancellation principle.

Cancellation Principle

$$\frac{ab}{ac} = \frac{b}{c}, \quad a \neq 0$$

This rule results from the fact that $\frac{a}{a} = 1$. Thus,

$$\frac{ab}{ac} = \frac{a}{a} \cdot \frac{b}{c} = 1 \cdot \frac{b}{c} = \frac{b}{c}$$

Once again we find that a rule for the arithmetic of fractions carries over to rational expressions.

EXAMPLE 2 *Factoring and Cancellation*

Simplify.

a. $\dfrac{x^2 - 4}{x^2 + 5x + 6}$ b. $\dfrac{3x^2(y-1)}{y+1} \div \dfrac{6x(y-1)^2}{(y+1)^3}$ c. $\dfrac{x^2 - x - 6}{3x - x^2}$

SOLUTION

a. $\dfrac{x^2 - 4}{x^2 + 5x + 6} = \dfrac{(x+2)(x-2)}{(x+3)(x+2)} = \dfrac{x-2}{x+3}, \quad x \neq -2$

b. $\dfrac{\dfrac{3x^2(y-1)}{y+1}}{\dfrac{6x(y-1)^2}{(y+1)^3}} = \dfrac{\dfrac{3x^2(y-1)}{y+1}}{\dfrac{6x(y-1)^2}{(y+1)^3}} \cdot \dfrac{(y+1)^3}{(y+1)^3} = \dfrac{3x^2(y-1)(y+1)^2}{6x(y-1)^2} = \dfrac{x(y+1)^2}{2(y-1)}, x \neq 0, y \neq -1$

c. $\dfrac{x^2 - x - 6}{3x - x^2} = \dfrac{(x-3)(x+2)}{x(3-x)} = \dfrac{(x-3)(x+2)}{-x(x-3)} = \dfrac{x+2}{-x}, \; x \neq 3 = -\dfrac{x+2}{x}, \; x \neq 3$ ■

Note that in Example 2c, we wrote $(3 - x)$ as $-(x - 3)$. This technique is often used to recognize factors that may be canceled.

Progress Check

Simplify.

a. $\dfrac{4 - x^2}{x^2 - x - 6}$ b. $\dfrac{8 - 2x}{y} \div \dfrac{x^2 - 16}{y}$

Answers

a. $\dfrac{2 - x}{x - 3}, \quad x \neq -2$ b. $-\dfrac{2}{x+4}, \quad x \neq 4, y \neq 0$

a. Only multiplicative factors can be canceled. Thus,

$$\frac{2x-4}{x} \neq 2-4$$

Since x is *not a factor* in the numerator, we may *not* perform cancellation.

b. Note that

$$\frac{y^2-x^2}{y-x} \neq y-x$$

To simplify correctly, write

$$\frac{y^2-x^2}{y-x} = \frac{(y+x)(y-x)}{y-x} = y+x, \quad y \neq x$$

1.5c Addition and Subtraction of Rational Expressions

Since the variables in rational expressions represent real numbers, the rules of arithmetic for addition and subtraction of fractions apply to rational expressions. When rational expressions have the same denominator, the addition and subtraction rules are as follows:

$$\frac{a}{c} + \frac{b}{c} = \frac{a+b}{c}$$

$$\frac{a}{c} - \frac{b}{c} = \frac{a-b}{c}$$

For example,

$$\frac{2}{x-1} - \frac{4}{x-1} + \frac{5}{x-1} = \frac{2-4+5}{x-1} = \frac{3}{x-1}$$

To add or subtract rational expressions with *different* denominators, we must first rewrite each rational expression as an equivalent one with the same denominator as the others. Although any common denominator will do, we will concentrate on finding the **least common denominator**, or **LCD**, of two or more rational expressions. We now outline the procedure and provide some examples.

EXAMPLE 3 *Least Common Denominator for Rational Numbers*

Find the LCD of the following three fractions:

$$\frac{1}{8} \qquad \frac{7}{90} \qquad \frac{3}{25}$$

SOLUTION

Method	Example			
Step 1. Factor the denominator of each fraction.	*Step 1.* $\frac{1}{2^3}$	$\frac{7}{2\cdot3^2\cdot5}$		$\frac{3}{5^2}$
Step 2. Find the different factors in the denominator and the highest power to which each factor occurs.	*Step 2.* Factor	Highest exponent		Final factors
	2	3		8
	3	2		9
	5	2		25
Step 3. The product of the final factors in *Step 2* is the LCD.	*Step 3.* The LCD is $2^3 \cdot 3^2 \cdot 5^2 = 8 \cdot 9 \cdot 25$.			

■

EXAMPLE 4 *Least Common Denominator for Rational Expressions*

Find the LCD of the following three rational expressions:

$$\frac{1}{x^3-x^2} \qquad \frac{-2}{x^3-x} \qquad \frac{3x}{x^2+2x+1}$$

SOLUTION

Method	Example			
Step 1. Factor the denominator of each fraction.	*Step 1.* $\frac{1}{x^2(x-1)}$	$\frac{-2}{x(x-1)(x+1)}$		$\frac{3x}{(x+1)^2}$
Step 2. Find the different factors in the denominator and the highest power to which each factor occurs.	*Step 2.* Factor	Highest exponent		Final factors
	x	2		x^2
	$x-1$	1		$x-1$
	$x+1$	2		$(x+1)^2$
Step 3. The product of the final factors in *Step 2* is the LCD.	*Step 3.* The LCD is $x^2(x-1)(x+1)^2$			

■

Find the LCD of the following fractions:

$$\frac{2a}{(3a^2+12a+12)b} \qquad \frac{-7b}{a(4b^2-8b+4)} \qquad \frac{3}{ab^3+2b^3}$$

Answer

$12ab^3(a+2)^2(b-1)^2$

Equivalent Fractions

The fractions $\frac{2}{5}$ and $\frac{6}{15}$ are said to be **equivalent** because we obtain $\frac{6}{15}$ by multiplying $\frac{2}{5}$ by $\frac{3}{3}$, which is the same as multiplying $\frac{2}{5}$ by 1. We also say that algebraic fractions are **equivalent fractions** if we can obtain one from the other by multiplying both the numerator and denominator by the same expression.

To add rational expressions, we must first determine the LCD and then convert each rational expression into an equivalent fraction with the LCD as its denominator. We can accomplish this conversion by multiplying the fraction by the appropriate equivalent of 1. We now outline the procedure and provide an example.

EXAMPLE 5 *Addition and Subtraction of Rational Expressions*

Simplify.

$$\frac{x+1}{2x^2} - \frac{2}{3x(x+2)}$$

SOLUTION

Method	Example
Step 1. Find the LCD.	*Step 1.* LCD $= 6x^2(x+2)$
Step 2. Multiply each rational expression by a fraction whose numerator and denominator are the same, and consist of all factors of the LCD that are missing in the denominator of the expression.	*Step 2.* $\frac{x+1}{2x^2} \cdot \frac{3(x+2)}{3(x+2)} = \frac{3x^2+9x+6}{6x^2(x+2)}$ $\frac{2}{3x(x+2)} \cdot \frac{(2x)}{(2x)} = \frac{4x}{6x^2(x+2)}$
Step 3. Add the rational expressions. Do not multiply out the denominators since it may be possible to cancel.	*Step 3.* $\frac{x+1}{2x^2} - \frac{2}{3x(x+2)} = \frac{3x^2+9x+6}{6x^2(x+2)} - \frac{4x}{6x^2(x+2)} = \frac{3x^2+5x+6}{6x^2(x+2)}$

■

Progress Check

Perform the indicated operations.

a. $\frac{x-8}{x^2-4} + \frac{3}{x^2-2x}$ b. $\frac{4r-3}{9r^3} - \frac{2r+1}{4r^2} + \frac{2}{3r}$

Answers

a. $\frac{x-3}{x(x+2)}$, $x \neq 2$ b. $\frac{6r^2+7r-12}{36r^3}$

1.5d Complex Fractions

At the beginning of this section, we said that we wanted to be able to simplify fractions such as

$$\frac{1-\frac{1}{x}}{\frac{1}{x^2}+\frac{1}{x}}$$

This is an example of a **complex fraction**, which is a fractional form with fractions in the numerator or denominator or both.

EXAMPLE 6 ***Division of Rational Expressions***

Simplify the following:

$$\frac{1-\frac{1}{x}}{\frac{1}{x^2}+\frac{1}{x}}$$

SOLUTION

Method	Example
Step 1. Find the LCD of all fractions in the numerator and denominator.	*Step 1.* The LCD of $\frac{1}{x}$ and $\frac{1}{x^2}$ is x^2.
Step 2. Multiply the numerator and denominator by the LCD. Since this is multiplication by 1, the result is an equivalent fraction.	*Step 2.* $\dfrac{\left(1-\frac{1}{x}\right)}{\left(\frac{1}{x^2}+\frac{1}{x}\right)}\cdot\dfrac{(x^2)}{(x^2)}=\dfrac{x^2-x}{1+x},\quad x\neq 0$ $=\dfrac{x(x-1)}{1+x},\quad x\neq 0$

■

Progress Check

Simplify.

a. $\dfrac{2+\frac{1}{x}}{1-\frac{2}{x}}$ b. $\dfrac{\frac{a}{b}+\frac{b}{a}}{\frac{1}{a}-\frac{1}{b}}$

Answers

a. $\dfrac{2x+1}{x-2},\quad x\neq 0$ b. $\dfrac{a^2+b^2}{b-a},\quad a\neq 0,\quad b\neq 0$

Exercise Set 1.5

Perform all possible simplifications in Exercises 1–20.

1. $\frac{x+4}{x^2-16}$

2. $\frac{y^2-25}{y+5}$

3. $\frac{x^2-8x+16}{x-4}$

4. $\frac{5x^2-45}{2x-6}$

5. $\frac{6x^2-x-1}{2x^2+3x-2}$

6. $\frac{2x^3+x^2-3x}{3x^2-5x+2}$

7. $\frac{2}{3x-6} \div \frac{3}{2x-4}$

8. $\frac{5x+15}{8} \div \frac{3x+9}{4}$

9. $\frac{25-a^2}{b+3} \cdot \frac{2b^2+6b}{a-5}$

10. $\frac{2xy^2}{x+y} \cdot \frac{x+y}{4xy}$

11. $\frac{x+2}{3y} \div \frac{x^2-2x-8}{15y^2}$

12. $\frac{3x}{x+2} \div \frac{6x^2}{x^2-x-6}$

13. $\frac{6x^2-x-2}{2x^2-5x+3} \cdot \frac{2x^2-7x+6}{3x^2+x-2}$

14. $\frac{6x^2+11x-2}{4x^2-3x-1} \cdot \frac{5x^2-3x-2}{3x^2+7x+2}$

15. $(x^2-4) \cdot \frac{2x+3}{x^2+2x-8}$

16. $(a^2-2a) \cdot \frac{a+1}{6-a-a^2}$

17. $(x^2-2x-15) \div \frac{x^2-7x+10}{x^2+1}$

18. $\frac{2y^2-5y-3}{y-4} \div (y^2+y-12)$

19. $\frac{x^2-4}{x^2+2x-3} \cdot \frac{x^2+3x-4}{x^2-7x+10} \cdot \frac{x+3}{x^2+3x+2}$

20. $\frac{x^2-9}{6x^2+x-1} \cdot \frac{2x^2+5x+2}{x^2+4x+3} \cdot \frac{x^2-x-2}{x^2-3x}$

In Exercises 21–30, find the LCD.

21. $\frac{4}{x}, \frac{x-2}{y}$

22. $\frac{x}{x-1}, \frac{x+4}{x+2}$

23. $\frac{5-a}{a}, \frac{7}{2a}$

24. $\frac{x+2}{x}, \frac{x-2}{x^2}$

25. $\frac{2b}{b-1}, \frac{3}{(b-1)^2}$

26. $\frac{2+x}{x^2-4}, \frac{3}{x-2}$

27. $\frac{4x}{x-2}, \frac{5}{x^2+x-6}$

28. $\frac{3}{y^2-3y-4}, \frac{2y}{y+1}$

29. $\frac{3}{x+1}, \frac{2}{x}, \frac{x}{x-1}$

30. $\frac{4}{x}, \frac{3}{x-1}, \frac{x}{x^2-2x+1}$

In Exercises 31–50, perform the indicated operations and simplify.

31. $\frac{8}{a-2} + \frac{4}{2-a}$

32. $\frac{x}{x^2-4} + \frac{2}{4-x^2}$

33. $\frac{x-1}{3} + 2$

34. $\frac{1}{x-1} + \frac{2}{x-2}$

35. $\frac{1}{a+2} + \frac{3}{a-2}$

36. $\frac{a}{8b} - \frac{b}{12a}$

37. $\frac{4}{3x} - \frac{5}{xy}$

38. $\frac{4x-1}{6x^3} + \frac{2}{3x^2}$

39. $\frac{5}{2x+6} - \frac{x}{x+3}$

40. $\frac{x}{x-y} - \frac{y}{x+y}$

41. $\frac{5x}{2x^2-18} + \frac{4}{3x-9}$

42. $\frac{4}{r} - \frac{3}{r+2}$

43. $\frac{1}{x-1} + \frac{2x-1}{(x-2)(x+1)}$

44. $\frac{2x}{2x+1} - \frac{x-1}{(2x+1)(x-2)}$

45. $\frac{2x}{x^2+x-2} + \frac{3}{x+2}$

46. $\frac{2}{x-2} + \frac{x}{x^2-x-6}$

47. $\frac{2x-1}{x^2+5x+6} - \frac{x-2}{x^2+4x+3}$

48. $\frac{2x-1}{x^3-4x} - \frac{x}{x^2+x-2}$

49. $\frac{2x}{x^2-1} + \frac{x+1}{x^2+3x-4}$

50. $\frac{2x}{x+2} + \frac{x}{x-2} - \frac{1}{x^2-4}$

In Exercises 51–66, simplify the complex fraction and perform all indicated operations.

51. $\frac{1+\frac{2}{x}}{1-\frac{3}{x}}$

52. $\frac{x-\frac{1}{x}}{2+\frac{1}{x}}$

53. $\frac{x+1}{1-\frac{1}{x}}$

54. $\frac{1-\frac{r^2}{s^2}}{1+\frac{r}{s}}$

55. $\frac{x^2-16}{\frac{1}{4}-\frac{1}{x}}$

56. $\frac{\frac{a}{a-b}-\frac{b}{a+b}}{a^2-b^2}$

57. $2-\frac{1}{1+\frac{1}{a}}$

58. $\frac{\frac{4}{x^2-4}+1}{\frac{x}{x^2+x-6}}$

59. $\dfrac{\frac{a}{b}-\frac{b}{a}}{\frac{1}{a}+\frac{1}{b}}$

60. $\dfrac{\frac{x}{x-2}-\frac{x}{x+2}}{\frac{2x}{x-2}+\frac{x^2}{x-2}}$

61. $3-\dfrac{2}{1-\frac{1}{1+x}}$

62. $2+\dfrac{3}{1+\frac{2}{1-x}}$

63. $\dfrac{y-\frac{1}{1-\frac{1}{y}}}{y+\frac{1}{1+\frac{1}{y}}}$

64. $1-\dfrac{1-\frac{1}{y}}{y-\frac{1}{y}}$

65. $1-\dfrac{1}{1+\frac{1}{1-\frac{1}{1+x}}}$

66. $1+\dfrac{1}{1-\frac{1}{1+\frac{1}{1+x}}}$

67. Combine and simplify.

a. $\dfrac{1}{R_1}+\dfrac{1}{R_2}+\dfrac{1}{R_3}+\dfrac{1}{R_4}$

b. $\dfrac{3}{c-d}+\dfrac{4d}{(c-d)^2}-\dfrac{5d^2}{(c-d)^3}$

c. $\dfrac{6}{k+3}+k-2$

68. Simplify the complex fractions.

a. $\dfrac{\frac{1}{(x+h)^2}-\frac{1}{x^2}}{h}$

b. $\dfrac{\frac{1}{x^2}-\frac{1}{9}}{x-3}$

69. Find the errors in the following and correct the statements.

a. $\dfrac{\frac{1}{b}}{\frac{1}{a}+\frac{1}{b}}=\dfrac{1}{\frac{1}{a}}$

b. $\dfrac{\frac{1}{b}}{\frac{1}{a}+\frac{1}{b}}=\left(\frac{1}{b}\right)\left(\frac{a}{1}+\frac{b}{1}\right)$

c. $\dfrac{a^2(3b+4a)}{a^2+b^2}=\dfrac{3b+4a}{b^2}$

d. $\dfrac{1-x}{1+x}=-1$

e. $(x^2-y^2)^2=x^4-y^4$

f. $\dfrac{a+b}{b}=a$

1.6 Integer Exponents

1.6a *Positive Integer Exponents*

In Section 1.3, we defined a^n for a real number a and a positive integer n as

$$a^n = \underbrace{a \cdot a \cdot \cdots \cdot a}_{n \text{ factors}}$$

and we showed that if m and n are positive integers then $a^m a^n = a^{m+n}$. The method we used to establish this rule was to write out the factors of a^m and a^n and count the total number of occurrences of a. The same method can be used to establish the rest of the properties in Table 1-9 when m and n are positive integers.

TABLE 1-9 Properties of Positive Integer Exponents, $m > 0$ and $n > 0$

Example	Property
$2^2 \cdot 2^3 = 2^5$ $4 \cdot 8 = 32$	$a^m a^n = a^{m+n}$
$(2^3)^2 = 2^{3 \cdot 2} = 2^6$ $8^2 = 64$	$(a^m)^n = a^{mn}$
$(2 \cdot 3)^2 = 2^2 \cdot 3^2$ $6^2 = 4 \cdot 9$	$(ab)^m = a^m b^m$
$\left(\frac{6}{2}\right)^2 = \frac{6^2}{2^2}$ $3^2 = \frac{36}{4}$	$\left(\frac{a}{b}\right)^m = \frac{a^m}{b^m}$
$\frac{2^5}{2^2} = 2^3$ $\frac{32}{4} = 8$	$\frac{a^m}{a^n} = a^{m-n}$ if $m > n$, $a \neq 0$
$\frac{2^2}{2^5} = \frac{1}{2^3}$ $\frac{4}{32} = \frac{1}{8}$	$\frac{a^n}{a^m} = \frac{1}{a^{m-n}}$ if $m > n$, $a \neq 0$
$\frac{5^2}{5^2} = \frac{25}{25} = 1$	$\frac{a^m}{a^m} = 1$ if $a \neq 0$

EXAMPLE 1 ***Multiplication with Positive Integer Exponents***

Simplify the following.

a. $(4a^2b^3)(2a^3b)$ b. $(2x^2y)^4$

SOLUTION

a. $(4a^2b^3)(2a^3b) = 4 \cdot 2 \cdot a^2a^3b^3b = 8a^5b^4$ b. $(2x^2y)^4 = 2^4(x^2)^4y^4 = 16x^8y^4$ ■

Progress Check

Simplify, using only positive exponents.

a. $(x^3)^4$ b. $x^4(x^2)^3$ c. $\frac{a^{14}}{a^8}$ d. $\frac{-2(x+1)^n}{(x+1)^{2n}}$

e. $(3ab^2)^3$ f. $\left(\frac{-ab^2}{c^3}\right)^3$

Answers

a. x^{12} b. x^{10} c. a^6 d. $-\frac{2}{(x+1)^n}$

e. $27a^3b^6$ f. $-\frac{a^3b^6}{c^9}$

1.6b *Zero and Negative Exponents*

We next expand our rules to include zero and negative exponents when the base is nonzero. We wish to define a^0 to be consistent with the previous rules for exponents. For example, applying the rule $a^m a^n = a^{m+n}$ yields

$$a^m a^0 = a^{m+0} = a^m$$

Dividing both sides by a^m, we obtain $a^0 = 1$. We therefore *define* a^0 for any nonzero real number by

$$a^0 = 1, \ a \neq 0$$

The same approach leads us to a definition of negative exponents. For consistency, if $m > 0$, we must have

$$a^m a^{-m} = a^{m-m} = a^0 = 1 \qquad \text{or} \qquad a^m a^{-m} = 1 \tag{1}$$

Division of both sides of Equation (1) by a^m suggests that we define a^{-m} as

$$a^{-m} = \frac{1}{a^m}, \ a \neq 0$$

Dividing Equation (1) by a^{-m}, we have

$$a^m = \frac{1}{a^{-m}}, \ a \neq 0$$

Thus, a^m and a^{-m} are reciprocals of one another. The rule for handling negative exponents can be expressed as follows:

A nonzero factor moves from numerator to denominator (or from denominator to numerator) by changing the sign of the exponent.

It is important to note that

0^0 is not defined. Furthermore, 0^{-m} is also not defined for $m > 0$.

We may also conclude that

$$a^0 \neq 0 \quad \text{and} \quad a^{-m} \neq 0, \quad m > 0$$

We summarize these results in Table 1-10.

TABLE 1-10 Properties of Integer Exponents, $a \neq 0$

Example	Property
$\left(-\frac{1}{2}\right)^0 = 1$	$a^0 = 1$
$8 = 2^3 = \frac{1}{2^{-3}}$	$a^m = \frac{1}{a^{-m}}$
$2^{-3} = \frac{1}{2^3} = \frac{1}{8}$	$a^{-m} = \frac{1}{a^m}$

EXAMPLE 2 *Operations with Integer Exponents*

Simplify the following, using only positive exponents:

a. $\dfrac{2}{(x-1)^0}$ b. $(x^2y^{-3})^{-5}$

c. $\dfrac{y^{3k+1}}{2y^{2k}}, \quad k > 0$

SOLUTION

a. $\dfrac{2}{(x-1)^0} = \dfrac{2}{1} = 2$ b. $(x^2y^{-3})^{-5} = (x^2)^{-5}(y^{-3})^{-5} = x^{-10}y^{15} = \dfrac{y^{15}}{x^{10}}$

c. $\dfrac{y^{3k+1}}{2y^{2k}} = \dfrac{y^{3k+1}y^{-2k}}{2} = \dfrac{y^{3k+1-2k}}{2} = \dfrac{1}{2}y^{k+1}$ ■

Progress Check

Simplify, using only positive exponents.

a. $x^{-2}y^{-3}$ b. $\dfrac{-3x^4y^{-2}}{9x^{-8}y^6}$ c. $\left(\dfrac{x^{-3}}{x^{-4}}\right)^{-1}$

Answers

a. $\dfrac{1}{x^2y^3}$ b. $-\dfrac{x^{12}}{3y^8}$ c. $\dfrac{1}{x}$

Do not confuse negative numbers and negative exponents.

a. $2^{-4} = \frac{1}{2^4}$

Note that $2^{-4} \neq -2^4$.

b. $(-2)^{-3} = \frac{1}{(-2)^3} = \frac{1}{-8} = -\frac{1}{8}$

Note that $(-2)^{-3} \neq \frac{1}{2^3} = \frac{1}{8}$.

1.6c Scientific Notation

One of the significant applications of integer exponents is that of **scientific notation.** This technique enables us to recognize the size of extremely large and extremely small numbers rather quickly and in a more concise form.

Consider the following examples for powers of 10:

$$\text{One thousand} = 1000 = 1 \times 10^3 = 10^3$$

$$\text{One thousandth} = 0.001 = \frac{1}{1000} = \frac{1}{10^3} = 10^{-3}$$

Reversing the procedure we obtain:

1. $10^2 = 1 \times 10^2 = 100.0$, namely, 1 with the decimal point *two* places to the *right* of it.
2. $10^{-2} = 10 \times 10^{-2} = \frac{1}{100} = 0.01$, namely, 1 with the decimal point *two* places to the *left* of it.

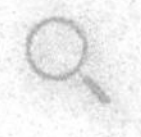

Scientific Notation

A number is written in scientific notation if it is of the format $\pm a \times 10^m$, where $1 \leq a < 10$ and m is some integer. If $a = 1$, it is generally omitted.

EXAMPLE 3 *Writing in Scientific Notation*

An angstrom (Å) equals 1 ten-billionth of a meter. Write this in scientific notation.

SOLUTION

1 Å = 0.0000000001 meters = 10^{-10} meters ■

EXAMPLE 4 *Writing in Scientific Notation*

One light-year is approximately 6 trillion miles. Write this in scientific notation.

SOLUTION

1 light-year ≈ 6,000,000,000,000 miles = 6×10^{12} miles ■

If $\pm a \times 10^m$ is the result of some calculations or measurements involving scientific notation, then the number of digits present in a are generally taken as the significant digits of the answer. For example, 6×10^{12} has one significant digit.

If we write a number in scientific notation with fewer significant digits than the original number presented, we must round the last significant digit used according to the following rule:

- Add 1 to the last significant digit if the digit following it in the original number is 5, 6, 7, 8, or 9.
- Leave the last significant digit alone if the digit following it in the original number is 0, 1, 2, 3, or 4.

EXAMPLE 5 *Writing in Scientific Notation*

The speed of light is approximately 186,282 miles per second. Write it in scientific notation with four significant digits.

SOLUTION

The speed of light $\approx 1.863 \times 10^5$ miles per second since 186,282 is rounded to 186,300. ■

EXAMPLE 6 *Writing in Scientific Notation*

There are 31,557,600 seconds in an average year (365.25 days). Write it in scientific notation with four significant digits.

SOLUTION

1 year $= 3.156 \times 10^7$ seconds. ■

EXAMPLE 7 *Calculations in Scientific Notation*

Find the number of miles in 1 light-year to four significant digits.

SOLUTION

One light-year is the number of miles light travels in 1 year.

$$\begin{aligned} \text{1 light-year} &= (1.863 \times 10^5 \text{ miles per second})(3.156 \times 10^7 \text{ seconds}) \\ &= 5.880 \times 10^{12} \text{ miles.} \end{aligned}$$ ■

Note that this becomes 6×10^{12} miles if we require only one significant digit. The number of significant digits of our answer equals the minimum number of significant digits involved in our calculations.

Graphing Calculator ALERT

To input scientific notation on your graphing calculator, you will use the [EE] (which is [2nd] [,]). For example, to input 5×10^4, key in [5] [EE] [4]. You will notice that the screen will show 5E4. When you hit [ENTER], you have 50000.

Sometimes computations are so large that the calculator reverts to scientific notation. For example, the computation in Example 7 will be done as:

1.863 [EE] 5 [×] 3.156 [EE] 7 [ENTER]

The output is 5.879628E12, which means our product is 5.879628×10^{12}.

Progress Check

1 ounce = 0.02834952 kilogram
Write this number in scientific notation using:

a. two significant digits

b. three significant digits

c. five significant digits

Answers

a. 2.8×10^{-2} b. 2.83×10^{-2} c. 2.8350×10^{-2}

Exercise Set 1.6

In Exercises 1–6, the right-hand side is incorrect. Find the correct term.

1. $x^2 \cdot x^4 = x^8$
2. $(y^2)^5 = y^7$
3. $\frac{b^6}{b^2} = b^3$
4. $\frac{x^2}{x^6} = x^4$
5. $(2x)^4 = 2x^4$
6. $\left(\frac{4}{3}\right)^4 = \frac{4}{3^4}$

In Exercises 7–64, use the rules for exponents to simplify. Write the answers using only positive exponents.

7. $\left(-\frac{1}{2}\right)^4\left(-\frac{1}{2}\right)^3$
8. $(x^m)^{3m}$
9. $(y^4)^{2n}$
10. $\frac{(-4)^6}{(-4)^{10}}$
11. $-\left(\frac{x}{y}\right)^3$
12. $-3r^2r^3$
13. $(x^3)^5 \cdot x^4$
14. $\frac{x^{12}}{x^8}$
15. $(-2x^2)^5$
16. $-(2x^2)^5$
17. $x^{3n} \cdot x^n$
18. $(-2)^m(-2)^n$
19. $\frac{x^n}{x^{n+2}}$
20. $\left(\frac{3x^3}{y^2}\right)^5$
21. $(-5x^3)(-6x^5)$
22. $(x^2)^3(y^2)^4(x^3)^7$
23. $\frac{(r^2)^4}{(r^4)^2}$
24. $[(3b+1)^5]^5$
25. $\left(\frac{3}{2}x^2y^3\right)^n$
26. $\frac{(-2a^2b)^4}{(-3ab^2)^3}$
27. $(2x+1)^3(2x+1)^7$
28. $\frac{y^3(y^3)^4}{(y^4)^6}$
29. $(-2a^2b^3)^{2n}$
30. $\left(-\frac{2}{3}a^2b^3c^2\right)^3$
31. $2^0 + 3^{-1}$
32. $(xy)^0 - 2^{-1}$
33. $\frac{3}{(2x^2+1)^0}$
34. $(-3)^{-3}$
35. $\frac{1}{3^{-4}}$
36. x^{-5}
37. $(-x)^3$
38. $-x^{-5}$
39. $\frac{1}{y^{-6}}$
40. $(2a)^{-6}$
41. $5^{-3}5^5$
42. $4y^5y^{-2}$
43. $(3^2)^{-3}$
44. $(x^{-2})^4$
45. $(x^{-3})^{-3}$
46. $[(x+y)^{-2}]^2$
47. $\frac{2^2}{2^{-3}}$
48. $\frac{x^8}{x^{-10}}$
49. $\frac{2x^4y^{-2}}{x^2y^{-3}}$
50. $(x^4y^{-2})^{-1}$
51. $(3a^{-2}b^{-3})^{-2}$
52. $\frac{1}{(2xy)^{-2}}$
53. $\left(-\frac{1}{2}x^3y^{-4}\right)^{-3}$
54. $\frac{(x^{-2})^2}{(3y^{-2})^3}$
55. $\frac{3a^5b^{-2}}{9a^{-4}b^2}$
56. $\left(\frac{x^3}{x^{-2}}\right)^2$
57. $\left(\frac{2a^2b^{-4}}{a^{-3}c^{-3}}\right)^2$
58. $\frac{2x^{-3}y^2}{x^{-3}y^{-3}}$
59. $(a-2b^2)^{-1}$
60. $\left(\frac{y^{-2}}{y^{-3}}\right)^{-1}$
61. $\frac{(a+b)^{-1}}{(a-b)^{-2}}$
62. $(a^{-1}+b^{-1})^{-1}$
63. $\frac{a^{-1}+b^{-1}}{a^{-1}-b^{-1}}$
64. $\left(\frac{a}{b}\right)^{-1}+\left(\frac{b}{a}\right)^{-1}$
65. Show that $\left(\frac{a}{b}\right)^{-n} = \left(\frac{b}{a}\right)^n$

Evaluate each expression in Exercises 66–69.

66. $(1.20^2)^{-1}$
67. $[(-3.67)^2]^{-1}$
68. $\left(\frac{7.65^{-1}}{7.65^2}\right)^2$
69. $\left(\frac{4.46^2}{4.46^{-1}}\right)^{-1}$

In Exercises 70–75, write each number using scientific notation.

70. 7000
71. 0.0091
72. 452,000,000,000
73. 23
74. 0.00000357
75. 0.8×10^{-3}

In Exercises 76–81, write each number without exponents.

76. 4.53×10^5
77. 8.93×10^{-4}
78. 0.0017×10^7
79. 145×10^3
80. 100×10^{-3}
81. 1253×10^{-6}
82. The dimensions of a rectangular field, measured in meters, are 4.1×10^3 by 3.75×10^5. Find the area of this field expressed in scientific notation.
83. The volume V of a spherical bubble of radius r is given by the formula
$$V = \frac{4}{3}\pi r^3$$
If we take the value of π to be 3.14, find the volume of a bubble, using scientific notation, if its radius is 0.09 inch.
84. Find the distance, expressed in scientific notation, that light travels in 0.000020 seconds if the speed of light is 1.86×10^5 miles per second.

85. The Republic of Singapore is said to have the highest population density of any country in the world. If its area is 270 square miles and its estimated population is 4,500,000, find the population density, that is, the approximate number of people per square mile, using scientific notation.

86. Scientists have suggested that the relationship between an animal's weight W and its surface area S is given by the formula

$$S = KW^{\frac{2}{3}}$$

where K is a constant chosen so that W, measured in kilograms, yields a value for S, measured in square meters. If the value of K for a horse is 0.10 and the horse weighs 350 kilograms, find the estimated surface area of the horse, using scientific notation.

87. Simplify the following:

a. $\dfrac{2^{n+3}+2^n+2^n}{4(2^{n+3}-2^{n+1})}$

b. $\dfrac{a(1-r^3)}{1-r}$

c. $\dfrac{9^{6m}}{3^{2m}}$

d. $\dfrac{8^4+8^4+8^4+8^4}{4^4}$

e. $\dfrac{(6\times 10^{-2})(2\times 10^{-3})}{3\times 10^8}$

88. Assuming a lifetime is 70 years, how much is that in seconds? Express your answer in scientific notation.

1.7 Rational Exponents and Radicals

1.7a *n*th Roots

Consider a square whose area is 25 cm^2 (square centimeters), and whose sides are of length a. We can then write

$$a^2 = 25$$

so that a is a number whose square is 25. We say that a is the *square root* of b if $a^2 = b$. Similarly, we say that a is a *cube root* of b if $a^3 = b$; and, in general, if n is a natural number, we say that

> a is an ***n*th root** of b if $a^n = b$

Thus, 5 is a square root of 25 since $5^2 = 25$, and -2 is a cube root of -8 since $(-2)^3 = -8$.

Since $(-5)^2 = 25$, we conclude that -5 is also a square root of 25. More generally, if $b > 0$ and a is a square root of b, then $-a$ is also a square root of b. If $b < 0$, there is no real number a such that $a^2 = b$, since the square of a real number is always nonnegative. (In Section 1.8, we introduce an extended number system in which there is a root when $b < 0$ and n is even.)

We would like to define rational exponents in a manner that is consistent with the rules for integer exponents. If the rule $(a^m)^n = a^{mn}$ is to hold, then we must have

$$(b^{1/n})^n = b^{n/n} = b$$

But a is an nth root of b if $a^n = b$. Then for every natural number n, we say that

> $b^{1/n}$ is an nth root of b

Principal *n*th Root

If n is even and b is positive, there are two numbers that are nth roots of b. For example,

$$4^2 = 16 \qquad \text{and} \qquad (-4)^2 = 16$$

There are then two candidates for $16^{1/2}$, namely 4 and -4. To avoid ambiguity we say that $16^{1/2} = 4$. That is, if n is even and b is positive, we always *choose the positive number a* such that $a^n = b$ is the nth root, and call it the *principal nth root* of b. Thus, $b^{1/n}$ denotes the principal nth root of b.

We summarize these results in Table 1-11.

TABLE 1-11 Properties of Powers $a^n = b$ and Roots $a = b^{1/n}$ for Integer $n > 0$

Example		Property
$2^3 = 8$	$(-2)^3 = -8$	Any power of a real number is a real number.
$8^{1/3} = 2$	$(-8)^{1/3} = -2$	The odd root of a real number is a real number.
$0^n = 0$	$0^{1/n} = 0$	A positive power or root of zero is zero.
$4^2 = 16$	$(-4)^2 = 16$	A positive number raised to an even power equals the negative of that number raised to the same even power.
$(16)^{1/2} = 4$		The principal root of a positive number is a positive number.
$(-4)^{1/2}$ is undefined in the real number system.		The even root of a negative number is not a real number.

EXAMPLE 1 ***Roots of a Real Number***

Evaluate.

a. $144^{1/2}$ b. $(-8)^{1/3}$ c. $(-25)^{1/2}$ d. $-\left(\frac{1}{16}\right)^{1/4}$

SOLUTION

a. $144^{1/2} = 12$ b. $(-8)^{1/3} = -2$ c. $(-25)^{1/2}$ is not a real number

d. $-\left(\frac{1}{16}\right)^{1/4} = -\frac{1}{2}$ ■

1.7b *Rational Exponents*

Now we are prepared to define $b^{m/n}$, where m is an integer (positive or negative), n is a positive integer, and $b > 0$ when n is even. We want the rules for exponents to hold for rational exponents as well. That is, we want to have

$$4^{3/2} = 4^{(1/2)(3)} = (4^{1/2})^3 = 2^3 = 8$$

and

$$4^{3/2} = 4^{(3)(1/2)} = (4^3)^{1/2} = (64)^{1/2} = 8$$

To achieve this consistency, we define $b^{m/n}$ for an integer m, a natural number n, and a real number b, by

$$b^{m/n} = (b^{1/n})^m = (b^m)^{1/n}$$

where b must be positive when n is even. With this definition, all the rules of exponents continue to hold when the exponents are rational numbers.

EXAMPLE 2 ***Operations with Rational Exponents***

Simplify.

a. $(-8)^{4/3}$ b. $x^{1/2} \cdot x^{3/4}$ c. $(x^{3/4})^2$ d. $(3x^{2/3}y^{-5/3})^3$

SOLUTION

a. $(-8)^{4/3} = [(-8)^{1/3}]^4 = (-2)^4 = 16$

b. $x^{1/2} \cdot x^{3/4} = x^{1/2+3/4} = x^{5/4}$

c. $(x^{3/4})^2 = x^{(3/4)(2)} = x^{3/2}$

d. $(3x^{2/3}y^{-5/3})^3 = 3^3 \cdot x^{(2/3)(3)}y^{(-5/3)(3)} = 27x^2y^{-5} = \frac{27x^2}{y^5}$ ■

Progress Check

Simplify. Assume all variables are positive real numbers.

a. $27^{4/3}$ b. $(a^{1/2}b^{-2})^{-2}$ c. $\left(\dfrac{x^{1/3}y^{2/3}}{z^{5/6}}\right)^{12}$

Answers

a. 81 b. $\dfrac{b^4}{a}$ c. $\dfrac{x^4y^8}{z^{10}}$

FOCUS *When Is a Proof Not a Proof?*

Books of mathematical puzzles love to include "proofs" that lead to false or contradictory results. Of course, there is always an incorrect step hidden somewhere in the proof. The error may be subtle, but a good grounding in the fundamentals of mathematics will enable you to catch it.

Examine the following "proof."

$$1 = 1^{1/2} \quad (1)$$
$$= [(-1)^2]^{1/2} \quad (2)$$
$$= (-1)^{2/2} \quad (3)$$
$$= (-1)^1 \quad (4)$$
$$= -1 \quad (5)$$

The result is obviously contradictory: we can't have $1 = -1$. Yet each step seems to be legitimate. Did you spot the flaw? The rule

$$(b^m)^{1/n} = b^{m/n}$$

used in going from Equation (2) to (3) does not apply when n is even and b is negative.

1.7c *Radicals*

The symbol $\sqrt{b}$ is an alternative way of writing $b^{1/2}$; that is, $\sqrt{b}$ denotes the nonnegative square root of b. The symbol $\sqrt{\ }$ is called a **radical sign**, and $\sqrt{b}$ is called the **principal square root** of b. Thus,

$$\sqrt{25} = 5 \qquad \sqrt{0} = 0 \qquad \sqrt{-25} \text{ is undefined}$$

In general, the symbol $\sqrt[n]{b}$ is an alternative way of writing $b^{1/n}$, the principal nth root of b. Of course, we must apply the same restrictions to $\sqrt[n]{b}$ that we established for $b^{1/n}$. In summary:

$$\sqrt[n]{b} = b^{1/n} = a \qquad \text{where } a^n = b$$

with these restrictions:

- if n is even and $b < 0$, $\sqrt[n]{b}$ is not a real number;
- if n is even and $b \geq 0$, $\sqrt[n]{b}$ is the *nonnegative* number a satisfying $a^n = b$.

Many students are accustomed to writing $\sqrt{4} = \pm 2$. This is incorrect since the symbol $\sqrt{\ }$ indicates the *principal* square root, which is nonnegative. Get in the habit of writing $\sqrt{4} = 2$. If you want to indicate *all* square roots of 4, write $\pm\sqrt{4} = \pm 2$.

In short, $\sqrt[n]{b}$ is the **radical form** of $b^{1/n}$. We can switch back and forth from one form to the other. For instance,

$$\sqrt[3]{7} = 7^{1/3} \qquad (11)^{1/5} = \sqrt[5]{11}$$

Finally, we treat the radical form of $b^{m/n}$ where m is an integer and n is a positive integer as follows:

$$b^{m/n} = (b^m)^{1/n} = \sqrt[n]{b^m}$$

and

$$b^{m/n} = (b^{1/n})^m = (\sqrt[n]{b})^m$$

Thus

$$\begin{aligned} 8^{2/3} &= (8^2)^{1/3} = \sqrt[3]{8^2} \\ &= (8^{1/3})^2 = (\sqrt[3]{8})^2 \end{aligned}$$

(Check that the last two expressions have the same value.)

EXAMPLE 3 *Radicals and Rational Exponents*

Change from radical form to rational exponent form or vice versa. Assume all variables are nonzero.

a. $(2x)^{-3/2}$, $x > 0$

b. $\dfrac{1}{\sqrt[7]{y^4}}$

c. $(-3a)^{3/7}$

d. $\sqrt{x^2 + y^2}$

SOLUTION

a. $(2x)^{-3/2} = \dfrac{1}{(2x)^{3/2}} = \dfrac{1}{\sqrt{8x^3}}$

b. $\dfrac{1}{\sqrt[7]{y^4}} = \dfrac{1}{y^{4/7}} = y^{-4/7}$

c. $(-3a)^{3/7} = \sqrt[7]{-27a^3}$

d. $\sqrt{x^2 + y^2} = (x^2 + y^2)^{1/2}$ ■

Change from radical form to rational exponent form or vice versa. Assume all variables are positive real numbers.

a. $\sqrt[4]{2rs^3}$ b. $(x+y)^{5/2}$ c. $y^{-5/4}$ d. $\dfrac{1}{\sqrt[4]{m^5}}$

Answers

a. $(2r)^{1/4}s^{3/4}$ b. $\sqrt{(x+y)^5}$ c. $\dfrac{1}{\sqrt[4]{y^5}}$ d. $m^{-5/4}$

Since radicals are just another way of writing exponents, the properties of radicals can be derived from the properties of exponents. In Table 1-12, n is a positive integer, a and b are real numbers, and all radicals are real numbers.

TABLE 1-12 Properties of Radicals

Example	Property
$\sqrt[3]{8^2} = (\sqrt[3]{8})^2 = 4$	$\sqrt[n]{b^m} = (\sqrt[n]{b})^m$
$\sqrt{4}\sqrt{9} = \sqrt{36} = 6$	$\sqrt[n]{a}\sqrt[n]{b} = \sqrt[n]{ab}$
$\dfrac{\sqrt[3]{8}}{\sqrt[3]{27}} = \sqrt[3]{\dfrac{8}{27}} = \dfrac{2}{3}$	$\dfrac{\sqrt[n]{a}}{\sqrt[n]{b}} = \sqrt[n]{\dfrac{a}{b}}$
$\sqrt[3]{(-2)^3} = -2$	$\sqrt[n]{a^n} = a$ if n is odd
$\sqrt{(-2)^2} = \lvert -2 \rvert = 2$	$\sqrt[n]{a^n} = \lvert a \rvert$ if n is even

Here are some examples using these properties.

EXAMPLE 4 *Operations with Radicals*

Simplify.

a. $\sqrt{18}$ b. $\sqrt[3]{-54}$ c. $2\sqrt[3]{8x^3y}$ d. $\sqrt{x^6}$

SOLUTION

a. $\sqrt{18} = \sqrt{9 \cdot 2} = \sqrt{9}\sqrt{2} = 3\sqrt{2}$

b. $\sqrt[3]{-54} = \sqrt[3]{(-27)(2)} = \sqrt[3]{-27}\sqrt[3]{2} = -3\sqrt[3]{2}$

c. $2\sqrt[3]{8x^3y} = 2\sqrt[3]{8}\,\sqrt[3]{x^3}\,\sqrt[3]{y} = 2(2)(x)\sqrt[3]{y} = 4x\sqrt[3]{y}$

d. $\sqrt{x^6} = \sqrt{x^2} \cdot \sqrt{x^2} \cdot \sqrt{x^2} = |x| \cdot |x| \cdot |x| = |x|^3$ ■

The properties of radicals state that

$$\sqrt{x^2} = |x|$$

It is a common error to write $\sqrt{x^2} = x$. This can lead to the conclusion that $\sqrt{(-6)^2} = -6$. Since the symbol $\sqrt{\ }$ represents the principal, or nonnegative, square root of a number, the result cannot be negative. It is therefore essential to write $\sqrt{x^2} = |x|$ (and, in fact, $\sqrt[n]{x^n} = |x|$ whenever n is even) unless we know that $x \geq 0$, in which case we can write $\sqrt{x^2} = x$.

Simplifying Radicals

A radical is said to be in **simplified form** when the following conditions are satisfied:

1. $\sqrt[n]{b^m}$ has $m < n$;
2. $\sqrt[n]{b^m}$ has no common factors between m and n;
3. A denominator is free of radicals.

The first two conditions can always be met by using the properties of radicals and by writing radicals in exponent form. For example,

$$\sqrt[3]{x^4} = \sqrt[3]{x^3 \cdot x} = \sqrt[3]{x^3}\sqrt[3]{x} = x\sqrt[3]{x}$$

and

$$\sqrt[6]{x^4} = x^{4/6} = x^{2/3} = \sqrt[3]{x^2}$$

The third condition can always be satisfied by multiplying the fraction by a properly chosen form of unity, a process called **rationalizing the denominator**. For example, to rationalize $\frac{1}{\sqrt{3}}$ we proceed as follows:

$$\frac{1}{\sqrt{3}} = \frac{1}{\sqrt{3}} \cdot \frac{\sqrt{3}}{\sqrt{3}} = \frac{\sqrt{3}}{\sqrt{3^2}} = \frac{\sqrt{3}}{3}$$

In this connection, a useful formula is

$$(\sqrt{m} + \sqrt{n})(\sqrt{m} - \sqrt{n}) = m - n$$

which we will apply in the following examples.

EXAMPLE 5 *Rationalizing Denominators*

Rationalize the denominator. Assume all variables denote positive numbers.

a. $\sqrt{\frac{x}{y}}$ b. $\frac{4}{\sqrt{5} - \sqrt{2}}$ c. $\frac{5}{\sqrt{x} + 2}$ d. $\frac{5}{\sqrt{x+2}}$

SOLUTION

a. $\sqrt{\frac{x}{y}} = \frac{\sqrt{x}}{\sqrt{y}} = \frac{\sqrt{x}}{\sqrt{y}} \cdot \frac{\sqrt{y}}{\sqrt{y}} = \frac{\sqrt{xy}}{\sqrt{y^2}} = \frac{\sqrt{xy}}{y}$

b. $\frac{4}{\sqrt{5} - \sqrt{2}} = \frac{4}{\sqrt{5} - \sqrt{2}} \cdot \frac{\sqrt{5} + \sqrt{2}}{\sqrt{5} + \sqrt{2}} = \frac{4(\sqrt{5} + \sqrt{2})}{5 - 2} = \frac{4}{3}(\sqrt{5} + \sqrt{2})$

c. $\frac{5}{\sqrt{x} + 2} = \frac{5}{\sqrt{x} + 2} \cdot \frac{\sqrt{x} - 2}{\sqrt{x} - 2} = \frac{5(\sqrt{x} - 2)}{x - 4}$

d. $\frac{5}{\sqrt{x+2}} = \frac{5}{\sqrt{x+2}} \cdot \frac{\sqrt{x+2}}{\sqrt{x+2}} = \frac{5\sqrt{x+2}}{x+2}$ ■

Rationalize the denominator. Assume all variables denote positive numbers.

a. $\dfrac{-9xy^3}{\sqrt{3xy}}$ b. $\dfrac{-6}{\sqrt{2}+\sqrt{6}}$ c. $\dfrac{4}{\sqrt{x}-\sqrt{y}}$

Answers

a. $-3y^2\sqrt{3xy}$ b. $\frac{3}{2}(\sqrt{2}-\sqrt{6})$ c. $\dfrac{4(\sqrt{x}+\sqrt{y})}{x-y}$

There are times in mathematics when it is necessary to **rationalize the numerator** instead of the denominator. Although this is in opposition to a simplified form, we illustrate this technique with the following example. Note that if an expression does not display a denominator, we assume a denominator of 1.

EXAMPLE 6 *Rationalizing Numerators*

Rationalize the numerator. Assume all variables denote positive numbers.

a. $\frac{4}{3}(\sqrt{5}+\sqrt{2})$ b. $\dfrac{x-\sqrt{3}}{x+4}$

c. $\sqrt{x}+4$ d. $\dfrac{\sqrt{x}-2}{x-4}$

SOLUTION

a. $\frac{4}{3}(\sqrt{5}+\sqrt{2})\cdot\dfrac{\sqrt{5}-\sqrt{2}}{\sqrt{5}-\sqrt{2}}=\dfrac{4}{3}\dfrac{(5-2)}{(\sqrt{5}-\sqrt{2})}=\dfrac{4}{\sqrt{5}-\sqrt{2}}$

See Example 5(b).

b. $\dfrac{x-\sqrt{3}}{x+4}\cdot\dfrac{x+\sqrt{3}}{x+\sqrt{3}}=\dfrac{x^2-3}{(x+4)(x+\sqrt{3})}$

c. $\dfrac{\sqrt{x}+4}{1}\cdot\dfrac{\sqrt{x}-4}{\sqrt{x}-4}=\dfrac{x-16}{\sqrt{x}-4}$

d. $\dfrac{\sqrt{x}-2}{x-4}\cdot\dfrac{\sqrt{x}+2}{\sqrt{x}+2}=\dfrac{x-4}{(x-4)(\sqrt{x}+2)}=\dfrac{1}{\sqrt{x}+2},\quad x\neq 4$ ■

EXAMPLE 7 *Simplified Forms with Radicals*

Write in simplified form. Assume all variables denote positive numbers.

a. $\sqrt[4]{y^5}$ b. $\sqrt{\dfrac{8x^3}{y}}$ c. $\sqrt[6]{\dfrac{x^3}{y^2}}$

SOLUTION

a. $\sqrt[4]{y^5} = \sqrt[4]{y^4 \cdot y} = \sqrt[4]{y^4}\sqrt[4]{y} = y\sqrt[4]{y}$

b. $\sqrt{\dfrac{8x^3}{y}} = \dfrac{\sqrt{(4x^2)(2x)}}{\sqrt{y}} = \dfrac{\sqrt{4x^2}\sqrt{2x}}{\sqrt{y}} = \dfrac{2x\sqrt{2x}}{\sqrt{y}} = \dfrac{2x\sqrt{2x}}{\sqrt{y}} \cdot \dfrac{\sqrt{y}}{\sqrt{y}} = \dfrac{2x\sqrt{2xy}}{y}$

c. $\sqrt[6]{\dfrac{x^3}{y^2}} = \dfrac{\sqrt[6]{x^3}}{\sqrt[6]{y^2}} = \dfrac{\sqrt{x}}{\sqrt[3]{y}} = \dfrac{\sqrt{x}}{\sqrt[3]{y}} \cdot \dfrac{\sqrt[3]{y^2}}{\sqrt[3]{y^2}} = \dfrac{\sqrt{x}\sqrt[3]{y^2}}{y}$ ■

Progress Check

Write in simplified form. Assume all variables denote positive numbers.

a. $\sqrt{75}$ b. $\sqrt{\dfrac{18x^6}{y}}$ c. $\sqrt[3]{ab^4c^7}$ d. $\dfrac{-2xy^3}{\sqrt[4]{32x^3y^5}}$

Answers

a. $5\sqrt{3}$ b. $\dfrac{3x^3\sqrt{2y}}{y}$ c. $bc^2\sqrt[3]{abc}$ d. $-\dfrac{y}{2}\sqrt[4]{8xy^3}$

1.7d *Operations with Radicals*

We can add or subtract expressions involving exactly the same radical forms. For example,

$$2\sqrt{2} + 3\sqrt{2} = 5\sqrt{2}$$

since

$$2\sqrt{2} + 3\sqrt{2} = (2+3)\sqrt{2} = 5\sqrt{2}$$

and

$$3\sqrt[3]{x^2y} - 7\sqrt[3]{x^2y} = -4\sqrt[3]{x^2y}$$

EXAMPLE 8 *Addition and Subtraction of Radicals*

Write in simplified form. Assume all variables denote positive numbers.

a. $7\sqrt{5} + 4\sqrt{3} - 9\sqrt{5}$ b. $\sqrt[3]{x^2y} - \frac{1}{2}\sqrt{xy} - 3\sqrt[3]{x^2y} + 4\sqrt{xy}$

SOLUTION

a. $7\sqrt{5} + 4\sqrt{3} - 9\sqrt{5} = -2\sqrt{5} + 4\sqrt{3}$

b. $\sqrt[3]{x^2y} - \frac{1}{2}\sqrt{xy} - 3\sqrt[3]{x^2y} + 4\sqrt{xy} = -2\sqrt[3]{x^2y} + \frac{7}{2}\sqrt{xy}$ ■

$$\sqrt{9} + \sqrt{16} \neq \sqrt{25}$$

You can perform addition only with identical radical forms. *Adding unlike radicals is one of the most common mistakes made by students in algebra!* You can easily verify that

$$\sqrt{9} + \sqrt{16} = 3 + 4 = 7$$

The product of $\sqrt[n]{a}$ and $\sqrt[m]{b}$ can be readily simplified only when $m = n$. Thus,

$$\sqrt[5]{x^2y} \cdot \sqrt[5]{xy} = \sqrt[5]{x^3y^2}$$

but

$$\sqrt[3]{x^2y} \cdot \sqrt[5]{xy}$$

cannot be readily simplified.

EXAMPLE 9 ***Multiplication of Radicals***

Multiply and simplify.

a. $2\sqrt[3]{xy^2} \cdot \sqrt[3]{x^2y^2}$ b. $\sqrt[5]{a^2b}\sqrt{ab}\sqrt[5]{ab^2}$

SOLUTION

a. $2\sqrt[3]{xy^2} \cdot \sqrt[3]{x^2y^2} = 2\sqrt[3]{x^3y^4} = 2xy\sqrt[3]{y}$ b. $\sqrt[5]{a^2b}\sqrt{ab}\sqrt[5]{ab^2} = \sqrt[5]{a^3b^3}\sqrt{ab}$ ■

Exercise Set 1.7

In Exercises 1–12, simplify, and write the answer using only positive exponents.

1. $16^{3/4}$
2. $(-125)^{-1/3}$
3. $(-64)^{-2/3}$
4. $c^{1/4}c^{-2/3}$
5. $\dfrac{2x^{1/3}}{x^{-3/4}}$
6. $\dfrac{y^{-2/3}}{y^{1/5}}$
7. $\left(\dfrac{x^{3/2}}{x^{2/3}}\right)^{1/6}$
8. $\dfrac{125^{4/3}}{125^{2/3}}$
9. $(x^{1/3}y^2)^6$
10. $(x^6y^4)^{-1/2}$
11. $\left(\dfrac{x^{15}}{y^{10}}\right)^{3/5}$
12. $\left(\dfrac{x^{18}}{y^{-6}}\right)^{2/3}$

In Exercises 13–18, write the expression in radical form.

13. $\left(\frac{1}{4}\right)^{2/5}$
14. $x^{2/3}$
15. $a^{3/4}$
16. $(-8x^2)^{2/5}$
17. $(12x^3y^{-2})^{2/3}$
18. $\left(\frac{8}{3}x^{-2}y^{-4}\right)^{-3/2}$

In Exercises 19–24, write the expression in exponent form.

19. $\sqrt[4]{8^3}$
20. $\sqrt[5]{3^2}$
21. $\dfrac{1}{\sqrt[5]{(-8)^2}}$
22. $\dfrac{1}{\sqrt[3]{x^7}}$
23. $\dfrac{1}{\sqrt[4]{\frac{4}{9}a^3}}$
24. $\sqrt[5]{(2a^2b^3)^4}$

In Exercises 25–33, evaluate the expression.

25. $\sqrt{\frac{4}{9}}$
26. $\sqrt{\frac{25}{4}}$
27. $\sqrt[4]{-81}$
28. $\sqrt[3]{\frac{1}{27}}$
29. $\sqrt{(5)^2}$
30. $\sqrt{\left(\frac{-1}{3}\right)^2}$
31. $\sqrt{\left(\frac{5}{4}\right)^2}$
32. $\sqrt{\left(-\frac{7}{2}\right)^2}$
33. $(14.43)^{3/2}$

In Exercises 34–36, provide a real value for each variable to demonstrate the result.

34. $\sqrt{x^2} \neq x$
35. $\sqrt{x^2+y^2} \neq x+y$
36. $\sqrt{x}\sqrt{y} \neq xy$

In Exercises 37–56, write the expression in simplified form. Every variable represents a positive number.

37. $\sqrt{48}$
38. $\sqrt{200}$
39. $\sqrt[3]{54}$
40. $\sqrt{x^8}$
41. $\sqrt[3]{y^7}$
42. $\sqrt[4]{b^{14}}$
43. $\sqrt[4]{96x^{10}}$
44. $\sqrt{x^5y^4}$
45. $\sqrt{x^5y^3}$
46. $\sqrt[3]{24b^{10}c^{14}}$
47. $\sqrt[4]{16x^8y^5}$
48. $\sqrt{20x^5y^7z^4}$
49. $\sqrt{\frac{1}{5}}$
50. $\dfrac{4}{3\sqrt{11}}$
51. $\dfrac{1}{\sqrt{3y}}$
52. $\sqrt{\frac{2}{y}}$
53. $\dfrac{4x^2}{\sqrt{2x}}$
54. $\dfrac{8a^2b^2}{2\sqrt{2b}}$
55. $\sqrt[3]{x^2y^7}$
56. $\sqrt[4]{48x^8y^6z^2}$

In Exercises 57–66, simplify and combine terms.

57. $2\sqrt{3}+5\sqrt{3}$
58. $4\sqrt[3]{11}-6\sqrt[3]{11}$
59. $3\sqrt{x}+4\sqrt{x}$
60. $3\sqrt{2}+5\sqrt{2}-2\sqrt{2}$
61. $2\sqrt{27}+\sqrt{12}-\sqrt{48}$
62. $\sqrt{20}-4\sqrt{45}+\sqrt{80}$
63. $\sqrt[3]{40}+\sqrt{45}-\sqrt[3]{135}+2\sqrt{80}$
64. $\sqrt{2abc}-3\sqrt{8abc}+\sqrt{\frac{abc}{2}}$
65. $2\sqrt{5}-(3\sqrt{5}+4\sqrt{5})$
66. $2\sqrt{18}-(3\sqrt{12}-2\sqrt{75})$

In Exercises 67–74, multiply and simplify.

67. $\sqrt{3}(\sqrt{3}+4)$
68. $\sqrt{8}(\sqrt{2}-\sqrt{3})$
69. $3\sqrt[3]{x^2y}\sqrt[3]{xy^2}$
70. $-4\sqrt[5]{x^2y^3}\sqrt[5]{x^4y^2}$
71. $(\sqrt{2}-\sqrt{3})^2$
72. $(\sqrt{8}-2\sqrt{2})(\sqrt{2}+2\sqrt{8})$
73. $(\sqrt{3x}+\sqrt{2y})(\sqrt{3x}-2\sqrt{2y})$
74. $(\sqrt[3]{2x}+3)(\sqrt[3]{2x}-3)$

In Exercises 75–86, rationalize the denominator.

75. $\dfrac{3}{\sqrt{2}+3}$
76. $\dfrac{-3}{\sqrt{7}-9}$
77. $\dfrac{-2}{\sqrt{3}-4}$
78. $\dfrac{3}{\sqrt{x}-5}$

79. $\frac{-3}{3\sqrt{a}+1}$ 80. $\frac{4}{2-\sqrt{2y}}$

81. $\frac{-3}{5+\sqrt{5y}}$ 82. $\frac{\sqrt{3}}{\sqrt{3}-5}$

83. $\frac{\sqrt{2}+1}{\sqrt{2}-1}$ 84. $\frac{\sqrt{5}+\sqrt{3}}{\sqrt{5}-\sqrt{3}}$

85. $\frac{\sqrt{6}+\sqrt{2}}{\sqrt{3}-\sqrt{2}}$ 86. $\frac{2\sqrt{a}}{\sqrt{2x}+\sqrt{y}}$

In Exercises 87–90, rationalize the numerator.

87. $\sqrt{12}-\sqrt{10}$ 88. $\frac{3-\sqrt{x}}{x-9}$

89. $\frac{\sqrt{x}-4}{16-x}$ 90. $\frac{2-\sqrt{x+1}}{3-x}$

In Exercises 91 and 92, provide real values for x and y and a positive integer value for n to demonstrate the result.

91. $\sqrt{x}+\sqrt{y} \neq \sqrt{x+y}$ 92. $\sqrt[n]{x^n+y^n} \neq x+y$

93. Find the step in the following "proof" that is incorrect. Explain.

$$2=\sqrt{4}=\sqrt{(-2)(-2)}=\sqrt{-2}\sqrt{-2}=-2$$

94. Prove that $|ab|=|a||b|$. (*Hint:* Begin with $|ab|=\sqrt{(ab)^2}$.)

95. Simplify the following.

a. $\sqrt{x\sqrt{x\sqrt{x}}}$ b. $(x^{1/2}-x^{-1/2})^2$

c. $\sqrt{1+x^2}-\frac{\sqrt{1+x^2}}{2}$

d. $\sqrt[5]{\frac{3^4+3^4+3^4}{5^4+5^4+5^4+5^4+5^4}}$

e. $\frac{5(1+x^2)^{1/2}-5x^2(1+x^2)^{-1/2}}{1+x^2}$

96. Write the following in simplest radical form.

a. $\sqrt{a^{-2}+c^{-2}}$ b. $\sqrt{1-\left(\frac{a}{c}\right)^2}$

c. $\sqrt{x+\frac{1}{x}+2}$

97. The frequency of an electrical circuit is given by

$$\frac{1}{2\pi}\sqrt{\frac{Lc_1c_2}{c_1+c_2}}$$

Make the denominator radical free. (*Hint:* Use the techniques for rationalizing the denominator.)

98. Use your calculator to find $\sqrt{0.4}$, $\sqrt{0.04}$, $\sqrt{0.004}$, $\sqrt{0.0004}$, and so on, until you see a pattern. Can you state a rule about the value of

$$\sqrt{\frac{a}{10^n}}$$

where a is a perfect square and n is a positive integer? Under what circumstances does this expression have an integer value? Test your rule for large values of n.

1.8 Complex Numbers

One of the central problems in algebra is to find solutions to a given polynomial equation. This problem will be discussed in later chapters of this book. For now, observe that there is no real number that satisfies a polynomial equation such as

$$x^2 = -4$$

since the square of a real number is always nonnegative.

To resolve this problem, mathematicians created a new number system built upon an **imaginary unit** i, defined by $i = \sqrt{-1}$. This number i has the property that when we square both sides of the equation we have $i^2 = -1$, a result that cannot be obtained with real numbers. By definition,

$$i = \sqrt{-1}$$
$$i^2 = -1$$

We also assume that i behaves according to all the algebraic laws we have already developed (with the exception of the rules for inequalities for real numbers). This allows us to simplify higher powers of i. Thus,

$$i^3 = i^2 \cdot i = (-1)i = -i$$
$$i^4 = i^2 \cdot i^2 = (-1)(-1) = 1$$

Now we may simplify i^n when n is any natural number. Since $i^4 = 1$, we seek the highest multiple of 4 that is less than or equal to n. For example,

$$i^5 = i^4 \cdot i = (1) \cdot i = i$$
$$i^{27} = i^{24} \cdot i^3 = (i^4)^6 \cdot i^3 = (1)^6 \cdot i^3 = i^3 = -i$$

EXAMPLE 1 ***Imaginary Unit* i**

Simplify.

a. i^{51}

b. $-i^{74}$

SOLUTION

a. $i^{51} = i^{48} \cdot i^3 = (i^4)^{12} \cdot i^3 = (1)^{12} \cdot i^3 = i^3 = -i$

b. $-i^{74} = -i^{72} \cdot i^2 = -(i^4)^{18} \cdot i^2 = -(1)^{18} \cdot i^2 = -(1)(-1) = 1$ ■

We may also write square roots of negative numbers in terms of i. For example,

$$\sqrt{-25} = i\sqrt{25} = 5i$$

and, in general, we define

$$\sqrt{-a} = i\sqrt{a} \quad \text{for} \quad a > 0$$

Any number of the form bi, where b is a real number, is called an **imaginary number**.

$$\sqrt{-4}\sqrt{-9} \neq \sqrt{36}$$

The rule $\sqrt{a} \cdot \sqrt{b} = \sqrt{ab}$ holds only when $a \geq 0$ and $b \geq 0$. Instead, write

$$\sqrt{-4}\sqrt{-9} = 2i \cdot 3i = 6i^2 = -6.$$

Having created imaginary numbers, we next combine real and imaginary numbers. We say that $a + bi$ is a **complex number** where a and b are real numbers. The number a is called the **real part** of $a + bi$, and b is called the **imaginary part**. The following are examples of complex numbers.

$$3 + 2i \qquad 2 - i \qquad -2i \qquad \frac{4}{5} + \frac{1}{5}i$$

Note that every real number a can be written as a complex number by choosing $b = 0$. Thus,

$$a = a + 0i$$

We see that the real number system is a subset of the complex number system. The desire to find solutions to every quadratic equation has led mathematicians to create a more comprehensive number system, which incorporates all previous number systems. We will show in a later chapter that complex numbers are all that we need to provide solutions to any polynomial equation.

EXAMPLE 2 *Complex Numbers $a + bi$*

Write as a complex number:

a. $-\frac{1}{2}$ b. $\sqrt{-9}$ c. $-1 - \sqrt{-4}$

SOLUTION

a. $-\frac{1}{2} = -\frac{1}{2} + 0i$ b. $\sqrt{-9} = i\sqrt{9} = 3i = 0 + 3i$

c. $-1 - \sqrt{-4} = -1 - i\sqrt{4} = -1 - 2i$ ■

Do not be concerned by the word "complex." You already have all the basic tools you need to tackle this number system. We will next define operations with complex numbers in such a way that the rules for the real numbers and the imaginary unit i continue to hold. We begin with equality and say that two complex numbers are equal if their real parts are equal and their imaginary parts are equal; that is,

$$a + bi = c + di \quad \text{if} \quad a = c \quad \text{and} \quad b = d$$

EXAMPLE 3 *Equality of Complex Numbers*

Solve the equation $x + 3i = 6 - yi$ for x and y.

SOLUTION

Equating the real parts, we have $x = 6$; equating the imaginary parts, $3 = -y$ or $y = -3$. ■

Complex numbers are added and subtracted by adding or subtracting the real parts and by adding or subtracting the imaginary parts.

Addition and Subtraction of Complex Numbers

$$(a + bi) + (c + di) = (a + c) + (b + d)i$$
$$(a + bi) - (c + di) = (a - c) + (b - d)i$$

Note that the sum or difference of two complex numbers is again a complex number.

EXAMPLE 4 *Addition and Subtraction of Complex Numbers*

Perform the indicated operations.

a. $(7 - 2i) + (4 - 3i)$ b. $14 - (3 - 8i)$

SOLUTION

a. $(7 - 2i) + (4 - 3i) = (7 + 4) + (-2 - 3)i = 11 - 5i$

b. $14 - (3 - 8i) = (14 - 3) + 8i = 11 + 8i$ ■

Progress Check

Perform the indicated operations.

a. $(-9 + 3i) + (6 - 2i)$ b. $7i - (3 + 9i)$

Answers

a. $-3 + i$ b. $-3 - 2i$

We now define multiplication of complex numbers in a manner that permits the commutative, associative, and distributive laws to hold, along with the definition $i^2 = -1$. We must have

$$\begin{aligned}(a + bi)(c + di) &= a(c + di) + bi(c + di)\\ &= ac + adi + bci + bdi^2\\ &= ac + (ad + bc)i + bd(-1)\\ &= (ac - bd) + (ad + bc)i\end{aligned}$$

The rule for multiplication is

Multiplication of Complex Numbers

$$(a + bi)(c + di) = (ac - bd) + (ad + bc)i$$

This result demonstrates that the product of two complex numbers is a complex number. It need not be memorized. Use the distributive law to form all the products and the substitution $i^2 = -1$ to simplify.

EXAMPLE 5 *Multiplication of Complex Numbers*

Find the product of $(2 - 3i)$ and $(7 + 5i)$.

SOLUTION

$$\begin{aligned}(2-3i)(7+5i) &= 2(7+5i) - 3i(7+5i)\\ &= 14 + 10i - 21i - 15i^2\\ &= 14 - 11i - 15(-1)\\ &= 29 - 11i\end{aligned}$$ ■

Progress Check

Find the product.

a. $(-3 - i)(4 - 2i)$ b. $(-4 - 2i)(2 - 3i)$

Answers

a. $-14 + 2i$ b. $-14 + 8i$

The complex number $a - bi$ is called the **complex conjugate**, or simply the **conjugate**, of the complex number $a + bi$. For example, $3 - 2i$ is the conjugate of $3 + 2i$, $4i$ is the conjugate of $-4i$, and 2 is the conjugate of 2. Forming the product $(a + bi)(a - bi)$, we have

$$\begin{aligned}(a+bi)(a-bi) &= a^2 - abi + abi - b^2i^2\\ &= a^2 + b^2 \quad \text{since } i^2 = -1\end{aligned}$$

Because a and b are real numbers, $a^2 + b^2$ is also a real number. We can summarize this result as follows:

The Complex Conjugate for Multiplication

The complex conjugate of $a + bi$ is $a - bi$. The product of a complex number and its conjugate is a real number.

$$(a+bi)(a-bi) = a^2 + b^2$$

Before we examine the quotient of two complex numbers, we consider the reciprocal of $a + bi$, namely, $\frac{1}{a+bi}$. This may be simplified by multiplying both numerator and denominator by the conjugate of the denominator.

$$\frac{1}{a+bi} = \left(\frac{1}{a+bi}\right)\left(\frac{a-bi}{a-bi}\right) = \frac{a-bi}{a^2+b^2} = \frac{a}{a^2+b^2} - \frac{b}{a^2+b^2}i$$

In general, the quotient of two complex numbers

$$\frac{a+bi}{c+di}$$

is simplified in a similar manner, that is, by multiplying both numerator and denominator by the conjugate of the denominator.

$$\frac{a+bi}{c+di} = \frac{a+bi}{c+di} \cdot \frac{c-di}{c-di}$$

$$= \frac{(ac+bd)+(bc-ad)i}{c^2+d^2}$$

$$= \frac{ac+bd}{c^2+d^2} + \frac{bc-ad}{c^2+d^2}i$$

Division of Complex Numbers

$$\frac{a+bi}{c+di} = \frac{ac+bd}{c^2+d^2} + \frac{bc-ad}{c^2+d^2}i, \quad c^2+d^2 \neq 0$$

This result demonstrates that the quotient of two complex numbers is a complex number. Instead of memorizing this formula for division, remember that quotients of complex numbers may be simplified by multiplying the numerator and denominator by the conjugate of the denominator.

EXAMPLE 6 *Division of Complex Numbers*

a. Write the quotient $\frac{-2+3i}{3-2i}$ in the form $a + bi$.

b. Write the reciprocal of $2 - 5i$ in the form $a + bi$.

SOLUTION

a. Multiplying numerator and denominator by the conjugate of the denominator, $3 + 2i$, we have

$$\frac{-2+3i}{3-2i} = \frac{-2+3i}{3-2i} \cdot \frac{3+2i}{3+2i} = \frac{-6-4i+9i+6i^2}{3^2+2^2} = \frac{-6+5i+6(-1)}{9+4}$$

$$= \frac{-12+5i}{13} = -\frac{12}{13} + \frac{5}{13}i$$

b. The reciprocal is $\frac{1}{2-5i}$. Multiplying both numerator and denominator by the conjugate $2 + 5i$, we have

$$\frac{1}{2-5i} \cdot \frac{2+5i}{2+5i} = \frac{2+5i}{2^2+5^2} = \frac{2+5i}{29} = \frac{2}{29} + \frac{5}{29}i$$

Verify that

$$(2-5i)\left(\frac{2}{29} + \frac{5}{29}i\right) = 1$$

■

Progress Check

Write the following in the form $a + bi$.

a. $\frac{4-2i}{5+2i}$ b. $\frac{1}{2-3i}$ c. $\frac{-3i}{3+5i}$

Answers

a. $\frac{16}{29} - \frac{18}{29}i$ b. $\frac{2}{13} + \frac{3}{13}i$ c. $-\frac{15}{34} - \frac{9}{34}i$

Exercise Set 1.8

Simplify in Exercises 1–9.

1. i^{60}
2. i^{27}
3. i^{83}
4. $-i^{54}$
5. $-i^{33}$
6. i^{-15}
7. i^{-84}
8. $-i^{39}$
9. $-i^{-25}$

In Exercises 10–21, write the number in the form $a + bi$.

10. 2
11. $-\frac{3}{4}$
12. -0.3
13. $\sqrt{-25}$
14. $-\sqrt{-5}$
15. $-\sqrt{-36}$
16. $-\sqrt{-18}$
17. $3 - \sqrt{-49}$
18. $-\frac{3}{2} - \sqrt{-72}$
19. $0.3 - \sqrt{-98}$
20. $-0.5 + \sqrt{-32}$
21. $-2 - \sqrt{-16}$

In Exercises 22–26, solve for x and y.

22. $(x + 2) + (2y - 1)i = -1 + 5i$
23. $(3x - 1) + (y + 5)i = 1 - 3i$
24. $\left(\frac{1}{2}x + 2\right) + (3y - 2)i = 4 - 7i$
25. $(2y + 1) - (2x - 1)i = -8 + 3i$
26. $(y - 2) + (5x - 3)i = 5$

In Exercises 27–42, compute the answer and write it in the form $a + bi$.

27. $2i + (3 - i)$
28. $-3i + (2 - 5i)$
29. $2 + 3i + (3 - 2i)$
30. $(3 - 2i) - \left(2 + \frac{1}{2}i\right)$
31. $-3 - 5i - (2 - i)$
32. $\left(\frac{1}{2} - i\right) + \left(1 - \frac{2}{3}i\right)$
33. $-2i(3 + i)$
34. $3i(2 - i)$
35. $i\left(-\frac{1}{2} + i\right)$
36. $\frac{i}{2}\left(\frac{4 - i}{2}\right)$
37. $(2 - i)(2 + i)$
38. $(5 + i)(2 - 3i)$
39. $(-2 - 2i)(-4 - 3i)$
40. $(2 + 5i)(1 - 3i)$
41. $(3 - 2i)(2 - i)$
42. $(4 - 3i)(2 + 3i)$

In Exercises 43–48, multiply by the conjugate and simplify.

43. $2 - i$
44. $3 + i$
45. $3 + 4i$
46. $2 - 3i$
47. $-4 - 2i$
48. $5 + 2i$

In Exercises 49–57, perform the indicated operations and write the answer in the form $a + bi$.

49. $\frac{2 + 5i}{1 - 3i}$
50. $\frac{1 + 3i}{2 - 5i}$
51. $\frac{3 - 4i}{3 + 4i}$
52. $\frac{4 - 3i}{4 + 3i}$
53. $\frac{3 - 2i}{2 - i}$
54. $\frac{2 - 3i}{3 - i}$
55. $\frac{2 + 5i}{3i}$
56. $\frac{5 - 2i}{-3i}$
57. $\frac{4i}{2 + i}$

In Exercises 58–64, find the reciprocal and write the answer in the form $a + bi$.

58. $3 + 2i$
59. $4 + 3i$
60. $\frac{1}{2} - i$
61. $1 - \frac{1}{3}i$
62. $-7i$
63. $-5i$
64. $\frac{3 - i}{3 + 2i}$

In Exercises 65–68, evaluate the polynomial $x^2 - 2x + 5$ for the given complex value of x.

65. $1 + 2i$
66. $2 - i$
67. $1 - i$
68. $1 - 2i$

69. Prove that the commutative law of addition holds for the set of complex numbers.
70. Prove that the commutative law of multiplication holds for the set of complex numbers.
71. Prove that $0 + 0i$ is the additive identity and $1 + 0i$ is the multiplicative identity for the set of complex numbers.
72. Prove that $-a - bi$ is the additive inverse of the complex number $a + bi$.
73. Prove the distributive property for the set of complex numbers.
74. For what values of x is $\sqrt{x - 3}$ a real number?
75. For what values of y is $\sqrt{2y - 10}$ a real number?
76. Perform the multiplications and simplify.
    a. $(x + yi)(x - yi)$
    b. $(1 - i)^5$
    c. $(1 - \sqrt{3})^4$
    d. $[x - (2 + 5i)][x - (2 - 5i)]$
77. *Mathematics in Writing*: Consider the addition and the multiplication of complex numbers. How does i differ from a variable like x? If you always treat i as though it is a variable, at what step in the procedures of addition or multiplication would you run into trouble?

Chapter Summary

Key Terms, Concepts, and Symbols

absolute value, | |	12	factoring	27	principal square root	55
algebraic expression	17	imaginary number	64	radical form	56
algebraic fraction	37	imaginary part	65	radical sign, $\sqrt{\ }$	55
algebraic operations	17	imaginary unit i	64	rational expression	37
base	17	inequalities	11	rational numbers	3
cancellation principle	38	inequality symbols, $<, >, \leq, \geq$	11	rationalize the numerator	59
coefficient	19	integers	3	rationalizing the denominator	58
complex conjugate	67	irrational numbers	3	real number line	10
complex fraction	42	irreducible polynomial	34	real numbers	3
complex number	65	leading coefficient	19	real part	65
constant	17	least common denominator (LCD)	39	scientific notation	48
constant term	19	like terms	20	set	2
degree of a monomial	19	member of a set, $\in$	2	set notation	2
degree of a polynomial	19	monomial	18	simplified form of a radical	58
Distance on the Real Number line	14	natural numbers	3	subset	2
element of a set, $\in$	2	nonnegative numbers	10	term	19
equality	5	nth root	53	value	17
equivalent fractions	41	origin	10	variable	17
evaluated	17	polynomial	18	zero polynomial	19
exponent	17	power	17		
factor	27	prime polynomial	34		

Key Ideas for Review

Topic	Page	Key Idea
Set	2	A set is a collection of objects or numbers.
The Set of Real Numbers	3	The set of real numbers is composed of the rational and irrational numbers. The rational numbers are those that can be written as the ratio of two integers, $\frac{p}{q}$, with $q \neq 0$; the irrational numbers cannot be written as the ratio of two integers.
Properties	5	The real number system satisfies a number of important properties, including: closure, commutativity, associativity, identities, inverses, distributivity
Equality	5	If two numbers are identical, we say that they are equal.
Properties	6	Equality satisfies these basic properties: reflexive property, symmetric property, transitive property, substitution property
Real Number Line	10	There is a one-to-one correspondence between the set of all real numbers and the set of all points on the real number line. That is, for every point on the line there is a real number, and for every real number there is a point on the line.

continues

Topic	Page	Key Idea
Inequalities	11	Algebraic statements using inequality symbols have geometric interpretations using the real number line. For example, $a < b$ says that a lies to the left of b on the real number line.
Operations	12	Inequalities can be operated on in the same manner as statements involving an equal sign with one important exception: when an inequality is multiplied or divided by a negative number, the direction of the inequality is reversed.
Absolute Value	12	Absolute value specifies distance independent of direction. Four important properties of absolute value are: $\lvert a\rvert \geq 0$ $\lvert a\rvert = \lvert -a\rvert$ $\lvert a-b\rvert = \lvert b-a\rvert$ $\lvert ab\rvert = \lvert a\rvert\lvert b\rvert$
Distance	14	The distance between points A and B whose coordinates are a and b, respectively, is given by $\overline{AB} = \lvert b-a\rvert$
Polynomials	18	Algebraic expressions of the form $a_nx^n + a_{n-1}x^{n-1} + \cdots + a_1x + a_0$ are called polynomials.
Operations	20	To add (subtract) polynomials, just add (subtract) like terms. To multiply polynomials, form all possible products, using the rule for exponents: $a^m a^n = a^{m+n}$
Factoring	27	A polynomial is said to be factored when it is written as a product of polynomials of lower degree.
Rational Expressions	37	Most of the rules of arithmetic for handling fractions carry over to rational expressions. For example, the LCD has the same meaning except that we deal with polynomials in factored form rather than with integers.
Exponents	45	The rules for positive integer exponents also apply to zero, negative integer exponents, and, in fact, to all rational exponents.
Scientific Notation	48	A number in scientific notation is of the form $\pm\, a \times 10^m$ where $1 \leq a < 10$ and m is some integer.
Radicals	55	Radical notation is another way of writing a rational exponent. That is, $\sqrt[n]{b} = b^{1/n}$
Principal n*th Root*	53	If n is even and b is positive, there are two real numbers a such that $b^{1/n} = a$. Under these circumstances, we insist that the nth root be positive. That is, $\sqrt[n]{b}$ is a positive number if n is even and b is positive. Thus $\sqrt{16} = 4$. Similarly, we must write $\sqrt{x^2} = \lvert x\rvert$ to ensure that the result is a positive number.
Simplifying	58	To be in simplified form, a radical must satisfy the following conditions: $\sqrt[n]{x^m}$ has $m < n$. $\sqrt[n]{x^m}$ has no common factors between m and n. The denominator has been rationalized.
Complex Numbers	64	Complex numbers were created because there were no real numbers that satisfy a polynomial equation such as $x^2 + 5 = 0$

continues

Topic	Page	Key Idea
Imaginary Unit i	64	Using the imaginary unit $i = \sqrt{-1}$, a complex number is of the form $a + bi$, where a and b are real numbers; the real part of $a + bi$ is a and the imaginary part of $a + bi$ is b.
Real Number System	65	The real number system is a subset of the complex number system.

Review Exercises

In Exercises 1–3, write each set by listing its elements within braces.

1. The set of natural numbers from -5 to 4, inclusive
2. The set of integers from -3 to -1, inclusive
3. The subset of $x \in S$, $S = \{0.5, 1, 1.5, 2\}$ such that x is an even integer

For Exercises 4–7, determine whether the statement is true (T) or false (F).

4. $\sqrt{7}$ is a real number.
5. -35 is a natural number.
6. -14 is not an integer.
7. 0 is an irrational number.

In Exercises 8–11, identify the property of the real number system that justifies the statement. All variables represent real numbers.

8. $3a + (-3a) = 0$
9. $(3 + 4)x = 3x + 4x$
10. $2x + 2y + z = 2x + z + 2y$
11. $9x \cdot 1 = 9x$

In Exercises 12–14, sketch the given set of numbers on a real number line.

12. The negative real numbers
13. The real numbers x such that $x > 4$
14. The real numbers x such that $-1 \le x < 1$
15. Find the value of $|-3| - |1 - 5|$.
16. Find $\overline{PQ}$ if the coordinates of P and Q are $\frac{9}{2}$ and 6, respectively.
17. A salesperson receives $7.25x + 0.15y$ dollars, where x is the number of hours worked and y is the number of miles driven. Find the amount due the salesperson if $x = 12$ hours and $y = 80$ miles.
18. Which of the following expressions are not polynomials?

    a. $-2xy^2 + x^2y$ b. $3b^2 + 2b - 6$

    c. $x^{-1/2} + 5x^2 - x$ d. $7.5x^2 + 3x - \frac{1}{2}x^0$

In Exercises 19 and 20, indicate the leading coefficient and the degree of each polynomial.

19. $-0.5x^7 + 6x^3 - 5$
20. $2x^2 + 3x^4 - 7x^5$

In Exercises 21–23, perform the indicated operations.

21. $(3a^2b^2 - a^2b + 2b - a) - (2a^2b^2 + 2a^2b - 2b - a)$
22. $x(2x - 1)(x + 2)$
23. $3x(2x + 1)^2$

In Exercises 24–29, factor each expression.

24. $2x^2 - 2$
25. $x^2 - 25y^2$
26. $2a^2 + 3ab + 6a + 9b$
27. $4x^2 + 19x - 5$
28. $x^8 - 1$
29. $27r^6 + 8s^6$

In Exercises 30–33, perform the indicated operations and simplify.

30. $\dfrac{14(y-1)}{3(x^2-y^2)} \cdot \dfrac{9(x+y)}{-7xy^2}$
31. $\dfrac{4-x^2}{2y^2} \div \dfrac{x-2}{3y}$
32. $\dfrac{x^2-2x-3}{2x^2-x} \div \dfrac{x^2-4x+3}{3x^3-3x^2}$
33. $\dfrac{a+b}{a+2b} \cdot \dfrac{a^2-4b^2}{a^2-b^2}$

In Exercises 34–37, find the LCD.

34. $\dfrac{-1}{2x^2}, \dfrac{2}{x^2-4}, \dfrac{3}{x-2}$
35. $\dfrac{4}{x}, \dfrac{5}{x^2-x}, \dfrac{-3}{(x-1)^2}$
36. $\dfrac{2}{(x-1)y}, \dfrac{-4}{y^2}, \dfrac{x+2}{5(x-1)^2}$
37. $\dfrac{y-1}{x^2(y+1)}, \dfrac{x-2}{2xy-2x}, \dfrac{3x}{4y^2+8y+4}$

In Exercises 38–41, perform the indicated operations and simplify.

38. $2 + \dfrac{4}{a^2-4}$
39. $\dfrac{3}{x^2-16} - \dfrac{2}{x-4}$
40. $\dfrac{\dfrac{3}{x+2} - \dfrac{2}{x-1}}{x-1}$
41. $x^2 + \dfrac{\dfrac{1}{x}+1}{x-\dfrac{1}{x}}$

In Exercises 42–50, simplify and express the answers using only positive exponents. All variables are positive numbers.

42. $(2a^2b^{-3})^{-3}$ 43. $2(a^2 - 1)^0$

44. $\left(\dfrac{x^3}{y^{-6}}\right)^{-4/3}$ 45. $\dfrac{x^{3+n}}{x^n}$

46. $\sqrt{80}$ 47. $\dfrac{2}{\sqrt{12}}$

48. $\sqrt{x^7y^5}$ 49. $\sqrt[4]{32x^8y^6}$

50. $\dfrac{\sqrt{x}}{\sqrt{x}+\sqrt{y}}$

51. Compute
$$\frac{(5.10\times10^7)(3.45\times10^{-2})}{7.10\times10^4}$$
to three decimal places and express the answer in scientific notation.

52. Rationalize the numerator for
$$\frac{\sqrt{x}-\sqrt{y}}{x-y}$$

In Exercises 53 and 54, perform the indicated operations. Simplify the answer.

53. $\sqrt[4]{x^2y^2} + 2\sqrt[4]{x^2y^2}$ 54. $(\sqrt{3}+\sqrt{5})^2$

55. Evaluate the given expressions using your calculator.

a. $\dfrac{12}{5} - \dfrac{3}{7}$ b. $|(-4)^3 - 5^6|$

c. $\sqrt{8}$ d. π^8

e. $\sqrt[5]{-27}$ f. $\dfrac{|2+\sqrt{3}|}{-6}$

g. $\sqrt[3]{4} + \sqrt{\dfrac{1}{8}}$ h. $\sqrt[10]{0.5}$

i. $\left(\dfrac{2}{3}\right)^4$ j. $9^{5/8}$

56. Solve for x and y:
$$(x-2) + (2y-1)i = -4 + 7i$$

57. Simplify i^{47}.

In Exercises 58–61, perform the indicated operations and write all answers in the form $a + bi$.

58. $2 + (6 - i)$ 59. $(2 + i)^2$

60. $(4 - 3i)(2 + 3i)$ 61. $\dfrac{4-3i}{2+3i}$

62. Perform the indicated operations.

a. Combine into one term with a common denominator
$$\frac{1}{a}+\frac{1}{b}+\frac{1}{c}$$

b. Simplify the quotient
$$\frac{\frac{1}{a}+\frac{1}{b}}{\frac{1}{c}+\frac{1}{d}}$$

63. Dan, at 200 pounds, wishes to reduce his weight to 180 pounds in time to attend his college reunion in 8 weeks. He learns that it takes 2400 calories per day to maintain his weight. A reduction of his caloric intake to 1900 calories per day will result in his losing weight at the rate of 1 pound per week. What should his daily caloric intake be to achieve this goal?

64. The executive committee of the student government association consists of a president, vice-president, secretary, and treasurer.

a. In how many ways can a committee of three persons be formed from among the executive committee members?

b. According to the by-laws, there must be at least three affirmative votes to carry a motion. If the president automatically has two votes, list all the minimal winning coalitions.

65. If 6 children can devour 6 hot dogs in $\frac{1}{10}$ of an hour, how many children would it take to devour 100 hot dogs in 6000 seconds?

66. A CD player costs a dealer \$16. If he wishes to make a profit of at least 25% of his cost, what must be the lowest selling price for the player?

67. Find the area of the shaded rectangle.

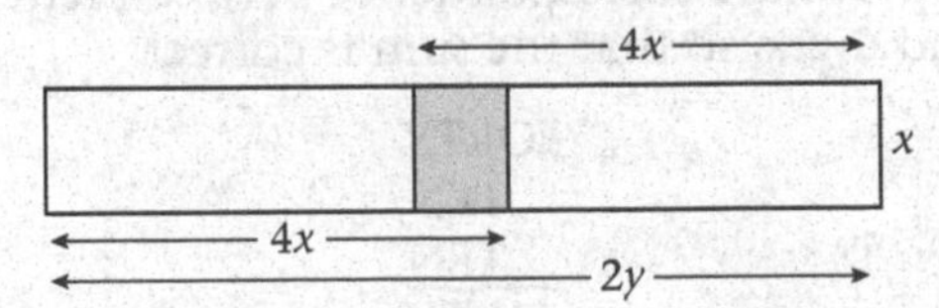

68. An open box is to be made from a 4 feet × 5 feet piece of tin by cutting out squares of equal size from the four corners and bending up the flaps to form sides. Find a formula for the volume in terms of s, the side of the square. Write the inequality that describes the restriction on s.

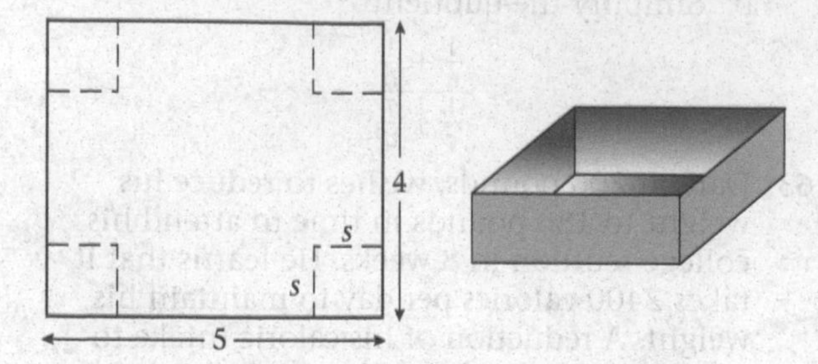

69. Compute the following products:

a. $(x - y)(x^2 + xy + y^2)$

b. $(x - y)(x^3 + x^2y + xy^2 + y^3)$

c. $(x - y)(x^4 + x^3y + x^2y^2 + xy^3 + y^4)$

70. Using Exercise 69, find a general formula that allows you to factor $x^n - y^n$, where n is a positive integer.

71. In ancient Alexandria, numbers were multiplied by using an abacus as follows:

$$\begin{aligned} 19 \times 28 &= (20 - 1)(30 - 2) \\ &= (20)(30) - (20)(2) - 30 + 2 \\ &= 600 - 40 - 30 + 2 \\ &= 532 \end{aligned}$$

Set up a comparable sequence of steps for 13 × 17.

72. Find two ways of grouping and then factoring $ac + ad - bc - bd$.

73. The following calculation represents a sum. If each letter represents a different digit, find the appropriate correspondence between letters and digits so that the sum is correct.

$$\begin{array}{r} \text{FORTY} \\ \text{TEN} \\ \text{TEN} \\ \hline \text{SIXTY} \end{array}$$

74. A natural number is said to be perfect if it is the sum of its divisors other than itself. For example, 6 is the first perfect number since $6 = 1 + 2 + 3$. Show that 28 is the second perfect number.

Every number of the form $2^{p-1}(2^p - 1)$, where $2^p - 1$ is prime, is an even perfect number. (Check your answer when $p = 2$.) Find the third and fourth even perfect numbers. The ancient Greeks could not find the fifth even perfect number. See if you can.

75. The speed of light is 3×10^8 meters per second. Write all answers using scientific notation.

a. How many seconds does it take an object traveling at the speed of light to go 1×10^{26} meters?

b. How many seconds are there in 1 year of 365 days?

c. Write the answer to part (a) in years. (This answer is the approximate age of the universe.)

76. Write $\sqrt{x + \sqrt{x + \sqrt{x}}}$ using exponents.

77. Determine if $(\sqrt{5} - \sqrt{24})^2$ and $(\sqrt{2} - \sqrt{3})^2$ have the same value.

78. The irrational number called the golden ratio

$$T = \frac{\sqrt{5} + 1}{2}$$

has properties that have intrigued artists, philosophers, and mathematicians through the ages. Show that T satisfies the identity

$$T = 1 + \frac{1}{T}$$

79. Rationalize the numerator in the following:

a. $\dfrac{\sqrt{x + h + 1} - \sqrt{x - 1}}{h}$

b. $\dfrac{\sqrt{3 + x} - \sqrt{3}}{x}$

80. In alternating-current theory, the current I (amps), voltage V (volts), and impedance Z (ohms) are treated as complex numbers. The formula relating these quantities is $V = IZ$. If $I = 2 - 3i$ amps and $Z = 6 + 2i$ ohms, find the voltage across this part of the circuit.

Review Test

In Problems 1 and 2, write each set by listing its elements within braces.

1. The set of positive, even integers less than 13
2. The subset of $x \in S$, $S = \{-1, 2, 3, 5, 7\}$, such that x is a multiple of 3

In Problems 3 and 4, determine whether the statement is true (T) or false (F).

3. -1.36 is an irrational number.
4. π is equal to $\frac{22}{7}$.

In Problems 5 and 6, identify the property of the real number system that justifies the statement. All variables represent real numbers.

5. $xy(z+1) = (z+1)xy$
6. $(-6)\left(-\frac{1}{6}\right) = 1$

In Problems 7 and 8, sketch the given set of numbers on a real number line.

7. The integers that are greater than -3 and less than or equal to 3
8. The real numbers x such that $-2 \le x < \frac{1}{2}$
9. Find the value of $|2-3| - |4-2|$.
10. Find $\overline{AB}$ if the coordinates of A and B are -6 and -4, respectively.
11. The area of a region is given by the expression $3x^2 - xy$. Find the area if $x = 5$ meters and $y = 10$ meters.
12. Evaluate the expression
$$\frac{-|y-2x|}{|xy|}$$
if $x = 3$ and $y = -1$.
13. Which of the following expressions are not polynomials?
    a. x^5 b. $5x^{-4}y + 3x^2 - y$
    c. $4x^3 + x$ d. $2x^2 + 3x^0$

In Problems 14 and 15, indicate the leading coefficient and the degree of each polynomial.

14. $-2.2x^5 + 3x^3 - 2x$
15. $14x^6 - 2x + 1$

In Problems 16 and 17, perform the indicated operations.

16. $3xy + 2x + 3y + 2 - (1 - y - x + xy)$
17. $(a+2)(3a^2 - a + 5)$

In Problems 18 and 19, factor each expression.

18. $8a^3b^5 - 12a^5b^2 + 16a^2b$
19. $4 - 9x^2$

In Problems 20 and 21, perform the indicated operations and simplify.

20. $\frac{m^4}{3n^2} \div \left(\frac{m^4}{9n} \cdot \frac{n}{2m^3}\right)$
21. $\frac{16-x^2}{x^2-3x-4} \cdot \frac{x-1}{x+4}$
22. Find the LCD of
$$\frac{-1}{2x^2} \qquad \frac{2}{4x^2-4} \qquad \frac{3}{x-2}$$

In Problems 23 and 24, perform the indicated operations and simplify.

23. $\frac{2x}{x^2-9} + \frac{5}{3x+9}$
24. $\frac{2-\frac{4}{x+1}}{x-1}$

In Problems 25–28, simplify and express the answers using only positive exponents.

25. $\left(\frac{x^{7/2}}{x^{2/3}}\right)^{-6}$
26. $\frac{y^{2n}}{y^{n-1}}$
27. $\frac{-1}{(x-1)^0}$
28. $(2a^2b^{-1})^2$

In Problems 29–31, perform the indicated operations.

29. $3\sqrt[3]{24} - 2\sqrt[3]{81}$
30. $(\sqrt{7}-5)^2$
31. $\frac{1}{2}\sqrt{\frac{xy}{4}} - \sqrt{9xy}$
32. For what values of x is $\sqrt{2-x}$ a real number?

In Problems 33–35, perform the indicated operations and write all answers in the form $a + bi$.

33. $(2-i) + (-3+i)$
34. $(5+2i)(2-3i)$
35. $\frac{5+2i}{2-i}$

Writing Exercises

1. Evaluate $(8)(1.4142)$ and $(8)(\sqrt{2})$. Are these results close to one another? Why?
2. Discuss the need for the complex number system.
3. Compare and contrast the properties of the complex numbers with those of the real numbers.
4. Discuss why division by zero is not permitted.

Chapter 1 Project

Polynomial expressions are used by physicists to study the motion of objects in free fall. Free fall means that the attraction of gravity is the only force operating on the object. In reality, other forces like air resistance play a role.

Take a look at Exercises 86 and 87 in Section 1.3 and Exercises 84–86 in Section 1.4. Set up a table for various planets or moons in our solar system, and use the Internet or other resources to find the data you need to write free-fall equations for objects on those worlds. (*Hint:* The value of a is all you need.) Here are some values to start you off:

Mars: $a = 3.72$

Earth: $a = 4.9$

The Moon: $a = 1.6$

All these values are in SI units, so the accelerations given above are in meters per second squared.

Try to redo the Exercises listed above for various planets. Write a paragraph explaining the problem described in the chapter opener.

2 Equations and Inequalities

The Internet is a short form of the word "internetworking." The Internet is a vast data network, with humble origins in the 1960s. A network, whether it connects computers or people, is just a way of facilitating communication. In a *full-mesh network*, elements are linked pairwise—that is, any two elements in the system are linked directly, without intermediary.

How many elements (users) could be linked in a full-mesh network with 190 two-way links? The answer to this problem is found by solving a quadratic equation (see the Chapter Project). This chapter will show you how.

Explore the Internet for its many mathematical offerings! Check out a site which is organized according to the Mathematics Subject Classification created by the American Mathematical Society. Look up graph theory to learn more about networking.

A major concern of algebra is the solution of equations. Does a given equation have a solution? Is it possible for an equation to have more than one solution? Is there a procedure for solving an equation? In this chapter we will explore the answers to these questions for polynomial equations of the first and second degree. We will also see that the ability to solve equations enables us to tackle a wide variety of applications and word problems.

Linear inequalities also play an important role in solving word problems. For example, if we are required to combine food products in such a way that a specified minimum daily requirement for various nutrients is provided, we need to use inequalities. Many important industries, including steel and petroleum, use computers daily to solve problems that involve thousands of inequalities. The solutions to such problems enable a company to optimize its "product mix" and its profitability.

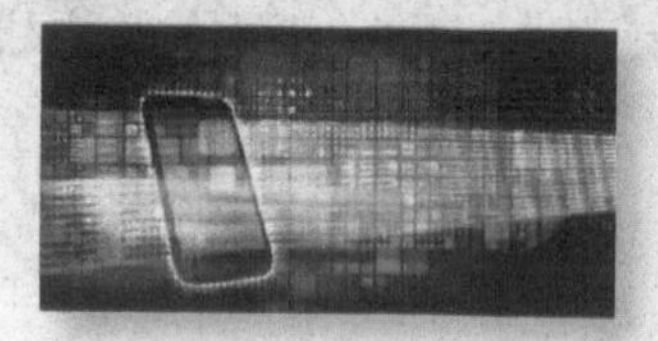

http://math-atlas.org

2.1 Linear Equations in One Unknown

2.1a Solving Equations

Expressions of the form

$$x - 2 = 0 \qquad x^2 - 9 = 0 \qquad 3(2x - 5) = 3$$

$$2x + 5 = \sqrt{x - 7} \qquad \frac{1}{2x + 3} = 5 \qquad x^3 - 3x^2 = 32$$

are examples of equations in the unknown x. An **equation** states that two algebraic expressions are equal. We refer to these expressions as the **left-hand side** and the **right-hand side** of the equation.

Our task is to find values of the unknown for which the equation is satisfied. These values are called **solutions** or **roots** of the equation, and the set of all solutions is called the **solution set**. For example, 2 is a solution of the equation $3x - 1 = 5$ since $3(2) - 1 = 5$. However, -2 is *not* a solution since $3(-2) - 1 \neq 5$.

Equations that do not have solutions in one number system may have solutions in a larger number system. For example, the equation $2x - 5 = 0$ has no integer solutions but does have a solution among the rational numbers, namely $\frac{5}{2}$. Similarly, the equation $x^2 = -4$ has no solutions among the real numbers but does have solutions if we consider complex numbers, namely $2i$ and $-2i$. The solution sets of these two equations are $\{\frac{5}{2}\}$ and $\{2i, -2i\}$, respectively.

Identities and Conditional Equations

We say that an equation is an **identity** if it is true for every real number for which both sides of the equation are defined. For example, the equation

$$x^2 - 1 = (x + 1)(x - 1)$$

is an identity because it is true for all real numbers. (Try any number and check that this equation holds.) The equation

$$x - 5 = 3$$

is only true when $x = 8$. (Try any number not equal to 8 and check that this equation does not hold.) An equation such as $x - 5 = 3$, which is not true for all values of x, is called a **conditional equation.**

When we say that we want to "solve an equation," we mean that we want to find *all* solutions or roots. If we can replace an equation with another, simpler equation that has the same solutions, we will have an approach to solving equations. Equations having the same solutions are called **equivalent equations.** For example, $3x - 1 = 5$ and $3x = 6$ are equivalent equations because it can be shown that $\{2\}$ is the solution set of both equations.

There are two important rules that allow us to replace an equation with an equivalent equation.

Equivalent Equations

The solutions of a given equation are not affected by the following operations:

1. addition (or subtraction) of the same number or expression on both sides of the equation
2. multiplication (or division) by the same number, different from 0, on both sides of the equation

EXAMPLE 1 *Solving Equations*

Solve $3x + 4 = 13$.

SOLUTION

We apply the preceding rules to this equation. The strategy is to isolate x, so we *subtract 4 from both sides of the equation.*

$$3x + 4 - 4 = 13 - 4$$
$$3x = 9$$

Dividing both sides by 3, we obtain the solution

$$x = 3$$

We check by substitution to make sure that 3 does, indeed, satisfy the original equation.

$$\begin{aligned}\text{left-hand side} &= 3x + 4 \qquad \text{right-hand side} = 13\\ &= 3(3) + 4\\ &= 13\end{aligned}$$

■

Although $x = 3$ is an equation that is *equivalent* to the original equation, in common usage we say that $3x + 4 = 13$ "has the solution $x = 3$."

When the given equation contains rational expressions, we eliminate fractions by first multiplying by the least common denominator of all fractions present. This technique is illustrated in Examples 2, 3, and 4.

EXAMPLE 2 *Solving Equations*

Solve the equation.

$$\frac{5}{6}x - \frac{4}{3} = \frac{3}{5}x + 1$$

SOLUTION

We first eliminate fractions by multiplying both sides of the equation by the LCD of all fractions, which is 30.

$$\left(\frac{5}{6}x - \frac{4}{3}\right)(30) = \left(\frac{3}{5}x + 1\right)(30)$$
$$25x - 40 = 18x + 30$$
$$7x = 70$$
$$x = 10$$

Verify that $x = 10$ is a solution of the original equation. ■

Progress Check

Solve and check.

a. $-\frac{2}{3}(x-5) = \frac{3}{2}(x+1)$ b. $\frac{1}{3}x + 2 - 3\left(\frac{x}{2} + 4\right) = 2\left(\frac{x}{4} - 1\right)$

Answers

a. $\frac{11}{13}$ b. $-\frac{24}{5}$

2.1b Solving Linear Equations

The equations we have solved are all of the first degree and involve only one unknown. Such equations are called **first-degree equations in one unknown**, or more simply, **linear equations.** The general form of such equations is

$$ax + b = 0$$

where a and b are any real numbers and $a \neq 0$. Let us see how to solve this equation.

$$ax + b = 0$$

$$ax + b - b = 0 - b \qquad \text{Subtract } b \text{ from both sides.}$$

$$ax = -b$$

$$\frac{ax}{a} = \frac{-b}{a} \qquad \text{Divide both sides by } a \neq 0.$$

$$x = -\frac{b}{a}$$

We verify that this is a solution.

$$a\left(-\frac{b}{a}\right) + b = 0$$

Furthermore, it can be shown that this is the only solution. We have thus obtained the following result:

Roots of a Linear Equation

The linear equation $ax + b = 0$, $a \neq 0$, has exactly one solution:

$$x = -\frac{b}{a}$$

Sometimes we are led to linear equations in the course of solving other equations. The following example illustrates this situation.

EXAMPLE 3 *Solving Equations*

Solve.

$$\frac{5x}{x+3} - 3 = \frac{1}{x+3}$$

SOLUTION

The LCD of all fractions is $x + 3$. Multiplying both sides of the equation by $x + 3$ to eliminate fractions, we obtain

$$5x - 3(x + 3) = 1$$

$$5x - 3x - 9 = 1$$

$$2x = 10$$

$$x = 5$$

Checking the solution, we have

$$\text{left-hand side} = \frac{5x}{x+3} - 3 \qquad \text{right-hand side} = \frac{1}{x+3}$$

$$= \frac{5(5)}{5+3} - 3 \qquad = \frac{1}{5+3}$$

$$= \frac{25}{8} - 3 \qquad = \frac{1}{8}$$

$$= \frac{25}{8} - \frac{24}{8}$$

$$= \frac{1}{8}$$

■

We said earlier that multiplication (or division) of both sides of an equation by any nonzero number results in an equivalent equation. What happens if we multiply or divide an equation by an expression that contains an unknown? In Example 3, this procedure worked and gave us a solution. However, this may not always be so since the answer we obtain may produce a zero denominator when substituted back into the original equation. Therefore, the following rule must be carefully observed:

Multiplying by an Unknown

Multiplication (or division) by the same expression on both sides of an equation may result in an equation that is *not* equivalent to the original equation. Always verify that the answer obtained to the subsequent equation is, indeed, a solution to the original equation.

EXAMPLE 4 *Equations with No Solution*

Solve and check.

$$\frac{8x+1}{x-2}+4=\frac{7x+3}{x-2}$$

SOLUTION

The LCD of all fractions is $x-2$. Multiplying both sides of the equation by $x-2$, we eliminate fractions and obtain

$$8x+1+4(x-2)=7x+3$$
$$8x+1+4x-8=7x+3$$
$$5x=10$$
$$x=2$$

Checking our answer, we find that $x=2$ is not a solution since substituting $x=2$ in the original equation yields a denominator of zero. We conclude that the given equation has no solution. ■

Progress Check

Solve and check.

a. $\frac{3}{x}-1=\frac{1}{2}-\frac{6}{x}$ b. $\frac{2x}{x+1}=1+\frac{2}{x+1}$

Answers

a. $x=6$ b. no solution

EXAMPLE 5 *Equations with No Solution*

Solve the equation $2x+1=2x-3$.

SOLUTION

Subtracting $2x$ from both sides, we have

$$2x+1-2x=2x-3-2x$$
$$1=-3$$

This equivalent equation is a contradiction, so we conclude that the given equation has no solution. ■

Exercise Set 2.1

In Exercises 1–4, determine whether the given statement is true (T) or false (F).

1. $x = -5$ is a solution of $2x + 3 = -7$.

2. $x = \frac{5}{2}$ is a solution of $3x - 4 = \frac{5}{2}$.

3. $x = \frac{6}{4-k}$, $k \neq 4$ is a solution of $kx + 6 = 4x$.

4. $x = \frac{7}{3k}$, $k \neq 0$ is a solution of $2kx + 7 = 5x$.

In Exercises 5–24, solve the given linear equation and check your answer.

5. $3x + 5 = -1$
6. $5r + 10 = 0$
7. $2 = 3x + 4$
8. $\frac{1}{2}s + 2 = 4$
9. $\frac{3}{2}t - 2 = 7$
10. $-1 = -\frac{2}{3}x + 1$
11. $0 = -\frac{1}{2}a - \frac{2}{3}$
12. $4r + 4 = 3r - 2$
13. $-5x + 8 = 3x - 4$
14. $2x - 1 = 3x + 2$
15. $-2x + 6 = -5x - 4$
16. $6x + 4 = -3x - 5$
17. $2(3b + 1) = 3b - 4$
18. $-3(2x + 1) = -8x + 1$
19. $4(x - 1) = 2(x + 3)$
20. $-3(x - 2) = 2(x + 4)$
21. $2(x + 4) - 1 = 0$
22. $3a + 2 - 2(a - 1) = 3(2a + 3)$
23. $-4(2x + 1) - (x - 2) = -11$
24. $3(a + 2) - 2(a - 3) = 0$

Solve for x in Exercises 25–28.

25. $kx + 8 = 5x$
26. $8 - 2kx = -3x$
27. $2 - k + 5(x - 1) = 3$
28. $3(2 + 3k) + 4(x - 2) = 5$

Solve and check in Exercises 29–44.

29. $\frac{x}{2} = \frac{5}{3}$
30. $\frac{3x}{4} - 5 = \frac{1}{4}$
31. $\frac{2}{x} + 1 = \frac{3}{x}$
32. $\frac{5}{a} - \frac{3}{2} = \frac{1}{4}$
33. $\frac{2y-3}{y+3} = \frac{5}{7}$
34. $\frac{1-4x}{1-2x} = \frac{9}{8}$
35. $\frac{1}{x-2} + \frac{1}{2} = \frac{2}{x-2}$
36. $\frac{4}{x-4} - 2 = \frac{1}{x-4}$
37. $\frac{2}{x-2} + \frac{2}{x^2-4} = \frac{3}{x+2}$
38. $\frac{3}{x-1} + \frac{2}{x+1} = \frac{5}{x^2-1}$
39. $\frac{x}{x-1} - 1 = \frac{3}{x+1}$
40. $\frac{2}{x-2} + 1 = \frac{x+2}{x-2}$
41. $\frac{4}{b} - \frac{1}{b+3} = \frac{3b+2}{b^2+2b-3}$
42. $\frac{3}{x^2-2x} + \frac{2x-1}{x^2+2x-8} = \frac{2}{x+4}$
43. $\frac{3r+1}{r+3} + 2 = \frac{5r-2}{r+3}$
44. $\frac{2x-1}{x-5} + 3 = \frac{3x-2}{5-x}$

In Exercises 45–48, indicate whether the equation is an identity (I) or a conditional equation (C).

45. $x^2 + x - 2 = (x + 2)(x - 1)$
46. $(x - 2)^2 = x^2 - 4x + 4$
47. $2x + 1 = 3x - 1$
48. $3x - 5 = 4x - x - 2 - 3$

In Exercises 49–54, write (T) if the equations within each exercise are all equivalent equations and (F) if they are not equivalent.

49. $2x - 3 = 5$ $\quad 2x = 8$ $\quad x = 4$
50. $5(x - 1) = 10$ $\quad x - 1 = 2$ $\quad x = 3$
51. $x(x - 1) = 5x$ $\quad x - 1 = 5$ $\quad x = 6$
52. $x = 5$ $\quad x^2 = 25$
53. $3(x^2 + 2x + 1) = -6$
$x^2 + 2x + 1 = -2$
$(x + 1)^2 = -2$
54. $(x + 3)(x - 1) = x^2 - 2x + 1$
$(x + 3)(x - 1) = (x - 1)^2$
$x + 3 = x - 1$

55. Write repeating decimal fractions as rational equivalents.

Example: $N = 0.1515\ldots = 0.\overline{15}$
$100N = 15.\overline{15}$
$-N = -0.\overline{15}$
$99N = 15$
$N = \frac{15}{99} = \frac{5}{33}$

a. $0.\overline{2}$ $\quad$ b. $0.\overline{123}$
c. $1.\overline{35}$ $\quad$ d. $0.\overline{9}$

56. Find the error in the following argument. Assume that $a = b \neq 0$.

$a^2 = ab$
$a^2 - b^2 = ab - b^2$
$(a + b)(a - b) = b(a - b)$
$a + b = b$
$b + b = b$ (since $a = b$)
$2b = b$
$2 = 1$ (since $b \neq 0$)

57. Solve

a. $\frac{w-c}{w-d} = \frac{c^2}{d^2}$ for w

b. $a^2 = \frac{a+c}{x} + c^2$ for x

c. $(a-y)(y+b) - c(y+c) = (c-y)(y+c) + ab$ for y

58. Solve for y.

a. $y + \frac{c}{y-3} = 3 + \frac{c}{y-3}$

b. $y + \frac{c}{y-3} = -3 + \frac{c}{y-3}$

59. The golden ratio is given by

$$T = \frac{1+\sqrt{5}}{2}$$

(See Chapter 1, Review Exercise 78.) Show that

$$T = 1 + \cfrac{1}{1 + \cfrac{1}{1 + \frac{1}{T}}}$$

2.2 Applications: From Words to Algebra

Many applied problems lead to linear equations. The challenge of applied problems is translating words into appropriate algebraic forms.

The steps listed here can guide you in solving word problems.

Step 1. Read the problem through the first time to get a general idea of what is being asked.

Step 2. Read the problem a second time to recognize what may be important in determining that which is to be found.

Step 3. If possible, estimate the solution to this problem, and then compare this estimate with your final answer.

Step 4. Let some algebraic symbol denote the quantity to be found.

Step 5. If possible, represent other quantities in the problem in terms of the algebraic symbol designated in *Step 4.*

Step 6. Find various relationships (equations or inequalities) in the problem.

Step 7. Use relationships established in *Step 6* to find the solution to the problem.

Step 8. Verify that your answer is, indeed, the solution to the problem.

The words and phrases in Table 2-1 may prove helpful in translating a word problem into an algebraic expression that can be solved.

TABLE 2-1 Translation of Words into Algebraic Expressions

Word or Phrase	Algebraic Symbol	Example	Algebraic Expression
Sum	$+$	Sum of two numbers	$a + b$
Difference	$-$	Difference of two numbers	$a - b$
		Difference of a number and 3	$x - 3$
Product	$\times$ or $\cdot$	Product of two numbers	$a \cdot b$, $(a)(b)$, or ab
Quotient	$\div$ or /	Quotient of two numbers	$\frac{a}{b}$, a/b, or $a \div b$
Exceeds		a exceeds b by 3	$a = b + 3$
More than		a is 3 more than b	or
More of		There are 3 more of a than of b.	$a - 3 = b$
Twice		Twice a number	$2x$
		Twice the difference of x and 3	$2(x - 3)$
		3 more than twice a number	$2x + 3$
		3 less than twice a number	$2x - 3$
Is or equals	$=$	The sum of a number and 3 is 15.	$x + 3 = 15$

EXAMPLE 1 *Prices and Discounts*

If you pay \$66 for a car radio after receiving a 25% discount, what was the price of the radio before the discount?

SOLUTION

Let $p =$ the price of the radio (in dollars) before the discount. Then

$$0.25p = \text{the amount discounted}$$

and the price of the radio after the discount is given by

$$p - 0.25p$$

Hence

$$p - 0.25p = 66$$
$$0.75p = 66$$
$$p = \frac{66}{0.75} = 88$$

The price of the radio was \$88 before the discount. ■

2.2a Coin Problems

When interpreting coin problems, always distinguish between the *number* of coins and the *value* of the coins. You may also find it helpful to use a chart, as in the following example:

EXAMPLE 2 *Coins*

A purse contains \$3.20 in quarters and dimes. If there are 3 more quarters than dimes, how many coins of each type are there?

SOLUTION

In this problem, we may let the unknown represent either the number of quarters or the number of dimes. Let

$$q = \text{the number of quarters}$$

Then

$$q - 3 = \text{the number of dimes}$$

since there are 3 more quarters than dimes.

Note that the number of coins times the value of each coin in cents is equal to the total value in cents using that particular coin.

	Number of coins ×	Value of each coin in cents =	Total value in cents using that coin
Quarters	q	25	$25q$
Dimes	$q - 3$	10	$10(q - 3)$

We know that

$$\begin{aligned} \text{total value} &= (\text{value of quarters}) + (\text{value of dimes}) \\ 320 &= 25q + 10(q - 3) \\ 320 &= 25q + 10q - 30 \\ 350 &= 35q \\ 10 &= q \end{aligned}$$

Then

$$\begin{aligned} q &= \text{number of quarters} = 10 \\ q - 3 &= \text{number of dimes} = 7 \end{aligned}$$

Now verify that the total value of all the coins is $3.20. ■

2.2b Simple Interest

Interest is the fee charged for borrowing money. In this section we will deal only with simple interest, which assumes the fee to be a fixed percentage r of the amount borrowed. We call the amount borrowed the **principal** and denote it by P.

If the principal P is borrowed at a simple annual interest rate r, then the interest due at the end of each year is Pr, and the total interest I due at the end of t years is

$$I = Prt$$

Consequently, if S is the total amount owed at the end of t years, then

$$S = P + I = P + Prt$$

since both the principal and interest are to be repaid. Thus, the basic formulas for simple interest calculations are

$$I = Prt$$
$$S = P + Prt$$

EXAMPLE 3 *Simple Interest*

A part of $7000 was borrowed at 6% simple annual interest and the remainder at 8%. If the total amount of interest due after 3 years is $1380, how much was borrowed at each rate?

SOLUTION

Let

$$s = \text{the amount borrowed at } 6\%$$

Then

$$7000 - s = \text{the amount borrowed at } 8\%$$

since the total amount is $7000. We can display the information in table form using the equation $I = Prt$.

	P ×	r ×	t =	Interest
6% Portion	s	0.06	3	$0.18s$
8% Portion	$7000 - s$	0.08	3	$0.24(7000 - s)$

Note that we write the rate r in its decimal form, so that $6\% = 0.06$ and $8\% = 0.08$.

Since the total interest of $1380 is the sum of the interest from the two portions, we have

$$\begin{aligned} 1380 &= 0.18s + 0.24(7000 - s) \\ 1380 &= 0.18s + 1680 - 0.24s \\ 0.06s &= 300 \\ s &= 5000 \end{aligned}$$

We conclude that $5000 was borrowed at 6% and $2000 was borrowed at 8%. ■

2.2c *Distance Problems (Uniform Motion)*

Here is the key to the solution of distance problems.

$$\text{Distance} = (\text{Rate})(\text{Time})$$

or

$$d = r \cdot t$$

The relationships that permit you to write an equation are sometimes obscured by the words. Here are some questions to ask as you set up a distance problem.

1. Are there two distances that are equal? (Will two objects have traveled the same distance? Is the distance on a return trip the same as the distance going?)
2. Is the sum (or difference) of two distances equal to a constant? (When two objects are traveling toward each other, they meet when the sum of the distances traveled by them equals the original distance between them.)

EXAMPLE 4 ***Travel***

Two trains leave New York for Chicago. The first train travels at an average speed of 60 mph. The second train, which departs an hour later, travels at an average speed of 80 mph. How long will it take the second train to overtake the first train?

SOLUTION

Since we are interested in the time the second train travels, we let

$t =$ the number of hours the second train travels

Then

$t + 1 =$ the number of hours the first train travels

since the first train departs 1 hour earlier. We display the information in table form using the equation $d = rt$.

	Rate	×	Time	=	Distance
First train	60		$t + 1$		$60(t + 1)$
Second train	80		t		$80t$

At the moment the second train overtakes the first, they must both have traveled the *same* distance. Thus,

$$\begin{aligned} 60(t+1) &= 80t \\ 60t + 60 &= 80t \\ 60 &= 20t \\ 3 &= t \end{aligned}$$

It takes the second train 3 hours to catch up with the first train. ■

2.2d Mixture Problems

One type of mixture problem involves mixing varieties of a commodity, say two or more types of coffee, to obtain a mixture with a desired value. If the commodity is measured in pounds, the relationships we need are as follows:

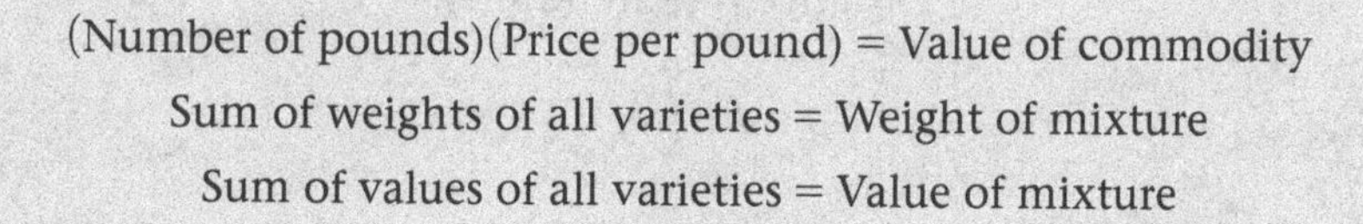

(Number of pounds)(Price per pound) = Value of commodity

Sum of weights of all varieties = Weight of mixture

Sum of values of all varieties = Value of mixture

EXAMPLE 5 *Mixtures*

How many pounds of Brazilian coffee worth \$10 per pound must be mixed with 20 pounds of Colombian coffee worth \$8 per pound to produce a mixture worth \$8.40 per pound?

SOLUTION

Let B = number of pounds of Brazilian coffee. We display all the information, using cents in place of dollars.

Type of coffee	Number of pounds ×	Price per pound	= Value (in cents)
Brazilian	B	1000	$1000B$
Colombian	20	800	16,000
Mixture	$B + 20$	840	$840(B + 20)$

Note that the weight of the mixture equals the sum of the weights of the Brazilian and Colombian coffees that make up the mixture. Since the value of the mixture is the sum of the values of the two types of coffee,

$$\begin{aligned} \text{value of mixture} &= (\text{value of Brazilian}) + (\text{value of Colombian}) \\ 840(B + 20) &= 1000B + 16{,}000 \\ 840B + 16{,}800 &= 1000B + 16{,}000 \\ 800 &= 160B \\ 5 &= B \end{aligned}$$

We must add 5 pounds of Brazilian coffee to make the required mixture. ■

2.2e Work Problems

Work problems typically involve two or more people or machines working on the same task. The key to these problems is to express the *rate of work per unit of time*, whether an hour, a day, a week, or some other unit. For example, if a machine can do a job in 5 days, then

$$\text{rate of machine} = \frac{1}{5} \text{ job per day}$$

If this machine is used for 2 days, it performs $2(\frac{1}{5}) = \frac{2}{5}$ of the job. In summary:

If a machine (or person) can complete a job in n days, then

$$\text{Rate of machine (or person)} = \frac{1}{n} \text{ job per day}$$

$$\text{Work done} = (\text{Rate})(\text{Time})$$

EXAMPLE 6 *Work*

Using a small mower, at 12 noon a student begins to mow a lawn, a job that would take 9 hours working alone. At 1 p.m. another student, using a tractor, joins the first student and they complete the job together at 3 p.m. How many hours would it take to do the job using only the tractor?

SOLUTION

Let t = number of hours to do the job by tractor alone. The small mower works from 12 noon to 3 p.m., or 3 hours. The tractor is used from 1 p.m. to 3 p.m., or 2 hours.

All the information can be displayed in table form.

	Rate	×	Time	=	Work done
Small mower	$\frac{1}{9}$		3		$\frac{3}{9} = \frac{1}{3}$
Tractor	$\frac{1}{t}$		2		$\frac{2}{t}$

Since

$$\text{Work done by small mower} + \text{Work done by tractor} = 1 \text{ Whole job}$$

$$\frac{1}{3} \quad + \quad \frac{2}{t} \quad = 1$$

To solve, multiply both sides by the LCD, which is $3t$.

$$\left(\frac{1}{3} + \frac{2}{t}\right)(3t) = 1(3t)$$

$$t + 6 = 3t$$

$$t = 3$$

Thus, by tractor alone, the job can be done in 3 hours. ■

2.2f Formulas

The circumference C of a circle is given by the formula

$$C = 2\pi r$$

where r is the radius of the circle. For every value of r, the formula gives us a value of C. If $r = 20$, we have

$$C = 2\pi(20) = 40\pi$$

It is sometimes convenient to be able to turn a formula around, that is, to be able to solve for a different variable. For example, if we want to express the radius of a circle in terms of the circumference, we have

$$C = 2\pi r$$

$$\frac{C}{2\pi} = \frac{2\pi r}{2\pi} \quad \text{Dividing by } 2\pi$$

$$\frac{C}{2\pi} = r$$

Now, given a value of C, we can determine a value of r.

EXAMPLE 7 ***Manipulation of Formulas***

If an amount P is borrowed at the simple annual interest rate r, then the amount S due at the end of t years is

$$S = P + Prt$$

Solve for P.

SOLUTION

$$P + Prt = S$$

$$P(1 + rt) = S \quad \text{Common factor } P$$

$$P = \frac{S}{1 + rt} \quad \text{Dividing both sides by } (1 + rt)$$

■

Exercise Set 2.2

In Exercises 1–3, let n represent the unknown. Translate from words to an algebraic expression or equation.

1. The number of blue chips is 3 more than twice the number of red chips.
2. The number of station wagons on a parking lot is 20 fewer than 3 times the number of sedans.
3. Five less than 6 times a number is 26.

In Exercises 4–41, translate from words to an algebraic problem and solve.

4. Janis is 3 years older than her sister. Thirty years from now the sum of their ages will be 111. Find the current ages of the sisters.
5. Luis is presently 12 years older than Carlos. Four years ago Luis was twice as old as Carlos. How old is each now?
6. The larger of two numbers is 3 more than twice the smaller. If their sum is 18, find the numbers.
7. Find three consecutive integers whose sum is 21.
8. A certain number is 5 less than another number. If their sum is 11, find the two numbers.
9. A resort guarantees that the average temperature over the period Friday, Saturday, and Sunday will be exactly 80°F, or else each guest pays only half price for the facilities. If the temperatures on Friday and Saturday were 90°F and 82°F, respectively, what must the temperature be on Sunday so that the resort does not lose half of its revenue?
10. A patient's temperature was taken at 6 a.m., 12 noon, 3 p.m., and 8 p.m. The first, third, and fourth readings were 102.5°, 101.5°, and 102°F, respectively. The nurse forgot to write down the second reading, but recorded that the average of the four readings was 101.5°F. What was the second temperature reading?
11. A 12-meter long steel beam is to be cut into two pieces so that one piece will be 4 meters longer than the other. How long will each piece be?
12. A rectangular field whose length is 10 meters longer than its width is to be enclosed with exactly 100 meters of fencing material. What are the dimensions of the field?
13. A vending machine contains $3.00 in nickels and dimes. If the number of dimes is 5 more than twice the number of nickels, how many coins of each type are there?
14. A wallet contains $460 in $5, $10, and $20 bills. The number of $5 bills exceeds twice the number of $10 bills by 4, and the number of $20 bills is 6 fewer than the number of $10 bills. How many bills of each type are there?
15. A movie theater charges $7.50 admission for an adult and $5 for a child. If 700 tickets are sold on a particular day and the total revenue received is $4500, how many tickets of each type are sold?
16. A student bought 23-cent, 41-cent, and 80-cent stamps with a total value of $31.50. If the number of 23-cent stamps is 2 more than the number of 41-cent stamps, and the number of 80-cent stamps is 5 more than one-half the number of 41-cent stamps, how many stamps of each denomination did the student obtain?
17. An amateur theater group is converting a classroom to an auditorium for a forthcoming play. The group sells $3, $5, and $6 tickets, and receives exactly $503 from the sale of tickets. If the number of $5 tickets is twice the number of $6 tickets, and the number of $3 tickets is 1 more than 3 times the number of $6 tickets, how many tickets of each type are there?
18. To pay for their child's college education, the parents invested $10,000, part in a certificate of deposit paying 8.5% annual interest, the rest in a mutual fund paying 7% annual interest. The annual income from the certificate of deposit is $200 more than the annual income from the mutual fund. How much money was put into each type of investment?
19. A bicycle store is closing out its entire stock of a certain brand of three-speed and ten-speed models. The profit on a three-speed bicycle is 11% of the sale price, and the profit on a ten-speed model is 22% of the sale price. If the entire stock is sold for $16,000 and the profit on the entire stock is 19%, how much is obtained from the sale of each type of bicycle?
20. A film shop carrying black-and-white film and color film has $4000 in inventory. The profit on black-and-white film is 12%, and the profit on color film is 21%. If all the film is sold, and if the profit on color film is $150 less than the profit on black-and-white film, how much was invested in each type of film?
21. A firm borrowed $12,000 at a simple annual interest rate of 8% for a period of 3 years. At the end of the first year, the firm found that its needs were reduced. The firm returned a portion of the original loan and retained the remainder until the end of the 3-year period. If the total interest paid was $1760, how much was returned at the end of the first year?

22. A finance company lent a certain amount of money to Firm A at 7% annual interest. An amount \$100 less than that lent to Firm A was lent to Firm B at 8%, and an amount \$200 more than that lent to Firm A was lent to Firm C at 8.5%. All loans were for one year. If the total annual income is \$126.50, how much was lent to each firm?

23. Two trucks leave Philadelphia for Miami. The first truck to leave travels at an average speed of 50 kilometers per hour. The second truck, which leaves 2 hours later, travels at an average speed of 55 kilometers per hour. How long does it take the second truck to overtake the first truck?

24. Jackie either drives or bicycles from home to school. Her average speed when driving is 36 mph, and her average speed when bicycling is 12 mph. If it takes her $\frac{1}{2}$ hour less to drive to school than to bicycle, how long does it take her to go to school, and how far is the school from her home?

25. Professors Roberts and Jones, who live 676 miles apart, are exchanging houses and jobs for the summer. They start out for their new locations at exactly the same time, and they meet after 6.5 hours of driving. If their average speeds differ by 4 mph, what are their average speeds?

26. Steve leaves school by moped for spring vacation. Forty minutes later his roommate, Frank, notices that Steve forgot to take his camera. So, Frank decides to try to catch up with Steve by car. If Steve's average speed is 25 mph and Frank averages 45 mph, how long does it take Frank to overtake Steve?

27. An express train and a local train start out from the same point at the same time and travel in opposite directions. The express train travels twice as fast as the local train. If after 4 hours they are 480 kilometers apart, what is the average speed of each train?

28. How many pounds of raisins worth \$3 per pound must be mixed with 10 pounds of peanuts worth \$2.40 per pound to produce a mixture worth \$2.80 per pound?

29. How many ounces of Ceylon tea worth \$1.50 per ounce and how many ounces of Formosa tea worth \$2.00 per ounce must be mixed to obtain a mixture of 8 ounces that is worth \$1.85 per ounce?

30. A copper alloy that is 40% copper is to be combined with a copper alloy that is 80% copper to produce 120 kilograms of an alloy that is 70% copper. How many kilograms of each alloy must be used?

31. A vat contains 27 gallons of water and 9 gallons of acetic acid. How many gallons of water must be evaporated if the resulting solution is to be 40% acetic acid?

32. A producer of packaged frozen vegetables wants to market mixed vegetables at \$1.20 per kilogram. How many kilograms of green beans worth \$1.00 per kilogram must be mixed with 100 kilograms of corn worth \$1.30 per kilogram and 90 kilograms of peas worth \$1.40 per kilogram to produce a satisfactory mixture?

33. A certain number is 3 times another. If the difference of their reciprocals is 8, find both numbers.

34. If $\frac{1}{3}$ is subtracted from 3 times the reciprocal of a certain number, the result is $\frac{25}{6}$. Find the number.

35. Computer A can carry out an engineering analysis in 6 hours, but computer B can do the same job in 4 hours. How long does it take to complete the job if both computers work together?

36. Maria can paint a certain room in 3 hours, Juanita in 4 hours, and Thuy in 2 hours. How long does it take to paint the room if they all work together?

37. A senior copy editor together with a junior copy editor can edit a book in 3 days. The junior editor, working alone, would take twice as long to complete the job as the senior editor would require if working alone. How long would it take each editor to complete the job by herself?

38. Hose A can fill a certain vat in 3 hours. After 2 hours of pumping, hose A is turned off. Hose B is then turned on and completes filling the vat in 3 more hours. How long would it take hose B to fill the vat alone?

39. A printing shop starts a job at 10 a.m. on press A. Using this press alone, it would take 8 hours to complete the job. At 2 p.m. press B is also turned on, and both presses together finish the job at 4 p.m. How long would it take press B to do the job alone?

40. A boat travels 20 kilometers upstream in the same time that it would take the same boat to travel 30 kilometers downstream. If the rate of the stream is 5 kilometers per hour, find the speed of the boat in still water.

41. An airplane flying against the wind travels 300 miles in the same time that it would take the same plane, flying the same speed, to travel 400 miles with the wind. If the wind speed is 20 mph, find the speed of the airplane in still air.

In Exercises 42–51 solve for the indicated variable in terms of the remaining variables.

42. $A = Pr$ for r

43. $C = 2\pi r$ for r

44. $V = \frac{1}{3}\pi r^2 h$ for h

45. $F = \frac{9}{5}C + 32$ for C

46. $S = \frac{1}{2}gt^2 + vt$ for v

47. $A = \frac{1}{2}h(b + b')$ for b

48. $A = P(1 + rt)$ for r

49. $\frac{1}{f} = \frac{1}{f_1} + \frac{1}{f_2}$ for f_2

50. $a = \frac{v_1 - v_0}{t}$ for v_0

51. $S = \frac{a - rL}{L - r}$ for L

52. Translate the following from words to an algebraic expression or equation, denoting the unknown by n.

a. The express train travels 5 mph faster than the local train.

b. The length of a rectangle is 7 inches more than its width.

c. The area of a triangle, if the altitude is twice the base.

d. The sum of 3 consecutive even numbers.

e. 15% of the amount by which a number exceeds 10,000.

53. If r and s represent two numbers, write the following:

a. twice the sum of the two numbers

b. 5% of the difference between the two numbers

c. 5 less than twice the second number

d. the ratio of the first to the second number

e. the sum of the squares of the two numbers

f. the average of the two numbers

g. 6 times the first number less 4 times the second number

54. Write formulas for each of the following:

a. the charge in cents for a long-distance telephone call lasting n minutes, n greater than 3, if the charge for the first 3 minutes is $1.20 and each additional minute costs 33 cents

b. the taxi fare for m miles, if the initial charge is $2.50 and the driver charges 70 cents for every $\frac{1}{5}$ mile traveled

c. the amount in an account at the end of a year, if simple interest is paid at the rate of 1.2%, and the account contains d dollars at the beginning of the year

d. the fine a company paid for dumping acid into the Mississippi River for d days, if the U.S. Environmental Protection Agency fined the company $150,000 plus $1000 per day until the company complied with the federal water pollution regulations

55. Find three consecutive even numbers such that twice the first plus 3 times the second is 4 times the third.

56. When exercising, Mary walks a distance to warm up, jogs $3\frac{1}{2}$ times as far as she walks, and sprints $3\frac{1}{3}$ times as far as she jogs. If she covers 4171 meters, find the distances that she walked, jogged, and sprinted.

57. A 10-quart radiator has 30% antifreeze. How much of the fluid should be drained and replaced with pure antifreeze to double the strength of the mixture?

58. There are two identical beakers in a chemistry laboratory, both filled to the same level. One contains sulfuric acid and the other contains water. First, one spoon of acid is put into the beaker with the water and mixed thoroughly. Then one spoon of this mixture is put back into the beaker with the acid. Is there more water in the acid or more acid in the water?

59. Two bicyclists leave cities A and B at the same time, heading toward each other. Their speeds are 20 mph and 30 mph, respectively. The distance between these cities is 100 miles. Simultaneously, a bird leaves city A, heading toward B, traveling at 40 mph. When it meets the bicyclist who left from B, it turns around and heads back toward A. When it subsequently meets the bicyclist who came from A, it turns around and heads back toward B, and so on. Find the total distance the bird will have flown by the time the two bicyclists meet.

60. To determine the number of deer in a forest, a conservationist catches 225 deer, tags them and then releases them. A week later, 102 deer are caught and, of those, 15 are found to be tagged. Assuming that the proportion of tagged deer in the second sample was the same as the proportion of all tagged deer in the total population, estimate the number of deer in the forest.

61. In a Tour de France bicycle race, Stefan averaged 20 mph for the first third of the race and 35 mph for the remainder. Enrique maintained a constant speed of 30 mph throughout the race. Of these two, who finished first?

In Exercises 62–65, solve for the indicated variable.

62. $I = \frac{E}{r + \frac{R}{n}}$ for n

63. $Wf = \left(\frac{W}{k} - 1\right)\left(\frac{1}{k}\right)$ for W

64. $W = \frac{2PR}{R - r}$ for r

65. $\frac{E}{c} = \frac{R + r}{r}$ for r

2.3 The Quadratic Equation

We now turn our attention to equations involving second-degree polynomials. A **quadratic equation** is an equation of the form

$$ax^2 + bx + c = 0, \qquad a \neq 0$$

where a, b, and c are real numbers. In this section we will explore techniques for solving this important class of equations. We will also show that there are several kinds of equations that can be transformed into quadratic equations and then solved.

2.3a *Solving by Factoring*

If we can factor the left-hand side of the quadratic equation

$$ax^2 + bx + c = 0, \qquad a \neq 0$$

into two linear factors, then we can solve the equation. For example, the quadratic equation

$$x^2 - 5x + 6 = 0$$

can be written as

$$(x - 2)(x - 3) = 0$$

since 0 is the only number with the following property:

If $ab = 0$, then $a = 0$ or $b = 0$

We can set each factor of the above quadratic equation equal to 0.

$$x - 2 = 0 \quad \text{or} \quad x - 3 = 0$$
$$x = 2 \quad \text{or} \quad x = 3$$

The solutions of the given quadratic equation are 2 and 3.

EXAMPLE 1 ***Solving by Factoring***

Solve the equation $2x^2 - 3x - 2 = 0$ by factoring.

SOLUTION

Factoring, we have

$$2x^2 - 3x - 2 = 0$$
$$(2x + 1)(x - 2) = 0$$

Since the product of the factors is 0, at least one factor must be 0. Setting each factor equal to 0, we have

$$2x + 1 = 0 \quad \text{or} \quad x - 2 = 0$$
$$x = -\frac{1}{2} \quad \text{or} \quad x = 2$$

■

EXAMPLE 2 ***Solving by Factoring***

Solve the equation $3x^2 + 5x - 2 = 0$ by factoring.

SOLUTION

Factoring, we have

$$(3x - 1)(x + 2) = 0$$

$$3x - 1 = 0 \quad \text{or} \quad x + 2 = 0$$

$$x = \frac{1}{3} \quad \text{or} \quad x = -2$$

■

EXAMPLE 3 ***Solving by Factoring***

Solve the equation $3x^2 - 4x = 0$ by factoring.

SOLUTION

Factoring, we have

$$3x^2 - 4x = 0$$

$$x(3x - 4) = 0$$

Setting each factor equal to zero,

$$x = 0 \quad \text{or} \quad x = \frac{4}{3}$$

■

When considering an equation with a common factor, such as

$$3x^2 - 4x = 0$$

$$x(3x - 4) = 0$$

always set each factor equal to zero. A common error of students is to divide both sides of the above equation by x.

$$\frac{x(3x-4)}{x} = \frac{0}{x}$$

concluding that

$$3x - 4 = 0$$

$$x = \frac{4}{3}$$

The only time this operation is permitted is if $x \neq 0$. If $x = 0$ were possible, then you would have "lost" this root of the original equation.

Solve each of the given equations by factoring.

a. $4x^2 - x = 0$ b. $3x^2 - 11x - 4 = 0$

Answers

a. $0, \frac{1}{4}$ b. $-\frac{1}{3}, 4$

One cannot always find "simple" factors to solve quadratic equations. For the most part, we will only attempt to use the factoring method for general quadratic equations with rational roots. In those cases, the factors only have integer coefficients.

Furthermore, there are some quadratic equations that cannot even be factored over the real numbers. Consider using the factoring method to solve

$$x^2 + x + 1 = 0$$

There do not exist any real numbers r and s such that

$$x^2 + x + 1 = (x + r)(x + s)$$

However, it can be written in this form if we permit r and s to be complex numbers. For this reason, it is necessary to develop solution techniques that are more powerful than factoring.

2.3b *Special Cases:* $x^2 - p = 0, x^2 + p = 0, a(x + h)^2 + c = 0$

There are certain quadratic equations that do not necessarily require the use of factoring when finding solutions. Because of their special form, we may use the method of taking roots.

EXAMPLE 4 *Special Cases*

Solve the equation $x^2 - 3 = 0$.

SOLUTION

We may write the original equation as

$$x^2 = 3$$

Taking the square root of both sides, we obtain

$$x = \sqrt{3} \quad \text{or} \quad x = -\sqrt{3}$$

Sometimes these solutions are written in the abbreviated form $x = \pm\sqrt{3}$.

Alternatively, if we recognize that the original equation can be factored as

$$x^2 - 3 = (x - \sqrt{3})(x + \sqrt{3}) = 0$$

we obtain the same results by setting each factor equal to 0. ■

For a positive number p, consider $x^2 - p = 0$ or, equivalently, $x^2 = p$. Taking the square root of both sides of this equation, we obtain $x = \pm\sqrt{p}$. Alternatively, we may factor

$$x^2 - p = (x - \sqrt{p})(x + \sqrt{p})$$

(Check this by multiplying the factors of the right-hand side of the equation.) This leads to the following result.

If $p > 0$ and $x^2 = p$, then $x = \pm\sqrt{p}$. Furthermore, the equation can be written in factored form as

$$x^2 - p = (x - \sqrt{p})(x + \sqrt{p}) = 0$$

EXAMPLE 5 *Special Cases*

Solve the equation $x^2 + 4 = 0$.

SOLUTION

We may write the original equation as

$$x^2 = -4$$

Taking the square root of both sides, we obtain

$$x = \pm 2i$$

Alternatively, if we recognize that the original equation can be factored as

$$x^2 + 4 = (x - 2i)(x + 2i) = 0$$

we obtain the same results by setting each factor equal to 0. ■

For a positive number p, consider $x^2 + p = 0$ or, equivalently, $x^2 = -p$. Taking the square root of both sides of this equation we obtain $x = \pm\sqrt{p}\,i$. Alternatively, we may factor

$$x^2 + p = (x - \sqrt{p}\,i)(x + \sqrt{p}\,i)$$

(Check by multiplying the factors of the right-hand side.)

This leads to the following result.

If $p > 0$ and $x^2 = -p$, then $x = \pm\sqrt{p}\,i$. Furthermore, the equation can be written in factored form as

$$x^2 + p = (x - \sqrt{p}\,i)(x + \sqrt{p}\,i) = 0$$

EXAMPLE 6 *Special Cases*

Solve the equation $2x^2 - 6 = 0$.

SOLUTION

$$\begin{aligned} 2x^2 - 6 &= 0 \\ 2x^2 &= 6 \\ x^2 &= 3 \\ x &= \pm\sqrt{3} \end{aligned}$$

■

EXAMPLE 7 ***Special Cases***

Solve the equation $2(x-1)^2 - 6 = 0$.

SOLUTION

$$\begin{aligned}2(x-1)^2 - 6 &= 0\\ 2(x-1)^2 &= 6\\ (x-1)^2 &= 3\\ x-1 &= \pm\sqrt{3}\\ x &= 1 \pm \sqrt{3}\end{aligned}$$

■

EXAMPLE 8 ***Special Cases***

Solve the equation $4(x+3)^2 + 20 = 0$.

SOLUTION

$$\begin{aligned}4(x+3)^2 + 20 &= 0\\ 4(x+3)^2 &= -20\\ (x+3)^2 &= -5\\ x+3 &= \pm\sqrt{5}\,i\\ x &= -3 \pm \sqrt{5}\,i\end{aligned}$$

■

Equations of the form $a(x+h)^2 + c = 0$ may be solved using the techniques of Examples 7 and 8 as follows:

$$\begin{aligned}a(x+h)^2 + c &= 0\\ a(x+h)^2 &= -c\\ (x+h)^2 &= -\frac{c}{a}\\ x+h &= \pm\sqrt{-\frac{c}{a}}\\ x &= -h \pm \sqrt{-\frac{c}{a}}\end{aligned}$$

Progress Check

Solve the given equation.

a. $5x^2 + 13 = 0$

b. $(2x-7)^2 - 5 = 0$

Answers

a. $\pm\frac{\sqrt{65}}{5}i$

b. $\frac{7 \pm \sqrt{5}}{2}$

We have seen that the solutions of a quadratic equation may be complex numbers, whereas the solution of a linear equation is a real number. In addition, quadratic equations appear to have two solutions. We will have more to say about these observations when we study the roots of polynomial equations in a later chapter.

We have shown that we can always find a solution to a quadratic equation of the form

$$a(x+h)^2 + c = 0 \tag{1}$$

A technique known as **completing the square** permits us to rewrite *any* quadratic equation in the form of Equation (1). Beginning with the expression $x^2 + dx$, we seek a constant h^2 to complete the square so that

$$x^2 + dx + h^2 = (x+h)^2$$

Expanding and solving, we have

$$x^2 + dx + h^2 = x^2 + 2hx + h^2$$
$$dx = 2hx$$
$$h = \frac{d}{2}$$
$$h^2 = \left(\frac{d}{2}\right)^2$$

so $h^2 = \left(\frac{d}{2}\right)^2$ is the amount to be added to $x^2 + dx$ to form a perfect square.

EXAMPLE 9 *Completing the Square*

Complete the square for each of the following.

a. $x^2 - 6x$ b. $x^2 + 3x$

SOLUTION

a. The coefficient of x is -6, so $h = -\frac{6}{2} = -3$ and $h^2 = 9$. Then

$$x^2 - 6x + 9 = (x-3)^2$$

b. The coefficient of x is 3, and $h^2 = \left(\frac{3}{2}\right)^2 = \frac{9}{4}$. Then

$$x^2 + 3x + \frac{9}{4} = \left(x + \frac{3}{2}\right)^2.$$ ■

We are now in a position to use this method to solve a quadratic equation.

EXAMPLE 10 ***Completing the Square***

Solve the quadratic equation $2x^2 - 10x + 1 = 0$ by completing the square.

SOLUTION

We outline and explain each step of the process below:

Method	Example
Step 1. Rewrite the equation with the constant term on the right-hand side.	*Step 1.* $2x^2 - 10x = -1$
Step 2. Factor out a, the coefficient of x^2.	*Step 2.* $2(x^2 - 5x) = -1$
Step 3. Divide both sides of the equation by a.	*Step 3.* $x^2 - 5x = -\frac{1}{2}$
Step 4. Find $h = \frac{d}{2}$ and $h^2 = \left(\frac{d}{2}\right)^2$, where d is the coefficient of x in *Step 3*.	*Step 4.* $h = \frac{-5}{2}$, $h^2 = \frac{25}{4}$
Step 5. Add h^2 to both sides of the equation.	*Step 5.* $x^2 - 5x + \frac{25}{4} = -\frac{1}{2} + \frac{25}{4}$
Step 6. Simplify.	*Step 6.* $\left(x - \frac{5}{2}\right)^2 = \frac{23}{4}$
Step 7. Solve for x.	*Step 7.* $x - \frac{5}{2} = \pm\sqrt{\frac{23}{4}}$ $x = \frac{5}{2} \pm \frac{\sqrt{23}}{2}$ $x = \frac{5 \pm \sqrt{23}}{2}$

■

Progress Check

Solve by completing the square.

a. $x^2 - 3x + 2 = 0$ b. $3x^2 - 4x + 2 = 0$

Answers

a. 1, 2 b. $\frac{2 \pm \sqrt{2}\,i}{3}$

2.3c The Quadratic Formula

We can apply the method of completing the square to the general quadratic equation

$$ax^2 + bx + c = 0, \qquad a \neq 0$$

Following the steps of the method as shown in the above table, we proceed as follows:

1. Move the constant term to the right-hand side. $ax^2 + bx = -c$
2. Factor out a, the coefficient of x^2. $a\left(x^2 + \frac{b}{a}x\right) = -c$
3. Divide both sides of the equation by a. $x^2 + \frac{b}{a}x = -\frac{c}{a}$
4. Find $h = \frac{d}{2}$ and $h^2 = \left(\frac{d}{2}\right)^2$, where d is the coefficient of x. $h = \frac{b}{2a}, \quad h^2 = \frac{b^2}{4a^2}$
5. Add h^2 to both sides of the equation. $x^2 + \frac{b}{a}x + \frac{b^2}{4a^2} = -\frac{c}{a} + \frac{b^2}{4a^2}$
6. Simplify.
$$\left(x + \frac{b}{2a}\right)^2 = \frac{b^2}{4a^2} - \frac{c(4a)}{a(4a)}$$
$$\left(x + \frac{b}{2a}\right)^2 = \frac{b^2 - 4ac}{4a^2}$$
7. Solve for x.
$$x + \frac{b}{2a} = \pm\sqrt{\frac{b^2 - 4ac}{4a^2}}$$
$$x = \frac{-b^2}{2a} \pm \frac{\sqrt{b^2 - 4ac}}{2a}$$
$$x = \frac{-b \pm \sqrt{b^2 - 4ac}}{2a}$$

The quadratic equation

$$ax^2 + bx + c = 0, \qquad a \neq 0$$

has *two* roots often written in the compact form

$$x = \frac{-b \pm \sqrt{b^2 - 4ac}}{2a}, \qquad a \neq 0$$

EXAMPLE 11 ***The Quadratic Formula***

Solve $2x^2 - 3x - 3 = 0$ by the quadratic formula.

SOLUTION

Since $a = 2$, $b = -3$ and $c = -3$, we have

$$x = \frac{-b \pm \sqrt{b^2 - 4ac}}{2a}$$

$$x = \frac{-(-3) \pm \sqrt{(-3)^2 - 4(2)(-3)}}{2(2)}$$

$$x = \frac{3 \pm \sqrt{33}}{4}$$

■

EXAMPLE 12 *The Quadratic Formula*

Solve $-5x^2 + 3x = 2$ by the quadratic formula.

SOLUTION

We first rewrite the given equation as $-5x^2 + 3x - 2 = 0$. Then $a = -5$, $b = 3$ and $c = -2$. Substituting into the quadratic formula, we have

$$x = \frac{-b \pm \sqrt{b^2 - 4ac}}{2a}$$

$$x = \frac{-3 \pm \sqrt{3^2 - 4(-5)(-2)}}{2(-5)}$$

$$x = \frac{-3 \pm \sqrt{-31}}{-10}$$

$$x = \frac{-3 \pm \sqrt{31}\,i}{-10}$$

(Show that this is equivalent to $x = \dfrac{3 \pm \sqrt{31}\,i}{10}$.) ■

Progress Check

Solve by the quadratic formula.

a. $x^2 - 8x = -10$ b. $4x^2 - 2x + 1 = 0$

Answers

a. $4 \pm \sqrt{6}$ b. $\dfrac{1 \pm \sqrt{3}\,i}{4}$

There are a number of errors that students make in using the quadratic formula.

a. To solve $x^2 - 3x = -4$, you must write the equation in the form

$$x^2 - 3x + 4 = 0$$

to properly identify a, b, and c. Note that $b = -3$, *not* 3.

b. The quadratic formula is

$$x = \frac{-b \pm \sqrt{b^2 - 4ac}}{2a}$$

Note that

$$x \neq -b \pm \frac{\sqrt{b^2 - 4ac}}{2a}$$

since the term $-b$ must also be divided by $2a$.

Now that there is a formula that works for any quadratic equation, it may be tempting to use it all the time. However, if you see an equation such as

$$x^2 - 15$$

it may be easier to obtain the answer: $x = \pm\sqrt{15}$. Similarly, when faced with

$$x^2 + 3x + 2 = 0$$

it may be faster to solve it if you see that

$$x^2 + 3x + 2 = (x + 1)(x + 2)$$

The method of completing the square is generally not used for solving quadratic equations once the quadratic formula is learned. The *technique* of completing the square is helpful in a variety of applications, and we will use it in a later chapter when we graph second-degree equations.

2.3d The Discriminant

By analyzing the quadratic formula

$$x = \frac{-b \pm \sqrt{b^2 - 4ac}}{2a}$$

we can learn a great deal about the roots of the quadratic equation

$$ax^2 + bx + c = 0, \qquad a \neq 0$$

The key to the analysis is the **discriminant** $b^2 - 4ac$ found under the radical.

- If $b^2 - 4ac$ is negative, we have the square root of a negative number, and the roots of the quadratic equation are complex numbers as conjugate pairs.
- If $b^2 - 4ac$ is positive, we have the square root of a positive number, and the roots of the quadratic equation are two different real numbers.
- If $b^2 - 4ac = 0$, then $x = -\frac{b}{2a}$, which we call a **double root** or **repeated root** of the quadratic equation. For example, if $x^2 - 10x + 25 = 0$, then the discriminant is 0 and $x = 5$. However,

  $$x^2 - 10x + 25 = (x - 5)(x - 5) = 0$$

  We call $x = 5$ a double root because the factor $x - 5$ is a double factor of $x^2 - 10x + 25 = 0$. This hints at the importance of the relationship between roots and factors, a relationship that we will explore in a later chapter. We summarize in the below table.

TABLE 2-2 Discriminant-Root Analysis

The quadratic equation $ax^2 + bx + c = 0$, $a \neq 0$, has exactly two roots, the nature of which are determined by the discriminant $b^2 - 4ac$.

Discriminant	Roots
Negative	Two complex roots as conjugate pairs
0	One real double root
Positive	Two different real roots

If the roots of the quadratic equation are real, and a, b, and c are rational numbers, the discriminant enables us to determine whether the roots are rational or irrational. Since $\sqrt{k}$ is a rational number only if k is a perfect square, we see that the quadratic formula produces a rational result only if $b^2 - 4ac$ is a perfect square.

EXAMPLE 13 *Discriminant-Root Analysis*

Without solving, determine the nature of the roots of the quadratic equation $3x^2 - 4x + 6 = 0$.

SOLUTION

We evaluate $b^2 - 4ac$ using $a = 3$, $b = -4$ and $c = 6$. Thus,

$$b^2 - 4ac = (-4)^2 - 4(3)(6) = 16 - 72 = -56$$

The discriminant is negative, so the equation has two complex roots. ■

EXAMPLE 14 *Discriminant-Root Analysis*

Without solving, determine the nature of the roots of the equation $2x^2 - 7x = -1$.

SOLUTION

We rewrite the equation in the standard form

$$2x^2 - 7x + 1 = 0$$

and then substitute $a = 2$, $b = -7$ and $c = 1$ in the discriminant. Therefore,

$$b^2 - 4ac = (-7)^2 - 4(2)(1) = 49 - 8 = 41$$

The discriminant is positive and is not a perfect square. Thus, the roots are real, unequal, and irrational. ■

Progress Check

Without solving, determine the nature of the roots of the quadratic equation by using the discriminant.

a. $4x^2 - 20x + 25 = 0$ b. $5x^2 - 6x = -2$

c. $10x^2 = x + 2$ d. $x^2 + x - 1 = 0$

Answers

a. a real, double root b. two complex roots

c. two real, rational roots d. two real, irrational roots

2.3e Forms Leading to Quadratics

Certain types of equations can be transformed into quadratic equations that can be solved by the methods discussed in this section. One form that leads to a quadratic equation is the **radical equation**, such as

$$x - \sqrt{x-2} = 4$$

which is solved in Example 15. To solve the equation, we isolate the radical and raise both sides to a suitable power. The following is the key to the solution of such equations.

> If P and Q are algebraic expressions, then the solution set of the equation
> $$P = Q$$
> is a subset of the solution set of the equation
> $$P^n = Q^n$$
> where n is a natural number.

This suggests that we can solve radical equations if we observe a precaution.

> If both sides of an equation are raised to the same power, the solutions of the resulting equation must be checked to see that they satisfy the original equation.

EXAMPLE 15 *Radical Equations*

Solve $x - \sqrt{x-2} = 4$.

SOLUTION

Step 1. If possible, isolate the radical on one side of the equation.	*Step 1.* $x - 4 = \sqrt{x-2}$
Step 2. Raise both sides of the equation to a suitable power to eliminate the radical. If necessary, go back to *Step 1*.	*Step 2.* Squaring both sides, we have $x^2 - 8x + 16 = x - 2$
Step 3. Solve for the unknown.	*Step 3.* $x^2 - 9x + 18 = 0$ $(x-3)(x-6) = 0$ $x = 3 \qquad x = 6$
Step 4. Check each solution by substituting in the *original* equation.	*Step 4.* LHS $= x - \sqrt{x-2}$ RHS $= 4$ Check $x = 3$ LHS $= 3 - \sqrt{3-2} = 2 \neq 4 =$ RHS Check $x = 6$ LHS $= 6 - \sqrt{6-2} = 6 - 2 = 4 =$ RHS

We will use the abbreviation LHS for left-hand side and RHS for right-hand side. We conclude that 6 is a solution of the original equation, and 3 is not a solution of the original equation. We say that 3 is an **extraneous solution** that was introduced when we raised both sides of the original equation to the second power. ■

Progress Check

Solve $x - \sqrt{1-x} = -5$.

Answer

-3

The equation in the next example contains more than one radical. Solving this equation requires that we square both sides *twice*.

EXAMPLE 16 *Radical Equations*

Solve $\sqrt{2x-4} - \sqrt{3x+4} = -2$.

SOLUTION

Before squaring, rewrite the equation so that we isolate one of the radicals on one side of the equation.

$$\sqrt{2x-4} = \sqrt{3x+4} - 2$$

$$2x - 4 = (3x + 4) - 4\sqrt{3x+4} + 4 \quad \text{Square both sides.}$$

$$-x - 12 = -4\sqrt{3x+4} \quad \text{Isolate the radical.}$$

$$x^2 + 24x + 144 = 16(3x + 4) \quad \text{Square both sides.}$$

$$x^2 - 24x + 80 = 0$$

$$(x - 20)(x - 4) = 0$$

$$x = 20 \qquad x = 4$$

Verify that both 20 and 4 are solutions of the original equation. ■

Progress Check

Solve $\sqrt{5x-1} - \sqrt{x+2} = 1$.

Answer

2

Although the equation

$$x^4 - x^2 - 2 = 0$$

is not a quadratic in the unknown x, it is a quadratic in the unknown x^2:

$$(x^2)^2 - (x^2) - 2 = 0$$

This may be seen more clearly by replacing x^2 with a new unknown u such that $u = x^2$. Substituting, we have

$$u^2 - u - 2 = 0$$

which is a quadratic equation in the unknown u. Solving, we find

$$(u+1)(u-2)=0$$
$$u=-1 \quad \text{or} \quad u=2$$

Since $x^2 = u$, we must next solve the equations

$$x^2=-1 \quad \text{and} \quad x^2=2$$
$$x=\pm i \qquad\qquad x=\pm\sqrt{2}$$

The original equation has four solutions: i, $-i$, $\sqrt{2}$, and $-\sqrt{2}$.

The technique we have used is called a **substitution of variable**. This is a powerful method that is commonly used in calculus.

Progress Check

Indicate an appropriate substitution of variable and solve each of the following equations.

a. $3x^4 - 10x^2 - 8 = 0$　　b. $4x^{2/3} + 7x^{1/3} - 2 = 0$

c. $\frac{2}{x^2} + \frac{1}{x} - 10 = 0$　　d. $\left(1+\frac{2}{x}\right)^2 - 8\left(1+\frac{2}{x}\right) + 15 = 0$

Answers

a. $u = x^2$; ± 2, $\pm\frac{\sqrt{6}i}{3}$　　b. $u = x^{1/3}$; $\frac{1}{64}$, -8

c. $u = \frac{1}{x}$; $-\frac{2}{5}$, $\frac{1}{2}$　　d. $u = 1 + \frac{2}{x}$; 1, $\frac{1}{2}$

Exercise Set 2.3

In Exercises 1–14, solve by factoring.

1. $x^2 - 3x + 2 = 0$
2. $x^2 - 6x + 8 = 0$
3. $x^2 + x - 2 = 0$
4. $3r^2 - 4r + 1 = 0$
5. $x^2 + 6x = -8$
6. $x^2 + 6x + 5 = 0$
7. $y^2 - 4y = 0$
8. $2x^2 - x = 0$
9. $2x^2 - 5x = -2$
10. $2s^2 - 5s - 3 = 0$
11. $t^2 - 4 = 0$
12. $4x^2 - 9 = 0$
13. $6x^2 - 5x + 1 = 0$
14. $6x^2 - x = 2$

In Exercises 15–24, solve the given equation.

15. $3x^2 - 27 = 0$
16. $4x^2 - 64 = 0$
17. $5y^2 - 25 = 0$
18. $6x^2 - 12 = 0$
19. $(2r + 5)^2 = 8$
20. $(3x - 4)^2 = -6$
21. $(3x - 5)^2 - 8 = 0$
22. $(4t + 1)^2 - 3 = 0$
23. $9x^2 + 64 = 0$
24. $81x^2 + 25 = 0$

In Exercises 25–36, solve by completing the square.

25. $x^2 - 2x = 8$
26. $t^2 - 2t = 15$
27. $2r^2 - 7r = 4$
28. $9x^2 + 3x = 2$
29. $3x^2 + 8x = 3$
30. $2y^2 + 4y = 5$
31. $2y^2 + 2y = -1$
32. $3x^2 - 4x = -3$
33. $4x^2 - x = 3$
34. $2x^2 + x = 2$
35. $3x^2 + 2x = -1$
36. $3u^2 - 3u = -1$

In Exercises 37–48, solve by the quadratic formula.

37. $2x^2 + 3x = 0$
38. $2x^2 + 3x + 3 = 0$
39. $5x^2 - 4x + 3 = 0$
40. $2x^2 - 3x - 2 = 0$
41. $5y^2 - 4y + 5 = 0$
42. $x^2 - 5x = 0$
43. $3x^2 + x - 2 = 0$
44. $2x^2 + 4x - 3 = 0$
45. $3y^2 - 4 = 0$
46. $2x^2 + 2x + 5 = 0$
47. $4u^2 + 3u = 0$
48. $4x^2 - 1 = 0$

In Exercises 49–58, solve by any method.

49. $2x^2 + 2x - 5 = 0$
50. $2t^2 + 2t + 3 = 0$
51. $3x^2 + 4x - 4 = 0$
52. $x^2 + 2x = 0$
53. $2x^2 + 5x + 4 = 0$
54. $2r^2 - 3r + 2 = 0$
55. $4u^2 - 1 = 0$
56. $x^2 + 2 = 0$
57. $4x^3 + 2x^2 + 3x = 0$
58. $4s^3 + 4s^2 - 15s = 0$

In Exercises 59–64, solve for the indicated variable in terms of the remaining variables.

59. $a^2 + b^2 = c^2$, for b
60. $s = \frac{1}{2}gt^2$, for t
61. $V = \frac{1}{3}\pi r^2 h$, for r
62. $A = \pi r^2$, for r
63. $s = \frac{1}{2}gt^2 + vt$, for t
64. $F = g\frac{m_1 m_2}{d^2}$, for d

Without solving, determine the nature of the roots of each quadratic equation in Exercises 65–80.

65. $x^2 - 2x + 3 = 0$
66. $3x^2 + 2x - 5 = 0$
67. $4x^2 - 12x + 9 = 0$
68. $2x^2 + x + 5 = 0$
69. $-3x^2 + 2x + 5 = 0$
70. $-3y^2 + 2y - 5 = 0$
71. $3x^2 + 2x = 0$
72. $4x^2 + 20x + 25 = 0$
73. $2r^2 = r - 4$
74. $3x^2 = 5 - x$
75. $3x^2 + 6 = 0$
76. $4x^2 - 25 = 0$
77. $6r = 3r^2 + 1$
78. $4x = 2x^2 + 3$
79. $12x = 9x^2 + 4$
80. $4s^2 = -4s - 1$

In Exercises 81–84, find a value or values of k for which the quadratic has a double root.

81. $kx^2 - 4x + 1 = 0$
82. $2x^2 + 3x + k = 0$
83. $x^2 - kx - 2k = 0$
84. $kx^2 - 4x + k = 0$

In Exercises 85–92, find the solution set.

85. $x + \sqrt{x+5} = 7$
86. $x - \sqrt{13-x} = 1$
87. $2x + \sqrt{x+1} = 8$
88. $3x - \sqrt{1+3x} = 1$
89. $\sqrt{3x+4} - \sqrt{2x+1} = 1$
90. $\sqrt{4-4x} - \sqrt{x+4} = 3$
91. $\sqrt{2x-1} + \sqrt{x-4} = 4$
92. $\sqrt{5x+1} + \sqrt{4x-3} = 7$

In Exercises 93–100, indicate an appropriate substitution of variable and solve each of the equations.

93. $3x^4 + 5x^2 - 2 = 0$
94. $2x^6 + 15x^3 - 8 = 0$
95. $\frac{6}{x^2} + \frac{1}{x} - 2 = 0$
96. $\frac{2}{x^4} - \frac{3}{x^2} - 9 = 0$
97. $2x^{2/5} + 5x^{1/5} + 2 = 0$
98. $3x^{4/3} - 4x^{2/3} - 4 = 0$
99. $2\left(\frac{1}{x}+1\right)^2 - 3\left(\frac{1}{x}+1\right) - 20 = 0$
100. $3\left(\frac{1}{x}-2\right)^2 + 2\left(\frac{1}{x}-2\right) - 1 = 0$

In Exercises 101 and 102, provide a proof of the following statements.

101. If r_1 and r_2 are the roots of the equation $ax^2 + bx + c = 0$, then (a) $r_1 r_2 = \frac{c}{a}$ and (b) $r_1 + r_2 = \frac{b}{a}$.
102. If a, b, and c are rational numbers, and the discriminant of the equation $ax^2 + bx + c = 0$ is positive, then the quadratic has either two rational roots or two irrational roots.

In Exercises 103–109, use the theorems of Exercise 101 to find a value or values of k that satisfies the indicated condition.

103. $kx^2 + 3x + 5 = 0$; sum of the roots is 6.

104. $2x^2 - 3kx - 2 = 0$; sum of the roots is -3.

105. $3x^2 - 10x + 2k = 0$; product of the roots is -4.

106. $2kx^2 + 5x - 1 = 0$; product of the roots is $\frac{1}{2}$.

107. $2x^2 - kx + 9 = 0$; one root is double the other.

108. $3x^2 - 4x + k = 0$; one root is triple the other.

109. $6x^2 - 13x + k = 0$; one root is the reciprocal of the other.

110. Show that if there is only one real root of the equation $ax^2 + bx + c = 0$, $a \neq 0$, then $b = \pm 2\sqrt{ac}$.

111. If $x^2 \leq 36$, is it true that $x \leq 6$? Prove this or give a counter example.

112. If $x^4 \geq 16$, is it true that $x \geq 2$? Prove this or give a counter example.

113. If $x \neq 0$, is $x > \frac{1}{x}$? Prove this or give a counter example.

114. Given a regular polygon having n sides, we can determine the number of diagonals D by the formula

$$D = \frac{n^2 - 3n}{2}$$

a. If a polygon has 65 diagonals, how many sides does the polygon have?

b. A polygon has 20 diagonals. How many sides does the polygon have?

c. A polygon has n sides. If the number of sides of the polygon is increased by one, by how much is the number of diagonals increased?

115. The sum of the first n positive integers is given by:

$$1 + 2 + 3 + 4 + \cdots + (n - 1) + n = \frac{n(n+1)}{2}$$

a. The sum of the first n integers is 55. What is n?

b. Find n if the sum of the first n integers is 36.

c. Find $2 + 3 + 4 + \cdots + n$.

116. The sum of the first n odd positive integers is given by: $1 + 3 + 5 + 7 + \cdots + (2n - 1) = n^2$

a. Find $5 + 7 + \cdots + (2n - 1)$.

b. Find $1 + 3 + 5 + \cdots + (2n - 3)$.

117. The sum of the first n odd positive integers is 121. (See Exercise 116.)

a. Find n.

b. Write out this sum explicitly.

118. The distance d between two points (x_1, y_1) and (x_2, y_2) is given by

$$d = \sqrt{(x_2 - x_1)^2 + (y_2 - y_1)^2}$$

If the distance between (2, 3) and $(x, 10)$ is 7, find x.

119. The number of two-way links n necessary to create a full-mesh network between x users is given by

$$\frac{x(x-1)}{2} = n$$

How many users can be connected in such a network with 190 links? 4950 links? (See chapter opener.)

120. Solve the following using the quadratic formula, by first storing the discriminant as D, then finding $\frac{-b \pm \sqrt{D}}{2a}$.

$$0.0001x^2 - 0.086x - 48.75 = 0$$

2.4 Applications of Quadratic Equations

As your knowledge of mathematical techniques and ideas grows, you will become capable of solving an ever wider variety of applied problems. In Section 2.2 we explored many types of word problems that lead to linear equations. We can now tackle a group of applied problems that lead to quadratic equations.

One word of caution: It is possible to arrive at a solution that makes no sense. For example, a negative solution that represents hours worked or the age of an individual is meaningless and must be rejected.

EXAMPLE 1 *Quadratic Equations and Word Problems*

The larger of two positive numbers exceeds the smaller by 2. If the sum of the squares of the two numbers is 74, find the two numbers.

SOLUTION

If we let

$$x = \text{the larger number}$$

then

$$x - 2 = \text{the smaller number}$$

The sum of the squares of the numbers is 74.

$$\begin{aligned}(\text{larger number})^2 + (\text{smaller number})^2 &= 74\\ x^2 + (x-2)^2 &= 74\\ x^2 + x^2 - 4x + 4 &= 74\\ 2x^2 - 4x - 70 &= 0\\ x^2 - 2x - 35 &= 0\\ (x+5)(x-7) &= 0\\ x &= 7 \qquad \text{Reject } x = -5.\end{aligned}$$

The numbers are then 7 and $(7 - 2) = 5$. Verify that the sum of the squares is indeed 74. ■

EXAMPLE 2 *Quadratic Equations and Word Problems*

The length of a pool is 3 times its width, and the pool is surrounded by a grass walk 4 feet wide. If the total area covered and enclosed by the walk is 684 square feet, find the dimensions of the pool.

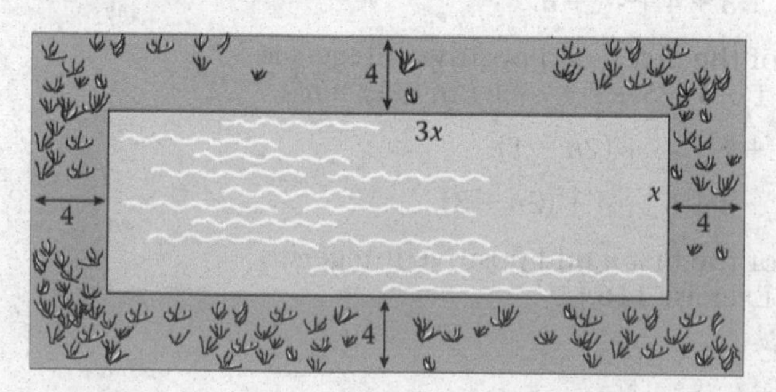

SOLUTION

Drawing diagrams is useful in solving geometric problems. For example, as shown above, if we let x = width of the pool, then $3x$ = length of the pool. The region enclosed by the walk has length $3x + 8$ and width $x + 8$. The total area is the product of the length and width, so

$$\begin{aligned} \text{length} \times \text{width} &= 684 \\ (3x + 8)(x + 8) &= 684 \\ 3x^2 + 32x + 64 &= 684 \\ 3x^2 + 32x - 620 &= 0 \\ (3x + 62)(x - 10) &= 0 \\ x &= 10 \qquad \text{Reject } x = -\frac{62}{3}. \end{aligned}$$

The dimensions of the pool are 10 feet by 30 feet. ■

EXAMPLE 3 *Quadratic Equations and Word Problems*

Working together, two cranes can unload a ship in 4 hours. The slower crane, working alone, requires 6 hours more than the faster crane to do the job. How long does it take each crane to do the job by itself?

SOLUTION

Let x = number of hours for the faster crane to do the job. Then $x + 6$ = number of hours for the slower crane to do the job. The rate of the faster crane is $\frac{1}{x}$, the portion of the whole job that it completes in 1 hour. Similarly, the rate of the slower crane is $\frac{1}{x+6}$. We display this information in a table.

	Rate	×	Time	=	Work done
Faster crane	$\frac{1}{x}$		4		$\frac{4}{x}$
Slower crane	$\frac{1}{x+6}$		4		$\frac{4}{x+6}$

When the two cranes work together, we must have

$$\left(\begin{array}{c}\text{work done}\\ \text{by fast crane}\end{array}\right) + \left(\begin{array}{c}\text{work done}\\ \text{by slow crane}\end{array}\right) = 1 \text{ whole job}$$

or

$$\frac{4}{x} + \frac{4}{x+6} = 1$$

To solve, we multiply by the LCD, $x(x + 6)$, obtaining

$$\begin{aligned} 4(x + 6) + 4x &= x^2 + 6x \\ 0 &= x^2 - 2x - 24 \\ 0 &= (x + 4)(x - 6) \\ x = -4 \quad \text{or} \quad x &= 6 \end{aligned}$$

The solution $x = -4$ is rejected because it makes no sense to speak of negative hours of work. Then

$x = 6$ is the number of hours in which the fast crane can do the job alone

$x + 6 = 12$ is the number of hours in which the slow crane can do the job alone ■

Exercise Set 2.4

1. Working together, computers A and B can complete a data-processing job in 2 hours. Computer A working alone can do the job in 3 hours less than computer B working alone. How long does it take each computer to do the job by itself?

2. A graphic designer and her assistant working together can complete an advertising layout in 6 days. The assistant working alone could complete the job in 16 more days than the designer working alone. How long would it take each person to do the job alone?

3. A roofer and his assistant working together can finish a roofing job in 4 hours. The roofer working alone could finish the job in 6 hours less than the assistant working alone. How long would it take each person to do the job alone?

4. A 16-inch by 20-inch mounting board is used to mount a photograph. How wide a uniform border is needed if the photograph occupies $\frac{3}{5}$ of the area of the mounting board?

5. The length of a rectangle exceeds twice its width by 4 feet. If the area of the rectangle is 48 square feet, find the dimensions.

6. The length of a rectangle is 4 centimeters less than twice its width. Find the dimensions if the area of the rectangle is 96 square centimeters.

7. The area of a rectangle is 48 square centimeters. If the length and width are each increased by 4 centimeters, the area of the newly formed rectangle is 120 square centimeters. Find the dimensions of the original rectangle.

8. The base of a triangle is 2 feet more than twice its altitude. If the area is 12 square feet, find the dimensions.

9. Find the width of a strip that has been mowed around a rectangular field 60 feet by 80 feet if $\frac{1}{2}$ the lawn has not yet been mowed.

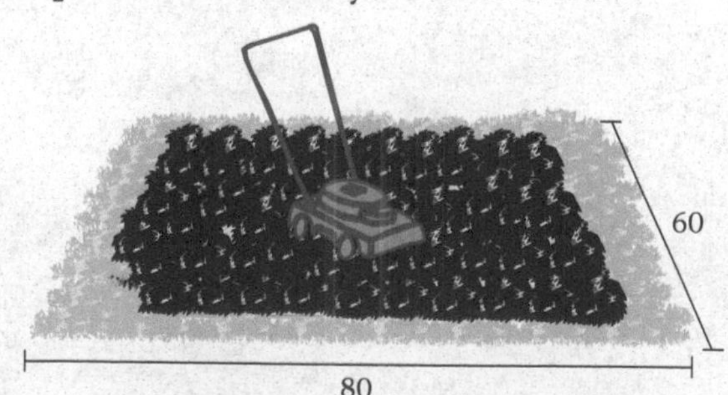

10. The sum of the reciprocals of two consecutive numbers is $\frac{7}{12}$. Find the numbers.

11. The sum of a number and its reciprocal is $\frac{26}{5}$. Find the number.

12. The difference of a number and its reciprocal is $\frac{35}{6}$. Find the number.

13. The smaller of two numbers is 4 less than the larger. If the sum of their squares is 58, find the numbers.

14. The sum of the reciprocals of two consecutive odd numbers is $\frac{8}{15}$. Find the numbers.

15. The sum of the reciprocals of two consecutive even numbers is $\frac{7}{24}$. Find the numbers.

16. A number of students rented a car for $160 for a one-week camping trip. If another student had joined the original group, each person's share of expenses would have been reduced by $8. How many students were in the original group?

17. An investor placed an order totaling $1200 for a certain number of shares of a stock. If the price of each share of stock was $2 more, the investor would get 30 fewer shares for the same amount of money. How many shares did the investor buy?

18. A fraternity charters a bus for a ski trip at a cost of $360. When 6 more students join the trip, each person's cost decreases by $2. How many students were in the original group of travelers?

19. A salesman worked a certain number of days to earn $192. If he had been paid $8 more per day, he would have earned the same amount of money in 2 fewer days. How many days did he work?

20. A freelance photographer worked a certain number of days for a newspaper to earn $480. If she had been paid $8 less per day, she would have earned the same amount in 2 more days. What was her daily rate of pay?

21. A wire 48 centimeters long is cut into two pieces. Each piece is bent to form a square. Where should the wire be cut so that the sum of the areas of the squares is equal to 80 square centimeters?

22. A circuit has a resistance of 25 ohms and a voltage of 110 volts. The power P in watts when a current I in amperes flows through the circuit is

$$P = 110I - I^2$$

If the power is 121 watts, find the current I.

The following are needed for Exercises 23 and 24: If an object is thrown with an initial speed of v_0, then the distance d the object travels in t seconds is

$$d = 16t^2 + v_0t$$

if the distance is measured in feet and

$$d = 4.9t^2 + v_0t$$

if the distance is measured in meters.

23. An object is thrown off the Gateway Arch in St. Louis with an initial velocity of 68 feet per second. The Gateway Arch is 630 feet tall. How long will it take for the object to reach the ground?

24. A warhead fired from an enemy ship in the Persian Gulf is a dud and travels only 100 meters before it hits the water. If it had an initial velocity of 489 meters per second, find the time from the initial launch of the warhead to impact.

25. A principal amount P is invested at a rate of $r \times 100\%$ per year. After 2 years the amount of the investment becomes $S = P(1 + r)^2$. Rachel invests $7000 at the start of her junior year in high school. What interest rate does her investment require so that she will have $9100 for her freshman tuition in college 2 years later?

26. The spread of a certain variety of ivy follows the investment model given in Exercise 25, $S = P(1 + r)^2$. Thirty square feet of ivy are planted. Two years later there are 75 square feet of ground cover. At what rate is the plant spreading?

27. According to Einstein's theory, relative to earth time, space travelers will not age as fast as those who remain on earth. In fact, if a space traveler could travel at the speed of light, relative to those on earth, the traveler would not age at all. The relationship between time on earth t_e, and time in space t_s is given by

$$t_s = t_e\sqrt{1 - \frac{v^2}{c^2}}$$

where v is the velocity of the space traveler and c is the speed of light. One of two brothers makes a round trip to Arcturus. The elapsed time of the trip for the brother on earth is 80.8 years, whereas the elapsed time for the traveling brother is 11.4 years. If the traveling brother maintained a constant speed for the entire trip, find this speed in terms of c.

28. There exists a gravitational force of attraction between all particles. Newton's law of gravitation defines the relationship between the force F exerted by a particle of mass m_1 on another particle of mass m_2 when the distance between the two particles is r.

$$F = \frac{Gm_1m_2}{r^2}$$

where $G = 6.67 \times 10^{-11}$ Nm^2/kg^2 is the universal gravitational constant. (N is the abbreviation for Newtons, a measure of force.) The gravitational force that attracts a 65-kilogram boy to a 50-kilogram girl is 8.67×10^{-7} N. How many meters apart are the boy and the girl?

29. Angular displacement θ is defined by:

$$\theta = \theta_0 + \omega_0 t + \frac{1}{2}\alpha_0 t^2$$

where θ_0, ω_0 and α_0 are the angular displacement, angular velocity, and angular acceleration, respectively, all at time $t = 0$.

a. Let $\theta = 11$ radians, $\theta_0 = 1$ radian, $\omega_0 = 3$ radians per second and $\alpha_0 = 2$ radians per second squared. Solve for t.

b. Find a general formula for t by using the quadratic formula.

30. The equilibrium point in economic theory is that price where demand equals supply. For the following supply and demand equations, find the equilibrium point.

a. $d = \dfrac{1500}{p}$, $s = 500p - 250$

b. $d = \dfrac{1400}{p}$, $s = 1200p - 3800$

31. In probability theory, a binomial random variable x can be approximated by a normal random variable z by the equation

$$z = \frac{x - np}{\sqrt{np(1-p)}}$$

where n is the number of trials in a binomial experiment and p is the probability of success in one trial of the experiment. If the number of trials is 16, $x = 1$ and $z = -2.48$, what is p?

32. An oil company has decided to replace two old cylindrical storage tanks with one new cylindrical storage tank constructed from a material guaranteed to keep the oil at a constant temperature. The old tanks were both 25 feet high, however, one tank had a radius of 12 feet whereas the other had a radius of 16 feet. The new tank will also be 25 feet high. Find the radius of the new tank if it is to hold the same amount of oil as both of the old tanks. (*Hint:* The volume of a cylinder is $V = \pi r^2 h$, where r is the radius and h is the height.)

33. The area of a circle is 10π. What is the radius?

34. The surface area of a cube is 294 square inches. What is the length of each edge of the cube?

35. It is believed that the most visually pleasing rectangle with length L and width W satisfies the following equation

$$\frac{L + W}{L} = \frac{L}{W}$$

a. What is L if $W = 5$?

b. Solve this equation for W in terms of L.

2.5 Linear and Quadratic Inequalities

Much of the terminology of equations carries over to inequalities. A **solution of an inequality** is a value of the unknown that satisfies the inequality, and the solution set is composed of all solutions. The properties of inequalities listed in Section 1.2 enable us to use the same procedures in solving inequalities as in solving equations *with one exception.*

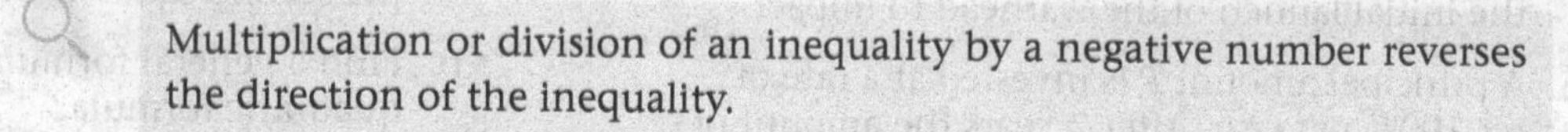

Multiplication or division of an inequality by a negative number reverses the direction of the inequality.

We will concentrate for now on solving a **linear inequality**, that is, an inequality in which the unknown appears only in the first degree.

EXAMPLE 1 *Linear Inequalities*

Solve the inequality $2x + 11 \geq 5x - 1$.

SOLUTION

We perform addition and subtraction for inequalities to collect terms in x just as we did for equations.

$$2x + 11 \geq 5x - 1$$
$$2x \geq 5x - 12$$
$$-3x \geq -12$$

We now divide both sides of the inequality by -3, a negative number, and therefore reverse the sense of the inequality.

$$\frac{-3x}{-3} \leq \frac{-12}{-3}$$
$$x \leq 4$$ ■

Progress Check

Solve the inequality $3x - 2 \geq 5x + 4$.

Answer

$x \leq -3$

Given the inequality

$$-2x \geq -6$$

it is a common error to conclude that dividing by -2 gives $x \leq -3$. Multiplication or division by a negative number changes the sense of the inequality, but the *signs* obey the usual rules of algebra. Thus,

$$-2x \geq -6$$

$$\frac{-2x}{-2} \leq \frac{-6}{-2} \quad \text{Reverse sense of the inequality.}$$

$$x \leq 3$$

There are three methods commonly used to describe subsets of the real numbers: graphs on a real number line, interval notation, and set notation. Since there will be occasions when we want to use each of these schemes, this is a convenient time to introduce them and to apply them to inequalities.

The **graph of an inequality** is the set of all points satisfying the inequality. The graph of the inequality $a \leq x < b$ is shown in Figure 2-1. The portion of the real number line that is in bold is the solution set of the inequality. The circle at point a is filled in to indicate that a is also a solution of the inequality; the circle at point b is left open to indicate that b is not a member of the solution set.

FIGURE 2-1
Graph of $a \leq x < b$

An **interval** is a set of numbers on the real number line that forms a line segment, a half line, or the entire real number line. The subset shown in Figure 2-1 is written in **interval notation** as $[a, b)$, where a and b are the **endpoints** of the interval. A bracket, [or], indicates that the endpoint is included, and a parenthesis, (or), indicates that the endpoint is not included. The interval $[a, b]$ is called a **closed interval** because both endpoints are included. The interval (a, b) is called an **open interval** because neither endpoint is included. Finally, the intervals $[a, b)$ and $(a, b]$ are called **half-open intervals.**

The set of all real numbers satisfying a given property P is written as

$$\{x \mid x \text{ satisfies property } P\}$$

which is read as "the set of all x such that x satisfies property P." This form, called **set notation**, provides a third means of designating subsets of the real number line. Thus, the interval $[a, b)$ shown in Figure 2-1 is written as

$$\{x \mid a \leq x < b\}$$

which indicates the x must satisfy the inequalities $x \geq a$ and $x < b$.

EXAMPLE 2 *Graphs of Finite Intervals*

Graph each of the given intervals on a real number line and indicate the same subset of the real number line in set notation.

a. $(-3, 2]$ b. $(1, 4)$ c. $[-4, -1]$

SOLUTION

a. 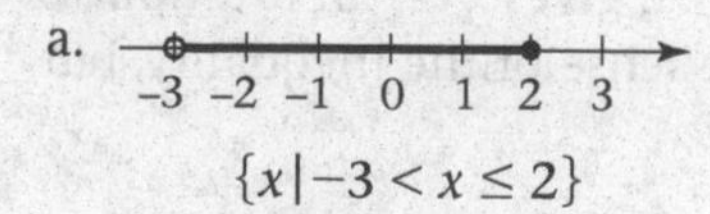

$\{x \mid -3 < x \leq 2\}$

b. 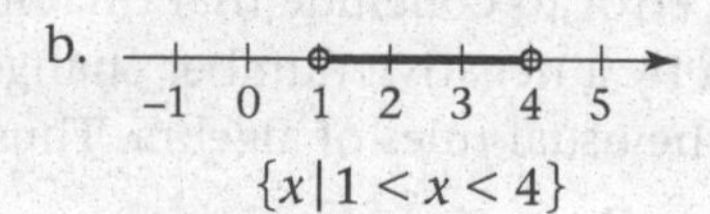

$\{x \mid 1 < x < 4\}$

c. -5 -4 -3 -2 -1 0 1

$\{x \mid -4 \leq x \leq -1\}$

■

To describe the inequality $x > 2$ or the inequality $x \leq 3$ in interval notation, we need to introduce the symbols ∞ and $-\infty$, read "**infinity**" and "minus infinity," respectively. The inequality $x > 2$ is then written as $(2, \infty)$, and the inequality $x \leq 3$ is written as $(-\infty, 3]$. They are graphed on a real number line as shown in Figure 2-2. Note that ∞ and $-\infty$ are symbols, not numbers, indicating that the intervals extend indefinitely. An interval using one of these symbols is called an **infinite interval.** The interval $(-\infty, \infty)$ designates the entire real number line. Square brackets must never be used around ∞ and $-\infty$ since they are not real numbers.

If the endpoint of the graph of an inequality is *not* specifically identified by an *open circle* or by a *filled-in circle,* then we assume that the graph continues forever in the direction where the endpoint is "missing." (See Figure 2-2.) (There are some texts that indicate that a graph continues forever in a particular direction by placing an arrow on the graph pointing in that direction. We shall *not* use this "arrow" notation in this text.)

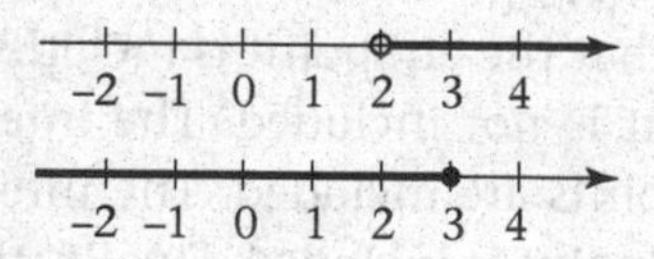

FIGURE 2-2
Graphs of Infinite Intervals

EXAMPLE 3 *Graphing and Solving Linear Inequalities*

Graph each inequality and write the solution set in interval notation.

a. $x \leq -2$ b. $x \geq -1$ c. $x < 3$

SOLUTION

a. 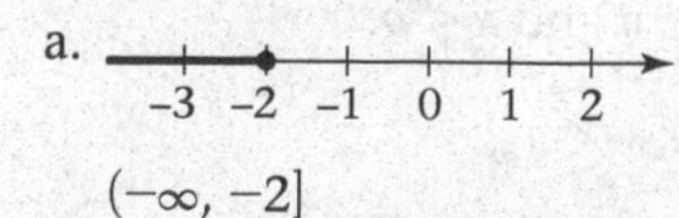

$(-\infty, -2]$

b. 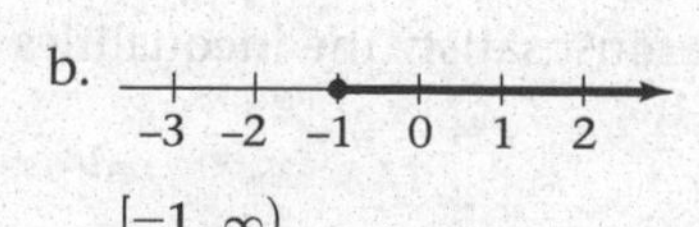

$[-1, \infty)$

c. -2 -1 0 1 2 3 4

$(-\infty, 3)$

■

EXAMPLE 4 *Graphing and Solving Linear Inequalities*

Solve the inequality.

$$\frac{x}{2} - 9 < \frac{1-2x}{3}$$

Graph the solution set, and write the solution set in both interval notation and set notation.

SOLUTION

To clear the inequality of fractions, we multiply both sides by the LCD of all fractions, which is 6.

$$\begin{aligned} 3x - 54 &< 2(1 - 2x) \\ 3x - 54 &< 2 - 4x \\ 7x &< 56 \\ x &< 8 \end{aligned}$$

We may write the solution set as $\{x \mid x < 8\}$ or as the infinite interval $(-\infty, 8)$. The graph of the solution set is shown below.

-1 0 1 2 3 4 5 6 7 8

■

EXAMPLE 5 *Linear Inequalities*

Solve the inequalities.

a. $\frac{2(x+1)}{3} < \frac{2x}{3} - \frac{1}{5}$ b. $2(x - 1) < 2x + 5$

SOLUTION

a. The LCD of all fractions is 15. Multiplying both sides of the inequality by 15, we obtain

$$\begin{aligned} 10(x + 1) &< 10x - 3 \\ 10x + 10 &< 10x - 3 \\ 10 &< -3 \end{aligned}$$

Our procedure has led to a contradiction, indicating that there is no solution to the inequality.

b. Expanding and simplifying leads to the inequality

$$-2 < 5$$

Since this inequality is true for all real values of x, we conclude that the solution set is the set of all real numbers. ■

Solve, and write the answers in interval notation.

a. $\frac{3x-1}{4} + 1 > 2 + \frac{x}{3}$ b. $\frac{2x-3}{2} \geq x + \frac{2}{5}$

Answers

a. $(3, \infty)$ b. no solution

EXAMPLE 6 *Inequalities and Word Problems*

A taxpayer may choose to pay a 20% tax on the gross income or a 25% tax on the gross income less \$4000. Above what income level should the taxpayer elect to pay at the 20% rate?

SOLUTION

If we let x = gross income, then the choice available to the taxpayer is

a. pay at the 20% rate on the gross income, that is, pay $0.20x$, or

b. pay at the 25% rate on the gross income less \$4000, that is, pay

$$0.25(x - 4000)$$

To determine when (a) produces a lower tax than (b), we must solve

$$0.20x < 0.25(x - 4000)$$
$$0.20x < 0.25x - 1000$$
$$-0.05x < -1000$$
$$x > \frac{-1000}{-0.05} = 20{,}000$$

The taxpayer should choose to pay at the 20% rate if the gross income is more than \$20,000. ■

Progress Check

A customer is offered the following choice of telephone services: unlimited local calls at a fixed \$20 monthly charge, or a base rate of \$8 per month plus \$0.06 per message unit. At what level of use does it cost less to choose the unlimited service?

Answer

Unlimited service costs less when the anticipated use exceeds 200 message units.

2.5a *Compound Inequalities*

We can solve **compound inequalities** such as

$$1 < 3x - 2 \leq 7$$

by operating on both inequalities at the same time.

$3 < 3x \leq 9$ Add +2 to each member.

$1 < x \leq 3$ Divide each member by 3.

The solution set is the half-open interval (1, 3].

Note that the statement of the compound inequality

$$1 < 3x - 2 \leq 7$$

actually represents three inequalities:

$$1 < 3x - 2$$
$$3x - 2 \leq 7$$
$$1 \leq 7$$

EXAMPLE 7 ***Compound Inequalities***

Solve the inequality $-3 \leq 1 - 2x < 6$, and write the answer in interval notation.

SOLUTION

Operating on this inequality, we have

$-4 \leq -2x < 5$ Add −1 to each member.

$2 \geq x > -\frac{5}{2}$ Divide each member by −2.

The solution set is the half-open interval $\left(-\frac{5}{2}, 2\right]$. ■

Progress Check

Solve the inequality $-5 < 2 - 3x < -1$, and write the answer in interval notation.

Answer

$\left(1, \frac{7}{3}\right)$

2.5b *Critical Value Method*

The **Critical Value Method** is an alternative approach to solving inequalities. In fact, we shall be relying upon this method throughout the remainder of the text.

> The **critical values** of an inequality are
> 1. those values for which either side of the inequality is not defined (such as a denominator equal to 0),
> 2. those values that are solutions to the equation obtained by replacing the inequality sign with an equal sign.

The critical values determine endpoints of intervals on the real number line. The inequality in question either satisfies all points in a given interval, or no points in a given interval. In order to find out in which intervals the inequality holds, we may test *any* point from each interval. We call such points **test points**. We follow this technique using the example in Table 2-3.

TABLE 2-3 Solving Inequalities by the Critical Value Method

Method	Example: $2x + 11 \geq 5x - 1$
Step 1. Find the critical values of the inequality. a. values where the inequality is not defined b. Replace the inequality sign by an equal sign and solve.	*Step 1.* a. Both sides of the inequality are defined everywhere. b. $2x + 11 = 5x - 1$ $12 = 3x$ $x = 4$
Step 2. Plot the critical values on the real number line.	*Step 2.* 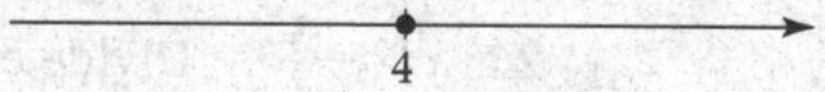
Step 3. Try a test point in each interval, and try each critical value.	*Step 3.* Interval: $x < 4$ Test point: $x = 0$ $2(0) + 11 \geq 5(0) - 1$ True Critical Value: $x = 4$ Test point: $x = 4$ $2(4) + 11 \geq 5(4) - 1$ True Interval: $x > 4$ Test point: $x = 5$ $2(5) + 11 \geq 5(5) - 1$ False
Step 4. Find the solution.	*Step 4.* $x \leq 4$

EXAMPLE 8 *Rational Expression Inequalities*

Solve the inequality.

$$\frac{x+1}{x-1} \geq 2$$

SOLUTION

The inequality is not defined where $x - 1 = 0$, that is, where $x = 1$. Solving the equation

$$\frac{x+1}{x-1} = 2$$

$$x + 1 = 2x - 2$$

$$x = 3$$

Therefore, the critical values are 1 and 3 as shown below.

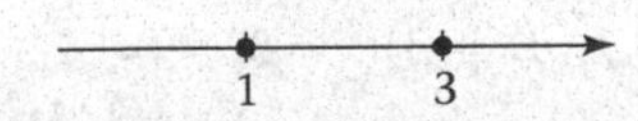

Interval, Critical Value	Test Point	Substitution	Verification
$x < 1$	$x = 0$	$\frac{0+1}{0-1} \geq 2$	False
$x = 1$	$x = 1$	$\frac{1+1}{1-1} \geq 2$	False
$1 < x < 3$	$x = 2$	$\frac{2+1}{2-1} \geq 2$	True
$x = 3$	$x = 3$	$\frac{3+1}{3-1} \geq 2$	True
$x > 3$	$x = 4$	$\frac{4+1}{4-1} \geq 2$	False

These results are summarized in Figure 2-3. Therefore, the solution set consists of all real numbers

$$\{x \mid 1 < x \leq 3\}$$

■

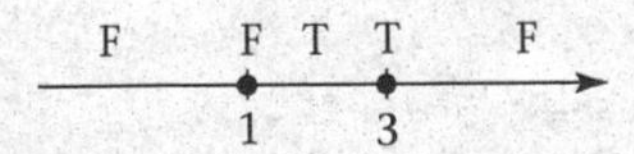

FIGURE 2-3
Summary of Critical Value Analysis

2.5c Second-Degree Inequalities

The Critical Value Method can also be applied to **second-degree inequalities.** This requires the solution of a quadratic equation rather than a linear equation.

EXAMPLE 9 *Quadratic Inequalities*

Solve the inequality $x^2 - 2x > 15$ and graph the solution.

SOLUTION

The inequality is defined everywhere. Solving the equation

$$\begin{aligned} x^2 - 2x &= 15 \\ x^2 - 2x - 15 &= 0 \\ (x+3)(x-5) &= 0 \\ x &= -3, 5 \end{aligned}$$

Therefore, the critical values are -3 and 5 as shown below.

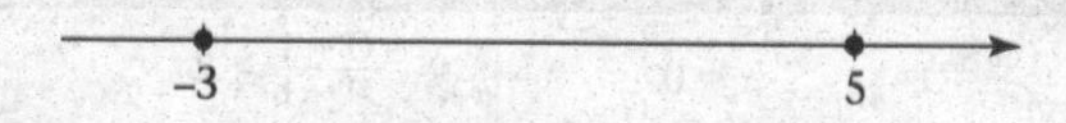

Interval, Critical Value	Test Point	Substitution	Verification
$x < -3$	$x = -5$	$(-5)^2 - 2(-5) > 15$	True
$x = -3$	$x = -3$	$(-3)^2 - 2(-3) > 15$	False
$-3 < x < 5$	$x = 0$	$0^2 - 2(0) > 15$	False
$x = 5$	$x = 5$	$(5)^2 - 2(5) > 15$	False
$x > 5$	$x = 6$	$(6)^2 - 2(6) > 15$	True

These results are summarized:

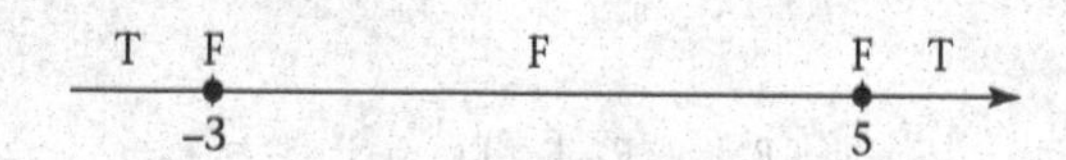

Therefore, the solution set is

$$\{x \mid x < -3 \quad \text{or} \quad x > 5\}$$

which consists of the real numbers in the open intervals $(-\infty, -3)$ and $(5, \infty)$. The solution set is shown below:

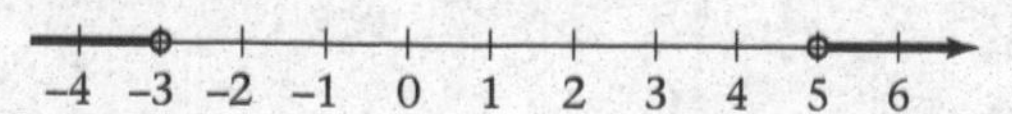

■

Solve the inequalities:

a. $2x^2 \geq 5x + 3$ b. $\frac{2x-3}{1-2x} \geq 0$

Answers

a. $\left\{x \middle| x \leq -\frac{1}{2} \text{ or } x \geq 3\right\}$ b. $\left\{x \middle| \frac{1}{2} < x \leq \frac{3}{2}\right\}$

EXAMPLE 10 *Polynomial Inequalities*

Solve the inequality $(x-2)(2x+5)(3-x) < 0$.

SOLUTION

The inequality is defined everywhere. Solving the equation

$$(x-2)(2x+5)(3-x) = 0$$

$$x = 2, -\frac{5}{2}, 3$$

Therefore the critical values are $-\frac{5}{2}$, 2, and 3 as shown below.

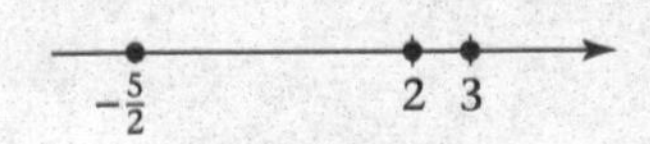

Interval, Critical Value	Test Point	Substitution	Verification
$x < -\frac{5}{2}$	$x = -3$	$(-3-2)(2(-3)+5)(3-(-3)) < 0$	False
$x = -\frac{5}{2}$	$x = -\frac{5}{2}$	$\left(-\frac{5}{2}-2\right)\left(2\left(-\frac{5}{2}\right)+5\right)\left(3-\left(-\frac{5}{2}\right)\right) < 0$	False
$-\frac{5}{2} < x < 2$	$x = 0$	$(0-2)(2(0)+5)(3-0) < 0$	True
$x = 2$	$x = 2$	$(2-2)(2(2)+5)(3-2) < 0$	False
$2 < x < 3$	$x = \frac{5}{2}$	$\left(\frac{5}{2}-2\right)\left(2\left(\frac{5}{2}\right)+5\right)\left(3-\frac{5}{2}\right) < 0$	False
$x = 3$	$x = 3$	$(3-2)(2(3)+5)(3-3) < 0$	False
$x > 3$	$x = 4$	$(4-2)(2(4)+5)(3-4) < 0$	True

These results are summarized as:

F F T F F F T

$-\frac{5}{2}$ 2 3

Therefore, the solution set consists of all real numbers

$$\left\{x \middle| -\frac{5}{2} < x < 2 \text{ or } x > 3\right\}$$

which consists of the real numbers in the open intervals $\left(-\frac{5}{2}, 2\right) \cup (3, \infty)$. ["∪" is the set notation for "or."] ■

Solve the inequality $(2y-9)(6-y)(y+5) \geq 0$.

Answers

$\left\{y \,\middle|\, y \leq -5 \text{ or } \frac{9}{2} \leq y \leq 6\right\}$ or $(-\infty, -5] \cup \left[\frac{9}{2}, 6\right]$

EXAMPLE 11 ***Quadratic Inequalities***

Solve the inequality $x^2 + 1 > 0$.

SOLUTION

The inequality is defined everywhere. The equation $x^2 = -1$ has no real roots. Thus, we have no critical values, and there is only one interval, namely, the entire real number line. If we choose our test point to be 0, we have

$$0^2 > -1$$

which is true. Therefore, the solution set consists of all real numbers. ■

Exercise Set 2.5

In Exercises 1–9, express the given inequality in interval notation.

1. $-5 \le x < 1$
2. $-4 < x \le 1$
3. $x > 9$
4. $x \le -2$
5. $-12 \le x \le -3$
6. $x \ge -5$
7. $3 < x < 7$
8. $x < 17$
9. $-6 < x \le -4$

In Exercises 10–18, express the given interval as an inequality.

10. $(-4, 3]$
11. $[5, 8]$
12. $(-\infty, -2]$
13. $(3, \infty)$
14. $[-3, 10)$
15. $(-\infty, 5]$
16. $(-2, -1)$
17. $[0, \infty)$
18. $(-5, 7)$

In Exercises 19–36, solve the inequality and graph the result.

19. $x + 4 < 8$
20. $x + 5 < 4$
21. $x + 3 < -3$
22. $x - 2 \le 5$
23. $x - 3 \ge 2$
24. $x + 5 \ge -1$
25. $2 < a + 3$
26. $-5 > b - 3$
27. $2y < -1$
28. $3x < 6$
29. $2x \ge 0$
30. $-\frac{1}{2}y \ge 4$
31. $2r + 5 < 9$
32. $3x - 2 > 4$
33. $3x - 1 \ge 2$
34. $\frac{-1}{2x+3} > 0$
35. $\frac{4}{5-3x} < 0$
36. $\frac{3}{3x-1} > 0$

Solve the given inequality in Exercises 37–60, and write the solution using interval notation.

37. $4x + 3 \le 11$
38. $\frac{1}{2}y - 2 \le 2$
39. $\frac{3}{2}x + 1 \ge 4$
40. $-5x + 2 > -8$
41. $4(2x + 1) < 16$
42. $3(3r - 4) \ge 15$
43. $2(x - 3) < 3(x + 2)$
44. $4(x - 3) \ge 3(x - 2)$
45. $3(2a - 1) > 4(2a - 3)$
46. $2(3x - 1) + 4 < 3(x + 2) - 8$
47. $\frac{2}{3}(x + 1) + \frac{5}{6} \ge \frac{1}{2}(2x - 1) + 4$
48. $\frac{1}{4}(3x + 2) - 1 \le -\frac{1}{2}(x - 3) + \frac{3}{4}$
49. $\frac{x-1}{3} + \frac{1}{5} < \frac{x+2}{5} - \frac{1}{3}$
50. $\frac{x}{5} - \frac{1-x}{2} > \frac{x}{2} - 3$
51. $3(x + 1) + 6 \ge 2(2x - 1) + 4$
52. $4(3x + 2) - 1 \le -2(x - 3) + 15$
53. $-2 < 4x \le 5$
54. $3 \le 6x < 12$
55. $-4 \le 2x + 2 \le -2$
56. $5 \le 3x - 1 \le 11$
57. $3 \le 1 - 2x < 7$
58. $5 < 2 - 3x \le 11$
59. $-8 < 2 - 5x \le 7$
60. $-10 < 5 - 2x < -5$

In Exercises 61–67, translate from words to an algebraic problem and solve.

61. A student has grades of 42 and 70 on the first two tests of the semester. If an average of 70 is required to obtain a C grade, what is the minimum score the student must achieve on the third exam to obtain a C?

62. A compact car can be rented from firm A for \$160 per week with no charge for mileage or from firm B for \$100 per week plus 20 cents for each mile driven. If the car is driven m miles, for what values of m does it cost less to rent from firm A?

63. An appliance salesperson is paid \$30 per day plus \$25 for each appliance sold. How many appliances must be sold for the salesperson's income to exceed \$130 per day?

64. A pension trust invests \$6000 in a bond that pays 5% simple interest per year. Additional funds are to be invested in a more speculative bond paying 9% simple interest per year, so that the return on the total investment will be at least 6%. What is the minimum amount that must be invested in the more speculative bond?

65. A book publisher spends \$38,000 on editorial expenses and \$12 per book for manufacturing and sales expenses in the course of publishing a psychology textbook. If the book sells for \$25, how many copies must be sold to show a profit?

66. If the area of a right triangle is not to exceed 80 square inches and the base is 10 inches, what values may be assigned to the altitude h?

67. A total of 70 meters of fencing material is available with which to enclose a rectangular area. If the width of the rectangle is 15 meters, what values can be assigned to the length L?

In Exercises 68–95, indicate the solution set of each inequality on a real number line.

68. $x^2 + 5x + 6 > 0$
69. $x^2 + 3x - 4 \le 0$
70. $2x^2 - x - 1 < 0$
71. $3x^2 - 4x - 4 \ge 0$
72. $4x - 2x^2 < 0$
73. $r^2 + 4r \ge 0$

74. $\frac{x+5}{x+3} \le 0$

75. $\frac{x-6}{x+4} \ge 0$

76. $\frac{2r+1}{r-3} \le 0$

77. $\frac{x-1}{2x-3} \ge 0$

78. $\frac{3s+2}{2s-1} \ge 0$

79. $\frac{4x+5}{x^2} \le 0$

80. $(x+2)(3x-2)(x-1) > 0$

81. $(x-4)(2x+5)(2-x) \le 0$

82. $x^2 + x - 6 > 0$

83. $x^2 - 3x - 10 \ge 0$

84. $2x^2 - 3x - 5 < 0$

85. $3x^2 - 4x - 4 \le 0$

86. $\frac{2r+3}{2r-1} < 0$

87. $\frac{3x+2}{2x-3} \ge 0$

88. $\frac{x-1}{x+1} \ge 0$

89. $\frac{2x-1}{x+2} \le 0$

90. $6x^2 + 8x + 2 \ge 0$

91. $2x^2 + 5x + 2 \le 0$

92. $(y-3)(2-y)(2y+4) \ge 0$

93. $(2x+5)(3x-2)(x+1) < 0$

94. $(x-3)(1+2x)(3x+5) > 0$

95. $(1-2x)(2x+1)(x-3) \le 0$

In Exercises 96–99, find the values of x for which the given expression has real values.

96. $\sqrt{(x-2)(x+1)}$

97. $\sqrt{(2x+1)(x-3)}$

98. $\sqrt{2x^2+7x+6}$

99. $\sqrt{2x^2+3x+1}$

100. A manufacturer of solar heaters finds that when x units are made and sold, the profit (in thousands of dollars) is given by $x^2 - 50x - 5000$. For what values of x will the firm show a loss?

101. A ball thrown directly upward from ground level at an initial velocity of 40 feet per second attains a height d given by $d = 40t - 16t^2$ after t seconds. During what time interval is the ball at a height of at least 16 feet?

102. A rectangle has length x and width $x - 4$.

a. Find the inequality that states that the perimeter of the rectangle must be at least 24 units, and solve for x.

b. Find the inequality that states that the area of the rectangle must be less than 12 square units, and solve for x.

103. Each of the two congruent sides of an isosceles triangle are 10 centimeters more than $\frac{1}{2}$ the length of the base. If the perimeter of the triangle is to be at most 100 centimeters, what is the maximum length of the base?

104. Morry's best time in the 70-meter track event is 9.5 seconds. Rob wants to beat Morry's best time. He runs the first half of the event at a speed of 7 meters per second. What is the maximum time Rob has left to run the second half of the event?

105. A carpet factory manufactures bolts of carpet 10 feet wide. A large bolt of carpet covers 8 linear feet more than a small bolt of carpet. If the large bolt of carpet covers at most 200 square feet of floor, what is the largest length of a small bolt?

106. Rob and Morry face each other at opposite ends of an 880-meter track. Rob runs this distance at 420 meters per minute, and Morry runs the same distance at 300 meters per minute. At the sound of the gun, the boys start running toward each other. Rob always arrives at a point P on the track before Morry. What is the farthest distance P could be from Rob's end of the track?

107. The power P in watts, total resistance R in ohms, and current I in amperes of a circuit are related by the equation

$$P = I^2R$$

The power output can be at most 1200 watts. What is the maximum current in the circuit if the total resistance is 48 ohms?

108. A string on a musical instrument has frequency f (hertz or vibrations per second) and is defined by

$$f = \frac{1}{2L}\sqrt{\frac{10^5 FL}{m}}$$

where F is the tension of the string in Newtons, L is the length of the string in centimeters and m is the mass of the string in grams. The D string on a violin has a length of 45 centimeters and a mass of 0.9 grams. What is the maximum frequency of the D string if the tension of the string can be at most 150 Newtons?

109. Suppose the pressure P in pounds and the volume V in cubic inches of a confined gas is

$$P = \frac{150}{V}$$

What range of volume corresponds to a range of pressure between 50 pounds per square inch and 75 pounds per square inch?

110. Marilyn called Y's Buy Oil company to have her oil tank filled. She was not exactly sure how many gallons of oil were needed. She was told that oil would cost \$2.079 per gallon if she purchased 150 gallons or more, and would cost \$2.179 per gallon if she purchased less than 150 gallons. For what number of gallons is it more cost effective for Marilyn to buy 150 gallons than to buy less than 150 gallons?

111. Michael earned 310 points before the final exam in his college algebra and trigonometry course. He must have at least 80% of a total of 600 points to get a B in this class. The final is worth 200 points. What is the lowest possible score Michael can get on his final exam and still get a B?

112. An economist hired by the Hiccup Seed Company has found that the company's profit, in hundred thousands of dollars, is

$$P = 6x^2 - 70x + 50$$

where x is the amount, in thousands, of seed packets sold. For what values of x does the Hiccup Seed Company make a profit?

113. The relationship between degrees Celsius and degrees Fahrenheit is given by

$$F = \frac{9}{5}C + 32$$

What temperature range in °F corresponds to −10°C to 20°C?

114. For what values of a is $a^2 > a$, provided $a > 0$?

115. Anthropologists use a ratio called the cephalic index in order to classify different genetic groupings. The cephalic index C requires that you measure the width W and length L of the top of a person's head. The index is as follows

$$C = \frac{100W}{L}$$

A Native American tribe in Oklahoma has a cephalic index of 67 ± 1. If the average length of their heads is 10 inches, what is the smallest and largest width of the heads of the members of this tribe?

116. Suppose you are setting up a full-mesh network for x users; and n, the number of two-way connections required to link all users pairwise, must be no greater than 132. For what range of x values can you set up your network?

$$\frac{x(x-1)}{2} = n$$

2.6 Absolute Value in Equations and Inequalities

In Section 1.2, we discussed the use of absolute value notation to indicate distance, and we provided this formal definition

$$|x| = \begin{cases} x \text{ when } x \geq 0 \\ -x \text{ when } x < 0 \end{cases}$$

The following example illustrates the application of this definition to the solution of equations involving absolute value.

EXAMPLE 1 ***Absolute Value in Equations***

Solve the equation $|2x - 7| = 11$.

SOLUTION

We apply the definition of absolute value and consider two cases.

Case 1. $2x - 7 \geq 0$

With the first part of the definition,

$$|2x - 7| = 2x - 7 = 11$$
$$2x = 18$$
$$x = 9$$

Case 2. $2x - 7 < 0$

With the second part of the definition,

$$|2x - 7| = -(2x - 7) = 11$$
$$-2x + 7 = 11$$
$$x = -2$$

Alternatively, we can solve $2x - 7 = 11$ to obtain $x = 9$ and $2x - 7 = -11$ to obtain $x = -2$. ■

Progress Check

Solve each equation and check the solution(s).

a. $|x + 8| = 9$ b. $|3x - 4| = 7$

Answers

a. $1, -17$ b. $\frac{11}{3}, -1$

When used in inequalities, absolute value notation plays an important and frequently used role in higher mathematics. To solve inequalities involving absolute value, we recall that $|x|$ is the distance between the origin and the point on the real number line corresponding to x. For $a > 0$, the solution set of the inequality $|x| < a$ is then seen to consist of all real numbers whose distance from the origin is less than a, that is, all real numbers in the open interval $(-a, a)$, shown in Figure 2-4. Similarly, if $|x| > a > 0$, the solution set consists of all real numbers whose distance from the origin is greater than a, that is, all points in the two infinite intervals $(-\infty, -a)$ or (a, ∞), shown in Figure 2-5. An alternative statement of the solution set is

$$\{x \mid x < -a \quad \text{or} \quad x > a\}$$

Of course, $|x| \leq a$ and $|x| \geq a$ include the endpoints a and $-a$, and the circles are filled in.

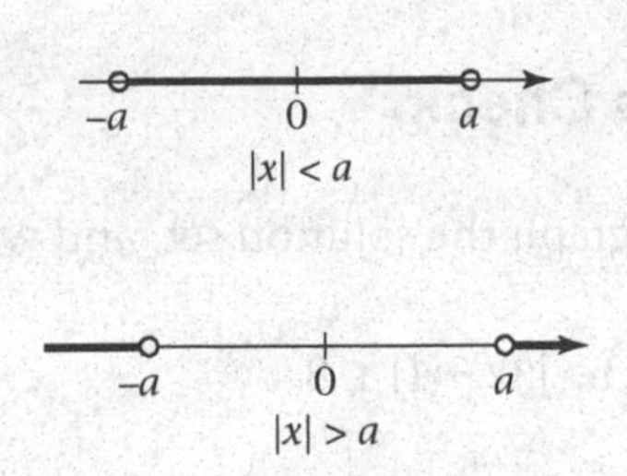

FIGURE 2-4
Graph of Solution Set for $|x| < a$

FIGURE 2-5
Graph of Solution Set for $|x| > a$

We will use the Critical Value Method for solving inequalities involving the absolute value in a manner similar to that found in Section 2.5.

EXAMPLE 2 *Absolute Value in Inequalities*

Solve the inequality $2|2x - 5| \leq 14$. Write the solution set in interval notation and graph the solution.

SOLUTION

The inequality is defined everywhere. Solving the equation

$$\begin{cases} 2|2x - 5| = 14 \\ |2x - 5| = 7 \end{cases}$$

$$\begin{aligned} 2x - 5 &= 7 \quad \text{or} \quad 2x - 5 = -7 \\ x &= 6 \quad \text{or} \quad x = -1 \end{aligned}$$

Therefore, the critical values are −1 and 6.

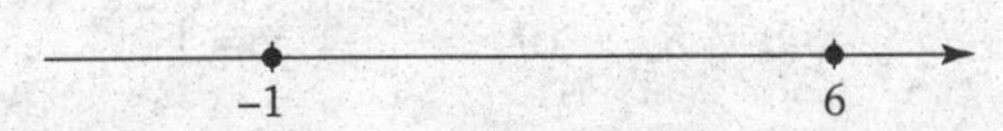

Interval, Critical Value	Test Point	Substitution	Verification
$x < -1$	$x = -2$	$2\|2(-2) - 5\| \leq 14$	False
$x = -1$	$x = -1$	$2\|2(-1) - 5\| \leq 14$	True
$-1 < x < 6$	$x = 0$	$2\|2(0) - 5\| \leq 14$	True
$x = 6$	$x = 6$	$2\|2(6) - 5\| \leq 14$	True
$x > 6$	$x = 7$	$2\|2(7) - 5\| \leq 14$	False

These results are summarized:

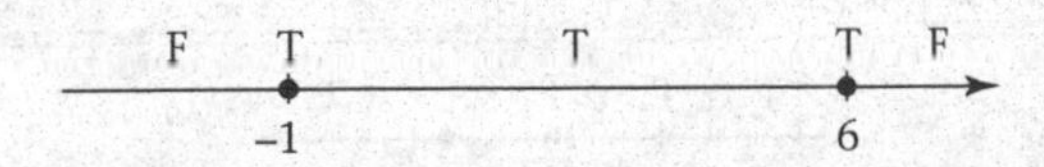

Therefore, the solution set consists of all real numbers in the closed interval $[-1, 6]$, and the graph of the solution set is shown below.

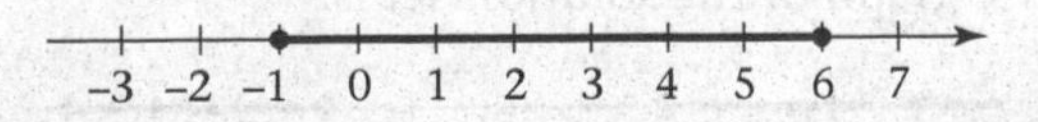

■

Progress Check

Solve each inequality, graph the solution set, and write the solution set in interval notation.

a. $|x| < 3$ b. $|3x - 1| \le 8$ c. $|x| < -2$

Answers

a. $(-3, 3)$ −3 0 3 b. $\left[-\frac{7}{3}, 3\right]$ −3 −2 −1 0 1 2 3 4

c. There is no solution since $|x|$ is always nonnegative and thus cannot be less than −2.

EXAMPLE 3 *Absolute Value in Inequalities*

Solve the inequality $|2x - 6| > 4$, write the solution set in interval notation, and graph the solution.

SOLUTION

The inequality is defined everywhere. Solving the equation

$$|2x - 6| = 4$$

$$2x - 6 = 4 \quad \text{or} \quad 2x - 6 = -4$$

$$x = 5 \quad \text{or} \quad x = 1$$

1 5

Interval, Critical Value	Test Point	Substitution	Verification
$x < 1$	$x = 0$	$\|2(0) - 6\| > 4$	True
$x = 1$	$x = 1$	$\|2(1) - 6\| > 4$	False
$1 < x < 5$	$x = 2$	$\|2(2) - 6\| > 4$	False
$x = 5$	$x = 5$	$\|2(5) - 6\| > 4$	False
$x > 5$	$x = 6$	$\|2(6) - 6\| > 4$	True

These results are summarized in Figure 2-6.

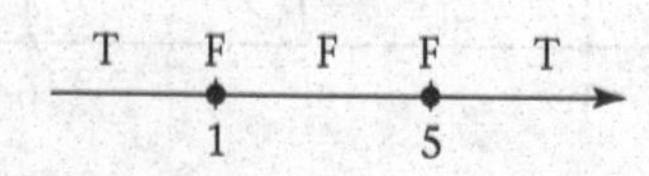

FIGURE 2-6
Summary of Critical Value Analysis

Therefore, the solution set consists of all real numbers in the infinite intervals $(-\infty, 1)$ or $(5, \infty)$. The graph of the solution set is:

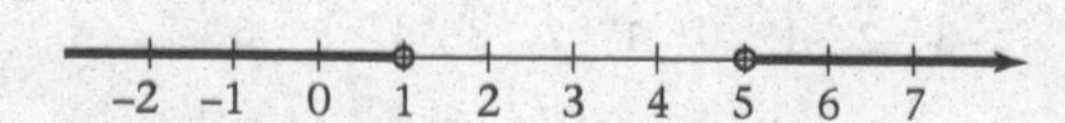

and the solution set is $(-\infty, 1) \cup (5, \infty)$. ■

Students sometimes write

$$1 > x > 5$$

This is a misuse of the inequality notation since it states that x is simultaneously less than 1 *and* greater than 5, which is impossible. What is usually intended is the pair of infinite intervals $(-\infty, 1)$ or $(5, \infty)$, and the inequalities must be written

$$x < 1 \quad \text{or} \quad x > 5,$$

written as $(-\infty, 1) \cup (5, \infty)$ in interval notation.

Two additional misuses of the inequality notation are

$$1 < x > 5 \quad \text{and} \quad 1 > x < 5$$

We summarize some facts concerning absolute values in equations and inequalities.

If $a > 0$, then:

- $|x| = a$ is equivalent to $x = \pm a$.
- $|x| < a$ is equivalent to $-a < x < a$.
- $|x| > a$ is equivalent to $x < -a$ or $x > a$.

Verify these results using the Critical Value Method.

Progress Check

Solve each inequality, write the solution set in interval notation, and graph the solution.

a. $|5x - 6| > 9$ b. $|2x - 2| \geq 8$

Answers

a. $\left(-\infty, -\frac{3}{5}\right) \cup (3, \infty)$ −4 −3 −2 −1 0 1 2 3 4 5 6

b. $(-\infty, -3] \cup [5, \infty)$ −4 −3 −2 −1 0 1 2 3 4 5 6

Exercise Set 2.6

In Exercises 1–9, solve and check.

1. $|x + 2| = 3$
2. $|r - 5| = \frac{1}{2}$
3. $|2x - 4| = 2$
4. $|5y + 1| = 11$
5. $|-3x + 1| = 5$
6. $|2t + 2| = 0$
7. $3|-4x - 3| = 27$
8. $\frac{1}{|x|} = 5$
9. $\frac{1}{|s-1|} = \frac{1}{3}$

In Exercises 10–15, solve the inequality and graph the solution set.

10. $|x + 3| < 5$
11. $|x + 1| > 3$
12. $|3x + 6| \leq 12$
13. $|4x - 1| > 3$
14. $|3x + 2| \geq -1$
15. $\left|\frac{1}{3} - x\right| < \frac{2}{3}$

In Exercises 16–24, solve the inequality, and write the solution set using interval notation.

16. $|x - 2| \leq 4$
17. $|x - 3| \geq 4$
18. $|2x + 1| < 5$
19. $\frac{|2x-1|}{4} < 2$
20. $\frac{|3x+2|}{2} \leq 4$
21. $\frac{|2x+1|}{3} < 0$
22. $\left|\frac{4}{3x-2}\right| < 1$
23. $\left|\frac{5-x}{3}\right| > 4$
24. $\left|\frac{2x+1}{3}\right| \leq 5$

In Exercises 25–28, solve for x.

25. $|2x + 1| - 3 = -2$
26. $3 - |2x + 4| = 1$
27. $2|3 - x| + 3 = 5$
28. $4 - 3|2x + 7| = -5$

In Exercises 29 and 30, x and y are real numbers.

29. Prove that $\left|\frac{x}{y}\right| = \frac{|x|}{|y|}$. (*Hint:* Consider four cases.)
30. Prove that $|x|^2 = x^2$.
31. A machine that packages 100 vitamin pills per bottle can make an error of 2 pills per bottle. If x is the number of pills in a bottle, write an inequality, using absolute value, that indicates a maximum error of 2 pills per bottle. Solve the inequality.
32. The weekly income of a worker in a manufacturing plant differs from \$500 by no more than \$50. If x is the weekly income, write an inequality, using absolute value, that expresses this relationship. Solve the inequality.
33. Express the statement $x > 6$ or $x < -6$ as a single inequality using absolute value.
34. Express the statement $-10 < x < 10$ as an inequality using absolute value.
35. Express the statement $x \geq 5$ or $x \leq 1$ as a single inequality using absolute value.
36. Express the statement $2d - 5 \leq x \leq 2d + 5$ as an inequality using absolute value.
37. Write an equation that states that x is 10 units from 4 on the real number line. Solve this equation.
38. Find all points x on the real number line such that x is 5 times as far from the origin as from 20.
39. Find all points x on the real number line such that x is 3 times as far from 4 as $2x$ is from 6.
40. Find the value of $\frac{|x|-x}{2}$ for
    a. $x \geq 0$
    b. $x < 0$
41. Chebyshev's Theorem from probability theory states that the probability that any random variable x will assume a value within k standard deviations σ of its mean μ is at least $1 - \frac{1}{k^2}$, or, equivalently,
$$P(|x - \mu| < k\sigma) \geq 1 - \frac{1}{k^2}$$
Solve $|x - \mu| < k\sigma$ for x.
42. In calculus, absolute value inequalities are used in the definitions of terms like *continuous function* and *limit*. Solve the following inequality for x when $\delta = 0.005$ and $x_0 = 0.001$:
$$|x - x_0| < \delta$$
43. Solve the following inequality for x when $L = 9$ and ε has each of the following values: 0.01, 0.001, 0.0001.
$$|(2x + 5) - L| < \varepsilon$$

Chapter Summary

Key Terms, Concepts, and Symbols

Key Ideas for Review

Topic	Page	Key Idea
Solutions of an Equation	78	A solution of an equation is a value that satisfies the equation.
Solution Process	78	To solve an equation, we generally form a succession of simpler, equivalent equations. We may add to or subtract from both sides of the equation any number or expression. We may also multiply both sides by any nonzero number. If we multiply the equation by an expression containing a variable, the answers must be substituted into the original equation to verify that they are solutions.
Linear Equations	80	The linear equation $ax + b = 0, \quad a \neq 0$ has precisely one solution $x = -\frac{b}{a}$
Completing the Square	99	$x^2 + dx + \frac{d^2}{4} = \left(x + \frac{d}{2}\right)^2$ Therefore, add $\frac{d^2}{4}$ to $x^2 + dx$ to "complete the square."
Quadratic Equations	94	The quadratic equation $ax^2 + bx + c = 0, \quad a \neq 0$ always has two solutions that are given by the quadratic formula $x = \frac{-b \pm \sqrt{b^2 - 4ac}}{2a}, \quad a \neq 0$ If $b = 0$, or if the quadratic equation can be factored, then faster solution methods are available.
Discriminant	103	The solutions or roots of a quadratic equation may be complex numbers. The expression $b^2 - 4ac$ under the radical of the quadratic formula is called the discriminant. Its value determines the nature of the roots of the quadratic equation.

continues

Topic	Page	Key Idea
Radical Equations	105	Radical equations often can be transformed into quadratic equations. Since the process involves raising both sides of an equation to a power, the answers must be checked to see that they satisfy the original equation.
Substitution of Variable	107	The method called *substitution of variable* can be used to transform certain equations into quadratic equations. This is a valuable technique that will be used in other chapters of this book.
Solutions of an Inequality	114	A solution of an inequality is a value that satisfies the inequality.
Solution Process	114	The permissible operations in solving an inequality are the same as those for solving equations with this proviso: multiplication or division by a negative number reverses the direction of the inequality.
Solution Set Representation	115	The solution set of an inequality can be represented by using set notation, interval notation, or a graph on the real number line.
Critical Value Method	120	The critical values of an inequality are as follows: 1. those values for which either side of the inequality is not defined (such as a denominator equal to 0) 2. those values that are solutions to the equation obtained by replacing the inequality sign with an equal sign An inequality can be solved by finding the critical values and checking a test point in each interval determined by those critical values.

Review Exercises

In Exercises 1–4, solve for x.

1. $3x - 5 = 3$
2. $2(2x - 3) - 3(x + 1) = -9$
3. $\dfrac{2 - x}{3 - x} = 4$
4. $k - 2 = 4kx$
5. The width of a rectangle is 4 centimeters less than twice its length. If the perimeter is 12 centimeters, find the measurement of each side.
6. A donation box contains coins consisting of dimes and quarters. The number of dimes is 4 more than twice the number of quarters. If the total value of the coins is \$2.65, how many coins of each type are there?
7. It takes 4 hours for a bush pilot in Australia to pick up mail at a remote village and return to home base. If the average speed going is 150 mph and the average speed returning is 100 mph, how far from the home base is the village?
8. Copying machines A and B, working together, can prepare enough copies of the annual report for the board of directors in 2 hours. Machine A, working alone, would require 3 hours to do the job. How long would it take machine B to do the job by itself?

In Exercises 9 and 10 indicate whether the statement is true (T) or false (F).

9. The equation $3x^2 = 9$ is an identity.
10. $x = 3$ is a solution of the equation $3x - 1 = 10$.
11. Solve $x^2 - x - 20 = 0$ by factoring.
12. Solve $6x^2 - 11x + 4 = 0$ by factoring.
13. Solve $x^2 - 2x + 6 = 0$ by completing the square.
14. Solve $2x^2 - 4x + 3 = 0$ by the quadratic formula.
15. Solve $3x^2 + 2x - 1 = 0$ by the quadratic formula.

In Exercises 16–18, solve for x.

16. $49x^2 - 9 = 0$
17. $kx^2 - 3\pi = 0$
18. $x^2 + x = 12$

In Exercises 19–21, determine the nature of the roots of the quadratic equation without solving.

19. $3r^2 = 2r + 5$
20. $4x^2 + 20x + 25 = 0$
21. $6y^2 - 2y = -7$

In Exercises 22–25, solve the given equation.

22. $\sqrt{x} + 2 = x$

23. $\sqrt{x+3} + \sqrt{2x-3} = 6$

24. $x^4 - 4x^2 + 3 = 0$

25. $\left(1-\frac{2}{x}\right)^2 - 8\left(1-\frac{2}{x}\right) + 15 = 0$

26. A charitable organization rented an auditorium for a meeting at a cost of \$420 and split the cost among the attendees. If 10 additional persons had attended the meeting, the cost per person would have decreased by \$1. How many persons actually attended?

27. Solve and graph $3 \le 2x + 1$.

28. Solve and graph $-4 < -2x + 1 \le 10$.

In Exercises 29–31, solve the inequality and express the solution set in interval notation.

29. $2(a + 5) > 3a + 2$

30. $\frac{-1}{2x-5} \le 0$

31. $\frac{2x}{3} + \frac{1}{2} \ge \frac{x}{2} - 1$

32. Solve $|3x + 2| = 7$ for x.

33. Solve and graph $|4x - 1| = 5$.

34. Solve and graph $|2x + 1| > 7$.

35. Solve $|2 - 5x| < 1$ and write the solution in interval notation.

36. Solve $|3x - 2| \ge 6$ and write the solution in interval notation.

37. Find the values of x for which $\sqrt{2x^2 - x - 6}$ has real values.

38. Using interval notation, write the solution set of the inequality $x^2 + 4x - 5 \le 0$.

39. Write the solution set for $\frac{2x+1}{x+5} \ge 0$ in interval notation.

40. Write the solution set for

$$(3 - x)(2x + 3)(x + 2) < 0$$

in interval notation.

41. A local school board is debating the question of whether or not to close an elementary school in its district. The board expects a larger than average number of local residents at this meeting. The typical meeting seats the residents in a rectangular formation of 10 rows, 15 seats to a row. In order to double the seating capacity with a new rectangular formation, the board decides to add an equal number of seats to each existing row and to add that same number of additional rows to the original formation. Find the number of chairs needed to add to each row.

42. A circle and a square have the same perimeter P. Show that the area of the circle is greater than the area of the square.

43. A rectangular window is to be cut into a door 6.5 feet tall so that the distance below the window is 6 inches less than twice the distance above the window. The window must be 18 inches wide and have an area of at least 324 square inches. What is the maximum distance between the top of the door and the top of the window?

44. In statistical theory, one can develop an interval estimate of the population mean μ with a known standard deviation σ by taking a sample of size n and finding the mean $\overline{x}$ of the sample. The 95% confidence interval is given by

$$\left|\frac{\mu - \overline{x}}{\frac{\sigma}{\sqrt{n}}}\right| < 1.96$$

Solve this inequality for μ.

45. The kinetic energy E of a particle depends on the particle's mass m and speed v and is defined as

$$E = \frac{1}{2}mv^2$$

a. A 10,000-kilogram bullet has kinetic energy of 5000 kilogram meters per second squared. Find the speed of the bullet.

b. If the mass of the bullet is halved, find the speed.

46. An open box is to be constructed from a piece of cardboard that is x inches by $x + 2$ inches. This is accomplished by cutting 2-inch squares from each corner of the cardboard and bending up the sides. If the volume of the box is to be 96 cubic inches, find the dimensions of the box.

Review Test

In Problems 1 and 2, solve for y.

1. $5 - 4y = 2$
2. $\dfrac{2+5y}{3y-1} = 6$

3. One side of a triangle is 2 meters shorter than the base, and the other side is 3 meters longer than half the base. If the perimeter is 15 meters, find the length of each side.
4. A trust fund invested a certain amount of money at 6.5% simple annual interest, a second amount \$200 more than the first amount at 7.5%, and a third amount \$300 more than twice the first amount at 9%. If the total annual income from these investments is \$1962, how much was invested at each rate?
5. Indicate whether the statement is true (T) or false (F): The equation $(2x-1)^2 = 4x^2 - 4x + 1$ is an identity.
6. Solve $x^2 - 5x = 14$ by factoring.
7. Solve $5x^2 - x + 4 = 0$ by completing the square.
8. Solve $12x^2 + 5x - 3 = 0$ by the quadratic formula.

In Problems 9 and 10, solve for x.

9. $(2x-5)^2 + 9 = 0$
10. $2 + \dfrac{1}{x} - \dfrac{3}{x^2} = 0$

In Problems 11 and 12, determine the nature of the roots of the quadratic equation without solving.

11. $6x^2 + x - 2 = 0$
12. $3x^2 - 2x = -6$

In Problems 13 and 14, solve the given equation.

13. $x - \sqrt{4-3x} = -8$
14. $3x^4 + 5x^2 - 2 = 0$

15. The area of a rectangle is 96 square meters. If the length and the width are each increased by 2 meters, the area of the newly formed rectangle is 140 square meters. Find the dimensions of the original rectangle.
16. Solve $-1 \le 2x + 3 < 5$ and graph the solution set.

In Problems 17 and 18, solve the inequality and express the solution set in interval notation.

17. $3(2a-1) - 4(a+2) \le 4$
18. $-2 \le 2 - x \le 6$
19. Solve $|4x - 1| = 9$.
20. Solve $|2x - 1| \le 5$ and graph the solution set.
21. Solve $|1 - 3x| > 5$ and write the solution in interval notation.
22. Find the values of x for which $\sqrt{3x^2 - 4x + 1}$ has real values.

In Problems 23–25, write the solution set in interval notation.

23. $-2x^2 + 3x - 1 \le 0$
24. $(x-1)(2-3x)(x+2) \le 0$
25. $\dfrac{2x-5}{x+1} > -\dfrac{1}{3}$

Writing Exercises

In Exercises 1–3, write in complete sentences the procedure that you follow in solving the following problems.

1. An automatic teller machine gives you \$150 in five and ten dollar bills. There are 2 more than twice as many five dollar bills as there are ten dollar bills. How many of each denomination are there?
2. A pound of raisins costs \$2.50 whereas a pound of chocolate bits costs \$4.00. If we want to make a one-pound mixture of these items to sell for \$3.25, how much of each item must be used?
3. Two students work in a library shelving books. Eric can shelve 50 books per hour and Steve can shelve 40 books per hour. If Steve starts work in the morning and then is relieved by Eric later in the day, how long did each student work if 355 books were shelved in an 8-hour day?
4. Why is it a good practice to check your answers? Give an example to show how not following this practice can lead to a wrong conclusion.

Chapter 2 Project

Have you ever considered the challenges involved in linking computers and databases across the world into a user-friendly network? The mathematics involved in any such project are very sophisticated. The type of network we discussed in this chapter is far simpler!

In Section 2.3, do Exercise 119, and in Section 2.5, do Exercise 116.

In a *central relay network,* all users are connected to one central point. Therefore, the number of users is exactly equal to the number of links necessary. For how many users would a central relay network actually require fewer links than a full-mesh network? The same number? Sketch a diagram to illustrate.

3 Functions

Try this experiment. Listen to a cricket chirping, and count the number of times it chirps in one minute. Divide that number by 4, then add 40. Now compare that result to the temperature in degrees Fahrenheit. Is your result close? It probably is.

This is one example of a *function.* For this experiment, the outside temperature in degrees Fahrenheit is treated as a function of the number of times a cricket chirps in fifteen seconds. This rule of thumb has actually been shown to have validity. Find out how by looking up Svante Arrhenius at http://scienceworld.wolfram.com/biography/.

We will look at this function more closely in several sections of this chapter. By looking at its graph, determining the slope, finding its inverse, etc., you will see how functions help us to predict events in the world around us. (See the Chapter Project.)

What is the result of increased fertilization on the growth of an azalea? If the minimum wage is increased, what will be the effect on the number of unemployed workers? When a submarine dives, can we calculate the water pressure against the hull at a given depth?

Each of the questions posed above seeks a relationship between phenomena. The search for relationships, or correspondence, is a central activity in our attempts to understand the universe; it is used in mathematics, engineering, the physical and biological sciences, the social sciences, business and economics.

The concept of a function has been developed as a means of organizing and assisting the study of relationships. Since graphs are powerful means of exhibiting relationships, we begin with a study of the Cartesian, or rectangular, coordinate system. We then formally define a function and offer a number of ways of viewing the concept of a function. Function notation will be introduced to provide a convenient means of writing functions.

We will also explore some special types of functional relationships (increasing and decreasing functions), the effect of combining functions in various ways and how functions can be used to describe certain processes.

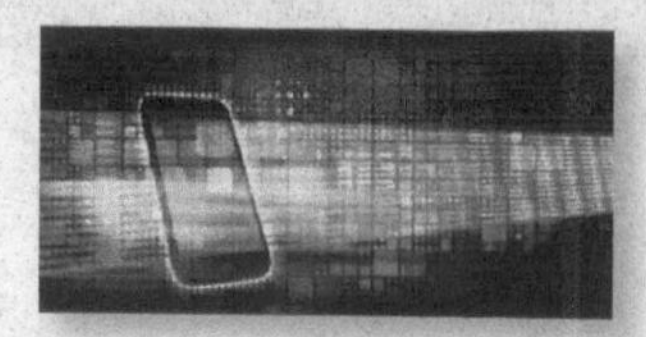

http://scienceworld.wolfram.com/biography

3.1 The Rectangular Coordinate System

In Chapter 1 we associated the system of real numbers with points on the real number line. That is, we saw that there is a one-to-one correspondence between the system of real numbers and points on the real number line.

We will now develop an analogous way to handle points in a plane. We begin by drawing a pair of perpendicular lines intersecting at a point called the **origin**. As shown in Figure 3-1, one of the lines, called the **x-axis**, is usually horizontal; and the other line, called the **y-axis**, is usually vertical.

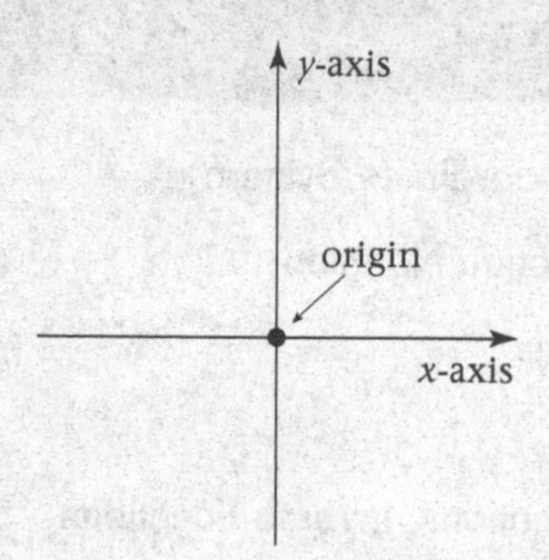

FIGURE 3-1
The Rectangular Coordinate System

If we think of the x-axis as a real number line, we may mark off some convenient unit of length, with positive numbers to the right of the origin and negative numbers to the left of the origin. Similarly, we may think of the y-axis as a real number line. Again, we may mark off a convenient unit of length (usually the same as the unit of length on the x-axis) with the upward direction representing positive numbers and the downward direction negative numbers. The x and y axes are called **coordinate axes**, and together they constitute a **rectangular,** or **Cartesian, coordinate system.** The coordinate axes divide the plane into four **quadrants,** which we label I, II, III, and IV as in Figure 3-2.

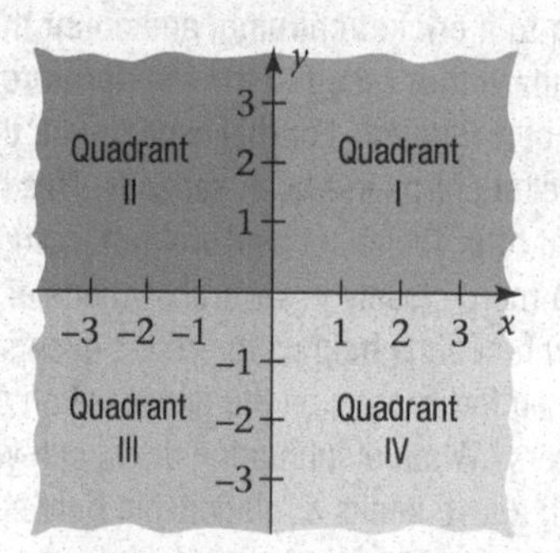

FIGURE 3-2
The Quadrants of the Rectangular Coordinate System

By using the coordinate axes, we can outline a procedure for labeling any point in the plane. Consider the point P that is 2 units to the right of the y-axis and 3 units above the x-axis as shown in Figure 3-3. This means that the line from P, perpendicular to the x-axis, meets the x-axis 2 units to the right of the origin. Furthermore, the line from P, perpendicular to the y-axis, meets the y-axis 3 units above the origin. We now give the point P the label (2, 3).

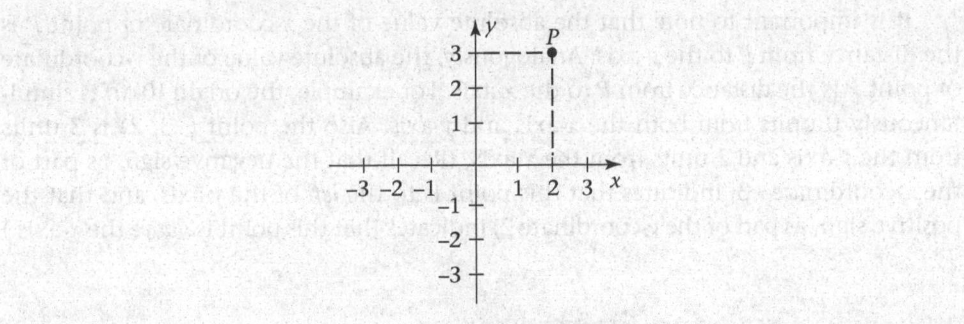

FIGURE 3-3
P with Coordinates (2, 3)

More generally, consider the point P in Figure 3-4. Starting from P, draw a line perpendicular to the x-axis, and let it meet the x-axis where x has the value a. Also starting from P, draw a line perpendicular to the y-axis, and let it meet the y-axis where y has the value b. Note the rectangle drawn in Figure 3-4, hence the name: *rectangular coordinates*. (The alternative name, "Cartesian," was given to honor the French mathematician and philosopher, René Descartes (1596–1650).) We say that the **coordinates** of P are given by the **ordered pair** (a, b). The term, "ordered pair," means that the order is significant; the ordered pair (a, b) is different from the ordered pair (b, a) when $a \neq b$.

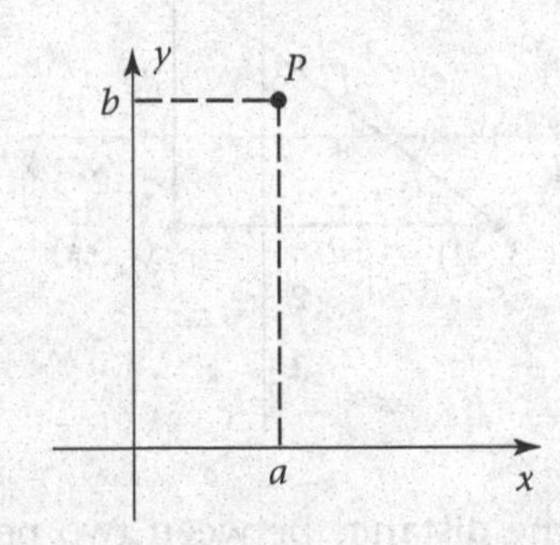

FIGURE 3-4
P with Coordinates (a, b)

The first number of the ordered pair (a, b) is called the **abscissa**, or ***x*-coordinate**, of P. The second number is called the **ordinate**, or ***y*-coordinate**, of P.

We have now developed a procedure for associating with each point P in a plane a unique ordered pair of real numbers (a, b) that we write as $P(a, b)$. Conversely, every ordered pair of real numbers (a, b) determines a unique point P in the plane. The point P is located at the intersection of two lines: one perpendicular to the x-axis at $x = a$ and one perpendicular to the y-axis at $y = b$. This establishes a one-to-one correspondence between the set of all points in the plane and the set of all ordered pairs of real numbers.

We have indicated a number of points in Figure 3-5. Note that all points on the x-axis have a y-coordinate of 0, and all points on the y-axis have an x-coordinate of 0.

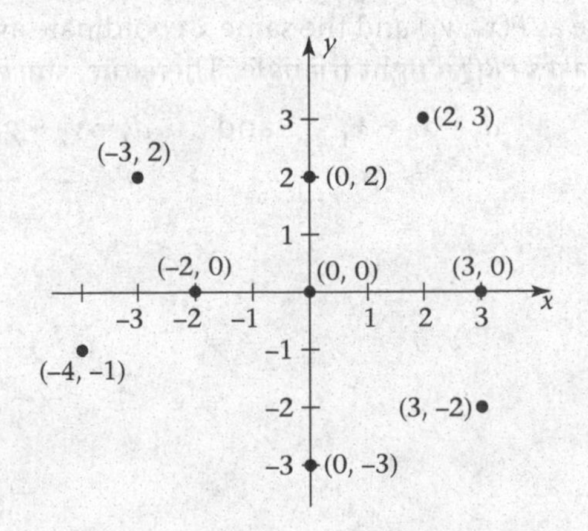

FIGURE 3-5
Points with Coordinates

It is important to note that the absolute value of the x-coordinate of point P is the distance from P to the y-axis. Analogously, the absolute value of the y-coordinate of point P is the distance from P to the x-axis. For example, the origin (0, 0) is simultaneously 0 units from both the x-axis and y-axis. Also the point (−3, 2) is 3 units from the y-axis and 2 units from the x-axis. (Recall that the negative sign, as part of the x-coordinate −3, indicates that this point is to the *left* of the y-axis, and that the positive sign, as part of the y-coordinate 2, indicates that this point is *above* the x-axis.)

3.1a *The Distance Formula*

Consider the problem of finding the distance between two points that are horizontal relative to each other, say (−3, −1) and (1, −1), as shown in Figure 3-6. We see that (−3, −1) is 3 units from the y-axis and (1, −1) is 1 unit from the other side of the y-axis. Therefore, since $1 > -3$, the distance between these points is given by $d_1 = 1 - (-3) = 4$.

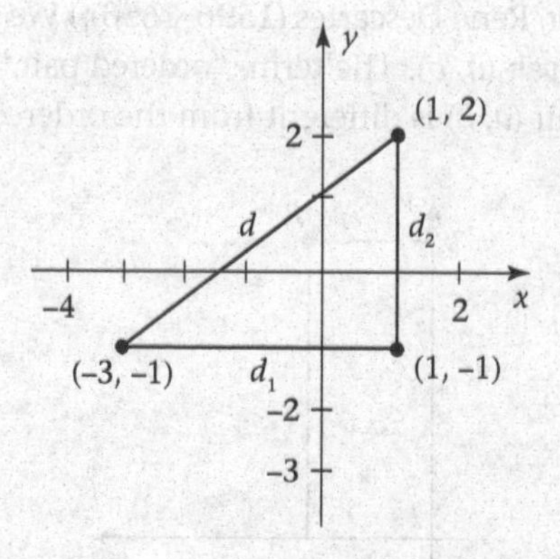

FIGURE 3-6
Finding the Distance Between (−3, −1) and (1, 2)

If we consider finding the distance between two points that are vertical relative to each other, say (1, −1) and (1, 2), as shown in Figure 3-6, we see that the first point is 1 unit from the x-axis, and the second point is 2 units from the other side of the x-axis. Therefore, since $2 > -1$, the distance between these points is given by $d_2 = 2 - (-1) = 3$.

A more complicated problem is to find the distance between (−3, −1) and (1, 2). (See Figure 3-6.) Since the triangle with vertices (−3, −1), (1, −1), and (1, 2) is a right triangle, we may use the Pythagorean Theorem. Thus

$$d^2 = d_1^2 + d_2^2 = 4^2 + 3^2 = 25$$

Therefore, $d = 5$, since distance is always nonnegative.

We can generalize this approach to derive a formula that gives the distance between any two points $P(x_1, y_1)$ and $Q(x_2, y_2)$, denoted $\overline{PQ}$. Choose R so that it has the same y-coordinate as $P(x_1, y_1)$ and the same x-coordinate as $Q(x_2, y_2)$, as shown in Figure 3-7. This makes PRQ a right triangle. Therefore, since $x_2 > x_1$ and $y_2 > y_1$,

$$d_1 = x_2 - x_1 \quad \text{and} \quad d_2 = y_2 - y_1$$

FIGURE 3-7
Deriving the Distance Formula

Hence, we can apply the Pythagorean Theorem.

$$d^2 = d_1^2 + d_2^2$$
$$d^2 = (x_2 - x_1)^2 + (y_2 - y_1)^2$$
$$d = \sqrt{(x_2 - x_1)^2 + (y_2 - y_1)^2}$$

because distance is always nonnegative.

Although points P and Q in Figure 3-7 are in quadrants III and I, respectively, the same result is true for any two points in the plane. Hence, we have

The Distance Formula
The distance $\overline{PQ}$ between the points $P(x_1, y_1)$ and $Q(x_2, y_2)$ in the plane is
$$\overline{PQ} = \sqrt{(x_2 - x_1)^2 + (y_2 - y_1)^2}$$

Use the distance formula to verify that $\overline{PQ} = \overline{QP}$.

EXAMPLE 1 *The Distance Formula*

Find the distance between the points $P(-2, -3)$ and $Q(1, 2)$.

SOLUTION
Using the distance formula, we have
$$\overline{PQ} = \sqrt{[1-(-2)]^2 + [2-(-3)]^2} = \sqrt{3^2 + 5^2} = \sqrt{34}$$ ■

Progress Check

Find the distance between the points $P(-3, 2)$ and $Q(4, -2)$.

Answer
$\sqrt{65}$

EXAMPLE 2 *Applications of the Distance Formula*

Show that the triangle with vertices $A(-2, 3)$, $B(3, -2)$, and $C(6, 1)$ is a right triangle.

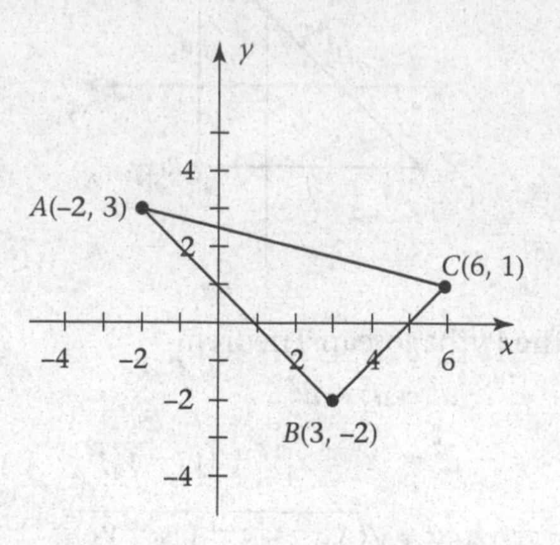

SOLUTION

We draw a diagram as shown above and compute the lengths of the three sides.

$$\overline{AB} = \sqrt{(3+2)^2+(-2-3)^2} = \sqrt{50}$$
$$\overline{BC} = \sqrt{(6-3)^2+(1+2)^2} = \sqrt{18}$$
$$\overline{AC} = \sqrt{(6+2)^2+(1-3)^2} = \sqrt{68}$$

If the Pythagorean Theorem holds, then triangle ABC is a right triangle. We see that

$$(\overline{AC})^2 = (\overline{AB})^2 + (\overline{BC})^2 \quad \text{since} \quad 68 = 50 + 18$$

and we conclude that triangle ABC is a right triangle whose hypotenuse is $\overline{AC}$. ■

3.1b *The Midpoint Formula*

Consider the line segment with endpoints $P(x_1, y_1)$ and $Q(x_2, y_2)$, as shown in Figure 3-8. Let $M(x_m, y_m)$ denote the midpoint of this segment. Since M is the midpoint of $\overline{PQ}$, then $c_1 = c_2$. From plane geometry, if $c_1 = c_2$, then $a_1 = a_2$ and $b_1 = b_2$. Since $x_2 > x_m > x_1$ and $y_2 > y_m > y_1$, $a_1 = x_m - x_1$, $a_2 = x_2 - x_m$, $b_1 = y_m - y_1$ and $b_2 = y_2 - y_m$. We may substitute into the equations as follows:

$$\begin{aligned} a_1 &= a_2 \\ x_m - x_1 &= x_2 - x_m \\ 2x_m &= x_1 + x_2 \\ x_m &= \frac{x_1 + x_2}{2} \end{aligned} \qquad\qquad \begin{aligned} b_1 &= b_2 \\ y_m - y_1 &= y_2 - y_m \\ 2y_m &= y_1 + y_2 \\ y_m &= \frac{y_1 + y_2}{2} \end{aligned}$$

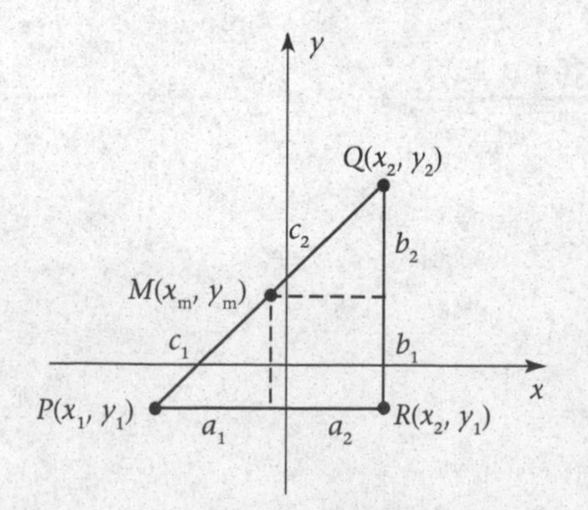

FIGURE 3-8
Deriving the Midpoint Formula

Although points P and Q in Figure 3-8 are in quadrants III and I, respectively, the same result is true for any two points in the plane. Hence, we have

The Midpoint Formula
The coordinates (x_m, y_m) of the midpoint M of the line segment with endpoints $P(x_1, y_1)$ and $Q(x_2, y_2)$ are given by

$$x_m = \frac{x_1 + x_2}{2} \qquad y_m = \frac{y_1 + y_2}{2}$$

Equivalently, we may say that the midpoint of a line segment is the average of the corresponding coordinates of its endpoints.

EXAMPLE 3 *The Midpoint Formula*

Find the coordinates (x_m, y_m) of the midpoint of the line segment with endpoints $P(2, 1)$ and $Q(6, 4)$.

SOLUTION
$P(x_1, y_1) = P(2, 1)$ and $Q(x_2, y_2) = Q(6, 4)$.

$$x_m = \frac{2+6}{2} = \frac{8}{2} = 4 \qquad y_m = \frac{1+4}{2} = \frac{5}{2}$$

The midpoint of $\overline{PQ}$ is $\left(4, \frac{5}{2}\right)$. ■

Progress Check

Find the coordinates of the midpoint of the line segment with endpoints $P(-2, -3)$ and $Q(1, 2)$.

Answer
$\left(-\frac{1}{2}, -\frac{1}{2}\right)$

EXAMPLE 4 *Applications of the Midpoint Formula*

Find the coordinates of the midpoint between $P(-4, 2)$ and the midpoint of the line segment with endpoints $P(-4, 2)$ and $Q(6, -2)$.

SOLUTION
The midpoint of $\overline{PQ}$ is

$$\left(\frac{-4+6}{2}, \frac{2-2}{2}\right) = (1, 0)$$

The midpoint of $(-4, 2)$ and $(1, 0)$ is

$$\left(\frac{-4+1}{2}, \frac{2+0}{2}\right) = \left(-\frac{3}{2}, 1\right)$$

(Check that $\left(-\frac{3}{2}, 1\right)$ divides the line segment $\overline{PQ}$ into two line segments whose lengths are in the ratio 1:3.) ■

If (1, 2) is the midpoint of $\overline{PQ}$, where the coordinates of P are (0, 1), find the coordinates of Q.

Answer
(2, 3)

3.1c *Graphs of Equations*

The **graph of an equation in two variables** x and y is the set of all points $P(x, y)$ whose coordinates satisfy the equation. We say that the ordered pair (a, b) is a **solution** of the equation if substituting a for x and b for y yields a true statement.

To graph $y = x^2 - 4$, an equation in the variables x and y, we proceed as follows: assign arbitrary values for x and compute the corresponding values of y. Thus, if $x = 3$, then $y = 3^2 - 4 = 5$, and the ordered pair (3, 5) is a solution of the equation. Table 3-1 shows a number of solutions. Next, we plot the points corresponding to these ordered pairs. Since the equation has an infinite number of solutions, the plotted points represent only a portion of the graph. We must plot enough points to feel reasonably certain of the curve, as shown in Figure 3-9(a). (Check intermediate values with a calculator.)

(Some texts indicate that a graph continues forever in a particular direction by placing an arrow on the graph pointing in that direction. We shall *not* use this "arrow" notation in this text.)

The abscissa of a point at which a graph meets the x-axis is called an ***x*-intercept**. Since the graph in Figure 3-9 meets the x-axis at points (2, 0) and (−2, 0), we see that 2 and −2 are the x-intercepts. Similarly, we define the ***y*-intercept** as the ordinate of a point at which the graph meets the y-axis. In Figure 3-9, the y-intercept is −4. Therefore, to find the x-intercept, set $y = 0$ and solve the resulting equation for x. Similarly, to find the y-intercept, set $x = 0$ and solve for y. Intercepts may be useful in sketching a graph.

TABLE 3-1 $y = x^2 - 4$

x	y
−3	5
−2	0
−1	−3
0	−4
1	−3
2	0
3	5

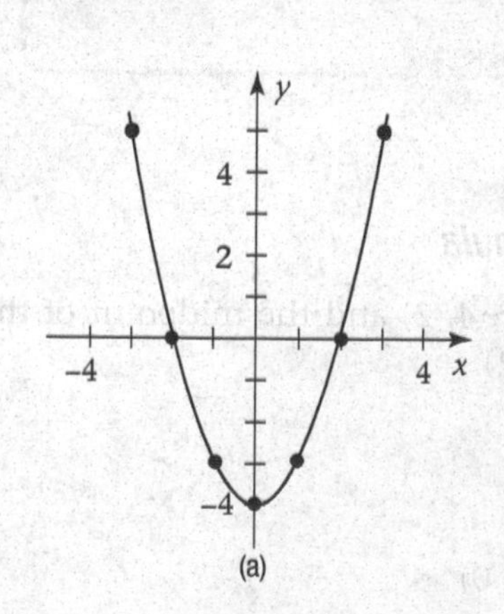

(a)

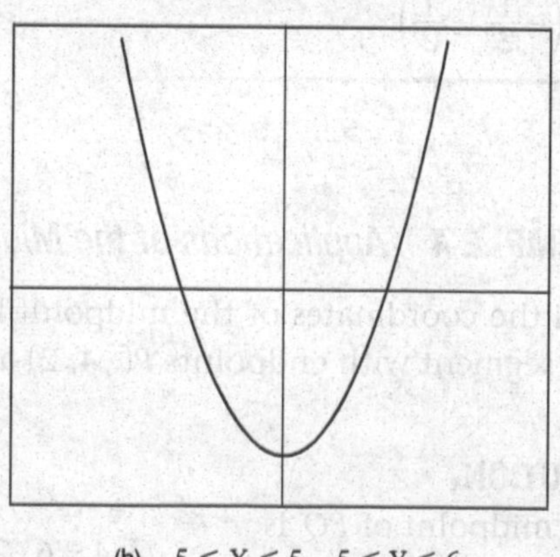
(b) $-5 \le X \le 5, -5 \le Y \le 6$

FIGURE 3-9 Graph of $y = x^2 - 4$, XSCL = 1, YSCL = 1

EXAMPLE 5 ***Graphs and Intercepts***

Sketch the graph of the equation $y = 2x + 1$. Determine the x- and y-intercepts, if any.

SOLUTION

We form a short table of values and sketch the graph. The graph appears to be a line that intersects the x-axis at $\left(-\frac{1}{2}, 0\right)$ and the y-axis at $(0, 1)$, implying that the x-intercept is $-\frac{1}{2}$ and the y-intercept is 1.

x	y
−2	−3
−1	−1
0	1
1	3
2	5

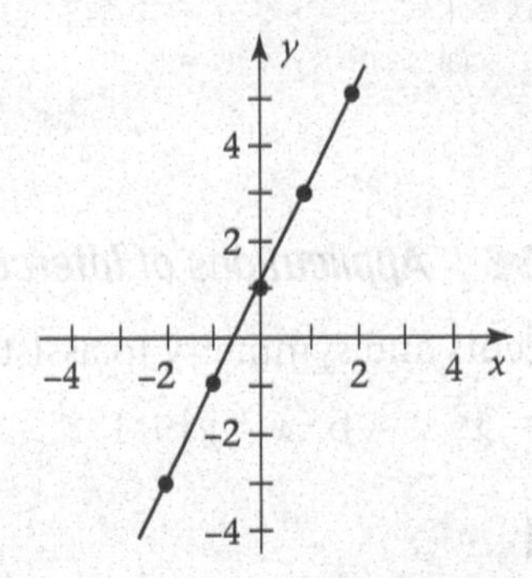

Alternatively, we can find the y-intercept algebraically by letting $x = 0$ so that

$$y = 2x + 1 = 2(0) + 1 = 1$$

and the x-intercept by letting $y = 0$ so that

$$y = 2x + 1$$
$$0 = 2x + 1$$
$$x = -\frac{1}{2}$$ ■

3.1d *Symmetry*

If we folded the graph of Figure 3-10(a) along the x-axis, the top and bottom portions would match exactly. This is what we mean when we speak of **symmetry** about the x-axis. We would like to develop a way of testing for symmetry that does not require examining the graph. We can then use this information to improve our sketching ability.

Returning to Figure 3-10(a), we see that every point typically labeled (x_1, y_1) on the portion of the curve above the x-axis is reflected in a point $(x_1, -y_1)$ that lies on the portion of the curve below the x-axis. Similarly, using the graph of Figure 3-10(b), we observe symmetry about the y-axis if every point (x_1, y_1) on the curve implies that $(-x_1, y_1)$ also lies on the curve. Finally, using the graph sketched in Figure 3-10(c), we see that symmetry occurs about the origin if every point (x_1, y_1) on the curve implies that $(-x_1, -y_1)$ also lies on the curve. We now summarize these results.

Tests for Symmetry

The graph of an equation is **symmetric with respect to the**

i. **x-axis** if replacing y with $-y$ results in an equivalent equation.
ii. **y-axis** if replacing x with $-x$ results in an equivalent equation.
iii. **origin** if replacing x with $-x$ and y with $-y$ results in an equivalent equation.

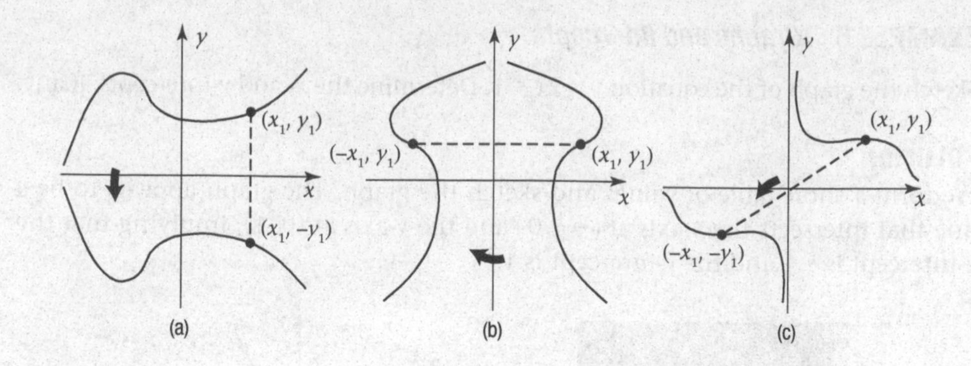

FIGURE 3-10
Symmetry with Respect to (a) x-axis, (b) y-axis, (c) origin

EXAMPLE 6 *Applications of Intercepts and Symmetry*

Use intercepts and symmetry to assist in graphing the equations.

a. $y = 1 - x^2$ b. $x = y^2 + 1$ c. $x^2 + y^2 = 25$

SOLUTION

a. To determine the intercepts, set $x = 0$ to yield $y = 1$ as the y-intercept. Setting $y = 0$, we have $x^2 = 1$ or $x = \pm 1$ as the x-intercepts.

To test for symmetry, replace x with $-x$ in the equation $y = 1 - x^2$ to obtain

$$y = 1 - (-x)^2 = 1 - x^2$$

Since the equation is unaltered, the curve is symmetric with respect to the y-axis. Now, replacing y with $-y$, we have

$$-y = 1 - x^2$$

which is *not* equivalent to the original equation. The curve is therefore not symmetric with respect to the x-axis. Finally, replacing x with $-x$ and y with $-y$ repeats the last result and shows that the curve is not symmetric with respect to the origin.

We can now form a table of values for $x \geq 0$ and use symmetry with respect to the y-axis to help sketch the graph of the equation.

x	y
0	1
1	0
2	−3
3	−8

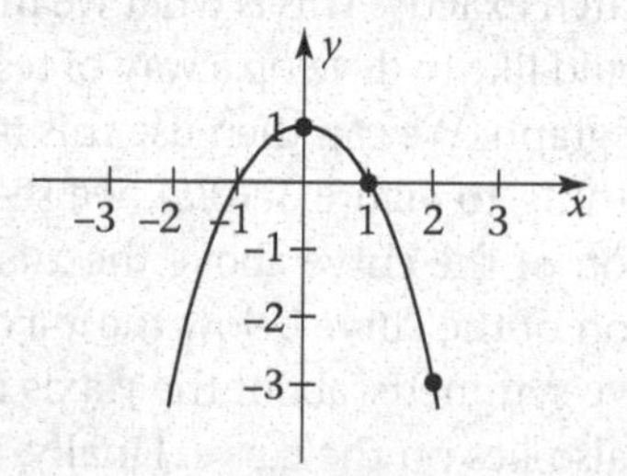

b. The y-intercepts occur where $x = 0$. Since this leads to the equation $y^2 = -1$, which has no real roots, there are no y-intercepts. Setting $y = 0$, we have $x = 1$ as the x-intercept.

Replacing x with $-x$ in the equation $x = y^2 + 1$ gives us

$$-x = y^2 + 1$$

which is *not* an equivalent equation. The curve is therefore not symmetric with respect to the y-axis. Replacing y with $-y$, we find that

$$x = (-y)^2 + 1 = y^2 + 1$$

which is the same as the original equation. Thus, the curve is symmetric with respect to the x-axis. Replacing x with $-x$ and y with $-y$ also results in the equation

$$-x = y^2 + 1$$

and demonstrates that the curve is not symmetric with respect to the origin. We next form the table of values by assigning nonnegative values to y and calculating the corresponding values of x from the equation. (Note that we have reversed the order of the table, putting y first and x second.) Symmetry enables us to sketch the lower half of the graph without plotting points.

x	y
0	1
1	2
2	5
3	10

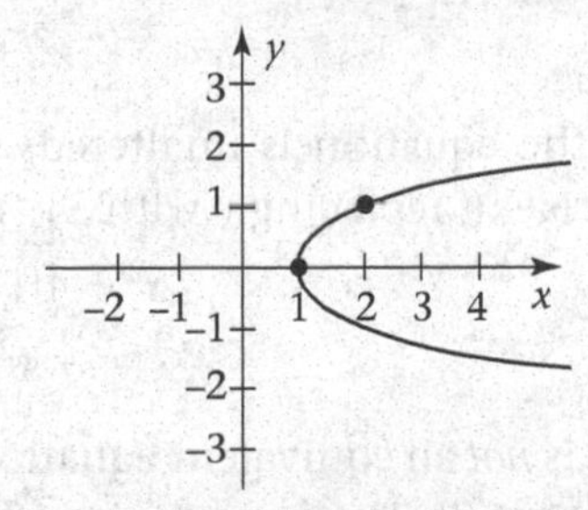

Solving the given equation for y yields $y = \pm\sqrt{x-1}$, which confirms the symmetry about the x-axis. The upper half of the curve is the graph of the equation $y = \sqrt{x-1}$, and the lower half is the graph of the equation $y = -\sqrt{x-1}$.

c. We find the y-intercepts when $x = 0$, that is, $y^2 = 25$ so $y = \pm 5$. The x-intercepts occur when $y = 0$, that is, $x^2 = 25$, hence $x = \pm 5$. Since x and y are raised to an even power, replacing x with $-x$, replacing y with $-y$, or replacing both x and y with $-x$ and $-y$, respectively, leaves the equation unchanged. Therefore, this curve is symmetric with respect to the x-axis, the y-axis, and the origin. If we only plot points in the first quadrant, the three symmetries will give us the complete graph. If $x^2 + y^2 = 25$, then $y = \pm\sqrt{25 - x^2}$. Then, for x and y in the first quadrant, we form the table of values shown below. (Approximate $\sqrt{24}$ and $\sqrt{21}$ with a calculator.) Symmetry enables us to sketch the remainder of the graph without plotting points.

x	y
0	5
1	$\sqrt{24}$
2	$\sqrt{21}$
3	4
4	3
5	0

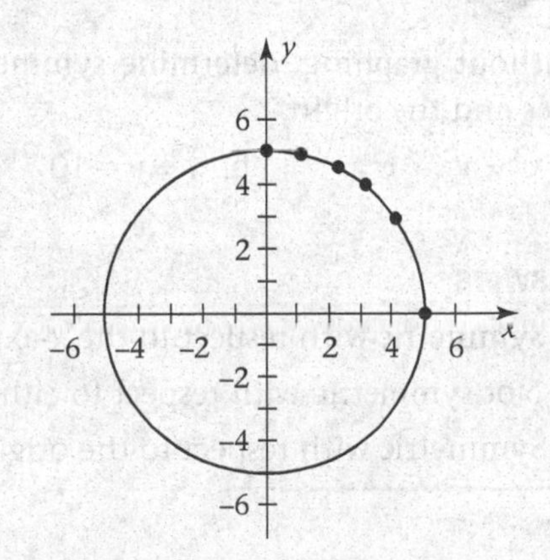

The upper half of the curve is the graph of $y = \sqrt{25 - x^2}$, and the lower half is the graph of $y = -\sqrt{25 - x^2}$. ■

EXAMPLE 7 *Determining Symmetry*

Without sketching the graph, determine symmetry with respect to the x-axis, the y-axis, and the origin.

a. $x^2 + 4y^2 - y = 1$ b. $xy = 5$ c. $y^2 = \frac{x^2+1}{x^2-1}$

SOLUTION

a. Replacing x with $-x$ in the equation, we have

$$(-x)^2 + 4y^2 - y = 1$$
$$x^2 + 4y^2 - y = 1$$

Since the equation is unaltered, the curve is symmetric with respect to the y-axis. Next, replacing y with $-y$, we have

$$x^2 + 4(-y)^2 - (-y) = 1$$
$$x^2 + 4y^2 + y = 1$$

which is *not* an equivalent equation. Replacing x with $-x$ and y with $-y$ repeats the last result. The curve is therefore not symmetric with respect to either the x-axis or the origin.

b. Replacing x with $-x$, we have $-xy = 5$, which is *not* an equivalent equation. Replacing y with $-y$, we again have $-xy = 5$. Thus, the curve is not symmetric with respect to either axis. However, replacing x with $-x$ and y with $-y$ gives us

$$(-x)(-y) = 5$$

which is equivalent to $xy = 5$. We conclude that the curve is symmetric with respect to the origin.

c. Since x and y both appear to the second power only, all tests lead to an equivalent equation. The curve is therefore symmetric with respect to both axes and the origin. ■

Progress Check

Without graphing, determine symmetry with respect to the coordinate axes and the origin.

a. $x^2 - y^2 = 1$ b. $x + y = 10$ c. $y = x + \frac{1}{x}$

Answers

a. Symmetric with respect to the x-axis, the y-axis, and the origin
b. Not symmetric with respect to either axis or the origin
c. Symmetric with respect to the origin only

Note that in Example 7(c) and in Progress Check (a) above, the curves are symmetric with respect to both the x-axis and y-axis, as well as the origin. In fact, we have the following rule:

A curve that is symmetric with respect to both coordinate axes is also symmetric with respect to the origin. However, a curve that is symmetric with respect to the origin need not be symmetric with respect to the coordinate axes.

The curve in Figure 3-10(c) illustrates this last point, namely, that it is symmetric with respect to the origin but not with respect to the coordinate axes.

Exercise Set 3.1

In each of Exercises 1 and 2, plot the given points on the same coordinate axes.

1. $(2, 3), (-3, -2), \left(-\frac{1}{2}, \frac{1}{2}\right), \left(0, \frac{1}{4}\right), \left(-\frac{1}{2}, 0\right), (3, -2)$
2. $(-3, 4), (5, -2), (-1, -3), \left(-1, \frac{3}{2}\right), (0, 1.5)$

In Exercises 3–8, find the distance between each pair of points and find the midpoint.

3. $(5, 4), (2, 1)$
4. $(-4, 5), (-2, 3)$
5. $(-1, -5), (-5, -1)$
6. $(-3, 0), (2, -4)$
7. $\left(\frac{2}{3}, \frac{3}{2}\right), (-2, -4)$
8. $\left(-\frac{1}{2}, 3\right), \left(-1, -\frac{3}{4}\right)$

In Exercises 9–12, find the length of the shortest side of the triangle determined by the three given points.

9. $A(6, 2), B(-1, 4), C(0, -2)$
10. $P(2, -3), Q(4, 4), R(-1, -1)$
11. $R\left(-1, \frac{1}{2}\right), S\left(-\frac{3}{2}, 1\right), T(2, -1)$
12. $F(-5, -1), G(0, 2), H(1, -2)$

In Exercises 13–16, determine if the given points form a right triangle. (*Hint*: A triangle is a right triangle if and only if the lengths of the sides satisfy the Pythagorean Theorem.)

13. $(1, -2), (5, 2), (2, 1)$
14. $(2, -3), (-1, -1), (3, 4)$
15. $(-4, 1), (1, 4), (4, -1)$
16. $(1, -1), (-6, 1), (1, 2)$

In Exercises 17–20, show that the points lie on the same line. (*Hint*: Three points are collinear if and only if the sum of the lengths of two sides equals the length of the third side.)

17. $(-1, 2), (1, 1), (5, -1)$
18. $(-1, -4), (1, 10), (0, 3)$
19. $(-1, 2), (1, 5), \left(-2, \frac{1}{2}\right)$
20. $(-1, -5), (1, 1), (-2, -8)$
21. Find the perimeter of the quadrilateral whose vertices are $(-2, -1), (-4, 5), (3, 5), (4, -2)$.
22. Show that the points $(-2, -1)$, $(2, 2)$, and $(5, -2)$ are the vertices of an isosceles triangle.
23. Show that the points $(9, 2)$, $(11, 6)$, $(3, 5)$, and $(1, 1)$ are the vertices of a parallelogram.
24. Show that the point $(-1, 1)$ is the midpoint of the line segment whose endpoints are $(-5, -1)$ and $(3, 3)$.
25. The points $A(1, 7)$, $B(4, 3)$ and $C(x, 5)$ determine a right triangle whose hypotenuse is AB. Find x. (*Hint*: There is more than one answer.)
26. The points $A(2, 6)$, $B(4, 6)$, $C(4, 8)$, and $D(x, y)$ form a rectangle. Find x and y.

In Exercises 27–32, determine the intercepts and sketch the graph of the given equation.

27. $y = 2x + 4$
28. $y = -2x + 5$
29. $y = \sqrt{x}$
30. $y = \sqrt{x-1}$
31. $y = |x + 3|$
32. $y = 2 - |x|$

In Exercises 33–38, determine the intercepts and use symmetry to assist in sketching the graph of the given equation.

33. $y = 3 - x^2$
34. $y = 3x - x^2$
35. $y = x^3 + 1$
36. $x = y^3 - 1$
37. $x = y^2 - 1$
38. $y = 3x$

In Exercises 39–44, use your graphing calculator to GRAPH the given equations in the indicated viewing rectangle. Set the XSCL and YSCL values appropriately. Use the TRACE command to estimate the intercepts of each graph.

39. $y = 7 - x$	$-5 \le X \le 15$	$-5 \le Y \le 15$
40. $y = 2x + 1$	$-5 \le X \le 5$	$-3 \le Y \le 3$
41. $y = \|x\| - x^2$	$-5 \le X \le 5$	$-3 \le Y \le 1$
42. $y = \sqrt{4-x}$	$-3 \le X \le 5$	$-1 \le Y \le 3$
43. $y = x^3 - x^2 - x$	$-5 \le X \le 5$	$-3 \le Y \le 3$
44. $y = 3x^2 + 4x$	$-5 \le X \le 5$	$-3 \le Y \le 3$

Without graphing, determine whether each curve in Exercises 45–59 is symmetric with respect to the x-axis, the y-axis, and the origin.

45. $3x + 2y = 5$
46. $y = 4x^2$
47. $y^2 = x - 4$
48. $x^2 - y = 2$
49. $y^2 = 1 + x^3$
50. $y = (x - 2)^2$
51. $y^2 = (x - 2)^2$
52. $y^2x + 2x = 4$
53. $y^2x + 2x^2 = 4x^2y$
54. $y^3 = x^2 - 9$
55. $y = \frac{x^2+4}{x^2-4}$
56. $y = \frac{1}{x^2+1}$
57. $y^2 = \frac{x^2+1}{x^2-1}$
58. $4x^2 + 9y^2 = 36$
59. $xy = 4$

60. A ladder leans against a wall. The foot of the ladder is 8 feet from the wall, and the top of the ladder is 6 feet from the floor. The foot of the ladder is pulled 1 foot away from the wall. How far will the top of the ladder slide down the wall? (*Hint*: Draw the coordinate axes and place the ladder up against the y-axis.)

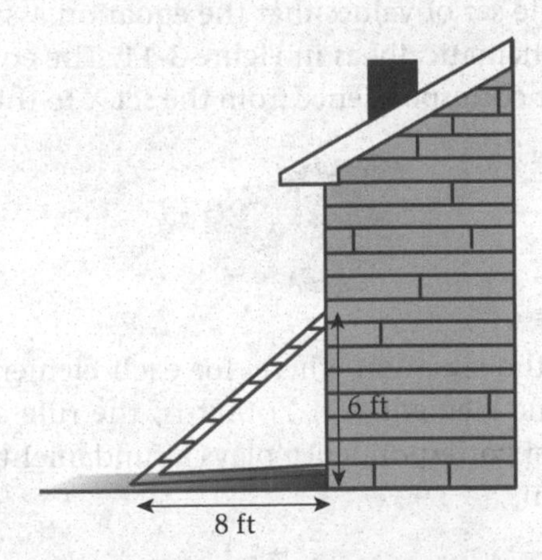

61. Show that the points $(0, 0)$, $(4, 3)$, and $(3, 4)$ form an isosceles triangle.

62. Graph the points $(0, 0)$, $(-3, 3)$, $(2, 8)$, and $(5, 5)$.

 a. Show that these points form a rectangle.

 b. Find the area of the rectangle.

63. Find the midpoint of the line segment whose endpoints are $(-3, 6)$ and $(5, -2)$.

64. A line segment has midpoint $(-3, -2)$. One endpoint of the line segment is $(1, -5)$. Find the other endpoint.

65. A line segment has midpoint $(2, 7)$. One endpoint has coordinates $(x, 10)$, and the other endpoint has coordinates $(6, y)$. Find the values of x and y.

66. The points $(0, 0)$, $(x, 5)$, and $(4, 4)$ form a triangle.

 a. Find x so that these three points form an isosceles triangle.

 b. Find the midpoint of the line segment whose endpoints are $(0, 0)$ and $(4, 4)$.

 c. The line segment from $(x, 5)$ to the midpoint from part (b) is the height of the triangle. Find the area of this triangle.

67. Prove that the midpoint of the hypotenuse of a right triangle is equidistant from the three vertices. Locate the right triangle as in the figure below.

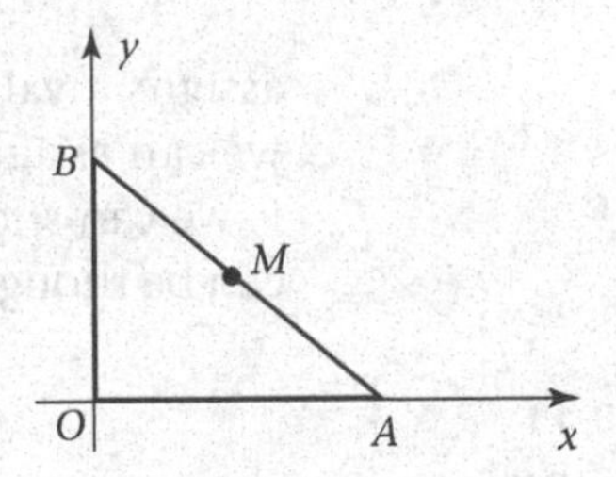

For Exercises 68–71, recall the definition of a median of a triangle: a line joining a vertex to the midpoint of the opposite side.

68. Prove that the medians from the equal angles of an isosceles triangle are of equal length. (*Hint*: Place the triangle so that its vertices are at the points $A(-a, 0)$, $B(a, 0)$, and $C(0, b)$.)

69. Show that the sum of the squares of the lengths of the medians of a triangle equals $\frac{3}{4}$ the sum of the squares of the lengths of the sides. (*Hint*: Place the triangle so that its vertices are at the points $(-a, 0)$, $(b, 0)$, and $(0, c)$.)

70. Prove that a triangle with two equal medians is isosceles.

71. Show that the midpoints of the sides of a rectangle are the vertices of a rhombus (a quadrilateral with four equal sides). (*Hint*: Place the rectangle so that its vertices are at the points $(0, 0)$, $(a, 0)$, $(0, b)$, and (a, b).)

72. Prove that the lengths of the diagonals of a rectangle are equal. (*Hint*: Place the rectangle so that its vertices are at the points $(0, 0)$, $(a, 0)$, $(0, b)$, and (a, b).)

3.2 Functions and Function Notation

The equation

$$y = 2x + 3$$

assigns a value to y for every value of x. If we let X denote the set of values that we can assign to x, and let Y denote the set of values that the equation assigns to y, we can show the correspondence schematically as in Figure 3-11. The equation can be thought of as a rule defining the correspondence from the set X to the set Y.

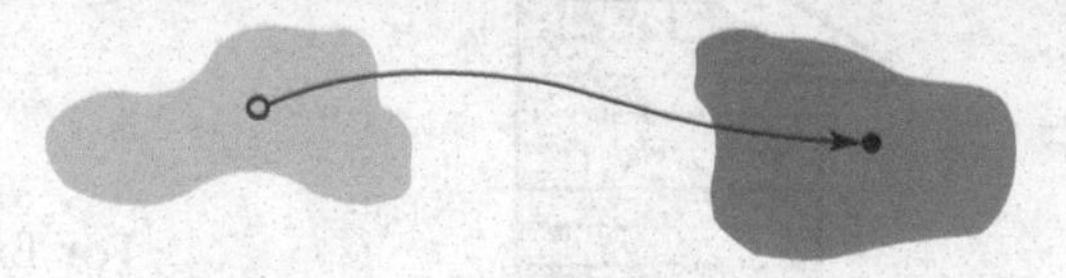

FIGURE 3-11
Correspondence Defined by $y = 2x + 3$

We are particularly interested in the situation where, for each element x in X, there corresponds one and only one element y in Y; that is, the rule assigns exactly one y for a given x. This type of correspondence plays a fundamental role in mathematics and is called a function.

Function, Domain, Image, and Range
A **function** is a rule that, for each x in a set X, assigns exactly one y in a set Y. The element y is called the **image** of x. The set X is called the **domain** of the function, and the set of all images is called the **range** of the function.

We can think of the rule defined by the equation $y = 2x + 3$ as a function machine as shown in Figure 3-12. Each time we drop a value of x from the domain into the input hopper, exactly one value of y falls out of the output chute. If we drop in $x = 5$, the function machine follows the rule and produces $y = 13$. Since we are free to choose the values of x that we drop into the machine, we call x the **independent variable**; the value of y that drops out depends upon the choice of x, so y is called the **dependent variable.** We say that the dependent variable is a function of the independent variable; that is, *the output is a function of the input.*

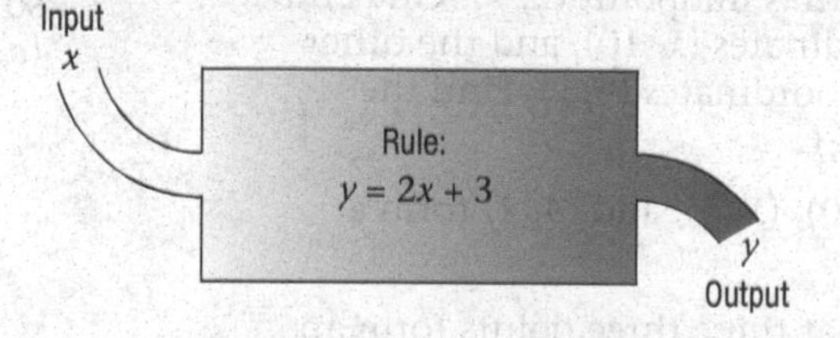

FIGURE 3-12
A Function Machine

Let us look at a few schematic presentations. The correspondence in Figure 3-13(a) is a function: for each x in X there is exactly one corresponding value of y in Y. The fact that y_1 is the image of both x_1 and x_2 does not violate the definition of a function. However, the correspondence in Figure 3-13(b) is not a function. Here x_1 has two images assigned to it, namely, y_1 and y_2, thus violating the definition of a function.

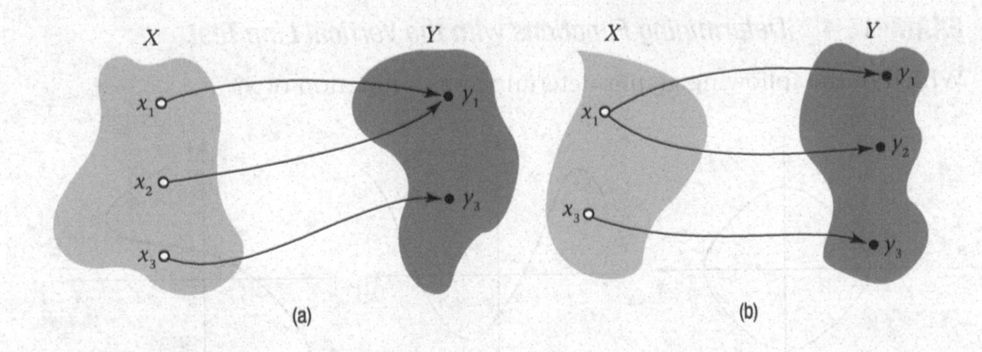

FIGURE 3-13
Correspondence:
(a) Is a Function,
(b) Is Not a Function

3.2a *Vertical Line Test*

The **graph of the function *f*** is defined as the graph of the equation $y = f(x)$. Therefore, it is possible to use the *graph of an equation* to test whether it determines a function. If we consider all vertical lines on the graph of Figure 3-14(a), we see that no vertical line intersects the graph at more than one point. This means that the correspondence used in sketching the graph assigns exactly one y-value for each x-value and therefore determines y as a function of x. If we consider all vertical lines on the graph of Figure 3-14(b), however, some vertical lines intersect the graph at two points. Since the correspondence graphed in Figure 3-14(b) assigns the values y_1 and y_2 to x_1, it does not determine y as a function of x. Thus, *not every equation or correspondence* in the variables x and y determines y as a function of x.

Vertical Line Test
A graph represents y as a function of x if, and only if, no vertical line meets the graph at more than one point.

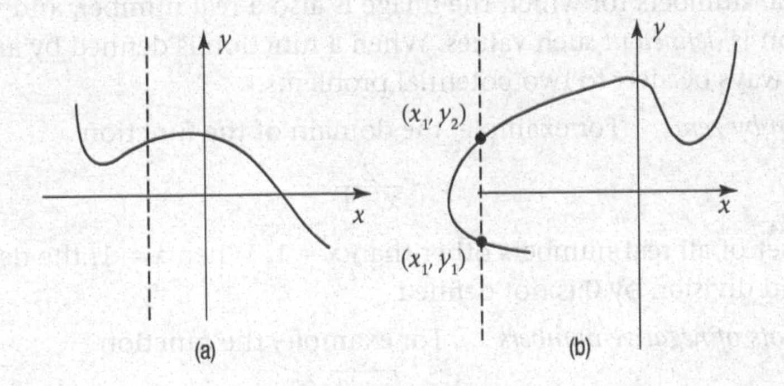

FIGURE 3-14
Vertical Line Test:
(a) Is a Function,
(b) Is Not a Function

EXAMPLE 1 ***Determining Functions with the Vertical Line Test***

Which of the following graphs determine y as a function of x?

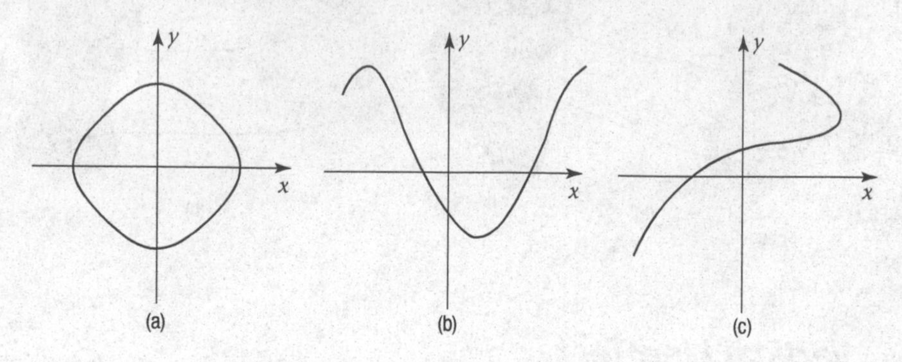

SOLUTION

a. This is not a function because some vertical line meets the graph in more than one point.

b. This is a function since the graph passes the vertical line test.

c. This is not a function since the graph fails the vertical line test. ■

3.2b *Domain and Range*

We have defined the domain of a function as the set of values assumed by the independent variable. In more advanced courses in mathematics, the domain may include complex numbers. In this book, we will restrict the domain of a function to those real numbers for which the image is also a real number, and we say that the function is *defined at* such values. When a function is defined by an equation, we must always be alert to two potential problems.

1. *Division by zero* For example, the domain of the function

$$y = \frac{2}{x-1}$$

is the set of all real numbers other than $x = 1$. When $x = 1$, the denominator is 0, and division by 0 is not defined.

2. *Even roots of negative numbers* For example, the function

$$y = \sqrt{x-1}$$

is defined only in the real number system for $x \geq 1$ since we exclude the square root of negative numbers. Hence the domain of the function consists of all real numbers $x \geq 1$.

In general, the range of a function is not as easily determined as its domain. The range is the set of all y-values that occur in the correspondence; that is, it is the set of all outputs of the function. For our purposes, it will suffice to determine the range by examining the graph of the function.

EXAMPLE 2 *Graphing and the Vertical Line Test*

Graph the equation $y = \sqrt{x}$. If the correspondence determines a function, find the domain and range.

SOLUTION

We obtain the graph of the equation by plotting points and connecting them to form a smooth curve.

x	y
0	0
1	1
4	2
9	3

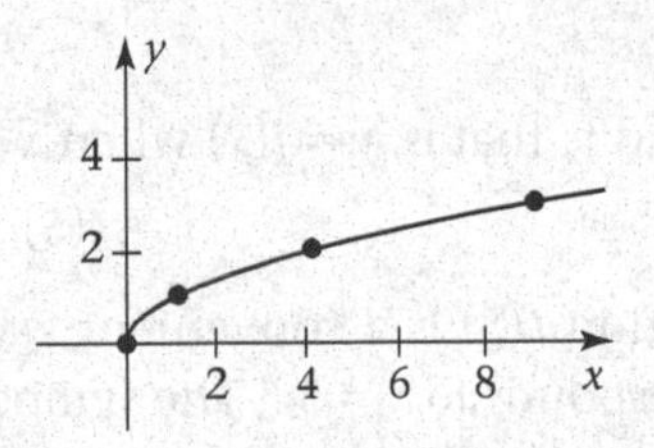

Graph of $y = \sqrt{x}$

Applying the vertical line test to the graph, we see that the equation determines a function. The domain of the function is the set $\{x \mid x \geq 0\}$, and the range is the set $\{y \mid y \geq 0\}$. ■

Progress Check

Graph the equation $y = x^2 - 4$, $-3 \leq x \leq 3$. If the correspondence determines a function, find the domain and range.

Answer

The graph is shown in Figure 3-15. The domain is $\{x \mid -3 \leq x \leq 3\}$; the range is $\{y \mid -4 \leq y \leq 5\}$.

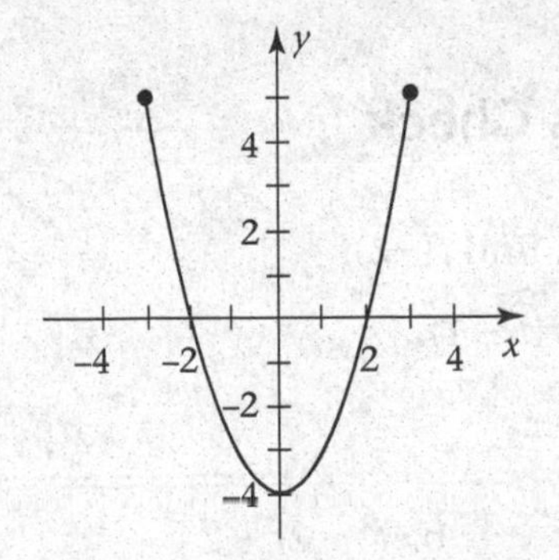

FIGURE 3-15
Graph of $y = x^2 - 4$

3.2c Function Notation

If we use the letter f to designate a function, then we denote the output corresponding to x by $f(x)$, which is read "f of x." Frequently, we use the letter y to denote the output corresponding to the input x. For example,

$$f(x) = 2x + 3$$

specifies a rule f for determining an output $f(x)$ for a given value of x. In other words, the symbol x can be thought of as holding a place. Equivalently, we might write

$$f(\quad) = 2(\quad) + 3$$

To find y, that is, $y = f(x)$ when $x = 5$, we substitute 5 for x and obtain

$$y = f(5) = 2(5) + 3 = 13$$

The notation $f(5)$ is a convenient way of specifying "the value of the function f that corresponds to $x = 5$." The symbol f represents the function or rule; the notation $f(x)$ or y represents the output produced by the rule. However, in common usage, the symbols f, $f(x)$, and y are often used interchangeably.

EXAMPLE 3 *Evaluating Functions*

a. If $f(x) = 2x^2 - 2x + 1$, find $f(-1)$.

b. If $f(t) = 3t^2 - 1$, find $f(2a)$.

SOLUTION

a. We substitute -1 for x.

$$f(-1) = 2(-1)^2 - 2(-1) + 1 = 5$$

b. We substitute $2a$ for t.

$$f(2a) = 3(2a)^2 - 1 = 3(4a^2) - 1 = 12a^2 - 1$$

■

Progress Check

a. If $f(u) = u^3 + 3u - 4$, find $f(-2)$.

b. If $f(t) = t^2 + 1$, find $f(t - 1)$.

Answers

a. -18 b. $t^2 - 2t + 2$

EXAMPLE 4 *Evaluating Functions*

Consider the function $f(x) = x^2 + 3$. Find the number (or numbers) whose image is

a. 3 b. 7 c. 2

SOLUTION

a. We seek values of x such that

$$f(x) = x^2 + 3 = 3$$

Solving for x, we obtain

$$x = 0$$

b. If $f(x) = x^2 + 3 = 7$, then

$$x^2 = 4$$
$$x = \pm 2$$

Note that $f(2) = 7$ and $f(-2) = 7$.

c. If $f(x) = x^2 + 3 = 2$, then

$$x^2 = -1$$

This has no solution for real values of x. Therefore, 2 is not the image of any element in the domain. ■

EXAMPLE 5 *Evaluating Functions*

Let the function f be defined by $f(x) = x^2 - 1$. Find the following:

a. $f(-2)$

b. $f(a)$

c. $f(a + h)$

d. $f(a + h) - f(a)$

e. $\dfrac{f(a+h)-f(a)}{h}$

SOLUTION

a. $f(-2) = (-2)^2 - 1 = 4 - 1 = 3$

b. $f(a) = a^2 - 1$

c. $f(a + h) = (a + h)^2 - 1 = a^2 + 2ah + h^2 - 1$

d.
$$\begin{aligned} f(a + h) - f(a) &= (a + h)^2 - 1 - (a^2 - 1) \\ &= a^2 + 2ah + h^2 - 1 - a^2 + 1 \\ &= 2ah + h^2 \end{aligned}$$

e. $\dfrac{f(a+h)-f(a)}{h} = \dfrac{2ah+h^2}{h} = \dfrac{h(2a+h)}{h} = 2a + h, \qquad h \neq 0$ ■

a. Note that $f(a+3) \neq f(a) + f(3)$. Function notation is not to be confused with the distributive law.

b. Note that $f(x^2) \neq f \cdot x^2$. The use of parentheses in function notation does *not* imply multiplication.

We may use letters other than f to designate a function as we see in the next example.

EXAMPLE 6 ***Functions and Word Problems***

A newspaper makes this offer to its advertisers: The first column inch will cost \$40, and each subsequent column inch will cost \$30. If T is the total cost of running an ad whose length is n column inches, and the minimum space is 1 column inch,

a. express T as a function of n;

b. find T when $n = 4$.

SOLUTION

a. After paying \$40 for the first inch, we must pay \$30 for each of the remaining $n - 1$ inches. Therefore, the equation

$$\begin{aligned} T &= 40 + 30(n-1) \\ &= 10 + 30n \end{aligned}$$

gives the correspondence between n and T. In function notation,

$$T(n) = 10 + 30n \qquad n \geq 1$$

b. When $n = 4$,

$$T(4) = 10 + 30(4) = 130$$

The total cost is \$130 for 4 column inches. ■

Exercise Set 3.2

In Exercises 1–6, graph the equation. If the graph determines y as a function of x, find the domain and use the graph to determine the range of the function.

1. $y = 2x - 3$
2. $y = x^2 + x, \quad -2 \le x \le 1$
3. $x = y + 1$
4. $x = y^2 - 1$
5. $y = \sqrt{x-1}$
6. $y = |x|$

In Exercises 7–12, determine the domain of the function defined by the given rule. Graph the function on your graphing calculator to verify.

7. $f(x) = \sqrt{2x-3}$
8. $f(x) = \sqrt{5-x}$
9. $f(x) = \dfrac{1}{\sqrt{x-2}}$
10. $f(x) = \dfrac{-2}{x^2+2x-3}$
11. $f(x) = \dfrac{\sqrt{x-1}}{x-2}$
12. $f(x) = \dfrac{x}{x^2-4}$

In Exercises 13–16, find the number (or numbers) whose image is 2.

13. $f(x) = 2x - 5$
14. $f(x) = x^2$
15. $f(x) = \dfrac{1}{x-1}$
16. $f(x) = \sqrt{x-1}$

In Exercises 17–23, determine the following if f is defined by $f(x) = 2x^2 + 5$.

17. $f(0)$
18. $f(-2)$
19. $f(a)$
20. $f(3x)$
21. $3f(x)$
22. $-f(x)$
23. $\dfrac{f(a+h)-f(a)}{h}$

In Exercises 24–29, determine the following if g is defined by $g(x) = x^2 + 2x$.

24. $g(-3)$
25. $g\left(\frac{1}{x}\right)$
26. $\dfrac{1}{g(x)}$
27. $g(-x)$
28. $g(a+h)$
29. $\dfrac{g(a+h)-g(a)}{h}$

In Exercises 30–34, determine the following if F is defined by

$$F(x) = \frac{x^2+1}{3x-1}$$

30. $F(-2.73)$ to two decimal places
31. $\dfrac{1}{F(x)}$
32. $F(-x)$
33. $2F(2x)$
34. $F(x^2)$

In Exercises 35–40, determine the following if r is defined by

$$r(t) = \frac{t-2}{t^2+2t-3}$$

35. $r(-8.27)$
36. $r(2.04)$
37. $r(2a)$
38. $2r(a)$
39. $r(a+1)$
40. $\dfrac{r(a+h)-r(a)}{h}$

41. If x dollars are borrowed at 7% simple annual interest, express the interest I at the end of 4 years as a function of x.
42. Express the area A of an equilateral triangle as a function of the length s of its side.
43. Express the diameter d of a circle as a function of its circumference C.
44. Express the perimeter P of a square as a function of its area A.
45. Container Corporation of America wants to manufacture a box with no top from a 10 inch by 12 inch piece of metal by cutting equal-sized squares from each corner and bending up the sides. Express the volume of the container as a function of its height.

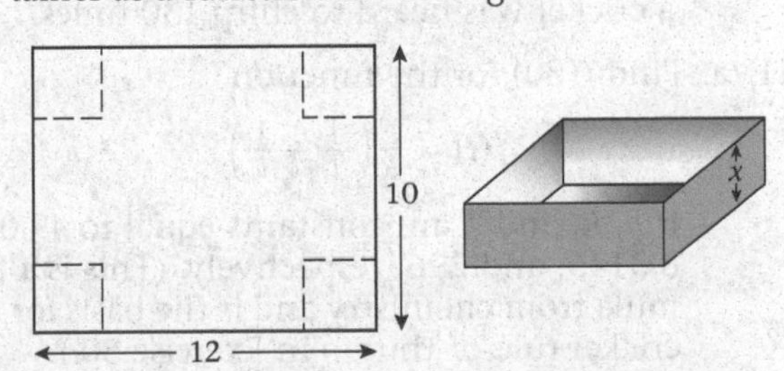

46. A rectangular box with a depth of 10 inches and a square base is to be manufactured by the Bold Box Company. The top of the box costs \$1.20 per square inch, the bottom costs \$.95 per square inch, and the sides cost \$.75 per square inch. Express the cost of manufacturing this box as a function of length.
47. The owners of an audio-video store have found that they can sell 400 HD televisions per month at \$300 per TV. For each \$5.00 drop in price, they can sell 10 more televisions.

    a. Express the gross sales as a function of each \$5.00 decrease in price.

    b. What would the gross sales of televisions be for ten \$5.00 decreases?

48. A power line runs due north. Westown is 5 miles due west from a point A on the power line. Another town, Westwood, is 12 miles west from a point B on the power line. Points A and B are 8 miles apart. A power company wants to locate a transformer between A and B. Express the sum of the distances from each town to the transformer as a function of the distance d from point A to the transformer.

49. One gram of hydrogen peroxide (H_2O_2) is made up of 6% hydrogen (H) and 94% oxygen (O).

    a. Express the number of grams of hydrogen needed in the formation of H_2O_2 as a function of the number of grams w of H_2O_2.

    b. Express the number of grams of oxygen needed in the formation of H_2O_2 as a function of the number of grams w of H_2O_2.

    c. How many grams of H and O are needed to form 250 grams of H_2O_2?

50. Read the paragraphs that open this chapter.

    a. Express the outside temperature T in degrees Fahrenheit as a function of the number of times n that a cricket is heard to chirp in one minute.

    b. What would the outside temperature be if a cricket was heard to chirp 130 times?

51. a. Find $f(80)$ for the function

$$f(t) = \frac{E}{R}\left(\frac{1}{T} - \frac{1}{t}\right)$$

if E, R, and T are constants equal to 49,000, 8.3145, and 296, respectively. (This is a formula from chemistry and is the basis for the cricket rule-of-thumb in Exercise 50.)

    b. Use your graphing calculator to set up a LIST of values for the constant t: {71, 76, 84, 89, 90}; then re-evaluate the function using the LIST.

52. *Mathematics in Writing:*

    a. Explain in your own words the phrase *domain of a function.*

    b. Give an example of a function whose domain excludes the number 2. Explain in a complete sentence.

    c. Give an example of a function whose domain excludes all real numbers less than 5. Explain in a complete sentence.

    d. How can your graphing calculator help you to determine the range of a function? Give an example.

3.3 Graphs of Functions

We have used the graph of an equation to help us find out when the equation determines a function. For example, the graph of the function f defined by the rule $f(x) = \sqrt{x}$ is the graph of the equation $y = \sqrt{x}$, which was sketched in Example 2 in Section 3.2.

There are times when different notation is used to identify points on the graph of the equation $y = f(x)$. One method is to identify specific points with numbered coordinates. For example, if $f(1) = 3$, then we would identify the point associated with this equation as (1, 3). More generally, if $f(x_1) = y_1$, then we may either identify this equation with the point (x_1, y_1) or with the point $(x_1, f(x_1))$ as shown in Figure 3-16.

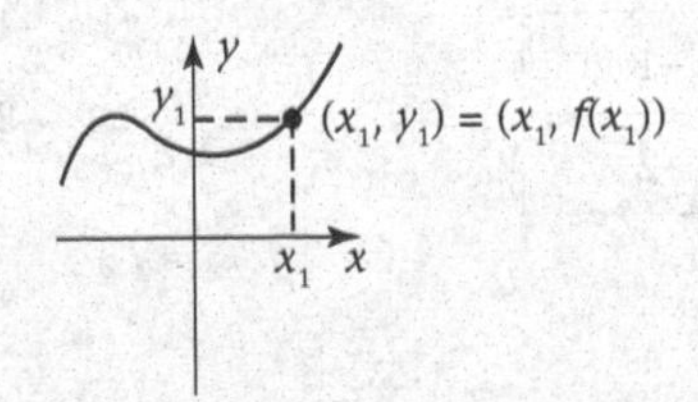

FIGURE 3-16
Points on Graphs of Functions

3.3a "Special" Functions and Their Graphs

There are a number of "special" functions that can be very useful in understanding many of the concepts presented here. Furthermore, these functions tend to arise quite frequently in many practical situations. Therefore, we will present a brief list of these functions, along with their graphs and information about symmetry, intercepts, domain, and range.

As you look at the graph of each function, see if you can anticipate the answers concerning the various characteristics mentioned above.

$f(x) = x$ Identity Function

The domain of f is the set of all real numbers. We form a table of values and use it to sketch the graph of $y = x$ in Figure 3-17. The graph is symmetric with respect to the origin. (Note that $-y = -x$ is equivalent to $y = x$.) The range of f is the set of all real numbers. The x-intercept and the y-intercept are both 0.

x	y
−2	−2
−1	−1
0	0
1	1
2	2

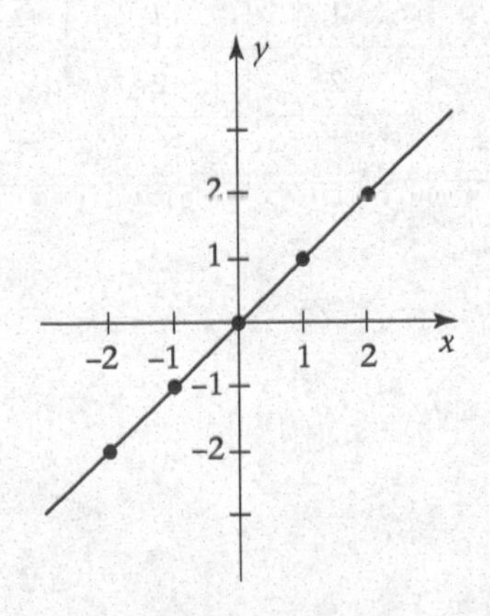

FIGURE 3-17
Graph of $y = x$

$f(x) = -x$ Negation Function

The domain of f is the set of all real numbers. A table of values is used to sketch the graph of $y = -x$ in Figure 3-18. The graph is symmetric with respect to the origin. (Note that $-y = x$ is equivalent to $y = -x$.) The range of f is the set of all real numbers. The x-intercept and the y-intercept are both 0.

FIGURE 3-18
Graph of $y = -x$

x	y
−2	2
−1	1
0	0
1	−1
2	−2

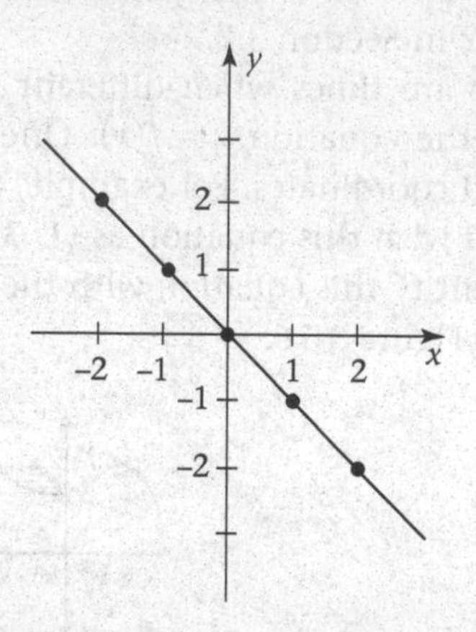

$f(x) = |x|$ Absolute Value Function

The domain of f is the set of all real numbers. A table of values allows us to sketch the graph in Figure 3-19. The graph is symmetric with respect to the y-axis. (Note that $y = |-x| = |x|$.) Since the graph lies on or above the x-axis, the range of f is the set of all nonnegative real numbers, that is, $\{y \mid y \geq 0\}$. The x-intercept and y-intercept are both 0. Note that we may also write

$$y = f(x) = |x| = \begin{cases} x \text{ if } x \geq 0 \\ -x \text{ if } x < 0 \end{cases}$$

Thus, $y = x$ if $x \geq 0$, and $y = -x$ if $x < 0$.

FIGURE 3-19
Graph of $y = |x|$

x	y
−2	2
−1	1
0	0
1	1
2	2

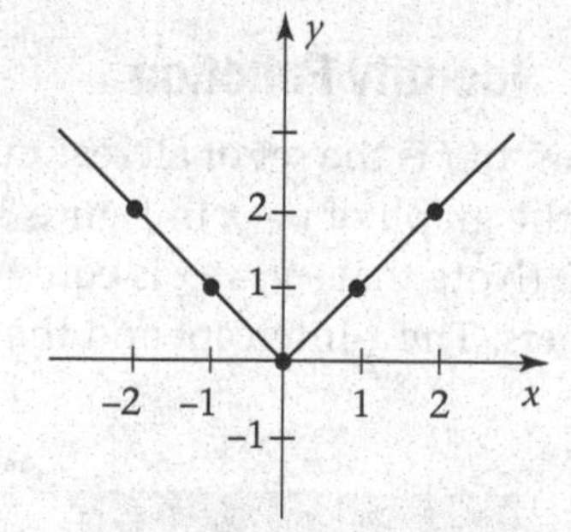

$f(x) = c$ Constant Function

The domain of f is the set of all real numbers. In fact, the value of f is the same for all values of x, as shown in Figure 3-20. The range of f is the set $\{c\}$. The graph is symmetric with respect to the y-axis. (Note that $y = c$ is unaltered when x is replaced by $-x$.) The y-intercept is c; there is no x-intercept.

x	y
−2	c
−1	c
0	c
1	c
2	c

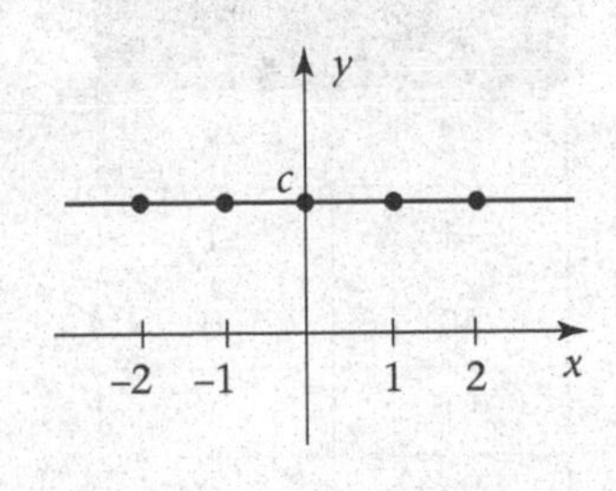

FIGURE 3-20
Graph of $y = c$, $c > 0$

$f(x) = x^2$ Squaring Function

The domain of f is the set of all real numbers. The graph in Figure 3-21 is called a **parabola** and illustrates the general shape of all second-degree polynomials. The graph of f is symmetric with respect to the y-axis. (Note that $y = (-x)^2 = x^2$.) Since the graph lies on or above the x-axis, the range of f is $\{y \mid y \geq 0\}$. Both the x-intercept and y-intercept are 0.

x	y
−2	4
−1	1
0	0
1	1
2	4

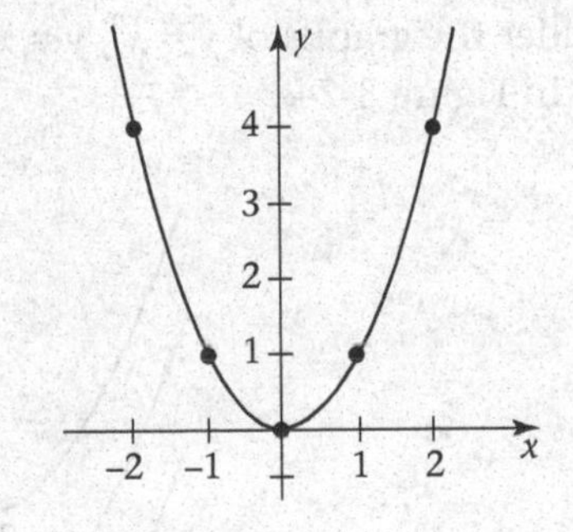

FIGURE 3-21
Graph of $y = x^2$

$f(x) = \sqrt{x}$ Square Root Function

Since $\sqrt{x}$ is not defined for $x < 0$, the domain is the set of nonnegative real numbers, that is, $\{x \mid x \geq 0\}$. The graph in Figure 3-22 always lies on or above the x-axis, so the range of f is $\{y \mid y \geq 0\}$. The graph is not symmetric with respect to either axis or the origin. Both the x-intercept and y-intercept are 0.

x	y
0	0
1	1
4	2
9	3

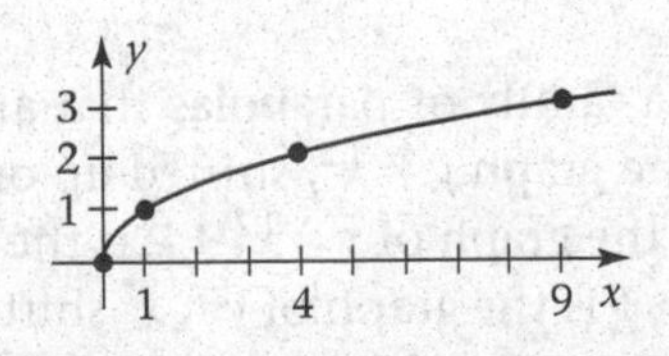

FIGURE 3-22
Graph of $y = \sqrt{x}$

$f(x) = x^3$ Cubic or Cubing Function

The domain is the set of all real numbers. From the graph in Figure 3-23, we see that the range is also the set of all real numbers. The graph is symmetric with respect to the origin. (Note that $-y = (-x)^3 = -x^3$ is equivalent to $y = x^3$.) Both the x-intercept and y-intercept are 0.

FIGURE 3-23
Graph of $y = x^3$

x	y
−2	−8
−1	−1
0	0
1	1
2	8

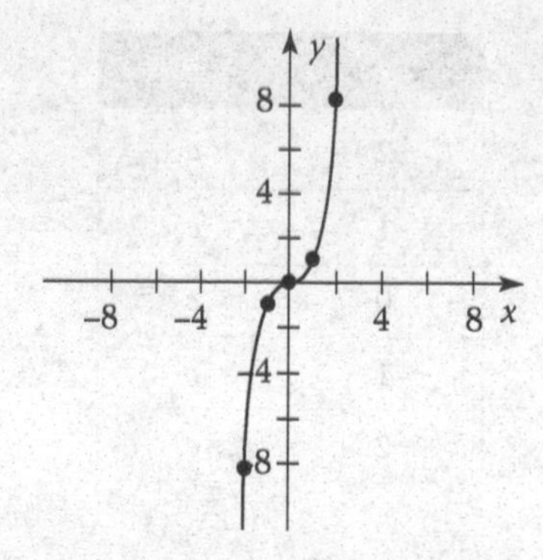

3.3b *Additional Graphing Techniques*

Although it is always possible to sketch a graph by finding many points, there are other methods that can make graph sketching a bit easier. We shall focus on specific transformations that create families of graphs. We will see how these transformations change graphs in predictable ways.

Consider the graphs of $y = x^2$, $y = x^2 + 1$, $y = x^2 + 2$, $y = x^2 - 1$, and $y = x^2 - 2$, as shown in Figure 3-24.

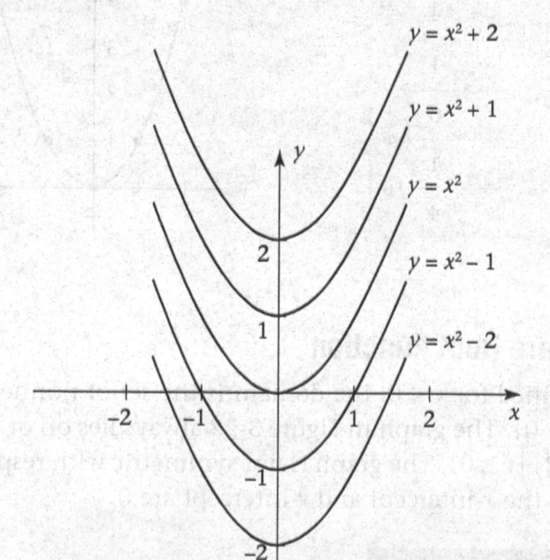

FIGURE 3-24
Vertical Shift of a Graph

We have a family of parabolas that are related as follows: each graph is an exact copy of the graph $y = x^2$, shifted up or down an appropriate number of units. For example, the graph of $y = x^2 + 2$ is the graph of $y = x^2$ shifted up 2 units; the graph of $y = x^2 - 1$ is the graph of $y = x^2$ shifted down 1 unit. This example illustrates the concept of vertical shifts.

Vertical Shift

If $p > 0$, the graph of $y = f(x) + p$ is the graph of $y = f(x)$ shifted up p units. Similarly, the graph of $y = f(x) - p$ is the graph of $y = f(x)$ shifted down p units.

Next, consider the graphs $y = x^3$, $y = (x - 1)^3$, $y = (x - 2)^3$, $y = (x + 1)^3$, and $y = (x + 2)^3$, as shown in Figure 3-25. We have a family of cubic functions that are related as follows: each graph is an exact copy of the graph of $y = x^3$, shifted to the right or left an appropriate number of units. For example, the graph of $y = (x - 2)^3$ is the graph of $y = x^3$ shifted 2 units to the right; the graph of $y = (x + 1)^3$ is the graph of $y = x^3$ shifted 1 unit to the left. This example illustrates the concept of horizontal shifts.

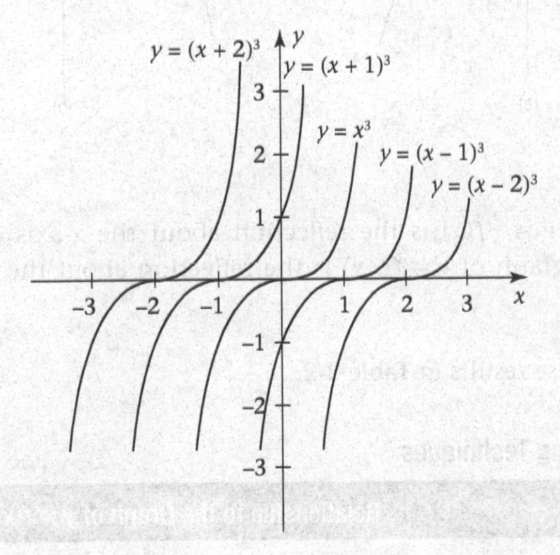

FIGURE 3-25
Horizontal Shift of a Graph

Horizontal Shift

If $p > 0$, the graph of $y = f(x - p)$ is the graph of $y = f(x)$ shifted p units to the right. Similarly, the graph of $y = f(x + p)$ is the graph of $y = f(x)$ shifted p units to the left.

Now, consider the graphs of $y = x^2$ and $y = -x^2$ as shown in Figure 3-26(a), or the graphs of $y = x^3$ and $y = -x^3$ as shown in Figure 3-26(b). We have two families of curves, parabolic and cubic. One member of a given family is the reflection about the x-axis of the other member of that family. This leads to the concept of **reflection**.

Note that $f(-x) = -x^3$ as well. This is equivalent to reflection about the y-axis of the graph of $y = x^3$. (In Figure 3-26(b), observe that the reflection of $y = x^3$ about either the x- or y-axis yields the same graph.)

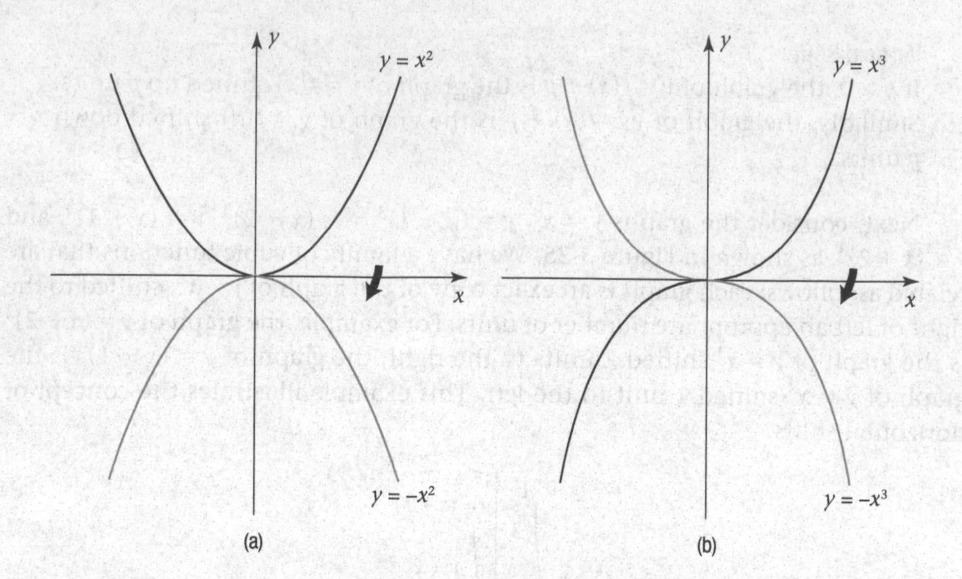

FIGURE 3-26
Reflection of a Graph About the x-axis

Reflection
The graph of $y = -f(x)$ is the reflection about the x-axis of the graph $y = f(x)$. The graph of $y = f(-x)$ is the reflection about the y-axis of the graph $y = f(x)$.

We summarize these results in Table 3-2.

TABLE 3-2 Graphing Techniques

Form	Relationship to the Graph of $y = f(x)$, $p > 0$
$y = f(x) + p$	Shift $f(x)$ p units up.
$y = f(x) - p$	Shift $f(x)$ p units down.
$y = f(x - p)$	Shift $f(x)$ p units to the right.
$y = f(x + p)$	Shift $f(x)$ p units to the left.
$y = -f(x)$	Reflect $f(x)$ about the x-axis.
$y = f(-x)$	Reflect $f(x)$ about the y-axis.

EXAMPLE 1 ***Applications of Graphing Techniques***

Sketch the graph of $y = \sqrt{x+2} + 1$.

SOLUTION

Take the graph of $y = \sqrt{x}$ and shift it 2 units to the left. Now take this graph and shift it up 1 unit.

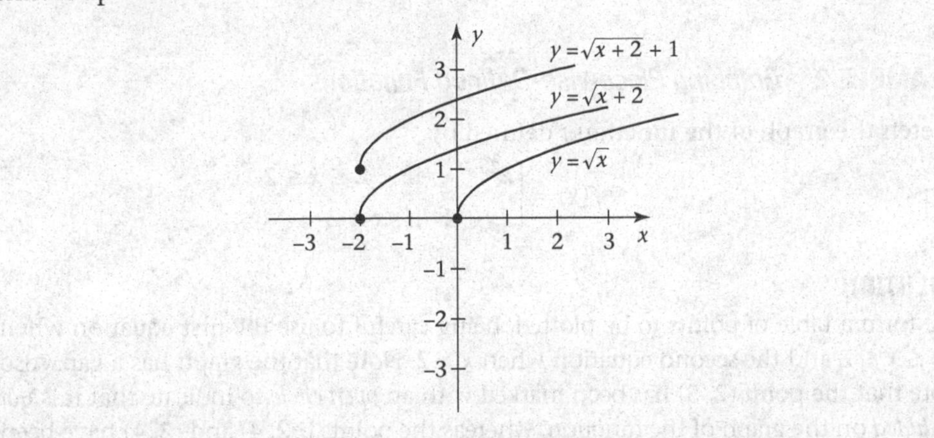

■

Consider the graphs of $y = x^2$, $y = 2x^2$, and $y = \frac{1}{2}x^2$ as shown in Figure 3-27. This is another family of parabolas. If we use the graph of $y = x^2$ as our reference point once again, we see that this graph will expand or contract depending upon whether the multiplying coefficient of x^2 is greater than 1 or less than 1.

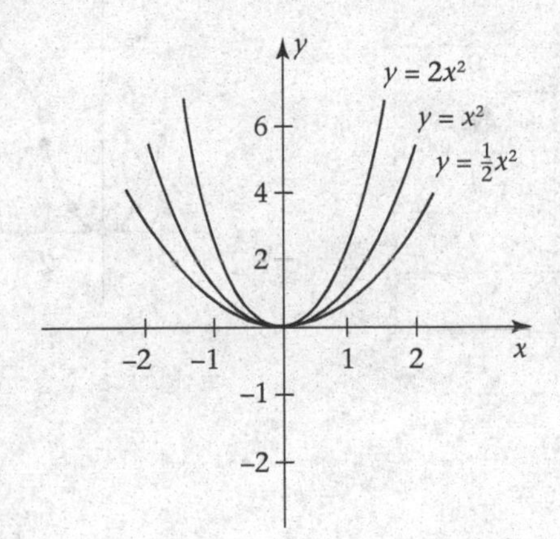

FIGURE 3-27
Stretching and Shrinking Graphs

Progress Check

Determine what must be done to the graph of $y = |x|$ in order to sketch the graph $y = -|x-1| + 3$.

Answer

Take the graph of $y = |x|$ and shift it 1 unit to the right. Now, reflect this graph about the x-axis. Finally, shift this reflected graph up 3 units. (As a check, try sketching this graph by plotting points.)

3.3c Piecewise-Defined Functions

Thus far we have defined each function by means of an equation. A function can also be defined by a table, by a graph, or by several equations. When a function is defined in different ways over different parts of its domain, it is said to be a **piecewise-defined function**. We illustrate this idea by several examples.

EXAMPLE 2 *Graphing Piecewise-Defined Functions*

Sketch the graph of the function f defined by

$$f(x) = \begin{cases} x^2 & \text{if } -2 \le x \le 2 \\ 2x+1 & \text{if } x > 2 \end{cases}$$

SOLUTION

We form a table of points to be plotted, being careful to use the first equation when $-2 \le x \le 2$ and the second equation when $x > 2$. Note that the graph has a gap. Also note that the point (2, 5) has been marked with an *open circle* to indicate that it is *not included* on the graph of the function, whereas the points (−2, 4) and (2, 4) have been marked with *filled-in circles* to indicate that these points *are included* on the graph.

x	y
−2	4
−1	1
0	0
1	1
2	4
3	7
4	9
5	11

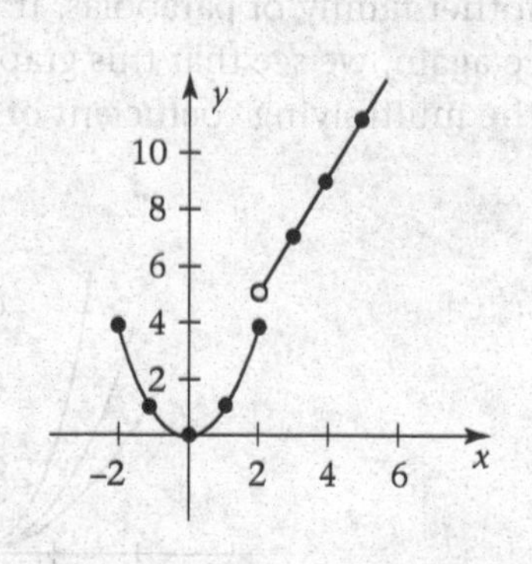

■

EXAMPLE 3 *Graphing with Absolute Value*

Sketch the graph of the function $f(x) = |x + 1|$.

SOLUTION

We apply the definition of absolute value to obtain

$$y = |x+1| = \begin{cases} x+1 & \text{if} \quad x+1 \ge 0 \\ -(x+1) & \text{if} \quad x+1 < 0 \end{cases}$$

or

$$y = \begin{cases} x+1 & \text{if} \quad x \ge -1 \\ -x-1 & \text{if} \quad x < -1 \end{cases}$$

From this example, we see that a function involving absolute value is usually a piecewise-defined function. As usual, we form a table of values, being careful to use $y = x + 1$ when $x \ge -1$ and $y = -x - 1$ when $x < -1$. It is a good idea to include the value of x in the table where the change occurs. The change in this example is at $x = -1$.

The graph consists of two rays or half-lines intersecting at $(-1, 0)$, as shown below.

x	y
−3	2
−2	1
−1	0
0	1
1	2
2	3
3	4

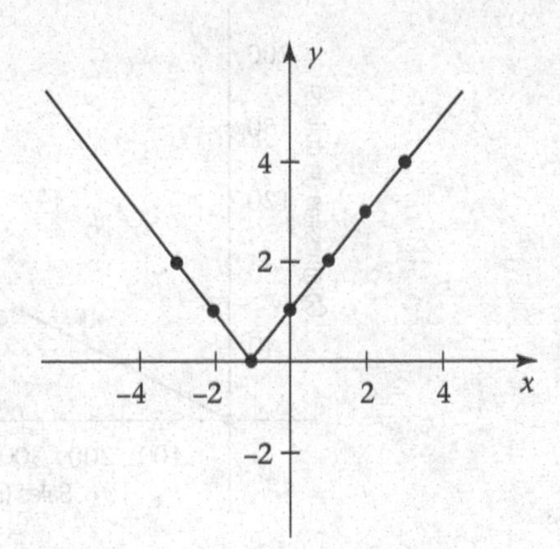

Alternatively, we may obtain the same result by shifting the graph of $y = |x|$ one unit to the left. See Table 3-2. ■

EXAMPLE 4 ***Piecewise-Defined Functions and Word Problems***

The commission earned by a door-to-door cosmetics salesperson is determined as shown in the accompanying table.

a. Express the commission C as a function of sales s.

b. Find the commission if the weekly sales are \$425.

c. Sketch the graph of the function.

Weekly Sales	Commission
less than \$300	20% of sales
\$300 or more but less than \$400	\$60 + 40% of sales over \$300
\$400 or more	\$100 + 60% of sales over \$400

SOLUTION

a. The function C can be described by three equations.

$$C(s) = \begin{cases} 0.20s & \text{if} \quad 0 \le s < 300 \\ 60 + 0.40(s - 300) & \text{if} \quad 300 \le s < 400 \\ 100 + 0.60(s - 400) & \text{if} \quad s \ge 400 \end{cases}$$

b. When $s = 425$, we must use the third equation and substitute to determine $C(425)$.

$$\begin{aligned} C(425) &= 100 + 0.60(425 - 400) \\ &= 100 + 0.60(25) \\ &= 115 \end{aligned}$$

The commission on sales of \$425 is \$115.

c. The graph of the function C consists of three line segments as shown below.

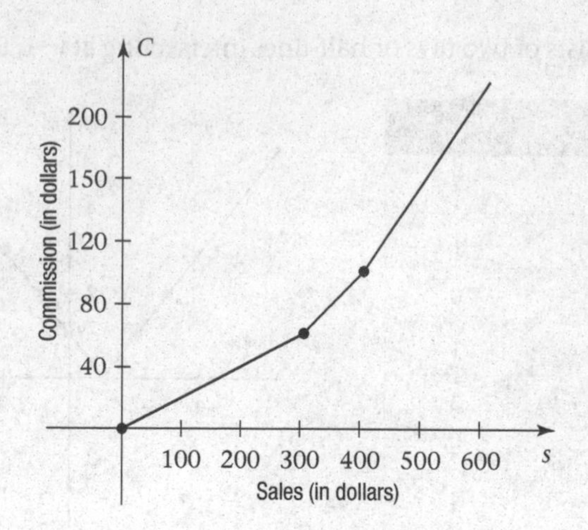

Progress Check

Let N denote the number of 50-cent stamps one can purchase with x dollars, assuming $0 \le x \le 1$. Express N as a function of x and sketch the graph.

Answer

$$N(x) = \begin{cases} 0 & \text{if} \quad 0 \le x < \frac{1}{2} \\ 1 & \text{if} \quad \frac{1}{2} \le x < 1 \\ 2 & \text{if} \quad x = 1 \end{cases}$$

The graph consists of two line segments and an isolated point:

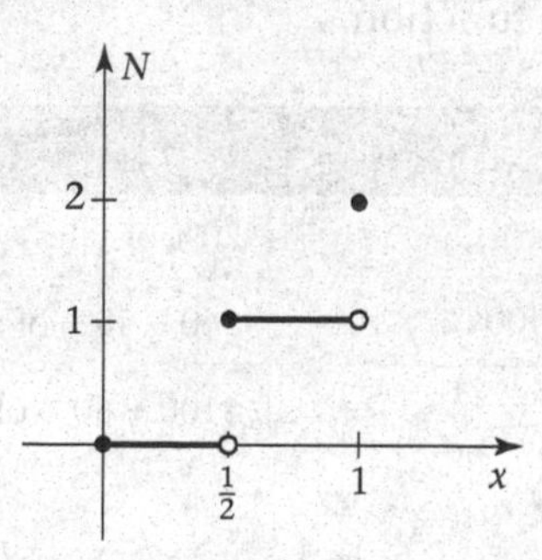

Number of 50¢ Stamps Purchased as a Function of Money

3.3d *Increasing and Decreasing Functions*

When we apply the terms **increasing** and **decreasing** to the graph of a function, we assume that we are viewing the graph from left to right. The line of Figure 3-28(a) is increasing, since the values of y increase as we move from left to right. Similarly, the graph in Figure 3-28(b) is decreasing, since the values of y decrease as we move from left to right. The graph in Figure 3-28(c) is a **constant**; hence it is neither increasing nor decreasing. One portion of the graph in Figure 3-28(d) is decreasing, and another is increasing. Observe that in Figure 3-28, the notation $f(x_1)$ and $f(x_2)$ represents the y-coordinates at $x = x_1$ and $x = x_2$, respectively. (See Figure 3-16.)

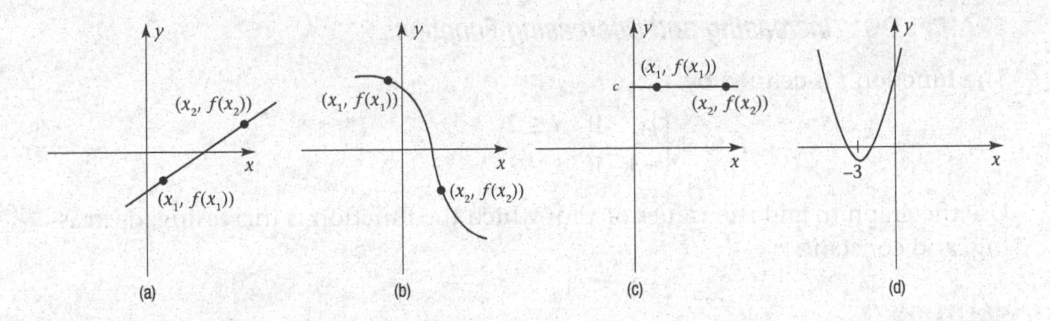

FIGURE 3-28
Functions: (a) Increasing, (b) Decreasing, (c) Constant, (d) Decreasing Then Increasing

In general, a function may increase over some intervals, decrease over others, and remain constant in still other intervals. Recall that there are four types of intervals to consider, namely, $[a, b]$, $(a, b]$, $[a, b)$, or (a, b). We define increasing, decreasing, and constant on an interval I.

Let x_1 and x_2 be any numbers in interval I in the domain of a function f. Then:

- f is increasing on I if $f(x_1) < f(x_2)$ whenever $x_1 < x_2$
- f is decreasing on I if $f(x_1) > f(x_2)$ whenever $x_1 < x_2$
- f is constant on I if $f(x_1) = f(x_2)$ for all x_1, x_2

Returning to Figure 3-28(d), note that the function is decreasing when $x \leq -3$ and increasing when $x \geq -3$. In other words, the function is decreasing on the interval $(-\infty, -3]$ and increasing on the interval $[-3, \infty)$. Observe that the point whose x-coordinate is -3 actually plays a dual role. The graph shows that the function has a minimum value at this point. It can be very useful, when sketching a graph, to find **turning points**, namely, those points where the graph changes from increasing to decreasing or from decreasing to increasing.

EXAMPLE 5 *Increasing and Decreasing Functions*

Use the graph of the function $f(x) = x^3 - 3x + 2$, shown in Figure 3-29, to determine where the function is increasing and where it is decreasing.

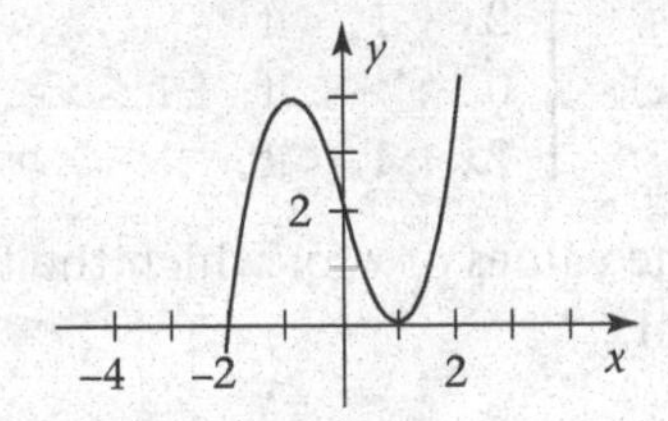

FIGURE 3-29
Graph of $y = x^3 - 3x + 2$

SOLUTION

From the graph, we see that there are turning points at $(-1, 4)$ and at $(1, 0)$. We conclude that

- f is increasing on the intervals $(-\infty, -1) \cup (1, \infty)$
- f is decreasing on the interval $(-1, 1)$ ■

EXAMPLE 6 *Increasing and Decreasing Functions*

The function f is defined by

$$f(x) = \begin{cases} |x| & \text{if } x \le 2 \\ -3 & \text{if } x > 2 \end{cases}$$

Use the graph to find the values of x for which the function is increasing, decreasing, and constant.

SOLUTION

Note that the piecewise-defined function f is composed of the absolute value function when $x \le 2$ and a constant function when $x > 2$. Therefore, we sketch the graph of f.

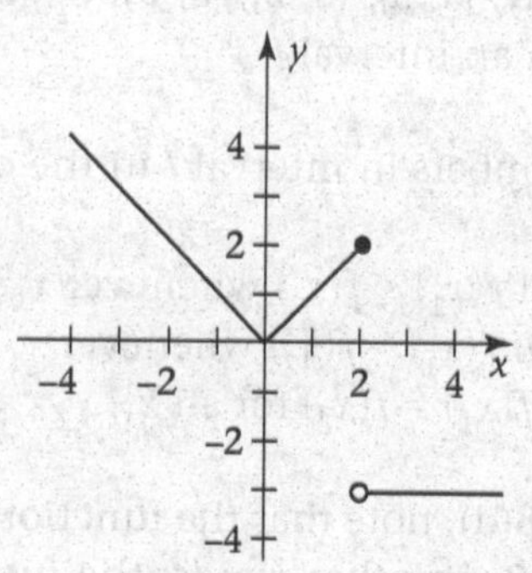

From this graph, we determine that

- f is decreasing on the interval $(-\infty, 0]$
- f is increasing on the interval $[0, 2]$
- f is constant and has value -3 on the interval $(2, \infty)$ ■

The function f is defined by

$$f(x) = \begin{cases} 2x + 1 & \text{if } \quad x < -1 \\ 0 & \text{if } \quad -1 \le x \le 3 \\ -2x + 1 & \text{if } \quad x > 3 \end{cases}$$

Use the graph to find the values of x for which the function is increasing, decreasing, and constant.

Answer

Increasing on the interval $(-\infty, -1)$, decreasing on $(3, \infty)$ and constant on $[-1, 3]$.

3.3e *Polynomial Functions*

The **polynomial function** of first degree

$$f(x) = ax + b$$

is called a **linear function.** We have already graphed a number of such functions in this chapter: $f(x) = 2x + 1$ (Example 5, Section 3.1), $f(x) = x$ (Figure 3-17), and $f(x) = -x$ (Figure 3-18). In each case, the graph appeared to be a **line.** We will prove in the next section that the graph of every linear function is indeed a line.

The polynomial function of second degree

$$f(x) = ax^2 + bx + c, \qquad a \neq 0$$

is called a **quadratic function.** We have graphed a few quadratic functions: $f(x) = x^2 - 4$ (Figure 3-9), $f(x) = 1 - x^2$ (Example 6a, Section 3.1), and $f(x) = x^2$ (Figure 3-21). The graph of the quadratic function is called a **parabola** and will be studied in detail in Chapter 5. For now, we offer an example for which a, b, and c are all nonzero.

EXAMPLE 7 *Graphs of Polynomials*

Sketch the graph of $f(x) = 2x^2 - 4x + 3$.

SOLUTION

We need to graph $y = 2x^2 - 4x + 3$. We form a table of values, plot the corresponding points, and connect them by a smooth curve.

x	y
−1	9
0	3
1	1
2	3
3	9

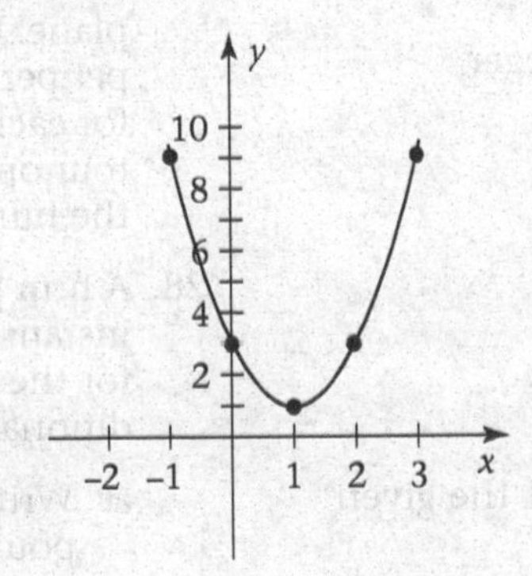

■

An investigation of polynomials of any degree reveals that they are all functions. The graphs of polynomials are always smooth curves. However, their shapes are not easily determined for degree greater than 2. We will take another look at this topic in the next chapter after learning more about the roots of polynomial equations. Note, however, that it may be difficult to graph polynomial functions accurately without results obtained by methods taught in calculus courses.

Exercise Set 3.3

In Exercises 1–16, sketch the graph of the function and state where it is increasing, decreasing, and constant.

1. $f(x) = 3x + 1$
2. $f(x) = 3 - 2x$
3. $f(x) = x^2 + 1$
4. $f(x) = x^2 - 4$
5. $f(x) = 9 - x^2$
6. $f(x) = 4x - x^2$
7. $f(x) = |2x + 1|$
8. $f(x) = |1 - x|$

9. $f(x) = \begin{cases} 2x & \text{if } x > -1 \\ -x - 1 & \text{if } x \le -1 \end{cases}$

10. $f(x) = \begin{cases} x + 1 & \text{if } x > 2 \\ 1 & \text{if } -1 \le x \le 2 \\ -x + 1 & \text{if } x < -1 \end{cases}$

11. $f(x) = \begin{cases} x & \text{if } x < 2 \\ 2 & \text{if } x \ge 2 \end{cases}$

12. $f(x) = \begin{cases} -x & \text{if } x \le -2 \\ x^2 & \text{if } -2 < x \le 2 \\ -x & \text{if } 3 \le x \le 4 \end{cases}$

13. $f(x) = \begin{cases} -x^2 & \text{if } -3 < x < 1 \\ 0 & \text{if } 1 \le x \le 2 \\ -3x & \text{if } x > 2 \end{cases}$

14. $f(x) = \begin{cases} 2 & \text{if } x \text{ is an integer} \\ -1 & \text{if } x \text{ is not an integer} \end{cases}$

15. $f(x) = \begin{cases} -2 & \text{if } x < -2 \\ -1 & \text{if } -2 \le x \le -1 \\ 1 & \text{if } x > -1 \end{cases}$

16. $f(x) = \begin{cases} \dfrac{x^2 - 1}{x - 1} & \text{if } x \ne 1 \\ 3 & \text{if } x = 1 \end{cases}$

In Exercises 17–24, sketch the graphs of the given functions on the same coordinate axes.

17. $f(x) = x^2,\quad g(x) = 2x^2,\quad h(x) = \frac{1}{2}x^2$
18. $f(x) = \frac{1}{2}x^2,\quad g(x) = \frac{1}{3}x^2,\quad h(x) = \frac{1}{4}x^2$
19. $f(x) = 2x^2,\quad g(x) = -2x^2$
20. $f(x) = x^2 - 2,\quad g(x) = 2 - x^2$
21. $f(x) = x^3,\quad g(x) = 2x^3$
22. $f(x) = \frac{1}{2}x^3,\quad g(x) = \frac{1}{4}x^3$
23. $f(x) = x^3,\quad g(x) = -x^3$
24. $f(x) = -2x^3,\quad g(x) = -4x^3$

25. The cell phone company charges a fee of $6.50 per month for the first 100 message units and an additional fee of $0.06 for each of the next 100 message units. A reduced rate of $0.05 is charged for each message unit after the first 200 units. Express the monthly charge C as a function of the number of message units u.

26. The annual dues of a union are as shown in the table.

Employees Annual Salary	Annual Dues
less than $8000	$60
$8000 or more but less than $15,000	$60 + 1% of the salary in excess of $8000
$15,000 or more	$130 + 2% of the salary in excess of $15,000

Express the annual dues d as a function of the salary S.

27. A tour operator who runs charter flights to Rome has established the following pricing schedule. For a group of no more than 100 people, the round trip fare per person is $300, with a minimum rental of $30,000 for the plane. For a group of more than 100, the fare per person for all passengers is reduced by $1 for each passenger in excess of 100. Write the tour operator's total revenue R as a function of the number of people x in the group.

28. A firm packages and ships 1-pound jars of instant coffee. The cost C of shipping is $2.50 for the first pound and 45 cents for each additional pound.

    a. Write C as a function of the weight w in pounds for $0 < w \le 30$.

    b. What is the cost of shipping a package containing 24 jars of instant coffee?

29. The daily rates of a car rental firm are $24 plus $0.15 per mile.

    a. Express the cost C of renting a car as a function of m, the number of miles traveled.

    b. What is the domain of the function?

    c. How much would it cost to rent a car for a 100-mile trip?

30. In a wildlife preserve, the population P of eagles depends on the population x of its basic food supply, rodents. Suppose that P is given by

$$P(x) = 0.002x + 0.004x^2$$

Find the eagle population when the rodent population is

a. 500 b. 2000

31. A parking lot in the center of the city of Philadelphia charges \$5.00 for any part of the first half-hour of parking. After the first half-hour, the rate is \$1.50 for any part of each additional half hour to a maximum rate of \$27.50 for the entire day.

   a. Express the parking fee F as a function of time t in hours.

   b. What does it cost to park if you enter the lot at 11:20 a.m. and leave at 1:05 p.m.?

   c. Sketch the graph of the function.

Graph functions (a) through (h) and then answer Exercises 32–36.

a. $f(x) = x^2 + 1$ b. $f(x) = 3x - \frac{1}{4}$

c. $f(x) = -2x - 1$ d. $f(x) = -3x^2$

e. $f(x) = \begin{cases} -x & \text{if} \quad x \le 0 \\ x & \text{if} \quad 0 < x \le 3 \\ 3 & \text{if} \quad x > 3 \end{cases}$

f. $f(x) = |x| - 2$ g. $f(x) = |x^2 - 9|$

h. $f(x) = \sqrt{1 - x}$

32. For each function, find the domain.

33. For each function, find the range.

34. For each function, find the intervals on which $f(x)$ is increasing.

35. For each function, find the intervals on which $f(x)$ is decreasing.

36. For each function, find the intervals on which $f(x)$ is neither increasing nor decreasing.

37. Graph the function in which the second coordinate is always −3 and the first coordinate may be any real number.

38. Graph the function in which the second coordinate is always 5 less than the first coordinate and the first coordinate may be any real number.

39. Graph the function whose second coordinate is always one more than the square of the first coordinate and whose first coordinate may be any real number.

40. Graph the function whose second coordinate is always the square root of the first coordinate and whose first coordinate may be any nonnegative real number.

The following is the graph of $y = f(x)$, associated with Exercises 41–44.

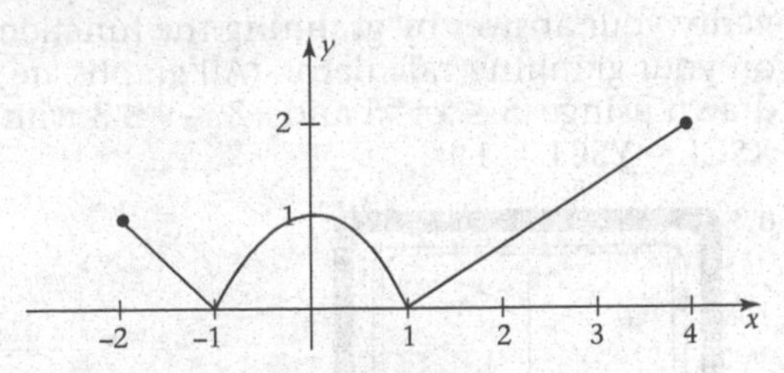

41. Sketch the graph of $y = 2 + f(x)$.

42. Sketch the graph of $y = f(x - 1)$.

43. Sketch the graph of $y = -f(x)$.

44. Sketch the graph of $y = f(-x)$.

The following is the graph of $y = f(x)$, associated with Exercises 45–49.

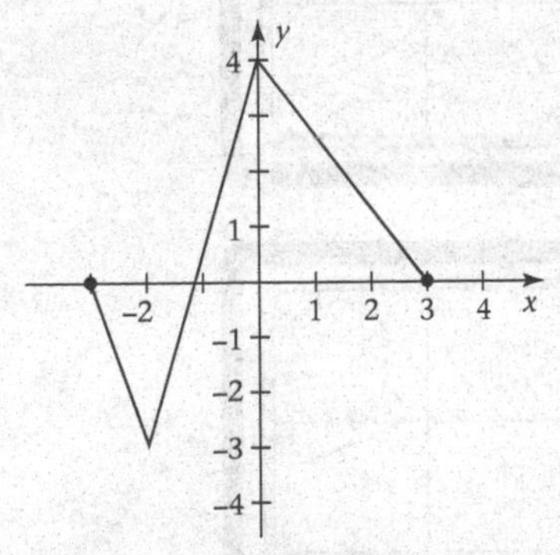

45. Sketch the graph of $y = f(x) - 3$.

46. Sketch the graph of $y = f(x + 2)$.

47. Sketch the graph of $y = -f(x)$.

48. Sketch the graph of $y = |f(x)|$.

49. Sketch the graph of $y = f(-x)$.

The following is the graph of $y = f(x)$, associated with Exercises 50–55.

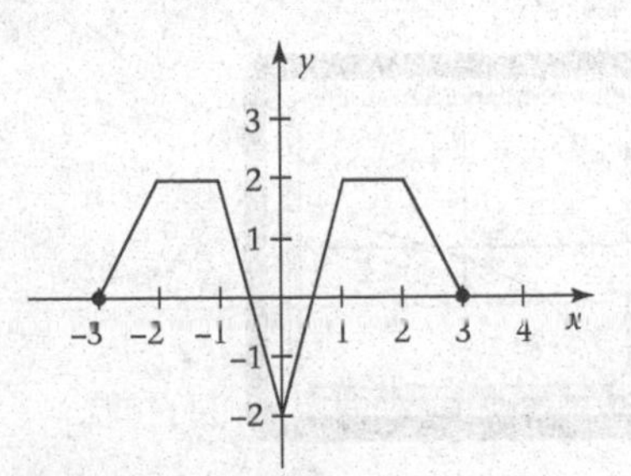

50. Sketch the graph of $y = f(x) + 5$.

51. Sketch the graph of $y = f(x + 1)$.

52. Sketch the graph of $y = -f(x)$.

53. Sketch the graph of $y = 5 - f(x)$.

54. Sketch the graph of $y = |f(x)|$.

55. Sketch the graph of $y = f(-x)$.

56. The following graphs are variations of the function $y = \sqrt{x}$. Determine each function and verify your answer by graphing the function on your graphing calculator. (All graphs are drawn using $-5 \leq x \leq 5$ and $-3 \leq y \leq 3$ with XSCL = YSCL = 1.)

a.

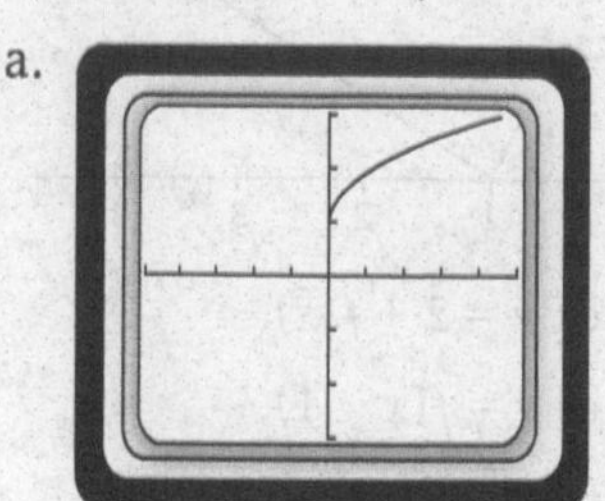

b.

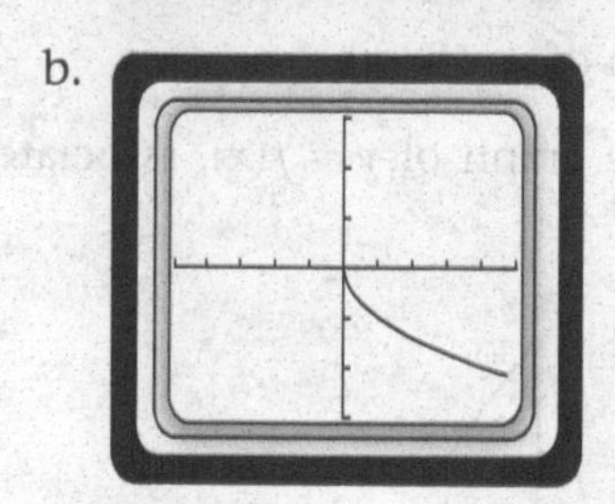

c.

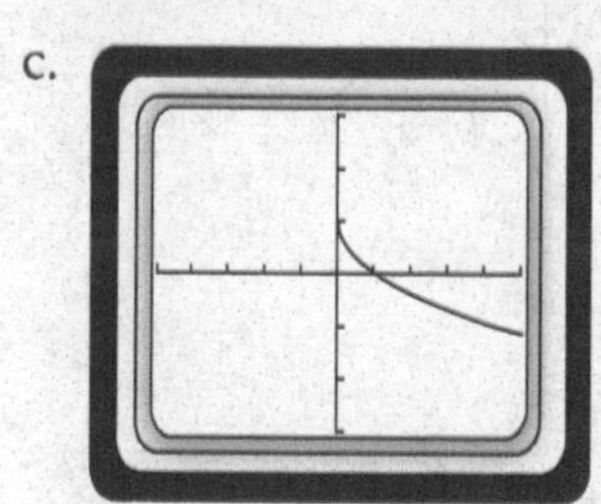

d.

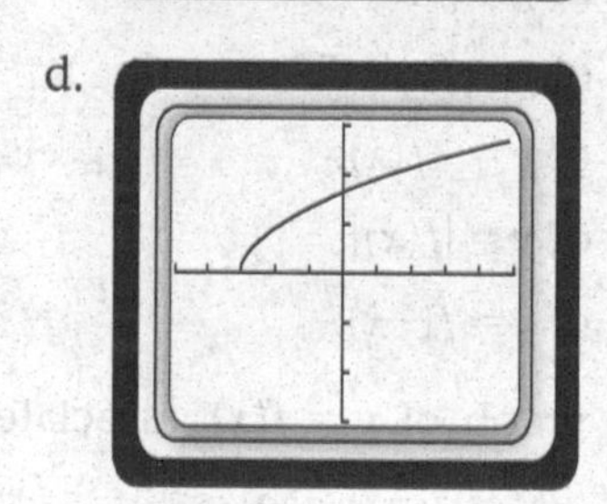

e.

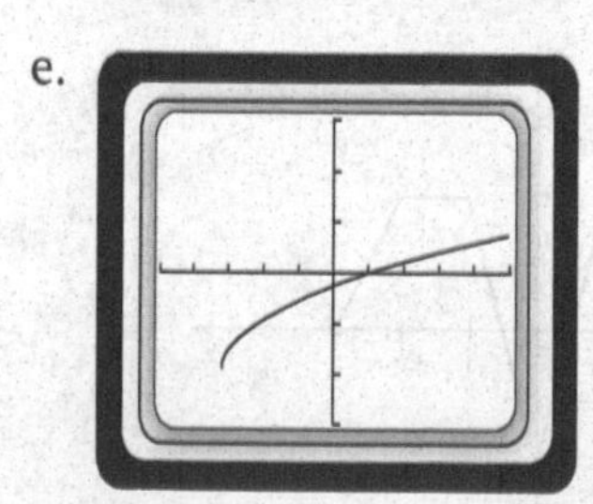

f.

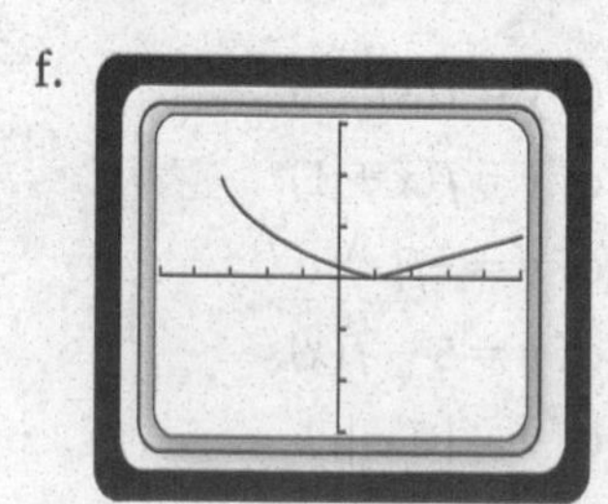

57. A ball is thrown upward from a bridge 96 feet over a river with an initial velocity of 80 feet per second. Its height after t seconds is given by $s = -16t^2 + 80t + 96$.

    a. Graph this function.

    b. From the graph, find the maximum height of the ball.

    c. At what time t after the ball is released does it reach its maximum height?

    d. Find the time when the ball reaches the water.

58. Johnny Appleseed currently has 20 winesap apple trees on one acre of ground. Each tree has an average yield of 40 bushels of apples. For every tree the farm adds, the yield per tree is one less bushel of apples due to overcrowding.

    a. Express the total number of bushels of apples as a function of the number of trees.

    b. Graph this function.

    c. From the graph, determine the number of trees that should be planted so that the yield of apples is a maximum.

    d. Find the maximum yield of apples, giving your answer in bushels.

59. *Mathematics in Writing*: Store the LIST $\{3, 4, 5, 6\}$ at L_1, then graph

$$y = 3x + L_1$$

Write a brief paragraph explaining your results.

3.4 Linear Functions

In the previous section, we said that the polynomial function of first degree

$$f(x) = ax + b$$

is called a linear function, and we observed that the graph of such a function *appears* to be a line. In this section, we will look at the property of a line that distinguishes it from all other curves. We will then develop equations for the line, and we will show that the graph of a linear function is indeed a line.

3.4a *Slope of a Line*

In Figure 3-30, we have drawn a line L that is not vertical. We have indicated the distinct points $P_1(x_1, y_1)$ and $P_2(x_2, y_2)$ on L. When moving from P_1 to P_2, the increment or change in the x-coordinate is $x_2 - x_1$ and in the y-coordinate, $y_2 - y_1$. Note that the increment $x_2 - x_1$ is not zero since L is not vertical.

If $P_3(x_3, y_3)$ and $P_4(x_4, y_4)$ are another pair of points on L, the increments $x_4 - x_3$ and $y_4 - y_3$ will, in general, be different from the increments obtained by using P_1 and P_2. However, since triangles P_1AP_2 and P_3BP_4 are similar, the corresponding sides are proportional to one another, that is,

$$\frac{y_2 - y_1}{x_2 - x_1} = \frac{y_4 - y_3}{x_4 - x_3}$$

This ratio is called the **slope of the line *L*** and is denoted by m.

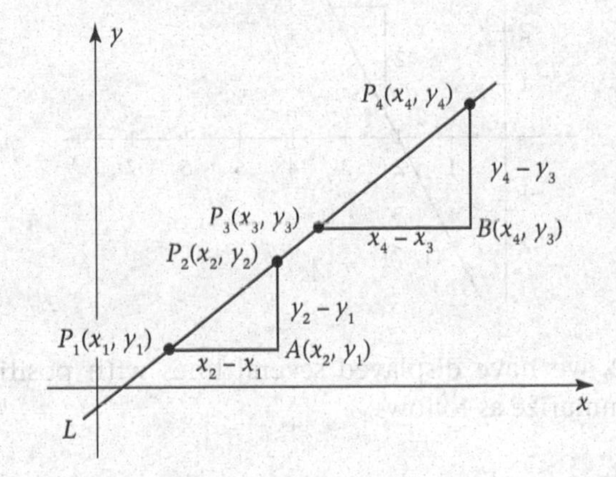

FIGURE 3-30
All Points on a Line Produce the Same Slope

Slope of a Line
The slope m of a line that is not vertical is given by

$$m = \frac{y_2 - y_1}{x_2 - x_1}$$

where $P_1(x_1, y_1)$ and $P_2(x_2, y_2)$ are any two distinct points on the line.

For a vertical line, $x_1 = x_2$, so $x_2 - x_1 = 0$. Since division by 0 is not defined, we say that a vertical line has no slope, or the slope of a vertical line does not exist.

The property of constant slope characterizes the line; that is, no other curve has this property.

EXAMPLE 1 *Finding the Slope of a Line*

Find the slope of the line that passes through the points (5, 6) and (1, −2).

SOLUTION

$$(x_1, y_1) = (5, 6) \quad \text{and} \quad (x_2, y_2) = (1, -2)$$

$$m = \frac{y_2 - y_1}{x_2 - x_1} = \frac{-2-6}{1-5} = \frac{-8}{-4} = 2$$

Verify that reversing the choice of P_1 and P_2 produces the same result for the slope m. Although we may choose either point as P_1 and the other as P_2, we must use this choice consistently once it has been made. ■

Slope is a means of measuring the steepness of a line. That is, slope specifies the number of units we must move up or down to reach the line after moving 1 unit to the left or right of the line. Specifically in Example 1 above, if we move 1 unit to the right of the line, we must move up 2 units to reach the line. Alternatively, if we move 1 unit to the left of the line, we must move down 2 units to reach the line. (See Figure 3-31.)

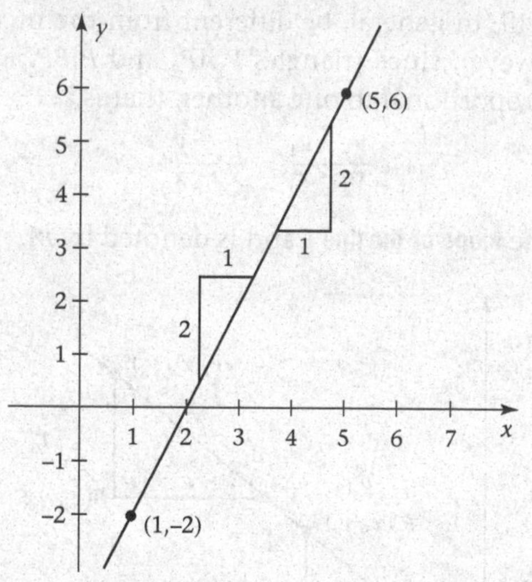

FIGURE 3-31
Graph of the Line for Example 1

In Figure 3-32, we have displayed several lines with positive and negative slopes. We can summarize as follows:

Let m be the slope of a line.

1. When $m > 0$, the line is the graph of an increasing function.
2. When $m < 0$, the line is the graph of a decreasing function.
3. When $m = 0$, the line is the graph of a constant function.
4. Slope does not exist for a vertical line, and a vertical line is not the graph of a function.

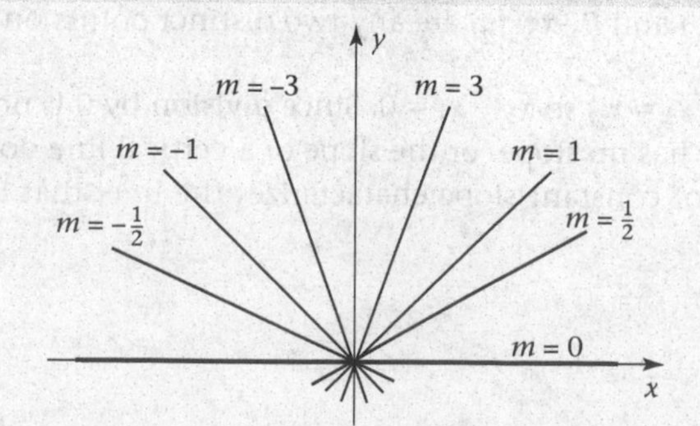

FIGURE 3-32
Lines with Different Slopes

Graphing Calculator ALERT

The function $y = x$ is graphed below in four different viewing rectangles. The slope *appears* to be different in each viewing rectangle. TRACE to locate two points on each line, and *compute* the slope to verify that each line has slope $m = 1$. *You cannot estimate the slope of a line on a graphing calculator "by eye."*

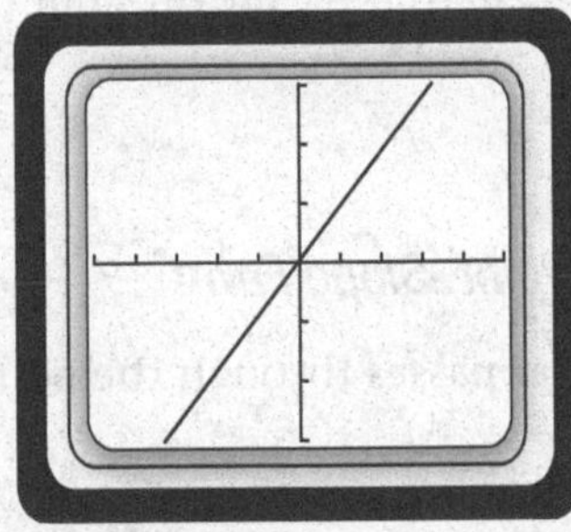

(a) EQUAL viewing rectangle
XSCL = 1, YSCL = 1

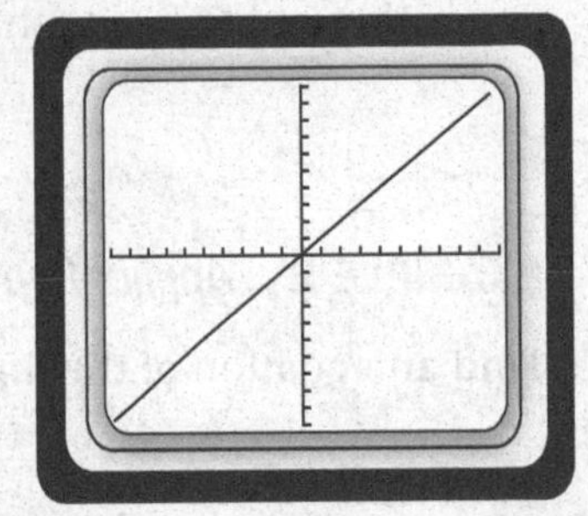

(b) TENS viewing rectangle
XSCL = 1, YSCL = 1

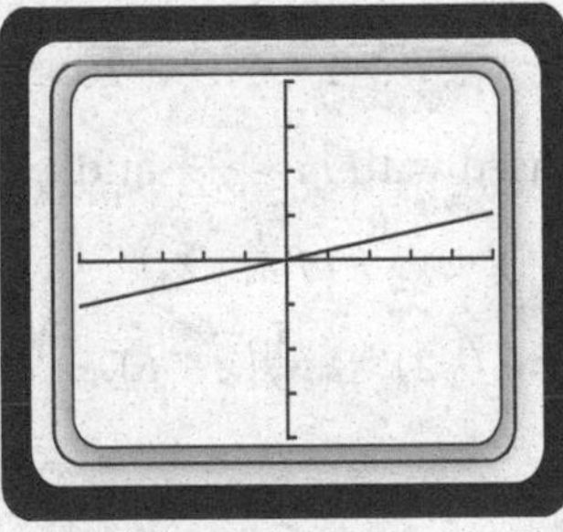

(c) $-5 \le X \le 5$ and $-20 \le X \le 20$
XSCL = 1, YSCL = 1

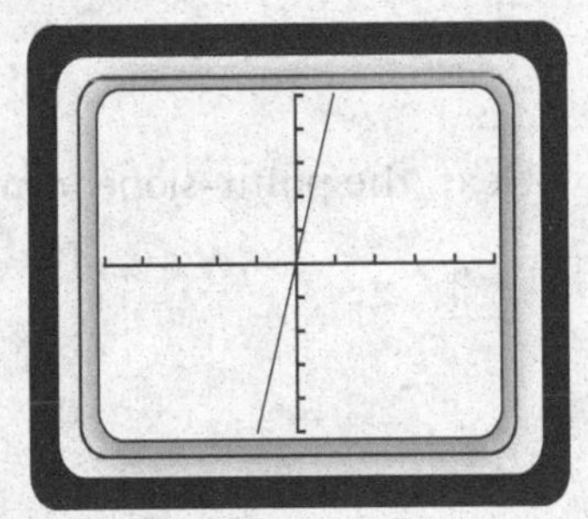

(d) $-20 \le X \le 20$ and $-5 \le Y \le 5$
XSCL = 5, YSCL = 1

3.4b *Equations of a Line*

We can apply the concept of slope to develop two important forms of the equation of a line. In Figure 3-33, the point $P_1(x_1, y_1)$ lies on a line L whose slope is assumed to be m. If $P(x, y)$ is any other point on L, then we may use P and P_1 to compute m. Therefore,

$$m = \frac{y - y_1}{x - x_1}$$

which can be written in the form

$$y - y_1 = m(x - x_1)$$

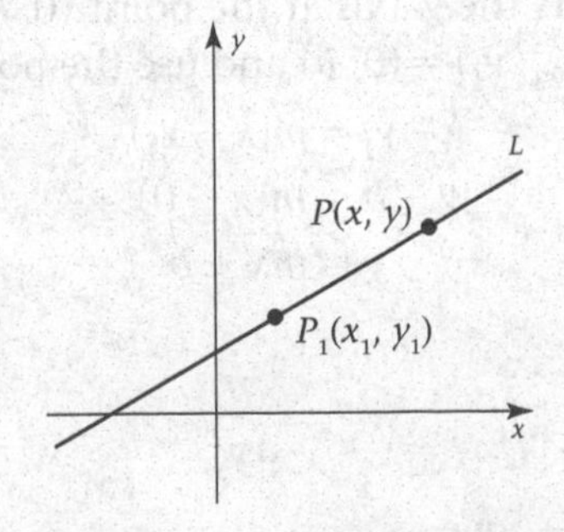

FIGURE 3-33
Graph of Line Using Point–Slope Form

Since (x_1, y_1) satisfies this equation, every point on L satisfies this equation. Conversely, any point satisfying this equation must lie on the line L, since there is only one line through $P_1(x_1, y_1)$ with slope m. This equation is called the **point–slope form** of a line.

Point–Slope Form

The equation

$$y - y_1 = m(x - x_1)$$

is that of the line with slope m that passes through the point (x_1, y_1).

EXAMPLE 2 ***Application of Point–Slope Form***

Find an equation of the line that passes through the points (6, −2) and (−4, 3).

SOLUTION

First we find the slope. Let $(x_1, y_1) = (6, -2)$ and $(x_2, y_2) = (-4, 3)$. Then

$$m = \frac{y_2 - y_1}{x_2 - x_1} = \frac{3-(-2)}{-4-6} = \frac{5}{-10} = -\frac{1}{2}$$

Next, the point–slope form is used with $m = -\frac{1}{2}$ and $(x_1, y_1) = (6, -2)$.

$$y - y_1 = m(x - x_1)$$

$$y - (-2) = -\frac{1}{2}(x - 6)$$

$$y = -\frac{1}{2}x + 1$$

Verify that using the point (−4, 3) and $m = -\frac{1}{2}$ in the point–slope form will also yield the same equation. ■

Progress Check

Find an equation of the line that passes through the points (−5, 0) and (2, −5).

Answer

$y = -\frac{5}{7}x - \frac{25}{7}$

There is another form of the equation of the line that is very useful. In Figure 3-34, the line L meets the y-axis at the point $(0, b)$ and is assumed to have slope m. Then, we can let $(x_1, y_1) = (0, b)$ and use the point–slope form:

$$y - y_1 = m(x - x_1)$$

$$y - b = m(x - 0)$$

$$y = mx + b$$

Recalling that b is the y-intercept, we call this equation the **slope–intercept form** of the line.

Slope–Intercept Form
The equation

$$y = mx + b$$

is that of the line with slope m and y-intercept b.

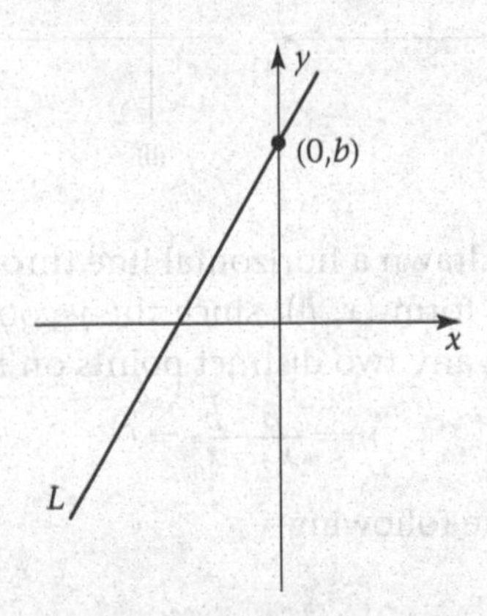

FIGURE 3-34
Graph of Line Using Slope–Intercept Form

The last result leads to the important conclusion mentioned in the introduction to this section. Since the graph of $y = mx + b$ is the graph of the function $f(x) = mx + b$, we have indeed shown that the *graph of a linear function is a line.*

EXAMPLE 3 ***Application of Slope–Intercept Form***

Find the slope and y-intercept of the line $y - 3x + 1 = 0$.

SOLUTION

The equation must be placed in the form $y = mx + b$. Solving for y gives

$$y = 3x - 1$$

and we find that $m = 3$ is the slope and $b = -1$ is the y-intercept. ■

Progress Check

Find the slope and y-intercept of the line $2y + x - 3 = 0$.

Answers

slope $= m = -\frac{1}{2}$; y-intercept $= b = \frac{3}{2}$

3.4c Horizontal and Vertical Lines

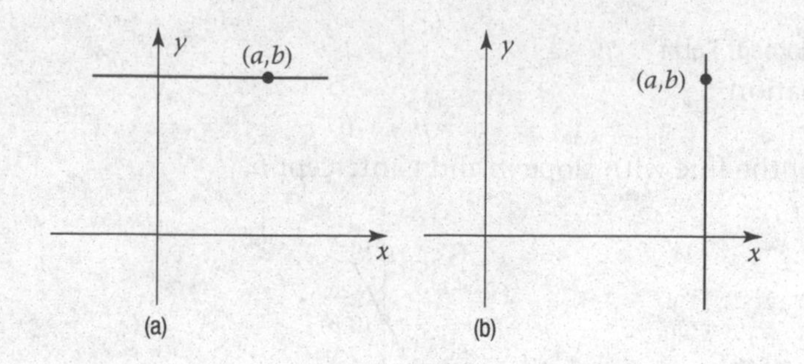

FIGURE 3-35
Horizontal and Vertical Lines

In Figure 3-35(a), we have drawn a horizontal line through the point (a, b). Every point on this line has the form (x, b), since the y-coordinate remains constant. If $P_1(x_1, b)$ and $P_2(x_2, b)$ are any two distinct points on the line, then the slope is

$$m = \frac{b - b}{x_2 - x_1} = 0$$

We have established the following:

Horizontal Lines
The equation

$$y = b$$

is that of the horizontal line through the point (a, b). The slope of a horizontal line is 0.

In Figure 3-35(b), every point on the vertical line through the point (a, b) has the form (a, y), since the x-coordinate remains constant. Calculating the slope using any two points $P_1(a, y_1)$ and $P_2(a, y_2)$ on the line produces

$$m = \frac{y_2 - y_1}{a - a} = \frac{y_2 - y_1}{0}$$

Since we cannot divide by 0, slope is not defined for a vertical line.

Vertical Lines
The equation

$$x = a$$

is that of the vertical line through the point (a, b). A vertical line has no slope.

Note that a vertical line is the only line that is *not* the graph of a function.

EXAMPLE 4 *Application of Horizontal and Vertical Line Forms*

Find the equations of the horizontal and vertical lines through $(-4, 7)$.

SOLUTION

The horizontal line has the equation $y = 7$. The vertical line has the equation $x = -4$.

■

Do not confuse "no slope" with "zero slope." A horizontal line has zero slope. A vertical line has no slope; in other words, its slope is undefined.

3.4d *General First-Degree Equation*

The **general first-degree equation** in x and y can always be written in the form

$$Ax + By + C = 0$$

where A, B and C are constants, and A and B are *not both* zero. We can rewrite this equation as

$$By = -Ax - C$$

If $B \neq 0$, the equation becomes

$$y = -\frac{A}{B}x - \frac{C}{B}$$

which we recognize as the equation of a line with slope $-\frac{A}{B}$ and y-intercept $-\frac{C}{B}$. If $B = 0$, the original equation becomes $Ax + C = 0$, whose graph is the vertical line $x = -\frac{C}{A}$. If $A = 0$, the original equation becomes $By + C = 0$, whose graph is the horizontal line $y = -\frac{C}{B}$.

The General First-Degree Equation

- The graph of the general first-degree equation is a line.
  $$Ax + By + C = 0, \qquad A \text{ and } B \text{ not both zero}$$
- If $A = 0$, the graph is a horizontal line.
- If $B = 0$, the graph is a vertical line.

3.4e *Parallel and Perpendicular Lines*

The concept of slope of a line can be used to determine when two lines are parallel or perpendicular. Since parallel lines have the same "steepness," we recognize that they must have the same slope.

Parallel Lines

Two lines with slopes m_1 and m_2 are parallel if, and only if,

$$m_1 = m_2$$

The criterion for perpendicular lines can be stated in this way.

Perpendicular Lines

Two lines with slopes m_1 and m_2 are perpendicular if, and only if,

$$m_1 m_2 = -1 \qquad \text{or equivalently,} \qquad m_2 = -\frac{1}{m_1}, \qquad m_1 \neq 0$$

These two theorems do not apply to vertical lines since the slope of a vertical line is undefined. The proofs of these theorems are geometric in nature and are outlined in Exercises 56 and 57.

EXAMPLE 5 ***Application of Parallel and Perpendicular Lines***

Given the line $y = 3x - 2$, find an equation of the line passing through the point $(-5, 4)$ that is (a) parallel to the given line and (b) perpendicular to the given line.

SOLUTION

We first note that the line $y = 3x - 2$ has slope $m_1 = 3$.

a. Every line parallel to the line $y = 3x - 2$ must have slope $m_2 = m_1 = 3$. Therefore, we seek a line with slope 3 that passes through the point $(-5, 4)$. Using the point–slope formula, we have

$$y - y_1 = m(x - x_1)$$
$$y - 4 = 3(x + 5)$$
$$y = 3x + 19$$

b. Since the line $y = 3x - 2$ has slope $m_1 = 3$, every line perpendicular to this line must have slope

$$m_2 = -\frac{1}{m_1} = -\frac{1}{3}$$

Then, the line we seek has slope $-\frac{1}{3}$ and passes through the point $(-5, 4)$. We can apply the point–slope formula once again to obtain

$$y - y_1 = m(x - x_1)$$
$$y - 4 = -\frac{1}{3}(x + 5)$$
$$y = -\frac{1}{3}x + \frac{7}{3}$$

The three lines are shown in Figure 3-36.

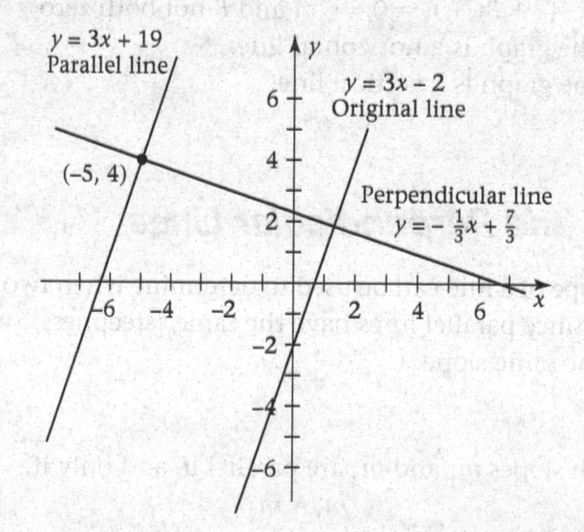

FIGURE 3-36

Graphs of Parallel and Perpendicular Lines

■

3.4f Summary

The slope m of a line can tell us a great deal about its graph. The slopes m_1 and m_2 of a pair of lines can be useful in determining some special relationship between their graphs. Various facts about the slope of a line are presented in Table 3-3.

TABLE 3-3 Slope of a Line

Slope(s)	Graph(s)	Example
$m > 0$	rising	y x
$m < 0$	falling	y x
$m = 0$	horizontal	y x
m undefined	vertical	y x
$m_1 = m_2$	parallel	y x
$m_1 m_2 = -1$	perpendicular	y x

Exercise Set 3.4

In Exercises 1–6, determine the slope of the line through the given points. State whether the line is the graph of an increasing function, a decreasing function, a constant function, or not a function.

1. (2, 3), (−1, −3)
2. (1, 2), (−2, 5)
3. (−2, 3), (0, 0)
4. (2, 4), (−3, 4)
5. $\left(\frac{1}{2}, 2\right), \left(\frac{3}{2}, 1\right)$
6. (−4, 1), (−1, −2)
7. Use slopes to show that the points $A(-1, -5)$, $B(1, -1)$, and $C(3, 3)$ are collinear, that is, they lie on the same line.
8. Use slopes to show that the points $A(-3, 2)$, $B(3, 4)$, $C(5, -2)$, and $D(-1, -4)$ are the vertices of a parallelogram.

In Exercises 9–12, determine an equation of the line with the given slope m that passes through the given point.

9. $m = 2$, (−1, 3)
10. $m = -\frac{1}{2}$, (1, −2)
11. $m = 3$, (0, 0)
12. $m = 0$, (−1, 3)

In Exercises 13–18, determine an equation of the line through the given points.

13. (2, 4), (−3, −6)
14. (−3, 5), (1, 7)
15. (0, 0), (3, 2)
16. (−2, 4), (3, 4)
17. $\left(-\frac{1}{2}, -1\right), \left(\frac{1}{2}, 1\right)$
18. (−8, −4), (3, −1)

In Exercises 19–24, determine an equation of the line with the given slope m and the given y-intercept b.

19. $m = 3, b = 2$
20. $m = -3, b = -3$
21. $m = 0, b = 2$
22. $m = -\frac{1}{2}, b = \frac{1}{2}$
23. $m = \frac{1}{3}, b = -5$
24. $m = -2, b = -\frac{1}{2}$

In Exercises 25–30, determine the slope m and y-intercept b of the given line.

25. $3x + 4y = 5$
26. $2x - 5y + 3 = 0$
27. $y - 4 = 0$
28. $x = -5$
29. $3x + 4y + 2 = 0$
30. $x = -\frac{1}{2}y + 3$

In Exercises 31–36, write an equation of (a) the horizontal line passing through the given point and (b) the vertical line passing through the given point.

31. (−6, 3)
32. (−5, −2)
33. (−7, 0)
34. (0, 5)
35. (9, −9)
36. $\left(-\frac{3}{2}, 1\right)$

In Exercises 37–40, determine the slope of (a) every line that is parallel to the given line and (b) every line that is perpendicular to the given line.

37. $y = -3x + 2$
38. $2y - 5x + 4 = 0$
39. $3y = 4x - 1$
40. $5y + 4x = -1$

In Exercises 41–44, determine an equation of the line through the given point that (a) is parallel to the given line; (b) is perpendicular to the given line.

41. (1, 3); $y = -3x + 2$
42. (−1, 2); $3y + 2x = 6$
43. (−3, 2); $3x + 5y = 2$
44. (−1, −3); $3y + 4x - 5 = 0$
45. The Celsius (C) and Fahrenheit (F) temperature scales are related by a linear equation. Water boils at 212°F or 100°C, and freezes at 32°F or 0°C.
    a. Write a linear equation expressing F in terms of C.
    b. What is the Fahrenheit temperature when the Celsius temperature is 20°?
46. The college bookstore sells a textbook that costs \$80 for \$94 and a textbook that costs \$84 for \$98.70. If the markup policy of the bookstore is linear, write a linear function that relates sales price S and cost C. What is the cost of a book that sells for \$105.75?
47. An appliance manufacturer finds that it had sales of \$200,000 five years ago and sales of \$600,000 this year. If the growth in sales is assumed to be linear, what will the sales amount be 5 years from now?
48. A product that cost \$2.50 three years ago sells for \$3 this year. If price increases are assumed to be linear, how much will the product cost 6 years from now?
49. Find a real number c such that $P(-2, 2)$ is on the line $3x + cy = 4$.
50. Find a real number c such that the line $cx - 5y + 8 = 0$ has x-intercept 4.
51. If the points (−2, −3) and (−1, 5) are on the graph of a linear function f, find $f(x)$.
52. If $f(1) = 4$ and $f(-1) = 3$ and the function f is linear, find $f(x)$.
53. Prove that the linear function $f(x) = ax + b$ is an increasing function if $a > 0$, a decreasing function if $a < 0$, and a constant function if $a = 0$.

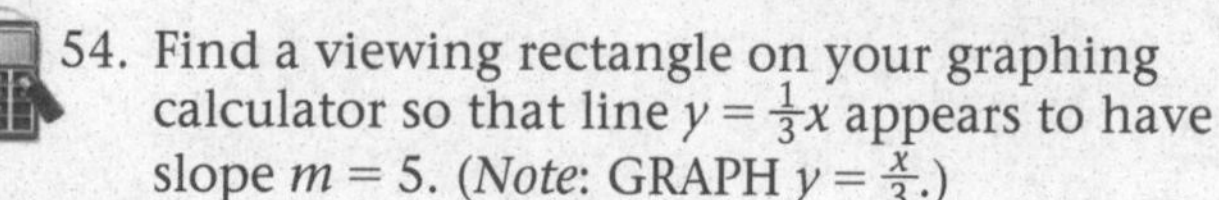

54. Find a viewing rectangle on your graphing calculator so that line $y = \frac{1}{3}x$ appears to have slope $m = 5$. (*Note*: GRAPH $y = \frac{x}{3}$.)

55. Prove that if two lines have the same slope, they are parallel.

56. In the accompanying figure, lines L_1 and L_2 are parallel. Points A and D are selected on lines L_1 and L_2, respectively. Lines parallel to the x-axis are constructed through A and D that intersect the y-axis at points B and E. Supply a reason for each of the steps in the following proof.

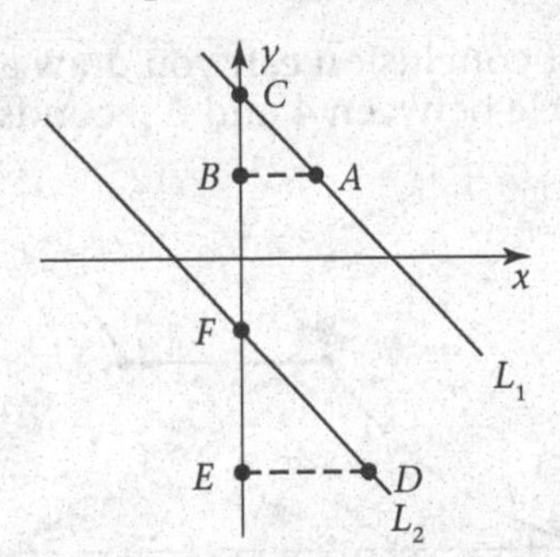

a. Angles ABC and DEF are equal.

b. Angles ACB and DFE are equal.

c. Triangles ABC and DEF are similar.

d. $\frac{\overline{CB}}{\overline{BA}} = \frac{\overline{FE}}{\overline{ED}}$

e. $m_1 = \frac{\overline{CB}}{\overline{BA}}, \quad m_2 = \frac{\overline{FE}}{\overline{ED}}$

f. $m_1 = m_2$

g. Parallel lines have the same slope.

57. In the accompanying figure, lines perpendicular to each other, with slopes m_1 and m_2, intersect at a point Q. A perpendicular line from Q to the x-axis intersects the x-axis at the point C. Supply a reason for each of the steps in the following proof.

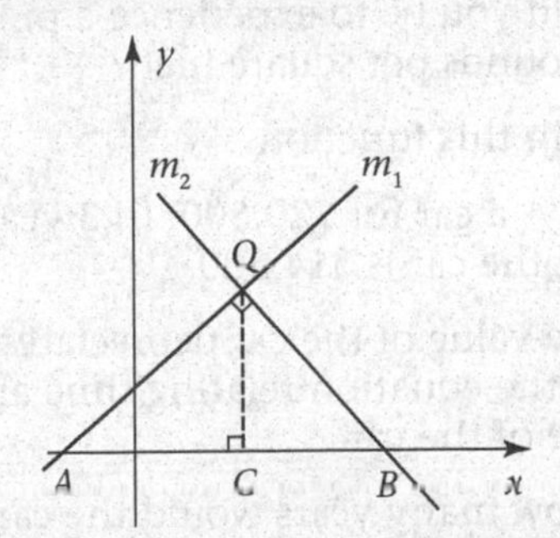

a. Angles CAQ and BQC are equal.

b. Triangles ACQ and QCB are similar.

c. $\frac{\overline{CQ}}{\overline{AC}} = \frac{\overline{CB}}{\overline{CQ}}$

d. $m_1 = \frac{\overline{CQ}}{\overline{AC}}, \quad m_2 = -\frac{\overline{CQ}}{\overline{CB}}$

e. $m_2 = -\frac{1}{m_1}$

58. Prove that if two lines have slopes m_1 and m_2 such that $m_2 = -\frac{1}{m_1}$, the lines are perpendicular.

59. If x_1 and x_2 are the abscissas of two distinct points on the graph of the function $y = f(x)$, show that the slope m of the line connecting the two points can be written as

$$m = \frac{f(x_2) - f(x_1)}{x_2 - x_1}$$

In Exercises 60–63,

a. Plot the data in the tables that follow.

b. If the graph is a line, find the linear relationship.

c. On your graphing calculator, GRAPH the line in an appropriate viewing rectangle for the problem and TRACE (approximately) to the points given in the table.

60. The table below shows the surface tension T in dynes per centimeter for pure water in contact with air at various temperatures C in °C

C	10	20	50
T	74	72	66

61. The table below gives the volume V of 1 gram of water at various temperatures T in °C.

T	0	50	75	100
V	1.00	1.01	1.02	1.04

62. The table shows the number N of earthquakes per year of different magnitudes M as measured on the Richter Scale.

M	7	5	4	2
N	14	1000	1493	2479

63. The profit P in dollars for selling G gallons of gasoline at a service station is shown in the following table.

G	2	10	50	85
P	−4	20	140	245

Exercises 64–70 require the following information.

The distance traveled by a moving object is a function of time. The average speed is defined as the ratio of the total distance traveled between two points to the elapsed time of travel, that is, *Average Speed* = $\frac{\textit{Distance Traveled}}{\textit{Elapsed Time}}$. It is also the slope of the straight line connecting the two points on the graph of the function.

64. What is the distance a car will travel in 12 minutes if it is going 50 mph?

65. An American Airlines jet leaves Raleigh-Durham at 10:05 a.m. nonstop to Fort Lauderdale. The jet arrives in Fort Lauderdale at 12:00 noon. If the distance between Raleigh-Durham and Fort Lauderdale is 683 air miles, find the average speed of the plane.

66. A particle is at $x = 2$ meters at $t = 0$ seconds, $x = -3$ meters at $t = 5$ seconds, and $x = 5$ meters at $t = 7$ seconds. Find the average speed during the intervals:

    a. $t = 0$ seconds to $t = 5$ seconds

    b. $t = 5$ seconds to $t = 7$ seconds

    c. $t = 0$ seconds to $t = 7$ seconds

67. An out-of-shape junior high school student is asked to run 0.25 miles during the last 5 minutes of a 50-minute gym class. The student starts running, but eventually ends up walking around the track. If the student's average speed is 2.5 miles per hour, will the student complete the full distance before the gym class ends?

The following two graphs show the position of a particle versus time. In Exercises 68 and 69, find the average speed during the given time intervals.

68. a. from $t = 0$ to $t = 2$

    b. from $t = 2$ to $t = 5$

    c. from $t = 5$ to $t = 7$

    d. from $t = 0$ to $t = 5$

    e. from $t = 0$ to $t = 7$

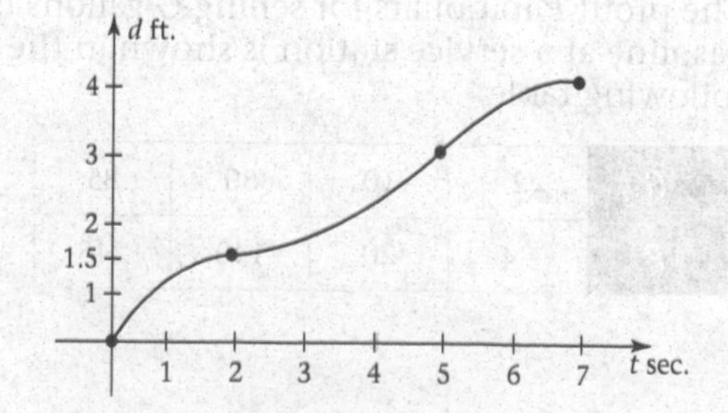

69. a. from $t = 0$ to $t = 4$

    b. from $t = 4$ to $t = 7$

    c. from $t = 7$ to $t = 10$

    d. from $t = 0$ to $t = 7$

    e. from $t = 4$ to $t = 10$

    f. from $t = 0$ to $t = 10$

    g. What conclusion can you draw about the particle between 4 and 7 seconds?

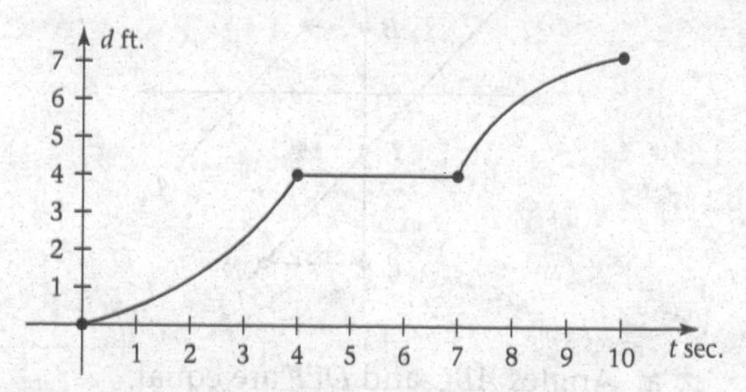

70. The speed of sound is 1100 feet per second. A clap of thunder occurs 5 miles from a school. How long will it take before the school children hear the thunder?

71. Assume that the pressure one experiences descending the depths of the ocean is linear.

    a. If the pressure is 15 pounds per square inch on the surface of the water and the pressure is 35 pounds per square inch 45 feet below the surface, find the equation giving the relationship between pressure p and depth d.

    b. What is the pressure 10,800 feet below the surface of the water?

    c. How far below the surface of the water would you be to experience a pressure of 99 pounds per square inch?

    d. Graph this function.

72. Katy buys a car for \$20,500. In 3 years the value of the car is \$14,500.

    a. If the value of the car depreciates linearly, find the equation relating time and the value of the car.

    b. In how many years would the car be worthless?

73. Refer to Katy's car purchase in Exercise 72. Suppose that 6 years after Katy purchased her car, the manufacturer stops producing the model that she purchased. This manufacturing decision causes the value of the car to begin to appreciate. After 8 years of ownership, Katy's car is now worth $900 more than it was worth after 6 years.

    a. If the appreciation of the car is again taken to be linear, express the relationship between the value of the car and time.

    b. Graph the relationship.

    c. What is the value of Katy's car after 10 years?

74. A veterinarian wants to create a new food mixture for cats with urinary tract disease. The precise level of ash in the feline's food is critical in maintaining the health of the cat. The veterinarian mixes two foods, *A* and *B*. *A* contains 25% ash and *B* contains 15% ash. The resulting mix must contain 20 pounds of ash.

    a. Write a linear equation relating the amount of food *A* and the amount of food *B* required to have the desired amount of ash.

    b. How much of food *A* would be needed if 50 pounds of food *B* were used in the mixture?

    c. Graph this function.

75. Refer to Exercise 50 in Section 3.2.

    a. Graph the function you found in part a.

    b. Find the slope of the line.

    c. What is the y-intercept of the line?

76. Refer to Exercises 45 and 75, in this section.

    a. Write a linear equation expressing the outside temperature in degrees Celsius in terms of n, the number of times a cricket chirps in one minute.

    b. Graph this linear relationship.

77. *Mathematics in Writing*: Explain in your own words how the slope of a line can help you to predict how the function outputs are changing with changes in the inputs. Are the graphs in Exercise 60 and 61 *rising* or *falling*? How can you tell? What is the importance of knowing whether a graph is rising or falling?

3.5 The Algebra of Functions; Inverse Functions

Functions such as

$$f(x) = x^2 \quad \text{and} \quad g(x) = x - 1$$

can be combined by the usual operations of addition, subtraction, multiplication, and division. Using these functions f and g, we can form

$$(f+g)(x) = f(x) + g(x) = x^2 + x - 1$$
$$(f-g)(x) = f(x) - g(x) = x^2 - (x-1) = x^2 - x + 1$$
$$(f \cdot g)(x) = f(x) \cdot g(x) = x^2(x-1) = x^3 - x^2$$
$$\left(\frac{f}{g}\right)(x) = \frac{f(x)}{g(x)} = \frac{x^2}{x-1}$$

In each case, we have combined two functions f and g to form a new function. Note, however, that the domain of the new function need not be the same as the domain of either of the original functions. The function formed by division in the above example has as its domain the set of all real numbers x except $x = 1$ since we cannot divide by 0. On the other hand, the original functions $f(x) = x^2$ and $g(x) = x - 1$ are both defined at $x = 1$.

EXAMPLE 1 *Algebra of Functions*

Given $f(x) = x - 4$ and $g(x) = x^2 - 4$, find the following:

a. $(f+g)(x)$

b. $(f-g)(x)$

c. $(f \cdot g)(x)$

d. $\left(\frac{f}{g}\right)(x)$

e. the domain of $\left(\frac{f}{g}\right)(x)$

SOLUTION

a. $(f+g)(x) = f(x) + g(x) = x - 4 + x^2 - 4 = x^2 + x - 8$

b. $(f-g)(x) = f(x) - g(x) = x - 4 - (x^2 - 4) = -x^2 + x$

c. $(f \cdot g)(x) = f(x) \cdot g(x) = (x-4)(x^2-4) = x^3 - 4x^2 - 4x + 16$

d. $\left(\frac{f}{g}\right)(x) = \frac{f(x)}{g(x)} = \frac{x-4}{x^2-4}$

e. The domain of $\left(\frac{f}{g}\right)(x)$ must exclude values of x for which $x^2 - 4 = 0$. Thus, the domain consists of the set of all real numbers except 2 and −2. ■

Given $f(x) = 2x^2$ and $g(x) = x^2 - 5x + 6$, find the following:

a. $(f+g)(x)$ b. $(f-g)(x)$

c. $(f \cdot g)(x)$ d. $\left(\frac{f}{g}\right)(x)$

e. the domain of $\left(\frac{f}{g}\right)(x)$

Answers

a. $3x^2 - 5x + 6$ b. $x^2 + 5x - 6$

c. $2x^4 - 10x^3 + 12x^2$ d. $\frac{2x^2}{x^2 - 5x + 6}$

e. The set of all real numbers except 2 and 3

3.5a *Composite Functions*

There is another important way in which two functions f and g can be combined to form a new function. In Figure 3-37, the function f takes the point x in set X and assigns it the value y in set Y. Then, the function g takes the point y in set Y and assigns it the value z in set Z. The net effect of this combination of f and g is a new function h, called the **composite function of *g* and *f***, denoted ***g* ∘ *f***. This function takes the point x in set X and assigns it the value z in set Z directly. We write the new function as

$$h(x) = (g \circ f)(x) = g[f(x)]$$

which is read "g of f of x." (The square bracket "[ ]" has been introduced in the notation regarding composite functions to form a more pronounced separation between g and f. We return to the use of parentheses "( )" when a "single substitution" is to take place.) Furthermore, if the notation $g[f(x)]$ is to make sense, then the domain of $g[f(x)]$ must consist only of those x in X for which $y = f(x)$ and $g(y) = g[f(x)]$ are both defined.

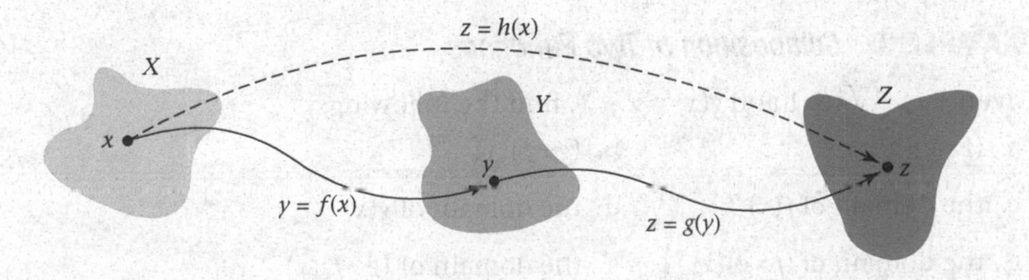

FIGURE 3-37
Correspondence for the Composition of Two Functions

EXAMPLE 2 *Evaluation of Composite Functions*

Given $f(x) = x^2$ and $g(x) = x - 1$, find the following:

a. $f[g(3)]$ b. $g[f(3)]$ c. $f[g(x)]$ d. $g[f(x)]$

SOLUTION

a. We begin by evaluating $g(3)$:

$$g(x) = x - 1$$
$$g(3) = 3 - 1 = 2$$

Therefore,

$$f[g(3)] = f(2)$$

Since

$$f(x) = x^2$$

then

$$f(2) = 2^2 = 4$$

Thus,

$$f[g(3)] = 4$$

b. Beginning with $f(3)$, we have

$$f(3) = 3^2 = 9$$

Then we find by substituting $f(3) = 9$ that

$$g[f(3)] = g(9) = 9 - 1 = 8$$

c. Since $g(x) = x - 1$, we make the substitution

$$f[g(x)] = f(x - 1) = (x - 1)^2 = x^2 - 2x + 1$$

d. Since $f(x) = x^2$, we make the substitution

$$g[f(x)] = g(x^2) = x^2 - 1$$

Note that in general, $f[g(x)] \neq g[f(x)]$, or equivalently $(f \circ g)(x) \neq (g \circ f)(x)$. ■

EXAMPLE 3 *Composition of Two Functions*

Given $f(x) = \sqrt{x} + 1$ and $g(x) = x - 1$, find the following:

a. $(f \circ g)(x)$ b. $(g \circ f)(x)$

c. the domain of $f(x)$ d. the domain of $g(x)$

e. the domain of $(f \circ g)(x)$ f. the domain of $(g \circ f)(x)$

SOLUTION

a. $(f \circ g)(x) = f[g(x)] = f(x - 1) = \sqrt{x - 1} + 1$

b. $(g \circ f)(x) = g[f(x)] = g(\sqrt{x} + 1) = (\sqrt{x} + 1) - 1 = \sqrt{x}$

c. If $f(x) = \sqrt{x} + 1$ is to be a real number, the domain is $\{x \mid x \geq 0\}$.

d. The domain of $g(x) = x - 1$ is the set of all real numbers.

e. If $(f \circ g)(x) = \sqrt{x-1} + 1$ is to be a real number, then we must have $x - 1 \geq 0$, so its domain is $\{x \mid x \geq 1\}$.

f. If $(g \circ f)(x) = \sqrt{x}$ is to be a real number, then its domain is $\{x \mid x \geq 0\}$.

(Verify that the only value for which $(f \circ g)(x) = (g \circ f)(x)$ in this example is $x = 1$.) ■

Progress Check

Given $f(x) = x^2 - 2x$ and $g(x) = 3x$, find the following:

a. $f[g(-1)]$ b. $g[f(-1)]$ c. $f[g(x)]$

d. $g[f(x)]$ e. $(f \circ g)(2)$ f. $(g \circ f)(2)$

Answers

a. 15 b. 9 c. $9x^2 - 6x$

d. $3x^2 - 6x$ e. 24 f. 0

EXAMPLE 4 ***Composition of Two Functions***

Find two functions $f(x)$ and $g(x)$, such that $h(x) = (f \circ g)(x)$ where $h(x) = (5x - 2)^8$.

SOLUTION

One way to examine $h(x) = (5x - 2)^8$ is to focus on what happens inside the parentheses, namely, $5x - 2$. After we find the value of $5x - 2$, we raise that value to the eighth power. Thus, if we let $g(x) = 5x - 2$ and $f(x) = x^8$, we obtain

$$(f \circ g)(x) = f[g(x)] = f[5x - 2] = (5x - 2)^8 = h(x)$$ ■

Progress Check

Find two functions $f(x)$ and $g(x)$, such that $h(x) = (f \circ g)(x)$ where

$$h(x) = \frac{2}{7 - x}$$

Answer

$f(x) = \frac{2}{x}$, $g(x) = 7 - x$

3.5b One-to-One Functions

An element in the range of a function may correspond to more than one element in the domain of the function. In Figure 3-38, we see that y in Y corresponds to both x_1 and x_2 in X. If we demand that every element in the domain be assigned to a *different* element of the range, then the function is called **one-to-one**. More formally:

One-to-One Function
A function f is one-to-one if $f(a) = f(b)$ only when $a = b$.

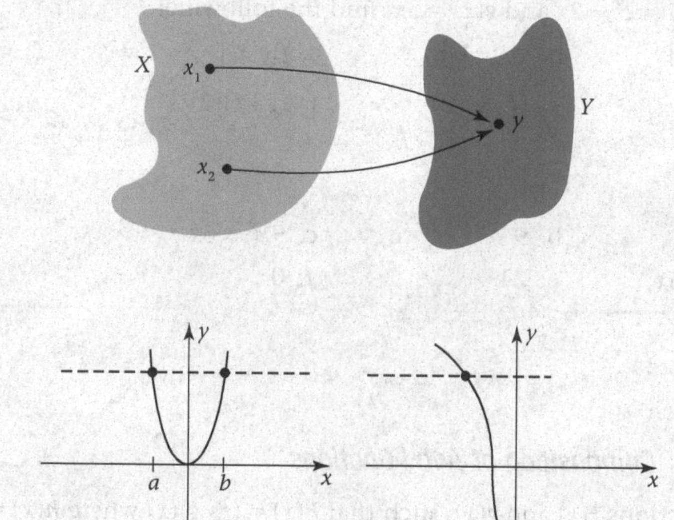

FIGURE 3-38
Correspondence of a Function That Is Not One-to-One

FIGURE 3-39
Horizontal Line Test: (a) Is Not One-to-One, (b) Is One-to-One

There is a simple means of determining if a function f is one-to-one by examining the graph of the function. In Figure 3-39(a), we see that a horizontal line meets the graph in more than one point. Thus, $f(a) = f(b)$ although $a \neq b$; hence the function is not one-to-one. On the other hand, no horizontal line meets the graph in Figure 3-39(b) in more than one point; the graph thus determines a one-to-one function. In summary, we have the following test.

Horizontal Line Test
If no horizontal line meets the graph of a function in more than one point, then the function is one-to-one.

EXAMPLE 5 ***Determining One-to-One Functions***

Which of the graphs below are graphs of one-to-one functions?

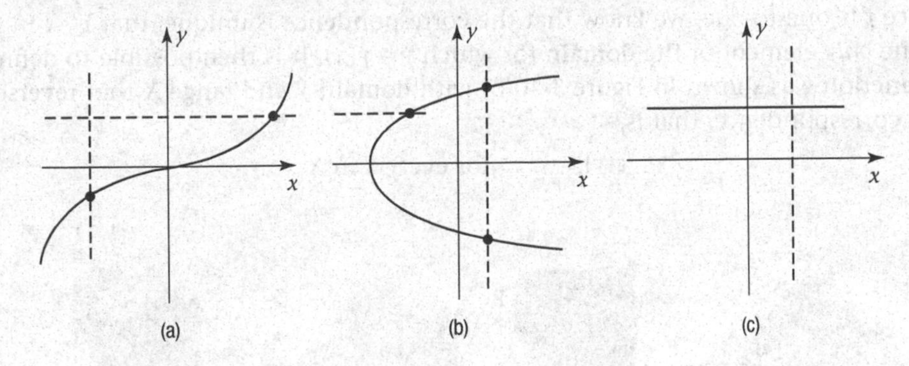

SOLUTION

a. No *vertical* line meets the graph in more than one point; hence, it is the graph of a function. No *horizontal* line meets the graph in more than one point; hence, it is the graph of a one-to-one function.

b. Although no *horizontal* line meets the graph in more than one point, at least one *vertical* line does meet the graph in more than one point. It is therefore not the graph of a function and consequently not the graph of a one-to-one function.

c. No *vertical* line meets the graph in more than one point; hence, it is the graph of a function. However, at least one *horizontal* line does meet the graph in more than one point. This is the graph of a function but not of a one-to-one function. ■

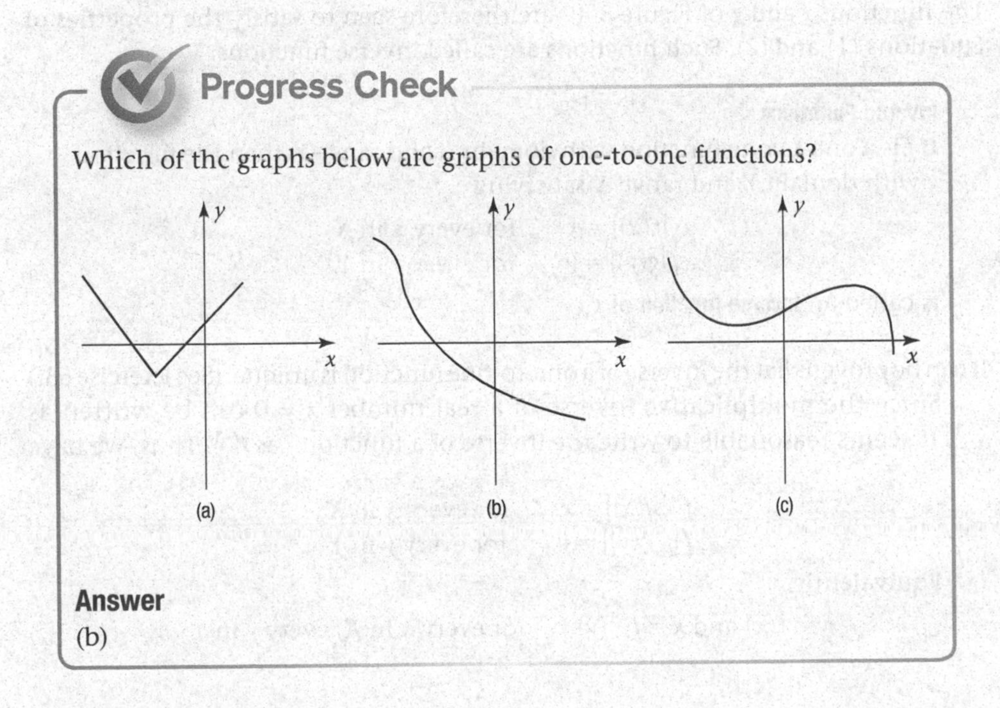

Progress Check

Which of the graphs below are graphs of one-to-one functions?

Answer

(b)

3.5c Inverse Functions

Suppose the function f in Figure 3-40(a) is a one-to-one function and that $y = f(x)$. Since f is one-to-one, we know that the correspondence is unique; that is, x in X is the *only* element of the domain for which $y = f(x)$. It is then possible to define a function g as shown in Figure 3-40(b) with domain Y and range X that reverses the correspondence, that is,

$$g(y) = x \qquad \text{for every } x \text{ in } X$$

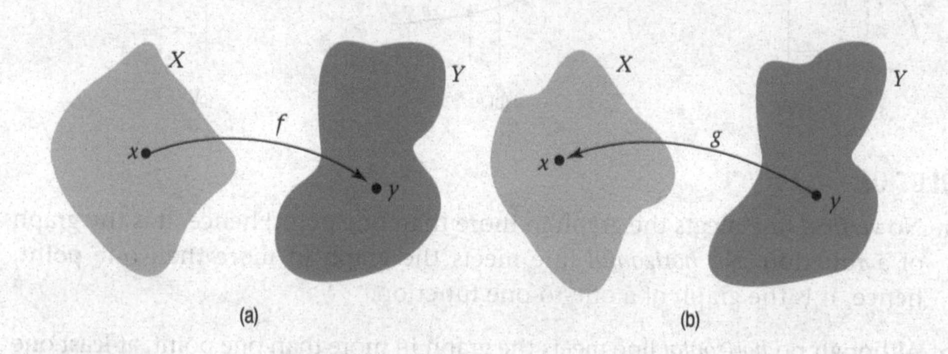

FIGURE 3-40
Correspondence of Inverse Functions

If we substitute $y = f(x)$, we have

$$g[f(x)] = x \qquad \text{for every } x \text{ in } X \tag{1}$$

Substituting $g(y) = x$ in the equation $f(x) = y$ yields

$$f[g(y)] = y \qquad \text{for every } y \text{ in } Y \tag{2}$$

The functions f and g of Figure 3-40 are therefore seen to satisfy the properties of Equations (1) and (2). Such functions are called inverse functions.

Inverse Functions

If f is a one-to-one function with domain X and range Y, then the function g with domain Y and range X satisfying

$$g[f(x)] = x \qquad \text{for every } x \text{ in } X$$
$$f[g(y)] = y \qquad \text{for every } y \text{ in } Y$$

is called an **inverse function** of f

It can be proven that the inverse of a one-to-one function is unique. (See Exercise 63.)

Since the multiplicative inverse of a real number $x \neq 0$ can be written as x^{-1}, it seems reasonable to write the inverse of a function f as f^{-1}. Thus, we have

$$f^{-1}[f(x)] = x \qquad \text{for every } x \text{ in } X$$
$$f[f^{-1}(y)] = y \qquad \text{for every } y \text{ in } Y$$

Equivalently,

$$y = f(x) \text{ and } x = f^{-1}(y) \qquad \text{for every } x \text{ in } X, \text{ every } y \text{ in } Y$$

Figure 3-41 is a graphical representation.

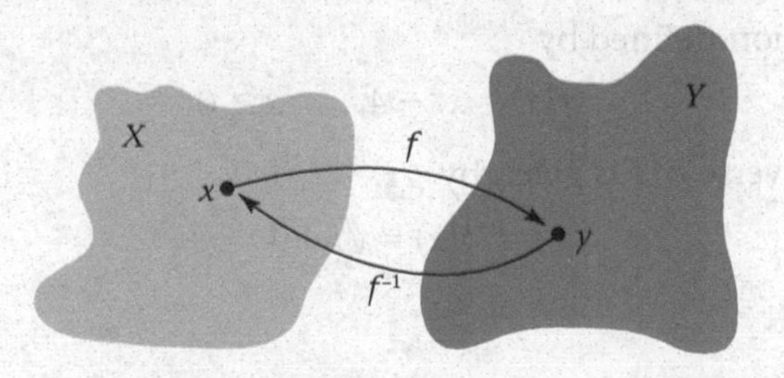

FIGURE 3-41
Correspondence for Inverse Functions and One-to-One Functions

We can define the inverse function of f only if f is one-to-one.

EXAMPLE 6 *Composition of Two Functions*

Given $f(x) = x + 5$ and $g(x) = x - 5$, find $(f \circ g)(x)$ and $(g \circ f)(x)$.

SOLUTION

$$(f \circ g)(x) = f[g(x)] = f[x - 5] = (x - 5) + 5 = x$$
$$(g \circ f)(x) = g[f(x)] = g[x + 5] = (x + 5) - 5 = x$$

■

In Example 6, we have shown that $f[g(x)] = x$ and $g[f(x)] = x$. Therefore, we may conclude that $f^{-1}(x) = x - 5$ and $g^{-1}(x) = x + 5$.

Progress Check

Given $f(x) = 5x$ and $g(x) = \frac{x}{5}$, find

a. $(f \circ g)(x)$ b. $(g \circ f)(x)$

c. Does $f^{-1}(x)$ exist? d. Does $g^{-1}(x)$ exist?

Answers

a. x b. x c. Yes d. Yes

EXAMPLE 7 *Verification of Inverse Functions*

Let f be the function defined by

$$f(x) = x^2 - 4, \qquad x \geq 0$$

Verify that the inverse of f is given by

$$f^{-1}(x) = \sqrt{x+4}$$

SOLUTION

We must verify that $f[f^{-1}(x)] = x$ and $f^{-1}[f(x)] = x$. Thus,

$$\begin{aligned} f[f^{-1}(x)] &= f(\sqrt{x+4}) \\ &= (\sqrt{x+4})^2 - 4 \\ &= x + 4 - 4 = x \end{aligned}$$

and

$$\begin{aligned} f^{-1}[f(x)] &= f^{-1}(x^2 - 4) \\ &= \sqrt{(x^2-4)+4} \\ &= \sqrt{x^2} = |x| \end{aligned}$$

Since $x \geq 0$,

$$f^{-1}[f(x)] = |x| = x$$

We have verified that the equations defining inverse functions hold, and we conclude that the inverse of f is as given. Verify the following: (a) the domain of f is the set of all real numbers in the interval $[0, \infty)$, and the range of f is the set of all real numbers in the interval $[-4, \infty)$. (b) The domain of f^{-1} is the range of f, and the range of f^{-1} is the domain of f. ■

It is sometimes possible to find an inverse by algebraic methods, as shown in the following example.

EXAMPLE 8 *Finding Inverse Functions*

Find the inverse function of $f(x) = 2x - 3$.

SOLUTION

If $f^{-1}(x)$ exists, then $y = f(x)$ implies that $x = f^{-1}(y)$. Thus,

$$y = f(x) = 2x - 3$$

and solving for x

$$\begin{aligned} 2x - 3 &= y \\ 2x &= y + 3 \\ x &= \frac{y+3}{2} \end{aligned}$$

But if $x = f^{-1}(y)$ and

$$x = \frac{y+3}{2}$$

then

$$f^{-1}(y) = \frac{y+3}{2}$$

We verify that

$$f^{-1}[f(x)] = f^{-1}(2x-3) = \frac{(2x-3)+3}{2} = \frac{2x}{2} = x$$

and

$$f[f^{-1}(x)] = f\left[\frac{x+3}{2}\right] = 2\left(\frac{x+3}{2}\right) - 3 = x + 3 - 3 = x$$ ■

In Table 3-4, we summarize a method for finding the inverse of a function when that inverse exists. We illustrate this method using an example to find the inverse of $f(x) = x^5 + 4$.

There are some inverses that cannot be found using the method from Table 3-4. Sometimes, the inverse must be "created." (See Section 6.3.)

TABLE 3-4 To Find the Inverse of a Function

Method	Example
Step 1. Given the function $f(x)$, set $y = f(x)$.	*Step 1.* $y = f(x) = x^5 + 4$.
Step 2. If possible, solve for x in terms of y, that is $x = g(y)$.	*Step 2.* $x = \sqrt[5]{y-4} = g(y)$.
Step 3. The function g is the inverse of f. Using the same variable we write $g(x) = f^{-1}(x)$.	*Step 3.* $g(y) = f^{-1}(y) = \sqrt[5]{y-4}$ Therefore, $f^{-1}(x) = \sqrt[5]{y-4}$

The inverse of a function that is not one-to-one does not exist. However, by restricting the domain of that function to where it is one-to-one, we are able to find the inverse of the restricted function

Consider the function $f(x)$ defined in the table below.

x	$f(x)$
1	4
2	5
3	4

Since $f(1) = 4$ and $f(3) = 4$ this function is not one-to-one. Let $g(x)$ denote the restriction of $f(x)$ to the domain $\{1, 2\}$.

x	$g(x)$
1	4
2	5

The function $g(x)$ is one-to-one, hence its inverse $g^{-1}(x)$ exists.

x	$g^{-1}(x)$
4	1
5	2

Consider $h(x) = x^2 - 4$. This function is not one-to-one since $h(2) = 0$ and $h(-2) = 0$. However, let us call $f(x)$ the restriction of $h(x)$ to $x \geq 0$, that is,

$$f(x) = x^2 - 4, \qquad x \geq 0$$

In Example 7, we verified that $f^{-1}(x) = \sqrt{x+4}$. Similarly, let us call $g(x)$ the restriction of $h(x)$ to $x \leq 0$, that is,

$$g(x) = x^2 - 4, \qquad x \leq 0$$

Verify that $g^{-1}(x) = -\sqrt{x+4}$. We show the graphs of $f(x)$, $g(x)$, and $h(x)$ in Figure 3-42.

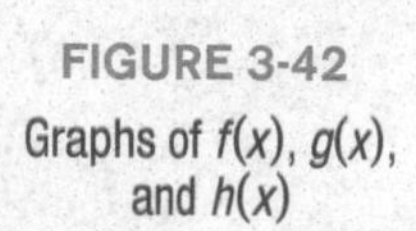

FIGURE 3-42

Graphs of $f(x)$, $g(x)$, and $h(x)$

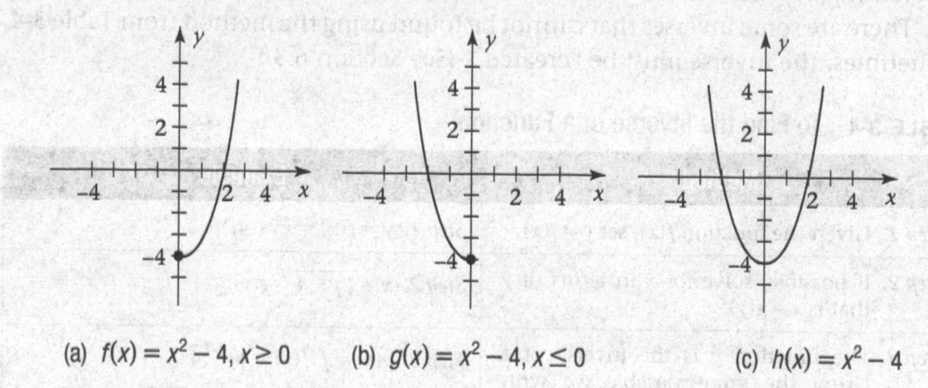

(a) $f(x) = x^2 - 4, x \geq 0$ (b) $g(x) = x^2 - 4, x \leq 0$ (c) $h(x) = x^2 - 4$

Progress Check

Given $f(x) = 3x + 5$, find f^{-1}

Answer

$f^{-1}(x) = \dfrac{x-5}{3}$

We may also think of the function f defined by $y = f(x)$ as the set of all ordered pairs $(x, f(x))$, where x assumes all values in the *domain* of f. Since the inverse function reverses the correspondence, the function f^{-1} is the set of all ordered pairs $(f(x), x)$, where $f(x)$ assumes all values in the *range* of f. With this approach, we see that the graphs of inverse functions are related in a specific manner. First, note that the points (a, b) and (b, a) in Figure 3-43(a) are located **symmetrically with respect to the graph of the line $y = x$.** That is, if we fold the paper along the line $y = x$, the two points will coincide. Furthermore, if (a, b) lies on the graph of the function f, then (b, a) must lie on the graph f^{-1}. Therefore,

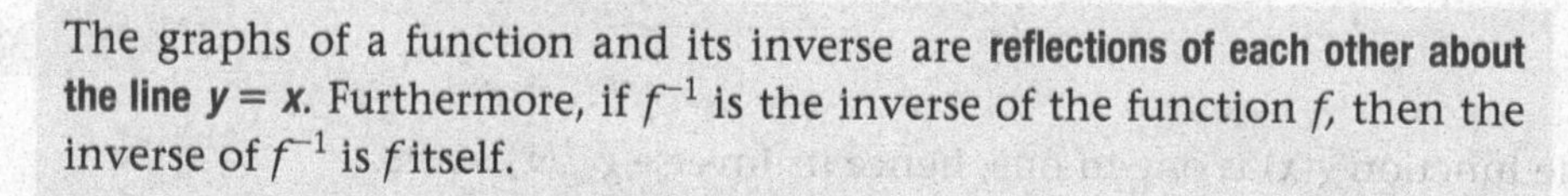

The graphs of a function and its inverse are **reflections of each other about the line $y = x$.** Furthermore, if f^{-1} is the inverse of the function f, then the inverse of f^{-1} is f itself.

In Figure 3-43(b), we have sketched the graphs of the functions from Example 7 on the same coordinate axes to demonstrate this relationship.

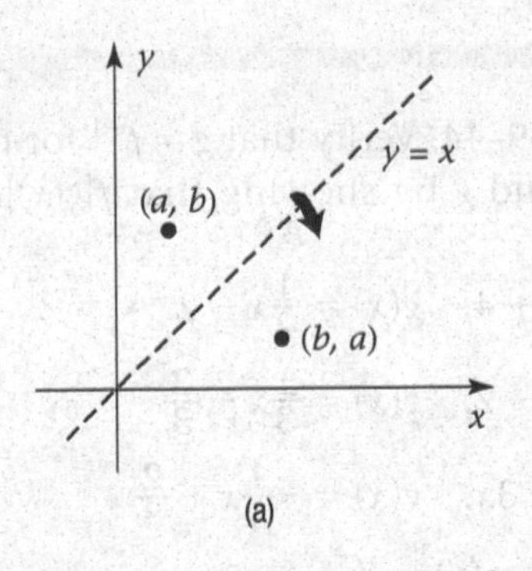

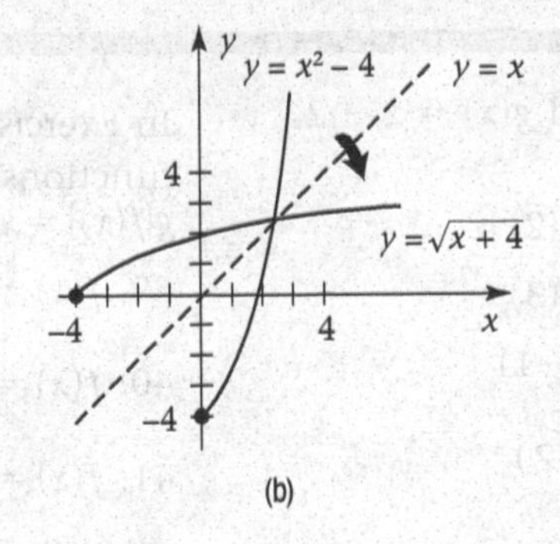

FIGURE 3-43

Graphs of Inverse Functions as Reflections About $y = x$

Although the notation f^{-1} for the inverse of f may have come from x^{-1}, the notation for the multiplicative inverse of x, it is important to avoid the possible confusion with the fact that

$$x^{-1} = \frac{1}{x}$$

a. In general,

$$f^{-1}(x) \neq \frac{1}{f(x)}$$

For example, if $g(x) = x - 1$, then

$$g^{-1}(x) \neq \frac{1}{x - 1}$$

In fact, using the methods of this section, we can show that

$$g^{-1}(x) = x + 1$$

b. The inverse function notation is *not* to be thought of as a power.

c. To avoid additional confusion, we may write

$$\frac{1}{f(x)} = [f(x)]^{-1}$$

Exercise Set 3.5

In Exercises 1–10, $f(x) = x^2 + 1$ and $g(x) = x - 2$. Determine the following:

1. $(f+g)(x)$
2. $(f+g)(2)$
3. $(f-g)(x)$
4. $(f-g)(3)$
5. $(f \cdot g)(x)$
6. $(f \cdot g)(-1)$
7. $\left(\frac{f}{g}\right)(x)$
8. $\left(\frac{f}{g}\right)(-2)$
9. the domains of f and of g
10. the domains of $\frac{f}{g}$ and of $\frac{g}{f}$

In Exercises 11–18, $f(x) = 2x + 1$ and $g(x) = 2x^2 + x$. Determine the following:

11. $(f \circ g)(x)$
12. $(g \circ f)(x)$
13. $(f \circ g)(2)$
14. $(g \circ f)(3)$
15. $(f \circ g)(x+1)$
16. $(f \circ f)(-2)$
17. $(g \circ f)(x-1)$
18. $(g \circ g)(x)$

In Exercises 19–24, $f(x) = x^2 + 4$ and $g(x) = \sqrt{x+2}$. Determine the following:

19. $(f \circ g)(x)$
20. $(g \circ f)(x)$
21. $(f \circ f)(-1)$
22. the domain of $(f \circ g)(x)$
23. the domain of $(g \circ f)(x)$
24. the domain of $(g \circ g)(x)$

In Exercises 25–28, determine $(f \circ g)(x)$ and $(g \circ f)(x)$.

25. $f(x) = x - 1, \quad g(x) = x + 2$
26. $f(x) = \sqrt{x+1}, \quad g(x) = x + 2$
27. $f(x) = \frac{1}{x+1}, \quad g(x) = \frac{1}{x-1}$
28. $f(x) = \frac{x+1}{x-1}, \quad g(x) = x$

In Exercises 29–38, write the given function $h(x)$ as a composite of two functions f and g so that $h(x) = (f \circ g)(x)$. (There may be more than one answer.)

29. $h(x) = x^2 + 3$
30. $h(x) = \frac{1}{x+2}$
31. $h(x) = (3x + 2)^8$
32. $h(x) = (x^3 + 2x^2 + 1)^{15}$
33. $h(x) = (x^3 - 2x^2)^{1/3}$
34. $h(x) = \left(\frac{x^2+2x}{x^3-1}\right)^{3/2}$
35. $h(x) = |x^2 - 4|$
36. $h(x) = |x^2 + x| - 4$
37. $h(x) = \sqrt{4-x}$
38. $h(x) = \sqrt{2x^2 - x + 2}$

In Exercises 39–44, verify that $g = f^{-1}$ for the given functions f and g by showing that $f[g(x)] = x$ and $g[f(x)] = x$.

39. $f(x) = 2x + 4, \quad g(x) = \frac{1}{2}x - 2$
40. $f(x) = 3x - 2, \quad g(x) = \frac{1}{3}x + \frac{2}{3}$
41. $f(x) = 2 - 3x, \quad g(x) = -\frac{1}{3}x + \frac{2}{3}$
42. $f(x) = x^3, \quad g(x) = \sqrt[3]{x}$
43. $f(x) = \frac{1}{x}, \quad g(x) = \frac{1}{x}$
44. $f(x) = \frac{1}{x-2}, \quad g(x) = \frac{1}{x} + 2$

In Exercises 45–52, find $f^{-1}(x)$. Sketch the graphs of $y = f(x)$ and $y = f^{-1}(x)$ on the same coordinate axes.

45. $f(x) = 2x + 3$
46. $f(x) = 3x - 4$
47. $f(x) = 3 - 2x$
48. $f(x) = \frac{1}{2}x + 1$
49. $f(x) = \frac{1}{3}x - 5$
50. $f(x) = 2 - \frac{1}{5}x$
51. $f(x) = x^3 + 1$
52. $f(x) = \frac{1}{x+1}$

53. Determine which functions are one-to-one.

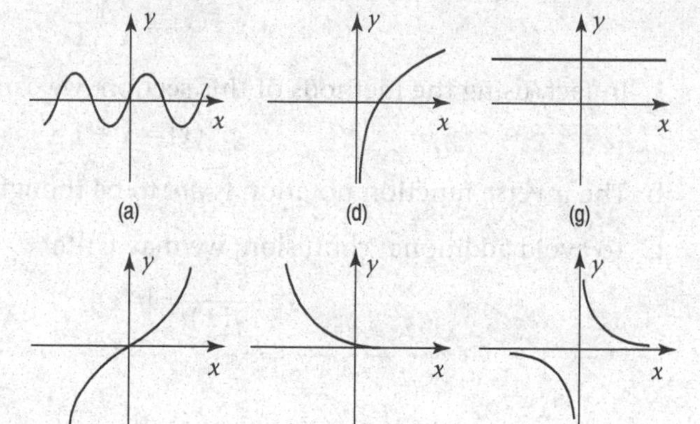

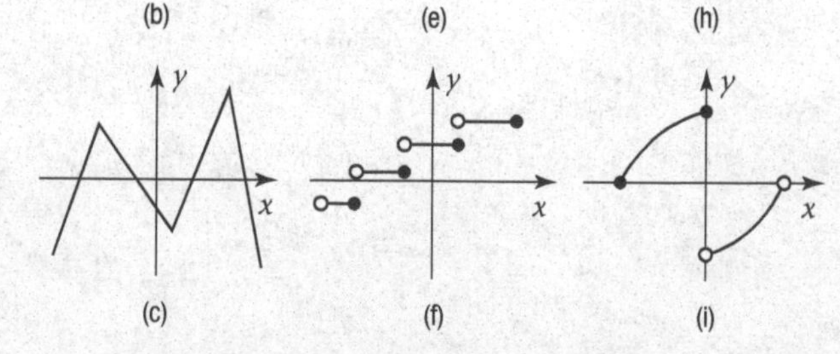

In Exercises 54–61, use the horizontal line test to determine whether the given function is a one-to-one function.

54. $f(x) = 2x - 1$
55. $f(x) = 3 - 5x$
56. $f(x) = x^2 - 2x + 1$
57. $f(x) = x^2 + 4x + 4$
58. $f(x) = -x^3 + 1$
59. $f(x) = x^3 - 2$

60. $f(x) = \begin{cases} 2x & \text{if} \quad x \le -1 \\ x^2 & \text{if} \quad -1 < x \le 0 \\ 3x - 1 & \text{if} \quad x > 0 \end{cases}$

61. $f(x) = \begin{cases} x^2 - 4x + 4 & \text{if} \quad x \le 2 \\ x & \text{if} \quad x > 1 \end{cases}$

62. Prove that a one-to-one function can have at most one inverse function. (*Hint*: Assume that the functions g and h are both inverses of the function f. Show that $g(x) = h(x)$ for all real values x in the range of f.)

63. Prove that the linear function $f(x) = ax + b$ is a one-to-one function if $a \neq 0$, and is not a one-to-one function if $a = 0$.

64. Find the inverse of the linear function $f(x) = ax + b$, $a \neq 0$.

65. Find two restrictions of the function $f(x) = |x + 3|$ that would allow it to have an inverse. Find the two inverses corresponding to these restrictions.

66. The boiling point of toluene is 230° on the Fahrenheit scale and 110° on the Celsius scale. The freezing point of toluene is −139°F and −95°C.

 a. Express degrees Celsius C as a function of degrees Fahrenheit F.

 b. Express degrees Fahrenheit F as a function of degrees Celsius C, by finding the inverse of the expression found in part (a).

67. The number of pairs q of sneakers manufactured in a day is a function of the number of working sewing machines m. A certain sneaker manufacturer finds the relationship to be as follows: $q = f(m) = 32m$. The total revenue r from selling these sneakers can be defined by: $r = g(q) = 30q$.

 a. Find $(g \circ f)(m)$.

 b. Interpret $(g \circ f)(m)$ in words.

68. The number of computers n a computer firm is willing to produce is a function of the price p_c it can get for selling each unit. This relationship is called the supply equation, and it is given as $n = f(p_c) = 20p_c - 5000$ for this computer manufacturer. The price per computer p_c is a function of the price of microchips p_m given as $p_c = g(p_m) = \frac{4000}{p_m}$.

 a. Find $(f \circ g)(p_m)$.

 b. Interpret $(f \circ g)(p_m)$ in words.

69. The Jazzy Jewel Company designs and manufactures gold earrings. The company's fixed costs, those independent of the level of production, are given by $C_F = f(p) = 12{,}500$. The variable costs, those directly related to the level of production, are a function of the price p of gold, and are given by $C_V = g(p) = 12p$.

 a. Find $(f + g)(p)$.

 b. Interpret $(f + g)(p)$ in words.

70. The Pirates Baseball Club Store's total costs C_T is a function of the number of items n stocked in the store, and is given by $C_T = f(n) = 5500 + 2.5n$. The revenue R earned by the Pirates Store is given by $R = g(n) = 22.5n$.

 a. Find $(g - f)(n)$.

 b. What relationship does $(g - f)(n)$ define?

 c. Find $(g - f)(500)$ and $(g - f)(250)$. Explain your results.

 d. Find the value of n that makes $(g - f)(n) = 0$.

 e. What information is gained by the Pirates Baseball Club Store by knowing the result of part (d)?

71. A ball is thrown off a cliff above a lake. At any time t in seconds, the distance d the ball is above the surface of the lake can be described by $r = f(t) = 112 + 96t - 16t^2$. The depth of the lake directly below the ball can be described by $d = g(t) = 1000 - 10t^2$.

 a. Find $(f + g)(t)$.

 b. Interpret $(f + g)(t)$.

 c. How many feet above the bottom of the lake is the ball after 4 seconds?

 d. When does the ball hit the water?

 e. How many feet above the bottom of the lake is the top of the cliff?

72. Refer to the function you defined in Section 3.2, Exercise 50.

 a. Find the inverse of this function.

 b. *Mathematics in Writing*: In a short paragraph, explain in your own words what the inverse you found in part enables you to do. How is its use different from that of the original function?

3.6 Direct and Inverse Variation

3.6a *Direct Variation*

There are two functional relationships that occur so frequently that they are given distinct names: direct and inverse variation. We say that two positive quantities *vary directly* if an increase in one causes a proportional increase in the other. In the table

x	1	2	3	4
y	3	6	9	12

we see that an increase in x causes a proportional increase in y. If we look at the ratios $\frac{y}{x}$, we see that

$$\frac{y}{x} = \frac{3}{1} = \frac{6}{2} = \frac{9}{3} = \frac{12}{4} = 3$$

or $y = 3x$. The ratio $\frac{y}{x}$ remains constant for all values of $x \neq 0$ and $y \neq 0$. This is an example of the principle of **direct variation**. This principle may be extended to include all real numbers through the following definition.

Principle of Direct Variation

If y varies directly as x, then $y = kx$ for some constant k.

As another example, when we say that y varies directly as the square of x, we mean that $y = kx^2$ for some constant k. The constant k is called the **constant of variation.**

EXAMPLE 1 ***Direct Variation***

Suppose that y varies directly as the cube of x and that $y = 24$ when $x = -2$. Write the appropriate equation, solve for the constant of variation k, and use this k to relate the variables.

SOLUTION

From the principle of direct variation, we know that the functional relationship is

$$y = kx^3 \quad \text{for some constant } k.$$

Substituting the values $y = 24$ and $x = -2$, we have

$$24 = k(-2)^3 = -8k$$
$$k = -3$$

Thus,

$$y = -3x^3$$

■

Progress Check

a. If P varies directly as the square of V, and $P = 64$ when $V = 16$, find the constant of variation.

b. The circumference C of a circle varies directly as the radius r. If $C = 25.13$ when $r = 4$, express C as a function of r, that is, use the constant of variation to relate the variables C and r.

Answers

a. $\frac{1}{4}$ b. $C = 6.2825r$

3.6b *Inverse Variation*

Two positive quantities are said to *vary inversely* if an increase in one causes a proportional decrease in the other. In the table

x	1	2	3	4
y	24	12	8	6

we see that an increase in x causes a proportional decrease in y. If we look at the product xy, we see that

$$xy = 1 \cdot 24 = 2 \cdot 12 = 3 \cdot 8 = 4 \cdot 6 = 24$$

or

$$y = \frac{24}{x}$$

In general, the principle of **inverse variation** may be extended to include all real numbers as follows.

Principle of Inverse Variation
If y varies inversely as x, then $y = \frac{k}{x}$ for some constant k.

Once again, k is called the constant of variation.

EXAMPLE 2 ***Inverse Variation***

Suppose that y varies inversely as x^2 and that $y = 10$ when $x = 10$. Write the appropriate equation, solve for the constant of variation k and use this k to relate the variables.

SOLUTION

The functional relationship is

$$y = \frac{k}{x^2} \qquad \text{for some constant } k.$$

Substituting $y = 10$ and $x = 10$, we have

$$10 = \frac{k}{(10)^2} = \frac{k}{100}$$
$$k = 1000$$

Thus,

$$y = \frac{1000}{x^2}$$ ■

Progress Check

If v varies inversely as the cube of w, and $v = 2$ when $w = -2$, find the constant of variation.

Answer

-16

3.6c *Joint Variation*

An equation of variation can involve more than two variables. We say that a quantity **varies jointly** as two or more other quantities if it varies directly as their product.

EXAMPLE 3 ***Joint Variation***

Express the following statement as an equation: P varies jointly as R, S, and the square of T.

SOLUTION

Since P must vary directly as RST^2, we have $P = kRST^2$ for some constant k. ■

EXAMPLE 4 ***Joint Variation and Word Problems***

A snow removal firm finds that the annual profit P varies jointly as the number of available plows p and the square of the total inches of snowfall s, and inversely as the price per gallon of gasoline g. If the profit is \$15,000 when the snowfall is 6 inches, 5 plows are used, and the price of gasoline is \$1.50 per gallon, express the profit P as a function of s, p, and g.

SOLUTION

We are given that

$$P = k\frac{ps^2}{g}$$

for some constant k. To determine k, we substitute $P = 15{,}000$, $p = 5$, $s = 6$, and $g = 1.5$. Thus,

$$15{,}000 = k\frac{(5)(6)^2}{1.5} = 120k$$

$$k = \frac{15{,}000}{120} = 125$$

Thus

$$P = 125\frac{ps^2}{g}$$

■

Exercise Set 3.6

1. In the following table, y varies directly with x.

x	2	3	4	6	8	12	
y	8	12	16	24		80	120

   a. Find the constant of variation.

   b. Write an equation showing that y varies directly with x.

   c. Fill in the blanks of the table.

2. In the accompanying table, y varies inversely with x.

x	1	2	3	6	9	12	15	18		
y	6	3	2	1	$\frac{2}{3}$				$\frac{1}{4}$	$\frac{1}{10}$

   a. Find the constant of variation.

   b. Write an equation showing that y varies inversely with x.

   c. Fill in the blanks of the table.

3. If y varies directly as x, and $y = -\frac{1}{4}$ when $x = 8$,

   a. find the constant of variation;

   b. find y when $x = 12$.

4. If C varies directly as the square of s, and $C = 12$ when $s = 6$,

   a. find the constant of variation;

   b. find C when $s = 9$.

5. If s varies directly as the square of t, and $s = 10$ when $t = 10$,

   a. find the constant of variation;

   b. find s when $t = 5$.

6. If V varies directly as the cube of T, and $V = 16$ when $T = 4$,

   a. find the constant of variation;

   b. find V when $T = 6$.

7. If y varies inversely as x, and $y = -\frac{1}{2}$ when $x = 6$,

   a. find the constant of variation;

   b. find y when $x = 12$.

8. If V varies inversely as the square of p, and $V = \frac{2}{3}$ when $p = 6$,

   a. find the constant of variation;

   b. find V when $p = 8$.

9. If K varies inversely as the cube of r, and $K = 8$ when $r = 4$,

   a. find the constant of variation;

   b. find K when $r = 5$.

10. If T varies inversely as the cube of u, and $T = 2$ when $u = 2$,

    a. find the constant of variation;

    b. find T when $u = 5$.

11. If M varies directly as the square of r and inversely as the square of s, and $M = 4$ when $r = 4$ and $s = 2$,

    a. write the appropriate equation relating M, r, and s;

    b. find M when $r = 6$ and $s = 5$.

12. If f varies jointly as u and v, and $f = 36$ when $u = 3$ and $v = 4$,

    a. write the appropriate equation connecting f, u, and v;

    b. find f when $u = 5$ and $v = 2$.

13. If T varies jointly as p and the cube of v, and inversely as the square of u, and $T = 24$ when $p = 3$, $v = 2$, and $u = 4$,

    a. write the appropriate equation connecting T, p, v, and u;

    b. find T when $p = 2$, $v = 3$, and $u = 36$.

14. If A varies jointly as the square of b and the square of c, and inversely as the cube of d, and $A = 18$ when $b = 4$, $c = 3$, and $d = 2$,

    a. write the appropriate equation relating A, b, c, and d;

    b. find A when $b = 9$, $c = 4$, and $d = 3$.

15. The distance s an object falls from rest in t seconds varies directly as the square of t. If an object falls 144 feet in 3 seconds,

    a. how far does it fall in 5 seconds?

    b. how long does it take to fall 784 feet?

16. In a certain state the income tax paid by a person varies directly as the income. If the tax is \$20 per month when the monthly income is \$1600, find the tax due when the monthly income is \$900.

17. The resistance R of a conductor varies inversely as the area A of its cross section. If $R = 20$ ohms when $A = 8$ square centimeters, find R when $A = 12$ square centimeters.

18. The pressure P of a certain enclosed gas varies directly as the temperature T and inversely as the volume V. Suppose that 300 cubic feet of gas exert a pressure of 20 pounds per square foot when the temperature is 500 K (absolute temperature measured on the Kelvin scale). What is the pressure of this gas when the temperature is lowered to 400 K and the volume is increased to 500 cubic feet?

19. The intensity of illumination I from a source of light varies inversely as the square of the distance d from the source. If the intensity is 200 candlepower when the source is 4 feet away,

    a. what is the intensity when the source is 6 feet away?

    b. how close should the source be to provide an intensity of 50 candlepower?

20. The weight of a body in space varies inversely as the square of its distance from the center of the earth. If a body weighs 400 pounds on the surface of the earth, how much does it weigh 1000 miles from the surface of the earth? Assume that the radius of the earth is 4000 miles.

21. The equipment cost of a printing job varies jointly as the number of presses and the number of hours that the presses are run. When 4 presses are run for 6 hours, the equipment cost is \$1200. If the equipment cost for 12 hours of running is \$3600, how many presses are being used?

22. The current I in a wire varies directly as the electromotive force E and inversely as the resistance R. In a wire whose resistance is 10 ohms, a current of 36 amperes is obtained when the electromotive force is 120 volts. Find the current produced when $E = 220$ volts and $R = 30$ ohms.

23. The illumination from a light source varies directly as the intensity of the source and inversely as the square of the distance from the source. If the illumination is 50 candlepower per square foot on a screen 2 feet away from a light source whose intensity is 400 candlepower, what is the illumination 4 feet away from a source whose intensity is 3840 candlepower?

24. If f varies directly as u and inversely as the square of v, what happens to f if both u and v are doubled?

25. The intensity of light from a point source varies inversely with the square of the distance from the source.

    a. Express the equation of intensity I as a function of distance d.

    b. If the intensity 5.5 meters from the source is 3.80 watts per square meter, what is the intensity 4 meters from the source?

26. Hooke's Law says that the distance a spring is stretched varies directly as the weight of the object attached to the spring. A 5-kilogram weight stretches a spring 35 centimeters.

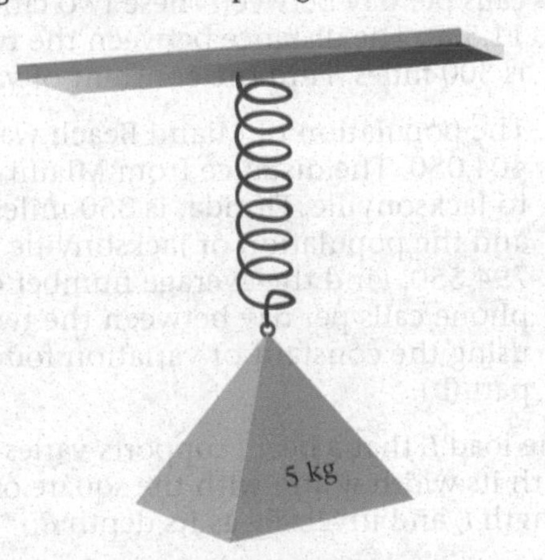

    a. Find the constant of variation.

    b. If a 3-kilogram weight is attached to the spring, how far will the spring be stretched?

27. Some economists theorize that the Dow Jones Industrial Average (an index of stock prices) varies inversely as the price of oil. The Dow Jones Average was 2520 when the price of oil was \$18.50 per barrel. After Iraq invaded Kuwait in 1990, the price of oil increased by \$13.50 per barrel. What should the Dow Jones Industrial Average be if this theory is correct?

28. The amount of pollution emitted into the air in a particular city varies directly with the number of cars driven in that city. If 35,000 cars cause 1988 tons of pollutants, how many cars would cause 2644.04 tons of pollutants to enter the atmosphere?

29. The volume V of gas varies directly as the temperature T and inversely as the pressure P. The volume of a particular gas is 245 cubic centimeters when the temperature is 56˚F and the pressure is 25 kilograms per cubic centimeter.

    a. Find the pressure of the gas when the volume is 210 cubic centimeters and the temperature is 45˚F.

    b. Find the temperature of the gas when the pressure is 40 kilograms per cubic centimeter and the volume is 305 cubic centimeters.

30. Sociologists developed a model to determine the average number of telephone calls per day between two cities. The model states that the average number of phone calls varies jointly with the population P_1 in one city and the population P_2 in the other city, and inversely as the square of the distance between the two cities.

    a. Write the equation of variation.

    b. When the population of Philadelphia was 1,488,400 and the population of Pittsburgh was 312,810, the average number of phone calls per day between these two cities was 11,500. The distance between the two cities is 300 miles. Find the constant of variation.

    c. The population of Miami Beach was 404,050. The distance from Miami Beach to Jacksonville, Florida, is 350 miles and the population of Jacksonville was 794,550. Find the average number of phone calls per day between the two cities, using the constant of variation found for part (b).

31. The load L that a beam supports varies jointly with its width w and with the square of its length l, and inversely as its depth d.

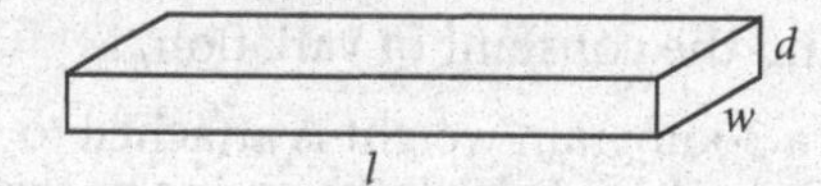

    a. Write the equation of variation.

    b. A 200 pound weight hangs safely from the center of a beam with dimensions 5 feet long by 3 inches wide by 4 inches deep. Find the constant of variation if all dimensions are measured in feet.

    c. What size weight would hang safely from a beam with dimensions 5 feet long by 6 inches wide by 2 inches deep?

    d. What is the effect on the load the beam can support if the width is doubled and the depth is halved?

    e. What is the effect on the load the beam can support if the length is doubled?

32. The intelligence quotient (IQ) varies directly as a person's mental age and varies inversely as a person's chronological age.

    a. Write the equation of variation.

    b. A child with 125 IQ has a mental age of 15 and a chronological age of 12. Find the constant of variation.

    c. If a 14-year-old child has an IQ of 135, what is the child's mental age?

Chapter Summary

Key Terms, Concepts, and Symbols

Key Ideas for Review

Topic	Page	Key Idea
Rectangular Coordinate System	140	In a rectangular coordinate system, every ordered pair of real numbers (a, b) corresponds to a point in the plane, and every point in the plane corresponds to an ordered pair of real numbers.
Distance Formula	142	The distance $\overline{PQ}$ between points $P(x_1, y_1)$ and $Q(x_2, y_2)$ is given by the distance formula $\overline{PQ} = \sqrt{(x_2 - x_1)^2 + (y_2 - y_1)^2}$
Midpoint Formula	144	The coordinates (x_m, y_m) of the midpoint M of the line segment with endpoints $P(x_1, y_1)$ and $Q(x_2, y_2)$ are given by $x_m = \frac{x_1 + x_2}{2}$ $y_m = \frac{y_1 + y_2}{2}$
Graphs of Equations	146	An equation in two variables can be graphed by plotting points that satisfy the equation and joining the points to form a smooth curve
Functions	154	A function is a rule that assigns exactly one element y of a set Y to each element x of a set X. The domain is the set of inputs, and the range is the set of outputs.
Domain	156	The domain of a function is the set of all real numbers for which the function is defined. Beware of division by zero and of even roots of negative numbers.

continues

Topic	Page	Key Idea
Range	156	The range of a function is the set of all outputs corresponding to the domain of that function.
Various Definitions	170	A function may be defined by an equation. However, sometimes a function may be defined by a table, a chart, or several equations. Moreover, not every equation defines a function.
Substitution	158	Function notation gives the definition of the function and also the value or expression at which to evaluate the function. If the function f is defined by $f(x) = x^2 + 2x$, then the notation $f(3)$ denotes the result of replacing the independent variable x by 3 wherever it appears. $f(x) = x^2 + 2x$ $f(3) = 3^2 + 2(3) = 15$
Vertical Line Test	155	A graph represents a function if and only if no vertical line meets the graph in more than one point.
Graphs	163	To graph $f(x)$, just graph $y = f(x)$.
Vertical Shift	167	If $p > 0$, the graph of $y = f(x) + p$ shifts the graph of $y = f(x)$ up p units, and the graph of $y = f(x) - p$ shifts the graph of $y = f(x)$ down p units.
Horizontal Shift	167	If $p > 0$, the graph of $y = f(x - p)$ shifts the graph of $y = f(x)$ p units to the right, and the graph of $y = f(x + p)$ shifts the graph of $y = f(x)$ p units to the left.
Reflection	168	The graph of $y = -f(x)$ is the reflection about the x-axis of the graph $y = f(x)$.
Piecewise-Defined Function	170	The graph of a function can have holes or gaps, and can be defined in "pieces."
Increasing, Decreasing, Constant	172	As we move from left to right, the graph of an increasing function rises, whereas the graph of a decreasing function falls. The graph of a constant function neither rises nor falls; it is horizontal.
Polynomials	175	Polynomials in one variable are all functions and have "smooth" curves as their graphs.
Line	179	The graph of the linear function $f(x) = ax + b$ is a line.
Slope	179	Any two points on a line can be used to find its slope m: $m = \frac{y_2 - y_1}{x_2 - x_1}$
Rising and Falling	180	Positive slope indicates that a line is rising; negative slope indicates that a line is falling.
Horizontal and Vertical	184	The slope of a horizontal line is 0; the slope of a vertical line is undefined.
Point–Slope Form	182	The point–slope form of a line is $y - y_1 = m(x - x_1)$.
Slope–Intercept Form	183	The slope–intercept form of a line is $y = mx + b$.
Horizontal and Vertical Forms	184	The equation of the horizontal line through the point (a, b) is $y = b$; the equation of the vertical line through the point (a, b) is $x = a$.
Graphs	185	The graphs of the linear function $f(x) = ax + b$ and the general first-degree equation $Ax + By + C = 0$ are always lines if A and B are not both zero.
Parallel	185	Parallel lines have the same slope.
Perpendicular	185	The slopes of perpendicular lines are negative reciprocals of each other, with the exception of horizontal and vertical lines.
Algebra of Functions	192	Functions can be combined by the usual operations of addition, subtraction, multiplication and division.
Composition	193	A composite function is a function of a function.

continues

Topic	Page	Key Idea
One-to-One	196	We say a function is one-to-one if every element of the range corresponds to precisely one element of the domain.
Horizontal Line Test	196	No horizontal line meets the graph of a one-to-one function in more than one point.
Inverse Function	198	The inverse f^{-1} of a function reverses the correspondence defined by the function f. The domain of f becomes the range of f^{-1}, and the range of f becomes the domain of f^{-1}.
Properties	198	A function f and its inverse f^{-1} satisfy $f^{-1}[f(x)] = x$ for all x in the domain of f; $f[f^{-1}(y)] = y$ for all y in the range of f
One-to-One	199	The inverse of a function f is defined only if f is one-to-one.
Graph	202	The graphs of a function and its inverse are reflections of one another about the line $y = x$.
Variation	206	Direct and inverse variation are functional relationships.
Direct Variation	206	We say that y varies directly as x if $y = kx$ for some constant k.
Inverse Variation	207	We say that y varies inversely as x if $y = \frac{k}{x}$ for some constant k.
Joint Variation	208	Joint variation is a term for direct variation involving more than two quantities.

Review Exercises

1. Find the distance between the points $(-4, -6)$ and $(2, -1)$.

2. Find the length of the longest side of the triangle whose vertices are $A(3, -4)$, $B(-2, -6)$, and $C(-1, 2)$.

In Exercises 3 and 4, sketch the graph of the given equation by forming a table of values.

3. $y = 1 - |x|$
4. $y = \sqrt{x-2}$

In Exercises 5 and 6, analyze the given equation for symmetry with respect to the x-axis, y-axis, and origin.

5. $y^2 = 1 - x^3$
6. $y^2 = \dfrac{x^2}{x^2-5}$

In Exercises 7 and 8, state if the graph determines y as a function of x.

7.

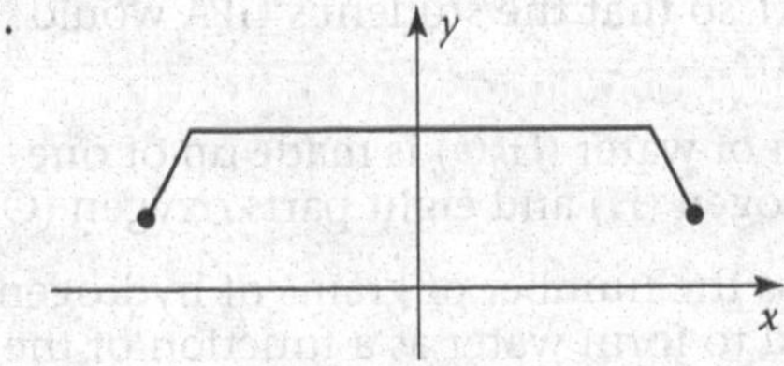

8.

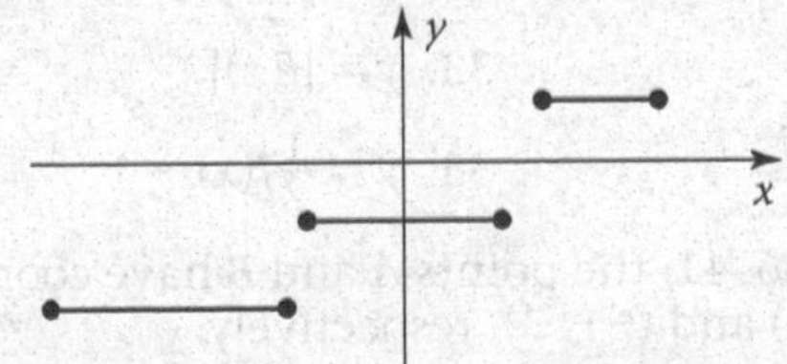

In Exercises 9 and 10 determine the domain of the given function.

9. $f(x) = \sqrt{3x-5}$
10. $f(x) = \dfrac{x}{x^2+2x+1}$

11. If $f(x) = \sqrt{x-1}$, find a real number whose image is 15.

12. If $f(t) = t^2 + 1$, find all real numbers whose image is 10.

In Exercises 13–22, sketch the graph of each function.

13. $y = 5x - 4$
14. $y = 3x^3 + 2$
15. $y = x - x^2$
16. $y = |x - x^2|$
17. $y = x - 3$
18. $y = |x - 3|$
19. $y = \dfrac{|x-3|}{x-3}$
20. $y = 2\sqrt{x} + 7$
21. $y = 2\sqrt{x+7}$
22. $y = \dfrac{5}{x^2+1}$

In Exercises 23–25, $f(x) = x^2 - x$. Evaluate the following:

23. $f(-3)$ 24. $f(y-1)$

25. $\dfrac{f(2+h)-f(2)}{h}$

Exercises 26–29 refer to the function f defined by

$$f(x) = \begin{cases} x-1 & \text{if} \quad x \le -1 \\ x^2 & \text{if} \quad -1 < x \le 2 \\ -2 & \text{if} \quad x > 2 \end{cases}$$

26. Sketch the graph of the function f.

27. Determine where the function f is increasing, decreasing, and constant.

28. Evaluate $f(-4)$. 29. Evaluate $f(4)$.

In Exercises 30–35, sketch the indicated graph, where the graph of $y = f(x)$ is

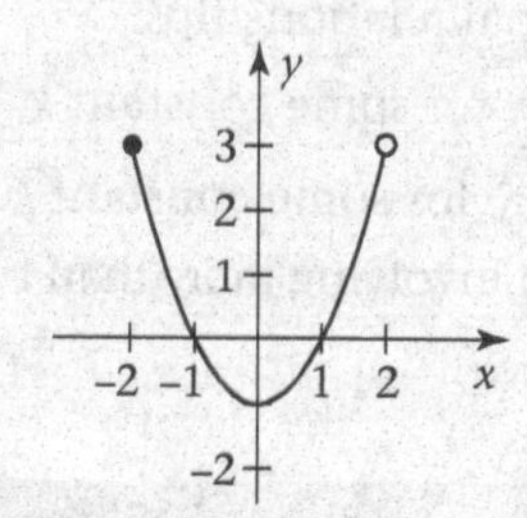

30. $y = f(x) + 3$ 31. $y = f(x+1)$

32. $y = -f(x)$ 33. $y = |f(x)|$

34. $y = 2f(x)$ 35. $y = \frac{1}{2}f(x) - 1$

In Exercises 36–41, the points A and B have coordinates $(-4, -6)$ and $(-1, 3)$, respectively.

36. Find the slope of the line through A and B.

37. Find an equation of the line through the points A and B.

38. Find an equation of the line through A that is parallel to the y-axis.

39. Find an equation of the horizontal line through B.

40. Find an equation of the line through A that is parallel to the line $4x - y - 3 = 0$.

41. Find an equation of the line through B that is perpendicular to the line $2y + x - 5 = 0$.

In Exercises 42–47, $f(x) = x + 1$ and $g(x) = x^2 - 1$. Determine the following:

42. $(f+g)(x)$ 43. $(f \cdot g)(-1)$

44. $\left(\dfrac{f}{g}\right)(x)$

45. the domain of $\left(\dfrac{f}{g}\right)(x)$

46. $(g \circ f)(x)$ 47. $(f \circ g)(2)$

In Exercises 48–51, $f(x) = \sqrt{x} - 2$ and $g(x) = x^2$. Determine the following:

48. $(f \circ g)(x)$ 49. $(g \circ f)(x)$

50. $(f \circ g)(-2)$ 51. $(g \circ f)(-2)$

In Exercises 52 and 53, $f(x) = 2x + 4$ and $g(x) = \frac{x}{2} - 2$.

52. Prove that f and g are inverse functions of each other.

53. Sketch the graphs of $y = f(x)$ and $y = g(x)$ on the same coordinate axes.

54. If R varies directly as q, and if $R = 20$ when $q = 5$, find R when $q = 40$.

55. If S varies inversely as the cube of t, and if $S = 8$ when $t = -1$, find S with $t = -2$.

56. P varies jointly as q and r and inversely as the square of t, and $P = -3$ when $q = 2$, $r = -3$ and $t = 4$. Find P when $q = -1$, $r = \frac{1}{2}$ and $t = 4$.

57. Jim has $10,000 to invest. He invests x dollars in tax-free municipal bonds yielding 8.5%. He invests 3 times the amount he invested in the bonds in certificates of deposit earning 9%. Jim puts the remainder of the money in a savings account at an interest rate of 5.25%.
    a. Express Jim's earned interest as a function of x.
    b. How much total interest does Jim earn if he invests $3000 in certificates of deposit?

58. A well-known college uses students' SAT scores to predict their grade point averages (GPA) at the end of their first year of college. The following table gives some representative SAT scores and GPAs from the linear model the college uses.

SAT	950	975	1050
GPA	2.00	2.10	2.40

    a. Express the GPA as a function of SAT scores.
    b. Find the inverse function of part (a).
    c. What score would a student have to get on the SAT so that the student's GPA would be 4.00?

59. One gram of water (H_2O) is made up of one part hydrogen (H) and eight parts oxygen (O).
    a. Express the number of grams of hydrogen needed to form water as a function of the number of grams w of water.
    b. Express the number of grams of oxygen needed to form water as a function of the number of grams w of water.
    c. How much hydrogen and how much oxygen are needed to form 315 grams of water?

60. Ohm's Law states that the resistance R in ohms of a conductor varies directly as the voltage V in volts and inversely as the current I in amperes. Electrical power P in watts varies jointly as the current I and the voltage V. Assume the constants of variation are equal to one.

    a. 120 volts operates a 1000-watt hairdryer. How much current does it draw and what is its resistance?

    b. In the United States the standard voltage is 120 volts, whereas in European countries the standard is 240 volts. Hairdryers produced in the United States now have a switch permitting the use of this appliance in foreign countries. How much current does the hairdryer in part (a) draw in England? (b) What is its resistance?

    c. The resistance of a light bulb is 240 ohms. The standard voltage required to light the light bulb is 120 volts. How much current does it draw and what is the power of the light bulb?

Review Test

1. Find the perimeter of the triangle whose vertices are $(2, 5)$, $(-3, 1)$ and $(-3, 4)$.
2. Use symmetry to assist in sketching the graph of the equation $y = 2x^2 - 1$.
3. Analyze the equation $y = \frac{1}{x^3}$ for symmetry with respect to the axes and origin.
4. Determine the domain of the function
$$f(x) = \frac{1}{\sqrt{x-1}}$$
5. If $f(x) = \sqrt{x-1}$, find a real number whose image is 4.
6. If $f(x) = 2x^2 + 3$, find $f(2t)$.

Exercises 7–10 refer to the function f defined by

$$f(x) = \begin{cases} 0 & \text{if} \quad x < -2 \\ |x| & \text{if} \quad -2 \le x \le 3 \\ x^2 - x & \text{if} \quad x > 3 \end{cases}$$

7. Determine where the function f is increasing, decreasing, and constant.
8. Evaluate $f(-5)$.
9. Evaluate $f(-2)$.
10. Sketch the graph of $y = \frac{1}{2}f(x + 1)$.
11. Find an equation of the line through the points $(-3, 5)$ and $(-5, 2)$.
12. Find an equation of the vertical line through the point $(-3, 4)$.
13. Find the slope m and y-intercept b of the line whose equation is $2y - x = 4$.
14. Find an equation of the line through the point $(4, -1)$ that is parallel to the x-axis.
15. Find an equation of the line that passes through the point $(-2, 3)$ and is perpendicular to the line $y - 3x - 2 = 0$.

In Exercises 16–18, $f(x) = \frac{1}{x-1}$ and $g(x) = x^2$. Find the following:

16. $(f - g)(2)$
17. $\left(\frac{g}{f}\right)(x)$
18. $(g \circ f)(3)$
19. Prove that $f(x) = -3x + 1$ and $g(x) = -\frac{1}{3}(x - 1)$ are inverse functions of each other.
20. If h varies directly as the cube of r, and $h = 2$ when $r = -\frac{1}{2}$, find h when $r = 4$.
21. T varies jointly as a and the square of b and inversely as the cube of c, and $T = 64$ when $a = -1$, $b = \frac{1}{2}$, and $c = 2$. Find T when $a = 2$, $b = 4$, and $c = -1$.

Cumulative Review Exercises: Chapters 1–3

In Exercises 1–5, determine if each expression is a polynomial.

1. $\sqrt{2}x^2 + 3x + \pi$
2. $2x^2 + \frac{3}{x} + 1$
3. $5x^2 + 3\sqrt{x} + 4$
4. $2x^2 + \frac{x}{3} + 1$
5. $-\frac{3}{4}$

In Exercises 6–8, simplify the given expressions.

6. $\sqrt[4]{a^8 b^6}$
7. $\frac{3}{7 - \sqrt{x}}$
8. $\left(\frac{a^2}{b^6}\right)^{1/2}\left(\frac{a^5}{b^{5/4}}\right)^{2/5}$

In Exercises 9 and 10, perform the indicated operations.

9. $\frac{3-x}{x^3+x} - \frac{x^2}{x^2+1}$
10. $\left(\frac{x^2-9}{2x^2+3x-2}\right)\left(\frac{2x-1}{3-x}\right)$

11. Solve for u and v: $(5 - u) + (7v + 2)i = 10 + 9i$.

In Exercises 12–14, perform the indicated operations.

12. $(3 - 2i)^2$
13. $(2 + 4i)(3 - 5i)$
14. $(2 + i)^4(3 - i)^2$

In Exercises 15–18, simplify the expression and write the answer using only positive exponents.

15. $x^{5/2}(x^{-2/3} - 1)$
16. $\sqrt{12} - \sqrt{75} + 2\sqrt{27}$
17. $\frac{2}{x^2 - x} - \frac{2x}{x - 1}$
18. $\frac{x^{-2} + 1}{x^{-1} - 3}$

In Exercises 19 and 20, factor completely.

19. $3x^3 + x^2 - 2x$
20. $(x - 1)^{2/3} + 2(x - 1)^{5/3}$
21. Rationalize the denominator:
$$\frac{2}{\sqrt{x} - \sqrt{2}}$$

In Exercises 22–25, solve the inequality.

22. $3x - 2 \le x + 3$
23. $|1 - 2h| > 2$
24. $2t^2 - 5t \ge 12$
25. $|x^2 - 3| \le 2$

In Exercises 26 and 27, find the domain of the function.

26. $f(x) = \frac{1}{\sqrt{2x - 1}}$
27. $f(x) = \frac{x}{x^2 - 1}$

28. Find the length of the hypotenuse of the right triangle whose vertices are (−2, 2), (4, 2), and (−2, −3).
29. Given $f(t) = 1 - t^2$:
    a. Find $f(2a - 1)$.
    b. Find all real numbers whose image is −15.
    c. Find $\frac{f(t+h) - f(t)}{h}$.
30. The function G is defined by
$$G(x) = \begin{cases} \frac{1}{2} & \text{if } x < -2 \\ x & \text{if } -2 \le x \le 2 \\ -x^2 & \text{if } x > 2 \end{cases}$$
    a. Find $G(0)$.
    b. Find $G(-3)$.
    c. Graph $G(x)$.
31. A chemist has several mixtures, each containing alcohol. She has x liters of mixture A with 5% alcohol, y liters of mixture B with 7% alcohol, and z liters of mixture C with $3\frac{1}{2}$% alcohol. Write an expression representing the total number of liters of alcohol in the three mixtures.

In Exercises 32–42, solve for x.

32. $2x^2 - x - 3 = 0$
33. $2x^2 - x + 19 = 15$
34. $\sqrt{x} + 12 = x$
35. $\frac{3 - 5x}{x + 2} = 4$
36. $6x^2 + 5x - 4 = 0$
37. $(3 - x)^2 - 16 = 0$
38. $x - \sqrt{-1 - 5x} = -3$
39. $-5 \le 1 - 4x \le 9$
40. $\left|\frac{x - 3}{2}\right| \ge 10$
41. $x^2 - 3x - 10 \le 0$
42. $\frac{3}{x - 2} \ge 1$

In Exercises 43 and 44, find the equation of the specified line.

43. The line is horizontal and passes through (2, −1).
44. The line is perpendicular to $x = 4y + 1$ and passes through $(\frac{1}{2}, 2)$.
45. Given the points $A(3, -2)$ and $B(-1, 2)$, find
    a. the slope of the line through the points A and B.
    b. an equation of the line through the points A and B.
    c. an equation of the line through B parallel to the y-axis.
    d. an equation of the line through the point $C(-4, -1)$ that is perpendicular to the line AB.
46. The hypotenuse of a right triangle has a length of 20 inches. If one leg of the triangle is 3 times the other leg, what is the length of the shortest side of the triangle?
47. Find the intercepts of the parabola $f(x) = -3x^2 - x + 2$.

48. Find the equation of the line whose x- and y-intercepts are both a.

49. Ron borrows $12,000 from his father (who does not charge interest) to buy a new car. He determines that if he increases his monthly payment by $50, he will have 8 fewer payments. What is Ron's new monthly payment, and how long will it take him to satisfy his debt?

50. Four towns are located as shown in the figure below. The lines connecting the points on the graph represent the roads between the four towns. Jack lives in Hometown and has to go to Workville everyday. He can choose either to go through town A, Route 1, or town B, Route 2. If he chooses to drive to Workville using Route 1, his average speed is 50 mph. His average speed using Route 2 is 30 mph. Which route should he choose so that he arrives at Workville in a minimum amount of time?

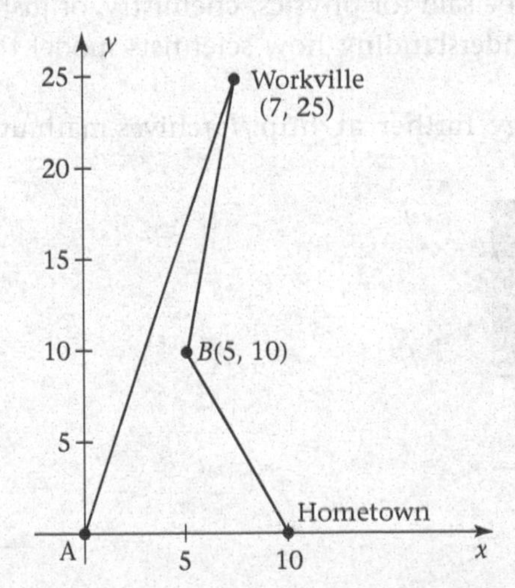

51. $P(x)$ represents the profit function of the Quark Textbook Company for sales of x copies of a book on advanced physics, where

$$P(x) = \begin{cases} \frac{1}{10}x^2 - 20x & \text{if} \quad 0 \leq x \leq 300 \\ -\frac{1}{10}x^2 + 120x - 24{,}000 & \text{if} \quad 300 < x \leq 1000 \end{cases}$$

a. What is the domain of $P(x)$?

b. Sketch the graph of $P(x)$.

c. Using the graph, determine:

i. the range of $P(x)$

ii. the number of books that maximizes $P(x)$

iii. the maximum profit

d. How many books does the Quark Textbook Company have to sell so that the company begins to make a profit?

52. The heat loss per hour through a glass window varies jointly as the area (A) of the window and the temperature difference (t_d) between the inside and outside temperatures, and varies inversely as the thickness (T) of the window.

a. A window measures 2 feet by 3 feet and is $\frac{1}{24}$ feet thick. The inside temperature is 68°F, and the outside temperature is 38°F. Find the constant of variation if 2592 calories of heat are conducted in 1 hour through the window.

b. The same window can conduct 3888 calories through it in 1 hour. What is the temperature difference?

Writing Exercises

1. Given the equation $y - y_1 = m(x - x_1)$, discuss how you would find the y-intercept.

2. Discuss how to use symmetry in sketching the graph of a function.

3. If we know the symmetries of the graph of a function that has an inverse, discuss the symmetries of the graph of the inverse function.

4. Discuss the relationship between the domain of $\frac{f}{g}$ with the domains of f and g.

5. Discuss in words the meaning of the following: The graph of a function is symmetric with respect to the x-axis, the y-axis, or the origin.

Chapter 3 Project

At the beginning of this chapter, we mentioned a function that relates the number of times a cricket chirps in one minute to the outside temperature in degrees Fahrenheit. Following is that function (you have seen it several times!).

$$f(n) = \frac{n}{4} + 40$$

What would the temperature be if you heard 36 chirps in one minute? 52? Use your graphing calculator to set up a TABLE, pairing the inputs (number of chirps) with the outputs (temperature) for 10 values of n.

For this chapter's project, look back at the following Exercises:

- Section 3.2, Exercise 50
- Section 3.4, Exercises 75 and 76
- Section 3.5, Exercise 72

Put these results together with your table, and answer the following question:

In your own words, explain how the concepts of *function, line, slope,* and *inverse* can be used to describe and predict phenomenon in the world around us.

Mathematics and biology go together well! The same could be said for physics, chemistry, or just about any branch of science. The concept of function is crucial to understanding how scientists model the real world with mathematics.

Explore the connections between mathematics and biology further at http://archives.math.utk.edu/mathbio/.

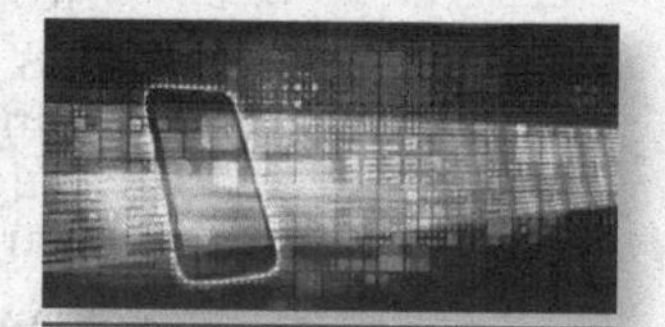

http://archives.math.utk.edu/mathbio/

4 Polynomial Functions

The United States Postal Service allows a package to have a combined length and girth of up to 130 inches. (The girth is the perimeter around the widest part of the package.) Suppose I created a package by cutting squares out of the sides of a rectangle of cardboard and then folding up the sides. How big should the squares be to give me the largest possible volume? How big should they be to maximize the volume and the combined length and girth? (You will be asked to answer these questions in this chapter's project.) The solutions can be found using polynomials (in this case, cubic or degree three polynomials).

The facts about package volume discussed here suggest how useful mathematics is in many careers, including packaging. Are you fascinated by numbers, patterns, and statistics? Learn more about them (and some careers in mathematics) at http://www.amstat.org, http://www.awm-math.org/, and http://fcit.usf.edu/math/websites/general.html.

In Section 3.3 we observed that the polynomial function

$$f(x) = ax + b \tag{1}$$

is called a linear function and that the polynomial function

$$g(x) = ax^2 + bx + c, \qquad a \neq 0 \tag{2}$$

is called a quadratic function. To facilitate the study of polynomial functions in general, we will use the notation

$$P(x) = a_n x^n + a_{n-1}x^{n-1} + \cdots + a_1 x + a_0, \qquad a \neq 0 \tag{3}$$

to represent a **polynomial function of degree *n***, where n is a nonnegative integer. Note that the subscript k of the coefficient in a_k is the same as the exponent in x^k. In general, the coefficients a_k may be real or complex numbers. Although this chapter will focus on real values for a_k, we will indicate which results hold true when the coefficients a_k are complex numbers.

If $a \neq 0$ in Equation (1), we set the polynomial function equal to zero and obtain the linear equation

$$ax + b = 0$$

which has precisely one solution,

$$x = -\frac{b}{a}$$

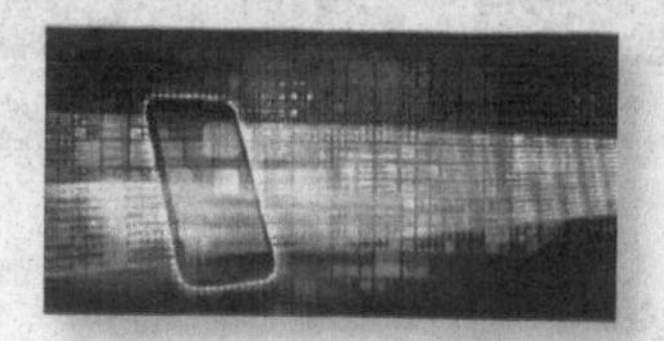

http://www.amstat.org, http://www.awm-math.org/, http://fcit.usf.edu/math/websites/general.html

If we set the polynomial function in Equation (2) equal to zero, we have the quadratic equation

$$ax^2 + bx + c = 0, \qquad a \neq 0$$

which has the two solutions given by the quadratic formula,

$$x = \frac{-b \pm \sqrt{b^2 - 4ac}}{2a}, \qquad a \neq 0$$

If we set the polynomial function in Equation (3) equal to zero, we have the **polynomial equation of degree *n***

$$a_n x^n + a_{n-1}x^{n-1} + \cdots + a_1 x + a_0 = 0, \qquad a_n \neq 0 \tag{4}$$

In this chapter, we study methods of finding solutions to Equation (4). Any complex number satisfying Equation (4) is called a *solution*, or **root**, of the polynomial equation $P(x) = 0$. Such values are also called the **zeros** of the polynomial $P(x)$.

Mathematicians, curious about polynomials, have considered many questions in this area. For example:

- Does a polynomial always have a zero?
- What is the total number of zeros of a polynomial of degree n?
- How many of the zeros of a polynomial are real numbers?
- What can one say about the zeros of a polynomial if all of its coefficients are integers?
- Is there a relationship between the zeros and the factors of a polynomial?
- Can we find a formula for expressing the zeros of a polynomial in terms of the coefficients of the polynomial?

We will explore some of these questions in the course of this chapter.

4.1 Quadratic Functions and Their Graphs

A function of the form

$$f(x) = ax^2 + bx + c \tag{1}$$

where a, b, and c are real numbers and $a \neq 0$, is called a **quadratic function**. By **completing the square**, it is always possible to rewrite Equation (1) in the form

$$f(x) = a(x - h)^2 + k \tag{2}$$

where h and k are constants. We demonstrate this process with a few examples.

EXAMPLE 1 *Completing the Square*

Write the quadratic function

$$f(x) = 2x^2 - 4x - 1$$

in the form of Equation (2).

SOLUTION

We complete the square in a manner analogous to that used in Section 2.3 for solving quadratic equations. Here, we will factor out the coefficient a of x^2, complete the square, and balance the equation as follows:

$$\begin{aligned} f(x) &= 2(x^2 - 2x \qquad) - 1 && \text{Factor out 2.} \\ &= 2(x^2 - 2x + 1) - 1 - 2 && \text{Complete the square and balance.} \\ &= 2(x - 1)^2 - 3 \end{aligned}$$

which is in the form of Equation (2) with $a = 2$, $h = 1$ and $k = -3$. ■

Write each quadratic function f in the form $f(x) = a(x - h)^2 + k$.

a. $f(x) = -3x^2 - 12x - 13$ b. $f(x) = 2x^2 - 2x + 3$

Answers

a. $f(x) = -3(x + 2)^2 - 1$ b. $f(x) = 2\left(x - \frac{1}{2}\right)^2 + \frac{5}{2}$

Be careful to balance the equation properly when completing the square. In Example 1, we wrote

$$\begin{aligned} f(x) &= 2(x^2 - 2x \quad\ ) - 1 \\ &= 2(x^2 - 2x + 1) - 1 - 2 \end{aligned}$$

We added +1 to the expression in parentheses that, due to the factor 2 in front of the parentheses, adds +2 to $f(x)$. We must balance by subtracting 2 as shown.

In Example 1 we completed the square to show that

$$f(x) = 2x^2 - 4x - 1 = 2(x - 1)^2 - 3$$

From Section 3.3, we know that this is the graph of the **parabola** $f(x) = 2x^2$ shifted 3 units downward and 1 unit to the right. In general, the graph of Equation (2) is that of the parabola $f(x) = ax^2$ shifted k units vertically and h units horizontally. Thus, the graph of Equation (2) is a parabola opening from the point (h, k), which is called the **vertex** of the parabola. If $a > 0$, the **parabola opens upward** from the vertex; if $a < 0$, the **parabola opens downward.** We can summarize these results in this way:

Graph of $f(x) = ax^2 + bx + c$

The quadratic function

$$f(x) = ax^2 + bx + c, \qquad a \neq 0$$

can be written in the form

$$f(x) = a(x - h)^2 + k$$

where h and k are constants. The graph is a parabola with vertex at (h, k), opening upward if $a > 0$ and downward if $a < 0$.

EXAMPLE 2 *Graphing Quadratic Functions*

Sketch the graphs of the following functions.

a. $f(x) = 2x^2 + 4x - 1$ b. $f(x) = -2x^2 + 4x$

SOLUTION

a. Completing the square in x, we have

$$\begin{aligned} f(x) &= 2x^2 + 4x - 1 \\ &= 2(x^2 + 2x \quad\ ) - 1 \\ &= 2(x^2 + 2x + 1) - 1 - 2 \\ &= 2(x + 1)^2 - 3 \end{aligned}$$

which is in the form $f(x) = a(x - h)^2 + k$ with $a = 2$, $h = -1$, and $k = -3$. The vertex of the parabola is at $(-1, -3)$, and the graph opens upward.

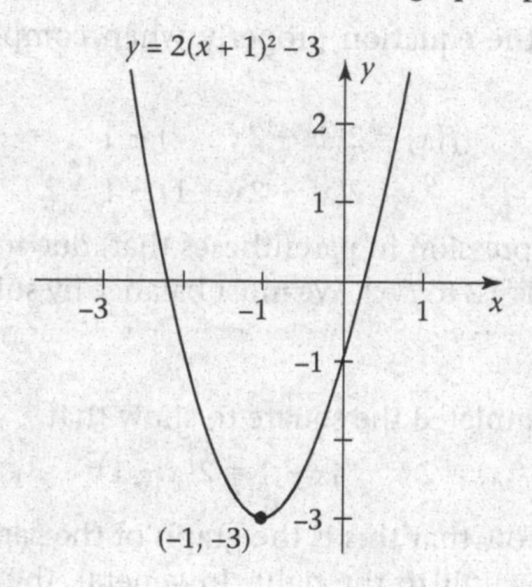

b. Completing the square,

$$\begin{aligned} f(x) &= -2x^2 + 4x \\ &= -2(x^2 - 2x \quad\ ) \\ &= -2(x^2 - 2x + 1) + 2 \\ &= -2(x - 1)^2 + 2 \end{aligned}$$

Here, $a = -2$, $h = 1$ and $k = 2$. The vertex of the parabola is at $(1, 2)$, and the parabola opens downward.

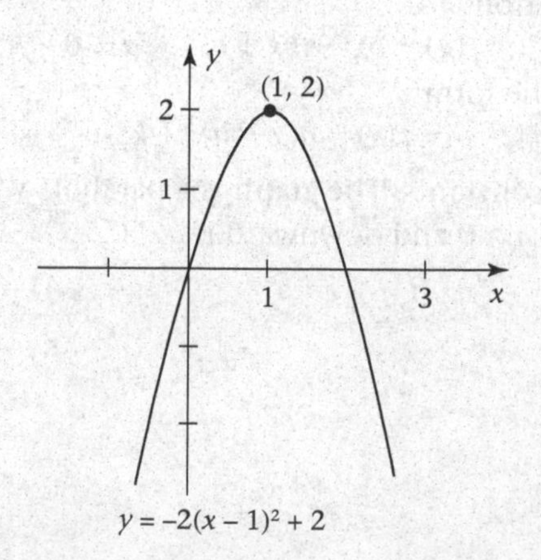

■

4.1a Intercepts and Roots

Since the graph of the quadratic function of Equation (1) is the graph of the equation

$$y = ax^2 + bx + c, \qquad a \neq 0$$

the graph intersects the x-axis at those points where $y = 0$. The x-intercepts are those points where $y = 0$, implying that $ax^2 + bx + c = 0$. The y-intercept is that point where $x = 0$, implying that $y = c$.

Intercepts of the Parabola

The x-intercepts of the parabola

$$f(x) = ax^2 + bx + c, \qquad a \neq 0$$

are the real roots of the quadratic equation

$$ax^2 + bx + c = 0$$

and are given by the quadratic formula. The y-intercept has coordinates $(0, c)$.

The discriminant of the quadratic equation tells us the number of real roots, and therefore, the number of **x-intercepts of the parabola.** Thus, there can be two different real roots, a **double root** or two complex roots. These graphs are shown in Figure 4-1 for the case when $a > 0$.

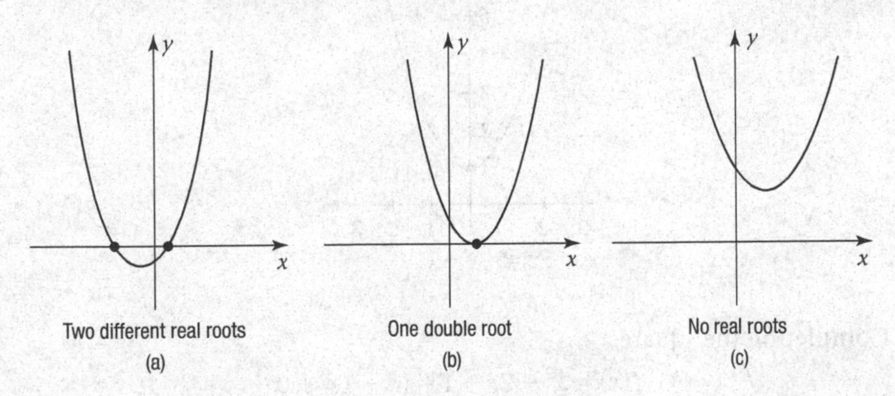

FIGURE 4-1
Roots, Intercepts, and Graphs of Quadratic Functions When $a > 0$

EXAMPLE 3 *Graphing Parabolas*

Find the vertex and all intercepts of each of the following parabolas. Sketch the graph.

a. $f(x) = -x^2 + 3x - 2$ b. $f(x) = x^2 + 1$ c. $f(x) = x^2 + 2x + 1$

SOLUTION

a. To find the vertex, we must complete the square

$$f(x) = -x^2 + 3x - 2 = -\left(x - \frac{3}{2}\right)^2 + \frac{1}{4}$$

The vertex is at $\left(\frac{3}{2}, \frac{1}{4}\right)$ and the parabola opens downward. Setting $y = f(x) = 0$, we see that

$$x^2 - 3x + 2 = 0$$
$$(x - 1)(x - 2) = 0$$

and the x-intercepts occur at $x = 1$ and $x = 2$. Finally, the y-intercept is $y = f(0) = -2$.

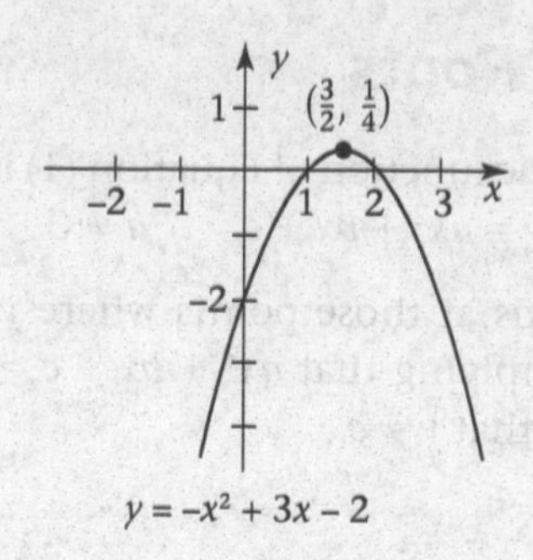

$y = -x^2 + 3x - 2$

b. Since

$$f(x) = x^2 + 1 = (x - 0)^2 + 1$$

the vertex is at (0, 1) and the parabola opens upward. To find the x-intercepts, we set $y = f(x) = 0$ so that

$$x^2 + 1 = 0$$

Since this quadratic equation has no real roots, there are no x-intercepts. To find the y-intercept, we set $x = 0$ and find that $y = f(0) = 1$.

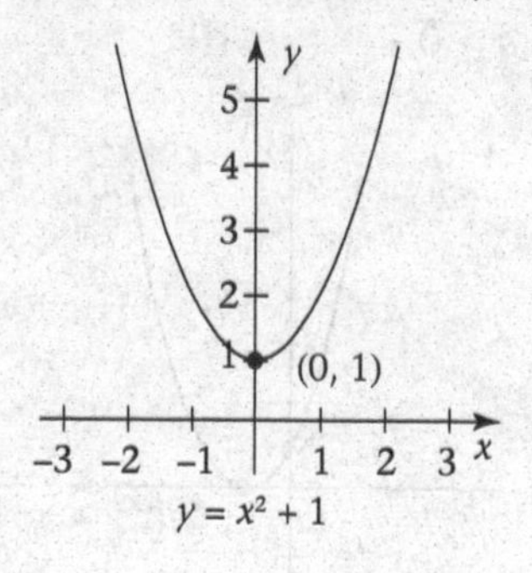

$y = x^2 + 1$

c. Completing the square,

$$f(x) = x^2 + 2x + 1 = (x + 1)^2 + 0$$

The vertex is at the point (−1, 0) and the parabola opens upward. Setting $f(x) = 0$, we see that

$$x^2 + 2x + 1 = (x + 1)^2 = 0$$

so that $x = -1$ is the only x-intercept. Finally, we set $x = 0$ and find that $y = f(0) = 1$ is the y-intercept.

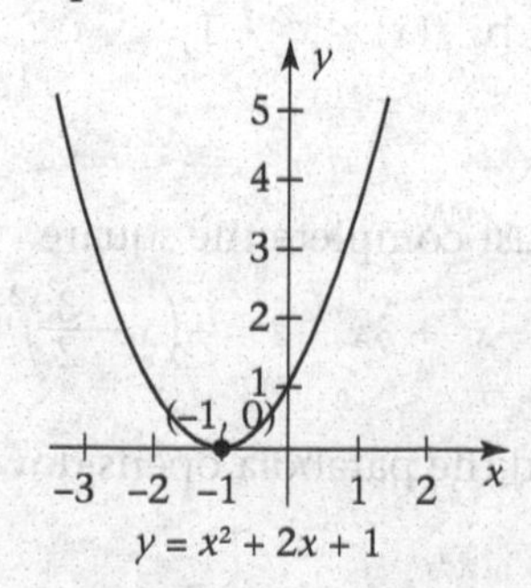

$y = x^2 + 2x + 1$

■

4.1b *Maximum and Minimum Values*

We know that the graph of the function

$$f(x) = ax^2 + bx + c, \qquad a \neq 0$$

is a parabola that opens from its vertex. If $a > 0$, the graph opens upward and the function takes on its **minimum** value at the vertex (see Figure 4-2(a)); if $a < 0$, the graph opens downward and the function takes on its **maximum** value at the vertex (see Figure 4-2(b)). Using the method for completing the square, we can find the coordinates of the vertex (h, k) from the general equation.

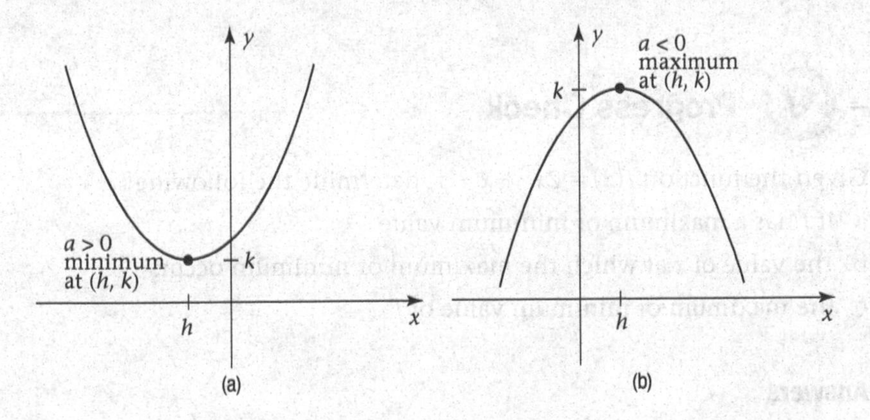

FIGURE 4-2
Maximum and Minimum for $f(x) = ax^2 + bx + c$

$$\begin{aligned} f(x) &= ax^2 + bx + c \\ &= a\left(x^2 + \frac{b}{a}x + \quad\right) + c \\ &= a\left(x^2 + \frac{b}{a}x + \left(\frac{b}{2a}\right)^2\right) + c - a\left(\frac{b}{2a}\right)^2 \\ &= a\left(x + \frac{b}{2a}\right)^2 + \left(c - \frac{b^2}{4a}\right) \end{aligned}$$

Since this last expression is of the form

$$f(x) = a(x - h)^2 + k$$

we see that

$$h = -\frac{b}{2a}$$

Therefore, the vertex has coordinates

$$\left(-\frac{b}{2a}, f\left(-\frac{b}{2a}\right)\right)$$

The maximum or minimum value of the quadratic function

$$f(x) = ax^2 + bx + c, \qquad a \neq 0$$

occurs at the vertex where

$$x = -\frac{b}{2a}$$

It is a maximum value if $a < 0$ and a minimum value if $a > 0$. The maximum or minimum value is

$$f\left(-\frac{b}{2a}\right)$$

EXAMPLE 4 ***Maximum and Minimum Values of Quadratic Functions***

Find the maximum or minimum value of the function $f(x) = -x^2 + 3x - 2$.

SOLUTION

For this quadratic function, $a = -1$, $b = 3$, and the vertex occurs at $x = -\frac{b}{2a} = \frac{3}{2}$. Since $a < 0$, the curve opens downward and the function attains a maximum value at the vertex. We find this maximum value by evaluating $f(x)$ when $x = \frac{3}{2}$, namely, $f(\frac{3}{2}) = \frac{1}{4}$. Conclusion: The maximum value of the function is $\frac{1}{4}$ and occurs when $x = \frac{3}{2}$. (See Example 2a.) ■

Progress Check

Given the function $f(x) = 2x^2 + x - 1$, determine the following:

a. if f has a maximum or minimum value

b. the value of x at which the maximum or minimum occurs

c. the maximum or minimum value of f

Answers

a. minimum b. $-\frac{1}{4}$ c. $-\frac{9}{8}$

EXAMPLE 5 ***Maximum and Minimum in Word Problems***

A rectangular region with one side against an existing building is to be fenced in. If 100 feet of fencing material are available, what dimensions will maximize the area of the region? What is the maximum area?

SOLUTION

Below, we indicate the lengths of the two parallel sides by x. Since we have 100 feet of material available, the length of the remaining side must be $100 - 2x$. The area A is a function of x and is given by

$$A(x) = x(100 - 2x) = -2x^2 + 100x$$

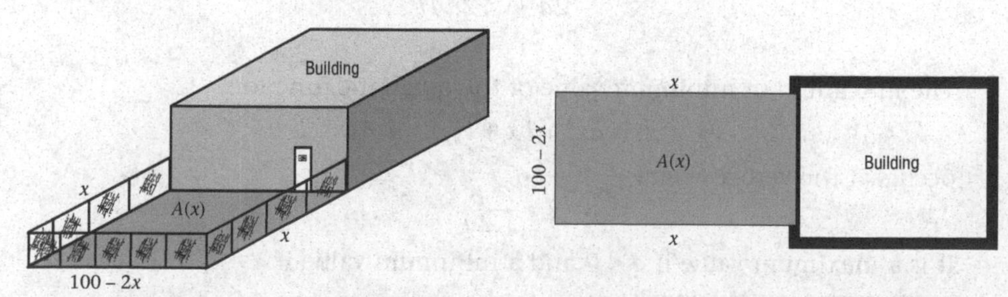

This quadratic function has a maximum (Why?) that is assumed when $x = -\frac{b}{2a} = 25$. The dimensions of the three sides are 25, 25, and 50 feet, and the maximum area is 1250 square feet. ■

EXAMPLE 6 ***Maximum and Minimum in Word Problems***

Find two numbers whose sum is 18 and whose product is a maximum.

SOLUTION

If one of the numbers is x, then the other number must be $18 - x$. The product P is a function of x such that

$$P(x) = x(18 - x) = -x^2 + 18x$$

The function has a maximum (Why?) at $x = -\frac{b}{2a} = 9$. The two numbers are 9 and 9, and the product is 81. ■

EXAMPLE 7 ***Maximum and Minimum in Word Problems***

A corporate vice-president finds that the cost C in dollars to produce and sell x units of the company's product is approximated by the function

$$C(x) = 3x^2 - 30x + 3000$$

If each unit sells for \$300, find the number of units manufactured and sold that will maximize profit.

SOLUTION

Since each unit sells for \$300, the revenue R from selling x units is

$$R(x) = 300x$$

The profit P is revenue minus cost, that is,

$$\begin{aligned} P(x) &= R(x) - C(x) \\ &= 300x - (3x^2 - 30x + 3000) \\ &= -3x^2 + 330x - 3000 \end{aligned}$$

Since P is a quadratic function and its leading coefficient is negative, P attains a maximum where $x = -\frac{b}{2a} = -\frac{330}{-6} = 55$. The company maximizes its profit if it manufactures and sells 55 units. (Verify that the maximum profit is \$6075.) ■

EXAMPLE 8 ***Applied Maximum and Minimum***

Which point on the graph of the function $f(x) = \sqrt{x}$ is closest to the point (2, 0)?

SOLUTION

In Figure 4-3, we use the symbol d to denote the distance from the point (2, 0) to a point $P(x, y)$ on the graph of $y = \sqrt{x}$. By the distance formula, we can write

$$d = \sqrt{(x-2)^2 + (y-0)^2} = \sqrt{x^2 - 4x + 4 + y^2}$$

Since point $P(x, y)$ lies on the graph of $y = \sqrt{x}$, we can substitute $\sqrt{x}$ for y and write d as a function of the variable x alone.

$$d(x) = \sqrt{x^2 - 4x + 4 + x} = \sqrt{x^2 - 3x + 4}$$

Since the square root function is an increasing function, d is minimum when d^2 is minimum; that is, the same value of x that provides a minimum for d provides a minimum for $d^2 = x^2 - 3x + 4$. This minimum occurs where $x = -\frac{b}{2a} = \frac{3}{2}$. The corresponding value of y is

$$y = \sqrt{x} = \sqrt{\frac{3}{2}} = \frac{\sqrt{6}}{2}$$

Therefore, the point we seek has coordinates $\left(\frac{3}{2}, \frac{\sqrt{6}}{2}\right)$.

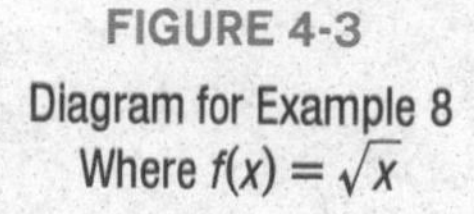

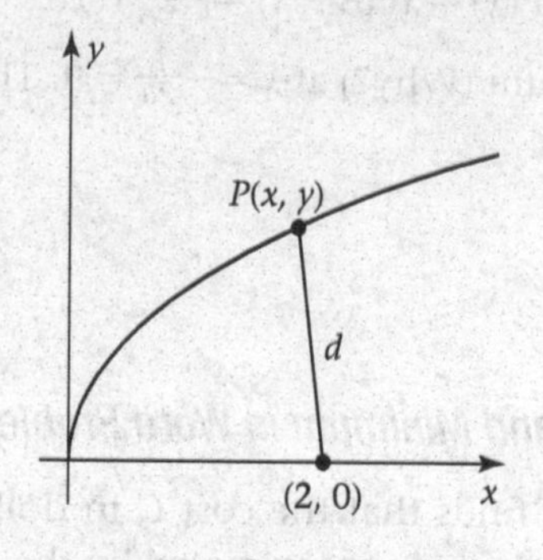

FIGURE 4-3
Diagram for Example 8
Where $f(x) = \sqrt{x}$

■

Exercise Set 4.1

In Exercises 1–10, write the quadratic function f in the form $2(x) = a(x-h)^2 + k$. Then, sketch the graph of the quadratic function.

1. $f(x) = x^2 - 6x + 10$
2. $f(x) = -x^2 - 2x - 3$
3. $f(x) = -2x^2 + 4x - 5$
4. $f(x) = 3x^2 + 12x + 14$
5. $f(x) = 2x^2 + 6x + 5$
6. $f(x) = -4x^2 + 4x$
7. $f(x) = -x^2 - x$
8. $f(x) = 3x^2 + 18x + 9$
9. $f(x) = -2x^2 + 5$
10. $f(x) = -\frac{1}{2}x^2 + 2x + 2$

In Exercises 11–18, find the vertex and all intercepts of the parabola. Sketch the graph.

11. $f(x) = 2x^2 - 4x$
12. $f(x) = -x^2 - 2x + 3$
13. $f(x) = -4x^2 + 4x - 1$
14. $f(x) = x^2 + 4x + 4$
15. $f(x) = \frac{1}{2}x^2 + 2x + 4$
16. $f(x) = -2x^2 + 2x - \frac{5}{2}$
17. $f(x) = -\frac{1}{2}x^2 + 3x - 4$
18. $f(x) = x^2 - x + 1$

In Exercises 19–26, use your graphing calculator to GRAPH the given functions in the default viewing rectangle. What is the relationship between the linear factors of the quadratic function and the x-intercepts of the graph?

19. $y = (x+1)(x-2)$
20. $y = -(x+1)(x-2)$
21. $y = x(x+5)$
22. $y = (x+5)^2$
23. $y = (x-3)(x-7)$
24. $y = 2(x-3)(x-7)$
25. $y = 0.3(x-3)(x-7)$
26. $y = -0.1(x-5)(x+5)$

In Exercises 27–30, the graphs represent quadratic functions of the form $f(x) = (x - r_1)(x - r_2)$ or $f(x) = -(x - r_1)(x - r_2)$. The graphs are drawn in the viewing rectangle $-10 \leq X \leq 10$ and $-20 \leq Y \leq 20$ with XSCL = 1 and YSCL = 5. For each function

a. Determine r_1 and r_2.

b. Write the function in the form $f(x) = ax^2 + bx + c$.

c. Use your graphing calculator to compare the graph of the function you found in part (b) with the graph shown below to verify your answer.

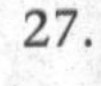

27. 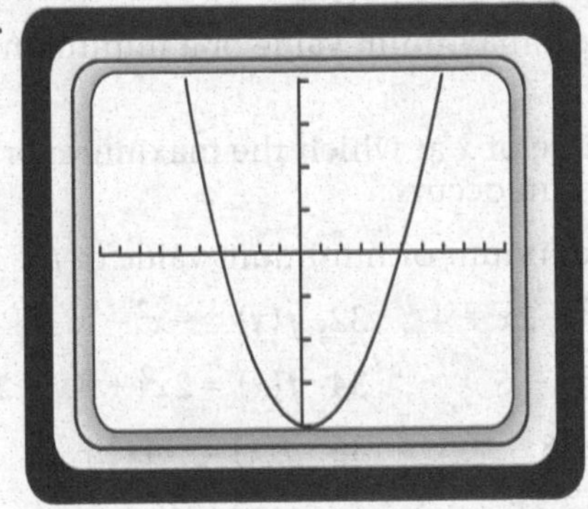

28. 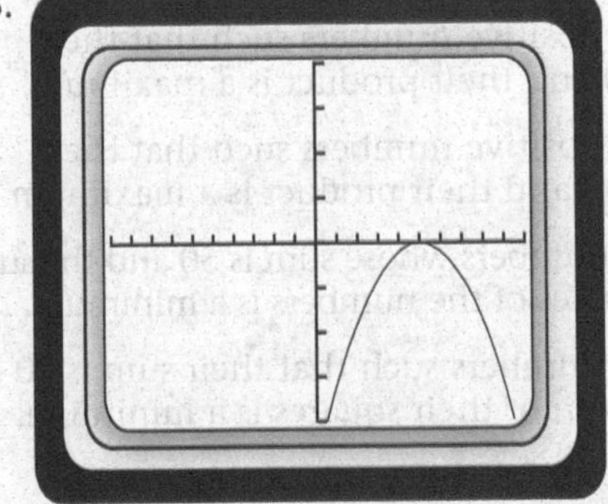

29. 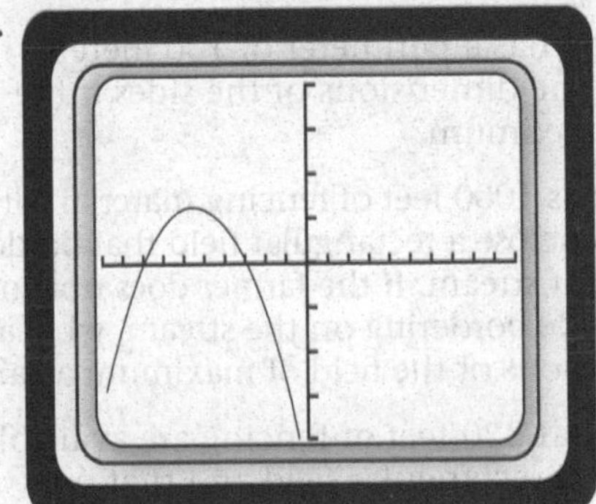

30. 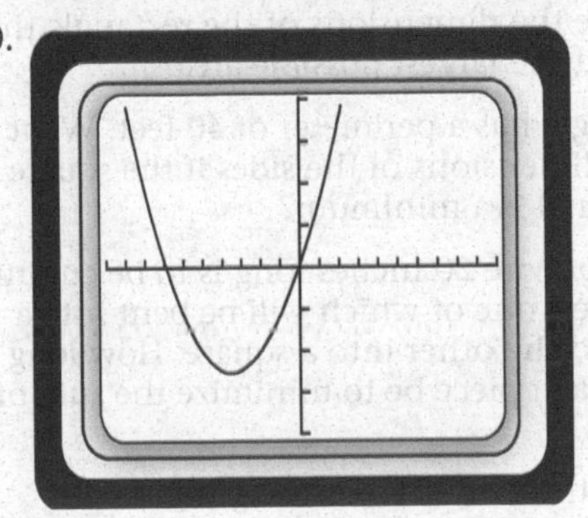

In Exercises 31–38, you are given a quadratic function. Determine the following:

a. if f has a maximum value or a minimum value

b. the value of x at which the maximum or minimum occurs

c. the maximum or minimum value of f

31. $f(x) = 3x^2 - 2x + 4$
32. $f(x) = -x^2 - x - 4$
33. $f(x) = -2x^2 - 5$
34. $f(x) = 2x^2 + 3x + 2$
35. $f(x) = x^2 + 5x$
36. $f(x) = -4x^2 + 3$
37. $f(x) = 2x^2 - \frac{1}{2}x - \frac{3}{2}$
38. $f(x) = \frac{2}{3}x^2 + x - 1$
39. Find two positive numbers such that their sum is 20 and their product is a maximum.
40. Find two positive numbers such that their sum is 100 and their product is a maximum.
41. Find two numbers whose sum is 50 and the sum of the squares of the numbers is a minimum.
42. Find two numbers such that their sum is 20 and the sum of their squares is a minimum.
43. Which point on the graph of the function $f(x) = \sqrt{x}$ is closest to the point (1, 0)?
44. A rectangle has a perimeter of 100 meters. What are the dimensions of the sides if the area is a maximum?
45. A farmer has 1000 feet of fencing material with which to enclose a rectangular field that borders on a straight stream. If the farmer does not enclose the side bordering on the stream, what are the dimensions of the field of maximum area?
46. Suppose that 320 feet of fencing are available to enclose a rectangular field and that one side of the field must be given double fencing. What are the dimensions of the rectangle that will yield the largest possible area?
47. A rectangle has a perimeter of 40 feet. What are the dimensions of the sides if the square of its diagonal is a minimum?
48. A piece of wire 20 inches long is to be cut into two pieces, one of which will be bent into a circle and the other into a square. How long should each piece be to minimize the sum of the areas?
49. A ball is thrown up from the ground with an initial velocity of 80 feet per second. The height s in feet can be expressed as a function of time t in seconds by

$$s(t) = 80t - 16t^2$$

When does the ball reach its maximum height? What is the maximum height?

50. At a rate of $40 per room, a 100-room motel is fully occupied each night. For each $1 increase in the room rate, 2 fewer rooms are rented. What increase in room rate will maximize revenue?
51. A movie theater finds that 300 people attend each performance at the current rate of $6 and that attendance decreases by 10 persons for each 25 cent increase in price. What increase yields the greatest gross revenue?
52. An apple grower finds that the average yield is 60 bushels per tree when 80 or fewer trees are planted in an orchard. Each additional tree decreases the average yield per tree by 2 bushels. How many trees will maximize the total yield of the orchard?
53. Show that the coordinates of the point on the line $y = mx + b$ that is closest to the origin are

$$\left(\frac{-bm}{1+m^2}, \frac{b}{1+m^2}\right)$$

54. A manufacturer believes that the net profit P is related to the advertising expenditure x (both in thousands of dollars) by

$$P(x) = 10 + 46x - \frac{1}{2}x^2$$

What advertising expenditure will maximize the profit?

55. A manufacturer is limited by the firm's production facilities to an output of at most 125 units per day. The daily cost C and revenue R from the production and sale of x units are given as

$$C(x) = 3x^2 - 750x + 100$$
$$R(x) = 50x - x^2$$

If all units produced are sold, how many units should be manufactured daily to maximize profit?

56. An athletic field is in the shape of a rectangle with semicircular ends of radius r. If the perimeter of the field is p units, show that the maximum area is attained when $r = \frac{p}{2\pi}$.

4.2 Graphs of Polynomial Functions of Higher Degree

We have already shown that the graph of the first-degree polynomial function

$$P(x) = ax + b$$

is always a line, and the graph of the second-degree polynomial function

$$P(x) = ax^2 + bx + c, \qquad a \neq 0$$

is always a parabola. However, for $n > 2$, the graph of the polynomial function

$$P(x) = a_n x^n + a_{n-1}x^{n-1} + \cdots + a_1 x + a_0, \qquad a_n \neq 0$$

is much more difficult to describe. Nevertheless, it is still possible to make some useful observations concerning the shape and nature of the graph of a polynomial of degree greater than 2.

4.2a *Continuity and the Intermediate Value Theorem*

We begin by exploring what the graph of a polynomial function can and cannot "do." The graphs in Figure 4-4 illustrate "typical" polynomials.

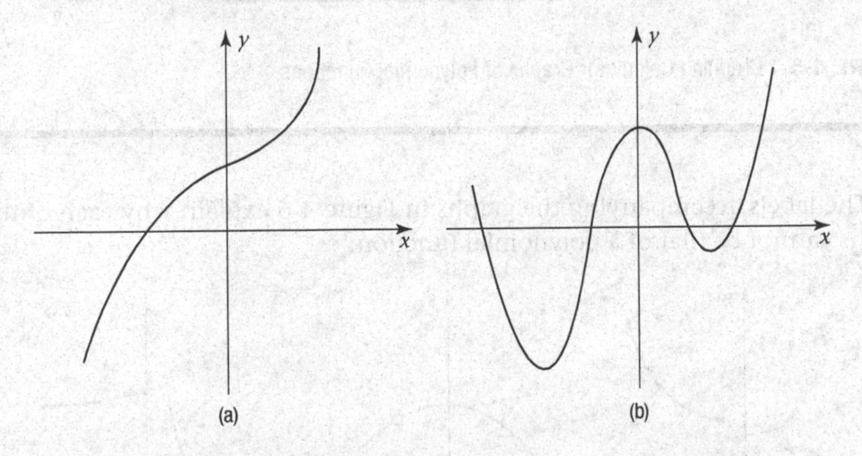

FIGURE 4-4
Graphs of Polynomial Functions

Continuity, Smoothness and Corners

- The graph of a polynomial function is **continuous**, which means that there are no "breaks" or "holes." (The graph can be drawn without lifting the pencil from the paper.)
- The graph of a polynomial function is **smooth**; it has no "corners" (also called **cusps**). In particular, no segment of the graph of a polynomial function of degree greater than 1 is a line.

Graphing Calculator ALERT

The properties of continuity and smoothness will not be evident on your graphing calculator display. You must train your eye to properly *interpret* the calculator display as a continuous, smooth graph (when the function is continuous and smooth). For example, consider the graphing calculator representations of the graphs in Figure 4-4(a) and Figure 4-4(b) as shown in Figure 4-5(a) and Figure 4-5(b), respectively.

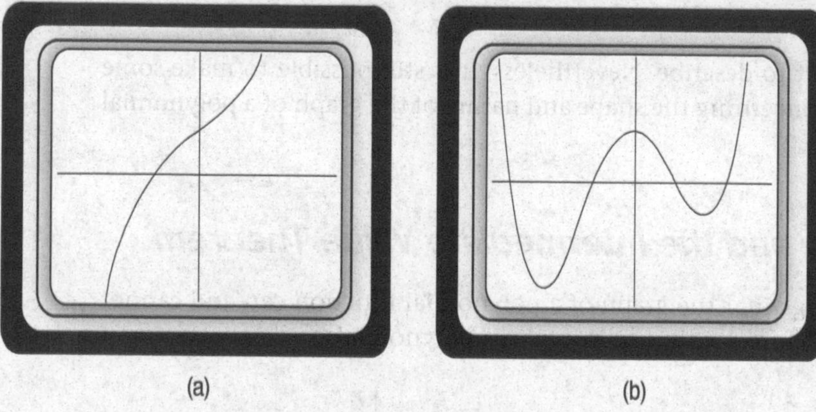

FIGURE 4-5 Graphing Calculator Graphs of Polynomial Functions

The labels accompanying the graphs in Figure 4-6 explain why each of these graphs cannot be that of a polynomial function.

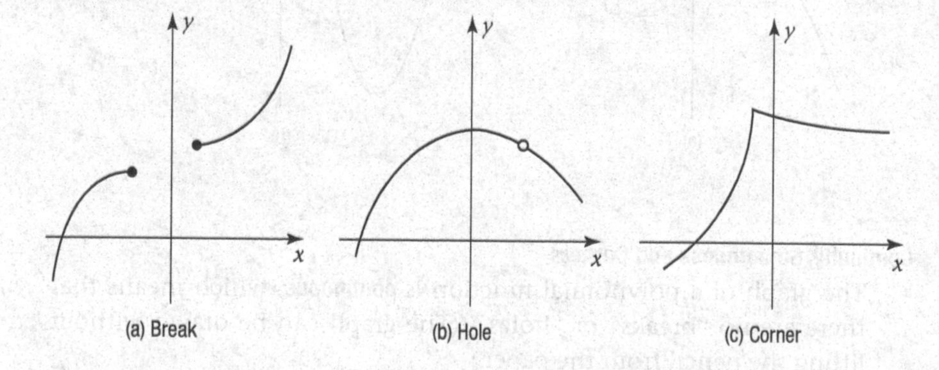

FIGURE 4-6 Graphs of Non-Polynomial Functions

The property of continuity has enabled us to plot points on the graph of a polynomial function and then to connect these points. In Figure 4-7, on the graph of $y = P(x)$, to get from the point $A(a, P(a))$ to $B(b, P(b))$ the graph must pass through every y-coordinate between $P(a)$ and $P(b)$. This result is formally known as the Intermediate Value Theorem for Polynomial Functions.

Intermediate Value Theorem for Polynomial Functions
If $a < b$ and P is a polynomial function such that $P(a) \neq P(b)$, then P takes on every value between $P(a)$ and $P(b)$ in the interval $[a, b]$.

This theorem is useful in finding the roots of the polynomial equation $P(x) = 0$. As shown in Figure 4-7, if we can find values a and b such that $P(a)$ and $P(b)$ are opposite in sign, then there must be at least one value c in the interval $[a, b]$ where $P(c) = 0$. This shows that there is at least one real root c such that $a < c < b$.

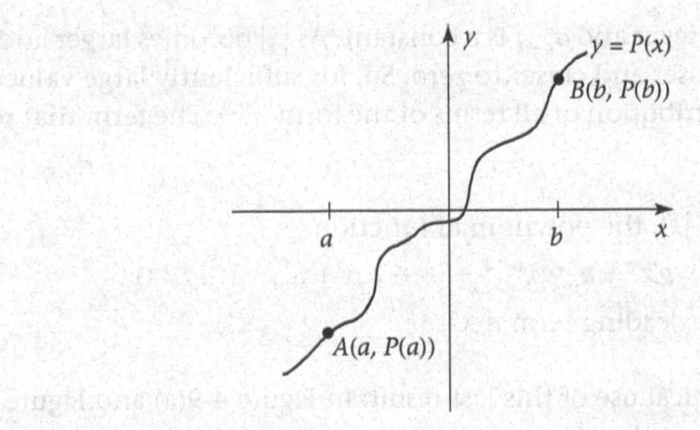

FIGURE 4-7
Intermediate Value Theorem for Polynomials

4.2b Turning Points

In Figure 4-8, A and B, called **turning points**, are those points at which the graph changes "direction," from rising to falling or from falling to rising. If we knew the number and location of the turning points of a polynomial graph, it would make it easier to sketch that graph. We shall give only a partial answer to these problems.

> **Turning Points**
> The graph of a polynomial function of degree n has at most $n - 1$ turning points.

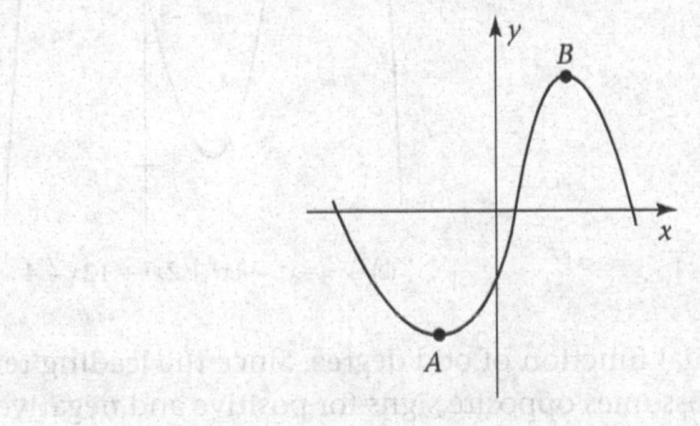

FIGURE 4-8
Turning Points

The polynomial function whose graph is shown in Figure 4-7 has no turning points. Therefore, the degree of this polynomial must be at least 1. Since the graph in Figure 4-8 has two turning points, the degree of this polynomial must be at least 3. Observe that the degree of the polynomial associated with Figure 4-4(b) must be at least 4.

4.2c Behavior for Large $|x|$

We can factor x^n out of

$$P(x) = a_n x^n + a_{n-1}x^{n-1} + \cdots + a_1 x + a_0$$

to obtain

$$P(x) = x^n\left(a_n + \frac{a_{n-1}}{x} + \frac{a_{n-2}}{x^2} + \cdots + \frac{a_1}{x^{n-1}} + \frac{a_0}{x^n}\right), \qquad x \neq 0$$

When the equation is written this way, it is possible to see how $P(x)$ behaves when $|x|$ assumes large values. (Observe that this may happen when x is positive or when x is negative.) Consider the expression

$$\frac{a_{n-k}}{x^k}$$

where k is a positive integer and a_{n-k} is a constant. As $|x|$ becomes larger and larger, this expression gets closer and closer to zero. So, for sufficiently large values of $|x|$, we can ignore the contribution of all terms of the form $\frac{a_{n-k}}{x^k}$. The term that remains is $a_n x^n$. In summary,

For large values of $|x|$, the polynomial function

$$P(x) = a_n x^n + a_{n-1} x^{n-1} + \cdots + a_1 x + a_0, \qquad a_n \neq 0$$

is dominated by its leading term $a_n x^n$.

We can make practical use of this last result. In Figure 4-9(a) and Figure 4-9(b), we have sketched the graphs of polynomial functions of degrees 3 and 4, respectively. Note that as $|x|$ increases, the **"ends" of the graph** of the third-degree polynomial function extend indefinitely in opposite directions, whereas the ends of the graph of the fourth-degree polynomial function extend indefinitely in the same direction.

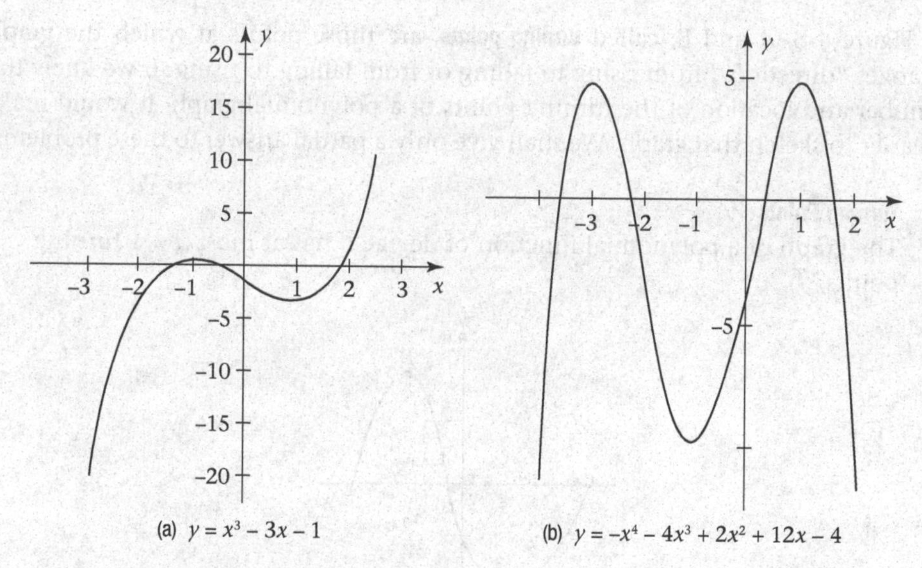

FIGURE 4-9
Polynomial Functions of Odd and Even Degree

(a) $y = x^3 - 3x - 1$

(b) $y = -x^4 - 4x^3 + 2x^2 + 12x - 4$

Consider any polynomial function of odd degree. Since the leading term $a_n x^n$ is of odd degree, this term assumes opposite signs for positive and negative values of x. When $|x|$ is large, this term dominates; hence, the polynomial assumes opposite signs for positive and negative values of x. Therefore, one end of the graph extends upward and the other downward.

If a polynomial has even degree, its leading term is also of even degree. Therefore, this term assumes the same sign for positive and negative values of x. When $|x|$ is large, this term dominates; hence, the polynomial assumes the same sign for positive and negative values of x. Thus, the ends of the graph both extend upward or both extend downward.

Graphs of Polynomial Functions

The graph of the polynomial function

$$P(x) = a_n x^n + a_{n-1} x^{n-1} + \cdots + a_1 x + a_0, \qquad a_n \neq 0$$

has these characteristics:

- If n is odd and $a_n > 0$, the graph extends upward if $|x|$ is large and $x > 0$, and extends downward if $|x|$ is large and $x < 0$; that is, the "right end" extends upward and the "left end" extends downward; if $a_n < 0$, the behavior is reversed.
- If n is even and $a_n > 0$, the graph extends upward at both ends; if $a_n < 0$, the graph extends downward at both ends.

EXAMPLE 1 ***Determining Behavior for Large $|x|$***

Without sketching the graph, determine the behavior of the graph of each of the following polynomial functions for large values of $|x|$.

a. $P(x) = -3x^5 + 26x^2 - 5$

b. $P(x) = -\frac{1}{2}x^6 + 209x^3 + 16x + 2$

SOLUTION

a. We need only concern ourselves with the leading term, $-3x^5$. Since the degree is odd and the leading coefficient is negative, the right end of the graph extends downward and the left end of the graph extends upward.

b. Since this is a polynomial of degree 6, the degree is even and the graph moves in the same direction at both ends. The negative leading coefficient indicates that both ends extend downward. ■

4.2d *Polynomials in Factored Form*

Given a polynomial function of degree n, we now know that the graph: (a) is continuous; (b) is smooth; (c) has at most $n - 1$ turning points; and (d) behaves in a predictable manner at the ends. We need some additional guidance to discover what happens when $|x|$ is not too large. We shall only consider polynomials that can be written as a product of linear factors. In this case, we are able to find the x-intercepts and to determine where the graph of the polynomial lies above the x-axis and where it lies below the x-axis. The following example illustrates the procedure.

EXAMPLE 2 ***Using Intercepts in Graphing***

Sketch the graph of the polynomial

$$P(x) = x^3 + x^2 - 6x$$

SOLUTION

Factoring, we find that

$$\begin{aligned} P(x) &= x(x^2 + x - 6) \\ &= x(x + 3)(x - 2) \end{aligned}$$

Since $P(x) = 0$ at $x = 0$, $x = -3$, and $x = 2$, these values are the x-intercepts of $P(x)$. We indicate these values on the real number line as shown in Figure 4-10.

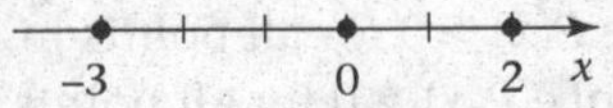

-3 0 2 x

FIGURE 4-10
x-intercepts of $P(x) = x^3 + x^2 - 6x$

We now utilize the Critical Value Method, which was presented in Section 2.5. (Specifically, refer to Example 10 of that section, which dealt with polynomial inequalities.) The x-intercepts presented above play the same role as the critical values. We consider the four intervals

$$(-\infty, -3),\ (-3, 0),\ (0, 2),\ \text{and}\ (2, \infty)$$

The polynomial $P(x)$ always has the same sign, provided the values of x are restricted to any one of these intervals. Thus, we can think of each interval as corresponding to part (if not all) of a solution set to some polynomial inequality. We choose a test point in each interval to determine the sign of the polynomial in that interval. To repeat, all test points within the same interval produce the same sign.

From interval $(-\infty, -3)$, choose test point $x = -4$. Since $P(-4) < 0$, $P(x) < 0$ for all points in $(-\infty, -3)$. From $(-3, 0)$, choose test point $x = -1$. Since $P(-1) > 0$, $P(x) > 0$ for all points in $(-3, 0)$. Choosing $x = 1$ from $(0, 2)$ implies that $P(1) < 0$, thus $P(x) < 0$ for all points in $(0, 2)$. Finally, choosing $x = 3$ from $(2, \infty)$ yields that $P(x) > 0$ for all points in $(2, \infty)$. We summarize these results in Figure 4-11. Plotting a few points, we obtain the graph of the polynomial.

FIGURE 4-11
Sign of $P(x)$ in Given Intervals

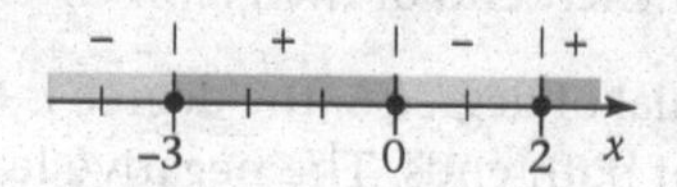

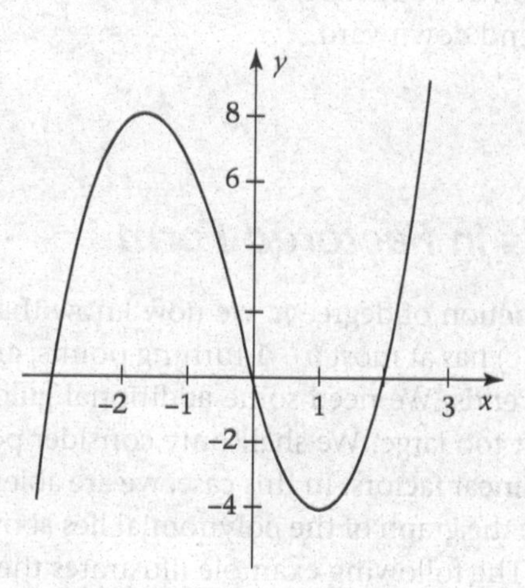

(Check that the right end extends upward and the left end extends downward from the analysis of $P(x)$ for large $|x|$.) ■

The sign of $P(x)$ in adjacent intervals does *not* always change from positive to negative or from negative to positive.

Consider the polynomial $P(x) = x^2$. This has only one x-intercept, $x = 0$. Thus, there are two intervals to examine: $(-\infty, 0)$ and $(0, \infty)$. For test point $x = -1$ from $(-\infty, 0)$, $P(-1) > 0$ and hence $P(x) > 0$ for all points in $(-\infty, 0)$. For test point $x = 1$ from $(0, \infty)$, $P(1) > 0$ and hence $P(x) > 0$ for all points in $(0, \infty)$.

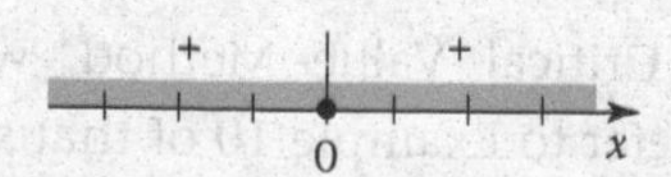

The graph of $P(x) = x^2$ is shown in Figure 4-12.

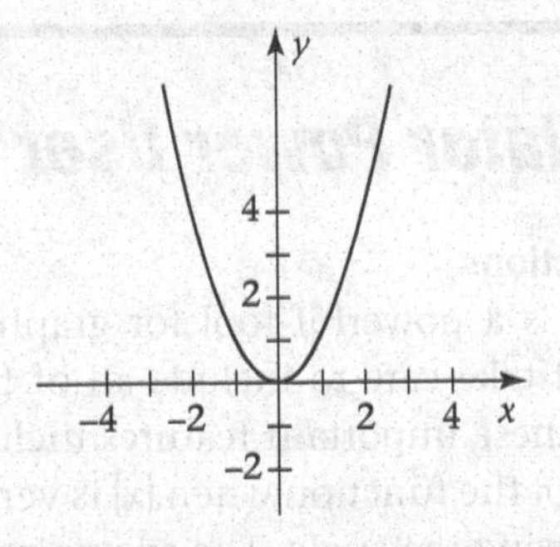

FIGURE 4-12
Graph of $P(x) = x^2$

Since we know that polynomials are continuous, if $P(x)$ is a polynomial with $P(a)$ and $P(b)$ of opposite sign where $a < b$, then $P(x)$ must have at least one x-intercept in the interval (a, b). Let r be such an x-intercept, that is, $P(r) = 0$, where $a < r < b$. Thus, r is a root of the equation $P(x) = 0$ with $a < r < b$.

EXAMPLE 3 ***Looking for Roots in Intervals***

Show that the equation $x^3 + 2x^2 + 5x + 1 = 0$ has a root in the interval $[-1, 0]$.

SOLUTION

Let $P(x) = x^3 + 2x^2 + 5x + 1$. Since $P(-1) = -3 < 0$ and $P(0) = 1 > 0$, $P(-1)$ and $P(0)$ are of opposite sign. Therefore, $P(x)$ has at least one x-intercept in $(-1, 0)$. Hence, the equation $x^3 + 2x^2 + 5x + 1 = 0$ must have at least one root in the interval $[-1, 0]$. ■

We have found the x-intercepts of a polynomial by putting the polynomial into factored form. We may also reverse the procedure; that is, we may construct a polynomial in factored form from a given set of x-intercepts.

EXAMPLE 4 ***Using x-Intercepts to Form a Polynomial***

Find a polynomial whose x-intercepts are 2, 4, and −1.

SOLUTION

If a polynomial has a factor of $x - 2$, then it has an x-intercept of 2. Similarly, if a polynomial has factors $x - 4$ and $x + 1$, then it has x-intercepts of 4 and −1. Therefore, the polynomial

$$P(x) = (x - 2)(x - 4)(x + 1)$$

has x-intercepts 2, 4, and −1. (Can you find another polynomial with the same x-intercepts?) ■

Graphing Calculator Power User's CORNER

Graphing Polynomial Functions

Your graphing calculator is a powerful tool for graphing polynomial functions. However, you must take care to include all of the important features of the function in your graph. For polynomials, these important features include the intercepts, the turning points of the functions, and the behavior of the function when $|x|$ is very large. It is frequently impossible to show all of this behavior in one viewing rectangle. The examples below show how several viewing rectangles can be used to provide complete information about a polynomial function.

EXAMPLE 5 *Graphing Polynomials with a Graphing Calculator*

Graph the function $y = 11x^3 - x - 5$ using the following information:

a. $-10 \le X \le 10$, $-10 \le Y \le 10$, XSCL = 1, and YSCL = 1

b. $-1 \le X \le 1$, $-6 \le Y \le -4$, XSCL = 1, and YSCL = 1

c. $-100 \le X \le 100$, $-11{,}000{,}000 \le Y \le 11{,}000{,}000$, XSCL = 10, and YSCL = 1,000,000

SOLUTION

a.

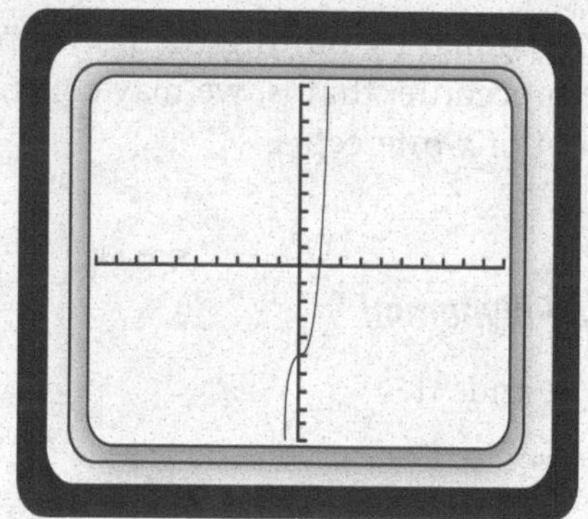

Observe that the x-intercept is approximately 0.808, and the y-intercept is -5.

b.

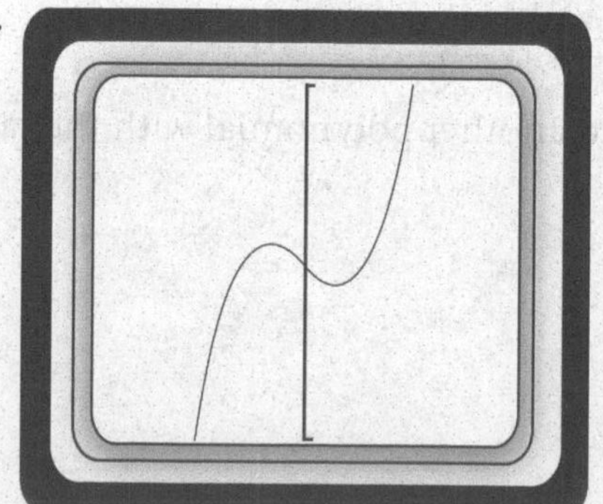

We see that the coordinates of the turning points are *approximately* $(0.174, -5.116)$ and $(-0.174, -4.884)$. Observe that these turning points are "hidden" in the default viewing rectangle.

c.

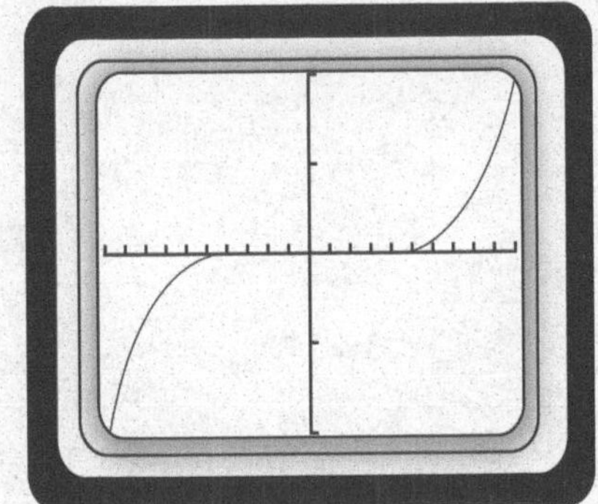

This viewing rectangle shows the behavior of y when $|x|$ is large. ■

It is important to note that the values you obtain when using TRACE on your graphing calculator are *approximate*. It is often necessary to ZOOM-IN to obtain more accurate results. Although the coordinates in the text are correct, they do not necessarily correspond to any values obtained when using the TRACE command.

EXAMPLE 6 ***Graphing Polynomials with a Graphing Calculator***

Graph the function $y = 2x^5 - 11x^4 + 20x^2 - 3$ using the following information:

a. $-10 \leq X \leq 10$, $-10 \leq Y \leq 10$, XSCL = 1, and YSCL = 1

b. $-10 \leq X \leq 10$, $-500 \leq Y \leq 100$, XSCL = 1, and YSCL = 100

c. $-100 \leq X \leq 100$, $-2 \cdot 10^{10} \leq Y \leq 2 \cdot 10^{10}$, XSCL = 10, and YSCL = 10^{10}

SOLUTION

a.

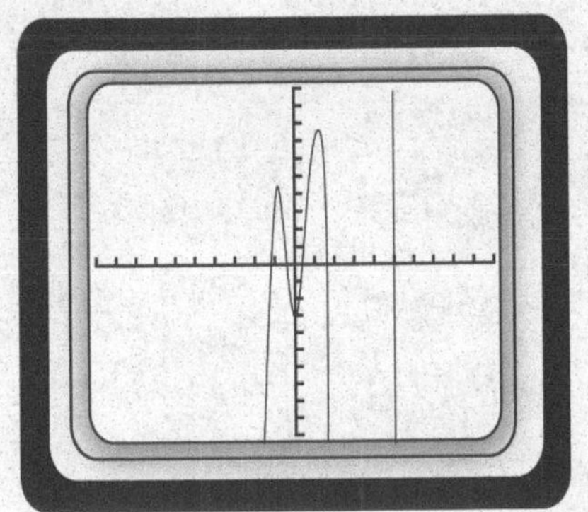

Observe that the x-intercepts are approximately −1.155, −0.408, 0.405, 1.537, and 5.121. (The x-intercept 5.121 may not be displayed on some graphing calculators.) The y-intercept is −3. The coordinates of the turning points are approximately (1.101, 8.316), (0, −3), and (−0.871, 4.839).

b.

We see that the largest x-intercept is approximately 5.121 and the coordinates of the turning point are approximately (4.170, −459.5).

c.

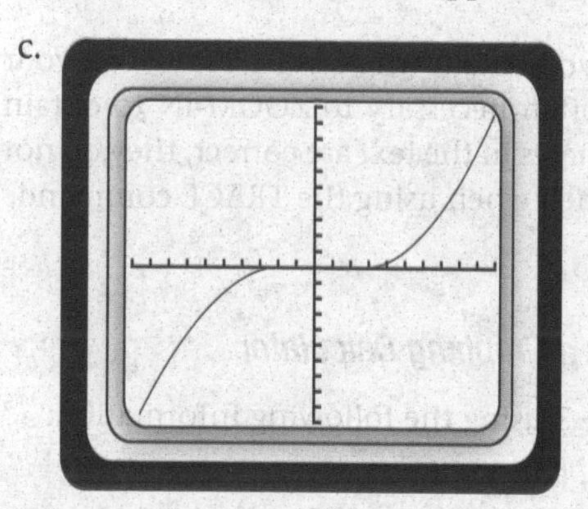

This viewing rectangle shows the behavior of y when $|x|$ is large. ■

Exercise Set 4.2

In Exercises 1–8, show that the equation has a root in the given interval.

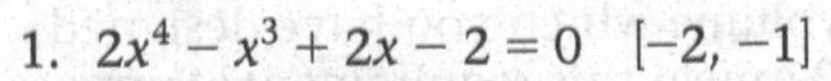

1. $2x^4 - x^3 + 2x - 2 = 0$ $[-2, -1]$
2. $3x^3 - 2x^2 + 5x + 4 = 0$ $[-1, 0]$
3. $x^5 - 3x^3 + x^2 - 3 = 0$ $[1, 2]$
4. $-x^4 + 3x^2 + 5 = 0$ $[2, 3]$
5. $x^6 - 3x^3 + x^2 - 2 = 0$ $[1, 2]$
6. $2x^5 - x^4 + 3x^2 - 6 = 0$ $[1, 2]$
7. $-2x^3 - x^2 + 3x - 4 = 0$ $[-2, -1]$
8. $2x^4 - 3x^3 + x^2 - 1 = 0$ $[-1, 0]$

In Exercises 9–16, without sketching, determine the behavior of the graph of the given polynomial function for large values of $|x|$. Use the letters U and D to indicate whether the graph extends upward or downward. Then, GRAPH $y = P(x)$ and $y =$ leading term of $P(x)$ together in each of the following viewing rectangles: (a) $-10 \le X \le 10$ and $-100 \le Y \le 100$; (b) $-10 \le X \le 10$ and $-10{,}000 \le Y \le 10{,}000$; and (c) $-10 \le X \le 10$ and $-100{,}000 \le Y \le 100{,}000$. For each $P(x)$ determine which viewing rectangle most clearly shows the polynomial function behaving like its leading term.

Polynomial Function	Leading Term	Large Values of $\lvert x\rvert$, $x > 0$	Large Values of $\lvert x\rvert$, $x < 0$
9. $P(x) = x^7 - 175x^3 + 23x^2$			
10. $P(x) = -3x^8 + 22x^4 + 3$			
11. $P(x) = -8x^3 + 17x^2 - 15$			
12. $P(x) = 14x^{12} - 5x^{11} + 3x - 1$			
13. $P(x) = -5x^{10} + 16x^7 + 5$			
14. $P(x) = 2x^5 - 11x^4 - 12x^3$			
15. $P(x) = 4x^8 - 10x^6 + x^3 - 8$			
16. $P(x) = -3x^9 + 6x^6 - 2x^5 + x$			

In Exercises 17–22, determine the x-intercepts and the intervals where $P(x) > 0$ and $P(x) < 0$. Sketch the graph of $P(x)$. Then, determine appropriate WINDOW values and GRAPH $P(x)$ on your graphing calculator. Be sure your graph shows the important features of each function.

17. $P(x) = (x - 3)(2x - 1)(x + 2)$
18. $P(x) = (2 - x)(x - 4)(x + 1)$
19. $P(x) = 2x^3 + 3x^2 - 5x$
20. $P(x) = x^4 - 5x^2 + 4$
21. $P(x) = x^4 - x^3 - 6x^2$
22. $P(x) = (2x + 5)(x - 1)(x + 1)(x - 3)$

In Exercises 23–28, determine a polynomial equation whose roots include the given values.

23. $2, -4, 4$
24. $5, -5, 1, -1$
25. $-1, -2, -3$
26. $-3, \sqrt{2}, -\sqrt{2}$
27. $4, 1 \pm \sqrt{3}$
28. $1, 2, 2 \pm \sqrt{2}$

In Exercises 29–33, use your graphing calculator to GRAPH the given function in the indicated viewing rectangle. Set the XSCL and YSCL values appropriately. Also, find the x- and y-intercepts.

29. $y = (x - 1)(x + 3)(x - 5)$
 $-10 \le X \le 10$ and $-30 \le Y \le 30$
30. $y = -(x + 8)(x - 2)$
 $-10 \le X \le 10$ and $-10 \le Y \le 25$
31. $y = -x(x - 5)(x + 5)(x + 8)$
 $-10 \le X \le 10$ and $-300 \le Y \le 600$
32. $y = (x + 20)(x + 10)(x - 5)$
 $-25 \le X \le 10$ and $-1250 \le Y \le 600$
33. $y = (x - 10)(x + 30)(x - 50)$
 $-100 \le X \le 100$ and $-30{,}000 \le Y \le 30{,}000$

In Exercises 34 and 35, we present graphs representing polynomial functions that are the product of linear factors. Graphs are drawn in the viewing rectangle $-10 \le X \le 10$ and $-500 \le Y \le 500$ with XSCL = 1 and YSCL = 100.

a. Write the function as a product of linear factors.

b. Use your graphing calculator to compare the graph of the function you found in part (a) with the graph shown below to verify your answer.

34.

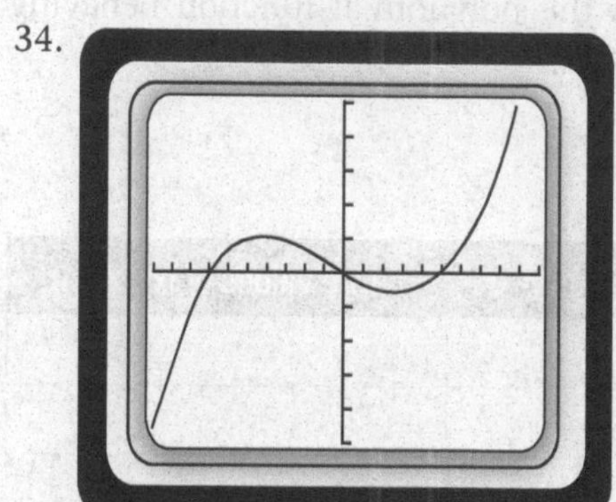

35.

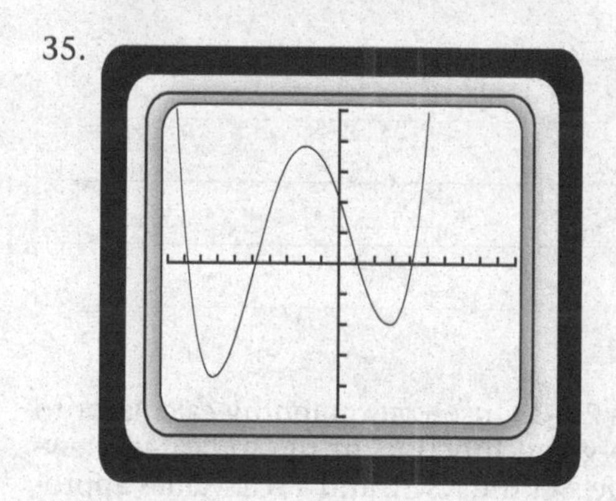

36. Construct a polynomial function to model the situation described in the chapter opener. A square piece of cardboard measuring 60 inches on a side is made into a rectangular box by cutting identical squares from the four corners and then folding up the flaps. Let x represent the length of the side of each discarded square. Write a formula for the volume of the box in terms of x. Put your formula in standard form.

37. Refer to Exercise 36. Graph your volume function, and select an appropriate viewing window. Find the maximum value of the function. What is the volume of the largest box that could be constructed in this way, and how large must the discarded squares be to obtain that volume?

38. The Post Office allows a package to have a maximum value of 130 inches for the sum of the length and girth. (The girth is the distance around the package.) Does the package of maximum volume which you have designed in Exercise 37 meet this requirement? Suppose you first cut one end from the original square of cardboard, leaving a rectangle measuring 60 inches by 40 inches. Repeat Exercises 36 and 37 for this situation. Would this package satisfy the postal requirement?

4.3 Polynomial Division and Synthetic Division

4.3a Polynomial Division

To find the zeros of a polynomial, it is necessary to divide the polynomial by a second polynomial. There is a procedure for polynomial division that parallels the long-division process of arithmetic. In arithmetic, if we divide an integer p by an integer $d \neq 0$, we obtain a quotient q and a remainder r. Thus we can write

$$\frac{p}{d} = q + \frac{r}{d} \tag{1}$$

where

$$0 \leq r < d \tag{2}$$

This result can also be written in the form

$$p = qd + r \tag{3}$$

For example,

$$\frac{7284}{13} = 560 + \frac{4}{13}$$

or

$$7284 = (560)(13) + 4$$

In the long-division process for polynomials, we divide the dividend $P(x)$ by the divisor $D(x) \neq 0$ to obtain a quotient $Q(x)$ and a remainder $R(x)$. We then have

$$\frac{P(x)}{D(x)} = Q(x) + \frac{R(x)}{D(x)} \tag{4}$$

where $R(x) = 0$ or where

$$\text{degree of } R(x) < \text{degree of } D(x) \tag{5}$$

This result can also be written as

$$P(x) = Q(x)D(x) + R(x) \tag{6}$$

Note that Equations (1) and (4) have the same form, as do Equations (3) and (6). Equation (2) requires that the remainder be less than the divisor, and the analogous requirement for polynomials in Equation (5) is that the *degree* of the remainder be less than the degree of the divisor.

We illustrate the long-division process for polynomials with an example.

EXAMPLE 1 *Polynomial Division*

Divide $3x^3 - 7x^2 + 1$ by $x - 2$.

SOLUTION

Step 1. Arrange the terms of both polynomials by descending powers of x. If a power is missing, write the term with a zero coefficient.	*Step 1.* $x - 2\overline{)\,3x^3 - 7x^2 + 0x + 1}$
Step 2. Divide the first term of the dividend by the first term of the divisor. The answer is written above the first term of the dividend.	*Step 2.* $\begin{array}{r} 3x^2 \\ x - 2\overline{)\,3x^3 - 7x^2 + 0x + 1} \end{array}$
Step 3. Multiply the divisor by the quotient obtained in *Step 2* and then subtract the product.	*Step 3.* $\begin{array}{r} 3x^2 \\ x - 2\overline{)\,3x^3 - 7x^2 + 0x + 1} \\ \underline{3x^3 - 6x^2} \\ -x^2 + 0x + 1 \end{array}$
Step 4. Repeat *Steps 2* and *3* until the remainder is zero or the degree of the remainder is less than the degree of the divisor.	*Step 4.* $\begin{array}{r} 3x^2 - x \quad - 2 = Q(x) \\ x - 2\overline{)\,3x^3 - 7x^2 + 0x + 1} \\ \underline{3x^3 - 6x^2} \\ -x^2 + 0x + 1 \\ \underline{-x^2 + 2x} \\ -2x + 1 \\ \underline{-2x + 4} \\ -3 = R(x) \end{array}$
Step 5. Write the answer in the form of Equation (4) or Equation (6)	*Step 5.* $P(x) = 3x^3 - 7x^2 + 1$ $= \underbrace{(3x^2 - x - 2)}_{Q(x)} \cdot \underbrace{(x - 2)}_{D(x)} + \underbrace{-3}_{R(x)}$

■

Progress Check

Divide $4x^2 - 3x + 6$ by $x + 2$.

Answer

$4x - 11 + \dfrac{28}{x+2}$

4.3b *Synthetic Division*

Our work in this chapter will frequently require division of a polynomial by a first-degree polynomial $x - r$, where r is a constant. Fortunately, there is a shortcut called **synthetic division** that simplifies this task. To demonstrate synthetic division, we do Example 1 again, writing only the coefficients.

$$\begin{array}{rrrrr}
 & 3 & -1 & -2 & \\ \cline{2-5}
-2) & 3 & -7 & 0 & 1 \\
 & 3 & -6 & & \\ \cline{2-5}
 & & -1 & 0 & 1 \\
 & & -1 & 2 & \\ \cline{2-5}
 & & & -2 & 1 \\
 & & & -2 & 4 \\ \cline{2-5}
 & & & & -3
\end{array}$$

Note that the boldface numerals are duplicated. We can use this to our advantage and simplify the process as follows:

$$\begin{array}{r|rrrr}
-2 & 3 & -7 & 0 & 1 \\
 & & -6 & 2 & 4 \\ \hline
 & 3 & -1 & -2 & \vert\,-3
\end{array}$$

coefficients of the quotient   remainder

In the third row we copied the leading coefficient (3) of the dividend, multiplied it by the divisor (−2), and wrote the result (−6) in the second row under the next coefficient. The numbers in the second column were subtracted to obtain $-7 - (-6) = -1$. The procedure is repeated until the third row is of the same length as the first row.

Since subtraction is more apt to produce errors than is addition, we can modify this process slightly. If the divisor is $x - r$, we write r instead of $-r$ in the box and use addition in each step instead of subtraction. Repeating our example, we have

$$\begin{array}{r|rrrr}
2 & 3 & -7 & 0 & 1 \\
 & & 6 & -2 & -4 \\ \hline
 & 3 & -1 & -2 & \vert\,-3
\end{array}$$

From Equations (5) and (6), we see that

$$P(x) = Q(x)D(x) + R(x)$$

where

$$\text{degree of } R(x) < \text{degree of } D(x)$$

Since we are only considering cases where the divisor $D(x)$ has degree 1, then the remainder $R(x)$ must have degree 0; that is, $R(x)$ is a constant.

EXAMPLE 2 *Synthetic Division*

Divide $4x^3 - 2x + 5$ by $x + 2$ using synthetic division.

SOLUTION

Step 1. If the divisor is $x - r$, write r in the box. Arrange the coefficients of the dividend by descending powers of x, supplying a zero coefficient for every missing power.	*Step 1.* $\begin{array}{r\|rrrr} -2 & 4 & 0 & -2 & 5 \end{array}$
Step 2. Copy the leading coefficient in the third row.	*Step 2.* $\begin{array}{r\|rrrr} -2 & 4 & 0 & -2 & 5 \\ & & & & \\ \hline & 4 & & & \end{array}$
Step 3. Multiply the last entry in the third row by the number in the box and write the result in the second row under the next coefficient. Add the numbers in that column.	*Step 3.* $\begin{array}{r\|rrrr} -2 & 4 & 0 & -2 & 5 \\ & & -8 & & \\ \hline & 4 & -8 & & \end{array}$
Step 4. Repeat *Step 3* until there is an entry in the third row for each entry in the first row. The last number in the third row is the remainder; the other numbers are the coefficients of the quotient in descending order.	*Step 4.* $\begin{array}{r\|rrrr} -2 & 4 & 0 & -2 & 5 \\ & & -8 & 16 & -28 \\ \hline & 4 & -8 & 14 & \|-23 \end{array}$ $\dfrac{4x^3 - 2x + 5}{x + 2} = 4x^2 - 8x + 14 - \dfrac{23}{x + 2}$

■

Progress Check

Use synthetic division to obtain the quotient $Q(x)$ and the constant remainder R when $2x^4 - 10x^2 - 23x + 6$ is divided by $x - 3$.

Answers

$Q(x) = 2x^3 + 6x^2 + 8x + 1; \quad R = 9$

a. Synthetic division can be used only when the divisor of the polynomial is a linear factor. Do not forget to write a zero for the coefficient of each missing term.

b. When dividing by $x - r$, place r in the box. For example, when the divisor is $x + 3$, place -3 in the box since $x + 3 = x - (-3)$. Similarly, when the divisor is $x - 3$, place $+3$ in the box since $x - 3 = x - (+3)$.

Exercise Set 4.3

In Exercises 1–10, use polynomial division to find the quotient $Q(x)$ and the remainder $R(x)$ when the first polynomial is divided by the second polynomial.

1. $x^2 - 7x + 12, \quad x - 5$
2. $x^2 + 3x + 3, \quad x + 2$
3. $2x^3 - 2x, \quad x^2 + 2x - 1$
4. $3x^3 - 2x^2 + 4, \quad x^2 - 2$
5. $3x^4 - 2x^2 + 1, \quad x + 3$
6. $x^5 - 1, \quad x^2 - 1$
7. $2x^3 - 3x^2, \quad x^2 + 2$
8. $3x^3 - 2x - 1, \quad x^2 - x$
9. $x^4 - x^3 + 2x^2 - x + 1, \quad x^2 + 1$
10. $2x^4 - 3x^3 - x^2 - x - 2, \quad x - 2$

In Exercises 11–20, use synthetic division to find the quotient $Q(x)$ and the constant remainder R when the first polynomial is divided by the second polynomial.

11. $x^3 - x^2 - 6x + 5, \quad x + 2$
12. $2x^3 - 3x^2 - 4, \quad x - 2$
13. $x^4 - 81, \quad x - 3$
14. $x^4 - 81, \quad x + 3$
15. $3x^3 - x^2 + 8, \quad x + 1$
16. $2x^4 - 3x^3 - 4x - 2, \quad x - 1$
17. $x^5 + 32, \quad x + 2$
18. $x^5 + 32, \quad x - 2$
19. $6x^4 - x^2 + 4, \quad x - 3$
20. $8x^3 + 4x^2 - x - 5, \quad x + 3$

4.4 The Remainder and Factor Theorems

4.4a *The Remainder Theorem*

From our work with the division process, we may surmise that division of a polynomial $P(x)$ by $x - r$ results in a quotient $Q(x)$ and a constant remainder R such that

$$P(x) = (x - r) \cdot Q(x) + R$$

Since this identity holds for all real values of x, it must hold when $x = r$. Consequently,

$$\begin{aligned} P(r) &= (r - r) \cdot Q(r) + R \\ &= 0 \cdot Q(r) + R \end{aligned}$$

or

$$P(r) = R$$

We have proved the Remainder Theorem.

Remainder Theorem
If a polynomial $P(x)$ is divided by $x - r$, the remainder is $P(r)$.

EXAMPLE 1 ***Applying the Remainder Theorem***

Determine the remainder when $P(x) = 2x^3 - 3x^2 - 2x + 1$ is divided by $x - 3$.

SOLUTION

By the Remainder Theorem, the remainder is $R = P(3)$. We then have

$$R = P(3) = 2(3)^3 - 3(3)^2 - 2(3) + 1 = 22$$

We can verify this result by using synthetic division.

$$\begin{array}{r|rrrr} 3 & 2 & -3 & -2 & 1 \\ & & 6 & 9 & 21 \\ \hline & 2 & 3 & 7 & \mathbf{22} \end{array}$$

The numeral in boldface is the remainder, so we have verified that $R = 22$. ■

Progress Check

Determine the remainder when $3x^2 - 2x - 6$ is divided by $x + 2$ using both the method of substitution and the method of synthetic division.

Answer
10

4.4b Factor Theorem

Assume that a polynomial $P(x)$ can be written as a product of polynomials.

$$P(x) = D(x)Q(x)$$

where both $D(x)$ and $Q(x)$ are of degree greater than 0. Since $D(x) \neq 0$,

$$\frac{P(x)}{D(x)} = \frac{D(x)Q(x)}{D(x)} = Q(x)$$

and we have the following definition:

The polynomial $D(x)$ is a **factor of a polynomial** $P(x)$ if division of $P(x)$ by $D(x)$ results in a remainder of zero.

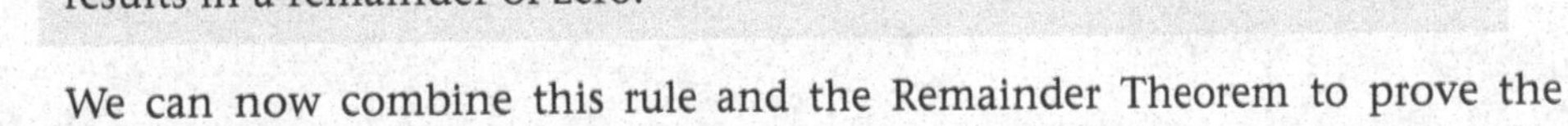

We can now combine this rule and the Remainder Theorem to prove the Factor Theorem.

Factor Theorem
A polynomial $P(x)$ has a factor $x - r$ if and only if $P(r) = 0$.

If $x - r$ is a factor of $P(x)$, then division of $P(x)$ by $x - r$ must result in a remainder of zero. By the Remainder Theorem, the remainder is $P(r)$, and hence $P(r) = 0$. Conversely, if $P(r) = 0$, then the remainder is zero and $P(x) = (x - r)Q(x)$ for some polynomial $Q(x)$ of degree one less than that of $P(x)$. By definition, $x - r$ is then a factor of $P(x)$.

EXAMPLE 2 *Applying the Factor Theorem*

Show that $x + 2$ is a factor of

$$P(x) = x^3 - x^2 - 2x + 8$$

SOLUTION
By the Factor Theorem, $x + 2 = x - (-2)$ is a factor if $P(-2) = 0$. Using synthetic division to evaluate $P(-2)$,

$$\begin{array}{r|rrrr} -2 & 1 & -1 & -2 & 8 \\ & & -2 & 6 & -8 \\ \hline & 1 & -3 & 4 & 0 \end{array}$$

we see that $P(-2) = 0$. Alternatively, we can evaluate

$$P(-2) = (-2)^3 - (-2)^2 - 2(-2) + 8 = 0$$

We conclude that $x + 2$ is a factor of $P(x)$. ■

Progress Check

Show that $x - 1$ is a factor of $P(x) = 3x^6 - 3x^5 - 4x^4 + 6x^3 - 2x^2 - x + 1$.

4.4c Summary

The following are equivalent statements for polynomial function $P(x)$ and real number r:

- r is a zero of $P(x)$.
- $x = r$ is a *root* of the equation $P(x) = 0$.
- $P(r) = 0$ (Remainder Theorem)
- $x - r$ is a *factor* of $P(x)$. (Factor Theorem)
- r is an x-intercept of the graph of the function $y = P(x)$.

Exercise Set 4.4

In Exercises 1–6, use the Remainder Theorem and synthetic division to find $P(r)$.

1. $P(x) = x^3 - 4x^2 + 1, \quad r = 2$
2. $P(x) = x^4 - 3x^2 - 5x, \quad r = -1$
3. $P(x) = x^5 - 2, \quad r = -2$
4. $P(x) = 2x^4 - 3x^3 + 6, \quad r = 2$
5. $P(x) = x^6 - 3x^4 + 2x^3 + 4, \quad r = -1$
6. $P(x) = x^6 - 2, \quad r = 1$

In Exercises 7–12, use the Remainder Theorem to determine the remainder when $P(x)$ is divided by $x - r$.

7. $P(x) = x^3 - 2x^2 + x - 3, \quad x - 2$
8. $P(x) = 2x^3 + x^2 - 5, \quad x + 2$
9. $P(x) = -4x^3 + 6x - 2, \quad x - 1$
10. $P(x) = 6x^5 - 3x^4 + 2x^2 + 7, \quad x + 1$
11. $P(x) = x^5 - 30, \quad x + 2$
12. $P(x) = x^4 - 16, \quad x - 2$

In Exercises 13–20, use the Factor Theorem to decide whether or not the first polynomial is a factor of the second polynomial.

13. $x - 2, \quad x^3 - x^2 - 5x + 6$
14. $x - 1, \quad x^3 + 4x^2 - 3x + 1$
15. $x + 2, \quad x^4 - 3x - 5$
16. $x + 1, \quad 2x^3 - 3x^2 + x + 6$
17. $x + 3, \quad x^3 + 27$
18. $x + 2, \quad x^4 + 16$
19. $x + 2, \quad x^4 - 16$
20. $x - 3, \quad x^3 + 27$

In Exercises 21–26, determine whether the given value of x is a root of the given polynomial equation.

21. $x^3 - 3x + 2 = 0, \quad x = -2$
22. $3x^2 - x + 1 = 0, \quad x = 2$
23. $-2 + x + 2x^2 - x^3 = 0, \quad x = -1$
24. $3x^2 + 2x - 1 = 0, \quad x = \frac{1}{3}$
25. $2x^2 + 4x - 1 = 0, \quad x = \frac{3}{2}$
26. $x^3 + 27 = 0, \quad x = -3$

In Exercises 27–32, determine all zeros of the polynomial function f.

27. $f(x) = (x + 1)(x - 2)$
28. $f(x) = (x - 1)^2(x + 3)$
29. $f(x) = (1 - x)(2x - 1)$
30. $f(x) = (2 - x)(x + 2)(1 + x)$
31. $f(x) = (1 - 2x)^2(1 + 2x)$
32. $f(x) = x^3(x - 2)^2$

In Exercises 33–36, use synthetic division to determine the value of k or r as requested.

33. Determine the values of r for which division of $x^2 - 2x - 1$ by $x - r$ has a remainder of 2.
34. Determine the values of r for which
$$\frac{x^2 - 6x - 1}{x - r}$$
has a remainder of −9.
35. Determine the values of k for which $x - 2$ is a factor of $x^3 - 3x^2 + kx - 1$.
36. Determine the values of k for which $2k^2x^3 + 3kx^2 - 2$ is divisible by $x - 1$.
37. Use the Factor Theorem to show that $x - 2$ is a factor of $P(x) = x^8 - 256$.
38. Use the Factor Theorem to show that $P(x) = 2x^4 + 3x^2 + 2$ has no factor of the form $x - r$, where r is a real number.
39. Use the Factor Theorem to show that $x - y$ is a factor of $x^n - y^n$, where n is a natural number.
40. *Mathematics in Writing*: Explain in your own words how and why the Remainder Theorem works to find a function value. Investigate whether the Remainder Theorem still yields correct function values when the input is a complex number. (If your graphing calculator has an i key, you can perform the operations very quickly!)

4.5 Factors and Zeros

4.5a Complex Numbers and Their Properties

We introduced the complex number system in Section 1.8. We then used this number system in Section 2.3 to provide solutions to quadratic equations. Recall that $z = a + bi$ is said to be a complex number where a and b are real numbers, and the imaginary unit $i = \sqrt{-1}$ has the property that $i^2 = -1$. We say that $a + bi$ is the *algebraic form* of z. We defined fundamental operations with complex numbers in the following way:

Equality: $a + bi = c + di$ if $a = c$ and $b = d$

Addition: $(a + bi) + (c + di) = (a + c) + (b + d)i$

Multiplication: $(a + bi)(c + di) = (ac - bd) + (ad + bc)i$

Conjugate: The conjugate of $a + bi$ is $a - bi$

Division: $\dfrac{a+bi}{c+di} = \dfrac{ac+bd}{c^2+d^2} + \dfrac{bc-ad}{c^2+d^2}i, \quad c^2 + d^2 \neq 0$

With this background, we can now explore further properties of the complex number system.

If we let $z = a + bi$, it is customary to write the conjugate $a - bi$ as $\bar{z}$. We will need the following properties of complex numbers and their **conjugates**.

Properties of Complex Conjugates

If z and w are complex numbers, then

1. $\bar{z} = \bar{w}$ if, and only if, $z = w$
2. $\bar{z} = z$ if, and only if, z is a real number
3. $\overline{z+w} = \bar{z} + \bar{w}$
4. $\overline{z \cdot w} = \bar{z} \cdot \bar{w}$
5. $\overline{z^n} = \bar{z}^n$, n is a positive integer

To prove Properties 1–5, let $z = a + bi$ and $w = c + di$. Properties 1 and 2 follow directly from the definition of equality of complex numbers. To prove Property 3, we note that $z + w = (a + c) + (b + d)i$. Then, by the definition of a complex conjugate,

$$\begin{aligned}\overline{z+w} &= (a+c) - (b+d)i \\ &= (a - bi) + (c - di) \\ &= \bar{z} + \bar{w}\end{aligned}$$

Properties 4 and 5 can be proved in a similar manner, although a proof of Property 5 requires the use of mathematical induction, a method to be discussed in a later chapter.

EXAMPLE 1 *Properties of Complex Conjugates*

If $z = 1 + 2i$ and $w = 3 - i$, verify that

a. $\overline{z+w} = \bar{z} + \bar{w}$ b. $\overline{z \cdot w} = \bar{z} \cdot \bar{w}$ c. $\overline{z^2} = \bar{z}^2$

SOLUTION

a. $z + w = (1 + 2i) + (3 - i)$ $\qquad \overline{z} = 1 - 2i$

$= 4 + i$ $\qquad \overline{w} = 3 + i$

$\overline{z+w} = 4 - i$ $\qquad \overline{z} + \overline{w} = (1 - 2i) + (3 + i)$

$= 4 - i$

Thus, $\overline{z+w} = \overline{z} + \overline{w}$.

b. $z \cdot w = (1 + 2i)(3 - i) = 5 + 5i$ $\qquad \overline{z} = 1 - 2i$

$\overline{z \cdot w} = 5 - 5i$ $\qquad \overline{w} = 3 + i$

$\overline{z} \cdot \overline{w} = (1 - 2i)(3 + i)$

$= 5 - 5i$

Thus, $\overline{z \cdot w} = \overline{z} \cdot \overline{w}$.

c. $z^2 = (1 + 2i)(1 + 2i)$ $\qquad \overline{z}^2 = (1 - 2i)(1 - 2i)$

$= -3 + 4i$ $\qquad = -3 - 4i$

$\overline{z^2} = -3 - 4i$

Thus, $\overline{z^2} = \overline{z}^2$. ■

Progress Check

If $z = 2 + 3i$ and $w = \frac{1}{2} - 2i$, verify the following:

a. $\overline{z+w} = \overline{z} + \overline{w}$ b. $\overline{z \cdot w} = \overline{z} \cdot \overline{w}$ c. $\overline{z^2} = \overline{z}^2$ d. $\overline{w^3} = \overline{w}^3$

4.5b *Factor Theorem*

By using the Factor Theorem, we can show that there is a close relationship between the factors and the zeros of the polynomial $P(x)$. By definition, r is a zero of $P(x)$ and is equivalent to $P(r) = 0$. But the Factor Theorem tells us that $P(r) = 0$ is equivalent to $x - r$ being a factor of $P(x)$. This leads to the following alternative statement of the Factor Theorem.

Factor Theorem
A polynomial $P(x)$ has a zero at $x = r$ if, and only if, $x - r$ is a factor of $P(x)$.

EXAMPLE 2 ***Application of the Factor Theorem***

Find a polynomial $P(x)$ of degree 3 whose zeros are -1, 1, and -2.

SOLUTION

By the Factor Theorem, $x + 1$, $x - 1$, and $x + 2$ are factors of $P(x)$. The product

$$P(x) = (x + 1)(x - 1)(x + 2) = x^3 + 2x^2 - x - 2$$

is a polynomial of degree 3 with the desired zeros. Note that multiplying $P(x)$ by any nonzero real number results in another polynomial with the same zeros. For example, the polynomial

$$5 \cdot P(x) = 5x^3 + 10x^2 - 5x - 10$$

also has -1, 1, and -2 as its zeros. Thus, the answer is not unique. ■

Find a polynomial $P(x)$ of degree 3 whose zeros are 2, 4, and -3.

Answer
$x^3 - 3x^2 - 10x + 24$

We began this chapter with the question, "Does a polynomial always have a zero?" The answer was supplied by Carl Friedrich Gauss (1777–1855), at age 22, in his doctoral dissertation. The importance of this theorem is reflected in its title.

The Fundamental Theorem of Algebra—Part 1
Every polynomial $P(x)$ of degree $n \geq 1$ has at least one zero among the complex numbers.

Note that the zero guaranteed by this theorem may be a real number since the real numbers are a subset of the complex number system. Furthermore, we may now see why it was necessary to create the complex numbers. No system beyond the complex numbers is required. (The proof of this theorem is beyond the scope of this text.)

In order to determine how many zeros a polynomial of degree n has, we must consider the following theorem:

Linear Factor Theorem
A polynomial $P(x)$ of degree $n \geq 1$ can be written as the product of n linear factors:

$$P(x) = a(x - r_1)(x - r_2) \cdots (x - r_n)$$

Note that a is the leading coefficient of $P(x)$ and that $r_1, r_2, \ldots, r_n$ may be complex numbers.

To prove this theorem, we first note that the Fundamental Theorem of Algebra guarantees the existence of a zero r_1. By the Factor Theorem, $x - r_1$ is a factor, and consequently,

$$P(x) = (x - r_1)Q_1(x) \tag{1}$$

where $Q_1(x)$ is a polynomial of degree $n - 1$. If $n - 1 \geq 1$ then $Q_1(x)$ must have a zero r_2. Thus,

$$Q_1(x) = (x - r_2)Q_2(x)$$

where $Q_2(x)$ is of degree $n - 2$. Substituting in Equation (1) for $Q_1(x)$, we have

$$P(x) = (x - r_1)(x - r_2)Q_2(x)$$

This process is repeated n times until $Q_n(x) = a$ is of degree 0. Hence,

$$P(x) = a(x - r_1)(x - r_2) \cdots (x - r_n) \tag{2}$$

Since a is the leading coefficient of the polynomial on the right side of Equation (2), it must also be the leading coefficient of $P(x)$.

EXAMPLE 3 ***Application of the Linear Factor Theorem***

Find the polynomial $P(x)$ of degree 3 that has the zeros -2, i, and $-i$, and satisfies $P(1) = -3$.

SOLUTION

Since -2, i, and $-i$ are zeros of $P(x)$, we may write

$$P(x) = a(x + 2)(x - i)(x + i)$$

To find the constant a, we use the condition $P(1) = -3$.

$$P(1) = -3 = a(1 + 2)(1 - i)(1 + i) = 6a$$

$$a = -\frac{1}{2}$$

Therefore,

$$P(x) = -\frac{1}{2}(x + 2)(x - i)(x + i)$$ ■

4.5c *Multiplicity of a Zero*

Recall that the zeros of a polynomial need not be different from one another. The polynomial

$$P(x) = x^2 - 2x + 1$$

can be written in factored form as

$$P(x) = (x - 1)(x - 1)$$

This shows that the zeros of $P(x)$ are 1 and 1. Since a zero is associated with a factor, and a factor may be repeated, we may have repeated zeros. If the factor $x - r$ appears k times, we say that r is a **zero of multiplicity *k*.**

We now establish an alternative form of the Fundamental Theorem of Algebra.

The Fundamental Theorem of Algebra—Part 2
If $P(x)$ is a polynomial of degree $n \geq 1$, then $P(x)$ has precisely n zeros among the complex numbers when a zero of multiplicity k is counted k times.

We can prove this theorem as follows: If we write $P(x)$ in the form of Equation (2), we see that $r_1, r_2, \ldots, r_n$ are zeros of the polynomial $P(x)$, hence there exist n zeros. If there is an additional zero r that is different from the zeros $r_1, r_2, \cdots, r_n$, then $r - r_1, r - r_2, \ldots, r - r_n$ are all different from 0. Substituting r for x in Equation (2) yields

$$P(r) = a(r - r_1)(r - r_2) \cdots (r - r_n)$$

which cannot equal 0, since the product of nonzero numbers cannot equal 0. Thus, $r_1, r_2, \ldots, r_n$ are zeros of $P(x)$, and there are no other zeros. We conclude that $P(x)$ has precisely n zeros.

EXAMPLE 4 *Polynomials in Factored Form*

Find all zeros of the polynomial

$$P(x) = \left(x - \frac{1}{2}\right)^3 (x + i)(x - 5)^4$$

SOLUTION

The distinct zeros of the polynomial are $\frac{1}{2}$, $-i$, and 5. Furthermore, $\frac{1}{2}$ is a zero of multiplicity 3, $-i$ is a zero of multiplicity 1, and 5 is a zero of multiplicity 4. ■

If we know that r is a zero of $P(x)$, we can write

$$P(x) = (x - r)Q(x)$$

We can then determine $Q(x)$ as the quotient

$$Q(x) = \frac{P(x)}{x - r}$$

This means that the degree of $Q(x)$ is one less than the degree of $P(x)$. Furthermore, if r_1 is a zero of $Q(x)$, then

$$P(r_1) = (r_1 - r)Q(r_1) = (r_1 - r) \cdot 0 = 0$$

Therefore, r_1 is also a zero of $P(x)$. We call $Q(x)$ the **quotient polynomial,** or **deflated polynomial,** of $P(x)$. Sometimes $Q(x) = 0$ is referred to as the **deflated equation** of the polynomial equation $P(x) = 0$.

EXAMPLE 5 *Finding Zeros of Polynomials*

If 4 is a zero of the polynomial $P(x) = x^3 - 8x^2 + 21x - 20$, find the other zeros.

SOLUTION

Since 4 is a zero of $P(x)$, $x - 4$ is a factor of $P(x)$. Therefore,

$$P(x) = (x - 4)Q(x)$$

To find the quotient polynomial, we compute

$$Q(x) = \frac{P(x)}{x - 4}$$

by synthetic division.

$$\begin{array}{r|rrrr} 4 & 1 & -8 & 21 & -20 \\ & & 4 & -16 & 20 \\ \hline & 1 & -4 & +5 & 0 \end{array}$$

$\underbrace{1 \quad -4 \quad +5}_{\text{coefficients of } Q(x)}$ $\quad$ 0 = remainder

The deflated equation is

$$x^2 - 4x + 5 = 0$$

Using the quadratic formula, we find the roots of the deflated equation to be $2 + i$ and $2 - i$. Therefore, the zeros of $P(x)$ are 4, $2 + i$, and $2 - i$. ■

If -2 is a zero of the polynomial $P(x) = x^3 - 7x - 6$, find the remaining zeros.

Answer

$-1, 3$

EXAMPLE 6 ***Finding Zeros and Factors of Polynomials***

If -1 is a zero of multiplicity 2 of $P(x) = x^4 + 4x^3 + 2x^2 - 4x - 3$, find the remaining zeros and write $P(x)$ as a product of linear factors.

SOLUTION

Since -1 is a double zero of $P(x)$, then $(x + 1)^2$ is a factor of $P(x)$. Therefore,

$$P(x) = (x + 1)^2 Q(x)$$

or

$$P(x) = (x^2 + 2x + 1)Q(x)$$

Using polynomial division, we divide both sides of this equation by $x^2 + 2x + 1$ to obtain

$$\begin{aligned} Q(x) &= \frac{x^4 + 4x^3 + 2x^2 - 4x - 3}{x^2 + 2x + 1} \\ &= x^2 + 2x - 3 \\ &= (x - 1)(x + 3) \end{aligned}$$

The roots of the deflated equation $Q(x) = 0$ are 1 and -3, and these are the remaining zeros of $P(x)$. By the Linear Factor Theorem,

$$P(x) = (x + 1)^2(x - 1)(x + 3)$$

■

Progress Check

If -2 is a zero of multiplicity 2 of $P(x) = x^4 + 4x^3 + 5x^2 + 4x + 4$, write $P(x)$ as a product of linear factors.

Answer

$P(x) = (x + 2)(x + 2)(x + i)(x - i)$

4.5d Conjugate Zeros

From the quadratic formula, if a quadratic equation with real coefficients has a complex root $a + bi$, $b \neq 0$, then the conjugate $a - bi$ is the other root. The following theorem extends this result to a polynomial of degree n with real coefficients.

Conjugate Zeros Theorem

If $P(x)$ is a polynomial of degree $n \geq 1$ with real coefficients, and if $a + bi$, $b \neq 0$, is a zero of $P(x)$, then the complex conjugate $a - bi$ is also a zero of $P(x)$.

To prove the Conjugate Zeros Theorem, we let $z = a + bi$ and make use of the Properties of Complex Conjugates developed earlier in this section. We may write

$$P(x) = a_n x^n + a_{n-1}x^{n-1} + \cdots + a_1 x + a_0$$

Since z is a zero of $P(x)$,

$$a_n z^n + a_{n-1}z^{n-1} + \cdots + a_1 z + a_0 = 0$$

From Property 1, we may take the conjugate of both sides of the equation to obtain

$$\overline{a_n z^n + a_{n-1}z^{n-1} + \cdots + a_1 z + a_0} = \overline{0} = 0$$

Properties 3 and 4 state that we may distribute the conjugate to sums and products respectively.

$$\overline{a_n z^n} + \overline{a_{n-1}z^{n-1}} + \cdots + \overline{a_1 z} + \overline{a_0} = 0$$

$$\overline{a_n}\,\overline{z^n} + \overline{a_{n-1}}\,\overline{z^{n-1}} + \cdots + \overline{a_1}\,\overline{z} + \overline{a_0} = 0$$

Since $a_0, a_1, \ldots, a_n$ are all real numbers, we know that $\overline{a_0} = a_0$, $\overline{a_1} = a_1, \ldots, \overline{a_n} = a_n$. Finally, using Property 5, we obtain

$$a_n \overline{z}^n + a_{n-1}\overline{z}^{n-1} + \cdots + a_1 \overline{z} + a_0 = 0$$

which establishes that $\overline{z}$ is a zero of $P(x)$.

EXAMPLE 7 *Application of the Conjugate Zeros Theorem*

Find a polynomial $P(x)$ with real coefficients of degree 3, whose zeros include -2 and $1 - i$.

SOLUTION

Since $1 - i$ is a zero of $P(x)$, it follows from the Conjugate Zeros Theorem that $1 + i$ is also a zero of $P(x)$. By the Factor Theorem, $x + 2$, $x - (1 - i)$, and $x - (1 + i)$ are factors of $P(x)$. Therefore,

$$\begin{aligned} P(x) &= (x + 2)[x - (1 - i)][x - (1 + i)] \\ &= (x + 2)(x^2 - 2x + 2) \\ &= x^3 - 2x + 4 \end{aligned}$$

■

Progress Check

Find a polynomial $P(x)$ with real coefficients of degree 4, whose zeros include i and $-3 + i$.

Answer

$P(x) = x^4 + 6x^3 + 11x^2 + 6x + 10$

The following is a corollary of the Conjugate Zeros Theorem.

> A polynomial $P(x)$ of degree $n \geq 1$ with real coefficients can be written as a product of linear and quadratic factors with real coefficients, where the quadratic factors have no real zeros.

By the Linear Factor Theorem, we may write

$$P(x) = a(x - r_1)(x - r_2) \cdots (x - r_n)$$

where $r_1, r_2, \ldots, r_n$ are the n zeros of $P(x)$. Some of these zeros may not be real numbers. Since the multiplicity of conjugate zeros must be the same (Why?), a complex zero $a + bi$, $b \neq 0$, may be paired with its conjugate $a - bi$ to provide the quadratic factor

$$[x - (a + bi)][x - (a - bi)] = x^2 - 2ax + a^2 + b^2 \tag{3}$$

which has real coefficients. Thus, a *quadratic* factor with real coefficients results from each pair of complex conjugate zeros; a *linear* factor with real coefficients results from each real zero. Furthermore, the discriminant of the quadratic factor in Equation (3) is $-4b^2$, which is always negative. Thus, this quadratic factor could never have real zeros.

4.5e *Polynomials with Complex Coefficients*

Although the definition of a polynomial, given at the beginning of this chapter, permits the coefficients to be complex numbers, we have limited our examples to polynomials with real coefficients. For completeness, we point out that both the Linear Factor Theorem and the Fundamental Theorem of Algebra hold for polynomials with complex coefficients.

On the other hand, the Conjugate Zeros Theorem may not hold if the polynomial $P(x)$ has complex coefficients. To see this, consider the polynomial

$$P(x) = x - (2 + i)$$

which has a complex coefficient and has the zero $2 + i$. The conjugate $\overline{2 + i} = 2 - i$ is *not* a zero of $P(x)$. Therefore, the Conjugate Zeros Theorem fails to apply.

EXAMPLE 8 *Polynomials with Complex Coefficients*

Find a polynomial $P(x)$ of degree 2 with complex coefficients that has the zeros -1 and $1 - i$.

SOLUTION

Since -1 is a zero of $P(x)$, $x + 1$ is a factor. Similarly, $x - (1 - i)$ is also a factor of $P(x)$. We can then write

$$\begin{aligned} P(x) &= (x + 1)[x - (1 - i)] \\ &= x^2 + ix - 1 + i \end{aligned}$$

which is a polynomial of degree 2 with complex coefficients with the required zeros. ■

Exercise Set 4.5

In Exercises 1–6, multiply by the conjugate and simplify.

1. $2-i$
2. $3+i$
3. $3+4i$
4. $2-3i$
5. $-4-2i$
6. $5+2i$

In Exercises 7–15, perform the indicated operations and write the answer in the form $a+bi$.

7. $\frac{2+5i}{1-3i}$
8. $\frac{1+3i}{2-5i}$
9. $\frac{3-4i}{3+4i}$
10. $\frac{4-3i}{4+3i}$
11. $\frac{3-2i}{2-i}$
12. $\frac{2-3i}{3-i}$
13. $\frac{2+5i}{3i}$
14. $\frac{5-2i}{-3i}$
15. $\frac{4i}{2+i}$

In Exercises 16–21, find the reciprocal and write the answer in the form $a+bi$.

16. $3+2i$
17. $4+3i$
18. $\frac{1}{2}-i$
19. $1-\frac{1}{3}i$
20. $-7i$
21. $-5i$
22. Prove that the multiplicative inverse of the complex number $a+bi$, a and b not both 0 is
$$\frac{a}{a^2+b^2}-\frac{b}{a^2+b^2}i$$
23. If z and w are complex numbers, prove that
$$\overline{z\cdot w}=\overline{z}\cdot\overline{w}$$
24. If z is a complex number, verify that $\overline{z^2}=\overline{z}^2$ and $\overline{z^3}=\overline{z}^3$.

In Exercises 25–30, find a polynomial $P(x)$ of lowest degree that has the indicated zeros.

25. $2, -4, 4$
26. $5, -5, 1, -1$
27. $-1, -2, -3$
28. $-3, \sqrt{2}, -\sqrt{2}$
29. $4, 1\pm\sqrt{3}$
30. $1, 2, 2\pm\sqrt{2}$

In Exercises 31–34, find the polynomial $P(x)$ of lowest degree that has the indicated zeros and satisfies the given condition.

31. $\frac{1}{2}, \frac{1}{2}, -2;\quad P(2)=3$
32. $3, 3, -2, 2;\quad P(4)=12$
33. $\sqrt{2}, -\sqrt{2}, 4;\quad P(-1)=5$
34. $\frac{1}{2}, -2, 5;\quad P(0)=5$

In Exercises 35–42, find the roots of the equation $P(x)=0$.

35. $(x-3)(x+1)(x-2)=0$
36. $(x-3)(x^2-3x-4)=0$
37. $(x+2)(x^2-16)=0$
38. $(x^2-x)(x^2-2x+5)=0$
39. $(x^2+3x+2)(2x^2+x)=0$
40. $(x^2+x+4)(x-3)^2=0$
41. $(x-5)^3(x+5)^2=0$
42. $(x+1)^2(x+3)^4(x-2)=0$

In Exercises 43–46, find a polynomial that has the indicated zeros and no others.

43. -2 of multiplicity 3
44. 1 of multiplicity 2, -4 of multiplicity 1
45. $\frac{1}{2}$ of multiplicity 2, -1 of multiplicity 2
46. -1 of multiplicity 2, 0 and 2 each of multiplicity 1

In Exercises 47–52, use the given root(s) to help in finding the remaining roots of the equation.

47. $x^3-3x-2=0;\quad -1$
48. $x^3-7x^2+4x+24=0;\quad 3$
49. $x^3-8x^2+18x-15=0;\quad 5$
50. $x^3-2x^2-7x-4=0;\quad -1$
51. $x^4+x^3-12x^2-28x-16=0;\quad -2$ (double root)
52. $x^4-2x^2+1=0;\quad 1$ (double root)

In Exercises 53–60, use your graphing calculator to GRAPH the given polynomial functions in the viewing rectangle $-10\le X\le 10$ and $-100\le Y\le 100$.

53. $y=(x+2)(x-3)$
54. $y=(x+2)(x-3)^2$
55. $y=(x+2)(x-3)^3$
56. $y=(x+2)^2(x-3)$
57. $y=(x+2)^3(x-3)$
58. $y=(x+2)^2(x-3)^2$
59. $y=(x+2)^3(x-3)^2$
60. $y=(x+2)^3(x-3)^3$

In Exercises 61–68, predict the shape of the following polynomial functions near $x=-2$ and near $x=3$. Determine appropriate values for YMIN and YMAX when $-10\le X\le 10$. Use your graphing calculator to GRAPH the functions to verify your predictions.

61. $y=(x+2)(x-3)^4$
62. $y=(x+2)(x-3)^5$
63. $y=(x+2)^4(x-3)$
64. $y=(x+2)^4(x-3)^2$
65. $y=(x+2)^4(x-3)^3$
66. $y=(x+2)^4(x-3)^4$
67. $y=(x+2)^4(x-3)^5$
68. $y=(x+2)^5(x-3)^5$

In Exercises 69–74, find a polynomial that has the indicated zeros and no others.

69. $1+3i, -2$

70. $1, -1, 2-i$

71. $1+i, 2-i$

72. $-2, 3, 1+2i$

73. -2 is a root of multiplicity 2, $3-2i$

74. 3 is a triple root, $-i$

In Exercises 75–80, use the given root(s) to help in writing the given polynomial as a product of linear and quadratic factors with real coefficients.

75. $x^3 - 7x^2 + 16x - 10;\ \ 3-i$

76. $x^3 + x^2 - 7x + 65;\ \ 2+3i$

77. $x^4 + 4x^3 + 13x^2 + 18x + 20;\ \ -1-2i$

78. $x^4 + 3x^3 - 5x^2 - 29x - 30;\ \ -2+i$

79. $x^5 + 3x^4 - 12x^3 - 42x^2 + 32x + 120;\ \ -3-i, -2$

80. $x^5 - 8x^4 + 29x^3 - 54x^2 + 48x - 16;\ \ 2+2i, 2$

81. Write a polynomial $P(x)$ with complex coefficients that has the zero $a+bi$, $b \neq 0$, and that does not have $a-bi$ as a zero.

82. Prove that a polynomial equation of degree 4 with real coefficients has four real roots, two real roots, or no real roots.

83. Prove that a polynomial equation of odd degree with real coefficients has at least one real root.

84. Prove that conjugate roots of a polynomial equation with real coefficients have the same multiplicity.

85. *Zeros and Extreme Values:* If you go on to study calculus, you will learn that the zeros of one function can often be used to find the maximum or minimum values of another function. For instance, the zeros of

$$12x^2 - 480x + 3600$$

can be used to find the answer to Exercise 37 in Section 4.2.

a. Find the zeros.

b. What is the relationship between the zeros of this function and the maximum value of the volume?

c. What is the significance of the second zero of the quadratic function?

86. Repeat Exercise 85 for this polynomial:

$$12x^2 - 400x + 2400$$

(These zeros will give you information about Exercise 38 in Section 4.2.)

87. Use your graphing calculator to find one root of

$$x^3 - 25x^2 + 187x - 363 = 0$$

between 0 and 10. Then use that root to help you find the other roots of the equation.

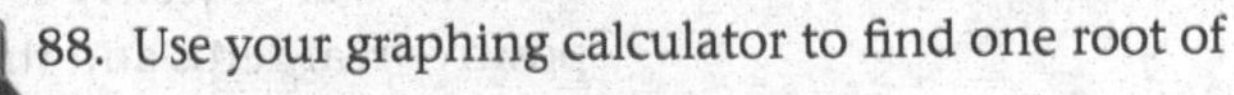

88. Use your graphing calculator to find one root of

$$4x^3 - 92x^2 - 932x - 1860 = 0$$

between -10 and 0. Then use that root to help you find the other roots of the equation.

4.6 Real, Complex, and Rational Zeros

In this section we will restrict our investigation to polynomials with real coefficients. Our objective is to obtain some information concerning the number of positive real zeros and the number of negative real zeros of such polynomials.

If the terms of a polynomial with real coefficients are written in descending order, then a **variation in sign** occurs whenever two successive terms have opposite signs. In determining the number of variations in sign, we ignore terms with zero coefficients. The polynomial

$$4x^5 - 3x^4 - 2x^2 + 1$$

has two variations in sign. The French mathematician René Descartes (1596–1650), who provided us with the foundations of analytic geometry, also gave us a theorem relating the nature of the real zeros of polynomials to the variations in sign. The proof of Descartes's Theorem is outlined in Exercises 45–50.

Descartes's Rule of Signs

If $P(x)$ is a polynomial with real coefficients and a nonzero constant term, then

a. the number of positive zeros is equal to the number of variations in sign of $P(x)$ less some nonnegative even number that could be 0;

b. the number of negative zeros is equal to the number of variations in sign of $P(-x)$ less some nonnegative even number that could be 0.

If it is determined that a polynomial of degree n has r real zeros, then the remaining $n - r$ zeros must consist of groups of conjugate pairs of complex numbers. Therefore, we see that $n - r$ must be an even number, possibly 0.

To apply Descartes's Rule of Signs to the polynomial

$$P(x) = 3x^5 + 2x^4 - x^3 + 2x - 3$$

we first note that there are three variations in sign as indicated. Thus, either there are three positive zeros or there is one positive zero. Next, we form $P(-x)$.

$$\begin{aligned} P(-x) &= 3(-x)^5 + 2(-x)^4 - (-x)^3 + 2(-x) - 3 \\ &= -3x^5 + 2x^4 + x^3 - 2x - 3 \end{aligned}$$

This can also be obtained by changing the signs of the coefficients of the odd-power terms. We see that $P(-x)$ has two variations in sign and conclude that $P(x)$ has either two negative zeros or no negative zeros. We summarize our results in Table 4-1.

TABLE 4-1 Analyzing the Nature of Zeros

$P(x) = 3x^5 + 2x^4 - x^3 + 2x - 3$			
Possible Positive Zeros	Possible Negative Zeros	Possible Complex Zeros	Total Number of Zeros
3	2	0	5
1	0	4	5
1	2	2	5

EXAMPLE 1 ***Application of Descartes's Rule of Signs***

Use Descartes's Rule of Signs to analyze the roots of the equation

$$2x^5 + 7x^4 + 3x^2 - 2 = 0$$

SOLUTION

Since

$$P(x) = 2x^5 + 7x^4 + \underbrace{3x^2 - 2}$$

has one variation in sign, there is precisely one positive zero. We calculate $P(-x)$ to be

$$P(-x) = \underbrace{-2x^5 + 7x^4} + \underbrace{3x^2 - 2}$$

There are two variations in sign, so $P(x)$ has either two negative zeros or no negative zeros. Since $P(x)$ has five zeros, the possibilities are as follows:

one positive zero, two negative zeros, two complex zeros
one positive zero, zero negative zeros, four complex zeros ■

Progress Check

Use Descartes's Rule of Signs to analyze the nature of the roots of the equation

$$x^6 + 5x^4 - 4x^2 - 3 = 0$$

Answers

one positive root, one negative root, four complex roots

4.6a *Rational Zeros*

When the coefficients of a polynomial are all integers, it is possible to search systematically for the *rational* zeros using the following theorem.

Rational Zero Theorem

If the coefficients of the polynomial

$$P(x) = a_n x^n + a_{n-1}x^{n-1} + \cdots + a_1 x + a_0, \quad a_n \neq 0$$

are all integers and $\frac{p}{q} \neq 0$ is a rational zero reduced to lowest terms, then

a. p is a factor of the constant term a_0, and
b. q is a factor of the leading coefficient a_n.

4.6b Proof of Rational Zero Theorem

Since $\frac{p}{q}$ is a zero of $P(x)$, $P\left(\frac{p}{q}\right) = 0$. Thus,

$$a_n\left(\frac{p}{q}\right)^n + a_{n-1}\left(\frac{p}{q}\right)^{n-1} + \cdots + a_1\left(\frac{p}{q}\right) + a_0 = 0 \tag{1}$$

Multiplying Equation (1) by q^n, we have

$$a_np^n + a_{n-1}p^{n-1}q + \cdots + a_1pq^{n-1} + a_0q^n = 0 \tag{2}$$

or

$$a_np^n + a_{n-1}p^{n-1}q + \cdots + a_1pq^{n-1} = -a_0q^n \tag{3}$$

Taking the common factor p out of the left-hand side of Equation (3) yields

$$p(a_np^{n-1} + a_{n-1}p^{n-2}q + \cdots + a_1q^{n-1}) = -a_0q^n$$

After dividing by p, we obtain

$$a_np^{n-1} + a_{n-0}p^{n-2}q + \cdots + a_1q^{n-1} = \frac{-a_0q^n}{p} \tag{4}$$

Since $a_1, a_2, \ldots, a_n, p$ and q are all integers, the left-hand side of Equation (4) is an integer; hence, the right-hand side is also an integer. Since $\frac{p}{q}$ is assumed to be reduced to lowest terms, p and q have no factors in common. Therefore, p must be a factor of a_0, proving part (a).

We can also rewrite Equation (2) in the form

$$q(a_{n-1}p^{n-1} + a_{n-2}p^{n-2}q + \cdots + a_1pq^{n-2} + a_0q^{n-1}) = -a_np^n$$

Thus, an argument similar to the preceding one now establishes part (b).

EXAMPLE 2 ***Rational Roots of a Polynomial Equation***

Find the rational roots of the equation

$$8x^4 - 2x^3 + 7x^2 - 2x - 1 = 0$$

SOLUTION

If $\frac{p}{q}$ is a rational root reduced to lowest terms, then p is a factor of 1 and q is a factor of 8. We can now list the possibilities:

possible numerators: ± 1 (the factors of 1)
possible denominators: $\pm 1, \pm 2, \pm 4, \pm 8$ (the factors of 8)
possible rational roots: $\pm 1, \pm\frac{1}{2}, \pm\frac{1}{4}, \pm\frac{1}{8}$

If $P(x) = 8x^4 - 2x^3 + 7x^2 - 2x - 1$, we see that $P(1) \neq 0$ and $P(-1) \neq 0$. Therefore, 1 and -1 are not roots. We may also use synthetic division to test if the other candidates are roots. Trying $\frac{1}{2}$ we have

$$\begin{array}{r|rrrrr} \frac{1}{2} & 8 & -2 & 7 & -2 & -1 \\ & & 4 & 1 & 4 & 1 \\ \hline & 8 & 2 & 8 & 2 & 0 \end{array}$$

which demonstrates that $\frac{1}{2}$ is a root. Similarly,

$$\begin{array}{r|rrrrr} -\frac{1}{4} & 8 & -2 & 7 & -2 & -1 \\ & & -2 & 1 & -2 & 1 \\ \hline & 8 & -4 & 8 & -4 & 0 \end{array}$$

which shows that $-\frac{1}{4}$ is also a root. Verify that none of the other possible rational roots produce a zero remainder. ■

Find the rational roots of the equation

$$9x^4 - 12x^3 + 13x^2 - 12x + 4 = 0$$

Answers

$\frac{2}{3}, \frac{2}{3}$

We can combine the Rational Zero Theorem and the quotient polynomial to give us more powerful methods of finding the zeros of a polynomial.

EXAMPLE 3 *Rational Roots and Deflated Equations*

Find the rational roots of the equation

$$8x^5 + 12x^4 + 14x^3 + 13x^2 + 6x + 1 = 0$$

SOLUTION

Since the coefficients of the polynomial are all integers, we may use the Rational Zero Theorem to list the possible rational roots:

possible numerators: ± 1 (the factors of 1)

possible denominators: $\pm 1, \pm 2, \pm 4, \pm 8$ (the factors of 8)

possible rational roots: $\pm 1, \pm\frac{1}{2}, \pm\frac{1}{4}, \pm\frac{1}{8}$

If $P(x) = 8x^5 + 12x^4 + 14x^3 + 13x^2 + 6x + 1$, we see that $P(1) \neq 0$, $P(-1) \neq 0$ and $P(\frac{1}{2}) \neq 0$. Therefore $1, -1$ and $\frac{1}{2}$ are not roots. Testing $-\frac{1}{2}$ by synthetic division yields

$$\begin{array}{r|rrrrrr} -\frac{1}{2} & 8 & 12 & 14 & 13 & 6 & 1 \\ & & -4 & -4 & -5 & -4 & -1 \\ \hline & 8 & 8 & 10 & 8 & 2 & 0 \end{array}$$

coefficients of quotient polynomial

Since the remainder is 0, $-\frac{1}{2}$ is a root. We now consider the deflated equation

$$8x^4 + 8x^3 + 10x^2 + 8x + 2 = 0$$

which has the same roots as

$$4x^4 + 4x^3 + 5x^2 + 4x + 1 = 0$$

While $1, -1$ and $\frac{1}{2}$ have been eliminated, we must still try $-\frac{1}{2}$ once again.

$$\begin{array}{r|rrrrr} -\frac{1}{2} & 4 & 4 & 5 & 4 & 1 \\ & & -2 & -1 & -2 & -1 \\ \hline & 4 & 2 & 4 & 2 & 0 \end{array}$$

coefficients of quotient polynomial

Observe that $-\frac{1}{2}$ is a root once again. This illustrates an important point: A rational root may be a multiple root! Applying the same technique to the resulting deflated equation

$$4x^3 + 2x^2 + 4x + 2 = 0$$

which has the same roots as

$$2x^3 + x^2 + 2x + 1 = 0$$

we obtain:

$$\begin{array}{r|rrrr} -\frac{1}{2} & 2 & 1 & 2 & 1 \\ & & -1 & 0 & -1 \\ \hline & 2 & 0 & 2 & \vert\ 0 \end{array}$$

coefficients of quotient polynomial

The final deflated equation is

$$2x^2 + 2 = 0 \quad \text{or} \quad x^2 + 1 = 0$$

which has the roots $\pm i$. Thus, the original equation has the rational roots

$$-\frac{1}{2}, -\frac{1}{2}, -\frac{1}{2}$$

Since the original equation had real coefficients, we could have used Descartes's Rule of Signs before trying different possible roots. We would have discovered that there are no positive real roots and 1, 3, or 5 negative real roots. ■

Progress Check

Find all zeros of the polynomial

$$P(x) = 9x^4 - 3x^3 + 16x^2 - 6x - 4$$

Answers

$\frac{2}{3}, -\frac{1}{3}, \pm\sqrt{2}i$

FOCUS on Solving Polynomial Equations

The quadratic formula provides us with the solutions of a polynomial equation of the second degree. How about polynomial equations of the third degree? of the fourth degree? of the fifth degree?

The search for formulas expressing the roots of polynomial equations in terms of the coefficients of the equations intrigued mathematicians for hundreds of years. A method for finding the roots of polynomial equations of degree 3 was published around 1535 and is known as Cardan's formula despite the possibility that Geronimo Cardano (also known as Cardan) stole the result from his friend Nicolo Tartaglia. Shortly afterward, a method that is attributed to Ferrari was published for solving polynomial equations of degree 4.
The next 250 years were spent in seeking formulas for the roots of polynomial equations of degree 5 or higher—without success. Finally, early in the nineteenth century, the Norwegian mathematician N. H. Abel and the French mathematician Evariste Galois proved that *no such formulas exist.* Galois's work on this problem was completed a year before his death, in a duel, at age 20. His proof, using the new concepts of group theory, was so advanced that his teachers wrote it off as being unintelligible gibberish.

Cardan's Formula

Cardano provided this formula for one root of the cubic equation:

$$x^3 + bx + c = 0$$

$$x = \sqrt[3]{\sqrt{\frac{b^3}{27} + \frac{c^2}{4}} - \frac{c}{2}} - \sqrt[3]{\sqrt{\frac{b^3}{27} + \frac{c^2}{4}} - \frac{c}{2}}$$

Try this formula for:

$$x^3 - x = 0$$

$$x^3 - 1 = 0$$

$$x^3 - 3x + 2 = 0$$

EXAMPLE 4 *Rational Zeros and Quotient Polynomials*

Write the polynomial

$$P(x) = 3x^4 + 2x^3 + 2x^2 + 2x - 1$$

as a product of linear and quadratic factors with real coefficients such that the quadratic factors have no real zeros.

SOLUTION

Since the coefficients of the polynomial are all integers, we may use the Rational Zero Theorem to obtain

possible numerators: ± 1 (factors of 1)

possible denominators: $\pm 1, \pm 3$ (factors of 3)

possible rational zeros: $\pm 1, \pm \frac{1}{3}$

Next, we note that $P(x)$ has real coefficients so that Descartes's Rule of Signs applies. There is one positive zero and one or three negative zeros. (Why?) If the positive zero is rational, it must be 1 or $\frac{1}{3}$. Since $P(1) \neq 0$, we evaluate $P(\frac{1}{3})$ using synthetic division.

$$\begin{array}{r|rrrr|r} \frac{1}{3} & 3 & 2 & 2 & 2 & -1 \\ & & 1 & 1 & 1 & 1 \\ \hline & 3 & 3 & 3 & 3 & 0 \end{array}$$

coefficients of quotient polynomial

Therefore, $\frac{1}{3}$ is a zero, and the deflated equation is

$$Q_1(x) = 3x^3 + 3x^2 + 3x + 3 = 0$$

which has the same roots as

$$Q_2(x) = x^3 + x^2 + x + 1 = 0$$

Since any zero of $Q_2(x)$ is also a zero of $P(x)$, and since we have removed the only positive zero, $Q_2(x)$ cannot have any positive zeros. (Verify that $Q_2(x)$ has no positive zeros and one or three negative zeros.) Since $Q_2(-1) = 0$, -1 is a zero. Using synthetic division, we obtain

$$\begin{array}{r|rrr|r} -1 & 1 & 1 & 1 & 1 \\ & & -1 & 0 & -1 \\ \hline & 1 & 0 & 1 & 0 \end{array}$$

coefficients of quotient polynomial

Therefore, we have the deflated equation

$$x^2 + 1 = 0$$

which has no real roots. (Why?) Thus,

$$P(x) = 3x^4 + 2x^3 + 2x^2 + 2x - 1 = 3\left(x - \frac{1}{3}\right)(x + 1)(x^2 + 1)$$

■

FOCUS on Transcendental Numbers

A real number that is a root of some polynomial equation with *integer* coefficients is said to be *algebraic*. We see that $\frac{2}{3}$ is algebraic since it is the root of the equation $3x - 2 = 0$. (Verify that all rational numbers are algebraic.) Also, $\sqrt{2}$ is algebraic since it satisfies the equation $x^2 - 2 = 0$. (Verify that the nth root of any positive integer is algebraic, where n is a positive integer.)

To show that a real number r is *not* algebraic, we must demonstrate that there is no polynomial equation with integer coefficients that has r as one of its roots. Although this appears to be an impossible task, it was performed in 1844 when Joseph Liouville exhibited specific examples of such numbers, called *transcendental* numbers. Subsequently, Georg Cantor (1845–1918) provided a more general proof of the existence of transcendental numbers.

The number π is a transcendental number; it is not a root of any polynomial equation with integer coefficients.

In Chapter 1 we discussed number systems and said that numbers such as $\sqrt{2}$ and $\sqrt{3}$ were irrational. The Rational Zero Theorem provides a direct means of verifying that this is indeed so.

EXAMPLE 5 *Application of the Rational Zero Theorem*

Prove that $\sqrt{3}$ is not a rational number.

SOLUTION

If we let $x = \sqrt{3}$, then $x^2 = 3$ or $x^2 - 3 = 0$. Let $P(x) = x^2 - 3$. By the Rational Zero Theorem, the only possible rational zeros of $P(x)$ are ± 1 and ± 3. Using substitution, we see that none of these numbers is a zero of $P(x)$, implying that $P(x)$ has *no* rational zeros. Since $P(\sqrt{3}) = 0$, $\sqrt{3}$ is a zero; hence, $\sqrt{3}$ cannot be a rational number. ■

Graphing Calculator Power User's CORNER

Analyzing Roots

We can combine the theory presented in this section with the power of the graphing calculator to obtain an efficient method for determining the nature of the roots of polynomials. Let us consider the function

$$P(x) = 2x^5 + 7x^4 + 3x^2 - 2$$

from Example 1. Using Descartes's Rule of Signs, we know that $P(x)$ has one positive root and either zero negative roots or two negative roots. Furthermore, the Rational Zero Theorem tells us that the only possible rational roots of $P(x)$ are ± 1, ± 2, and $\pm\frac{1}{2}$. We graph $P(x)$ in the default viewing rectangle as shown in Figure 4-13. We see that $P(x)$ has one positive root between 0 and 1, and two negative roots, one between 0 and -1, the other between -3 and -4. Thus, the only possible rational roots of $P(x)$ are $\pm\frac{1}{2}$. Using a calculator or synthetic division, we verify that neither $\pm\frac{1}{2}$ is a root of $P(x)$. Thus, we conclude that $P(x)$ has three irrational roots and two complex roots. We will approximate the irrational roots in Section 4.7.

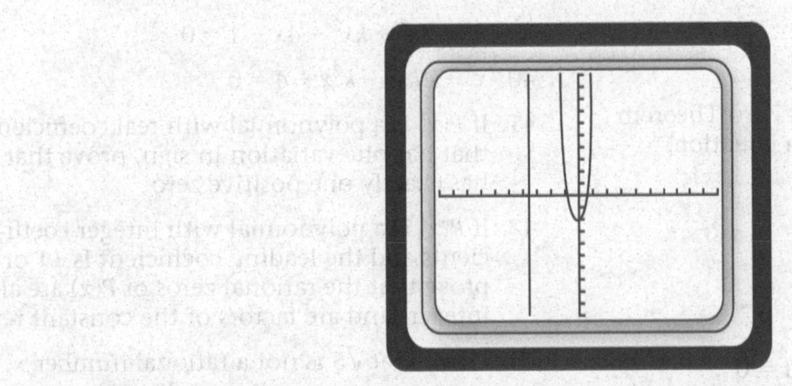

FIGURE 4-13 Graph of $y = 2x^5 + 7x^4 + 3x^2 - 2$, $-10 \le X \le 10$, $-10 \le Y \le 10$, XSCL = 1, YSCL = 1

Exercise Set 4.6

In Exercises 1–12, use Descartes's Rule of Signs to analyze the nature of the roots of the given equation. List all possibilities.

1. $3x^4 - 2x^3 + 6x^2 + 5x - 2 = 0$
2. $2x^6 + 5x^5 + x^3 - 6 = 0$
3. $x^6 + 2x^4 + 4x^2 + 1 = 0$
4. $3x^3 - 2x + 2 = 0$
5. $x^5 - 4x^3 + 7x - 4 = 0$
6. $2x^3 - 5x^2 + 8x - 2 = 0$
7. $5x^3 + 2x^2 + 7x - 1 = 0$
8. $x^5 + 6x^4 - x^3 - 2x - 3 = 0$
9. $x^4 - 2x^3 + 5x^2 + 2 = 0$
10. $3x^4 - 2x^3 - 1 = 0$
11. $x^8 + 7x^3 + 3x - 5 = 0$
12. $x^7 + 3x^5 - x^3 - x + 2 = 0$

In Exercises 13–22, use the Rational Zero Theorem to find all rational roots of the given equation.

13. $x^3 - 2x^2 - 5x + 6 = 0$
14. $3x^3 - x^2 - 3x + 1 = 0$
15. $6x^4 - 7x^3 - 13x^2 + 4x + 4 = 0$
16. $36x^4 - 15x^3 - 26x^2 + 3x + 2 = 0$
17. $5x^6 - x^5 - 5x^4 + 6x^3 - x^2 - 5x + 1 = 0$
18. $16x^4 - 16x^3 - 29x^2 + 32x - 6 = 0$
19. $4x^4 - x^3 + 5x^2 - 2x - 6 = 0$
20. $6x^4 + 2x^3 + 7x^2 + x + 2 = 0$
21. $2x^5 - 13x^4 + 26x^3 - 22x^2 + 24x - 9 = 0$
22. $8x^5 - 4x^4 + 6x^3 - 3x^2 - 2x + 1 = 0$

In Exercises 23–30, use the Rational Zero Theorem and quotient polynomials to find all roots of the given equation.

23. $4x^4 + x^3 + x^2 + x - 3 = 0$
24. $x^4 + x^3 + x^2 + 3x - 6 = 0$
25. $5x^5 - 3x^4 - 10x^3 + 6x^2 - 40x + 24 = 0$
26. $12x^4 - 52x^3 + 75x^2 - 16x - 5 = 0$
27. $6x^4 - x^3 - 5x^2 + 2x = 0$
28. $2x^4 - \frac{3}{2}x^3 + \frac{11}{2}x^2 + \frac{23}{2}x + \frac{5}{2} = 0$
29. $2x^4 - x^3 - 28x^2 + 30x - 8 = 0$
30. $12x^4 + 4x^3 - 17x^2 + 6x = 0$

In Exercises 31–36, use Descartes's Rule of Signs, the Rational Zero Theorem, and quotient polynomials to find all roots of the given equation.

31. $x^4 - 6x^3 + 10x^2 - 6x + 9 = 0$
32. $2x^4 - 3x^3 + 5x^2 - 6x + 2 = 0$
33. $x^4 - 6x^2 + 8 = 0$
34. $x^4 - 4x^3 + 7x^2 - 6x + 2 = 0$
35. $4x^4 + 4x^3 - 3x^2 - 4x - 1 = 0$
36. $x^5 + x^4 - 7x^3 - 11x^2 - 8x - 12 = 0$

In Exercises 37–40, find the integer value(s) of k for which the given equation has rational roots, and find the roots. (*Hint*: Use synthetic division.)

37. $x^3 + kx^2 + kx + 2 = 0$
38. $x^4 - 4x^3 - kx^2 + 6kx + 9 = 0$
39. $x^4 - 3x^3 + kx^2 - 4x - 1 = 0$
40. $x^3 - 4kx^2 - k^2x + 4 = 0$
41. If $P(x)$ is a polynomial with real coefficients that has one variation in sign, prove that $P(x)$ has exactly one positive zero.
42. If $P(x)$ is a polynomial with integer coefficients and the leading coefficient is +1 or −1, prove that the rational zeros of $P(x)$ are all integers and are factors of the constant term.
43. Prove that $\sqrt{5}$ is not a rational number.
44. If p is a prime, prove that $\sqrt{p}$ is not a rational number.
45. Prove that if $P(x)$ is a polynomial with real coefficients and r is a positive zero of $P(x)$, then the quotient polynomial
$$Q(x) = \frac{P(x)}{(x-r)}$$
has at least one fewer variation in sign than $P(x)$. (*Hint*: Assume the leading coefficient of $P(x)$ to be positive, and use synthetic division to obtain $Q(x)$. Note that the coefficients of $Q(x)$ remain positive at least until there is a variation in sign in $P(x)$.)
46. Prove that if $P(x)$ is a polynomial with real coefficients, the number of positive zeros is not greater than the number of variations in sign in $P(x)$. (*Hint*: Let $r_1, r_2, \ldots, r_k$ be the positive zeros of $P(x)$, and let
$$P(x) = (x - r_1)(x - r_2) \cdots (x - r_k)\, Q(x)$$
Use the result of Exercise 45 to show that $Q(x)$ has at least k fewer variations in sign than does $P(x)$.)

47. Prove that if $r_1, r_2, \ldots, r_k$ are positive numbers, then

$$P(x) = (x - r_1)(x - r_2) \cdots (x - r_k)$$

has alternating signs. (*Hint*: Use the result of Exercise 46.)

48. Prove that the number of variations in sign of a polynomial with real coefficients is even if the first and last coefficients have the same sign, and is odd if they are of opposite sign.

49. Prove that if the number of positive zeros of the polynomial $P(x)$ with real coefficients is less than the number of variations in sign, it is less by an even number. (*Hint*: Write $P(x)$ as a product of linear factors corresponding to the positive and negative zeros, and of quadratic factors corresponding to complex zeros. Apply the results of Exercises 47 and 48.)

50. Prove that the positive zeros of $P(-x)$ correspond to the negative zeros of $P(x)$; that is, prove that if $a > 0$ is a zero of $P(-x)$, then $-a$ is a zero of $P(x)$.

4.7 Approximation of the Zeros of Polynomial Functions

There are many techniques available for approximating the real zeros of polynomial functions. We shall discuss two methods that are suitable as an introduction to this topic. They will require the use of a calculator.

4.7a *Approximating Roots by Successive Digits*

Suppose we seek a real root of the polynomial equation $P(x) = 0$, and we can find values a and b such that $P(a)$ and $P(b)$ are of opposite sign. We can deduce from the Intermediate Value Theorem that there must be at least one value c in the interval (a, b) where $P(c) = 0$. Thus, c is a real root and $a < c < b$. Our problem is to get a better estimate for c.

Consider the following example with the equation

$$P(x) = 2x^4 + 4x^3 - x^2 - 10x - 10 = 0$$

Since $P(0) < 0$ and $P(3) > 0$, there must be at least one root in the interval $(0, 3)$. Let us form Table 4-2.

TABLE 4-2 $P(x)$ **in the Interval [0, 3]**

x	0	1	2	3
$P(x)$	−10	−15	30	221

Since $P(1)$ is negative and $P(2)$ is positive, there is a root in the interval $(1, 2)$. To improve our estimate, we start at $x = 1$ and increase x by tenths to form Table 4-3. Note that we only record the *sign* of $P(x)$. Also, we stop as soon as we record a change in sign. Since $P(1.5)$ is negative and $P(1.6)$ is positive, we isolate the root in the interval $(1.5, 1.6)$.

TABLE 4-3 $P(x)$ **in the Interval [1, 2]**

x	1.0	1.1	1.2	1.3	1.4	1.5	1.6	1.7	1.8	1.9	2.0
$P(x)$	−	−	−	−	−	−	+				

The process just illustrated may be repeated any number of times, providing another decimal place of accuracy at each stage. We take this one stage further as shown in Table 4-4.

TABLE 4-4 $P(x)$ **in the Interval [1.5, 1.6]**

x	1.50	1.51	1.52	1.53	1.54	1.55	1.56	1.57	1.58	1.59	1.60
$P(x)$	−	−	−	−	−	−	−	−	−	+	

At this point we can conclude that there is a root in the interval $(1.58, 1.59)$, and thus the first three digits of the root are 1.58.

Progress Check

Use the method of successive digits to find the root of the equation

$$x^3 + x^2 - 3x - 3 = 0$$

in the interval [1, 2] accurate to two decimal places.

Answer

1.73

4.7b *Approximating Roots by Bisection*

Many of the sophisticated methods for approximating roots require a knowledge of calculus and are not suitable for this text. There is, however, a technique known as **bisection** that does not require calculus, and can be of use in many circumstances.

Suppose we are told that the polynomial equation

$$P(x) = x^4 + x^3 - 5x^2 - 6x - 6 = 0$$

has a root in the interval [2, 3]. We *bisect* the interval and evaluate the polynomial at the endpoints and midpoint of the interval [2, 3] and find

$$P(2) < 0 \qquad P(2.5) > 0 \qquad P(3) > 0$$

By the Intermediate Value Theorem, we have isolated the root in the interval (2, 2.5). Repeating the process, we find that

$$P(2) < 0 \qquad P(2.25) < 0 \qquad P(2.5) > 0$$

which narrows the interval containing the root to (2.25, 2.5). We can repeat this process until we obtain a result of desired accuracy.

It is not necessarily clear that one method is better than the other in all situations. However, the method of bisection is well suited for computer usage. It is a technique for finding approximate solutions that can be applied to a large variety of problems.

Progress Check

Find a root of the equation $x^5 + x^4 + x + 2 = 0$ in the interval [−2, −1] to two decimal places by the method of bisection.

Answer

−1.27

Exercise Set 4.7

In Exercises 1–12, find a root of the polynomial equation in the stated interval by the method of successive digits to two decimal places.

1. $2x^4 - x^3 + 2x - 2 = 0 \quad [-2, -1]$
2. $3x^3 - 2x^2 + 5x + 4 = 0 \quad [-1, 0]$
3. $x^5 - 3x^3 + x^2 - 3 = 0 \quad [1, 2]$
4. $-x^4 + 3x^2 + 5 = 0 \quad [2, 3]$
5. $x^6 - 3x^3 + x^2 - 2 = 0 \quad [1, 2]$
6. $2x^5 - x^4 + 3x^2 - 6 = 0 \quad [1, 2]$
7. $-2x^3 - x^2 + 3x - 4 = 0 \quad [-2, -1]$
8. $2x^4 - 3x^3 + x^2 - 1 = 0 \quad [-1, 0]$
9. $2x^5 - 3x^2 - 5 = 0 \quad [1, 2]$
10. $2x^3 + 3x^2 - 2x + 4 = 0 \quad [-3, -2]$
11. $2x^5 - x^4 + x^2 - 3 = 0 \quad [1, 2]$
12. $2x^3 - 2x^2 + 3x - 2 = 0 \quad [0, 1]$

In Exercises 13–24, repeat Exercises 1–12 using the method of bisection.

Chapter Summary

Key Terms, Concepts, and Symbols

Term	Page
bisection method	275
completing the square	222
complex conjugate	254
continuous function	233
cusp	233
deflated equation	258
deflated polynomial	258
double root	225
"ends" of a graph	236
factor of a polynomial	251
maximum	227
minimum	227
multiplicity of a zero	257
parabola	223
parabola opening downward	223
parabola opening upward	223
polynomial equation of degree n	222
polynomial function of degree n	221
quadratic function	222
quotient polynomial	258
root of an equation	222
smooth	233
synthetic division	247
turning point	235
variation in sign	264
vertex of a parabola	223
x-intercepts of a parabola	225
$\overline{z}$	254
zero of a polynomial	222
zero of multiplicity k	257

Key Ideas for Review

Topic	Page	Key Idea
Quadratic Functions and Graphs	222	The quadratic function $f(x) = ax^2 + bx + c = 0, \quad a \neq 0$ can be written in the form $f(x) = a(x - h)^2 + k$ where h and k are constants. The graph is a parabola with vertex at (h, k), opening upward if $a > 0$ and downward if $a < 0$.
x-intercepts	225	The discriminant of the quadratic function is $b^2 - 4ac$.

$b^2 - 4a$	x-intercepts
positive	2
zero	1
negative	0

Topic	Page	Key Idea
Maximum, Minimum	227	The maximum or minimum of a quadratic function occurs at $x = -\dfrac{b}{2a}$
Polynomial Functions	233	The polynomial function of degree n has the form $P(x) = a_n x^n + a_{n-1}x^{n-1} + \cdots + a_1 x + a_0, \quad a_n \neq 0$
Zeros and Roots	234	The zeros of the polynomial function $P(x)$ are the roots of the polynomial equation $P(x) = 0$.
Graphs	233	The graph of a polynomial function of degree n is continuous and smooth, and has at most $n - 1$ turning points.
Intermediate Value Theorem	234	If P is a polynomial function and $a < b$ with $P(a) \neq P(b)$, then P assumes every value between $P(a)$ and $P(b)$ in the interval $[a, b]$.
Lead Term Dominance	236	For large values of $\lvert x \rvert$, a polynomial function is dominated by its lead term; that is, the lead term determines the sign of the function.
"Ends" of a Graph	236	For a polynomial of odd degree, the "ends" of the graph extend indefinitely in opposite directions. For a polynomial of even degree, both ends of the graph extend upward or both extend downward.

continues

Topic	Page	Key Idea
Polynomial Division	245	Polynomial division results in a quotient and a remainder, both of which are polynomials. The degree of the remainder is less than the degree of the divisor.
Synthetic Division	247	Synthetic division is a quick way to divide a polynomial by a first-degree polynomial $x - r$, where r is a real constant.
Remainder Theorem	250	If a polynomial $P(x)$ is divided by $x - r$, the remainder is $P(r)$.
Factor Theorem	251	A polynomial $P(x)$ has a zero at $x = r$ if, and only if, $x - r$ is a factor of $P(x)$.
Fundamental Theorem of Algebra	256	If $P(x)$ is a polynomial of degree $n \geq 1$, then $P(x)$ has precisely n zeros among the complex numbers when a zero of multiplicity k is counted k times.
Deflated Equation	258	If r is a real root of polynomial equation $P(x) = 0$, then the roots of $Q(x) = \frac{P(x)}{x - r} = 0$ are the other roots of $P(x) = 0$.
Conjugate Zeros Theorem	260	If $a + bi$, $b \neq 0$, is a zero of the polynomial $P(x)$ with real coefficients, then the conjugate of $a + bi$, $\overline{a+bi} = a - bi$, is also a zero of $P(x)$.
Descartes's Rule of Signs	264	If $P(x)$ is a polynomial with real coefficients and a nonzero constant term, then a. the number of positive zeros equals the number of variations in sign of $P(x)$ less some nonnegative even number that could be zero; b. the number of negative zeros equals the number of variations in sign of $P(-x)$ less some nonnegative even number that could be zero.
Rational Zero Theorem	265	If $\frac{p}{q}$ is a rational zero in lowest terms of the polynomial $P(x)$ with integer coefficients, then p is a factor of the constant term a_0 of $P(x)$, and q is a factor of the leading coefficient a_n of $P(x)$.
Listing Possible Rational Zeros	266	If $P(x)$ has integer coefficients, then the Rational Zero Theorem enables us to list all possible rational zeros of $P(x)$. Recall that r is a zero if and only if $P(r) = 0$.
Approximating Roots	275	The methods of successive digits or bisection can be used to approximate the roots of a polynomial equation.

Review Exercises

In Exercises 1 and 2, find the vertex and all intercepts of the parabola.

1. $f(x) = -x^2 - 4x$
2. $f(x) = x^2 - 5x + 7$

In Exercises 3 and 4, determine (a) if f has a maximum or minimum value; (b) the value of x at which the maximum or minimum occurs; (c) the maximum or minimum value of f.

3. $f(x) = 2x^2 - x + 1$
4. $f(x) = -x^2 - 3x - 1$

In Exercises 5 and 6, determine the behavior of the graph of the given polynomial function for large values of $|x|$.

5. $P(x) = -2x^5 + 27x^2 + 100$
6. $P(x) = 4x^3 - 10{,}000$

In Exercises 7 and 8, use synthetic division to find the quotient $Q(x)$ and the constant remainder R when the first polynomial is divided by the second polynomial.

7. $2x^3 + 6x - 4, \quad x - 1$
8. $x^4 - 3x^3 + 2x - 5, \quad x + 2$

In Exercises 9 and 10, use synthetic division to find $P(2)$ and $P(-1)$.

9. $P(x) = 7x^3 - 3x^2 + 2$
10. $P(x) = x^5 - 4x^3 + 2x$

In Exercises 11 and 12, use the Factor Theorem to show that the second polynomial is a factor of the first polynomial.

11. $2x^4 + 4x^3 + 3x^2 + 5x - 2, \quad x + 2$
12. $2x^3 - 5x^2 + 6x - 2, \quad x - \frac{1}{2}$

In Exercises 13–15, write the given quotient in the form $a + bi$.

13. $\frac{3-2i}{4+3i}$ 14. $\frac{2+i}{-5i}$ 15. $\frac{-5}{1+i}$

In Exercises 16–18, write the reciprocal of the given complex number in the form $a + bi$.

16. $1 + 3i$ 17. $-4i$ 18. $2 - 5i$

In Exercises 19–21, find a polynomial of lowest degree that has the indicated zeros.

19. $-3, -2, -1$ 20. $3, \pm\sqrt{-3}$

21. $-2, \pm\sqrt{3}, 1$

In Exercises 22–24, find a polynomial that has the indicated zeros and no others.

22. $\frac{1}{2}$ of multiplicity 2, -1 of multiplicity 2

23. $i, -i$, each of multiplicity 2

24. -1 of multiplicity 3, 3 of multiplicity 1

In Exercises 25–27, use the given root to assist in finding the remaining roots of the equation.

25. $2x^3 - x^2 - 13x - 6 = 0;\ \ -2$

26. $x^3 - 2x^2 - 9x + 4 = 0;\ \ 4$

27. $2x^4 - 15x^3 + 34x^2 - 19x - 20 = 0;\ \ -\frac{1}{2}$

In Exercises 28–31, use Descartes's Rule of Signs to determine the maximum number of positive and negative real roots of the given equation.

28. $x^4 - 2x - 1 = 0$

29. $x^5 - x^4 + 3x^3 - 4x^2 + x - 5 = 0$

30. $x^3 - 5 = 0$ 31. $3x^4 - 2x^2 + 1 = 0$

In Exercises 32–34, find all the rational roots of the given equation.

32. $6x^3 - 5x^2 - 33x - 18 = 0$

33. $6x^4 - 7x^3 - 19x^2 + 32x - 12 = 0$

34. $x^4 + 3x^3 + 2x^2 + x - 1 = 0$

In Exercises 35 and 36, find all roots of the given equation.

35. $6x^3 + 15x^2 - x - 10 = 0$

36. $2x^4 - 3x^3 - 10x^2 + 19x - 6 = 0$

Review Test

1. Find the vertex and intercepts of the parabola whose equation is $y = 3x^2 - 2x + 1$. Find the maximum or minimum value on the graph of the parabola.

Exercises 2 and 3 refer to the polynomial function

$$P(x) = -2x^9 + 3x^6 + 200$$

2. Describe the behavior of the graph of $P(x)$ for large values of $|x|$, $x > 0$.
3. Describe the behavior of the graph of $P(x)$ for large values of $|x|$, $x < 0$.
4. Find the quotient and remainder when $2x^4 - x^2 + 1$ is divided by $x^2 + 2$.
5. Use synthetic division to find the quotient and remainder when $3x^4 - x^3 - 2$ is divided by $x + 2$.
6. If $P(x) = x^3 - 2x^2 + 7x + 5$, use synthetic division to find $P(-2)$.
7. Determine the remainder when $4x^5 - 2x^4 - 5$ is divided by $x + 2$.
8. Use the Factor Theorem to show that $x - 3$ is a factor of $2x^4 - 9x^3 + 9x^2 + x - 3$.

In Exercises 9 and 10, find a polynomial of lowest degree that has the indicated zeros.

9. $-2, 1, 3$ 10. $-1, 1, 3 \pm \sqrt{2}$

In Exercises 11 and 12, find the roots of the given equation.

11. $(x^2 + 1)(x - 2) = 0$

12. $(x + 1)^2(x^2 - 3x - 2) = 0$

In Exercises 13–15, find a polynomial that has the indicated zeros and no others.

13. -3 of multiplicity 2, 1 of multiplicity 3

14. $-\frac{1}{4}$ of multiplicity 2, $i, -i, 1$

15. $i, 1 + i$

In Exercises 16 and 17, use the given root to help in finding the remaining roots of the equation.

16. $4x^3 - 3x + 1 = 0,\ \ -1$

17. $x^4 - x^2 - 2x + 2 = 0,\ \ 1$

18. If $2 + i$ is a root of $x^3 - 6x^2 + 13x - 10 = 0$, write the polynomial as a product of linear and quadratic factors with real coefficients.

In Exercises 19 and 20, determine the maximum number of roots of specified type for the given equation.

19. $2x^5 - 3x^4 + 1 = 0;$ positive real roots

20. $3x^4 + 2x^3 - 2x^2 - 1 = 0;$ negative real roots

In Exercises 21 and 22, find all rational roots of the given equation.

21. $6x^3 - 17x^2 + 14x + 3 = 0$

22. $2x^5 - x^4 - 4x^3 + 2x^2 + 2x - 1 = 0$

23. Find all roots of the equation

$$3x^4 + 7x^3 - 3x^2 + 7x - 6 = 0$$

Writing Exercises

1. Make up a word problem which uses a quadratic function and requires finding a maximum value.

2. Discuss under what conditions a polynomial function has a maximum (minimum).

3. When using the method of bisection for approximating the root of a polynomial, one has to decide when to stop the procedure. Describe criteria which can be used to make this decision.

4. Given a polynomial, describe a procedure for determining the number and type of roots (real, complex, rational, irrational). Justify each step in your explanation.

Chapter 4 Project

Cubic polynomials can be used to model volumes, as you saw in this chapter (Section 4.2, Exercises 36–38; Section 4.5, Exercises 85–86). You also took a look ahead to see how the zeros of polynomials can sometimes be used to find the maximum (or minimum) values of other polynomials. Review those exercises now.

Make sketches that show the method described in those exercises. Label them carefully.

Now, set up a general volume function for packages constructed in this way, which have the maximum allowable value for the sum of length and girth, according to postal requirements. Use L and W to represent the length and width of the original cardboard rectangle. Let x represent the length of the side of each discarded square, as before. Start by explaining in your own words why

$$(L - 2x) + 2(W - 2x) + 2x = 130 \tag{1}$$

Simplify this formula.

Suppose you always start with a square of cardboard, so that $L = W$. Replace L with W in formula (1), and then solve for x. Write a volume function for a package designed in this way. (The independent variable for this function will be W. You may simply substitute your expression which gives x in terms of W for x in the volume formula you derived in Section 4.2, Exercise 36.)

Interpret this polynomial, using any of the techniques you have learned in this chapter.

What is the largest possible volume of a package designed to meet these specifications?

5 Rational Functions and Conic Sections

How does a business make decisions about how many units of their products to produce in order to make a profit? Suppose you were a manufacturer of graphing calculators. Your fixed or start-up cost is, let's say, \$10,000. Before you produce a single calculator, you are already spending money! Now let's assume each calculator costs you \$50 to manufacture. What is the average cost to you of producing your first hundred calculators? What formula will give you the average cost of the first x calculators?

The formula you need is a rational function of x. Take a look at this chapter's project at the end of the chapter. The intersecting worlds of business, finance, and economics generally all rely upon mathematics, and especially the mathematics of functions. Investigate the history of functions and the rest of mathematics at a site hosted by the British Society for the History of Mathematics, http://www.dcs.warwick.ac.uk/bshm/resources.html.

In this chapter, we are going to investigate two types of relationships. The first is derived from the polynomial functions considered in Chapter 3. Specifically, we will study a function that is the quotient of two polynomials. Although sums, differences, and products of two polynomials still produce polynomials, quotients in general tend to be more complicated. Additionally, we will examine the various curves that arise when one considers the intersection of a plane and a cone. These curves will be the graphs of an important class of relationships that are not necessarily functions.

In 1637, the French philosopher and scientist, René Descartes, developed an idea that combined the techniques of algebra with those of geometry. He created a new field of study called **analytic geometry**, an area in which one can apply the methods and equations of algebra to the solution of problems in geometry and also obtain geometric representations of algebraic equations. In Chapter 3, we observed some applications of these methods in deriving the distance and midpoint formulas. Analytic geometry is also well suited to the study of conic sections, since geometry plays such an important role here as well.

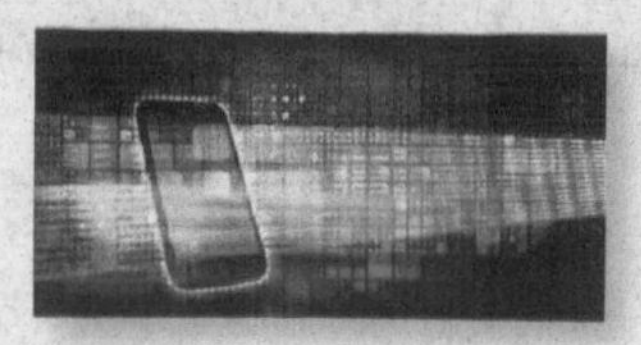

http://www.dcs.warwick.ac.uk/bshm/resources.html

5.1 Rational Functions and Their Graphs

We can apply our knowledge of polynomial functions to the study of a function of the form

$$f(x) = \frac{P(x)}{Q(x)}$$

where $P(x)$ and $Q(x)$ are polynomials and $Q(x) \neq 0$. This is called a **rational function**. (You may recall that a *rational number* is the quotient of two integers.) We will assume that the polynomials $P(x)$ and $Q(x)$ have no common factors, and we will call such a rational function **irreducible**. (Show that $P(x)$ and $Q(x)$ have no common zeros.) We will also assume that $Q(x)$ is of degree 1 or higher. (If $Q(x)$ were of degree 0, it would actually be a constant; hence $f(x)$ would be a polynomial.)

5.1a *Domain and Intercepts*

Since $P(x)$ and $Q(x)$ are polynomials, they are both defined for all real values of x. The function f can have "problems" only where the denominator is zero. Consequently, the domain of f consists of all real numbers except those for which $Q(x) = 0$.

To find the y-intercepts of the function f, set x equal to 0 and evaluate $y = f(0)$. Should $Q(0) = 0$, then $f(0)$ is undefined and there are no y-intercepts. ($f(x)$ has at most one y-intercept.)

To find the x-intercepts, we note that $y = f(x)$ can be 0 only if the numerator $P(x)$ is zero. Therefore, the x-intercepts correspond to the roots of the polynomial equation $P(x) = 0$. Since we have assumed that f is irreducible, if r is such that $P(r) = 0$, then $Q(r) \neq 0$. In other words, $P(x)$ and $Q(x)$ have no common zeros.

EXAMPLE 1 ***Domain and Intercepts***

Find the domain and intercepts of each irreducible rational function.

a. $f(x) = \dfrac{x+1}{x-1}$ b. $g(x) = \dfrac{x^3 + 2x^2 - 3x}{x^2 - 4}$ c. $h(x) = \dfrac{x^2 - 9}{x^2 + 1}$

SOLUTION

a. The denominator is 0 when $x = 1$. Thus, the domain of f is the set of all real numbers except $x = 1$.

To find the y-intercept, we set $x = 0$ and find $y = f(0) = -1$. To find the x-intercepts, we set the numerator equal to 0 and find that $y = f(x) = 0$ when $x = -1$. Summarizing, the y-intercept is $(0, -1)$, and the x-intercept is $(-1, 0)$.

b. The denominator is 0 when $x = 2$ and when $x = -2$. The domain of g is then the set of all real numbers except $x = \pm 2$.

To find the y-intercept, we set $x = 0$ and find $y = g(0) = 0$. To find the x-intercepts, we set the numerator equal to 0 and find that

$$x^3 + 2x^2 - 3x = 0$$
$$x(x^2 + 2x - 3) = 0$$
$$x(x-1)(x+3) = 0$$

has the solutions $x = 0$, $x = 1$, and $x = -3$. Summarizing, the y-intercept is $(0, 0)$, and the x-intercepts are $(0, 0)$, $(1, 0)$, and $(-3, 0)$.

c. Since the denominator $x^2 + 1$ can never be zero (Why?), the domain of h is the set of all real numbers.

To find the y-intercept, we set $x = 0$ and find $y = h(0) = -9$. To find the x-intercepts, we set the numerator $x^2 - 9 = 0$ and find that $x = \pm 3$. Summarizing, the y-intercept is $(0, -9)$, and the x-intercepts are $(3, 0)$ and $(-3, 0)$. ■

Progress Check

Find the domain and intercepts of each rational function.

a. $S(x) = \dfrac{x-3}{2x^2 - 3x - 2}$ b. $T(x) = \dfrac{5}{x^4 + x^2 + 5}$

Answers

a. *Domain:* all real numbers except $x = -\frac{1}{2}$, $x = 2$; y-intercept: $\left(0, \frac{3}{2}\right)$; x-intercept: $(3, 0)$.

b. *Domain:* all real numbers; y-intercept: $(0, 1)$; no x-intercept.

5.1b Graphing $\frac{k}{x}$ and $\frac{k}{x^2}$

We begin the study of graphs of rational functions by considering examples in which the numerator is a constant.

EXAMPLE 2 *Rational Functions with Constant Numerators*

Sketch the graph of the function

$$f(x) = \frac{1}{x}$$

SOLUTION

Domain: The denominator is 0 when $x = 0$. Thus, the domain of f is the set of all real numbers except $x = 0$.

Intercepts: There are no x-intercepts or y-intercepts.

Symmetry: The graph of f is symmetric with respect to the origin since the equation remains unchanged when x and y are replaced by $-x$ and $-y$, respectively. Therefore, we only need to plot those points corresponding to positive values of x.

x	$y = \frac{1}{x}$
$\frac{1}{1000}$	1000
$\frac{1}{100}$	100
$\frac{1}{10}$	10
1	1
2	$\frac{1}{2}$
4	$\frac{1}{4}$

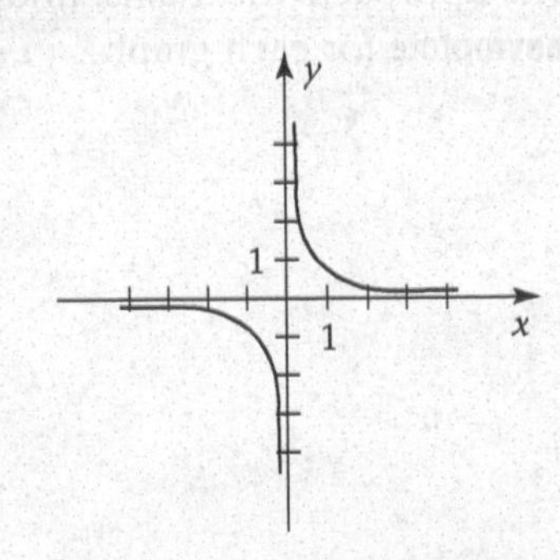

■

EXAMPLE 3 *Rational Functions with Constant Numerators*

Sketch the graph of the function

$$f(x) = \frac{1}{x^2}$$

SOLUTION

Domain: The denominator is 0 when $x = 0$. Thus, the domain of f is the set of all real numbers except $x = 0$.

Intercepts: There are no x-intercepts or y-intercepts.

Symmetry: The graph of f is symmetric with respect to the y-axis since the equation remains unchanged when x is replaced by $-x$. Therefore, we only need consider positive values of x.

x	$y = \frac{1}{x^2}$
$\frac{1}{1000}$	1,000,000
$\frac{1}{100}$	10,000
$\frac{1}{10}$	100
1	1
2	$\frac{1}{4}$
4	$\frac{1}{16}$

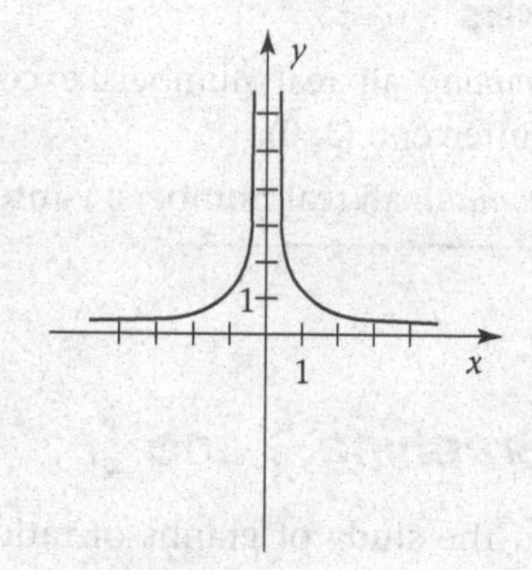

■

5.1c Asymptotes

The graphs in Example 2 and Example 3 illustrate an important concept: the graphs appear to approach specific horizontal and vertical lines without ever touching them. Such lines play an important role in the graphs of many functions, and we may define them in the following intuitive way.

> A line is said to be an **asymptote** of a graph if the graph gets closer and closer to the line as we move farther and farther out along the line.

Note the behavior of the graphs in Example 2 and Example 3 as x gets closer and closer to 0. Both graphs approach the y-axis, and we say that the line $x = 0$ (the y-axis) is a **vertical asymptote** for each graph. Similarly, as $|x|$ gets extremely large, both graphs approach the x-axis, and we say that the line $y = 0$ (the x-axis) is a **horizontal asymptote** for each graph.

EXAMPLE 4 *Using Asymptotes in Graphing*

Sketch the graph of the rational function

$$F(x) = \frac{1}{(x-1)}$$

SOLUTION

If we compare $F(x) = \frac{1}{x-1}$ with the function from Example 2, $f(x) = \frac{1}{x}$, we see that

$$F(x) = f(x-1)$$

From Section 3.3, we may observe that the graph of $F(x)$ is that of $f(x)$ shifted 1 unit to the right, as shown below. Note that the vertical asymptote has also been shifted or *translated* 1 unit to the right. (Shifting the graph to the right or left leaves the horizontal asymptote unchanged.)

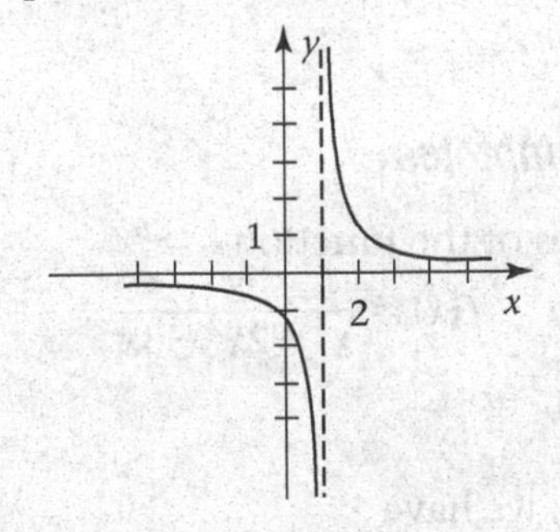

■

EXAMPLE 5 *Using Asymptotes in Graphing*

Sketch the graph of the rational function

$$F(x) = \frac{3}{(x+2)^2}$$

SOLUTION

If we compare

$$F(x) = \frac{3}{(x+2)^2}$$

with the function from Example 3, $f(x) = \frac{1}{x^2}$, we see that

$$F(x) = 3f(x+2)$$

From Section 3.3, we may observe that the graph of $F(x)$ is that of $f(x)$ shifted 2 units to the left and stretched by a factor of 3, as shown below. Note that the vertical asymptote has also been translated 2 units to the left.

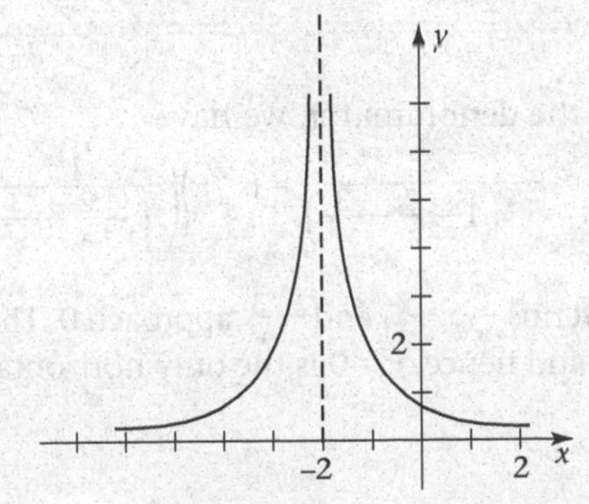

■

Since the asymptotes play an important role in graphing rational functions, it is useful to find a procedure for locating them. The graphs in each of the preceding figures of this chapter indicate that the functions *increase without bound,* **approaching** ∞, or *decrease without bound,* **approaching** $-\infty$, as the curve approaches a vertical asymptote. Note that in these cases, the absolute value of the denominator of the quotient gets closer and closer to 0. The following theorem provides the means for finding all vertical asymptotes.

Vertical Asymptote Theorem

The graph of the rational function

$$f(x) = \frac{P(x)}{Q(x)}$$

has a vertical asymptote at $x = r$ if r is a real root of $Q(x)$ but not of $P(x)$.

EXAMPLE 6 *Vertical Asymptotes*

Find the vertical asymptotes of the function

$$f(x) = \frac{2}{x^3 - 2x^2 - 3x}$$

SOLUTION

Factoring the denominator, we have

$$f(x) = \frac{2}{x(x+1)(x-3)}$$

Therefore, $x = 0$, $x = -1$, and $x = 3$ are the vertical asymptotes of $f(x)$. ■

To find the horizontal asymptotes of a function, we must examine the behavior of that function as x approaches ∞ and as x approaches $-\infty$, that is, as $|x|$ *increases without bound.* Recall the expression

$$\frac{k}{x^n}$$

where k is a constant and n is a positive integer. This expression becomes very small as $|x|$ becomes very large. In other words, $\frac{k}{x^n}$ approaches 0 as $|x|$ approaches ∞.

EXAMPLE 7 *Horizontal Asymptotes*

Find the horizontal asymptotes of the function

$$f(x) = \frac{2}{x^3 - 2x^2 - 3x}$$

SOLUTION

If we factor out x^3 from the denominator, we have

$$f(x) = \frac{2}{x^3\left(1 - \frac{2}{x} - \frac{3}{x^2}\right)} = \left(\frac{2}{x^3}\right)\left(\frac{1}{1 - \frac{2}{x} - \frac{3}{x^2}}\right)$$

As $|x|$ approaches ∞, the terms $\frac{2}{x^3}$, $-\frac{2}{x}$, and $-\frac{3}{x^2}$ approach 0. Therefore, $f(x)$ approaches 0 as $|x|$ approaches ∞; and hence, $y = 0$ is the only horizontal asymptote. ■

EXAMPLE 8 *Using Asymptotes in Graphing*

Sketch the graph of the function

$$f(x) = \frac{2}{x^3 - 2x^2 - 3x}$$

SOLUTION

Since we can write

$$f(x) = \frac{2}{x(x+1)(x-3)}$$

we determine that the critical values of $f(x)$ are -1, 0, and 3. (See the Critical Value Method in Section 2.5)

Interval	Test Point	Substitution	Sign
$x < -1$	$x = -2$	$f(-2) < 0$	$-$
$-1 < x < 0$	$x = -\frac{1}{2}$	$f(-\frac{1}{2}) > 0$	$+$
$0 < x < 3$	$x = 1$	$f(1) < 0$	$-$
$x > 3$	$x = 4$	$f(4) > 0$	$+$

These results are summarized as follows:

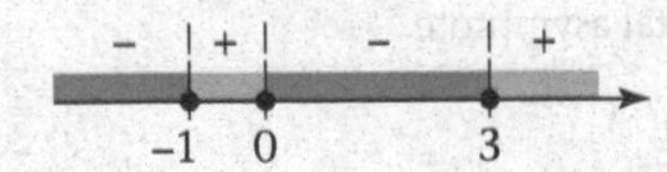

If $x < -1$ and x approaches -1, then $f(x)$ approaches $-\infty$. If $-1 < x < 0$ and x approaches -1, then $f(x)$ approaches ∞. Using a similar analysis at each point corresponding to a vertical asymptote, we may draw the following partial graphs of $f(x)$ as shown in Figure 5-1.

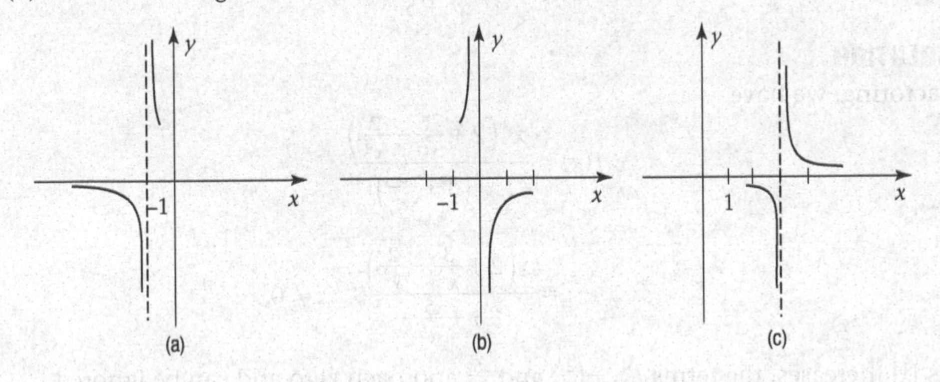

FIGURE 5-1
Partial Graphs of
$f(x) = \dfrac{2}{x(x+1)(x-3)}$

From Example 7, we have that $y = 0$ is the horizontal asymptote. Combining this observation with the portions of the graph of $f(x)$ sketched in Figure 5 1 leads to the graph of $f(x)$ sketched below.

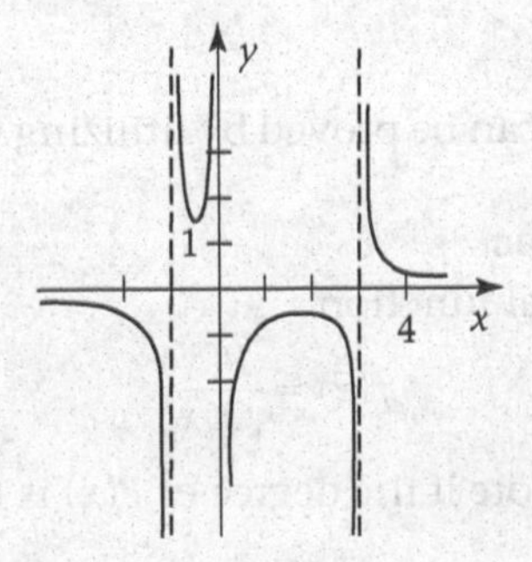

■

EXAMPLE 9 *Horizontal Asymptotes*

Find the horizontal asymptotes of the function

$$f(x) = \frac{2x^2 - 5}{3x^2 + 2x - 4}$$

SOLUTION

We illustrate the steps of the procedure as follows:

Step 1. Factor out the highest power of x found in the numerator, and factor out the highest power of x found in the denominator.	*Step 1.* $f(x) = \dfrac{x^2\left(2 - \frac{5}{x^2}\right)}{x^2\left(3 + \frac{2}{x} - \frac{4}{x^2}\right)}$
Step 2. Since we are interested in large values of $\lvert x\rvert$, we may cancel common factors in the numerator and denominator.	*Step 2.* $f(x) = \dfrac{2 - \frac{5}{x^2}}{3 + \frac{2}{x} - \frac{4}{x^2}},\ x \neq 0$
Step 3. As $\lvert x\rvert$ increases, terms of the form $\frac{k}{x^n}$ approach 0 and may be ignored.	*Step 3.* The terms $-\frac{5}{x^2}$, $\frac{2}{x}$ and $-\frac{4}{x^2}$ approach 0 as $\lvert x\rvert$ approaches ∞.
Step 4. If $f(x)$ approaches some number c, then $y = c$ is a horizontal asymptote. Otherwise, there is no horizontal asymptote.	*Step 4.* Ignoring these terms, we have $y = \frac{2}{3}$ as the horizontal asymptote.

■

EXAMPLE 10 *Horizontal Asymptotes*

Find the horizontal asymptotes of the function

$$f(x) = \frac{2x^3 + 3x - 2}{x^2 + 5}$$

SOLUTION

Factoring, we have

$$f(x) = \frac{x^3\left(2 + \frac{3}{x^2} - \frac{2}{x^3}\right)}{x^2\left(1 + \frac{5}{x^2}\right)}$$

$$= \frac{x\left(2 + \frac{3}{x^2} - \frac{2}{x^3}\right)}{1 + \frac{5}{x^2}}, \quad x \neq 0$$

As $\lvert x\rvert$ increases, the terms $\frac{3}{x^2}$, $-\frac{2}{x^3}$, and $\frac{5}{x^2}$ approach zero and can be ignored. Therefore, as $\lvert x\rvert$ increases, $f(x)$ approaches $2x$. However, as $\lvert x\rvert$ approaches ∞, $f(x)$ does not approach some number c. In fact, $\lvert y\rvert$ approaches ∞ as $\lvert x\rvert$ approaches ∞. Thus, there is no horizontal asymptote. ■

The following theorem can be proved by utilizing the procedure of Example 9.

Horizontal Asymptote Theorem

The graph of the rational function

$$f(x) = \frac{P(x)}{Q(x)}$$

has a horizontal asymptote if the degree of $P(x)$ is less than or equal to the degree of $Q(x)$.

Note that the graph of a rational function may have many vertical asymptotes but at most one horizontal asymptote. A more specific version of the Horizontal Asymptote Theorem can be found in Exercises 28 and 29.

Progress Check

Determine the horizontal asymptote of the graph of each function.

a. $f(x) = \dfrac{x-1}{2x^2+1}$ b. $g(x) = \dfrac{4x^2-3x+1}{-3x^2+1}$

c. $h(x) = \dfrac{3x^3-x+1}{2x^2-1}$

Answers

a. $y=0$ b. $y=-\frac{4}{3}$

c. no horizontal asymptote

5.1d *Sketching Graphs*

We now summarize the information that can be gathered in preparation for sketching the graph of a rational function:

- symmetry with respect to the axes and the origin
- x-intercepts, y-intercepts
- vertical asymptotes
- horizontal asymptotes
- brief table of values including points near the vertical asymptotes

EXAMPLE 11 ***Graphing Rational Functions***

Sketch the graph of

$$f(x) = \frac{x^2}{x^2-1}$$

SOLUTION

Symmetry: Replacing x with $-x$ results in the same equation, establishing symmetry with respect to the y-axis.

Intercepts: Setting $x = 0$, we obtain the y-intercept $y = 0$. Setting $y = f(x) = 0$ yields the x-intercept $x = 0$. Therefore, the point (0, 0) is both the x- and y-intercept.

Vertical asymptotes: Setting the denominator equal to zero, we find that $x = 1$ and $x = -1$ are vertical asymptotes of the graph of f.

Horizontal asymptotes: We note that

$$f(x) = \frac{x^2}{x^2\left(1-\frac{1}{x^2}\right)} = \frac{1}{1-\frac{1}{x^2}}, \quad x \neq 0$$

As $|x|$ gets larger and larger, $\frac{1}{x^2}$ approaches 0 and the values of $f(x)$ approach 1. Thus, $y = 1$ is the horizontal asymptote.

We determine the critical values of $f(x)$ to be 0 and ± 1. Our analysis of the behavior of $f(x)$ in the various intervals yields

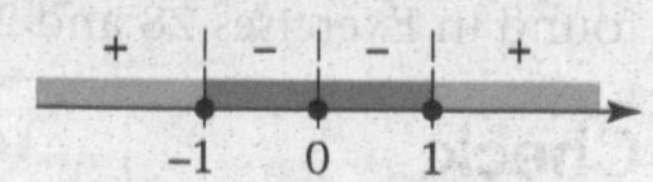

Following the technique used in Example 8 and plotting a few points, we sketch the graph:

x	y
$\frac{1}{2}$	-0.33
$\frac{3}{4}$	-1.29
$\frac{5}{4}$	2.78
$\frac{3}{2}$	1.80
2	1.33

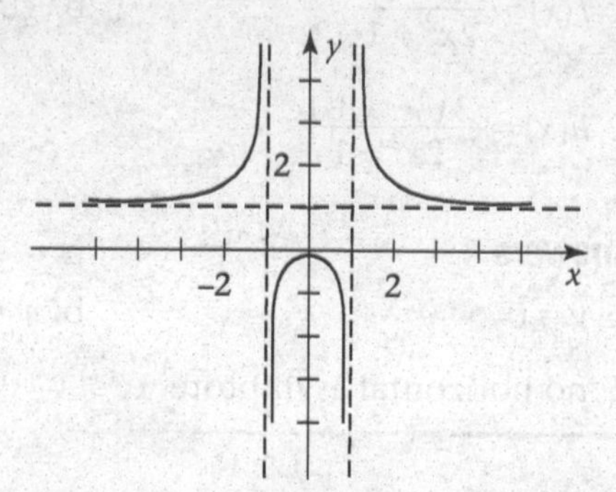

■

Progress Check

Find the horizontal and vertical asymptotes, and the x- and y-intercepts. Sketch the graph of the function

$$f(x) = \frac{x^2 - x - 6}{x^2 - 2x}$$

Answers

The horizontal asymptote is $y = 1$. The vertical asymptotes are $x = 0$ and $x = 2$. There is no y-intercept, and the x-intercepts are $(3, 0)$ and $(-2, 0)$. We sketch the graph of $f(x)$:

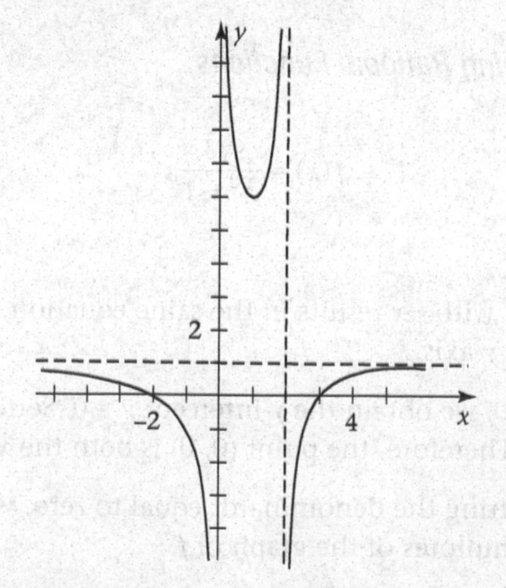

5.1e *Reducible Rational Functions*

We conclude this section with an example of a **reducible rational function**, that is, one in which the numerator and denominator have a factor in common other than a constant. Reducible rational functions are often used to illustrate functions that have **"holes"** in their graphs. Such functions are not continuous at these holes.

EXAMPLE 12 *Graphing Reducible Rational Functions*

Sketch the graph of the function

$$f(x) = \frac{x^2 - 1}{x - 1}$$

SOLUTION

We observe that

$$f(x) = \frac{x^2 - 1}{x - 1} = \frac{(x+1)(x-1)}{x-1} = x + 1, \quad x \neq 1$$

Thus, the graph of the function $f(x)$ coincides with the line $y = x + 1$, with the exception that $f(x)$ is undefined at $x = 1$. We sketch the graph of $f(x)$ below.

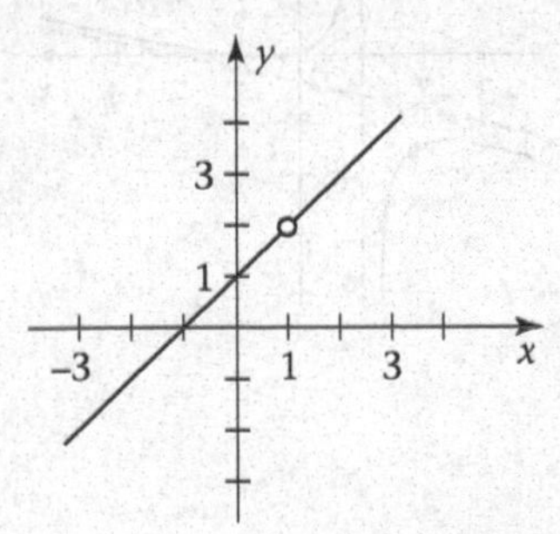

■

Progress Check

Sketch the graph of the function

$$f(x) = \frac{4 - x^2}{x + 2}$$

Answer

The graph of $f(x)$ is:

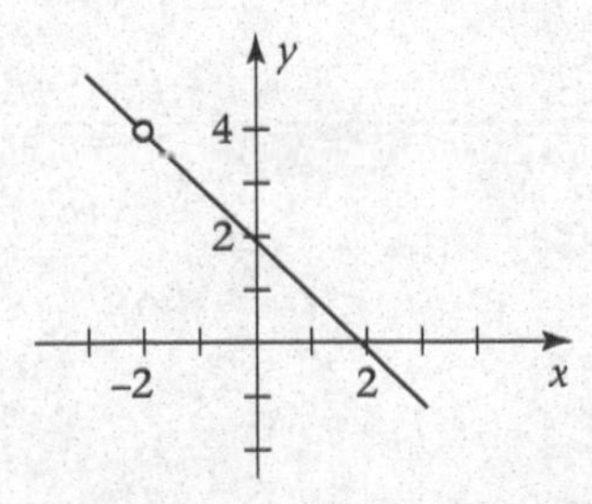

In discussing graphs and their asymptotes, we did not consider the possibility of an **oblique**, or **slanted, asymptote**. This is an asymptote that is neither horizontal nor vertical. A rational function has an oblique asymptote if the degree of the numerator is one more than the degree of the denominator.

EXAMPLE 13 *Finding Oblique Asymptotes*

Find the oblique asymptote of

$$f(x) = \frac{x^2 - x}{x + 1}$$

SOLUTION

Since the degree of the numerator − degree of the denominator = 2 − 1 = 1 there is an oblique asymptote. After we perform the division, we obtain

$$f(x) = \frac{x^2 - x}{x + 1} = x - 2 + \frac{2}{x + 1}$$

As $|x|$ gets larger and larger, $\frac{2}{x+1}$ approaches 0 and $f(x)$ gets closer and closer to its oblique asymptote, the line

$$y = x - 2$$

as shown in the following figure:

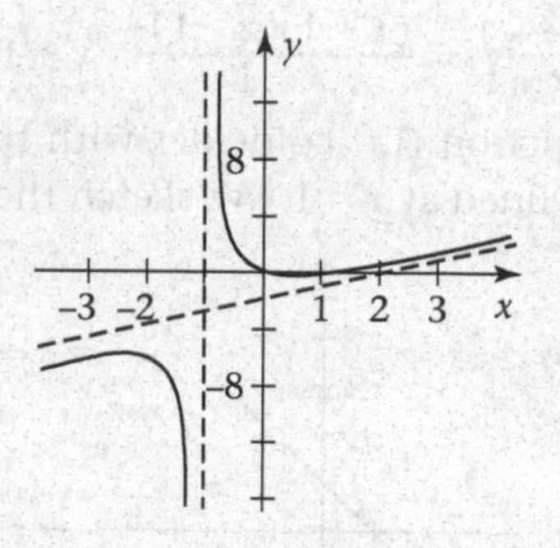

■

Exercise Set 5.1

In Exercises 1–6, determine the domain and intercepts of the given function.

1. $f(x) = \dfrac{x^2}{x-1}$
2. $f(x) = \dfrac{x-1}{x^2+x-2}$
3. $g(x) = \dfrac{x^2+1}{x^2-2x}$
4. $g(x) = \dfrac{x^2+2}{x^2-2}$
5. $F(x) = \dfrac{x^2-3}{x^2+3}$
6. $T(x) = \dfrac{3x+2}{2x^3-x^2-x}$

In Exercises 7–21, determine the vertical and horizontal asymptotes of the graph of the given function. Sketch the graph.

7. $f(x) = \dfrac{1}{x-4}$
8. $f(x) = \dfrac{-2}{x-3}$
9. $f(x) = \dfrac{3}{x+2}$
10. $f(x) = \dfrac{-1}{(x-1)^2}$
11. $f(x) = \dfrac{1}{(x+1)^2}$
12. $f(x) = \dfrac{-1}{x^2+1}$
13. $f(x) = \dfrac{x+2}{x-2}$
14. $f(x) = \dfrac{x}{x+2}$
15. $f(x) = \dfrac{2x^2+1}{x^2-4}$
16. $f(x) = \dfrac{x^2+1}{x^2+2x-3}$
17. $f(x) = \dfrac{x^2+2}{2x^2-x-6}$
18. $f(x) = \dfrac{x^2-1}{x+2}$
19. $f(x) = \dfrac{x^2}{4x-4}$
20. $f(x) = \dfrac{x-1}{2x^3-2x}$
21. $f(x) = \dfrac{x^3+4x^2+3x}{x^2-25}$

In Exercises 22–27, determine the domain and sketch the graph of the reducible function.

22. $f(x) = \dfrac{x^2-25}{2x-10}$
23. $f(x) = \dfrac{2x^2-8}{x+2}$
24. $f(x) = \dfrac{2x^2+2x-12}{3x-6}$
25. $f(x) = \dfrac{x^2+2x-8}{2x^2-8x+8}$
26. $f(x) = \dfrac{x+2}{x^2-x-6}$
27. $f(x) = \dfrac{2x}{x^2+x}$

In Exercises 28 and 29, $f(x) = \dfrac{P(x)}{Q(x)}$ is an irreducible rational function. Provide a proof for the stated theorem.

28. If the polynomials $P(x)$ and $Q(x)$ are of the same degree, then there is a horizontal asymptote $y = k$, where k is the ratio of the leading coefficients of $P(x)$ and $Q(x)$.

29. If the degree of $P(x)$ is less than the degree of $Q(x)$, then there is a horizontal asymptote at $y = 0$.

30. Suppose you are a manufacturer of graphing calculators. Your fixed or start-up cost is $10,000. Each calculator costs you $50 to manufacture.
    a. What is the average cost to you of producing your first 100 calculators?
    b. What formula will give you the average cost of the first x calculators?

31. The average cost of producing n units of a product is given by
$$A(n) = \frac{C(n)}{n}$$
Given the cost function
$$C(n) = 0.01n^3 - 0.1n^2 + 100n + 1000$$
find
    a. The average cost of producing n units
    b. The average cost as a function of x, if the number of units produced is $n = x - 2$

    c. GRAPH the average cost formulas you found in parts a and b, and use the graphs to determine when the cost is increasing or decreasing. (You will need to find an appropriate viewing WINDOW.)

5.2 The Circle

The conic sections provide us with an opportunity to illustrate the power of analytic geometry. We shall see that a geometric figure defined as a set of points can often be described analytically by an algebraic equation. Furthermore, we can start with an algebraic equation and use graphing procedures to study the properties of the curve, as well as the equation.

First, consider how the term "conic section" originates. If we pass a plane through a cone at various angles, as shown in Figure 5-2, the intersections are called **conic sections**. In exceptional cases, the intersection of a plane and a cone may be a point, a line or a pair of lines.

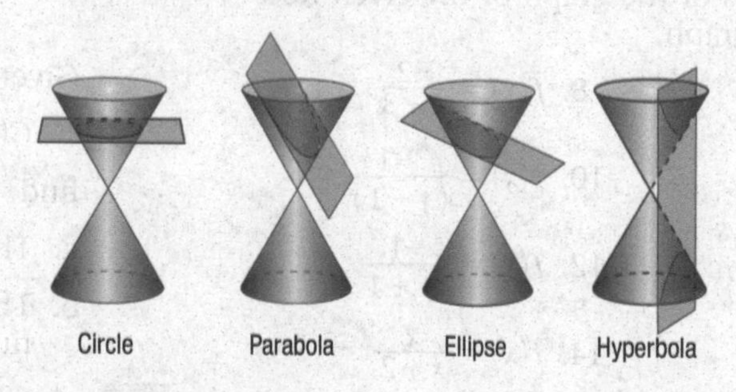

FIGURE 5-2
Examples of Conic Sections

We begin with the geometric definition of a circle.

> **Definition of a Circle**
> A **circle** is the set of all points in a plane that are at a given distance from a fixed point in the plane. The fixed point is called the **center** of the circle, and the given distance is called the **radius**.

Recall from Section 3.1 the formula for the length d of the line segment joining points $P_1(x_1, y_1)$ and $P_2(x_2, y_2)$ as

$$d = \sqrt{(x_2 - x_1)^2 + (y_2 - y_1)^2}$$

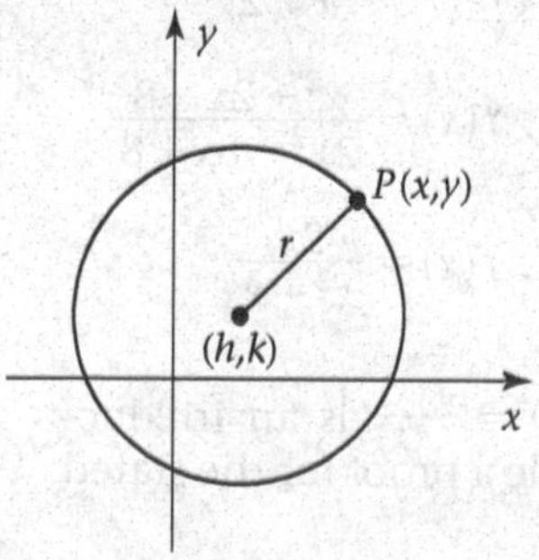

FIGURE 5-3
Deriving the Equation of a Circle

Using the methods of analytic geometry, we place the center at a point (h, k) as shown in Figure 5-3. If $P(x, y)$ is a point on the circle, then by the distance formula, the distance from P to the center (h, k) is

$$\sqrt{(x-h)^2 + (y-k)^2}$$

Since this distance is equal to the radius r, we can write

$$\sqrt{(x-h)^2 + (y-k)^2} = r$$

or

$$(x-h)^2 + (y-k)^2 = r^2$$

Since $P(x, y)$ is any point on the circle, we say that

Standard Form of the Equation of a Circle

$$(x-h)^2 + (y-k)^2 = r^2$$

is the standard form of the equation of the circle with center (h, k) and radius r.

Note the special case of

$$x^2 + y^2 = r^2$$

which is the equation of the circle of radius r, centered at the origin.

EXAMPLE 1 *Finding the Equation of a Circle*

Write the equation of the circle with center at $(2, -5)$ and radius 3.

SOLUTION

Substituting $h = 2$, $k = -5$, and $r = 3$ into the equation

$$(x-h)^2 + (y-k)^2 = r^2$$

yields

$$(x-2)^2 + (y+5)^2 = 9$$

■

EXAMPLE 2 *Finding the Center and Radius of a Circle*

Find the center and radius of the circle whose equation is

$$(x+1)^2 + (y-3)^2 = 4$$

SOLUTION

If we compare this equation with the standard form

$$(x-h)^2 + (y-k)^2 = r^2$$

we have

$$h = -1 \qquad k = 3 \qquad r = 2$$

The center is at $(-1, 3)$, and the radius is 2. ■

Progress Check

Find the center and radius of the circle whose equation is

$$\left(x - \frac{1}{2}\right)^2 + (y+5)^2 = 15$$

Answers

center $\left(\frac{1}{2}, -5\right)$, radius $\sqrt{15}$

5.2a General Form

If we are given the equation of a circle in the general form

$$Ax^2 + Ay^2 + Dx + Ey + F = 0, \qquad A \neq 0$$

in which the coefficients of x^2 and y^2 are the same, we may rewrite the equation in standard form. The process involves completing the square in each variable.

Recall from Section 2.3 that if we have the expression

$$x^2 + dx$$

we add $\left(\frac{d}{2}\right)^2$ to form

$$x^2 + dx + \frac{d^2}{2} = \left(x + \frac{d}{2}\right)^2$$

For example, starting with the expressions

$$x^2 + 4x \qquad \text{and} \qquad y^2 - 10y$$

we complete the squares in this way:

$$x^2 + 4x + 4 = (x+2)^2 \qquad \text{and} \qquad y^2 - 10y + 25 = (y-5)^2$$

EXAMPLE 3 *Standard Form of the Equation of a Circle*

Write the equation of the circle $2x^2 + 2y^2 - 12x + 16y - 31 = 0$ in standard form.

SOLUTION

Grouping the terms in x and y and factoring produces

$$2(x^2 - 6x) + 2(y^2 + 8y) = 31$$

Completing the square in both x and y, we have

$$2(x^2 - 6x + 9) + 2(y^2 + 8y + 16) = 31 + 18 + 32$$
$$2(x-3)^2 + 2(y+4)^2 = 81$$

Note that the quantities 18 and 32 were added to the right-hand side to maintain equality. The last equation can be written as

$$(x-3)^2 + (y+4)^2 = \frac{81}{2}$$

This is the standard form of the equation of the circle with center at (3, −4) and radius

$$r = \sqrt{\frac{81}{2}} = \frac{9\sqrt{2}}{2}$$

■

Progress Check

Write the equation of the circle $4x^2 + 4y^2 - 8x + 4y = 103$ in standard form, and determine the center and radius.

Answers

$(x-1)^2 + \left(y + \frac{1}{2}\right)^2 = 27$, center $\left(1, -\frac{1}{2}\right)$, radius $3\sqrt{3}$

EXAMPLE 4 ***Standard Form of the Equation of a Circle***

Write the equation $3x^2 + 3y^2 - 6x + 15 = 0$ in standard form.

SOLUTION

Regrouping, we have

$$3(x^2 - 2x) + 3y^2 = -15$$

We then complete the square in x and y:

$$3(x^2 - 2x + 1) + 3y^2 = -15 + 3$$
$$3(x - 1)^2 + 3y^2 = -12$$
$$(x - 1)^2 + y^2 = -4$$

Since $r^2 = -4$ is impossible, the graph of the equation is not a circle. (Note that the left-hand side of the equation in standard form is a sum of squares and hence nonnegative. However, the right-hand side is negative.) Thus, there are no real values of x and y that satisfy the equation. This is an example of an equation that does not have a graph. ■

EXAMPLE 5 ***Standard Form of the Equation of a Circle***

Find an equation of the circle that has its center at $C(-1, 2)$ and that passes through the point $P(3, 4)$.

SOLUTION

Since the distance from the center to any point on the circle determines the radius, we can use the distance formula to find

$$r = \overline{PC} = \sqrt{20}$$

Then we can write the equation of the circle in standard form as

$$(x + 1)^2 + (y - 2)^2 = 20$$ ■

Progress Check

Write the equation $x^2 + y^2 - 12y + 36 = 0$ in standard form, and analyze its graph.

Answers

The standard form is $x^2 + (y - 6)^2 = 0$. The equation is that of a "circle" with center at $(0, 6)$ and radius 0. The "circle" is actually the point $(0, 6)$.

Exercise Set 5.2

In Exercises 1–8, find an equation of the circle with center at (h, k) and radius r.

1. $(h, k) = (2, 3), \quad r = 2$
2. $(h, k) = (-3, 0), \quad r = 3$
3. $(h, k) = (-2, -3), \quad r = \sqrt{5}$
4. $(h, k) = (2, -4), \quad r = 4$
5. $(h, k) = (0, 0), \quad r = 3$
6. $(h, k) = (0, -3), \quad r = 2$
7. $(h, k) = (-1, 4), \quad r = 2\sqrt{2}$
8. $(h, k) = (2, 2), \quad r = 2$

In Exercises 9–16, find the center and radius of the circle with the given equation.

9. $(x-2)^2 + (y-3)^2 = 16$
10. $(x+2)^2 + y^2 = 9$
11. $(x-2)^2 + (y+2)^2 = 4$
12. $\left(x+\frac{1}{2}\right)^2 + (y-2)^2 = 8$
13. $(x+4)^2 + \left(y+\frac{3}{2}\right)^2 = 18$
14. $x^2 + (y-2)^2 = 4$
15. $\left(x-\frac{1}{3}\right)^2 + y^2 = -\frac{1}{9}$
16. $(x-1)^2 + \left(y-\frac{1}{2}\right)^2 = 3$

In Exercises 17–24, write the equation of each given circle in standard form, and determine the center and radius if it exists.

17. $x^2 + y^2 + 4x - 8y + 4 = 0$
18. $x^2 + y^2 - 2x + 6y - 15 = 0$
19. $2x^2 + 2y^2 - 6x - 10y + 6 = 0$
20. $2x^2 + 2y^2 + 8x - 12y - 8 = 0$
21. $2x^2 + 2y^2 - 4x - 5 = 0$
22. $4x^2 + 4y^2 - 2y + 7 = 0$
23. $3x^2 + 3y^2 - 12x + 18y + 15 = 0$
24. $4x^2 + 4y^2 + 4x + 4y - 4 = 0$

In Exercises 25–36, write the given equation in standard form, and determine if the graph of the equation is a circle, a point, or neither.

25. $x^2 + y^2 - 6x + 8y + 25 = 0$
26. $x^2 + y^2 + 4x + 6y + 5 = 0$
27. $x^2 + y^2 + 3x - 5y + 7 = 0$
28. $x^2 + y^2 - 4x - 6y - 13 = 0$
29. $2x^2 + 2y^2 - 12x - 4 = 0$
30. $2x^2 + 2y^2 + 4x - 4y + 25 = 0$
31. $2x^2 + 2y^2 - 6x - 4y - 2 = 0$
32. $2x^2 + 2y^2 - 10y + 6 = 0$
33. $3x^2 + 3y^2 + 12x - 4y - 20 = 0$
34. $x^2 + y^2 + x + y = 0$
35. $4x^2 + 4y^2 + 12x - 20y + 38 = 0$
36. $4x^2 + 4y^2 - 12x - 36 = 0$

37. Find the area of the circle whose equation is
$$x^2 + y^2 - 2x + 4y - 4 = 0$$
38. Find the circumference of the circle whose equation is
$$x^2 + y^2 - 6x + 8 = 0$$
39. Show that the circles whose equations are
$$x^2 + y^2 - 4x + 9y - 3 = 0$$
and
$$3x^2 + 3y^2 - 12x + 27y - 27 = 0$$
are concentric (same center).
40. Find an equation of the circle that has its center at $(3, -1)$ and passes through the point $(-2, 2)$.
41. Find an equation of the circle that has its center at $(-5, 2)$ and passes through the point $(-3, 4)$.
42. The two points $(-2, 4)$ and $(4, 2)$ are the endpoints of the diameter of a circle. Write the equation of the circle in standard form.
43. The two points $(3, 5)$ and $(7, -3)$ are the endpoints of the diameter of a circle. Write the equation of the circle in standard form.
44. Use simultaneous equations to determine the equation of the circle passing through the points $(1, 2)$, $(-1, 2)$, and $(0, 3)$, given that the general form of a circle is
$$Ax^2 + Ay^2 + Dx + Ey + F = 0, \quad A \neq 0$$
45. Convert your solution to Exercise 44 to standard form, and state the center and radius of the circle.
46. Find the equation of the path of a satellite in a circular orbit 250,000 miles above a planet whose radius is 75,000 miles.

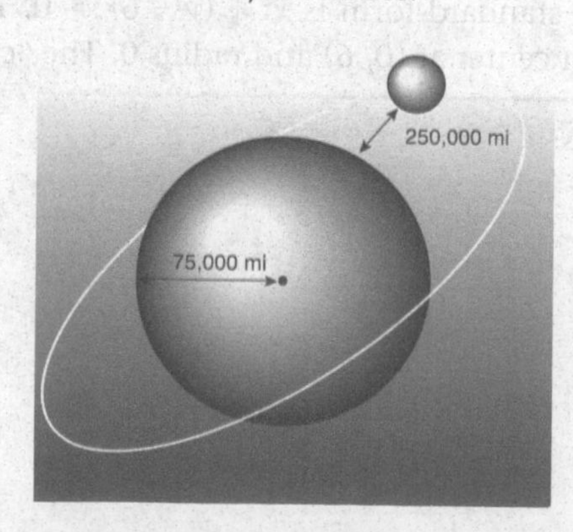

5.3 The Parabola

We begin our study of the parabola with the geometric definition

Definition of a Parabola

A **parabola** is the set of all points in a plane that are equidistant from a given point and a given line, both in the plane. The given point is called the **focus**, and the given line is called the **directrix.**

In Figure 5-4, all points P_i on the parabola are equidistant from the focus F and the directrix D; that is, $\overline{P_iF} = \overline{P_iQ_i}$. The line through the focus that is perpendicular to the directrix is called the **axis of the parabola.** This line is also called the **axis of symmetry** since the parabola is symmetric with respect to it. The point V in Figure 5-4, where the parabola intersects its axis, is called the **vertex** of the parabola. The vertex is the point from which the parabola opens. Note that the vertex is the point on the parabola that is closest to the directrix.

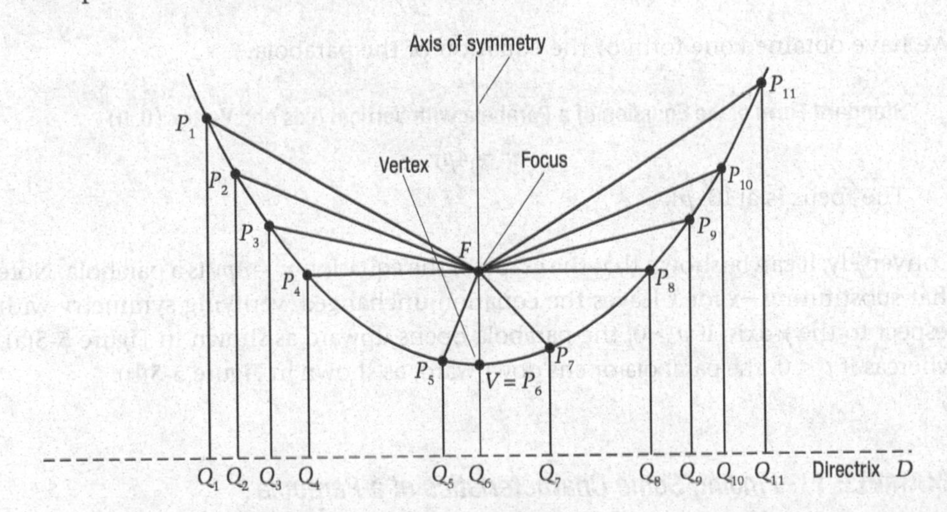

FIGURE 5-4
Directrix, Focus, Vertex, and Axis of a Parabola

We can apply the methods of analytic geometry to find an equation of a parabola. For a parabola with a vertical axis, choose that axis to be the y-axis, and take the origin as its vertex, as shown in Figure 5-5. Since the vertex is on the parabola, it is equidistant from the focus and the directrix. Thus, if the coordinates of the focus F are $(0, p)$, then the equation of the directrix is $y = -p$. We let $P(x, y)$ be any point on the parabola, and we equate the distance from P to the focus F with the distance from P to the directrix D. Using the distance formula,

$$\overline{PF} = \overline{PQ}$$

$$\sqrt{(x-0)^2+(y-p)^2} = \sqrt{(x-x)^2+(y+p)^2}$$

FIGURE 5-5
Deriving the Equation of a Parabola

(a) $p > 0$ (b) $p < 0$

Squaring both sides,

$$x^2 + y^2 - 2py + p^2 = y^2 + 2py + p^2$$
$$x^2 = 4py$$

We have obtained one form of the equation of the parabola.

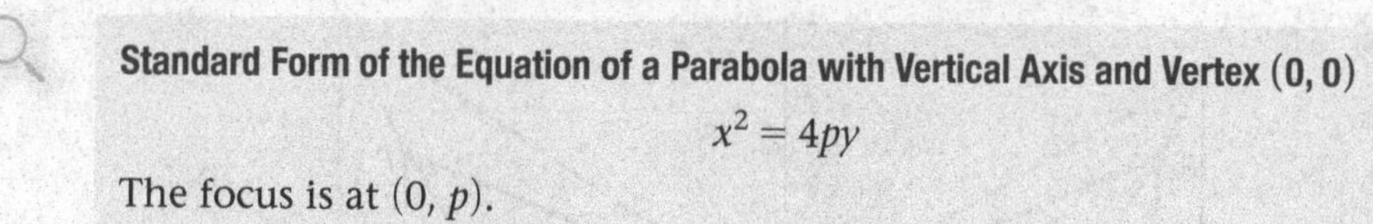

Standard Form of the Equation of a Parabola with Vertical Axis and Vertex (0, 0)

$$x^2 = 4py$$

The focus is at $(0, p)$.

Conversely, it can be shown that the graph of the equation $x^2 = 4py$ is a parabola. Note that substituting $-x$ for x leaves the equation unchanged, verifying symmetry with respect to the y-axis. If $p > 0$, the parabola opens upward as shown in Figure 5-5(a), whereas if $p < 0$, the parabola opens downward, as shown in Figure 5-5(b).

EXAMPLE 1 *Finding Some Characteristics of a Parabola*

Determine the focus and directrix of the parabola $x^2 = 8y$, and sketch its graph.

SOLUTION

The equation of the parabola is of the form

$$x^2 = 4py = 8y$$

so $p = 2$. The equation of the directrix is $y = -p = -2$, and the focus is at $(0, p) = (0, 2)$. Since $p > 0$, the parabola opens upward. The graph of the parabola is shown below.

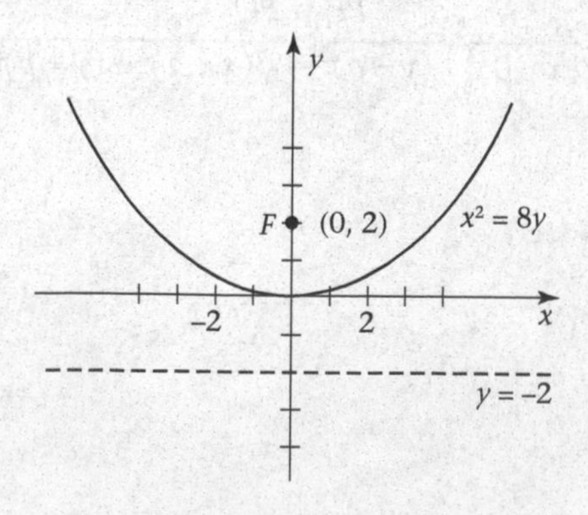

■

Determine the focus and directrix of the parabola $x^2 = -3y$.

Answers

focus at $\left(0, -\frac{3}{4}\right)$, directrix $y = \frac{3}{4}$

EXAMPLE 2 ***Finding the Equation of a Parabola***

Find the equation of the parabola with vertex at (0, 0) and focus at $\left(0, -\frac{3}{2}\right)$.

SOLUTION

Since the focus is at $(0, p)$, we have $p = -\frac{3}{2}$. The equation of the parabola is

$$x^2 = 4py$$
$$x^2 = 4\left(-\frac{3}{2}y\right)$$
$$x^2 = -6y$$

■

Progress Check

Find the equation of the parabola with vertex at (0, 0) and focus at (0, 3).

Answer

$x^2 = 12y$

If we place the parabola as shown in Figure 5-6, we can proceed as we did for a parabola with a vertical axis and vertex (0, 0) to obtain following result.

Standard Form of the Equation of a Parabola with Horizontal Axis and Vertex (0, 0)

$$y^2 = 4px$$

The focus is at $(p, 0)$.

Note that substituting $-y$ for y leaves this equation unchanged, verifying symmetry with respect to the x-axis. If $p > 0$, the parabola opens to the right as shown in Figure 5-6(a); but if $p < 0$, the parabola opens to the left as shown in Figure 5-6(b).

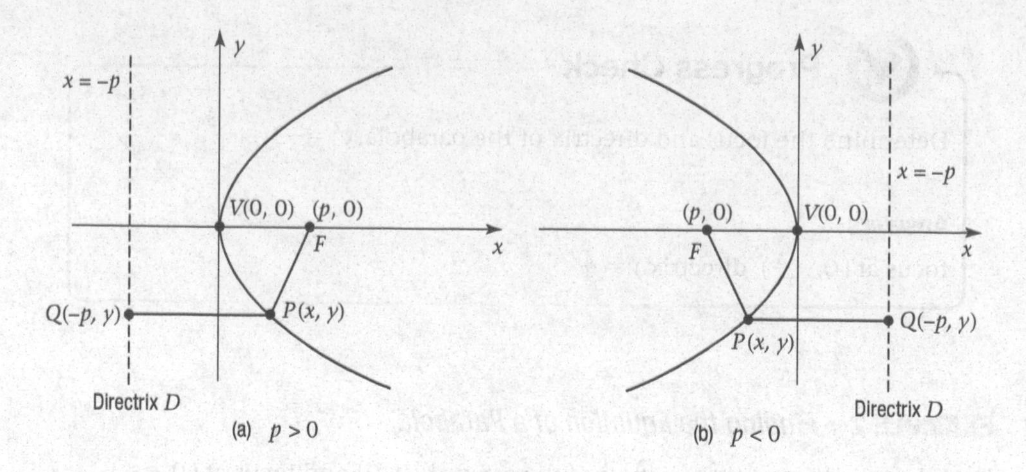

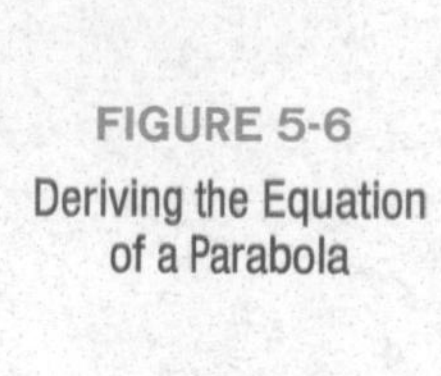

FIGURE 5-6 Deriving the Equation of a Parabola

FOCUS with a Parabolic Shape

The properties of the parabola are used in the design of some important devices. For example, by rotating a parabola about its axis, we obtain a *parabolic reflector*, a shape used in the headlight of an automobile. The light source (the bulb) is placed at the focus of the parabola. The headlight is coated with a reflecting material, and the rays of light bounce back in lines that are parallel to the axis of the parabola, permitting a headlight to disperse light in front of the automobile where it is needed.

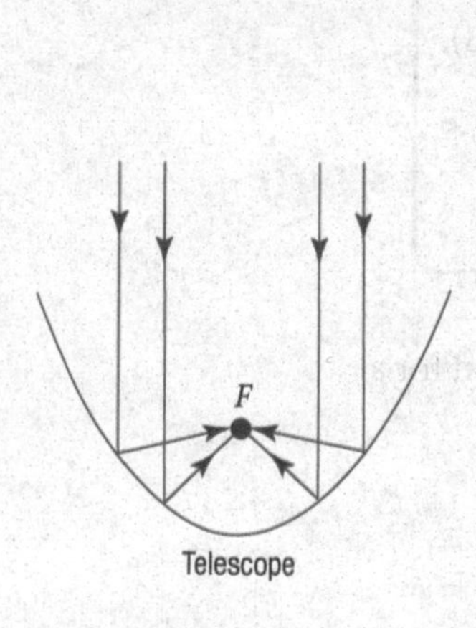

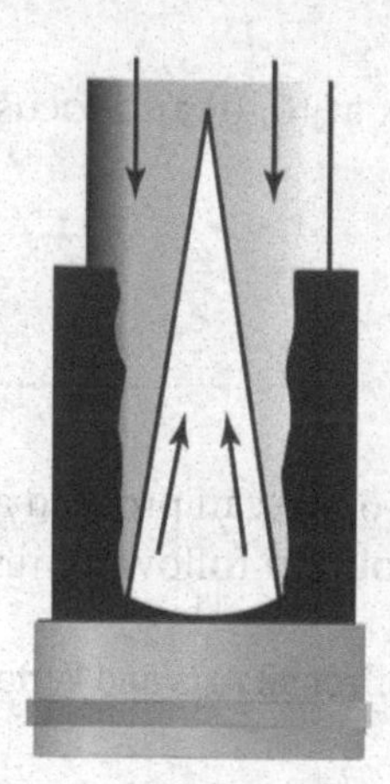

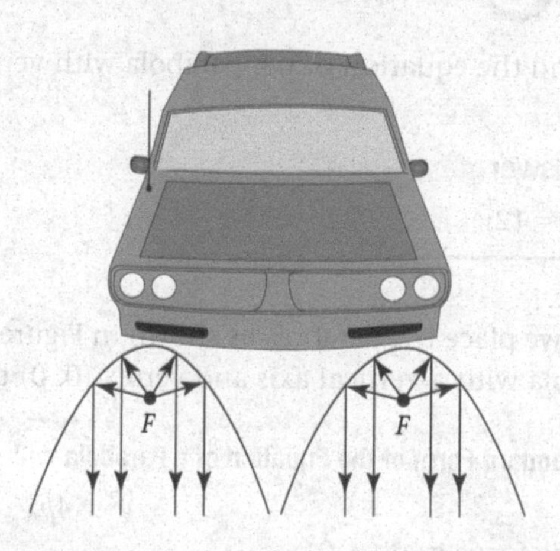

A reflecting telescope reverses the use of these same properties. Here, the rays of light from a distant star, which are nearly parallel to the axis of the parabola, are reflected by the mirror to the focus. The eyepiece is placed at the focus, where the rays of light are gathered.

EXAMPLE 3 *Finding the Equation of a Parabola*

Find the equation of the parabola with vertex at (0, 0) and directrix $x = \frac{1}{2}$.

SOLUTION

The directrix is $x = -p$, so $p = -\frac{1}{2}$. The equation of the parabola is then

$$y^2 = 4px$$
$$y^2 = 4\left(-\frac{1}{2}\right)x$$
$$y^2 = -2x$$

■

EXAMPLE 4 *Finding the Equation of a Parabola*

Find the equation of the parabola that has its axis on the x-axis, its vertex at (0, 0), and passes through the point (−2, 3).

SOLUTION

Since the axis of the parabola is the x-axis, the equation of the parabola is $y^2 = 4px$. The parabola passes through the point (−2, 3), so the coordinates of this point must satisfy the equation of the parabola. Thus,

$$y^2 = 4px$$
$$(3)^2 = 4p(-2)$$
$$4p = -\frac{9}{2}$$

and the equation of the parabola is

$$y^2 = 4px$$
$$y^2 = -\frac{9}{2}x$$

■

Progress Check

Find the equation of the parabola that has its axis on the y-axis, its vertex at (0, 0), and passes through the point (1, −2).

Answer

$x^2 = -\frac{1}{2}y$

5.3a Vertex at (h, k)

It is also possible to determine an equation of a parabola when the vertex is at some arbitrary point (h, k). The form of the equation depends on whether the axis of the parabola is parallel to the x-axis or to the y-axis. The situations are summarized in Table 5-1. Note that if the point (h, k) is the origin, then $h = k = 0$, and we arrive at the equations we derived previously. Thus, in all cases, the sign of the constant p determines the direction in which the parabola opens. Furthermore, an equation of a parabola can always be written in one of the standard forms shown in Table 5-1.

TABLE 5-1 Standard Forms of the Equation of a Parabola with Vertex (h, k)

Equation	Axis	Directrix	Direction of Opening
$(x-h)^2 = 4p(y-k)$	$x = h$	$y = k - p$	Up if $p > 0$ Down if $p < 0$
$(y-k)^2 = 4p(x-h)$	$y = k$	$x = h - p$	Right if $p > 0$ Left if $p < 0$

Note that these changes in the equations of the parabola are similar to the change that occurs in the equation of the circle when the center is moved from the origin to a point (h, k). In both cases, x is replaced by $x - h$ and y is replaced by $y - k$.

EXAMPLE 5 Finding Some Characteristics of a Parabola

Determine the vertex, axis, and the direction in which the parabola opens.

$$\left(x - \frac{1}{2}\right)^2 = -3(y + 4)$$

SOLUTION

Comparison of the equation with the standard form

$$(x-h)^2 = 4p(y-k)$$

yields $h = \frac{1}{2}$, $k = -4$, and $p = -\frac{3}{4}$. The axis of the parabola is found by setting the square term equal to 0.

$$\left(x - \frac{1}{2}\right)^2 = 0$$

$$x = \frac{1}{2}$$

Thus, the vertex is at $(h, k) = (\frac{1}{2}, -4)$, the axis is $x = \frac{1}{2}$, and the parabola opens downward since $p < 0$. ■

Progress Check

Determine the vertex, axis, and the direction in which the parabola opens.

$$3(y+1)^2 = 12\left(x - \frac{1}{3}\right)$$

Answers

vertex $(\frac{1}{3}, -1)$, axis $y = -1$, opens to the right

EXAMPLE 6 ***Finding Some Characteristics of a Parabola***

Locate the vertex and the axis of symmetry of each of the given parabolas. Sketch the graph.

a. $x^2 + 2x - 2y - 3 = 0$ b. $y^2 - 4y + x + 1 = 0$

SOLUTION

a. We complete the square in x:

$$x^2 + 2x = 2y + 3$$
$$x^2 + 2x + 1 = 2y + 3 + 1$$
$$(x + 1)^2 = 2(y + 2)$$

The vertex of the parabola is at $(-1, -2)$ and the axis is $x = -1$.

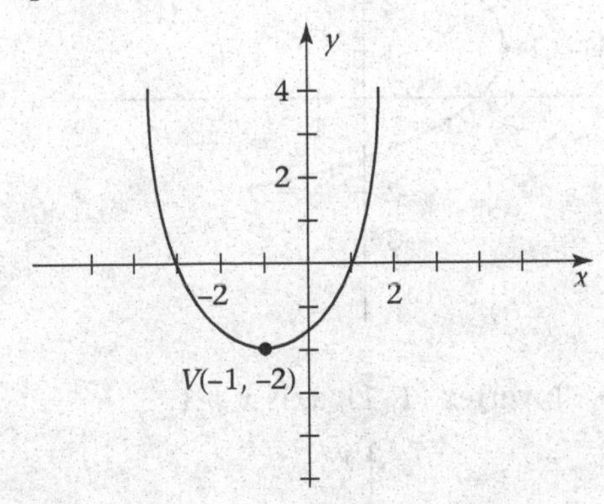

b. We complete the square in y:

$$y^2 - 4y = -x - 1$$
$$y^2 - 4y + 4 = -x - 1 + 4$$
$$(y - 2)^2 = -(x - 3)$$

The vertex of the parabola is at $(3, 2)$ and the axis is $y = 2$.

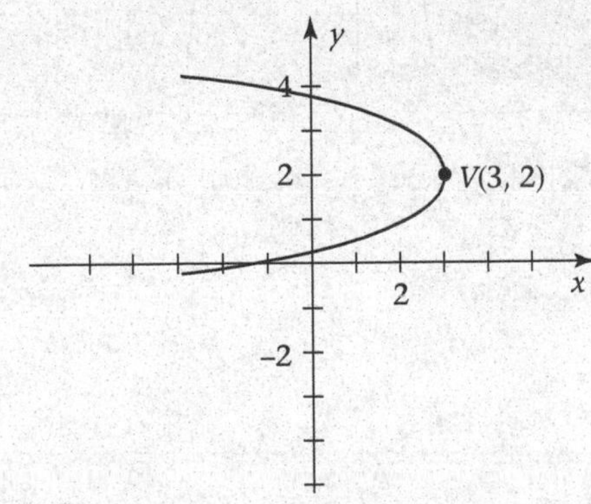

■

Progress Check

Write the equation of the parabola in standard form. Locate the vertex and the axis, and sketch the graph.

a. $y^2 - 2y - 2x - 5 = 0$ b. $x^2 - 2x + 2y - 1 = 0$

Answers

a. $(y-1)^2 = 2(x+3)$, vertex $(-3, 1)$, axis $y = 1$.

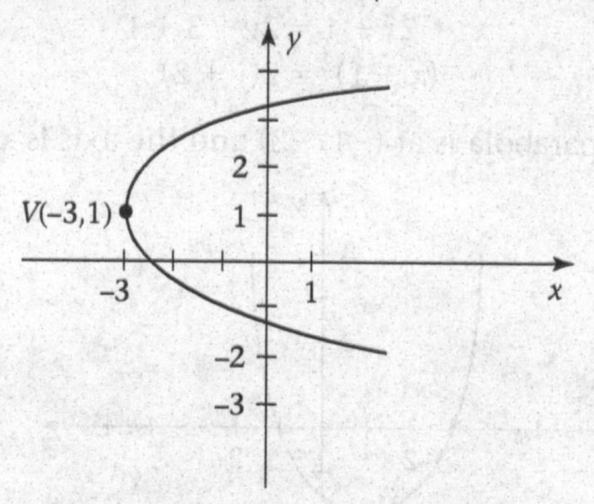

b. $(x-1)^2 = -2(y-1)$, vertex $(1, 1)$, axis $x = 1$.

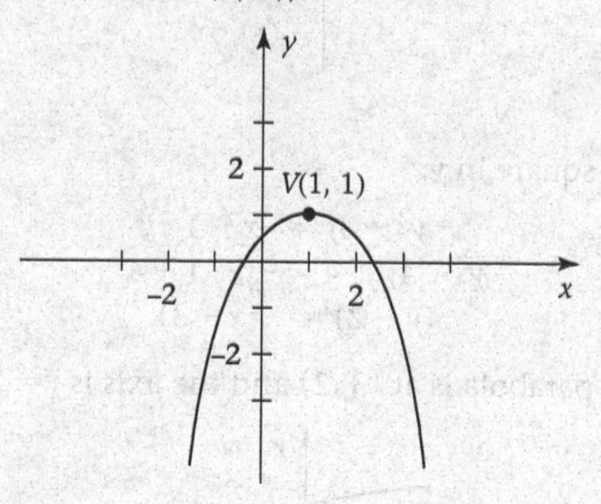

Exercise Set 5.3

In Exercises 1–8, determine the focus and directrix of the given parabola, and sketch the graph.

1. $x^2 = 4y$
2. $x^2 = -4y$
3. $y^2 = 2x$
4. $y^2 = -\frac{3}{2}x$
5. $x^2 + 5y = 0$
6. $2y^2 - 3x = 0$
7. $y^2 - 12x = 0$
8. $x^2 - 9y = 0$

In Exercises 9–20, determine the equation of the parabola that has its vertex at the origin and satisfies the given conditions.

9. Focus at $(1, 0)$
10. Focus at $(0, -3)$
11. Directrix $x = -\frac{3}{2}$
12. Directrix $y = \frac{5}{2}$
13. Axis is the y-axis, and the parabola passes through the point $(4, -2)$.
14. Axis is the x-axis, and the parabola passes through the point $(2, 1)$.
15. Axis is the x-axis and $p = -\frac{5}{4}$.
16. Axis is the y-axis and $p = 2$.
17. Focus at $(-1, 0)$ and directrix $x = 1$.
18. Focus at $(0, -\frac{5}{2})$ and directrix $y = \frac{5}{2}$.
19. Axis is the x-axis, and the parabola passes through the point $(4, 2)$.
20. Axis is the y-axis, and the parabola passes through the point $(2, 4)$.

In Exercises 21–24, determine in which direction each parabola opens.

21. $4x^2 + y = 0$
22. $4x^2 - y = 0$
23. $2x + y^2 = 0$
24. $2x - 5y^2 = 0$

In Exercises 25–44, write the equation in standard form. Determine the vertex, axis, and the direction in which each parabola opens. Sketch the graph.

25. $x^2 - 2x - 3y + 7 = 0$
26. $x^2 + 4x + 2y - 2 = 0$
27. $y^2 - 8y + 2x + 12 = 0$
28. $y^2 + 6y - 3x + 12 = 0$
29. $x^2 - x + 3y + 1 = 0$
30. $y^2 + 2y - 4x - 3 = 0$
31. $y^2 - 10y - 3x + 24 = 0$
32. $x^2 + 2x - 5y - 19 = 0$
33. $x^2 - 3x - 3y + 1 = 0$
34. $y^2 + 4y + x + 3 = 0$
35. $y^2 + 6y + \frac{1}{2}x + 7 = 0$
36. $x^2 + 2x - 3y + 19 = 0$
37. $x^2 + 2x + 2y + 3 = 0$
38. $y^2 - 6y + 2x + 17 = 0$
39. $x^2 - 4x - 2y + 2 = 0$
40. $y^2 + 2x - 4y + 6 = 0$
41. $2x^2 + 16x + y + 34 = 0$
42. $2x^2 - y + 3 = 0$
43. $y^2 + 2x + 2 = 0$
44. $y^2 + 3x - 2y - 5 = 0$

45. A stuntman has agreed to perform his famous human cannonball routine for the "Greatest Show in the Galaxy" carnival. The stuntman determines that his path will follow the parabola $y = 40x - x^2$, with the units measured in feet. His assistant has a 30 square foot circular net to catch his landing. The assistant places the net so that its center is 45 feet from the cannon.

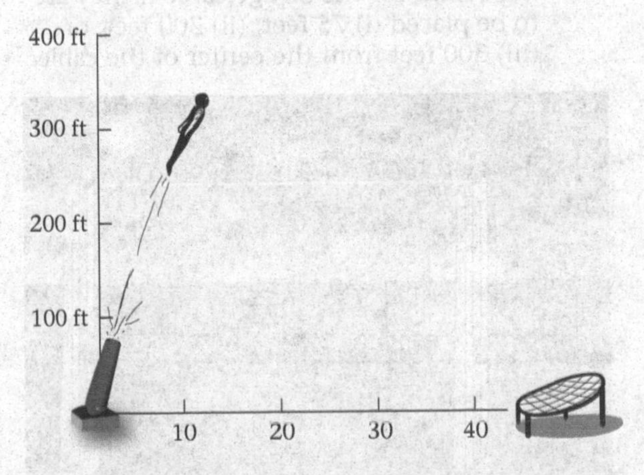

a. Sketch the path the stuntman will travel.
b. Determine if the net will "catch" the stuntman.

46. Consider a large parabolic reflector in the shape of a headlight used by a construction crew late at night. The lamp is designed in such a way that its light source is placed at the focus of the parabola, which lies $3\frac{1}{4}$ inches from its vertex. Suppose the lamp is to be $16\frac{1}{8}$ inches deep.

a. Sketch a graph of the parabola.
b. How wide will the reflector be?
c. How far will the rim of the lamp be from the light source?

47. A police helicopter has a searchlight that is designed so that its light source lies at the focal point of a parabolic reflector that is 6 inches from its vertex. Suppose the searchlight is designed to be 25 inches deep.

    a. Find the width of the searchlight.

    b. How far will the light source be from the rim of the searchlight?

48. A cable from a suspension bridge in the shape of a parabola hangs from two 75-feet towers, which are 650 feet apart. The lowest part of the cable hangs 17 feet from the roadway of the bridge.

    a. Assume that the bridge cable will be supported by vertical beams. How long should such a support beam be if it is located 125 feet from the center of the span of the bridge?

    b. Determine the length of vertical supporting beams for the bridge cable if they are to be placed (i) 75 feet, (ii) 200 feet, or (iii) 300 feet from the center of the cable.

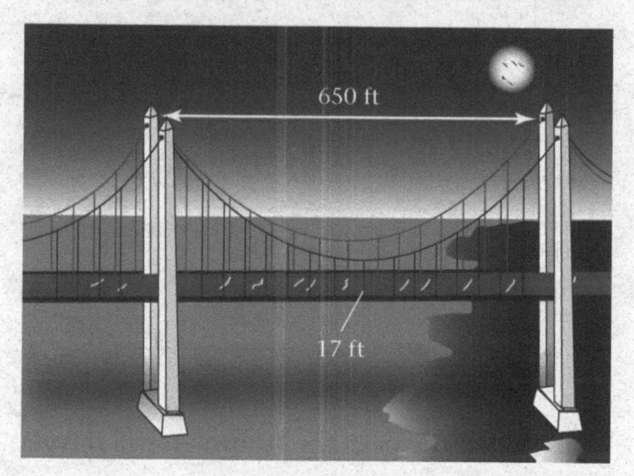

5.4 The Ellipse and Hyperbola

5.4a The Ellipse

The geometric definition of an ellipse is as follows.

Definition of an Ellipse
An **ellipse** is the set of all points in a plane, the sum of whose distances from two fixed points in the plane is a constant. Each fixed point is called a **focus** of the ellipse.

An ellipse may be constructed in the following way. Place a thumbtack at each of the foci F_1 and F_2, and attach one end of a string to each of the thumbtacks. Hold a pencil tightly against the string, as shown in Figure 5-7, and move the pencil. The point P describes an ellipse since the sum of the distances from P to the foci is always a constant, the length of the string.

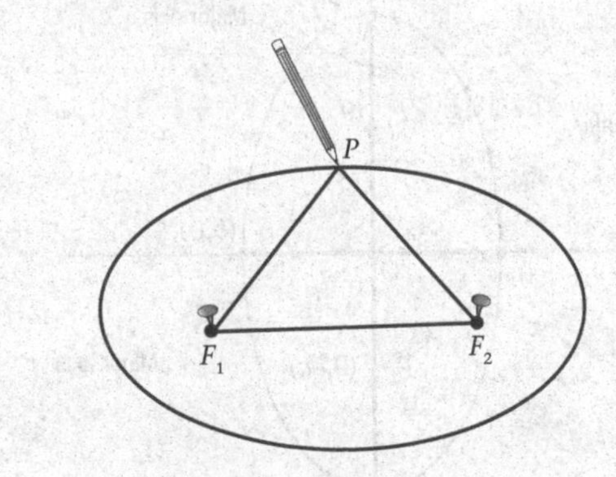

FIGURE 5-7
The Foci of an Ellipse

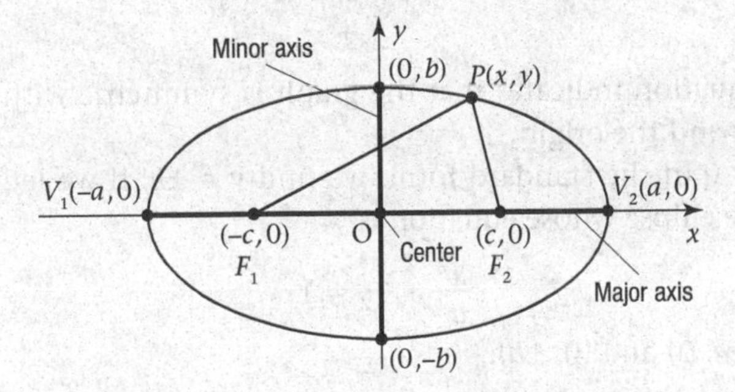

FIGURE 5-8
Deriving the Equation of an Ellipse

The **equation of the ellipse** is in standard form if the two fixed points are on either the x-axis or the y-axis and they are equidistant from the origin. In Figure 5-8, we have placed these foci F_1 and F_2 on the x-axis. The line segment joining the foci is called the **major axis** of the ellipse, and the midpoint of the major axis is called the **center** of the ellipse. The line segment through the center and perpendicular to the major axis is called the **minor axis**. The points at which the ellipse intersects the major axis are called the **vertices** (V_1 and V_2 in Figure 5-8) of the ellipse. Note that the length of the major axis is $2a$, the length of the minor axis is $2b$, and the center of the ellipse is the origin. By placing the point P at either V_1 or V_2, we see that "the length of the string" (see Figure 5-7) is $2a$. (Why?) Therefore, we have that

$$\overline{F_1P} + \overline{F_2P} = 2a$$

which is to be true for all points P on the ellipse. The equation of an ellipse in standard form is shown (see Exercise 35) to be as follows:

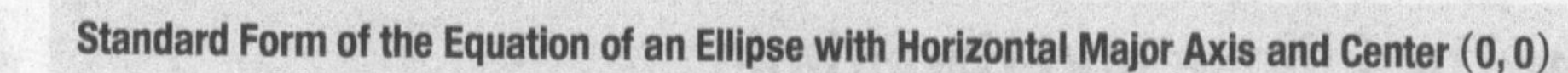

Standard Form of the Equation of an Ellipse with Horizontal Major Axis and Center (0, 0)

$$\frac{x^2}{a^2}+\frac{y^2}{b^2}=1, \quad a>b$$

If we place foci F_1 and F_2 on the y-axis, symmetrically about the origin as shown in Figure 5-9, we can obtain the other standard form of the equation of an ellipse. Combining the above results, we have the following:

Standard Form of the Equation of an Ellipse with Center (0, 0)

$$\frac{x^2}{a^2}+\frac{y^2}{b^2}=1, \quad a\neq b$$

If $a>b$, the major axis is horizontal. If $a<b$, the major axis is vertical.

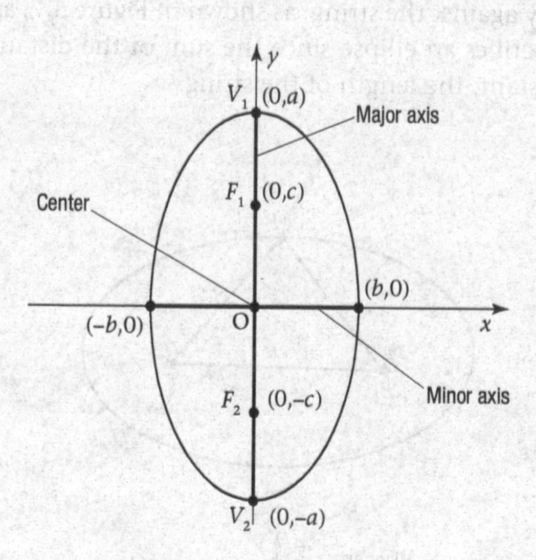

FIGURE 5-9
Deriving the Equation of an Ellipse

Note that the equation indicates that the graph is symmetric with respect to the x-axis, the y-axis and the origin.

If we let $x=0$ in the standard form, we find $y=\pm b$; if we let $y=0$, we find $x=\pm a$. Thus, the ellipse whose equation is

$$\frac{x^2}{a^2}+\frac{y^2}{b^2}=1$$

has intercepts $(\pm a, 0)$ and $(0, \pm b)$.

EXAMPLE 1 *Using the Equation of an Ellipse*

Find the intercepts and sketch the graph of the ellipse whose equation is

$$\frac{x^2}{4}+\frac{y^2}{25}=1$$

SOLUTION

Setting $x=0$ and solving for y yields the y-intercepts ± 5; setting $y=0$ and solving for x yields the x-intercepts ± 2.

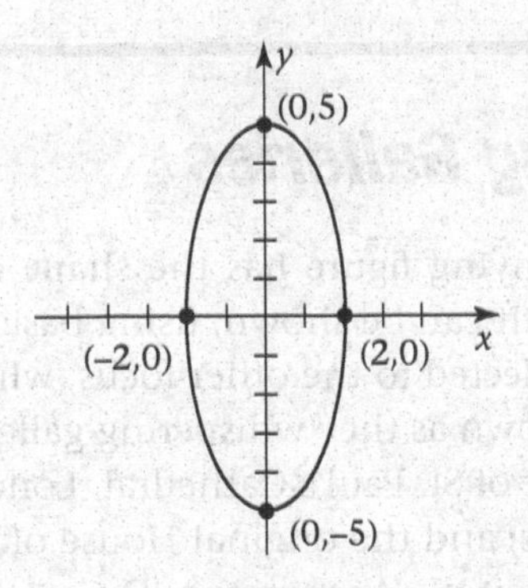

EXAMPLE 2 ***Using the Equation of an Ellipse***

Write the equation of the ellipse in standard form and determine the intercepts.

a. $4x^2 + 3y^2 = 12$ b. $9x^2 + y^2 = 10$

SOLUTION

a. Dividing by 12 to make the right-hand side equal to 1, we have

$$\frac{x^2}{3} + \frac{y^2}{4} = 1$$

The x-intercepts are $(\pm\sqrt{3}, 0)$; the y-intercepts are $(0, \pm 2)$.

b. Dividing by 10 we have

$$\frac{9x^2}{10} + \frac{y^2}{10} = 1$$

But this is *not* standard form. However, if we write

$$\frac{9x^2}{10} \quad \text{as} \quad \frac{x^2}{\frac{10}{9}}$$

then

$$\frac{x^2}{\frac{10}{9}} + \frac{y^2}{10} = 1$$

is the standard form of an ellipse. The intercepts are

$$\left(\frac{\pm\sqrt{10}}{3}, 0\right) \quad \text{and} \quad (0, \pm\sqrt{10})$$ ■

Progress Check

Write the equation of each ellipse in standard form and determine the intercepts.

a. $2x^2 + 3y^2 = 6$ b. $3x^2 + y^2 = 5$

Answers

a. $\frac{x^2}{3} + \frac{y^2}{2} = 1$; $(\pm\sqrt{3}, 0)$, $(0, \pm\sqrt{2})$ b. $\frac{x^2}{\frac{5}{3}} + \frac{y^2}{5} = 1$; $\left(\frac{\pm\sqrt{15}}{3}, 0\right)$, $(0, \pm\sqrt{5})$

FOCUS on Whispering Galleries

The domed roof in the accompanying figure has the shape of an ellipse that has been rotated about its major axis. It can be shown, using basic laws of physics, that a sound uttered at one focus is reflected to the other focus, where it is clearly heard. This property of such rooms is known as the "whispering gallery effect."

Famous whispering galleries include the dome of St. Paul's Cathedral, London; St. John Lateran, Rome; the Salle des Carlatides in the Louvre, Paris; and the original House of Representatives (now the National Statuary Hall in the United States Capitol), Washington, D.C.

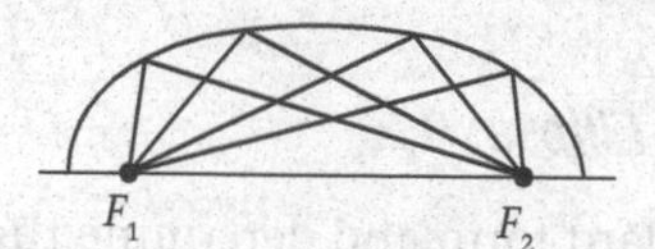

5.4b The Hyperbola

The geometric definition of a hyperbola is as follows.

Definition of a Hyperbola

A **hyperbola** is the set of all points in a plane, the difference of whose distances from two fixed points in the plane is a positive constant. Each fixed point is called a **focus** of the hyperbola.

The **equation of the hyperbola** is in standard form if the two fixed points are on either the x-axis or the y-axis, and they are equidistant from the origin. In Figure 5-10(a), we have placed these foci, F_1 and F_2, on the x-axis, and in Figure 5-10(b), on the y-axis. The line through the foci is called the **transverse axis**, and the midpoint of this axis is called the **center** of the hyperbola. The line through the center and perpendicular to the transverse axis is called the **conjugate axis**. The points at which the hyperbola intersects the transverse axis are called the **vertices** (V_1 and V_2 in Figure 5-10(a) and Figure 5-10(b)) of the hyperbola. The two separate parts of the hyperbola are called its **branches**.

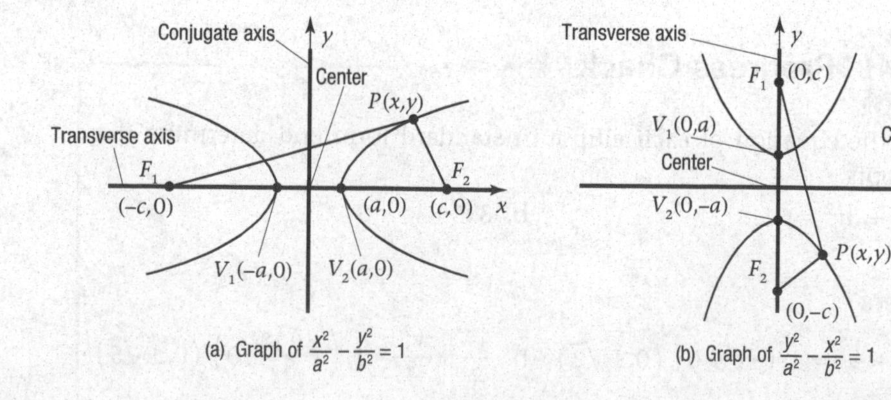

FIGURE 5-10 Deriving the Equation of a Hyperbola

Consider placing the point P at either V_1 or V_2. If the distance $\overline{PF_1} > \overline{PF_2}$ (see Figure 5-10(a)), we observe (Why?) that

$$\overline{PF_1} - \overline{PF_2} = 2a$$

Alternatively, if $\overline{PF_1} < \overline{PF_2}$ (see Figure 5-10(b)), we observe (Why?) that

$$\overline{PF_2} - \overline{PF_1} = 2a$$

Therefore

$$\left|\overline{PF_2} - \overline{PF_1}\right| = 2a$$

is true for all points P on the hyperbola. The equations of hyperbola in standard form can be shown (see Exercise 36) to be as follows.

Standard Forms of the Equation of a Hyperbola with Center (0, 0)

$$\frac{x^2}{a^2} - \frac{y^2}{b^2} = 1 \qquad \text{foci on the } x\text{-axis} \qquad (1)$$

$$\frac{y^2}{a^2} - \frac{x^2}{b^2} = 1 \qquad \text{foci on the } y\text{-axis} \qquad (2)$$

These equations indicate that the graphs are symmetric with respect to the x-axis, the y-axis, and the origin.

Letting $y = 0$, we see that the x-intercepts of the graph of Equation (1) are $x = \pm a$. Letting $x = 0$, we find there are no y-intercepts since the equation $y^2 = -b^2$ has no real roots. (See Figure 5-10(a)) Similarly, the graph of Equation (2) has y-intercepts of $\pm a$ and no x-intercepts. (See Figure 5-10(b))

EXAMPLE 3 *Using the Equation of a Hyperbola*

Find the intercepts and sketch the graph of each equation.

a. $\frac{x^2}{9} - \frac{y^2}{4} = 1$ b. $\frac{y^2}{4} - \frac{x^2}{3} = 1$

SOLUTION

a. When $y = 0$, we have $x^2 = 9$ or $x = \pm 3$. The x-intercepts are $(3, 0)$ and $(-3, 0)$. With the assistance of a few plotted points, we can sketch the graph:

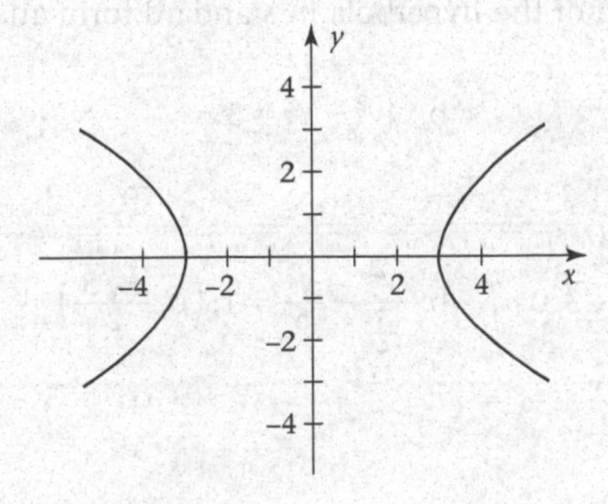

b. When $x = 0$, we have $y^2 = 4$ or $y = \pm 2$. The y-intercepts are (0, 2) and (0, −2). Plotting a few points, we can sketch the graph:

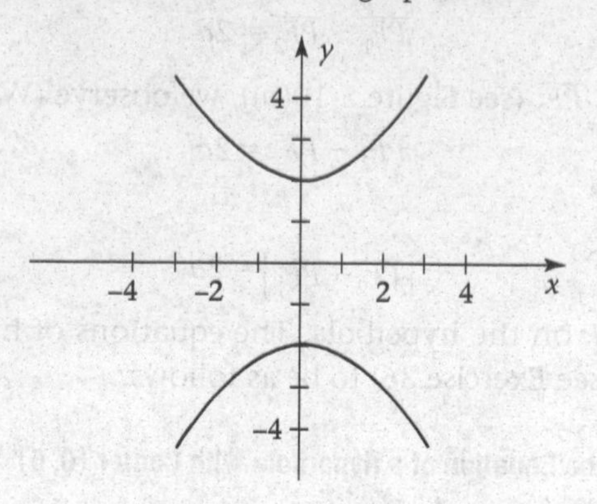

■

EXAMPLE 4 *Using the Equation of a Hyperbola*

Write the equation of the hyperbola $9x^2 - 5y^2 = 10$ in standard form and determine the intercepts.

SOLUTION

Dividing by 10, we have

$$\frac{9x^2}{10} - \frac{y^2}{2} = 1$$

Rewriting the equation in standard form, we have

$$\frac{x^2}{\frac{10}{9}} + \frac{y^2}{2} = 1$$

The x-intercepts are $\left(\frac{\pm\sqrt{10}}{3}, 0\right)$; there are no y-intercepts. ■

Progress Check

Write the equation of the hyperbola in standard form and determine the intercepts.

a. $2x^2 - 5y^2 = 6$

b. $4y^2 - x^2 = 5$

Answers

a. $\frac{x^2}{3} - \frac{y^2}{\frac{6}{5}} = 1$; $(\pm\sqrt{3}, 0)$

b. $\frac{y^2}{\frac{5}{4}} - \frac{x^2}{5} = 1$; $\left(0, \frac{\pm\sqrt{5}}{2}\right)$

5.4c Asymptotes of a Hyperbola

There is a feature of a hyperbola that enables us to sketch its branches without having to plot many points. By examining the corresponding values of y for values of x that are very far from the origin, we can show that a hyperbola has two **asymptotes.** (There is a discussion concerning asymptotes in Section 5.1.) Solving

$$\frac{x^2}{a^2} - \frac{y^2}{b^2} = 1$$

for y, we obtain

$$\begin{aligned} y &= \pm\frac{b}{a}\sqrt{x^2 - a^2} \\ &= \pm\frac{b}{a}\sqrt{x^2\left(1 - \frac{a^2}{x^2}\right)} \\ &= \pm\frac{b}{a}x\sqrt{1 - \frac{a^2}{x^2}} \end{aligned}$$

For large values of $|x|$, the term $\frac{a^2}{x^2}$ is very close to 0, in which case y approaches the asymptotes

$$y = \pm\frac{b}{a}x$$

The graph of this hyperbola with its asymptotes is shown in Figure 5-11. Note that the rectangle with vertices $(\pm a, \pm b)$ is such that the lines that contain the diagonals are the asymptotes.

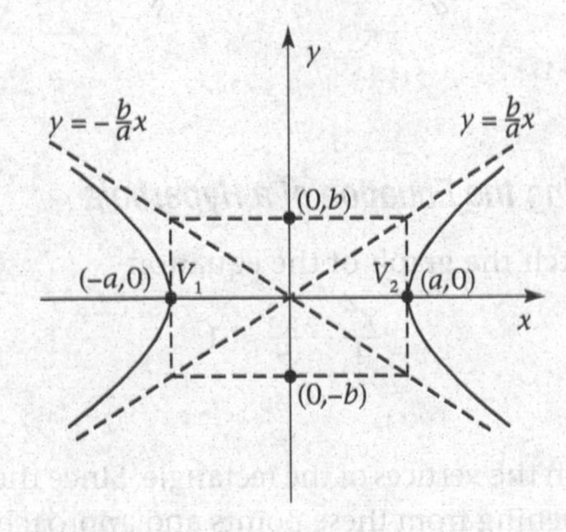

FIGURE 5-11
Graph of $\frac{x^2}{a^2} - \frac{y^2}{b^2} = 1$ with its Asymptotes

A similar argument may be used with the hyperbola

$$\frac{y^2}{a^2} - \frac{x^2}{b^2} = 1$$

to obtain a different pair of asymptotes. We summarize the results.

Asymptotes of a Hyperbola

$\frac{x^2}{a^2} - \frac{y^2}{b^2} = 1$ has asymptotes $y = \pm\frac{b}{a}x$

containing the diagonals of the rectangle with vertices $(\pm a, \pm b)$.

$\frac{y^2}{a^2} - \frac{x^2}{b^2} = 1$ has asymptotes $y = \pm\frac{a}{b}x$

containing the diagonals of the rectangle with vertices $(\pm b, \pm a)$.

EXAMPLE 5 *Analyzing the Equation of a Hyperbola*

Using asymptotes, sketch the graph of the equation

$$25x^2 - 4y^2 = 100$$

SOLUTION

We first write the standard form of the equation by dividing both sides by 100, obtaining

$$\frac{x^2}{4} - \frac{y^2}{25} = 1$$

The points $(\pm 2, \pm 5)$ form the vertices of the rectangle. Using the fact that $(\pm 2, 0)$ are intercepts, we sketch the graph opening from these points and approaching the asymptotes.

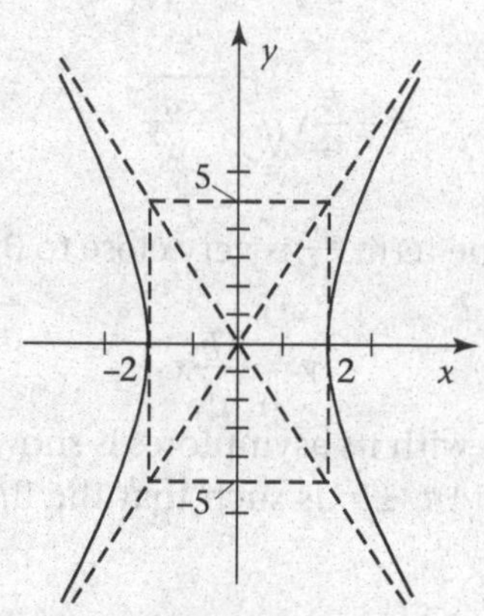

■

EXAMPLE 6 *Analyzing the Equation of a Hyperbola*

Using asymptotes, sketch the graph of the equation

$$\frac{y^2}{4} - \frac{x^2}{9} = 1$$

SOLUTION

The points $(\pm 3, \pm 2)$ form the vertices of the rectangle. Since the intercepts are $(0, \pm 2)$, we sketch the graph opening from these points and approaching the asymptotes.

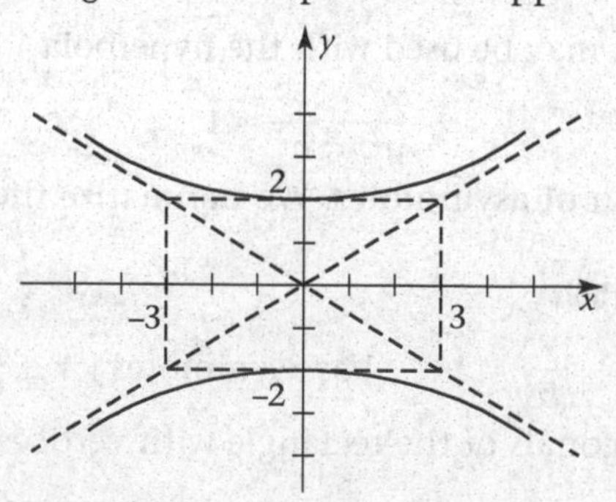

■

Progress Check

Using asymptotes, sketch the graph of the equation

$$4x^2 - 9y^2 = 144$$

Answer

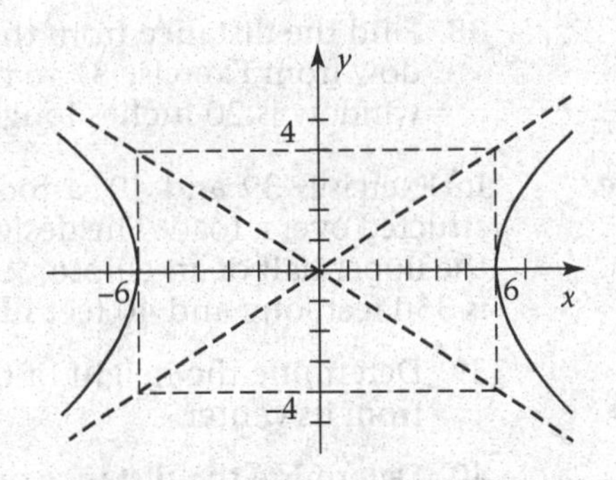

Exercise Set 5.4

In Exercises 1–6, find the intercepts and sketch the graph of the ellipse.

1. $\frac{x^2}{25} + \frac{y^2}{4} = 1$
2. $\frac{x^2}{4} + \frac{y^2}{16} = 1$
3. $\frac{x^2}{8} + \frac{y^2}{4} = 1$
4. $\frac{x^2}{12} + \frac{y^2}{18} = 1$
5. $\frac{x^2}{16} + \frac{y^2}{25} = 1$
6. $\frac{x^2}{1} + \frac{y^2}{3} = 1$

In Exercises 7–16, write the equation of the ellipse in standard form and determine the intercepts.

7. $4x^2 + 9y^2 = 36$
8. $16x^2 + 9y^2 = 144$
9. $4x^2 + 16y^2 = 16$
10. $25x^2 + 4y^2 = 100$
11. $4x^2 + 16y^2 = 4$
12. $8x^2 + 4y^2 = 32$
13. $8x^2 + 6y^2 = 24$
14. $5x^2 + 6y^2 = 50$
15. $36x^2 + 8y^2 = 9$
16. $5x^2 + 4y^2 = 45$

In Exercises 17–22, find the intercepts and sketch the graph of the hyperbola.

17. $\frac{x^2}{25} - \frac{y^2}{16} = -1$
18. $\frac{y^2}{9} - \frac{x^2}{4} = 1$
19. $\frac{x^2}{36} - \frac{y^2}{1} = 1$
20. $\frac{y^2}{49} - \frac{x^2}{25} = 1$
21. $\frac{x^2}{6} - \frac{y^2}{8} = 1$
22. $\frac{y^2}{8} - \frac{x^2}{10} = -1$

In Exercises 23–28, write the equation of the hyperbola in standard form and determine the intercepts.

23. $16x^2 - y^2 = 64$
24. $4x^2 - 25y^2 = 100$
25. $4y^2 - 4x^2 = 1$
26. $2x^2 - 3y^2 = 6$
27. $4x^2 - 5y^2 = 20$
28. $25y^2 - 16x^2 = 400$

In Exercises 29–34, use the asymptotes and intercepts to sketch the graph of the hyperbola.

29. $16x^2 - 9y^2 = 144$
30. $16y^2 - 25x^2 = 400$
31. $y^2 - x^2 = 9$
32. $25x^2 - 9y^2 = 225$
33. $\frac{x^2}{25} - \frac{y^2}{36} = 1$
34. $y^2 - 4x^2 = 4$

35. Derive the standard form of the equation of an ellipse from the geometric definition of an ellipse. (*Hint*: In Figure 5-8, let $P(x, y)$ be any point on the ellipse and let $F_1(-c, 0)$ and $F_2(c, 0)$ be the foci. Note that the point $V_2(a, 0)$ lies on the ellipse and that $\overline{V_2F_1} + \overline{V_2F_2} = 2a$. Thus, the sum of the distances $\overline{PF_1} + \overline{PF_2}$ must also equal $2a$. Use the distance formula, simplify, and substitute $b^2 = a^2 - c^2$.)

36. Derive the standard form of the equation of a hyperbola from the geometric definition of a hyperbola. (*Hint*: Proceed in a manner similar to that of Exercise 35 and refer to Figure 5-10.)

37. An architect decides to install an elliptical window in the sun room of a house. The window is to be 36 inches long and 9 inches high. Find the vertical height of the window 5 inches horizontally from the window's center.

38. Find the distance from the center of the window from Exercise 37 to the place where the window is 20 inches long.

In Exercises 39 and 40, a footbridge is to be constructed over a road. The design of the footbridge is the upper half of an ellipse. Suppose the footbridge is 350 feet long and 40 feet high.

39. Determine the height of the bridge 125 feet from its center.

40. Determine the distance on the ground from the center of the bridge to the point at which the bridge is 30 feet high.

41. Find the equation of the ellipse, center at the origin, vertex at (3, 0) and passing through $(\sqrt{5}, \frac{4}{3})$.

42. Find the equation of the ellipse, center at the origin, vertex at (4, 0), and passing through $(\frac{16}{5}, -3)$.

43. Using Exercise 35 and find the equation of the ellipse, center at the origin, focus at (0, −4), and vertex at (0, 5).

44. Using Exercise 35, find the equation of the ellipse, center at the origin, focus at (2, 0), and vertex at (3, 0).

45. Find the equation of the hyperbola, center at the origin, vertex at (3, 0), and passing through $(5, \frac{4}{3})$.

46. Find the equation of the hyperbola, center at the origin, vertex at (0, 4), and with asymptotes $y = \pm\frac{4}{3}x$.

5.5 Translation of Axes

In Figure 5-12, two sets of coordinate axes are displayed. There is the standard x- and y-coordinate axes with origin O; and there is another set of coordinate axes, x' and y', that are parallel to the x-axis and y-axis, respectively.

Consider shifting the x-axis vertically to the x'-axis and shifting the y-axis horizontally to the y'-axis. This process is called **translation of axes**, and we say that the x- and y-axes have been translated.

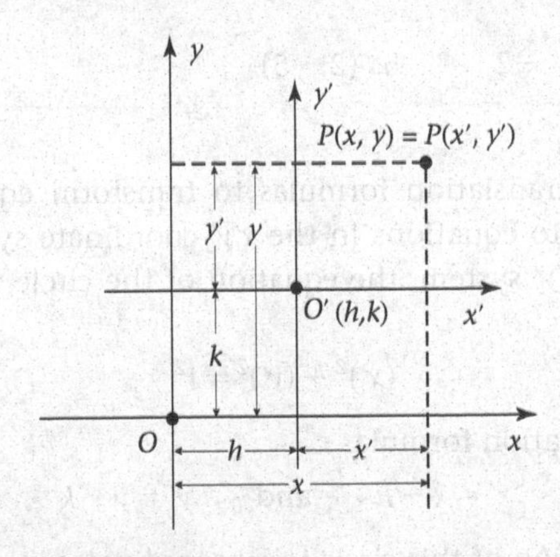

FIGURE 5-12
Deriving the Formulas for Translation of Axes

A point P in the plane has coordinates (x, y) with respect to the xy-coordinate system and (x', y') with respect to the $x'y'$-coordinate system. Suppose that O', the origin of the $x'y'$ system, has coordinates (h, k) in the xy system. We can obtain the following formulas relating x to x' and y to y'.

Translation of Axes Formulas

$$x' = x - h \quad \text{and} \quad y' = y - k$$

EXAMPLE 1 ***Using the Formulas for Translation of Axes***

The origin O' of the $x'y'$-coordinate system is at $(-2, 4)$.

a. Express x' and y' in terms of x and y, respectively.

b. Find the $x'y'$-coordinates of the point P whose xy-coordinates are $(4, -6)$.

SOLUTION

a. Substituting $h = -2$ and $k = 4$, we obtain the translation formulas

$$x' = x + 2 \qquad y' = y - 4$$

b. Substituting $x = 4$ and $y = -6$ yields

$$x' = 4 + 2 = 6 \qquad y' = -6 - 4 = -10$$

The $x'y'$-coordinates of P are $(6, -10)$. ■

The origin of O' of the $x'y'$-coordinate system is at $(-1, -2)$.

a. Express x and y in terms of x' and y', respectively.

b. Find the xy-coordinates of the point P whose $x'y'$-coordinates are $(3, -3)$.

Answers

a. $x = x' - 1, y = y' - 2$ b. $(2, -5)$

We can use the translation formulas to transform equations given in the xy-coordinate system to equations in the $x'y'$-coordinate system, and vice versa. For example, in the $x'y'$ system, the equation of the circle with center at O' and radius r is

$$(x')^2 + (y')^2 = r^2$$

Substituting the translation formulas

$$x' = x - h \quad \text{and} \quad y' = y - k$$

we find that the equation of this circle in xy-coordinates is

$$(x - h)^2 + (y - k)^2 = r^2$$

which is precisely the standard form of the equation of the circle with center at (h, k) and radius r, as discussed in Section 5.2.

EXAMPLE 2 *Analyzing and Sketching a Conic Section*

Discuss and sketch the graph of the equation

$$x^2 - 4x + y^2 + 2y + 1 = 0$$

SOLUTION

We group the terms in x and the terms in y

$$(x^2 - 4x \quad) + (y^2 + 2y \quad) = -1$$

Completing the square in each variable

$$(x^2 - 4x + 4) + (y^2 + 2y + 1) = -1 + 4 + 1 = 4$$

(Note that the equation is balanced by adding 4 + 1 to the right-hand side.) We then have

$$(x - 2)^2 + (y + 1)^2 = 4$$

which is the equation of a circle with center at $(2, -1)$ and radius 2. In terms of the $x'y'$ coordinate system with origin O' at $(2, -1)$, the equation becomes

$$(x')^2 + (y')^2 = 4$$

We see that the equation and analysis are simplified by translating the axes to the point $(h, k) = (2, -1)$.

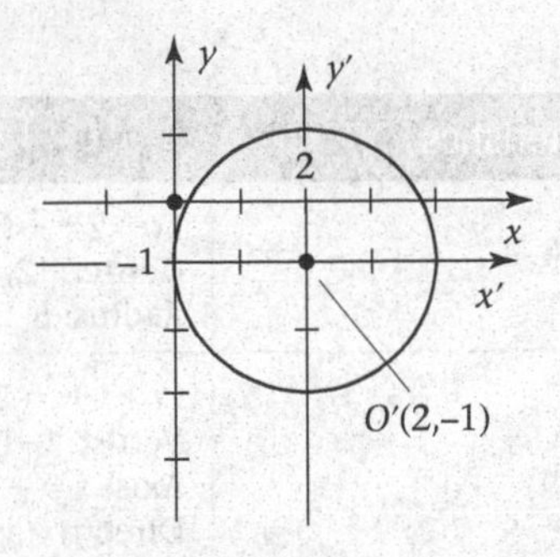

■

The technique of translation of axes can be applied to each of the conic sections. The results can be summarized as shown in Table 5-2. If we write the equation of a conic section in standard form, we can perform a translation of axes to the origin $O'(h, k)$ and then analyze and sketch the graph in the simplified form relative to the $x'y'$-coordinate system.

EXAMPLE 3 ***Analyzing and Sketching a Conic Section***

Sketch the graph of the equation

$$x^2 + 4x - 4y^2 + 24y - 48 = 0$$

SOLUTION

Write the equation as

$$(x^2 + 4x \qquad) - 4(y^2 - 6y \qquad) = 48$$

and complete the square in both x and y.

$$(x^2 + 4x + 4) - 4(y^2 - 6y + 9) = 48 + 4 - 36$$
$$(x + 2)^2 - 4(y - 3)^2 = 16$$

Letting

$$x' = x + 2 \quad \text{and} \quad y' = y - 3$$

we have

$$(x')^2 - 4(y')^2 = 16$$

and divide through by 16 to obtain

$$\frac{(x')^2}{16} - \frac{(y')^2}{4} = 1$$

On the $x'y'$-coordinate system this is seen to be the equation of a hyperbola with center at $O'(-2, 3)$, $a = 4$, $b = 2$. Using the intercepts and asymptotes, the graph is sketched:

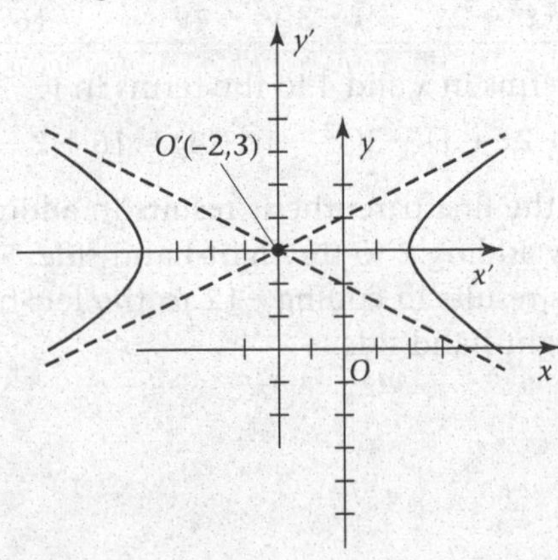

■

TABLE 5-2 Standard Forms of Conic Sections

Conic Section	Standard Form	Characteristics	Example
Circle	$(x-h)^2+(y-k)^2=r^2$	Center: (h, k) Radius: r	$(x-2)^2+(y+4)^2=25$ Center: $(2, -4)$ Radius: 5
Parabola	$(x-h)^2=4p(y-k)$	Vertex: (h, k) Axis: $x=h$ Directrix: $y=k-p$ $p>0$: Opens up $p<0$: Opens down	$(x+1)^2=2(y-3)$ Vertex: $(-1, 3)$ Axis: $x=-1$ Directrix: $y=\frac{5}{2}$ Opens up
	$(y-k)^2=4p(x-h)$	Vertex: (h, k) Axis: $y=k$ Directrix: $x=h-p$ $p>0$: Opens right $p<0$: Opens left	$(y+4)^2=-4(x+5)$ Vertex: $(-5, -4)$ Axis: $y=-4$ Directrix: $x=-4$ Opens left
Ellipse	$\frac{(x-h)^2}{a^2}+\frac{(y-k)^2}{b^2}=1$	Center: (h, k) Endpoints of axes of ellipse: $(h\pm a, k)$, $(h, k\pm b)$	$\frac{(x-1)^2}{4}+\frac{(y+2)^2}{9}=1$ Center: $(1, -2)$ Endpoints of axes of ellipse: $(3, -2)$, $(-1, -2)$, $(1, 1)$, $(1, -5)$
Hyperbola	$\frac{(x-h)^2}{a^2}-\frac{(y-k)^2}{b^2}=1$	Center: (h, k) Vertices: $(h\pm a, k)$ Asymptotes: $y-k=\pm\frac{b}{a}(x-h)$ Opens left and right	$\frac{(x+3)^2}{25}-\frac{(y-2)^2}{16}=1$ Center: $(-3, 2)$ Vertices: $(2, 2)$, $(-8, 2)$ Asymptotes: $y-2=\pm\frac{4}{5}(x+3)$
	$\frac{(y-k)^2}{a^2}-\frac{(x-h)^2}{b^2}=1$	Center: (h, k) Vertices: $(h, k\pm a)$ Asymptotes: $y-k=\pm\frac{a}{b}(x-h)$ Opens up and down	$\frac{(y-5)^2}{4}-\frac{(x+1)^2}{25}=1$ Center: $(-1, 5)$ Vertices: $(-1, 7)$, $(-1, 3)$ Asymptotes: $y-5=\pm\frac{2}{5}(x+1)$ Opens up and down

To complete the square in the equation

$$2(x^2+2x\qquad)-3(y^2-4y\qquad)=16$$

we must add 1 to the terms in x and 4 to the terms in y:

$$2(x^2+2x+1)-3(y^2-4y+4)=16+2-12$$

Note that adding 1 in the first parenthesis results in adding 2 to the left-hand side and is balanced by adding 2 to the right-hand side. Similarly, adding 4 in the second parenthesis results in adding -12 to the left-hand side, and this is also balanced on the right-hand side.

Progress Check

a. Show that the graph of the equation

$$4x^2 + 16x + y^2 + 2y + 13 = 0$$

is an ellipse.

b. Show that the graph of the equation

$$4y^2 - 24y - 25x^2 - 50x - 89 = 0$$

is a hyperbola.

EXAMPLE 4 ***Analyzing and Sketching a Conic Section***

Sketch the graph of the equation

$$x^2 - 4x - 4y - 4 = 0$$

SOLUTION

Write the equation as

$$(x^2 - 4x \qquad) = 4y + 4$$

and complete the square in x.

$$(x^2 - 4x + 4) = 4y + 4 + 4$$

$$(x - 2)^2 = 4y + 8 = 4(y + 2)$$

Letting

$$x' = x - 2 \qquad \text{and} \qquad y' = y + 2$$

we obtain

$$(x')^2 = 4y'$$

In the $x'y'$-coordinate system, this is seen to be the equation of a parabola with vertex at $O'(2, -2)$ and $p = 1$. The graph is sketched:

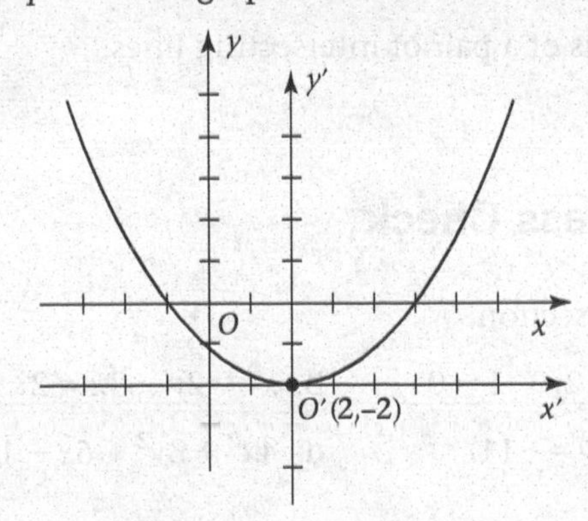

■

Progress Check

Show that the graph of the equation

$$y^2 + 4y - 6x + 22 = 0$$

is a parabola.

It can be shown that the graph of the equation

$$Ax^2 + Cy^2 + Dx + Ey + F = 0$$

is a conic section or a degenerate form of a conic section, such as a point, a line, a pair of lines, or no graph.

EXAMPLE 5 *Identifying the Conic Section*

Identify the graph of the equation

$$x^2 - 4x + y^2 - 2y + 9 = 0$$

SOLUTION

We rewrite the equation as

$$(x^2 - 4x \qquad) + (y^2 - 2y \qquad) = -9$$

and complete the square in both x and y.

$$(x^2 - 4x + 4) + (y^2 - 2y + 1) = -9 + 4 + 1$$
$$(x - 2)^2 + (y - 1)^2 = -4$$

Since the sum of two squares is nonnegative, this equation has no graph. ■

EXAMPLE 6 *Identifying the Conic Section*

Identify the graph of the equation

$$y^2 - x^2 = 0$$

SOLUTION

Solving for y,

$$y = \pm x$$

These are the equations of a pair of intersecting lines. ■

Progress Check

Identify the conic section.

a. $\frac{x^2}{5} - 3y^2 - 2x + 2y - 4 = 0$

b. $x^2 - 2y - 3x = 2$

c. $x^2 + y^2 - 4x - 6y = -11$

d. $4x^2 + 3y^2 + 6x - 10 = 0$

e. $x^2 - 2x - 3 = 0$

Answers

a. hyperbola

b. parabola

c. circle

d. ellipse

e. pair of parallel lines

We present some facts in Table 5-3 that may help to identify the various conic sections.

TABLE 5-3 Identifying Conic Sections

$Ax^2 + Cy^2 + Dx + Ey + F = 0$	Conic Section	Remarks
$A = 0$ or $C = 0$	Parabola	Second degree in one variable, first degree in the other variable
$A = C \neq 0$	Circle	Coefficients A and C are the same. *Caution:* Complete the square to obtain the standard form and check that radius $r > 0$.
$A \neq C$, $AC > 0$	Ellipse	A and C are unequal but have the same sign. *Caution:* Complete the square and check that the right-hand side is a positive constant.
$AC < 0$	Hyperbola	A and C have opposite signs.

Exercise Set 5.5

In Exercises 1–4, the origin O' of the $x'y'$-coordinate system is at $(-1, 4)$. Find the $x'y'$-coordinates of the point whose xy-coordinates are given.

1. $(0, 0)$
2. $(-2, 1)$
3. $(4, 3)$
4. $(-6, -2)$

In Exercises 5–8, the origin O' of the $x'y'$-coordinate system is at $(-3, 4)$. Find the xy-coordinates of the point whose $x'y'$-coordinates are given.

5. $(0, 0)$
6. $(-2, 1)$
7. $(4, 3)$
8. $(-6, -2)$

In Exercises 9–18, sketch the graph of the given equation. Then, determine appropriate WINDOW values, and check your answer using your graphing calculator.

9. $36x^2 - 100y^2 + 216x + 99 = 0$
10. $x^2 - 4y^2 + 10x - 16y + 25 = 0$
11. $x^2 + 4x - y + 5 = 0$
12. $2x^2 - 12x + y + 21 = 0$
13. $y^2 + 2x + 15 = 0$
14. $16x^2 + 4y^2 + 12y - 7 = 0$
15. $x^2 + 4y^2 + 10x - 8y + 13 = 0$
16. $9x^2 + 25y^2 - 36x + 50y - 164 = 0$
17. $x^2 + 9y^2 - 54y + 72 = 0$
18. $x^2 - y^2 + 4x + 8y - 11 = 0$

In Exercises 19–42, identify the conic section.

19. $y^2 - 8x + 6y + 17 = 0$
20. $4x^2 + 4y^2 - 12x + 16y - 11 = 0$
21. $4x^2 + y^2 + 24x - 4y + 24 = 0$
22. $4x^2 - y^2 - 40x - 4y + 80 = 0$
23. $x^2 - y^2 + 6x + 4y - 4 = 0$
24. $2x^2 + y^2 - 4x + 4y - 12 = 0$
25. $25x^2 - 16y^2 + 210x + 96y + 656 = 0$
26. $x^2 - 10x + 8y + 1 = 0$
27. $2x^2 + y - x + 3 = 0$
28. $4y^2 - x^2 + 2x - 3y + 5 = 0$
29. $4x^2 + 4y^2 - 2x + 3y - 4 = 0$
30. $3x^2 + 6y^2 - 2x + 8 = 0$
31. $36x^2 - 4y^2 + x - y + 2 = 0$
32. $x^2 + y^2 - 6x + 4y + 13 = 0$
33. $16x^2 + 4y^2 - 2y + 3 = 0$
34. $2y^2 - 3x + y + 4 = 0$
35. $x^2 + y^2 - 4x - 2y + 8 = 0$
36. $x^2 + y^2 - 2x - 2y + 6 = 0$
37. $4x^2 + 9y^2 - x + 2 = 0$
38. $3x^2 + 3y^2 - 3x + y = 0$
39. $4x^2 - 9y^2 + 2x + y + 3 = 0$
40. $x^2 + y^2 + 6x - 2y + 10 = 0$
41. $x^2 + y^2 - 4x + 4 = 0$
42. $4x^2 + y^2 = 32$

Chapter Summary

Key Terms, Concepts, and Symbols

analytic geometry	281	conjugate axis	312	minor axis	309
approaching ∞	286	directrix	299	oblique asymptote	291
approaching $-\infty$	286	ellipse	309	parabola	299
asymptote	284	equation of circle	295	radius of a circle	294
asymptotes of a hyperbola	315	equation of ellipse	309	rational function	282
axes of ellipse	309	equation of hyperbola	312	reducible rational function	291
axes of hyperbola	312	equation of parabola	300, 301	slanted asymptote	291
axis of parabola	299	foci of ellipse	309	translation of axes	319
axis of symmetry	299	foci of hyperbola	312	transverse axis	312
branches of hyperbola	312	focus of parabola	299	vertex of parabola	299
center of a circle	294	"holes"	291	vertical asymptote	284
center of ellipse	309	horizontal asymptote	284	vertices of ellipse	309
center of hyperbola	312	hyperbola	312	vertices of hyperbola	312
circle	294	irreducible rational function	282		
conic sections	294	major axis	309		

Key Ideas for Review

Topic	Page	Key Idea
Horizontal Asymptote	284	A rational function has a unique horizontal asymptote if the degree of the numerator is less than or equal to the degree of the denominator.
Vertical Asymptote	284	An irreducible rational function has a vertical asymptote corresponding to each zero of the denominator.
Graphing Rational Functions	289	Determine the intercepts, symmetry, and horizontal and vertical asymptotes of a rational function before attempting to sketch its graph.
Reducible Rational Functions	291	The graph of a reducible rational function has a "hole" corresponding to each unique common factor of the numerator and denominator.
Analytic Geometry	281	Analytic geometry applies algebraic techniques to the study of geometry. Theorems from plane geometry can be proved using these methods.
Conic Sections	294	The conic sections represent the possible intersections of a plane and a cone. In general, a conic section can be a circle, parabola, ellipse, or hyperbola. In special cases, these may reduce to a point, a line, two lines, or no graph. Each conic section has a geometric definition that can be used to derive a second-degree equation in two variables. The graph of this equation corresponds to that particular conic section.
Circle	294	A circle is the set of all points in a plane that are a given distance from a fixed point in the plane. The standard equation is $(x-h)^2 + (y-k)^2 = r^2$
Parabola	299	A parabola is the set of all points in a plane that are equidistant from a given point and a given line, both in the plane. The standard equation is $(x-h)^2 = 4p(y-k)$ or $(y-k)^2 = 4p(x-h)$

continues

Topic	Page	Key Idea
Ellipse	309	An ellipse is the set of all points in a plane, the sum of whose distances from two fixed points in the plane is a constant. The standard equation is $$\frac{(x-h)^2}{a^2}+\frac{(y-k)^2}{b^2}=1$$
Hyperbola	312	A hyperbola is the set of all points in a plane, the difference of whose distances from two fixed points in the plane is a positive constant. The standard equation is $$\frac{(x-h)^2}{a^2}-\frac{(y-k)^2}{b^2}=1 \quad \text{or} \quad \frac{(y-k)^2}{a^2}-\frac{(x-h)^2}{b^2}=1$$
Translation of Axes	319	It is possible to consider a graph of an equation with respect to a new coordinate system with axes x' and y', parallel to the standard x- and y-coordinate axes, respectively. If the origin of the new coordinate system has coordinates (h, k) relative to the standard xy-coordinate system, then the relationships between the two coordinate systems is given by $$x'=x-h \quad \text{and} \quad y'=y-k$$

Review Exercises

In Exercises 1–3, sketch the graph of the given function.

1. $f(x)=\dfrac{x}{x+1}$

2. $f(x)=\dfrac{x^2}{x+1}$

3. $f(x)=\dfrac{x^2+2}{x^2-1}$

4. Write an equation of the circle whose center is at $(-5, 2)$ and whose radius is 4.

5. Write an equation of the circle whose center is at $(-3, 3)$ and whose radius is 2.

In Exercises 6–11, determine the center and radius of the circle with the given equation.

6. $(x-2)^2+(y+3)^2=9$

7. $\left(x+\frac{1}{2}\right)^2+(y-4)^2=\frac{1}{9}$

8. $x^2+y^2+4x-6y=-10$

9. $2x^2+2y^2-4x+4y=-3$

10. $x^2+y^2-6y+3=0$

11. $x^2+y^2-2x-2y=8$

In Exercises 12 and 13, determine the vertex and axis of the given parabola. Sketch the graph.

12. $(y+5)^2=4\left(x-\frac{3}{2}\right)$

13. $(x-1)^2=2-y$

In Exercises 14–19, determine the vertex, axis, and direction of the given parabola.

14. $y^2+3x+9=0$

15. $y^2+4y+x+2=0$

16. $2x^2-12x-y+16=0$

17. $x^2+4x+2y+5=0$

18. $y^2-2y-4x+1=0$

19. $x^2+6x+4y+9=0$

In Exercises 20 and 21, determine the focus and directrix of the given parabola, and sketch the graph.

20. $x^2=-\frac{2}{3}y$

21. $3y^2+2x=0$

In Exercises 22 and 23, determine the equation of the parabola with its vertex at the origin that satisfies the given conditions.

22. directrix $y=\frac{7}{4}$

23. axis the y-axis; parabola passing through the point $\left(1, \frac{5}{2}\right)$

In Exercises 24–29, write the given equation in standard form and determine the intercepts.

24. $9x^2-4y^2=36$

25. $9x^2+y^2=9$

26. $5x^2+7y^2=35$

27. $9x^2-16y^2=144$

28. $3x^2+4y^2=9$

29. $3y^2-5x^2=20$

In Exercises 30 and 31 use the intercepts and asymptotes of the hyperbola to sketch the graph.

30. $4x^2 - 4y^2 = 1$

31. $9y^2 - 4x^2 = 36$

32. Find the equation of the hyperbola with center at the origin that has a vertex at (2, 0) and passes through the point (3, 1).

In Exercises 33–42, identify the conic section whose equation is given.

33. $2x^2 - 4x + y^2 = 0$

34. $4y + x^2 - 2x = 1$

35. $x^2 - 2x + y^2 - 4y = -6$

36. $-x^2 - 4x + y^2 + 4 = 0$

37. $y^2 + 2x - 4 = x^2 - 2x$

38. $x^2 - 4x + 2y = 6 + y^2$

39. $2y^2 + 6y - 3x + 2 = 0$

40. $6x^2 - 7y^2 - 5x + 6y = 0$

41. $2x^2 + y^2 + 12x - 2y + 17 = 0$

42. $9x^2 + 4y^2 = -36$

Review Test

1. Sketch the graph of the function
$$f(x) = \frac{2x}{x^2 - 1}$$

2. Write an equation of the circle of radius 6 whose center is at (2, −3).

In Exercises 3 and 4, determine the center and radius of the circle.

3. $x^2 + y^2 - 2x + 4y = -1$

4. $x^2 - 4x + y^2 = 1$

In Exercises 5 and 6, determine the vertex and axis of the parabola. Sketch the graph.

5. $x^2 + 6x + 2y + 7 = 0$

6. $y^2 - 4x - 4y + 8 = 0$

In Exercises 7 and 8, determine the vertex, axis, and the direction in which the parabola opens.

7. $x^2 - 6x + 2y + 5 = 0$

8. $y^2 + 8y - x + 14 = 0$

In Exercises 9–11, write the given equation in standard form and determine the intercepts.

9. $x^2 + 4y^2 = 4$

10. $4y^2 - 9x^2 = 36$

11. $4x^2 - 4y^2 = 1$

In Exercises 12 and 13, use the intercepts and asymptotes of the hyperbola to sketch its graph.

12. $9x^2 - y^2 = 9$

13. $4y^2 - x^2 = 1$

In Exercises 14–17, identify the conic section.

14. $5y^2 - 4x^2 - 6x + 2 = 0$

15. $3x^2 - 5x + 6y = 3$

16. $x^2 + y^2 + 2x - 2y - 2 = 0$

17. $x^2 + 9y^2 - 4x + 6y + 4 = 0$

Writing Exercises

1. We define the eccentricity of the ellipse $\frac{x^2}{a^2} + \frac{y^2}{b^2} = 1$ as $\frac{\sqrt{a^2 - b^2}}{a}$. Describe what happens to the shape of this ellipse when its eccentricity varies from 0 to 1.

2. We define the eccentricity of the hyperbola $\frac{x^2}{a^2} - \frac{y^2}{b^2} = 1$ as $\frac{\sqrt{a^2 + b^2}}{a}$. Describe what happens to the shape of this hyperbola when its eccentricity increases.

3. Discuss how circles, parabolas, and ellipses are used in everyday life.

4. Suppose you know that the rational function $f(x)$ has the line $y = c$ as a horizontal asymptote. Discuss what can be deduced from this information.

5. Suppose you know that the rational function $f(x)$ has the line $x = c$ as a vertical asymptote. Discuss what can be deduced from this information.

6. The acronym for long-distance radio navigation is LORAN. Research how this system uses hyperbolas to determine the location of ships.

Chapter 5 Project

To determine how many of each item to produce, how many should sell, and how much to charge, businesses must make use of functions which model cost, revenue, and profit. The independent variable for these functions could be the number of units produced. We could then express the profit as a difference function, an algebraic combination of cost and revenue (see Section 3.5 to review the algebra of functions).

What is the profit function? Review Section 5.1, Exercises 30 and 31. Suppose the revenue function for producing graphing calculators is

$$R(x) = 120x$$

and the cost is given by

$$C(x) = 50x + 10{,}000$$

where R and C are in dollars and x represents the number of graphing calculators produced. Assume that every calculator produced sells (a very favorable assumption!).

Answer the following questions:

- How much does each calculator sell for?
- What is the average cost function? What does its graph look like? Why?
- What is the average profit function?
- Set up a TABLE in your graphing calculator, with one column for the number of calculators produced, one for the cost, one for revenue, and the last column for profit. Use increments of 100 units.
- Is the average profit function an algebraic combination of the average revenue and average cost function? Why or why not?

6 Exponential and Logarithmic Functions

On December 26, 2004, off the west coast of Northern Sumatra (in Indonesia), an earthquake occurred that was felt in India, Myanmar, and Thailand, causing landslides, mud volcanoes, and tsunamis. The magnitude of the earthquake was measured at 9.0 on a special scale created in 1935 by Charles F. Richter at the California Institute of Technology, using data from seismograph stations in Southern California.

The Richter scale uses *logarithms* to compare the intensity of a given earthquake to the "standard intensity." Each integer step represents a tenfold increase in amplitude and about 31 times as much energy released. Look at the end of this chapter for a project dealing with this and other earthquakes.

The story of measuring earthquakes has become a lot more complicated since Richter's time, including newer concepts like body-wave and surface-wave magnitude; but the Richter scale continues to be used. Discover the details of the mathematical methods employed at http://quake.usgs.gov/.

Thus far in our study of algebra, we have dealt primarily with functions that are polynomials or some combination of polynomials. In this chapter, we introduce another family: the exponential and logarithmic functions. These functions are inverses of each other.

Exponential functions are needed to help analyze the behavior of various phenomena found in biology, chemistry, physics and economics. For example, we shall observe how an exponential function can describe the growth of bacteria, as well as the increase in funds subject to compound interest.

Logarithms can be viewed as another way to write exponents. Historically, logarithms were used to simplify calculations. Prior to the availability of calculators, a device called a slide rule was used to perform various calculations. The design of the slide rule was based upon logarithms and their properties. Although the need for manipulating logarithms in calculations has diminished, the logarithmic functions are basic to developing more sophisticated mathematical ideas.

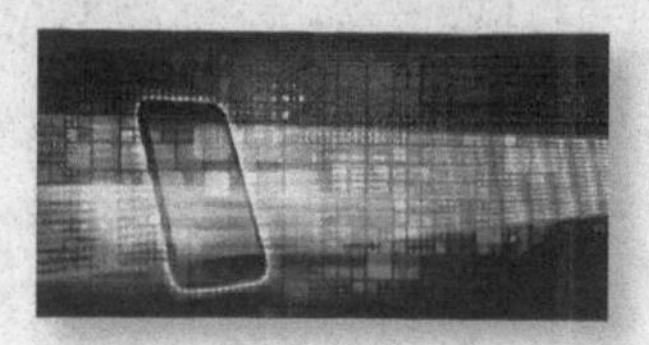

http://quake.usgs.gov/

6.1 A Brief Review of Inverse Functions

In Section 3.5, we introduced the concept of an inverse function. Since the exponential and logarithmic functions are inverses of each other, we are presenting a summary of facts concerning inverse functions. We want to keep the properties of inverse functions in mind in order to help us understand the development of the exponential and logarithmic functions.

Recall that we first consider functions that are one-to-one.

A function f is **one-to-one** if $f(a) = f(b)$ only when $a = b$.

EXAMPLE 1 *One-to-One Functions*

Show that $f(x) = \frac{1}{2}x + 1$ is one-to-one.

SOLUTION

Since $f(a) = \frac{1}{2}a + 1$ and $f(b) = \frac{1}{2}b + 1$, we set

$$f(a) = f(b)$$
$$\frac{1}{2}a + 1 = \frac{1}{2}b + 1$$
$$\frac{1}{2}a = \frac{1}{2}b$$
$$a = b$$

Therefore, $f(x)$ is one-to-one. ■

EXAMPLE 2 *One-to-One Functions*

Show that $f(x) = x^2 + 1$ is not one-to-one.

SOLUTION

Setting

$$f(a) = f(b)$$
$$a^2 + 1 = b^2 + 1$$
$$a^2 = b^2$$

However, this does not imply that $a = \text{b}$. For example, $2 \neq -2$, yet $f(2) = 5 = f(-2)$. Therefore, $f(x)$ is not one-to-one. ■

We may determine if a function is one-to-one by examining its graph.

Horizontal-Line Test

If no horizontal line meets the graph of a function in more than one point, then the function is one-to-one.

EXAMPLE 3 *Determining One-to-One Functions*

Show that $f(x) = \frac{1}{2}x + 1$ is one-to-one.

SOLUTION

The graph of $f(x) = \frac{1}{2}x + 1$ is shown in Figure 6-1. Any horizontal line meets $f(x)$ in at most one point. Therefore, $f(x)$ is one-to-one.

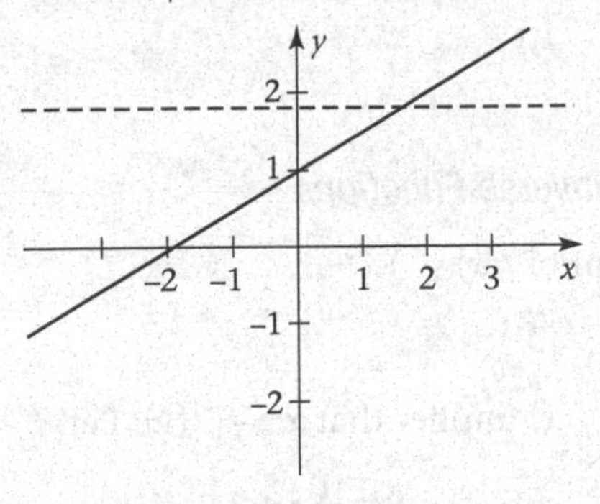

FIGURE 6-1
Graph of $f(x) = \frac{1}{2}x + 1$

■

EXAMPLE 4 *Determining One-to-One Functions*

Show that $f(x) = x^2 + 1$ is a function, but not a one-to-one function.

SOLUTION

The graph of $f(x) = x^2 + 1$ is shown in Figure 6-2. Note that any vertical line meets $f(x)$ in at most one point, and hence $f(x)$ is a function. However, in Figure 6-2 we see a horizontal line that meets $f(x)$ in two points. Therefore, $f(x)$ is not one-to-one.

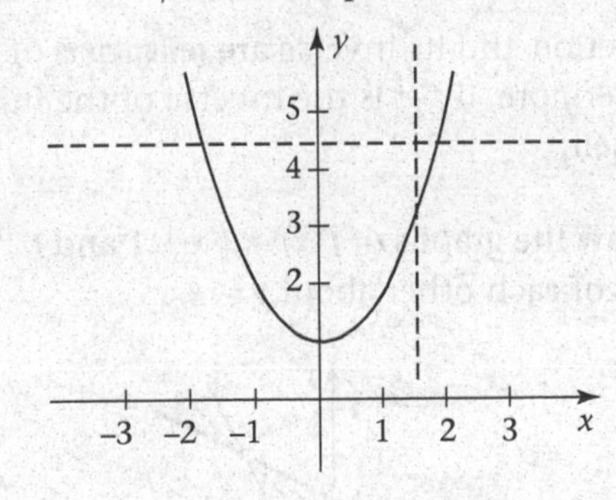

FIGURE 6-2
Graph of $f(x) = x^2 + 1$

■

We now recall the definition of an inverse function.

Inverse Functions

If f is a one-to-one function with domain X and range Y, then the function g with domain Y and range X satisfying

$$g[f(x)] = x \quad \text{for every } x \text{ in } X$$

$$f[g(y)] = y \quad \text{for every } y \text{ in } Y$$

is called an **inverse function** of f.

As an alternative, we may write

$$f^{-1}[f(x)] = x \quad \text{for every } x \text{ in } X$$
$$f[f^{-1}(y)] = y \quad \text{for every } y \text{ in } Y$$

Equivalently,

$$y = f(x) \text{ and } x = f^{-1}(y) \text{ for every } x \text{ in } X, \text{ every } y \text{ in } Y$$

EXAMPLE 5 *Finding Inverse Functions*

Find the inverse function of $f(x) = \frac{1}{2}x + 1$.

SOLUTION

If $f^{-1}(x)$ exists, then $y = f(x)$ implies that $x = f^{-1}(y)$. Let

$$y = \frac{1}{2}x + 1$$

Solving for x, we obtain

$$x = 2y - 2$$

and

$$f^{-1}(y) = 2y - 2$$

Writing this as a function of x,

$$f^{-1}(x) = 2x - 2$$

(Verify that $f^{-1}[f(x)] = x$ and $f[f^{-1}(y)] = y$ for all x in X and all y in Y.) ■

The graphs of a function and its inverse are **reflections** of each other about the line $y = x$. Furthermore, if f^{-1} is the inverse of the function f, then the inverse of f^{-1} is f itself.

In Figure 6-3, we draw the graphs of $f(x) = \frac{1}{2}x + 1$ and $f^{-1}(x) = 2x - 2$. Observe that they are reflections of each other about $y = x$.

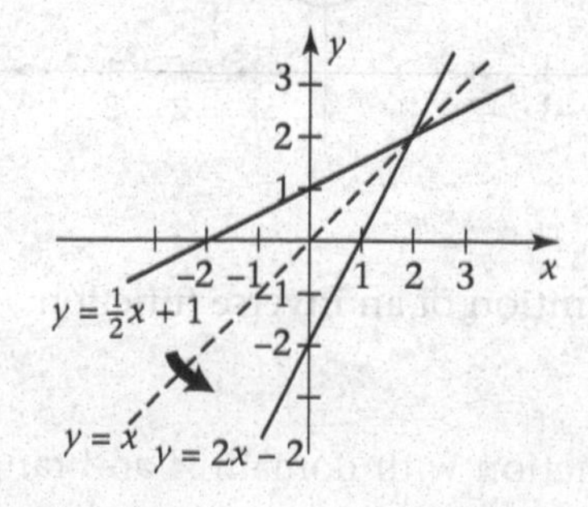

FIGURE 6-3
Graphs of Inverse Functions as Reflections About $y = x$

Exercise Set 6.1

1. Let $F(x) = 2x - 2$. Find
   a. $F^{-1}[F(4)]$
   b. $(F \circ F^{-1})(25)$
   c. $F^{-1}[F(y+1)]$
2. Let $f(x) = x^3 + 2x + 3$ and assume the domain of f^{-1} to be the set of all real numbers. Find
   a. $f[f^{-1}(-3)]$
   b. $(f^{-1} \circ f)(\sqrt{3})$
   c. $f[f^{-1}(2a-1)]$

In Exercises 3–6, determine whether the given function $f(x)$ is one-to-one. If it is, find its inverse $f^{-1}(x)$, and sketch the graphs of $y = f(x)$ and $y = f^{-1}(x)$ on the same coordinate axes.

3. $f(x) = \sqrt{x}$
4. $f(x) = |x|$
5. $f(x) = \frac{1}{x}$
6. $f(x) = -1$
7. Let $f(x) = x^3 + 2x + 3$. Find $f^{-1}(3)$.
8. Let $H(x) = 2x^3 + 5x - 2$. Find $H^{-1}(-2)$.
9. A function G is defined by the following table:

x	−10	−5	0	5	10
$G(x)$	17	20	−4	3	10

   a. Is G a one-to-one function? Why?
   b. Find $G^{-1}(3)$.
   c. Find $G^{-1}(-4)$.
   d. Find $G^{-1}(x)$.
10. A function H is defined by the following table.

x	−4	−1	0	2	3
$H(x)$	0	2	−1	0	1

   a. Does H have an inverse? Why?
   b. Find $H(-1)$.
   c. Find $H(1)$.
11. Find two restrictions of the function $f(x) = x^2$ that allow it to have an inverse.
12. Find the inverse function for $f(x) = x^2 + 3$ on $[0, \infty)$.
13. Sketch the inverse function for $f(x) = x^4$ restricted to $(-\infty, 0]$.
14. Let $f(x) = -2x + 5$. Find $f^{-1}(x)$ and then sketch both functions on the same graph.

6.2 Exponential Functions

The function $f(x) = 2^x$ is quite different from any of the functions we have considered thus far. Previously, we defined functions using the basic algebraic operations (addition, subtraction, multiplication, division, powers, and roots). However, $f(x) = 2^x$ has a variable in the exponent and does not fall into the class of algebraic functions. Rather, it is our first example of an exponential function.

An **exponential function** has the form

$$f(x) = a^x, \qquad a > 0, \qquad a \neq 1$$

where the real constant a is called the **base**, and the independent variable x may assume any real value.

6.2a *Graphs of Exponential Functions*

To get a better understanding of exponential functions, we shall consider some examples.

EXAMPLE 1 *Graphing an Exponential Function*

Sketch the graph of $f(x) = 2^x$.

SOLUTION

Let $y = 2^x$, and make a table of x and y values. After plotting these points, we sketch the smooth curve as shown below. Note that the x-axis is a horizontal asymptote; that is, the curve gets closer and closer to the x-axis as x approaches $-\infty$.

x	$y = 2^x$
−4	$\frac{1}{16}$
−3	$\frac{1}{8}$
−2	$\frac{1}{4}$
−1	$\frac{1}{2}$
0	1
1	2
2	4
3	8

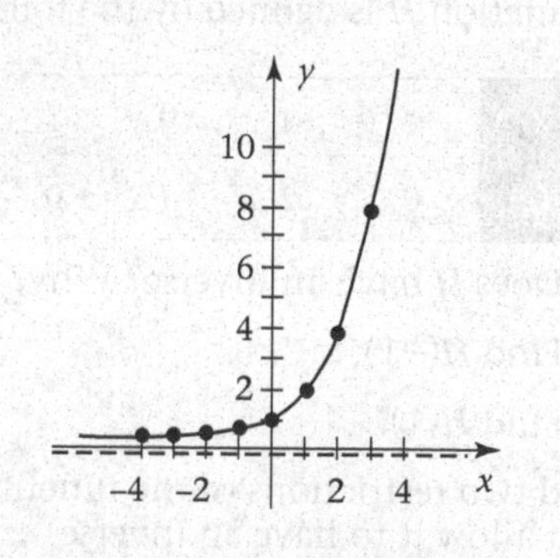

■

Observe that when we sketched the graph of $f(x) = 2^x$, we obtained some values of 2^x without giving a complete explanation of our computations. Specifically, we have not explained how to find 2^x when x is irrational. For example, how do we calculate $2^{\sqrt{2}}$? For our purposes, we shall think of $2^{\sqrt{2}}$ as that value that 2^x approaches as x gets closer and closer to $\sqrt{2}$, such as $2^{1.4}$, $2^{1.41}$, $2^{1.414}$, ... A precise definition is given in more advanced mathematics courses, noting that the same laws of exponents hold for irrational exponents as well as for rational exponents.

The exponential function 2^x grows very rapidly for large values of x compared with x^2. For example, if $x = 2$, they both yield the same answer, 4. However, if $x = 20$

$$2^{20} = 1{,}048{,}576 \qquad \text{while} \qquad (20)^2 = 400$$

Previously, we considered functions in which a *variable is raised to a constant power.* It is possible to confuse this with exponential functions in which *a constant is raised to a variable power.* For example,

Exponential Functions	Nonexponential Functions
2^x	x^2
$(10)^x$	x^{10}
$\left(\frac{1}{2}\right)^x$	$x^{1/2}$
$(\sqrt{2})^x$	$x^{\sqrt{2}}$
π^x	x^{π}

EXAMPLE 2 ***Graphing an Exponential Function***

Sketch the graph of $f(x) = \left(\frac{1}{2}\right)^x = 2^{-x}$.

SOLUTION

After finding the coordinates of some points, as listed in the table, we sketch the graph:

x	$y = 2^{-x}$
−3	8
−2	4
−1	2
0	1
1	$\frac{1}{2}$
2	$\frac{1}{4}$
3	$\frac{1}{8}$
4	$\frac{1}{16}$

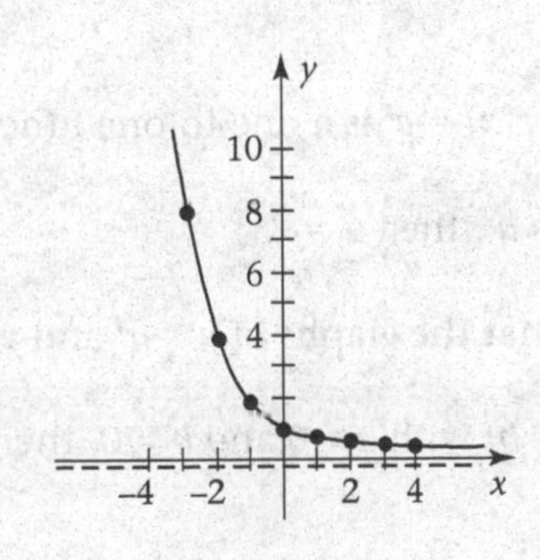

■

Note that the graph of $y = 2^{-x}$ is a reflection about the y-axis of the graph $y = 2^x$.

6.2b Properties of Exponential Functions

Consider the graphs of $y = 2^x$, $y = (\frac{1}{2})^x$, $y = 3^x$, and $y = (\frac{1}{3})^x$ all sketched on the same coordinate system as shown in Figure 6-4. The graphs of $y = (\frac{1}{2})^x$ and $y = (\frac{1}{3})^x$ are typical of the graph of $y = a^x$ where $0 < a < 1$. Similarly, the graphs of $y = 2^x$ and $y = 3^x$ are typical of the graph of $y = a^x$ for $a > 1$. We state various properties in Table 6-1.

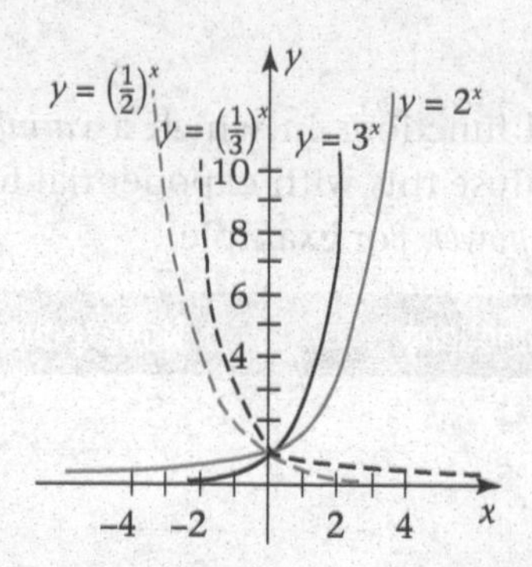

FIGURE 6-4
Graph of Various Exponential Functions

TABLE 6-1 Properties of Exponential Functions

$f(x) = a^x, \quad a > 0, \quad a \neq 1$
1. The graph of $f(x) = a^x$ always passes through the points (0, 1) and (1, a).
2. The graph of $f(x) = (\frac{1}{a})^x = a^{-x}$ is the reflection about the y-axis of the graph of $f(x) = a^x$.
3. The graph of $f(x) = a^x$ has the horizontal asymptote $y = 0$.
4. The domain of $f(x) = a^x$ consists of the set of real numbers; the range is the set of positive numbers.
5. $f(x) = a^x$ is increasing if $a > 1$; $f(x) = a^x$ is decreasing if $0 < a < 1$.
6. $f(x) = a^x$ is a one-to-one function since it passes the horizontal-line test.
7. Let $0 < a < b$. If $x > 0$, then $a^x < b^x$; if $x < 0$, then $a^x > b^x$.

Since $f(x) = a^x$ is a one-to-one function, we have the following result:

If $a^u = a^v$, then $u = v$.

Observe that the graphs of $y = a^x$ and $y = b^x$ intersect only at $x = 0$.

If $a^u = b^u$ with $a > 0$ and $b > 0$, then $u = 0$ or $a = b$.

EXAMPLE 3 *Solving Exponential Equations*

Solve for x.

a. $3^{10} = 3^{5x}$ b. $2^7 = (x-1)^7$ if $x > 1$ c. $3^{3x} = 9^{x-1}$

SOLUTION

a. Since $a^u = a^v$ implies $u = v$, we have

$$10 = 5x$$
$$x = 2$$

b. Since $a^u = b^u$ with $a > 0$, $b > 0$, and $u \neq 0$ implies $a = b$, we have

$$2 = x - 1$$
$$x = 3$$

c. Rewriting $3^{3x} = 9^{x-1}$ with the same bases, we have

$$3^{3x} = (3^2)^{x-1}$$
$$3^{3x} = 3^{2x-2}$$

Since $a^u = a^v$ implies $u = v$, we have

$$3x = 2x - 2$$
$$x = -2$$

■

Progress Check

Solve for x.

a. $2^8 = 2^{x+1}$ b. $4^{2x+1} = 4^{11}$ c. $8^{x+1} = 2$

Answers

a. 7 b. 5 c. $-\frac{2}{3}$

6.2c *The Number e*

There is an irrational number that was first designated by the letter e by the Swiss mathematician Leonhard Euler (1707–1783). The number e is the value that the expression

$$\left(1 + \frac{1}{m}\right)^m$$

approaches as m gets larger and larger. We will approximate this expression for different values of m, as shown in Table 6-2.

TABLE 6-2 Approximating the Value of *e*

m	1	2	10	100	1000	10,000	100,000	1,000,000
$\left(1+\frac{1}{m}\right)^m$	2.0	2.25	2.5937	2.7048	2.7169	2.7181	2.71827	2.71828

Alternatively, we write the following:

As m approaches ∞, $\left(1+\frac{1}{m}\right)^m$ approaches $e \approx 2.71828$.

Although the definition of the number e appears to be rather unnatural, the function $f(x) = e^x$ is called the **natural exponential function**. This function plays a most important role in mathematics.

The graphs of $f(x) = e^x$ and $f(x) = e^{-x}$ are shown in Figure 6-5. Since $2 < e < 3$, the graph of $y = e^x$ falls between the graphs of $y = 2^x$ and $y = 3^x$.

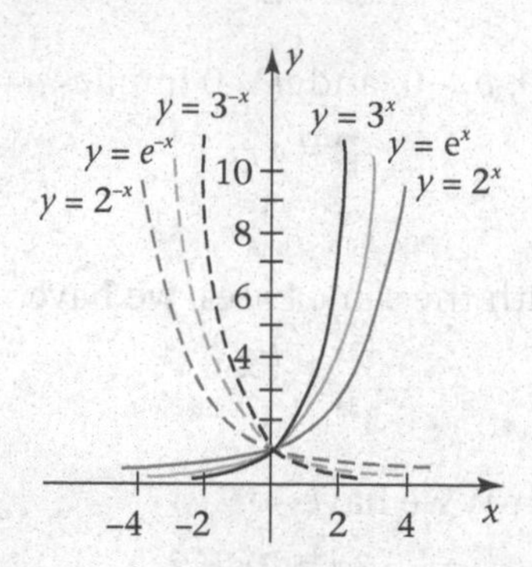

FIGURE 6-5
The Natural Exponential Function with Other Exponential Functions

Graphing Calculator ALERT

To evaluate e^x on a graphing calculator, you will use [2nd] [LN]. To compute e to 9 decimal places, press the buttons [2nd] [LN] 1. (Note that on some models you will need to close the parentheses. On others, you may not need to do that.)

6.2d Applications

There are many applications involving the use of exponential functions. We shall consider models associated with business, biology, and physics. Specifically, we shall investigate a mechanism for computing compound interest earned on an investment, a model that can represent population growth under most favorable environmental conditions and, finally, a model that simulates the process of radioactive decay.

Compound Interest

Suppose you invest P dollars at an interest rate r that is paid to you at the end of a period of t years. Then, the amount A available to you is calculated by the formula for simple interest:

$$A = P + Prt = P(1 + rt)$$

Most banks, however, advertise that they pay **compound interest**; that is, each successive payment includes interest on the previously accrued interest.

We use Table 6-3 to explore the effect of annual compounding over a period of 3 years, first with an investment of \$100 compounded annually at a rate of 10% and then in general. We discover that the amount of interest paid increases each year since the principal at the start of the next year has increased. Further, the amount A available at the end of each year suggests the following formula:

$$A = P(1 + r)^t$$

TABLE 6-3 Effect of Compound Interest over Three Years

Year	Starting Amount	Ending Amount	Starting Amount	Ending Amount
1	100.00	$100(1 + 0.1) = 110.00$	P	$P(1 + r)$
2	110.00	$110(1 + 0.1) = 121.00$	$P(1 + r)$	$P(1 + r)^2$
3	121.00	$121(1 + 0.1) = 133.10$	$P(1 + r)^2$	$P(1 + r)^3$

EXAMPLE 4 *Compound Interest*

Suppose that \$5000 is invested at an interest rate of 7.5% compounded annually. Find the value of the investment after 4 years.

SOLUTION

To apply the formula for compound interest, we must express the interest rate as a decimal, that is, $7.5\% = 0.075$.

We are given $P = 5000$, $r = 0.075$, and $t = 4$. Substituting,

$$\begin{aligned} A &= P(1 + r)^t \\ &= 5000(1 + 0.075)^4 \approx \$6677.35 \end{aligned}$$

■

What happens if interest is compounded more frequently than once per year? Many banks advertise that they pay interest compounded semiannually, quarterly, or daily. We can modify the compound interest formula to handle any **compounding period**, the time period between successive additions of interest. For example, suppose that compounding takes place semiannually, that is, twice a year at an *annual* rate r. Then, the interest rate applied to each period is assumed to be $\frac{r}{2}$. In addition, if the investment is held for t years, then there are $2t$ compounding periods. The amount available after t years is then

$$A = P\left(1 + \frac{r}{2}\right)^{2t}$$

We may now generalize to the situation where compounding takes place k times per year. The interest rate applied to each period is $\frac{r}{k}$ and there are a total of kt compounding periods. The compound interest formula can then be stated the following way:

Compound Interest Formula

If P dollars are invested at an annual interest rate r compounded k times annually, then the amount A available at the end of t years is

$$A = P\left(1 + \frac{r}{k}\right)^{kt}$$

EXAMPLE 5 ***Compound Interest***

Suppose that \$6000 is invested at an annual interest rate of 8%. What will the value of the investment be after 3 years if

a. interest is compounded quarterly?

b. interest is compounded semiannually?

SOLUTION

a. We are given $P = 6000$, $r = 0.08$, $k = 4$ and $t = 3$. Thus,

$$A = P\left(1 + \frac{r}{k}\right)^{kt} = 6000(1 + 0.02)^{12} \approx \$7609.45$$

b. We have $P = 6000$, $r = 0.08$, $k = 2$ and $t = 3$. Therefore

$$A = 6000(1 + 0.04)^6 \approx \$7591.91$$

■

Note that the interest obtained when compounding quarterly is greater than the interest obtained when compounding semiannually.

Progress Check

Suppose that \$5000 is invested at an annual interest rate of 6% compounded semiannually. What is the value of the investment after 12 years?

Answer

\$10,163.97

Continuous Compounding

When P, r, and t, are held fixed and the frequency of compounding is increased, the return on the investment is increased. Table 6-4 displays the approximate increase in the value of an investment at the end of 1 year using various compounding periods.

TABLE 6-4 **Investment of $P = \$1000$ at Rate $r = 8.0\%$ for One Year**

Compounding Period	Compounding Frequency k	Amount A	Additional Interest
annual	1	\$1080.00	
quarter	4	\$1082.43	\$2.43
month	12	\$1083.00	\$0.57
day	365	\$1083.28	\$0.28
hour	8760	\$1083.29	\$0.01

Although the value of the investment increases as we increase the compounding frequency, the additional benefit appears to diminish. In fact, the result in Table 6-4 is almost the same to two decimal places for daily and hourly compounding. (Show that the amount A continues to increase slightly if we consider more than two decimal places. In fact, show that this increase continues for a compounding period of 1 minute and 1 second.) This observation leads us to suspect that there may be a limiting value that is approached as we increase the compounding frequency. Consider the limit of this compounding process as we let the compounding period approach zero and the compounding frequency get larger and larger. The limit is called **continuous compounding**.

Recall the compound interest formula

$$A = P\left(1+\frac{r}{k}\right)^{kt}$$

If we let $m = \frac{k}{r}$, we obtain

$$A = P\left(1+\frac{1}{m}\right)^{mkt} = P\left[\left(1+\frac{1}{m}\right)^{m}\right]^{rt}$$

Letting the compounding frequency k become larger and larger is equivalent to letting m get larger and larger. Earlier in this section, we observed that

As m approaches ∞, $\left(1+\frac{1}{m}\right)^m$ approaches e.

Thus, we obtain the following result:

Continuous Compounding

If P dollars are invested at an interest rate r compounded continuously, then the amount A available at the end of t years is

$$A = Pe^{rt} \qquad (1)$$

EXAMPLE 6 *Continuous Compounding*

Suppose that \$20,000 is invested at an annual interest rate of 7% compounded continuously. What is the value of the investment after 4 years?

SOLUTION

We have $P = 20{,}000$, $r = 0.07$, and $t = 4$. Substituting into Equation (1), we obtain

$$\begin{aligned} A &= Pe^{rt} \\ &= 20{,}000e^{0.07(4)} = 20{,}000e^{0.28} \\ &\approx 26{,}462.596 \end{aligned}$$

The amount available after 4 years is \$26,462.60. ■

Progress Check

Suppose that \$10,000 is invested at an annual interest rate of 10% compounded continuously. What is the value of the investment after 6 years?

Answer

The investment value is approximately \$18,221

By solving Equation (1) for P, we can determine the principal P that must be invested at continuous compounding to have a certain amount A at some future time.

EXAMPLE 7 ***Continuous Compounding***

Suppose that a principal P is to be invested at continuous compound interest of 8% per year to yield $10,000 in 5 years. Approximately how much should be invested?

SOLUTION

Using Equation (1) with $A = 10{,}000$, $r = 0.08$ and $t = 5$, we have

$$A = Pe^{rt}$$
$$10{,}000 = Pe^{0.08(5)} = Pe^{0.40}$$
$$P = \frac{10{,}000}{e^{0.40}}$$
$$= 10{,}000e^{-0.40}$$
$$\approx 6703$$

Thus, approximately $6703 should be invested initially. ■

Progress Check

Approximately how much money should a 35-year-old woman invest now at continuous compound interest of 10% per year to obtain the sum of $20,000 upon her retirement at age 65?

Answer

Approximately $996 should be invested now.

Exponential Growth

Biologists have observed that under extremely favorable environmental conditions, populations appear to increase following the **exponential growth model.**

Exponential Growth Model

$$Q(t) = q_0e^{kt}, \qquad q_0 > 0, \qquad k > 0$$

This model predicts the quantity Q (size or biomass) of a population that is present at time t. Both q_0 and k are constants specific to the particular population in question. Note that when $t = 0$, we have

$$Q(0) = q_0e^0 = q_0$$

which says that q_0 is the initial quantity. (It is customary to use subscript 0 to denote an initial value.) The constant k is called the **growth constant.**

EXAMPLE 8 ***Exponential Growth Model—Bacteria in a Culture***

The number of bacteria in a culture after t hours is described by the exponential growth model

$$Q(t) = 50e^{0.7t}$$

a. Find the initial number of bacteria q_0 in the culture.

b. How many bacteria are in the culture after 10 hours?

SOLUTION

a. To find q_0 we need to evaluate $Q(t)$ at $t = 0$.

$$Q(0) = 50e^{0.7(0)} = 50e^0 = 50 = q_0$$

Thus, there are initially 50 bacteria in the culture.

b. The number of bacteria in the culture after 10 hours is given by $Q(10)$.

$$Q(10) = 50e^{0.7(10)} = 50e^7 \approx 54,832$$

Thus, there are approximately 54,832 bacteria after 10 hours. ■

Progress Check

The number of bacteria in a culture after t minutes is described by the exponential growth model $Q(t) = q_0e^{0.005t}$. If there were 100 bacteria present initially, how many bacteria will be present after 1 hour has elapsed?

Answer

Approximately 135 bacteria will be present.

EXAMPLE 9 ***Exponential Growth Model—U.S. Population***

Statistics indicate that the U.S. population, in 2017, is growing at an average rate of 1.08% per year. According to the 2014 National Projections provided by the United States Census Bureau at www.census.gov/population, the U.S. population in the year 2016 was approximately 324 million. Assuming an exponential growth model, find the approximate population of the United States in the year 2020.

SOLUTION

We let $k = 1.08\% = 0.0108$ and $q_0 = 324$. Then the exponential growth model

$$Q(t) = q_0e^{kt}$$

becomes

$$Q(t) = 324e^{0.0108t}$$

For the year 2020, $t = 4$, so we find that

$$Q(4) = 324e^{0.0108(4)} = 324e^{0.0432} \approx 338$$

Therefore, we project a U.S. population of approximately 338 million persons for the year 2020. ■

Exponential Decay

Radioactive elements decay in a manner that is described by the **exponential decay model:**

Exponential Decay Model

$$Q(t) = q_0 e^{-kt}, \qquad q_0 > 0, \qquad k > 0$$

This model predicts the quantity Q (mass) of a particular radioactive material remaining after time t has elapsed. As in the previous model, q_0 represents the initial quantity. The constant k is called the **decay constant.**

We use the term **half-life** to describe the time it takes for half of the atoms of a radioactive element to break down or decay. Table 6-5 displays the approximate half-lives of a number of elements. Half-lives can vary from a fraction of a second to billions of years. The long half-life of certain radioactive elements, obtained as by-products of nuclear processes, has caused environmentalists to warn us of the potential hazards of nuclear wastes.

TABLE 6-5 Half-Lives of Radioactive Elements

Radioactive Element	Half-Life	Radioactive Element	Half-Life
Iridium 198	1 minute	Radon 222	4 days
Polonium 210	4 months	Radium 226	1620 years
Uranium 238	4.5 billion years	Thorium 232	14 billion years

FOCUS on Radioactivity in the News

On April 25, 1986, a nuclear disaster of unprecedented proportions took place at Chernobyl in the former Soviet Union. A fire occurred within a building housing a nuclear reactor, resulting in an explosion that sent radioactive material into the atmosphere. The air and soil of the surrounding farmlands were seriously contaminated.

Soil tests have indicated the presence of cesium 137, a radioactive element that has a half-life of 37 years. Scientists believe that the level of contamination in the soil must be reduced to 1/125th of its current reading before the region can again be used for farming. This will take 7 half-lives (since $2^7 = 128$) or approximately 259 years.

On September 13, 1988, the *New York Times* headline read

MAJOR RADON PERIL IS DECLARED BY U.S. IN CALL FOR TESTS

Radon, a colorless, invisible gas, is released by the breakdown of uranium in the earth's crust. Outdoors, the gas dissipates rapidly and is considered to be harmless. Indoors, however, it may accumulate to dangerous levels, especially in newer homes that have been constructed very tightly to conserve energy. Since the half-life of radon is only 3.8 days, homes showing high concentrations of radon may be salvaged by the installation of ventilation systems.

EXAMPLE 10 *Exponential Decay Model*

A radioactive substance has a decay rate of 5% per hour. If 500 grams are present initially, how much of the substance remains after 4 hours?

SOLUTION
We let $k = 5\% = 0.05$ and $q_0 = 500$. Then the exponential decay model

$$Q(t) = q_0e^{-kt}$$

becomes

$$Q(t) = 500e^{-0.05t}$$

After 4 hours,

$$Q(4) = 500e^{-0.05(4)} = 500e^{-0.2} \approx 409.37$$

Thus, there remain approximately 409.37 grams of the substance. ■

EXAMPLE 11 *Exponential Decay Model—Half-Life of Radium 226*

Radium 226 has an approximate half-life of 1620 years. Show that the quantity Q of radium 226 present after t years is given by

$$Q(t) = q_02^{(-t/1620)}$$

SOLUTION
If q_0 represents the initial quantity, then $\frac{q_0}{2}$ is the quantity present when $t = 1620$. Substituting in the exponential decay model

$$Q(t) = q_0e^{-kt}$$

$$\frac{q_0}{2} = q_0e^{-1620k}$$

We can write this last equation as

$$\frac{1}{2} = e^{-1620k}$$

Taking reciprocals, we obtain

$$2 = e^{1620k} = (e^k)^{1620}$$

and so

$$e^k = 2^{(1/1620)}$$

Substituting in the original equation, we have

$$Q(t) = q_0e^{-kt} = q_0(e^k)^{-t} = q_0(2^{(1/1620)})^{-t} = q_02^{(-t/1620)}$$ ■

Progress Check

The number of grams Q of a certain radioactive substance present after t seconds is given by the exponential decay model $Q(t) = q_0e^{-0.4t}$. If 200 grams of the substance are present initially, find how much remains after 0.1 minute.

Answer
Approximately 18.14 grams remain.

Exercise Set 6.2

In Exercises 1–12, sketch the graph of the given function f.

1. $f(x) = 4^x$
2. $f(x) = 4^{-x}$
3. $f(x) = 10^x$
4. $f(x) = 10^{-x}$
5. $f(x) = 2^{x+1}$
6. $f(x) = 2^{x-1}$
7. $f(x) = 2^{|x|}$
8. $f(x) = 2^{-|x|}$
9. $f(x) = 2^{2x}$
10. $f(x) = 3^{-2x}$
11. $f(x) = e^{x+1}$
12. $f(x) = e^{-2x}$

In Exercises 13–20, solve for x.

13. $2^x = 2^3$
14. $2^{x-1} = 2^4$
15. $3^x = 9^{x-2}$
16. $2^x = 8^{x+2}$
17. $2^{3x} = 4^{x+1}$
18. $3^{4x} = 9^{x-1}$
19. $e^{x-1} = e^3$
20. $e^{x-1} = 1$

In Exercises 21–24, solve for a if $a > \frac{1}{2}$ and $x \neq 0$.

21. $(a+1)^x = (2a-1)^x$
22. $(2a+1)^x = (a+4)^x$
23. $(a+1)^x = (2a)^x$
24. $(2a+3)^x = (3a+1)^x$

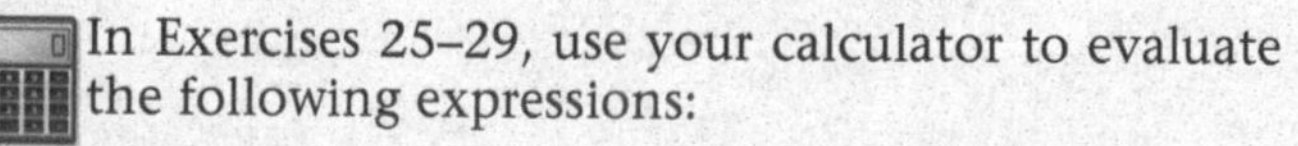
In Exercises 25–29, use your calculator to evaluate the following expressions:

25. e^2
26. $6e^{-0.02}$
27. $\dfrac{e^3 + e^{-3}}{2}$
28. $\dfrac{1}{1+e^{-5}}$
29. $\dfrac{5}{6-3e^{-8}}$

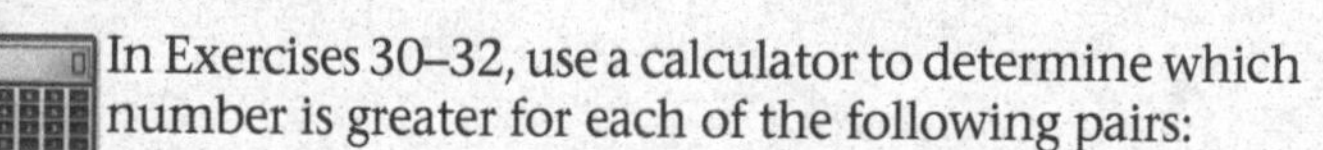
In Exercises 30–32, use a calculator to determine which number is greater for each of the following pairs:

30. 2^π, π^2
31. 3^π, π^3
32. e^π, π^e

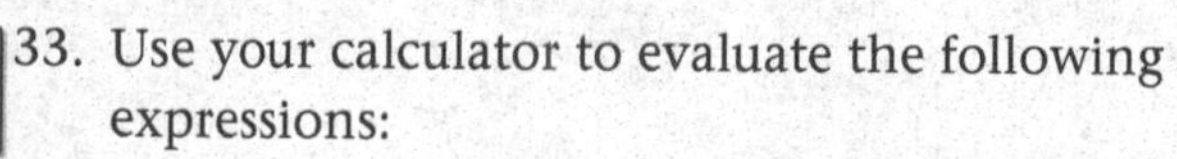
33. Use your calculator to evaluate the following expressions:
    a. $500(1+0.075)^5$
    b. $500\left(1+\dfrac{0.075}{2}\right)^{2(5)}$
    c. $500\left(1+\dfrac{0.075}{4}\right)^{4(5)}$
    d. $500\left(1+\dfrac{0.075}{12}\right)^{12(5)}$
    e. $500\left(1+\dfrac{0.075}{365}\right)^{365(5)}$
    f. $500\left(1+\dfrac{0.075}{365(24)}\right)^{365(24)(5)}$
    g. $500e^{0.07(5)}$
    h. What do the computations in parts (a) to (g) represent?

34. GRAPH $y = \left(1+\frac{1}{x}\right)^x$ in the following viewing rectangles to visualize that as $x \to \infty$, $\left(1+\frac{1}{x}\right)^x \to 2.71828$:
    a. $-10 \le X \le 10$ $\quad -10 \le Y \le 10$
    b. $-100 \le X \le 100$ $\quad 0 \le Y \le 5$
    c. $-1000 \le X \le 1000$ $\quad 0 \le Y \le 5$
    d. $-1{,}000{,}000 \le X \le 1{,}000{,}000$ $\quad 0 \le Y \le 5$

35. GRAPH $y = (1+x)^{1/x}$ in the following viewing rectangles to visualize that as $x \to 0$, $(1+x)^{1/x} \to 2.71828$:
    a. $-10 \le X \le 10$ $\quad -10 \le Y \le 10$
    b. $-1 \le X \le 1$ $\quad 0 \le Y \le 5$
    c. $-0.1 \le X \le 0.1$ $\quad 0 \le Y \le 5$
    d. $-0.00001 \le X \le 0.00001$ $\quad 0 \le Y \le 5$

    (On some graphing calculators you may find that $y = 1$ when x is close to 0. This is a result of the internal round-off error of the calculator.)

36. What point is common to the graph of all exponential functions?
37. What symmetry do you observe when you compare the graph of $y = \left(\frac{1}{4}\right)^x$ with $y = \left(\frac{1}{4}\right)^{-x}$?
38. If \$5000 is invested at 3.2% interest compounded monthly, how much would you have after 54 months?
39. Find the interest received if \$10,000 is invested at 5.1% for 2 years, compounded continuously.
40. Suppose we have an investment of \$10,000 that is earning 2.1% interest compounded continuously.
    a. How long would it take to double the initial investment?
    b. How long would it take to triple the initial investment?
    c. What would be the answers to parts (a) and (b) if your initial investment were \$20,000?
41. If a bacteria colony doubles its number at the end of each day, and there were 10,000 at the end of day 2, find q_0 and the number at the end of a week. Use $Q(t) = q_0 2^{kt}$.
42. If the half-life of a radioactive substance is 25 minutes, what fraction of the substance remains after 125 minutes? Use $Q(t) = q_0 2^{-kt}$.

43. The number of bacteria in a culture after t hours is described by the exponential growth model $Q(t) = 200e^{0.25t}$.

a. What is the initial number of bacteria in the culture?

b. Find the number of bacteria in the culture after 20 hours.

c. Complete the following table:

t	1	4	8	10
Q				

44. The number of bacteria in a culture after t hours is described by the exponential growth model $Q(t) = q_0e^{0.01t}$. If there were 400 bacteria present initially, how many bacteria will be present after 2 days?

45. At the beginning of 2000, the world population was approximately 6.0 billion. Suppose that the population is described by an exponential growth model, and that the rate of growth is 1.2% per year. Give the approximate world population by the beginning of the year 2050.

46. The number of grams of potassium 42 present after t hours is given by the exponential decay model $Q(t) = q_0e^{-0.055t}$. If 400 grams of the substance is present initially, how much remains after 10 hours?

47. A radioactive substance has a decay rate of 4% per hour. If 1000 grams are present initially, how much of the substance remains after 10 hours?

48. An investor purchases a $12,000 savings certificate paying 2.25% annual interest compounded semiannually. Find the amount received when the savings certificate is redeemed at the end of 8 years.

49. The parents of a newborn infant place $10,000 in an investment that pays 3% annual interest compounded quarterly. What sum is available at the end of 18 years to finance the child's college education?

50. A widow is offered a choice of two investments. Investment A pays 2% annual interest compounded quarterly, and investment B pays 2.5% compounded annually. Which investment yields a greater return?

51. A firm intends to replace some of its present computers in 5 years. The treasurer suggests that $25,000 be set aside in an investment paying 4% compounded monthly. What sum will be available for the purchase of the new computers?

52. If $5000 is invested at an annual interest rate of 2% compounded continuously, how much is available after 5 years?

53. If $100 is invested at an annual interest rate of 1.5% compounded continuously, how much is available after 10 years?

54. A principal P is to be invested at continuous compound interest of 1.9% to yield $50,000 in 20 years. What is the approximate value of P to be invested?

55. A 40-year-old executive plans to retire at age 65. How much should be invested at 6.5% annual interest compounded continuously to provide the sum of $250,000 upon retirement?

56. Investment A offers 8% annual interest compounded semiannually, and investment B offers 8% annual interest compounded continuously. If $1000 were invested in each, what would be the approximate difference in value after 10 years?

57. If $1000 is deposited in a savings account that earns interest at an annual rate of 3.5% compounded continuously, what is the value of the account at the end of 8 years?

58. A trust fund is being set up by a single payment so that at the end of 25 years there will be $100,000 in the fund. If interest is compounded continuously at a rate of 8%, how much should be paid into the fund initially?

59. In the late 1970s, the minimum wage in the United States was approximately $2.30 per hour. Since then, assume that the minimum wage has increased according to the exponential function $y(t) = 2.30 \times 1.95^{t/20}$, where t is the number of years after 1975.

a. Find the minimum wage predicted by this model for 2017.

b. How does this prediction in part (a) compare to the minimum wage of $7.25 per hour in 2017? Was this a good prediction model for the long term?

60. An investment of $10,000 earns interest at an annual rate of 10% compounded continuously. After t years, its value S is given by $S = 10{,}000e^{0.1t}$. Find the value of this investment after 25 years.

61. Write a PROGRAM in your calculator that will enable the user to enter the principal, interest rate, and time for an investment, and will return the value of the investment. You may be able to find such a program at your calculator manufacturer's website.

62. *Mathematics in Writing:* Under what circumstances is an exponential function increasing (rising) or decreasing? Compare Exercise 2 in this section with the following:

$$f(x) = \left(\frac{1}{4}\right)^x$$

What conclusions can you draw?

6.3 Logarithmic Functions

6.3a Logarithms as Exponents

We noted in the previous section that the exponential function $f(x) = a^x$, $a > 0$, $a \neq 1$ is a one-to-one function. Therefore, it has an inverse, and it is called the **logarithmic function**. Specifically, if

$$f(x) = a^x$$

then we write the inverse using the notation

$$f^{-1}(x) = \log_a x$$

which is read as "log of x to the base a" or "log base a of x." Note that the base a of the exponential function becomes the **base** of the corresponding logarithmic function. Thus, if

$$y = a^x$$

we are able to solve this equation for x in terms of y. The solution is

$$x = \log_a y$$

Consider the case where $a = 2$, namely,

$$f(x) = 2^x \quad \text{and} \quad f^{-1}(x) = \log_2 x$$

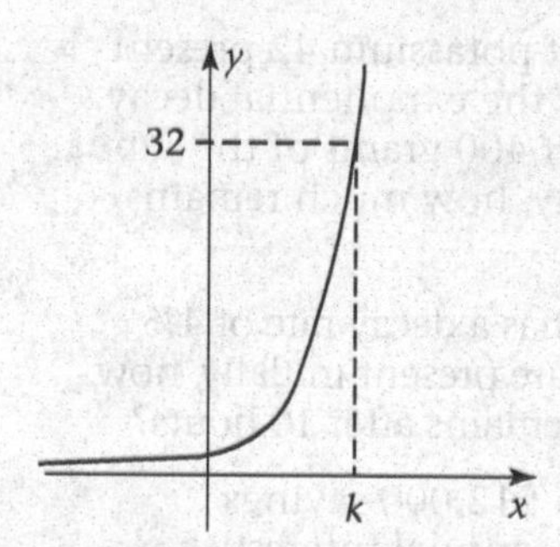

FIGURE 6-6
Graph of $y = 2^x$

Recall that the range of f is the set of all positive numbers and that $f(x)$ is one-to-one. Thus, for any positive number that we choose, there is a unique point on the graph of $f(x)$ whose y-coordinate is that positive number. For example, if we choose the y-coordinate to be 32, as shown in Figure 6-6, and we denote the x-coordinate by k, we have that

$$f(k) = 32$$

or equivalently

$$2^k = 32 \tag{1}$$

The x-coordinate k is the *logarithm of 32 to the base 2*, which we write as

$$k = \log_2 32 \tag{2}$$

In this example $k = 5$, that is, $\log_2 32 = 5$.

In other words, if we raise 2 to the exponent 5, we obtain 32. Equivalently, if we raise 2 to the exponent $\log_2 32$, we obtain 32. More generally, $\log_a x$ is the exponent to which a must be raised to obtain x. Equations (1) and (2) demonstrate that a *logarithm is an exponent* and can be generalized to provide the following definition:

Logarithmic Function Base *a*

$$y = \log_a x \quad \text{if and only if} \quad x = a^y$$

When no base is indicated, the notation **log *x*** is interpreted to mean $\log_{10} x$ which is also called the **common logarithm.**

Common Logarithm

$$\log x = \log_{10} x$$

The notation **ln *x*** is used to indicate the logarithm to the base e. Since $\ln x$ is the inverse of the natural exponential function e^x, it is called the **natural logarithm.**

Natural Logarithm

$$\ln x = \log_e x$$

The exponential form $x = a^y$ and the logarithmic form $y = \log_a x$ are two ways of expressing the same relationship among x, y, and a. Further, it is always possible to convert one form into the other. A natural question, then, is why create a logarithmic form when we already have an equivalent exponential form. One reason is to allow us to switch an equation from the form $x = a^y$ to a form in which y is a function of x. We will also demonstrate that the logarithmic function has some very useful properties.

EXAMPLE 1 *Logarithmic to Exponential Form*

Write in exponential form.

a. $\log_3 9 = 2$ b. $\log_2 \frac{1}{8} = -3$ c. $\log_{16} 4 = \frac{1}{2}$ d. $\ln 7.39 \approx 2$

SOLUTION

We change from the logarithmic form $\log_a x = y$ to the equivalent exponential form $a^y = x$.

a. $3^2 = 9$ b. $2^{-3} = \frac{1}{8}$ c. $16^{1/2} = 4$ d. $e^2 \approx 7.39$ ■

Progress Check

Write in exponential form.

a. $\log_4 64 = 3$

b. $\log_{10}\left(\frac{1}{10{,}000}\right) = -4$

c. $\log_{25} 5 = \frac{1}{2}$

d. $\ln 0.3679 \approx -1$

Answers

a. $4^3 = 64$

b. $10^{-4} = \frac{1}{10{,}000}$

c. $25^{1/2} = 5$

d. $e^{-1} \approx 0.3679$

EXAMPLE 2 ***Exponential to Logarithmic Form***

Write in logarithmic form.

a. $36 = 6^2$ b. $7 = \sqrt{49}$ c. $\frac{1}{100} = 10^{-2}$ d. $0.1353 \approx e^{-2}$

SOLUTION

Since $y = \log_a x$ if and only if $x = a^y$, the logarithmic forms are

a. $\log_6 36 = 2$ b. $\log_{49} 7 = \frac{1}{2}$ c. $\log \frac{1}{100} = -2$ d. $\ln 0.1353 \approx -2$ ■

Progress Check

Write in logarithmic form.

a. $1000 = 10^3$ b. $6 = 36^{1/2}$

c. $\frac{1}{7} = 7^{-1}$ d. $20.09 \approx e^3$

Answers

a. $\log 1000 = 3$ b. $\log_{36} 6 = \frac{1}{2}$

c. $\log_7 \frac{1}{7} = -1$ d. $\ln 20.09 \approx 3$

6.3b *Graphs of the Logarithmic Functions*

To sketch the graph of the logarithmic function $\log_a x$ we can take advantage of the fact that it is the inverse of the exponential function a^x. We know from our earlier work with inverse functions that the curves are reflections about the line $y = x$. This enables us to sketch the graphs shown in Figure 6-7 for $a > 1$.

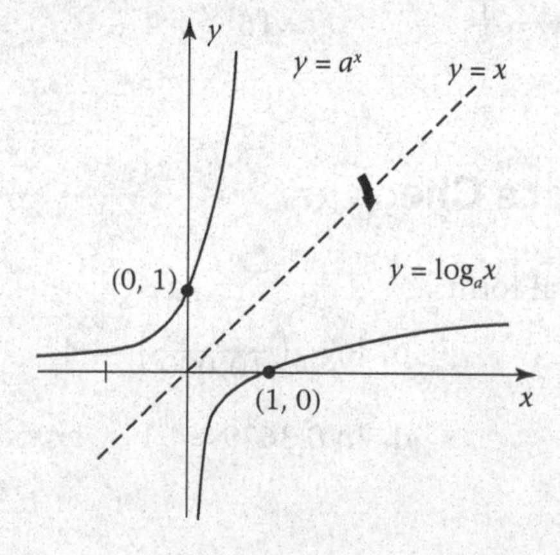

FIGURE 6-7
The Logarithmic Function

The graphs in Figure 6-7 demonstrate that the exponential and logarithmic functions for a base $a > 1$ are both increasing functions. However, the rate of growth for these two functions is dramatically different. The exponential function continues to rise more steeply, accelerating its rate of growth, whereas the logarithmic function continues to flatten out, decelerating its rate of growth.

The next example offers an alternative means for sketching the graph of a logarithmic function.

EXAMPLE 3 ***Graphing Logarithmic Functions***

Sketch the graph of $y = \log_2 x$.

SOLUTION

To sketch the graph of a logarithmic function, we convert to the equivalent exponential form. Thus, to sketch the graph of $y = \log_2 x$, we form a table of values for the equivalent exponential equation $x = 2^y$.

y	−3	−2	−1	0	1	2	3
$x = 2^y$	1/8	1/4	1/2	1	2	4	8

We can now plot these points and sketch a smooth curve, as shown below. Note that the y-axis is a vertical asymptote. We have included the graph of $y = 2^x$ to stress that the graphs of a pair of inverse functions are reflections of each other about the line $y = x$.

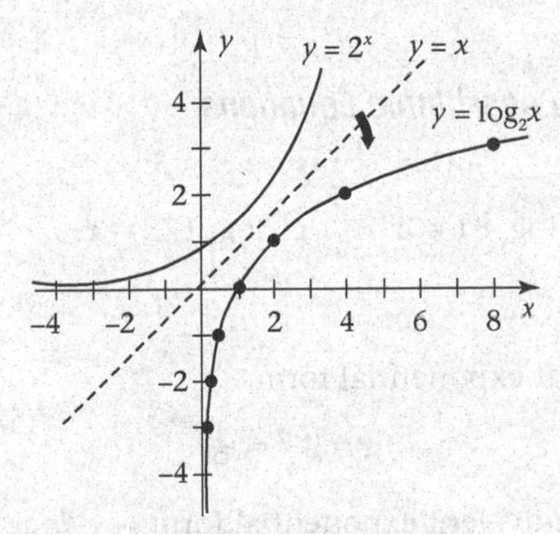

■

EXAMPLE 4 ***Logarithms with Base Less Than 1***

Sketch the graphs of $y = \log_3 x$ and $y = \log_{1/3} x$ on the same coordinate axes.

SOLUTION

We could set up a table of values for each function and then sketch the graphs. However, by switching from the logarithmic form

$$y = \log_{1/3} x$$

to the exponential form

$$x = \left(\frac{1}{3}\right)^y = 3^{-y}$$

and then back again to logarithmic form

$$\log_3 x = -y$$

or

$$y = -\log_3 x$$

we can conclude that

$$y = \log_{1/3} x = -\log_3 x$$

This demonstrates that the graph of $y = \log_{1/3} x$ is the reflection of the graph of $y = \log_3 x$ about the x-axis:

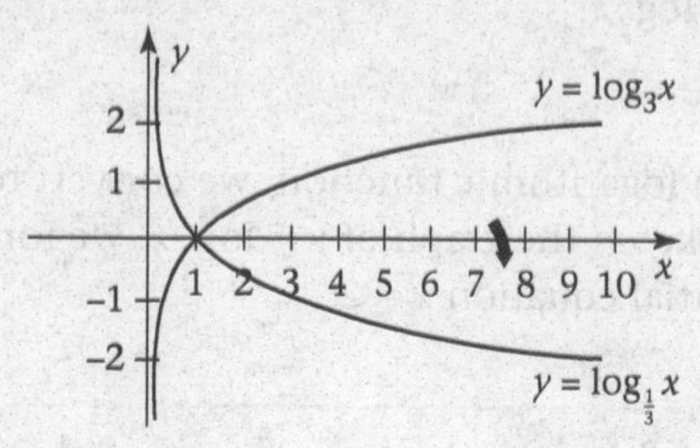

■

6.3c Logarithmic Equations and Calculators

Logarithmic equations can often be solved by changing them into equivalent exponential forms.

EXAMPLE 5 *Solving Logarithmic Equations*

Solve for x.

a. $\log_3 x = -2$ b. $\log_x 81 = 4$ c. $\log_5 125 = x$

SOLUTION

a. Using the equivalent exponential form,

$$x = 3^{-2} = \frac{1}{9}$$

b. Changing to the equivalent exponential form,

$$x^4 = 81 = 3^4$$

Since x is the base of a logarithm, $x > 0$. Hence $a^u = b^u$ with $a > 0$, $b > 0$, and $u \neq 0$ implies that

$$x = 3$$

c. In exponential form we have

$$5^x = 125$$

Writing 125 to the base 5, we have

$$5^x = 5^3$$

and since $a^u = a^v$ implies $u = v$, we conclude that

$$x = 3$$

■

Solve for x.

a. $\log_x 1000 = 3$ b. $\log_2 x = 5$ c. $x = \log_7 \frac{1}{49}$

Answers

a. 10 b. 32 c. −2

EXAMPLE 6 ***Calculators and Logarithms***

Use a calculator to solve for x to four decimal places.

a. $\ln x = -0.75$

b. $\log x = 1.25$

SOLUTION

a. $\ln x = -0.75$
$$x = e^{-0.75}$$
$$x \approx 0.4724$$

b. $\log x = 1.25$
$$x = 10^{1.25}$$
$$x \approx 17.7828$$

■

6.3d *Logarithmic Identities*

If $f(x) = a^x$, then $f^{-1}(x) = \log_a x$. Recall that inverse functions have the property that

$$f[f^{-1}(x)] = x \quad \text{and} \quad f^{-1}[f(x)] = x$$

Substituting $f(x) = a^x$ and $f^{-1}(x) = \log_a x$, we have

$$f[f^{-1}(x)] = x \qquad f^{-1}[f(x)] = x$$
$$f(\log_a x) = x \qquad f^{-1}(a^x) = x$$
$$a^{\log_a x} = x \qquad \log_a a^x = x$$

These two identities are useful in simplifying expressions.

$$a^{\log_a x} = x$$
$$\log_a a^x = x$$

The following pair of identities can be established by converting to the equivalent exponential form.

$$\log_a a = 1$$
$$\log_a 1 = 0$$

EXAMPLE 7 ***Working with Logarithmic Identities***

Evaluate.

a. $8^{\log_8 5}$ b. $\log 10^{-3}$ c. $\log_7 7$ d. $\log_4 1$

SOLUTION

a. 5 b. −3 c. 1 d. 0 ■

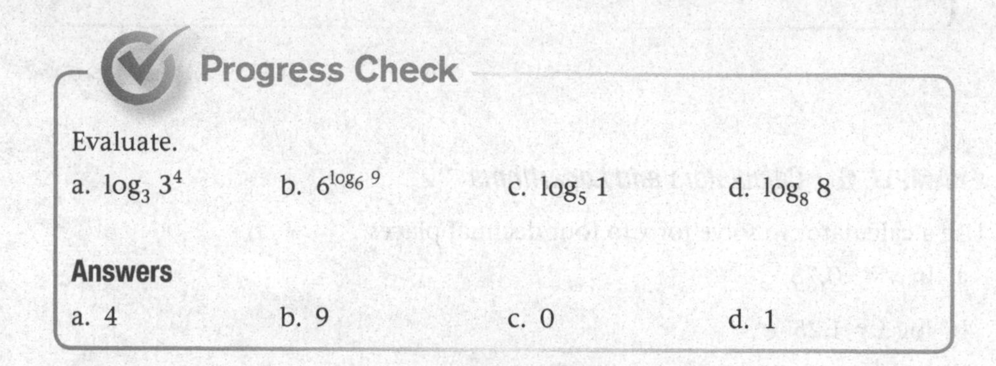

Progress Check

Evaluate.

a. $\log_3 3^4$ b. $6^{\log_6 9}$ c. $\log_5 1$ d. $\log_8 8$

Answers

a. 4 b. 9 c. 0 d. 1

6.3e *Properties of Logarithmic Functions*

The graph of $y = \log_2 x$ in Example 3 and the graph of $y = \log_{1/3} x$ in Example 4 are typical of the graph of $y = \log_a x$ for $a > 1$ and $0 < a < 1$, respectively. We state the following properties in Table 6-6.

TABLE 6-6 Properties of the Logarithmic Function

$f(x) = \log_a x, \quad a > 0, \quad a \neq 1$
1. The graph of $f(x) = \log_a x$ always passes through the points (1, 0) and $(a, 1)$ since $\log_a 1 = 0$ and $\log_a a = 1$.
2. The graph of $f(x) = \log_{1/a} x$ is the reflection about the x-axis of the graph of $f(x) = \log_a x$.
3. The domain of $f(x) = \log_a x$ consists of the set of all positive numbers; the range is the set of all real numbers.
4. $f(x) = \log_a x$ is increasing if $a > 1$; $f(x) = \log_a x$ is decreasing if $0 < a < 1$.
5. $f(x) = \log_a x$ is a one-to-one function since it passes the horizontal-line test.

These results are in accord with what we anticipate for a pair of inverse functions. As expected, the domain of the logarithmic function is the range of the corresponding exponential function, and vice versa.

Since $f(x) = \log_a x$ is a one-to-one function, we have the following result:

If $\log_a u = \log_a v$, then $u = v$.

Observe that the graphs of $y = \log_a x$ and $y = \log_b x$ intersect only at $x = 1$.

If $\log_a u = \log_b u$, either $u = 1$ or $a = b$.

EXAMPLE 8 ***Solving Logarithmic Equations***

Solve for x.

a. $\log_5 (x + 1) = \log_5 25$ b. $\log_{x-1} 31 = \log_5 31$

SOLUTION

a. Since $\log_a u = \log_a v$ implies $u = v$, then

$$x + 1 = 25$$
$$x = 24$$

b. Since $\log_a u = \log_b u$ and $u \neq 1$, then $a = b$. Hence,

$$x - 1 = 5$$
$$x = 6$$

■

Progress Check

Solve for x.

a. $\log_2 x^2 = \log_2 9$ b. $\log_7 14 = \log_{2x} 14$

Answers

a. 3, −3 b. $\frac{7}{2}$

Summary of Logarithmic Identities and Properties

Identities	Properties
$a^{\log x} = x$	$\log_a u = \log_a v$ implies $u = v$
$\log_a a^x = x$	If $\log_a u = \log_b u$, either $u = 1$ or $a = b$.
$\log_a a = 1$	
$\log_a 1 = 0$	

FOCUS on Measuring an Earthquake

The Loma Prieta Earthquake of 1989

On October 17, 1989, at 5:04 p.m. a major earthquake with a reading of 7.1 on the Richter scale struck the San Francisco area. Its epicenter was located 75 miles south of San Francisco between Santa Cruz and San Jose. About 63 people died, and damage was estimated to be approximately 6 billion dollars. The earthquake forced the cancellation of the third game of baseball's World Series that was due to start at Candlestick Park in San Francisco at 5:30 p.m.

This earthquake was the fourth worst in U.S. history, as far as loss of life is concerned, since the 1906 San Francisco earthquake (3000+ dead, 7.9 on the Richter scale), the 1933 Long Beach earthquake (120 dead, 6.3 on the Richter scale), and the 1964 Alaskan earthquake (114 dead, 9.2 on the Richter scale).

On the Richter scale, the magnitude M of an earthquake is defined as

$$M = \log \frac{I}{I_0}$$

where I_0 is a constant that represents a standard intensity and I is the intensity of the earthquake being measured. The Richter scale is a means of measuring a given earthquake against a "standard earthquake" of intensity I_0.

What does 3.0 on the Richter scale mean? Substituting $M = 3$ in the above equation, we have

$$3 = \log \frac{I}{I_0}$$

or, in the equivalent exponential form,

$$1000 = \frac{I}{I_0}$$

Solving for I,

$$I = 1000\, I_0$$

which states that an earthquake with a Richter scale reading of 3.0 is $10^3 = 1000$ times as intense as the standard. Therefore, an earthquake registering 8.0 on the Richter scale has an intensity $10^8 = 100{,}000{,}000$ times that of the standard.

Based on past experience, the following damages can be expected from an earthquake of various Richter scale readings:

Reading	Damage
2.0	Not noticed
4.5	Some damage in a very limited area
6.0	Hazardous serious damage with destruction of buildings in a limited area
7.0	Felt over a wide area with significant damage
8.0	Great damage
9.5	Maximum recorded

Exercise Set 6.3

In Exercises 1–12, write each equation in exponential form.

1. $\log_2 4 = 2$
2. $\log_5 125 = 3$
3. $\log_9 \frac{1}{81} = -2$
4. $\log_{64} 4 = \frac{1}{3}$
5. $\ln 20.09 \approx 3$
6. $\ln \frac{1}{7.39} \approx -2$
7. $\log_{10} 1000 = 3$
8. $\log_{10} \frac{1}{1000} = -3$
9. $\ln 1 = 0$
10. $\log_{10} 0.01 = -2$
11. $\log_3 \frac{1}{27} = -3$
12. $\log_{125} \frac{1}{5} = -\frac{1}{3}$

In Exercises 13–26, write each equation in logarithmic form.

13. $25 = 5^2$
14. $27 = 3^3$
15. $10{,}000 = 10^4$
16. $\frac{1}{100} = 10^{-2}$
17. $\frac{1}{8} = 2^{-3}$
18. $\frac{1}{27} = 3^{-3}$
19. $1 = 2^0$
20. $1 = e^0$
21. $6 = \sqrt{36}$
22. $2 = \sqrt[3]{8}$
23. $64 = 16^{3/2}$
24. $81 = 27^{4/3}$
25. $\frac{1}{3} = 27^{-1/3}$
26. $\frac{1}{2} = 16^{-1/4}$

In Exercises 27–44, solve for x without using a calculator.

27. $\log_5 x = 2$
28. $\log_{16} x = \frac{1}{2}$
29. $\log_{25} x = -\frac{1}{2}$
30. $\log_{1/2} x = 3$
31. $\ln x = 2$
32. $\ln x = -3$
33. $\ln x = -\frac{1}{2}$
34. $\log_4 64 = x$
35. $\log_5 \frac{1}{25} = x$
36. $\log_x 4 = \frac{1}{2}$
37. $\log_x \frac{1}{8} = -\frac{1}{3}$
38. $\log_3 (x-1) = 2$
39. $\log_5 (x+1) = 3$
40. $\log_2 (x-1) = \log_2 10$
41. $\log_{x+1} 24 = \log_3 24$
42. $\log_3 x^3 = \log_3 64$
43. $\log_{x+1} 17 = \log_4 17$
44. $\log_{3x} 18 = \log_4 18$

In Exercises 45–64, evaluate the expression without using a calculator.

45. $3^{\log_3 6}$
46. $2^{\log_2 (2/3)}$
47. $e^{\ln 2}$
48. $e^{\ln (1/2)}$
49. $\log_5 5^3$
50. $\log_4 4^{-2}$
51. $\log_8 8^{1/2}$
52. $\log_{64} 64^{-1/3}$
53. $\log_7 49$
54. $\log_7 \sqrt{7}$
55. $\log_5 5$
56. $\ln e^{x^2}$
57. $\ln 1$
58. $\log_4 1$
59. $\log_2 \frac{1}{4}$
60. $\log_{16} 4$
61. $\log 10{,}000$
62. $e^{\ln(x+1)}$
63. $\ln e^2$
64. $\ln e^{-2/3}$

In Exercises 65–70, use your calculator to evaluate the following expressions:

65. $\log \frac{8}{5}$
66. $\frac{\ln 8}{5}$
67. $\frac{\ln 8}{\ln 5}$
68. $\frac{\log 8}{\log 5}$
69. $\frac{\log\left(\frac{1}{2}\right)}{-0.0006}$
70. $2^{\ln 0.015}$

In Exercises 71–78, sketch the graph of each given function.

71. $f(x) = \log_4 x$
72. $f(x) = \log_{1/2} x$
73. $f(x) = \log 2x$
74. $f(x) = \frac{1}{2} \log x$
75. $f(x) = \ln \frac{x}{2}$
76. $f(x) = \ln 3x$
77. $f(x) = \log_3(x-1)$
78. $f(x) = \log_3 (x+1)$

In Exercises 79–86, determine the domain of the given function. Sketch the graph.

79. $f(x) = \ln (1-x)$
80. $f(x) = \ln (1-x)^2$
81. $f(x) = \log \frac{x}{x-1}$
82. $f(x) = \log \frac{\sqrt{x-1}}{x}$
83. $f(x) = \ln 2^x$
84. $f(x) = \log e^{-x}$
85. $f(x) = \log(-\sqrt{x})$
86. $f(x) = \ln x^3$
87. The Davis National Bank pays 6% interest compounded quarterly. How long will it take for a deposit to triple in value?
88. Suppose interest is compounded continuously at an annual rate of 5%.
    a. How long will it take for the principal to triple?
    b. How long will it take for the principal to quadruple?
89. If interest is compounded continuously, at what annual rate will a principal of $100 triple in 20 years?
90. Fechner's Logarithmic Law in psychology is used to measure notable changes in sensory response based upon equal increases of a particular sensory stimulus. The law is given by the equation $S = k \log_{10} I$ where S is the magnitude of the sensory experience, I is the physical intensity, and k is a scaling constant. Determine the physical intensity of a sensory experiment if the recorded magnitude of sensation is 0.5005 and the constant $k = 10$.

Exercises 91–93 refer to the following table of values:

x	0	1	2	3
y	2	6	18	54

91. Plot the points (x, z) where $z = \ln y$.

92. Find the equation of the line relating x and z.

93. Find the equation relating x and y, written in the form $y = ab^x$.

94. The Richter scale discussed in the chapter opener is described by the model

$$M = \log_{10} \frac{I}{S}$$

where I is the intensity of the earthquake measured in microns (the amplitude of a seismograph reading taken from 100 km from the epicenter), and S represents a standard intensity of 1 micron.

a. Convert the Richter model to an exponential equation.

b. Given any two earthquakes with intensities I_1 and I_2 with Richter scale intensities M_1 and M_2, find a formula for the ratio of the intensities.

95. Use your results from Exercise 94b to find the ratio of the intensity of the Indonesia earthquake to each of the following earthquakes (all data found at http://bvtlab.com/53aJ8):

a. November 28, 2004, in Hokkaido, Japan. $M = 7.0$

b. June 9, 1994, in northern Bolivia. $M = 8.2$

c. March 6, 1988, at the Gulf of Alaska. $M = 7.8$

6.4 Fundamental Properties of Logarithms

Logarithms are an important computational device because of three fundamental properties. For any positive numbers x and y,

Property 1. $\log_a xy = \log_a x + \log_a y$

Property 2. $\log_a\left(\frac{x}{y}\right) = \log_a x - \log_a y$

Property 3. $\log_a x^n = n \log_a x$, n a real number

These properties can be proved by using equivalent exponential forms. To prove the first property, $\log_a xy = \log_a x + \log_a y$, we let

$$\log_a x = u \qquad \text{and} \qquad \log_a y = v$$

Then the equivalent exponential forms are

$$a^u = x \qquad \text{and} \qquad a^v = y$$

Multiplying the left-hand and right-hand sides of these equations, respectively, we have

$$a^u a^v = xy$$

or

$$a^{u+v} = xy$$

Substituting a^{u+v} for xy in $\log_a xy$ we have

$$\begin{aligned} \log_a xy &= \log_a a^{u+v} \\ &= u + v \quad \text{since } \log_a a^x = x \end{aligned}$$

Substituting for u and v.

$$\log_a xy = \log_a x + \log_a y$$

Properties 2 and *3* can be established in much the same way.

It is these properties of logarithms that convert the operations of multiplication, division, and exponentiation to addition, subtraction, and multiplication, respectively. We next demonstrate the use of these properties.

EXAMPLE 1 *Simplifying Logarithmic Expressions*

Write in terms of simpler logarithmic forms.

a. $\log_{10}[(225)(478)]$ b. $\log_8 \frac{422}{735}$ c. $\log_2 2^5$ d. $\log_a \frac{xy}{z}$

SOLUTION

a. $\log_{10}[(225)(478)] = \log_{10} 225 + \log_{10} 478$

b. $\log_8 \frac{422}{735} = \log_8 422 - \log_8 735$

c. $\log_2 2^5 = 5 \log_2 2 = 5 \cdot 1 = 5$

d. $\log_a \frac{xy}{z} = \log_a x + \log_a y - \log_a z$ ■

Progress Check

Write in terms of simpler logarithmic forms.

a. $\log_4 [(1.47)(22.3)]$ b. $\log_a \frac{x-1}{\sqrt{x}}$

Answers

a. $\log_4 1.47 + \log_4 22.3$ b. $\log_a (x-1) - \frac{1}{2} \log_a x$

EXAMPLE 2 ***Applying the Properties of Logarithms***

Prove that

$$\log_a \frac{1}{x} = -\log_a x$$

SOLUTION

Note that

$$\log_a \frac{1}{x} = \log_a 1 - \log_a x$$

$$\log_a \frac{1}{x} = -\log_a x$$

since $\log_a 1 = 0$. ■

The result of Example 2 is useful in simplifying logarithmic forms.

6.4a Simplifying Logarithms

The next example illustrates rules that speed the handling of logarithmic forms.

EXAMPLE 3 ***Simplifying Complex Logarithmic Expressions***

Write

$$\log_a \frac{(x-1)^{-2}(y+2)^3}{\sqrt{x}}$$

in terms of simpler logarithmic forms.

SOLUTION

Step 1. Rewrite the expression so that each factor has a positive exponent.	*Step 1.* $\log_a \frac{(x-1)^{-2}(y+2)^3}{\sqrt{x}} = \log_a \frac{(y+2)^3}{(x-1)^2 x^{1/2}}$
Step 2. Apply *Property 1* and *Property 2* for logarithms of products and quotients. Each factor in the numerator will yield a term with a plus sign. Each factor in the denominator will yield a term with a minus sign.	*Step 2.* $= \log_a (y+2)^3 - \log_a (x-1)^2 - \log_a x^{1/2}$
Step 3. Apply *Property 3* to simplify.	*Step 3.* $= 3 \log_a (y+2) - 2 \log_a (x-1) - \frac{1}{2} \log_a x$

■

Progress Check

Simplify

$$\log_a \frac{(2x-3)^{1/2}(y+2)^{-2/3}}{z^4}$$

Answer

$\frac{1}{2}\log_a(2x-3) - \frac{2}{3}\log_a(y+2) - 4\log_a z$

EXAMPLE 4 ***Applying the Properties of Logarithms***

If $\log_a 1.5 = r$, $\log_a 2 = s$, and $\log_a 5 = t$, find the following:

a. $\log_a 7.5$

b. $\log_a \left[(1.5)^3 \sqrt[5]{\frac{2}{5}}\right]$

SOLUTION

a. Since

$$\begin{aligned} 7.5 &= (1.5)(5) \\ \log_a 7.5 &= \log_a[(1.5)(5)] \\ &= \log_a 1.5 + \log_a 5 && \textit{Property 1} \\ &= r + t && \text{Substitution} \end{aligned}$$

b. $\log_a \left[(1.5)^3 \sqrt[5]{\frac{2}{5}}\right]$

$$\begin{aligned} &= \log_a (1.5)^3 + \log_a \left(\frac{2}{5}\right)^{1/5} && \textit{Property 1} \\ &= 3\log_a 1.5 + \frac{1}{5}\log_a \frac{2}{5} && \textit{Property 3} \\ &= 3\log_a 1.5 + \frac{1}{5}[\log_a 2 - \log_a 5] && \textit{Property 2} \\ &= 3r + \frac{1}{5}(s-t) && \text{Substitution} \end{aligned}$$

■

Progress Check

If $\log_a 2 = 0.43$ and $\log_a 3 = 0.68$, find the following:

a. $\log_a 18$

b. $\log_a \sqrt[3]{\frac{9}{2}}$

Answers

a. 1.79

b. 0.31

a. Note that

$$\log_a (x + y) \neq \log_a x + \log_a y$$

Property 1 tells us that

$$\log_a (x \cdot y) = \log_a x + \log_a y$$

None of the three properties permits simplification of $\log_a (x + y)$.

b. Note that

$$\log_a x^n \neq (\log_a x)^n$$

By *Property 3,*

$$\log_a x^n = n \log_a x$$

We can also apply the properties of logarithms to combine terms involving logarithms.

EXAMPLE 5 *Combining Logarithmic Expressions*

Write as a single logarithm.

$$2 \log_a x - 3 \log_a (x + 1) + \log_a \sqrt{x - 1}$$

SOLUTION

$$2 \log_a x - 3 \log_a (x + 1) + \log_a \sqrt{x - 1}$$

$$= \log_a x^2 - \log_a (x + 1)^3 + \log_a \sqrt{x - 1} \qquad \textit{Property 3}$$

$$= \log_a x^2\sqrt{x - 1} - \log_a (x + 1)^3 \qquad \textit{Property 1}$$

$$= \log_a \frac{x^2\sqrt{x - 1}}{(x + 1)^3} \qquad \textit{Property 2}$$

■

Progress Check

Write as a single logarithm.

$$\frac{1}{3}[\log_a (2x - 1) - \log_a (2x - 5)] + 4 \log_a x$$

Answer

$$\log_a x^4 \sqrt[3]{\frac{2x - 1}{2x - 5}}$$

a. Note that

$$\frac{\log_a x}{\log_a y} \neq \log_a (x - y)$$

Property 2 tells us that

$$\log_a \frac{x}{y} = \log_a x - \log_a y$$

None of the three properties permits simplification of

$$\frac{\log_a x}{\log_a y}$$

b. The expressions

$$\log_a x + \log_b x$$

and

$$\log_a x - \log_b x$$

cannot be simplified. Logarithms with different bases do not readily combine except in special cases.

6.4b *Change of Base*

Sometimes it is convenient to write a logarithm that is given in terms of a base a in terms of another base b, that is, to convert $\log_a x$ to $\log_b x$. (As always, we require a and b to be positive real numbers different from 1.)

To compute $\log_b x$ given $\log_a x$, let $y = \log_b x$. The equivalent exponential form is then

$$b^y = x$$

Taking logarithms to the base a of both sides of this equation, we have

$$\log_a b^y = \log_a x$$

We now apply the fundamental properties of logarithms developed earlier in this section. By *Property 3*,

$$y \log_a b = \log_a x$$

Solving for y,

$$y = \frac{\log_a x}{\log_a b}$$

Since $y = \log_b x$, we have

Change of Base Formula

$$\log_b x = \frac{\log_a x}{\log_a b}, \qquad a > 0, a \neq 1 \qquad b > 0, b \neq 1$$

EXAMPLE 6 ***Change of Base***

Use the [log] key on your calculator to compute $\log_2 27$.

SOLUTION

We use the change of base formula

$$\log_b x = \frac{\log_a x}{\log_a b}$$

with $b = 2$, $a = 10$, and $x = 27$. Then

$$\log_2 27 = \frac{\log 27}{\log 2} \approx 4.7549$$

■

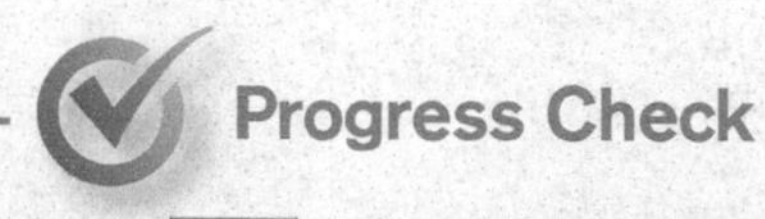

Progress Check

Use the [ln] key on your calculator to find:

a. $\log_5 16$ b. $\log_{10} e$

Answers

a. Approximately 1.7227 b. Approximately 0.4343

In Table 6-7, we summarize the rules we have encountered for manipulating logarithms as well as the common errors to be avoided.

TABLE 6-7 Logarithms

Properties	Warnings
$\log_a xy = \log_a x + \log_a y$	$\log_a (x + y) \neq \log_a x + \log_a y$
$\log_a \frac{x}{y} = \log_a x - \log_a y$	$\frac{\log_a x}{\log_a y} \neq \log_a (x - y)$
$\log_a x^n = n \log_a x$ $\log_a \frac{1}{x} = -\log_a x$	$\log_a x^n \neq (\log_a x)^n$
$\log_b x = \frac{\log_a x}{\log_a b}$	

Exercise Set 6.4

In Exercises 1–20, write each expression in terms of simpler logarithmic forms.

1. $\log_{10}[(120)(36)]$
2. $\log_6 \frac{187}{39}$
3. $\log_3(3^4)$
4. $\log_3(4^3)$
5. $\log_a(2xy)$
6. $\ln(4xyz)$
7. $\log_a \frac{x}{yz}$
8. $\ln \frac{2x}{y}$
9. $\ln x^5$
10. $\log_3 y^{2/3}$
11. $\log_a(x^2y^3)$
12. $\log_a(xy)^3$
13. $\log_a \sqrt{xy}$
14. $\log_a \sqrt[3]{xy^4}$
15. $\ln(x^2y^3z^4)$
16. $\log_a(xy^3z^2)$
17. $\ln(\sqrt{x}\sqrt[3]{y})$
18. $\ln(\sqrt[3]{xy^2}\sqrt[4]{z})$
19. $\log_a\left(\frac{x^2y^3}{z^4}\right)$
20. $\ln \frac{x^4y^2}{z^{1/2}}$

In Exercises 21–30, if $\log 2 \approx 0.30$, $\log 3 \approx 0.47$, and $\log 5 \approx 0.70$, evaluate.

21. $\log 6$
22. $\log \frac{2}{3}$
23. $\log 9$
24. $\log \sqrt{5}$
25. $\log 12$
26. $\log \frac{6}{5}$
27. $\log \frac{15}{2}$
28. $\log 0.3$
29. $\log \sqrt{7.5}$
30. $\log \sqrt[4]{30}$

In Exercises 31–44, write each expression as a single logarithm.

31. $2 \log x + \frac{1}{2} \log y$
32. $3 \log_a x - 2 \log_a z$
33. $\frac{1}{3} \ln x + \frac{1}{3} \ln y$
34. $\frac{1}{3} \ln x - \frac{2}{3} \ln y$
35. $\frac{1}{3} \log_a x + 2 \log_a y - \frac{3}{2} \log_a z$
36. $\frac{2}{3} \log_a x + 2 \log_a y - 2 \log_a z$
37. $\frac{1}{2}(\log_a x + \log_a y)$
38. $\frac{2}{3}(4 \ln x - 5 \ln y)$
39. $\frac{1}{3}(2 \ln x + 4 \ln y) - 3 \ln z$
40. $\ln x - \frac{1}{2}(3 \ln x + 5 \ln y)$
41. $\frac{1}{2} \log_a(x-1) - 2 \log_a(x+1)$
42. $2 \log_a(x+2) - \frac{1}{2}(\log_a y + \log_a z)$
43. $3 \log_a x - 2 \log_a(x-1) + \frac{1}{2} \log_a \sqrt[3]{x+1}$
44. $4 \ln(x-1) + \frac{1}{2} \ln(x+1) - 3 \ln y$

A calculator is used to compute $\ln 10 \approx 2.3026$, $\ln 6 \approx 1.7918$, and $\ln 3 \approx 1.0986$. In Exercises 45–50, use the first value to find the required value.

45. $\ln 17 \approx 2.8332$; find $\log 17$
46. $\ln 22 \approx 3.0910$; find $\log_6 22$
47. $\ln 141 \approx 4.9488$; find $\log_3 141$
48. $\ln 78 \approx 4.3567$; find $\log_6 78$
49. $\ln 245 \approx 5.5013$; find $\log 245$
50. $\ln 7 \approx 1.9459$; find $\log_3 7$
51. Given $\log 2 \approx 0.301$ and $\log 3 \approx 0.477$, use the rules of logarithms to find
    a. $\log 18$
    b. $\log \frac{4}{9}$
    c. $\log 5$
52. Use the rules of logarithms to find
    a. $\ln \frac{1}{\sqrt{e}}$
    b. $\ln 3e - \ln 3$
    c. $e^{(\ln 2 - \ln 3)}$
    d. $\log 2 + \log 50$
    e. $e^{-2 \ln 3}$

In Exercises 53–56, use the change of base formula to evaluate the following logarithms on your calculator:

53. $\log_5 10$
54. $\log_3 \frac{1}{2}$
55. $\log_2 \sqrt{7}$
56. $\log_{1/6} 13$

In Exercises 57 and 58, use the change of base formula to graph the following logarithmic functions on your calculator viewing rectangle:

57. $y = \log_5 x$
58. $y = \log_{1/4} x$
59. Refer back to Exercises 94 and 95 in Section 6.3. On May 22, 1960, there was an earthquake in Chile that was about 3.162 times as intense as the Indonesia earthquake. What was its magnitude on the Richter scale?
60. Set up a table including the intensities of various earthquakes, and the ratio of their intensities to that of the Indonesian earthquake. Use data from http://bvtlab.com/53aJ8.
61. *Mathematics in Writing*: How are logarithms useful in comparing the intensities of earthquakes? How might the Richter scale be confusing to someone who did not understand that it is a logarithmic, rather than linear, scale?

6.5 Exponential and Logarithmic Equations

The properties of exponentials and logarithms mentioned in the previous sections can be combined with the following suggestions to solve exponential and logarithmic equations:

- When solving an exponential equation, consider taking logarithms of both sides of the equation.
- When solving a logarithmic equation, consider forming a single logarithm on one side of the equation, and then converting this equation to the equivalent exponential form.

EXAMPLE 1 *Solving an Exponential Equation*

Solve $3^{2x-1} = 17$.

SOLUTION

Consider taking logarithms to the base 10 of both sides of the equation.

$$\log 3^{2x-1} = \log 17$$

Then by *Property 3,*

$$(2x-1)\log 3 = \log 17$$

$$2x - 1 = \frac{\log 17}{\log 3}$$

$$2x = 1 + \frac{\log 17}{\log 3}$$

$$x = \frac{1}{2} + \frac{\log 17}{2\log 3}$$

$$x \approx 1.7895$$

Note that we could have taken logarithms to *any* base in solving this equation. (Verify that the answer is the same if we had used ln instead of log.) ■

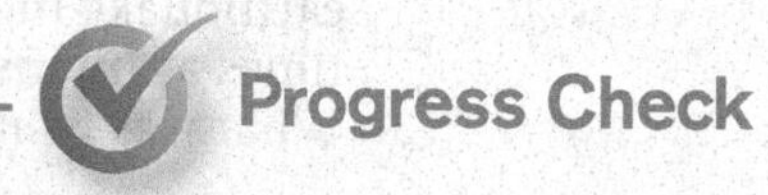

Solve $2^{x+1} = 3^{2x-3}$ and express your answer in terms of common logarithms.

Answer

$$\frac{\log 2 + 3\log 3}{2\log 3 - \log 2}$$

EXAMPLE 2 ***Solving an Exponential Equation***

Solve the equation $5e^{2-x} = 3$.

SOLUTION

Since the equation contains the base e, we take logarithms to the base e of both sides of the equation.

$$\ln(5e^{2-x}) = \ln 3$$
$$\ln 5 + \ln e^{2-x} = \ln 3 \qquad \textit{Property 1}$$
$$2 - x = \ln 3 - \ln 5$$
$$x = 2 + \ln 5 - \ln 3$$
$$x = 2 + \ln \frac{5}{3} \qquad \textit{Property 2}$$
$$x \approx 2.5108$$

(Verify the answer using log instead of ln.) ■

Solve the equation $5 - 2e^{3x-1} = 0$.

Answer

$\frac{1}{3}\left(1 + \ln\frac{5}{2}\right)$

EXAMPLE 3 ***Solving a Logarithmic Equation***

Solve for x: $\log(2x + 8) = 1 + \log(x - 4)$.

SOLUTION

Rewrite the equation in the form

$$\log(2x + 8) - \log(x - 4) = 1$$

We can now apply *Property 2* to form a single logarithm.

$$\log \frac{2x+8}{x-4} = 1$$

Converting to the equivalent exponential form, we have

$$\frac{2x+8}{x-4} = 10^1 = 10$$

$$2x + 8 = 10x - 40$$
$$x = 6$$

■

Solve $\log x - \frac{1}{2} = -\log 3$.

Answer

$\frac{\sqrt{10}}{3}$

EXAMPLE 4 *Solving a Logarithmic Equation*

Solve for x: $\log_2 x = 3 - \log_2 (x + 2)$.

SOLUTION

Rewriting the equation with a single logarithm, we have

$$\log_2 x + \log_2 (x + 2) = 3$$

$$\log_2 [x(x + 2)] = 3 \quad \text{Why?}$$

$$x(x + 2) = 2^3 = 8 \quad \text{Equivalent exponential form}$$

$$x^2 + 2x - 8 = 0$$

$$(x - 2)(x + 4) = 0 \quad \text{Factor}$$

$$x = 2 \text{ or } x = -4$$

The "solution" $x = -4$ must be rejected since the original equation contains $\log_2 x$. Note that the domain of the logarithmic function does not include negative values. ■

Solve for x: $\log_3 (x - 8) = 2 - \log_3 x$.

Answer

$x = 9$

FOCUS on Dating the Latest Ice Age

All organic forms of life contain radioactive carbon 14. In 1947, the chemist Willard Libby, who won the Nobel Prize in chemistry in 1960, found that the percentage of carbon 14 in the atmosphere equals the percentage found in the living tissues of all organic forms of life. When an organism dies, it stops replacing carbon 14 in its living tissues. Yet the carbon 14 continues decaying at the rate of 0.012% per year. By measuring the amount of carbon 14 in the remains of an organism, it is possible to estimate fairly accurately when the organism died.

$$Q(t) = q_0 e^{-kt}$$
$$0.254 q_0 = q_0 e^{-0.00012t}$$
$$0.254 = e^{-0.00012t}$$
$$\ln 0.254 = \ln e^{-0.00012t}$$
$$-1.3704 \approx -0.00012t$$
$$t \approx 11{,}420$$

In the late 1940s, radioactive carbon dating was used to date the last ice sheet to cover the North American and European continents. Remains of trees in the Two Creeks Forest in northern Wisconsin were found to have lost 74.6%of their carbon 14 content. The remaining carbon 14, therefore, was 25.4% of the original quantity q0 that was present when the descending ice sheet felled the trees. The accompanying computations use the general equation of an exponential decay model to find the age t of the wood. Conclusion: The latest ice age occurred approximately 11,420 years before the measurements were taken.

EXAMPLE 5 *Exponential Growth Model*

In a certain country, population is increasing at an annual rate of 2.5%. If we assume an exponential growth model, in how many years will the population double?

SOLUTION

The exponential growth model

$$Q(t) = q_0 e^{0.025t}$$

describes the population Q as a function of time t. Since the initial population is $Q(0) = q_0$, we seek the time t required for the population to double or become $2q_0$. We wish to solve the equation

$$Q(t) = 2q_0 = q_0 e^{0.025t}$$

for t. We then have

$$2q_0 = q_0 e^{0.025t}$$
$$2 = e^{0.025t} \qquad \text{Divide by } q_0$$
$$\ln 2 = \ln e^{0.025t} \qquad \text{Take natural logarithms of both sides}$$
$$\ln 2 = 0.025t \qquad \text{Since } \ln e^x = x$$
$$t = \frac{\ln 2}{0.025} \approx 27.7$$

or approximately 28 years. ■

EXAMPLE 6 ***Continuous Compounding***

A trust fund invests \$8000 at an annual interest rate of 8% compounded continuously. How long does it take for the initial investment to grow to \$12,000?

SOLUTION

From Section 6.2, we have the equation

$$A = Pe^{rt}$$

where A is the amount available at the end of t years, having initially invested the principal P at a continuously compounded interest rate r. If $A = 12{,}000$, $P = 8000$, and $r = 8\% = 0.08$, we must find t. Therefore,

$$12{,}000 = 8000e^{0.08t}$$

$$\frac{12{,}000}{8{,}000} = e^{0.08t}$$

$$e^{0.08t} = 1.5$$

Taking natural logarithms of both sides, we have

$$0.08t = \ln 1.5$$

$$t = \frac{\ln 1.5}{0.08}$$

$$\approx 5.07$$

It takes approximately 5.07 years for the initial \$8000 to grow to \$12,000. ■

Progress Check

The number of bacteria in a culture after t minutes is described by the exponential growth model $Q(t) = q_0e^{0.1t}$. How long does it take for the number of bacteria to double?

Answer

Approximately 6.93 minutes

Exercise Set 6.5

In Exercises 1–31, solve for x.

1. $5^x = 18$
2. $2^x = 24$
3. $2^{x-1} = 7$
4. $3^{x-1} = 12$
5. $3^{2x} = 46$
6. $2^{2x-1} = 56$
7. $5^{2x-5} = 564$
8. $3^{3x-2} = 23.1$
9. $3^{x-1} = 2^{2x-1}$
10. $4^{2x-1} = 3^{2x-3}$
11. $2^{-x} = 15$
12. $3^{-x+2} = 103$
13. $4^{-2x+1} = 12$
14. $3^{-3x+2} = 2^{-x}$
15. $e^x = 18$
16. $e^{x-1} = 2.3$
17. $e^{2x+3} = 30$
18. $e^{-3x+2} = 40$
19. $\log x + \log 2 = 3$
20. $\log x - \log 3 = 2$
21. $\log_x(3 - 5x) = 1$
22. $\log_x(8 - 2x) = 2$
23. $\log x + \log(x - 3) = 1$
24. $\log x + \log(x + 21) = 2$
25. $\log(3x + 1) - \log(x - 2) = 1$
26. $\log(7x - 2) - \log(x - 2) = 1$
27. $\log_2 x = 4 - \log_2(x - 6)$
28. $\log_2(x - 4) = 2 - \log_2 x$
29. $\log_2(x + 4) = 3 - \log_2(x - 2)$
30. $y = \dfrac{e^x + e^{-x}}{2}$
31. $y = \dfrac{e^x - e^{-x}}{2}$
32. Suppose that the world population is increasing at an annual rate of 1.2%. If we assume an exponential growth model, in how many years will the population double?
33. Suppose that the population of a certain city is increasing at an annual rate of 3%. If we assume an exponential growth model, in how many years will the population triple?
34. The population P of a certain city t years from now is given by
$$P = 20{,}000e^{0.05t}$$
How many years from now will the population be 50,000?
35. Potassium 42 has a decay rate of approximately 5.5% per hour. Assuming an exponential decay model, in how many hours will the original quantity of potassium 42 have been halved?
36. Consider an exponential decay model given by
$$Q = q_0e^{-0.4t}$$
where t is in weeks. How many weeks does it take for Q to decay to $\frac{1}{4}$ of its original amount?
37. How long does it take an amount of money to double if it is invested at a rate of 8% per year compounded semiannually?
38. At what rate of annual interest, compounded semiannually, should a certain amount of money be invested so that it will double in 8 years?
39. The number N of radios that an assembly line worker can assemble daily after t days of training is given by
$$N = 60 - 60e^{-0.04t}$$
After how many days of training does the worker assemble 40 radios daily?
40. The quantity Q in grams of a radioactive substance that is present after t days of decay is given by
$$Q = 400e^{-kt}$$
If $Q = 300$ when $t = 3$, find k, the decay rate.
41. A person on an assembly line produces P items per day after t days of training, where
$$P = 400(1 - e^{-t})$$
How many days of training will it take this person to be able to produce 300 items per day?
42. Suppose that the number N of mopeds sold when x thousands of dollars are spent on advertising is given by
$$N = 4000 + 1000 \ln(x + 2)$$
How much advertising money must be spent to sell 6000 mopeds?

In Exercises 43–45, solve for y.

43. $x^2 + x = \ln(y + 1)$
44. $\dfrac{1}{2x + 5} = \dfrac{1}{2}\ln(y - 3)^4$
45. $\ln x + 3 = \ln(y + 1) + \ln(y - 1)$

Chapter Summary

Key Terms, Concepts, and Symbols

Key Ideas for Review

Topic	Page	Key Idea
One-to-One	332	We say a function is one-to-one if every element of the range corresponds to precisely one element of the domain.
Horizontal Line Test	332	No horizontal line meets the graph of a one-to-one function in more than one point.
Inverse Function	333	The inverse of a function, f^{-1}, reverses the correspondence defined by the function f. The domain of f becomes the range of f^{-1}, and the range of f becomes the domain of f^{-1}.
Properties	333	A function f and its inverse f^{-1} satisfy $f^{-1}[f(x)] = x$ for all x in the domain of f; $f[f^{-1}(y)] = y$ for all y in the range of f
One-to-One	333	The inverse of a function f is defined only if f is one-to-one.
Inverse	334	The inverse of the inverse of a function is the function itself.
Graph	334	The graphs of a function and its inverse are reflections of each other about the line $y = x$.
Exponential Function	336	An exponential function has the form $f(x) = a^x, \quad a > 0, \quad a \neq 1$
Domain	338	The domain of $f(x) = a^x$ is the set of all real numbers.
Range	338	The range of $f(x) = a^x$ is the set of all positive numbers.
Increasing, Decreasing	338	The graph of $f(x) = a^x$ is increasing if $a > 1$ and decreasing if $0 < a < 1$.
One-to-One	338	The exponential function is one-to-one.
Equality, Same Base	338	If $a^x = a^y$, then $x = y$.
Equality, Different Bases	338	If $a^x = b^x$ with $a > 0$ and $b > 0$, then $x = 0$ or $a = b$.
e	339	As m approaches ∞, $(1 + \frac{1}{m})^m$ approaches $e \approx 2.71828$.

continues

Topic	Page	Key Idea
Applications	340	There are many applications involving the use of exponential functions, including models associated with business, biology, and physics.
Compound Interest	341	$A = P\left(1 + \frac{r}{k}\right)^{kt}$
Continuous Compounding	343	$A = Pe^{rt}$
Exponential Growth Model	344	$Q(t) = q_0 e^{kt}, \quad q_0 > 0, \quad k > 0$
Exponential Decay Model	346	$Q(t) = q_0 e^{-kt}, \quad q_0 > 0, \quad k > 0$
Logarithmic Function	350	The logarithmic function has the form $f(x) = \log_a x, \quad a > 0, \ a \neq 1$ It is the inverse of the exponential function a^x, and both functions have base a.
Special Cases	351	The common logarithm is $\log x = \log_{10} x$, and the natural logarithm is $\ln x = \log_e x$.
Logarithms as Exponents	351	The logarithmic function $y = \log_a x$ and the exponential form $x = a^y$ are two ways of expressing the same relationship. In short, logarithms are exponents. Consequently, it is always possible to convert one form into the other.
Identities	355	The following identities are useful in simplifying expressions and in solving equations: $a^{\log_a x} = x$ $\log_a a^x = x$ $\log_a a = 1$ $\log_a 1 = 0$
Domain	356	The domain of $f(x) = \log_a x$ is the set of all positive numbers.
Range	356	The range of $f(x) = \log_a x$ is the set of all real numbers.
Increasing, Decreasing	356	The graph of $f(x) = \log_a x$ is increasing if $a > 1$ and decreasing if $0 < a < 1$.
One-to-One	356	The logarithmic function is one-to-one.
Equality, Same Base	356	If $\log_a x = \log_a y$, then $x = y$.
Equality, Different Bases	356	If $\log_a x = \log_b x$, either $x = 1$ or $a = b$.
Fundamental Properties	361	The fundamental properties of logarithms are as follows: *Property 1.* $\log_a xy = \log_a x + \log_a y$ *Property 2.* $\log_a \frac{x}{y} = \log_a x - \log_a y$ *Property 3.* $\log_a x^n = n \log_a x$
Change of Base	365	The change of base formula is $\log_b x = \frac{\log_a x}{\log_a b}$

Review Exercises

In Exercises 1 and 2, determine whether the given function $f(x)$ is one-to-one. If it is, find its inverse $f^{-1}(x)$, and sketch the graphs of $y = f(x)$ and $y = f^{-1}(x)$ on the same coordinate axes.

1. $f(x) = \frac{x}{3} - 2$
2. $f(x) = \frac{|x|}{3} - 2$
3. Sketch the graph of $f(x) = (\frac{1}{3})^x$. Label the point $(-1, f(-1))$.
4. Solve $2^{2x} = 8^{x-1}$ for x.
5. Solve $(2a + 1)^x = (3a - 1)^x$ for a if $a > \frac{1}{3}$ and $x \neq 0$.
6. The amount of \$8000 is invested in a certificate paying 12% annual interest compounded semiannually. What amount is available at the end of 4 years?

In Exercises 7–10, write each logarithmic form in exponential form and vice versa.

7. $27 = 9^{3/2}$
8. $\log_{64} 8 = \frac{1}{2}$
9. $\log_2 \frac{1}{8} = -3$
10. $6^0 = 1$

In Exercises 11–14, solve for x.

11. $\log_x 16 = 4$
12. $\log_5 \frac{1}{125} = x - 1$
13. $\ln x = -4$
14. $\log_3 (x + 1) = \log_3 27$

In Exercises 15–18, evaluate the given expression.

15. $\log_3 3^5$
16. $\ln e^{-1/3}$
17. $\log_3 \frac{1}{3}$
18. $e^{\ln 3}$
19. Sketch the graph of $f(x) = \log_3 x + 1$.

In Exercises 20–23, write the given expression in terms of simpler logarithmic forms.

20. $\log_a \frac{\sqrt{x-1}}{2x}$
21. $\log_a \frac{x(2-x)^2}{(y+1)^{1/2}}$
22. $\ln[(x + 1)^4(y - 1)^2]$
23. $\log \sqrt[5]{\frac{y^2 z}{z+3}}$

In Exercises 24–33, use your calculator to evaluate the following expressions:

24. e^{-5}
25. $\ln \sqrt{2}$
26. $\sqrt[3]{5 - e^3}$
27. $2500\left(1+\frac{0.08}{4}\right)^{80}$
28. $2500e^{1.6}$
29. $\frac{\log 5}{\log 2}$
30. $\log_3 4$
31. $e^{5 \ln 2}$
32. 2^e
33. $\log(5 - \pi)$

34. A father has two sons, Cliff and Joe. Both wish to contribute a portion of their savings to their father's construction firm that they will one day inherit. Cliff and Joe both have different accounts. Cliff originally deposited \$8000 in his bank account where 9% interest was compounded quarterly, when he was 12 years old. Joe, however, deposited \$7000 in his bank account where 11% interest was compounded semiannually, when he was 14 years old. Who has the potential to contribute more to their father's business if Cliff is now 21 years old and Joe is now 23 years old? By how much?
35. Veterinary medicine sometimes uses an antibiotic called chloramphenecol to suppress blood cell production. Suppose a patient's red blood cell count drops to 3.0×10^6 RBC/cm^3 after therapy with this antibiotic. Assume that cell production can be stimulated at 50 cells exponentially per day, that is, 50^t, t is time measured in days. How long would it be before the patient's count was back to a normal range of 5.0×10^6 RBC/cm^3? (Note that red blood cells are produced from bone marrow and not from other blood cells.)
36. The number of bacteria in a certain culture is 320,000 at 4 p.m. At 6 p.m. on the same day, the count is 640,000. Assuming that the population grows exponentially, find the count at 10 p.m. that day.
37. With reference to Exercise 36, suppose there were 150,000 cells at 4 p.m. and 225,000 at 6 p.m. What will the count be at 2 a.m. the next morning?
38. Lou has \$1000 he wishes to invest in a certificate of deposit at bank A. The terms are 5% compounded continuously for 5 years. At bank B the terms are 5.05% compounded semiannually for 5 years. Which is the better investment?
39. A laboratory medical experiment involves testing the absorption rate of an AIDS antidote drug, once that drug has been injected into the circulatory system. Suppose $C_F = C_0 e^{-kt}$, where C_F = the final concentration, C_0 = the initial concentration, k = a predetermined constant, and t = time measured in minutes.
    a. Determine t if $C_F = 1000$ mg, $C_0 = 10{,}000$ mg, and $k = 2$.
    b. Plot C_F for $k = 1, 2, 3$ on the same set of axes, where $t \geq 0$.

Exercises 40 and 41 refer to the following:

Noise is related to the intensity of sound waves as the waves travel through a medium, such as air, water, etc. The intensity I is the energy carried by a wave per unit time, measured in watts per square meter. The loudness B of sound intensity levels is rated on a logarithmic scale and is measured in decibels. The equation relating these variables is

$$B = 10 \log \frac{I}{I_0}$$

where I_0 is the intensity of some reference level, usually taken as the minimum intensity audible to the human ear, 1.0×10^{-12} watts per square meter.

40. What is the loudness level of a radio whose intensity is 1.0×10^{-8} watts per square meter?

41. A single mosquito 1 meter from a person makes a sound close to the threshold of human hearing. What is the loudness level of 1000 mosquitos at such a distance?

Review Test

In Exercises 1 and 2, find $f^{-1}(x)$, the inverse of $f(x)$, and sketch the graphs of $y = f(x)$ and $y = f^{-1}(x)$ on the same coordinate axes.

1. $f(x) = -4x + 2$
2. $f(x) = -\frac{1}{x}$

3. Sketch the graph of $f(x) = 2^{x+1}$. Label the point $(1, f(1))$.

4. Solve $\left(\frac{1}{2}\right)^x = \left(\frac{1}{4}\right)^{2x+1}$

In Exercises 5 and 6, convert from logarithmic form to exponential form or vice versa.

5. $\log_3 \frac{1}{9} = -2$
6. $64 = 16^{3/2}$

In Exercises 7 and 8, solve for x.

7. $\log_x 27 = 3$
8. $\log_6\left(\frac{1}{36}\right) = 3x + 1$

In Exercises 9 and 10, evaluate the given expression.

9. $\ln e^{5/2}$
10. $\log_5 \sqrt{5}$

In Exercises 11 and 12, write the given expression in terms of simpler logarithmic forms.

11. $\log_a \frac{x^3}{y^2 z}$
12. $\log \frac{x^2 \sqrt{2y-1}}{y^3}$

In Exercises 13 and 14, use the value $\log 2 \approx 0.3$ to evaluate the given expression.

13. $\log 5$
14. $\log 2\sqrt{2}$

In Exercises 15 and 16, write the given expression as a single logarithm.

15. $2 \log x - 3 \log(y + 1)$

16. $\frac{2}{3}[\log_a (x + 3) - \log_a (x - 3)]$

17. The number of bacteria in a culture is described by the exponential growth model

$$Q(t) = q_0 e^{0.02t}$$

Approximately how many hours are required for the number of bacteria to double?

18. Suppose that \$500 is invested in a certificate at an annual interest rate of 12% compounded monthly. What is the value of the investment after 6 months?

In Exercises 19 and 20, solve for x.

19. $\log x - \log 2 = 2$

20. $\log_4 (x - 3) = 1 - \log_4 x$

In Exercises 21–24, use the values $\log 2 \approx 0.30$, $\log 3 \approx 0.48$, and $\log 7 \approx 0.85$ to evaluate the given expression.

21. $\log 14$
22. $\log 3.5$
23. $\log \sqrt{6}$
24. $\log 0.7$

In Exercises 25–28, write the given expression as a single logarithm.

25. $\frac{1}{3} \log_a x - \frac{1}{2} \log_a y$

26. $\frac{4}{3}[\log x + \log(x - 1)]$

27. $\ln 3x + 2\left(\ln y - \frac{1}{2} \ln z\right)$

28. $2 \log_a (x + 2) - \frac{3}{2} \log_a (x + 1)$

In Exercises 29 and 30, use the values $\log 32 \approx 1.5$, $\log 8 \approx 0.9$, and $\log 5 \approx 0.7$ to find the requested value.

29. $\log_8 32$
30. $\log_5 32$

31. A substance is known to have a decay rate of 6% per hour. Approximately how many hours are required for the remaining quantity to be half of the original quantity?

In Exercises 32–34, solve for x.

32. $2^{3x-1} = 14$

33. $2 \log x - \log 5 = 3$

34. $\log(2x - 1) = 2 + \log(x - 2)$

Cumulative Review Exercises: Chapters 4–6

1. Given $P(x) = 2x^3 + 5x^2 + 3$,
   a. find $P(-2)$.
   b. find the quotient and remainder when $P(x)$ is divided by $x + 1$.
2. Find all rational roots of the equation
$$5x^3 - x^2 + 20x - 4 = 0$$
3. The graph of the function
$$f(x) = 2x^3 - x^2 - 6x + 3$$
crosses the x-axis at $x = \frac{1}{2}$. Find the other x-intercepts, if any.
4. Given the function
$$f(x) = \frac{2x^2 + 6x}{x^2 - 9}$$
   a. determine the domain of f.
   b. find all asymptotes of the graph of f.
   c. sketch the graph of f.

In Exercises 5–7, solve for x.

5. $27^{x-1} = 9^{3-x}$
6. $\log_x (2x + 3) = 2$
7. $\log_2 (\log_3 x) = 1$
8. A radioactive isotope is known to decay exponentially. If the half-life of the isotope is 5 years, how long will it take for 9 grams to decay to 3 grams?
9. Sketch the graph of $f(x) = 2^x - 1$.
10. If $\log 2 = s$, $\log 3 = t$, and $\log 5 = u$, express the following in terms of s, t, and u:
    a. $\log 18$
    b. $\log \frac{5}{4}$
    c. $\log 1.5$
    d. $\log \sqrt{7.5}$
11. A certificate of deposit is purchased paying interest at the rate of 8% per year, compounded continuously. If the certificate matures in 6 months at a value of $1000, how much was invested initially?
12. Simplify
$$\ln \frac{x^2\sqrt{x+1}}{x-1}$$
13. Suppose an amount of money P is placed in an account where interest is compounded continuously for 15 years. What interest rate would be necessary in order for the value of the account to quadruple?
14. A $6000 investment is deposited in a local bank where interest is compounded quarterly at a rate of 1.5%. Find the amount in the account after 4 years. Also, what is the amount of the compound interest earned during the 4 years?
15. A father has an 8-year-old son for whom he is setting up a fund. He wants to have enough money accumulated in the fund to pay for a $20,000 college tuition, to start when his son turns 18. If interest is compounded continuously at an annual rate of 2.1%, how much money should the father initially deposit into the fund?
16. A firm wishes to deposit $75,000 in a bank account for future investment use. The firm estimates it will need 4 times its original deposit to have an effect on the stock market. If the going rate is 5% interest compounded continuously, how long will it take for the original investment to quadruple?
17. Maria will attend a university in the fall. Over the summer she will earn approximately $4000. She wishes to save about half of that for future college expenses. Bond A, maturing in three years, yields an interest rate of 7% compounded continuously and can be purchased for $1000. Bond B, also maturing in three years, yields an interest rate of 9% compounded continuously and can be purchased for $950. Which bond offers Katy the better investment?
18. Newton's law of cooling states that
$$T = (T_0 - T_R)e^{-kt} + T_R$$
where T_0 is the original temperature of the object in question, t is the elapsed time, T_R is the room temperature, k is the constant of decay, and T is the temperature at time t. Suppose a cup of coffee has a temperature of 150°F. It is placed in a room whose temperature is 80°F. If the temperature of the coffee is 90°F 2 hours later, find the constant of decay.

Writing Exercises

1. Write a brief report illustrating how exponential and logarithmic functions occur in everyday life.
2. Discuss how the invention of logarithms enabled seventeenth century astronomers in Europe to simplify their mathematical computations.
3. Discuss the environmental impact of the exponential growth of the world's population.
4. A newscaster stated that the 1989 earthquake in San Francisco, with a reading on the Richter scale of 7.1, was twice as powerful as a subsequent earthquake whose reading on the Richter scale was 3.05. Comment on the newscaster's understanding of the Richter scale.
5. Write a report on Napier's "bones."
6. Suppose you borrow $6000 at simple interest of 5% for one year. The bank gives you $5700. You must repay the loan in one payment when the loan is due. Explain in words how you would find the true rate of interest.
7. Explain how a slide rule works.

Chapter 6 Project

In 1865 in Memphis, Tennessee, there was an earthquake that measured 5.0 on the Richter scale. Memphis is part of a region that the United States Geological Survey refers to as the New Madrid Seismic Zone. Many buildings in this part of the country were never designed with large earthquakes in mind, but seismologists consider the likelihood of a large earthquake in this region within the next fifty years to be high. Tennessee schools have had Earthquake Awareness Week since 1995. (This information comes from http://www.cusec.org/.)

What is the likelihood of a large earthquake in your area? What precautions exist, if any, against the possibility of damage or loss of life from seismic activity?

For this project, first refer back to Exercises 94 and 95 in Section 6.3, and Exercises 59, 60, and 61 in Section 6.4. On your graphing calculator, display the graph of $y = \log x$ and set the viewing window large enough so that you can TRACE a point on the graph all the way up to $y = 9.0$. What is the range of x values? Write a brief essay comparing this graph to the Richter scale and describing how the y values are related to the x values. (*Hint:* The independent variable represents the ratio I/S.)

Copy your graph, and label the points on the graph which give information about the earthquakes in Bolivia, Indonesia, Memphis, Alaska, and Japan.

7 The Trigonometric Functions

The word "trigonometry" derives from the Greek, meaning "measurement of triangles." Historical interest in trigonometry was strongly motivated by the development of such fields as navigation, astronomy, and surveying. Hipparchus of Nicaea (190–125 BC) is often referred to as the "father of trigonometry."

We begin this chapter with the classical approach to trigonometry, that is, examining the relationships between the sides and angles of right triangles. We then follow with the more modern approach that emphasizes the concept of a function, as introduced in Chapter 3. Here, the trigonometric functions are viewed as functions of real numbers whose ranges are related to points on a circle. Thus, they are also called *circular functions*.

After examining the graphs of the trigonometric functions, we consider the possibility of inverses of these functions. We conclude the chapter with some applications.

7.1 Angles and Their Measurement

7.1a Definition of an Angle

In plane geometry, an angle is usually formed by the sides of a triangle. In trigonometry, we need to introduce a more general concept of an angle. Here, an angle is unlimited in magnitude, and it can be either positive or negative.

An **angle** is the geometric shape formed by two rays, or half-lines, with a common endpoint. For our purposes, we wish to define an angle as the result of a rotation of a ray about its endpoint. In Figure 7-1, the **initial side** is rotated about its endpoint O until it coincides with the **terminal side** to form the angle α. When the initial side coincides with the positive x-axis with its endpoint at the origin, the angle generated is said to be in **standard position**. In Figure 7-2(a), the x-axis, called the **initial side**, rotates in a counterclockwise direction until it coincides with the terminal side, forming the angle α. In this case we say that α is a **positive angle**. In Figure 7-2(b), the ray has been rotated in a clockwise direction to form the angle β. In this case, we say that β is a **negative angle**. If the terminal side coincides with a coordinate axis, the angle is called a **quadrantal angle**. In all other cases, the angle is said to lie in the same quadrant as its terminal side. Figure 7-3 displays an angle in each of the four quadrants. Note that the designation depends only upon the quadrant in which the terminal side lies and not upon the direction of rotation.

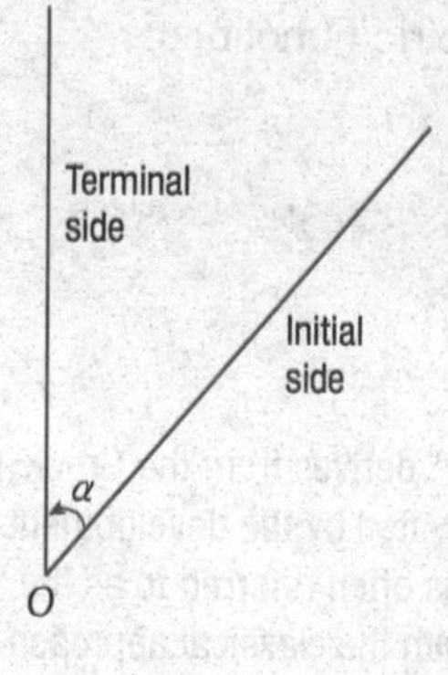

FIGURE 7-1
Forming an Angle

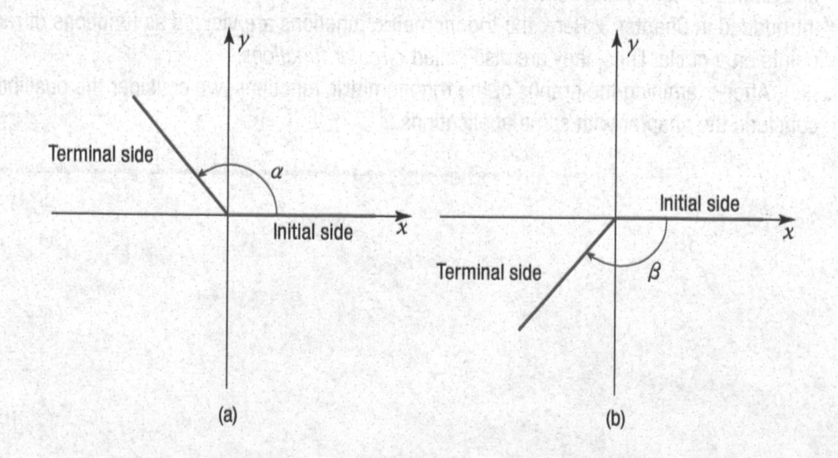

FIGURE 7-2
Positive and Negative Angles in Standard Position

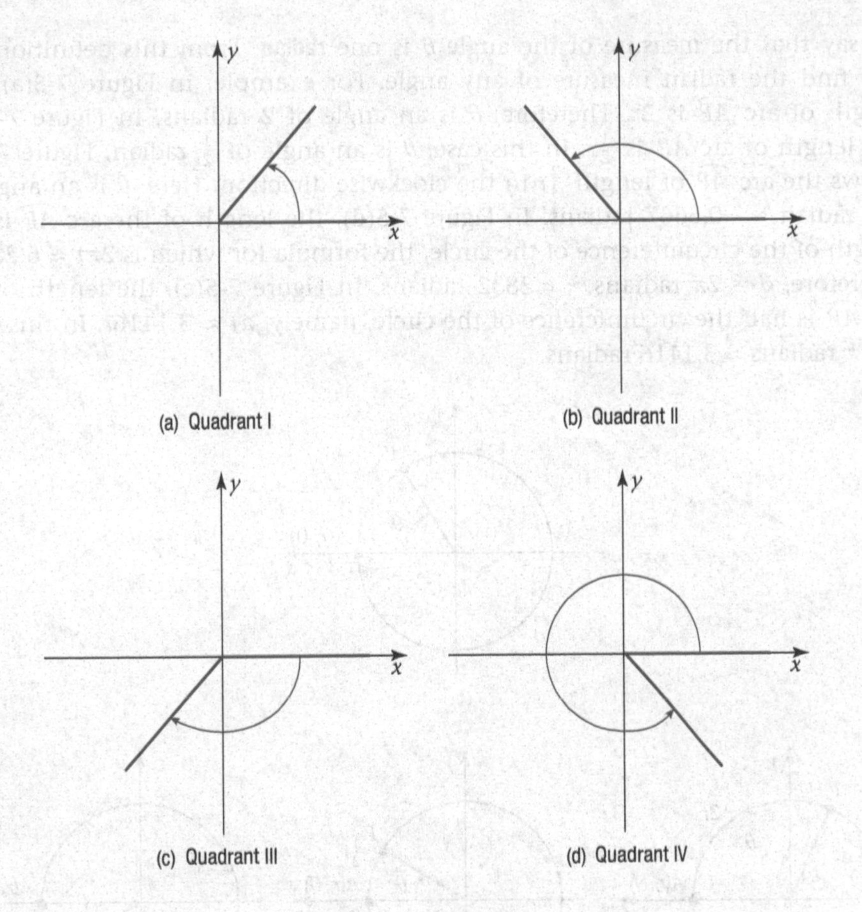

FIGURE 7-3
An Angle in Each Quadrant

7.1b *Angular Measurement: Degrees and Radians*

There are two commonly used units for measuring angles. An angle is said to have a measure of one **degree**, written 1°, if the angle is formed by rotating the initial side $\frac{1}{360}$ of a complete rotation in a counterclockwise direction. It follows that an angle obtained by a complete rotation of the initial side has a measure of 360°, and an angle obtained by one-fourth of a complete rotation has a measure of $\frac{1}{4}(360°) = 90°$. One degree is subdivided into 60 **minutes**, written 60′, and one minute is subdivided into 60 **seconds**, written 60″. For example, the notation 14°24′18″ is read *14 degrees, 24 minutes and 18 seconds*. This is equivalent to 14.405°, since

$$\begin{aligned} 14°24'18'' &= 14°24' + \left(\frac{18}{60}\right)' = 14°24' + 0.3' \\ &= 14°24.3' = 14° + \left(\frac{24.3}{60}\right)^{\circ} \\ &= 14° + 0.405° = 14.405° \end{aligned}$$

An angle in standard position greater than 0° and less than 90° lies in the first quadrant and is called an **acute angle**. (See Figure 7-3(a).) An angle greater than 90° and less than 180° lies in the second quadrant and is called an **obtuse angle**. (See Figure 7-3(b).) An angle measuring 90° is a quadrantal angle and is called a **right angle**. Angles measuring 0°, 180°, 270°, 360°, and −90° are also examples of quadrantal angles.

To define the second unit of angular measurement, consider a circle centered at the origin with radius r, as shown in Figure 7-4. Choose point P on the circle, counterclockwise from A, so that the arc $\widehat{AP}$ has length equal to r. Point P thus determines the angle θ. As the length or measure of the arc $\widehat{AP}$ is *1 radius*, we say that the measure of the angle θ is one **radian**. From this definition, we can find the radian measure of any angle. For example, in Figure 7-5(a), the length of arc $\widehat{AP}$ is $2r$. Therefore, θ is an angle of 2 radians. In Figure 7-5(b), the length of arc $\widehat{AP}$ is $\frac{1}{2}r$. In this case, θ is an angle of $\frac{1}{2}$ radian. Figure 7-5(c) shows the arc $\widehat{AP}$ of length $\frac{2}{3}r$ in the clockwise direction. Here, θ is an angle of $-\frac{2}{3}$ radian ≈ -0.6667 radians. In Figure 7-5(d), the length of the arc $\widehat{AP}$ is the length of the circumference of the circle, the formula for which is $2\pi r \approx 6.2832r$. Therefore, $\theta = 2\pi$ radians ≈ 6.2832 radians. In Figure 7-5(e), the length of the arc $\widehat{AP}$ is half the circumference of the circle, namely, $\pi r \approx 3.1416r$. In this case, $\theta = \pi$ radians ≈ 3.1416 radians.

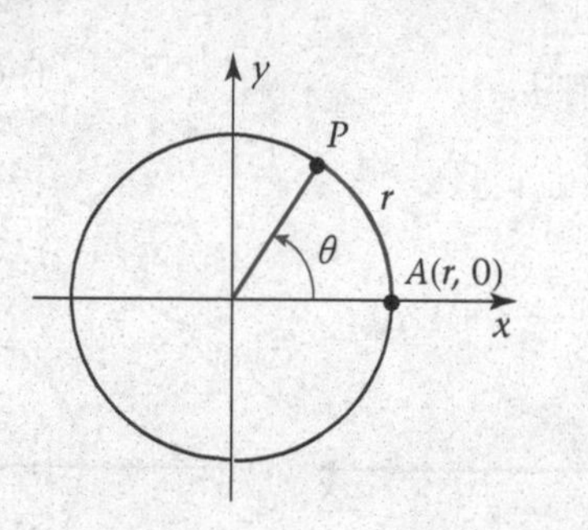

FIGURE 7-4
An Angle of 1 Radian

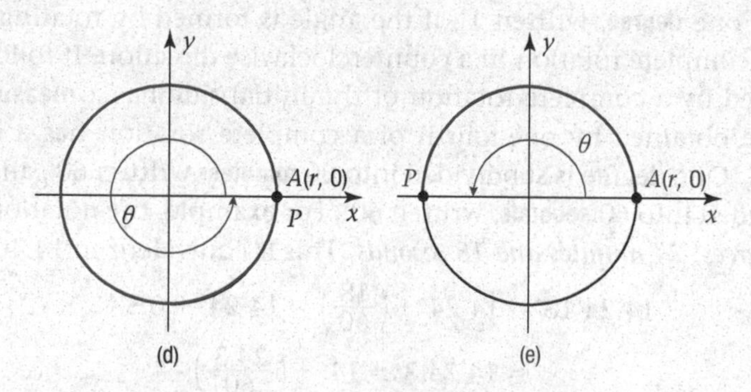

FIGURE 7-5
Measuring Angles in Radians

7.1c Angle Conversion

An angle of 360° traces a complete revolution in the counterclockwise direction. As we have just observed in the discussion concerning Figure 7-5(d), that same angle has a measure of 2π radians. Therefore, 2π radians = 360° or

$$\pi \text{ radians} = 180° \quad (1)$$

This relationship enables us to transform angular measure from radians to degrees and vice versa. One way to handle such conversions for any angle θ is by establishing a proportion.

$$\frac{\text{radian measure of angle } \theta}{\pi \text{ radians}} = \frac{\text{degree measure of angle } \theta}{180°}$$

Alternatively, we can solve Equation (1) to provide the conversion formulas

$$1 \text{ radian} = \left(\frac{180}{\pi}\right)^{\circ} \quad \text{and} \quad 1° = \frac{\pi}{180} \text{ radians}$$

Equivalently,

$$1 \text{ radian} \approx 57.29578° \quad \text{and} \quad 1° \approx 0.0174533 \text{ radians}$$

EXAMPLE 1 ***Angle Conversion***

Convert 150° to radian measure.

SOLUTION

With θ representing the radian measure of the angle, we establish the proportion

$$\frac{\theta}{\pi} = \frac{150°}{180°}$$

Solving, we have

$$\theta = \frac{150\pi}{180} = \frac{5\pi}{6}$$

Alternatively, since $1° = \frac{\pi}{180}$ radians, we have

$$150° = 150\left(\frac{\pi}{180}\right) = \frac{5\pi}{6} \text{ radians}$$

Thus, $\theta = \frac{5\pi}{6}$ radians. ■

Progress Check

Convert the following from degree to radian measure.

a. −210°　　b. 390°

Answers

a. $-\frac{7\pi}{6}$ radians　　b. $\frac{13\pi}{6}$ radians

EXAMPLE 2 ***Angle Conversion***

Convert $\frac{2\pi}{3}$ radians to degree measure.

SOLUTION

With θ denoting the degree measure of the angle, we establish the proportion

$$\frac{\frac{2\pi}{3}}{\pi} = \frac{\theta}{180^\circ}$$

Solving, we have

$$\theta = \frac{2}{3}(180^\circ) = 120^\circ$$

Alternatively, since 1 radian $= \frac{180}{\pi}$ degrees,we have

$$\frac{2\pi}{3} \text{ radians} = \frac{2\pi}{3}\left(\frac{180}{\pi}\right)^\circ = 120^\circ$$

Thus, $\theta = 120^\circ$. ■

Progress Check

Convert the following from radian measure to degrees.

a. $\frac{9\pi}{2}$ radians b. $-\frac{4\pi}{3}$ radians

Answers

a. 810° b. −240°

7.1d *"Special" Angles*

There are certain angles that we will use frequently in the examples and exercises throughout this chapter. It may prove helpful to verify the conversions shown in Table 7-1. Figure 7-6 displays some angles in standard position and shows both the radian and degree measure.

TABLE 7-1 Radians and Degrees

Radians	0	$\frac{\pi}{6}$	$\frac{\pi}{4}$	$\frac{\pi}{3}$	$\frac{\pi}{2}$	$\frac{2\pi}{3}$	$\frac{3\pi}{4}$	$\frac{5\pi}{6}$	π
Degrees	0°	30°	45°	60°	90°	120°	135°	150°	180°

Radians	$\frac{7\pi}{6}$	$\frac{5\pi}{4}$	$\frac{4\pi}{3}$	$\frac{3\pi}{2}$	$\frac{5\pi}{3}$	$\frac{7\pi}{4}$	$\frac{11\pi}{6}$	2π
Degrees	210°	225°	240°	270°	300°	315°	330°	360°

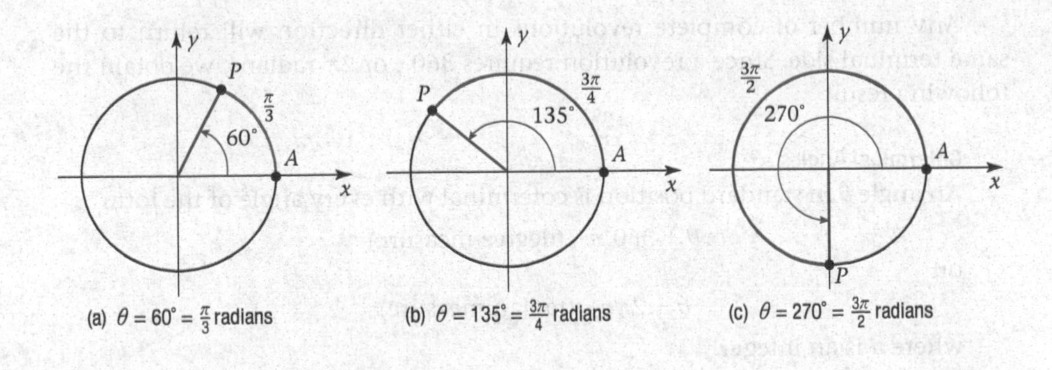

FIGURE 7-6
Degree and Radian Measure

EXAMPLE 3 *Finding Quadrants*

If the angle θ is in standard position, determine the quadrant in which the given angle lies.

a. $\theta = 200^\circ$ b. $\theta = \frac{7\pi}{4}$ radians

SOLUTION

Figure 7-7 shows the quadrantal angles in standard position.

a. Since $\theta = 200^\circ$ is between 180° and 270°, θ lies in quadrant III.

b. Since $\theta = \frac{7\pi}{4}$ radians is between $\frac{3\pi}{2}$ and 2π, θ lies in quadrant IV.

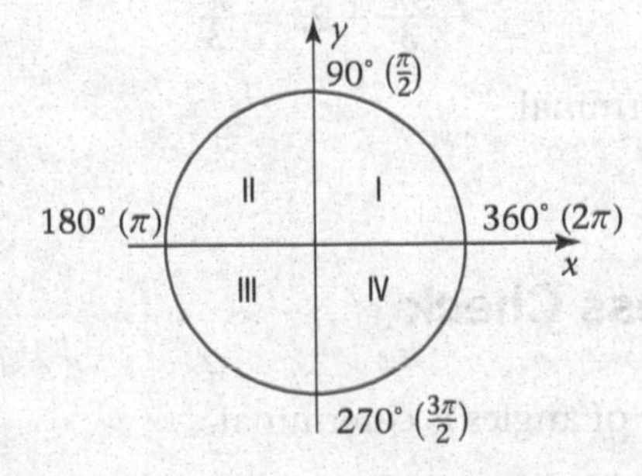

FIGURE 7-7
Quadrants and Angles

7.1e *The Reference Angle*

Since a complete revolution about a circle returns to the initial position, different angles in standard position may have the same terminal side. For example, angles of 30° and 390° in standard position have the same terminal side, as shown in Figure 7-8(a). Similarly, the angles of 45° and −315° shown in Figure 7-8(b) have the same terminal side. Such angles are said to be **coterminal**.

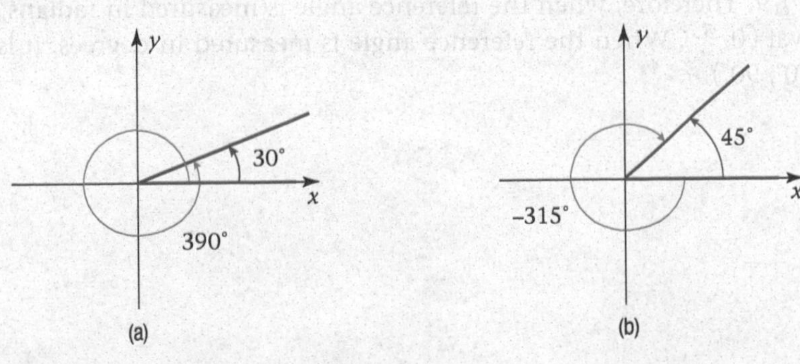

FIGURE 7-8
Coterminal Angles

Any number of complete revolutions in either direction will return to the same terminal side. Since a revolution requires 360°, or 2π radians, we obtain the following result

Coterminal Angles

An angle θ in standard position is coterminal with every angle of the form

$$\theta + 360^\circ n \quad \text{(degree measure)}$$

or

$$\theta + 2\pi n \quad \text{(radian measure)}$$

where n is an integer.

EXAMPLE 4 *Finding Coterminal Angles*

Find a first quadrant angle that is coterminal with an angle of

a. 410° b. $-\frac{5\pi}{3}$ radians

SOLUTION

a. Since 410° − 360° = 50°, it follows that an angle of 50° is coterminal with an angle of 410°.

b. Since

$$-\frac{5\pi}{3} + 2\pi = \frac{\pi}{3}$$

$\frac{\pi}{3}$ and $-\frac{5\pi}{3}$ are coterminal. ■

Progress Check

Show that each pair of angles is coterminal.

a. −265° and 95° b. $\frac{22\pi}{3}$ and $\frac{4\pi}{3}$ radians

For an angle in standard position that is not a quadrantal angle, it is convenient to define an acute angle that is called the **reference angle.**

The reference angle θ' associated with the angle θ is the acute angle formed by the terminal side of θ and the x-axis.

If θ lies in quadrant I and it is an acute angle, then $\theta' = \theta$. Other cases are illustrated in Figure 7-9. Therefore, when the reference angle is measured in radians, it is in the interval $(0, \frac{\pi}{2})$. When the reference angle is measured in degrees, it is in the interval (0°, 90°).

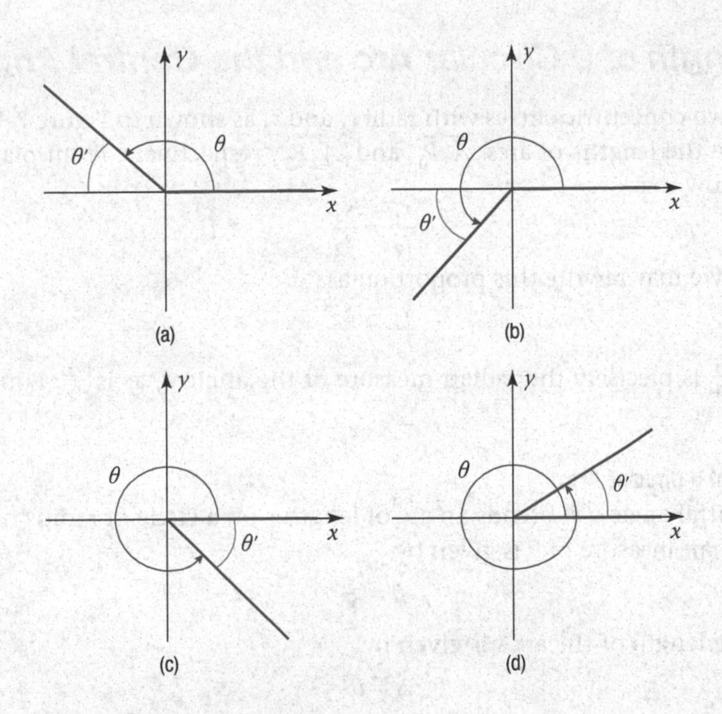

FIGURE 7-9
The Reference Angle

EXAMPLE 5 ***Reference Angle Computation***

Find the reference angle θ' if

a. $\theta = 240^\circ$ b. $\theta = \frac{5\pi}{3}$ radians

SOLUTION

a. Since $\theta = 240^\circ$ lies in the third quadrant, the reference angle is

$$\theta' = 240^\circ - 180^\circ = 60^\circ$$

b. Since $\theta = \frac{5\pi}{3}$ radians lies in the fourth quadrant, the reference angle is

$$\theta' = 2\pi - \frac{5\pi}{3} = \frac{\pi}{3}$$

■

Progress Check

Find the reference angle θ' if

a. $\theta = 160^\circ$ b. $\theta = \frac{4\pi}{3}$ radians

Answers

a. 20° b. $\frac{\pi}{3}$ radians

7.1f Length of a Circular Arc and the Central Angle

Consider two concentric circles with radii r_1 and r_2 as shown in Figure 7-10. We let s_1 and s_2 be the lengths of arcs $\widehat{A_1P_1}$ and $\widehat{A_2P_2}$, respectively. From plane geometry, we know that

$$\frac{r_1}{r_2} = \frac{s_1}{s_2}$$

Therefore, we may rewrite this proportion as

$$\frac{s_1}{r_1} = \frac{s_2}{r_2}$$

The ratio $\frac{s_1}{r_1}$ is precisely the radian measure of the angle θ, as is $\frac{s_2}{r_2}$. Equivalently, we have

Length of a Circular Arc

If a central angle θ subtends an arc of length s on a circle of radius r, then the radian measure of θ is given by

$$\theta = \frac{s}{r}$$

and the length of the arc s is given by

$$s = r\theta$$

Note that if the length of the arc s is equal to the length of one radius r, then the measure of θ is one radian.

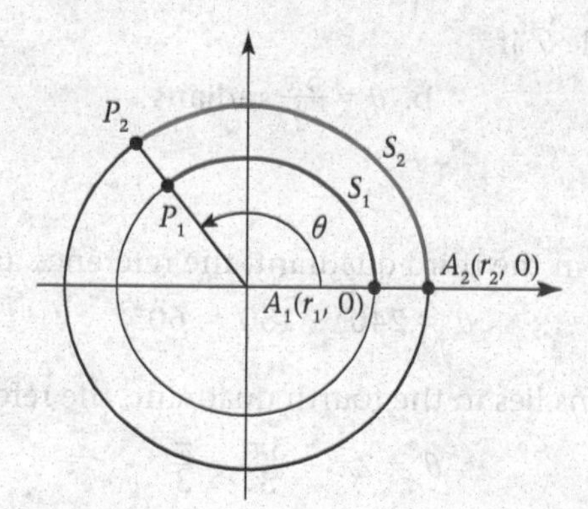

FIGURE 7-10
Radian Measure with Any Circle

EXAMPLE 6 *The Central Angle Formula*

A central angle θ subtends an arc of length 12 inches on a circle whose radius is 6 inches. Find the radian measure of the central angle.

SOLUTION

We have $s = 12$ and $r = 6$, so that $\theta = \frac{s}{r} = \frac{12}{6} = 2$ radians.

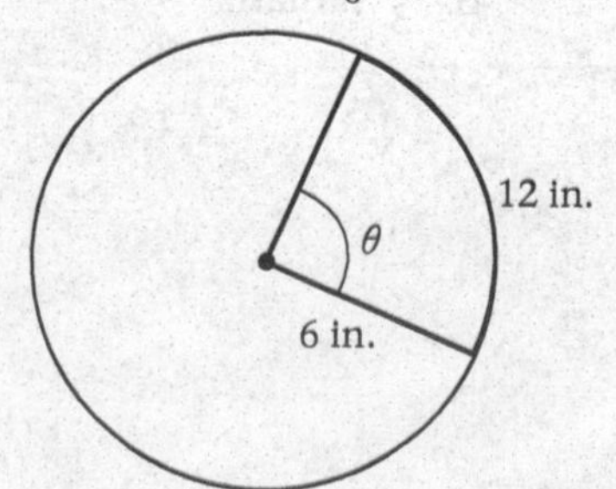

■

The formula

$$\theta = \frac{s}{r}$$

can only be applied if the angle θ is in radian measure.

EXAMPLE 7 ***The Central Angle Formula***

A designer has to place the word "ALMONDS" on a can using equally spaced letters, as shown below. For good visibility, the letters must cover a sector of the circle having a 90° central angle. If the base of the can is a circle of radius 2 inches, as shown, what is the maximum width of each letter, ignoring the spacing between letters?

(a)

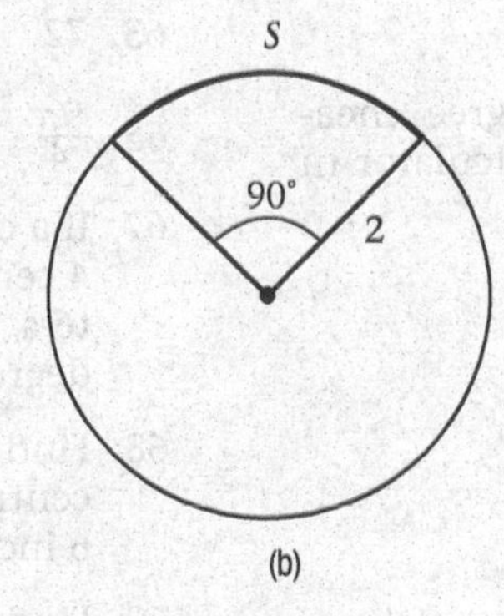

(b)

SOLUTION

Since $\theta = 90° = \frac{\pi}{2}$ radians, the arc has length

$$s = r\theta = 2\left(\frac{\pi}{2}\right) = \pi$$

Each of the seven letters can occupy $\frac{1}{7}$ of this arc, namely, $\frac{\pi}{7}$ inches. ■

Exercise Set 7.1

In Exercises 1–18, the angle θ is in standard position.

Determine the quadrant in which the angle lies.

1. $\theta = 313^\circ$
2. $\theta = 182^\circ$
3. $\theta = 14^\circ$
4. $\theta = 227^\circ$
5. $\theta = 141^\circ$
6. $\theta = -167^\circ$
7. $\theta = -345^\circ$
8. $\theta = 555^\circ$
9. $\theta = 618^\circ$
10. $\theta = -428^\circ$
11. $\theta = -195^\circ$
12. $\theta = 730^\circ$
13. $\theta = \frac{7\pi}{8}$
14. $\theta = -\frac{3\pi}{5}$
15. $\theta = -\frac{8\pi}{3}$
16. $\theta = \frac{3\pi}{8}$
17. $\theta = \frac{13\pi}{3}$
18. $\theta = \frac{9\pi}{5}$

In Exercises 19–34, convert from degree measure to radian measure. Do not use a calculator in Exercises 19–30.

19. 30°
20. 200°
21. -150°
22. -330°
23. 75°
24. 570°
25. -450°
26. -570°
27. 135°
28. 405°
29. 120°
30. 90°
31. 45.22°
32. 196.54°
33. $123^\circ 20'$
34. $87^\circ 12' 30''$

In Exercises 35–48, convert from radian measure to degree measure. Do not use a calculator in Exercises 35–46.

35. $\frac{\pi}{4}$
36. $\frac{\pi}{3}$
37. $\frac{3\pi}{2}$
38. $\frac{5\pi}{6}$
39. $-\frac{\pi}{2}$
40. $-\frac{7\pi}{12}$
41. $\frac{4\pi}{3}$
42. 3π
43. $\frac{5\pi}{2}$
44. -5π
45. $-\frac{5\pi}{3}$
46. $\frac{9\pi}{2}$
47. 1.72
48. 24.98

In Exercises 49–54, for each pair of angles, write T if they are coterminal and F if they are not coterminal.

49. $30^\circ, 390^\circ$
50. $50^\circ, -310^\circ$
51. $45^\circ, -45^\circ$
52. $120^\circ, \frac{14\pi}{3}$
53. $\frac{\pi}{2}, \frac{7\pi}{2}$
54. $-60^\circ, 760^\circ$

In Exercises 55–66, for each given angle, find the reference angle.

55. 130°
56. $\frac{5\pi}{6}$
57. -20°
58. 25°
59. -455°
60. 700°
61. $\frac{12\pi}{5}$
62. $\frac{5\pi}{4}$
63. 72°
64. $-\frac{2\pi}{3}$
65. $\frac{9\pi}{4}$
66. $\frac{5\pi}{3}$

67. If a central angle θ subtends an arc of length 4 centimeters on a circle of radius 7 centimeters, find the measure of θ in radians and in degrees (to two decimal places).

68. Find the length of an arc subtended by a central angle of $\frac{\pi}{5}$ radians on a circle of radius 6 inches.

69. Find the radius of a circle if a central angle of $\frac{2\pi}{3}$ radians subtends an arc of 4 meters.

70. In a circle of radius 150 centimeters, what is the length of an arc subtended by a central angle of 45°?

71. A subcompact car uses a tire whose radius is 13 inches. How far has the car moved when the tire completes one rotation? How many rotations are completed when the tire has traveled 1 mile?

72. A builder intends to place 7 equally spaced homes on a semicircular plot as shown in the accompanying figure. If the circle has a diameter of 400 feet, what is the distance between any two adjacent homes?

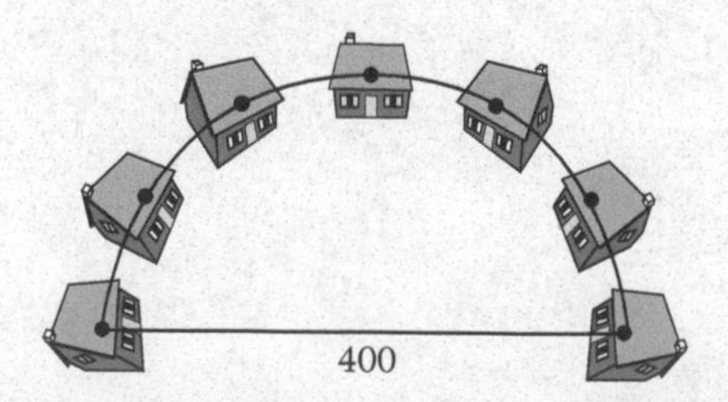

73. How many ribs are there in an umbrella if the length of each rib is 1.5 feet and the arc between two adjacent ribs measures $\frac{3\pi}{10}$ feet?

74. In a circle of radius r, a central angle of d degrees subtends an arc of length s. Establish a formula for s in terms of the variables r and d.

75. The minute hand on a clock is 3 inches long. Through what distance does the tip of the minute hand move between the hours of 4:10 p.m. and 4:22 p.m.?

76. A steam roller, whose wheel has a diameter of 2 feet, has just completed a pass over a driveway that is 30 feet in length. How many rotations of the roller were required?

77. The diameter of the earth is approximately 7900 miles. What is the difference in latitude between Lexington, Kentucky, and Jacksonville, Florida, if one city is 500 miles due north of the other? (*Hint:* Latitude is expressed in degrees.)

7.2 Right Triangle Trigonometry

Historically, trigonometry developed as a study of triangles. We begin by investigating various properties of right triangles. In Figure 7-11, we show right triangle ABC.

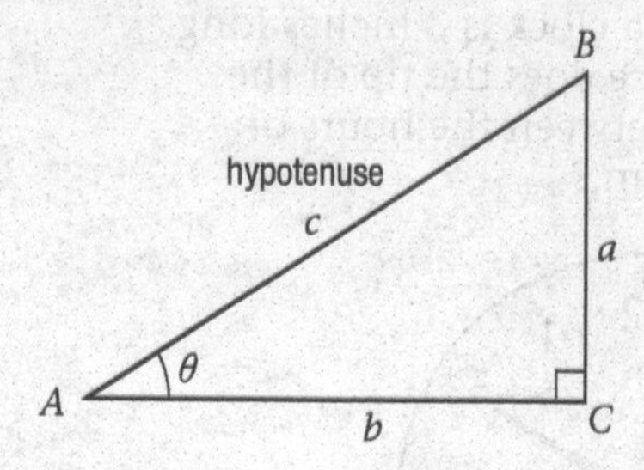

FIGURE 7-11
Right Triangle *ABC*

The letters A, B, and C play a dual role. They identify the vertices of this triangle, and they also identify the angles formed by the sides of the triangle at the respective vertices. Furthermore, in Figure 7-11, θ and A refer to the same angle.

From plane geometry, we know that the sum of the angles of a triangle equals 180°. Since angle C is a right angle (90°), we have that

$$\text{angle } A + \text{angle } B = 90^\circ$$

From this we conclude that both A and B are acute angles. The length of the side opposite the right angle C is called the **hypotenuse.** It will be called $\overline{AB}$, c, or *hypotenuse.* With our focus on angle θ, we identify the two legs or other two sides of triangle ABC. The length of the side opposite θ will be called $\overline{BC}$, a, or *opposite*. The length of the side adjacent to θ will be called $\overline{AC}$, b, or *adjacent*. The terms "opposite" and "adjacent" are names that may only be applied when our focus is directed toward a particular angle. In Figure 7-12(a), we focus on angle A and note the labels on the legs of triangle ABC as described above. In Figure 7-12(b), our focus is on angle B and the labels on the legs are reversed from those of Figure 7-12(a).

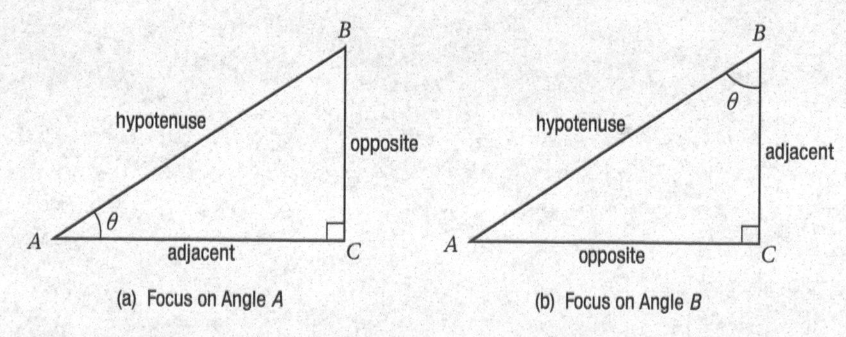

FIGURE 7-12
Right Triangle *ABC*

Now consider the two right triangles shown in Figure 7-13. If $\theta = \theta'$, then these two triangles must be similar. (Why?) The similarity of these triangles implies that their corresponding sides are proportional, that is,

$$\frac{a}{a'} = \frac{b}{b'} = \frac{c}{c'}$$

FIGURE 7-13
Similar Right Triangles ABC and $A'B'C'$

Alternatively, we may write

$$\frac{a}{b} = \frac{a'}{b'}, \quad \frac{a}{c} = \frac{a'}{c'}, \quad \frac{b}{c} = \frac{b'}{c'}$$

This indicates that the ratios of the corresponding sides are equal, hence, *independent* of the "size" of the right triangle. Because of this independence, consider all possible ratios of two sides of triangle ABC:

$$\frac{a}{c}, \quad \frac{b}{c}, \quad \frac{a}{b}, \quad \frac{b}{a}, \quad \frac{c}{b}, \quad \frac{c}{a}$$

These six ratios are used to define the six **trigonometric functions: sine, cosine, tangent, cotangent, secant,** and **cosecant.** We present the definitions of these functions in Table 7-2, which also contains triangle ABC.

7.2a Definition of sin θ, cos θ, tan θ, cot θ, sec θ, csc θ

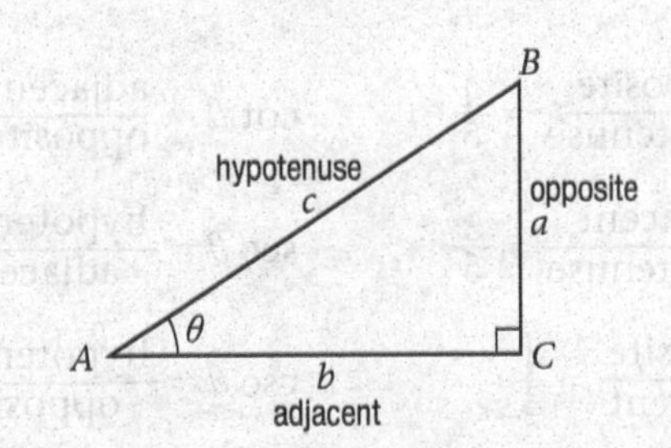

TABLE 7-2 Definition of Trigonometric Functions

Function Name	Notation and Definition
sine	$\sin\theta = \frac{\text{opposite}}{\text{hypotenuse}} = \frac{a}{c}$
cosine	$\cos\theta = \frac{\text{adjacent}}{\text{hypotenuse}} = \frac{b}{c}$
tangent	$\tan\theta = \frac{\text{opposite}}{\text{adjacent}} = \frac{a}{b}$
cotangent	$\cot\theta = \frac{\text{adjacent}}{\text{opposite}} = \frac{b}{a}$
secant	$\sec\theta = \frac{\text{hypotenuse}}{\text{adjacent}} = \frac{c}{b}$
cosecant	$\csc\theta = \frac{\text{hypotenuse}}{\text{opposite}} = \frac{c}{a}$

Before we consider some examples, we wish to recall the Pythagorean Theorem as applied to the right triangle in Table 7-2.

$$a^2 + b^2 = c^2$$

or equivalently,

$$(opposite)^2 + (adjacent)^2 = (hypotenuse)^2$$

Whereas this formula may be helpful in solving problems, it will also be useful in finding important relationships among the trigonometric functions.

EXAMPLE 1 ***Right Triangle Trigonometry***

Find the values of the trigonometric functions of the angle θ in Figure 7-14.

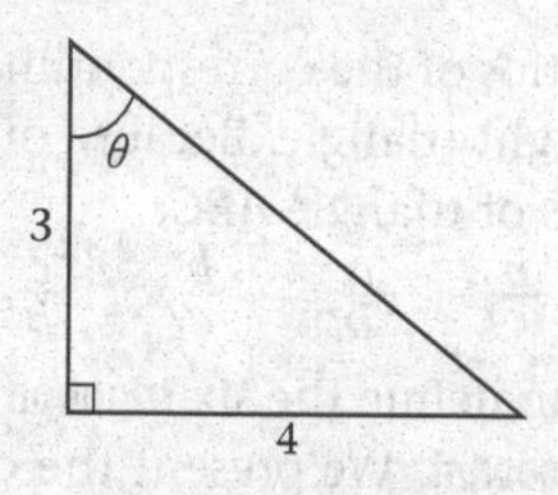

FIGURE 7-14
Diagram for Example 1

SOLUTION

Using the Pythagorean Theorem, the

$$\text{hypotenuse} = \sqrt{3^2+4^2} = \sqrt{25} = 5$$

Therefore,

$$\sin\theta = \frac{\text{opposite}}{\text{hypotenuse}} = \frac{4}{5} \qquad \cot\theta = \frac{\text{adjacent}}{\text{opposite}} = \frac{3}{4}$$

$$\cos\theta = \frac{\text{adjacent}}{\text{hypotenuse}} = \frac{3}{5} \qquad \sec\theta = \frac{\text{hypotenuse}}{\text{adjacent}} = \frac{5}{3}$$

$$\tan\theta = \frac{\text{opposite}}{\text{adjacent}} = \frac{4}{3} \qquad \csc\theta = \frac{\text{hypotenuse}}{\text{opposite}} = \frac{5}{4}$$

■

If we examine Table 7-2, we observe that the six trigonometric functions are related in a number of ways. For example,

$$\sin\theta = \frac{\text{opposite}}{\text{hypotenuse}} \quad \text{and} \quad \csc\theta = \frac{\text{hypotenuse}}{\text{opposite}} \tag{1}$$

$$\cos\theta = \frac{\text{adjacent}}{\text{hypotenuse}} \quad \text{and} \quad \sec\theta = \frac{\text{hypotenuse}}{\text{adjacent}} \tag{2}$$

$$\tan\theta = \frac{\text{opposite}}{\text{adjacent}} \quad \text{and} \quad \cot\theta = \frac{\text{adjacent}}{\text{opposite}} \tag{3}$$

Thus, we see the functions in (1) as reciprocals of one another. The functions in (2) are similarly reciprocals, as are the functions in (3).

Furthermore, since

$$\sin\theta = \frac{\text{opposite}}{\text{hypotenuse}} \quad \text{and} \quad \cos\theta = \frac{\text{adjacent}}{\text{hypotenuse}}$$

we have that

$$\frac{\sin\theta}{\cos\theta} = \frac{\frac{\text{opposite}}{\text{hypotenuse}}}{\frac{\text{adjacent}}{\text{hypotenuse}}} = \frac{\text{opposite}}{\text{adjacent}} = \tan\theta \tag{4}$$

and similarly,

$$\frac{\cos\theta}{\sin\theta} = \cot\theta \tag{5}$$

If an equation is true for *all* values in the domain of the variable, then the equation is called an **identity**. The functional relationships in (1), (2), (3), (4) and (5) are true for any θ, where $0^\circ < \theta < 90^\circ$. Therefore, we may refer to these equations as **trigonometric identities**. These identities are summarized in Table 7-3.

TABLE 7-3 Trigonometric Identities for $0^\circ < \theta < 90^\circ$

$\csc\theta = \dfrac{1}{\sin\theta}$	$\sin\theta = \dfrac{1}{\csc\theta}$
$\sec\theta = \dfrac{1}{\cos\theta}$	$\cos\theta = \dfrac{1}{\sec\theta}$
$\cot\theta = \dfrac{1}{\tan\theta}$	$\tan\theta = \dfrac{1}{\cot\theta}$
$\dfrac{\sin\theta}{\cos\theta} = \tan\theta$	$\dfrac{\cos\theta}{\sin\theta} = \cot\theta$

Verify the calculations in Example 1 using these identities.

EXAMPLE 2 *Right Triangle Trigonometry*

Find the values of the trigonometric functions of 45°.

SOLUTION

Since the sum of the two acute angles in a right triangle must be 90°, if one angle is 45°, then the other must be 45° as well. Two equal angles in a triangle imply that the triangle is isosceles. Since the size of the triangle does not matter when evaluating the six trigonometric functions, we choose the two equal legs of our isosceles right triangle to be of length 1, as shown in Figure 7-15. From the Pythagorean Theorem, we see that the

$$\text{hypotenuse} = \sqrt{1^2 + 1^2} = \sqrt{2}$$

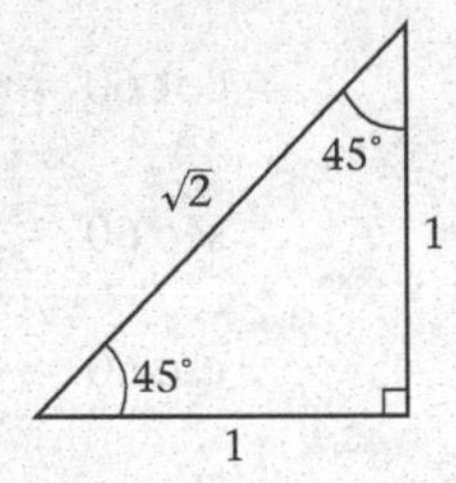

FIGURE 7-15
Diagram for Example 2

Therefore,

$$\sin 45^\circ = \frac{1}{\sqrt{2}} = \frac{\sqrt{2}}{2} \qquad \cot 45^\circ = 1$$

$$\cos 45^\circ = \frac{\sqrt{2}}{2} \qquad \sec 45^\circ = \sqrt{2}$$

$$\tan 45^\circ = 1 \qquad \csc 45^\circ = \sqrt{2}$$

■

EXAMPLE 3 ***Right Triangle Trigonometry***

Find the values of the trigonometric functions of 30° and 60°.

SOLUTION

Consider equilateral triangle ABD with each side of length 2, as shown below. Draw the angle bisector of angle B and let it intersect AD at point C.

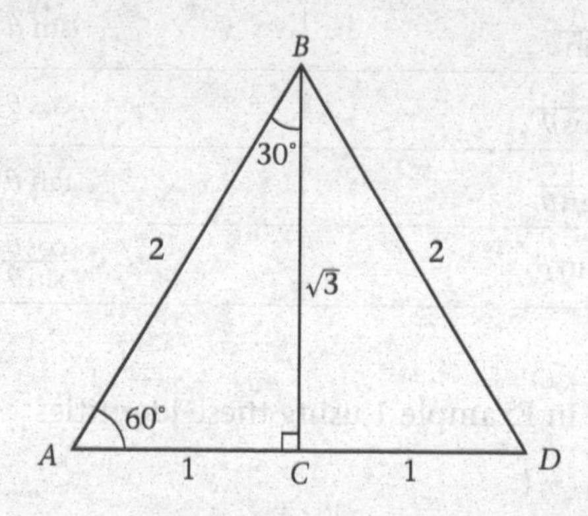

From plane geometry, we know that $\overline{AC} = \overline{CD} = 1$ and that BC is perpendicular to AD. Therefore, ABC is a right triangle, where angle A is 60° and angle ABC is 30°. From the Pythagorean Theorem, we calculate

$$\overline{BC} = \sqrt{2^2 - 1^2} = \sqrt{3}$$

Therefore,

$$\sin 30^\circ = \frac{1}{2} \qquad \cot 30^\circ = \sqrt{3}$$

$$\cos 30^\circ = \frac{\sqrt{3}}{2} \qquad \sec 30^\circ = \frac{2}{\sqrt{3}} = \frac{2\sqrt{3}}{3}$$

$$\tan 30^\circ = \frac{1}{\sqrt{3}} = \frac{\sqrt{3}}{3} \qquad \csc 30^\circ = 2$$

and

$$\sin 60^\circ = \frac{\sqrt{3}}{2} \qquad \cot 60^\circ = \frac{\sqrt{3}}{3}$$

$$\cos 60^\circ = \frac{1}{2} \qquad \sec 60^\circ = 2$$

$$\tan 60^\circ = \sqrt{3} \qquad \csc 60^\circ = \frac{2\sqrt{3}}{3}$$

■

Since we have already presented formulas for evaluating $\tan\theta$, $\cot\theta$, $\sec\theta$, and $\csc\theta$ in terms of $\sin\theta$ and $\cos\theta$, we summarize in Table 7-4 the values of $\sin\theta$ and $\cos\theta$ for the "special angles" of Example 2 and Example 3.

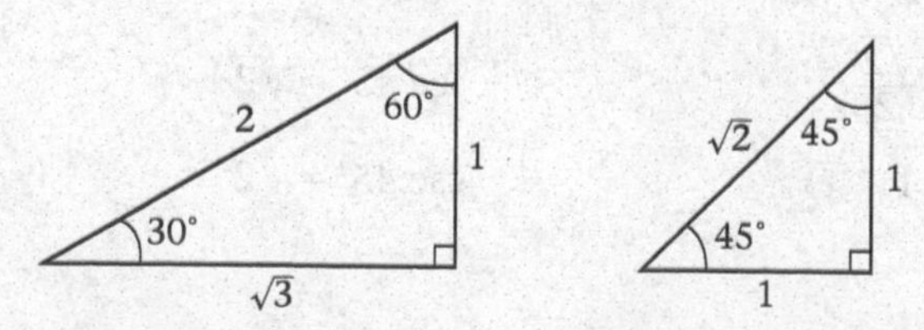

TABLE 7-4 sin θ and cos θ for 30°, 45°, and 60°

θ in Degrees	θ in Radians	sin θ	cos θ
30°	$\frac{\pi}{6}$	$\frac{1}{2}$	$\frac{\sqrt{3}}{2}$
45°	$\frac{\pi}{4}$	$\frac{\sqrt{2}}{2}$	$\frac{\sqrt{2}}{2}$
60°	$\frac{\pi}{3}$	$\frac{\sqrt{3}}{2}$	$\frac{1}{2}$

Progress Check

Verify that each of the following triangles is similar to one of the triangles found in Table 7-4. Show that the corresponding sides are proportional. Furthermore, verify that the entries for sin θ and cos θ for the following triangles are identical to those found in Table 7-4.

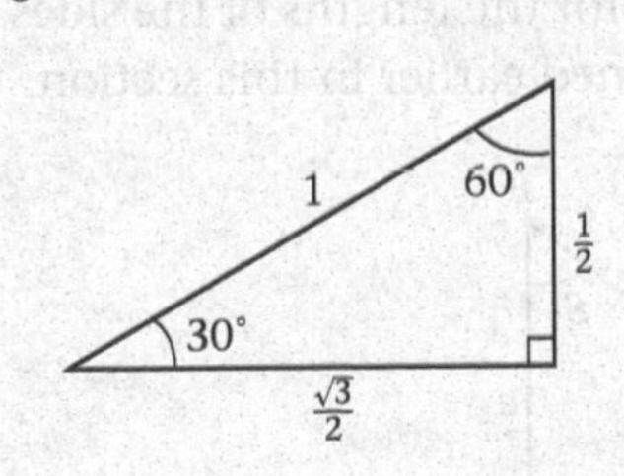

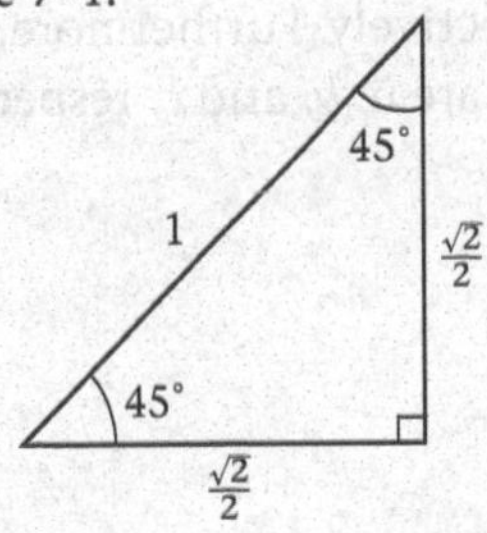

If we examine the values of the trigonometric functions in Example 3, we observe that the same six numbers appear in the answers for 30° and for 60°, although in a different order. We will now establish the relationships causing these repetitions by considering triangle ABC in Figure 7-16.

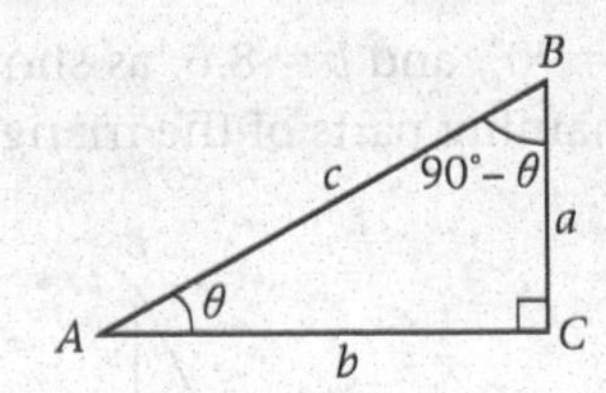

FIGURE 7-16
Right Triangle *ABC*

$$\sin\theta = \frac{a}{c} = \cos(90^\circ - \theta) \tag{6}$$

$$\cos\theta = \frac{b}{c} = \sin(90^\circ - \theta) \tag{7}$$

$$\tan\theta = \frac{a}{b} = \cot(90^\circ - \theta) \tag{8}$$

$$\cot\theta = \frac{b}{a} = \tan(90^\circ - \theta) \tag{9}$$

$$\sec\theta = \frac{c}{b} = \csc(90^\circ - \theta) \tag{10}$$

$$\csc\theta = \frac{c}{a} = \sec(90^\circ - \theta) \tag{11}$$

Angles θ and $90^\circ - \theta$ are called **complementary** since their sum is 90°. In relationship (6), we see that the cosine of the angle complementary to θ is the sine of θ. In (7), the sine of the angle complementary to θ is the cosine of θ. Therefore, we say that sine and cosine are **cofunctions**. The origin of the prefix "co" used in cosine and cofunction is from the first two letters of the word "complement." In a similar manner, tangent and cotangent are cofunctions from (8) and (9), as are secant and cosecant from (10) and (11).

Verify the calculations in Examples 2 and 3 using the cofunction relationships.

7.2b *Solving a Triangle*

The expression "to **solve a triangle**" is used to indicate that we seek all parts of the triangle, that is, the length of each side and the measure of each angle. For any right triangle, given any two sides, or given one side and an acute angle, it is always possible to solve the triangle. We will standardize our notation as shown in Figure 7-17, namely, we identify the measure of the angles at A, B, and C by α, β, and γ, respectively. Furthermore, the notation for the lengths of the sides opposite A, B, and C are a, b, and c, respectively, as defined earlier in this section.

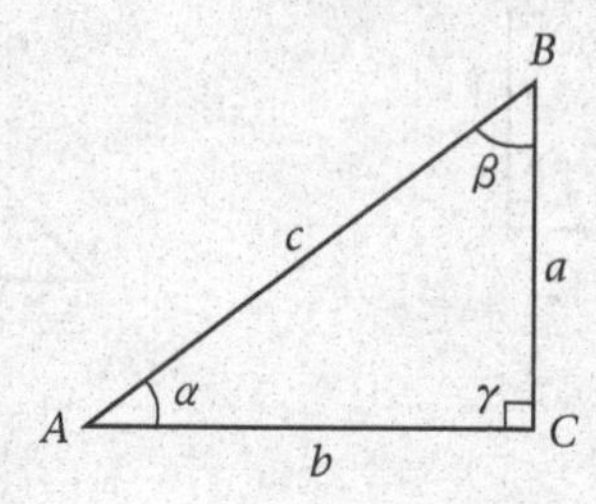

FIGURE 7-17
Notation for Parts of a Triangle

EXAMPLE 4 *Solving a Triangle*

In Triangle ABC, $\beta = 27^\circ$, $\gamma = 90^\circ$, and $b = 8.6$, as shown in Figure 7-18. Find approximate values for the remaining parts of the triangle using a calculator.

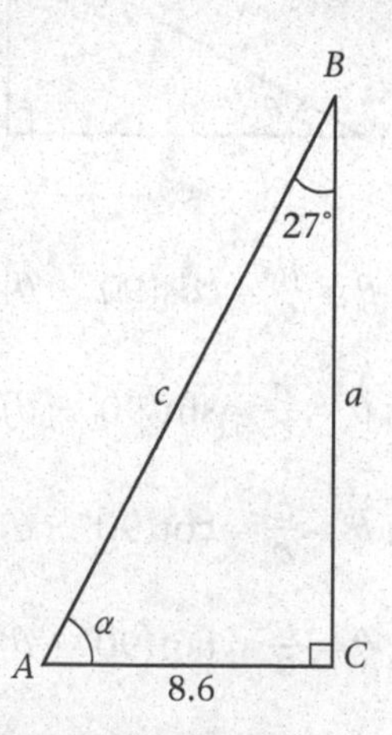

FIGURE 7-18
Triangle for Example 4

SOLUTION

Since $\alpha + 27^\circ = 90^\circ$, $\alpha = 63^\circ$. To find c, we may consider

$$\sin 27^\circ = \frac{8.6}{c}$$

in which case

$$c = \frac{8.6}{\sin 27^\circ} \approx \frac{8.6}{0.4540} \approx 18.94$$

To find a, consider

$$\tan 27^\circ = \frac{8.6}{a}$$

in which case

$$a = \frac{8.6}{\tan 27^\circ} \approx 16.88$$

(Verify the answers to a and c using trigonometric functions and 63°.) ■

Progress Check

In triangle ABC, $\alpha = 64^\circ$, $\gamma = 90^\circ$, and $b = 24.7$. Solve the triangle.

Answer

$\beta = 26^\circ$, $a = 50.64$, $c = 56.34$

Exercise Set 7.2

In Exercises 1–8, find the values of the trigonometric functions of the angle θ.

1.

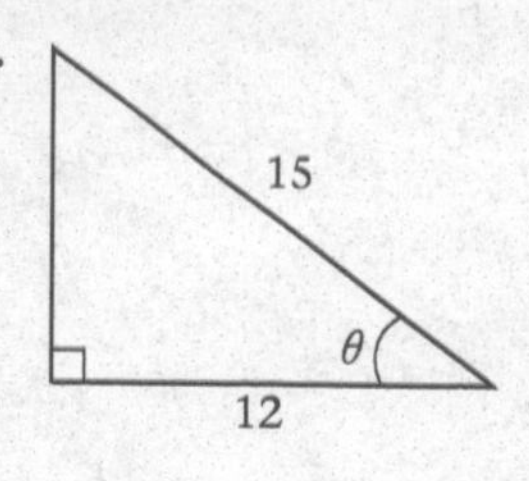

2.

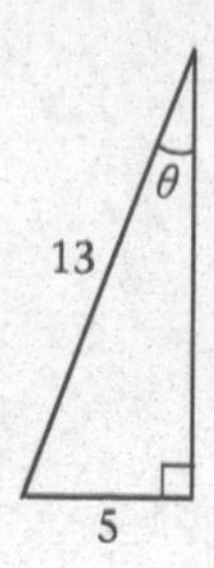

3.

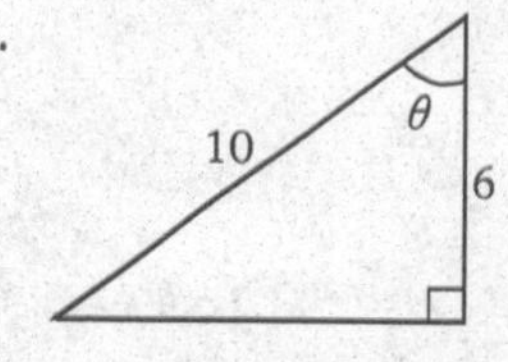

4.

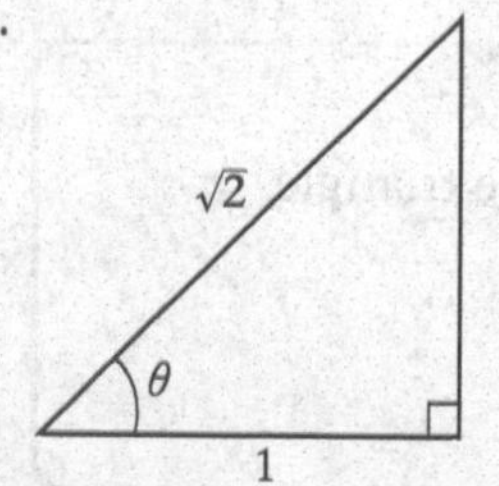

5.

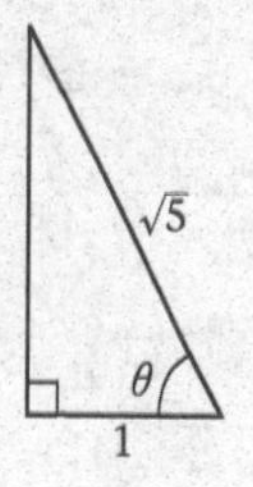

6.

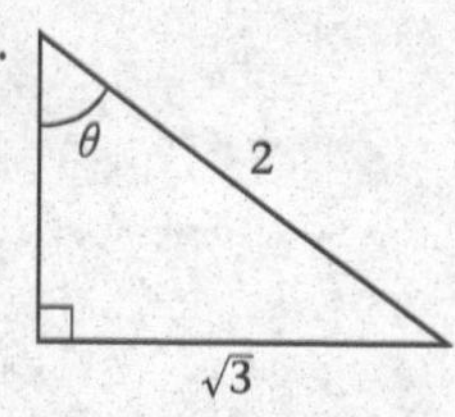

7.

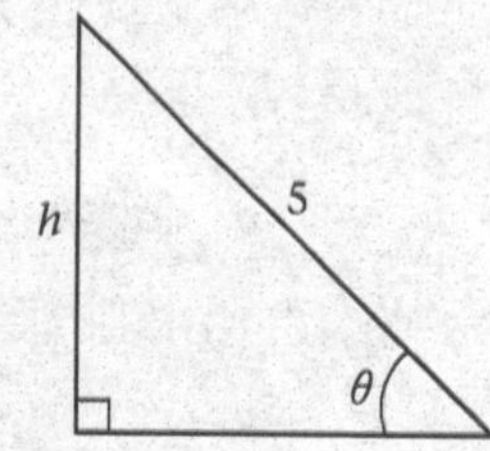

8.

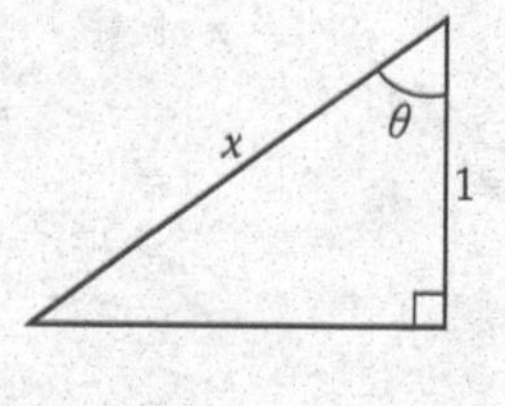

In Exercises 9–14, express the length h using a trigonometric function of the angle θ.

9.

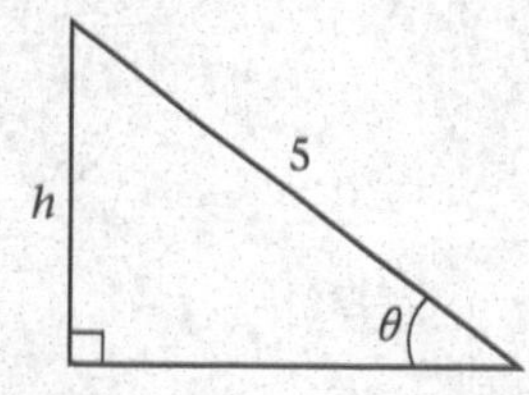

10.

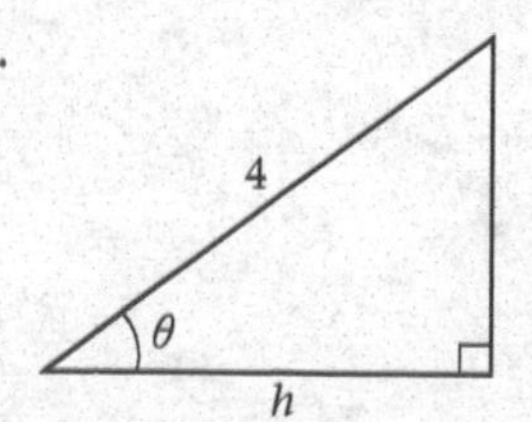

11.

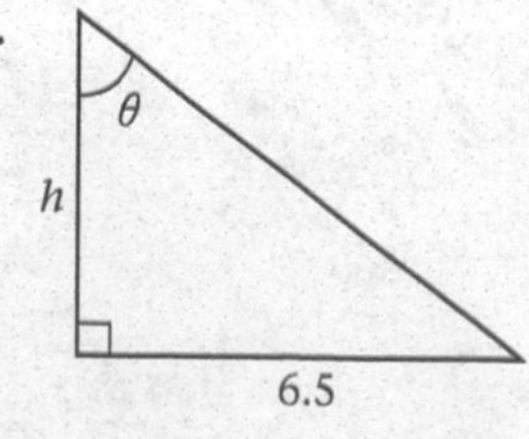

12.

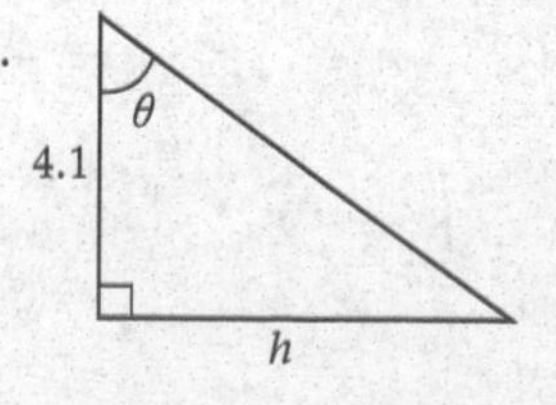

13.

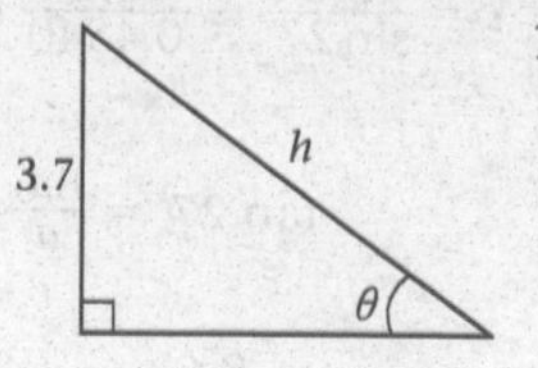

14.

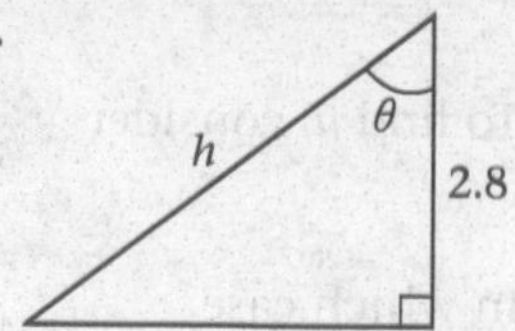

In Exercises 15–20, we are given triangle ABC with $\gamma = 90°$. Find approximate values for the remaining parts of the triangle.

15. $b = 40$, $\beta = 40°$; find c.
16. $a = 22$, $\alpha = 36°$; find b.
17. $a = 75$, $\beta = 22°$; find b.
18. $b = 60$, $\alpha = 53°$; find c.
19. $a = 25$, $\beta = 42°30'$; find c.
20. $b = 50$, $\alpha = 36°20'$; find a.

7.3 The Trigonometric Functions

In this section, we extend the definitions of the six trigonometric functions to include angles that may take on any value. However, it is important that these extensions retain their previous meaning for angles between 0° and 90°. Furthermore, since the trigonometric functions can all be written in terms of sine and cosine, we shall specifically focus on these two functions.

Let us begin by considering the coordinate axes with point P having coordinates (x, y) in the first quadrant, as shown in Figure 7-19. We denote the distance from the origin O to point P by r, and we form right triangle OPQ, where the coordinates of Q are $(x, 0)$. This enables us to calculate any trigonometric functions of θ. Specifically, observe that

$$\cos\theta = \frac{x}{r} \quad \text{and} \quad \sin\theta = \frac{y}{r}$$

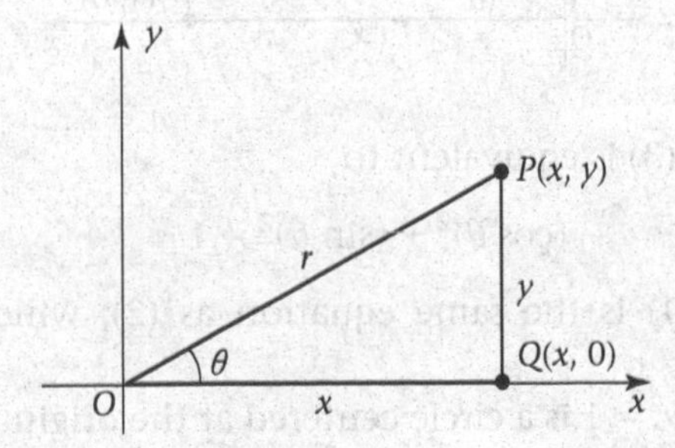

FIGURE 7-19
Right Triangle *OPQ*

Furthermore, the Pythagorean Theorem states that

$$x^2 + y^2 = r^2 \tag{1}$$

Dividing both sides by r^2, we obtain

$$\left(\frac{x}{r}\right)^2 + \left(\frac{y}{r}\right)^2 = 1$$

In this case, Equation (1) is equivalent to

$$(\cos\theta)^2 + (\sin\theta)^2 = 1 \tag{2}$$

Now, suppose we fix the value of θ and change the value of r to r', so that r' is the distance from O to $P'(x', y')$, as shown in Figure 7-20. If Q' has coordinates $(x', 0)$, then triangle $OP'Q'$ is similar to triangle OPQ. For triangle $OP'Q'$,

$$\cos\theta = \frac{x'}{r'} \quad \text{and} \quad \sin\theta = \frac{y'}{r'}$$

However, as we already discussed in Section 7.2,

$$\frac{x'}{r'} = \frac{x}{r} \quad \text{and} \quad \frac{y'}{r'} = \frac{y}{r}$$

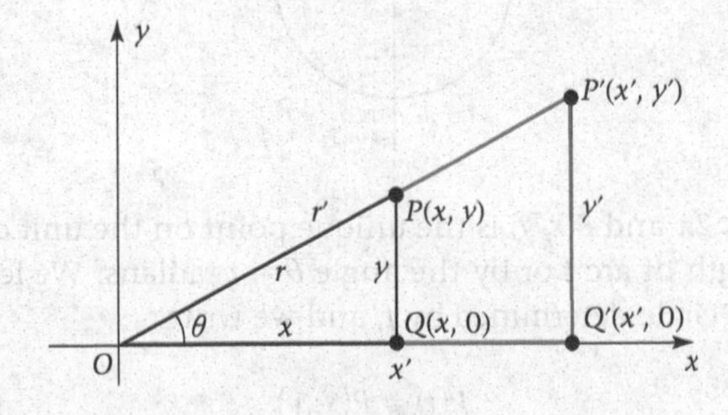

FIGURE 7-20
Right Triangles *OPQ* and *OP'Q'*

Therefore, since the "size of the triangle," or equivalently, the value for r, does not matter in determining $\sin\theta$ or $\cos\theta$, we choose $r = 1$, as shown in Figure 7-21. We observe that

$$\cos\theta = x \quad \text{and} \quad \sin\theta = y$$

Alternatively, we may say that the coordinates of point P are $(\cos\theta, \sin\theta)$. Furthermore, the Pythagorean Theorem states that

$$x^2 + y^2 = 1 \tag{3}$$

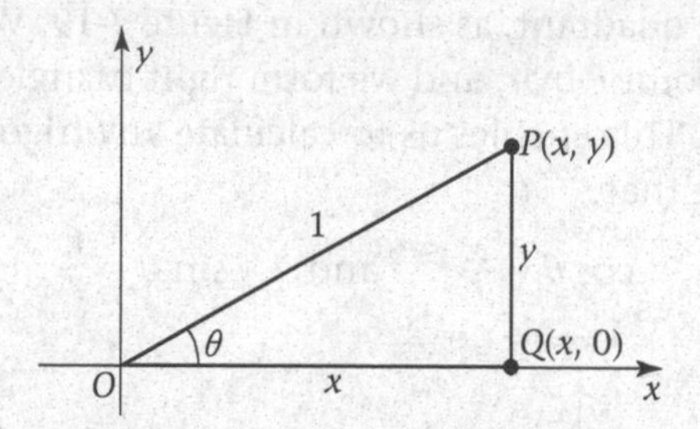

FIGURE 7-21
Right Triangle *OPQ*

In this case, Equation (3) is equivalent to

$$(\cos\theta)^2 + (\sin\theta)^2 = 1 \tag{4}$$

Note that Equation (4) is the same equation as (2), which we obtained from Figure 7-19 when $\overline{OP} = r$.

The graph of $x^2 + y^2 = 1$ is a circle centered at the origin with radius 1. We call this graph the **unit circle**. (Refer to Sections 3.1 and 5.2 for additional information concerning the equation of a circle and its symmetries.) Since the coordinates of the unit circle in the first quadrant correspond to $(\cos\theta, \sin\theta)$, it might seem reasonable to extend the definitions of $\cos\theta$ and $\sin\theta$ using $(\cos\theta, \sin\theta)$ for points on the unit circle in the remaining three quadrants as well. In fact, that is precisely what we will do. (Because the trigonometric functions are so closely related to the unit circle, they are sometimes referred to as **circular functions**.)

The circumference of the unit circle is $2\pi r = 2\pi(1) = 2\pi$. Choose a value t with $0 \le t < 2\pi$, and sweep out $\widehat{QP}$ of length t in the counterclockwise direction, as shown in Figure 7-22. If θ is measured in radians, then we may use the formula relating θ to the length of the arc s subtended by θ.

$$\theta = \frac{s}{r} = \frac{t}{1} = t$$

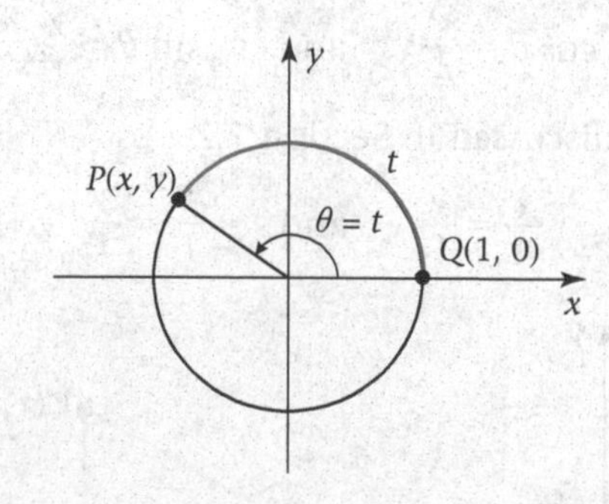

FIGURE 7-22
Choosing t on the Unit Circle, $0 \le t < 2\pi$

Therefore, $0 \le \theta < 2\pi$ and $P(x, y)$ is the unique point on the unit circle determined either by the length of arc t or by the angle $\theta = t$ radians. We let $P(t)$ denote the point on the unit circle determined by t, and we write

$$P(t) = P(x, y)$$

to indicate that the rectangular coordinates of the point on the unit circle are (x, y). We now define

$$\cos t = x \qquad \text{and} \qquad \sin t = y \qquad \text{for } 0 \le t < 2\pi$$

Equivalently,

$$P(t) = P(x, y) = P(\cos t, \sin t) \qquad \text{for } 0 \le t < 2\pi$$

EXAMPLE 1 *Finding Coordinates of Points on the Unit Circle*

Find the coordinates of the point $P(t)$ on the unit circle for each of the following.

a. $t = 0$ b. $t = \frac{\pi}{2}$ c. $t = \pi$ d. $t = \frac{3\pi}{2}$

SOLUTION

a. When $t = 0$, $P(0)$ and $Q(1, 0)$ represent the same point. Therefore, as we see in Figure 7-23(a),

$$P(0) = P(1, 0)$$

b. Since

$$\frac{\frac{\pi}{2}}{2\pi} = \frac{1}{4}$$

$\frac{\pi}{2}$ represents one-quarter of the circumference of the circle. Therefore, as we see in Figure 7-23(b),

$$P\left(\frac{\pi}{2}\right) = P(0, 1)$$

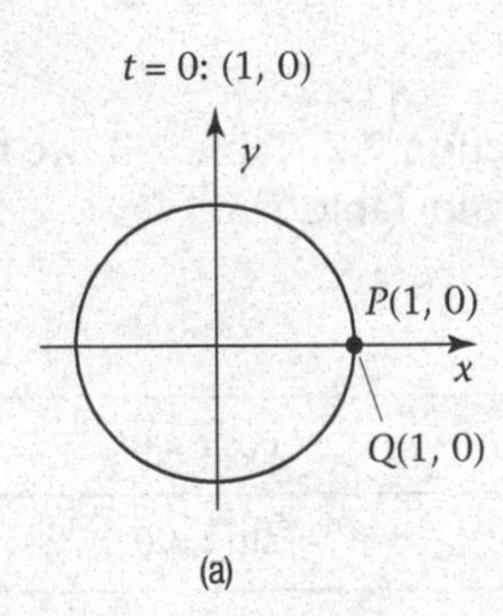

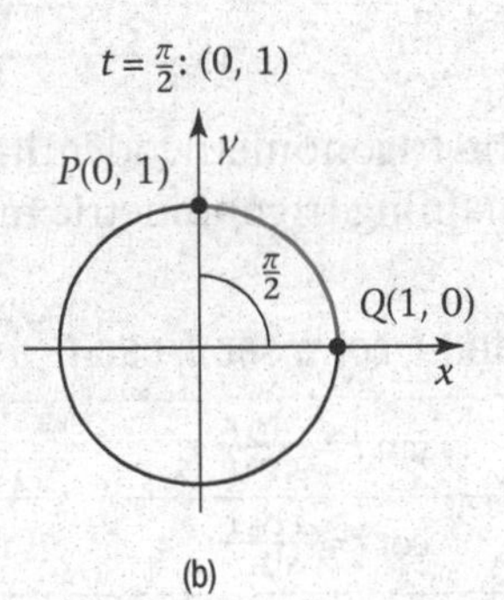

FIGURE 7-23
Coordinates of Points on the Unit Circle

c. Since

$$\frac{\pi}{2\pi} = \frac{1}{2}$$

π represents one-half of the circumference of the circle. As we see in Figure 7-24(a),

$$P(\pi) = P(-1, 0)$$

d. Since

$$\frac{\frac{3\pi}{2}}{2\pi} = \frac{3}{4}$$

$\frac{3\pi}{2}$ represents three-quarters of the circumference of the circle. As we see in Figure 7-24(b),

$$P\left(\frac{3\pi}{2}\right) = P(0, -1)$$

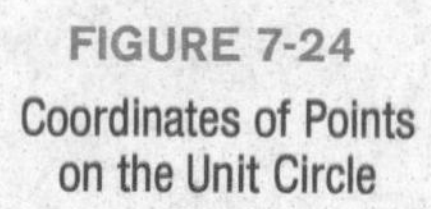

FIGURE 7-24
Coordinates of Points on the Unit Circle

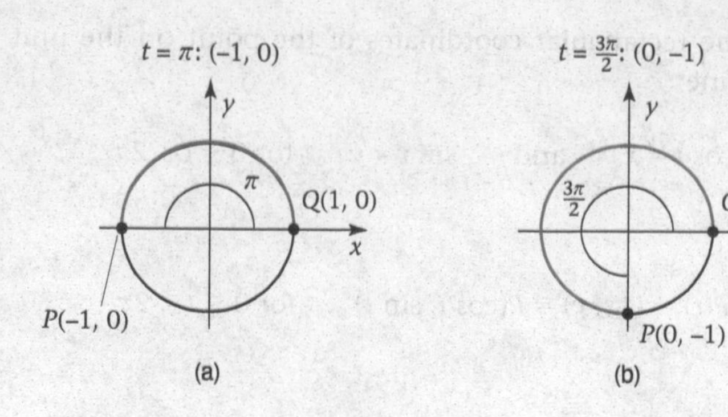

■

EXAMPLE 2 ***Finding cos t and sin t***

Find cos t and sin t for the following values of t.

a. $t = 0$ b. $t = \frac{\pi}{2}$ c. $t = \pi$ d. $t = \frac{3\pi}{2}$

SOLUTION

From Example 1, we know that $P(0) = (1, 0)$, $P(\frac{\pi}{2}) = (0, 1)$, $P(\pi) = (-1, 0)$, and $P(\frac{3\pi}{2}) = (0, -1)$. Since we have $x = \cos t$ and $y = \sin t$ on the unit circle, it follows that

t	cos t	sin t
0	1	0
$\frac{\pi}{2}$	0	1
π	−1	0
$\frac{3\pi}{2}$	0	−1

■

Using the trigonometric identities from Section 7.2, Table 7-3, we now define the four remaining trigonometric functions of t in Table 7-5.

TABLE 7-5 **tan t, cot t, sec t, csc t**

$\tan t = \frac{\sin t}{\cos t}$	$\cos t \neq 0$
$\cot t = \frac{\cos t}{\sin t}$	$\sin t \neq 0$
$\sec t = \frac{1}{\cos t}$	$\cos t \neq 0$
$\csc t = \frac{1}{\sin t}$	$\sin t \neq 0$

EXAMPLE 3 *Finding* tan *t*, cot *t*, sec *t*, *and* csc *t*

Find tan t, cot t, sec t, and csc t for the following values of t.

a. $t = 0$ b. $t = \frac{\pi}{2}$ c. $t = \pi$ d. $t = \frac{3\pi}{2}$

SOLUTION

Using the results of Example 2, we obtain

t	tan t	cot t	sec t	csc t
0	0	undefined	1	undefined
$\frac{\pi}{2}$	undefined	0	undefined	1
π	0	undefined	−1	undefined
$\frac{3\pi}{2}$	undefined	0	undefined	−1

(Verify that tan $\frac{\pi}{2}$ produces an error message on your calculator.) ■

Our goal is to define sin t and cos t for *all* values of t. To accomplish this, let us consider $t \geq 2\pi$. We sweep out the arc of length t in the counterclockwise direction, or equivalently, the angle $\theta = t$ radians, making as many revolutions as are needed. Since each revolution sweeps out an arc of length 2π and returns to the initial point $Q(1, 0)$, we can "ignore" all multiples of 2π. For example, if $t = \frac{5\pi}{2}$, we can write

$$t = \frac{5\pi}{2} = 2\pi + \frac{\pi}{2}$$

Sweeping out an arc of length 2π brings us back to $Q(1, 0)$. Then, we sweep out an arc of length $\frac{\pi}{2}$, arriving at the point $P(0, 1)$. To illustrate this, we use $\theta = \frac{5\pi}{2}$ in Figure 7-25. Therefore, $P(0, 1)$ corresponds to both $t = \frac{5\pi}{2}$ and $t = \frac{\pi}{2}$, or equivalently,

$$P\left(\frac{5\pi}{2}\right) = P\left(\frac{\pi}{2}\right) = P(0, 1)$$

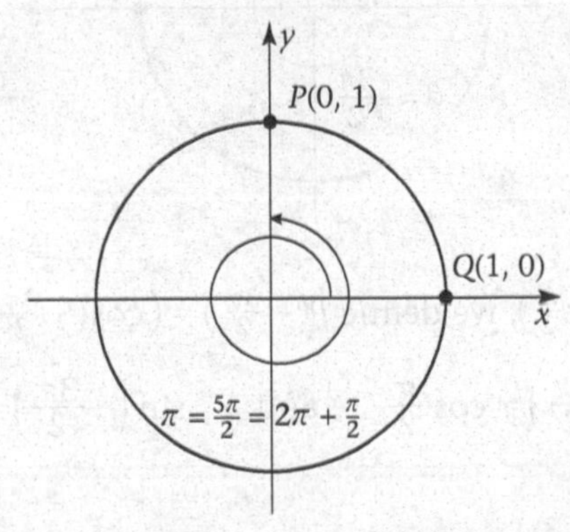

FIGURE 7-25
The Angle $\frac{5\pi}{2}$ on the Unit Circle

Since $P(\frac{\pi}{2}) = (\cos\frac{\pi}{2}, \sin\frac{\pi}{2})$, we define $P(\frac{5\pi}{2}) = (\cos\frac{5\pi}{2}, \sin\frac{5\pi}{2})$ so that

$$\cos\frac{5\pi}{2} = \cos\frac{\pi}{2} \quad \text{and} \quad \sin\frac{5\pi}{2} = \sin\frac{\pi}{2}$$

Similarly, if $t = \frac{19\pi}{4}$, we can write

$$\frac{19\pi}{4} = 2\pi + 2\pi + \frac{3\pi}{4}$$

Therefore, as shown in Figure 7-26,

$$P\left(\frac{19\pi}{4}\right) = P\left(\frac{3\pi}{4}\right)$$

and hence

$$\left(\cos\frac{19\pi}{4}, \sin\frac{19\pi}{4}\right) = \left(\cos\frac{3\pi}{4}, \sin\frac{3\pi}{4}\right)$$

or

$$\cos\frac{19\pi}{4} = \cos\frac{3\pi}{4} \quad \text{and} \quad \sin\frac{19\pi}{4} = \sin\frac{3\pi}{4}$$

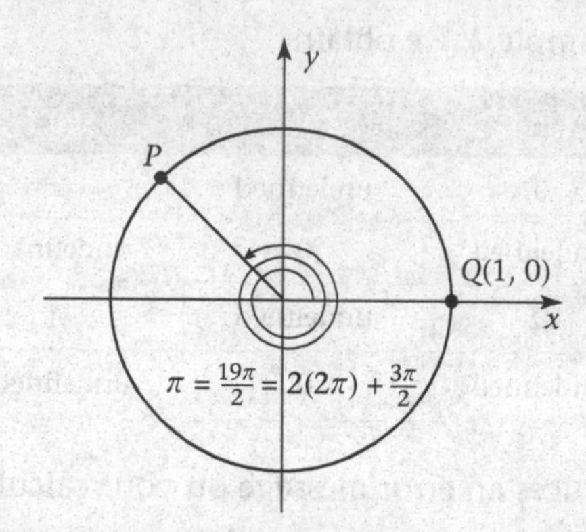

FIGURE 7-26
The Angle $\frac{19\pi}{2}$ on the Unit Circle

Finally, if we let $t < 0$, we sweep out the arc of length $|t|$ in the *clockwise* direction, or equivalently, the angle $\theta = t$. For example, if $t = -\frac{3\pi}{2}$, sweeping out an arc of length $\frac{3\pi}{2}$ in the clockwise direction takes us to $P(0, 1)$. We illustrate this in Figure 7-27 using $\theta = -\frac{3\pi}{2}$ and the arc of length $\frac{3\pi}{2}$. Therefore, $P(0, 1)$ corresponds to $t = -\frac{3\pi}{2}$ as well as to $t = \frac{\pi}{2}$, or equivalently,

$$P\left(-\frac{3\pi}{2}\right) = P\left(\frac{\pi}{2}\right) = P(0, 1)$$

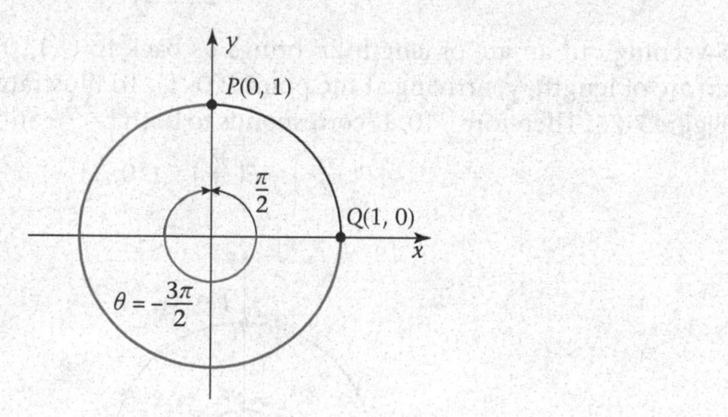

FIGURE 7-27
The Angles $\frac{\pi}{2}$ and $\frac{-3\pi}{2}$ on the Unit Circle

Since $P(\frac{\pi}{2}) = (\cos\frac{\pi}{2}, \sin\frac{\pi}{2})$, we define $P(-\frac{3\pi}{2}) = (\cos(-\frac{3\pi}{2}), \sin(-\frac{3\pi}{2}))$ so that

$$\cos\left(-\frac{3\pi}{2}\right) = \cos\frac{\pi}{2} \quad \text{and} \quad \sin\left(-\frac{3\pi}{2}\right) = \sin\frac{\pi}{2}$$

We have established the following:

For every real number t, there is a unique point $P(t)$ on the unit circle that can be found by sweeping out an arc of length $|t|$ in

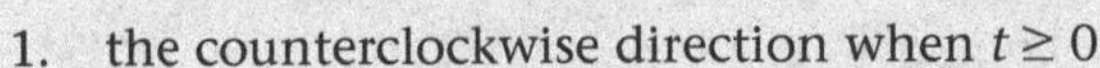

1. the counterclockwise direction when $t \geq 0$
2. the clockwise direction when $t < 0$

Equivalently, the point $P(t)$ is uniquely determined by the angle $\theta = t$ radians. Furthermore, if

$$P(t) = P(x, y)$$

then we define

$$\cos t = x \quad \text{and} \quad \sin t = y \quad \text{for any number } t$$

Equivalently,

$$P(t) = P(x, y) = P(\cos t, \sin t) \quad \text{for any number } t$$

Note that $P(t)$, $\cos t$ and $\sin t$ are not one-to-one functions since

$$P\left(\frac{\pi}{2}\right) = P\left(\frac{5\pi}{2}\right) = P\left(-\frac{3\pi}{2}\right)$$

$$\cos\frac{\pi}{2} = \cos\frac{5\pi}{2} = \cos\left(-\frac{3\pi}{2}\right)$$

$$\sin\frac{\pi}{2} = \sin\frac{5\pi}{2} = \sin\left(-\frac{3\pi}{2}\right)$$

Since the circumference of the unit circle is 2π, or equivalently, since there are 2π radians in a circle, every complete revolution *in either direction* returns to the point $Q(1, 0)$. Therefore,

For every integer n,

$$P(t) = P(t + 2\pi n)$$

$$\sin t = \sin(t + 2\pi n)$$

$$\cos t = \cos(t + 2\pi n)$$

EXAMPLE 4 *Finding Points on the Unit Circle*

Find the point on the unit circle corresponding to each of the following.

a. $t = 6\pi$ b. $t = 5\pi$ c. $t = -\frac{11\pi}{2}$

SOLUTION

a. Since $t = 6\pi = 3(2\pi)$, we must complete three revolutions, returning to the point on the unit circle $(1, 0)$. Thus, $P(6\pi) = P(1, 0)$.

b. If we write

$$t = 5\pi = 2(2\pi) + \pi$$

then we only need to sweep out an arc of length π to arrive at the point $(-1, 0)$. Thus, $P(5\pi) = P(-1, 0)$.

c. We write this in the form

$$t = -\frac{11\pi}{2} = -2(2\pi) - \frac{3\pi}{2}$$

Discarding the multiples of 2π, we sweep out an arc of length $\frac{3\pi}{2}$ in a clockwise direction. This takes us to the point $(0, 1)$. Thus, $P\left(-\frac{11\pi}{2}\right) = P(0, 1)$. ■

Progress Check

Find the point on the unit circle corresponding to

a. $t = -\frac{15\pi}{2}$ b. $t = -3\pi$ c. $t = -\frac{\pi}{2}$

Answers

a. $P(0, 1)$ b. $P(-1, 0)$ c. $P(0, -1)$

We now extend the definitions of tan t, cot t, sec t, and csc t to all values of t that are consistent with the previous definitions. In Table 7-6, we present these definitions, including the corresponding domains for each of the six trigonometric functions. Note that the domains corresponding to tan t, cot t, sec t, and csc t can be found as a consequence of the results of Example 3 and the fact that $\sin t = \sin(t + 2\pi n)$ and $\cos t = \cos(t + 2\pi n)$ for any integer n.

TABLE 7-6 Domains of the Trigonometric Functions

Function	Domain
$\sin t$	All values of t
$\cos t$	All values of t
$\tan t = \frac{\sin t}{\cos t}$	All values of $t \neq \frac{\pi}{2} + n\pi$ for any integer n
$\cot t = \frac{\cos t}{\sin t}$	All values of $t \neq n\pi$ for any integer n
$\sec t = \frac{1}{\cos t}$	All values of $t \neq \frac{\pi}{2} + n\pi$ for any integer n
$\csc t = \frac{1}{\sin t}$	All values of $t \neq n\pi$ for any integer n

Graphing Calculator Power User's CORNER

A Common Source of Confusion

When students first begin working with trigonometric functions, one common mistake is to assume that to find the value of the cosecant of an angle one should type 2nd sin. THIS IS NOT THE CASE. Typing 2nd sin finds the arcsine—the inverse of the sine function. Similarly, 2nd cos and 2nd tan find the arccosine and arctangent.

In order to find the value of the cosecant of an angle such as 2.09 radians, begin by considering the sine of that angle. Using the calculator, you would type sin 2.09 ENTER and get .868.

Bearing in mind that the cosecant is the reciprocal of the sine, you find the cosecant of the same angle by typing 1/ sin 2.09 ENTER to get the correct answer of 1.152.

When working with secant, type 1/ cos and then the angle measure.

When working with cotangent, type 1/ tan and then the angle measure.

Practice Problems

Use your calculator to find the following. (All angles are in radians unless otherwise specified.)

1. sin .785 csc .785
2. cos 2.749 sec 2.749
3. tan 5.760 cot 5.760

Exercise Set 7.3

In Exercises 1–8, determine the quadrant in which t lies.

1. $t = \frac{4\pi}{3}$
2. $t = -\frac{\pi}{6}$
3. $t = \frac{5\pi}{6}$
4. $t = \frac{5\pi}{4}$
5. $t = -\frac{19\pi}{18}$
6. $t = \frac{3\pi}{4}$
7. $t = -\frac{7\pi}{6}$
8. $t = \frac{10\pi}{6}$

In Exercises 9–20, for each given real number s, find a real number t in the interval $[0, 2\pi)$ so that the point on the unit circle $P(t) = P(s)$.

9. 4π
10. $\frac{13\pi}{2}$
11. $\frac{15\pi}{7}$
12. $-\frac{25\pi}{4}$
13. $-\frac{21\pi}{2}$
14. $-\frac{11\pi}{2}$
15. $\frac{41\pi}{6}$
16. $\frac{11\pi}{2}$
17. -9π
18. 7π
19. $\frac{27\pi}{5}$
20. $-\frac{22\pi}{3}$

In Exercises 21–24, plot the approximate positions of the points on the unit circle.

21. $P(7\pi)$, $P\left(\frac{4\pi}{3}\right)$, $P\left(\frac{5\pi}{2}\right)$, $P\left(-\frac{7\pi}{4}\right)$, $P\left(-\frac{13\pi}{6}\right)$
22. $P\left(\frac{11\pi}{2}\right)$, $P\left(-\frac{3\pi}{4}\right)$, $P\left(\frac{11\pi}{6}\right)$, $P\left(-\frac{10\pi}{3}\right)$, $P\left(\frac{33\pi}{4}\right)$
23. $P(-10)$, $P(8)$, $P(3.3)$, $P(-4)$, $P(1.7)$
24. $P(14)$, $P(-0.5)$, $P(-8)$, $P(6)$, $P(3)$

In Exercises 25–36, find the point on the unit circle corresponding to each of the following.

25. $t = 3\pi$
26. $t = 12\pi$
27. $t = \frac{7\pi}{2}$
28. $t = \frac{15\pi}{2}$
29. $t = \frac{23\pi}{2}$
30. $t = -3\pi$
31. $t = -18\pi$
32. $t = -\frac{7\pi}{2}$
33. $t = -\frac{15\pi}{2}$
34. $t = -\frac{23\pi}{2}$
35. $t = \frac{17\pi}{2}$
36. $t = -\frac{17\pi}{2}$

In Exercises 37–44, $P = P(x, y) = P(t)$ is a point on the unit circle. Find $\sin t$ and $\cos t$.

37. $P\left(-\frac{3}{5}, \frac{4}{5}\right)$
38. $P\left(-\frac{4}{5}, \frac{3}{5}\right)$
39. $P\left(\frac{\sqrt{3}}{2}, -\frac{1}{2}\right)$
40. $P\left(-\frac{1}{2}, -\frac{\sqrt{3}}{2}\right)$
41. $P\left(-\frac{\sqrt{2}}{2}, \frac{\sqrt{2}}{2}\right)$
42. $P\left(\frac{\sqrt{2}}{2}, -\frac{\sqrt{2}}{2}\right)$
43. $P\left(\frac{\sqrt{15}}{4}, -\frac{1}{4}\right)$
44. $P\left(\frac{1}{4}, \frac{\sqrt{15}}{4}\right)$

In Exercises 45–52, find $\tan t$, $\cot t$, $\sec t$, and $\csc t$ for the points on the unit circle given in Exercises 37–44, respectively.

7.4 Special Values and Properties of Trigonometric Functions

In the previous section, we showed that for every real number t, there is a unique point

$$P(t) = (\cos t, \sin t)$$

on the unit circle. We have found the point $P(t)$ for $t = 0, \frac{\pi}{2}, \pi,$ and $\frac{3\pi}{2}$. However, it is not easy to determine the point $P(t)$ for every given value of t. In this section, we shall find $P(t)$ by using right triangle trigonometry.

Since the unit circle is symmetric with respect to the x-axis, the y-axis, and the origin, these symmetries will be quite useful in evaluating the extensions of the trigonometric functions discussed previously. For example, if $\theta = \frac{\pi}{3}$ radians, the coordinates on the unit circle are $\left(\frac{1}{2}, \frac{\sqrt{3}}{2}\right)$. (See Table 7-4.) Using the symmetries of the circle, we see that $\left(-\frac{1}{2}, \frac{\sqrt{3}}{2}\right)$, $\left(-\frac{1}{2}, -\frac{\sqrt{3}}{2}\right)$, and $\left(\frac{1}{2}, -\frac{\sqrt{3}}{2}\right)$ are also points on the unit circle, as shown in Figure 7-28. However, what values of θ correspond to these points? Consider the unit circle in Figure 7-29. Since the angle swept out by the arc counterclockwise from Q to $\left(\frac{1}{2}, \frac{\sqrt{3}}{2}\right)$ is $\frac{\pi}{3}$, using plane geometry, we calculate that the angle swept out by the arc counterclockwise from Q to $\left(-\frac{1}{2}, \frac{\sqrt{3}}{2}\right)$ is $\frac{2\pi}{3}$. Then $\cos\frac{2\pi}{3} = -\frac{1}{2}$ and $\sin\frac{2\pi}{3} = \frac{\sqrt{3}}{2}$. In a similar fashion, we calculate $\cos\frac{4\pi}{3} = -\frac{1}{2}$, $\sin\frac{4\pi}{3} = -\frac{\sqrt{3}}{2}$, $\cos\frac{5\pi}{3} = \frac{1}{2}$, and $\sin\frac{5\pi}{3} = -\frac{\sqrt{3}}{2}$. (Verify this.)

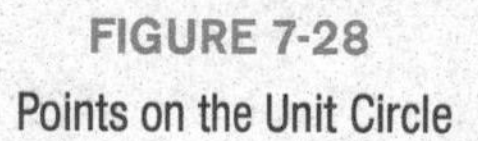
FIGURE 7-28
Points on the Unit Circle

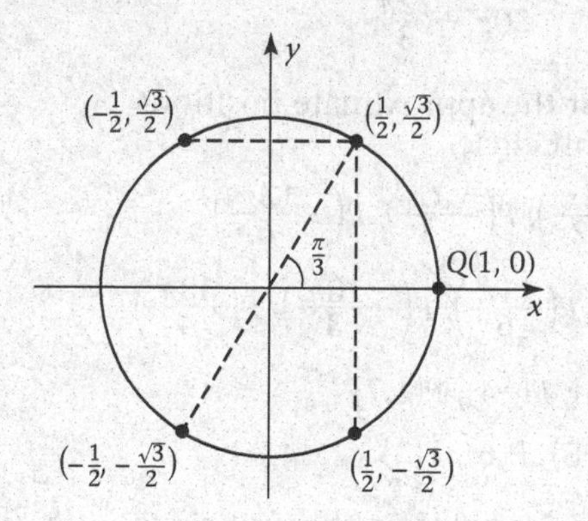

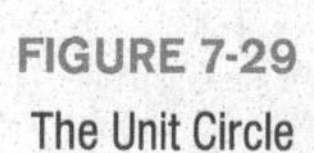
FIGURE 7-29
The Unit Circle

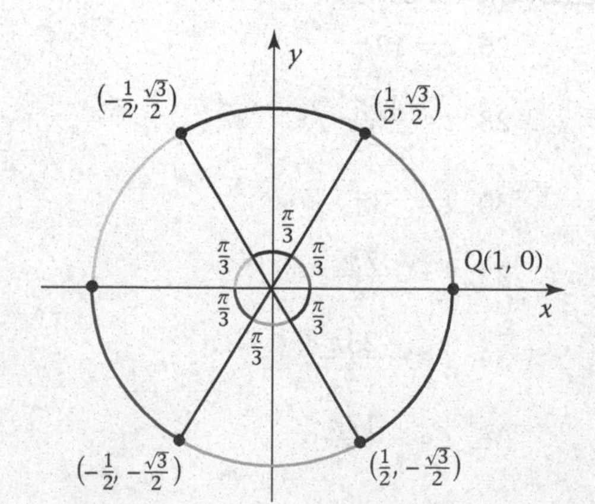

Recall the definition of a reference angle from Section 7.1 with associated Figure 7-9, which we present here again as Figure 7-30. The reference angle θ' is acute.

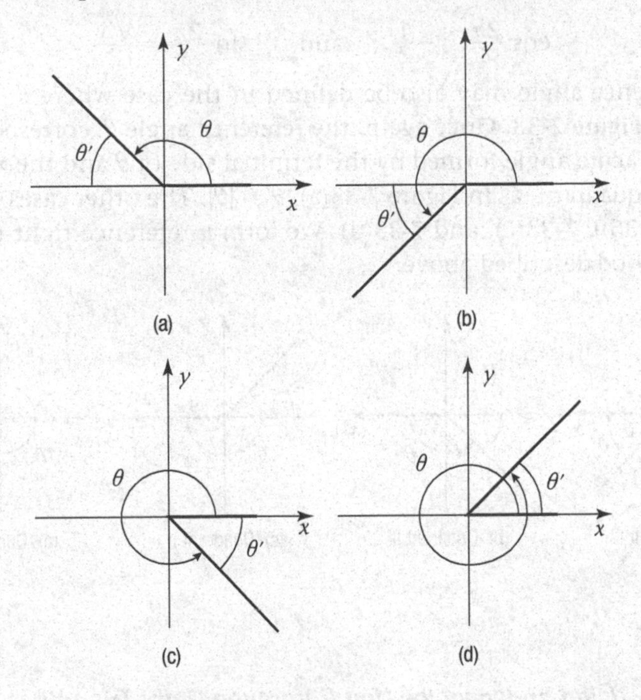

FIGURE 7-30
Reference Angles

Therefore, we form a **reference right triangle**, having one leg on the x-axis, hypotenuse of length r, and an acute angle θ', as shown in Figure 7-31. The reference right triangle is called a **unit reference right triangle** if $r = 1$.

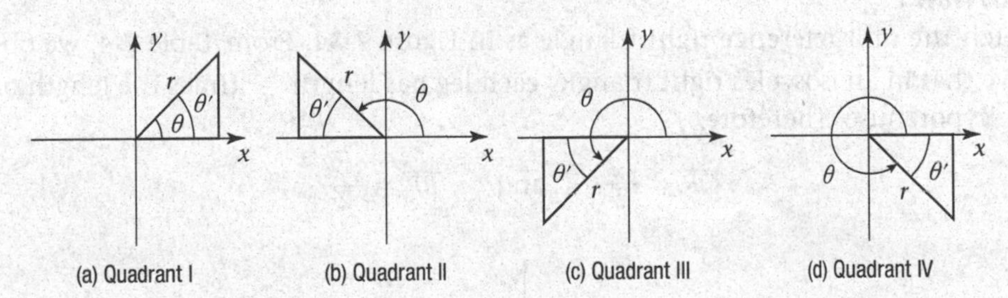

FIGURE 7-31
Reference Right Triangles

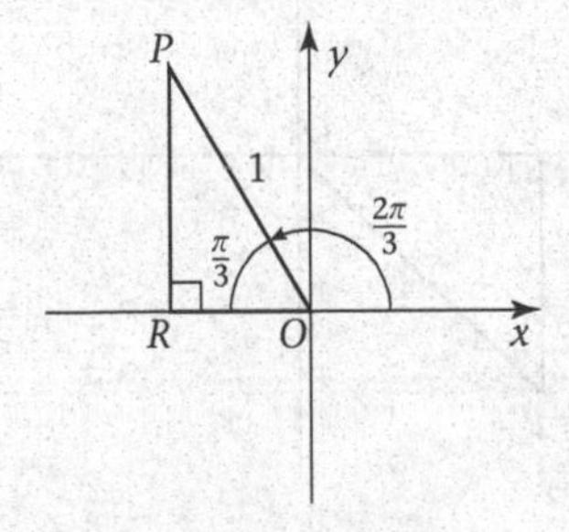

FIGURE 7-32
The Unit Reference Right Triangle for $\theta = \frac{2\pi}{3}$

Using the unit reference right triangle, we shall calculate $\cos\frac{2\pi}{3}$ and $\sin\frac{2\pi}{3}$. In Figure 7-32, we have the corresponding right triangle. From Table 7-4 in Section 7.2, we observe that in a 30°, 60°, 90° right triangle, or equivalently, a $\frac{\pi}{6}$, $\frac{\pi}{3}$, $\frac{\pi}{2}$ radian right triangle, the side opposite the 30°, or $\frac{\pi}{6}$ radian, angle has length $\frac{1}{2}$ the length of the hypotenuse. Furthermore, the side opposite the 60°, or $\frac{\pi}{3}$ radian, angle has length $\frac{\sqrt{3}}{2}$ times the length of the hypotenuse. From this we determine that

$$\overline{OR} = \frac{1}{2} \quad \text{and} \quad \overline{RP} = \frac{\sqrt{3}}{2}$$

Now we are able to find the coordinates of P, namely, $\left(-\frac{1}{2}, \frac{\sqrt{3}}{2}\right)$ from which we see that

$$\cos \frac{2\pi}{3} = -\frac{1}{2} \quad \text{and} \quad \sin \frac{2\pi}{3} = \frac{\sqrt{3}}{2}$$

The reference angle may also be defined in the case where $\theta < 0$. Consider the angles in Figure 7-33. Once again, the reference angle θ' corresponding to the angle θ is the acute angle formed by the terminal side of θ and the x-axis. If θ lies in the fourth quadrant, as in Figure 7-33(a), $\theta' = |\theta|$. The other cases are illustrated in Figures 7-33(b), 7-33(c), and 7-33(d). We form a reference right triangle using the same method described above.

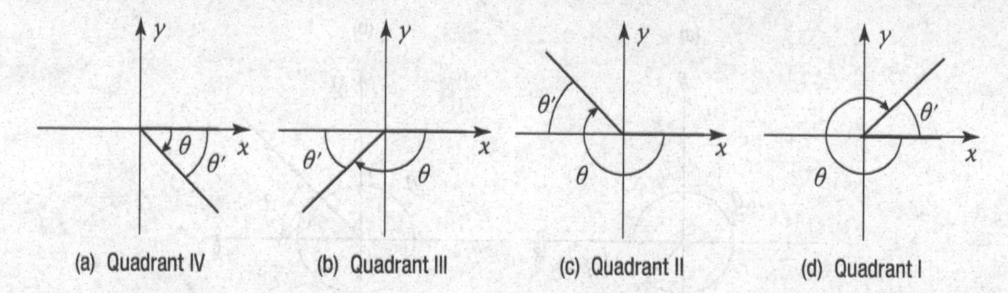

FIGURE 7-33
Reference Angles

EXAMPLE 1 *Using an Isosceles Unit Reference Right Triangle*

Find $\cos\left(-\frac{3\pi}{4}\right)$ and $\sin\left(-\frac{3\pi}{4}\right)$.

SOLUTION

Sketch the unit reference right triangle as in Figure 7-34. From Table 7-4, we observe that in an isosceles right triangle, each leg has length $\frac{\sqrt{2}}{2}$ times the length of the hypotenuse. Therefore,

$$\overline{OR} = \frac{\sqrt{2}}{2} \quad \text{and} \quad \overline{RP} = \frac{\sqrt{2}}{2}$$

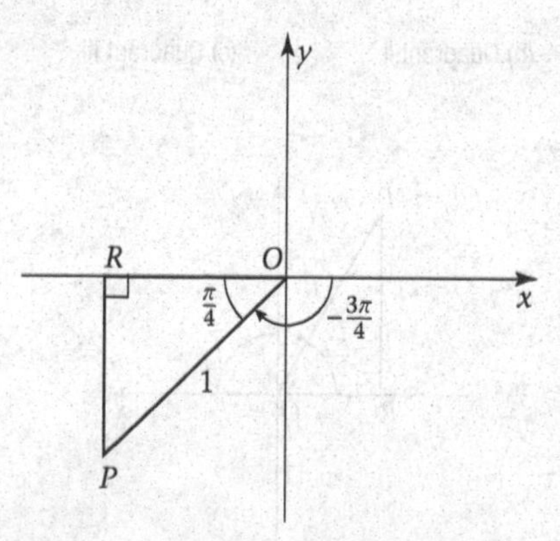

FIGURE 7-34
The Unit Reference Right Triangle for $\theta = -\frac{3\pi}{4}$

We are now able to find the coordinates of P, namely, $\left(-\frac{\sqrt{2}}{2}, -\frac{\sqrt{2}}{2}\right)$ from which we obtain

$$\cos\left(-\frac{3\pi}{4}\right) = -\frac{\sqrt{2}}{2} \quad \text{and} \quad \sin\left(-\frac{3\pi}{4}\right) = -\frac{\sqrt{2}}{2}$$ ■

EXAMPLE 2 *Using a Unit Reference Right Triangle*

Find $\tan \frac{5\pi}{6}$.

SOLUTION

Using the properties of a 30°, 60°, 90° right triangle, sketch the triangle as shown in Figure 7-35. We find that $\overline{OR} = \frac{\sqrt{3}}{2}$ and $\overline{RP} = \frac{1}{2}$. Therefore, the coordinates of P are $\left(-\frac{\sqrt{3}}{2}, \frac{1}{2}\right)$, hence, $\cos \frac{5\pi}{6} = -\frac{\sqrt{3}}{2}$ and $\sin \frac{5\pi}{6} = \frac{1}{2}$. Since $\tan t = \frac{\sin t}{\cos t}$, we have

$$\tan \frac{5\pi}{6} = \frac{\sin \frac{5\pi}{6}}{\cos \frac{5\pi}{6}} = \frac{\frac{1}{2}}{-\frac{\sqrt{3}}{2}} = -\frac{1}{\sqrt{3}} = -\frac{\sqrt{3}}{3}$$

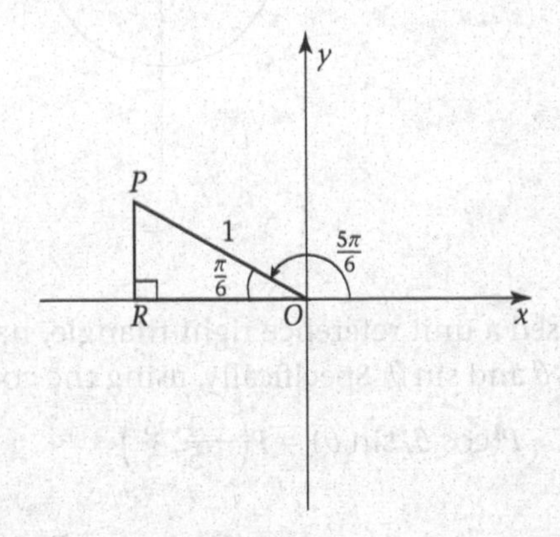

FIGURE 7-35
The Unit Reference Right Triangle for $\theta = \frac{5\pi}{6}$

Progress Check

Find $\csc\left(-\frac{4\pi}{3}\right)$.

Answer

$\frac{2\sqrt{3}}{3}$

EXAMPLE 3 *Using a Unit Reference Right Triangle*

Find $\cos \theta$ and $\sin \theta$ if the point $(-4, 3)$ lies on the terminal side of θ.

SOLUTION

The reference angle of θ is θ'. Using the Pythagorean Theorem, we find that the length of the hypotenuse is

$$\sqrt{(-4)^2 + (3)^2} = \sqrt{16+9} = \sqrt{25} = 5$$

As shown in the sketch below, the line from the origin to $P'(-4, 3)$ intersects the unit circle at $P\left(-\frac{4}{5}, \frac{3}{5}\right)$ since

$$\overline{OP} = \frac{1}{5}\overline{OP'} = \frac{1}{5}(5) = 1$$

Therefore,

$$\cos \theta = -\frac{4}{5} \quad \text{and} \quad \sin \theta = \frac{3}{5}$$

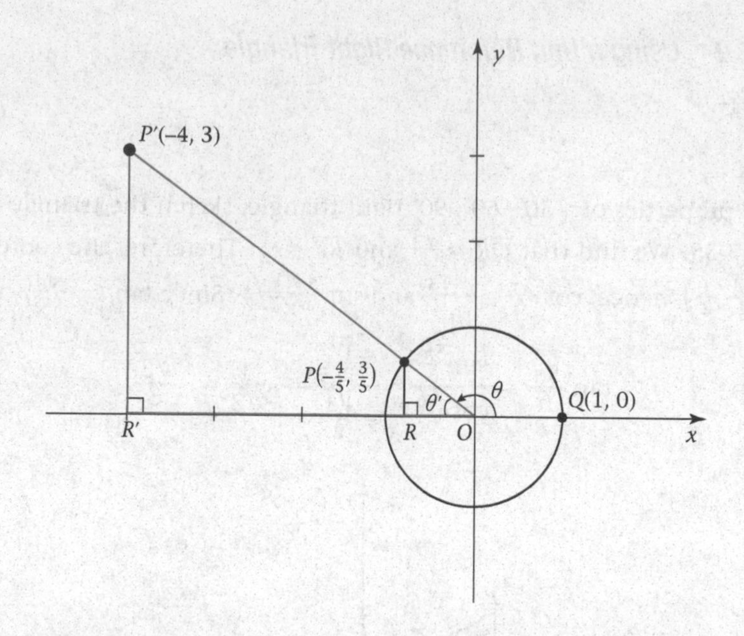

In Example 3, we used a unit reference right triangle, namely, triangle *OPR* in the drawing, to find $\cos\theta$ and $\sin\theta$. Specifically, using the coordinates of *P*, we had

$$P(\cos\theta, \sin\theta) = P\left(-\frac{4}{5}, \frac{3}{5}\right) \tag{1}$$

Hence

$$\cos\theta = -\frac{4}{5} \qquad \text{and} \qquad \sin\theta = \frac{3}{5}$$

Is it possible to find $\cos\theta$ and $\sin\theta$ from the coordinates of P'? Since right triangle *OPR* and right triangle $OP'R'$ both have reference angle θ', they are similar. Thus, their corresponding sides are proportional. Since $\overline{OP} = 1$ and $\overline{OP'} = 5$,

$$\overline{OP'} = 5\overline{OP}$$

in which case

$$\overline{OR'} = 5\overline{OR} \qquad \text{and} \qquad \overline{P'R'} = 5\overline{PR}$$

Thus, the coordinates of P' are 5 times the coordinates of *P*, respectively. From Equation (1),

$$P'(5\cos\theta, 5\sin\theta) = P'\left(5\left(-\frac{4}{5}\right), 5\left(\frac{3}{5}\right)\right)$$

$$P'(5\cos\theta, 5\sin\theta) = P'(-4, 3) \tag{2}$$

Hence

$$5\cos\theta = -4 \qquad \text{and} \qquad 5\sin\theta = 3$$

Therefore, using the coordinates of P' from Equation (2), we are able to obtain

$$\cos\theta = -\frac{4}{5} \qquad \text{and} \qquad \sin\theta = \frac{3}{5}$$

In general, one may use a reference right triangle to find $\cos\theta$ and $\sin\theta$, not only a unit reference right triangle. Thus, as shown in Figure 7-36, if *P* has coordinates (x, y) and $r = \sqrt{x^2 + y^2} \neq 0$, then

$$P(r\cos\theta, r\sin\theta) = P(x, y) \tag{3}$$

Hence

$$r\cos\theta = x \qquad \text{and} \qquad r\sin\theta = y$$

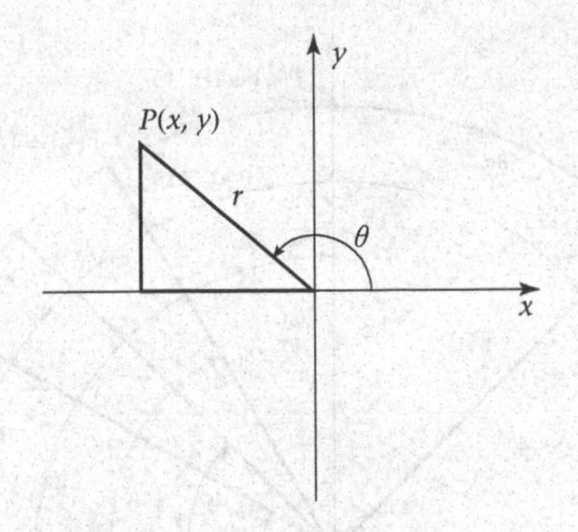

FIGURE 7-36
The Reference Right Triangle for θ

Therefore, using the coordinates of P from Equation (3), we are able to obtain

$$\cos\theta = \frac{x}{r} \quad \text{and} \quad \sin\theta = \frac{y}{r}$$

To summarize:

> If P has coordinates (x, y) with $r = \sqrt{x^2+y^2} \neq 0$, then
> $$P(x, y) = P(r\cos\theta, r\sin\theta)$$
> and
> $$\cos\theta = \frac{x}{r}, \qquad \sin\theta = \frac{y}{r}$$

Observe that these are the same equations for $\cos\theta$ and $\sin\theta$ presented at the beginning of Section 7.3.

In Table 7-7, we summarize some special values of $\cos\theta$ and $\sin\theta$. We illustrate some of these values in Figure 7-37.

TABLE 7-7 Special Values of $\cos\theta$ and $\sin\theta$

θ	$\cos\theta$	$\sin\theta$	θ	$\cos\theta$	$\sin\theta$
0	1	0	π	-1	0
$\frac{\pi}{6}$	$\frac{\sqrt{3}}{2}$	$\frac{1}{2}$	$\frac{7\pi}{6}$	$-\frac{\sqrt{3}}{2}$	$-\frac{1}{2}$
$\frac{\pi}{4}$	$\frac{\sqrt{2}}{2}$	$\frac{\sqrt{2}}{2}$	$\frac{5\pi}{4}$	$-\frac{\sqrt{2}}{2}$	$-\frac{\sqrt{2}}{2}$
$\frac{\pi}{3}$	$\frac{1}{2}$	$\frac{\sqrt{3}}{2}$	$\frac{4\pi}{3}$	$-\frac{1}{2}$	$-\frac{\sqrt{3}}{2}$
$\frac{\pi}{2}$	0	1	$\frac{3\pi}{2}$	0	-1
$\frac{2\pi}{3}$	$-\frac{1}{2}$	$\frac{\sqrt{3}}{2}$	$\frac{5\pi}{3}$	$\frac{1}{2}$	$-\frac{\sqrt{3}}{2}$
$\frac{3\pi}{4}$	$-\frac{\sqrt{2}}{2}$	$\frac{\sqrt{2}}{2}$	$\frac{7\pi}{4}$	$\frac{\sqrt{2}}{2}$	$-\frac{\sqrt{2}}{2}$
$\frac{5\pi}{6}$	$-\frac{\sqrt{3}}{2}$	$\frac{1}{2}$	$\frac{11\pi}{6}$	$\frac{\sqrt{3}}{2}$	$-\frac{1}{2}$
			2π	1	0

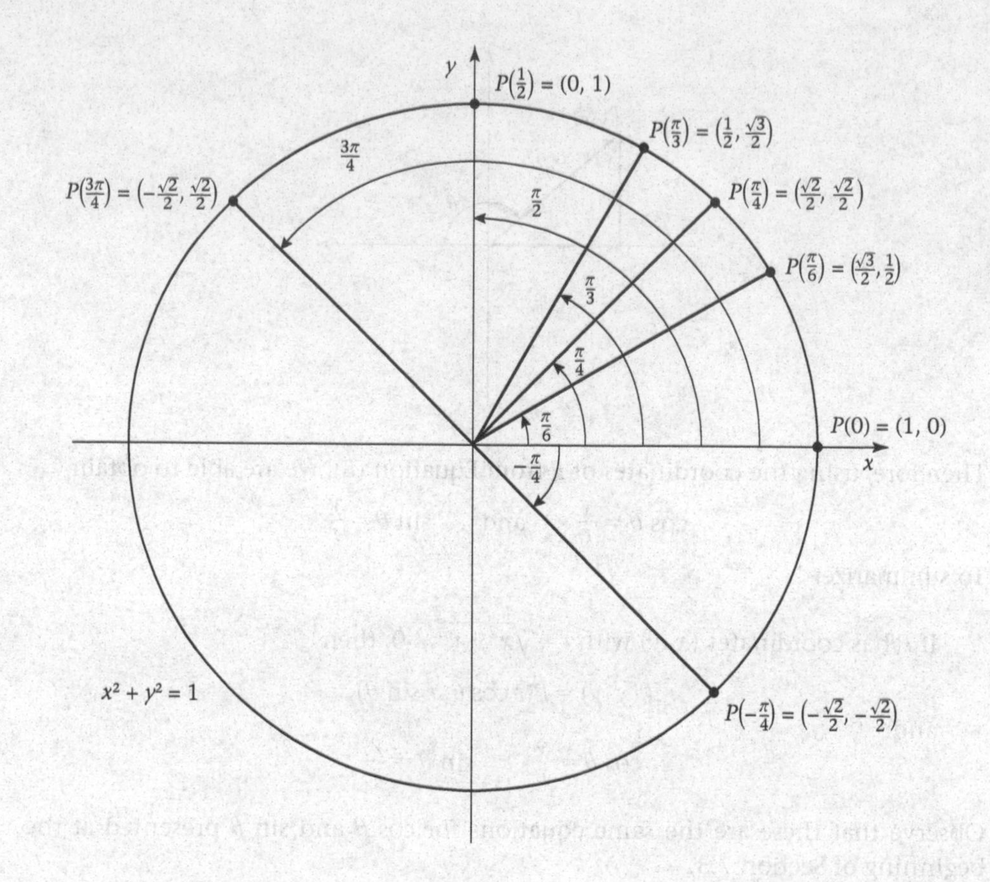

FIGURE 7-37
Special Values of cos θ and sin θ and the Unit Circle

7.4a *Properties of the Trigonometric Functions*

Consider Figure 7-38. Let point P be any point on the unit circle. If $P(x, y)$ is in the first or fourth quadrant, then $x > 0$, whereas if $P(x, y)$ is in the second or third quadrant, then $x < 0$. Since $\cos t = x$, we conclude that $\cos t > 0$ if t is in quadrant I or IV and $\cos t < 0$ if t is in quadrant II or III. Similarly, $y > 0$ if $P(x, y)$ is in quadrant I or II, whereas $y < 0$ if $P(x, y)$ is in quadrant III or IV. Since $\sin t = y$, we may write similar inequalities for $\sin t$. We present these results in Figure 7-39.

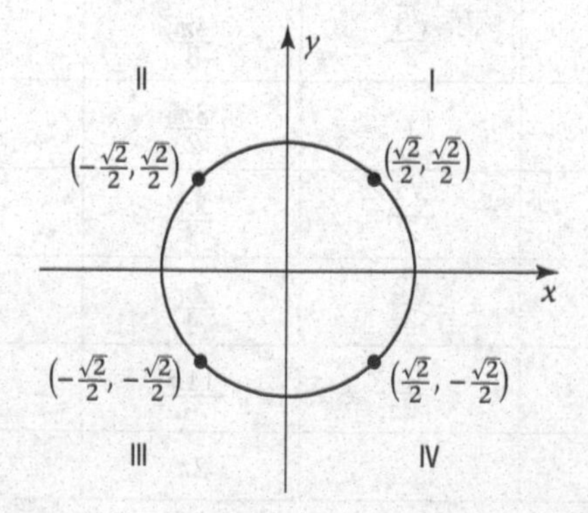

FIGURE 7-38
Points on the Unit Circle

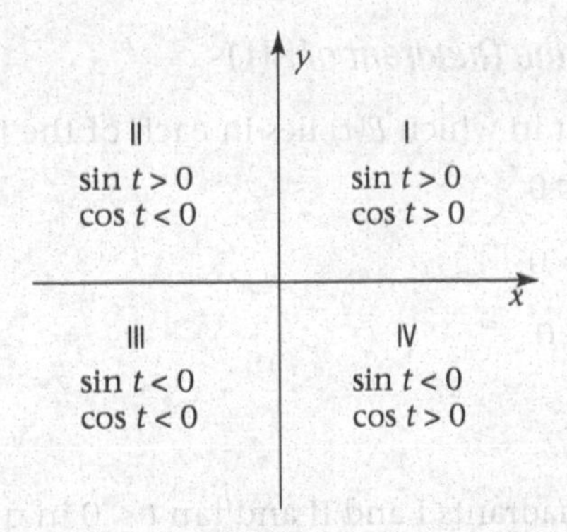

FIGURE 7-39
Signs of sin t and cos t

Since $\tan t = \frac{\sin t}{\cos t}$ if sin t and cos t have the same *sign*, then $\tan t > 0$. If sin t and cos t have opposite signs, then $\tan t < 0$. Using this with the other definitions presented in Table 7-6, Section 7.3, we summarize the signs for all the trigonometric functions in Figure 7-40. We have presented two columns in each quadrant of Figure 7-40. The order of the inequalities appearing in the first column of a given quadrant is the same as the order of the corresponding reciprocals appearing in the second column of the same quadrant. Therefore, we focus on the column listing sin t, cos t, and tan t. Additionally, we present Figure 7-41, which identifies those quadrants where sin t, cos t, and tan t are positive. Hence, by inference, any of these three functions omitted in a particular quadrant is negative in that quadrant.

II		I	
$\sin t > 0$	$\csc t > 0$	$\sin t > 0$	$\csc t > 0$
$\cos t < 0$	$\sec t < 0$	$\cos t > 0$	$\sec t > 0$
$\tan t < 0$	$\cot t < 0$	$\tan t > 0$	$\cot t > 0$
III		**IV**	
$\sin t < 0$	$\csc t < 0$	$\sin t < 0$	$\csc t < 0$
$\cos t < 0$	$\sec t < 0$	$\cos t > 0$	$\sec t > 0$
$\tan t > 0$	$\cot t > 0$	$\tan t < 0$	$\cot t < 0$

FIGURE 7-40
Signs of the Trigonometric Functions

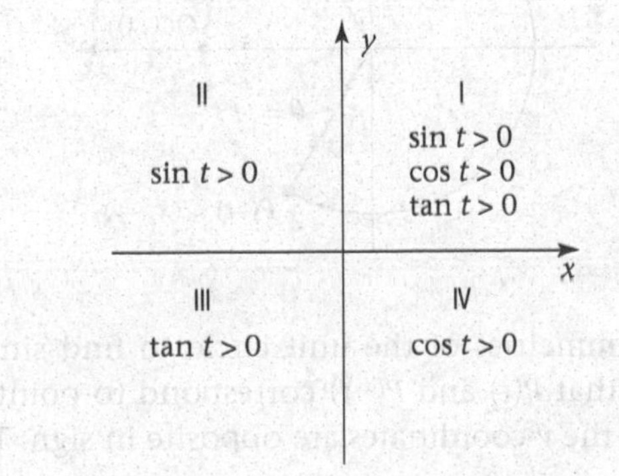

FIGURE 7-41
Signs of sin t, cos t, and tan t

EXAMPLE 4 *Finding the Quadrant of* P(t)

Determine the quadrant in which $P(t)$ lies in each of the following.

a. $\sin t > 0$ and $\tan t < 0$

b. $\sin t < 0$ and $\cos t > 0$

c. $\sec t > 0$ and $\csc t < 0$

SOLUTION

a. Since $\sin t > 0$ in quadrants I and II and $\tan t < 0$ in quadrants II and IV, both conditions hold only in quadrant II.

b. Since $\sin t < 0$ in quadrants III and IV and $\cos t > 0$ in quadrants I and IV, both conditions apply only in quadrant IV.

c. Since $\sec t > 0$ in quadrants I and IV and $\csc t < 0$ in quadrants III and IV, both conditions apply only in quadrant IV. ■

Progress Check

Determine the quadrant in which $P(t)$ lies in each of the following:

a. $\cos t < 0$ and $\tan t > 0$ b. $\cos t < 0$ and $\sin t > 0$

c. $\tan t < 0$ and $\csc t > 0$

Answers

a. quadrant III b. quadrant II c. quadrant II

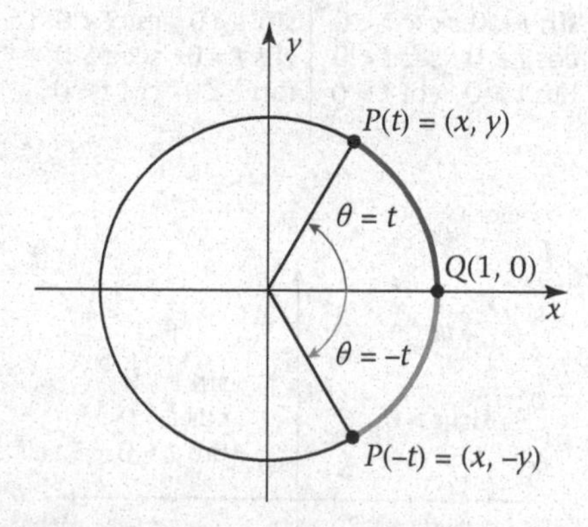

FIGURE 7-42
P(*t*) and *P*(−*t*) on a Unit Circle

We can use the symmetries of the unit circle to find $\sin(-t)$ and $\cos(-t)$. In Figure 7-42, we see that $P(t)$ and $P(-t)$ correspond to points having the same x-coordinates, whereas the y-coordinates are opposite in sign. Then

$$\sin t = y \quad \text{and} \quad \sin(-t) = -y$$

so

$$\sin(-t) = -\sin t$$

Similarly,

$$\cos t = x \quad \text{and} \quad \cos(-t) = x$$

so

$$\cos(-t) = \cos t$$

Using the definitions from Table 7-6, we obtain

$$\sin(-t) = -\sin t$$
$$\cos(-t) = \cos t$$
$$\tan(-t) = -\tan t$$
$$\cot(-t) = -\cot t$$
$$\sec(-t) = \sec t$$
$$\csc(-t) = -\csc t$$

A function f is said to be an **even function** if $f(-x) = f(x)$. We see that cosine and secant are even functions. A function f is said to be an **odd function** if $f(-x) = -f(x)$. The remaining four trigonometric functions are odd functions. In the case of polynomial functions, observe that $f(x) = x^2$ is an even function, but $f(x) = x^3$ is an odd function. From Section 3.3, we know that the graph of an even function is symmetric about the y-axis, and the graph of an odd function is symmetric with respect to the origin.

EXAMPLE 5 ***Even and Odd Functions***

Find $\sin\left(-\frac{\pi}{4}\right)$ and $\cos\left(-\frac{\pi}{3}\right)$.

SOLUTION

$$\sin\left(-\frac{\pi}{4}\right) = -\sin\frac{\pi}{4} = -\frac{\sqrt{2}}{2} \qquad \cos\left(-\frac{\pi}{3}\right) = \cos\frac{\pi}{3} = \frac{1}{2}$$

■

Progress Check

Verify the results of Example 1 and the first Progress Check in this section using the formulas for $\sin(-t)$ and $\cos(-t)$.

7.4b *Identities*

Trigonometry often involves the use of identities, which are equations that are true for *all* values which may be assumed by the variable. Identities are useful in simplifying equations and in providing alternative forms for computations. Although we will devote an entire section to this topic in the next chapter, there are some basic identities that we present at this time. First we review the identities from Table 7-6, Section 7.3.

$$\tan t = \frac{\sin t}{\cos t} \qquad \sec t = \frac{1}{\cos t}$$

$$\cot t = \frac{\cos t}{\sin t} \qquad \csc t = \frac{1}{\sin t}$$

Next, observe that the coordinates (x, y) of every point on the unit circle satisfy the equation

$$x^2 + y^2 = 1$$

Since $x = \cos t$ and $y = \sin t$, we obtain

$$(\cos t)^2 + (\sin t)^2 = 1$$

(See Equation (4) of Section 7.3.)

Since powers of trigonometric functions, such as $(\sin t)^n$ and $(\cos t)^n$, occur quite frequently, a special notation is used. For example, we write

$$(\sin t)^n = \sin^n t \qquad \text{and} \qquad (\cos t)^n = \cos^n t \qquad \text{provided } n \neq -1$$

Using the notation and rearranging terms, the identity becomes

$$\sin^2 t + \cos^2 t = 1$$

We may also use this identity in the alternative forms

$$\sin^2 t = 1 - \cos^2 t$$

$$\cos^2 t = 1 - \sin^2 t$$

Do not confuse

$$\sin t^2 \qquad \text{and} \qquad \sin^2 t$$

We have defined $\sin^2 t$ by

$$\sin^2 t = (\sin t)^2$$

which indicates that we find $\sin t$ and then square the result. But $\sin t^2$ indicates that we are to square t and *then* find the sine of the argument t^2.

EXAMPLE 6 ***Applying the Trigonometric Identities***

If $\cos t = \frac{3}{5}$ and t is in quadrant IV, find $\sin t$ and $\tan t$.

SOLUTION

Using the identity $\sin^2 t + \cos^2 t = 1$, we have

$$\sin^2 t + \left(\frac{3}{5}\right)^2 = 1$$

$$\sin^2 t = 1 - \frac{9}{25} = \frac{16}{25}$$

$$\sin t = \pm\frac{4}{5}$$

Since t is in quadrant IV, $\sin t$ must be negative so that $\sin t = -\frac{4}{5}$. Then

$$\tan t = \frac{\sin t}{\cos t} = \frac{-\frac{4}{5}}{\frac{3}{5}} = -\frac{4}{3}$$

■

Progress Check

If $\sin t = \frac{12}{13}$ and t is in quadrant II, find the following:

a. $\cos t$ b. $\tan t$

Answers

a. $-\frac{5}{13}$ b. $-\frac{12}{5}$

EXAMPLE 7 ***Proving an Identity***

Show that $1 + \tan^2 x = \sec^2 x$.

SOLUTION

We will use the trigonometric identities to transform the left-hand side of the equation into the right-hand side. Since $\tan x = \frac{\sin x}{\cos x}$, we have

$$1 + \tan^2 x = 1 + \frac{\sin^2 x}{\cos^2 x}$$

$$= \frac{\cos^2 x + \sin^2 x}{\cos^2 x}$$

Since $\cos^2 x + \sin^2 x = 1$,

$$1 + \tan^2 x = \frac{1}{\cos^2 x} = \sec^2 x$$ ■

Progress Check

Use identities to transform the expression

$$\tan t \cos t + \sin t + \frac{1}{\csc t}$$

into 3 sin t.

You cannot verify an identity by checking to see that it "works" for one or more values of the variable as these values could turn out to be solutions to a conditional equation. (See Section 2.1.) You must show that an equation is true for *all* values in the domain of its variable to prove that it is an identity.

Exercise Set 7.4

In Exercises 1–12, determine the reference angle for each angle.

1. 250°
2. -130°
3. -330°
4. 125°
5. $\frac{6\pi}{5}$
6. $\frac{9\pi}{5}$
7. 335°
8. -10°
9. -47°
10. 110°
11. $\frac{15\pi}{7}$
12. $-\frac{3\pi}{5}$

In Exercises 13–24, use the rectangular coordinates of $P(t)$ to find $\sin t$ and $\cos t$.

13. $t=\frac{5\pi}{3}$
14. $t=\frac{3\pi}{4}$
15. $t=-5\pi$
16. $t=-\frac{5\pi}{4}$
17. $t=\frac{7\pi}{4}$
18. $t=\frac{7\pi}{6}$
19. $t=\frac{2\pi}{3}$
20. $t=\frac{5\pi}{6}$
21. $t=-\frac{\pi}{3}$
22. $t=-\frac{5\pi}{6}$
23. $t=\frac{5\pi}{4}$
24. $t=-11\pi$

In Exercises 25–40, without using a calculator, find the values of the six trigonometric functions for each argument.

25. 135°
26. 300°
27. -30°
28. -300°
29. $\frac{\pi}{3}$
30. $\frac{\pi}{6}$
31. $\frac{\pi}{4}$
32. $\frac{\pi}{2}$
33. $\frac{5\pi}{6}$
34. $\frac{4\pi}{3}$
35. $\frac{3\pi}{2}$
36. $\frac{7\pi}{4}$
37. $\frac{3\pi}{4}$
38. $-\frac{11\pi}{6}$
39. $-\frac{5\pi}{4}$
40. $-\frac{7\pi}{6}$

In Exercises 41–56, find the rectangular coordinates of the given point.

41. $P(5\pi)$
42. $P\left(\frac{5\pi}{2}\right)$
43. $P\left(-\frac{\pi}{4}\right)$
44. $P\left(-\frac{3\pi}{2}\right)$
45. $P\left(\frac{5\pi}{4}\right)$
46. $P(8\pi)$
47. $P\left(\frac{4\pi}{3}\right)$
48. $P\left(\frac{2\pi}{3}\right)$
49. $P\left(-\frac{2\pi}{3}\right)$
50. $P\left(-\frac{19\pi}{3}\right)$
51. $P\left(\frac{19\pi}{6}\right)$
52. $P\left(\frac{17\pi}{6}\right)$
53. $P\left(-\frac{5\pi}{6}\right)$
54. $P\left(-\frac{11\pi}{6}\right)$
55. $P\left(\frac{19\pi}{3}\right)$
56. $P\left(\frac{25\pi}{3}\right)$

In Exercises 57–64, determine both a positive and a negative real number t, $|t| < 2\pi$, for which $P(t)$ has the following rectangular coordinates.

57. $(-1, 0)$
58. $(0, -1)$
59. $\left(-\frac{\sqrt{2}}{2}, \frac{\sqrt{2}}{2}\right)$
60. $\left(\frac{\sqrt{2}}{2}, -\frac{\sqrt{2}}{2}\right)$
61. $\left(-\frac{\sqrt{3}}{2}, \frac{1}{2}\right)$
62. $\left(-\frac{1}{2}, \frac{\sqrt{3}}{2}\right)$
63. $\left(\frac{1}{2}, -\frac{\sqrt{3}}{2}\right)$
64. $\left(-\frac{\sqrt{3}}{2}, -\frac{1}{2}\right)$
65. Given $P(t)=\left(\frac{3}{5}, \frac{4}{5}\right)$ use the symmetries of the circle to find
   a. $P(t+\pi)$
   b. $P\left(t-\frac{\pi}{2}\right)$
   c. $P(-t)$
   d. $P(-t-\pi)$
66. Given $P(t)=\left(-\frac{4}{5}, -\frac{3}{5}\right)$, use the symmetries of the circle to find
   a. $P(t-\pi)$
   b. $P\left(t+\frac{\pi}{2}\right)$
   c. $P(-t)$
   d. $P(-t+\pi)$
67. If the point (a, b) is on the unit circle, show that $(a, -b)$, $(-a, b)$, and $(-a, -b)$ also lie on the unit circle.
68. If the point (a, b) is on the unit circle, show that (b, a), $(b, -a)$, $(-b, a)$, and $(-b, -a)$ also lie on the unit circle.

In Exercises 69–82, find the quadrant in which $P(t)$ lies if the following conditions hold.

69. $\sin t > 0$, $\cos t < 0$
70. $\sin t < 0$, $\tan t > 0$
71. $\cos t < 0$, $\tan t > 0$
72. $\tan t < 0$, $\sin t > 0$
73. $\sin t < 0$, $\cos t < 0$
74. $\tan t < 0$, $\cos t < 0$
75. $\sec t < 0$, $\sin t < 0$
76. $\tan t < 0$, $\sec t < 0$
77. $\csc t > 0$, $\sec t < 0$
78. $\sin t < 0$, $\cot t > 0$
79. $\sec t < 0$, $\cot t > 0$
80. $\cot t < 0$, $\sin t > 0$
81. $\sec t < 0$, $\csc t < 0$
82. $\csc t < 0$, $\cot t > 0$

In Exercises 83–94, find the value of the trigonometric function when t is replaced by $-t$. (For example, given $\sin t = \frac{1}{2}$ find $\sin(-t)$).

83. $\tan t = \frac{3}{2}$
84. $\sin t = 1$
85. $\tan t = 1$
86. $\cos t = -1$

87. $\tan t = \frac{\sqrt{2}}{2}$

88. $\cos t = \frac{\sqrt{3}}{2}$

89. $\cos t = -\frac{\sqrt{3}}{2}$

90. $\sin t = -\frac{1}{2}$

91. $\tan t = \sqrt{3}$

92. $\sin t = \frac{1}{2}$

93. $\sin t = \frac{\sqrt{3}}{2}$

94. $\tan t = \frac{\sqrt{3}}{3}$

95. If the point $\left(\frac{12}{13}, \frac{5}{13}\right)$ on the unit circle corresponds to the real number t, use the symmetries of the circle to find the coordinates of the point on the unit circle corresponding to the real number

a. $t + \pi$

b. $t - \frac{\pi}{2}$

b. $-t$

d. $-t - \pi$

96. If the point $\left(-\frac{4}{5}, -\frac{3}{5}\right)$ on the unit circle corresponds to the real number t, use the symmetries of the circle to find the coordinates of the point on the unit circle corresponding to the real number

a. $t - \pi$

b. $t + \frac{\pi}{2}$

c. $-t$

d. $-t + \pi$

97. The point $P(a, b)$ on the unit circle corresponding to the real number t lies in quadrant II. Find the values of

a. $\sin\left(t + \frac{\pi}{2}\right)$

b. $\cos\left(t + \frac{\pi}{2}\right)$

In Exercises 98–100, the points $P(t) = P(a, b)$ and $P\left(t \pm \frac{\pi}{2}\right) = P'(a', b')$, as shown in the accompanying figure.

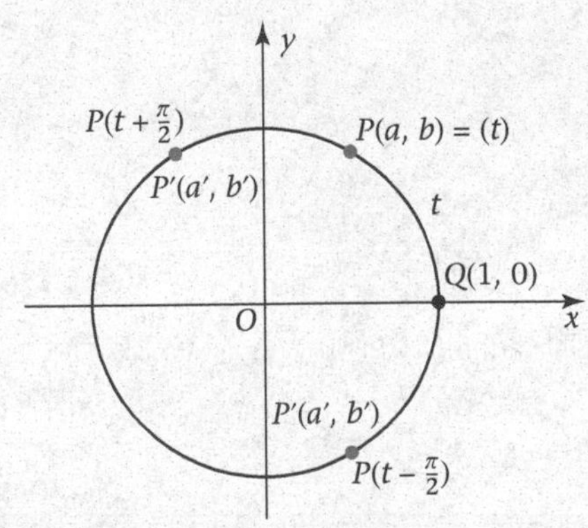

98. Show that $\frac{b'}{a'} = -\frac{a}{b}$. (*Hint*: Show that the lines OP and OP' are perpendicular and then determine their slopes.)

99. Show that $b' = \pm a$ and $a' = \pm b$. (*Hint*: The radii OP and OP' are equal in length. Use the distance formula combined with the result of Exercise 98 above to substitute alternately for b' and for a'.)

100. Show that either (i) $(a', b') = (-b, a)$ or (ii) $(a', b') = (b, -a)$. (*Hint*: Begin with the result of Exercise 99 and apply the result of Exercise 98.)

In Exercises 101–108, find the indicated value under the given conditions.

101. If $\sin t = \frac{3}{5}$ and t is in quadrant II, find $\tan t$.

102. If $\tan t = -\frac{3}{4}$ and t is in quadrant II, find $\cos t$.

103. If $\cos t = -\frac{5}{13}$ and t is in quadrant III, find $\sin t$.

104. If $\sin t = -\frac{5}{13}$ and t is in quadrant III, find $\tan t$.

105. If $\cos t = \frac{4}{5}$ and $\sin t < 0$, find $\sin t$.

106. If $\tan t = \frac{12}{5}$ and $\cos t < 0$, find $\sin t$.

107. If $\sin t = -\frac{3}{5}$ and $\tan t < 0$, find $\cos t$.

108. If $\tan t = -\frac{5}{12}$ and $\sin t > 0$, find $\sin t$.

In Exercises 109–120, find the values of the trigonometric functions of the angle θ if the point P lies on the terminal side of θ.

109. $P(-5, 12)$

110. $P(3, -4)$

111. $P(-1, -1)$

112. $P(1, 2)$

113. $P(-8, 6)$

114. $P(12, 5)$

115. $P(12, -5)$

116. $P(-1, \sqrt{3})$

117. $P(-12, -5)$

118. $P(-3, 4)$

119. $P(-2, 1)$

120. $P(-2, -1)$

In Exercises 121–126, use a calculator and the polynomial approximations

$$\sin t \approx t - \frac{t^3}{6} + \frac{t^5}{120} - \frac{t^7}{5040}$$

$$\cos t \approx 1 - \frac{t^2}{2} + \frac{t^4}{24} - \frac{t^6}{720}$$

to test the accuracy of the approximation.

121. $\sin 0.80$

122. $\cos 1.10$

123. $\sin(-0.20)$

124. $\cos(-0.75)$

125. $\tan 0.1$

126. $\tan(-1.2)$

127. Using the polynomial approximation for $\sin t$ given above, show that sine is an odd function, that is, $\sin(-t) = -\sin(t)$.

128. Using the polynomial approximation for $\cos t$ given above, show that cosine is an even function, that is, $\cos(-t) = \cos(t)$.

In Exercises 129–138, use trigonometric identities to transform the first expression into the second.

129. $\tan t \cos t$, $\sin t$

130. $\frac{\cos t}{\sin t}$, $\frac{1}{\tan t}$

131. $\frac{1 - \sin^2 t}{\sin t}$, $\frac{\cos t}{\tan t}$

132. $\tan t \sin t + \cos t$, $\frac{1}{\cos t}$

133. $\cos t\left(\frac{1}{\cos t} - \cos t\right)$, $\sin^2 t$

134. $\dfrac{1-\cos^2 t}{\sin t}$, $\sin t$

135. $\dfrac{1-\cos^2 t}{\cos^2 t}$, $\tan^2 t$

136. $\dfrac{\cos^2 t}{1-\sin t}$, $1+\sin t$

137. $(\sin t-\cos t)^2$, $1-2\sin t\cos t$

138. $\dfrac{1}{1-\sin t}+\dfrac{1}{1+\sin t}$, $\dfrac{2}{\cos^2 t}$

7.5 Graphs of the Trigonometric Functions

7.5a Periodic Functions

In Section 7.3, we observed a certain repetitive nature to the trigonometric functions. This repetition gives us an additional method for graphing these functions. Thus, once we know the structure of the pattern that repeats itself we may use that knowledge to draw the graph over any specified region. This characteristic makes these functions especially useful in describing cyclic phenomena in a wide variety of applications.

A function f is **periodic** if there exists a positive number c, such that

$$f(x + c) = f(x)$$

for all x in the domain of f. The least number $c > 0$ for which f is periodic is called the **period** of f.

Recall from Section 7.3,

For every integer n,

$$P(t + 2\pi n) = P(t)$$
$$\sin (t + 2\pi n) = \sin t$$
$$\cos (t + 2\pi n) = \cos t$$

In particular, for $n = 1$,

$$\sin (t + 2\pi) = \sin t \quad \text{and} \quad \cos (t + 2\pi) = \cos t$$

Therefore, $\sin t$ and $\cos t$ are periodic functions. Furthermore, they both have period 2π. (See Exercises 49 and 50.) Because of the definitions of $\tan t$, $\cot t$, $\sec t$, and $\csc t$ in terms of $\sin t$ and $\cos t$, we note that these four trigonometric functions are also periodic. However, as we will see, not all of them have period 2π.

7.5b Graphs of Sine and Cosine

The definitions of $\sin t$ and $\cos t$ as the coordinates of points on the unit circle $x^2 + y^2 = 1$ tell us that they must satisfy the inequalities

$$-1 \leq \sin t \leq 1$$
$$-1 \leq \cos t \leq 1$$

Therefore, we indicate these minimum and maximum values on the y-axis as shown in Figure 7-43. Since these functions have a period of 2π, we initially sketch their graphs over the interval $[0, 2\pi]$.

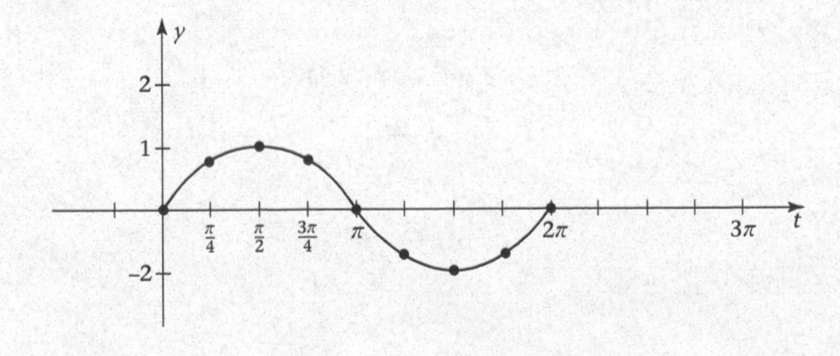

FIGURE 7-43
Graph of $y = \sin t$ for $0 \leq t \leq 2\pi$

We form Table 7-8 using the values from Table 7-7 to two decimal places. Since we are letting t denote the independent variable, we use the label t for the horizontal axis. Verify the values in Table 7-8 using your calculator.

TABLE 7-8 Plotting sin t and cos t

t	$\sin t$	$\cos t$
0	0	1
$\frac{\pi}{4}$	0.71	0.71
$\frac{\pi}{2}$	1	0
$\frac{3\pi}{4}$	0.71	−0.71
π	0	−1
$\frac{5\pi}{4}$	−0.71	−0.71
$\frac{3\pi}{2}$	−1	0
$\frac{7\pi}{4}$	−0.71	0.71
2π	0	1

In Figure 7-44, we repeat the graph of Figure 7-43 for adjacent intervals of width 2π.

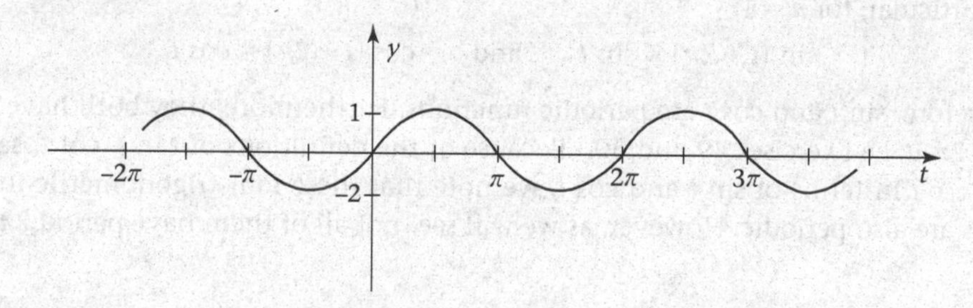

FIGURE 7-44
Graph of $y = \sin t$

We sketch $y = \cos t$ in Figure 7-45 using the values from Table 7-8 and the same procedure that we used to graph $y = \sin t$.

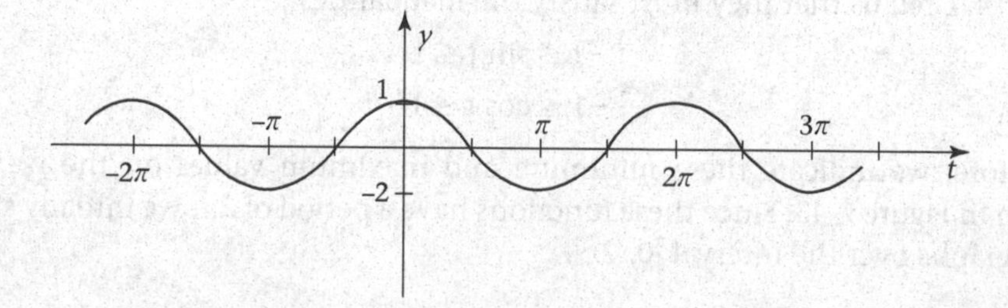

FIGURE 7-45
Graph of $y = \cos t$

7.5c Graphs of Tangent, Cotangent, Secant and Cosecant

Since sin t and cos t both have period 2π, one might suspect that tan t has period 2π as well. However, consider Figure 7-46. In Figure 7-46(a),

$$\tan(t+\pi) = \frac{\sin(t+\pi)}{\cos(t+\pi)} = \frac{-y}{-x} = \frac{y}{x} = \frac{\sin t}{\cos t} = \tan t$$

whereas in Figure 7-46(b),

$$\tan(t+\pi) = \frac{\sin(t+\pi)}{\cos(t+\pi)} = \frac{-y}{x} = \frac{y}{-x} = \frac{\sin t}{\cos t} = \tan t$$

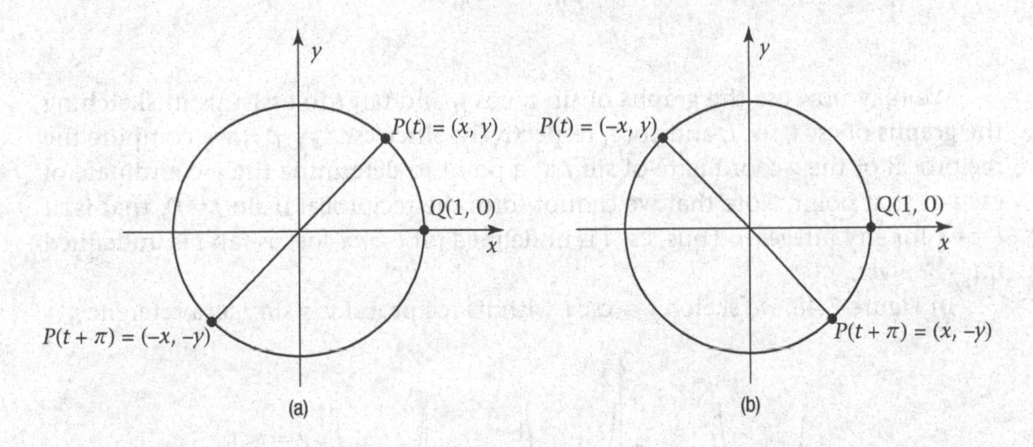

FIGURE 7-46
Circle Symmetry

Thus,

$$\tan(t+\pi) = \tan t$$

Since there are no other numbers c with $0 < c < \pi$ such that $\tan(t+c) = \tan t$ (See Exercise 51.), tan t has period π.

It turns out to be convenient to plot tan t in the interval $(-\frac{\pi}{2}, \frac{\pi}{2})$ which is of length π. First, we form Table 7-9 using special values of t as well as a calculator to two decimal places. Since tan t is undefined at $-\frac{\pi}{2}$ and $\frac{\pi}{2}$ we need to carefully consider the behavior of the graph *near* these values of t. That is why Table 7-9 contains the values -1.5 and -1.57 near $-\frac{\pi}{2}$, and 1.5 and 1.57 near $\frac{\pi}{2}$. As t gets closer and closer to $\frac{\pi}{2}$ with $t < \frac{\pi}{2}$, we say that tan t **approaches infinity**. Similarly, as t gets closer and closer to $-\frac{\pi}{2}$ with $t > -\frac{\pi}{2}$, we say that tan t **approaches negative infinity**. These considerations lead us to the graph of tan t as shown in Figure 7-47. Note the asymptotes in this graph, two of which are found at $t = \pm\frac{\pi}{2}$.

TABLE 7-9 Plotting tan t

t	tan t	t	tan t
$-\frac{\pi}{2}$	undefined	$\frac{\pi}{6}$	0.58
−1.57	−1255.77	$\frac{\pi}{4}$	1
−1.5	−14.10	$\frac{\pi}{3}$	1.73
$-\frac{\pi}{3}$	−1.73	1.5	14.10
$-\frac{\pi}{4}$	−1	1.57	1255.77
$-\frac{\pi}{6}$	−0.58	$\frac{\pi}{2}$	undefined
0	0		

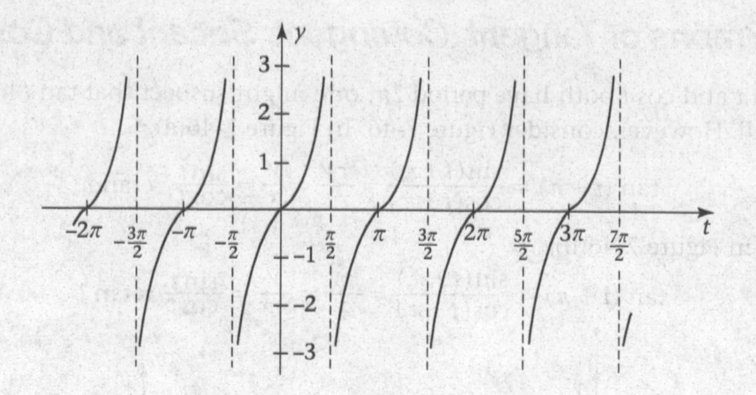

FIGURE 7-47
Graph of $y = \tan t$

We may now use the graphs of sin t, cos t, and tan t to assist us in sketching the graphs of csc t, sec t, and cot t, respectively. Since $\csc t = \frac{1}{\sin t}$, we compute the reciprocal of the y-coordinate of sin t at a point to determine the y-coordinate of csc t at that point. Note that we cannot form the reciprocal if $\sin t = 0$, that is, if $t = n\pi$ for any integer n. Thus, csc t is undefined for $t = n\pi$ just as tan t is undefined if $t = \frac{\pi}{2} + n\pi$.

In Figure 7-48, we sketch $y = \csc t$ with its reciprocal $y = \sin t$ as a reference.

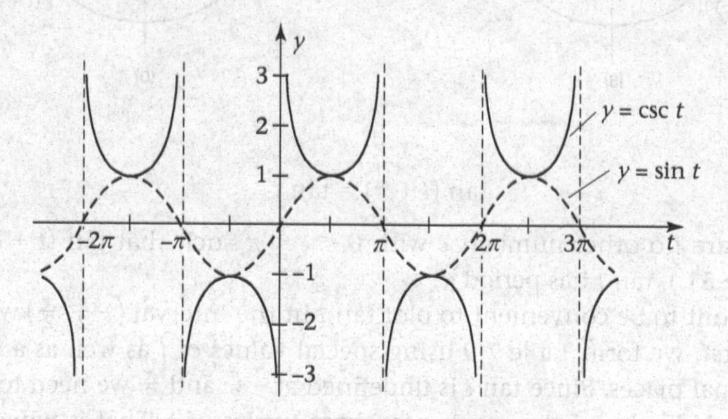

FIGURE 7-48
Graph of $y = \csc t$

Similarly in Figure 7-49, we sketch $y = \sec t$ with reference $y = \cos t$, and lastly, in Figure 7-50, we sketch $y = \cot t$ with its reference $y = \tan t$.

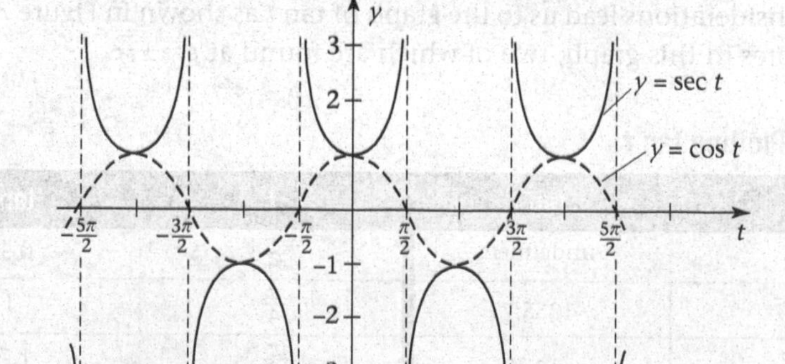

FIGURE 7-49
Graph of $y = \sec t$

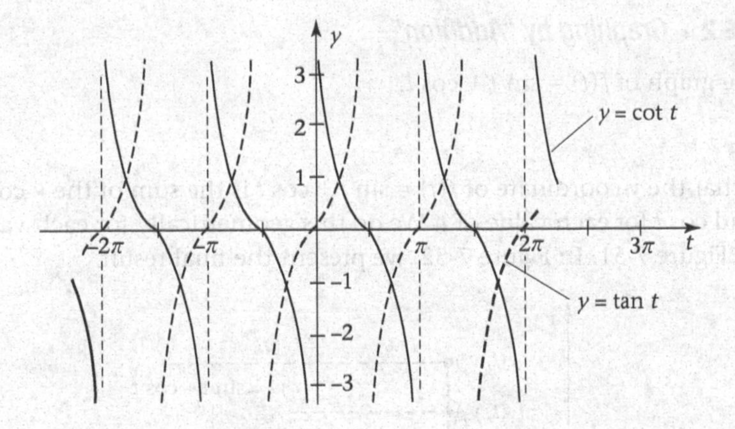

FIGURE 7-50
Graph of $y = \cot t$

We summarize the period, domain, and range for the six trigonometric functions, as shown in Table 7-10.

TABLE 7-10 Domain and Range for Trigonometric Functions

Function	Period	Domain	Range
$\sin t$	2π	all t	$[-1, 1]$
$\cos t$	2π	all t	$[-1, 1]$
$\tan t$	π	$t \neq \frac{\pi}{2} + n\pi$	$(-\infty, \infty)$
$\csc t$	2π	$t \neq n\pi$	$(-\infty, -1], [1, \infty)$
$\sec t$	2π	$t \neq \frac{\pi}{2} + n\pi$	$(-\infty, -1], [1, \infty)$
$\cot t$	π	$t \neq n\pi$	$(-\infty, \infty)$

EXAMPLE 1 ***Graphing by "Addition"***

Sketch the graph of $f(t) = 1 + \sin t$.

SOLUTION

Rather than form a table of values and plot points, we note that the graph of $f(t) = 1 + \sin t$ is that of the graph of $f(t) = \sin t$ shifted up one unit. (See Section 3.3.)

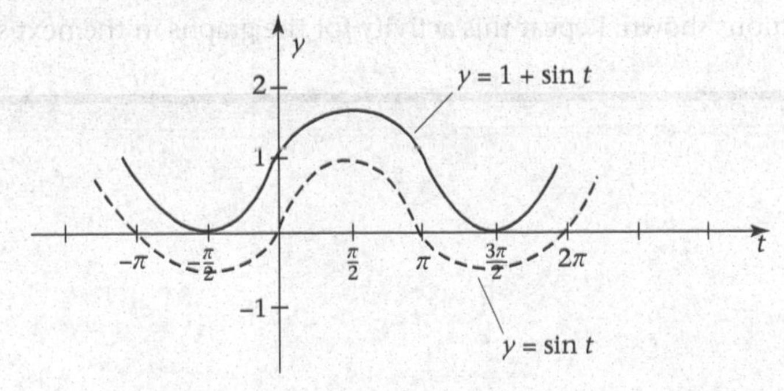

■

EXAMPLE 2 *Graphing by "Addition"*

Sketch the graph of $f(t) = \sin t + \cos t$.

SOLUTION

We note that the y-coordinate of $f(t) = \sin t + \cos t$ is the sum of the y-coordinates of $\sin t$ and $\cos t$ for each value of t. We do this geometrically for each value of t as shown in Figure 7-51. In Figure 7-52, we present the final result.

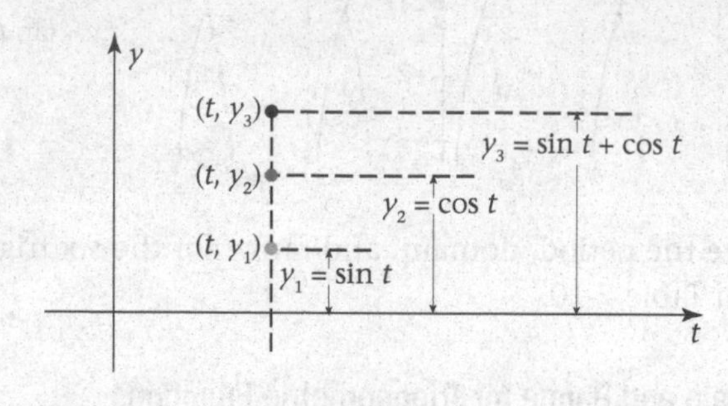

FIGURE 7-51
Adding y-coordinates Geometrically

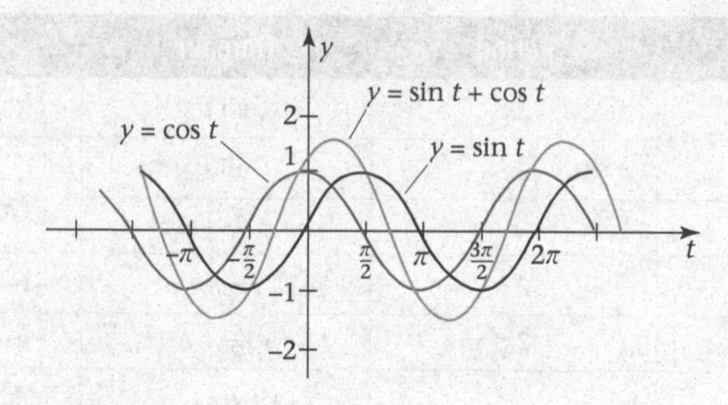

FIGURE 7-52
Graph of $y = \sin t + \cos t$

■

Graphing Calculator Power User's CORNER

When graphing trigonometric functions, your calculator *must* be set to RADIAN mode. (*Note*: If your calculator has built-in trigonometric RANGE settings and your calculator is set to degree mode, you will get what appears to be a correct graph. However, using TRACE will reveal incorrect function values.)

For each graph in this section, set the WINDOW values on your calculator to match those in the figure and GRAPH the function or functions shown. Repeat this activity for the graphs in the next section.

Exercise Set 7.5

In Exercises 1–38, use the graph of the particular trigonometric function to find the values of t in the interval $[0, 2\pi)$ that satisfy the given equation.

1. $\cos t = 0$
2. $\sin t = 1$
3. $\tan t = 1$
4. $\cos t = -1$
5. $\sin t = \frac{\sqrt{2}}{2}$
6. $\cos t = \frac{\sqrt{3}}{2}$
7. $\cos t = -\frac{\sqrt{3}}{2}$
8. $\sin t = -1$
9. $\tan t = \sqrt{3}$
10. $\sin t = \frac{3}{2}$
11. $\sin t = \frac{\sqrt{3}}{2}$
12. $\tan t = \frac{\sqrt{3}}{3}$
13. $\sin t = -\frac{1}{2}$
14. $\cos t = -\frac{\sqrt{2}}{2}$
15. $\sin t = 2$
16. $\cos t = \frac{\sqrt{2}}{2}$
17. $\sin t = \frac{1}{2}$
18. $\cos t = \frac{3}{2}$
19. $\sec t = 1$
20. $\sec t = -1$
21. $\csc t = -2$
22. $\csc t = 0$
23. $\cot t = 1$
24. $\cot t = \sqrt{3}$
25. $\cot t = -1$
26. $\cot t = \frac{\sqrt{3}}{3}$
27. $\sec t = \sqrt{2}$
28. $\csc t = -\sqrt{2}$
29. $\cot t = -\sqrt{3}$
30. $\csc t = \frac{2\sqrt{3}}{3}$
31. $\sin t = \frac{1}{2}$, $\sec t < 0$
32. $\tan t = \sqrt{3}$, $\csc t < 0$
33. $\sec t = -2$, $\csc t > 0$
34. $\csc t = -2$, $\cot t > 0$
35. $\csc t = -\sqrt{2}$, $\sec t < 0$
36. $\sec t = \sqrt{2}$, $\cot t > 0$
37. $\cot t = -1$, $\sec t < 0$
38. $\cot t = \sqrt{3}$, $\csc t < 0$

In Exercises 39–48, sketch the graph of the given function. Then, determine appropriate WINDOW values and GRAPH the functions on your graphing calculator. Be sure your calculator is in RADIAN mode. Replace the variable t with the variable x.

39. $f(t) = 1 + \cos t$
40. $f(t) = -1 + \sin t$
41. $f(t) = 2 \sin t$
42. $f(t) = \frac{1}{2} \cos t$
43. $f(t) = \sin t + \frac{1}{2} \cos t$
44. $f(t) = 2 \sin t + \cos t$
45. $f(t) = \sin t - \cos t$
46. $f(t) = \sin(-t) + \cos t$
47. $f(t) = t + \sin t$
48. $f(t) = -t + \cos t$
49. Prove that the period of the sine function is 2π. (*Hint*: Assume $\sin(t + c) = \sin t$, $0 < c < 2\pi$, for all t. By letting $t = 0$, show that $\sin c = 0$ and, consequently, that $c = \pi$. Finally, conclude that $\sin(t + \pi) = \sin t$ does not hold for $t = \frac{\pi}{2}$.)
50. Prove that the period of the cosine function is 2π.
51. Prove that the period of the tangent function is π.
52. Verify that $\sin(-t) = -\sin t$ by using the graph of the sine function.
53. Verify that $\cos(-t) = \cos t$ by using the graph of the cosine function.
54. Determine the domain and range of the functions in Exercises 39–48 by examining the graph of each function.
55. Use the identity

$$\tan t = \frac{\sin t}{\cos t}$$

to determine the vertical asymptotes of the graph of $\tan t$.

7.6 Graphs: Amplitude, Period, and Phase Shift

Our objective in this section is to sketch graphs, such as $f(x) = A \sin (Bx + C)$ and $f(x) = A \cos (Bx + C)$, where A, B, and C are real numbers with $A \neq 0$ and $B \neq 0$. (We suggest reviewing Section 3.3 at this time.) Note that we now use the symbol x to denote the independent variable rather than the symbol t, which was used in the previous sections of this chapter. The use of the independent variable x in this context should not be confused with the x-coordinate of the point on the unit circle, $P(t) = P(x, y)$.

7.6a *Amplitude*

Since the sine function and the cosine function both have a maximum value of 1 and a minimum value of −1, the functions $f(x) = A \sin x$ and $f(x) = A \cos x$ have a maximum value of $|A|$ and a minimum value of $-|A|$. We define

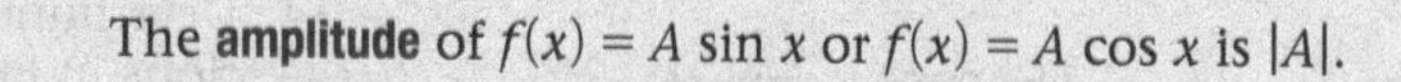

The **amplitude** of $f(x) = A \sin x$ or $f(x) = A \cos x$ is $|A|$.

The multiplier A acts as a vertical "stretching" factor when $|A| > 1$, and as a vertical "shrinking" factor when $|A| < 1$.

EXAMPLE 1 ***Amplitude of f(x) = A sin x***

Sketch the graphs of $y = 2\sin x$ and $y = \frac{1}{2}\sin x$ on the same coordinate axes.

SOLUTION

The graph of $y = 2\sin x$ has an amplitude of 2. Its maximum value is 2 and its minimum value is −2. Similarly, the graph of $y = \frac{1}{2}\sin x$ has an amplitude of $\frac{1}{2}$ with maximum and minimum values of $\frac{1}{2}$ and $-\frac{1}{2}$, respectively.

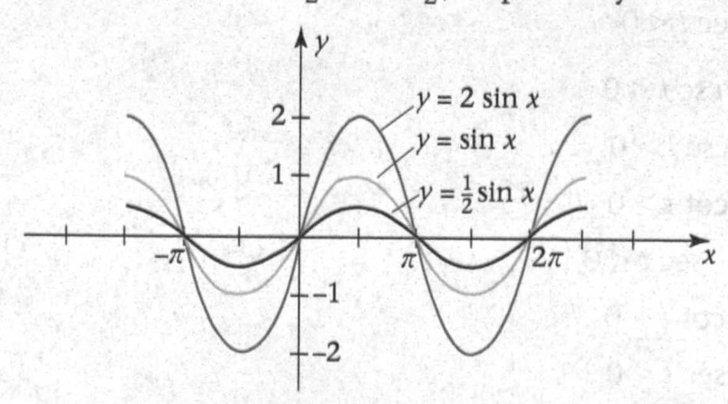

■

EXAMPLE 2 ***Graphing with a Negative "Stretching" Factor***

Sketch the graph of $f(x) = -3 \cos x$.

SOLUTION

The graph of $y = -3 \cos x$ has an amplitude of 3 with maximum and minimum values of 3 and −3, respectively. This graph is said to be a **reflection** about the x-axis of the graph $y = 3 \cos x$.

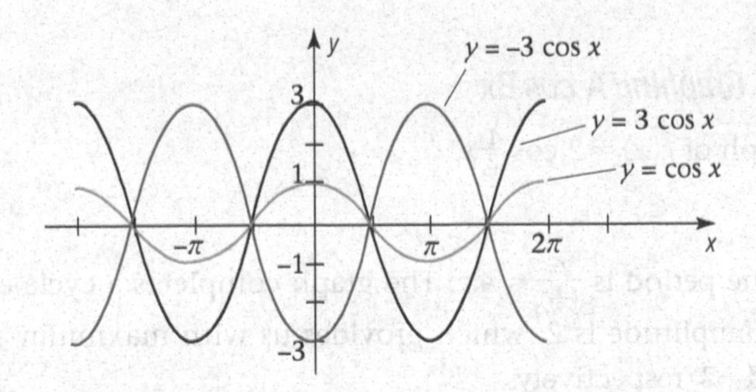

7.6b Period

We now seek to determine the period of functions such as $f(x) = A \sin Bx$ or $f(x) = A \cos Bx$. Although the argument of the functions has been generalized from x to Bx from our earlier presentation, the amplitude remains unchanged, that is, the amplitude of either function is $|A|$. Now, since $y = \sin x$ and $y = \cos x$ both have period 2π, we say that either function completes one **cycle** as x varies from 0 to 2π. Similarly, $f(x) = A \sin Bx$ or $f(x) = A \cos Bx$ completes one cycle as Bx varies from 0 to 2π if $B > 0$, or as $-Bx$ varies from 0 to 2π if $B < 0$. We may combine both cases by writing, "as $|B|x$ varies from 0 to 2π." This leads to the equation

$$|B|x = 0 \qquad \text{in which case} \qquad x = 0$$

and to the equation

$$|B|x = 2\pi \qquad \text{in which case} \qquad x = \frac{2\pi}{|B|}$$

Therefore, both $f(x) = A \sin Bx$ and $f(x) = A \cos Bx$ complete one cycle as x varies from 0 to $\frac{2\pi}{|B|}$, or equivalently,

The **period** of $f(x) = A \sin Bx$ or $f(x) = A \cos Bx$ is $\frac{2\pi}{|B|}$.

The multiplier B acts as a horizontal stretching factor if $0 < |B| < 1$ and as a horizontal shrinking factor if $|B| > 1$.

EXAMPLE 3 *The Period of sin* Bx

Sketch the graph of $f(x) = \sin 2x$.

SOLUTION

Since $B = 2$, the period is $\frac{2\pi}{2} = \pi$. Therefore, the graph completes a cycle every π units.

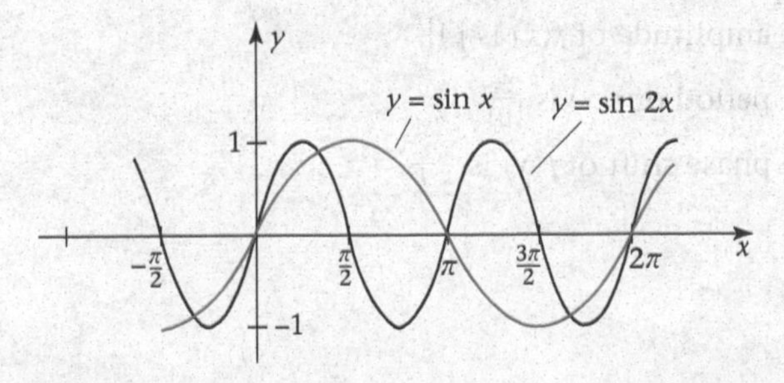

■

EXAMPLE 4 *Graphing* A *cos* Bx

Sketch the graph of $f(x) = 2 \cos \frac{1}{2}x$.

SOLUTION

Since $B = \frac{1}{2}$, the period is $\frac{2\pi}{\left(\frac{1}{2}\right)} = 4\pi$. The graph completes a cycle every 4π units. Note that the amplitude is 2, which provides us with maximum and minimum values of 2 and -2, respectively.

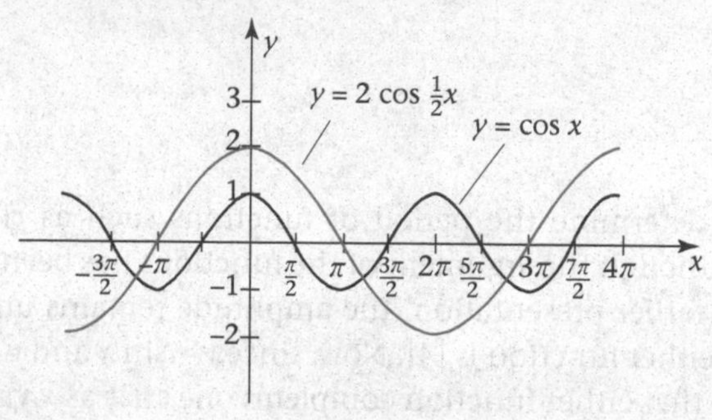

■

7.6c Phase Shift

We now examine the more general functions $f(x) = A \sin (Bx + C)$ and $f(x) = A \cos (Bx + C)$. As before, generalizing the argument of the trigonometric function does not affect the amplitude, which is $|A|$. Since $y = \sin x$ and $y = \cos x$ both complete one cycle as x varies from 0 to 2π, this leads to the equation

$$Bx + C = 0 \quad \text{in which case} \quad x = -\frac{C}{B}$$

and to the equation

$$Bx + C = 2\pi \quad \text{in which case} \quad x = \frac{2\pi - C}{B}$$

The value $-\frac{C}{B}$ is called the **phase shift**. Consider the function

$$f(x) = A \sin (Bx + C).$$

If $-\frac{C}{B} > 0$, shifting the graph of $y = A \sin Bx$ to the right $-\frac{C}{B}$ units yields the graph of $y = A \sin (Bx + C)$. If $-\frac{C}{B} < 0$, shifting the graph of $y = A \sin Bx$ to the left $\frac{C}{B}$ units yields the graph of $y = A \sin (Bx + C)$. Similar statements can be made about $y = A \cos Bx$ and $y = A \cos (Bx + C)$.

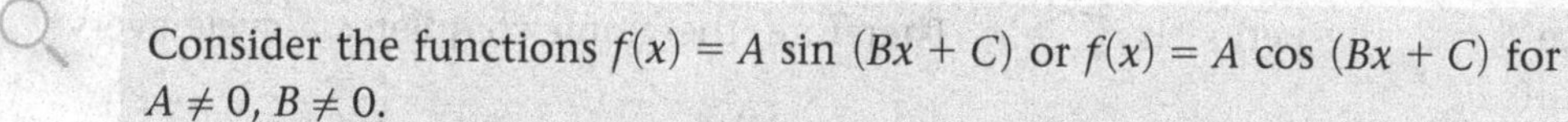

Consider the functions $f(x) = A \sin (Bx + C)$ or $f(x) = A \cos (Bx + C)$ for $A \neq 0, B \neq 0$.

a. The amplitude of $f(x)$ is $|A|$.

b. The period of $f(x)$ is $\frac{2\pi}{|B|}$.

c. The phase shift of $f(x)$ is $-\frac{C}{B}$.

EXAMPLE 5 *Graphing with a Phase Shift*

Sketch the graph $f(x) = 3\sin(2x - \pi)$.

SOLUTION

Step 1. Determine A, B, and C.	*Step 1.* Since $f(x) = 3\sin(2x - \pi) = A\sin(Bx + C)$ $A = 3$, $B = 2$, and $C = -\pi$
Step 2. Determine the amplitude, period, and phase shift.	*Step 2.* amplitude $= \lvert A \rvert = 3$ period $= \frac{2\pi}{B} = \frac{2\pi}{2} = \pi$ phase shift $= -\frac{C}{B} = \frac{\pi}{2}$ or, $2x - \pi = 0$ yields $x = \frac{\pi}{2}$ as the phase shift.
Step 3. Analyze the effect of the phase shift on the point $(0, 0)$.	*Step 3.* A phase shift of $\frac{\pi}{2}$ causes the cycle to "begin" at $\left(\frac{\pi}{2}, 0\right)$ rather than at $(0, 0)$.
Step 4. Use the period to determine the values of x at which a cycle is complete.	*Step 4.* Adding the period π to the phase shift $\frac{\pi}{2}$, we have $x = \frac{\pi}{2} + \pi = \frac{3\pi}{2}$ The graph completes a cycle in the interval $\left[\frac{\pi}{2}, \frac{3\pi}{2}\right]$
Step 5. Using the amplitude, sketch the graph.	*Step 5.* Recalling that the amplitude is 3, sketch the graph.

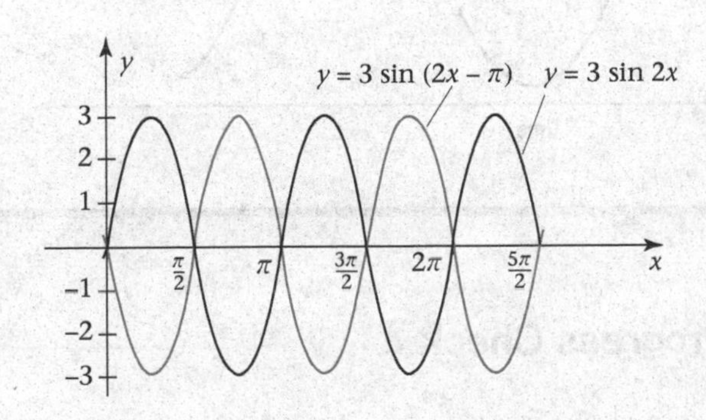

■

FOCUS *on Predator-Prey Interaction*

In the natural world, we frequently find that two plant or animal species interact in their environment in such manner that one species, the prey, serves as the primary food supply for the second species, the predator. Examples of such interaction are the relationships between trees (prey) and insects (predators) and between rabbits (prey) and lynxes (predators). As the population of the prey increases, the additional food supply results in an increase in the population of the predators. More predators consume more food, so the population of the prey decreases, which, in turn, leads, to a decrease in the population of the predators. The reduction in the predator population results in an increase in the number of prey and the cycle starts all over again.

The accompanying figure, adapted from *Mathematics: Ideas and Applications,* by Daniel D. Benice, Academic Press, 1978 (used with permission), shows the interaction between lynx and rabbit populations. Both curves demonstrate periodic behavior and can be described by trigonometric functions.

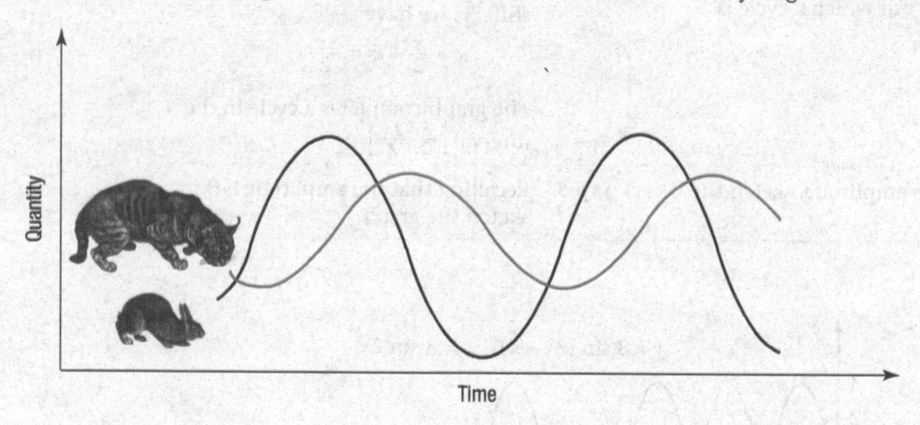

Progress Check

If $f(x) = 2\cos\left(2x + \frac{\pi}{2}\right)$, find the amplitude, period, and phase shift of f. Sketch the graph of the function.

Answers

The amplitude = 2, the period = π, and the phase shift = $-\frac{\pi}{4}$.

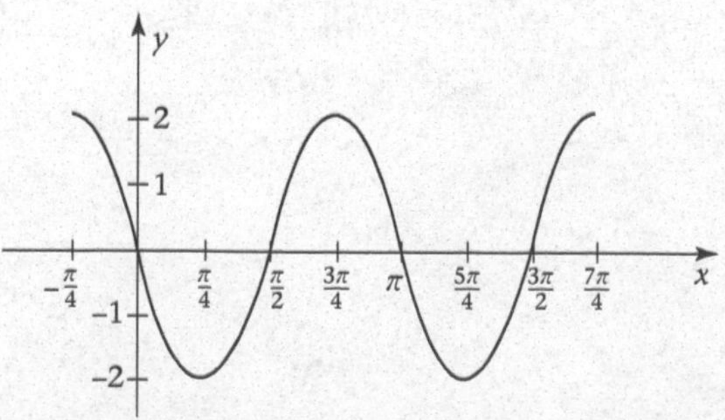

Exercise Set 7.6

In Exercises 1–12, determine the amplitude and period, and sketch the graph of each of the following functions.

1. $f(x) = 3 \sin x$
2. $f(x) = \frac{1}{4} \cos x$
3. $f(x) = \cos 4x$
4. $f(x) = \sin \frac{x}{4}$
5. $f(x) = -2 \sin 4x$
6. $f(x) = -\cos \frac{x}{4}$
7. $f(x) = 2 \cos \frac{x}{3}$
8. $f(x) = 4 \sin 4x$
9. $f(x) = \frac{1}{4} \sin \frac{x}{4}$
10. $f(x) = \frac{1}{2} \cos \frac{x}{4}$
11. $f(x) = -3 \cos 3x$
12. $f(x) = -2 \sin 3x$

In Exercises 13–20, for each given function, determine the amplitude, period, and phase shift. Sketch the graph of the function.

13. $f(x) = 2 \sin (x - \pi)$
14. $f(x) = \frac{1}{2} \cos \left(x + \frac{\pi}{2}\right)$
15. $f(x) = 3 \cos (2x - \pi)$
16. $f(x) = 4 \sin \left(x + \frac{\pi}{4}\right)$
17. $f(x) = \frac{1}{3} \sin \left(3x + \frac{3\pi}{4}\right)$
18. $f(x) = 2 \cos \left(2x + \frac{\pi}{2}\right)$
19. $f(x) = 2 \cos \left(\frac{x}{4} - \pi\right)$
20. $f(x) = 6 \sin \left(\frac{x}{2} + \frac{\pi}{2}\right)$

In Exercises 21–24, find a trigonometric function with the given graph. Verify your answer with your graphing calculator.

21.

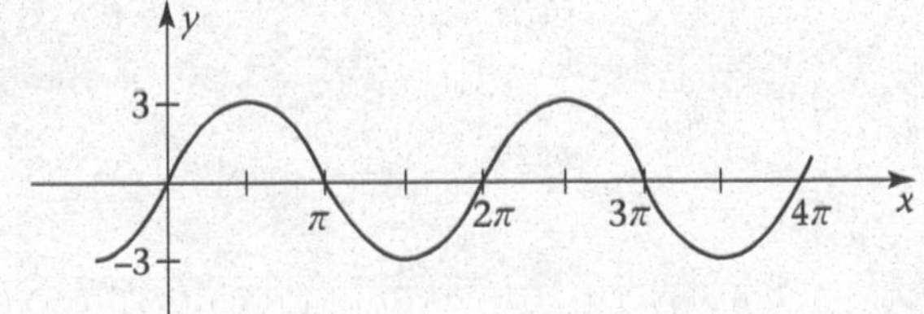

22.

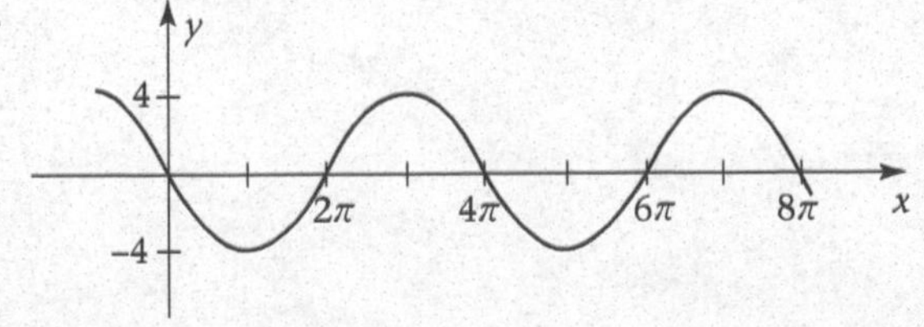

23.

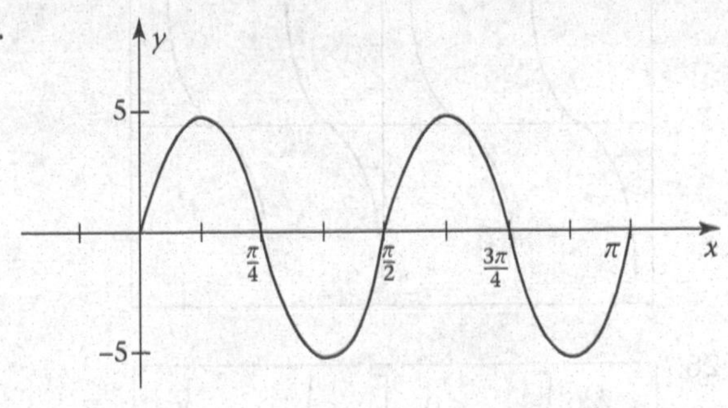

24.

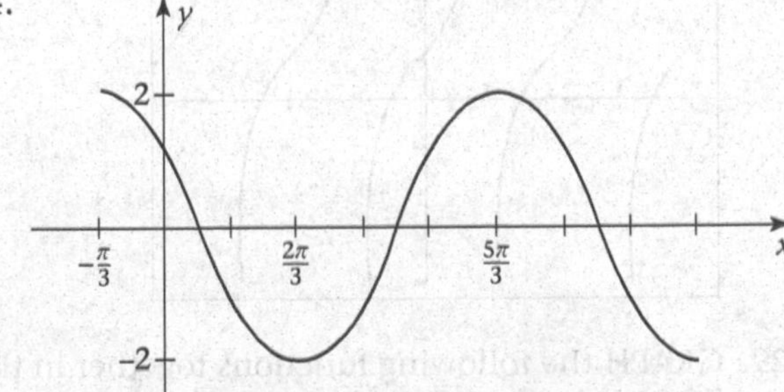

In Exercises 25–28, the graphs are variations of the function $y = \sin x$ or $y = \tan x$. Determine A, B, and C, where $y = A \sin (Bx + C)$ or $y = A \tan (Bx + C)$, for each graph below. Verify your answer by GRAPHing the function on your graphing calculator. (All graphs are drawn in the viewing rectangle $-6.28 \leq X \leq 6.28$, $-4 \leq Y \leq 4$, with XSCL = 1.57 and YSCL = 1.)

25.

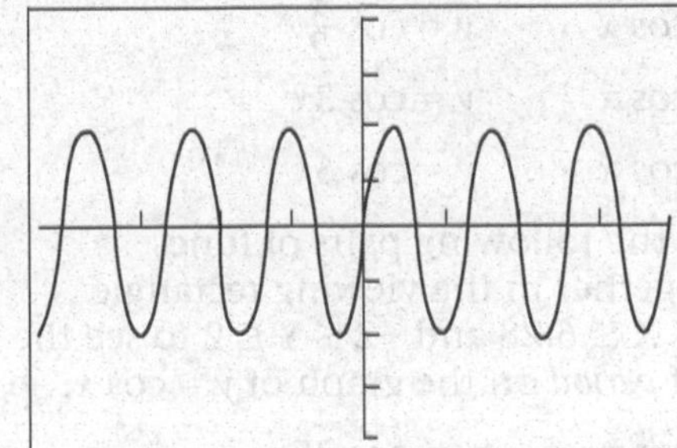

26.

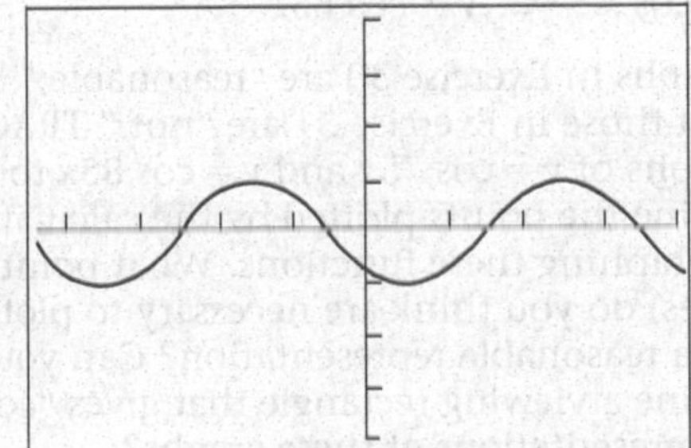

27.

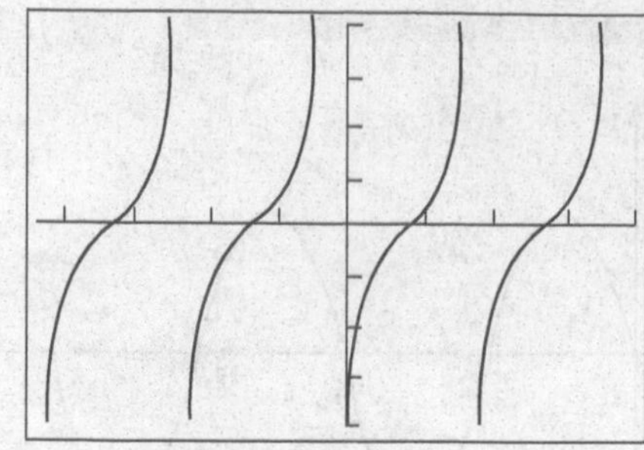

28.

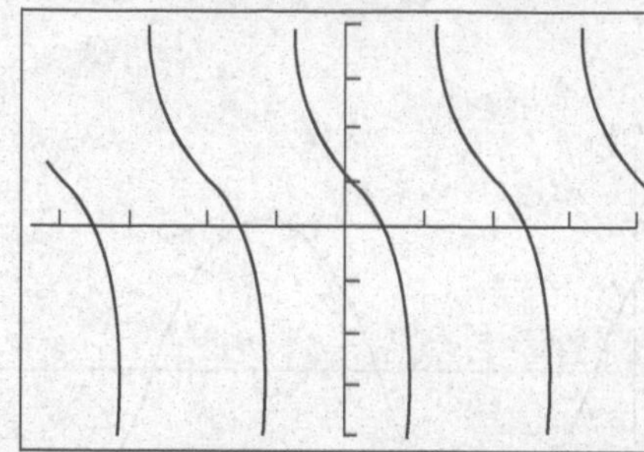

29. GRAPH the following functions together in the ZOOM standard viewing rectangle to see the effect of *amplitude* on the graph of $y = \sin x$.

a. $y = \sin x$ b. $y = 3 \sin x$

c. $y = 8 \sin x$ d. $y = -5 \sin x$

30. GRAPH the following pairs of functions together in the viewing rectangle $-6.28 \le X \le 6.28$ and $-2 \le Y \le 2$ to see the effect of *period* on the graph of $y = \cos x$.

a. $y = \cos x$ $y = \cos \frac{x}{2}$

b. $y = \cos x$ $y = \cos 3x$

c. $y = \cos x$ $y = \cos 5x$

31. GRAPH the following pairs of functions together in the viewing rectangle $-6.28 \le X \le 6.28$ and $-2 \le Y \le 2$ to see the effect of *period* on the graph of $y = \cos x$.

a. $y = \cos x$ $y = \cos 75x$

b. $y = \cos x$ $y = \cos 85x$

32. The graphs in Exercise 30 are "reasonable," but some of those in Exercise 31 are "not." TRACE the graphs of $y = \cos 75x$ and $y = \cos 85x$ to determine the points plotted by the calculator when graphing these functions. What points (x values) do you think are necessary to plot to obtain a reasonable representation? Can you determine a viewing rectangle that gives "correct" representations of these graphs?

In Exercises 33–35, GRAPH the following pairs of functions together in the viewing rectangle $-6.28 \le X \le 6.28$ and $-2 \le Y \le 2$ to see the effect of *phase shift* on the graph of $y = \sin x$. TRACE to determine the x-intercept of each shifted graph.

33. $y = \sin x$ $y = \sin (x - \pi)$

34. $y = \sin x$ $y = \sin \left(x + \frac{\pi}{3}\right)$

35. $y = \sin 2x$ $y = \sin \left(2x + \frac{\pi}{3}\right)$

36. GRAPH the following functions together in the viewing rectangle $-6.28 \le X \le 6.28$ and $-2 \le Y \le 2$.

a. $y = \sin x$ b. $y = x - \frac{x^3}{3!}$

c. $y = x - \frac{x^3}{3!} + \frac{x^5}{5!}$

d. $y = x - \frac{x^3}{3!} + \frac{x^5}{5!} - \frac{x^7}{7!} + \frac{x^9}{9!} - \frac{x^{11}}{11!}$

37. The polynomials in Exercise 36 are called Taylor Polynomials for $\sin x$. Note that the graphs of the Taylor Polynomials with higher degree coincide with the graph of $y = \sin x$ over a larger interval of the x-axis. Can you determine the smallest Taylor Polynomial for $\sin x$ that coincides with $y = \sin x$ on the interval $[-2\pi, 2\pi]$?

38. Taylor Polynomials for $\cos x$ are

$$1 - \frac{x^2}{2!}$$

$$1 - \frac{x^2}{2!} + \frac{x^4}{4!}$$

$$1 - \frac{x^2}{2!} + \frac{x^4}{4!} - \frac{x^6}{6!}$$

and so on. Compare the graphs of these polynomials with the graph of $y = \cos x$.

7.7 The Inverse Trigonometric Functions

Inverse functions were introduced in Section 3.5 and also discussed in Section 6.1. Furthermore, they were used to define the logarithmic function in Section 6.3. We have seen that if f is a one-to-one function whose domain is the set X and whose range is the set Y, then the inverse function f^{-1} reverses the correspondence, that is,

$$f^{-1}(y) = x \qquad \text{for all } y \text{ in } Y$$

if and only if

$$f(x) = y \qquad \text{for all } x \text{ in } X$$

Using this definition, we saw that the following identities characterize inverse functions.

$$f^{-1}[f(x)] = x \qquad \text{for all } x \text{ in } X$$
$$f[f^{-1}(y)] = y \qquad \text{for all } y \text{ in } Y$$

If we attempt to find an inverse of the sine function, we have an immediate problem. Since sine is a periodic function, it is not a one-to-one function and has no inverse. However, we can resolve this problem by defining a function that agrees with the sine function, but over a restricted domain. That is, we would like to find an interval such that $y = \sin x$ is one-to-one and y assumes all values between -1 and 1 over this interval. If we define the function f by

$$f(x) = \sin x, \qquad -\frac{\pi}{2} \le x \le \frac{\pi}{2}$$

then f assumes all real values in the interval $[-1, 1]$ as shown in Figure 7-53. We see that f is an increasing function and, therefore, one-to-one. Consequently, f has an inverse, and we are led to the following definition.

The inverse sine function, denoted by **arcsin**, or **sin^{-1}**, is defined by

$$\sin^{-1} x = y \qquad \text{if and only if} \qquad \sin y = x$$

where $-1 \le x \le 1$ and $-\frac{\pi}{2} \le y \le \frac{\pi}{2}$.

In words, the inverse sine of x takes on the value y, where the domain is $-1 \le x \le 1$ and the range is $-\frac{\pi}{2} \le y \le \frac{\pi}{2}$. Sometimes $y = \sin^{-1} x$ is read, "y equals the arcsine of x" or "y is the angle whose sine is x."

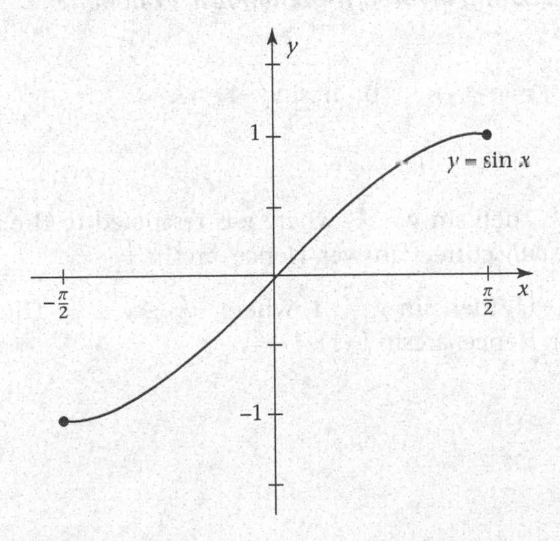

FIGURE 7-53
Graph of $y = \sin x$, $-\frac{\pi}{2} \le x \le \frac{\pi}{2}$

When we defined $\sin^n t = (\sin t)^n$, we said that this definition does not hold when $n = -1$. This allowed us to reserve the notation $\sin^{-1}$ for the inverse sine function. Therefore, $\sin^{-1} x$ is not to be confused with

$$(\sin x)^{-1} = \frac{1}{\sin x}$$

The notation arcsin and $\sin^{-1}$ are both in common use, and we will use them interchangeably. (Recall from our presentation of the unit circle that the coordinates of $P(t)$ are (x, y), where $y = \sin t$. Thus, t determines an *arc whose sine is y*, that is, $t = \arcsin y$.)

We now form a table of values and sketch the graph of $y = \sin^{-1} x$, as shown in Figure 7-54. Observe that the restricted graph of $y = \sin x$ from Figure 7-53 and the graph of $y = \sin^{-1} x$ from Figure 7-54 are reflections of one another about the line $y = x$.

x	y
-1	$-\frac{\pi}{2}$
$-\frac{\sqrt{3}}{2}$	$-\frac{\pi}{3}$
$-\frac{1}{2}$	$-\frac{\pi}{6}$
0	0
$\frac{1}{2}$	$\frac{\pi}{6}$
$\frac{\sqrt{3}}{2}$	$\frac{\pi}{3}$
1	$\frac{\pi}{2}$

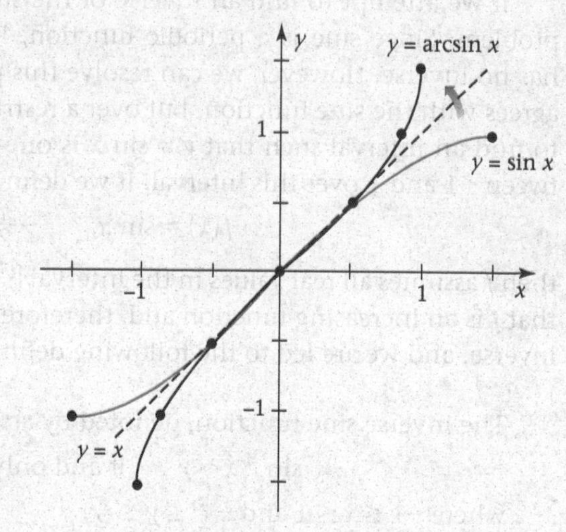

FIGURE 7-54
Graph of $y = \arcsin x$

EXAMPLE 1 *Evaluating Inverse Trigonometric Functions*

Find.

a. $\arcsin \frac{1}{2}$

b. $\arcsin(-1)$

SOLUTION

a. If $y = \arcsin \frac{1}{2}$, then $\sin y = \frac{1}{2}$, where y is restricted to the interval $[-\frac{\pi}{2}, \frac{\pi}{2}]$. Thus, $\frac{\pi}{6}$ is the *only* correct answer. Hence, $\arcsin \frac{1}{2} = \frac{\pi}{6}$.

b. If $y = \arcsin(-1)$, then $\sin y = -1$, where $-\frac{\pi}{2} \le y \le \frac{\pi}{6}$. Thus, $-\frac{\pi}{2}$ is the *only* correct answer. Hence, $\arcsin(-1) = -\frac{\pi}{2}$. ■

EXAMPLE 2 *Evaluating the Inverse Trigonometric Functions*

Evaluate $\sin^{-1}(\cos \frac{\pi}{6})$.

SOLUTION

Since $\cos \frac{\pi}{6} = \frac{\sqrt{3}}{2}$, we have

$$\sin^{-1}\left(\cos \frac{\pi}{6}\right) = \sin^{-1}\left(\frac{\sqrt{3}}{2}\right)$$

We let

$$y = \sin^{-1}\left(\frac{\sqrt{3}}{2}\right)$$

Then

$$\sin y = \frac{\sqrt{3}}{2} \qquad \text{where} \qquad -\frac{\pi}{2} \le y \le \frac{\pi}{2}$$

and

$$y = \frac{\pi}{3}$$

is the *only* solution. Therefore, $\sin^{-1}(\cos \frac{\pi}{6}) = \frac{\pi}{3}$ ■

Progress Check

Find.

a. $\sin^{-1}\left(-\frac{\sqrt{3}}{2}\right)$ b. $\arcsin\left(\tan \frac{5\pi}{4}\right)$

Answers

a. $-\frac{\pi}{3}$ b. $\frac{\pi}{2}$

We may use a similar approach to define the inverse cosine function. If we define the function f by

$$f(x) = \cos x, \qquad 0 \le x \le \pi$$

then f assumes all real values in the interval {−1, 1] as shown in Figure 7-55. We see that f is a decreasing function and, hence, one-to-one. Therefore, f has an inverse, and we have the following definition.

The inverse cosine function, denoted by **arccos**, or **$\cos^{-1}$**, is defined by

$$\cos^{-1} x = y \qquad \text{if and only if} \qquad \cos y = x$$

where $-1 \le x \le 1$ and $0 \le y \le \pi$.

We form a table of values and sketch the graph of $y = \cos^{-1} x$, as shown in Figure 7-56. Note that the graphs in Figure 7-55 and Figure 7-56 are reflections of each other about the line $y = x$.

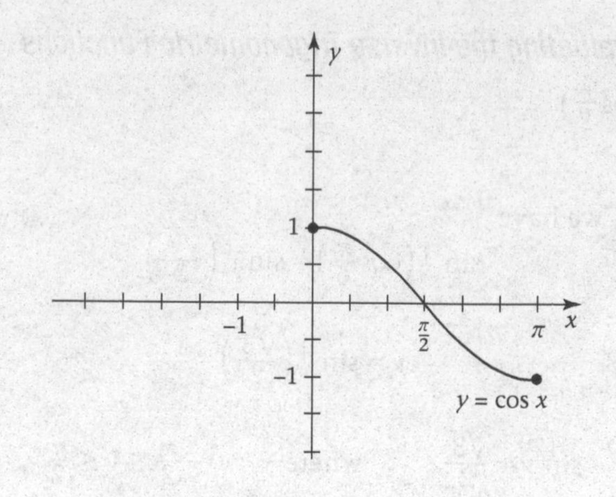

FIGURE 7-55
Graph of $y = \cos x$, $0 \le x \le \pi$

x	y
1	0
$\frac{\sqrt{3}}{2}$	$\frac{\pi}{6}$
$\frac{1}{2}$	$\frac{\pi}{3}$
0	$\frac{\pi}{2}$
$-\frac{1}{2}$	$\frac{2\pi}{3}$
$-\frac{\sqrt{3}}{2}$	$\frac{5\pi}{6}$
−1	π

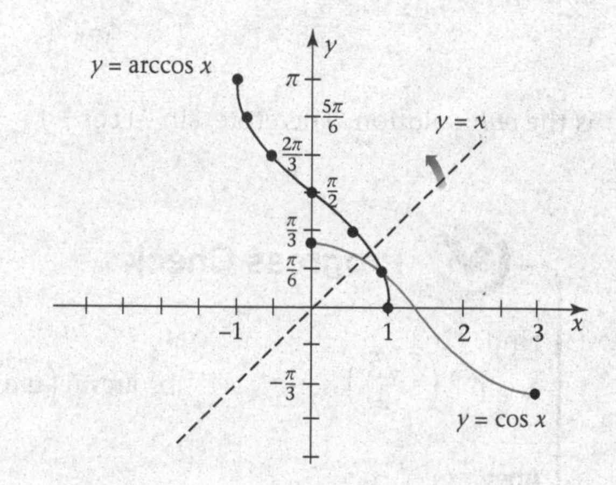

FIGURE 7-56
Graph of $y = \arccos x$

EXAMPLE 3 *Evaluating Inverse Trigonometric Functions*

Find.

a. $\cos^{-1}\left(-\frac{1}{2}\right)$ b. $\arccos\left(\sin\frac{\pi}{2}\right)$

SOLUTION

a. If $y = \cos^{-1}(-\frac{1}{2})$, then $\cos y = -\frac{1}{2}$, where y is restricted to the interval $[0, \pi]$. Consequently, $y = \frac{2\pi}{3}$ is the *only* correct answer. Thus, $\cos^{-1}(-\frac{1}{2}) = \frac{2\pi}{3}$.

b. Since $\sin\frac{\pi}{2} = 1$, we let $y = \arccos 1$. Then $\cos y = 1$, where $0 \le y \le \pi$. Therefore, $y = 0$ is the *only* correct answer. Thus, $\arccos(\sin\frac{\pi}{2}) = 0$. ■

We define the inverse tangent function with an appropriate restriction to the tangent function. Let

$$f(x) = \tan x, \qquad -\frac{\pi}{2} < x < \frac{\pi}{2}$$

Note that f assumes all real values, as shown in Figure 7-57. Here, f is increasing, so that it is one-to-one and has an inverse.

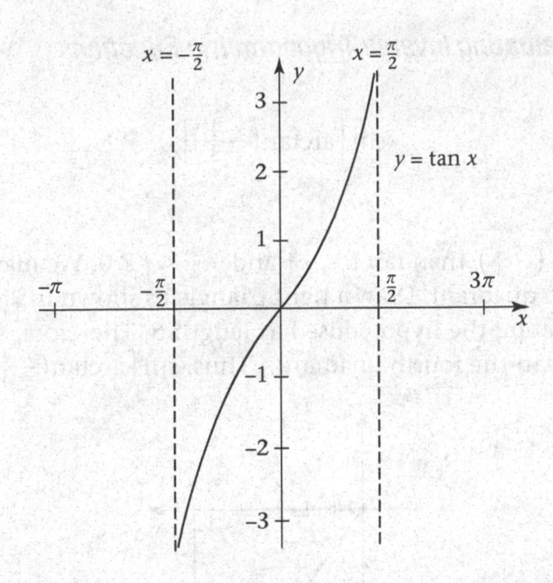

FIGURE 7-57
Graph of $y = \tan x$, $-\frac{\pi}{2} < x < \frac{\pi}{2}$

The inverse tangent function, denoted by **arctan**, or $\tan^{-1}$, is defined by

$$\tan^{-1} x = y \quad \text{if and only if} \quad \tan y = x$$

where $-\infty < x < \infty$ and $-\frac{\pi}{2} < y < \frac{\pi}{2}$.

Proceeding as before, we sketch the graph of $y = \tan^{-1} x$ in Figure 7-58.

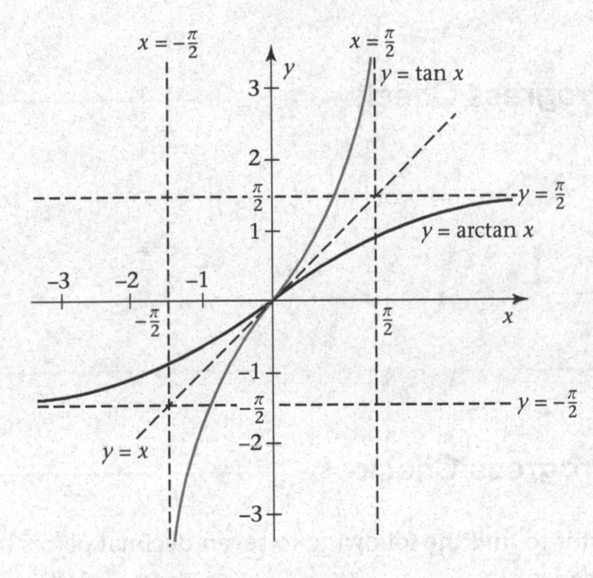

FIGURE 7-58
Graph of $y = \arctan x$

EXAMPLE 4 *Evaluating Inverse Trigonometric Functions*

Find.

a. $\tan^{-1} \sqrt{3}$ b. $\tan^{-1}(\cos \pi)$

SOLUTION

a. If $y = \tan^{-1} \sqrt{3}$, then $\tan y = \sqrt{3}$. Since $-\frac{\pi}{2} < y < \frac{\pi}{2}$, we must have $y = \frac{\pi}{3}$. Thus, $\tan^{-1} \sqrt{3} = \frac{\pi}{3}$.

b. Since $\cos \pi = -1$, we let $y = \tan^{-1}(-1)$. Then $\tan y = -1$, where $-\frac{\pi}{2} < y < \frac{\pi}{2}$. Therefore, $y = -\frac{\pi}{4}$. Thus, $\tan^{-1}(\cos \pi) = -\frac{\pi}{4}$. ■

EXAMPLE 5 *Evaluating Inverse Trigonometric Functions*

Find.

$$\sin\left[\arctan\left(-\tfrac{4}{3}\right)\right]$$

SOLUTION

If we let $t = \arctan\left(-\frac{4}{3}\right)$, then $\tan t = -\frac{4}{3}$ and $-\frac{\pi}{2} < t < 0$. We may think of t as an angle in the fourth quadrant. Draw a right triangle as shown in Figure 7-59. By the Pythagorean Theorem, the hypotenuse has length 5. Therefore, $\sin t = -\frac{4}{5}$. (Note that $\sin t < 0$ if t is in the fourth quadrant.) Thus, $\sin\left[\arctan\left(-\frac{4}{3}\right)\right] = -\frac{4}{5}$.

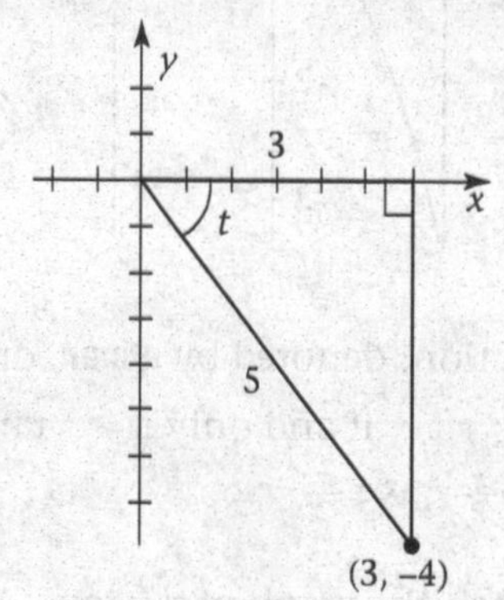

FIGURE 7-59
$t = \arctan\left(\frac{-4}{3}\right)$

■

Progress Check

Find

$$\cot\left[\sin^{-1}\left(-\tfrac{5}{13}\right)\right]$$

Answer

$-\frac{12}{5}$

Progress Check

Use a calculator to find the following to seven decimal places in radians.

a. $\sin^{-1}(-0.725)$ b. $\sec(\arcsin(-0.429))$

Answers

a. -0.8110344 b. 1.1070464

EXAMPLE 6 *Applying the Inverse Trigonometric Functions*

Find the solutions of $3 \sin x = 1$ in the interval $[0, \frac{\pi}{2}]$ to seven decimal places.

SOLUTION

First we obtain

$$\sin x = \frac{1}{3}$$

and then

$$x = \sin^{-1} \frac{1}{3} \approx 0.3398369$$

■

EXAMPLE 7 *Applying the Inverse Trigonometric Functions*

Find the solutions of $2 \sin x = \sqrt{2}$ in the interval $[\frac{\pi}{2}, \pi]$.

SOLUTION

Since

$$\sin x = \frac{\sqrt{2}}{2}$$

we have

$$x = \sin^{-1} \frac{\sqrt{2}}{2} = \frac{\pi}{4}$$

This is *not* the solution since $\frac{\pi}{4}$ does not lie in the interval $[\frac{\pi}{2}, \pi]$, namely, the second quadrant. However, observe that

$$\sin x = \frac{\sqrt{2}}{2}$$

is positive, and $\sin x$ is positive in both the first and second quadrants. Since $\frac{3\pi}{4}$ is in the second quadrant and has reference angle $\frac{\pi}{4}$, we see that $\frac{3\pi}{4}$ is the solution. Thus, $x = \frac{3\pi}{4}$. ■

EXAMPLE 8 *Applying the Inverse Trigonometric Functions*

Find the solutions of $5 \cos^2 x - 3 = 0$ in the interval $[0, \pi]$.

SOLUTION

We treat the equation as quadratic in cosine.

$$5 \cos^2 x = 3$$

$$\cos x = \pm\sqrt{\frac{3}{5}} = \pm\frac{\sqrt{15}}{5}$$

Thus

$$x = \cos^{-1} \frac{\sqrt{15}}{5} \quad \text{or} \quad x = \cos^{-1}\left(-\frac{\sqrt{15}}{5}\right)$$

Since $x = \cos^{-1} \frac{\sqrt{15}}{5} \approx 0.6847192$ and $x = \cos^{-1}\left(-\frac{\sqrt{15}}{5}\right) = 2.4568735$, both values lie in the interval $[0, \pi]$. Therefore, they are both solutions of the given equation. ■

Find the solutions of $2 \sin^2 x + 2 \sin x - 1 = 0$ that are in the interval $[-\frac{\pi}{2}, \frac{\pi}{2}]$.

Answer

$$\sin^{-1}\left(\frac{-1+\sqrt{3}}{2}\right) \approx 0.3747344$$

Note that $\sin^{-1}\left(\frac{-1-\sqrt{3}}{2}\right)$ does not exist.

It is important to remember the domain and range for each inverse trigonometric function. For example, given

$$y = \tan^{-1}(-1)$$

students often write $y = \frac{3\pi}{4}$, which is incorrect. Although $\tan \frac{3\pi}{4} = -1$, the correct answer is $y = -\frac{\pi}{4}$ since the answer must come from the interval $(-\frac{\pi}{2}, \frac{\pi}{2})$.

Although it is possible to define **csc**$^{-1}$ x, **sec**$^{-1}$ x, and **cot**$^{-1}$ x by similarly restricting the functions $\csc x$, $\sec x$, and $\cot x$, respectively, so that they are one-to-one, we will not discuss these functions here.

Exercise Set 7.7

In Exercises 1–18, evaluate the given expression without using a calculator.

1. $\sin^{-1}\left(-\frac{1}{2}\right)$
2. $\arccos \frac{\sqrt{3}}{2}$
3. $\arctan \sqrt{3}$
4. $\tan^{-1} 0$
5. $\arcsin\left(-\frac{\sqrt{2}}{2}\right)$
6. $\cos^{-1}(-1)$
7. $\arccos\left(-\frac{\sqrt{3}}{2}\right)$
8. $\tan^{-1}\left(\frac{\sqrt{3}}{3}\right)$
9. $\sin^{-1}(-1)$
10. $\arctan 1$
11. $\cos^{-1} 0$
12. $\sin^{-1}\left(-\frac{\sqrt{3}}{2}\right)$
13. $\cos^{-1} 1$
14. $\arcsin \frac{\sqrt{2}}{2}$
15. $\arctan(-1)$
16. $\sin^{-1} 0$
17. $\cos^{-1}\left(-\frac{1}{2}\right)$
18. $\arcsin \frac{1}{2}$

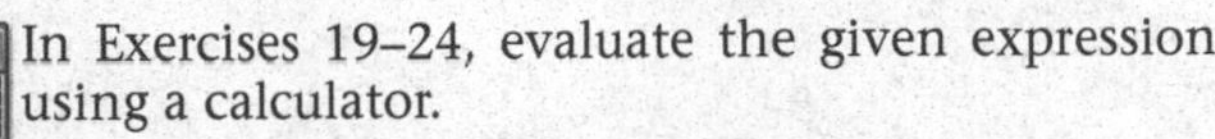

In Exercises 19–24, evaluate the given expression using a calculator.

19. $\sin^{-1} 0.3709$
20. $\arctan 1.398$
21. $\cos^{-1}(-0.7648)$
22. $\tan^{-1}(-3.010)$
23. $\arcsin 0.9636$
24. $\arccos(-0.921)$

In Exercises 25–36, evaluate the given expression without using a calculator.

25. $\sin(\arctan 1)$
26. $\cos\left[\arcsin\left(-\frac{1}{2}\right)\right]$
27. $\tan^{-1}\left(\cos\frac{\pi}{2}\right)$
28. $\sin^{-1}(\sin 0.62)$
29. $\cos^{-1}\left(\sin\frac{9\pi}{4}\right)$
30. $\tan(\sin^{-1} 0)$
31. $\cos^{-1}\left(\cos\frac{2\pi}{3}\right)$
32. $\sin^{-1}\left(\cos\frac{\pi}{6}\right)$
33. $\tan\left[\sin^{-1}\left(-\frac{5}{13}\right)\right]$
34. $\sin\left[\arctan\left(-\frac{12}{5}\right)\right]$
35. $\cos\left[\sin^{-1}\left(\frac{4}{5}\right)\right]$
36. $\tan\left[\cos^{-1}\left(-\frac{3}{5}\right)\right]$

In Exercises 37–44, solve the given equation on the given interval.

37. $\sin x = \frac{\sqrt{3}}{2}$, $\quad x \in \left[0, \frac{\pi}{2}\right]$
38. $\cos x = -\frac{1}{2}$, $\quad x \in [0, \pi]$
39. $\tan x = \frac{1}{\sqrt{3}}$, $\quad x \in \left[0, \frac{\pi}{2}\right]$
40. $2 \tan x = 2$, $\quad x \in [0, 2\pi]$
41. $2 \sin x = -1$, $\quad x \in [\pi, 2\pi]$
42. $\cos x = \frac{\sqrt{3}}{2}$, $\quad x \in [\pi, 3\pi]$
43. $\tan x = \sqrt{3}$, $\quad x \in \left[\frac{\pi}{2}, \frac{3\pi}{2}\right]$
44. $\tan x = -1$, $\quad x \in [-\pi, 2\pi]$

In Exercises 45–50, use the inverse trigonometric functions to express the solutions of the given equation exactly on the given interval.

45. $7 \sin^2 x - 1 = 0$, $\quad x \in \left[-\frac{\pi}{2}, \frac{\pi}{2}\right]$
46. $6 \cos^2 y - 5 = 0$, $\quad y \in [0, \pi]$
47. $12 \cos^2 x - \cos x - 1 = 0$, $\quad x \in [0, \pi]$
48. $2 \tan^2 t + 4 \tan t - 3 = 0$, $\quad t \in \left[-\frac{\pi}{2}, \frac{\pi}{2}\right]$
49. $9 \sin^2 t - 12 \sin t + 4 = 0$, $\quad t \in \left[-\frac{\pi}{2}, \frac{\pi}{2}\right]$
50. $3 \cos^2 x - 7 \cos x - 6 = 0$, $\quad x \in [0, \pi]$

In Exercises 51–54, provide a value for x to show that the equation is not an identity.

51. $\sin^{-1} x = \frac{1}{\sin x}$
52. $(\sin^{-1} x)^2 + (\cos^{-1} x)^2 = 1$
53. $\sin^{-1}(\sin x) = x$
54. $\arccos(\cos x) = x$

In Exercises 55–58, use a calculator to assist in finding all solutions in the indicated interval.

55. $2 \cos^2 x + \cos x = 2$, $\quad [0, \pi]$
56. $2 \tan^2 x = 3$, $\quad \left[0, \frac{\pi}{2}\right]$
57. $\sin^2 x - 2 \sin x - 2 = 0$, $\quad \left[-\frac{\pi}{2}, \frac{\pi}{2}\right]$
58. $9 \cos^2 x + 3 \cos x = 2$, $\quad [0, \pi]$

7.8 Applications

In Section 7.2, we introduced the expression "to solve a triangle," meaning to find the length of each side and the measure of each angle. In this section, we return to using the units of degrees to measure the size of the angles.

Many applied problems involve right triangles. We are now prepared to use our ability in solving triangles to tackle a variety of problems.

EXAMPLE 1 *Solving a Triangle*

In triangle ABC, $\gamma = 90^\circ$, $a = 12.8$, and $b = 22.5$. Find approximate values for the remaining parts of the triangle.

SOLUTION

The parts of the triangle are displayed in Figure 7-60. Using angle β, we have

$$\tan \beta = \frac{12.8}{22.5} \approx 0.5689$$

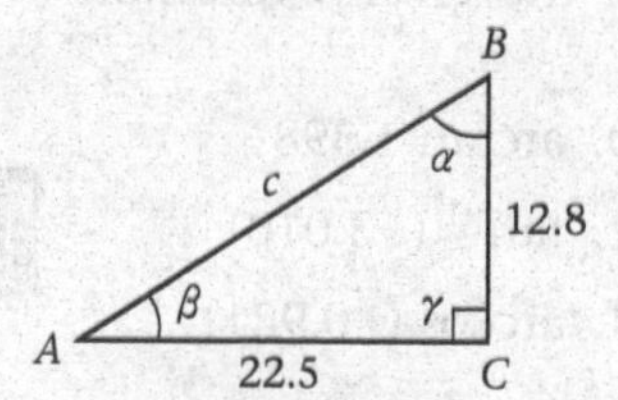

FIGURE 7-60
Diagram for Example 1

Solving for β,

$$\beta = \tan^{-1}\left(\frac{12.8}{22.5}\right) \approx 29.64^\circ$$

Sometimes, we want the answer in degrees and minutes.

$$(0.64^\circ)(60') \approx 38'$$

Therefore, $\beta = 29^\circ 38'$. Since $\alpha + \beta + \gamma = 180^\circ$, $\alpha \approx 60^\circ 22'$.

We can find c using the Pythagorean Theorem.

$$c = \sqrt{a^2 + b^2} = \sqrt{(12.8)^2 + (22.5)^2}$$

$$= \sqrt{163.84 + 506.25} = \sqrt{670.09} \approx 25.9$$

(Verify these values by finding $\sin \alpha$ two ways.) ■

Progress Check

In triangle ABC, $\gamma = 90^\circ$, $a = 17.4$ and $b = 38.2$. Solve the triangle.

Answer

$a \approx 24^\circ 29'$ $\quad b \approx 65^\circ 31'$ $\quad c \approx 42.0$

EXAMPLE 2 ***Applying Trigonometry***

A ladder leaning against a building makes an angle of 35° with the ground. If the bottom of the ladder is 5 meters from the building, how long is the ladder? To what height does it rise along the building?

SOLUTION

In Figure 7-61, we seek the length d of the ladder and the height h along the building. Using right triangle trigonometry,

$$\cos 35^\circ = \frac{5}{d} \qquad \text{and} \qquad \tan 35^\circ = \frac{h}{5}$$

$$d = \frac{5}{\cos 35^\circ} \qquad\qquad h = 5 \tan 35^\circ$$

$$d \approx 6.1 \text{ meters} \qquad \text{and} \qquad h \approx 3.5 \text{ meters}$$

The ladder is about 6.1 m long and it rises about 3.5 m along the building.

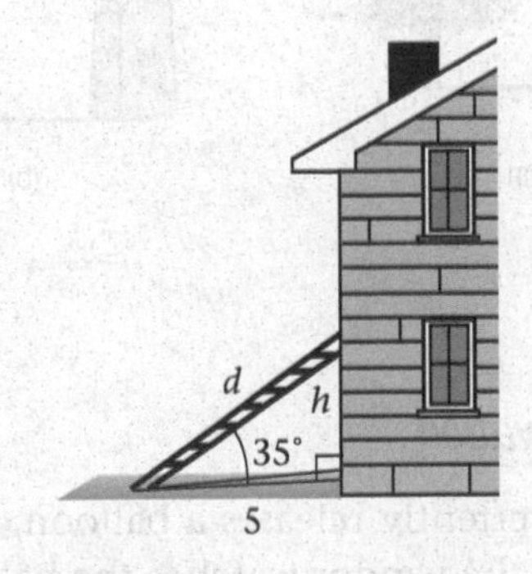

FIGURE 7-61
Diagram for Example 2

■

The string of a kite makes an angle of 32°30′ with the ground. If 125 meters of string have been let out, how high is the kite?

Answer

Approximately 67 meters

7.8a Elevation and Depression

There are two terms that will occur frequently in our word problems. The **angle of elevation** is the angle between the horizontal and the line of sight when looking up. In Figure 7-62(a), θ is the angle of elevation of the top T of a tree from a point x meters from the base of the tree.

The **angle of depression** is the angle between the horizontal and the line of sight when looking down. In Figure 7-62(b), θ is the angle of depression of a boat B as seen from a lighthouse L.

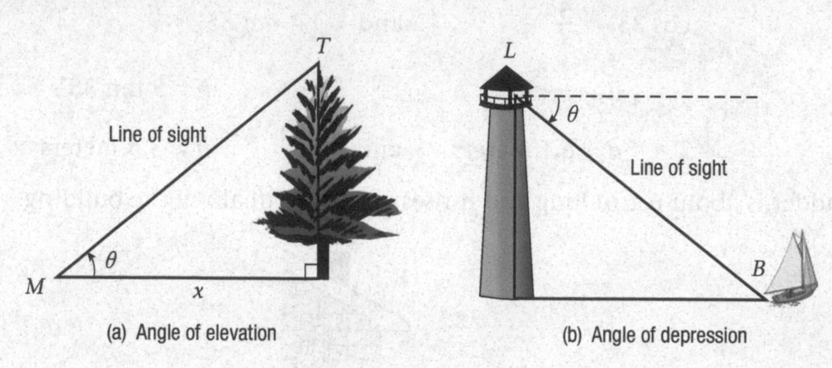

FIGURE 7-62
Angles of Elevation and Depression

EXAMPLE 3 *Angle of Elevation*

A vendor of balloons inadvertently releases a balloon, which rises straight up. A child standing 50 feet from the vendor watches the balloon rise. When the angle of elevation of the balloon reaches 44°, how high is the balloon?

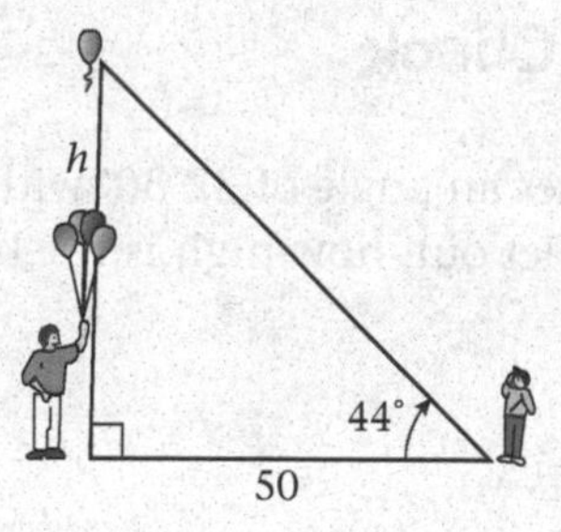

FIGURE 7-63
Diagram for Example 3

SOLUTION

We seek the height h in Figure 7-63. Thus,

$$\tan 44^\circ = \frac{h}{50}$$

$$h = 50 \tan 44^\circ$$

$$h \approx 48$$

The balloon has risen approximately 48 feet. ■

EXAMPLE 4 *Angle of Depression*

A forest ranger is in a tower 65 feet above the ground. If the ranger spots a fire at an angle of depression of 6°30′, how far is the fire from the base of the tower, assuming level terrain?

SOLUTION

We need to find the distance d as shown in the diagram below. Since $\theta + 6°30' = 90°$, $\theta = 83°30'$.

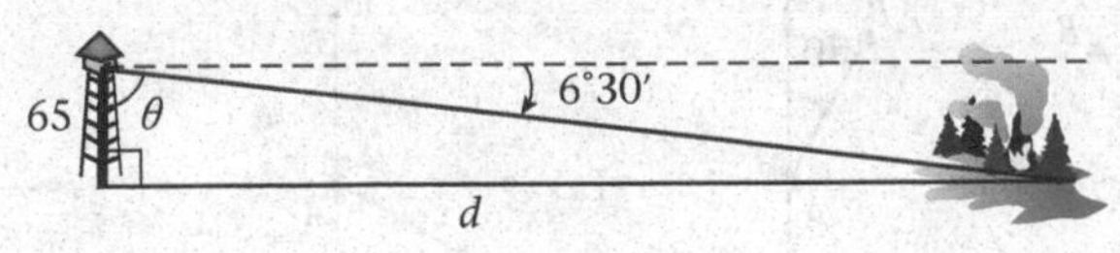

Then

$$\tan\theta = \frac{d}{65}$$
$$d = 65 \tan 83°30'$$
$$d = 65 \tan 83.5°$$
$$d \approx 570$$

The fire is approximately 570 feet from the base of the tower. ■

EXAMPLE 5 *Applying Trigonometry*

A mathematics professor walks toward the university clock tower on the way to her office and decides to find the height of the clock above ground. She determines the angle of elevation to be 30° and, after proceeding an additional 60 feet toward the base of the tower, finds the angle of elevation to be 40°. What is the height of the clock?

SOLUTION

As shown below, we seek to determine h. From triangle ACD,

$$\tan 30° = \frac{h}{d+60} \quad \text{or} \quad h = (d+60)(\tan 30°)$$

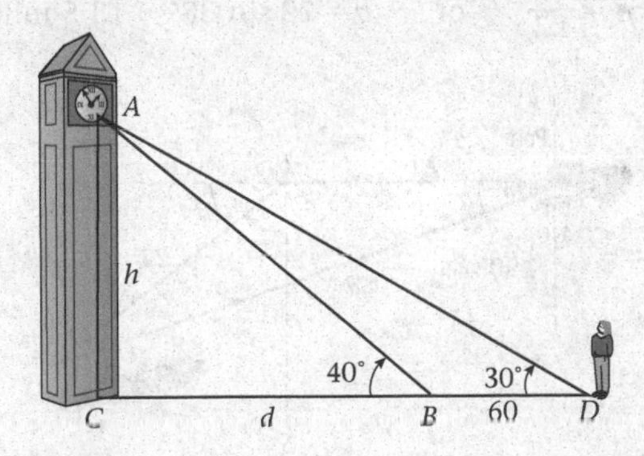

From triangle ACB,

$$\tan 40° = \frac{h}{d} \quad \text{or} \quad d = \frac{h}{\tan 40°}$$

Substituting for d in the first expression, we obtain

$$h = \left(\frac{h}{\tan 40°} + 60\right)\tan 30°$$
$$h \approx 0.6881\, h + 34.6410$$
$$0.3119\, h \approx 34.641$$
$$h \approx 111.1$$

The height of the clock is approximately 111.1 feet. ■

7.8b Navigation and Surveying

In navigation and surveying, directions are often given by **bearings**, which specify an acute angle and its direction from the north-south line. In Figure 7-64(a), the bearing of point B from point A is N 40° E, that is, 40° east of north. In Figure 7-64(b), the bearing of point B from point A is N 30° W. In Figure 7-64(c) and Figure 7-64(d), it is S 60° W and S 20° E, respectively.

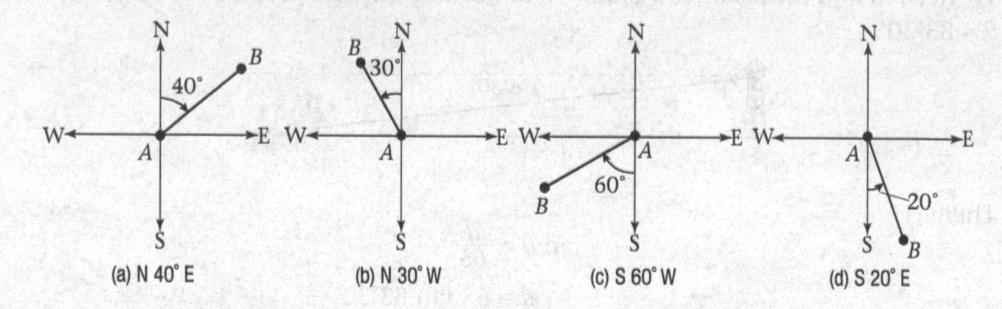

FIGURE 7-64
Bearings

EXAMPLE 6 *Bearings*

A ship leaves port at 10 a.m. and heads due east at a rate of 22 mph. At 11 a.m. the course is changed to S 52° E. Find the distance and bearing of the ship from the point of departure at noon.

SOLUTION

The ship travels due east from port, point A, reaches B at 11 a.m. and changes direction, arriving at E at noon, as shown in the diagram. Since the ship travels at 22 mph, $\overline{AB}$ and $\overline{BE}$ are each 22 miles in length. Further, since angle EBS has a measure of 52°, we find angle $\beta = 38°$. From right triangle BCE,

$$\cos\beta = \frac{e}{22} \quad \text{or} \quad e = 22\cos 38° \approx 17.3 \text{ miles}$$

$$\sin\beta = \frac{b}{22} \quad \text{or} \quad b = 22\sin 38° \approx 13.5 \text{ miles}$$

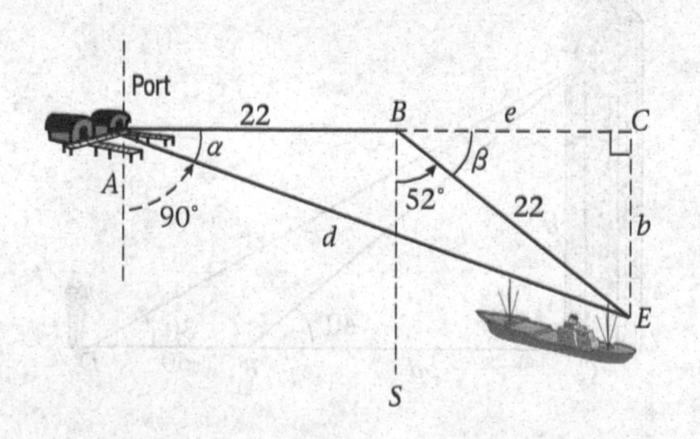

We now know two sides of right triangle ACE, namely

$$\overline{AC} = 22 + e \approx 22 + 17.3 = 39.3$$

$$\overline{CE} = b \approx 13.5$$

From the Pythagorean Theorem,

$$d = \sqrt{(\overline{AC})^2 + (\overline{CE})^2} \approx \sqrt{(39.3)^2 + (13.5)^2} = \sqrt{1726.74} \approx 41.6$$

To find the bearing,

$$\tan \alpha = \frac{\overline{CE}}{\overline{AC}} \approx \frac{13.5}{39.3}$$

so

$$\alpha = 19^\circ$$

Therefore, the ship is approximately 41.6 miles from port at a bearing of approximately S 71° E. ■

Exercise Set 7.8

In Exercises 1 and 2, find the required part of triangle ABC if $\gamma = 90^\circ$.

1. If $a = 12$ and $b = 16$, find α.
2. If $a = 5$ and $b = 15$, find β.
3. A ladder 20 feet in length touches a wall at a point 16 feet above the ground. Find the angle the ladder makes with the ground.
4. A monument is 550 feet high. What is the length of the shadow cast by the monument when the sun is 64° above the horizon?

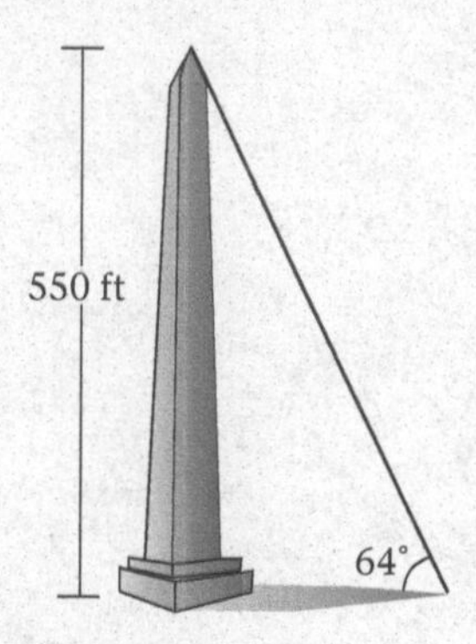

5. Find the angle of elevation of the sun when a tower 45 meters in height casts a horizontal shadow 25 meters in length.
6. A technician positioned on an oil-drilling rig 120 feet above the water spots a boat at an angle of depression of 16°. How far is the boat from the rig?
7. A mountainside hotel is located 8000 feet above sea level. From the hotel, a trail leads farther up the mountain to an inn at an elevation of 10,400 feet. If the trail has an angle of inclination of 18° (that is, the angle of elevation of the inn from the hotel is 18°), find the distance along the trail from the hotel to the inn.

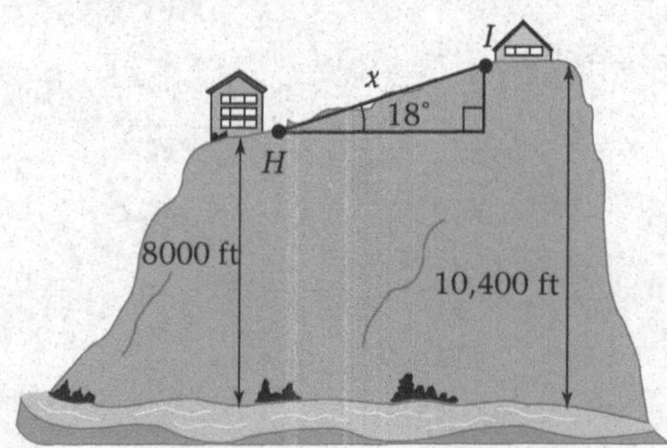

8. A hill is known to be 200 meters high. A surveyor standing on the ground finds the angle of elevation of the top of the hill to be 42°50′. Find the distance from the surveyor's feet to a point directly below the top of the hill.
9. An observer is 425 meters from a launching pad when a rocket is launched vertically. If the angle of elevation of the rocket at its apogee (highest point) is 66°20′, how high does the rocket rise?

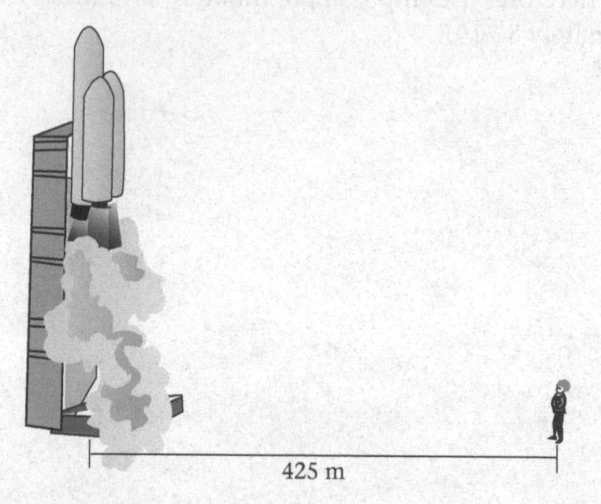

10. An airplane pilot wants to climb from an altitude of 6000 feet to an altitude of 16,000 feet. If the plane climbs at an angle of 9° with a constant speed of 22,000 feet per minute, how long will it take to reach the increased altitude?
11. A rectangle is 16 inches long and 13 inches wide. Find the measure of the angles formed by a diagonal with the sides.
12. The sides of an isosceles triangle are 15, 15, and 26 centimeters. Find the measures of the angles of the triangle. (*Hint*: The altitude of an isosceles triangle bisects the base.)
13. The side of a regular pentagon is 22 centimeters. Find the radius of the circle circumscribed about the pentagon. (*Hint*: The radii from the center of the circumscribed circle to any two adjacent vertices of the regular pentagon form an isosceles triangle. The altitude of an isosceles triangle bisects the base.)

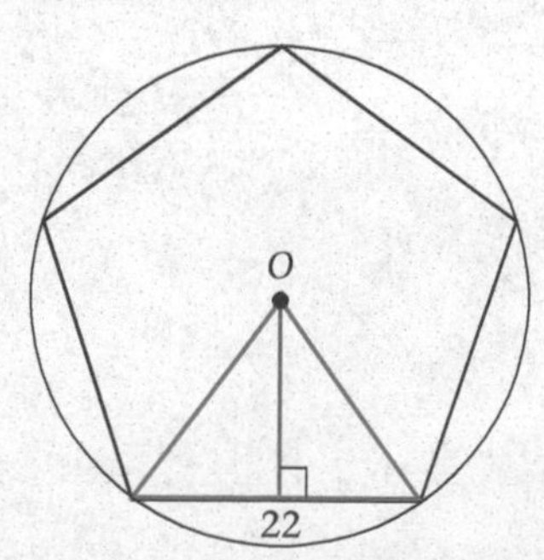

14. To determine the width of a river, markers are placed at each side of the river in line with the base of a tower that rises 23.4 meters above the ground. From the top of the tower, the angles of depression of the markers are 58°20′ and 11°40′. Find the width of the river.

15. The angle of elevation of the top of building B from the base of building A is 29°. From the top of building A, the angle of depression of the base of building B is 15°. If building B is 110 feet high, find the height of building A.

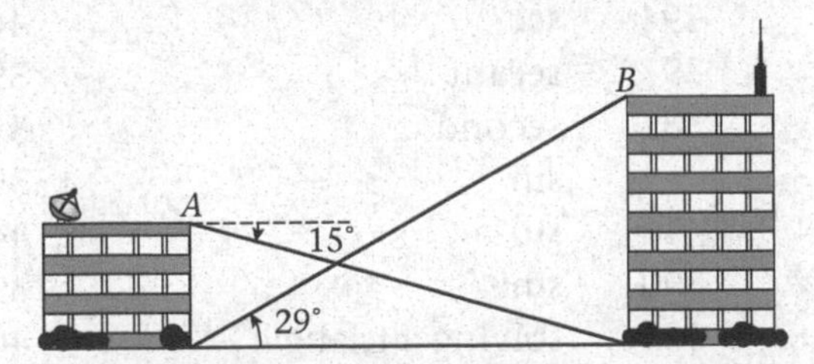

16. A ship leaves port at 2 p.m. and heads due east at a rate of 40 kilometers per hour. At 4 p.m. the course is changed to N 32° E. Find the distance and bearing of the ship from the point of departure at 6 p.m.

17. An attendant in a lighthouse receives a request for aid from a stalled craft located 15 miles due east of the lighthouse. The attendant contacts a second boat located 14 miles from the lighthouse at a bearing of N 23° W. What is the distance of the rescue ship from the stalled craft?

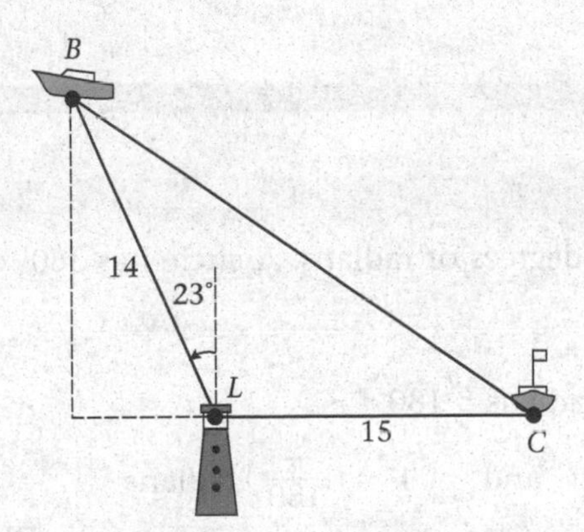

Chapter Summary

Key Terms, Concepts, and Symbols

Term	Page
acute angle	383
amplitude	434
angle	382
angle of depression	452
angle of elevation	452
approaches infinity	429
approaches negative infinity	429
arccos	443
arcsin	441
arctan	445
bearings	454
circular functions	404
cofunctions	400
complementary angles	400
cos	395
$\cos^{-1}$	443
cosecant	395
cosine	395
cot	395
$\cot^{-1}$	448
cotangent	395
coterminal	387
csc	395
$\csc^{-1}$	448
cycle	435
degree	383
even function	421
hypotenuse	394
identity	397
initial side of an angle	382
inverse trigonometric function	441
length of a circular arc	390
minute	383
negative angle	382
obtuse angle	383
odd function	421
period	427, 435
periodic function	427
phase shift	436
positive angle	382
$P(t)$	404
$P(x, y)$	404
quadrantal angle	382
radian	384
reference angle	388
reference right triangle	413
reflection	434
right angle	383
sec	395
$\sec^{-1}$	448
secant	395
second	383
sin	395
$\sin^{-1}$	441
sine	395
solving a triangle	400
standard position of an angle	382
tan	395
$\tan^{-1}$	445
tangent	395
terminal side of an angle	382
trigonometric functions	395
trigonometric identities	397
unit circle	404
unit reference right triangle	413

Key Ideas for Review

Topic	Page	Key Idea
Measurements of Angles	383	Angles are usually measured in degrees or radians. A circle has 360° or 2π radians.
Conversion Formulas	385	Since $$\pi \text{ radians} = 180^\circ$$ $$1 \text{ radian} = \left(\frac{180}{\pi}\right)^\circ \quad \text{and} \quad 1^\circ = \left(\frac{\pi}{180}\right) \text{ radians}$$
Reference Angle	387	The reference angle θ' associated with the angle θ is the acute angle formed by the terminal side of θ and the x-axis.
Right Triangle Trigonometry	394	Right triangle trigonometry relates a trigonometric function of an angle θ of a right triangle to the ratio of the lengths of two of its three sides: the *hypotenuse*, *opposite*, and *adjacent*. $$\sin\theta = \frac{\text{opposite}}{\text{hypotenuse}} \qquad \csc\theta = \frac{\text{hypotenuse}}{\text{opposite}}$$ $$\cos\theta = \frac{\text{adjacent}}{\text{hypotenuse}} \qquad \sec\theta = \frac{\text{hypotenuse}}{\text{adjacent}}$$ $$\tan\theta = \frac{\text{opposite}}{\text{adjacent}} \qquad \cot\theta = \frac{\text{adjacent}}{\text{opposite}}$$

hypotenuse
opposite
θ
adjacent

continues

Topic	Page	Key Idea
"Special" Right Triangles	398	Two triangles that can be helpful in determining special values of trigonometric functions are 2, 60°, 1, 30°, $\sqrt{3}$ $\sqrt{2}$, 45°, 1, 45°, 1
Points on the Unit Circle	405	For every real number t, there is a unique point $P(t)$ on the unit circle. If the rectangular coordinates of $P(t)$ are (x, y), we write $P(t) = P(x, y)$ We defined $\cos t = x$ and $\sin t = y$ so that $P(t) = P(\cos t, \sin t)$ Furthermore, $P(t) = P(t + 2\pi n)$ where n is any integer.
Points Not at the Origin	417	If P has coordinates (x, y) with $r = \sqrt{x^2 + y^2} \neq 0$, then $P(x, y) = P(r \cos \theta, r \sin \theta)$ and $\cos \theta = \frac{x}{r}, \; \sin \theta = \frac{y}{r}$
Trigonometric or Circular Functions	406	If $P(t)$ is a point on the unit circle, then $\tan t = \frac{\sin t}{\cos t} \qquad \cot t = \frac{\cos t}{\sin t}$ $\sec t = \frac{1}{\cos t} \qquad \csc t = \frac{1}{\sin t}$
Periodicity	427	All trigonometric functions are periodic. Both $\sin t$ and $\cos t$ have period 2π, whereas $\tan t$ has period π.
Signs	419	$\sin t$, $\cos t$, and $\tan t$, are all positive in quadrant I, $\sin t$ is positive in quadrant II, $\tan t$ is positive in quadrant III, and $\cos t$ is positive in quadrant IV.
Even and Odd Functions	421	Sine and tangent are odd functions, whereas cosine is an even function. $\sin(-t) = -\sin t$ $\cos(-t) = \cos t$ $\tan(-t) = -\tan t$
Identity	422	$\sin^2 t + \cos^2 t = 1$

continues

Topic	Page	Key Idea
Graphs of Sine and Cosine	427	The graph of $y = \sin x$ for $0 \le x \le 2\pi$ is 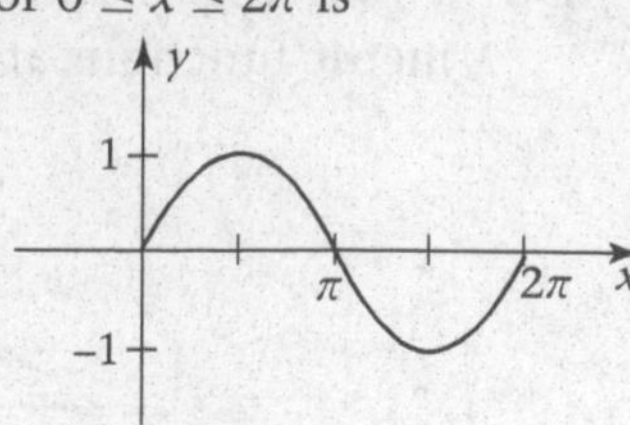 The graph of $y = \cos x$ for $0 \le x \le 2\pi$ is 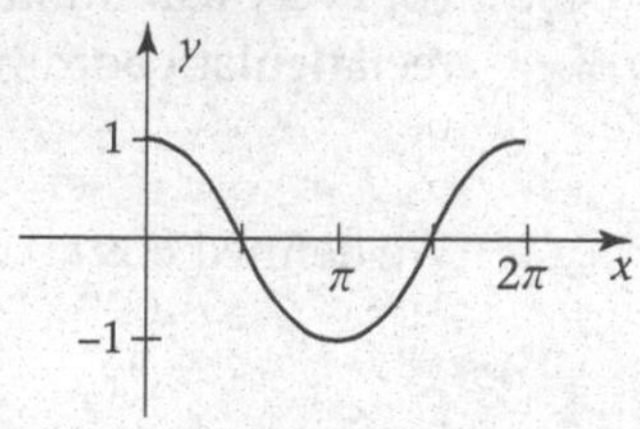
Amplitude, Period, and Phase Shift	436	To sketch the graph of $f(x) = A \sin (Bx + c)$ or $f(x) = A \cos (Bx + C)$ for $A \neq 0$, $B \neq 0$, note that 1. The amplitude is $\lvert A \rvert$. 2. The period is $\frac{2\pi}{\lvert B \rvert}$. 3. The phase shift is $-\frac{C}{B}$.
Inverse Trigonometric Functions	441	To define inverse trigonometric functions, it is necessary to restrict the domain of the trigonometric functions so that the resulting function is one-to-one.

Review Exercises

In Exercises 1–4, convert from degree measure to radian measure or from radian measure to degree measure.

1. $-60°$
2. $\frac{3\pi}{2}$
3. $-\frac{5\pi}{12}$
4. $45°$

In Exercises 5–7, determine if the pair of angles are coterminal.

5. $100°, \frac{5\pi}{9}$
6. $\frac{4\pi}{3}, 480°$
7. $\frac{5\pi}{4}, -135°$

In Exercises, 8–11, determine the quadrant in which t or θ lies.

8. $t = \frac{11\pi}{6}$
9. $\theta = -220°$
10. $\theta = 490°$
11. $t = -\frac{11\pi}{3}$

In Exercises 12–14, find the reference angle of the given angle.

12. $310°$
13. $-185°$
14. $405°$

15. If a central angle θ subtends an arc of length 14 centimeters on a circle whose radius is 10 centimeters, find the radian measure of θ.

16. A central angle of $\frac{2\pi}{3}$ radians subtends an arc of length $\frac{5\pi}{2}$ centimeters. Find the radius of the circle.

In Exercises 17–20, replace each given real number t by t', $0 \le t' < 2\pi$, so that $P(t') = P(t)$.

17. $\frac{9\pi}{2}$
18. $-\frac{15\pi}{2}$
19. -6π
20. $\frac{23\pi}{3}$

In Exercises 21–23, express the required trigonometric function as a ratio of the given parts of the right triangle ABC with $\gamma = 90°$.

21. If $a = 5$ and $b = 12$, find $\sin \alpha$.
22. If $a = 3$ and $c = 5$, find $\tan \beta$.
23. If $a = 4$ and $b = 7$, find $\sec \alpha$.

In Exercises 24–27, the point P lies on the terminal side of the angle θ. Find the value of the required trigonometric function without using a calculator.

24. $P(-\sqrt{3},1)$, $\csc\theta$
25. $P(\sqrt{2},-\sqrt{2})$, $\cot\theta$
26. $P(-1,-\sqrt{3})$, $\cos\theta$
27. $P(\sqrt{2},\sqrt{2})$, $\sin\theta$

In Exercises 28–32, find the rectangular coordinates of the given point without using a calculator.

28. $P\left(\frac{7\pi}{6}\right)$
29. $P\left(-\frac{8\pi}{3}\right)$
30. $P\left(\frac{5\pi}{6}\right)$
31. $P\left(-\frac{7\pi}{4}\right)$
32. $P\left(\frac{11\pi}{6}\right)$

In Exercises 33–36, determine the value of the indicated trigonometric function without using a calculator.

33. $\sin\frac{2\pi}{3}$
34. $\sec\left(-\frac{5\pi}{4}\right)$
35. $\tan\frac{5\pi}{6}$
36. $\csc\left(-\frac{\pi}{6}\right)$

In Exercises 37–41, $P(t) = \left(\frac{4}{5},-\frac{3}{5}\right)$. Use the symmetries of the circle to find the rectangular coordinates of the given point.

37. $P(t-\pi)$
38. $P\left(t+\frac{\pi}{2}\right)$
39. $P(-t)$
40. $P\left(t-\frac{\pi}{2}\right)$
41. $P(-t-\pi)$

In Exercises 42–45, find a value of t in the interval $[0, 2\pi)$ satisfying the given conditions.

42. $\sin t = -\frac{\sqrt{2}}{2}$, $P(t)$ in quadrant III
43. $\cos t = \frac{\sqrt{3}}{2}$, $P(t)$ in quadrant IV
44. $\cot t = \frac{\sqrt{3}}{3}$, $P(t)$ in quadrant I
45. $\sec t = -2$, $P(t)$ in quadrant II

In Exercises 46 and 47, find the quadrant in which t lies if the following conditions hold.

46. $\sin t < 0$ and $\cos t > 0$
47. $\sin(-t) > 0$ and $\tan t > 0$

In Exercises 48–51, use the trigonometric identities

$$\sin^2 t + \cos^2 t = 1 \qquad \tan t = \frac{\sin t}{\cos t}$$

to find the indicated value under the given conditions.

48. If $\cos t = \frac{3}{5}$ and $P(t)$ is in quadrant IV, find $\cot t$.
49. If $\sin t = -\frac{4}{5}$ and $\tan t > 0$, find $\sec t$.
50. If $\sin t = \frac{12}{13}$ and $\cos t < 0$, find $\tan t$.
51. If $\cos t = -\frac{5}{13}$ and $\tan t < 0$, find $\csc t$.

In Exercises 52 and 53, use the trigonometric identities to transform the first expression into the second.

52. $(\sin t)(\sec t)$, $\tan t$
53. $\frac{\sin t}{\cos^2 t}$, $(\tan t)(\sec t)$

In Exercises 54 and 55, use a calculator to evaluate the given expression.

54. $\cos 3.71 - \sin 1.44$
55. $\tan(-2.74)$

In Exercises 56 and 57, sketch the graph of the given function.

56. $f(x) = 1 - \sin x$
57. $f(x) = 2\sin\left(\frac{x}{2}+\pi\right)$

In Exercises 58–60, determine the amplitude, period and phase shift for each given function. Sketch the graph.

58. $f(x) = -\cos(2x-\pi)$
59. $f(x) = 4\sin\left(-x+\frac{\pi}{2}\right)$
60. $f(x) = -2\sin\left(\frac{x}{3}+\frac{\pi}{3}\right)$

In Exercises 61–63, evaluate the given expression.

61. $\arcsin\left(-\frac{1}{2}\right)$
62. $\tan(\cos^{-1} 1)$
63. $\tan(\tan^{-1} 5)$

64. Use the inverse cosine function to express the exact solutions of the equation
$$5\cos^2 x - 4 = 0$$

In Exercises 65–68, find the required part of triangle ABC with $\gamma = 90°$.

65. If $a = 50$ and $b = 60$, find α.
66. If $a = 40$ and $\beta = 20°$, find b.
67. If $a = 20$ and $\alpha = 52°$, find c.
68. If $b = 15$ and $\alpha = 25°$, find c.

69. A ladder 6 meters in length leans against a vertical wall. If the ladder makes an angle of 65° with the ground, find the height that the ladder reaches above the ground.

70. Find the angle of elevation of the sun when a tree 25 meters in height casts a horizontal shadow 10 meters in length.

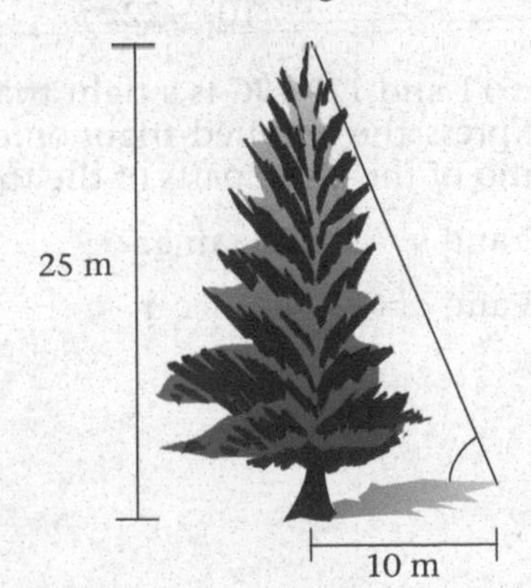

71. A rectangle is 22 centimeters long and 16 centimeters wide. Find the measure of the smaller angle formed by the diagonal with a side.

In Exercises 72 and 73, the following graphs are variations of the function $y = \cos x$. Determine A, B, and C, where $y = A \cos(Bx + C)$ for each graph below. Verify your answer by GRAPHing the function on your graphing calculator. (All graphs are drawn in the viewing rectangle $-6.28 \le X \le 6.28$, $-4 \le Y \le 4$, with XSCL = 1.57 and YSCL = 1.)

72.

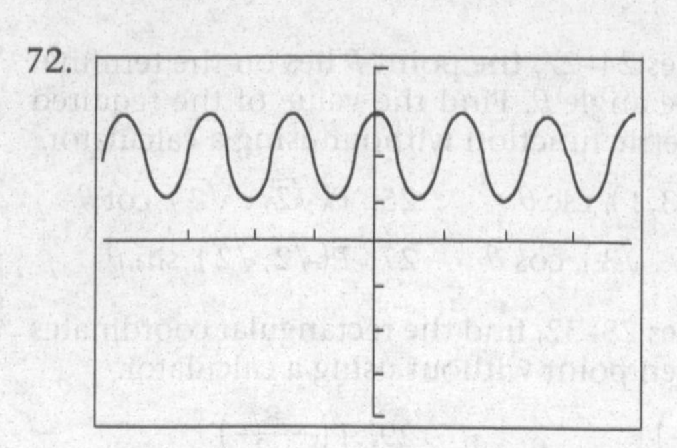

73.

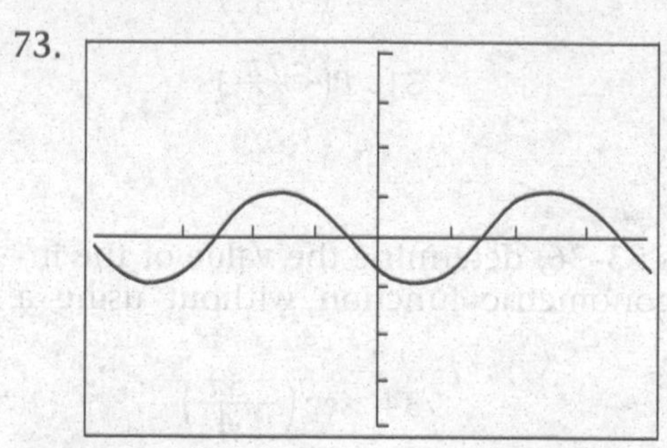

Review Test

In Exercises 1–3, convert from degree measure to radian measure or from radian measure to degree measure.

1. $\frac{5\pi}{3}$ 2. -200°

3. 75°

In Exercises 4 and 5, find the angle θ, $0^\circ \le \theta < 360^\circ$, that is coterminal with the given angle.

4. -25° 5. $\frac{17\pi}{4}$

In Exercises 6 and 7, find the reference angle of the given angle.

6. 160° 7. $\frac{7\pi}{4}$

8. If a central angle θ subtends an arc of length 12 inches on a circle whose radius is 15 inches, find the radian measure of θ.

In Exercises 9 and 10, replace the given real number t by t', $0 \le t' < 2\pi$, so that $P(t') = P(t)$.

9. $\frac{19\pi}{3}$ 10. -22π

In Exercises 11 and 12, ABC is a right triangle with $\gamma = 90^\circ$. Express the required trigonometric function as a ratio of the given parts of the triangle.

11. If $a = 7$ and $b = 5$, find $\tan \alpha$.

12. If $b = 5$ and $c = 15$, find $\sec \alpha$.

In Exercises 13–15, the point P lies on the terminal side of the angle θ. Find the value of the required trigonometric function without using a calculator.

13. $P(-\sqrt{2}, \sqrt{2})$, $\cot \theta$ 14. $P(0, -5)$, $\sin \theta$

15. $P(2, 2\sqrt{3})$, $\sec \theta$

In Exercises 16 and 17, find the rectangular coordinates of the given point.

16. $P\left(\frac{29\pi}{6}\right)$ 17. $P\left(-\frac{\pi}{3}\right)$

In Exercises 18–20, $P(t) = \left(-\frac{5}{13}, \frac{12}{13}\right)$. Use the symmetries of the circle to find the rectangular coordinates of the given point.

18. $P(t + \pi)$ 19. $P\left(t - \frac{\pi}{2}\right)$

20. $P(-t)$

In Exercises 21 and 22, determine the value of the indicated trigonometric function without using a calculator.

21. $\cos \frac{7\pi}{3}$ 22. $\csc\left(-\frac{2\pi}{3}\right)$

In Exercises 23 and 24, find a value of t in the interval $[0, 2\pi)$ satisfying the given conditions.

23. $\tan t = 1$, $P(t)$ in quadrant III

24. $\sec t = \sqrt{2}$, $P(t)$ in quadrant IV

In Exercises 25 and 26, use the trigonometric identities

$$\sin^2 t + \cos^2 t = 1 \qquad \tan t = \frac{\sin t}{\cos t}$$

to find the indicated value under the given conditions.

25. If $\cos t = -\frac{12}{13}$ and $\tan t > 0$, find $\sin t$.

26. If $\sin t = \frac{3}{5}$ and $P(t)$ is in quadrant II, find $\sec t$.

27. Use the trigonometric identities given for Exercises 25 and 26 to transform

$$1 - \tan x \qquad \text{to} \qquad \frac{\cos x - \sin x}{\cos x}$$

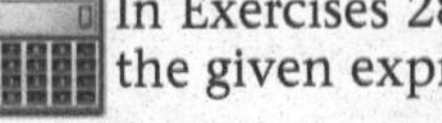

In Exercises 28 and 29, use a calculator to evaluate the given expression.

28. $\tan(-3.68)$

29. $\cos 1.15 - \sin 0.72$

30. Sketch the graph of the function f defined by

$$f(x) = x + \cos x$$

In Exercises 31 and 32, determine the amplitude, period, and phase shift of each given function.

31. $f(x) = -2\cos(\pi - x)$

32. $f(x) = 2\sin\left(\frac{x}{2} - \frac{\pi}{2}\right)$

In Exercises 33 and 34, evaluate the given expression.

33. $\tan^{-1}(-\sqrt{3})$

34. $\cos\left(\sin^{-1}\frac{\sqrt{3}}{2}\right)$

35. Use the inverse tangent function to express the exact solutions of the equation

$$6\tan^2 x - 13\tan x + 6 = 0$$

where $x \in \left(-\frac{\pi}{2}, \frac{\pi}{2}\right)$

In Exercises 36–38, find the required part of triangle ABC with $\gamma = 90°$.

36. If $a = 25$ and $c = 30$, find α.

37. If $b = 20$ and $\alpha = 32°$, find c.

38. If $a = 15$ and $b = 20$, find β.

39. From the top of a cliff 100 meters in height, the angle of depression of the entrance to a castle is 36°. Find the distance of the castle from the base of the cliff.

8 Analytic Trigonometry

Much of the language and terminology of algebra carries over to trigonometry. For example, we have seen that algebraic expressions involve variables, constants, and algebraic operations. **Trigonometric expressions** involve these same elements but also permit trigonometric functions of variables and constants. They also allow algebraic operations upon these trigonometric functions. Thus,

$$x + \sin x \qquad \sin x + \tan x \qquad \frac{1 - \cos x}{\sec^2 x}$$

are all examples of trigonometric expressions.

The distinction between an identity and an equation also carries over to trigonometry. Thus, a **trigonometric identity** is true for all values that may be assumed by the variable, but a **trigonometric equation** is true only for certain values of the variable, called *solutions*. (Note that the solutions of a trigonometric equation may be expressed as real numbers or as angles.) As usual, the set of all solutions of a trigonometric equation is called the *solution set*.

8.1 Trigonometric Identities and Their Verification

8.1a Fundamental Identities

In Section 7.4, we established the identity

$$\sin^2 t + \cos^2 t = 1 \tag{1}$$

If $\cos t \neq 0$, we may divide both sides of Equation (1) by $\cos^2 t$ to obtain

$$\frac{\sin^2 t}{\cos^2 t} + \frac{\cos^2 t}{\cos^2 t} = \frac{1}{\cos^2 t}$$

or

$$\tan^2 t + 1 = \sec^2 t \tag{2}$$

Similarly, if $\sin t \neq 0$, dividing Equation (1) by $\sin^2 t$ yields

$$\frac{\sin^2 t}{\sin^2 t} + \frac{\cos^2 t}{\sin^2 t} = \frac{1}{\sin^2 t}$$

or

$$\cot^2 t + 1 = \csc^2 t \tag{3}$$

Observe that $\tan t$ and $\sec t$ are undefined for exactly those values of t for which $\cos t = 0$. Similarly, $\cot t$ and $\csc t$ are undefined for those values of t for which $\sin t = 0$. It follows that Equations (2) and (3) are also identities. These identities, together with those discussed in Section 7.4, are called the **fundamental identities**. We summarize them in Table 8-1.

TABLE 8-1 Fundamental Identities

$\tan t = \frac{\sin t}{\cos t}$	$\sin^2 t + \cos^2 t = 1$	$\tan t = \frac{1}{\cot t}$
$\cot t = \frac{\cos t}{\sin t}$	$\tan^2 t + 1 = \sec^2 t$	$\sin^2 t = 1 - \cos^2 t$
$\csc t = \frac{1}{\sin t}$	$\cot^2 t + 1 = \csc^2 t$	$\cos^2 t = 1 - \sin^2 t$
$\sec t = \frac{1}{\cos t}$	$\sin t = \frac{1}{\csc t}$	$\tan^2 t = \sec^2 t - 1$
$\cot t = \frac{1}{\tan t}$	$\cos t = \frac{1}{\sec t}$	$\cot^2 t = \csc^2 t - 1$

The fundamental identities can be used to simplify trigonometric expressions as well as to verify trigonometric identities. Such manipulations may enable us to see relationships that would otherwise be obscured.

The preferred method of verifying an identity is to transform one side of the equation into the other. Although we will use this method whenever practical, we recognize that it is also acceptable to transform each side independently, with the hope of arriving at the same expression and then showing that the process can be reversed. This is often the technique used when both sides of the equation involve complicated expressions.

Unfortunately, we cannot present a specific set of steps that will "always work" to transform one side into the other. In fact, there are often many ways to handle a given identity. We will, however, demonstrate different techniques that we have found to be of value. We list the following suggestions on how to proceed, with examples to follow.

1. Factoring may help to simplify an expression.
2. It is often helpful to write all of the trigonometric functions in terms of sine and cosine.
3. Consider beginning with the more complicated expression and perform some of the indicated operations.
4. If you have the ratio of two trigonometric functions, it may be worthwhile to multiply both numerator and denominator by some trigonometric expression to obtain forms such as $1 - \sin^2 \theta$, $1 - \cos^2 \theta$, or $\sec^2 \theta - 1$, which can be further simplified.

EXAMPLE 1 *Using Trigonometric Identities*

Simplify the expression $\sin^2 x + \sin^2 x \tan^2 x$.

SOLUTION

We begin by noting that $\sin^2 x$ appears in both terms, which suggests that we factor.

$$
\begin{aligned}
\sin^2 x + \sin^2 x \tan^2 x &= \sin^2 x\,(1 + \tan^2 x) && \text{Factoring} \\
&= \sin^2 x \sec^2 x && 1 + \tan^2 x = \sec^2 x \\
&= \sin^2 x \left(\frac{1}{\cos^2 x}\right) && \sec x = \frac{1}{\cos x} \\
&= \tan^2 x && \frac{\sin x}{\cos x} = \tan x
\end{aligned}
$$

■

Progress Check

Simplify the expression $\dfrac{\csc\theta}{1+\cot^2\theta}$.

Answer

$\sin \theta$

EXAMPLE 2 *Verifying an Identity*

Verify the identity

$$\sin \alpha - \sin^2 \alpha = \frac{1-\sin\alpha}{\csc\alpha}$$

SOLUTION

$$
\begin{aligned}
\sin \alpha - \sin^2 \alpha &= \sin \alpha\,(1 - \sin \alpha) \\
&= \frac{1-\sin\alpha}{\csc\alpha}
\end{aligned}
$$

■

Verify the identity

$$\frac{\sin^2 y - 1}{1 - \sin y} = -1 - \sin y$$

When verifying identities, we are trying to show that equality holds. Since adding, subtracting, multiplying, and dividing both sides on an equation are properties that only hold for equality, we cannot use those when verifying identities.

EXAMPLE 3 ***Verifying an Identity***

Verify the identity $\cos x \tan x \csc x = 1$.

SOLUTION

$$\begin{aligned}\cos x \tan x \csc x &= \cos x \left(\frac{\sin x}{\cos x}\right)\left(\frac{1}{\sin x}\right)\\ &= 1\end{aligned}$$

■

Verify the identity $\sin x \sec x = \tan x$.

EXAMPLE 4 ***Verifying an Identity***

Verify the identity

$$\frac{1}{1-\sin x} + \frac{1}{1+\sin x} = 2\sec^2 x$$

SOLUTION

We begin with the left-hand side, combining fractions.

$$\begin{aligned}\frac{1}{1-\sin x} + \frac{1}{1+\sin x} &= \frac{1+\sin x + 1 - \sin x}{(1-\sin x)(1+\sin x)}\\ &= \frac{2}{1-\sin^2 x}\\ &= \frac{2}{\cos^2 x}\\ &= 2\sec^2 x\end{aligned}$$

■

Verify the identity $\cos x + \tan x \sin x = \sec x$.

EXAMPLE 5 ***Verifying an Identity***

Verify the identity

$$\frac{\cos\theta}{1-\sin\theta} = \sec\theta + \tan\theta$$

SOLUTION

To obtain the expression $1 - \sin^2\theta$ in the denominator, we multiply numerator and denominator by $1 + \sin\theta$.

$$\begin{aligned}
\frac{\cos\theta}{1-\sin\theta} &= \left(\frac{\cos\theta}{1-\sin\theta}\right)\left(\frac{1+\sin\theta}{1+\sin\theta}\right) \\
&= \frac{\cos\theta(1+\sin\theta)}{1-\sin^2\theta} \\
&= \frac{\cos\theta(1+\sin\theta)}{\cos^2\theta} \\
&= \frac{1+\sin\theta}{\cos\theta} \\
&= \frac{1}{\cos\theta} + \frac{\sin\theta}{\cos\theta} \\
&= \sec\theta + \tan\theta
\end{aligned}$$

■

Progress Check

Verify the identity

$$\frac{1+\cos t}{\sin t} + \frac{\sin t}{1+\cos t} = 2\csc t$$

EXAMPLE 6 ***Verifying an Identity***

Verify the identity

$$\frac{\cot u - \tan u}{\sin u \cos u} = \csc^2 u - \sec^2 u$$

SOLUTION

We transform both sides of the equation by writing all trigonometric functions in terms of sine and cosine. For the left-hand side, we have

$$\begin{aligned}
\frac{\cot u - \tan u}{\sin u \cos u} &= \frac{\dfrac{\cos u}{\sin u} - \dfrac{\sin u}{\cos u}}{\sin u \cos u} \\
&= \frac{\cos^2 u - \sin^2 u}{\sin^2 u \cos^2 u}
\end{aligned}$$

and for the right-hand side, we have

$$\begin{aligned}
\csc^2 u - \sec^2 u &= \frac{1}{\sin^2 u} - \frac{1}{\cos^2 u} \\
&= \frac{\cos^2 u - \sin^2 u}{\sin^2 u \cos^2 u}
\end{aligned}$$

We have successfully transformed both sides of the equation into the same expression. Since all the steps are reversible, we have verified the identity. ■

Progress Check

Verify the identity

$$\frac{\sin x + \cos x}{\tan^2 x - 1} = \frac{\cos^2 x}{\sin x - \cos x}$$

Exercise Set 8.1

In Exercises 1–46, verify each of the identities.

1. $\csc\gamma - \cos\gamma\cot\gamma = \sin\gamma$
2. $\cot x\sec x = \csc x$
3. $\sec v + \tan v = \dfrac{1+\sin v}{\cos v}$
4. $\cos\theta + \tan\theta\sin\theta = \sec\theta$
5. $\sin\alpha\sec\alpha = \tan\alpha$
6. $\sec\beta - \cos\beta = \sin\beta\tan\beta$
7. $3 - \sec^2 x = 2 - \tan^2 x$
8. $1 - 2\sin^2 t = 2\cos^2 t - 1$
9. $\dfrac{\sec^2 y}{\tan y} = \tan y + \cot y$
10. $\dfrac{\sin x + \cos x}{\cos x} = 1 + \tan x$
11. $\dfrac{\sin u}{\csc u} + \dfrac{\cos u}{\sec u} = 1$
12. $\dfrac{\tan^2\alpha}{1+\sec\alpha} = \sec\alpha - 1$
13. $\dfrac{\sec^2\theta - 1}{\sec^2\theta} = \sin^2\theta$
14. $\sin^4 x + 2\sin^2 x\cos^2 x + \cos^4 x = 1$
15. $\cos\gamma + \cos\gamma\tan^2\gamma = \sec\gamma$
16. $\dfrac{1}{\tan u + \cot u} = \cos u\sin u$
17. $\dfrac{\sec w\sin w}{\tan w + \cot w} = \sin^2 w$
18. $(1 - \cos^2\beta)(1 + \cot^2\beta) = 1$
19. $(\sin\alpha + \cos\alpha)^2 + (\sin\alpha - \cos\alpha)^2 = 2$
20. $\dfrac{1+\tan^2 u}{\csc^2 u} = \tan^2 u$
21. $\sec^2 v + \cos^2 v = \dfrac{\sec^4 v + 1}{\sec^2 v}$
22. $\sin^2\theta - \tan^2\theta = -\tan^2\theta\sin^2\theta$
23. $\dfrac{\sin^2\alpha}{1+\cos\alpha} = 1 - \cos\alpha$
24. $\cot x\sin^2 x = \cos x\sin x$
25. $\dfrac{\cos t}{1+\sin t} = \dfrac{1-\sin t}{\cos t}$
26. $\dfrac{\sin\beta}{1+\cos\beta} + \dfrac{1+\cos\beta}{\sin\beta} = 2\csc\beta$
27. $\csc^2\theta - \dfrac{\cos^2\theta}{\sin^2\theta} = 1$
28. $\dfrac{\cos^2 u}{1-\sin u} = 1 + \sin u$
29. $\dfrac{\cot y}{1+\cot^2 y} = \sin y\cos y$
30. $\dfrac{1+\tan^2 x}{\tan^2 x} = \csc^2 x$
31. $\cos(-t)\csc(-t) = -\cot t$
32. $\sin(-\theta)\sec(-\theta) = -\tan\theta$
33. $\dfrac{\sec x + \csc x}{1+\tan x} = \csc x$
34. $\dfrac{\sec u}{\sec u - 1} = \dfrac{1}{1-\cos u}$
35. $\dfrac{1+\tan x}{1+\cot x} = \dfrac{\sec x}{\csc x}$
36. $(\tan u + \sec u)^2 = \dfrac{1+\sin u}{1-\sin u}$
37. $\dfrac{1-\sin t}{1+\sin t} = (\sec t - \tan t)^2$
38. $2\csc^2\theta - \csc^4\theta = 1 - \cot^4\theta$
39. $\dfrac{\sin^2 w}{\cos^4 w + \cos^2 w\sin^2 w} = \tan^2 w$
40. $\dfrac{\sin z + \tan z}{1+\cos z} = \tan z$
41. $\dfrac{\sec\gamma - \csc\gamma}{\sec\gamma + \csc\gamma} = \dfrac{\tan\gamma - 1}{\tan\gamma + 1}$
42. $\dfrac{\cot x - 1}{1 - \tan x} = \dfrac{\csc x}{\sec x}$
43. $\dfrac{\tan\gamma - \sin\gamma}{\tan\gamma} = \dfrac{\sin^2\gamma}{1+\cos\gamma}$
44. $\cos^4 u - \sin^4 u = \cos^2 u - \sin^2 u$
45. $\dfrac{\csc x}{1+\csc x} - \dfrac{\csc x}{1-\csc x} = 2\sec^2 x$
46. $\sin^3\theta + \cos^3\theta = (1 - \sin\theta\cos\theta)(\sin\theta + \cos\theta)$

In Exercises 47–52, show that each of the equations is not an identity by finding a value of the variable for which the equation is not true.

47. $\sin x = \sqrt{1-\cos^2 x}$
48. $\tan x = \sqrt{\sec^2 x - 1}$
49. $(\sin t + \cos t)^2 = \sin^2 t + \cos^2 t$
50. $\sin\theta + \cos\theta = \sec\theta + \csc\theta$
51. $\sqrt{\cos^2 x} = \cos x$
52. $\sqrt{\cot^2 x} = \cot x$

8.2 The Addition and Subtraction Formulas

The identities that we verified in the examples and exercises of Section 8.1 were, in general, of no special significance. We were primarily interested in demonstrating manipulation with the fundamental identities. There are, however, many trigonometric identities that are indeed of importance. These identities are called **trigonometric formulas.**

Our objective in this section is to develop the **addition formulas** for sin $(s + t)$, cos $(s + t)$, and tan $(s + t)$, as well as the **subtraction formulas** for sin $(s - t)$, cos $(s - t)$, and tan $(s - t)$.

We begin with the derivation of cos $(s - t)$. For convenience, assume that s, t, and $s - t$ are all positive and less than 2π. Let $P_1 = P(s)$, $Q_1 = P(t)$, and $P_2 = P(s - t)$ be the points on the unit circle determined by s, t, and $s - t$, respectively, as shown in Figure 8-1. Then $\overset{\frown}{Q_2P_2Q_1P_1} = s$, $\overset{\frown}{Q_2P_2Q_1} = t$, and $\overset{\frown}{Q_2P_2} = s - t$. From Section 7.3, we know that the coordinates of P_1, Q_1, and P_2 are $P_1 = P(\cos s, \sin s)$, $Q_1 = P(\cos t, \sin t)$, and $P_2 = P(\cos (s - t), \sin (s - t))$. Since the length of the arcs $\overset{\frown}{Q_1P_1}$ and $\overset{\frown}{Q_2P_2}$ are both equal (length of $s - t$), the length of the chords $\overline{Q_1P_1}$ and $\overline{Q_2P_2}$ are also equal. By the distance formula, we have

$$\overline{Q_1P_1} = \overline{Q_2P_2}$$

$$\sqrt{(\cos s - \cos t)^2 + (\sin s - \sin t)^2} = \sqrt{[\cos(s-t) - 1]^2 + [\sin(s-t) - 0]^2}$$

Squaring both sides and rearranging terms, we have

$$\sin^2 s + \cos^2 s + \sin^2 t + \cos^2 t - 2\cos s \cos t - 2 \sin s \sin t$$
$$= \sin^2 (s - t) + \cos^2 (s - t) - 2\cos (s - t) + 1$$

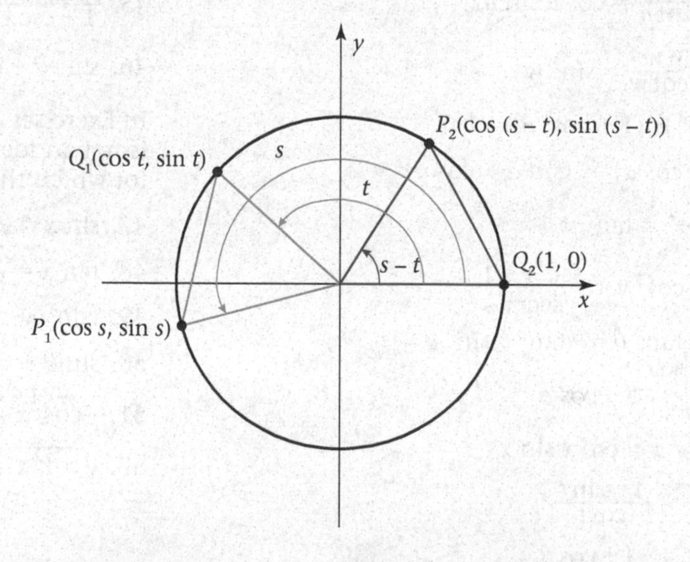

FIGURE 8-1
The Derivation of cos $(s - t)$

Since $\sin^2 s + \cos^2 s = 1$, $\sin^2 t + \cos^2 t = 1$ and $\sin^2 (s - t) + \cos^2 (s - t) = 1$, we have

$$2 - 2\cos s \cos t - 2 \sin s \sin t = 2 - 2\cos (s - t)$$

Solving for cos $(s - t)$ yields the formula

$$\cos (s - t) = \cos s \cos t + \sin s \sin t \qquad (1)$$

We obtain the formula for cos $(s + t)$ by writing

$$s + t = s - (-t)$$

Therefore

$$\begin{aligned}\cos(s+t) &= \cos(s-(-t))\\ &= \cos s\cos(-t) + \sin s\sin(-t)\end{aligned}$$

Since $\cos(-t) = \cos t$ and $\sin(-t) = -\sin t$,

$$\cos(s+t) = \cos s\cos t - \sin s\sin t \qquad (2)$$

EXAMPLE 1 *Using the Subtraction Formula*

Find $\cos 15^\circ$ without using a calculator.

SOLUTION

Since $15^\circ = 45^\circ - 30^\circ$, we may use the formula for $\cos(s-t)$ to obtain

$$\begin{aligned}\cos 15^\circ &= \cos(45^\circ - 30^\circ)\\ &= \cos 45^\circ\cos 30^\circ + \sin 45^\circ \sin 30^\circ\\ &= \frac{\sqrt{2}}{2}\cdot\frac{\sqrt{3}}{2} + \frac{\sqrt{2}}{2}\cdot\frac{1}{2}\\ &= \frac{\sqrt{6}+\sqrt{2}}{4}\end{aligned}$$

■

Redo Example 1 using $15^\circ = 60^\circ - 45^\circ$.

EXAMPLE 2 *Using the Addition Formula*

Find the exact value of $\cos\frac{5\pi}{12}$.

SOLUTION

We note that $\frac{5\pi}{12} = \frac{2\pi}{12} + \frac{3\pi}{12} = \frac{\pi}{6} + \frac{\pi}{4}$. Then

$$\begin{aligned}\cos\frac{5\pi}{12} &= \cos\left(\frac{\pi}{6}+\frac{\pi}{4}\right)\\ &= \cos\frac{\pi}{6}\cos\frac{\pi}{4} - \sin\frac{\pi}{6}\sin\frac{\pi}{4}\\ &= \frac{\sqrt{3}}{2}\cdot\frac{\sqrt{2}}{2} - \frac{1}{2}\cdot\frac{\sqrt{2}}{2}\\ &= \frac{\sqrt{6}-\sqrt{2}}{4}\end{aligned}$$

■

Redo Example 2 using the identity $\frac{5\pi}{12} = \frac{9\pi}{12} - \frac{4\pi}{12}$.

We may now establish the following relationships:

Cofunctions

$$\cos\left(\frac{\pi}{2}-t\right)=\sin t \qquad (3)$$

$$\sin\left(\frac{\pi}{2}-t\right)=\cos t \qquad (4)$$

$$\tan\left(\frac{\pi}{2}-t\right)=\cot t \qquad (5)$$

As we stated in Section 7.2, sine and cosine are **cofunctions**, as are tangent and cotangent, and secant and cosecant. We now verify them for all real numbers or angles here.

Using the subtraction formula for cosine, we have

$$\begin{aligned}\cos\left(\frac{\pi}{2}-t\right)&=\cos\frac{\pi}{2}\cos t+\sin\frac{\pi}{2}\sin t\\&=0\cdot\cos t+1\sin t\\&=\sin t\end{aligned}$$

which establishes the identity in Equation (3). Replacing t with $\frac{\pi}{2}-t$ in this identity yields

$$\begin{aligned}\cos\left[\frac{\pi}{2}-\left(\frac{\pi}{2}-t\right)\right]&=\sin\left(\frac{\pi}{2}-t\right)\\\cos t&=\sin\left(\frac{\pi}{2}-t\right)\end{aligned}$$

which establishes the identity in Equation (4). The third identity follows from the definition of tangent and from Equations (3) and (4).

$$\tan\left(\frac{\pi}{2}-t\right)=\frac{\sin\left(\frac{\pi}{2}-t\right)}{\cos\left(\frac{\pi}{2}-t\right)}=\frac{\cos t}{\sin t}=\cot t$$

We will now prove the identity in Equation (6). (We leave the proof of Equation (7) as an exercise.)

$$\sin(s+t)=\sin s\cos t+\cos s\sin t \qquad (6)$$

$$\sin(s-t)=\sin s\cos t-\cos s\sin t \qquad (7)$$

From Equation (3),

$$\begin{aligned}\sin(s+t)&=\cos\left[\frac{\pi}{2}-(s+t)\right]\\&=\cos\left[\left(\frac{\pi}{2}-s\right)-t\right]\\&=\cos\left(\frac{\pi}{2}-s\right)\cos t+\sin\left(\frac{\pi}{2}-s\right)\sin t\\&=\sin s\cos t+\cos s\sin t\end{aligned}$$

We conclude with the proof of the identity in Equation (8). (We leave the proof of Equation (9) as an exercise.)

$$\tan(s+t)=\frac{\tan s+\tan t}{1-\tan s\tan t} \qquad (8)$$

$$\tan(s-t)=\frac{\tan s-\tan t}{1+\tan s\tan t} \qquad (9)$$

$$\tan(s+t) = \frac{\sin(s+t)}{\cos(s+t)}$$
$$= \frac{\sin s \cos t + \cos s \sin t}{\cos s \cos t - \sin s \sin t}$$
$$= \frac{\left(\frac{\sin s}{\cos s} \cdot \frac{\cos t}{\cos t}\right) + \left(\frac{\cos s}{\cos s} \cdot \frac{\sin t}{\cos t}\right)}{\left(\frac{\cos s}{\cos s} \cdot \frac{\cos t}{\cos t}\right) - \left(\frac{\sin s}{\cos s} \cdot \frac{\sin t}{\cos t}\right)}$$
$$= \frac{\tan s + \tan t}{1 - \tan s \tan t}$$

EXAMPLE 3 ***Applying the Addition Formula***

Show that $\sin\left(x + \frac{3\pi}{2}\right) = -\cos x$.

SOLUTION

$$\sin\left(x + \frac{3\pi}{2}\right) = \sin x \cos \frac{3\pi}{2} + \cos x \sin \frac{3\pi}{2}$$
$$= (\sin x)\, 0 + (\cos x)(-1)$$
$$= -\cos x$$

■

Progress Check

Verify that $\tan(x - \pi) = \tan x$.

EXAMPLE 4 ***Using the Addition Formula***

Given $\sin \alpha = -\frac{4}{5}$, with α an angle in quadrant III, and $\cos \beta = -\frac{5}{13}$, with β an angle in quadrant II, use the addition formula to find $\sin(\alpha + \beta)$ and the quadrant in which $\alpha + \beta$ lies.

SOLUTION

The addition formula

$$\sin(\alpha + \beta) = \sin \alpha \cos \beta + \cos \alpha \sin \beta$$

requires that we know $\sin \alpha$, $\cos \alpha$, $\sin \beta$, and $\cos \beta$. Using the fundamental identity $\sin^2 \alpha + \cos^2 \alpha = 1$, we have

$$\cos^2 \alpha = 1 - \sin^2 \alpha = 1 - \frac{16}{25} = \frac{9}{25}$$

Taking the square root of both sides, we must have $\cos \alpha = -\frac{3}{5}$ since α is in quadrant III. Similarly,

$$\sin^2 \beta = 1 - \cos^2 \beta = 1 - \frac{25}{169} = \frac{144}{169}$$

Taking the square root of both sides, we must have $\sin \beta = \frac{12}{13}$ since β is in quadrant II. Thus,

$$\sin(\alpha + \beta) = \left(-\frac{4}{5}\right)\left(-\frac{5}{13}\right) + \left(-\frac{3}{5}\right)\left(\frac{12}{13}\right)$$
$$= \frac{20}{65} - \frac{36}{65} = -\frac{16}{65}$$

Since $\sin(\alpha + \beta)$ is negative, $\alpha + \beta$ lies in either quadrant III or quadrant IV. However, the sum of an angle that lies in quadrant III and an angle that lies in quadrant II cannot lie in quadrant III. Thus, $\alpha + \beta$ lies in quadrant IV. ■

Given $\cos \alpha = -\frac{4}{5}$, with α in quadrant III, and $\cos \beta = \frac{3}{5}$, with β in quadrant I, find $\cos(\alpha - \beta)$ and the quadrant in which $\alpha - \beta$ lies.

Answers

$-\frac{24}{25}$, quadrant II

Exercise Set 8.2

In Exercises 1–6, show that the given equation is not an identity. (*Hint*: For each equation, find values of s and t for which the equation is not true.)

1. $\cos(s - t) = \cos s - \cos t$
2. $\sin(s + t) = \sin s + \sin t$
3. $\sin(s - t) = \sin s - \sin t$
4. $\cos(s + t) = \cos s + \cos t$
5. $\tan(s + t) = \tan s + \tan t$
6. $\tan(s - t) = \tan s - \tan t$

In Exercises 7–22, use the addition and subtraction formulas to find exact values.

7. $\cos\left(\frac{\pi}{6}+\frac{\pi}{4}\right)$
8. $\sin\left(\frac{\pi}{6}-\frac{\pi}{4}\right)$
9. $\sin\left(\frac{\pi}{4}+\frac{\pi}{3}\right)$
10. $\cos\left(\frac{\pi}{3}-\frac{\pi}{4}\right)$
11. $\cos(30^\circ + 180^\circ)$
12. $\tan(60^\circ + 300^\circ)$
13. $\tan(300^\circ - 60^\circ)$
14. $\sin(270^\circ - 45^\circ)$
15. $\sin\frac{11\pi}{12}$
16. $\tan\frac{7\pi}{12}$
17. $\cos\frac{7\pi}{12}$
18. $\tan 75^\circ$
19. $\sin\frac{7\pi}{6}$
20. $\cos\frac{5\pi}{6}$
21. $\tan 15^\circ$
22. $\tan 165^\circ$

In Exercises 23–28, write the given expression in terms of cofunctions of complementary angles.

23. $\sin 47^\circ$
24. $\cos 78^\circ$
25. $\tan\frac{\pi}{6}$
26. $\tan 84^\circ$
27. $\cos\frac{\pi}{3}$
28. $\sin 72^\circ 30'$
29. If $\sin t = -\frac{3}{5}$, with t in quadrant III, find $\sin\left(\frac{\pi}{2} - t\right)$.
30. If $\cos t = -\frac{5}{13}$, with t in quadrant II, find $\sin(t - \pi)$.
31. If $\tan\theta = \frac{4}{3}$ and angle θ lies in quadrant III, find $\tan\left(\theta + \frac{\pi}{4}\right)$.
32. If $\sec\theta = \frac{5}{3}$ and angle θ lies in quadrant I, find $\sin\left(\theta + \frac{\pi}{6}\right)$.
33. If $\cos t = 0.4$, with t in quadrant IV, find $\tan(t + \pi)$.
34. If $\sec\alpha = 1.2$ and angle α lies in quadrant IV, find $\tan(\alpha - \pi)$.
35. If $\sin s = \frac{3}{5}$ and $\cos t = -\frac{12}{13}$, with s in quadrant II and t in quadrant III, find $\sin(s + t)$.
36. If $\sin s = -\frac{4}{5}$ and $\csc t = \frac{13}{5}$, with s in quadrant IV and t in quadrant II, find $\cos(s - t)$.
37. If $\cos\alpha = \frac{5}{13}$ and $\tan\beta = -2$, with angle α in quadrant I and angle β in quadrant II, find $\tan(\alpha + \beta)$.
38. If $\sec\alpha = \frac{5}{3}$ and $\cot\beta = \frac{15}{8}$, with angle α in quadrant IV and angle β in quadrant III, find $\tan(\alpha - \beta)$.

In Exercises 39–54, prove each of the following identities by transforming the left-hand side of the equation into the expression on the right-hand side.

39. $\sin 2\alpha = 2\sin\alpha\cos\alpha$
40. $\cos 2t = \cos^2 t - \sin^2 t$
41. $\tan 2\alpha = \dfrac{2\tan\alpha}{1-\tan^2\alpha}$
42. $\sin(x + y)\sin(x - y) = \sin^2 x - \sin^2 y$
43. $\cos(x - y)\cos(x + y) = \cos^2 x\cos^2 y - \sin^2 x\sin^2 y$
44. $\dfrac{\sin(s+t)}{\sin(s-t)} = \dfrac{\tan s + \tan t}{\tan s - \tan t}$
45. $\csc\left(t+\frac{\pi}{2}\right) = \sec t$
46. $\tan(\alpha + 90^\circ) = -\cot\alpha$
47. $\tan\left(x+\frac{\pi}{4}\right) = \dfrac{1+\tan x}{1-\tan x}$
48. $\csc(t - \pi) = -\csc t$
49. $\cot(s - t) = \dfrac{1+\tan s\tan t}{\tan s - \tan t}$
50. $\cot(u + v) = \dfrac{\cot u\cot v - 1}{\cot u + \cot v}$
51. $\sin(s + t) + \sin(s - t) = 2\sin s\cos t$
52. $\cos(s + t) + \cos(s - t) = 2\cos s\cos t$
53. $\dfrac{\sin(x+h)-\sin x}{h} = \sin x\left(\dfrac{\cos h - 1}{h}\right) + \cos x\left(\dfrac{\sin h}{h}\right)$
54. $\dfrac{\cos(x+h)-\cos x}{h} = \cos x\left(\dfrac{\cos h - 1}{h}\right) - \sin x\left(\dfrac{\sin h}{h}\right)$

8.3 Double-Angle and Half-Angle Formulas

8.3a Double-Angle Formulas

Our initial objective in this section is to derive expressions for sin $2t$, cos $2t$, and tan $2t$ in terms of trigonometric functions of t. We will establish the following **double-angle formulas.**

$$\sin 2t = 2 \sin t \cos t \quad (1)$$

$$\cos 2t = \cos^2 t - \sin^2 t \quad (2)$$

$$\tan 2t = \frac{2 \tan t}{1 - \tan^2 t} \quad (3)$$

To establish Equation (1), we rewrite $2t$ as $(t + t)$ and use the addition formula.

$$\begin{aligned} \sin 2t &= \sin (t + t) \\ &= \sin t \cos t + \cos t \sin t \\ &= 2 \sin t \cos t \end{aligned}$$

We proceed in the same manner to prove Equation (2).

$$\begin{aligned} \cos 2t &= \cos (t + t) \\ &= \cos t \cos t - \sin t \sin t \\ &= \cos^2 t - \sin^2 t \end{aligned}$$

Using the addition formula for the tangent function yields a proof of Equation (3).

$$\begin{aligned} \tan 2t &= \tan (t + t) \\ &= \frac{\tan t + \tan t}{1 - \tan t \tan t} \\ &= \frac{2 \tan t}{1 - \tan^2 t} \end{aligned}$$

EXAMPLE 1 ***Using the Double-Angle Formulas***

If $\cos t = -\frac{3}{5}$ and $P(t)$ is in quadrant II, evaluate sin $2t$ and cos $2t$. In which quadrant does $P(2t)$ lie?

SOLUTION

We first find sin t by use of the fundamental identity $\sin^2 t + \cos^2 t = 1$. Thus,

$$\sin^2 t + \frac{9}{25} = 1$$

$$\sin^2 t = \frac{16}{25}$$

Since $P(t)$ is in quadrant II, sin t must be positive. Therefore,

$$\sin t = \frac{4}{5}$$

Applying the double-angle formulas with $\cos t = -\frac{3}{5}$ and $\sin t = \frac{4}{5}$, we have

$$\sin 2t = 2 \sin t \cos t = 2\left(\frac{4}{5}\right)\left(-\frac{3}{5}\right) = -\frac{24}{25}$$

$$\cos 2t = \cos^2 t - \sin^2 t = \frac{9}{25} - \frac{16}{25} = -\frac{7}{25}$$

Since sin $2t$ and cos $2t$ are both negative, we conclude that $P(2t)$ lies in quadrant III. ■

Progress Check

If $\sin \theta = \frac{5}{13}$ and θ is in quadrant I, evaluate $\sin 2\theta$ and $\tan 2\theta$.

Answers

$\sin 2\theta = \frac{120}{169}$, $\tan 2\theta = \frac{120}{119}$

EXAMPLE 2 ***Using the Addition and Double-Angle Formulas***

Express $\sin 3t$ in terms of $\sin t$ and $\cos t$.

SOLUTION

We write $3t$ as $(2t + t)$. Then

$$\begin{aligned}\sin 3t &= \sin (2t + t)\\ &= \sin 2t \cos t + \cos 2t \sin t\\ &= 2 \sin t \cos t \cos t + (\cos^2 t - \sin^2 t) \sin t\\ &= 2 \sin t \cos^2 t + \sin t \cos^2 t - \sin^3 t\\ &= 3 \sin t \cos^2 t - \sin^3 t\end{aligned}$$

■

Progress Check

Express $\cos 3t$ in terms of $\sin t$ and $\cos t$.

Answer

$\cos 3t = \cos^3 t - 3 \sin^2 t \cos t$

If we begin with the formula for $\cos 2t$ and use the fundamental identity $\cos^2 t = 1 - \sin^2 t$, we obtain

$$\begin{aligned}\cos 2t &= \cos^2 t - \sin^2 t\\ &= (1 - \sin^2 t) - \sin^2 t\\ &= 1 - 2 \sin^2 t\end{aligned}$$

Similarly, replacing $\sin^2 t$ by $1 - \cos^2 t$ yields

$$\begin{aligned}\cos 2t &= \cos^2 t - \sin^2 t\\ &= \cos^2 t - (1 - \cos^2 t)\\ &= 2 \cos^2 t - 1\end{aligned}$$

We then have two additional formulas for $\cos 2t$.

$$\cos 2t = 1 - 2 \sin^2 t \qquad (4)$$

$$\cos 2t = 2 \cos^2 t - 1 \qquad (5)$$

EXAMPLE 3 ***Using Double-Angle Formulas in Verifying an Identity***

Verify the identity

$$\frac{1-\cos 2\alpha}{2\sin\alpha\cos\alpha} = \tan\alpha$$

SOLUTION

Substituting $\cos 2\alpha = 1 - 2\sin^2\alpha$, we have

$$\begin{aligned}\frac{1-\cos 2\alpha}{2\sin\alpha\cos\alpha} &= \frac{1-(1-2\sin^2\alpha)}{2\sin\alpha\cos\alpha}\\ &= \frac{2\sin^2\alpha}{2\sin\alpha\cos\alpha}\\ &= \frac{\sin\alpha}{\cos\alpha}\\ &= \tan\alpha\end{aligned}$$

■

Verify the identity

$$\frac{1+\cos 2\theta}{\sin 2\theta} = \cot\theta$$

Note that

$$\frac{\sin 2t}{2} \neq \sin t$$

From Equation (1),

$$\frac{\sin 2t}{2} = \frac{2\sin t\cos t}{2} = \sin t\cos t$$

8.3b *Half-Angle Formulas*

If we begin with the alternative forms for $\cos 2t$ given in Equations (4) and (5), we can obtain the following expressions for $\sin^2 t$ and $\cos^2 t$. The expressions are often used in calculus.

$$\sin^2 t = \frac{1-\cos 2t}{2} \qquad (6)$$

$$\cos^2 t = \frac{1+\cos 2t}{2} \qquad (7)$$

We will use the identities in Equations (6) and (7) to derive formulas for $\sin\frac{t}{2}$, $\cos\frac{t}{2}$, and $\tan\frac{t}{2}$. Substituting $s = 2t$ into Equations (6) and (7), we obtain

$$\sin^2\frac{s}{2} = \frac{1-\cos s}{2}$$

$$\cos^2\frac{s}{2} = \frac{1-\cos s}{2}$$

Replacing s with t and solving, we have

$$\sin\frac{t}{2} = \pm\sqrt{\frac{1-\cos t}{2}} \qquad (8)$$

$$\cos\frac{t}{2} = \pm\sqrt{\frac{1+\cos t}{2}} \qquad (9)$$

The appropriate sign to use in Equations (8) and (9) depends on the quadrant in which $P\left(\frac{t}{2}\right)$ is located. Thus, $\sin\frac{t}{2}$ is positive if $P\left(\frac{t}{2}\right)$ lies in quadrant I or II. Similarly, we choose the positive root for $\cos\frac{t}{2}$ in Equation (9) if $P\left(\frac{t}{2}\right)$ lies in quadrant I or IV.

Using the identity

$$\tan\frac{t}{2} = \frac{\sin\frac{t}{2}}{\cos\frac{t}{2}}$$

we obtain

$$\tan\frac{t}{2} = \pm\sqrt{\frac{1-\cos t}{1+\cos t}} \qquad (10)$$

Formulas (8), (9), and (10) are known as the **half-angle formulas.**

EXAMPLE 4 *Applying the Half-Angle Formulas*

Find the exact values of sin 22.5° and cos 112.5°.

SOLUTION

Applying the half-angle formulas with $22.5^\circ = \frac{45^\circ}{2}$, we have

$$\begin{aligned}
\sin 22.5^\circ &= \sin\frac{45^\circ}{2} \\
&= \sqrt{\frac{1-\cos 45^\circ}{2}} = -\sqrt{\frac{1-\frac{\sqrt{2}}{2}}{2}} \\
&= \sqrt{\frac{2-\sqrt{2}}{2}}
\end{aligned}$$

Note that we chose the positive square root since 22.5° is in the first quadrant and the sine function is positive in the first quadrant. Similarly,

$$\begin{aligned}
\cos 112.5^\circ &= \cos\frac{225^\circ}{2} \\
&= -\sqrt{\frac{1+\cos 225^\circ}{2}} \\
&= -\sqrt{\frac{1-\cos 45^\circ}{2}} = -\sqrt{\frac{1-\frac{\sqrt{2}}{2}}{2}} \\
&= -\sqrt{\frac{2-\sqrt{2}}{2}}
\end{aligned}$$

The negative square root was selected since 112.5° is in the second quadrant and the cosine function is negative in quadrant II. ■

Use the half-angle formulas to evaluate $\tan \frac{3\pi}{8}$.

Answer

$\sqrt{2}+1$

EXAMPLE 5 *Applying the Half-Angle Formulas*

If $\sin \theta = -\frac{3}{5}$ and θ is in quadrant III, evaluate $\cos \frac{\theta}{2}$.

SOLUTION

We first evaluate $\cos \theta$ by using the identity

$$\cos^2 \theta = 1 - \sin^2 \theta = 1 - \frac{9}{25} = \frac{16}{25}$$

Since θ is in quadrant III, $\cos \theta$ is negative. Thus, $\cos \theta = -\frac{4}{5}$. Note that since $180^\circ < \theta < 270^\circ$, we see that $90^\circ < \frac{\theta}{2} < 135^\circ$. Thus, $\frac{\theta}{2}$ is in quadrant II and $\cos \frac{\theta}{2}$ is negative. We can now employ the half-angle formula

$$\cos \frac{\theta}{2} = -\sqrt{\frac{1+\cos\theta}{2}}$$

$$= -\sqrt{\frac{1-\frac{4}{5}}{2}}$$

$$= -\frac{\sqrt{10}}{10}$$

■

Progress Check

If $\tan \alpha = \frac{3}{4}$ and α is in quadrant III, evaluate $\tan \frac{\alpha}{2}$.

Answer

-3

Exercise Set 8.3

In Exercises 1–12, use the given conditions to determine the value of the specified trigonometric function.

1. If $\sin u = \frac{3}{5}$ and $P(u)$ is in quadrant II, find $\cos 2u$.
2. If $\cos x = -\frac{5}{13}$ and $P(x)$ is in quadrant III, find $\sin 2x$.
3. If $\sec \alpha = -2$ and α is in quadrant II, find $\sin 2\alpha$.
4. If $\tan \theta = \frac{4}{3}$ and θ is in quadrant I, find $\cos 2\theta$.
5. If $\csc t = -\frac{17}{8}$ and $P(t)$ is in quadrant IV, find $\tan 2t$.
6. If $\cot \beta = \frac{3}{4}$ and β is in quadrant III, find $\cot 2\beta$.
7. If $\sin 2\alpha = -\frac{4}{5}$ and 2α is in quadrant IV, find $\sin 4\alpha$.
8. If $\sec 5x = -\frac{13}{12}$ and $P(5x)$ is in quadrant III, find $\tan 10x$.
9. If $\cos \frac{\theta}{2} = \frac{8}{17}$ and $\frac{\theta}{2}$ is acute, find $\cos \theta$.
10. If $\csc \frac{t}{2} = -\frac{13}{5}$ and $P(\frac{t}{2})$ is in quadrant IV, find $\cos t$.
11. If $\sin 42^\circ \approx 0.67$, find $\cos 84^\circ$.
12. If $\cos 77^\circ \approx 0.22$, find $\cos 154^\circ$.

In Exercises 13–18, use the half-angle formulas to find exact values for each of the following.

13. $\sin 15^\circ$
14. $\cos 75^\circ$
15. $\tan \frac{\pi}{8}$
16. $\sec \frac{5\pi}{8}$
17. $\csc 165^\circ$
18. $\cot \frac{7\pi}{12}$

In Exercises 19–26, use the given conditions to determine the exact value of the specified trigonometric function.

19. If $\sin \theta = -\frac{4}{5}$ and θ is in quadrant IV, find $\cos \frac{\theta}{2}$.
20. If $\cos \theta = \frac{3}{5}$ and θ is in quadrant I, find $\sin \frac{\theta}{2}$.
21. If $\sec t = -3$ and $P(t)$ is in quadrant II, find $\sin \frac{t}{2}$.
22. If $\tan x = \frac{4}{3}$ and $P(x)$ is in quadrant III, find $\cos \frac{x}{2}$.
23. If $\cot \beta = \frac{3}{4}$ and β is in quadrant III, find $\tan \frac{\beta}{2}$.
24. If $\csc \alpha = \frac{13}{5}$ and α is in quadrant II, find $\tan \frac{\alpha}{2}$.
25. If $\cos 4x = \frac{1}{3}$ and $P(4x)$ is in quadrant IV, find $\cos 2x$.
26. If $\sec 6\alpha = -\frac{13}{12}$ and 6α is in quadrant III, find $\sin 3\alpha$.

In Exercises 27–46, verify the identity.

27. $\sin 50x = 2 \sin 25x \cos 25x$
28. $(\sin \theta + \cos \theta)^2 = 1 + \sin 2\theta$
29. $\tan 2y = \dfrac{2 \cot y}{\csc^2 y - 2}$
30. $2 \sin^2 2t + \cos 4t = 1$
31. $\sin 4\alpha = 4 \sin \alpha \cos^3 \alpha - 4 \sin^3 \alpha \cos \alpha$
32. $\cos 4\beta = 1 - 8 \sin^2 \beta \cos^2 \beta$
33. $\cos 2u = \dfrac{1 - \tan^2 u}{1 + \tan^2 u}$
34. $\sin 2\theta = \dfrac{2 \tan \theta}{1 + \tan^2 \theta}$
35. $\sin \frac{t}{2} \cos \frac{t}{2} = \frac{\sin t}{2}$
36. $\tan \frac{y}{2} = \csc y - \cot y$
37. $\sin \alpha - \cos \alpha \tan \frac{\alpha}{2} = \tan \frac{\alpha}{2}$
38. $\dfrac{1 - \cos 2\beta}{1 + \cos 2\beta} = \tan^2 \beta$
39. $\cos^4 x - \sin^4 x = \cos 2x$
40. $\dfrac{\sin 2t}{\sin t} - \dfrac{\cos 2t}{\cos t} = \sec t$
41. $\dfrac{2 \tan \alpha}{1 + \tan^2 \alpha} = \sin 2\alpha$
42. $\cos^2 \frac{x}{2} = \dfrac{\tan x + \sin x}{2 \tan x}$
43. $\sec 2t = \dfrac{\sec^2 t}{2 - \sec^2 t}$
44. $\cos 2t + \cot 2t = \cot 2t (\sin t + \cos t)^2$
45. $\tan \frac{t}{2} = \dfrac{1 - \cos t}{\sin t}$
46. $\tan \frac{t}{2} = \dfrac{\sin t}{1 + \cos t}$

In Exercises 47–50, find the requested value.

47. $\sin\left(2 \arccos \frac{3}{5}\right)$
48. $\cos\left(2 \sin^{-1} \frac{3}{5}\right)$
49. $\tan\left(2 \arcsin \frac{5}{13}\right)$
50. $\cos\left(2 \arctan \frac{12}{5}\right)$

8.4 The Product-Sum Formulas

The objective of this section is to derive formulas that can transform sums of sines and cosines into products of sines and cosines, and vice versa. We use the word "sum" in a more general way to include the word "difference," since subtraction can be thought of as adding a negative quantity.

The following formulas express a product as a sum.

Product-Sum Formulas

$$\sin s \cos t = \frac{\sin(s+t)+\sin(s-t)}{2} \qquad (1)$$

$$\cos s \sin t = \frac{\sin(s+t)-\sin(s-t)}{2} \qquad (2)$$

$$\cos s \cos t = \frac{\cos(s+t)+\cos(s-t)}{2} \qquad (3)$$

$$\sin s \sin t = \frac{\cos(s-t)-\cos(s+t)}{2} \qquad (4)$$

To prove the identity in Equation (1), we begin with the right-hand side of the equation.

$$\frac{\sin(s+t)+\sin(s-t)}{2} = \frac{(\sin s \cos t + \cos s \sin t)+(\sin s \cos t - \cos s \sin t)}{2}$$

$$= \frac{2 \sin s \cos t}{2}$$

$$= \sin s \cos t$$

The proofs of the identities in Equations (2), (3), and (4) are very similar.

EXAMPLE 1 *Applying the Product-Sum Formulas*

Express sin $4x$ cos $3x$ as a sum or a difference.

SOLUTION

Applying Equation (1), we obtain

$$\sin 4x \cos x = \frac{\sin(4x+3x)+\sin(4x-3x)}{2}$$

$$= \frac{\sin 7x + \sin x}{2}$$ ■

Progress Check

Express sin $5x$ sin $2x$ as a sum or as a difference.

Answer

$\frac{1}{2}(\cos 3x - \cos 7x)$

EXAMPLE 2 *Applying the Product-Sum Formulas*

Evaluate the product $\cos\frac{5\pi}{8}\cos\frac{3\pi}{8}$ by using a product-sum formula.

SOLUTION

Using Equation (3), we have

$$\cos\frac{5\pi}{8}\cos\frac{3\pi}{8} = \frac{1}{2}\left[\cos\left(\frac{5\pi}{8}+\frac{3\pi}{8}\right)+\cos\left(\frac{5\pi}{8}-\frac{3\pi}{8}\right)\right]$$

$$= \frac{1}{2}\left[\cos\pi+\cos\frac{\pi}{4}\right]$$

$$= \frac{1}{2}\left[-1+\frac{\sqrt{2}}{2}\right] = \frac{\sqrt{2}-2}{4}$$

■

Progress Check

Evaluate $\cos\frac{\pi}{3}\sin\frac{\pi}{6}$ by a product-sum formula.

Answer

$\frac{1}{4}$

The following formulas express a sum as a product.

Sum-Product Formulas

$$\sin s + \sin t = 2\sin\frac{s+t}{2}\cos\frac{s-t}{2} \tag{5}$$

$$\sin s - \sin t = 2\cos\frac{s+t}{2}\sin\frac{s-t}{2} \tag{6}$$

$$\cos s + \cos t = 2\cos\frac{s+t}{2}\cos\frac{s-t}{2} \tag{7}$$

$$\cos s - \cos t = -2\sin\frac{s+t}{2}\sin\frac{s-t}{2} \tag{8}$$

To prove the identity in Equation (5), we begin with the right-hand side and apply Equation (1). Then

$$2\sin\frac{s+t}{2}\cos\frac{s-t}{2} = 2\left\{\frac{1}{2}\left[\sin\left(\frac{s+t}{2}+\frac{s-t}{2}\right)+\sin\left(\frac{s+t}{2}-\frac{s-t}{2}\right)\right]\right\}$$

$$= \sin s + \sin t$$

This establishes Equation (5). The other identities are established in a similar fashion.

EXAMPLE 3 ***Applying the Product-Sum Formulas***

Express $\sin 5x - \sin 3x$ as a product.

SOLUTION

Using Equation (6), we have

$$\begin{aligned}\sin 5x - \sin 3x &= 2\cos\frac{5x+3x}{2}\sin\frac{5x-3x}{2}\\ &= 2\cos 4x \sin x\end{aligned}$$ ■

Progress Check

Express $\cos 6x + \cos 2x$ as a product.

Answer

$2\cos 4x\cos 2x$

EXAMPLE 4 ***Applying the Product-Sum Formulas***

Evaluate $\cos\frac{5\pi}{12} - \cos\frac{\pi}{12}$ using a product-sum-formula.

SOLUTION

Using Equation (8), we have

$$\begin{aligned}\cos\frac{5\pi}{12} - \cos\frac{\pi}{12} &= -2\sin\frac{\pi}{4}\sin\frac{\pi}{6}\\ &= -2\left(\frac{\sqrt{2}}{2}\right)\frac{1}{2} = -\frac{\sqrt{2}}{2}\end{aligned}$$ ■

Progress Check

Evaluate

$$\sin\frac{11\pi}{12} - \sin\frac{5\pi}{12}$$

using a product-sum formula.

Answer

$-\frac{\sqrt{2}}{2}$

Exercise Set 8.4

In Exercises 1–8, express each product as a sum or difference.

1. $2 \sin 5\alpha \cos \alpha$
2. $-3 \cos 6x \sin 2x$
3. $\sin 3x \sin (-2x)$
4. $\cos 7t \cos (-3t)$
5. $-2 \cos 2\theta \cos 5\theta$
6. $\sin \frac{5\theta}{2} \sin \frac{\theta}{2}$
7. $\cos (\alpha + \beta) \cos (\alpha - \beta)$
8. $-\sin 2u \cos 4u$

In Exercises 9–12, evaluate each product by using a product-sum formula.

9. $\cos \frac{7\pi}{8} \sin \frac{5\pi}{8}$
10. $\cos \frac{\pi}{3} \cos \frac{\pi}{6}$
11. $\sin 120^\circ \cos 60^\circ$
12. $\sin \frac{13\pi}{12} \sin \frac{11\pi}{12}$

In Exercises 13–20, express each sum or difference as a product.

13. $\sin 5x + \sin x$
14. $\cos 8t - \cos 2t$
15. $\cos 2\theta + \cos 6\theta$
16. $\sin 5\alpha - \sin 7\alpha$
17. $\sin (\alpha + \beta) + \sin (\alpha - \beta)$
18. $\cos \frac{x}{2} - \cos \frac{3x}{2}$
19. $\sin 7x - \sin 3x$
20. $\cos 5\theta + \cos 3\theta$

In Exercises 21–24, evaluate each sum by using a product-sum formula.

21. $\cos 75^\circ + \cos 15^\circ$
22. $\sin \frac{5\pi}{12} + \sin \frac{\pi}{12}$
23. $\cos \frac{3\pi}{4} - \cos \frac{\pi}{4}$
24. $\sin \frac{13\pi}{12} - \sin \frac{5\pi}{12}$

In Exercises 25–34, verify the identities.

25. $\sin 40^\circ + \sin 20^\circ = \cos 10^\circ$
26. $\cos 70^\circ - \cos 10^\circ = -\sin 40^\circ$
27. $\frac{\sin 5\theta - \sin 3\theta}{\cos 3\theta - \cos 5\theta} = \cot 4\theta$
28. $\frac{\cos 7x - \cos x}{\sin 7x + \sin x} = -\tan 3x$
29. $\frac{\sin t - \sin s}{\cos t - \cos s} = -\cot \frac{s+t}{2}$
30. $\frac{\sin s + \sin t}{\cos s + \cos t} = \tan \frac{s+t}{2}$
31. $\frac{\sin 50^\circ - \sin 10^\circ}{\cos 50^\circ - \cos 10^\circ} = -\sqrt{3}$
32. $2 \sin \left(\theta + \frac{\pi}{4}\right) \sin \left(\theta - \frac{\pi}{4}\right) = -\cos 2\theta$
33. $\frac{\cot x - \tan x}{\cos x + \tan x} = \cos 2x$
34. $\cos 6x \cos 2x + \sin^2 4x = \cos^2 2x$
35. Express $\sin ax \cos bx$ as a sum.
36. Express $\cos ax \cos bx$ as a sum.
37. Prove the product-sum formulas given in Equations (2), (3), and (4).
38. Prove the product-sum formulas given in Equations (6), (7), and (8).

8.5 Trigonometric Equations

Thus far, this chapter has dealt exclusively with trigonometric identities. We now seek to solve trigonometric equations that may be true for some values of the variable but not for all values.

We have seen that algebraic equations may have just one or two solutions. The situation is quite different with trigonometric equations. Since trigonometric functions are periodic by nature, if we find one solution, there must be an infinite number of solutions. To deal with this situation, we first seek all solutions t such that $0 \leq t < 2\pi$. Then, for every integer n, $t + 2\pi n$ is also a solution. The following example illustrates this procedure for finding the solution set.

EXAMPLE 1 *Solving a Trigonometric Equation*

Find all solutions of the equation $\cos t = 0$.

SOLUTION

The only values in the interval $[0, 2\pi)$ for which $\cos t = 0$ are $\frac{\pi}{2}$ and $\frac{3\pi}{2}$. Then every solution is included among those values of t such that

$$t = \frac{\pi}{2} + 2\pi n \qquad \text{or} \qquad t = \frac{3\pi}{2} + 2\pi n, \qquad n \text{ an integer}$$

Since $\frac{3\pi}{2} = \frac{\pi}{2} + \pi$, the solution set can be written in a more compact form as

$$t = \frac{\pi}{2} + \pi n, \qquad n \text{ an integer}$$

■

Factoring provides an important technique for solving trigonometric equations. If we can write the equation in the form $P(x)Q(x) = 0$, we can find the solutions by setting $P(x) = 0$ and $Q(x) = 0$.

It may also be helpful to think of substituting a new variable for some trigonometric expression. Thus, the equation

$$4 \sin^2 x + 3 \sin x - 1 = 0$$

can be viewed as a quadratic in u

$$4u^2 + 3u - 1 = 0$$

by substituting $u = \sin x$.

EXAMPLE 2 *Restricting the Solutions of a Trigonometric Equation*

Find all solutions of the equation $2 \cos^2 t - \cos t - 1 = 0$ in the interval $[0, 2\pi)$.

SOLUTION

Factoring the left side of the equation yields

$$(2 \cos t + 1)(\cos t - 1) = 0$$

Setting each factor equal to 0, we have

$$2 \cos t + 1 = 0 \qquad \text{or} \qquad \cos t - 1 = 0$$

so that

$$\cos t = -\frac{1}{2} \qquad \text{or} \qquad \cos t = 1$$

The solutions of $\cos t = -\frac{1}{2}$ in the interval $[0, 2\pi)$ are $t = \frac{2\pi}{3}$ and $t = \frac{4\pi}{3}$. The only solution of $\cos t = 1$ in the interval $[0, 2\pi)$ is $t = 0$. Thus, the solutions of the original problem are

$$t = \frac{2\pi}{3}, \qquad t = \frac{4\pi}{3}, \qquad \text{and} \qquad t = 0$$

■

Progress Check

Find all solutions of the equation $2\sin^2 t - 3\sin t + 1 = 0$ in the interval $[0, 2\pi)$.

Answers

$\frac{\pi}{6}, \frac{5\pi}{6}, \frac{\pi}{2}$

If the solutions of the trigonometric equation are angles, the answer may be given in either radians or degrees.

EXAMPLE 3 *Expressing Solutions in Radians and Degrees*

Find all solutions of the equation $\tan\theta\cos^2\theta - \tan\theta = 0$.

SOLUTION

Factoring the left side yields

$$(\tan\theta)(\cos^2\theta - 1) = 0$$

Setting each factor equal to 0,

$$\tan\theta = 0 \qquad \text{or} \qquad \cos^2\theta = 1$$

so that

$$\tan\theta = 0, \qquad \cos\theta = 1, \qquad \text{or} \qquad \cos\theta = -1$$

These equations yield the following solutions in the interval $[0, 2\pi)$.

$$\tan\theta = 0: \quad \theta = 0 \quad \text{or} \quad \theta = \pi$$
$$\cos\theta = 1: \quad \theta = 0$$
$$\cos\theta = -1: \quad \theta = \pi$$

The solutions of the original equation are

$$\theta = 0 + 2\pi n \qquad \text{and} \qquad \theta = \pi + 2\pi n, \qquad n \text{ an integer}$$

which can be expressed more compactly as

$$\theta = \pi n, \qquad n \text{ an integer}$$

In degree measure, the solutions are

$$\theta = 180^\circ n, \qquad n \text{ an integer}$$

■

EXAMPLE 4 *Expressing Restricted Solutions in Radians and Degrees*

Find all solutions of the equation $\sin 2\theta - 3 \sin \theta = 0$ in the interval $[0, 2\pi)$ and when $0^\circ \leq \theta < 360^\circ$.

SOLUTION

Using the identity $\sin 2\theta = 2 \sin \theta \cos \theta$, we have

$$2 \sin \theta \cos \theta - 3 \sin \theta = 0$$
$$\sin \theta (2 \cos \theta - 3) = 0$$
$$\sin \theta = 0 \quad \text{or} \quad 2 \cos \theta - 3 = 0$$
$$\sin \theta = 0 \quad \text{or} \quad \cos \theta = \frac{3}{2}$$

The equation $\cos \theta = \frac{3}{2}$ has no solutions. The solutions of $\sin \theta = 0$ are $\theta = 0$ and $\theta = \pi$. Therefore, the solutions of the original equation are

$$\theta = 0 \quad \text{and} \quad \theta = \pi.$$

Equivalently, in degree measure, the solutions are

$$\theta = 0^\circ \quad \text{and} \quad \theta = 180^\circ.$$ ■

Progress Check

Find all solutions of the equation $\cos 2\theta + \cos \theta = 0$ in both radians and degrees.

Answers

$\frac{\pi}{3} + 2\pi n, \pi + 2\pi n, \frac{5\pi}{3} + 2\pi n$ or $60^\circ + 360^\circ n, 180^\circ + 360^\circ n, 300^\circ + 360^\circ n$

Equations involving multiple angles can often be solved by using a substitution of variable. The following example shows what may occur when seeking solutions in the interval $[0, 2\pi)$.

EXAMPLE 5 *Substitution of a Variable in Trigonometric Equations*

Find all solutions of the equation $\cos 3x = 0$ in the interval $[0, 2\pi)$.

SOLUTION

We are given

$$\cos 3x = 0, \qquad 0 \leq x < 2\pi$$

Substituting $t = 3x$, we obtain

$$\cos t = 0, \qquad 0 \leq \frac{t}{3} < 2\pi$$

or

$$\cos t = 0, \qquad 0 \leq t < 6\pi$$

Note that we seek solutions of $\cos t = 0$ in the interval $[0, 6\pi)$ rather than $[0, 2\pi)$. The solutions are then

$$t = \frac{\pi}{2}, \frac{3\pi}{2}, \frac{5\pi}{2}, \frac{7\pi}{2}, \frac{9\pi}{2}, \frac{11\pi}{2}$$

Since $x = \frac{t}{3}$, we obtain

$$x = \frac{\pi}{6}, \frac{\pi}{2}, \frac{5\pi}{6}, \frac{7\pi}{6}, \frac{3\pi}{2}, \frac{11\pi}{6}$$ ■

When you perform a substitution of variable, you must remember to go back and express the answers in terms of the original variable.

EXAMPLE 6 ***Substitution of Variable***

Find all solutions of the equation

$$3\tan^2 x + \tan x - 1 = 0$$

in the interval $[0, \pi)$.

SOLUTION

Since the equation does not factor easily, consider $u = \tan x$. Then we obtain

$$3u^2 + u - 1 = 0$$

From the quadratic formula,

$$u = \frac{-1 \pm \sqrt{13}}{6}$$

so that

$$\tan x = \frac{-1 \pm \sqrt{13}}{6}$$

Therefore,

$$\tan x \approx 0.4342586 \qquad \text{and} \qquad \tan x \approx -0.7675919$$

in which case

$$x \approx \tan^{-1} 0.4342586 \approx 0.4096865$$

and

$$x \approx \tan^{-1}(-0.7675919) \approx -0.6546652$$

Although the first value for x is in the interval $[0, \pi)$, the second value for x is not. In fact, this second value is in the interval $(-\frac{\pi}{2}, \frac{\pi}{2})$, the range of $\tan^{-1} x$. Since the period of $\tan x$ is π,

$$\begin{aligned} x &\approx -0.6546652 + \pi \\ &\approx 2.4869275 \end{aligned}$$

is also a solution. Observe that this new value for x is in the interval $[0, \pi)$. Thus the solutions in the given interval are $x \approx 0.4097$ and $x \approx 2.4869$. ■

Exercise Set 8.5

In Exercises 1–20, find all solutions of the given equation in the interval $[0, 2\pi)$ and on $[0°, 360°)$.

1. $2 \sin \theta - 1 = 0$
2. $2 \cos \theta + 1 = 0$
3. $\cos \alpha + 1 = 0$
4. $\cot \gamma + 1 = 0$
5. $4 \cos^2 \alpha = 3$
6. $\tan^2 \theta = 3$
7. $3 \tan^2 \alpha = 1$
8. $2 \cos^2 \alpha - 1 = 0$
9. $2 \sin^2 \beta = \sin \beta$
10. $\sin \alpha = \cos \alpha$
11. $2 \cos^2 \theta - 3 \cos \theta + 1 = 0$
12. $2 \sin^2 \theta - \sin \theta - 1 = 0$
13. $\sin 5\theta = 1$
14. $\tan 3\beta = -\sqrt{3}$
15. $2 \sin^2 \alpha - 3 \cos \alpha = 0$
16. $2 \cos^2 \theta - 1 = \sin \theta$
17. $\csc 2\theta = 2$
18. $\cos^2 2\alpha = \frac{1}{4}$
19. $\sin^2 \beta + 3 \cos \beta - 3 = 0$
20. $2 \cos^2 \theta \tan \theta - \tan \theta = 0$

In Exercises 21–38, find all the solutions of the given equation.

21. $3 \tan^2 x - 1 = 0$
22. $2 \sin^2 y - 1 = 0$
23. $3 \cot^2 \theta - 1 = 0$
24. $1 - 4 \cos^2 t = 0$
25. $\sec 2u - 2 = 0$
26. $\tan 3x - 1 = 0$
27. $\sin 4x = 0$
28. $\cos 5t = -1$
29. $4 \cos^2 2t - 3 = 0$
30. $\csc^2 2x - 2 = 0$
31. $\sin 2t + 2 \cos t = 0$
32. $\sin 2t + 3 \cos t = 0$
33. $\cos 2t + \sin t = 0$
34. $2 \cos 2t + 2 \sin t = 0$
35. $\tan^2 x - \tan x = 0$
36. $\sec^2 x - 3 \sec x + 2 = 0$
37. $2 \sin^2 x + 3 \sin x - 2 = 0$
38. $2 \cos^2 x - 5 \cos x - 3 = 0$

In Exercises 39–42, solve the equations on the interval $[0, 2\pi)$ and state the solutions to two decimal places.

39. $5 \sin^2 x - \sin x - 2 = 0$
40. $\sec^2 y - 5 \sec y + 6 = 0$
41. $3 \tan^2 u + 5 \tan u + 1 = 0$
42. $\cos^2 t - 2 \sin t + 3 = 0$

In Exercises 43–46, find the approximate solutions of the given equation in the interval $[0, 2\pi)$ by finding the point(s) of intersection of appropriate graphs on your graphing calculator.

43. $\cos x = x$
44. $\sin x = \cos x$
45. $\tan x = 8 - \frac{1}{2}x^2$
46. $3 \cos \frac{x}{2} = x^2 - 3$

Chapter Summary

Key Terms, Concepts, and Symbols

Key Ideas for Review

Topic	Page	Key Idea
Trigonometric Identity	465	A trigonometric identity is true for all values that may be assumed by the variable.
Fundamental Identities	466	The fundamental identities are trigonometric identities that are directly related to the definitions of the trigonometric functions.
Verification of Identities	467	The fundamental identities can be used to verify other trigonometric identities. The techniques frequently used to verify identities include: 1. factoring 2. writing trigonometric functions in terms of sine and cosine. 3. performing some of the indicated operations and simplifying complicated expressions 4. multiplying numerator and denominator of some fraction involving trigonometric functions by some trigonometric expression to obtain forms such as $1 - \sin^2\theta$, $1 - \cos^2\theta$, or $\sec^2\theta - 1$, which can be further simplified.
Trigonometric Formulas	472	Some of the most useful trigonometric formulas are the following.
Addition Formulas	472	$\sin(s+t) = \sin s \cos t + \cos s \sin t$ $\cos(s+t) = \cos s \cos t - \sin s \sin t$ $\tan(s+t) = \dfrac{\tan s + \tan t}{1 - \tan s \tan t}$
Double-Angle Formulas	478	$\sin 2t = 2 \sin t \cos t$ $\cos 2t = \cos^2 t - \sin^2 t$ $\tan 2t = \dfrac{2\tan t}{1 - \tan^2 t}$
Half-Angle Formulas	481	$\sin \dfrac{t}{2} = \pm\sqrt{\dfrac{1-\cos t}{2}}$ $\cos \dfrac{t}{2} = \pm\sqrt{\dfrac{1+\cos t}{2}}$ $\tan \dfrac{t}{2} = \pm\sqrt{\dfrac{1-\cos t}{1+\cos t}}$
Trigonometric Equations	488	Since the trigonometric functions are periodic, a trigonometric equation has either no solution or an infinite number of solutions.
Restricted Solutions	488	Sometimes the solutions to a trigonometric equation are restricted to a specific interval.

Review Exercises

In Exercises 1–3, verify the given identity.

1. $\sin\theta\sec\theta + \tan\theta = 2\tan\theta$
2. $\dfrac{\cos^2 x}{1-\sin x} = 1 + \sin x$
3. $\sin\alpha + \sin\alpha\cot^2\alpha = \csc\alpha$

In Exercises 4–7, determine the exact value of the given expression by using the addition formulas.

4. $\sin\left(\frac{\pi}{6}+\frac{\pi}{4}\right)$
5. $\cos(45^\circ + 90^\circ)$
6. $\tan\left(\frac{\pi}{3}+\frac{\pi}{4}\right)$
7. $\sin\frac{7\pi}{12}$

In Exercises 8–11, write the given expression in terms of cofunctions of complementary angles.

8. $\csc 15^\circ$
9. $\cos 23^\circ$
10. $\sin\frac{\pi}{8}$
11. $\tan\frac{2\pi}{7}$
12. If $\cos\theta = -\frac{12}{13}$ and $0 \le \theta \le 180^\circ$ find $\sin(\pi-\theta)$.
13. If $\sec\alpha = \frac{5}{4}$ and α lies in quadrant IV, find $\csc\left(\alpha+\frac{\pi}{3}\right)$.
14. If $\sin t = -\frac{3}{5}$ and $P(t)$ is in quadrant III, find $\tan(t+\pi)$.
15. If $\cos\alpha = -\frac{12}{13}$ and $\tan\beta = -\frac{5}{2}$, with angles α and β in quadrant II, find $\tan(\alpha+\beta)$.
16. If $\sin x = \frac{3}{5}$ and $\csc y = \frac{13}{12}$, with $P(x)$ in quadrant II and $P(y)$ in quadrant I, find $\cos(x-y)$.
17. If $\csc u = -\frac{5}{4}$ and $P(u)$ is in quadrant IV, find $\cos 2u$.
18. If $\tan\alpha = -\frac{3}{4}$ and $0 \le \alpha \le 180^\circ$ find $\sin 2\alpha$.
19. If $\sin 2t = \frac{3}{5}$ and $P(2t)$ is in quadrant I, find $\sin 4t$.
20. If $\sin\theta = 0.5$ and $\frac{\pi}{2} \le \theta \le \pi$ find $\sin 2\theta$.
21. If $\cos\frac{\theta}{2} = \frac{12}{13}$ and θ is acute, find $\sin\theta$.
22. If $\sin\alpha = -\frac{3}{5}$ and α is in quadrant III, find $\cos\frac{\alpha}{2}$.
23. If $\cot t = -\frac{4}{3}$ and $P(t)$ is in quadrant IV, find $\tan\frac{t}{2}$.
24. If $\cos 4x = \frac{2}{3}$ and $P(4x)$ is in quadrant IV, find $\cos 2x$.
25. Find the exact value of $\cos 15^\circ$ by using a half-angle formula.
26. Find the exact value of $\sin\frac{\pi}{8}$ by using a half-angle formula.
27. Find the exact value of $\tan 112.5^\circ$ by using a half-angle formula.

In Exercises 28–30, verify the given identity.

28. $\cos 30x = 1 - 2\sin^2 15x$
29. $\frac{1}{2}\sin 2y = \dfrac{\sin y}{\sec y}$
30. $\tan\frac{\alpha}{2} = \dfrac{1-\cos\alpha}{\sin\alpha}$
31. Express $\sin\frac{3\alpha}{2}\sin\frac{\alpha}{2}$ as a sum or difference.
32. Express $\cos 3x - \cos x$ as a product.
33. Evaluate $\sin 75^\circ \sin 15^\circ$ by using a product-sum formula.
34. Evaluate $\cos\frac{3\pi}{4} + \cos\frac{\pi}{4}$ by using a product-sum formula.

In Exercises 35–37, find all solutions of the given equation in the interval $[0, 2\pi)$. Express the answers in radian measure.

35. $2\cos^2\alpha - 1 = 0$
36. $2\sin\theta\cos\theta = 0$
37. $\sin 2t - \sin t = 0$

In Exercises 38–40, find all solutions of the given equation. Express the answers in degree measure.

38. $\cos^2\alpha - 2\cos\alpha = 0$
39. $\tan 3x + 1 = 0$
40. $4\sin^2 2t = 3$

41. Find the approximate solutions of $x\sin x = 10 - x^2$ in the interval $[0, 2\pi)$ by finding the points of intersection of appropriate graphs on your graphing calculator.

Review Test

1. Verify the identity $4 - \tan^2 x = 5 - \sec^2 x$.

In Exercises 2 and 3, determine exact values of the given expressions by using the addition formulas.

2. $\cos(270^\circ + 30^\circ)$
3. $\tan\left(\frac{\pi}{4} - \frac{\pi}{3}\right)$
4. Write $\sin 47^\circ$ in terms of its cofunction.
5. If $\cos\theta = \frac{4}{5}$ and θ lies in quadrant IV, find $\sin(\theta - \pi)$.
6. If $\sin x = -\frac{5}{13}$ and $\tan y = \frac{8}{3}$ with angles x and y in quadrant III, find $\tan(x - y)$.
7. If $\sin v = -\frac{12}{13}$ and $P(v)$ is in quadrant IV, find $\cos 2v$.
8. If $\cos 2\alpha = -\frac{4}{5}$ and 2α is in quadrant II, find $\cos 4\alpha$.
9. If $\csc\alpha = -2$ and α is in quadrant III, find $\cos\frac{\alpha}{2}$.
10. Find the exact value of $\tan 15^\circ$ by using a half-angle formula.
11. Verify the identity
$$\sin\frac{x}{4} = 2\sin\frac{x}{8}\cos\frac{x}{8}$$
12. Express $\sin 2x + \sin 3x$ as a product.
13. Express $\sin 150^\circ - \sin 30^\circ$ by using a product-sum formula.
14. Find all solutions of the equation $4\sin^2\alpha = 3$ in the interval $[0, 2\pi)$. Express the answers in radian measure.
15. Find all solutions of the equation $\sin^2\theta - \cos^2\theta = 0$ and express the answers in degree measure.

9 Applications of Trigonometry

In Chapter 7, we investigated the relationships of trigonometric functions pertaining to right triangles. In this chapter, we develop relationships with trigonometric functions that are more general in nature because they apply to all triangles.

We also introduce a way of blending the study of trigonometry and complex numbers. This technique allows us to solve a larger class of polynomial equations.

In addition, we investigate a different method of identifying points in the plane. With rectangular coordinates, we locate points by moving horizontally and vertically, forming the shape of a rectangle. Using polar coordinates, we choose a radius of specific length starting from the origin and sweep out a specified angle from the positive *x*-axis, forming the shape of a sector of a circle.

Finally, we conclude with a presentation of vectors. This permits us to address additional applications of mathematics; more specifically, those oriented toward physics. Furthermore, we use both rectangular and polar coordinates to develop this material.

9.1 Law of Sines

In Section 7.2, we studied the trigonometry of a right triangle. In this and the next section, we will examine an **oblique triangle**, a triangle that does not contain a right angle.

It is always possible to solve an oblique triangle by using right triangles. Consider triangle *ABC* in either Figure 9-1(a) or 9-1(b). In both cases, we drop the perpendicular from *C* to the opposite side of the triangle and analyze right triangles *ACD* and *BCD*. However, we can derive a more simplified approach to this problem that will yield two results: the *law of sines* and the *law of cosines*. We now state the law of sines, using the notation introduced in Section 7.2 and illustrated in Figure 9-1. We denote the angles of triangle *ABC* by α, β, and γ, with opposite sides a, b, and c, respectively.

FIGURE 9-1

The Law of Sines

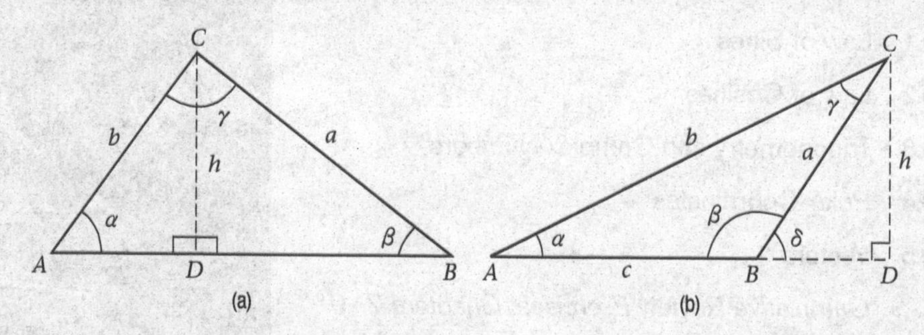

The Law of Sines

In triangle *ABC*,

$$\frac{\sin \alpha}{a} = \frac{\sin \beta}{b} = \frac{\sin \gamma}{c}$$

or alternatively,

$$\frac{a}{\sin \alpha} = \frac{b}{\sin \beta} = \frac{c}{\sin \gamma}$$

To prove the law of sines, we examine the two cases shown in Figure 9-1: (a) where all angles are acute, and (b) where there is an obtuse angle.

9.1a Case 1

In triangle *ACD*,

$$\sin \alpha = \frac{h}{b} \quad \text{or} \quad h = b \sin \alpha$$

In triangle *BCD*,

$$\sin \beta = \frac{h}{a} \quad \text{or} \quad h = a \sin \beta$$

Equating the expressions for h yields

$$b \sin \alpha = a \sin \beta$$

which can be written as

$$\frac{\sin \alpha}{a} = \frac{\sin \beta}{b} \quad \text{or} \quad \frac{a}{\sin \alpha} = \frac{b}{\sin \beta}$$

To complete the proof, we drop a perpendicular from *A* to *BC* and use a similar argument to obtain

$$\frac{\sin \beta}{b} = \frac{\sin \gamma}{c} \quad \text{or} \quad \frac{b}{\sin \beta} = \frac{c}{\sin \gamma}$$

9.1b *Case 2*

In triangle *ACD*,

$$\sin\alpha = \frac{h}{b} \qquad \text{or} \qquad h = b\sin\alpha$$

In triangle *BCD*, $\delta = 180^\circ - \beta$ so

$$\sin\delta = \sin(180^\circ - \beta) = \frac{h}{a} \qquad \text{or} \qquad h = a\sin(180^\circ - \beta)$$

Using the formula for the sine of the difference of two angles, we obtain

$$\begin{aligned} h = a\sin(180^\circ - \beta) &= a(\sin 180^\circ \cos\beta - \cos 180^\circ \sin\beta) \\ &= a\sin\beta \end{aligned}$$

Substituting, we obtain

$$b\sin\alpha = a\sin\beta$$

We now proceed as in *Case 1* to conclude the proof.

Observe that the proof of *Case 1* holds if $\gamma = 90^\circ$. In fact, consider right triangle *ABC* with $\gamma = 90^\circ$. The law of sines yields

$$\frac{\sin\alpha}{a} = \frac{\sin\beta}{b} = \frac{\sin\gamma}{c} = \frac{\sin 90^\circ}{c} = \frac{1}{c}$$

Although the law of sines may be applied to any triangle, certain minimum information is necessary before we can solve a triangle using the law of sines.

Applying the Law of Sines

The law of sines may be used when the known parts of a triangle are

1. one side and two angles (SAA)
2. two sides and the angle between them (SAS)
3. two sides and an angle opposite one of these sides (SSA)

EXAMPLE 1 ***Using the Law of Sines***

In triangle *ABC*, $\alpha = 38^\circ$, $\beta = 64^\circ$, and $c = 24$. Find approximate values for the remaining parts of the triangle.

SOLUTION

Using the information given, we sketch a triangle:

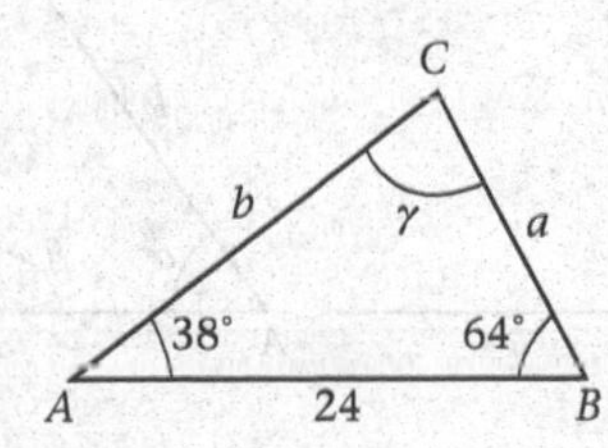

Since α and β are known,

$$\gamma = 180^\circ - (\alpha + \beta) = 180^\circ - (38^\circ + 64^\circ) = 78^\circ$$

Applying the law of sines,

$$\frac{a}{\sin\alpha} = \frac{c}{\sin\gamma}0$$

$$\frac{a}{\sin 38^\circ} = \frac{24}{\sin 78^\circ}$$

$$a = \frac{24\sin 38^\circ}{\sin 78^\circ} \approx 15.1$$

Similarly, from

$$\frac{b}{\sin\beta} = \frac{c}{\sin\gamma}$$

we obtain

$$\frac{b}{\sin 64^\circ} = \frac{24}{\sin 78^\circ}$$

$$b = \frac{24\sin 64^\circ}{\sin 78^\circ} \approx 22.05$$ ■

9.1c Unique and Ambiguous Cases

When the given parts of a triangle are two sides and an angle opposite one of them, the situation is not straightforward since a *unique* triangle is not always determined. In Figure 9-2, we have constructed angle α and side b and then used a compass to construct a side of length a with an endpoint at C. In Figure 9-2(a), no triangle exists satisfying the given conditions. Figure 9-2(b) shows that we may obtain a right triangle. In Figure 9-2(c), there is a possibility that two triangles will satisfy the given conditions. Lastly, Figure 9-2(d) shows that exactly one acute triangle may be possible.

Although there are inequalities that can be used to determine which of the four cases applies to a given set of conditions (see Exercise 23), we prefer to let the results lead us to the appropriate answer. Assume that sides a and b and angle α of triangle ABC are known and that we use the law of sines to determine angle β. These are the results that correspond to the possibilities of Figure 9-2.

a. $\sin\beta > 1$. Since $|\sin\theta| \le 1$ for all θ, there is no angle β satisfying the given conditions. This corresponds to the illustration in Figure 9-2(a).
b. $\sin\beta = 1$. Then $\beta = 90^\circ$, and the given parts determine a unique right triangle as shown in Figure 9-2(b).
c. $0 < \sin\beta < 1$. There are two possible choices for β, which is why this is called the **ambiguous case**. Since the sine function is positive in quadrants I and II, one choice will be an acute angle and one will be an obtuse angle, as shown in Figure 9-2(c).
d. $0 < \sin\beta < 1$. There are two possible choices for β, but the obtuse angle does not form a triangle, since $\alpha + \beta > 180^\circ$, as shown in Figure 9-2(d).

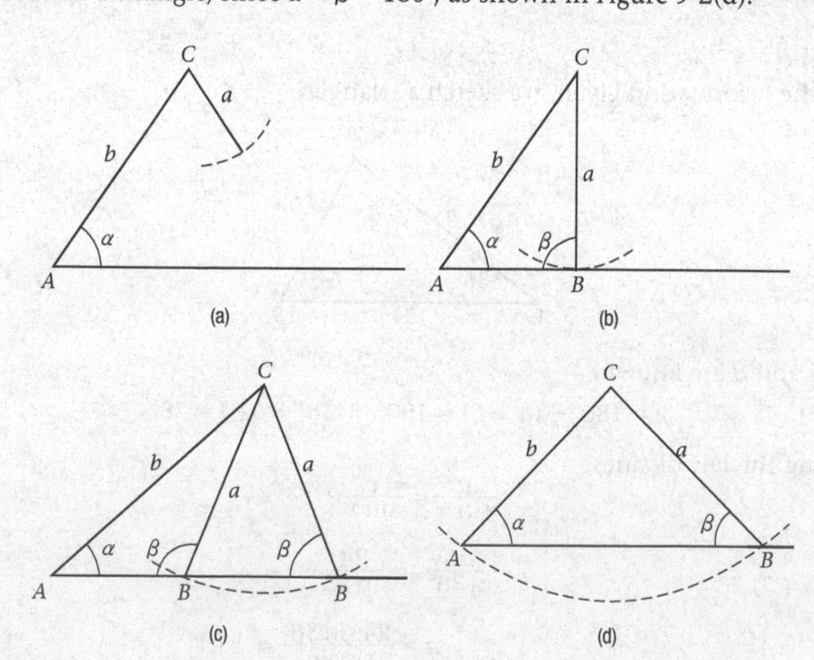

FIGURE 9-2
Construction Given Two Sides and an Angle

EXAMPLE 2 ***The Law of Sines (No Solution)***

In triangle ABC, $\alpha = 60^\circ$, $a = 5$, and $b = 7$. Find angle β.

SOLUTION
Using the law of sines,

$$\frac{\sin\alpha}{a} = \frac{\sin\beta}{b}$$

$$\sin\beta = \frac{b\sin\alpha}{a} = \frac{7\sin 60^\circ}{5} \approx 1.2$$

Since the sine function has a maximum value of 1, there is no angle β such that $\sin\beta = 1.2$. Hence, there is no triangle with the given values. This example corresponds to Figure 9-2(a). ■

EXAMPLE 3 ***The Law of Sines (Ambiguous Case)***

In triangle ABC, $a = 5$, $b = 8$, and $\alpha = 22^\circ$. Find the remaining angles of the triangle.

SOLUTION
Using the law of sines,

$$\frac{\sin\alpha}{a} = \frac{\sin\beta}{b}$$

$$\sin\beta = \frac{b\sin\alpha}{a} = \frac{8\sin 22^\circ}{5} \approx 0.5994$$

We find that $\beta \approx 36.82^\circ \approx 36^\circ 49'$ or $\beta \approx 180^\circ - 36^\circ 49' = 143^\circ 11'$. Since $\gamma = 180^\circ - \alpha - \beta$, we have

$$\alpha = 22^\circ, \qquad \beta \approx 36^\circ 49', \qquad \gamma \approx 121^\circ 11'$$

or

$$\alpha = 22^\circ, \qquad \beta \approx 143^\circ 11', \qquad \gamma \approx 14^\circ 49'$$

This is an example of an ambiguous case, which corresponds to Figure 9-2(c). ■

EXAMPLE 4 ***The Law of Sines (Unique Solution)***

In triangle ABC, $a = 9$, $b = 6$, and $\alpha = 35^\circ$. Find angles β and γ.

SOLUTION
Applying the law of sines,

$$\frac{\sin\alpha}{a} = \frac{\sin\beta}{b}$$

$$\sin\beta = \frac{b\sin\alpha}{a} = \frac{6\sin 35^\circ}{9} \approx 0.3824$$

We find that $\beta \approx 22^\circ 29'$ or $\beta \approx 180^\circ - 22^\circ 29' = 157^\circ 31'$. Since $\gamma = 180^\circ - \alpha - \beta$, we have

$$\alpha = 35^\circ, \qquad \beta \approx 22^\circ 29', \qquad \gamma \approx 122^\circ 31'$$

However, $\beta \approx 157^\circ 31'$ is impossible since $\alpha + \beta > 180$. This example corresponds to Figure 9-2(d). ■

Exercise Set 9.1

In Exercises 1–12, use the law of sines to approximate the required part(s) of triangle ABC. Give both solutions if more than one triangle satisfies the given conditions.

1. If $\alpha = 25°$, $\beta = 82°$, and $a = 12.4$, find b.
2. If $\alpha = 74°$, $\gamma = 36°$, and $c = 6.8$, find a.
3. If $\beta = 23°$, $\gamma = 47°$, and $a = 9.3$, find c.
4. If $\alpha = 46°$, $\beta = 88°$, and $c = 10.5$, find b.
5. If $\alpha = 42°20'$, $\gamma = 78°40'$, and $b = 20$, find a.
6. If $\beta = 16°30'$, $\gamma = 84°40'$, and $a = 15$, find c.
7. If $\alpha = 65°$, $a = 25$, and $b = 30$, find β.
8. If $\beta = 32°$, $b = 20$, and $c = 14$, find α and γ.
9. If $\gamma = 30°$, $a = 12.6$, and $c = 6.3$, find b.
10. If $\beta = 64°$, $a = 10$, and $b = 8$, find c.
11. If $\gamma = 45°$, $b = 7$, and $c = 6$, find a.
12. If $\alpha = 64°$, $a = 11$, and $b = 12$, find β and γ.
13. Points A and B are chosen on opposite sides of a rock quarry. A point C is 160 meters from B, and the measures of angles BAC and ABC are found to be 95° and 47°, respectively. Find the width of the quarry.
14. A tunnel is to be dug between points A and B on opposite sides of a hill. A point C is chosen that is 150 meters from A and 180 meters from B. If angle ABC measures 54°, find the length of the tunnel.
15. A ski lift 750 meters in length rises to the top of a mountain at an angle of inclination of 40°. A second lift is to be built whose base is in the same horizontal plane as the initial lift. If the angle of elevation of the second lift is 45°, what is the length of the second lift?

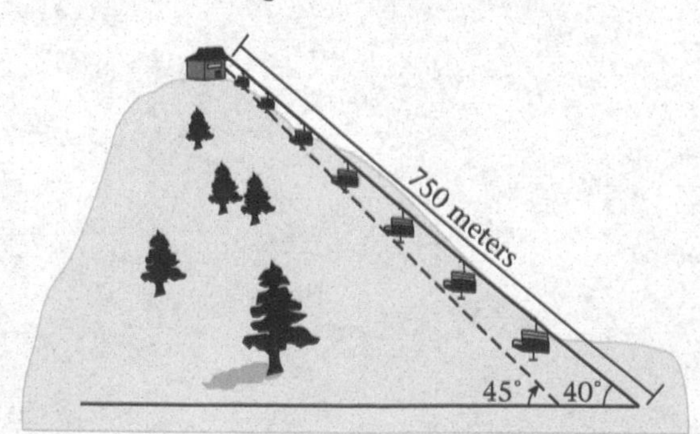

16. A tree leans away from the sun at an angle of 9° from the vertical. The tree casts a shadow 20 meters in length when the angle of elevation of the sun is 62°. Find the height of the tree.

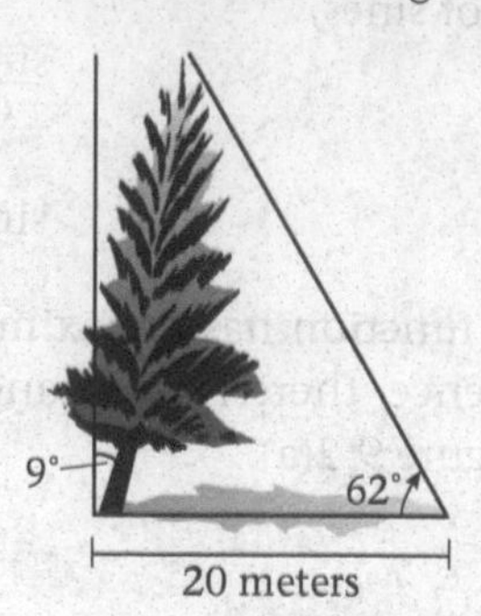

17. A ship is sailing due north at a rate of 22 miles per hour. At 2 p.m. a lighthouse is seen at a bearing of N 15° W. At 4 p.m. the bearing of the same lighthouse is S 65° W. Find the distance of the ship from the lighthouse at 2 p.m.
18. A plane leaves airport A and flies at a bearing of N 32° E. A few moments later, the plane is spotted from airport B at a bearing of N 56° W. If airport B lies 15 miles due east of airport A, find the distance of the plane from airport B at the moment it is spotted.
19. A guy wire attached to the top of a vertical pole has an angle of inclination of 65° with the ground. From a point 10 meters farther from the pole, the angle of elevation of the top of the pole is 45°. Find the height of the pole.

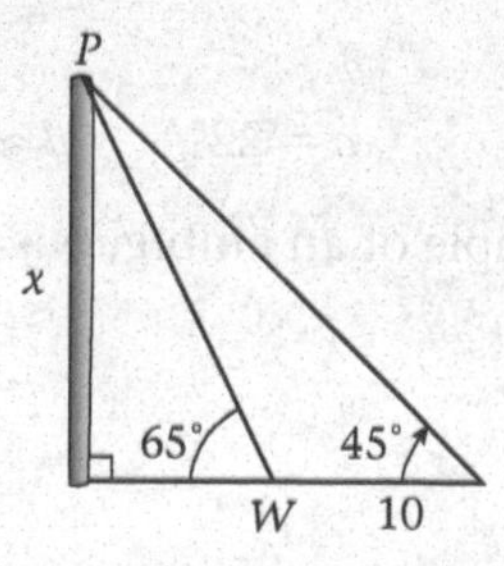

20. At 5 p.m. a sailor on board a ship sailing at a rate of 18 miles per hour spots an island due east of the ship. The ship maintains a bearing of N 26° E. At 6 p.m. the sailor finds the bearing of the island to be S 37° E. Find the distance of the island from the ship at 6 p.m.

21. The short side of a parallelogram and the shorter diagonal measure 80 centimeters and 100 centimeters, respectively. If the angle between the longer side and the shorter diagonal is 43°, find the length of the longer side.

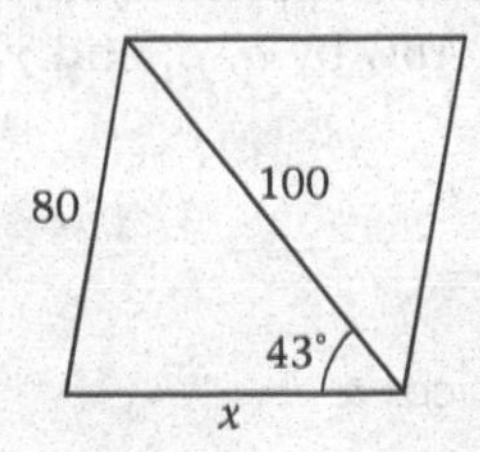

22. An archaeological mound is discovered in a jungle in Central America. To determine the height of the mound, a point A is chosen from which the angle of elevation of the top of the mound is found to be 31°. A second point B is chosen on a line with A and the base of the mound, 30 meters closer to the base of the mound. If the angle of elevation of the top of the mound from point B is 39°, find the height of the mound.

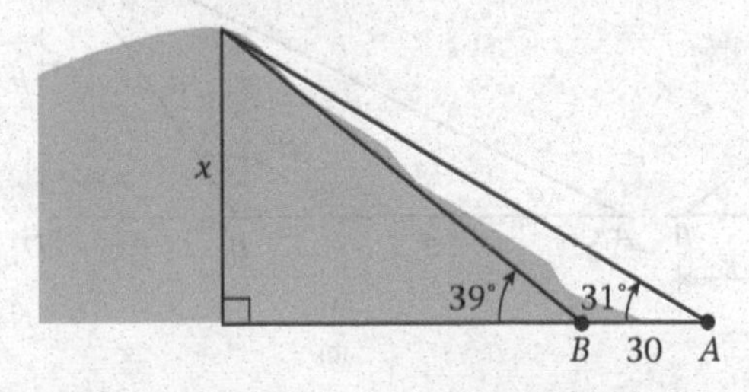

23. In a triangle, sides of length a and b and an angle α are given. Prove the following.

   a. If $b \sin \alpha > a$, there is no triangle with the given parts.

   b. If $b \sin \alpha = a$, the parts determine a right triangle.

   c. If $b \sin \alpha < a < b$, there are two triangles with the given parts.

   d. If $b \leq a$, there is one acute triangle with the given parts.

9.2 Law of Cosines

In the previous section, we applied the law of sines to an oblique triangle. We will now state and prove the law of cosines, which also applies to an oblique triangle. Once again, we denote the angles of triangle ABC by α, β, and γ, with opposite sides a, b, and c, respectively.

The Law of Cosines
In triangle ABC,

$$a^2 = b^2 + c^2 - 2bc \cos \alpha \tag{1}$$

$$b^2 = a^2 + c^2 - 2ac \cos \beta \tag{2}$$

$$c^2 = a^2 + b^2 - 2ab \cos \gamma \tag{3}$$

Consider triangle ABC in Figure 9-3: (a) where all angles are acute and (b) where there is an obtuse angle. In both cases, we construct the perpendicular CD to AB, forming right triangles ACD and BCD.

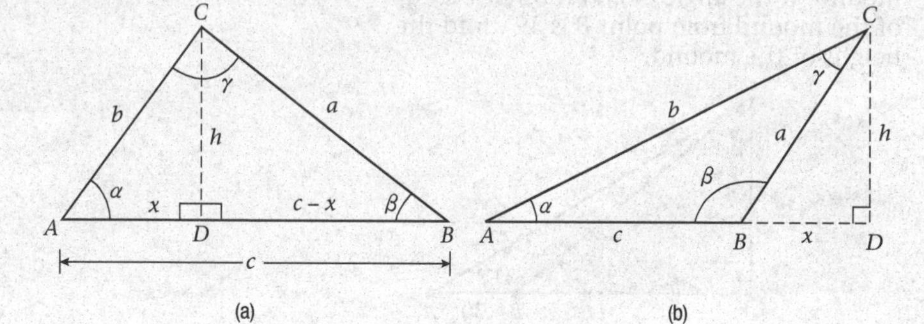

FIGURE 9-3
The Law of Cosines

9.2a Case 1

Applying the Pythagorean Theorem to right triangles ACD and BCD yields

$$b^2 = x^2 + h^2 \qquad \text{and} \qquad a^2 = (c - x)^2 + h^2$$

Expanding, we have

$$a^2 = c^2 - 2cx + x^2 + h^2 = c^2 - 2cx + b^2 \tag{4}$$

Also, in triangle ACD,

$$\cos \alpha = \frac{x}{b} \qquad \text{or} \qquad x = b \cos \alpha$$

Substituting this value for x in Equation (4) establishes Equation (1), namely,

$$a^2 = b^2 + c^2 - 2bc \cos \alpha$$

9.2b Case 2

In triangle *ACD*,

$$\sin \alpha = \frac{h}{b} \quad \text{or} \quad h = b \sin \alpha$$

and

$$\cos \alpha = \frac{c + x}{b} \quad \text{or} \quad x = b \cos \alpha - c$$

Applying the Pythagorean Theorem to triangle *BCD* yields

$$a^2 = x^2 + h^2$$

Substituting the values for x and h from above, we obtain

$$\begin{aligned} a^2 &= (b \sin \alpha)^2 + (b \cos \alpha - c)^2 \\ &= b^2 \sin^2 \alpha + b^2 \cos^2 \alpha - 2bc \cos \alpha + c^2 \\ &= b^2 (\sin^2 \alpha + \cos^2 \alpha) + c^2 - 2bc \cos \alpha \\ &= b^2 + c^2 - 2bc \cos \alpha \end{aligned}$$

which establishes Equation (1) once again.

A similar argument can be used to establish the other two forms given in Equations (2) and (3).

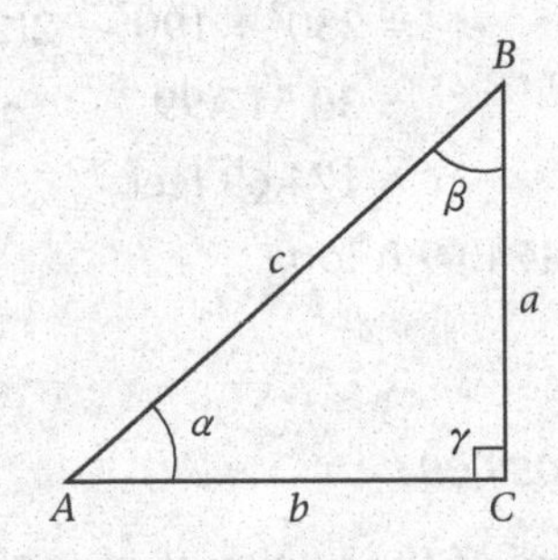

FIGURE 9-4
Right Triangle *ABC*

The law of cosines can be viewed as a generalization of the Pythagorean Theorem. Consider right triangle *ABC* with $\gamma = 90^\circ$, as shown in Figure 9-4. From Equation (3), we have

$$c^2 = a^2 + b^2 - 2ab \cos \gamma$$

However, $\cos \gamma = \cos 90^\circ = 0$, and we obtain

$$c^2 = a^2 + b^2$$

which is the Pythagorean Theorem.

We have seen that the law of cosines may be applied to any triangle. However, certain minimum information must be given before we can solve a triangle using the law of cosines.

Applying the Law of Cosines

The law of cosines may be used when

1. three sides of a triangle are known (*SSS*)
2. two sides of a triangle are known and the measure of the angle formed by those sides is known (*SAS*)

EXAMPLE 1 *Applying the Law of Cosines*

Highway engineers, who are to dig a tunnel through a small mountain, wish to determine the length of the tunnel. Points A and B are chosen as the endpoints of the tunnel, as shown in Figure 9-5. Point C is then selected, and the distances from A and from B are found to be 190 feet, and 230 feet, respectively. If angle ACB measures 48°, find the approximate length of the tunnel.

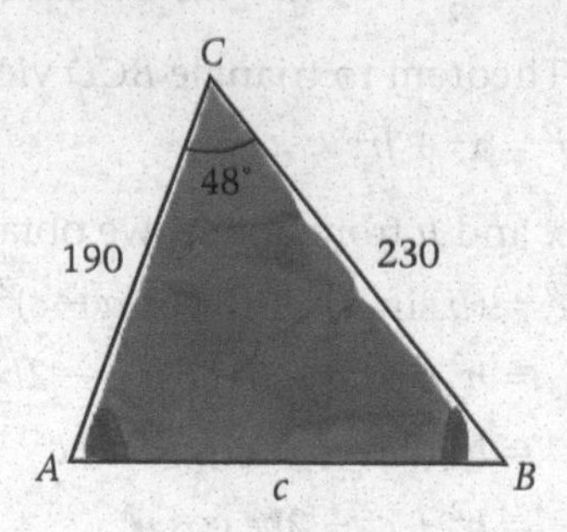

FIGURE 9-5
Diagram for Example 1

SOLUTION

Applying the law of cosines,

$$\begin{aligned} c^2 &= a^2 + b^2 - 2ab \cos \gamma \\ &= 230^2 + 190^2 - 2(230)(190) \cos 48^\circ \\ &\approx 30{,}517.99 \\ c &\approx 174.69 \text{ feet} \end{aligned}$$

the tunnel is approximately 174.69 ft long. ■

EXAMPLE 2 *The Law of Cosines*

Find the approximate measure of the angles of triangle ABC if $a = 150$, $b = 100$, and $c = 75$.

SOLUTION

Substituting into the equation

$$\begin{aligned} a^2 &= b^2 + c^2 - 2bc \cos \alpha \\ 150^2 &= 100^2 + 75^2 - 2(100)(75) \cos \alpha \\ 22{,}500 &= 10{,}000 + 5625 - 15{,}000 \cos \alpha \\ \cos \alpha &\approx -0.4583 \end{aligned}$$

Since $\cos \alpha$ is negative, angle α must lie in the second quadrant and is an obtuse angle. Therefore,

$$\alpha \approx 117.28^\circ \approx 117^\circ 17'$$

Similarly,

$$\begin{aligned} b^2 &= a^2 + c^2 - 2ac \cos \beta \\ 100^2 &= 150^2 + 75^2 - 2(150)(75) \cos \beta \\ 10{,}000 &= 22{,}500 + 5625 - 22{,}500 \cos \beta \\ \cos \beta &\approx 0.8056 \\ \beta &\approx 36^\circ 20' \end{aligned}$$

Since $\alpha + \beta + \gamma = 180^\circ$,

$$\gamma \approx 180^\circ - (117^\circ 17' + 36^\circ 20') = 26^\circ 23'$$

(Verify that $c^2 = a^2 + b^2 - 2ab \cos \gamma$.) ■

We conclude this section recalling a result from plane geometry.

In any triangle, the smallest angle is opposite the smallest side and the largest angle is opposite the largest side.

This theorem provides a quick check if the solution of the triangle in question is reasonable.

Exercise Set 9.2

In Exercises 1–10, use the law of cosines to approximate the required part of triangle ABC.

1. If $a = 10$, $b = 15$, and $c = 21$, find β.
2. If $a = 5$, $b = 12$, and $c = 15$, find γ.
3. If $a = 25$, $c = 30$, and $\beta = 28°30'$, find b.
4. If $b = 20$, $c = 13$, and $\alpha = 19°10'$, find a.
5. If $a = 10$, $b = 12$, and $\gamma = 108°$, find c.
6. If $a = 30$, $c = 40$, and $\beta = 122°$, find b.
7. If $b = 6$, $a = 7$, and $\gamma = 68°$, find α.
8. If $a = 6$, $b = 15$, and $c = 16$, find β.
9. If $a = 9$, $b = 12$, and $c = 15$, find γ.
10. If $a = 11$, $c = 15$, and $\beta = 33°$, find γ.
11. The sides of a parallelogram measure 25 centimeters and 40 centimeters, and the longer diagonal measures 50 centimeters. Find the approximate measure of the smaller angle of the parallelogram.
12. The sides of a parallelogram measure 40 inches and 70 inches, and one of the angles is 108°. Find the approximate length of each diagonal of the parallelogram.
13. A ship leaves port at 9 a.m. and travels due west at a rate of 15 miles per hour. At 11 a.m. the ship changes direction to S 32° W. What is the distance and bearing of the ship from port at 1 p.m.?
14. A ship leaves from port A intending to travel direct to port B, a distance of 25 kilometers. After travelling 12 kilometers, the captain finds that his course has been in error by 10°. How far is the ship from port B?
15. Two trains leave Pennsylvania Station in New York City at 2 p.m. and travel in directions that differ by 55°. If the trains travel at constant rates of 50 miles per hour and 80 miles per hour, respectively, what is the distance between them at 2:30 p.m.?
16. Hurricane David has left a telephone pole in a nonvertical position. Workmen place a 30-foot ladder at a point 10 feet from the base of the pole. If the ladder touches the pole at a point 26 feet up the pole, find the angle the pole makes with the ground.

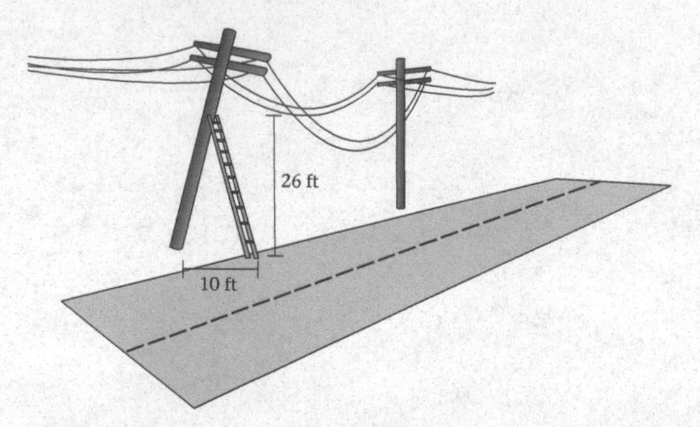

17. Find the approximate perimeter of triangle ABC if $a = 20$, $b = 30$, and $\gamma = 37°$.
18. A hill makes an angle of 10° with the horizontal. An antenna 50 feet in height is erected at the top of the hill and a guy wire is run to a point 30 feet from the base of the antenna. What is the length of the guy wire?

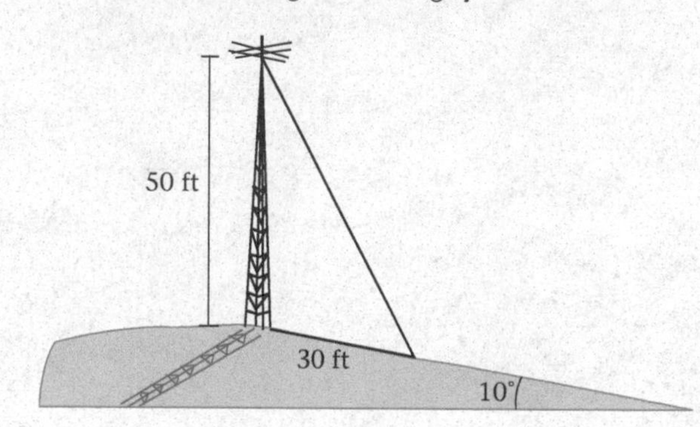

19. Prove the following in triangle ABC.
    a. $a^2 + b^2 + c^2 = 2(bc \cos \alpha + ac \cos \beta + ab \cos \gamma)$
    b. $\dfrac{\cos \alpha}{a} + \dfrac{\cos \beta}{b} + \dfrac{\cos \gamma}{c} = \dfrac{a^2 + b^2 + c^2}{2abc}$
20. Prove that if
$$\frac{\cos \beta}{a} = \frac{\cos \alpha}{b}$$
triangle ABC is either a right triangle or an isosceles triangle.
21. Prove that the area of triangle ABC is $\frac{1}{2} ab \sin \gamma$ if the area of a triangle is
$$\tfrac{1}{2}(\text{length of base})(\text{length of altitude})$$
22. Prove that the area of triangle ABC is
$$\sqrt{s(s-a)(s-b)(s-c)}$$
where $s = \frac{1}{2}(a+b+c)$. This is known as Heron's Formula. (*Hint*: Use Exercise 21 and the law of cosines.)

9.3 Trigonometry and Complex Numbers

9.3a *The Complex Plane*

We make the association between the complex number $2 + 3i$ and the point in the plane whose coordinates are (2, 3). Figure 9-6(a) shows the geometric representation of several complex numbers. In fact, as shown in Figure 9-6(b), to any complex number $a + bi$, we may associate the point (a, b) in the complex plane.

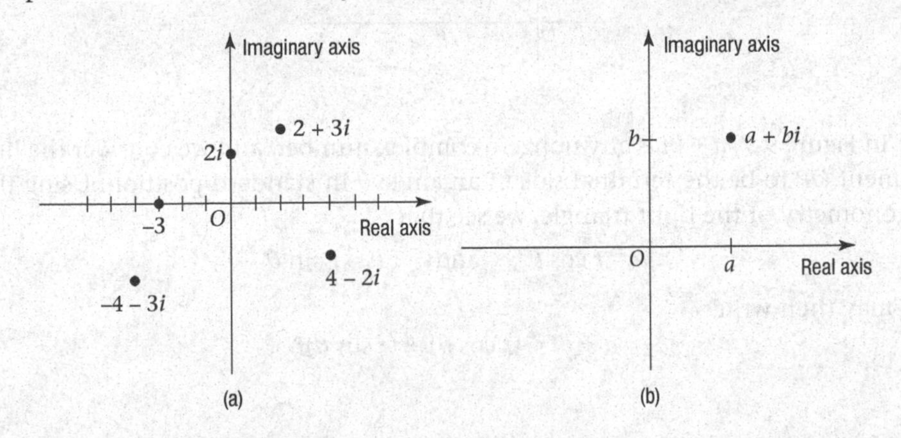

FIGURE 9-6 The Complex Plane

Conversely, every point (a, b) in the complex plane can be viewed as representing the complex number $a + bi$. When a rectangular coordinate system is used to represent complex numbers, it is called a **complex plane** and the x- and y-axes are called the **real axis** and the **imaginary axis**, respectively.

We can extend the concept of absolute value to complex numbers. Since $|x|$ represents the distance on a real number line from the origin to a point that corresponds to x, it would be consistent to define the **absolute value** $|a + bi|$ as the distance from the origin to the point corresponding to $a + bi$. Applying the distance formula, we are led to the following definition.

> The absolute value of a complex number $a + bi$ is denoted by $|a + bi|$ and is defined by $|a + bi| = \sqrt{a^2 + b^2}$.

The quantity $|a + bi|$ is also called the **modulus** of $a + bi$.

EXAMPLE 1 ***Absolute Value of Complex Numbers***

Find the absolute value of each of the following complex numbers.

a. $2 - 3i$ b. $4i$ c. -2

SOLUTION

Applying the definition of absolute value,

a. $|2 - 3i| = \sqrt{4 + 9} = \sqrt{13}$

b. $|4i| = \sqrt{0 + 16} = 4$

c. $|-2| = \sqrt{4 + 0} = 2$

Observe that $|-2|$ produces the same result, whether we view -2 as a complex number or as a real number. ■

The representation of a complex number as a point in a coordinate plane can be used to link complex numbers with the trigonometry of the right triangle.

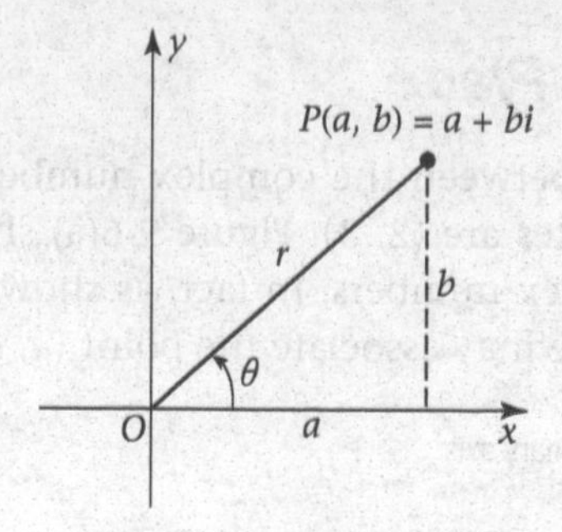

FIGURE 9-7
Polar Form

In Figure 9-7, $a + bi$ is any nonzero complex number, and we consider the line segment OP to be the terminal side of an angle θ in standard position. Using the trigonometry of the right triangle, we see that

$$a = r\cos\theta \qquad \text{and} \qquad b = r\sin\theta$$

We may then write

$$a + bi = (r\cos\theta) + (r\sin\theta)i$$

or

$$a + bi = r(\cos\theta + i\sin\theta) \tag{1}$$

where $r = \overline{OP} = |a + bi| = \sqrt{a^2 + b^2}$. If $a + bi = 0$, then $r = 0$, and θ may assume any value. We also note that $\tan\theta = \frac{b}{a}$.

Equation (1) is known as the **trigonometric form**, or *polar form,* of a complex number. Since we have an infinite number of choices for the angle θ, the polar form of a complex number is not unique. We call r the **modulus** and θ the **argument** of the complex number $r(\cos\theta + i\sin\theta)$. If $0 \le \theta < 360°$, or, equivalently, $0 \le \theta < 2\pi$, then θ is called the **principal argument**.

EXAMPLE 2 *Converting to Trigonometric Form*

Write the complex number $-2 + 2i$ in trigonometric form.

SOLUTION

The geometric representation is shown in Figure 9-8. The modulus of $-2 + 2i$ is

$$r = |-2 + 2i| = \sqrt{4 + 4} = 2\sqrt{2}$$

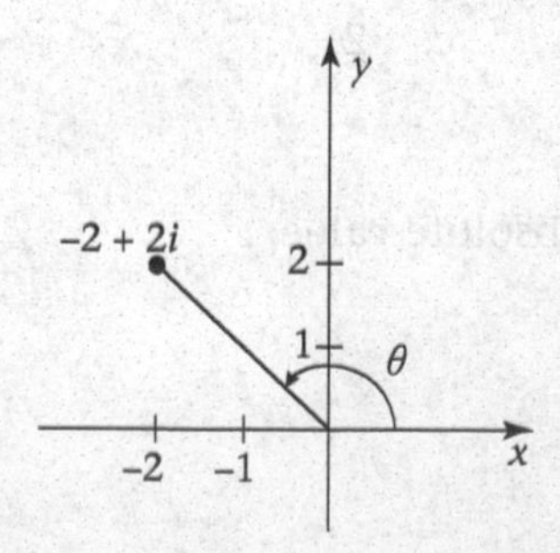

FIGURE 9-8
Trigonometric Form of $-2 + 2i$

The principal argument θ is an angle in the second quadrant such that

$$\tan\theta = \frac{2}{-2} = -1$$

Thus $\theta = 135°$, and using the trigonometric form of a complex number in Equation (1), we have

$$-2 + 2i = 2\sqrt{2}(\cos 135° + i \sin 135°)$$ ■

Progress Check

Write the complex number $1 - \sqrt{3}i$ in trigonometric form.

Answer

$2(\cos 300° + i \sin 300°)$

EXAMPLE 3 ***Converting from Trigonometric Form***

Write the complex number $2\sqrt{3}(\cos 150° + i \sin 150°)$ in the form $a + bi$.

SOLUTION

We substitute $\cos 150° = -\frac{\sqrt{3}}{2}$ and $\sin 150° = \frac{1}{2}$. Thus,

$$2\sqrt{3}(\cos 150° + i \sin 150°) = 2\sqrt{3}\left(-\frac{\sqrt{3}}{2} + \frac{1}{2}i\right)$$
$$= -3 + \sqrt{3}i$$ ■

Progress Check

Write the complex number

$$\sqrt{2}\left(\cos\frac{\pi}{4} + i\sin\frac{\pi}{4}\right)$$

in the form $a + bi$.

Answer

$1 + i$

9.3b Multiplication and Division

Now that we have introduced the trigonometric form of a complex number, we wish to examine what happens with this form when we multiply and divide complex numbers.

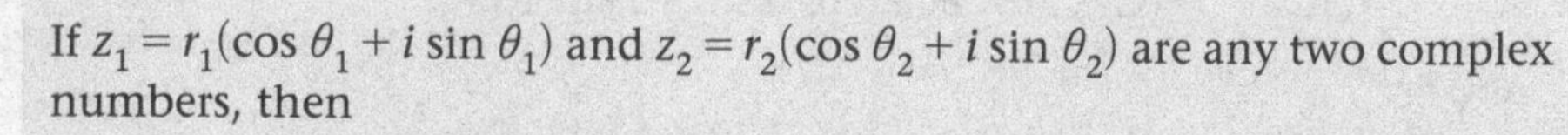

If $z_1 = r_1(\cos\theta_1 + i\sin\theta_1)$ and $z_2 = r_2(\cos\theta_2 + i\sin\theta_2)$ are any two complex numbers, then

$$\begin{aligned} z_1z_2 &= r_1(\cos\theta_1 + i\sin\theta_1)\cdot r_2(\cos\theta_2 + i\sin\theta_2) \\ &= r_1r_2[\cos(\theta_1+\theta_2) + i\sin(\theta_1+\theta_2)] \end{aligned} \quad (2)$$

and

$$\frac{z_1}{z_2} = \frac{r_1(\cos\theta_1 + i\sin\theta_1)}{r_2(\cos\theta_2 + i\sin\theta_2)} = \frac{r_1}{r_2}[\cos(\theta_1-\theta_2) + i\sin(\theta_1-\theta_2)] \quad (3)$$

Note that the rule for multiplication requires the multiplication of the moduli and the addition of the arguments. To prove this, we see that

$$\begin{aligned} &r_1(\cos\theta_1 + i\sin\theta_1)\cdot r_2(\cos\theta_2 + i\sin\theta_2) \\ &= r_1r_2[(\cos\theta_1\cos\theta_2 - \sin\theta_1\sin\theta_2) + i(\sin\theta_1\cos\theta_2 + \cos\theta_1\sin\theta_2)] \\ &= r_1r_2[\cos(\theta_1+\theta_2) + i\sin(\theta_1+\theta_2)] \end{aligned}$$

where the last step results from the addition and subtraction formulas.

The rule for division requires the division of moduli and the subtraction of the arguments. The proof is left as an exercise.

EXAMPLE 4 *Product of Complex Numbers*

Find the product of the complex numbers $1 + i$ and $-2i$ by

a. writing the numbers in trigonometric form, and

b. multiplying the numbers algebraically.

SOLUTION

a. The trigonometric forms of these complex numbers are

$$1 + i = \sqrt{2}(\cos 45^\circ + i\sin 45^\circ)$$

and

$$-2i = 2(\cos 270^\circ + i\sin 270^\circ)$$

Multiplying, we have

$$\begin{aligned} &\sqrt{2}(\cos 45^\circ + i\sin 45^\circ)\cdot 2(\cos 270^\circ + i\sin 270^\circ) \\ &= 2\sqrt{2}(\cos 315^\circ + i\sin 315^\circ) \\ &= 2\sqrt{2}\left(\frac{\sqrt{2}}{2} - i\frac{\sqrt{2}}{2}\right) \\ &= 2 - 2i \end{aligned}$$

b. Multiplying algebraically,

$$(1+i)(-2i) = -2i - 2i^2 = -2i + 2 = 2 - 2i$$

■

Progress Check

Express the complex numbers $1+\sqrt{3}\,i$ and $1-\sqrt{3}\,i$ in trigonometric form and find their product.

Answers

$2(\cos 60^\circ + i\sin 60^\circ)$, $2(\cos 300^\circ + i\sin 300^\circ)$, 4

EXAMPLE 5 ***Quotient of Complex Numbers***

Write the quotient $\frac{1+i}{-2i}$ in the form $a+bi$ by

a. writing the numbers in trigonometric form, and

b. using the conjugate of the denominator.

SOLUTION

a. From Example 4,

$$\frac{1+i}{-2i} = \frac{\sqrt{2}(\cos 45^\circ + i\sin 45^\circ)}{2(\cos 270^\circ + i\sin 270^\circ)} = \frac{\sqrt{2}}{2}[(\cos(-225^\circ) + i\sin(-225^\circ))]$$

$$= \frac{\sqrt{2}}{2}(\cos 135^\circ + i\sin 135^\circ) = \frac{\sqrt{2}}{2}\left(-\frac{\sqrt{2}}{2} + i\frac{\sqrt{2}}{2}\right) = -\frac{2}{4} + \frac{2}{4}i$$

$$= -\frac{1}{2} + \frac{1}{2}i$$

b. $$\frac{1+i}{-2i}\left(\frac{2i}{2i}\right) = \frac{2i+2i^2}{-4i^2} = \frac{2i-2}{4} = -\frac{1}{2} + \frac{1}{2}i$$ ■

Progress Check

Write the quotient $\frac{1+\sqrt{3}i}{1-\sqrt{3}i}$ in trigonometric form and then in the form $a+bi$.

Answer

$\cos(-240^\circ) + i\sin(-240^\circ) = -\frac{1}{2} + \frac{\sqrt{3}}{2}i$

9.3c De Moivre's Theorem

Since repeated multiplication is equivalent to exponentiation to a positive integer, we might anticipate the following result when a complex number in trigonometric form is raised to a power. The theorem that states this result is credited to Abraham De Moivre, a French-born English mathematician.

De Moivre's Theorem
If $z = r(\cos\theta + i\sin\theta)$ is a complex number and n is a natural number, then

$$z^n = [r(\cos\theta + i\sin\theta)]^n = r^n(\cos n\theta + i\sin n\theta)$$

We verify this theorem for $n = 2$ and $n = 3$. By Equation (2),

$$\begin{aligned}[r(\cos\theta + i\sin\theta)]^2 &= r(\cos\theta + i\sin\theta)\cdot r(\cos\theta + i\sin\theta)\\ &= r^2[\cos(\theta+\theta) + i\sin(\theta+\theta)]\\ &= r^2(\cos 2\theta + i\sin 2\theta)\end{aligned}$$

and

$$\begin{aligned}[r(\cos\theta + i\sin\theta)]^3 &= r^2(\cos 2\theta + i\sin 2\theta)\cdot r(\cos\theta + i\sin\theta)\\ &= r^3(\cos 3\theta + i\sin 3\theta)\end{aligned}$$

EXAMPLE 6 *Applying De Moivre's Theorem*

Evaluate $(1 - i)^{10}$.

SOLUTION

Writing $1 - i$ in trigonometric form, we have

$$1 - i = \sqrt{2}(\cos 315^\circ + i\sin 315^\circ)$$

and

$$(1 - i)^{10} = [\sqrt{2}(\cos 315^\circ + i\sin 315^\circ)]^{10}$$

Applying De Moivre's Theorem,

$$\begin{aligned}(1 - i)^{10} &= (\sqrt{2})^{10}[\cos 3150^\circ + i\sin 3150^\circ]\\ &= 32[\cos 270^\circ + i\sin 270^\circ]\\ &= 32[0 + i(-1)] = -32i\end{aligned}$$

■

Progress Check

Evaluate $(\sqrt{3} + i)^6$.

Answer

-64

9.3d *n*th Root of a Complex Number

Recall that a real number a is said to be an nth root of a real number b if $a^n = b$ for a positive integer n. In a similar manner, we say that the complex number u is an ***n*th root** of the nonzero complex number z if $u^n = z$. If we express u and z in trigonometric form as

$$u = s(\cos\phi + i\sin\phi) \qquad z = r(\cos\theta + i\sin\theta) \tag{4}$$

we can then apply De Moivre's Theorem to obtain

$$u^n = s^n(\cos n\phi + i\sin n\phi) = r(\cos\theta + i\sin\theta) \tag{5}$$

Since the two complex numbers u^n and z are equal, they are represented by the same point in the complex plane. Hence, the moduli must be equal, since the modulus is the distance of the point from the origin. Therefore, $s^n = r$ or

$$s = \sqrt[n]{r}$$

Since $z \neq 0$, we know that $r \neq 0$. We may therefore divide Equation (5) by r to obtain

$$\cos n\phi + i\sin n\phi = \cos\theta + i\sin\theta$$

By the definition of equality of complex numbers, we must have

$$\cos n\phi = \cos\theta \qquad \sin n\phi = \sin\theta$$

Since both sine and cosine are periodic functions with period 2π, we conclude that

$$n\phi = \theta + 2\pi k$$

or

$$\phi = \frac{\theta + 2\pi k}{n}$$

where k is an integer. Substituting for s and for ϕ in the trigonometric form of u given in Equation (4), we have proved the following theorem.

The *n*th Roots of a Complex Number

Let $z = r(\cos\theta + i\sin\theta)$ be any nonzero complex number and n any positive integer. Then, the n distinct roots $u_0, u_1, \ldots, u_{n-1}$ of z are

$$u_k = \sqrt[n]{r}\left[\cos\left(\frac{\theta + 2\pi k}{n}\right) + i\sin\left(\frac{\theta + 2\pi k}{n}\right)\right]$$

where $k = 0, 1, 2, \ldots, n - 1$

Note that when k exceeds $n - 1$, we repeat a previous root. For example, when $k = n$, the angle is

$$\frac{\theta + 2\pi n}{n} = \frac{\theta}{n} + 2\pi = \frac{\theta}{n}$$

which is the same result that is obtained when $k = 0$.

EXAMPLE 7 *The Cube Roots of a Complex Number*

Find the cube roots of $-8i$.

SOLUTION

For $z = a + bi = 0 - 8i$, we have $a = 0$, $b = -8$, and thus

$$r = \sqrt{a^2+b^2} = \sqrt{0^2+(-8)^2} = 8$$

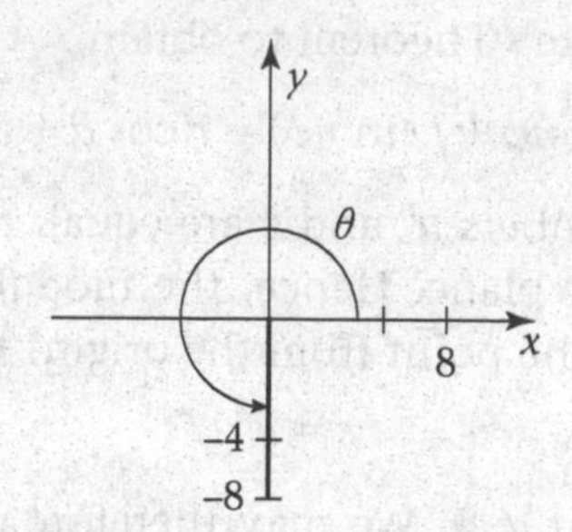

FIGURE 9-9
Trigonometric Form of $-8i$

From Figure 9-9, we see that $\theta = 270^\circ$. Then, in trigonometric form,

$$-8i = 8(\cos 270^\circ + i \sin 270^\circ)$$

With $r = 8$, $\theta = 270^\circ$, and $n = 3$, the cube roots are

$$\sqrt[3]{8}\left[\cos\left(\frac{270^\circ + 360^\circ k}{3}\right) + i \sin\left(\frac{270^\circ + 360^\circ k}{3}\right)\right]$$

for $k = 0, 1, 2$. Substituting for each value of k we have

$$2(\cos 90^\circ + i \sin 90^\circ) = 2i$$

$$2(\cos 210^\circ + i \sin 210^\circ) = -\sqrt{3} - i$$

$$2(\cos 330^\circ + i \sin 330^\circ) = \sqrt{3} - i$$

■

When $z = 1$, we call the n distinct nth roots of z the ***n*th roots of unity.**

EXAMPLE 8 *The Fourth Roots of Unity*

Find the four fourth roots of unity.

SOLUTION

In trigonometric form,

$$1 = 1(\cos 0^\circ + i \sin 0^\circ)$$

so that $r = 1$, $\theta = 0^\circ$, and $n = 4$. The fourth roots are then given by

$$\sqrt[4]{1}\left[\cos\left(\frac{0^\circ + 360^\circ k}{4}\right) + i \sin\left(\frac{0^\circ + 360^\circ k}{4}\right)\right]$$

for $k = 0, 1, 2, 3$. Substituting these values for k yields

$$\cos 0^\circ + i \sin 0^\circ = 1$$

$$\cos 90^\circ + i \sin 90^\circ = i$$

$$\cos 180^\circ + i \sin 180^\circ = -1$$

$$\cos 270^\circ + i \sin 270^\circ = -i$$

See Figure 9-10.

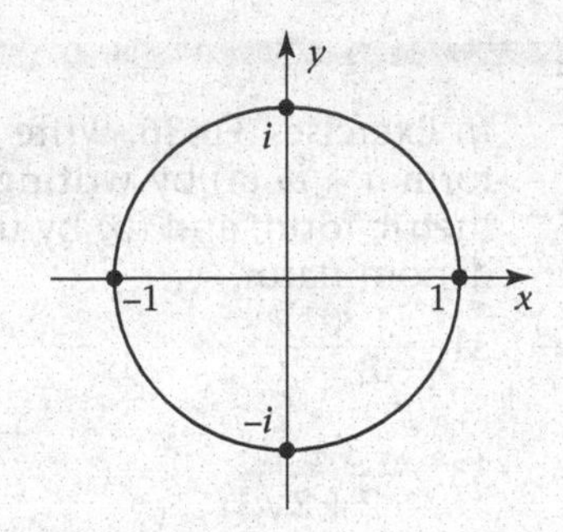

FIGURE 9-10
The Fourth Roots of Unity

■

Progress Check

Find the two square roots of $\frac{\sqrt{3}}{2} - \frac{1}{2}i$. Express the answers in trigonometric form.

Answers

$\cos 165^\circ + i \sin 165^\circ$, $\cos 345^\circ + i \sin 345^\circ$

Exercise Set 9.3

In Exercises 1–6, find the absolute value of each given complex number.

1. $3-2i$
2. $-7+6i$
3. $1+i$
4. $\frac{1}{2}+\frac{1}{2}i$
5. $-6-2i$
6. $3-i$

In Exercises 7–14, express the given complex number in trigonometric form.

7. $3-3i$
8. $2+2i$
9. $\sqrt{3}-i$
10. $-2-2\sqrt{3}i$
11. $-1+i$
12. $-2i$
13. -4
14. $3i$

In Exercises 15–20, convert the given complex number from trigonometric form to the algebraic form $a+bi$.

15. $4(\cos 180^\circ + i\sin 180^\circ)$
16. $\frac{1}{2}\left(\cos\frac{\pi}{2}+i\sin\frac{\pi}{2}\right)$
17. $\sqrt{2}(\cos 135^\circ + i\sin 135^\circ)$
18. $2(\cos 120^\circ + i\sin 120^\circ)$
19. $5\left(\cos\frac{3\pi}{2}+i\sin\frac{3\pi}{2}\right)$
20. $4(\cos 240^\circ + i\sin 240^\circ)$

In Exercises 21–24, find the product of the given complex numbers. Express the answers in trigonometric form.

21. $2(\cos 150^\circ + i\sin 150^\circ)\cdot 3(\cos 210^\circ + i\sin 210^\circ)$
22. $3(\cos 120^\circ + i\sin 120^\circ)\cdot 3(\cos 150^\circ + i\sin 150^\circ)$
23. $2\left(\cos\frac{\pi}{5}+i\sin\frac{\pi}{5}\right)\cdot\left(\cos\frac{\pi}{4}+i\sin\frac{\pi}{4}\right)$
24. $3\left(\cos\frac{4\pi}{7}+i\sin\frac{4\pi}{7}\right)\cdot 4\left(\cos\frac{5\pi}{3}+i\sin\frac{5\pi}{3}\right)$

In Exercises 25–30, express the given complex numbers in trigonometric form, compute the product, and write the answer in the form $a+bi$.

25. $1-i,\ 2i$
26. $-\sqrt{3}+i,\ -2$
27. $-2+2\sqrt{3}i,\ 3+3i$
28. $1-\sqrt{3}i,\ 1+\sqrt{3}i$
29. $5,\ -2-2i$
30. $-4i,\ -3i$

In Exercises 31–36, write the given quotient in the form $a+bi$ (a) by writing the numbers in trigonometric form, and (b) by using the conjugate of the denominator.

31. $\frac{5}{-2-2i}$
32. $\frac{2}{-\sqrt{3}+i}$
33. $\frac{3+3i}{-2+2\sqrt{3}i}$
34. $\frac{3i}{-4i}$
35. $\frac{2i}{1-i}$
36. $\frac{1-\sqrt{3}i}{1+\sqrt{3}i}$

In Exercises 37–42, use De Moivre's Theorem to express the given number in the form $a+bi$.

37. $(-2+2i)^6$
38. $(\sqrt{3}-i)^{10}$
39. $(1-i)^9$
40. $(-1+\sqrt{3}i)^{10}$
41. $(-1-i)^7$
42. $(-\sqrt{2}+\sqrt{2}i)^6$

In Exercises 43–46, find the indicated roots of the given complex number. Express the answer in the indicated form.

43. The fourth roots of -16 in algebraic form $a+bi$.
44. The square roots of -25 in trigonometric form.
45. The square roots of $1-\sqrt{3}i$ in trigonometric form.
46. The four fourth roots of unity in algebraic form.

In Exercises 47–50, determine all roots of the given equation.

47. $z^3+8=0$
48. $z^3+125=0$
49. $z^4-16=0$
50. $z^4+16=0$

51. Prove $\frac{r_1(\cos\theta_1+i\sin\theta_1)}{r_2(\cos\theta_2+i\sin\theta_2)} = \frac{r_1}{r_2}[\cos(\theta_1-\theta_2)+i\sin(\theta_1-\theta_2)]$

52. Prove that the sum of the nth roots of unity is 0 for every positive integer n.

9.4 Polar Coordinates

Thus far, we have represented a point P in the plane by the ordered pair (x, y), where x and y represent the location of P relative to the x-axis and y-axis, respectively. In this section we discuss the polar coordinate system, another way of representing points in the plane.

We start with a point O, called the *origin* or **pole**, draw a fixed ray, called the **polar axis**, with an endpoint at O, and select a unit of length for measuring distance. For any point P in the plane other than O, we draw a ray from O to P, as shown in Figure 9-11. If $r = \overline{OP}$, the length of the segment OP, and θ is the angle between the polar axis and the ray OP, then r and θ are called the **polar coordinates** of P, and P is denoted by $P(r, \theta)$, or simply by (r, θ). As usual, the angle θ is positive if it is measured in a counterclockwise direction and negative if measured in a clockwise direction. The angle θ may be expressed in degrees or radians.

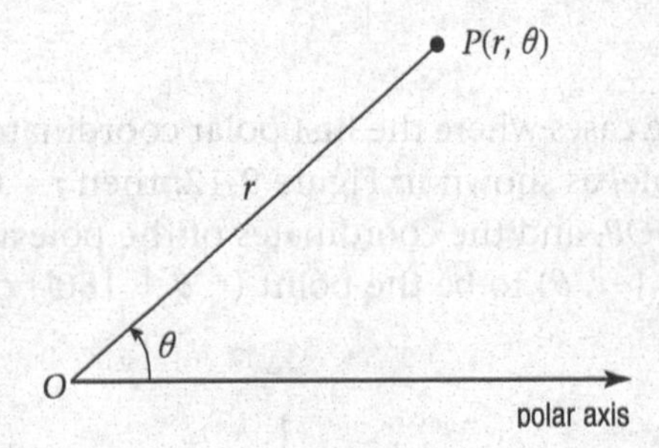

FIGURE 9-11
The Pole and Polar Axis

EXAMPLE 1 *Plotting Polar Coordinates*

Plot the points with the given polar coordinates.

a. $P(3, 45°)$ b. $P\left(3, \frac{\pi}{6}\right)$ c. $P(2, -120°)$ d. $P\left(2, -\frac{\pi}{6}\right)$

SOLUTION

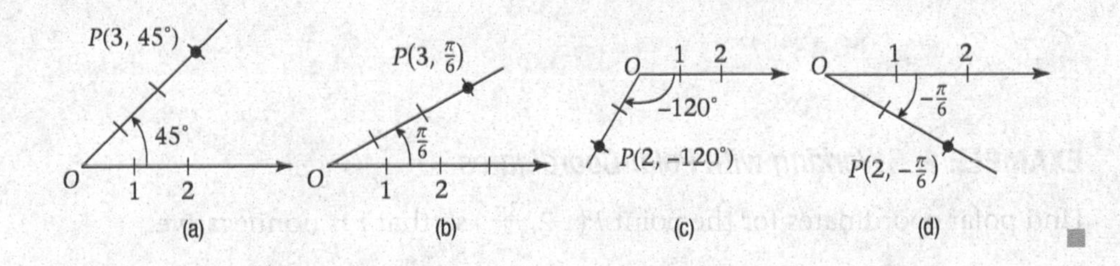

■

There is an important difference between the polar coordinates and the rectangular coordinates of a point P. The rectangular coordinates of P are unique, but the polar coordinates are not. In fact, P has infinitely many polar coordinates since $(r, \theta + 360°n)$ or $(r, \theta + 2\pi n)$ designates the same point for all integer values of n. For example, $(2, 10°)$ and $(2, 370°)$ designate the same point.

Progress Check

Give three other polar coordinate representations for the point (3, −30°)

Answers

Some possibilities are (3, 330°), (3, −390°), (3, 690°)

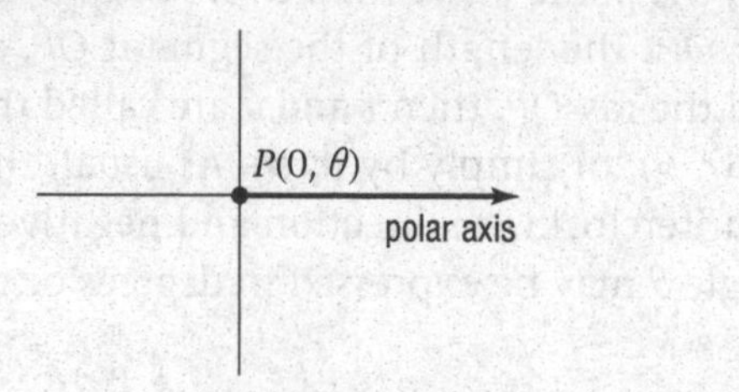

FIGURE 9-12
Point P at the Origin

Let us examine the cases where the first polar coordinate is zero or negative. If P is at the origin or pole, as shown in Figure 9-12, then $r = 0$ and any angle θ can be assigned to the ray OP, and the coordinates of the pole are $(0, \theta)$. If $r > 0$, then we consider the point $(-r, \theta)$ to be the point $(r, \theta + 180°)$ or $(r, \theta + \pi)$, as shown in Figure 9-13.

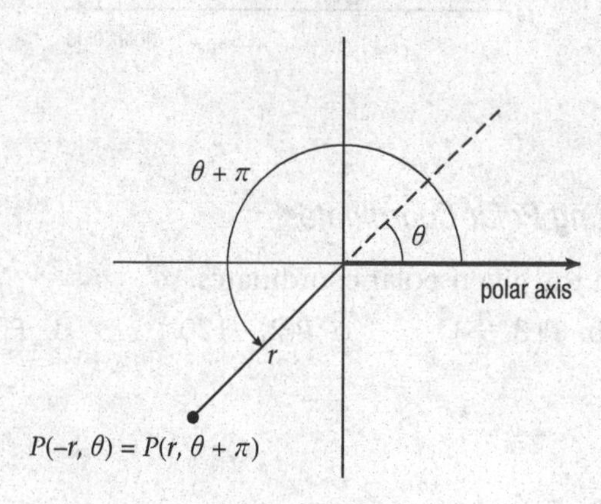

FIGURE 9-13
Point P at $(-r, \theta)$

EXAMPLE 2 *Working with Polar Coordinates*

Find polar coordinates for the point $P\left(-2, \frac{3\pi}{4}\right)$ so that r is nonnegative.

SOLUTION

By definition, we have

$$P\left(-2, \frac{3\pi}{4}\right) = P\left(2, \frac{3\pi}{4} + \pi\right) = P\left(2, \frac{7\pi}{4}\right)$$

as shown in Figure 9-14. Moreover, the given point can also be designated as $P\left(2, \frac{7\pi}{4} + 2\pi n\right)$ for all integer values of n.

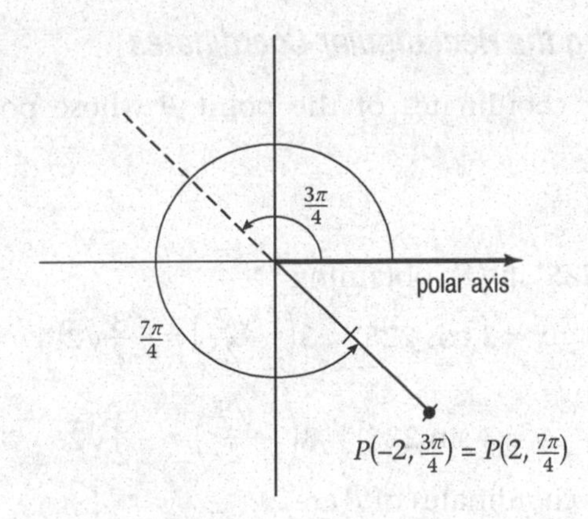

FIGURE 9-14
Point P at $\left(-2, \frac{3\pi}{4}\right)$

Progress Check

Which of the following polar coordinates represent the point $(-3, -45^\circ)$?

a. $(3, 45^\circ)$ b. $(3, 135^\circ)$ c. $(-3, 315^\circ)$ d. $(3, 225^\circ)$

Answers

b. and c.

9.4a Rectangular and Polar Coordinates

It is often convenient to be able to convert between polar coordinates and rectangular coordinates. To obtain the relationship between the two coordinate systems, we let the polar axis coincide with the positive x-axis and the pole with the origin of the rectangular system. From Figure 9-15, we obtain the following relations.

Conversion Equations

$$x = r\cos\theta \qquad y = r\sin\theta \tag{1}$$

$$\tan\theta = \frac{y}{x} \qquad r = \sqrt{x^2 + y^2} \tag{2}$$

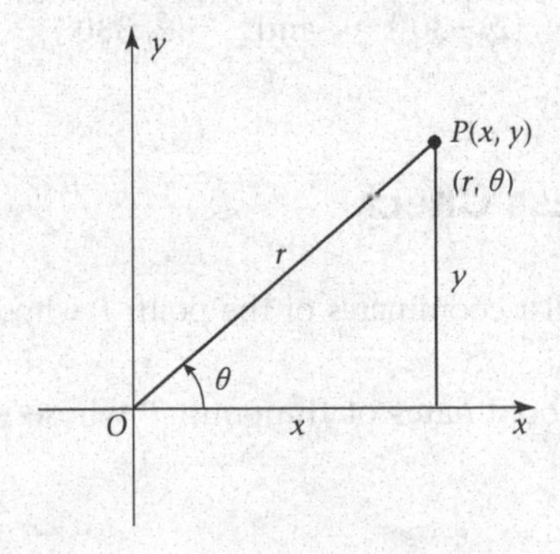

FIGURE 9-15
Polar and Rectangular Coordinates

EXAMPLE 3 *Finding the Rectangular Coordinates*

Find the rectangular coordinates of the point P whose polar coordinates are $(3, 225°)$.

SOLUTION

We let $r = 3$ and $\theta = 225°$ in (1), obtaining

$$x = 3\cos 225° = 3\left(-\frac{\sqrt{2}}{2}\right) = -\frac{3}{2}\sqrt{2}$$

$$y = 3\sin 225° = 3\left(-\frac{\sqrt{2}}{2}\right) = -\frac{3}{2}\sqrt{2}$$

Thus, the rectangular coordinates of P are

$$\left(-\frac{3}{2}\sqrt{2}, -\frac{3}{2}\sqrt{2}\right)$$

■

EXAMPLE 4 *Finding the Polar Coordinates*

Find polar coordinates of the point P whose rectangular coordinates are $(\sqrt{3}, -1)$.

SOLUTION

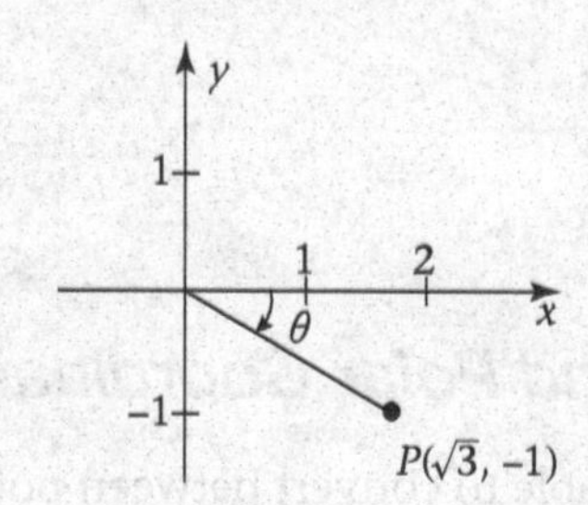

FIGURE 9-16
Diagram for Example 4

The given point is shown in Figure 9-16. Using the equations in (2) with $x = \sqrt{3}$ and $y = -1$, we have

$$r = \sqrt{x^2 + y^2} = \sqrt{4} = 2$$

$$\tan\theta = \frac{y}{x} = \frac{-1}{\sqrt{3}} = -\frac{\sqrt{3}}{3}$$

Since P lies in the fourth quadrant, θ may be taken as $-30°$ or as $330°$. Thus, two possible polar coordinate representations for P are

$$(2, -30°) \qquad \text{and} \qquad (2, 330°)$$

■

Progress Check

a. Find the rectangular coordinates of the point P whose polar coordinates are $\left(2, \frac{\pi}{3}\right)$.

b. Find the polar coordinates of the point P whose rectangular coordinates are $(-1, -1)$.

Answers

a. $(1, \sqrt{3})$

b. $\left(\sqrt{2}, \frac{5\pi}{4}\right)$ (not unique)

9.4b Polar Equations

An equation in the variables r and θ is called a **polar equation.** A *solution* of a polar equation is an ordered pair (a, b) that satisfies the polar equation when $r = a$ and $\theta = b$ are substituted into the equation. Just as the graph of an equation in x and y is obtained by plotting all the solutions of the equation on the rectangular coordinate system, the *graph of a polar equation* is obtained by plotting all the solutions of the equation on a **polar coordinate system.** When sketching the graph of a polar equation, it is convenient to superimpose the rectangular system on the polar coordinate system so that the origin coincides with the pole and the positive x-axis coincides with the polar axis. When sketching a graph in polar coordinates, it may be helpful to use an outline of the polar coordinate system such as appears in Figure 9-17.

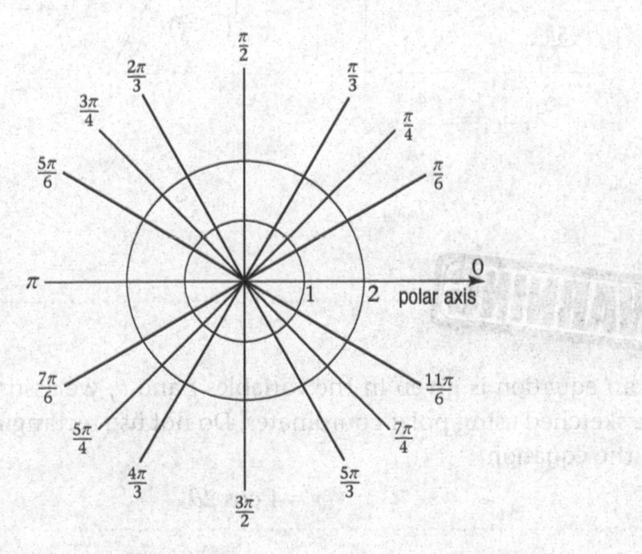

FIGURE 9-17
The Polar Coordinate System

EXAMPLE 5 *Sketching a Polar Equation*

Sketch the graph of the polar equation $r = 4 \cos \theta$.

SOLUTION

We form a table of values as shown below. Plotting the points in this table, we seem to obtain a circle of radius 2, with center at $(2, 0)$, as shown below. As θ takes values from π to 2π, we trace over the same graph a second time. (In Example 6, we shall show that the graph of this polar equation is indeed a circle of radius 2, with center at $(2, 0)$ in the xy-coordinate system.)

θ	r
0	4
$\frac{\pi}{6}$	$2\sqrt{3}$
$\frac{\pi}{4}$	$2\sqrt{2}$
$\frac{\pi}{3}$	2
$\frac{\pi}{2}$	0
$\frac{2\pi}{3}$	-2
$\frac{3\pi}{4}$	$-2\sqrt{2}$
$\frac{5\pi}{6}$	$-2\sqrt{3}$
π	-4

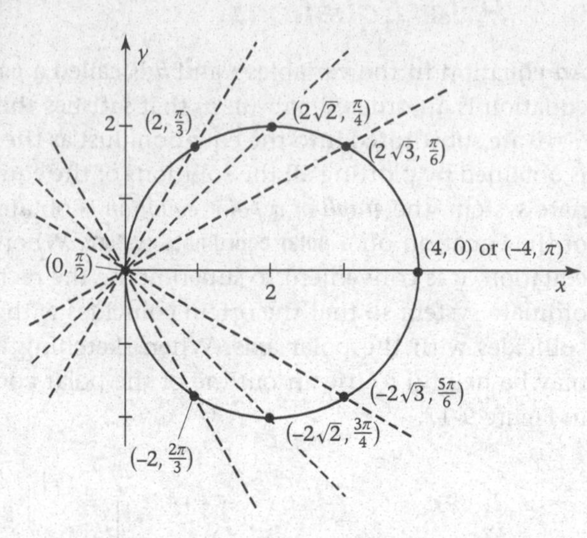

■

When an equation is given in the variables r and θ, we assume that the graph is to be sketched using polar coordinates. Do not use rectangular coordinates to sketch the equation

$$r = 4 \cos 2\theta$$

The conversion formulas enable us to transform polar equations to equations in x and y, and vice versa.

EXAMPLE 6 *Transforming a Polar Equation*

Transform the polar equation $r = 4 \cos \theta$ to an equation in x and y.

SOLUTION

This equation was plotted in Example 5. Multiplying both sides of the given equation by r, we have

$$r^2 = 4r \cos \theta$$

Substituting $x^2 + y^2 = r^2$,

$$x^2 + y^2 = 4x$$

Completing the square in the previous equation, we have

$$(x - 2)^2 + y^2 = 4$$

whose graph is a circle of radius 2 with center at $(2, 0)$. This result verifies our solution of Example 5. ■

EXAMPLE 7 *Sketching a Polar Equation*

Sketch the graph of the polar equation $r = 1 + \sin\theta$.

SOLUTION

We form a table of values as shown below. Plotting these points, we obtain the heart-shaped curve, which is called a **cardioid**.

θ	0	$\frac{\pi}{6}$	$\frac{\pi}{3}$	$\frac{\pi}{2}$	$\frac{2\pi}{3}$	$\frac{5\pi}{6}$	π	$\frac{7\pi}{6}$	$\frac{4\pi}{3}$	$\frac{3\pi}{2}$	$\frac{5\pi}{3}$	$\frac{11\pi}{6}$	2π
r	1	$\frac{3}{2}$	$1+\frac{\sqrt{3}}{2}$	2	$\frac{3}{2}$	$1+\frac{\sqrt{3}}{2}$	1	$\frac{1}{2}$	$1-\frac{\sqrt{3}}{2}$	0	$1-\frac{\sqrt{3}}{2}$	$\frac{1}{2}$	1

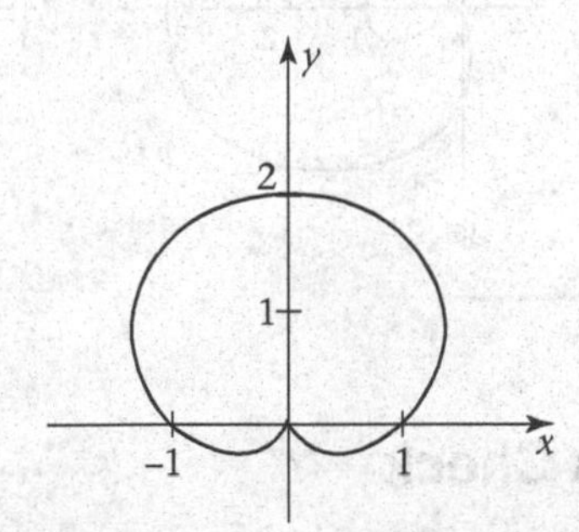

■

EXAMPLE 8 *Sketching a Polar Equation*

Sketch the graph of the polar equation $r = 2\sin 2\theta$.

SOLUTION

Instead of using a table of values, we proceed as follows. When $\theta = 0$, $r = 0$. As θ increases from 0 to $\frac{\pi}{4}$, $\sin 2\theta$ increases from 0 to 1, so r increases from 0 to 2. As θ increases from $\frac{\pi}{4}$ to $\frac{\pi}{2}$ $\sin 2\theta$ decreases from 1 to 0, so r decreases from 2 to 0. Plotting these points, we obtain the loop shown in Figure 9-18(a). Similarly, as θ increases from $\frac{\pi}{2}$ to $\frac{3\pi}{4}$, r decreases from 0 to -2, and as θ increases from $\frac{3\pi}{4}$ to π, we find that r increases from -2 to 0. These points yield the second loop, shown in Figure 9-18(b). As θ increases from π to 2π, we obtain two additional loops. The final figure, shown in Figure 9-18(c) is called a **four-leaved rose**.

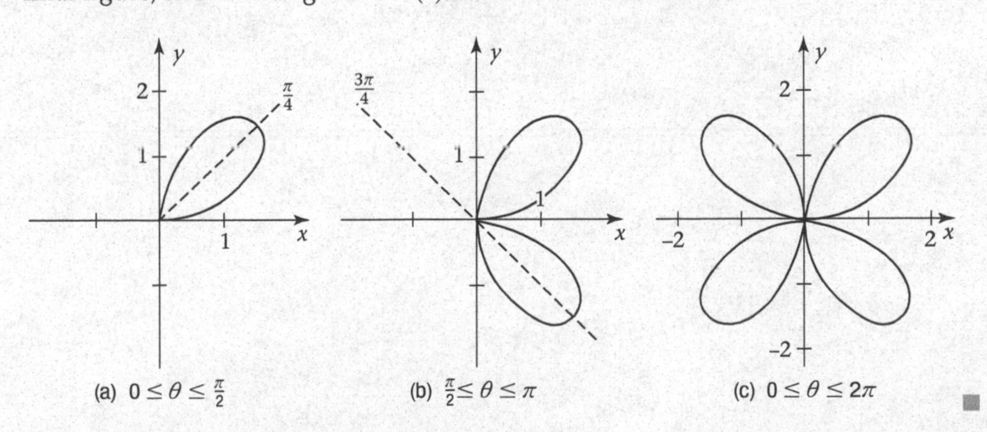

FIGURE 9-18
Graph of $r = 2\sin\theta$

■

Progress Check

Sketch the graph of the given polar equation.

a. $r = 2 \sin \theta$ b. $r = 1 + 2 \cos \theta$ c. $r = \cos 3\theta$

Answers

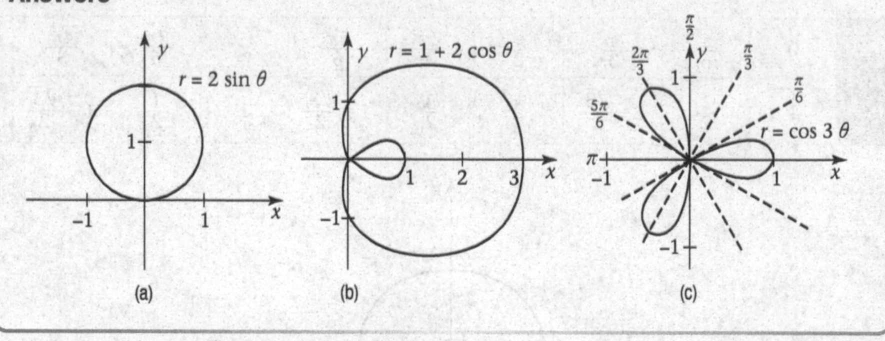

Progress Check

Transform the rectangular equation $x = 5$ to a polar equation.

Answer

$r \cos \theta = 5$

Exercise Set 9.4

1. Plot the points with the given polar coordinates.

   a. $(4, 30^\circ)$ b. $(-2, 60^\circ)$

   c. $\left(5, \frac{\pi}{4}\right)$ d. $\left(-3, -\frac{\pi}{2}\right)$

2. Plot the points with the given polar coordinates.

   a. $(2, 225^\circ)$ b. $(-4, -150^\circ)$

   c. $\left(5, \frac{2\pi}{3}\right)$ d. $\left(-4, +\frac{3\pi}{4}\right)$

3. For each point given in polar coordinates, give two other polar coordinate representations.

   a. $(6, 135^\circ)$ b. $(-2, 120^\circ)$

   c. $\left(4, \frac{5\pi}{6}\right)$ d. $\left(-4, \frac{-7\pi}{4}\right)$

4. For each point given in polar coordinates, give two other polar coordinate representations.

   a. $(4, 315^\circ)$ b. $(-3, -150^\circ)$

   c. $\left(1, \frac{11\pi}{6}\right)$ d. $\left(-1, -\frac{3\pi}{2}\right)$

5. For each point given in polar coordinates, give a polar coordinate representation with $r \geq 0$.

   a. $(-2, 30^\circ)$ b. $(-4, -60^\circ)$

   c. $\left(-3, \frac{2\pi}{3}\right)$ d. $\left(-1, -\frac{7\pi}{6}\right)$

6. For each point given in polar coordinates, give a polar coordinate representation with $r \geq 0$.

   a. $(-2, 60^\circ)$ b. $(-3, 45^\circ)$

   c. $\left(-5, \frac{7\pi}{6}\right)$ d. $\left(-4, \frac{-2\pi}{3}\right)$

7. Which of the following polar coordinates represent the point $(2, -30^\circ)$?

   a. $(-2, 150^\circ)$ b. $(-2, 330^\circ)$

   c. $(2, 330^\circ)$ d. $(2, 510^\circ)$

8. Which of the following polar coordinates represent the point $\left(4, \frac{2\pi}{3}\right)$?

   a. $\left(4, \frac{5\pi}{3}\right)$ b. $\left(-4, -\frac{5\pi}{3}\right)$

   c. $\left(4, -\frac{4\pi}{3}\right)$ d. $\left(4, -\frac{2\pi}{3}\right)$

9. Find the rectangular coordinates of the points with the given polar coordinates.

   a. $(5, 330^\circ)$ b. $(2, 270^\circ)$

   c. $\left(4, \frac{\pi}{6}\right)$ d. $\left(-3, -\frac{2\pi}{3}\right)$

10. Find the rectangular coordinates of the points with the given polar coordinates.

    a. $(1, 315^\circ)$ b. $(3, 150^\circ)$

    c. $\left(2, \frac{3\pi}{4}\right)$ d. $\left(-4, -\frac{4\pi}{3}\right)$

11. Find polar coordinates of the points with the given rectangular coordinates.

    a. $(-2, 2)$ b. $(1, -\sqrt{3})$

    c. $(\sqrt{3}, 1)$ d. $(-4, -4)$

12. Find polar coordinates of the points with the given rectangular coordinates.

    a. $(-1, -\sqrt{3})$ b. $(-\sqrt{3}, 1)$

    c. $(3, -3)$ d. $(3, 3)$

In Exercises 13–26, sketch the graph of the given polar equation.

13. $\theta = 45^\circ$ 14. $\theta = \frac{2\pi}{3}$

15. $r = 3$ 16. $r = 1 - \sin\theta$

17. $r = 2\sin\theta$ 18. $r = 2$

19. $r = -3\cos\theta$ 20. $r = 2 + 4\sin\theta$

21. $r = 3\sin 5\theta$ 22. $r = 3\cos 4\theta$

23. $r^2 = 1 + \sin 2\theta$ 24. $r^2 = 4\sin 2\theta$

25. $r = \theta, \theta \geq 0$ 26. $r\theta = 2$

In Exercises 27–32, transform the given polar equation to an equation in x and y.

27. $r = 4$ 28. $\theta = \frac{\pi}{4}$

29. $r + 2\sin\theta = 0$ 30. $r = 3\sec\theta$

31. $r\cos\theta = 2$ 32. $r = 2\tan\theta$

In Exercises 33–38, transform the given rectangular equation to a polar equation.

33. $x^2 + y^2 = 25$ 34. $y = 3$

35. $x = -5$ 36. $y^2 = 2x$

37. $y = 3x$ 38. $x^2 - y^2 = 9$

In Exercises 39–42, use your graphing calculator to graph each polar equation. How does changing the set of 0 values affect the shape of the graphs?

39. a. $r = \cos 3\theta$ $0 \leq \theta \leq \pi$

    b. $r = \cos 3\theta$ $0 \leq \theta \leq 2\pi$

40. a. $r = \cos 4\theta$ $0 \leq \theta \leq \pi$

    b. $r = \cos 4\theta$ $0 \leq \theta \leq 2\pi$

41. a. $r = 2\theta$ $0 \leq \theta \leq 10$

    b. $r = 2\theta$ $-10 \leq \theta \leq 10$

42. a. $r = 3 + \cos\theta$ $0 \leq \theta \leq 2\pi$

    b. $r = 3 + 3\cos\theta$ $0 \leq \theta \leq 2\pi$

    c. $r = 3 + 5\cos\theta$ $0 \leq \theta \leq 2\pi$

9.5 Vectors

Throughout this text, we have demonstrated a number of applications of mathematics to various disciplines. Physics is a branch of science that has required the development of many new areas of mathematics. It is within this context that we focus our attention on the subject of vectors.

In previous sections, our applications have been directed to quantities that were measured in units, such as dollars, feet, seconds, pounds, miles per hour and so on. The magnitude or number associated with any of these units is called a **scalar**, and the number together with its units is called a **scalar quantity**. There are times, however, when we are concerned with more than just magnitude. Consider a wind blowing south at 40 miles per hour or the force exerted by a person to lift up a child weighing 20 pounds. In both of these examples, we are given magnitude and direction. We use a **directed line segment**, namely, a line segment with an arrowhead, to illustrate these examples in Figure 9-19. We now define a **vector**, or **vector quantity**, as an object that has both magnitude and direction. A vector is always represented in a figure as a directed line segment, where the *length* of the segment equals the **magnitude** of the vector and the *arrowhead* on the segment points in the **direction** of the vector.

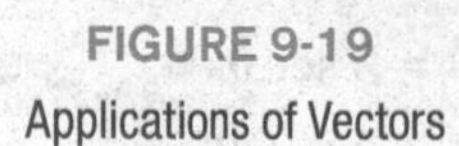

FIGURE 9-19
Applications of Vectors

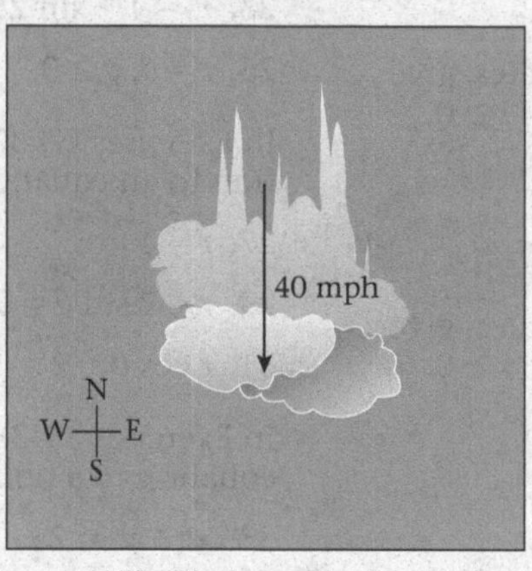

(a) Wind blowing south at 40 mph

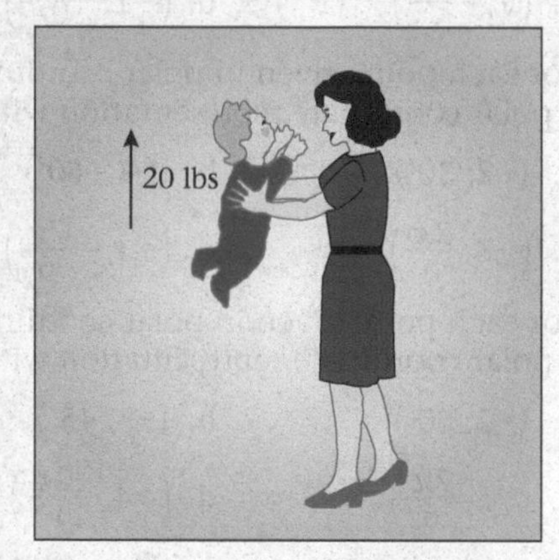

(b) Person lifting a 20 lb child

We identify vectors using $\vec{v}$ or $\overrightarrow{PQ}$ as shown in Figure 9-20. Note that the order of the letters $\overrightarrow{PQ}$ indicates the direction of the vector. We call the first letter P the **initial point**, or **tail**, and the second letter Q, the **terminal point**, or **head.**

FIGURE 9-20
Vector Terminology

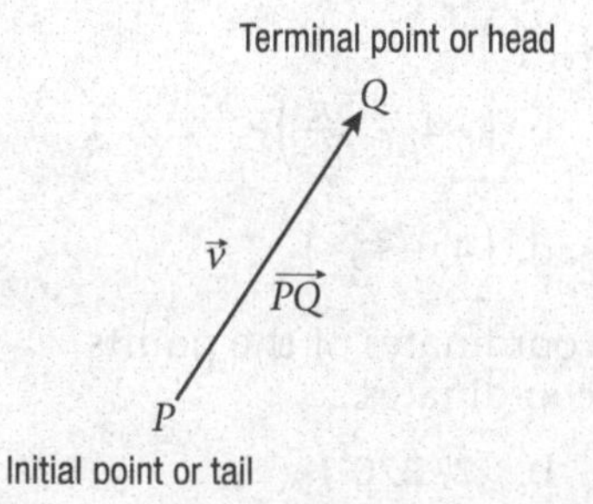

Consider the vectors in Figure 9-21. Note the difference between $\overrightarrow{PQ}$ and $\overrightarrow{QP}$. They both have the same length, but point in opposite directions. Whereas the length of the line segment PQ is denoted $\overline{PQ}$, the magnitude of the vector $\overrightarrow{PQ}$ is denoted $\|\overrightarrow{PQ}\|$. Observe that

$$\|\overrightarrow{PQ}\| = \overline{PQ} = \overline{QP} = \|\overrightarrow{QP}\|$$

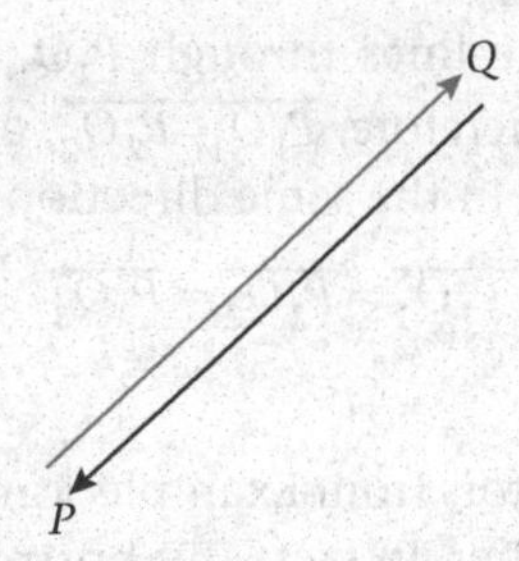

FIGURE 9-21
Two Vectors with the Same Magnitude

We say that two vectors are **equal** if they have the same magnitude and direction. Thus, in Figure 9-22, even though $\overrightarrow{PQ}$ and $\overrightarrow{P'Q'}$ have different initial points and terminal points,

$$\overrightarrow{P'Q'} = \overrightarrow{PQ}$$

or equivalently,

$$\vec{v} = \vec{v'}$$

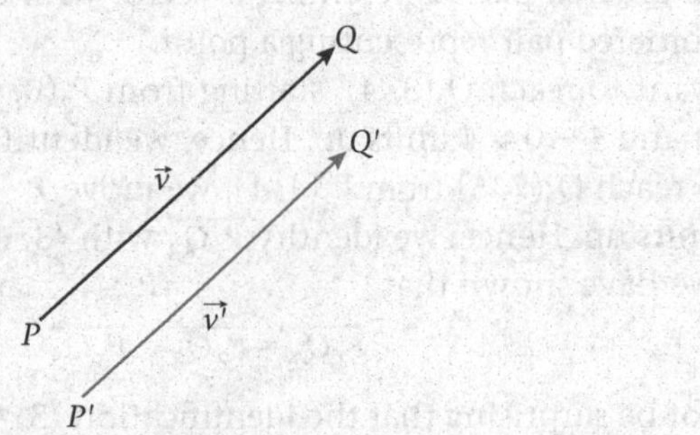

FIGURE 9-22
Equal Vectors

Therefore, all vectors that are parallel to $\overrightarrow{PQ}$ with the same magnitude, pointing in the same direction, are equal. Note that $\overrightarrow{PQ}$ and $\overrightarrow{QP}$ in Figure 9-21 are parallel with the same magnitude, but they are *not* equal.

We now consider vectors in a coordinate system.

EXAMPLE 1 *Relationships Between Vectors*

Determine if $\overrightarrow{P_1Q_1} = \overrightarrow{P_2Q_2} = \overrightarrow{P_3Q_3}$, where the vectors are shown below.

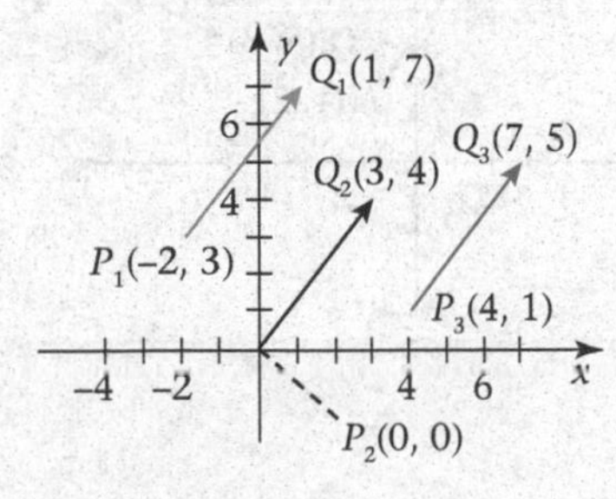

SOLUTION

We find that

$$\|\overrightarrow{P_1Q_1}\| = \overline{P_1Q_1} = \sqrt{(1-(-2))^2+(7-3)^2} = 5$$
$$\|\overrightarrow{P_2Q_2}\| = \overline{P_2Q_2} = \sqrt{(3-0)^2+(4-0)^2} = 5$$
$$\|\overrightarrow{P_3Q_3}\| = \overline{P_3Q_3} = \sqrt{(7-4)^2+(5-1)^2} = 5$$

Furthermore, the slopes of the lines through P_1Q_1, through P_2Q_2, and through P_3Q_3 are all $\frac{4}{3}$. (Verify this.) Therefore, $\overrightarrow{P_1Q_1}$, $\overrightarrow{P_2Q_2}$, and $\overrightarrow{P_3Q_3}$ are all parallel, with the same magnitude, pointing in the same direction. Hence,

$$\overrightarrow{P_1Q_1} = \overrightarrow{P_2Q_2} = \overrightarrow{P_3Q_3}$$

■

Observe that the three vectors from Example 1 are combinations of horizontal and vertical shifts of one another. In fact, any horizontal or vertical shift of these vectors produces vectors that are equal to the original ones. Among the many representations of this vector, we focus on one whose initial point is the origin, and we say that such a vector is in standard position. In Example 1, the vector in **standard position** is $\overrightarrow{P_2Q_2}$.

We shall now outline a procedure to identify an ordered pair with each of these vectors. If we want to reach $Q_1(1, 7)$, starting from $P_1(-2, 3)$, we move $1 - (-2) = 3$ units to the right and $7 - 3 = 4$ units up. Hence, we identify $\overrightarrow{P_1Q_1}$ with $\langle 3, 4 \rangle$. (We enclose this ordered pair representing a vector with angle brackets to distinguish it from an ordered pair representing a point.)

If we want to reach $Q_2(3, 4)$, starting from $P_2(0, 0)$, we move $3 - 0 = 3$ units to the right and $4 - 0 = 4$ units up. Hence, we identify $\overrightarrow{P_2Q_2}$ with $\langle 3, 4 \rangle$. Finally, if we want to reach $Q_3(7, 5)$ from $P_3(4, 1)$, we move $7 - 4 = 3$ units to the right and $5 - 1 = 4$ units up. Hence, we identify $\overrightarrow{P_1Q_3}$ with $\langle 3, 4 \rangle$.

Since we have shown that

$$\overrightarrow{P_1Q_1} = \overrightarrow{P_2Q_2} = \overrightarrow{P_3Q_3}$$

it should not be surprising that the identification $\langle 3, 4 \rangle$ is the same for each of the three vectors.

EXAMPLE 2 *Vector Notation*

Find the ordered pair identified with $\overrightarrow{P_1Q_1}$ and $\overrightarrow{P_2Q_2}$, where the coordinates of P_1, Q_1, P_2, and Q_2 are shown.

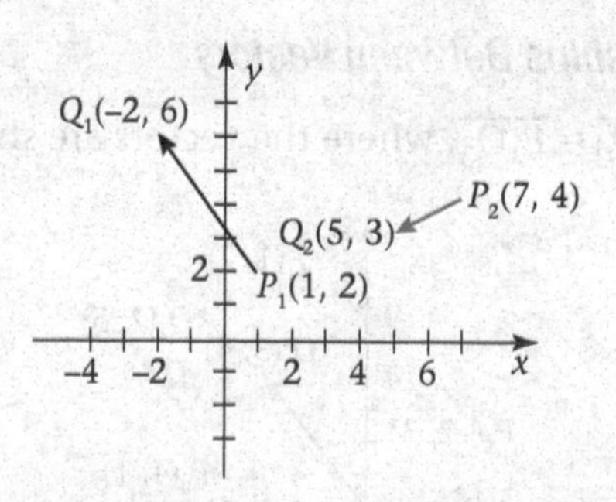

SOLUTION

To reach $Q_1(-2, 6)$ from $P_1(1, 2)$, we move $-2 - 1 = -3$ units to the right, which is equivalent to moving 3 units to the left, and we move $6 - 2 = 4$ units up. Therefore, we identify $\overrightarrow{P_1Q_1}$ with $\langle -3, 4 \rangle$.

To reach $Q_2(5, 3)$ from $P_2(7, 4)$, we move $5 - 7 = -2$ units to the right, meaning 2 units to the left, and $3 - 4 = -1$ units up, which is equivalent to moving 1 unit down. Therefore, we identify $\overrightarrow{P_2Q_2}$ with $\langle -2, -1 \rangle$. ■

For the more general case, consider Figure 9-23:

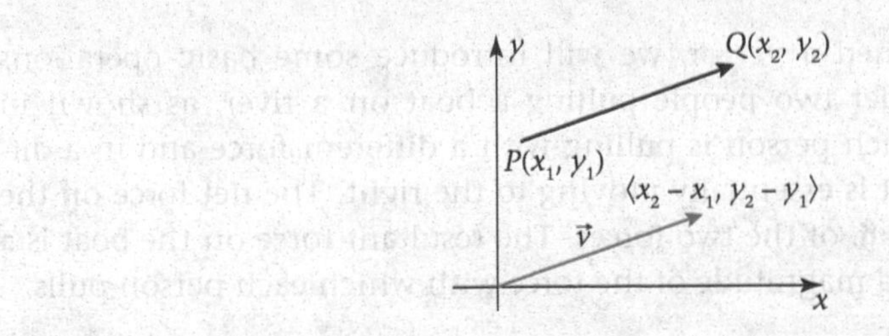

FIGURE 9-23
Vectors in a Coordinate System

If $\overrightarrow{PQ}$ is the vector from $P(x_1, y_1)$ to $Q(x_2, y_2)$, then we identify the ordered pair $\langle x_2 - x_1, y_2 - y_1 \rangle$ with $\overrightarrow{PQ}$ and we write

$$\overrightarrow{PQ} = \langle x_2 - x_1, y_2 - y_1 \rangle$$

The numbers $x_2 - x_1$ and $y_2 - y_1$ are called the **components** of the vector $\overrightarrow{PQ}$. Furthermore, the magnitude of $\overrightarrow{PQ}$,

$$\|\overrightarrow{PQ}\| = \sqrt{(x_2 - x_1)^2 + (y_2 - y_1)^2}$$

Note that $\vec{v}$ in Figure 9-23 is the representation of $\overrightarrow{PQ}$ in standard position. Verify that $\|\vec{v}\| = \|\overrightarrow{PQ}\|$.

Since we may always choose a representative of any vector to be the one in standard position, consider $\vec{v}$ in Figure 9-24. The vector $\vec{v} = \langle a, b \rangle$ has magnitude $\|\vec{v}\| = \sqrt{a^2 + b^2}$.

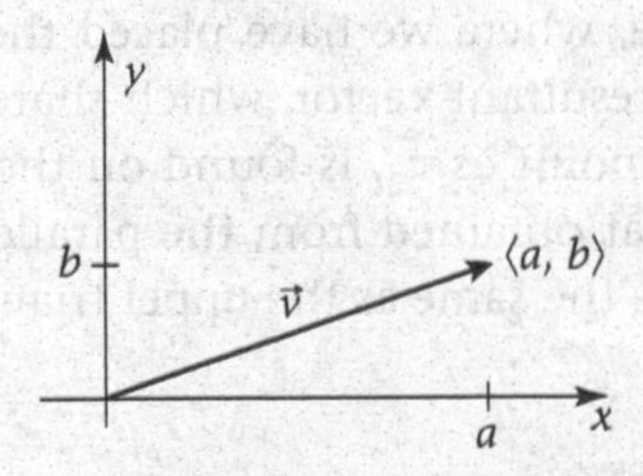

FIGURE 9-24
Vectors in a Standard Position

Recalling the discussion following Example 1, we have the following definition for the equality of vectors.

Suppose $\overrightarrow{P_1Q_1} = \langle a_1, b_1 \rangle$ and $\overrightarrow{P_2Q_2} = \langle a_2, b_2 \rangle$. Then

$$\overrightarrow{P_1Q_1} = \overrightarrow{P_2Q_2} \text{ is equivalent to } a_1 = a_2 \text{ and } b_1 = b_2$$

that is, two vectors are equal if and only if their corresponding components are equal.

9.5a *Vector Addition*

Now that we have defined a vector, we will introduce some basic operations between vectors. Consider two people pulling a boat on a river, as shown in Figure 9-25. Although each person is pulling with a different force and in a different direction, the boat is essentially moving to the right. The net force on the boat is the **sum**, or **resultant**, of the two forces. The resultant force on the boat is a function of the angle and magnitude of the force with which each person pulls.

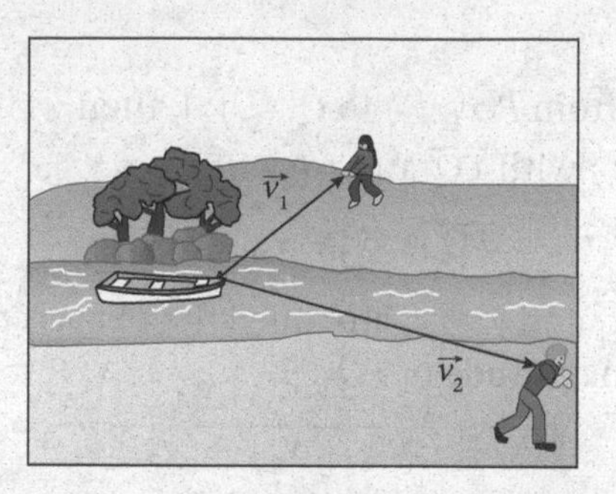

FIGURE 9-25
Sum of Two Forces

We focus on the vectors of the previous illustration, as shown in Figure 9-26. Note that in Figure 9-26(a) we have formed a parallelogram by drawing lines parallel to the original vectors. The vector $\vec{v_3}$ on the diagonal of the parallelogram is the sum of vectors $\vec{v_1}$ and $\vec{v_2}$. It represents the resultant force on the boat. This method for addition is sometimes referred to as the **parallelogram law**. In Figure 9-26(b), we have formed a triangle, where we have placed the initial point of $\vec{v_2}$ at the terminal point of $\vec{v_1}$. The resultant vector, which shares the same initial point as $\vec{v_1}$ and the same terminal point as $\vec{v_2}$, is found on the third side of the triangle. It is the same vector as that obtained from the parallelogram law. Note that the triangle in Figure 9-26(b) is the same as the upper triangle in the parallelogram in Figure 9-26(a).

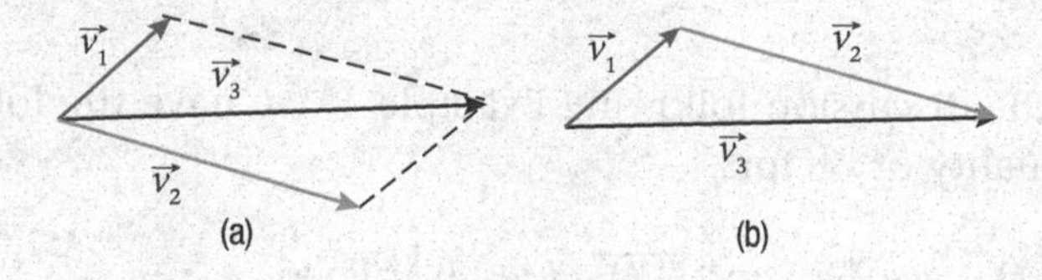

FIGURE 9-26
Sum of Two Vectors

We now consider vectors in a coordinate system:

Vector Addition

If $\vec{v_1} = \langle a_1, b_1 \rangle$ and $\vec{v_2} = \langle a_2, b_2 \rangle$, then the sum

$$\vec{v_1} + \vec{v_2} = \vec{v_3} \text{ is equivalent to } \langle a_3, b_3 \rangle = \langle a_1 + a_2, b_1 + b_2 \rangle$$

Thus, the sum of two vectors is obtained by adding their corresponding components.

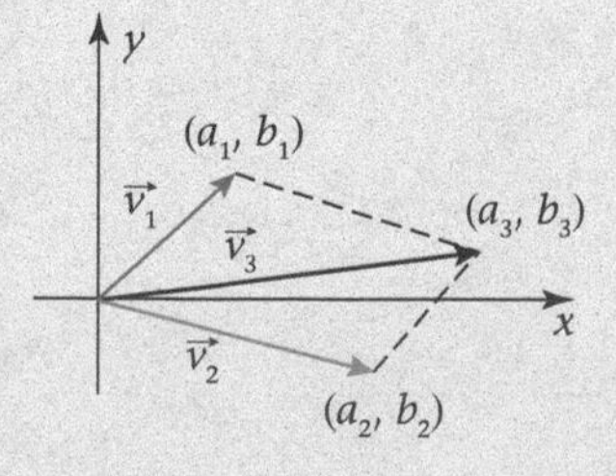

EXAMPLE 3 ***Sum of Two Vectors***

If $\vec{v_1} = \langle 2, 3\rangle$ and $\vec{v_2} = \langle 4, -5\rangle$, find $\vec{v_1} + \vec{v_2}$. Sketch $\vec{v_1}$, $\vec{v_2}$, and $\vec{v_1} + \vec{v_2}$.

SOLUTION

$\vec{v_1} + \vec{v_2} = \langle 2, 3\rangle + \langle 4, -5\rangle = \langle 6, -2\rangle$.

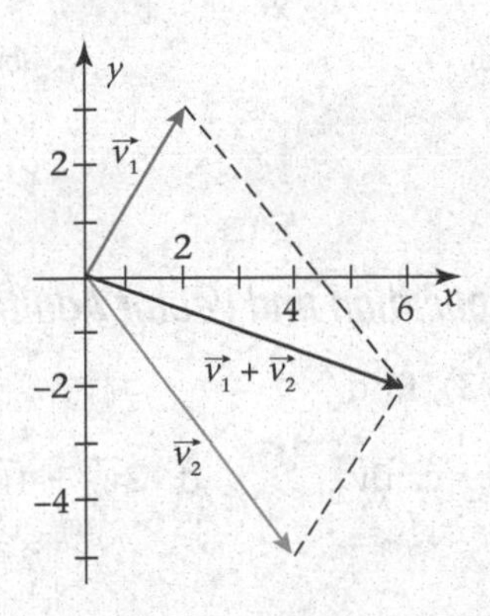

■

Progress Check

If $\vec{v_1} = \langle -1, 1\rangle$ and $\vec{v_2} = \langle 5, 3\rangle$, find $\vec{v_1} + \vec{v_2}$. Sketch $\vec{v_1}$, $\vec{v_2}$, and $\vec{v_1} + \vec{v_2}$.

Answer

$\langle 4, 4\rangle$

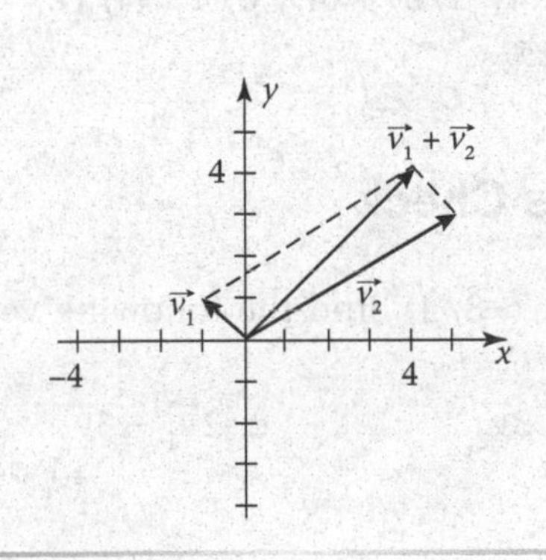

9.5b *Scalar Multiplication*

Consider the special case where two people pull with equal force on the same object in the same direction, or equivalently, suppose we add $\vec{v} + \vec{v}$, where $\vec{v} = \langle a, b\rangle$.

$$\vec{v} + \vec{v} = \langle a, b\rangle + \langle a, b\rangle = \langle 2a, 2b\rangle$$

If we write $\vec{v} + \vec{v} = 2\vec{v}$, then we have

$$2\vec{v} = \langle 2a, 2b\rangle$$

We show $\vec{v}$ and $2\vec{v}$ in Figure 9-27. This result motivates the following definition.

Scalar Multiplication of a Vector

If $\vec{v} = \langle a, b\rangle$, then for any scalar k

$$k\vec{v} = k\langle a, b\rangle = \langle ka, kb\rangle$$

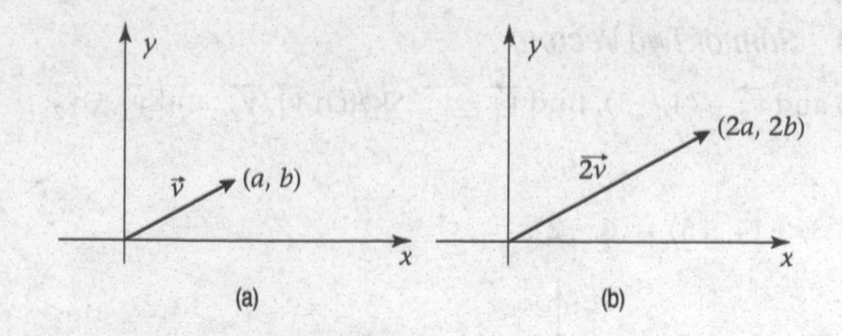

FIGURE 9-27
Sum of Two Equal Vectors

EXAMPLE 4 *Scalar Multiplication and Vector Addition*

If $\vec{v_1} = \langle 2, -4 \rangle$ and $\vec{v_2} = \langle -1, 3 \rangle$, find

a. $\frac{1}{2}\vec{v_1}$ b. $-\vec{v_2}$ c. $0\vec{v_2}$ d. $2\vec{v_1} + 4\vec{v_2}$

SOLUTION

a. $\frac{1}{2}\vec{v_1} = \frac{1}{2}\langle 2, -4 \rangle = \langle 1, -2 \rangle$

b. $-\vec{v_2} = -1\langle -1, 3 \rangle = \langle 1, -3 \rangle$

c. $0\vec{v_2} = 0\langle -1, 3 \rangle = \langle 0, 0 \rangle$

d. $2\vec{v_1} + 4\vec{v_2} = 2\langle 2, -4 \rangle + 4\langle -1, 3 \rangle = \langle 4, -8 \rangle + \langle -4, 12 \rangle = \langle 0, 4 \rangle$ ■

Progress Check

If $\vec{v_1} = \langle 1, 2 \rangle$ and $\vec{v_2} = \langle -3, 4 \rangle$, find the following vectors and sketch the resulting vectors.

a. $-5\vec{v_1}$ b. $3\vec{v_2}$ c. $2\vec{v_1} - 3\vec{v_2}$

Answers

a. $\langle -5, -10 \rangle$ b. $\langle -9, 12 \rangle$ c. $\langle 11, -8 \rangle$

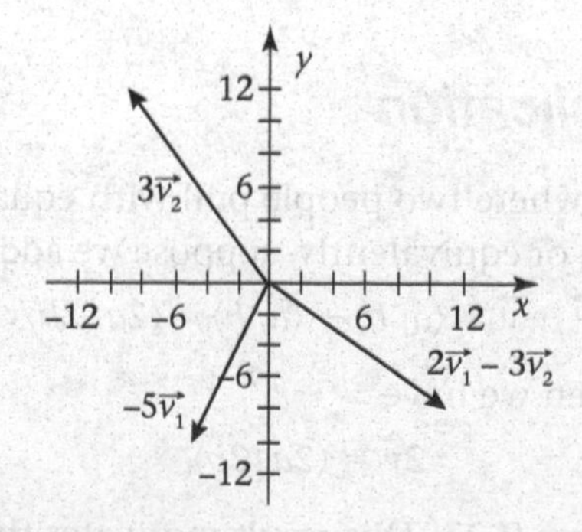

If $\vec{v} = \langle a, b\rangle$, we have found that $\|\vec{v}\| = \sqrt{a^2+b^2}$. We calculate

$$\|k\vec{v}\| = \|k\langle a,b\rangle\| = \|\langle ka,kb\rangle\| = \sqrt{k^2a^2+k^2b^2} = \sqrt{k^2}\sqrt{a^2+b^2} = |k|\|\vec{v}\|.$$

Therefore, if we know the length of the vector $\vec{v}$, then $k\vec{v}$ is $|k|$ times the length of $\vec{v}$.

We can verify this with the results from Example 4. From (a),

$$\|\vec{v_1}\| = \|\langle 2,-4\rangle\| = \sqrt{4+16} = \sqrt{20} = 2\sqrt{5}$$

$$\left\|\frac{1}{2}\vec{v_1}\right\| = \|\langle 1,-2\rangle\| = \sqrt{1+4} = \sqrt{5} = \left|\frac{1}{2}\right|(2\sqrt{5})$$

In Figure 9-28(a) and Figure 9-28(b), we show $\vec{v_1}$ and $\frac{1}{2}\vec{v_1}$, respectively. From (b),

$$\|\vec{v_2}\| = \|\langle -1,3\rangle\| = \sqrt{1+9} = \sqrt{10}$$

$$\|-\vec{v_2}\| = \|\langle 1,-3\rangle\| = \sqrt{1+9} = \sqrt{10} = |-1|(\sqrt{10})$$

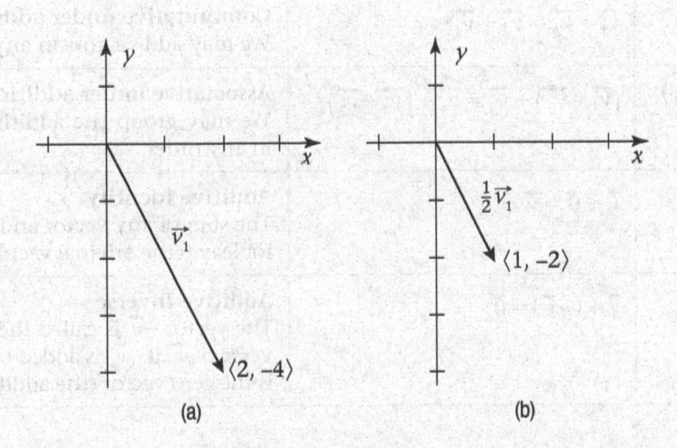

FIGURE 9-28
Scalar Multiplication of Vectors

In Figure 9-29, we show $\vec{v_2}$ and $-\vec{v_2}$. From (c),

$$\|0\vec{v_1}\| = \|\langle 0,0\rangle\| = \sqrt{0+0} = 0 = |0|(\sqrt{10})$$

For $k \neq 0$, we show that $\vec{v}$ and $k\vec{v}$ are parallel. Their directions are the same if $k > 0$ and opposite if $k < 0$.

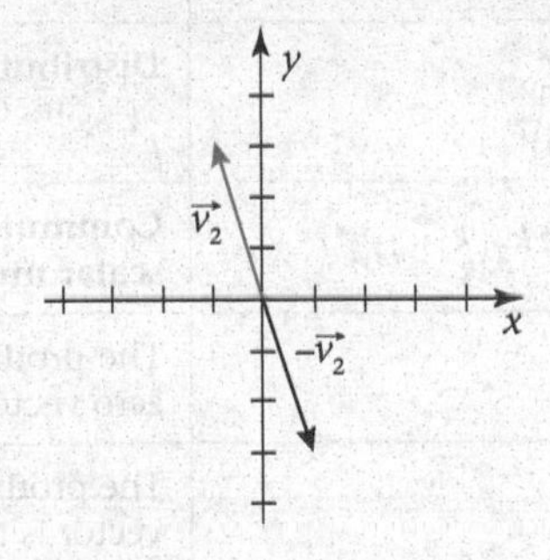

FIGURE 9-29
$\vec{v_2}$ and $-\vec{v_2}$

9.5c Properties of Vectors

Since a vector is an ordered pair of real numbers, many properties of vectors are inherited from the properties of real numbers. Following Table 1-1 and Table 1-3 in Chapter 1, we list the properties of vectors here in Table 9-1 and Table 9-2. However, first we define

$$\vec{0} = \langle 0, 0 \rangle$$

which is called the **zero vector**.

TABLE 9-1 Properties of Vectors

Example	Algebraic Expression	Property
$\langle 1, 2\rangle + \langle 3, 4\rangle$ is a vector.	$\vec{v_1} + \vec{v_2}$ is a vector.	**Closure under addition** The sum of two vectors is a vector.
$\langle 1, 2\rangle + \langle 3, 4\rangle = \langle 3, 4\rangle + \langle 1, 2\rangle$	$\vec{v_1} + \vec{v_2} = \vec{v_2} + \vec{v_1}$	**Commutative under addition** We may add vectors in any order.
$(\langle 1, 2\rangle + \langle 3, 4\rangle) + \langle 5, 6\rangle = \langle 1, 2\rangle + (\langle 3, 4\rangle + \langle 5, 6\rangle)$	$(\vec{v_1} + \vec{v_2}) + \vec{v_3} = \vec{v_1} + (\vec{v_2} + \vec{v_3})$	**Associative under addition** We may group the addition of vectors in any order.
$\langle 1, 2\rangle + \langle 0, 0\rangle = \langle 1, 2\rangle$	$\vec{v} + \vec{0} = \vec{v}$	**Additive Identity** The sum of any vector and the zero vector leaves the original vector unchanged.
$\langle 1, 2\rangle + \langle -1, -2\rangle = \langle 0, 0\rangle$	$\vec{v} + (-\vec{v}) = \vec{0}$	**Additive Inverse** The vector $-\vec{v}$ is called the **additive inverse** of $\vec{v}$. If $-\vec{v}$ is added to $\vec{v}$, the result is the zero vector (the additive identity).

TABLE 9-2 Properties of Scalar Multiplication of Vectors

Example	Algebraic Expression	Property
$1\langle 1, 2\rangle = \langle 1, 2\rangle$	$1\vec{v} = \vec{v}$	**Scalar multiplicative identity**
$2(\langle 1, 2\rangle + \langle 3, 4\rangle) = 2\langle 1, 2\rangle + 2\langle 3, 4\rangle$ $(2 + 3)\langle 1, 2\rangle = 2\langle 1, 2\rangle + 3\langle 1, 2\rangle$	$k(\vec{v_1} + \vec{v_2}) = k\vec{v_1} + k\vec{v_2}$ $(k_1 + k_2)\vec{v} = k_1\vec{v} + k_2\vec{v}$	**Distributive for scalars**
$(2 \cdot 3)\langle 1, 2\rangle = 2\langle 3, 6\rangle = 3\langle 2, 4\rangle$	$(k_1, k_2)\vec{v} = k_1(k_2\vec{v}) = k_2(k_1\vec{v})$	**Commutative and associative under scalar multiplications**
$0\langle 1, 2\rangle = \langle 0, 0\rangle$	$0\vec{v} = \vec{0}$	The product of 0 and any vector is the zero vector.
$3\langle 0, 0\rangle = \langle 0, 0\rangle$	$k\vec{0} = \vec{0}$	The product of any scalar with the zero vector is the zero vector

We define subtraction of two vectors $\vec{a} - \vec{b}$ as $\vec{a} + (-1)\vec{b}$. Therefore, we have

Vector Subtraction

If $\vec{v_1} = \langle a_1, b_1\rangle$ and $\vec{v_2} = \langle a_2, b_2\rangle$, then

$$\vec{v_1} - \vec{v_2} = \langle a_1 - a_2, b_1 - b_2\rangle$$

that is, we subtract vectors by subtracting their corresponding components.

9.5d Unit Vectors and Linear Combinations

If a vector $\vec{v}$ has magnitude $\|\vec{v}\| = 1$, then $\vec{v}$ is called a **unit vector**. For example,

$$\vec{v} = \left\langle \frac{1}{2}, \frac{\sqrt{3}}{2} \right\rangle$$

is a unit vector since

$$\|\vec{v}\| = \left\| \left\langle \frac{1}{2}, \frac{\sqrt{3}}{2} \right\rangle \right\| = \sqrt{\left(\frac{1}{2}\right)^2 + \left(\frac{\sqrt{3}}{2}\right)^2} = \sqrt{\frac{1}{4} + \frac{3}{4}} = \sqrt{1} = 1$$

We define the **standard unit vectors**

$$\vec{i} = \langle 1, 0 \rangle \quad \text{and} \quad \vec{j} = \langle 0, 1 \rangle$$

as shown in Figure 9-30.

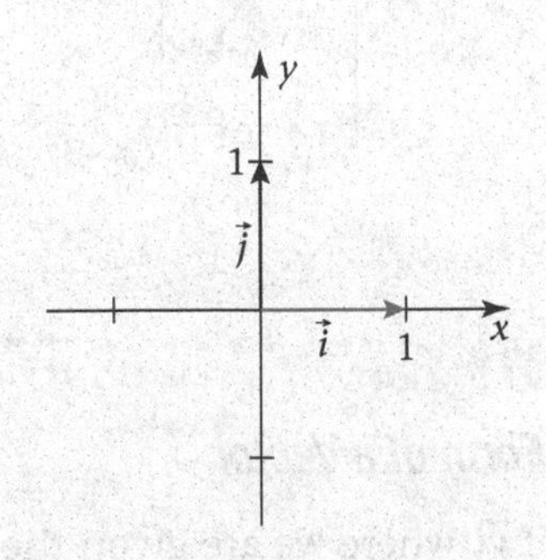

FIGURE 9-30
Standard Unit Vectors

Since $\langle a, b \rangle = a\langle 1, 0 \rangle + b\langle 0, 1 \rangle$,

> Any vector $\vec{v} = \langle a, b \rangle$ can be written in the unique form
> $$\vec{v} = a\vec{i} + b\vec{j}$$

The **horizontal component** of $\vec{v}$ is a and the **vertical component** of $\vec{v}$ is b, as shown in Figure 9-31. The vector $\vec{v}$ is said to be a **linear combination** of $\vec{i}$ and $\vec{j}$ if there are scalars a and b such that

$$\vec{v} = a\vec{i} + b\vec{j}$$

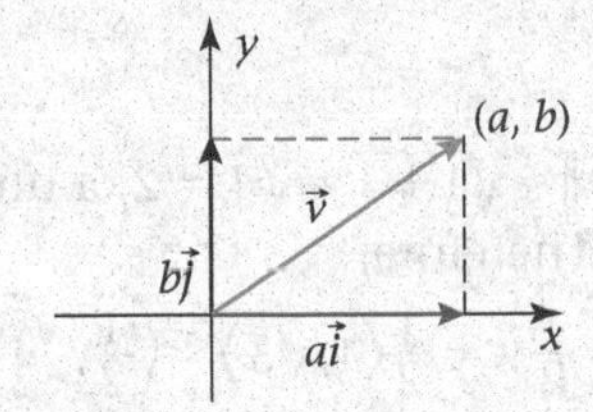

FIGURE 9-31
The Decomposition of $\vec{v}$ into Horizontal and Vertical Components

Do not confuse

$$\vec{i} = \langle 1, 0 \rangle \quad \text{with} \quad i = \sqrt{-1}$$

EXAMPLE 5 *Decomposition of a Vector*

Find the decomposition of $\vec{v} = \langle 2, -3 \rangle$ into its horizontal and vertical components.

SOLUTION

$$\vec{v} = 2\langle 1, 0 \rangle - 3\langle 0, 1 \rangle = 2\vec{i} - 3\vec{j}$$

as shown in Figure 9-32.

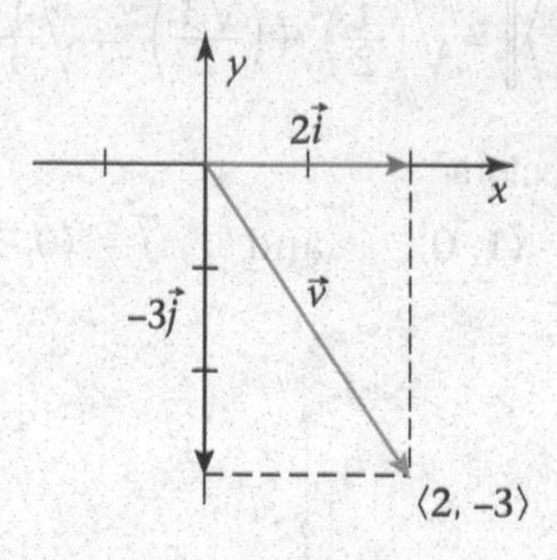

FIGURE 9-32
The Decomposition of $\langle 2, -3 \rangle$

■

EXAMPLE 6 *Component Form of a Vector*

Find the component form of $\vec{v}$, where we are given the linear combination $\vec{v} = 2\vec{i} - 3\vec{j}$.

SOLUTION

$$\vec{v} = 2\vec{i} - 3\vec{j} = 2\langle 1, 0 \rangle - 3\langle 1, 0 \rangle = \langle 2, 0 \rangle + \langle 0, -3 \rangle = \langle 2, -3 \rangle$$

■

EXAMPLE 7 *Unit Vectors*

Find a unit vector $\vec{u}$ in the same direction as $\vec{v} = \langle 1, \sqrt{3} \rangle$.

SOLUTION

Since the magnitude of $\vec{v}$, $\|\vec{v}\| = \sqrt{1+3} = \sqrt{4} = 2$, a unit vector in the same direction as $\vec{v}$ is one-half as long. Therefore,

$$\vec{u} = \frac{1}{2}\vec{v} = \frac{1}{2}\langle 1, \sqrt{3} \rangle = \left\langle \frac{1}{2}, \frac{\sqrt{3}}{2} \right\rangle$$

■

Progress Check

Find the component form and the decomposition of the vector that is the resultant of $2\langle 3, -1 \rangle - 2\langle 4, 2 \rangle - 3\langle -2, 0 \rangle$.

Furthermore, find the unit vector in the same direction as the resultant.

Answers

$\langle 4, -6 \rangle = 4\vec{i} - 6\vec{j}$; $\frac{2\sqrt{13}}{13}\vec{i} - \frac{3\sqrt{13}}{13}\vec{j}$

9.5e *Trigonometry and Vectors*

The **direction angle** of vector $\vec{v}$ is the angle that $\vec{v}$ makes with the positive x-axis, measured in the counterclockwise direction. In other words, it is the angle measured in the standard way when considering polar coordinates. Consider $\vec{v} = \langle a, b\rangle$ as shown in Figure 9-33.

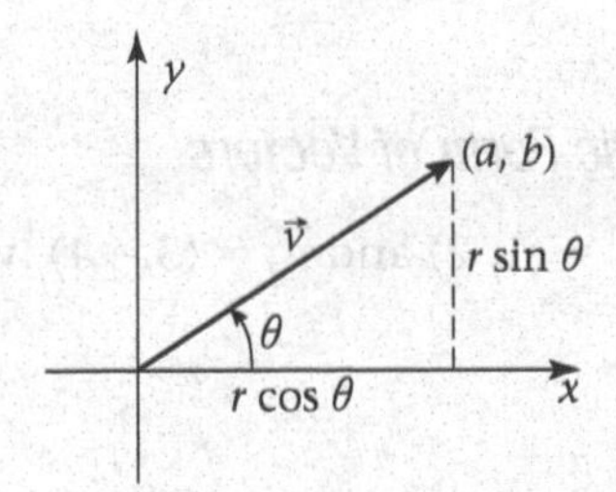

FIGURE 9-33
Trigonometric Form of Vectors

If we identify the rectangular coordinates (a, b) with the polar coordinates (r, θ), then r is the length of $\vec{v}$, that is,

$$r = \|\vec{v}\|$$

and θ is called the direction angle of $\vec{v}$. From Figure 9-33 we observe that $a = r\cos\theta$ and $b = r\sin\theta$. We summarize these results as follows:

Trigonometric Form of Vectors

If $\vec{v} = \langle a, b\rangle$ and $\|\vec{v}\| = r$ with direction angle θ, then

$$\vec{v} = \langle r\cos\theta, r\sin\theta\rangle, \qquad a^2 + b^2 = r^2 \qquad \text{and} \qquad \tan\theta = \frac{b}{a}$$

(Compare these results with the material in Section 9.4.)

EXAMPLE 8 *Trigonometric Form of Vectors*

Find the magnitude and direction angle of $\vec{v} = \left\langle \frac{1}{2}, \frac{\sqrt{3}}{2} \right\rangle$.

SOLUTION

We have already shown in Example 7 that $\|\vec{v}\| = 1$. Observe the angle θ in Figure 9-34. Since θ is in the first quadrant and

$$\tan\theta = \frac{\frac{\sqrt{3}}{2}}{\frac{1}{2}} = \sqrt{3}$$

we find that $\theta = \frac{\pi}{3}$.

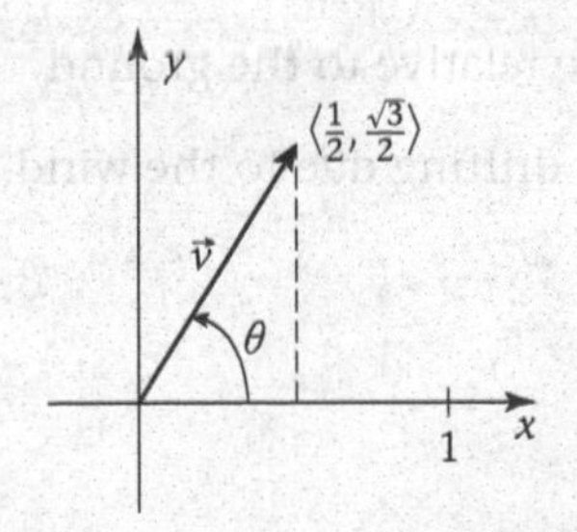

FIGURE 9-34
Vector in Example 8

■

If $\vec{v} = \langle a, b \rangle = a\vec{i} + b\vec{j}$ and $\vec{v} = \langle r\cos\theta, r\sin\theta \rangle$, then

$a = r\cos\theta$ is the horizontal component of $\vec{v}$

and

$b = r\sin\theta$ is the vertical component of $\vec{v}$

EXAMPLE 9 *Trigonometric Form of Vectors*

Find the angle θ between $\vec{v_1} = \langle 1, 2 \rangle$ and $\vec{v_2} = \langle 3, -4 \rangle$, where $0^\circ \le \theta \le 180^\circ$.

SOLUTION

We sketch $\vec{v_1}$ and $\vec{v_2}$.

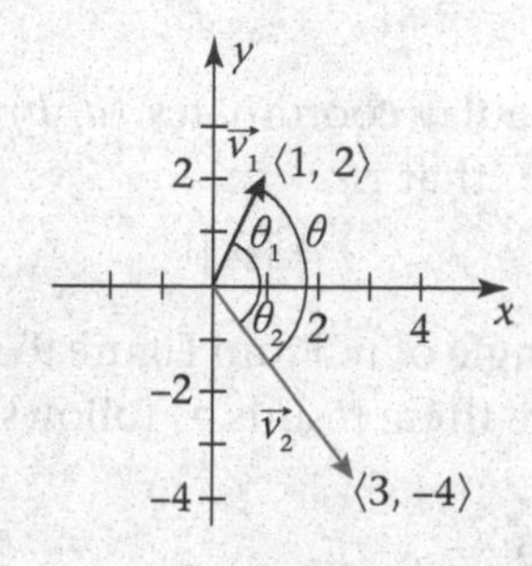

Since (1, 2) is in the first quadrant, $\vec{v_2}$ has direction angle

$$\tan^{-1}\left(\frac{2}{1}\right) \approx 63.43^\circ$$

Therefore, $\theta_1 \approx 63.43^\circ$. Since (3, −4) is in the fourth quadrant, $\vec{v_2}$ has direction angle

$$\tan^{-1}\left(\frac{-4}{3}\right) \approx 306.87^\circ$$

We observe that $\theta_2 \approx 360^\circ - 306.87^\circ = 53.13^\circ$. Since $\theta = \theta_1 + \theta_2$, we have

$$\theta \approx 63.43^\circ + 53.13^\circ = 116.56^\circ$$

■

EXAMPLE 10 *Applications of Vectors*

An airplane is flying N 49° E at 250 miles per hour, and the wind is blowing S 26° E at 50 miles per hour, as shown below. Find

a. the velocity vector representing the true heading of the airplane

b. the speed of the airplane relative to the ground

c. the angle the airplane is drifting due to the wind.

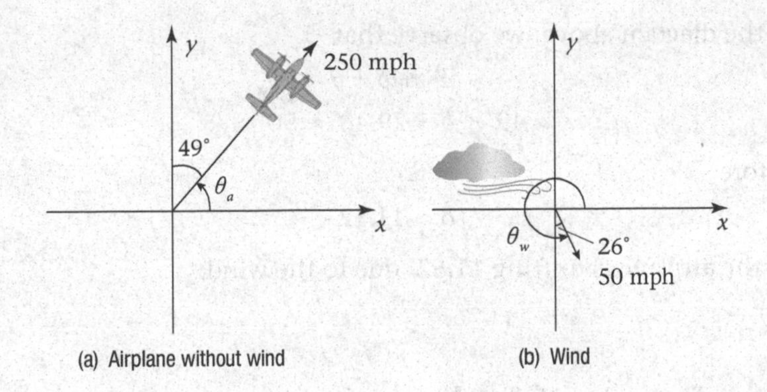

SOLUTION

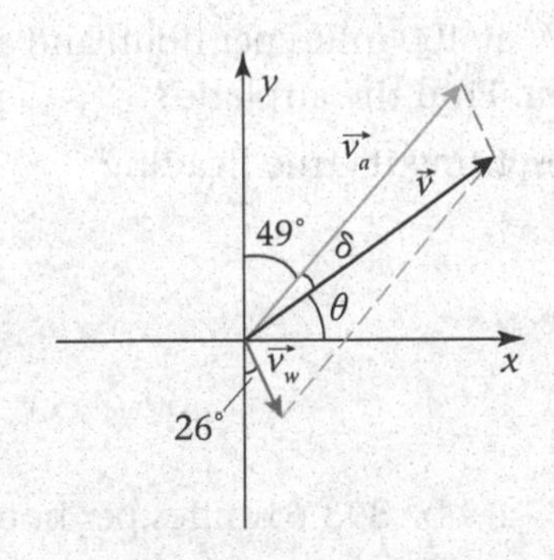

a. As shown above, let $\vec{v}_a$ and $\vec{v}_w$ denote the vectors of the airplane and wind, respectively. Let $\vec{v}$ be the vector denoting the true heading, and let δ be the drift angle. Then, for the airplane without wind, $\theta_a = 90^\circ - 49^\circ = 41^\circ$, and the wind has $\theta_w = 270^\circ + 26^\circ = 296^\circ$.

Therefore,

$$\vec{v}_a = \langle 250 \cos \theta_a, 250 \sin \theta_a \rangle = \langle 250 \cos 41^\circ, 250 \sin 41^\circ \rangle$$
$$\vec{v}_a \approx \langle 188.68, 164.01 \rangle$$

and

$$\vec{v}_w = \langle 50 \cos \theta_w, 50 \sin \theta_w \rangle = \langle 50 \cos 296^\circ, 50 \sin 296^\circ \rangle$$
$$\vec{v}_w \approx \langle 21.92, -44.94 \rangle$$

Thus, the vector of the true heading is

$$\vec{v} = \vec{v}_a + \vec{v}_w \approx \langle 210.60, 119.07 \rangle$$

b. The ground speed of the plane is

$$\|\vec{v}\| \approx \sqrt{(210.60)^2 + (119.07)^2} \approx 241.93 \text{ miles per hour}$$

c. We can find the direction angle θ of $\vec{v}$ from the trigonometric form of the vector. Since $\vec{v}$ is heading into the first quadrant,

$$\theta \approx \tan^{-1} \frac{119.07}{210.60} \approx 29.48^\circ$$

From the diagram above, we observe that

$$49^\circ + \delta + \theta = 90^\circ$$
$$49^\circ + \delta + 29.48^\circ \approx 90^\circ$$

Therefore

$$\delta \approx 11.52^\circ$$

Thus, the airplane is drifting 11.52° due to the wind. ■

Progress Check

A plane is flying S 20° W at 400 miles per hour, and a wind is blowing due east at 20 miles per hour. Find the airplane's

a. velocity vector representing its true heading
b. ground speed
c. drift angle

Answers

a. $\langle -116.81, -375.88 \rangle$ b. 393.61 miles per hour c. 2.74°

Exercise Set 9.5

In Exercises 1–6, sketch the following vectors.

1. $\langle 2, 5\rangle$
2. $\langle -3, 4\rangle$
3. $\langle -5, -1\rangle$
4. $\left\langle 2, -\frac{1}{2}\right\rangle$
5. $\langle -4, 1\rangle$
6. $\left\langle -\frac{3}{2}, -2\right\rangle$

In Exercises 7–12, find the vector $\overrightarrow{PQ}$ in component form $\langle a, b\rangle$, where the coordinates of P and Q are given. Furthermore, find the magnitude $\|\overrightarrow{PQ}\|$.

7. $P(0, 0), Q(1, 2)$
8. $P(1, 2), Q(3, 4)$
9. $P(-1, 3), Q(2, -1)$
10. $P(-2, -3), Q(4, 5)$
11. $P(4, -1), Q(-7, -3)$
12. $P(3, -4), Q(6, -1)$

In Exercises 13–20, $\vec{v_1} = \langle 1, -1\rangle$, $\vec{v_2} = \langle -3, 2\rangle$, $\vec{v_3} = \langle \frac{1}{2}, -\frac{1}{3}\rangle$, $\vec{v_4} = 7\vec{i} - 3\vec{j}$, and $\vec{v_5} = -\vec{i} + 4\vec{j}$. Simplify the following, writing your answer in component form $\langle a, b\rangle$.

13. $2\vec{v_1} + 3\vec{v_2}$
14. $-\vec{v_1} + \vec{v_2}$
15. $-2\vec{v_1} + \vec{v}2 - 6\vec{v_3}$
16. $3\vec{v_1} - 4\vec{v_2} + 8\vec{v_3}$
17. $2\vec{v_4} - 3\vec{v_5}$
18. $-\vec{v_4} + 4\vec{v_5}$
19. $\vec{v_1} + 2\vec{v_2} + 12\vec{v_3} - \vec{v_4} + \vec{v_5}$
20. $-\vec{v_1} - 2\vec{v_2} + 3\vec{v_3} + \vec{v_4} - 2\vec{v_5}$

In Exercises 21–24, find $\vec{v_1} + \vec{v_2}$, $\vec{v_1} - \vec{v_2}$, $2\vec{v_1}$, and $-\vec{v_2}$, and write your answers as linear combinations of the standard unit vectors. In addition, sketch the resulting vectors.

21. $\vec{v_1} = \langle 1, -1\rangle, \vec{v_2} = \langle -3, 2\rangle$
22. $\vec{v_1} = \langle 2, 5\rangle, \vec{v_2} = \langle -5, -1\rangle$
23. $\vec{v_1} = \langle -4, 1\rangle, \vec{v_2} = \langle -3, -2\rangle$
24. $\vec{v_1} = \langle -1, -3\rangle, \vec{v_2} = \langle 4, -2\rangle$

In Exercises 25–32, find the unit vectors $\vec{u_1}$ and $\vec{u_2}$, where $\vec{u_1}$ points in the same direction as $\vec{v}$ and $\vec{u_2}$ points in the opposite direction from $\vec{v}$.

25. $\vec{v} = \langle 2, 0\rangle$
26. $\vec{v} = \langle 0, 4\rangle$
27. $\vec{v} = \langle 5, 1\rangle$
28. $\vec{v} = \langle 2, 4\rangle$
29. $\vec{v} = \langle -3, 2\rangle$
30. $\vec{v} = \langle 4, -1\rangle$
31. $\vec{v} = -3i - j$
32. $\vec{v} = -2i + 6j$

In Exercises 33–38, find the vector $\vec{v}$ in component form $\langle a, b\rangle$ given the magnitude $\|\vec{v}\|$ and the direction angle θ. Sketch the resulting vectors.

33. $\|\vec{v}\| = 3, \theta = 60°$
34. $\|\vec{v}\| = 4, \theta = 150°$
35. $\|\vec{v}\| = 2, \theta = 300°$
36. $\|\vec{v}\| = 1, \theta = 210°$
37. $\|\vec{v}\| = 8, \theta = 315°$
38. $\|\vec{v}\| = 4, \theta = 240°$

In Exercises 39–44, find the angle θ between the given vectors $\vec{v_1}$ and $\vec{v_2}$, where $0° \le \theta \le 180°$.

39. $\vec{v_1} = \langle 3, -1\rangle, \vec{v_2} = \langle 2, -3\rangle$
40. $\vec{v_1} = \langle 1, 2\rangle, \vec{v_2} = \langle 1, 4\rangle$
41. $\vec{v_1} = \langle -1, -2\rangle, \vec{v_2} = \langle -2, -3\rangle$
42. $\vec{v_1} = \langle 1, 3\rangle, \vec{v_2} = \langle -1, 3\rangle$
43. $\vec{v_1} = \langle 2, -5\rangle, \vec{v_2} = \langle -2, -5\rangle$
44. $\vec{v_1} = \langle -2, 3\rangle, \vec{v_2} = \langle -2, 4\rangle$

In Exercises 45 and 46, find the sum of the vectors, written in component form $\langle a, b\rangle$, find the direction angle, and sketch the vectors and their sum.

45. $\vec{v_1} = \langle 1, -1\rangle, \vec{v_2} = \langle 2, 0\rangle$
46. $\vec{v_1} = \langle 4, -1\rangle, \vec{v_2} = -3\vec{i} + \vec{j}, \vec{v_3} = \langle 2, -1\rangle$
47. A plane is flying N 20° W at 280 miles per hour. A wind is blowing due south at 40 miles per hour. Find the airplane's
    a. velocity vector representing its true heading
    b. ground speed
    c. drift angle
48. A ship is heading S 10° E at 40 miles per hour. A current is moving S 15° W at 10 miles per hour. Find the ship's
    a. velocity vector representing its true heading
    b. speed relative to the land
    c. drift angle
49. Find the solution to Exercise 47 using the law of cosines and the law of sines.
50. Find the solution to Exercise 48 using the law of cosines and the law of sines.

Chapter Summary

Key Terms, Concepts, and Symbols

Key Ideas for Review

Topic	Page	Key Idea
Law of Sines	498	The law of sines is useful in solving any triangle when the known parts are *SAA*, *ASA*, or *SSA*. $$\frac{\sin\alpha}{a} = \frac{\sin\beta}{b} = \frac{\sin\gamma}{c}$$
Law of Cosines	504	The law of cosines is useful in solving any triangle when the known parts are *SSS* or *SAS*. $$a^2 = b^2 + c^2 - 2bc\cos\alpha$$ $$b^2 = a^2 + c^2 - 2ac\cos\beta$$ $$c^2 = a^2 + b^2 - 2ab\cos\gamma$$
Trigonometric Form of $a + bi$	510	The complex number $a + bi$ can be associated with the point $P(a, b)$ in the complex plane. The trigonometric or polar form of $a + bi$ is $$r(\cos\theta + i\sin\theta)$$ where r is the distance from the origin O to P and θ is the measure of the angle in standard position whose terminal side is OP.
Multiplication and Division	512	If $z_1 = r_1(\cos\theta_1 + i\sin\theta_1)$ and $z_2 = r_2(\cos\theta_2 + i\sin\theta_2)$, then $$z_1z_2 = r_1r_2(\cos(\theta_1+\theta_2) + i\sin(\theta_1+\theta_2))$$ and $$\frac{z_1}{z_2} = \frac{r_1}{r_2}(\cos(\theta_1-\theta_2) + i\sin(\theta_1-\theta_2))$$

continues

Topic	Page	Key Idea
De Moivre's Theorem	514	If $z = r(\cos\theta + i\sin\theta)$, then for natural number n $$z^n = r^n(\cos n\theta + i\sin n\theta)$$
nth Roots of a Complex Number	515	If $z = r(\cos\theta + i\sin\theta) \neq 0$ and n is a positive integer, then the n distinct roots of z are $$u_k = \sqrt[n]{r}\left[\cos\left(\frac{\theta+2\pi k}{n}\right)+i\sin\left(\frac{\theta+2\pi k}{n}\right)\right]$$ where $k = 0, 1, 2, \ldots, n-1$.
Polar Coordinates	519	If P is a point in the plane with rectangular coordinates (x, y), then P has polar coordinates (r, θ), where r is the distance from the origin O to P and θ is the measure of the angle in standard position whose terminal side is OP. Specifically, the relationships are $$x = r\cos\theta \qquad y = r\sin\theta$$ $$\tan\theta = \frac{y}{x} \qquad r = \sqrt{x^2+y^2}$$
Vectors	528	A vector $\vec{v}$ is a directed line segment from $P(x_1, y_1)$ to $Q(x_2, y_2)$ with components $$\vec{v} = \langle x_2 - x_1, y_2 - y_1\rangle$$ and magnitude $$\lVert\vec{v}\rVert = \sqrt{(x_2-x_1)^2+(y_2-y_1)^2}$$
Vector Addition and Scalar Multiplication	532 533	If $\vec{v_1} = \langle a_1, b_1\rangle$ and $\vec{v_2} = \langle a_2, b_2\rangle$, then • $\vec{v_1} + \vec{v_2} = \langle a_1 + a_2, b_1 + b_2\rangle$ • $k\vec{v_1} = \langle ka_1, kb_1\rangle$
Trigonometric Form of Vectors	539	If $\vec{v} = \langle a + b\rangle$ with $\lVert\vec{v}\rVert = r$ and direction angle θ, then • $\vec{v} = \langle r\cos\theta, r\sin\theta\rangle$ • $a^2 + b^2 = r^2$ • $\tan\theta = \frac{b}{a}$

Review Exercises

In Exercises 1–4, use the law of sines or the law of cosines to approximate the required part of triangle ABC.

1. If $a = 12$, $b = 7$, and $c = 15$, find α.
2. If $a = 20$, $b = 15$, and $\alpha = 55°$, find β.
3. If $a = 10$, $\alpha = 38°$, and $\beta = 22°$, find c.
4. If $b = 8$, $c = 12$, and $\alpha = 35°$, find a.

In Exercises 5–7, determine the absolute value of the given complex number.

5. $2 - i$
6. $-3 + 2i$
7. $-4 - 5i$

In Exercises 8–11, convert from trigonometric to algebraic form and vice versa.

8. $-3 + 3i$
9. $2(\cos 90° + i\sin 90°)$
10. $\sqrt{2}(\cos 315° + i\sin 315°)$
11. -2

In Exercises 12–14, find the indicated product or quotient. Express the answer in trigonometric form.

12. $4(\cos 22° + i\sin 22°) \cdot (6\cos 15° + i\sin 15°)$
13. $\dfrac{5(\cos 71° + i\sin 71°)}{3(\cos 50° + i\sin 50°)}$
14. $2(\cos 210° + i\sin 210°) \cdot (\cos 240° + i\sin 240°)$

In Exercises 15 and 16, use De Moivre's Theorem to express the given number in the form $a + bi$.

15. $(3 - 3i)^5$

16. $[2(\cos 90^\circ + i \sin 90^\circ)]^3$

17. Express the two square roots of -9 in trigonometric form.

18. Determine all roots of the equation $z^3 - 1 = 0$.

In Exercises 19 and 20, sketch the graph of the given polar equation.

19. $r = \sec\theta \tan\theta$

20. $r = |\theta|$

21. Transform the equation $r = 3 \csc\theta$ to an equation in x and y.

22. Transform the equation $x^2 + 9y^2 = 9$ to a polar equation.

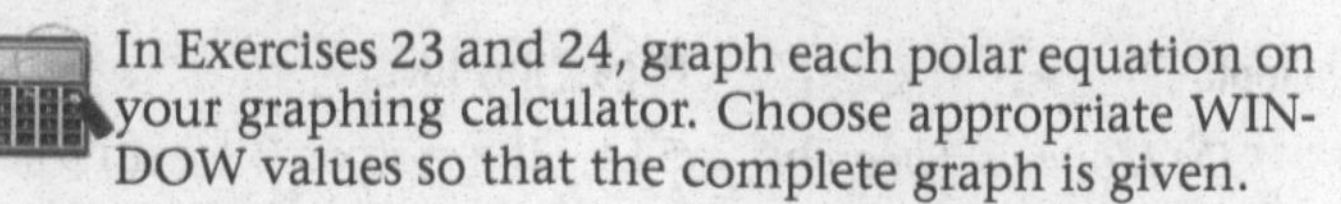

In Exercises 23 and 24, graph each polar equation on your graphing calculator. Choose appropriate WINDOW values so that the complete graph is given.

23. a. $r = 3 \sin\theta$ $\quad 0 \le \theta \le 2\pi$
    b. $r = -3 \cos\theta$ $\quad 0 \le \theta \le 2\pi$
    c. $r = -3 \sin\theta$ $\quad 0 \le \theta \le 2\pi$
    d. $r = 3 \cos\theta$ $\quad 0 \le \theta \le 2\pi$

24. a. $r = 2\sqrt{\cos(2\theta)}$ $\quad 0 \le \theta \le 2\pi$
    b. $r = -2\sqrt{\cos(2\theta)}$ $\quad 0 \le \theta \le 2\pi$

In Exercises 25–27, we are given $\vec{v_1} = \langle -1, 3 \rangle$, $\vec{v_2} = \langle 4, 2 \rangle$, and $\vec{v_3} = \langle -2, -3 \rangle$.

25. Find $3\vec{v_1} - 2\vec{v_2} + \vec{v_3}$ in terms of the standard unit vectors.

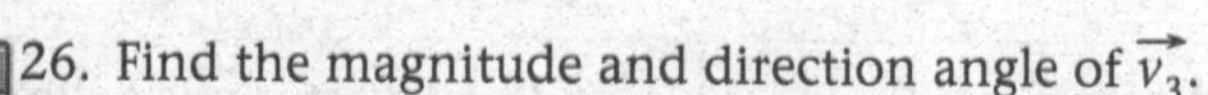

26. Find the magnitude and direction angle of $\vec{v_3}$.

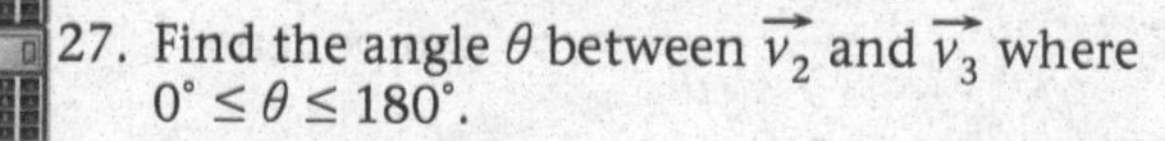

27. Find the angle θ between $\vec{v_2}$ and $\vec{v_3}$ where $0^\circ \le \theta \le 180^\circ$.

Review Test

In Exercises 1–4, find the required part of triangle ABC.

1. If $a = 2$, $b = 4$, and $c = 5$, find α.
2. If $b = 10$, $\alpha = 15^\circ$, and $\beta = 28^\circ$, find c.
3. If $b = 10$, $c = 13$, and $\alpha = 54^\circ$, find β.
4. If $a = 5$, $c = 9$, and $\beta = 36^\circ$, find b.

In Exercises 5 and 6, determine the absolute value.

5. $3 - 4i$ $\qquad$ 6. $-2 + 2i$

In Exercises 7–10, find the indicated product or quotient. Express the answer in trigonometric form.

7. $\frac{1}{2}(\cos 14^\circ + i \sin 14^\circ) \cdot 10(\cos 72^\circ + i \sin 72^\circ)$

8. $\dfrac{3(\cos 85^\circ + i \sin 85^\circ)}{6(\cos 8^\circ + i \sin 8^\circ)}$

9. $\left(\cos\frac{\pi}{5} + i \sin\frac{\pi}{5}\right) \cdot 5\left(\cos\frac{\pi}{8} + i \sin\frac{\pi}{8}\right)$

10. $\dfrac{\frac{1}{2}\left(\cos\frac{\pi}{10} + i \sin\frac{\pi}{10}\right)}{\frac{1}{4}\left(\cos\frac{\pi}{7} + i \sin\frac{\pi}{7}\right)}$

In Exercises 11 and 12, use De Moivre's Theorem to express the given form as $a + bi$.

11. $\left[\frac{1}{5}(\cos 120^\circ + i \sin 120^\circ)\right]^4$

12. $(-2i)^6$

In Exercises 13 and 14, express the roots in trigonometric form.

13. $z^3 + 1 = 0$ $\qquad$ 14. $z^3 - 27 = 0$

15. Sketch the graph of the polar equation $r = 4 \csc\theta$.

16. Transform the equation $r = \sec^2\frac{\theta}{2}$ to an equation in x and y.

17. Transform the equation $x^2 + xy + y^2 = 1$ to a polar equation.

18. Find the angle θ between $\vec{v_1} = \langle 2, -5 \rangle$ and $\vec{v_2} = \langle -4, 1 \rangle$, where $0^\circ \le \theta \le 180^\circ$.

Cumulative Review Exercises: Chapter 7–9

1. Write each of the following trigonometric functions in terms of an angle t, where t is in the first quadrant. Your answers may use only the forms $\sin t$, $-\sin t$, $\cos t$, $-\cos t$, $\tan t$ and $-\tan t$.

   a. $\sin\left(-\frac{9\pi}{5}\right)$ b. $\cos\frac{8\pi}{3}$

   c. $\tan\left(-\frac{11\pi}{7}\right)$

2. Evaluate without using a calculator.

   a. $\sec\left(-\frac{37\pi}{3}\right)$ b. $\cot\frac{23\pi}{6}$

   c. $\sin\left(-\frac{29\pi}{6}\right)$

3. Sketch the graph of $f(x) = 3\sin\left(2x + \frac{\pi}{2}\right)$.

4. Find the values of t in the interval $[0, 2\pi]$ that satisfy the given equation.

   a. $\cos t = -\frac{\sqrt{2}}{2}$ b. $\csc t = 2$

   c. $\tan t = -\sqrt{3}$

5. Evaluate

   a. $\tan\left(\cos^{-1}\frac{3}{5}\right)$ b. $\sec\left(\sin^{-1}\frac{5}{13}\right)$

   c. $\cos^{-1}\left(\cot\left(-\frac{\pi}{4}\right)\right)$

6. Find t in the interval $[0, 2\pi]$ if

   a. $\tan t = \sqrt{3}$ and $\cos t < 0$

   b. $\sec t = 2$ and $\tan t < 0$

   c. $\cos t = -\sqrt{3}$ and $\tan t > 0$

7. Prove.

   a. $\sec t \cot t = \csc t$

   b. $\frac{1}{\sec t - \tan t} = \sec t + \tan t$

8. Determine the amplitude, period and phase shift.

   a. $f(x) = -\sin\left(\frac{x}{2} + \frac{\pi}{2}\right)$

   b. $f(x) = 3\cos\left(2x - \frac{\pi}{2}\right)$

9. Solve and express the exact solution in terms of inverse trigonometric functions.

   a. $3\sin^2 x = 2$

   b. $2\cos^2 x - \cos x - 3 = 0$

10. Use a calculator to find all solutions in the interval $[0, 360°]$.

    $$\tan^2 x - 4\tan x + 2 = 0$$

11. Find the reference angle for the given angle.

    a. $265°$ b. $-\frac{7\pi}{3}$

    c. $-120°$

12. Convert from radians to degrees or degrees to radians.

    a. $5°$ b. $-\frac{\pi}{6}$

    c. 2

13. A right triangle has legs of length 3 and 5. Let θ be the angle between the leg of length 5 and the hypotenuse. Evaluate.

    a. $\sec\theta$ b. $\sin\left(\frac{\pi}{2} - \theta\right)$

    c. $\cot\theta$

14. To bring the seat of a ferris wheel to a point at which it can be serviced, the ferris wheel must go through a central angle of 30°. The distance covered is 20 feet. What is the diameter of the ferris wheel?

15. Find the angle of elevation of the sun when a building 50 feet in height casts a horizontal shadow 30 feet in length.

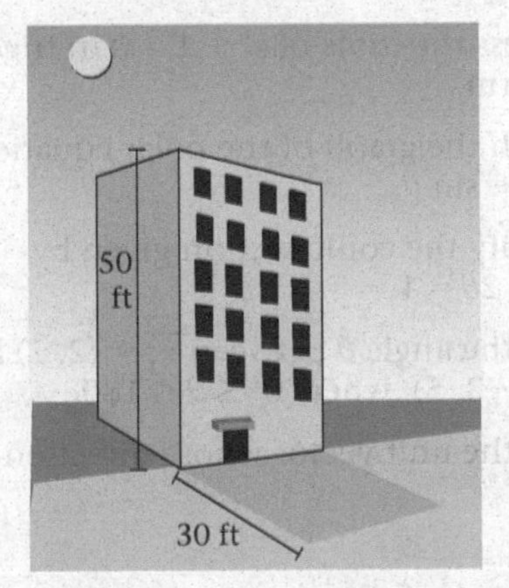

16. The lengths of the sides of a rectangle are 6 and 9. What is the degree measure of the angle between the shorter side and the diagonal?

17. Use the law of sines or the law of cosines to approximate the required part of triangle ABC.

    a. If $a = 5$, $b = 12$ and $c = 8$, find γ.

    b. If $a = 15$, $b = 18$ and $\alpha = 50°$, find γ.

18. Verify the identity

    $$\frac{\tan^2 t - 1}{\tan^2 t + 1} = 1 - 2\cos^2 t$$

19. Write the following in terms of trigonometric functions of θ alone.

    a. $\cos\left(\frac{\pi}{6} - \theta\right)$ b. $\tan\left(\theta - \frac{\pi}{4}\right)$

20. Find the exact values without the use of a calculator.

a. $\cos\left(-\frac{\pi}{8}\right)$ b. $\tan 255^\circ$

21. Write in terms of the cofunction.

a. $\cot 62^\circ$ b. $\sin\frac{\pi}{12}$

22. Given $\sin t = -\frac{3}{5}$ and $\cos t > 0$, find

a. $\sin\frac{t}{2}$ b. $\tan 2t$

23. Express as a product: $\cos 40^\circ + \cos 20^\circ$.

24. Express as a sum: $\sin 2t \cos 3t$.

25. Find all solutions of the equation $\sin\theta - \cos 2\theta = 0$ in the interval $[0, 2\pi]$.

26. Determine the absolute value of the complex number.

a. $3 + 2i$ b. $6 - 3i$

27. Write the complex number in trigonometric form.

a. $-3i$ b. $4 - 4i$

28. Simplify the quotient

$$\frac{35(\cos 210^\circ + i\sin 210^\circ)}{7(\cos 180^\circ + i\sin 180^\circ)}$$

29. Express the roots of $z^4 + 1 = 0$ in trigonometric form.

30. Sketch the graph of the polar equation $r = 1 - \sin\theta$.

31. Identify the conic section given by $r^2\cos 2\theta = 1$.

32. Find the angle θ between $\overrightarrow{v_1} = \langle 2, 7\rangle$ and $\overrightarrow{v_2} = \langle -3, 5\rangle$, where $0^\circ \le \theta \le 180^\circ$.

33. Find the unit vector whose direction angle is 230°.

10 Systems of Equations and Inequalities

Mathematics students become accustomed to seeing equations. However, if you look at the world around you, you rarely see equations! What you do see is relationships which can be modeled by equations. One way to do this is by polynomial curve fitting. This is a process whereby we find the equation of a polynomial which passes through any point we are given (perhaps these points represent paired data, like heights and weights). The process of finding an equation in standard form for a polynomial that passes through a given set of points requires that we solve a system of equations. You will get some practice doing this in this chapter's project.

Continue your investigation of mathematics on the Web by checking out the following site: www.cut-the-knot.org/.

Many problems in business and engineering require the solution of systems of equations and inequalities. In fact, systems of linear equations and inequalities occur with such frequency that mathematicians and computer scientists have devoted considerable energy to devising methods for their solution. With the aid of large-scale computers, it is possible to solve systems involving thousands of equations or inequalities that contain thousands of variables.

We begin with the study of the methods of substitution and elimination, methods that are applicable to all types of systems. We then introduce graphical methods for solving systems of linear inequalities and apply this technique to linear programming problems, a type of optimization problem.

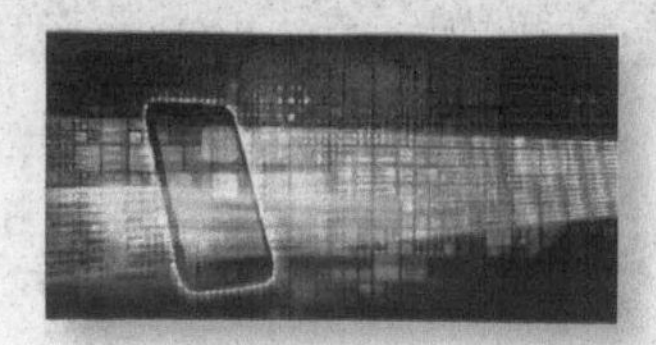

www.cut-the-knot.org/

10.1 Systems of Equations in Two Unknowns

Consider the following problem:

> A pile of 9 coins consists of nickels and quarters. If the total value of the coins is \$1.25, how many of each type of coin are there?

Although we could solve this problem using the methods discussed in Chapter 2, we are going to focus on a somewhat different approach. Let

$$x = \text{the number of nickels}$$

and

$$y = \text{the number of quarters}$$

The requirements can then be expressed as

$$\begin{aligned} x + \quad y &= \quad 9 \\ 5x + 25y &= 125 \end{aligned}$$

This is an example of a **system of equations**, and we seek values of x and y that satisfy *both* equations. If $x = a$ and $y = b$ satisfy both equations, then the ordered pair (a, b) is called a **solution** of the system. Thus,

$$x = 5 \qquad \text{and} \qquad y = 4$$

or equivalently

$$(5, 4)$$

is a solution since substituting into the above system yields

$$\begin{aligned} 5 + \quad 4 &= \quad 9 \\ 5(5) + 25(4) &= 125 \end{aligned}$$

10.1a *Solving by Substitution*

If we can use one of the equations of a system to express one variable in terms of the other variable, then we can **substitute** this expression into the other equation.

EXAMPLE 1 ***Solving by Substitution***

Solve the system of equations.

$$\begin{aligned} x^2 + y^2 &= 25 \\ x + y &= -1 \end{aligned}$$

SOLUTION

From the second equation we have

$$y = -1 - x$$

Substituting this expression for y into the first equation, we obtain

$$\begin{aligned} x^2 + (-1 - x)^2 &= 25 \\ x^2 + 1 + 2x + x^2 &= 25 \\ 2x^2 + 2x - 24 &= 0 \\ x^2 + x - 12 &= 0 \\ (x + 4)(x - 3) &= 0 \end{aligned}$$

which yields $x = -4$ and $x = 3$. Substituting these values for x in the second equation, we obtain the corresponding values of y.

$$x = -4: \quad -4 + y = -1 \qquad x = 3: \quad 3 + y = -1$$
$$y = 3 \qquad\qquad\qquad\qquad y = -4$$

Thus, $x = -4$, $y = 3$ and $x = 3$, $y = -4$ are solutions of the system of equations. ■

10.1b Solving by Graphing

The coordinates of every point on the graph of an equation must satisfy that equation. If we sketch the graphs of a pair of equations on the same coordinate axes, it follows that the *points of intersection* must satisfy *both* equations. Thus, we have a graphical means of solving a system of equations.

Consider the equations of Example 1. The graph of the first equation is a circle of radius 5, centered at the origin. The graph of the second equation is a line. We sketch both curves on the same set of axes, as shown in Figure 10-1. Note the solutions $(-4, 3)$ and $(3, -4)$ are the points of intersection of these graphs.

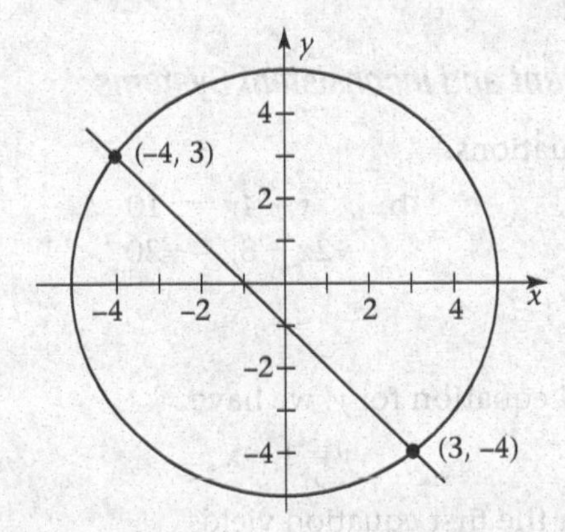

FIGURE 10-1
Diagram for Example 1

The expression for x or y obtained from an equation *must not be substituted into the same equation*. From the first equation of the system

$$x + 2y = -1$$
$$3x^2 + y = 2$$

we obtain

$$x = -1 - 2y$$

Substituting in the *same* equation results in

$$(-1 - 2y) + 2y = -1$$
$$-1 = -1$$

The substitution $x = -1 - 2y$ must be made into the *second* equation.

Solve the system of equations.

a. $x^2 + 3y^2 = 12$
$x + 3y = 6$

b. $x^2 + y^2 = 34$
$x - y = 2$

Answers

a. $x = 3, y = 1; x = 0, y = 2$

b. $x = -3, y = -5; x = 5, y = 3$

It is also possible for a system of equations to have no solution or an infinite number of solutions. The following terminology is used to distinguish these situations.

Consistent and Inconsistent Systems

- A **consistent** system of equations has one or more solutions.
- An **inconsistent** system of equations has no solutions.

EXAMPLE 2 *Consistent and Inconsistent Systems*

Solve the system of equations.

a. $x^2 - 2x - y + 3 = 0$
$x + y - 1 = 0$

b. $x + 4y = 10$
$-2x - 8y = -20$

SOLUTION

a. Solving the second equation for y, we have

$$y = 1 - x$$

and substituting in the first equation yields

$$x^2 - 2x - (1 - x) + 3 = 0$$
$$x^2 - x + 2 = 0$$

Since the discriminant of this quadratic equation is negative, the equation has no real roots. However, any solution of the system of equations must satisfy this quadratic equation. We can therefore conclude that the system is inconsistent. The graphs of the equations are a parabola and a line that do not intersect, as shown in Figure 10-2.

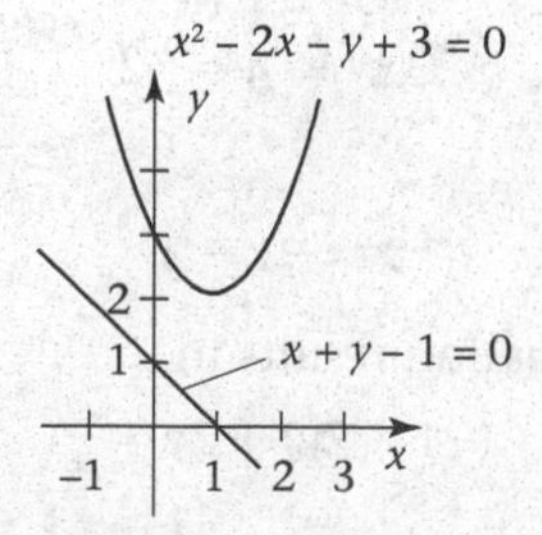

FIGURE 10-2
Sketch of the System in Example 2(a)

b. Solving the first equation for x, we have

$$x = 10 - 4y$$

and substituting in the second equation gives

$$\begin{aligned} -2(10-4y)-8y &= -20 \\ -20+8y-8y &= -20 \\ -20 &= -20 \end{aligned}$$

The substitution procedure has resulted in an identity, indicating that any solution of the first equation also satisfies the second equation. Since there are an infinite number of ordered pairs (a, b) with $x = a$ and $y = b$ satisfying the first equation, the system is consistent. (Verify that the graphs of these two equations are identical.) ■

Progress Check

Solve by substitution.

a. $3x^2 - y^2 = 7$
$-9x + 3y = -2$

b. $-5x + 2y = -4$
$\frac{5}{2}x - y = 2$

Answers

a. no solution

b. any point on the line $-5x + 2y = -4$

10.1c *Systems of Linear Equations*

A system consisting only of equations that are of the first degree in x and y is called a **system of linear equations**, or simply a **linear system**. When we graph a linear system of two equations on the same set of coordinate axes, there are three possibilities.

Consistent and Inconsistent Systems

1. The two lines intersect at a point, as shown in Figure 10-3(a). The system is consistent and has a unique solution, namely, the point of intersection.
2. The two equations are different forms of the same line, as shown in Figure 10-3(b). The system is consistent and has an infinite number of solutions, namely, all points on the line.
3. The two lines are parallel, as shown in Figure 10-3(c). Since the lines do not intersect, the system is inconsistent and has no solution.

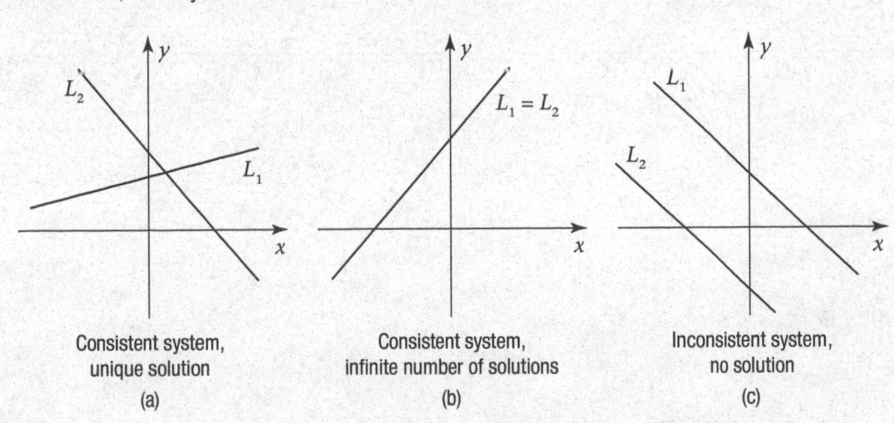

FIGURE 10-3
Possible Graphs for a Pair of Linear Equations

10.1d *Solving by Elimination*

The **method of elimination** seeks to combine the equations of a system in such a way as to **eliminate** one of the unknowns.

EXAMPLE 3 *Solving by Elimination*

Solve the system.

$$\begin{aligned} x + y &= 9 \\ 5x + 25y &= 125 \end{aligned}$$

SOLUTION

Step 1. Multiply each equation by a constant so that the coefficients of either x or y differ only in sign.	*Step 1.* Multiply the first equation by -5 and the second equation by 1 so that the coefficients of x are -5 and 5. $-5x - 5y = -45$ $5x + 25y = 125$
Step 2. Add the equations. At least one unknown has been "eliminated" in the resulting equation.	*Step 2.* $20y = 80$
Step 3. Solve the resulting equation in one unknown, if possible.	*Step 3.* $y = 4$
Step 4. Substitute into either of the *original* equations to solve for the second unknown.	*Step 4.* Substitute $y = 4$ into the first equation of the original system. $x + y = 9$ $x + 4 = 9$ $x = 5$
Step 5. Check in both equations.	*Step 5.* Verify that the solution $x = 5, \quad y = 4$ satisfies $x + y = 9$ $5 + 4 = 9$ ✓ $5x + 25y = 125$ $5(5) + 25(4) = 25 + 100 = 125$ ✓

■

EXAMPLE 4 *Solving by Elimination*

Solve the system.

$$\begin{aligned} 4x^2 + 9y^2 &= 36 \\ -9x^2 + 18y^2 &= 4 \end{aligned}$$

SOLUTION

Step 1. Multiply each equation by a constant so that the coefficients of either x^2 or y^2 differ only in sign.	*Step 1.* Multiply the first equation by −2 and the second equation by 1 so that the coefficients of y^2 are −18 and 18. $-8x^2 - 18y^2 = -72$ $-9x^2 + 18y^2 = 4$
Step 2. Add the equations. At least one unknown has been "eliminated" in the resulting equation.	*Step 2.* $-17x^2 = -68$
Step 3. Solve the resulting equation in one unknown, if possible.	*Step 3.* $x^2 = 4$ $x = \pm 2$
Step 4. Substitute into either of the *original* equations to solve for the second unknown.	*Step 4.* Substitute $x = 2$ into the first equation of the original system. $4x^2 + 9y^2 = 36$ $4(2)^2 + 9y^2 = 36$ $y = \pm\frac{2}{3}\sqrt{5}$ Substituting $x = -2$ yields the same values for y.
Step 5. Check in both equations.	*Step 5.* Verify that the solutions $x = 2,\ y = \frac{2}{3}\sqrt{5}$ $x = 2,\ y = -\frac{2}{3}\sqrt{5}$ $x = -2,\ y = \frac{2}{3}\sqrt{5}$ $x = -2,\ y = -\frac{2}{3}\sqrt{5}$ satisfy both equations.

The graphs of these equations (ellipse and hyperbola) for Example 4 are shown in Figure 10-4.

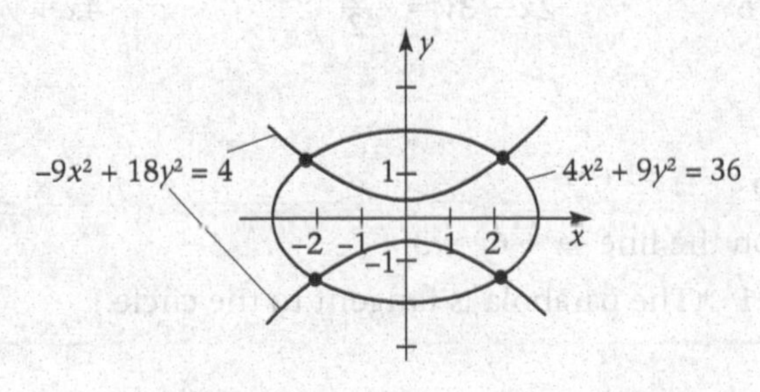

FIGURE 10-4
Graph for Example 4

■

EXAMPLE 5 *Solving by Elimination*

Solve the system.

a. $2x^2 - 3y^2 = 9$
$x^2 + y^2 = 4$

b. $5x + 6y = 4$
$-10x - 12y = -8$

SOLUTION

a. Adding -2 times the second equation to the first equation yields

$$-5y^2 = 1 \quad \text{or} \quad y^2 = -\frac{1}{5}$$

Since this equation has no real solutions, the graphs of the given system do not intersect; the system is inconsistent, as shown below.

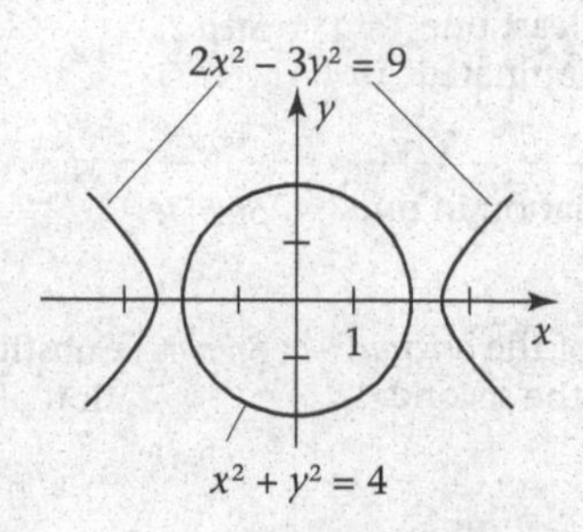

b. Multiplying the first equation by 2, we have

$$\begin{array}{r} 10x + 12y = \ \ 8 \\ -10x - 12y = -8 \\ \hline 0x + \ \ 0y = \ \ 0 \end{array}$$

We conclude that the equations represent the same line and that the solution set consists of all points on the line $5x + 6y = 4$. ■

Progress Check

Solve by elimination.

a. $x - y = 2$
$3x - 3y = -6$

b. $4x + 6y = 3$
$-2x - 3y = -\frac{3}{2}$

c. $x^2 - 4x + y^2 - 4y = 1$
$x^2 - 4x + y = -5$

Answers

a. no solution

b. all points on the line $4x + 6y = 3$

c. $x = 2, y = -1$ (The parabola is tangent to the circle.)

Exercise Set 10.1

In Exercises 1–10, find approximate solutions of the given system by graphing.

1. $x + y = 1$, $x - y = 3$
2. $x - y = 1$, $x + y = 5$
3. $3x - y = 4$, $6x - 2y = -8$
4. $x^2 + 4y^2 = 32$, $x + 2y = 0$
5. $xy = -4$, $4x - y = 8$
6. $4x^2 + y^2 = 4$, $x^2 - y^2 = 9$
7. $4x^2 + 9y^2 = 72$, $4x - 3y^2 = 0$
8. $2y^2 - x^2 = -1$, $4y^2 + x^2 = 25$
9. $x^2 + y^2 = 1$, $y^2 - 3x^2 = 5$
10. $3x^2 + 8y^2 = 21$, $x^2 + 4y^2 = 10$

In Exercises 11–20, solve the system of equations by the method of substitution.

11. $x + y = 1$, $x - y = 3$
12. $x + 2y = 8$, $3x - 4y = 4$
13. $x^2 + y^2 = 13$, $2x - y = 4$
14. $x^2 + 4y^2 = 32$, $x + 2y = 0$
15. $y^2 - x = 0$, $y - 4x = -3$
16. $xy = -4$, $4x - y = 8$
17. $x^2 - 2x + y^2 = 3$, $2x + y = 4$
18. $4x^2 + y^2 = 4$, $x - y = 3$
19. $xy = 1$, $x - y + 1 = 0$
20. $\frac{1}{2}x - \frac{3}{2}y = 4$, $\frac{3}{2}x + y = 1$

In Exercises 21–30, solve the system of equations by the method of elimination.

21. $x + 2y = 1$, $5x + 2y = 13$
22. $x - 4y = -7$, $2x + 3y = -8$
23. $25y^2 - 16x^2 = 400$, $9y^2 - 4x^2 = 36$
24. $x^2 - y^2 = 3$, $x^2 + y^2 = 5$
25. $2x^2 + 3y^2 = 30$, $x - y^2 = -1$
26. $x^2 + y^2 + 2y = 9$, $y - 2x = 4$
27. $2x + y = 4$, $-6x - 3y = -8$
28. $2x + 3y = -2$, $-3x - 5y = 4$
29. $y^2 - x^2 = -5$, $3y^2 + x^2 = 21$
30. $x^2 + 4y^2 = 25$, $4x^2 + y^2 = 25$

In Exercises 31–42, determine whether the system is consistent or inconsistent. If the system is consistent, find all the solutions.

31. $2x + 2y = 6$, $3x + 3y = 6$
32. $2x + y = 2$, $3x - y = 8$
33. $y^2 - 8x^2 = 9$, $y^2 + 3x^2 = -31$
34. $4y^2 + 3x^2 = 24$, $3y^2 - 2x^2 = 35$
35. $3x + 3y = 9$, $2x + 2y = -6$
36. $x - 4y = -7$, $2x - 8y = -4$
37. $3x - y = 18$, $\frac{3}{2}x - \frac{1}{2}y = 9$
38. $2x + y = 6$, $x + \frac{1}{2}y = 3$
39. $x^2 - 3xy - 2y^2 - 2 = 0$, $x - y - 2 = 0$
40. $3x^2 + 8y^2 = 20$, $x^2 + 4y^2 = 10$
41. $2x - 2y = 4$, $x - y = 8$
42. $2x - 3y = 8$, $4x - 6y = 16$

10.2 Applications: Word Problems

In the beginning of the previous section, we saw that it is possible to formulate and solve a word problem using two variables and a system of linear equations rather than using one variable and a single linear equation. We now consider a number of word problems using this approach. (For additional practice, see Section 2.2 and try to formulate and solve those problems using two variables and a system of linear equations.)

EXAMPLE 1 *Applying Linear Systems*

If 3 vitamin pills and 4 herbal supplements cost 69 cents, and 5 vitamin pills and 2 herbal supplements cost 73 cents, what is the cost of each type of pill?

SOLUTION

Using two variables, we let

x = the cost in cents of each vitamin pill
y = the cost in cents of each herbal supplement pill

Then

$$\begin{aligned} 3x + 4y &= 69 \\ 5x + 2y &= 73 \end{aligned}$$

We multiply the second equation by -2 and add to eliminate y.

$$\begin{aligned} 3x + 4y &= 69 \\ -10x - 4y &= -146 \\ \hline -7x \quad\quad &= -77 \\ x &= 11 \end{aligned}$$

Substituting into the first equation, we have

$$\begin{aligned} 3(11) + 4y &= 69 \\ 4y &= 36 \\ y &= 9 \end{aligned}$$

Thus, each vitamin pill costs 11 cents, and each herbal supplement pill costs 9 cents. ■

EXAMPLE 2 *Applying Linear Systems*

Swimming downstream, a swimmer can cover 2 kilometers in 15 minutes. The return trip upstream requires 20 minutes. What is the rate of the swimmer and of the current in kilometers per hour? (The rate of the swimmer is the swimmer's speed if there was no current.)

SOLUTION

Let

x = the rate of the swimmer in kilometers per hour
y = the rate of the current in kilometers per hour

For swimming downstream, the rate of the current is added to the rate of the swimmer, so $x + y$ is the rate downstream. On the other hand, $x - y$ is the rate for swimming upstream. We display the information we have in the table, expressing time in hours.

Swimming	Rate	×	Time	=	Distance
Downstream	$x+y$		$\frac{1}{4}$		$\frac{1}{4}(x+y)$
Upstream	$x-y$		$\frac{1}{3}$		$\frac{1}{3}(x-y)$

Since distance upstream = distance downstream = 2 kilometers,

$$\frac{1}{4}(x+y) = 2$$
$$\frac{1}{3}(x-y) = 2$$

or equivalently,

$$x + y = 8$$
$$x - y = 6$$

Solving, we have

$x = 7$ rate of the swimmer
$y = 1$ rate of the current

Thus, the rate of the swimmer is 7 kilometers per hour, and the rate of the current is 1 kilometer per hour. ■

EXAMPLE 3 *Applying Linear Systems*

The sum of a two-digit number and its units digit is 64, and the sum of the number and its tens digit is 62. Find the number.

SOLUTION

The basic idea in solving digit problems is to note that if we let

$$t = \text{tens digit}$$

and

$$u = \text{units digit}$$

then

$$10t + u = \text{the two-digit number}$$

Then "the sum of a two-digit number and its units digit is 64" translates into

$$(10t + u) + u = 64 \quad \text{or} \quad 10t + 2u = 64$$

Also, "the sum of the number and its tens digit is 62" becomes

$$(10t + u) + t = 62 \quad \text{or} \quad 11t + u = 62$$

Solving, we find that $t = 5$ and $u = 7$. Hence, the number is 57. ■

10.2a Applications in Business and Economics: Break-Even Analysis

An important problem faced by a manufacturer is that of determining the **level of production**, that is, the number of units of the product to be manufactured during a given time period—a day, a week, or a month. Suppose that

$$C = 400 + 2x \tag{1}$$

is the total cost in thousands of dollars of producing x units of the product. Thus, after setting up production at a cost of \$400,000, the manufacturer has an additional cost of \$2000 to make each unit.

Furthermore, suppose that

$$R = 4x \tag{2}$$

is the total revenue in thousands of dollars when x units of the product are sold. In other words, a revenue of \$4000 is earned from the sale of each unit. If all units that are manufactured are sold, the total profit P is the difference between total revenue and total cost.

$$\begin{aligned} P &= R - C \\ &= 4x - (400 + 2x) \\ &= 2x - 400 \end{aligned}$$

The value of x for which $R = C$, where the profit is zero, is called the **break-even point**. Therefore, at that level of production and sales, the manufacturer neither makes money nor loses money. To find the break-even point, we set $R = C$. Using Equations (1) and (2), we obtain

$$\begin{aligned} 400 + 2x &= 4x \\ x &= 200 \end{aligned}$$

Thus, the break-even point is 200 units.

The break-even point can also be obtained graphically as follows. Observe that Equations (1) and (2) are linear equations. The break-even point is the x-coordinate of the point where the two lines intersect. Figure 10-5 shows the lines and their point of intersection (200, 800). When 200 units of the product are made, the cost (\$800,000) is exactly equal to the revenue, and the profit is \$0. If $x > 200$, then $R > C$, so the manufacturer is making a profit. If $x < 200$, then $R < C$ and the manufacturer is losing money.

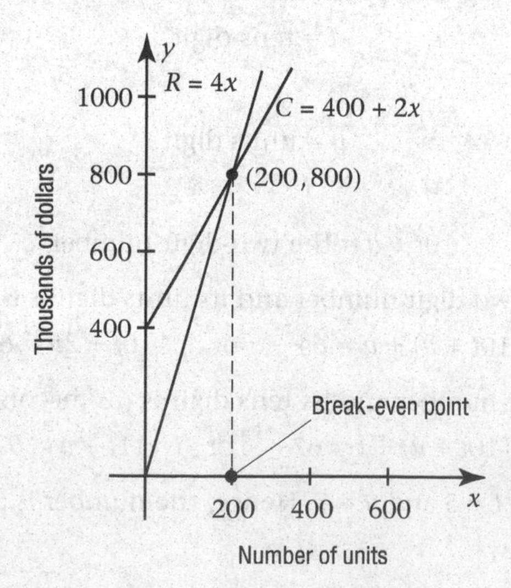

FIGURE 10-5
Break-Even Analysis

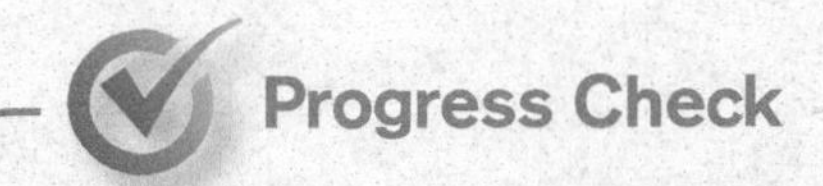

A producer of photographic developer finds that the total weekly cost of producing x liters of developer is given in dollars by $C = 5500 + 3.90x$. The manufacturer sells the product at $4.40 per liter.

a. What is the total revenue received when x liters of developer are sold?

b. Find the break-even point graphically.

c. What is the total revenue received at the break-even point?

Answers

a. $R = 4.40x$ b. 11,000 liters c. $48,400

10.2b Applications in Business and Economics: Supply and Demand

A manufacturer of a product may set any price p (in dollars) for each unit of the product. If the price is too high, few people may buy the product. If the price is too low, so many people may want the product that the producer may not be able to satisfy demand. Thus, in setting price, the manufacturer must take into consideration the demand for the product as well as the supply.

Let S be the number of units that the manufacturer is willing to supply at the price p. Call S the **supply**. Generally, the value of S increases as p increases, that is, the manufacturer is willing to supply more of the product as the price p increases. Let D be the number of units of the product that consumers are willing to buy at the price p. Call D the **demand**. Generally, the value of D decreases as p increases; that is, consumers are willing to buy fewer units of the product as the price rises. For example, suppose that S and D are given by

$$S = 2p + 3 \quad (3)$$

$$D = -p + 12 \quad (4)$$

Equations (3) and (4) are linear equations, so they are equations of lines. (See Figure 10-6.) The price at which supply S and demand D are equal is called the **equilibrium price**. At this price, every unit that is supplied is purchased. Thus, there is neither a surplus nor a shortage. In Figure 10-6 the equilibrium price is $p = 3$. At this price, the number of units supplied equals the number of units demanded and is found by substituting into Equation (3): $S = 2(3) + 3 = 9$. This value can also be obtained by finding the ordinate at the point of intersection in Figure 10-6.

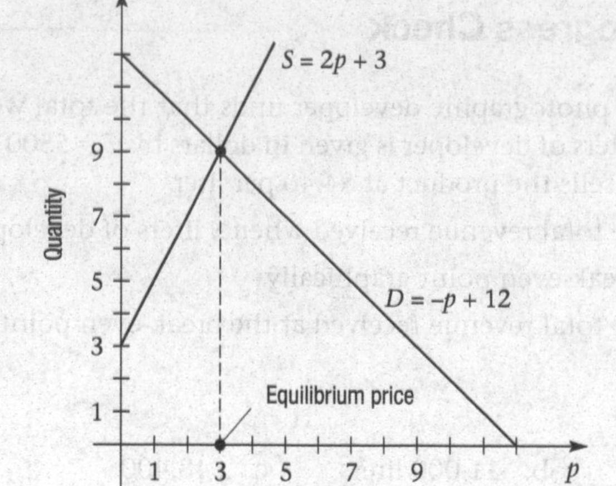

FIGURE 10-6
Break-Even Analysis

If we are in an economic system in which there is pure competition, the law of supply and demand states that the selling price of a product is its equilibrium price. That is, if the selling price was higher than the equilibrium price, consumers' reduced demand would leave the manufacturer with an unsold surplus. To sell this surplus, the manufacturer would be forced to reduce the selling price. If the selling price was below the equilibrium price, the increased demand would cause a shortage of the product, leading the manufacturer to raise the selling price.

In actual practice, the marketplace rarely operates under pure competition. Mathematical analysis of more realistic economic systems requires the use of more sophisticated equations.

EXAMPLE 4 ***Applying Linear Systems***

Suppose that supply and demand for ballpoint pens are given by

$$S = p + 5$$
$$D = -p + 7$$

a. Find the equilibrium price.

b. Find the number of pens sold at that price.

SOLUTION

a. Figure 10-7 illustrates the graphical solution. The equilibrium price is $p = 1$. (Verify that algebraic methods yield the same solution.)

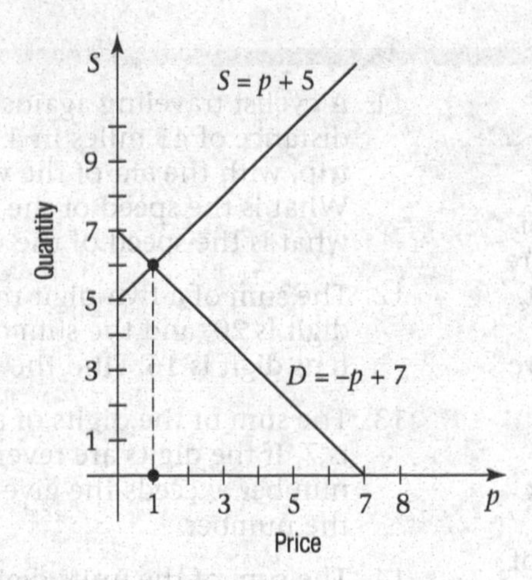

FIGURE 10-7
Graph of System for Example 4

b. When $p = 1$, the number of pens sold is $S = 1 + 5 = 6$, the value of the ordinate at the point of intersection. ■

Progress Check

Suppose that supply and demand for radios are given by

$$S = 3p + 120$$
$$D = -p + 200$$

a. Find the equilibrium price.

b. Find the number of radios sold at that price.

Answers

a. 20 b. 180

Exercise Set 10.2

1. A pile of 40 coins consists of nickels and dimes. If the value of the coins is $2.75, how many of each type of coin are there?
2. An automatic vending machine in the post office sells a packet of 20 stamps for $8.60. There are 34-cent and 46-cent stamps in this packet. If there is no additional cost for using this machine, how many of each type of stamp are in the packet?
3. A photography store sells sampler A consisting of 6 rolls of color film and 4 rolls of black and white film for $82. It also sells sampler B, consisting of 4 rolls of color film and 6 rolls of black and white film for $68. What is the cost per roll of each type of film?
4. A hardware store sells power pack A, consisting of 4 D cell batteries and 2 C cell batteries for $12.60, and power pack B, consisting of 6 D cell batteries and 4 C cell batteries for $20.80. What is the price of each type of battery?
5. A fund manager invested $60,000 in two types of bonds, A and B. Bond A, which is safer than bond B, pays annual interest of 5%, whereas bond B pays annual interest of 8%. If the total annual return on both investments is $3600, how much was invested in each type of bond?
6. A trash removal company carries waste material in sealed containers weighing 4 kilograms and 3 kilograms. On a certain trip the company carries 30 containers weighing a total of 100 kilograms. How many of each type of container are there?
7. A paper firm makes rolls of paper 12 inches wide and 15 inches wide by cutting a sheet that is 180 inches wide. Suppose that a total of 14 rolls of paper are to be cut without any waste. How many of each type of roll will be made?
8. An animal-feed producer mixes two types of grain, A and B. Each unit of grain A contains 2 grams of fat and 80 calories, and each unit of grain B contains 3 grams of fat and 60 calories. If the producer wants the final product to provide 18 grams of fat and 480 calories, how much of each type of grain should be used?
9. A supermarket mixes coffee that sells for $7.60 per pound with coffee that sells for $10.00 per pound to obtain 24 pounds of coffee selling for $8.00 per pound. How much of each type of coffee should be used?
10. An airplane flying against the wind covers a distance of 3000 kilometers in 6 hours. The return trip, with the aid of the wind, takes 5 hours. What is the speed of the airplane in still air, and what is the speed of the wind?
11. A cyclist traveling against the wind covers a distance of 45 miles in 4 hours. The return trip, with the aid of the wind, takes 3 hours. What is the speed of the cyclist in still air, and what is the speed of the wind?
12. The sum of a two-digit number and its units digit is 20, and the sum of the number and its tens digit is 16. Find the number.
13. The sum of the digits of a two-digit number is 7. If the digits are reversed, the resulting number exceeds the given number by 9. Find the number.
14. The sum of the units digit and 3 times the tens digit of a two-digit number is 14, and the sum of the tens digit and twice the units digit is 18. Find the number.
15. A health food shop mixes nuts and raisins into a snack pack. How many pounds of nuts, selling for $7.50 per pound, and how many pounds of raisins, selling for $2.50 per pound, must be mixed to produce a 50-pound mixture selling for $4.10 per pound?
16. A movie theater charges $8.00 admission for an adult and $5.75 for a child. On a particular day 1200 tickets were sold and the total revenue received was $8700. How many tickets of each type were sold?
17. A Segway dealer selling a model A and a model B Segway has $180,000 in inventory. The profit on a model A Segway is 12%, and the profit on a model B Segway is 18%. If the profit on the entire stock is 16%, how much was invested in each model?
18. A water bill is determined as follows: There is a flat monthly service charge and a uniform rate for each hundred cubic feet of water used. Suppose that the use of 40 hundred cubic feet of water resulted in a bill of $130 and a use of 50 hundred cubic feet was billed at $157.50. Find the monthly service charge and the uniform rate for each 100 cubic feet.
19. A certain epidemic is treated by a combination of the drugs Epiline I and Epiline II. Suppose that each unit of Epiline I contains 1 milligram of factor X and 2 milligrams of factor Y, whereas each unit of Epiline II contains 2 milligrams of factor X and 3 milligrams of factor Y. Successful treatment of the disease calls for 13 milligrams of factor X and 22 milligrams of factor Y. How many units of Epiline I and Epiline II should be administered to a patient?

20. **(Break-even analysis)** An animal-feed manufacturer finds that the weekly cost of making x kilograms of feed is given in dollars by $C = 10{,}000 + 4x$ and that the revenue received from selling the feed is given by $R = 6x$.
    a. Find the break-even point graphically and algebraically.
    b. What is the total revenue at the break-even point?

21. **(Break-even analysis)** A small manufacturer of a new solar device finds that the annual cost of making x units is given in dollars by $C = 240{,}000 + 75x$. Each device sells for $125.
    a. What is the total revenue received when x devices are sold?
    b. Find the break-even point graphically and algebraically.
    c. What is the total revenue received at the break-even point?

22. **(Supply and demand)** A manufacturer of calculators finds that the supply and demand are given by

$$S = 750p + 40{,}000$$
$$D = -400p + 132{,}000$$

    a. Find the equilibrium price.
    b. What is the number of calculators sold at this price?

23. **(Supply and demand)** A manufacturer of motor scooters finds that the supply and demand are given by

$$S = 5p + 2000$$
$$D = -2p + 16{,}000$$

    a. Find the equilibrium price.
    b. What is the number of scooters sold at this price?

24. Find the dimensions of a rectangle with an area of 30 square feet and a perimeter of 22 feet.

25. Find two numbers whose product is 20 and whose sum is 9.

26. Find two numbers the sum of whose squares is 65 and whose sum is 11.

27. A pile of 34 coins worth $4.10 consists of nickels and quarters. Find the number of each type of coin.

28. Car A can travel 20 kilometers per hour faster than car B. If car A travels 240 kilometers in the same time that car B travels 200 kilometers, what is the speed of each car?

29. How many pounds of nuts worth $4.00 per pound and how many pounds of raisins worth $1.50 per pound must be mixed to obtain a mixture of 2 pounds that is worth $3.00 per pound?

30. A part of $8000 was invested at an annual interest of 7% and the remainder at 8%. If the total interest received at the end of 1 year is $590, how much was invested at each rate?

31. During World War II, the owner of a service station sold 1325 gallons of gasoline and collected 200 ration tickets. If type A ration tickets are used to purchase 10 gallons of gasoline and type B are used to purchase 1 gallon of gasoline, how many of each type of ration ticket did the station collect?

32. A bank is paying 6% annual interest on 1-year certificates, and treasury notes are paying 3% annual interest. An investor wishes to receive $2400 interest at the end of 1 year by investing a total of $60,000. How much was invested at each rate?

33. The sum of the squares of the length and the width of a rectangle is 100 square meters. If the area of the rectangle is 48 square meters, find the length of each side of the rectangle.

34. Recall that the standard form for a linear polynomial is

$$y = a_1x + a_0$$

Suppose you want to model the following data by a linear function: (1, −3), (3, 5). Proceed by substituting the given values for x and y into the standard form. This will give you two equations with the two unknowns a_1, a_0. Solve this system, then use your solution to write the linear polynomial that fits this data.

In Exercises 35–40, redo Exercises 13–18 in Exercise Set 3.4 using the method of polynomial curve fitting described in Exercise 34 of this section.

10.3 Systems of Linear Equations in Three Unknowns

The method of substitution and the method of elimination can both be applied to systems of linear equations in three unknowns and, more generally, to systems of linear equations in any number of unknowns. There is yet another method, ideally suited for computers, which we will now apply to solving linear systems in three unknowns.

10.3a *Gaussian Elimination and Triangular Form*

In solving equations, we found it convenient to transform an equation into an equivalent equation having the same solution set. Similarly, we can attempt to transform a system of equations into another system, called an **equivalent system**, that has the same solution set. In particular, the objective of **Gaussian Elimination** is to transform a linear system into an equivalent system in triangular form, such as

$$\begin{aligned} 3x - y + 3z &= -11 \\ 2y + z &= 2 \\ 2z &= -4 \end{aligned}$$

A linear system is in **triangular form** when the only nonzero coefficient of x appears in the first equation, the only nonzero coefficients of y appear in the first and second equations, and so on.

Note that when a linear system is in triangular form, the last equation immediately yields the value of an unknown. In our example, we see that

$$\begin{aligned} 2z &= -4 \\ z &= -2 \end{aligned}$$

Substituting $z = -2$ into the second equation yields

$$\begin{aligned} 2y + (-2) &= 2 \\ y &= 2 \end{aligned}$$

Finally, substituting $z = -2$ and $y = 2$ into the first equation yields

$$\begin{aligned} 3x - (2) + 3(-2) &= -11 \\ 3x &= -3 \\ x &= -1 \end{aligned}$$

Therefore, when a linear system is in triangular form, the process of **back-substitution** allows us to solve the system quickly. The challenge is to find a means of transforming a linear system into this form.

Operations that transform a system of linear equations into an equivalent system are as follows:

1. Interchange any two equations.
2. Multiply an equation by a nonzero constant.
3. Replace an equation by the sum of itself plus a constant times another equation.

Using these operations we can demonstrate the method of Gaussian Elimination.

EXAMPLE 1 ***Solving a Linear System by Gaussian Elimination***

Solve the linear system

$$\begin{aligned} 2y - z &= -5 \\ x - 2y + 2z &= 9 \\ 2x - 3y + 3z &= 14 \end{aligned}$$

SOLUTION

Step 1. If necessary, interchange equations to obtain a nonzero coefficient for x in the first equation.	*Step 1.* Interchanging the first two equations yields $$\begin{aligned} x - 2y + 2z &= 9 \\ 2y - z &= -5 \\ 2x - 3y + 3z &= 14 \end{aligned}$$
Step 2. Replace the second equation with the sum of itself and an appropriate multiple of the first equation, so that the result has a zero coefficient for x.	*Step 2.* The coefficient of x in the second equation is already 0
Step 3. Replace the third equation with the sum of itself and an appropriate multiple of the first equation, so that the result has a zero coefficient for x.	*Step 3.* Replace the third equation with the sum of itself plus −2 times the first equation. $$\begin{aligned} x - 2y + 2z &= 9 \\ 2y - z &= -5 \\ y - z &= -4 \end{aligned}$$
Step 4. Apply the procedures of *Steps 1–3* to the second and third equations and the coefficients of y and z, respectively.	*Step 4.* Replace the third equation with the sum of itself and $-\frac{1}{2}$ times the second equation. $$\begin{aligned} x - 2y + 2z &= 9 \\ 2y - z &= -5 \\ -\tfrac{1}{2}z &= -\tfrac{3}{2} \end{aligned}$$
Step 5. The system is now in triangular form. The solution is obtained by back-substitution.	*Step 5.* From the third equation, $$\begin{aligned} -\tfrac{1}{2}z &= -\tfrac{3}{2} \\ z &= 3 \end{aligned}$$ Substituting this value of z into the second equation, we have $$\begin{aligned} 2y - (3) &= -5 \\ y &= -1 \end{aligned}$$ Substituting for y and for z into the first equation, we have $$\begin{aligned} x - 2(-1) + 2(3) &= 9 \\ x + 8 &= 9 \\ x &= 1 \end{aligned}$$ The solution is $x = 1$, $y = -1$, $z = 3$.

■

WARNING

When solving a system of linear equations, we must be sure to align the unknowns. For example, to solve the system

$$\begin{aligned} -z + 2y &= -5 \\ -2y + x + 2z &= \ \ 9 \\ 2x - 3y + 3z &= 14 \end{aligned}$$

we rewrite it as

$$\begin{aligned} 2y - \ \ z &= -5 \\ x - 2y + 2z &= \ \ 9 \\ 2x - 3y + 3z &= 14 \end{aligned}$$

Observe that this is the linear system in Example 1.

Progress Check

Solve by Gaussian Elimination.

a. $\begin{aligned} 2x - 4y + 2z &= \ \ 1 \\ 3x + \ \ y + 3z &= \ \ 5 \\ x - \ \ y - 2z &= -8 \end{aligned}$ b. $\begin{aligned} -2x + 3y - 12z &= -17 \\ 3x - \ \ y - 15z &= \ \ 11 \\ -x + 5y + \ \ 3z &= \ -9 \end{aligned}$

Answers

a. $x = -\frac{3}{2}, y = \frac{1}{2}, z = 3$ b. $x = 5, y = -1, z = \frac{1}{3}$

10.3b *Consistent and Inconsistent Systems*

The graph of a linear equation in three unknowns is a plane in three-dimensional space. A system of three linear equations in three unknowns corresponds to three planes. If the planes intersect in a point P, as shown in Figure 10-8(a), then the coordinates of the point P are the unique solution of the system; and this can be found by Gaussian Elimination. In Figure 10-8(b), the coordinates of *each* point of the line L is a solution of the system. One case where there is no solution is shown in Figure 10-8(c). (See the presentation in Figure 10-3.)

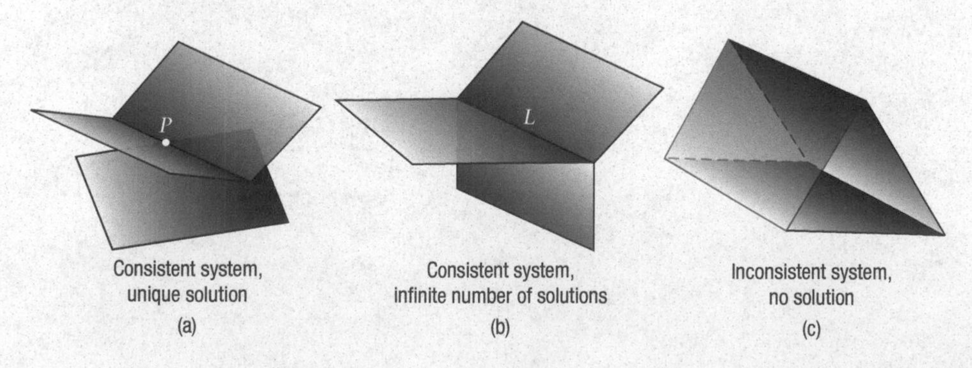

FIGURE 10-8
Possible Situations with Three Linear Equations and Three Unknowns

Consistent and Inconsistent Systems

- A **consistent system** of equations has one or more solutions. (See Figures 10-3(a), 10-3(b), 10-8(a), and 10-8(b).) It has an infinite number of solutions if Gaussian Elimination yields an equation of the form
$$0x + 0y + 0z = 0$$
- An **inconsistent system** of equations has no solutions. (See Figures 10-3(c) and 10-8(c).) If Gaussian Elimination yields an equation of the form
$$0x + 0y + 0z = c, \qquad c \neq 0$$

then the system is inconsistent.

EXAMPLE 2 *Consistent System*

Solve the linear system.

$$\begin{aligned} x - 2y + 2z &= -4 \\ x + y - 7z &= 8 \\ -x - 4y + 16z &= -20 \end{aligned}$$

SOLUTION

Replacing the second equation by itself minus the first equation, and replacing the third equation by itself plus the first equation, we have

$$\begin{aligned} x - 2y + 2z &= -4 \\ 3y - 9z &= 12 \\ -6y + 18z &= -24 \end{aligned}$$

Replacing the third equation of the last system by itself plus 2 times the second equation results in the system

$$\begin{aligned} x - 2y + 2z &= -4 \\ 3y - 9z &= 12 \\ 0x + 0y + 0z &= 0 \end{aligned}$$

The last equation indicates that the system is consistent. It has an infinite number of solutions. If we solve the second equation of the last system for y, we have

$$y = 3z + 4$$

Now solving the first equation for x, we have

$$x = 2y - 2z - 4$$

Substituting for y,

$$\begin{aligned} x &= 2(3z + 4) - 2z - 4 \\ x &= 4z + 4 \end{aligned}$$

The equations

$$\begin{aligned} x &= 4z + 4 \\ y &= 3z + 4 \end{aligned}$$

yield a solution of the original system for every real value of z. For example, if $z = 0$, then $x = 4$, $y = 4$, $z = 0$ satisfies the original system; if $z = -2$, then $x = -4$, $y = -2$, $z = -2$ is another solution. ■

Progress Check

a. Verify that the linear system

$$\begin{aligned} x - 2y + z &= 3 \\ 2x + y - 2z &= -1 \\ -x - 8y + 7z &= 5 \end{aligned}$$

is inconsistent.

b. Verify that the linear system

$$\begin{aligned} 2x + y + 2z &= 1 \\ x - 4y + 7z &= -4 \\ x - y + 3z &= -1 \end{aligned}$$

has an infinite number of solutions.

Exercise Set 10.3

In Exercises 1–18, solve by Gaussian Elimination. Indicate if the system is inconsistent or has an infinite number of solutions.

1. $$\begin{aligned} x + 2y + 3z &= -6 \\ 2x - 3y - 4z &= 15 \\ 3x + 4y + 5z &= -8 \end{aligned}$$

2. $$\begin{aligned} 2x + 3y + 4z &= -12 \\ x - 2y + z &= -5 \\ 3x + y + 2z &= 1 \end{aligned}$$

3. $$\begin{aligned} x + y + z &= 1 \\ x + y - 2z &= 3 \\ 2x + y + z &= 2 \end{aligned}$$

4. $$\begin{aligned} 2x - y + z &= 3 \\ x - 3y + z &= 4 \\ -5x - 2z &= -5 \end{aligned}$$

5. $$\begin{aligned} x + y + z &= 2 \\ x - y + 2z &= 3 \\ 3x + 5y + 2z &= 6 \end{aligned}$$

6. $$\begin{aligned} x + y + z &= 0 \\ x + y &= 3 \\ y + z &= 1 \end{aligned}$$

7. $$\begin{aligned} x + 2y + z &= 7 \\ x + 2y + 3z &= 11 \\ 2x + y + 4z &= 12 \end{aligned}$$

8. $$\begin{aligned} 4x + 2y - z &= 5 \\ 3x + 3y + 6z &= 1 \\ 5x + y - 8z &= 8 \end{aligned}$$

9. $$\begin{aligned} x + y + z &= 2 \\ x + 2y + z &= 3 \\ x + y - z &= 2 \end{aligned}$$

10. $$\begin{aligned} x + y - z &= 2 \\ x + 2y + z &= 3 \\ x + y + 4z &= 3 \end{aligned}$$

11. $$\begin{aligned} 2x + y + 3z &= 8 \\ -x + y + z &= 10 \\ x + y + z &= 12 \end{aligned}$$

12. $$\begin{aligned} 2x - 3z &= 4 \\ x + 4y - 5z &= -6 \\ 3x + 4y - z &= -2 \end{aligned}$$

13. $$\begin{aligned} x + 3y + 7z &= 1 \\ 3x - y - 5z &= 9 \\ 2x + y + z &= 4 \end{aligned}$$

14. $$\begin{aligned} 2x - y + z &= 2 \\ 3x + y + 2z &= 3 \\ x + y - z &= -1 \end{aligned}$$

15. $$\begin{aligned} x - 2y + 3z &= -2 \\ x - 5y + 9z &= 4 \\ 2x - y &= 6 \end{aligned}$$

16. $$\begin{aligned} x + 2y - 2z &= 8 \\ 5y - z &= 6 \\ -2x + y + 3z &= -2 \end{aligned}$$

17. $$\begin{aligned} z - 2y + x &= -5 \\ z + 2x &= -10 \\ y - z &= 15 \end{aligned}$$

18. $$\begin{aligned} -3z + 2y &= 4 \\ 2z + x &= -2 \\ -8y + x + 14z &= -18 \end{aligned}$$

19. A special low-calorie diet consists of dishes A, B, and C. Each unit of A has 2 grams of fat, 1 gram of carbohydrate and 3 grams of protein. Each unit of B has 1 gram of fat, 2 grams of carbohydrate, and 1 gram of protein. Each unit of C has 1 gram of fat, 2 grams of carbohydrate, and 3 grams of protein. The diet must provide exactly 10 grams of fat, 14 grams of carbohydrate, and 18 grams of protein. How much of each dish should be used?

20. A furniture manufacturer makes chairs, coffee tables, and dining room tables. Each chair requires 2 minutes of sanding, 2 minutes of staining, and 4 minutes of varnishing. Each coffee table requires 5 minutes of sanding, 4 minutes of staining, and 3 minutes of varnishing. Each dining room table requires 5 minutes of sanding, 4 minutes of staining, and 6 minutes of varnishing. The sanding bench is available 6 hours per day, the staining bench 5 hours per day, and the varnishing bench 6 hours per day. How many of each type of furniture can be made if all facilities are used to capacity?

21. A manufacturer produces 40-inch, 46-inch, and 52-inch television sets that require assembly, testing, and packing. Each 40-inch set requires 45 minutes to assemble, 30 minutes to test, and 10 minutes to package. Each 46-inch set requires 1 hour to assemble, 45 minutes to test, and 15 minutes to package. Each 52-inch set requires 1.5 hours to assemble, 1 hour to test, and 15 minutes to package. If the assembly line operates for 17.75 hours per day, the test facility is used for 12.5 hours per day, and the packing equipment is used for 3.75 hours per day, how many of each type of set can be produced?

22. Recall that a second degree polynomial can be written as
$$y = a_2x^2 + a_1x + a_0$$
Suppose you want to model the following data by a quadratic function: (0, 1), (3, 16), and (−1, 8). Proceed by substituting the given values for x and y into the standard form. This will give you three equations with three unknowns a_2, a_1, a_0. Solve this system, then use your solution to write the quadratic polynomial that fits this data.

23. Repeat Exercise 22 for the following data points: (1, 11), (2, 24), (−2, −4).

10.4 Applications: Partial Fractions

Students of algebra are taught how to add fractions, such as

$$\frac{2}{x-1}+\frac{3}{x+2}=\frac{5x+1}{(x-1)(x+2)}$$

Surprisingly, there is an important application of integral calculus that requires the *inverse* procedure. This procedure requires that we solve a system of linear equations.

Our objective in this section is to write a rational function $\frac{P(x)}{Q(x)}$ as a sum of fractions, each of which is called a **partial fraction**. Now we need only consider the situation where $\frac{P(x)}{Q(x)}$ is a **proper fraction**, that is, where the degree of $P(x)$ is less than the degree of $Q(x)$. If it is not a proper fraction, then it is called an **improper fraction**, and we divide through by $Q(x)$. For example,

$$\frac{2x^3-x^2-x+1}{x^2-1}$$

is an improper fraction. We use long division to divide through by x^2-1 as follows:

$$\begin{array}{rll}
 & 2x\ -1 & \text{quotient} \\
x^2-1 \,\big) & 2x^3-x^2-x+1 & \\
 & 2x^3 \quad\ -2x & \\
\hline
 & -x^2+x+1 & \\
 & -x^2 \quad\ +1 & \\
\hline
 & x & \text{remainder}
\end{array}$$

Therefore

$$\frac{2x^3-x^2-x+1}{x^2-1}=2x-1+\frac{x}{x^2-1}$$

We can now work with the proper fraction $\frac{x}{x^2-1}$.

The procedure for partial-fraction decomposition of the rational function $\frac{P(x)}{Q(x)}$ begins with factoring the denominator $Q(x)$. As we showed in Chapter 4 (as a corollary to the Conjugate Zeros Theorem), every polynomial with real coefficients can be written as a product of linear and quadratic factors with real coefficients, such that the quadratic factors have no real zeros. Such quadratic factors are said to be *irreducible* over the reals. For example,

$$Q(x)=x^3-x^2+2x-2=(x-1)(x^2+2)$$

where the quadratic factor x^2+2 cannot be decomposed into linear factors with real coefficients since the polynomial x^2+2 has no real zeros.

Having factored $Q(x)$ in this manner, we collect the repeated factors so that $Q(x)$ is the product of distinct factors of the forms

$$(ax+b)^m \qquad \text{and} \qquad (ax^2+bx+c)^n$$

The following rules tell us the format of the partial-fraction decomposition of $\frac{P(x)}{Q(x)}$.

Rules for Partial-Fraction Decomposition

Rule 1. Linear Factors

For each distinct factor of the form $(ax + b)^m$ in the denominator $Q(x)$, introduce the sum of m partial fractions

$$\frac{A_1}{ax+b}+\frac{A_2}{(ax+b)^2}+\dots+\frac{A_m}{(ax+b)^m}$$

where $A_1, A_2, \dots, A_m$ are constants.

Rule 2. Quadratic Factors

For each distinct factor of the form $(ax^2 + bx + c)^n$ in the denominator $Q(x)$, introduce the sum of n partial fractions

$$\frac{A_1x+B_1}{ax^2+bx+c}+\frac{A_2x+B_2}{(ax^2+bx+c)^2}+\dots+\frac{A_nx+B_n}{(ax^2+bx+c)^n}$$

where $A_1, A_2, \dots, A_n$ and $B_1, B_2, \dots, B_n$ are constants.

We must now find a method to determine the value of these constants. The constants used in *Rule 1* are different from the constants used in *Rule 2.*

EXAMPLE 1 *Partial-Fraction Decomposition: Linear Factors*

Find the partial-fraction decomposition of

$$\frac{x+5}{x^2+x-2}$$

SOLUTION

Factoring the denominator gives us

$$\frac{x+5}{x^2+x-2}=\frac{x+5}{(x+2)(x-1)}$$

The denominator consists of linear factors. By *Rule 1,* each linear factor introduces one term.

factor	$x+2$	$x-1$
term	$\frac{A}{x+2}$	$\frac{B}{x-1}$

The partial-fraction decomposition is then

$$\frac{x+5}{(x+2)(x-1)}=\frac{A}{x+2}+\frac{B}{x-1}$$

To solve for the constants A and B, we clear fractions by multiplying both sides by $(x + 2)(x - 1)$ to yield

$$x+5=A(x-1)+B(x+2)$$

This last equation may be solved for A and B by substituting different values for x. Sometimes, we may choose a value for x that will cause an unknown to "disappear." Setting $x = 1$ gives us

$$6=3B \quad \text{or} \quad B=2$$

Setting $x = -2$ yields

$$3=-3A \quad \text{or} \quad A=-1$$

Substituting these values for A and B, we arrive at the partial-fraction decomposition

$$\frac{x+5}{(x+2)(x-1)}=\frac{-1}{x+2}+\frac{2}{x-1}$$

Verify that this is indeed an *identity.* Furthermore, verify that you obtain the same solution for A and B using different values of x. ■

EXAMPLE 2 *Repeated Linear Factors*

Find the partial-fraction decomposition of

$$\frac{x^2-2x-9}{x(x+3)^2}$$

SOLUTION

The denominator is already in factored form and consists of the linear factor x and the repeated linear factor $x + 3$. By *Rule 1*, the factor x introduces one term

$$\frac{A}{x}$$

and the factor $x + 3$ introduces two terms ($m = 2$)

$$\frac{B}{x+3} \quad \text{and} \quad \frac{C}{(x+3)^2}$$

The partial-fraction decomposition is then

$$\frac{x^2-2x-9}{x(x+3)^2} = \frac{A}{x} + \frac{B}{x+3} + \frac{C}{(x+3)^2}$$

Clearing fractions by multiplying both sides by $x(x + 3)^2$ yields

$$x^2 - 2x - 9 = A(x + 3)^2 + Bx(x + 3) + Cx \qquad (1)$$

Setting $x = -3$ in Equation (1) allows us to solve for C.

$$6 = -3C \quad \text{or} \quad C = -2$$

Setting $x = 0$ in Equation (1) allows us to solve for A.

$$-9 = 9A \quad \text{or} \quad A = -1$$

Expanding the right-hand side of Equation (1) and collecting terms in the powers of x, we have

$$x^2 - 2x - 9 = (A + B)x^2 + (6A + 3B + C)x + 9A \qquad (2)$$

Substituting $A = -1$ and $C = -2$ in Equation (2) yields

$$x^2 - 2x - 9 = (B - 1)x^2 + (3B - 8)x - 9 \qquad (3)$$

Since the solution must be an identity, the coefficients of the same powers of x on both sides of Equation (3) must be equal. Equating the coefficients of the terms in x^2, we have

$$1 = B - 1 \quad \text{or} \quad B = 2$$

The same result is obtained by equating the coefficients of the terms in x. Alternatively, we may substitute another value for x in Equation (3) and obtain an equation in B yielding the same result, $B = 2$.

Therefore, the partial-fraction decomposition is

$$\frac{x^2-2x-9}{x(x+3)^2} = -\frac{1}{x} + \frac{2}{x+3} - \frac{2}{(x+3)^2}$$ ■

For the rational function

$$\frac{2x-1}{x^2(x^2+1)}$$

the factor x^2 in the denominator must be treated as the repeated *linear* factor $(x - 0)^2$. The partial-fraction decomposition has the structure

$$\frac{2x-1}{x^2(x^2+1)} = \frac{A}{x} + \frac{B}{x^2} + \frac{Cx+D}{x^2+1}$$

EXAMPLE 3 *Linear and Quadratic Factors*

Find the partial-fraction decomposition of

$$\frac{x^2 - 5x + 1}{2x^3 - x^2 + 2x = 1}$$

SOLUTION

First we factor the denominator. By the factor theorem, $x - r$ is a factor of

$$Q(x) = 2x^3 - x^2 + 2x - 1$$

if and only if r is a zero of $Q(x)$. The Rational Zero Theorem tells us that the possible rational roots are $1, -1, \frac{1}{2}$, and $-\frac{1}{2}$. Using the condensed form of synthetic division to test these possible roots

	2	−1	2	−1
1	2	1	3	2
−1	2	−3	5	−6
$\frac{1}{2}$	2	0	2	0

coefficients of quotient polynomial

we see that $Q\left(\frac{1}{2}\right) = 0$, so $x = \frac{1}{2}$ is a zero of $Q(x)$. Consequently $\left(x - \frac{1}{2}\right)$ is a factor of $Q(x)$, and $2x^2 + 2 = 0$ is seen to be the deflated equation. Since $2x^2 + 2 = 2(x^2 + 1)$, we may write

$$Q(x) = \left(x - \frac{1}{2}\right)(2)(x^2 + 1)$$

To avoid fractions, multiply the first two factors to obtain

$$Q(x) = (2x - 1)(x^2 + 1)$$

By *Rule 1*, the factor $2x - 1$ introduces a term of the form

$$\frac{A}{2x - 1}$$

and by *Rule 2*, the factor $x^2 + 1$ introduces a term of the form

$$\frac{Bx + C}{x^2 + 1}$$

Then

$$\frac{x^2 - 5x + 1}{(2x - 1)(x^2 + 1)} = \frac{A}{2x - 1} + \frac{Bx + C}{x^2 + 1}$$

Multiplying by $(2x - 1)(x^2 + 1)$, we have

$$x^2 - 5x + 1 = A(x^2 + 1) + (Bx + C)(2x - 1)$$

Setting $x = \frac{1}{2}$, we obtain

$$-\frac{5}{4} = \frac{5}{4}A \quad \text{or} \quad A = -1$$

Setting $x = 0$ yields

$$1 = A - C$$

Since $A = -1$, we have

$$1 = -1 - C$$
$$C = -2$$

Setting $x = 1$, we obtain

$$-3 = 2A + B + C$$

Since $A = -1$ and $C = -2$,

$$-3 = -2 + B - 2$$
$$B = 1$$

Therefore, the partial-fraction decomposition is

$$\frac{x^2 - 5x + 1}{2x^3 - x^2 + 2x - 1} = \frac{-1}{2x - 1} + \frac{x - 2}{x^2 + 1}$$ ■

EXAMPLE 4 ***Linear and Repeated Quadratic Factors***

Find the partial-fraction decomposition of

$$\frac{x^2 - 2x}{(x+2)(x^2+4)^2}$$

SOLUTION

The linear factor $x + 2$ introduces one term of the form

$$\frac{A}{x+2}$$

The quadratic factor $x^2 + 4$ has no real roots. By *Rule 2,* this irreducible quadratic factor introduces two terms ($n = 2$) of the form

$$\frac{Bx + C}{x^2 + 4} \quad \text{and} \quad \frac{Dx + E}{(x^2+4)^2}$$

We then have to solve

$$\frac{x^2 - 2x}{(x+2)(x^2+4)^2} = \frac{A}{x+2} + \frac{Bx + C}{x^2 + 4} + \frac{Dx + E}{(x^2+4)^2}$$

for values of A, B, C, D, and E that will produce an identity. Multiplying by $(x + 2)(x^2 + 4)^2$ yields

$$x^2 - 2x = A(x^2 + 4)^2 + (Bx + C)(x + 2)(x^2 + 4) + (Dx + E)(x + 2) \tag{4}$$

Setting $x = -2$ enables us to solve for A.

$$8 = 64A \quad \text{or} \quad A = \frac{1}{8}$$

A methodical way to solve for B, C, D, and E is to successively equate the coefficients of the powers x^4, x^3, and x^2, in the left-hand and right-hand sides of Equation (4).

$$\text{Coefficients of } x^4\text{: } 0 = A + B \quad \text{or} \quad B = -A = -\frac{1}{8}$$

$$\text{Coefficients of } x^3\text{: } 0 = 2B + C \quad \text{or} \quad C = -2B = \frac{1}{4}$$

$$\text{Coefficients of } x^2\text{: } 1 = 8A + 4B + 2C + D$$
$$1 = 1 - \frac{1}{2} + \frac{1}{2} + D$$
$$D = 0$$

To find E, we may equate the coefficients of x or the constant terms. Choosing the latter approach,

$$0 = 16A + 8C + 2E$$
$$0 = 2 + 2 + 2E$$
$$E = -2$$

Alternatively, we may substitute four other values for x in Equation (4) obtaining four linear equations in B, C, D, and E. This also yields the same solution. (Verify this by letting $x = 0$, $x = 1$, $x = 2$, and $x = 3$.)

The partial-fraction decomposition is

$$\frac{x^2 - 2x}{(x+2)(x^2+4)^2} = \frac{\frac{1}{8}}{x+2} + \frac{-\frac{1}{8}x + \frac{1}{4}}{x^2+4} + \frac{-2}{(x^2+4)^2}$$ ■

Exercise Set 10.4

Find the partial-fraction decomposition of each of the following:

1. $\dfrac{2x-11}{(x+2)(x-3)}$

2. $\dfrac{1}{x^2+3x+2}$

3. $\dfrac{3x-2}{6x^2-5x+1}$

4. $\dfrac{2x+1}{x^2-1}$

5. $\dfrac{x^2+x+2}{x^3-x}$

6. $\dfrac{3x-14}{(x-3)(x^2-4)}$

7. $\dfrac{3x-2}{x^3+2x^2}$

8. $\dfrac{4x^2-5x+2}{2x^3-x^2}$

9. $\dfrac{x^2-x+2}{(x-1)(x+1)^2}$

10. $\dfrac{3x-1}{x(x-1)^3}$

11. $\dfrac{1-2x}{x^3+4x}$

12. $\dfrac{x^2-2x+1}{x^3+2x^2+2x}$

13. $\dfrac{2x^3-x^2+x}{(x^2+3)^2}$

14. $\dfrac{x^2+2x+10}{x^3-2x^2+2x-4}$

15. $\dfrac{-x}{x^3-2x^2-4x-1}$

16. $\dfrac{x^3-2x^2+1}{x^4+2x^2+1}$

17. $\dfrac{x^4-x^2-9}{(x+1)(x^2+2)^2}$

18. $\dfrac{-x^3-x^2+5x+1}{(2x-1)(x^2+1)^2}$

19. $\dfrac{x^4+x^3+x^2+3x-2}{(x+1)(x^2+1)}$

20. $\dfrac{x^3-5x+5}{(x-1)^3}$

21. *Mathematics in Writing*: Suppose you are given a rational expression and asked to rewrite it as the sum of several other rational expressions, each one having a numerator which is constant. Is it always possible to do this? Under what circumstances would this task be easy? Difficult?

10.5 Systems of Linear Inequalities

10.5a Graphing Linear Inequalities

When we draw the graph of a linear equation, say

$$y = 2x - 1$$

we see that the graph of the line divides the plane into two regions, which we call **half-planes**. (See Figure 10-9.) If we replace the equal sign in $y = 2x - 1$ with any of the symbols, $<$, $>$, $\leq$, or $\geq$, we have a **linear inequality in two variables**. By the **graph of a linear inequality**, such as

$$y < 2x - 1$$

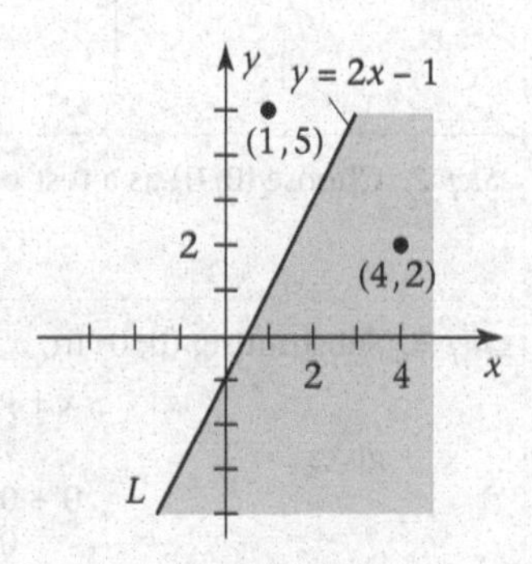

FIGURE 10-9
Half-planes Formed by the Graph of a Linear Equation

we mean the set of all points whose coordinates satisfy the inequality. Thus, the point (4, 2) lies on the graph of $y < 2x - 1$, since

$$2 < (2)(4) - 1$$
$$2 < 7$$

is true. However, the point (1, 5) does *not* lie on the graph of $y < 2x - 1$ since

$$5 < (2)(1) - 1$$
$$5 < 1$$

is not true. Since the coordinates of every point on the line L in Figure 10-9 satisfy the *equation* $y = 2x - 1$, we see that the coordinates of those points in the half-plane below the line must satisfy the *inequality* $y < 2x - 1$. Similarly, the coordinates of those points in the half-plane above the line satisfy the *inequality* $y > 2x - 1$. This suggests that the graph of a linear inequality in two variables is a half-plane, and it leads to a method for graphing inequalities.

EXAMPLE 1 *Graphing a Linear Inequality*

Sketch the graph of the inequality $x + y \geq 1$.

SOLUTION

Step 1. Replace the inequality sign with an equal sign and plot the line. a. If the inequality is $\leq$ or $\geq$, plot a solid line. (Points on the line satisfy the inequality.) b. If the inequality is $<$ or $>$, plot a dashed line. (Points on the line do not satisfy the inequality.)	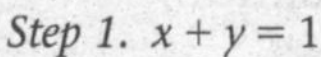 *Step 1.* $x + y = 1$ 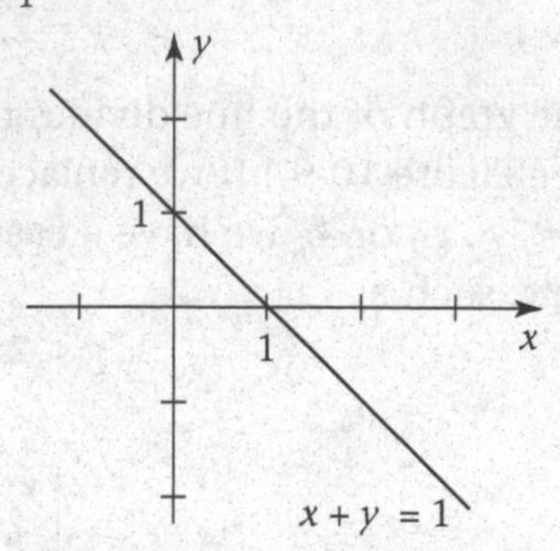
Step 2. Choose any point that is not on the line as a test point. If the origin is not on the line, it may be a convenient choice.	*Step 2.* Choose (0, 0) as a test point.
Step 3. Substitute the coordinates of the test point into the inequality. a. If the test point satisfies the inequality, then the coordinates of every point in the half-plane that contains the test point satisfy the inequality. b. If the test point does not satisfy the inequality, then the half-plane that does not contain the test point will satisfy the inequality.	*Step 3.* Substituting (0, 0) in $x + y \geq 1$ gives $0 + 0 \geq 1$ (?) $0 \geq 1$ which is false. Since (0, 0) is in the half-plane below the line and does not satisfy the inequality, all points above the line satisfy the inequality.
Step 4. Shade the half-plane that satisfied the inequality.	*Step 4.* 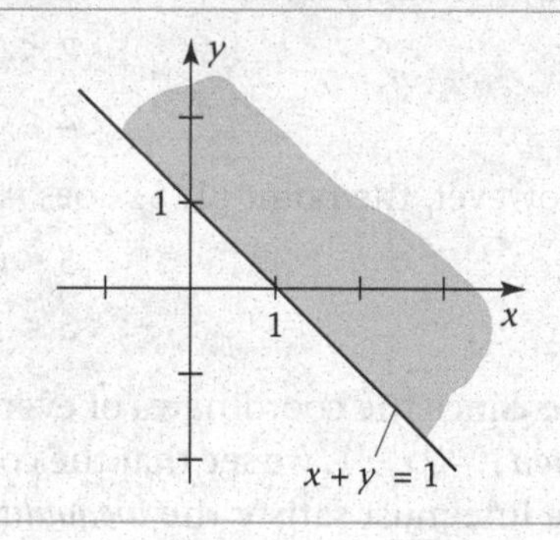

■

EXAMPLE 2 ***Graphing a Linear Inequality***

Sketch the graph of the inequality $2x - 3y > 6$.

SOLUTION

We first graph the line $2x - 3y = 6$. We draw a dashed or broken line to indicate that $2x - 3y = 6$ is not part of the graph. Since (0, 0) is not on the line, we can use it as a test point.

$$2x - 3y > 6$$
$$2(0) - 3(0) > 6 \quad (?)$$
$$0 - 0 > 6 \quad (?)$$
$$0 > 6$$

is false. Since (0, 0) is in the half-plane above the line, the graph consists of the half-plane below the line.

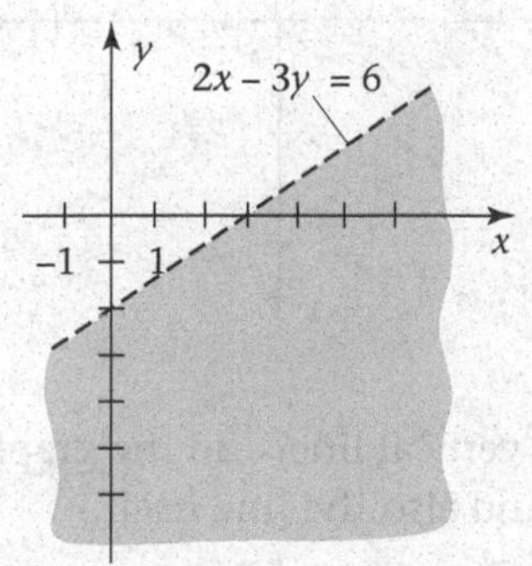

■

Progress Check

Graph the inequalities.

a. $y \le 2x + 1$ b. $y + 3x > -2$ c. $y \ge -x + 1$

Answers

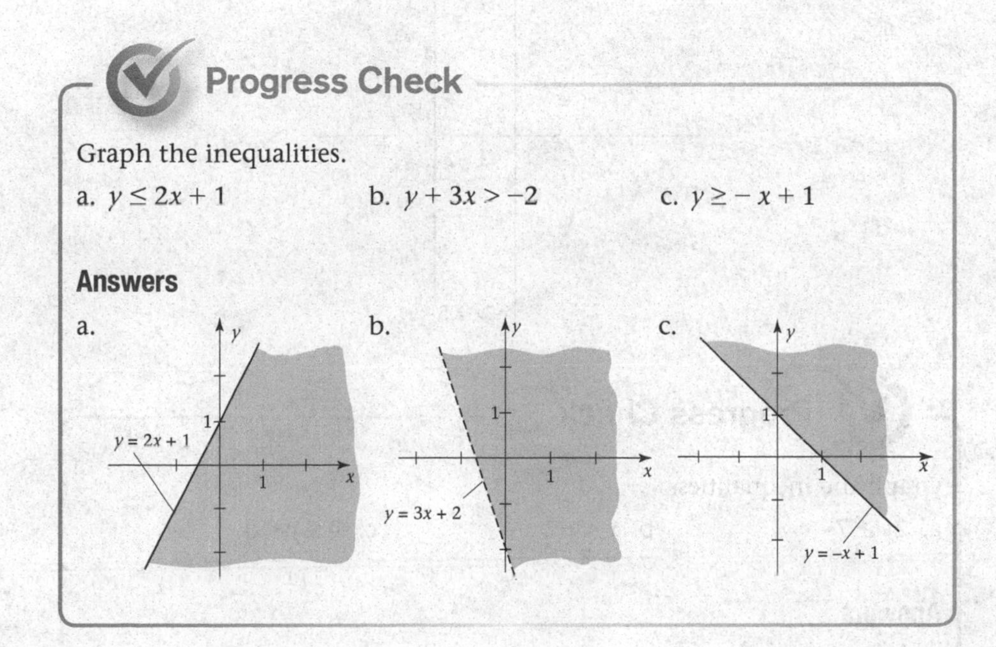

EXAMPLE 3 ***Graphing a Linear Inequality***

Graph the inequalities.

a. $y > x$ b. $2x \geq 5$

SOLUTION

a. Since the origin lies on the line $y = x$, we choose another test point, say (0, 1) above the line. Since (0, 1) satisfies the inequality, the graph of the inequality is the half-plane above the line.

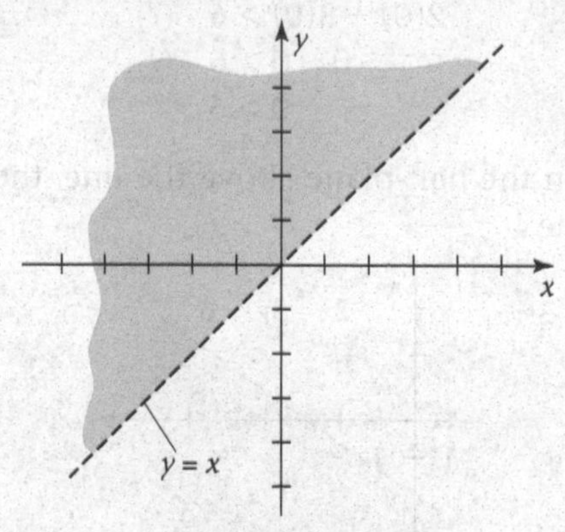

b. The graph of $2x = 5$ is a vertical line, and the graph of $2x \geq 5$ is the half-plane to the right of the line and also the line itself.

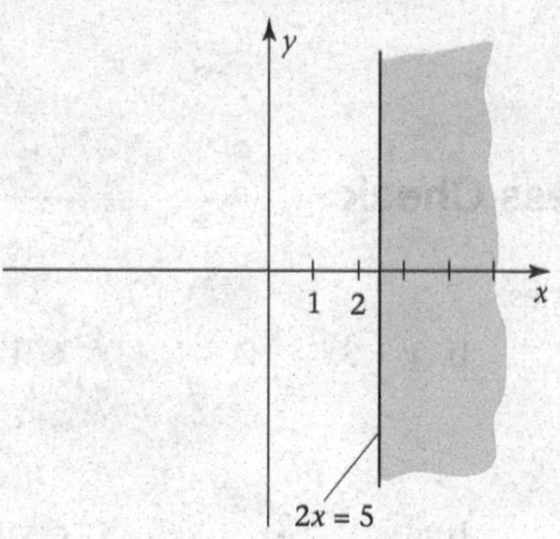

■

Progress Check

Graph the inequalities.

a. $2y \geq 7$ b. $x < -2$ c. $1 \leq y < 3$

Answers

a. b. c.

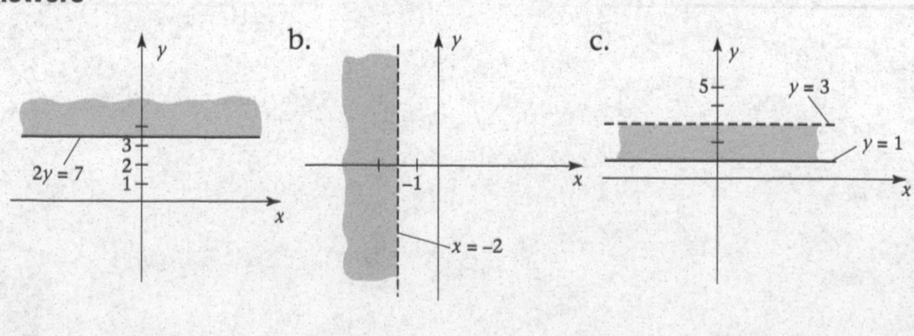

10.5b Systems of Linear Inequalities

We may also consider **systems of linear inequalities** in two variables x and y. Examples of such systems are

$$\begin{cases} 2x - 3y > 6 \\ x + 2y < 2 \end{cases} \quad \text{and} \quad \begin{cases} 2x - 5y \leq 12 \\ 2x + y \leq 18 \\ x \geq 0 \\ y \geq 0 \end{cases}$$

The **solution of a system of linear inequalities** consists of all ordered pairs (a, b) such that the substitution $x = a$, $y = b$ satisfies *all* the inequalities. Thus, the ordered pair $(0, -3)$ is a solution of the system

$$\begin{aligned} 2x - 3y &> 6 \\ x + 2y &< 2 \end{aligned}$$

since it satisfies both inequalities.

$$\begin{aligned} (2)(0) - 3(-3) &= 9 > 6 \\ 0 + (2)(-3) &= -6 < 2 \end{aligned}$$

We can graph the solution set of a system of linear inequalities by graphing the solution set of each inequality and marking that portion of the graph that satisfies *all* the inequalities.

EXAMPLE 4 *Graphing a System of Linear Inequalities*

Graph the solution set of the system.

$$\begin{aligned} 2x - 3y &\leq 2 \\ x + y &\leq 6 \end{aligned}$$

SOLUTION

In the graph below, we show the solution set of each of the inequalities. The darkest region indicates those points that satisfy both inequalities and is, therefore, the solution set of the system of inequalities.

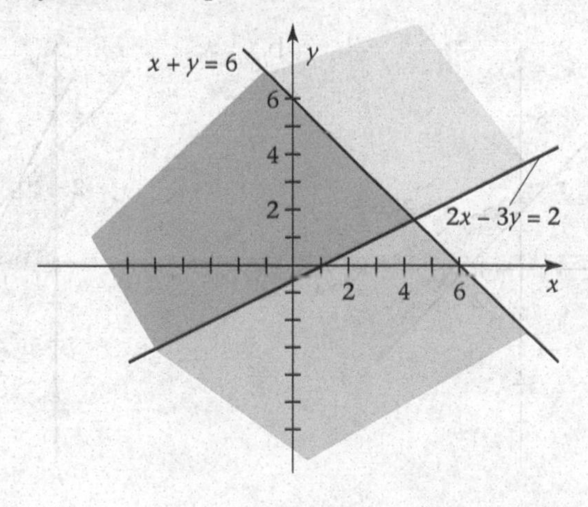

■

EXAMPLE 5 *Graphing a System of Linear Inequalities*

Graph the solution set of the system.

$$x + y < 2$$
$$2x + 3y \geq 9$$
$$x \geq 1$$

SOLUTION

Since there are no points satisfying *all* the inequalities, we conclude that the system is inconsistent and has no solutions.

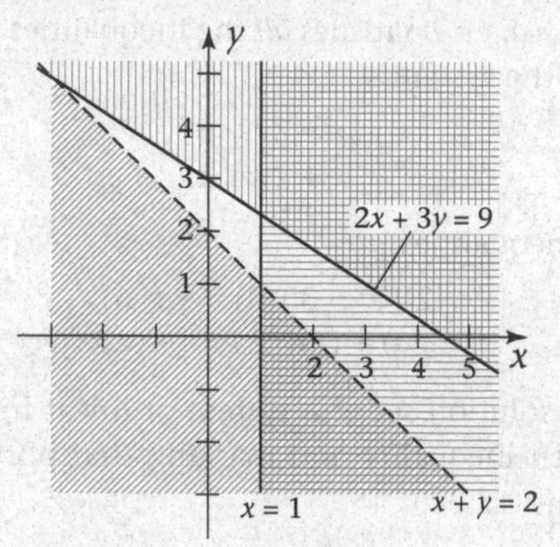

■

Progress Check

Graph the solution set of the given system.

a. $x + y \geq 3$
$x + 2y < 8$

b. $2x + y \leq 4$
$x + y \leq 3$
$x \geq 0$
$y \geq 0$

Answers

a.

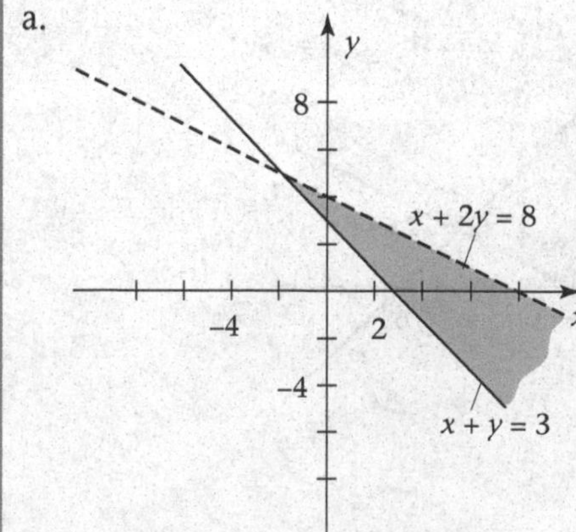

b.

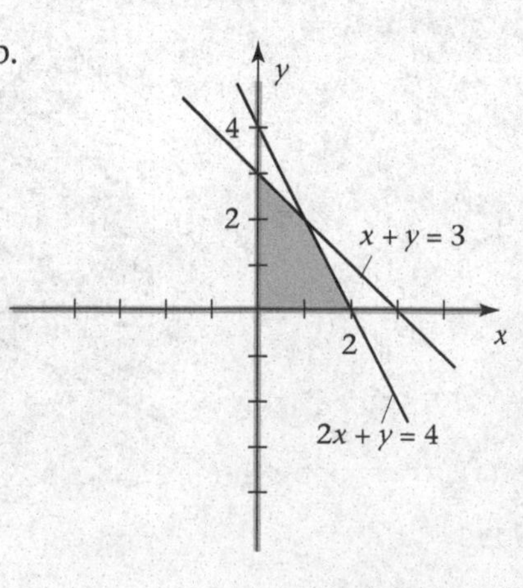

EXAMPLE 6 *Applying Systems of Linear Inequalities*

A dietitian at a university is planning a menu that consists of two primary foods, A and B, whose nutritional contents are shown in the table. The dietitian insists that the meal provide at most 12 units of fat, at least 2 units of carbohydrate and at least 1 unit of protein. If x and y represent the number of grams of food types A and B, respectively, write a system of linear inequalities expressing the restrictions. Graph the solution set.

Nutritional Content in Units per Gram

	Fat	Carbohydrate	Protein
A	2	2	0
B	3	1	1

SOLUTION

The number of units of fat contained in the meal is $2x + 3y$, so x and y must satisfy the inequality

$$2x + 3y \leq 12 \qquad \text{fat requirement}$$

Similarly, the requirements for carbohydrate and protein result in the inequalities

$$2x + y \geq 2 \qquad \text{carbohydrate requirement}$$
$$y \geq 1 \qquad \text{protein requirement}$$

We must also have $x \geq 0$, since negative quantities of food type A would make no sense. The system of linear inequalities is then

$$\begin{aligned} 2x + 3y &\leq 12 \\ 2x + y &\geq 2 \\ x &\geq 0 \\ y &\geq 0 \end{aligned}$$

Thus, the solution is the region:

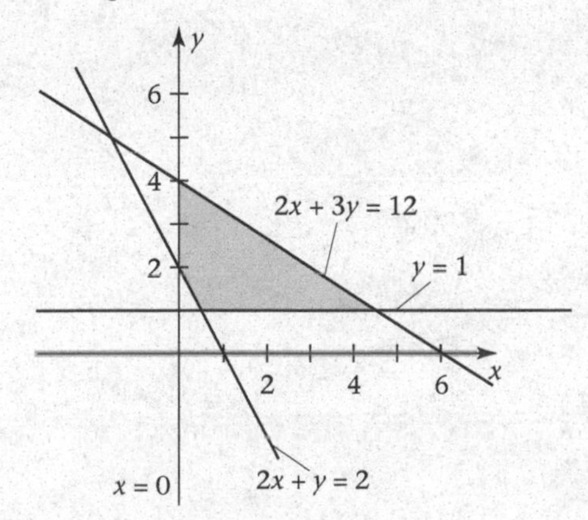

■

Graphing Calculator Power User's CORNER

Solving Systems of Inequalities

Some graphing calculators draw inequality function graphs ($y \geq f(x)$, etc.), shading the appropriate portion of the plane. This capability differs for each model, and you should consult your owner's manual for specific details. Some calculators automatically shade the intersection of the regions you have graphed. On some calculators, you need to shade each region with a different resolution to see the intersection of several regions. (Using resolutions that have no common factors leads to the best results.) Finally, if neither of these approaches works on your calculator, you can shade the other side of each inequality so that the solution set of the given system of equations is the portion of the plane that remains unshaded.

Exercise Set 10.5

In Exercises 1–18, graph the solution set of the given inequality.

1. $y \le x + 2$
2. $y \ge x + 3$
3. $y > x - 4$
4. $y < x - 5$
5. $y \le 4 - x$
6. $y \ge 2 - x$
7. $y > x$
8. $y \le 2x$
9. $3x - 5y > 15$
10. $2y - 3x < 12$
11. $x \le 4$
12. $3x > -2$
13. $y > -3$
14. $5y \le 25$
15. $x > 0$
16. $y < 0$
17. $-2 \le x \le 3$
18. $-6 < y < -2$

19. A steel producer makes two types of steel, regular and special. A ton of regular steel requires 2 hours in the open-hearth furnace and a ton of special steel requires 5 hours. Let x and y denote the number of tons of regular and special steel made per day, respectively. If the open-hearth furnace is available at most 15 hours per day, find the inequalities that must be satisfied by x and y, and graph the solution set.

20. A patient is placed on a diet that restricts caloric intake to 1500 calories per day. The patient plans to eat x ounces of cheese, y slices of bread, and z apples on the first day of the diet. If cheese contains 100 calories per ounce, bread 110 calories per slice, and apples 80 calories each, write an inequality that must be satisfied by x, y, and z.

In Exercises 21–32, graph the solution set of the system of linear inequalities.

21. $2x - y \le 3$, $2x + 3y \ge -3$
22. $x - y \le 4$, $2x + y \ge 6$
23. $3x - y \ge -7$, $3x + y \le -2$
24. $3x - 2y > 1$, $2x + 3y \le 18$
25. $3x - 2y \ge -5$, $4x - y \le 10$, $y \ge 2$
26. $2x - y \ge -3$, $x + y \le 5$, $y \ge 1$
27. $2x - y \le 5$, $x + 2y \ge 1$, $x \ge 0$, $y \ge 0$
28. $-x + 3y \le 2$, $4x + 3y \le 18$, $x \ge 0$, $y \ge 0$
29. $3x + y \le 6$, $x - 2y \le -1$, $x \ge 2$
30. $x - y \ge -2$, $x + y \ge -5$, $y \ge 0$
31. $3x - 2y \le -6$, $8x + 3y \le 24$, $5x + 4y \ge 20$, $x \ge 0$, $y \ge 0$
32. $2x + 3y \ge 18$, $x + 3y \ge 12$, $4x + 3y \ge 24$, $x \ge 0$, $y \ge 0$

33. A farmer has 10 quarts of milk and 15 quarts of cream that he will use to make ice cream and yogurt. Each quart of ice cream requires 0.4 quarts of milk and 0.2 quarts of cream, and each quart of yogurt requires 0.2 quarts of milk and 0.4 quarts of cream. Graph the set of points representing the possible production of ice cream and yogurt.

34. A coffee packer uses Jamaican and Colombian coffee to prepare a mild blend and a strong blend. Each pound of mild blend contains 0.5 pounds of Jamaican coffee and 0.5 pounds of Colombian coffee, and each pound of the strong blend requires 0.25 pounds of Jamaican coffee and 0.75 pounds of Colombian coffee. The packer has available 100 pounds of Jamaican coffee and 125 pounds of Colombian coffee. Graph the set of points representing the possible production of the two blends.

35. A trust fund of $1,000,000, established to provide university scholarships, must adhere to certain restrictions.
    a. No more than half of the fund may be invested in common stocks.
    b. No more than $350,000 may be invested in preferred stocks.
    c. No more than $600,000 may be invested in all types of stocks.
    d. The amount invested in common stock cannot be more than twice the amount invested in preferred stocks.

    Graph the solution set representing the possible investments in common and preferred stocks.

36. An institution serves a luncheon consisting of two dishes, A and B, whose nutritional content in grams per unit served is given in the accompanying table.

	Fat	Carbohydrate	Protein
A	1	1	2
B	2	1	6

The luncheon is to provide no more than 10 grams of fat, no more than 7 grams of carbohydrate, and at least 6 grams of protein. Graph the solution set of possible quantities of dishes A and B.

In Exercises 37–40, write the inequality whose solution is graphed. (Each WINDOW is using ZOOM standard).

37.

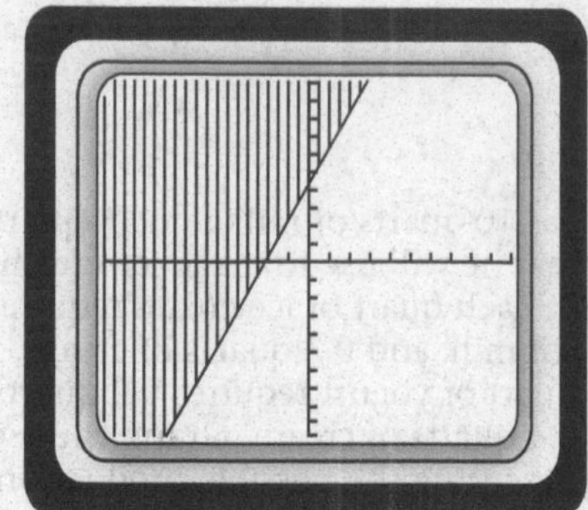

38.

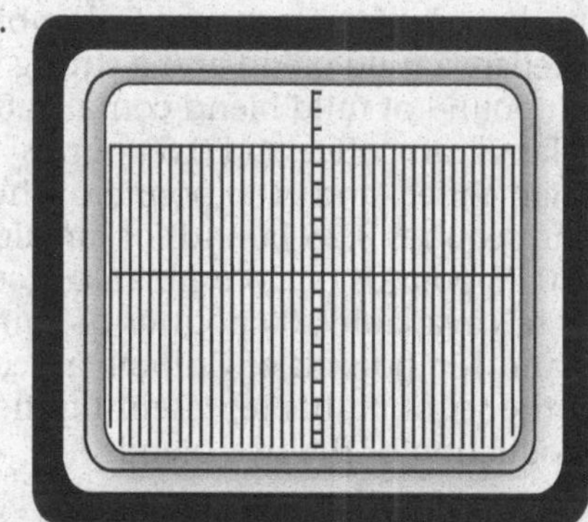

39.

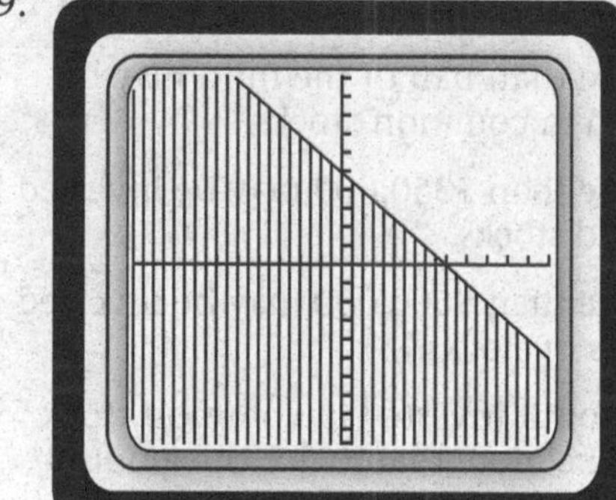

40.

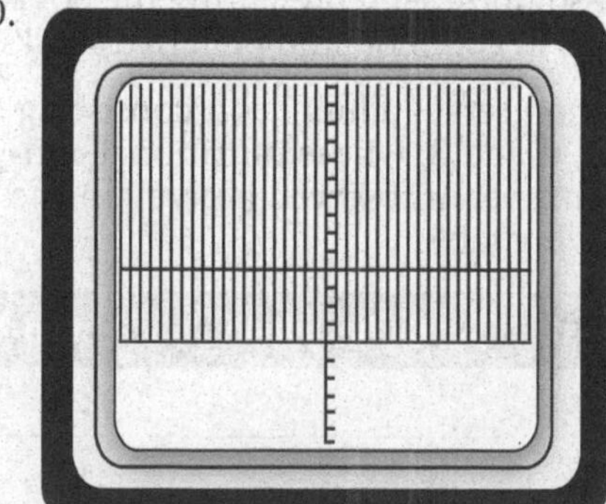

10.6 Linear Programming

Consider the following problem.

A lot is zoned for an apartment building to consist of no more than 40 apartments, totaling no more than 45,000 square feet. A builder is planning to construct two types of apartments. A one-bedroom apartment will require 1000 square feet and will rent for $800 per month. A two-bedroom apartment will require 1500 square feet and will rent for $1120 per month. If all available apartments can be rented, how many apartments of each type should be built to maximize the builder's monthly rental revenue?

Let x denote the number of one-bedroom units and y the number of two-bedroom units. The accompanying table displays the information given in the problem.

	Number of Units	Square Feet	Rental ($)
one-bedroom	x	1,000	800
two-bedroom	y	1,500	1120
Total	40	45,000	z

Using the methods of the previous section, we can translate the **constraints**, or requirements, of the variables x and y into a system of inequalities. The total number of apartments is $x + y$, so we have

$$x + y \leq 40 \qquad \text{number of units constraint}$$

Since each one-bedroom apartment occupies 1000 square feet of space, x apartments will occupy $1000x$ square feet of space. Similarly, the two-bedroom apartments will occupy $1500y$ square feet of space. The total amount of space is $1000x + 1500y$, so we must have

$$1000x + 1500y \leq 45{,}000 \qquad \text{square footage constraint}$$

Moreover, since x and y denote the number of apartments to be built, we must have $x \geq 0$, $y \geq 0$. Thus, we have obtained the following system of inequalities:

$$\begin{aligned} x + y &\leq 40 && \text{number of units constraint} \\ 1000x + 1500y &\leq 45{,}000 && \text{square footage constraint} \\ x &\geq 0 && \text{need for number of apartments} \\ y &\geq 0 && \text{to be nonnegative} \end{aligned}$$

We graph the solution set of this system of linear inequalities in Figure 10-10.

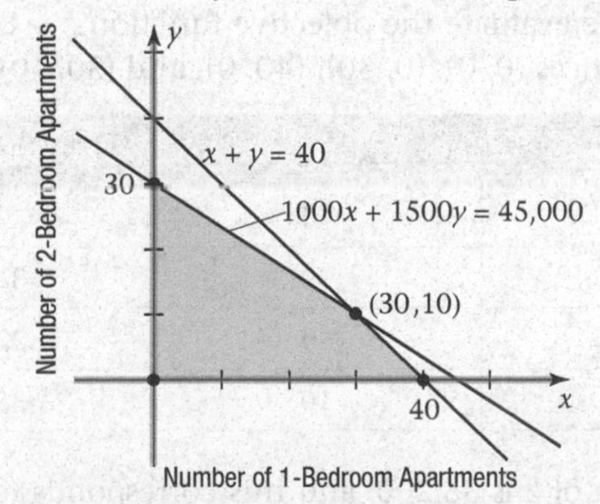

FIGURE 10-10
Solution Set of the System of Linear Inequalities

However, the problem as stated asks that we *maximize* the monthly rental $z = 800x + 1120y$. It is this requirement to **optimize**, that is, to seek a maximum or a minimum value of a linear expression, that characterizes a linear programming problem.

Linear Programming Problem

A **linear programming problem** seeks the optimal (either the largest or the smallest) value of a linear expression called the **objective function** while satisfying constraints that can be formulated as a system of linear inequalities.

Returning to our apartment builder, we can state the linear programming problem in this way:

$$\begin{aligned} \text{maximize} \quad & z = 800x + 1120y \\ \text{subject to} \quad & x + y \leq 40 \\ & 1000x + 1500y \leq 45{,}000 \\ & x \geq 0 \\ & y \geq 0 \end{aligned}$$

The coordinates of each point of the solution set shown in Figure 10-10 are a **feasible solution**; that is, the coordinates give us ordered pairs (a, b) that satisfy the system of linear inequalities. However, which points provide us with values of x and y that maximize the rental income z? For example, the points (40, 0) and (15, 20) are feasible solutions yielding these results for z:

x	y	$z = 800x + 1120y$
40	0	32,000
15	20	34,400

Building 15 one-bedroom and 20 two-bedroom units certainly yields a higher rental revenue than building 40 one-bedroom units. However, is there a solution that yields an even greater value for z?

Before providing the key to solving linear programming problems, we first note that the solution set is bounded by lines. We use the term **vertex** to denote an intersection point of any two boundary lines. We are then ready to state the following theorem:

Fundamental Theorem of Linear Programming

If a linear programming problem has an optimal solution, that solution occurs at a vertex of the set of feasible solutions.

With this result, the builder need only examine the vertices of the solution set of Figure 10-10, rather than considering each of the infinite number of feasible solutions available. We evaluate the objective function $z = 800x + 1120y$ for the coordinates of the vertices (0, 0), (0, 30), (40, 0), and (30, 10).

x	y	$z = 800x + 1120y$
0	0	0
0	30	33,600
40	0	32,000
30	10	35,200

Since the largest value of z is 35,200, and this corresponds to $x = 30$, $y = 10$, the builder finds that the optimal strategy is to build 30 one-bedroom units and 10 two-bedroom units.

Linear programming problems occur in real-life situations with great frequency. In certain industries these problems can involve thousands of variables and hundreds of constraints. Obviously, the method of graphical solution we presented for two variables cannot be used in those situations. A solution method known as the simplex algorithm was first devised by George Dantzig in 1947. Despite the sophistication of this approach, the number of calculations required becomes unmanageably large for hand computation even with a relatively small number of constraints. Fortunately, the discovery of the simplex algorithm occurred at the time electronic computers made their initial appearance. Since then, industries, such as oil refining and steel production, have used linear programming to determine the optimum use of their facilities.

We can now illustrate the steps in solving a linear programming problem.

EXAMPLE 1 *Linear Programming*

Solve the linear programming problem

$$\begin{aligned} \text{minimize} \quad & z = x - 4y \\ \text{subject to} \quad & x + 2y \le 10 \\ & -x + 4y \le 8 \\ & x \ge 0 \\ & y \ge 1 \end{aligned}$$

SOLUTION

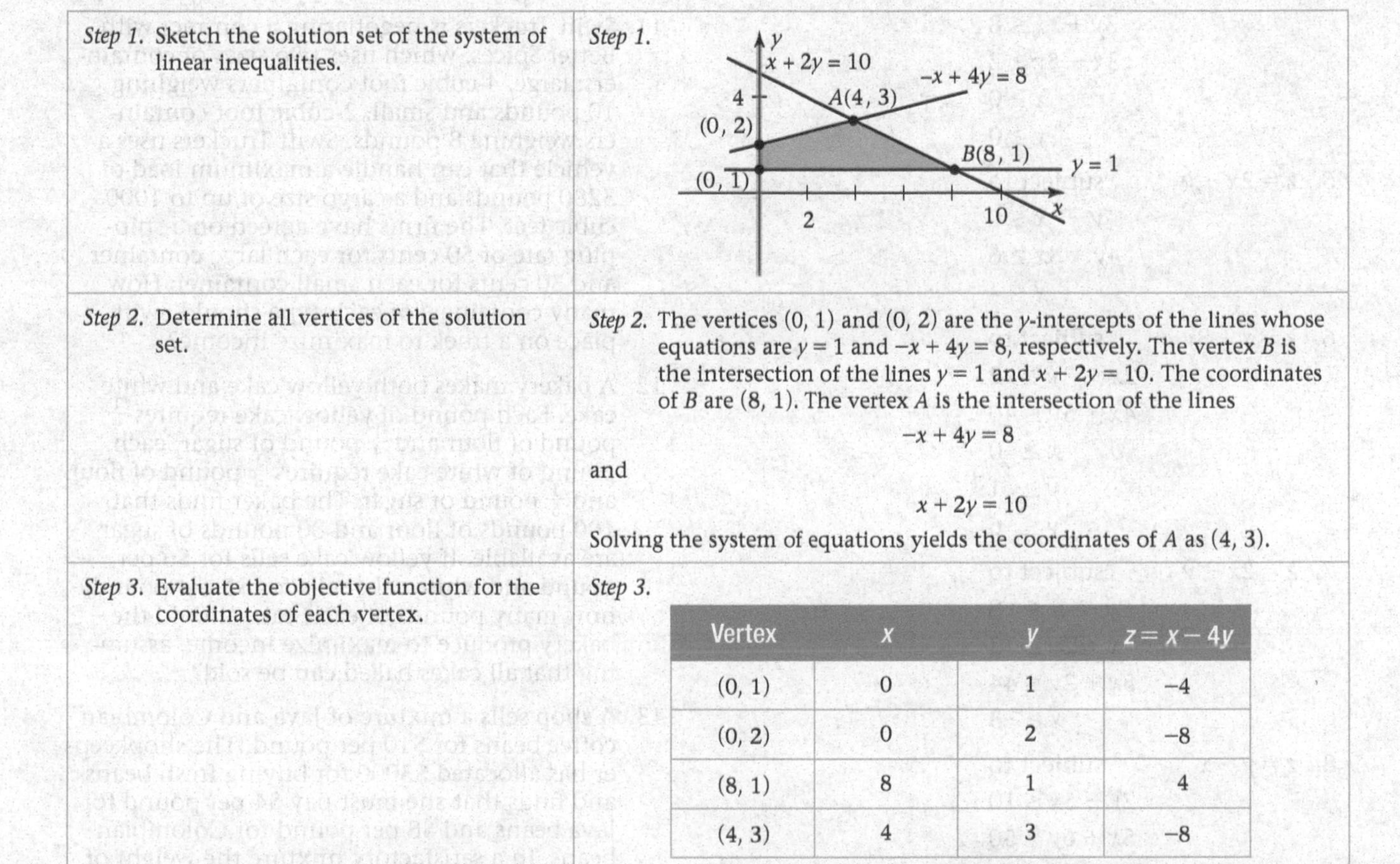

Step 1. Sketch the solution set of the system of linear inequalities.

Step 1.

Step 2. Determine all vertices of the solution set.

Step 2. The vertices (0, 1) and (0, 2) are the y-intercepts of the lines whose equations are $y = 1$ and $-x + 4y = 8$, respectively. The vertex B is the intersection of the lines $y = 1$ and $x + 2y = 10$. The coordinates of B are (8, 1). The vertex A is the intersection of the lines

$$-x + 4y = 8$$

and

$$x + 2y = 10$$

Solving the system of equations yields the coordinates of A as (4, 3).

Step 3. Evaluate the objective function for the coordinates of each vertex.

Step 3.

Vertex	x	y	$z = x - 4y$
(0, 1)	0	1	−4
(0, 2)	0	2	−8
(8, 1)	8	1	4
(4, 3)	4	3	−8

Step 4. The point or points providing the optimal value of the objective function are solutions of the linear programming problem.

Step 4. The minimum value of the objective function is −8, which occurs at the vertices (0, 2) an (4, 3). Thus, $x = 0$, $y = 2$ and $x = 4$, $y = 3$ are both solutions of the linear programming problem.

■

Exercise Set 10.6

In Exercises 1–8, find the minimum value and the maximum value of the linear expression, subject to the given constraints. Indicate coordinates of the vertices at which the minimum and maximum values occur.

1. $z = x - \frac{1}{2}y$ subject to
$3x - y \geq 1$
$x \geq 0$
$x \leq 5$
$y \geq 0$

2. $z = 2x + y$ subject to
$x + y \leq 4$
$x \geq 1$
$y \geq 2$

3. $z = \frac{1}{2}x - 2y$ subject to
$x + 2y \leq 6$
$3y - 2x \leq 2$
$x \geq 0$
$y \geq 0$

4. $z = 0.2x + 0.8y$ subject to
$3y + x \leq 8$
$3x - 5y \geq 2$
$x \geq 0$
$y \geq 0$

5. $z = 2x - y$ subject to
$y - x \leq 0$
$4y + 3x \geq 6$
$x \leq 4$

6. $z = x + 3y$ subject to
$2x + y \geq 2$
$4x + 5y \leq 40$
$x \geq 0$
$y \geq 1$
$y \leq 6$

7. $z = 2x - y$ subject to
$2y - x \leq 8$
$x + 2y \geq 12$
$5x + 2y \leq 44$
$x \geq 3$

8. $z = y - x$ subject to
$2y - 5x \leq 10$
$5x + 6y \leq 50$
$5x + y \leq 20$
$x \geq 0$
$y \geq 1$

9. A firm has budgeted \$1500 for display space at a toy show. Two types of display booths are available. "Preferred space" costs \$18 per square foot, with a minimum rental of 60 square feet, and "regular space" costs \$12 per square foot, with a minimum rental of 30 square feet. It is estimated that there will be 120 visitors for each square foot of "preferred space" and 60 visitors for each square foot of "regular space." How should the firm allot its budget to maximize the number of potential clients that will visit the booths?

10. A company manufactures a desktop computer and a laptop computer. To meet existing orders, it must schedule at least 50 desktop computers for the next production cycle and can produce no more than 150 desktop computers. The manufacturing facilities are adequate to produce no more than 300 laptop computers, but the total number of computers that can be produced cannot exceed 400. The profit on each desktop computer is \$310, and on each laptop computer is \$275. Find the number of computers of each type that should be manufactured to maximize profit.

11. Swift Truckers is negotiating a contract with Better Spices, which uses two sizes of containers: large, 4-cubic foot containers weighing 10 pounds and small, 2-cubic foot containers weighing 8 pounds. Swift Truckers uses a vehicle that can handle a maximum load of 3280 pounds and a cargo size of up to 1000 cubic feet. The firms have agreed on a shipping rate of 50 cents for each large container and 30 cents for each small container. How many containers of each type should Swift place on a truck to maximize income?

12. A bakery makes both yellow cake and white cake. Each pound of yellow cake requires $\frac{1}{4}$ pound of flour and $\frac{1}{4}$ pound of sugar; each pound of white cake requires $\frac{1}{3}$ pound of flour and $\frac{1}{5}$ pound of sugar. The baker finds that 100 pounds of flour and 80 pounds of sugar are available. If yellow cake sells for \$6 per pound and white cake sells for \$5 per pound, how many pounds of each cake should the bakery produce to maximize income, assuming that all cakes baked can be sold?

13. A shop sells a mixture of Java and Colombian coffee beans for \$10 per pound. The shopkeeper has allocated \$3000 for buying fresh beans and finds that she must pay \$4 per pound for Java beans and \$8 per pound for Colombian beans. In a satisfactory mixture the weight of Colombian beans is at least the weight of Java beans and no more than twice the weight of the Java beans. How many pounds of each type of coffee bean should be ordered to maximize the profit if all the mixture can be sold?

14. A pension fund plans to invest up to $500,000 in U.S. Treasury bonds yielding 5% interest per year and corporate bonds yielding 8% interest per year. The fund manager is told to invest a minimum of $250,000 in the Treasury bonds and a minimum of $100,000 in the corporate bonds, with no more than $\frac{1}{4}$ of the total investment to be in corporate bonds. How much should the manager invest in each type of bond to achieve a maximum amount of annual interest? What is the maximum interest?

15. A farmer intends to plant crops A and B on all or part of a 100-acre field. Seed for crop A costs $30 per acre, and labor and equipment costs $100 per acre. For crop B, seed costs $50 per acre, and labor and equipment cost $150 per acre. The farmer cannot spend more than $4500 for seed and $12,000 for labor and equipment. If the income per acre is $300 for crop A and $400 for crop B, how many acres of each crop should be planted to maximize total income?

16. The farmer in Exercise 15 finds that a worldwide surplus in crop B reduces the income to $250 per acre, but the income for crop A remains steady at $300 per acre. How many acres of each crop should be planted to maximize total income?

17. In preparing food for the college cafeteria, a dietitian combines volume pack A and volume pack B. Each pound of volume pack A costs $2.50 and contains 4 units of carbohydrate, 3 units of protein, and 5 units of fat. Each pound of volume pack B costs $1.50 and contains 3 units of carbohydrate, 4 units of protein, and 1 unit of fat. If minimum monthly requirements are 60 units of carbohydrates, 52 units of protein, and 42 units of fat, how many pounds of each food pack does the dietitian use to minimize costs?

18. A lawn service uses a riding mower that cuts a 5000-square foot area per hour and a smaller mower that cuts a 3000-square foot area per hour. Surprisingly, each mower uses $\frac{1}{2}$ gallon of gasoline per hour. Near the end of the day, the supervisor finds that both mowers are empty and that there remains 0.6 gallons of gasoline in the storage cans. Suppose at least 4000 square feet of lawn must still be mowed. If the cost of operating the riding mower is $9 per hour and the cost of operating the smaller mower is $5 per hour, how much of the remaining gasoline should be allocated to each mower to do the job at the least possible cost?

Chapter Summary

Key Terms, Concepts, and Symbols

Key Ideas for Review

Topic	Page	Key Idea
Solving Systems of Equations	550	To solve a system of equations, you must find a value for each unknown that satisfies all equations of the system.
Method of Substitution	550	The method of substitution involves solving an equation for one unknown and substituting the result into another equation.
Method of Elimination	554	The method of elimination involves multiplying an equation by a constant so that when it is added to a second equation, an unknown is eliminated.
Gaussian Elimination	566	Gaussian Elimination is a systematic way of transforming a linear system to triangular form. A linear system in triangular form may be solved by back-substitution.
Consistent Systems	568	A consistent system of equations has one or more solutions.
Inconsistent Systems	568	An inconsistent system has no solutions.
Applications	558	It is often more natural to set up word problems using two or more unknowns.
Partial Fractions	572	A proper fraction can always be written as a sum of partial fractions whose denominators are of the form $(ax+b)^m$ and $(ax^2+bx+c)^n$
Graphing Linear Systems	553	The graph of a pair of linear equations in two unknowns is two lines, not necessarily distinct. 1. If the lines intersect at a point, the system is consistent; and the coordinates of the point of intersection is the unique solution of the system of linear equations. 2. If the lines are the same, the system is consistent; and any point on the line is a solution of the system of linear equations. 3. If the lines are parallel and distinct, the system is inconsistent; and the system of linear equations has no solution.

continues

Topic	Page	Key Idea
Solving a System of Linear Inequalities	583	The solution of a system of linear inequalities can be found graphically as the region that is the intersection of all regions satisfying each inequality.
Linear Programming Problems	589	To solve a linear programming problem, it is only necessary to consider the coordinates of the vertices of the region of feasible solutions.

Review Exercises

In Exercises 1 and 2, solve the given system by graphing.

1. $2x + 3y = 2$, $4x + 5y = 3$
2. $-y^2 + x = 1$, $x + y = 7$

In Exercises 3–8, solve the given system by the method of substitution.

3. $-x + 6y = -11$, $2x + 5y = 5$
4. $2x - 4y = -14$, $-x - 6y = -5$
5. $2x + y = 0$, $x - 3y = \frac{7}{4}$
6. $x^2 + y^2 = 25$, $x + 3y = 5$
7. $x^2 - 4y^2 = 9$, $y - 2x = 0$
8. $y^2 - 4x = 0$, $y^2 + x - 2y = 12$

In Exercises 9–14, solve the given system by the method of elimination.

9. $x + 4y = 17$, $2x - 3y = -21$
10. $5x - 2y = 14$, $-x - 3y = 4$
11. $-3x + y = -13$, $2x - 3y = 11$
12. $7x - 2y = -20$, $3x - y = -9$
13. $y^2 = 2x - 1$, $x - y = 2$
14. $x^2 + y^2 = 9$, $-x^2 + y = 3$

15. The sum of a two-digit number and its tens digit is 49. If we reverse the digits of the number, the resulting number is 9 more than the original number. Find the number.
16. The sum of the digits of a two-digit number is 9. The sum of the number and its units digit is 74. Find the number.
17. Five pounds of hamburger and 4 pounds of steak cost \$55, and 3 pounds of hamburger and 7 pounds of steak cost \$79. Find the cost per pound of hamburger and of steak.
18. An airplane flying with a tail wind can complete a journey of 3500 kilometers in 5 hours. Flying the reverse direction, the plane completes the same trip in 7 hours. What is the speed of the plane in still air?
19. A manufacturer of faucets finds that the supply S and demand D are related to price p as follows:
$$S = 3p + 2$$
$$D = -2p + 17$$
Find the equilibrium price and the number of faucets sold at that price.
20. An auto repair shop finds that its monthly expenditure in dollars is given by $C = 34{,}500 + 40x$, where x is the total number of hours worked by all employees. If the revenue received in dollars is given by $R = 100x$, find the break-even point in number of work hours and the total revenue received at that point.

In Exercises 21–24, use Gaussian Elimination to solve the given linear system.

21. $-3x - y + z = 12$, $2x + 5y - 2z = -9$, $-x + 4y + 2z = 15$
22. $3x + 2y - z = -8$, $2x + 3z = 5$, $x - 4y = -4$
23. $5x - y + 2z = 10$, $-2x + 3y - z = -7$, $3x + 2z = 7$
24. $x + 4y = 4$, $-x + 3z = -4$, $2x + 2y - z = \frac{41}{6}$

In Exercises 25–28, solve by any method.

25. $2x + 3y = 6$, $3x - y = -13$
26. $x + 2y = 0$, $-x + 4y = 5$
27. $2x + 3y - z = -4$, $x - 2y + 2z = -6$, $2x - 3z = 5$
28. $2x + 2y - 3z = -4$, $3y - z = -4$, $4x - y + z = 4$

In Exercises 29–31, find the partial-fraction decomposition of the given rational function.

29. $\dfrac{8 - x}{2x^2 + 3x - 2}$
30. $\dfrac{3x^3 + 5x - 1}{(x^2 + 1)^2}$
31. $\dfrac{2x^3 - 3x^2 + 4x - 2}{(x - 1)^2}$

In Exercises 32–37, graph the solution set of the linear inequality or system of linear inequalities.

32. $x - 2y \le 5$

33. $2x + y > 4$

34. $2x + 3y \le 2$
$x - y \ge 1$

35. $x - 2y \ge 4$
$2x - y \le 2$

36. $2x + 3y \le 6$
$x \ge 0$
$y \ge 1$

37. $2x + y \le 4$
$2x - y \le 3$
$x \ge 0$
$y \ge 0$

In Exercises 38 and 39, solve the given linear programming problem.

38. maximize $z = 5y - x$
subject to $8y - 3x \le 36$
$6x + y \le 30$
$y \ge 1$
$x \ge 0$

39. minimize $z = x + 4y$
subject to $4x - y \ge 8$
$4x + y \le 24$
$5y + 4x \ge 32$

Review Test

1. Solve the linear system by graphing.
$3x - y = -17$
$x + 2y = -1$

In Exercises 2 and 3, solve the given system by the method of substitution.

2. $2x + y = 4$
$3x - 2y = -15$

3. $y^2 - 5x = 0$
$y^2 - x^2 = 6$

In Exercises 4 and 5, solve the given linear system by the method of elimination.

4. $x - 2y = 7$
$3x + 4y = -9$

5. $x^2 + y^2 = 25$
$4x^2 - y^2 = 20$

6. The sum of the digits of a two-digit number is 11. If the sum of the number and its tens digit is 41, find the number.

7. An elegant men's shop is having a post-Christmas sale. All shirts are reduced to one low price, and all ties are reduced to an even lower price. A customer purchases 3 ties and 7 shirts, paying $570. Another customer selects 5 ties and 3 shirts and pays $430. What is the sale price of each tie and of each shirt?

8. A school cafeteria manager finds that the weekly cost of operation is $6000 plus $3.00 for every meal served. If the average meal produces a revenue of $5.00, find the number of meals served that results in zero profit.

9. Solve by Gaussian Elimination.
$3x + 2y - z = -4$
$x - y + 3z = 12$
$2x - y - 2z = -20$

Solve Exercises 10 and 11 by any method.

10. $-3x + 2y = -1$
$6x - y = -1$

11. $3x + y - 2z = 8$
$3y - 4z = 14$
$3x + \frac{1}{2}y + z = 1$

12. Find the partial-fraction decomposition of
$$\frac{x-12}{x^2+x-6}$$

In Exercises 13 and 14, graph the solution set of the system of linear inequalities.

13. $2x - 3y \ge 6$
$3x + y \le 3$

14. $2x + y \le 4$
$2x - 5y \le 5$
$y \ge 1$

15. Solve the linear programming problem.
maximize $z = x + 4y$
subject to $4x - y \ge 8$
$4x + y \le 24$
$4x + 5y \ge 32$

Writing Exercises

1. Using complete sentences, write the procedure that you follow in solving the following problem using two unknowns: A stamp machine dispenses 41¢ stamps and 26¢ stamps. When you deposit a five dollar bill, you receive a packet containing 50% more 41¢ stamps than 26¢ stamps with $1.25 change.

2. Write a report on the origins of linear programming and its connection to World War II.

3. Describe an example of linear programming in everyday life.

4. Construct an example of a linear programming problem in two variables. Describe in words the geometric method used to solve it.

5. What are the limitations of solving linear programming problems geometrically?

6. Describe the method of elimination for solving a system of three linear equations in three unknowns.

Chapter 10 Project

In the chapter opener you read about polynomial curve fitting and how it can be used to model (that is, design equations for) data from the real world. Solving systems of equations, both "two-by-two" (two equations and two unknowns) and "three-by-three," was necessary to complete this task. Review Exercises 34–40 in Section 10.2 and Exercises 22–23 in Section 10.3.

Now use the method outlined in those exercises to model the following data with a quadratic polynomial:

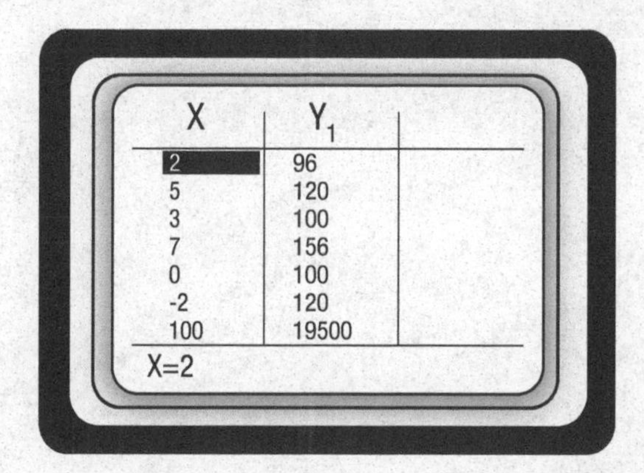

X	Y_1	
2	96	
5	120	
3	100	
7	156	
0	100	
-2	120	
100	19500	

X=2

(This table gives you many more data points than are needed. Use the "extra" data points to check your answer.)

Not all data perfectly matches a polynomial function. Your graphing calculator probably has regression capabilities. This means that you can enter your own data, select a type of function, and your calculator will give you the coefficients to complete the function. Try it!

11 Matrices, Linear Systems, and Determinants

Have you recently sent or received a picture of a friend by email or cell phone? Or perhaps you plan on watching a DVD with friends tonight. The pictures you will see are digital images, which are made up of pixels. For a black-and-white image, each pixel has a value representing the gray-level intensity. If we replace each pixel in the image with its value, a number, we get a rectangular array that looks like this:

$$\begin{bmatrix} 5 & 6 & 8 \\ 4 & 0 & 1 \\ 2 & 1 & 10 \end{bmatrix}$$

This array is called a matrix. By multiplying each entry by 3, we would increase the contrast. Other matrix operations (Section 11.2) can be applied to alter the image in other ways. The chapter project explores some possibilities.

The material on matrices and determinants presented in this chapter serves as an introduction to linear algebra, a mathematical subject that is used in the natural sciences, business and economics, and the social sciences. Since methods involving matrices may require millions of numerical computations, computers have played an important role in expanding the use of matrix techniques to a wide variety of practical problems.

Our study of matrices and determinants will focus on their application to the solution of systems of linear equations. We will see that the method of Gaussian Elimination, studied in the previous chapter, can be readily implemented using matrices. We will show that matrix notation provides a convenient means for writing linear systems and that the inverse of a matrix enables us to solve such a system. Determinants will also provide us with an additional technique, known as Cramer's Rule, for the solution of certain linear systems.

It should be emphasized that this material is a very brief introduction to matrices and determinants. Their properties and applications are both extensive and important.

Explore some different topics like this and others on the Mathematical Association of America website: http://www.maa.org/press/periodicals. There you will find many uses of different concepts in mathematics.

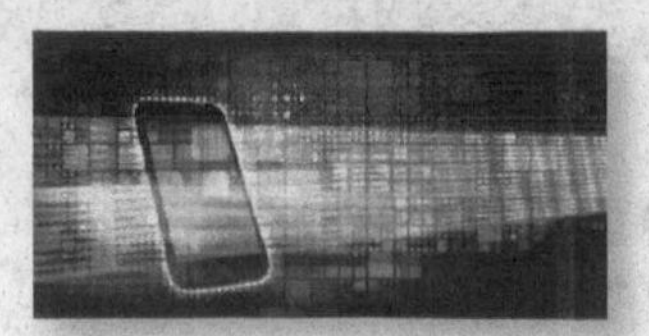

http://www.maa.org/press/periodicals

11.1 Matrices and Linear Systems

11.1a Definitions

We have already studied several methods for solving a linear system, such as

$$\begin{aligned} 2x + 3y &= -7 \\ 3x - y &= 17 \end{aligned}$$

This system can be displayed by a **matrix**, which is a rectangular array of mn real numbers arranged in m horizontal rows and n vertical columns. The numbers are called the **entries**, or **elements**, of the matrix and are enclosed within brackets. Thus,

$$A = \begin{bmatrix} 2 & 3 & -7 \\ 3 & -1 & 17 \end{bmatrix} \begin{matrix} \leftarrow \\ \leftarrow \end{matrix} \text{ rows}$$

$$\uparrow \quad \uparrow \quad \uparrow$$

columns

is a matrix consisting of two rows and three columns, whose entries are obtained from the two given equations. In general, a matrix of m rows and n columns is said to be of **dimension *m* by *n***, written $m \times n$. The matrix A is seen to be of dimension 2×3. If the numbers of rows and columns of a matrix are both equal to n, the matrix is called a **square matrix of order *n*.**

EXAMPLE 1 *Dimension of a Matrix*

a. $A = \begin{bmatrix} -1 & 4 \\ 1 & -2 \end{bmatrix}$

is a 2×2 matrix. Since matrix A has two rows and two columns, it is a square matrix of order 2.

b. $B = \begin{bmatrix} 4 & -5 \\ -2 & 1 \\ 3 & 0 \end{bmatrix}$

has three rows and two columns and is a 3×2 matrix.

c. $C = [-8 \quad 6 \quad 1]$

is a 1×3 matrix and is called a **row matrix** since it has precisely one row.

d. $D = \begin{bmatrix} 2 \\ -4 \end{bmatrix}$

is a 2×1 matrix and is called a **column matrix** since it has exactly one column. ■

11.1b Subscript Notation

There is a convenient way of denoting a general $m \times n$ matrix, using "double subscripts."

$$A = \begin{bmatrix} a_{11} & a_{12} & \cdots & a_{1j} & \cdots & a_{1n} \\ a_{21} & a_{22} & \cdots & a_{2j} & \cdots & a_{2n} \\ \vdots & \vdots & & \vdots & & \vdots \\ a_{i1} & a_{i2} & \cdots & a_{ij} & \cdots & a_{in} \\ \vdots & \vdots & & \vdots & & \vdots \\ a_{m1} & a_{m2} & \cdots & a_{mj} & \cdots & a_{mn} \end{bmatrix} \begin{matrix} \leftarrow \text{first row} \\ \leftarrow \text{second row} \\ \\ \leftarrow i\text{th row} \\ \\ \leftarrow m\text{th row} \end{matrix}$$

↑ first column, ↑ second column, ↑ jth column, ↑ nth column

Thus, a_{ij} is the entry in the ith row and jth column of the matrix A. It is customary to write $A = [\mathbf{a}_{ij}]$ to indicate that a_{ij} is the entry in row i and column j of matrix A.

EXAMPLE 2 *Matrix Dimension and Element Notation*

Let

$$A = \begin{bmatrix} 3 & -2 & 4 & 5 \\ 9 & 1 & 2 & 0 \\ -3 & 2 & -4 & 8 \end{bmatrix}$$

Matrix A is of dimension 3×4. The element a_{12} is found in the first row and second column and is seen to be -2. Similarly, we see that $a_{31} = -3$, $a_{33} = -4$ and $a_{34} = 8$. ■

Progress Check

Let

$$B = \begin{bmatrix} 4 & 8 & 1 \\ 2 & -5 & 3 \\ -8 & 6 & -4 \\ 0 & 1 & -1 \end{bmatrix}$$

Find the following:

a. b_{11} b. b_{23} c. b_{31} d. b_{42}

Answers

a. 4 b. 3 c. -8 d. 1

11.1c Coefficient and Augmented Matrices

If we begin with the system of linear equations

$$\begin{aligned} 2x + 3y &= -7 \\ 3x - y &= 17 \end{aligned}$$

the matrix

$$\begin{bmatrix} 2 & 3 \\ 3 & -1 \end{bmatrix}$$

in which the first column is formed from the coefficients of x and the second column is formed from the coefficients of y, is called the **coefficient matrix**. The matrix

$$\left[\begin{array}{cc|c} 2 & 3 & -7 \\ 3 & -1 & 17 \end{array}\right]$$

which includes the column consisting of the right-hand sides of the equations separated by a dashed line, is called the **augmented matrix**. Note that the unknowns should always be aligned when forming the coefficient and augmented matrices.

EXAMPLE 3 *Linear Systems and the Augmented Matrix*

Write a system of linear equations that corresponds to the augmented matrix.

$$\left[\begin{array}{ccc|c} -5 & 2 & -1 & 15 \\ 0 & -2 & 1 & -7 \\ \frac{1}{2} & 1 & -1 & 3 \end{array}\right]$$

SOLUTION

We attach the unknown x to the first column, the unknown y to the second column, and the unknown z to the third column. The resulting system is

$$\begin{aligned} -5x + 2y - z &= 15 \\ -2y + z &= -7 \\ \tfrac{1}{2}x + y - z &= 3 \end{aligned}$$

■

Now that we have seen how a matrix can be used to represent a system of linear equations, we next proceed to show how operations on that matrix can yield the solution of the system. These "matrix methods" are simply a streamlining of the methods already studied in the previous chapter.

In Section 10.3, we used three elementary operations to transform a system of linear equations into triangular form. When applying the same procedures to a matrix, we speak of rows, columns, and elements instead of equations, unknowns, and coefficients. The three elementary operations that yield an equivalent system now become the **elementary row operations**.

Elementary Row Operations

The following elementary row operations transform an augmented matrix into another augmented matrix. These augmented matrices correspond to equivalent linear systems.

1. Interchange any two rows.
2. Multiply each element of any row by a constant $k \neq 0$.
3. Replace each element of a given row by the sum of itself plus k times the corresponding element of any other row.

The method of **Gaussian Elimination**, introduced in Section 10.3, can now be restated in terms of matrices. By use of elementary row operations we seek to transform an augmented matrix into a matrix for which $a_{ij} = 0$ when $i > j$. The resulting matrix has the following appearance for a system of three linear equations in three unknowns:

$$\left[\begin{array}{ccc|c} * & * & * & * \\ 0 & * & * & * \\ 0 & 0 & * & * \end{array}\right]$$

Since this matrix represents a linear system in triangular form, back-substitution provides a solution of the original system. We will illustrate the process with an example.

EXAMPLE 4 *Elementary Row Operations and Gaussian Elimination*

Solve the system.

$$\begin{aligned} x - y + 4z &= 4 \\ 2x + 2y - z &= 2 \\ 3x - 2y + 3z &= -3 \end{aligned}$$

SOLUTION

We describe and illustrate the steps of the procedure.

Step 1. Form the augmented matrix.	*Step 1.* The augmented matrix is $\left[\begin{array}{ccc\|c} 1 & -1 & 4 & 4 \\ 2 & 2 & -1 & 2 \\ 3 & -2 & 3 & -3 \end{array}\right]$
Step 2. If necessary, interchange rows to make sure that a_{11}, the first element of the first row, is nonzero. We call a_{11} the **pivot element** and row 1 the **pivot row**.	*Step 2.* We see that $a_{11} = 1 \neq 0$. The pivot element is a_{11} and is shown in blue.
Step 3. Arrange to have 0 as the first element of every row below row 1. This is done by replacing row 2, row 3, and so on by the sum of itself and an appropriate multiple of row 1.	*Step 3.* To make $a_{21} = 0$, replace row 2 by the sum of itself and (−2) times row 1. To make $a_{31} = 0$, replace row 3 by the sum of itself and (−3) times row 1. $\left[\begin{array}{ccc\|c} 1 & -1 & 4 & 4 \\ 0 & 4 & -9 & -6 \\ 0 & 1 & -9 & -15 \end{array}\right]$
Step 4. Repeat the process defined by *Steps 2* and *3*, allowing row 2, row 3, and so on to play the role of the first row. Thus, row 2, row 3, and so on serve as the pivot rows, with a_{22} the pivot element of row 2, a_{33}, the pivot element of row 3, and so on.	*Step 4.* Since $a_{22} = 4 \neq 0$, it serves as the next pivot element and is shown in blue. To make $a_{32} = 0$, replace row 3 by the sum of itself and $\left(-\frac{1}{4}\right)$ times row 2. $\left[\begin{array}{ccc\|c} 1 & -1 & 4 & 4 \\ 0 & 4 & -9 & -6 \\ 0 & 0 & -\frac{27}{4} & -\frac{27}{2} \end{array}\right]$
Step 5. The corresponding linear system is in triangular form. Solve by back-substitution.	*Step 5.* The third row of the final matrix yields $\frac{27}{4}z = -\frac{27}{2}$, $z = 2$. Substituting $z = 2$, we obtain from the second row of the final matrix $4y - 9z = -6$, $4y - 9(2) = -6$, $y = 3$. Substituting $y = 3$ and $z = 2$, we obtain from the first row of the final matrix $x - y + 4z = 4$, $x - 3 + 4(2) = 4$, $x = -1$. The solution is $x = -1$, $y = 3$, $z = 2$.

■

Solve the linear system by matrix methods.

$$\begin{aligned} 2x + 4y - z &= 0 \\ x - 2y - 2z &= 2 \\ -5x - 8y + 3z &= -2 \end{aligned}$$

Answer

$x = 6, y = -2, z = 4$

Note that we described the process of Gaussian Elimination in a manner that applies to any augmented matrix that is $n \times (n + 1)$. Thus, Gaussian Elimination may be used on any system of n linear equations in n unknowns that has a unique solution.

It is also permissible to perform elementary row operations in ways to simplify the arithmetic. For example, you may wish to interchange rows or multiply a row by a constant to obtain a pivot element equal to 1. We will illustrate these ideas with an example.

EXAMPLE 5 *Elementary Row Operations and Gaussian Elimination*

Solve by matrix methods.

$$\begin{aligned} 2y + 3z &= 4 \\ 4x + y + 8z + 15w &= -14 \\ x - y + 2z &= 9 \\ -x - 2y - 3z - 6w &= 10 \end{aligned}$$

SOLUTION

We begin with the augmented matrix and perform a sequence of elementary row operations. The pivot element is shown in blue.

Write the augmented matrix. Note that $a_{11} = 0$.

$$\left[\begin{array}{cccc|c} 0 & 2 & 3 & 0 & 4 \\ 4 & 1 & 8 & 15 & -14 \\ 1 & -1 & 2 & 0 & 9 \\ -1 & -2 & -3 & -6 & 10 \end{array}\right]$$

Interchange rows 1 and 3 so that $a_{11} = 1$.

$$\left[\begin{array}{cccc|c} 1 & -1 & 2 & 0 & 9 \\ 4 & 1 & 8 & 15 & -14 \\ 0 & 2 & 3 & 0 & 4 \\ -1 & -2 & -3 & -6 & 10 \end{array}\right]$$

To make $a_{21} = 0$, replace row 2 by the sum of itself and (-4) times row 1. To make $a_{41} = 0$, replace row 4 by the sum of itself and row 1.

$$\left[\begin{array}{cccc|c} 1 & -1 & 2 & 0 & 9 \\ 0 & 5 & 0 & 15 & -50 \\ 0 & 2 & 3 & 0 & 4 \\ 0 & -3 & -1 & -6 & 19 \end{array}\right]$$

Multiply row 2 by $\frac{1}{5}$ so that $a_{22} = 1$.

$$\left[\begin{array}{cccc|c} 1 & -1 & 2 & 0 & 9 \\ 0 & 1 & 0 & 3 & -10 \\ 0 & 2 & 3 & 0 & 4 \\ 0 & -3 & -1 & -6 & 19 \end{array}\right]$$

To make $a_{32} = 0$, replace row 3 by the sum of itself and (−2) times row 2. To make $a_{42} = 0$, replace row 4 by the sum of itself and 3 times row 2.

$$\left[\begin{array}{cccc|c} 1 & -1 & 2 & 0 & 9 \\ 0 & 1 & 0 & 3 & -10 \\ 0 & 0 & 3 & -6 & 24 \\ 0 & 0 & -1 & 3 & -11 \end{array}\right]$$

Interchange rows 3 and 4 so that the next pivot is $a_{33} = -1$.

$$\left[\begin{array}{cccc|c} 1 & -1 & 2 & 0 & 9 \\ 0 & 1 & 0 & 3 & -10 \\ 0 & 0 & -1 & 3 & -11 \\ 0 & 0 & 3 & -6 & 24 \end{array}\right]$$

To make $a_{43} = 0$, replace row 4 by the sum of itself and 3 times row 3.

$$\left[\begin{array}{cccc|c} 1 & -1 & 2 & 0 & 9 \\ 0 & 1 & 0 & 3 & -10 \\ 0 & 0 & -1 & 3 & -11 \\ 0 & 0 & 0 & 3 & -9 \end{array}\right]$$

The last row of the matrix indicates that

$$3w = -9$$
$$w = -3$$

The remaining unknowns are found by back-substitution.

Third Row of Final Matrix	Second Row of Final Matrix	First Row of Final Matrix
$-z + 3w = -11$	$y + 3w = -10$	$x - y + 2z = 9$
$-z + 3(-3) = -11$	$y + 3(-3) = -10$	$x - (-1) + 2(2) = 9$
$z = 2$	$y = -1$	$x = 4$

The solution is $x = 4$, $y = -1$, $z = 2$, $w = -3$. ■

11.1d Gauss-Jordan Elimination

There is an important variant of Gaussian Elimination known as **Gauss-Jordan Elimination.** The objective of this variant is to transform a linear system into a form that yields a solution without back-substitution. For a 3 × 3 system that has a unique solution, the final matrix and equivalent linear system look like this:

$$\left[\begin{array}{ccc|c} 1 & 0 & 0 & c_1 \\ 0 & 1 & 0 & c_2 \\ 0 & 0 & 1 & c_3 \end{array}\right] \qquad \begin{aligned} x + 0y + 0z &= c_1 \\ 0x + y + 0z &= c_2 \\ 0x + 0y + z &= c_3 \end{aligned}$$

The solution is then seen to be $x = c_1$, $y = c_2$ and $z = c_3$.

The execution of the Gauss-Jordan Method is essentially the same as that of Gaussian Elimination with these exceptions:

1. The pivot elements are always required to be equal to 1.
2. All elements in a column other than the pivot element are forced to be 0.

These objectives are accomplished by the use of elementary row operations as illustrated in the following example.

EXAMPLE 6 *Gauss-Jordan Elimination*

Solve the linear system by the Gauss-Jordan Method.

$$\begin{aligned} x - 3y + 2z &= 12 \\ 2x + y - 4z &= -1 \\ x + 3y - 2z &= -8 \end{aligned}$$

SOLUTION

We begin with the augmented matrix. At each stage, the pivot element is shown in blue and is used to force all elements in that column other than the pivot element itself to be zero.

The pivot element is a_{11}.

$$\left[\begin{array}{ccc|c} 1 & -3 & 2 & 12 \\ 2 & 1 & -4 & -1 \\ 1 & 3 & -2 & -8 \end{array}\right]$$

To make $a_{21} = 0$, replace row 2 by the sum of itself and -2 times row 1. To make $a_{31} = 0$, replace row 3 by the sum of itself and -1 times row 1.

$$\left[\begin{array}{ccc|c} 1 & -3 & 2 & 12 \\ 0 & 7 & -8 & -25 \\ 0 & 6 & -4 & -20 \end{array}\right]$$

Replace row 2 by the sum of itself and -1 times row 3 to yield the next pivot, $a_{22} = 1$.

$$\left[\begin{array}{ccc|c} 1 & -3 & 2 & 12 \\ 0 & 1 & -4 & -5 \\ 0 & 6 & -4 & -20 \end{array}\right]$$

To make $a_{12} = 0$, replace row 1 by the sum of itself and 3 times row 2. To make $a_{32} = 0$, replace row 3 by the sum of itself and -6 times row 2.

$$\left[\begin{array}{ccc|c} 1 & 0 & -10 & -3 \\ 0 & 1 & -4 & -5 \\ 0 & 0 & 20 & 10 \end{array}\right]$$

Multiply row 3 by $\frac{1}{20}$ so that $a_{33} = 1$.

$$\left[\begin{array}{ccc|c} 1 & 0 & -10 & -3 \\ 0 & 1 & -4 & -5 \\ 0 & 0 & 1 & \frac{1}{2} \end{array}\right]$$

To make $a_{13} = 0$, replace row 1 by the sum of itself and 10 times row 3. To make $a_{23} = 0$, replace row 2 by the sum of itself and 4 times row 3.

$$\left[\begin{array}{ccc|c} 1 & 0 & 0 & 2 \\ 0 & 1 & 0 & -3 \\ 0 & 0 & 1 & \frac{1}{2} \end{array}\right]$$

We see the solution directly from the final matrix: $x = 2$, $y = -3$, and $z = \frac{1}{2}$. ■

Graphing Calculator Power User's CORNER

Reduced Row Echelon Form

Your graphing calculator can take an augmented matrix and return the reduced row echelon form required by Gauss-Jordan Elimination. As you will see in Example 7, the dashed line customarily found in an augmented matrix does not appear. Your graphing calculator is a powerful tool for solving systems of equations. However, you must be able to interpret the information it gives you. Recall the cases in which there are either infinitely many solutions, or no solution.

EXAMPLE 7 *Gauss-Jordan Elimination Using the Graphing Calculator*

Consider the system

$$\begin{aligned} 0.03x + \quad y - 0.07z &= 0.89 \\ x - 0.01y + 0.12z &= 1.23 \\ 1.02x - 1.02y + \quad z &= 2 \end{aligned}$$

The augmented matrix for this system is

$$\begin{bmatrix} 0.03 & 1 & -0.07 & 0.89 \\ 1 & -0.01 & 0.12 & 1.23 \\ 1.02 & -1.02 & 1 & 2 \end{bmatrix}$$

After entering this matrix into the graphing calculator and naming it A, we select reduced row echelon form from the MATRIX MATH menu:

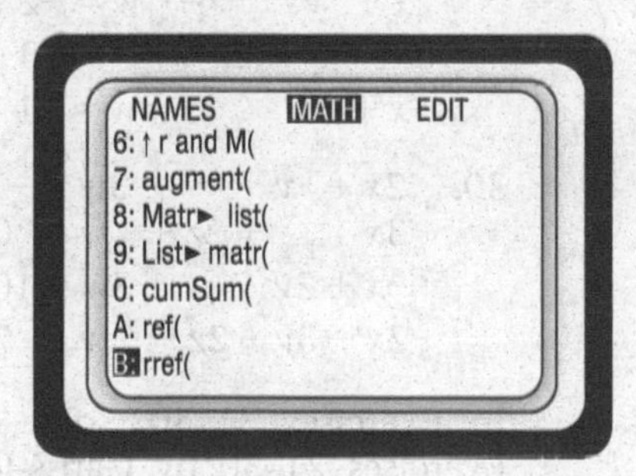

Here is the result:

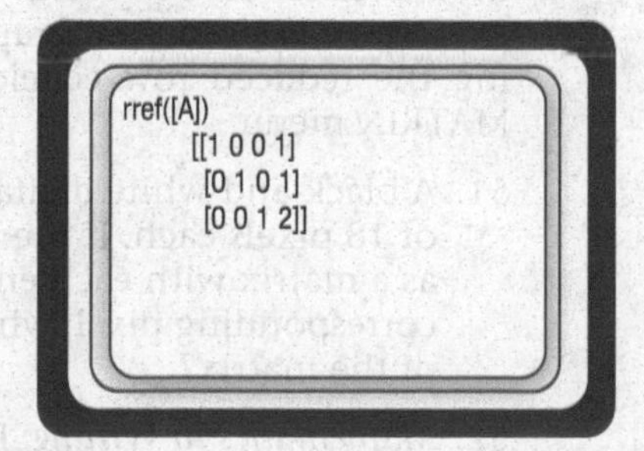

The solution is (1, 1, 2). ■

Exercise Set 11.1

In Exercises 1–6, state the dimension of each matrix.

1. $\begin{bmatrix} 3 & -1 \\ 2 & 4 \end{bmatrix}$

2. $[1 \ \ 2 \ \ 3 \ \ -1]$

3. $\begin{bmatrix} 4 & 2 & 3 \\ 5 & -1 & 4 \\ 2 & 3 & 6 \\ -8 & -1 & 2 \end{bmatrix}$

4. $\begin{bmatrix} -1 \\ 3 \\ 2 \end{bmatrix}$

5. $\begin{bmatrix} 4 & 2 & 1 \\ 3 & 1 & 5 \\ -4 & -2 & 3 \end{bmatrix}$

6. $\begin{bmatrix} 3 & -1 & 2 & 6 \\ 2 & 8 & 4 & 1 \end{bmatrix}$

7. Given

$$A = \begin{bmatrix} 3 & -4 & -2 & 5 \\ 8 & 7 & 6 & 2 \\ 1 & 0 & 9 & -3 \end{bmatrix}$$

find

a. a_{12} b. a_{22} c. a_{23} d. a_{34}

8. Given

$$B = \begin{bmatrix} -5 & 6 & 8 \\ 4 & 1 & 3 \\ 0 & 2 & -6 \\ -3 & 9 & 7 \end{bmatrix}$$

find

a. b_{13} b. b_{21} c. b_{33} d. b_{42}

In Exercises 9–12, write the coefficient matrix and the augmented matrix for each given linear system.

9. $\begin{aligned} 3x - 2y &= 12 \\ 5x + y &= -8 \end{aligned}$

10. $\begin{aligned} 3x - 4y &= 15 \\ 4x - 3y &= 12 \end{aligned}$

11. $\begin{aligned} \tfrac{1}{2}x + y + z &= 4 \\ 2x - y - 4z &= 6 \\ 4x + 2y - 3z &= 8 \end{aligned}$

12. $\begin{aligned} 2x + 3y - 4z &= 10 \\ -3x + y &= 12 \\ 5x - 2y + z &= -8 \end{aligned}$

In Exercises 13–16, write the linear system whose augmented matrix is given.

13. $\left[\begin{array}{cc|c} \frac{3}{2} & 6 & -1 \\ 4 & 5 & 3 \end{array}\right]$

14. $\left[\begin{array}{cc|c} 4 & 0 & 2 \\ -7 & 8 & 3 \end{array}\right]$

15. $\left[\begin{array}{ccc|c} 1 & 1 & 3 & -4 \\ -3 & 4 & 0 & 8 \\ 2 & 0 & 7 & 6 \end{array}\right]$

16. $\left[\begin{array}{ccc|c} 4 & 8 & 3 & 12 \\ 1 & -5 & 3 & -14 \\ 0 & 2 & 7 & 18 \end{array}\right]$

In Exercises 17–20, the augmented matrix corresponding to a linear system has been transformed to the given matrix by elementary row operations. Find a solution of the original linear system.

17. $\left[\begin{array}{ccc|c} 1 & 2 & 0 & 3 \\ 0 & 1 & -2 & 4 \\ 0 & 0 & 1 & 2 \end{array}\right]$

18. $\left[\begin{array}{ccc|c} 1 & 0 & 2 & -1 \\ 0 & 1 & 3 & 2 \\ 0 & 0 & 1 & 5 \end{array}\right]$

19. $\left[\begin{array}{ccc|c} 1 & -2 & 1 & 3 \\ 0 & 1 & 3 & 2 \\ 0 & 0 & 1 & -4 \end{array}\right]$

20. $\left[\begin{array}{ccc|c} 1 & -4 & 2 & -4 \\ 0 & 1 & 3 & -2 \\ 0 & 0 & 1 & 5 \end{array}\right]$

In Exercises 21–30, solve the given linear system by applying Gaussian Elimination to the augmented matrix.

21. $\begin{aligned} x - 2y &= -4 \\ 2x + 3y &= 13 \end{aligned}$

22. $\begin{aligned} 2x + y &= -1 \\ 3x - y &= -7 \end{aligned}$

23. $\begin{aligned} x + y + z &= 4 \\ 2x - y + 2z &= 11 \\ x + 2y + 2z &= 6 \end{aligned}$

24. $\begin{aligned} x - y + z &= -5 \\ 3x + y + 2z &= -5 \\ 2x - y - z &= -2 \end{aligned}$

25. $\begin{aligned} 2x + y - z &= 9 \\ x - 2y + 2z &= -3 \\ 3x + 3y + 4z &= 11 \end{aligned}$

26. $\begin{aligned} 2x + y - z &= -2 \\ -2x - 2y + 3z &= 2 \\ 3x + y - z &= -4 \end{aligned}$

27. $\begin{aligned} -x - y + 2z &= 9 \\ x + 2y - 2z &= -7 \\ 2x - y + z &= -9 \end{aligned}$

28. $\begin{aligned} 4x + y - z &= -1 \\ x - y + 2z &= 3 \\ -x + 2y - z &= 0 \end{aligned}$

29. $\begin{aligned} x + y - z + 2w &= 0 \\ 2x + y - w &= -2 \\ 3x + 2z &= -3 \\ -x + 2y + 3w &= 1 \end{aligned}$

30. $\begin{aligned} 2x + y - 3w &= -7 \\ 3x + 2z + w &= 0 \\ -x + 2y + 3w &= 10 \\ -2x - 3y + 2z - w &= 7 \end{aligned}$

In Exercises 31–40, solve the linear systems of Exercises 21–30 by Gauss-Jordan Elimination applied to the augmented matrix.

In Exercises 41–50, solve the linear systems of Exercises 21–30 in your graphing calculator by using the reduced row echelon option under your MATRIX menu.

51. A black-and-white digital image has 30 rows of 18 pixels each. If the image is represented as a matrix with each entry the value of the corresponding pixel, what are the dimensions of the matrix?

52. *Mathematics in Writing:* In your own words, describe the difference between Gaussian elimination and Gauss-Jordan elimination. Which do you prefer? Why?

11.2 Matrix Operations and Applications

Now that we have defined a matrix, we can define various operations with matrices. First we begin with the definition of equality.

Equality of Matrices
Two matrices are equal if they are of the same dimension and their corresponding entries are equal.

EXAMPLE 1 *Matrix Equality*

Solve for all unknowns.

$$\begin{bmatrix} -2 & 2x & 9 \\ y-1 & 3 & -4s \end{bmatrix} = \begin{bmatrix} z & 6 & 9 \\ -4 & r & 7 \end{bmatrix}$$

SOLUTION
Equating corresponding elements, we must have

$$\begin{aligned} -2 &= z \quad \text{or} \quad z = -2 \\ 2x &= 6 \quad \text{or} \quad x = 3 \\ y - 1 &= -4 \quad \text{or} \quad y = -3 \\ 3 &= r \quad \text{or} \quad r = 3 \\ -4s &= 7 \quad \text{or} \quad s = -\frac{7}{4} \end{aligned}$$

■

Matrix addition can be performed only when the matrices are of the same dimension.

Matrix Addition
The sum $A + B$ of two $m \times n$ matrices A and B is the $m \times n$ matrix obtained by adding the corresponding elements of A and B.

EXAMPLE 2 *Matrix Addition*

Given the following matrices,

$$A = [2 \quad -3 \quad 4] \qquad B = [5 \quad 3 \quad 2]$$

$$C = \begin{bmatrix} 1 & 6 & -1 \\ -2 & 4 & 5 \end{bmatrix} \qquad D = \begin{bmatrix} 16 & 2 & 9 \\ 4 & -7 & -1 \end{bmatrix}$$

find (if possible):

a. $A + B$ b. $A + D$ c. $C + D$

SOLUTION

a. Since A and B are both 1×3 matrices, they can be added, giving

$$A + B = [2+5 \quad -3+3 \quad 4+2] = [7 \quad 0 \quad 6]$$

b. Matrices A and D are not of the same dimension and cannot be added.

c. C and D are both 2×3 matrices. Thus,

$$C + D = \begin{bmatrix} 1+16 & 6+2 & -1+9 \\ -2+4 & 4+(-7) & 5+(-1) \end{bmatrix} = \begin{bmatrix} 17 & 8 & 8 \\ 2 & -3 & 4 \end{bmatrix}$$

■

A matrix is a way of writing the information displayed in a table. For example, Table 11-1 displays the current inventory of the Quality TV Company at its various outlets.

TABLE 11-1 Inventory of Television Sets

TV Sets	Boston	Miami	Chicago
17 inch	140	84	25
19 inch	62	17	48

The same data is displayed by the matrix S, where we understand the columns to represent the cities and the rows to represent the sizes of the television sets.

$$S = \begin{bmatrix} 140 & 84 & 25 \\ 62 & 17 & 48 \end{bmatrix}$$

If the matrix

$$M = \begin{bmatrix} 30 & 46 & 15 \\ 50 & 25 & 60 \end{bmatrix}$$

specifies the number of sets of each size received at each outlet the following month, then the matrix

$$T = S + M = \begin{bmatrix} 170 & 130 & 40 \\ 112 & 42 & 108 \end{bmatrix}$$

gives the revised inventory.

Suppose the salespeople at each outlet are told that half of the revised inventory is to be placed on sale. To determine the number of sets of each size to be placed on sale, we need to multiply each element of the matrix T by 0.5. When working with matrices, we call a real number such as 0.5 a **scalar** and define **scalar multiplication** as follows.

Scalar Multiplication
To multiply a matrix A by a scalar c, multiply each entry of A by c.

EXAMPLE 3 *Scalar Multiplication*

The matrix Q

$$Q = \begin{array}{c} \\ \end{array}\begin{matrix} \text{Regular} & \text{Unleaded} & \text{Premium} \\ \end{matrix}$$

$$Q = \begin{bmatrix} 130 & 250 & 60 \\ 110 & 180 & 40 \end{bmatrix} \begin{matrix} \text{City A} \\ \text{City B} \end{matrix}$$

(columns: Regular, Unleaded, Premium)

shows the quantity (in thousands of gallons) of the principal types of gasolines stored by a refiner at two different locations. It is decided to increase the quantity of each type of gasoline stored at each site by 10%. Use scalar multiplication to determine the desired inventory levels.

SOLUTION

To increase each entry of matrix Q by 10%, we compute the scalar product $1.1Q$.

$$1.1Q = 1.1\begin{bmatrix} 130 & 250 & 60 \\ 110 & 180 & 40 \end{bmatrix} = \begin{bmatrix} 1.1(130) & 1.1(250) & 1.1(60) \\ 1.1(110) & 1.1(180) & 1.1(40) \end{bmatrix} = \begin{bmatrix} 143 & 275 & 66 \\ 121 & 198 & 44 \end{bmatrix}$$ ■

We denote $A + (-1)B$ by $A - B$ and refer to this as the *difference* of A and B.

Matrix Subtraction

The difference $A - B$ of two $m \times n$ matrices A and B is the $m \times n$ matrix obtained by subtracting each entry of B from the corresponding entry of A.

EXAMPLE 4 *Matrix Subtraction*

Using the matrices C and D of Example 2, find $C - D$.

SOLUTION

By definition,

$$C - D = \begin{bmatrix} 1-16 & 6-2 & -1-9 \\ -2-4 & 4-(-7) & 5-(-1) \end{bmatrix} = \begin{bmatrix} -15 & 4 & -10 \\ -6 & 11 & 6 \end{bmatrix}$$ ■

11.2a Matrix Multiplication

We will use the Quality TV Company again, this time to help us arrive at a definition of matrix multiplication. Suppose

$$\begin{array}{cccc} & \text{Boston} & \text{Miami} & \text{Chicago} \\ B = & \begin{bmatrix} 60 & 85 & 70 \\ 40 & 100 & 20 \end{bmatrix} & & \begin{array}{l} \text{17 inch} \\ \text{19 inch} \end{array} \end{array}$$

is a matrix representing the number of television sets in stock at the end of the year. Further, suppose the cost of each 17-inch set is \$80 and the cost of each 19-inch set is \$125. To find the total cost of the inventory at each outlet, we multiply the number of 17-inch sets by \$80, the number of 19-inch sets by \$125, and add the two products. If we let

$$A = [80 \quad 125]$$

be the cost matrix, we seek to define the product

$$AB = [80 \quad 125] \begin{bmatrix} 60 & 85 & 70 \\ 40 & 100 & 20 \end{bmatrix}$$

so that the result is a matrix displaying the total inventory cost at each outlet. We need to calculate for

the Boston outlet	$(80)(60) + (125)(40) = 9800$
the Miami outlet	$(80)(85) + (125)(100) = 19{,}300$
the Chicago outlet	$(80)(70) + (125)(20) = 8100$

The total inventory cost at each outlet can then be displayed by the 1×3 matrix

$$C = [9800 \quad 19{,}300 \quad 8100]$$

which is the product of A and B. Thus,

$$\begin{aligned} AB &= [80 \quad 125] \begin{bmatrix} 60 & 85 & 70 \\ 40 & 100 & 20 \end{bmatrix} \\ &= [(80)(60) + (125)(40) \quad (80)(85) + (125)(100) \quad (80)(70) + (125)(20)] \\ &= [9800 \quad 19{,}300 \quad 8100] = C \end{aligned}$$

This example illustrates the process for multiplying a 1×2 matrix times a 2×3 matrix. The general definition of matrix multiplication utilizes the same basic idea. That is, multiplication of matrices requires calculating sums of products. In this example, the first matrix had two columns and the second matrix had two rows. If we denote the elements of the first matrix as

$$A = [a_{11} \quad a_{12}]$$

and the elements of the second matrix as

$$B = \begin{bmatrix} b_{11} & b_{12} & b_{13} \\ b_{21} & b_{22} & b_{23} \end{bmatrix}$$

then matrix multiplication requires that we calculate

$$[a_{11}b_{11} + a_{12}b_{21} \quad a_{11}b_{12} + a_{12}b_{22} \quad a_{11}b_{13} + a_{12}b_{23}]$$

If we denote the elements of this product by

$$C = [c_{11} \quad c_{12} \quad c_{13}]$$

then we see that

$$c_{1k} = a_{11}b_{1k} + a_{12}b_{2k} \qquad \text{for } k = 1, 2, 3$$

Matrix Multiplication

The product of AB of $m \times n$ matrix $A = [a_{ij}]$ and the $n \times r$ matrix $B = [b_{jk}]$ is obtained by forming the $m \times r$ matrix $C = [c_{ik}]$, where

$$c_{ik} = a_{i1}b_{1k} + a_{i2}b_{2k} + \cdots + a_{in}b_{nk}$$

for $i = 1, 2, \ldots, m$ and $k = 1, 2, \ldots, r$.

It is important to note that the product AB only exists if the number of columns of A equals the number of rows of B. See Figure 11-1.

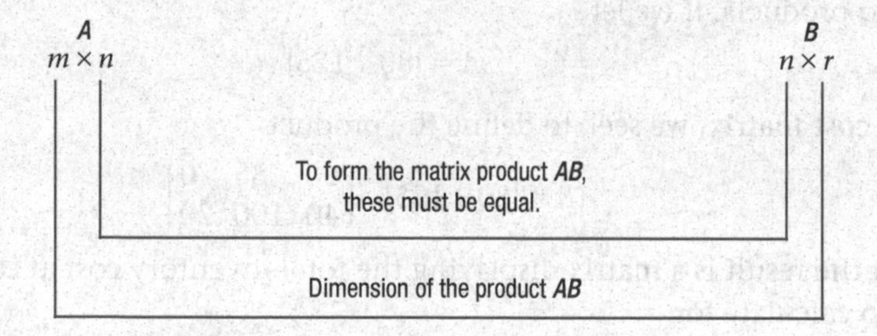

FIGURE 11-1
Dimension of the Product Matrix

EXAMPLE 5 *Matrix Multiplication*

Find the product AB if

$$A = \begin{bmatrix} 2 & 1 \\ 3 & 5 \end{bmatrix} \qquad B = \begin{bmatrix} 4 & -6 & -2 & 4 \\ 2 & 0 & 1 & -5 \end{bmatrix}$$

SOLUTION

$$AB = \begin{bmatrix} (2)(4)+(1)(2) & (2)(-6)+(1)(0) & (2)(-2)+(1)(1) & (2)(4)+(1)(-5) \\ (3)(4)+(5)(2) & (3)(-6)+(5)(0) & (3)(-2)+(5)(1) & (3)(4)+(5)(-5) \end{bmatrix}$$

$$= \begin{bmatrix} 10 & -12 & -3 & 3 \\ 22 & -18 & -1 & -13 \end{bmatrix}$$

■

Progress Check

Find the product AB if

$$A = \begin{bmatrix} -2 & -1 & 2 \\ 4 & 3 & 1 \end{bmatrix} \qquad B = \begin{bmatrix} 5 & -4 \\ 3 & 1 \\ -1 & 0 \end{bmatrix}$$

Answer

$$AB = \begin{bmatrix} -15 & 7 \\ 28 & -13 \end{bmatrix}$$

EXAMPLE 6 ***Matrix Multiplication***

Given the matrices

$$A = \begin{bmatrix} 1 & -1 \\ 2 & 3 \end{bmatrix} \quad B = \begin{bmatrix} 5 & -3 \\ -2 & 2 \end{bmatrix} \quad C = \begin{bmatrix} 3 & -1 & -2 \\ 1 & 0 & 4 \end{bmatrix} \quad D = \begin{bmatrix} 1 \\ 2 \\ 3 \end{bmatrix}$$

a. Show that $AB \neq BA$.

b. Determine the dimension of CD.

SOLUTION

a. $AB = \begin{bmatrix} (1)(5)+(-1)(-2) & (1)(-3)+(-1)(2) \\ (2)(5)+(3)(-2) & (2)(-3)+(3)(2) \end{bmatrix} = \begin{bmatrix} 7 & -5 \\ 4 & 0 \end{bmatrix}$

$BA = \begin{bmatrix} (5)(1)+(-3)(2) & (5)(-1)+(-3)(3) \\ (-2)(1)+(2)(2) & (-2)(-1)+(2)(3) \end{bmatrix} = \begin{bmatrix} -1 & -14 \\ 2 & 8 \end{bmatrix}$

Since the corresponding elements of AB and BA are not equal, $AB \neq BA$.

b. The product of a 2×3 matrix and a 3×1 matrix is a 2×1 matrix. ■

Progress Check

If possible, find the dimension of CD and of CB, using the matrices of Example 6.

Answers

2×1; not defined

We saw in Example 6 that $AB \neq BA$; that is, the commutative law does not hold for matrix multiplication. However, the associative law $A(BC) = (AB)C$ does hold when the dimensions of A, B, and C permit us to find the necessary products.

Progress Check

Verify that $A(BC) = (AB)C$ for the matrices of A, B, and C of Example 6.

11.2b Matrices and Linear Systems

Matrix multiplication provides a convenient shorthand for writing a linear system. For example, the linear system

$$\begin{aligned} 2x - y - 2z &= 3 \\ 3x + 2y + z &= -1 \\ x + y - 3z &= 14 \end{aligned}$$

can be expressed as

$$AX = B$$

where

$$A = \begin{bmatrix} 2 & -1 & -2 \\ 3 & 2 & 1 \\ 1 & 1 & -3 \end{bmatrix} \qquad X = \begin{bmatrix} x \\ y \\ z \end{bmatrix} \qquad B = \begin{bmatrix} 3 \\ -1 \\ 14 \end{bmatrix}$$

To verify this, form the matrix product AX and then apply the definition of matrix equality to the matrix equation $AX = B$.

EXAMPLE 7 *Matrices and Linear Systems*

Write the linear system $AX = B$ if

$$A = \begin{bmatrix} -2 & 3 \\ 1 & 4 \end{bmatrix} \quad X = \begin{bmatrix} x \\ y \end{bmatrix} \quad B = \begin{bmatrix} 16 \\ -3 \end{bmatrix}$$

SOLUTION

Equating corresponding elements of the matrix equation $AX = B$ yields

$$\begin{aligned} -2x + 3y &= 16 \\ x + 4y &= -3 \end{aligned}$$

■

Exercise Set 11.2

1. For what values of a, b, c, and d are the matrices A and B equal?

$$A = \begin{bmatrix} a & b \\ 6 & -2 \end{bmatrix} \quad B = \begin{bmatrix} 3 & -4 \\ c & d \end{bmatrix}$$

2. For what values of a, b, c, and d are the matrices A and B equal?

$$A = \begin{bmatrix} a+b & 2c \\ a & c-d \end{bmatrix} \quad B = \begin{bmatrix} -1 & 6 \\ 5 & 10 \end{bmatrix}$$

In Exercises 3–18, the following matrices are given:

$$A = \begin{bmatrix} 2 & 3 & 1 \\ -3 & 4 & 1 \end{bmatrix} \quad B = \begin{bmatrix} 2 & -1 \\ 3 & 2 \\ 4 & 1 \end{bmatrix}$$

$$C = \begin{bmatrix} 1 & 2 & 3 \\ 4 & -1 & 2 \\ 3 & 2 & 5 \end{bmatrix} \quad D = \begin{bmatrix} -3 & 2 \\ 4 & 1 \end{bmatrix}$$

$$E = \begin{bmatrix} 1 & -3 & 2 \\ 3 & 2 & 4 \\ 1 & 1 & 2 \end{bmatrix} \quad F = \begin{bmatrix} 1 & 3 \\ -2 & 4 \end{bmatrix}$$

$$G = \begin{bmatrix} -2 & 4 & 2 \\ 1 & 0 & 3 \end{bmatrix}$$

In Exercises 3–18, if possible, compute the indicated matrix.

3. $C + E$
4. $C - E$
5. $2A + 3G$
6. $3G - 4A$
7. $A + F$
8. $2B - D$
9. AB
10. BA
11. $CB + D$
12. $EB - FA$
13. $DF + AB$
14. $AC + 2DG$
15. $DA + EB$
16. $FG + B$
17. $2GE - 3A$
18. $AB + FG$

19. If

$$A = \begin{bmatrix} -2 & 3 \\ 2 & -3 \end{bmatrix} \quad B = \begin{bmatrix} -1 & 3 \\ 2 & 0 \end{bmatrix}$$

$$C = \begin{bmatrix} -4 & -3 \\ 0 & -4 \end{bmatrix}$$

show that $AB = AC$.

20. If

$$A = \begin{bmatrix} 1 & 2 \\ 3 & 2 \end{bmatrix} \quad \text{and} \quad B = \begin{bmatrix} 2 & -1 \\ -3 & 4 \end{bmatrix}$$

show that $AB \neq BA$.

21. If

$$A = \begin{bmatrix} -2 & 3 \\ 2 & -3 \end{bmatrix} \quad \text{and} \quad B = \begin{bmatrix} 3 & 6 \\ 2 & 4 \end{bmatrix}$$

show that

$$AB = \begin{bmatrix} 0 & 0 \\ 0 & 0 \end{bmatrix}$$

22. If

$$A = \begin{bmatrix} 0 & 1 \\ 1 & 0 \end{bmatrix}$$

show that

$$A \cdot A = \begin{bmatrix} 1 & 0 \\ 0 & 1 \end{bmatrix}$$

23. If

$$I = \begin{bmatrix} 1 & 0 & 0 \\ 0 & 1 & 0 \\ 0 & 0 & 1 \end{bmatrix} \quad \text{and} \quad A = \begin{bmatrix} a_{11} & a_{12} & a_{13} \\ a_{21} & a_{22} & a_{23} \\ a_{31} & a_{32} & a_{33} \end{bmatrix}$$

show that $AI = A$ and $IA = A$.

24. Pesticides are sprayed on plants to eliminate harmful insects. However, some of the pesticide is absorbed by the plant, and the pesticide is then absorbed by herbivores (plant-eating animals, such as cows) when they eat the plants that have been sprayed. Suppose that we have three pesticides and four plants, and that the amounts of pesticide absorbed by the different plants are given by the matrix

	Plant 1	Plant 2	Plant 3	Plant 4
Pesticide 1	3	2	4	3
Pesticide 2	6	5	2	4
Pesticide 3	4	3	1	5

$$A = \begin{bmatrix} 3 & 2 & 4 & 3 \\ 6 & 5 & 2 & 4 \\ 4 & 3 & 1 & 5 \end{bmatrix}$$

where a_{ij} denotes the amount of pesticide i in milligrams that has been absorbed by plant j. Thus, plant 4 has absorbed 5 milligrams of pesticide 3. Now suppose that we have three herbivores and that the numbers of plants eaten by these animals are given by the matrix

	Herbivore 1	Herbivore 2	Herbivore 3
Plant 1	18	30	20
Plant 2	12	15	10
Plant 3	16	12	8
Plant 4	6	4	12

$$B = \begin{bmatrix} 18 & 30 & 20 \\ 12 & 15 & 10 \\ 16 & 12 & 8 \\ 6 & 4 & 12 \end{bmatrix}$$

How much of pesticide 2 has been absorbed by herbivore 3?

25. What does entry a_{23} in the matrix product AB of Exercise 24 represent?

In Exercises 26 29, find the matrices A, X, and B so that the matrix equation $AX = B$ is equivalent to the given linear system.

26. $\begin{aligned} 7x - 2y &= 6 \\ -2x + 3y &= -2 \end{aligned}$

27. $\begin{aligned} 3x + 4y &= -3 \\ 3x - y &= 5 \end{aligned}$

28. $\begin{aligned} 5x + 2y - 3z &= 4 \\ 2x - \tfrac{1}{2}y + z &= 10 \\ x + y - 5z &= -3 \end{aligned}$

29. $\begin{aligned} 3x - y + 4z &= 5 \\ 2x + 2y + \tfrac{3}{4}z &= -1 \\ x - \tfrac{1}{4}y + z &= \tfrac{1}{2} \end{aligned}$

In Exercises 30–33, write the linear system that is represented by the matrix equation $AX = B$.

30. $A = \begin{bmatrix} 2 & -1 \\ -3 & 4 \end{bmatrix}$ $X = \begin{bmatrix} x \\ y \end{bmatrix}$ $B = \begin{bmatrix} -2 \\ 10 \end{bmatrix}$

31. $A = \begin{bmatrix} 1 & -5 \\ 4 & 3 \end{bmatrix}$ $X = \begin{bmatrix} x_1 \\ x_2 \end{bmatrix}$ $B = \begin{bmatrix} 0 \\ 2 \end{bmatrix}$

32. $A = \begin{bmatrix} 1 & 7 & -2 \\ 3 & 6 & 1 \\ -4 & 2 & 0 \end{bmatrix}$ $X = \begin{bmatrix} x \\ y \\ z \end{bmatrix}$ $B = \begin{bmatrix} 3 \\ -3 \\ 2 \end{bmatrix}$

33. $A = \begin{bmatrix} 4 & 5 & -2 \\ 0 & 3 & -1 \\ 0 & 0 & 2 \end{bmatrix}$ $X = \begin{bmatrix} x_1 \\ x_2 \\ x_3 \end{bmatrix}$ $B = \begin{bmatrix} 2 \\ -5 \\ 4 \end{bmatrix}$

34. The $m \times n$ matrix all of whose elements are zero is called the **zero matrix** and is denoted by 0. Show that $A + 0 = A$ for every $m \times n$ matrix A.

35. The square matrix of order n, such that $a_{ij} = 1$ and $a_{ij} = 0$ when $i \neq j$, is called the **identity matrix** of order n and is denoted by I_n. (*Note*: The definition indicates that the diagonal elements are all equal to 1 and all elements of the diagonal are 0.) Show that $AI_n = I_nA$ for every square matrix A of order n.

36. The matrix B, each of whose entries is the negative of the corresponding entry of matrix A, is called the **additive inverse** of the matrix A. Show that $A + B = 0$ where 0 is the zero matrix. (See Exercise 34.)

37. A square black-and-white digital image with 9 pixels may be represented as a matrix, like matrix A in Exercise 23. Suppose the image has 4 bits per pixel. Each bit has a value of 0 or 1. Each entry in A must be an integer between 0 (darkest black) to $2^4 - 1$, or 15 (whitest white). (*Note*: The integers from 0 to 15 represent 16 possible values.)

Suppose $A = \begin{bmatrix} 0 & 4 & 6 \\ 5 & 0 & 1 \\ 7 & 2 & 3 \end{bmatrix}$

The *contrast* is increased by multiplying each entry by a scaling factor. Find the matrix $2A$, representing an image with increased contrast.

38. The *digital negative image* of an image is found by subtracting each element of the image matrix from its maximum possible value. The *i,j* entry of the matrix N for the digital negative of A in Exercise 37 is

$$n_{ij} = 15 - a_{ij}$$

Find the matrix N.

39. We can add one image to another and represent the resulting image by the matrix sum of the image matrix for each. Find the matrix for the image that results from adding the image represented by A to its negative N. Describe the image qualitatively. What would it look like?

11.3 Inverses of Matrices

If $a \neq 0$, then the linear equation $ax = b$ can be solved by multiplying both sides by the reciprocal of a. Thus, we obtain $x = \left(\frac{1}{a}\right)b$. It would be nice if we could multiply both sides of the matrix equation $AX = B$ by the "reciprocal of A." Unfortunately, a matrix has *no* reciprocal. However, we shall discuss a notion that, for a square matrix, provides an analogue of the reciprocal of a real number and will enable us to solve the linear system in a manner distinct from the Gauss-Jordan Method discussed earlier in this chapter.

In this section we confine our attention to square matrices. The $n \times n$ matrix

$$I_n = \begin{bmatrix} 1 & 0 & 0 & \cdots & 0 \\ 0 & 1 & 0 & \cdots & 0 \\ \cdot & \cdot & \cdot & \cdots & \cdot \\ \cdot & \cdot & \cdot & \cdots & \cdot \\ \cdot & \cdot & \cdot & \cdots & \cdot \\ 0 & 0 & 0 & \cdots & 1 \end{bmatrix}$$

that has 1 for each entry on the main diagonal and 0 elsewhere is called the **identity matrix**. Examples of identity matrices follow:

$$I_2 = \begin{bmatrix} 1 & 0 \\ 0 & 1 \end{bmatrix} \quad I_3 = \begin{bmatrix} 1 & 0 & 0 \\ 0 & 1 & 0 \\ 0 & 0 & 1 \end{bmatrix} \quad I_4 = \begin{bmatrix} 1 & 0 & 0 & 0 \\ 0 & 1 & 0 & 0 \\ 0 & 0 & 1 & 0 \\ 0 & 0 & 0 & 1 \end{bmatrix}$$

If A is any $n \times n$ matrix, we can show that

$$AI_n = I_nA = A$$

(See Exercise 35, Section 11.2.) Thus, I_n is the matrix analogue of the real number 1.

An $n \times n$ matrix A is called **invertible**, or **nonsingular**, if we can find an $n \times n$ matrix B such that

$$AB = BA = I_n$$

The matrix B is called an **inverse** of A.

EXAMPLE 1 ***Verifying Inverses***

Show that A and B are inverses of one another where

$$A = \begin{bmatrix} 2 & 1 \\ 3 & 2 \end{bmatrix} \quad \text{and} \quad B = \begin{bmatrix} 2 & -1 \\ -3 & 2 \end{bmatrix}$$

SOLUTION

Since

$$AB = BA = \begin{bmatrix} 1 & 0 \\ 0 & 1 \end{bmatrix}$$

we conclude that A is an invertible matrix and that B is an inverse of A. (Verify the above equation.) Note that if B is an inverse of A, then A is an inverse of B. ■

It can be shown that if an $n \times n$ matrix A has an inverse, it can have only one inverse. We denote the inverse of A by A^{-1}. Thus, we have

$$AA^{-1} = I_n \quad \text{and} \quad A^{-1}A = I_n$$

Note that the products AA^{-1} and $A^{-1}A$ yield the *identity matrix* I_n, whereas the products $aa^{-1} = a\left(\frac{1}{a}\right)$ and $a^{-1}a = \left(\frac{1}{a}\right)a$ yield the *identity element* 1 for any real number $a \neq 0$. For this reason, A^{-1} may be thought of as the *matrix analogue* of the reciprocal $\frac{1}{a}$.

Verify that the matrices

$$A = \begin{bmatrix} 4 & 5 \\ 5 & 2 \end{bmatrix} \quad \text{and} \quad B = \begin{bmatrix} -1 & \frac{5}{2} \\ 1 & -2 \end{bmatrix}$$

are inverses of each other.

If $a \neq 0$ is a real number, then a^{-1} has the property that $aa^{-1} = a^{-1}a = 1$. Since $a^{-1} = \frac{1}{a}$, we may refer to a^{-1} as the inverse, *or* reciprocal, of a. Although the matrix A^{-1} is the inverse of the $n \times n$ matrix A, since $AA^{-1} = A^{-1}A = I_n$, it cannot be referred to as the reciprocal of A, since *matrix division is not defined.*

We now develop a practical method for finding the inverse of an invertible matrix. Suppose we want to find the inverse of the matrix

$$A = \begin{bmatrix} 1 & 3 \\ 2 & 5 \end{bmatrix}$$

Let the inverse be denoted by

$$B = \begin{bmatrix} b_1 & b_2 \\ b_3 & b_4 \end{bmatrix}$$

Then we must have

$$AB = I_2 \tag{1}$$

and

$$BA = I_2 \tag{2}$$

Equation (1) now becomes

$$\begin{bmatrix} 1 & 3 \\ 2 & 5 \end{bmatrix} \begin{bmatrix} b_1 & b_2 \\ b_3 & b_4 \end{bmatrix} = \begin{bmatrix} 1 & 0 \\ 0 & 1 \end{bmatrix}$$

or

$$\begin{bmatrix} b_1 + 3b_3 & b_2 + 3b_4 \\ 2b_1 + 5b_3 & 2b_2 + 5b_4 \end{bmatrix} = \begin{bmatrix} 1 & 0 \\ 0 & 1 \end{bmatrix}$$

Since two matrices are equal if, and only if, their corresponding entries are equal, we have

$$\begin{aligned} b_1 + 3b_3 &= 1 \\ 2b_1 + 5b_3 &= 0 \end{aligned} \tag{3}$$

and

$$\begin{aligned} b_2 + 3b_4 &= 0 \\ 2b_2 + 5b_4 &= 1 \end{aligned} \tag{4}$$

We solve the linear systems (3) and (4) by Gauss-Jordan Elimination. We begin with the augmented matrices of the linear systems and perform a sequence of elementary row operations as follows:

	(3)	(4)
Write the augmented matrices of (3) and (4).	$\left[\begin{array}{cc\|c} 1 & 3 & 1 \\ 2 & 5 & 0 \end{array}\right]$	$\left[\begin{array}{cc\|c} 1 & 3 & 0 \\ 2 & 5 & 1 \end{array}\right]$
To make $a_{21} = 0$, replace row 2 with the sum of itself and -2 times row 1.	$\left[\begin{array}{cc\|c} 1 & 3 & 1 \\ 0 & -1 & -2 \end{array}\right]$	$\left[\begin{array}{cc\|c} 1 & 3 & 0 \\ 0 & -1 & 1 \end{array}\right]$
Multiply row 2 by -1 to obtain $a_{22} = 1$.	$\left[\begin{array}{cc\|c} 1 & 3 & 1 \\ 0 & 1 & 2 \end{array}\right]$	$\left[\begin{array}{cc\|c} 1 & 3 & 0 \\ 0 & 1 & -1 \end{array}\right]$
To make $a_{12} = 0$, replace row 1 with the sum of itself and -3 times row 2.	$\left[\begin{array}{cc\|c} 1 & 0 & -5 \\ 0 & 1 & 2 \end{array}\right]$	$\left[\begin{array}{cc\|c} 1 & 0 & 3 \\ 0 & 1 & -1 \end{array}\right]$

Thus, $b_1 = -5$ and $b_3 = 2$ is the solution of (3), and $b_2 = 3$ and $b_4 = -1$ is the solution of (4). Check that

$$B = \begin{bmatrix} -5 & 3 \\ 2 & -1 \end{bmatrix}$$

also satisfies the requirement $BA = I_2$ of Equation (2).

Observe that the linear systems (3) and (4) have the same coefficient matrix (which is also the same as the original matrix A) and that an identical sequence of elementary row operations was performed in the Gauss-Jordan Elimination. This suggests that we can solve the systems at *the same time.* We write the coefficient matrix A and next to it list the right-hand sides of (3) and (4) to obtain the matrix

$$\left[\begin{array}{cc|cc} 1 & 3 & 1 & 0 \\ 2 & 5 & 0 & 1 \end{array}\right] \tag{5}$$

Note that the columns to the right of the dashed line in (5) form the identity matrix I_2. Performing the same sequence of elementary row operations on matrix (5) as we did on matrices (3) and (4) yields

$$\left[\begin{array}{cc|cc} 1 & 0 & -5 & 3 \\ 0 & 1 & 2 & -1 \end{array}\right] \tag{6}$$

Then A^{-1} is the matrix to the right of the dashed line in (6).

The procedure outlined for the 2×2 matrix A applies in general. Thus, we have the following method for finding the inverse of an invertible $n \times n$ matrix A.

Computing A^{-1}

Step 1. Form the $n \times 2n$ matrix $[A \mid I_n]$ by adjoining the identity matrix I_n to the given matrix A.

Step 2. Apply elementary row operations to the matrix $[A \mid I_n]$ to transform the matrix A to I_n.

Step 3. The final matrix is of the form $[I_n \mid B]$ where B is A^{-1}.

EXAMPLE 2 *Computing Inverses*

Find the inverse of

$$A = \begin{bmatrix} 1 & 2 & 3 \\ 2 & 5 & 7 \\ 1 & 1 & 1 \end{bmatrix}$$

SOLUTION

We form the 3×6 matrix $[A \mid I_3]$ and transform it by elementary row operations to the form $[I_3 \mid A^{-1}]$. The pivot element at each stage is shown in **blue**.

Write matrix A augmented by I_3.

$$\left[\begin{array}{ccc|ccc} \mathbf{1} & 2 & 3 & 1 & 0 & 0 \\ 2 & 5 & 7 & 0 & 1 & 0 \\ 1 & 1 & 1 & 0 & 0 & 1 \end{array}\right]$$

To make $a_{21} = 0$, replace row 2 with the sum of itself and -2 times row 1. To make $a_{31} = 0$, replace row 3 by the sum of itself and -1 times row 1.

$$\left[\begin{array}{ccc|ccc} 1 & 2 & 3 & 1 & 0 & 0 \\ 0 & \mathbf{1} & 1 & -2 & 1 & 0 \\ 0 & -1 & -2 & -1 & 0 & 1 \end{array}\right]$$

To make $a_{12} = 0$, replace row 1 with the sum of itself and -2 times row 2. To make $a_{32} = 0$, replace row 3 with the sum of itself and row 2.

$$\left[\begin{array}{ccc|ccc} 1 & 0 & 1 & 5 & -2 & 0 \\ 0 & 1 & 1 & -2 & 1 & 0 \\ 0 & 0 & -1 & -3 & 1 & 1 \end{array}\right]$$

Multiply row 3 by -1.

$$\left[\begin{array}{ccc|ccc} 1 & 0 & 1 & 5 & -2 & 0 \\ 0 & 1 & 1 & -2 & 1 & 0 \\ 0 & 0 & \mathbf{1} & 3 & -1 & -1 \end{array}\right]$$

To make $a_{13} = 0$, replace row 1 with the sum of itself and -1 times row 3. To make $a_{23} = 0$, replace row 2 with the sum of itself and -1 times row 3.

$$\left[\begin{array}{ccc|ccc} 1 & 0 & 0 & 2 & -1 & 1 \\ 0 & 1 & 0 & -5 & 2 & 1 \\ 0 & 0 & 1 & 3 & -1 & -1 \end{array}\right]$$

The final matrix is of the form $[I_3 \vdots A^{-1}]$ that is,

$$A^{-1} = \begin{bmatrix} 2 & -1 & 1 \\ -5 & 2 & 1 \\ 3 & -1 & -1 \end{bmatrix}$$

■

We now have a practical method for finding the inverse of an invertible matrix, but we do not know whether a given square matrix *has* an inverse. It can be shown that if the preceding procedure is carried out with the matrix $[A \mid I_n]$ and we arrive at a point at which all possible candidates for the next pivot element are zero, then the matrix is not invertible; and we may stop our calculations.

EXAMPLE 3 ***Computing Inverses***

Find the inverse of

$$A = \begin{bmatrix} 1 & 2 & 6 \\ 0 & 0 & 2 \\ -3 & -6 & -9 \end{bmatrix}$$

SOLUTION

We begin with $[A \mid I_3]$.

$$\left[\begin{array}{ccc|ccc} 1 & 2 & 6 & 1 & 0 & 0 \\ 0 & 0 & 2 & 0 & 1 & 0 \\ -3 & -6 & -9 & 0 & 0 & 1 \end{array}\right]$$

To make $a_{31} = 0$, replace row 3 by the sum of itself and 3 times row 1.

$$\left[\begin{array}{ccc|ccc} 1 & 2 & 6 & 1 & 0 & 0 \\ 0 & 0 & 2 & 0 & 1 & 0 \\ 0 & 0 & 9 & 3 & 0 & 1 \end{array}\right]$$

Note that $a_{22} = a_{32} = 0$ in the last matrix. We cannot perform any elementary row operations upon rows 2 and 3 that will produce a nonzero pivot element for a_{22}. We conclude that the matrix A does not have an inverse. ■

Progress Check

Show that the matrix A is not invertible.

$$A = \begin{bmatrix} 1 & 2 & -3 \\ 3 & 2 & 1 \\ 5 & 6 & -5 \end{bmatrix}$$

11.3a *Solving Linear Systems*

Consider a linear system of n equations in n unknowns.

$$\begin{aligned} a_{11}x_1 + a_{12}x_2 + \cdots + a_{1n}x_n &= b_1 \\ a_{21}x_1 + a_{22}x_2 + \cdots + a_{2n}x_n &= b_2 \\ \vdots \qquad \vdots \qquad\quad \vdots \quad &\quad \vdots \\ a_{n1}x_1 + a_{n2}x_2 + \cdots + a_{nn}x_n &= b_n \end{aligned} \tag{7}$$

As has already been pointed out in Section 11.2 of this chapter, we can write the linear system (7) in matrix form as

$$AX = B \tag{8}$$

where

$$A = \begin{bmatrix} a_{11} & a_{12} & \cdots & a_{1n} \\ a_{21} & a_{22} & \cdots & a_{2n} \\ \vdots & \vdots & & \vdots \\ a_{n1} & a_{n2} & \cdots & a_{nn} \end{bmatrix} \qquad X = \begin{bmatrix} x_1 \\ x_2 \\ \vdots \\ x_n \end{bmatrix} \qquad B = \begin{bmatrix} b_1 \\ b_2 \\ \vdots \\ b_n \end{bmatrix}$$

Suppose now that the coefficient matrix A is invertible so that we can compute A^{-1}. Multiplying both sides of (8) by A^{-1}, we have

$$\begin{aligned} A^{-1}(AX) &= A^{-1}B \\ (A^{-1}A)X &= A^{-1}B && \text{Associative law} \\ I_nX &= A^{-1}B && A^{-1}A = I_n \\ X &= A^{-1}B && I_nX = X \end{aligned}$$

Thus, we have the following result:

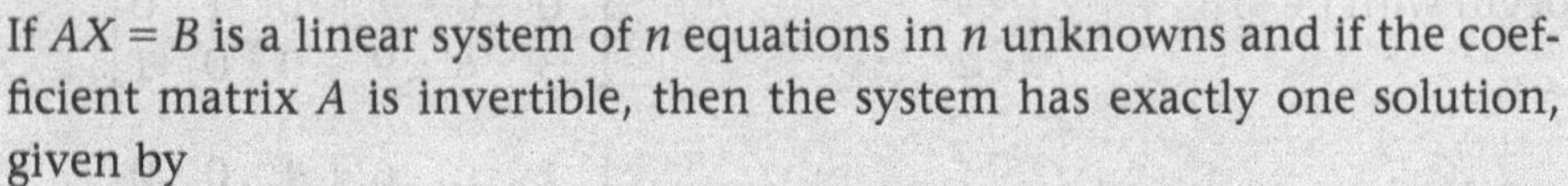

If $AX = B$ is a linear system of n equations in n unknowns and if the coefficient matrix A is invertible, then the system has exactly one solution, given by

$$X = A^{-1}B$$

Since matrix multiplication is not commutative, be careful to write the solution to the system $AX = B$ as $X = A^{-1}B$ and *not* $X = BA^{-1}$.

EXAMPLE 4 ***Solving a System of Linear Equations Using Inverses***

Solve the linear system by finding the inverse of the coefficient matrix.

$$\begin{aligned} x + 2y + 3z &= -3 \\ 2x + 5y + 7z &= 4 \\ x + y + z &= 5 \end{aligned}$$

SOLUTION

The coefficient matrix

$$A = \begin{bmatrix} 1 & 2 & 3 \\ 2 & 5 & 7 \\ 1 & 1 & 1 \end{bmatrix}$$

is the matrix whose inverse was obtained in Example 2 as

$$A^{-1} = \begin{bmatrix} 2 & -1 & 1 \\ -5 & 2 & 1 \\ 3 & -1 & -1 \end{bmatrix}$$

Since

$$B = \begin{bmatrix} -3 \\ 4 \\ 5 \end{bmatrix}$$

we obtain the solution of the given system as

$$X = A^{-1}B = \begin{bmatrix} 2 & -1 & 1 \\ -5 & 2 & 1 \\ 3 & -1 & -1 \end{bmatrix} \begin{bmatrix} -3 \\ 4 \\ 5 \end{bmatrix} = \begin{bmatrix} -5 \\ 28 \\ -18 \end{bmatrix}$$

Thus, $x = -5$, $y = 28$, $z = -18$. ■

Solve the linear system by finding the inverse of the coefficient matrix.

$$\begin{aligned} x - 2y + z &= 1 \\ x + 3y + 2z &= 2 \\ -x + z &= -11 \end{aligned}$$

Answer

$x = 7, y = 1, z = -4$

The inverse of the coefficient matrix is especially useful when we need to solve a number of linear systems

$$AX = B_1, AX = B_2, \ldots, AX = B_k$$

where the coefficient matrix is the same, and the right-hand side changes.

EXAMPLE 5 ***Solving a System of Linear Equations Using Inverses***

A steel producer makes two types of steel, regular and special. A ton of regular steel requires 2 hours in the open-hearth furnace and 5 hours in the soaking pit; a ton of special steel requires 2 hours in the open-hearth furnace and 3 hours in the soaking pit. How many tons of each type of steel can be manufactured daily if

a. the open-hearth furnace is available 8 hours per day and the soaking pit is available 15 hours per day?

b. the open-hearth furnace is available 9 hours per day and the soaking pit is available 15 hours per day?

SOLUTION

Let

x = the number of tons of regular steel to be made
y = the number of tons of special steel to be made

Then the total amount of time required in the open-hearth furnace is

$$2x + 2y$$

Similarly, the total amount of time required in the soaking pit is

$$5x + 3y$$

If we let b_1 and b_2 denote the number of hours that the open-hearth furnace and the soaking pit are available per day, respectively, then we have

$$\begin{aligned} 2x + 2y &= b_1 \\ 5x + 3y &= b_2 \end{aligned}$$

or

$$\begin{bmatrix} 2 & 2 \\ 5 & 3 \end{bmatrix} \begin{bmatrix} x \\ y \end{bmatrix} = \begin{bmatrix} b_1 \\ b_2 \end{bmatrix}$$

Then

$$\begin{bmatrix} x \\ y \end{bmatrix} = \begin{bmatrix} 2 & 2 \\ 5 & 3 \end{bmatrix}^{-1} \begin{bmatrix} b_1 \\ b_2 \end{bmatrix}$$

Verify that the inverse of the coefficient matrix is

$$\begin{bmatrix} 2 & 2 \\ 5 & 3 \end{bmatrix}^{-1} = \begin{bmatrix} -\frac{3}{4} & \frac{1}{2} \\ \frac{5}{4} & -\frac{1}{2} \end{bmatrix}$$

a. If $b_1 = 8$ and $b_2 = 15$, then

$$\begin{bmatrix} x \\ y \end{bmatrix} = \begin{bmatrix} -\frac{3}{4} & \frac{1}{2} \\ \frac{5}{4} & -\frac{1}{2} \end{bmatrix} \begin{bmatrix} 8 \\ 15 \end{bmatrix} = \begin{bmatrix} \frac{3}{2} \\ \frac{5}{2} \end{bmatrix}$$

That is, $\frac{3}{2}$ tons of regular steel and $\frac{5}{2}$ tons of special steel can be manufactured daily.

b. If $b_1 = 9$ and $b_2 = 15$, then

$$\begin{bmatrix} x \\ y \end{bmatrix} = \begin{bmatrix} -\frac{3}{4} & \frac{1}{2} \\ \frac{5}{4} & -\frac{1}{2} \end{bmatrix} \begin{bmatrix} 9 \\ 15 \end{bmatrix} = \begin{bmatrix} \frac{3}{4} \\ \frac{15}{4} \end{bmatrix}$$

That is, $\frac{3}{4}$ tons of regular steel and $\frac{15}{4}$ tons of special steel can be manufactured daily. ■

FOCUS on Coded Messages

Cryptography is the study of methods for encoding and decoding messages. One of the very simplest techniques for doing this involves the use of the inverse of a matrix.

A	B	C	D	E	F	G
↕	↕	↕	↕	↕	↕	↕
1	2	3	4	5	6	7
H	I	J	K	L	M	N
↕	↕	↕	↕	↕	↕	↕
8	9	10	11	12	13	14
O	P	Q	R	S	T	U
↕	↕	↕	↕	↕	↕	↕
15	16	17	18	19	20	21
V	W	X	Y	Z		
↕	↕	↕	↕	↕		
22	23	24	25	26		

Suppose that Leslie and Ronnie are drug enforcement agents in the New York City police department and that Leslie has infiltrated a major drug operation. To avoid detection, the agents communicate with each other by using coded messages. First, they agree to attach a different number to every letter of the alphabet. For example, they let A be 1, B be 2, and so on, as shown in the accompanying table. Suppose that on Thursday, Ronnie wants to send Leslie the message

STRIKE MONDAY

to indicate that the police will raid the drug operation on the following Monday. Substituting for each letter, Ronnie sends the message

$$19, 20, 18, 9, 11, 5, 13, 15, 14, 4, 1, 25 \qquad (1)$$

Unfortunately, this simple code can be easily cracked by analyzing the frequency of letters in the English alphabet. A much better method involves the use of matrices.

First, Ronnie breaks the message (1) into four 3×1 matrices

$$X_1 = \begin{bmatrix} 19 \\ 20 \\ 18 \end{bmatrix} \qquad X_2 = \begin{bmatrix} 9 \\ 11 \\ 5 \end{bmatrix} \qquad X_3 = \begin{bmatrix} 13 \\ 15 \\ 14 \end{bmatrix} \qquad X_4 = \begin{bmatrix} 4 \\ 1 \\ 25 \end{bmatrix}$$

Sometime ago, Ronnie and Leslie had jointly selected an invertible 3×3 matrix such as

$$A = \begin{bmatrix} 1 & 1 & 2 \\ 1 & 1 & 1 \\ 1 & 0 & 1 \end{bmatrix}$$

which no one else knows. Ronnie now forms the 3×1 matrices

$$AX_1 = \begin{bmatrix} 75 \\ 57 \\ 37 \end{bmatrix} \qquad AX_2 = \begin{bmatrix} 30 \\ 25 \\ 14 \end{bmatrix} \qquad AX_3 = \begin{bmatrix} 56 \\ 42 \\ 27 \end{bmatrix} \qquad AX_4 = \begin{bmatrix} 55 \\ 30 \\ 29 \end{bmatrix}$$

and sends the message

$$75, 57, 37, 30, 25, 14, 56, 42, 27, 55, 30, 29 \qquad (2)$$

To decode the message, Leslie uses the inverse of matrix A,

$$A^{-1} = \begin{bmatrix} -1 & 1 & 1 \\ 0 & 1 & -1 \\ 1 & -1 & 0 \end{bmatrix}$$

and forms

$$A^{-1}\begin{bmatrix} 75 \\ 57 \\ 37 \end{bmatrix} = X_1 \quad A^{-1}\begin{bmatrix} 30 \\ 25 \\ 14 \end{bmatrix} = X_2 \quad A^{-1}\begin{bmatrix} 56 \\ 42 \\ 27 \end{bmatrix} = X_3 \quad A^{-1}\begin{bmatrix} 55 \\ 30 \\ 29 \end{bmatrix} = X_4$$

which, of course, is the original message (1) and which can be understood by using the accompanying table.

If Leslie responds with the message

$$33, 21, 16, 52, 39, 14, 66, 47, 28, 52, 38, 23$$

what is Ronnie being told?

Exercise Set 11.3

In Exercises 1–4, determine whether the matrix B is the inverse of the matrix A.

1. $A = \begin{bmatrix} 2 & \frac{1}{2} \\ -1 & 3 \end{bmatrix}$ $\quad B = \begin{bmatrix} 1 & -1 \\ 2 & 4 \end{bmatrix}$

2. $A = \begin{bmatrix} 3 & -1 \\ -2 & 2 \end{bmatrix}$ $\quad B = \begin{bmatrix} \frac{1}{2} & \frac{1}{4} \\ \frac{1}{2} & \frac{3}{4} \end{bmatrix}$

3. $A = \begin{bmatrix} 1 & 2 & 2 \\ -1 & 3 & 0 \\ 0 & 2 & 1 \end{bmatrix}$ $\quad B = \begin{bmatrix} 3 & 2 & -6 \\ 1 & 1 & -2 \\ -2 & -2 & 5 \end{bmatrix}$

4. $A = \begin{bmatrix} 1 & 0 & -2 \\ 2 & 1 & 3 \\ -4 & 1 & 2 \end{bmatrix}$ $\quad B = \begin{bmatrix} 1 & 2 & -2 \\ -2 & -4 & 1 \\ 0 & 1 & -1 \end{bmatrix}$

In Exercises 5–10, find the inverse of the given matrix.

5. $\begin{bmatrix} -1 & 5 \\ 2 & -4 \end{bmatrix}$

6. $\begin{bmatrix} 2 & 0 \\ -1 & 2 \end{bmatrix}$

7. $\begin{bmatrix} -1 & 1 \\ -2 & 1 \end{bmatrix}$

8. $\begin{bmatrix} 2 & 1 & 0 \\ 1 & 1 & 0 \\ 1 & 1 & 1 \end{bmatrix}$

9. $\begin{bmatrix} 1 & -2 & 3 \\ -1 & 3 & -4 \\ 0 & 5 & -4 \end{bmatrix}$

10. $\begin{bmatrix} 1 & 1 & 0 \\ 1 & 0 & 0 \\ 1 & 2 & 2 \end{bmatrix}$

In Exercises 11–18, find the inverse, if possible.

11. $\begin{bmatrix} 1 & 3 \\ -1 & 4 \end{bmatrix}$

12. $\begin{bmatrix} 6 & -4 \\ 9 & -6 \end{bmatrix}$

13. $\begin{bmatrix} 1 & 1 & 3 \\ 2 & -8 & -4 \\ -1 & 2 & 0 \end{bmatrix}$

14. $\begin{bmatrix} 8 & 7 & -1 \\ -5 & -5 & 1 \\ -4 & -4 & 1 \end{bmatrix}$

15. $\begin{bmatrix} 2 & 0 \\ 0 & -3 \end{bmatrix}$

16. $\begin{bmatrix} -1 & 0 & 0 \\ 0 & 4 & 0 \\ 0 & 0 & 2 \end{bmatrix}$

17. $\begin{bmatrix} 1 & 0 & -1 \\ 2 & 1 & 0 \\ 0 & 1 & 1 \end{bmatrix}$

18. $\begin{bmatrix} 1 & 0 & -3 & 0 \\ 0 & 1 & 0 & 0 \\ -1 & 0 & 4 & 0 \\ 2 & 0 & -6 & 1 \end{bmatrix}$

In Exercises 19–24, solve the given linear system by finding the inverse of the coefficient matrix.

19. $\begin{aligned} 2x + y &= 5 \\ x - 3y &= 6 \end{aligned}$

20. $\begin{aligned} 2x - 3y &= -5 \\ 3x + y &= -13 \end{aligned}$

21. $\begin{aligned} 3x + y - z &= 2 \\ x - 2y &= 8 \\ 3y + z &= -8 \end{aligned}$

22. $\begin{aligned} 3x + y - z &= 10 \\ 2x - y + z &= -1 \\ -x + y - 2z &= 5 \end{aligned}$

23. $\begin{aligned} 2x - y + 3z &= -11 \\ 3x - y + z &= -5 \\ x + y + z &= -1 \end{aligned}$

24. $\begin{aligned} 2x + 3y - 2z &= 13 \\ 4x + 2y + z &= 3 \\ y - z &= 5 \end{aligned}$

In Exercises 25–34, solve the linear systems in Exercises 21–30, Exercise Set 11.1, by finding the inverse of the coefficient matrix.

35. Solve the linear systems $AX = B_1$ and $AX = B_2$, given

$$A^{-1} = \begin{bmatrix} 3 & -2 & 4 \\ 2 & -1 & 0 \\ 0 & 4 & 1 \end{bmatrix}$$

$$B_1 = \begin{bmatrix} 1 \\ -1 \\ 5 \end{bmatrix} \quad B_2 = \begin{bmatrix} 4 \\ 3 \\ -2 \end{bmatrix}$$

36. Solve the linear systems $AX = B_1$ and $AX = B_2$, given

$$A^{-1} = \begin{bmatrix} 1 & 0 & -1 \\ 1 & 2 & 0 \\ -1 & -1 & 3 \end{bmatrix}$$

$$B_1 = \begin{bmatrix} 2 \\ -3 \\ 2 \end{bmatrix} \quad B_2 = \begin{bmatrix} 4 \\ -3 \\ -5 \end{bmatrix}$$

37. Show that the matrix

$$\begin{bmatrix} a & b & c \\ 0 & 0 & 0 \\ d & e & f \end{bmatrix}$$

is not invertible.

38. A trustee decides to invest \$500,000 in two mortgages, which yield 4% and 8% per year, respectively. How should the \$500,000 be invested in the two mortgages if the total annual interest is to be

a. \$30,000? b. \$40,000? c. \$50,000?

(*Hint*: Some of these investment objectives cannot be attained.)

39. Many graphing calculators can find the inverse of a matrix, just by entering the name of the matrix you have stored and then hitting the inverse key. The display looks like this:

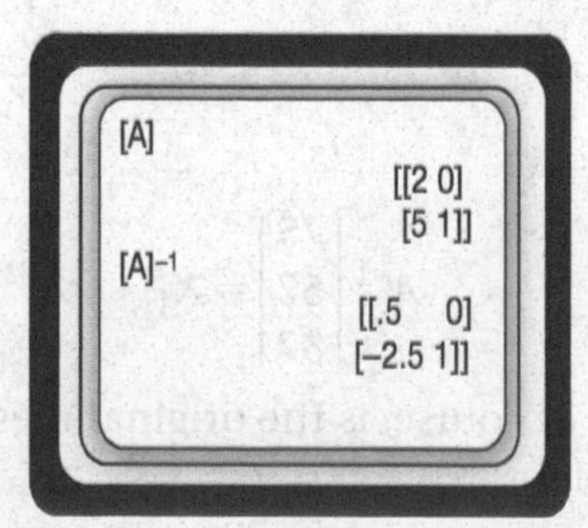

Use this method to find the inverse of the coefficient matrix for the system of equations in Example 5 in Section 11.1.

40. Now use the inverse you found in Exercise 39 to solve the system, verifying the solution given at the end of the example.

41. *Mathematics in Writing*: Explain in your own words how the inverse of a matrix is used to solve a system of equations. How is this process similar to the method for solving a linear equation in one unknown, discussed in Section 2.1? How is it different?

11.4 Determinants

In this section, we will define a determinant and develop manipulative skills for evaluating determinants. We will then show that determinants have important applications and can be used to solve linear systems.

Associated with every square matrix A is a number called the **determinant** of A, denoted by $|A|$. If A is 1×1, that is, if $A = [a_{11}]$, then we define $|A| = a_{11}$. If A is the 2×2 matrix

$$A = \begin{bmatrix} a_{11} & a_{12} \\ a_{21} & a_{22} \end{bmatrix}$$

then $|A|$ is said to be a *determinant of second order* and is defined by the rule

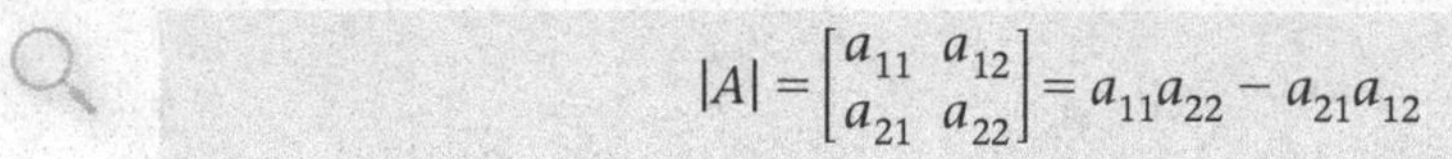

$$|A| = \begin{bmatrix} a_{11} & a_{12} \\ a_{21} & a_{22} \end{bmatrix} = a_{11}a_{22} - a_{21}a_{12}$$

EXAMPLE 1 ***Determinant of Second Order***

Compute the real number represented by

$$\begin{vmatrix} 4 & -5 \\ 3 & -1 \end{vmatrix}$$

SOLUTION

We apply the rule for a determinant of second order.

$$\begin{vmatrix} 4 & -5 \\ 3 & -1 \end{vmatrix} = (4)(-1) - (3)(-5) = 11$$ ■

Progress Check

Compute the real number represented by

a. $\begin{vmatrix} -6 & 2 \\ -1 & -2 \end{vmatrix}$ b. $\begin{vmatrix} \frac{1}{2} & \frac{1}{4} \\ -4 & -2 \end{vmatrix}$

Answers

a. 14 b. 0

11.4a Minors and Cofactors

Consider the 3 × 3 matrix

$$A = \begin{bmatrix} a_{11} & a_{12} & a_{13} \\ a_{21} & a_{22} & a_{23} \\ a_{31} & a_{32} & a_{33} \end{bmatrix}$$

The **minor** of an element a_{ij} is the determinant of the matrix remaining after deleting the row and column in which the element a_{ij} appears. Given the matrix

$$\begin{bmatrix} 4 & 0 & -2 \\ 1 & -6 & 7 \\ -3 & 2 & 5 \end{bmatrix}$$

the minor of the element in row 2, column 3 is

$$\begin{vmatrix} 4 & 0 & -2 \\ 1 & -6 & 7 \\ -3 & 2 & 5 \end{vmatrix} = \begin{vmatrix} 4 & 0 \\ -3 & 2 \end{vmatrix} = 8 - 0 = 8$$

The **cofactor** of the element a_{ij} is the minor of the element a_{ij} multiplied by $(-1)^{i+j}$. Since $(-1)^{i+j}$ is +1 if $i+j$ is even and −1 if $i+j$ is odd, we see that the cofactor is the minor with a sign attached. The cofactor attaches the sign to the minor according to this pattern:

$$\begin{bmatrix} + & - & + & - & \cdots \\ - & + & - & + & \cdots \\ + & - & + & - & \cdots \\ - & + & - & + & \cdots \\ \vdots & \vdots & \vdots & \vdots & \vdots \end{bmatrix}$$

EXAMPLE 2 *Determining Cofactors*

Find the cofactor of each element in the first row of the matrix.

$$\begin{bmatrix} -2 & 0 & 12 \\ -4 & 5 & 3 \\ 7 & 8 & -6 \end{bmatrix}$$

SOLUTION

The cofactors are

$$(-1)^{1+1}\begin{vmatrix} -2 & 0 & 12 \\ -4 & 5 & 3 \\ 7 & 8 & -6 \end{vmatrix} = \begin{vmatrix} 5 & 3 \\ 8 & -6 \end{vmatrix} = -30 - 24 = -54$$

$$(-1)^{1+2}\begin{vmatrix} -2 & 0 & 12 \\ -4 & 5 & 3 \\ 7 & 8 & -6 \end{vmatrix} = -\begin{vmatrix} -4 & 3 \\ 7 & -6 \end{vmatrix} = -(24 - 21) = -3$$

$$(-1)^{1+3}\begin{vmatrix} -2 & 0 & 12 \\ -4 & 5 & 3 \\ 7 & 8 & -6 \end{vmatrix} = \begin{vmatrix} -4 & 5 \\ 7 & 8 \end{vmatrix} = -32 - 35 = -67$$

■

Find the cofactor of each entry in the second column of the matrix.

$$\begin{bmatrix} 16 & -9 & 3 \\ -5 & 2 & 0 \\ -3 & 4 & -1 \end{bmatrix}$$

Answers

cofactor of −9 is −5; cofactor of 2 is −7; cofactor of 4 is −15

The cofactor is the key to the process of evaluating determinants of any order.

Expansion by Cofactors

To evaluate a determinant, form the sum of the products obtained by multiplying each entry of any row or any column by its cofactor. This process is called **expansion by cofactors.**

Consider the matrix

$$A = \begin{bmatrix} 4 & -5 \\ 3 & -1 \end{bmatrix}$$

and choose the second column. The cofactor of

$$a_{12} = -5 \quad \text{is} \quad (-1)^{1+2}\begin{bmatrix} 4 & -5 \\ 3 & -1 \end{bmatrix} = -3$$

and the cofactor of

$$a_{22} = -1 \quad \text{is} \quad (-1)^{2+2}\begin{bmatrix} 4 & -5 \\ 3 & -1 \end{bmatrix} = 4$$

Therefore

$$|A| = (-5)(-3) + (-1)(4) = 15 - 4 = 11$$

Note that the above is an alternative method for Example 1. In fact, verify the formula given for a determinant of order 2 at the beginning of this section using the method of expansion by cofactors, using any row or any column.

Let us illustrate the process for a 3×3 matrix.

EXAMPLE 3 *Expansion by Cofactors*

Evaluate the determinant of the matrix

$$\begin{bmatrix} -2 & 7 & 2 \\ 6 & -6 & 0 \\ 4 & 10 & -3 \end{bmatrix}$$

using the method of expansion by cofactors.

SOLUTION

Step 1. Choose a row or column about which to expand. (In general, a row or column containing zeros simplifies the work.	*Step 1.* We expand about column 3.
Step 2. Expand about the cofactors of the chosen row or column by multiplying each entry of the row or column by its cofactor. Repeat the procedure until all determinants are of order 2.	*Step 2.* The expansion about column 3 is $(2)(-1)^{1+3}\begin{vmatrix} 6 & -6 \\ 4 & 10 \end{vmatrix}$ $+(0)(-1)^{2+3}\begin{vmatrix} -2 & 7 \\ 4 & 10 \end{vmatrix}$ $+(-3)(-1)^{3+3}\begin{vmatrix} -2 & 7 \\ 6 & -6 \end{vmatrix}$
Step 3. Evaluate the cofactors and form the sum indicated in *Step 2.*	*Step 3.* Using the rule for evaluating a determinant of order 2, we have $(2)(1)[(6)(10)-(4)(-6)]+0+(-3)(1)[(-2)(-6)-(6)(7)]$ $=2(60+24)-3(12-42)$ $=258$

Observe that it was unnecessary for us to calculate the cofactor corresponding to the 0 element in column 3. We only did it here to reinforce the method of finding cofactors. ■

Note that expansion by cofactors of *any row or any column* produces the same result. This property of determinants can be used to simplify the arithmetic. The best choice of a row or column about which to expand is generally the one that has the most zero entries. If an entry is zero, the entry times its cofactor is also zero, so we do not have to evaluate that cofactor.

Progress Check

Find the determinant of the matrix in Example 3 by expanding about the second row.

Answer

258

EXAMPLE 4 *Expansion by Cofactors*

Verify the rule for evaluating the determinant of the matrix of order 3.

$$\begin{vmatrix} a_{11} & a_{12} & a_{13} \\ a_{21} & a_{22} & a_{23} \\ a_{31} & a_{32} & a_{33} \end{vmatrix} = a_{11}a_{22}a_{33} - a_{11}a_{32}a_{23} - a_{12}a_{21}a_{33} + a_{12}a_{31}a_{23} + a_{13}a_{21}a_{32} - a_{13}a_{31}a_{22}$$

SOLUTION

Expanding about the first row, we have

$$\begin{aligned} \begin{vmatrix} a_{11} & a_{12} & a_{13} \\ a_{21} & a_{22} & a_{23} \\ a_{31} & a_{32} & a_{33} \end{vmatrix} &= a_{11}\begin{vmatrix} a_{22} & a_{23} \\ a_{32} & a_{33} \end{vmatrix} - a_{12}\begin{vmatrix} a_{21} & a_{23} \\ a_{31} & a_{33} \end{vmatrix} + a_{13}\begin{vmatrix} a_{21} & a_{22} \\ a_{31} & a_{32} \end{vmatrix} \\ &= a_{11}(a_{22}a_{33} - a_{32}a_{23}) - a_{12}(a_{21}a_{33} - a_{31}a_{23}) + a_{13}(a_{21}a_{32} - a_{31}a_{22}) \\ &= a_{11}a_{22}a_{33} - a_{11}a_{32}a_{23} - a_{12}a_{21}a_{33} + a_{12}a_{31}a_{23} + a_{13}a_{21}a_{32} - a_{13}a_{31}a_{22} \end{aligned}$$

(Verify this answer using any other column or row.) ■

Progress Check

Show that the determinant of the matrix is equal to zero.

$$\begin{vmatrix} a & b & c \\ a & b & c \\ d & e & f \end{vmatrix}$$

The process of expanding by cofactors works for determinants of any order. If we apply the method to a determinant of order 4, we produce determinants of order 3; applying the method again results in determinants of order 2.

EXAMPLE 5 *Expansion by Cofactors*

Evaluate the determinant of the matrix.

$$\begin{vmatrix} -3 & 5 & 0 & -1 \\ 1 & 2 & 3 & -3 \\ 0 & 4 & -6 & 0 \\ 0 & -2 & 1 & 2 \end{vmatrix}$$

SOLUTION

Expanding about the cofactors of the first column, we have

$$\begin{aligned} \begin{vmatrix} -3 & 5 & 0 & -1 \\ 1 & 2 & 3 & -3 \\ 0 & 4 & -6 & 0 \\ 0 & -2 & 1 & 2 \end{vmatrix} &= -3\begin{vmatrix} 2 & 3 & -3 \\ 4 & -6 & 0 \\ -2 & 1 & 2 \end{vmatrix} - 1\begin{vmatrix} 5 & 0 & -1 \\ 4 & -6 & 0 \\ -2 & 1 & 2 \end{vmatrix} \\ &= -3\left[-4\begin{vmatrix} 3 & -3 \\ 1 & 2 \end{vmatrix} - 6\begin{vmatrix} 2 & -3 \\ -2 & 2 \end{vmatrix}\right] - 1\left[-1\begin{vmatrix} 4 & -6 \\ -2 & 1 \end{vmatrix} + 2\begin{vmatrix} 5 & 0 \\ 4 & -6 \end{vmatrix}\right] \\ &= -3[(-4)(9) - 6(-2)] - 1[(-1)(-8) + 2(-30)] \\ &= -3[-36 + 12] - 1[8 - 60] \\ &= -3(-24) - 1(-52) = 124 \end{aligned}$$

■

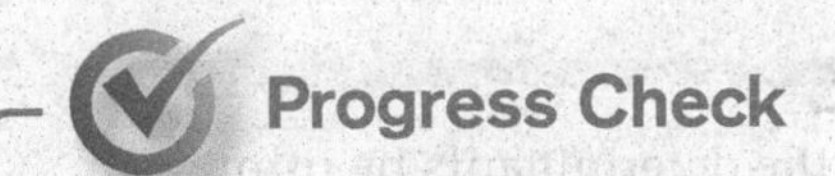

Evaluate the determinant of the matrix.

$$\begin{vmatrix} 0 & -1 & 0 & 2 \\ 3 & 0 & 4 & 0 \\ 0 & 5 & 0 & -3 \\ 1 & 0 & 1 & 0 \end{vmatrix}$$

Answer

7

Exercise Set 11.4

In Exercises 1–6, evaluate the determinant of the given matrix.

1. $\begin{vmatrix} 2 & -3 \\ 4 & 5 \end{vmatrix}$ 2. $\begin{vmatrix} 3 & 4 \\ -1 & 2 \end{vmatrix}$

3. $\begin{vmatrix} -4 & 1 \\ 0 & 2 \end{vmatrix}$ 4. $\begin{vmatrix} 2 & 2 \\ 3 & 3 \end{vmatrix}$

5. $\begin{vmatrix} 0 & 0 \\ 1 & 3 \end{vmatrix}$ 6. $\begin{vmatrix} -4 & -1 \\ -2 & 3 \end{vmatrix}$

In Exercises 7–10, let

$$A = \begin{bmatrix} 3 & -1 & 2 \\ 4 & 1 & -3 \\ 5 & -2 & -0 \end{bmatrix}$$

7. Compute the minor of each of the following elements:
   a. a_{11} b. a_{23} c. a_{31} d. a_{33}
8. Compute the minor of each of the following elements:
   a. a_{12} b. a_{22} c. a_{23} d. a_{32}
9. Compute the cofactor of each of the following elements:
   a. a_{11} b. a_{23} c. a_{31} d. a_{33}
10. Compute the cofactor of each of the following elements:
   a. a_{12} b. a_{22} c. a_{23} d. a_{32}

In Exercises 11–20, evaluate the determinant of the given matrix.

11. $\begin{vmatrix} 4 & -2 & 5 \\ 5 & 2 & 0 \\ 2 & 0 & 4 \end{vmatrix}$ 12. $\begin{vmatrix} 4 & 1 & 2 \\ 0 & 2 & 3 \\ 0 & 0 & -4 \end{vmatrix}$

13. $\begin{vmatrix} -1 & 2 & 0 \\ 3 & 4 & 1 \\ 6 & 5 & 2 \end{vmatrix}$ 14. $\begin{vmatrix} -1 & 3 & 2 \\ 0 & 7 & 7 \\ 2 & 1 & 3 \end{vmatrix}$

15. $\begin{vmatrix} 0 & -1 & 0 & 3 \\ 0 & 1 & 2 & 1 \\ 2 & -2 & 2 & 3 \\ 3 & 3 & 1 & 0 \end{vmatrix}$ 16. $\begin{vmatrix} 0 & 0 & 2 & 3 \\ -1 & 1 & -2 & 3 \\ 0 & 2 & 2 & 1 \\ 3 & 1 & 3 & 0 \end{vmatrix}$

17. $\begin{vmatrix} 2 & 1 & -3 & 1 \\ 2 & 0 & 3 & -5 \\ 1 & -1 & 2 & 2 \\ 0 & -1 & 1 & 3 \end{vmatrix}$ 18. $\begin{vmatrix} 2 & 2 & 1 & 0 \\ 1 & 0 & -1 & -1 \\ -3 & 3 & 2 & 1 \\ 1 & -5 & 2 & 3 \end{vmatrix}$

19. $\begin{vmatrix} 0 & 0 & 2 & 4 \\ 0 & 1 & 2 & 1 \\ 5 & 1 & 3 & 3 \\ 3 & 3 & 1 & 0 \end{vmatrix}$ 20. $\begin{vmatrix} 3 & 2 & 0 & -1 \\ -2 & 3 & 1 & 0 \\ -2 & -2 & 4 & 4 \\ 1 & -5 & 2 & 3 \end{vmatrix}$

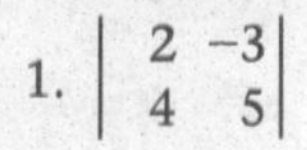

21. Finding the determinants by using your graphing calculator's MATRIX menu, investigate what happens to the determinant of the matrix in Exercise 17 if you change the matrix in the following ways:
   a. Interchange row 2 and row 3.
   b. Interchange row 2 and row 3, then the new row 3 and row 4.

   What can you conclude from these results?

11.5 Properties of Determinants

In general, the computations required to evaluate the determinant of a matrix can get rather time-consuming as the dimension of the matrix becomes quite large. Therefore, it may be worthwhile to consider alternative methods that may reduce the number of operations involved. We have already observed that if an element of a matrix equals zero, then we need not evaluate the corresponding cofactor since the product of the two is also zero. Thus, we will examine methods to enable us to obtain more zero entries in a matrix whose determinant is equal to that of the original matrix under consideration.

In Section 11.1, we presented the elementary row operations:

1. Interchange any two rows.
2. Multiply each element of any row by a constant $k \neq 0$.
3. Replace each element of a given row by the sum of itself plus k times the corresponding element of any other row.

We have observed that these operations are important in transforming one matrix into another matrix. We wish to explore what effect these operations have on the determinant of the original matrix compared with the determinant of the transformed matrix. We also wish to examine the determinant of some special matrices:

1. a matrix with a row of zeros
2. a matrix where two rows are identical
3. a matrix where the rows and columns are interchanged, called the **transpose** of the original matrix

Let

$$A = \begin{bmatrix} 5 & 0 & -1 \\ 4 & -6 & 0 \\ -2 & 1 & 2 \end{bmatrix}$$

Expanding about the cofactors of the first row, we find that

$$|A| = -52$$

Interchanging rows 1 and 2 of A, we have

$$\begin{vmatrix} 4 & -6 & 0 \\ 5 & 0 & -1 \\ -2 & 1 & 2 \end{vmatrix} = 52$$

Multiplying the second row of A by $\frac{1}{2}$, we obtain

$$\begin{vmatrix} 5 & 0 & -1 \\ 2 & -3 & 0 \\ -2 & 1 & 2 \end{vmatrix} = -26$$

Adding 2 times row 1 to row 3, we find that

$$\begin{vmatrix} 5 & 0 & -1 \\ 4 & -6 & 0 \\ 8 & 1 & 0 \end{vmatrix} = -52$$

If we replace the second row of A with 0 elements, we have

$$\begin{vmatrix} 5 & 0 & -1 \\ 0 & 0 & 0 \\ -2 & 1 & 2 \end{vmatrix} = 0$$

If we replace row 3 of A with row 2, we obtain

$$\begin{vmatrix} 5 & 0 & -1 \\ 4 & -6 & 0 \\ 4 & -6 & 0 \end{vmatrix} = 0$$

Taking the transpose of A, where we interchange the rows and columns of A, or equivalently, replacing a_{ij} with a_{ji} we find that

$$\begin{vmatrix} 5 & 4 & -2 \\ 0 & -6 & 1 \\ -1 & 0 & 2 \end{vmatrix} = -52$$

(Verify the calculations of the previous determinants, expanding by any row or any column.) These examples suggest the properties of determinants shown in Table 11-2.

Note that the determinant of a matrix expanded by cofactors yields the same answer, whether the expansion uses a particular row or a particular column. This fact allows us to replace the word "row" by the word "column" and obtain the same property. If a row or column has all zero entries, then expansion by cofactors about this zero row or column produces a determinant of 0. If two rows or two columns are identical, then we may add −1 times one to the other to produce a row or column of zeros, respectively.

TABLE 11-2 Properties of Determinants

Properties
1. Interchange any two rows of A or interchange any two columns of A, and call the new matrix B. Then $\lvert B\rvert = -\lvert A\rvert$
2. Multiply each element of any row of A or any column of A by a constant k, and call the new matrix B. Then $\lvert B\rvert = k\lvert A\rvert$
3. Add k times one row to any other row or k times one column to any other column and call the new matrix B. Then $\lvert B\rvert = \lvert A\rvert$
4. If A has a row or column with 0 elements or if A has two identical rows or two identical columns then $\lvert A\rvert = 0$
5. Take the transpose of A, where we replace a_{ij} with a_{ji}, so that the rows become columns and the columns become rows. If we call the new matrix B, then $\lvert B\rvert = \lvert A\rvert$

EXAMPLE 1 *Using Properties of Determinants*

Evaluate the determinant.

$$\begin{vmatrix} 0 & 2 & 3 & 0 \\ 4 & 1 & 8 & 15 \\ 1 & -1 & 2 & 0 \\ -1 & -2 & -3 & -6 \end{vmatrix}$$

SOLUTION

To make $a_{21} = 0$, replace row 2 by the sum of itself and (−4) times row 3. To make $a_{41} = 0$, replace row 4 by the sum of itself and row 3.

$$\begin{vmatrix} 0 & 2 & 3 & 0 \\ 0 & 5 & 0 & 15 \\ 1 & -1 & 2 & 0 \\ 0 & -3 & -1 & -6 \end{vmatrix}$$

Now expand the determinant by the cofactors of the first column, obtaining

$$\begin{vmatrix} 2 & 3 & 0 \\ 5 & 0 & 15 \\ -3 & -1 & -6 \end{vmatrix}$$

We factor out 5 from the second row to obtain

$$5\begin{vmatrix} 2 & 3 & 0 \\ 1 & 0 & 3 \\ -3 & -1 & -6 \end{vmatrix}$$

To make $a_{23} = 0$, replace column 3 by the sum of itself and (-3) times column 1.

$$5\begin{vmatrix} 2 & 3 & -6 \\ 1 & 0 & 0 \\ -3 & -1 & 3 \end{vmatrix}$$

Expand this determinant by the cofactors of the second row, obtaining

$$-5\begin{vmatrix} 3 & -6 \\ -1 & 3 \end{vmatrix}$$

Evaluating this last 2×2 determinant, we have

$$-5(9 - 6) = -15$$

■

Evaluate the determinant.

$$\begin{vmatrix} 4 & 0 & 0 & 3 \\ 2 & 4 & 5 & 8 \\ -2 & 1 & 0 & 2 \\ -4 & -1 & -2 & -3 \end{vmatrix}$$

Answer

-10

Exercise Set 11.5

In Exercises 1–6, evaluate the determinant of the given matrix.

1. $\begin{vmatrix} 2 & 2 & 4 \\ 3 & 8 & 1 \\ 1 & 1 & 2 \end{vmatrix}$

2. $\begin{vmatrix} 0 & 1 & 3 \\ 2 & 5 & -1 \\ 4 & 2 & -2 \end{vmatrix}$

3. $\begin{vmatrix} 3 & 2 & 1 & 0 \\ -1 & -3 & -1 & 0 \\ 0 & 0 & 2 & 2 \\ 4 & 1 & 3 & 3 \end{vmatrix}$

4. $\begin{vmatrix} -1 & 2 & 4 & 0 \\ 3 & -2 & -3 & 0 \\ 0 & 4 & 2 & 5 \\ 0 & -3 & 1 & 4 \end{vmatrix}$

5. $\begin{vmatrix} 2 & -3 & 2 & -4 \\ 0 & 4 & -1 & 9 \\ 0 & 1 & 2 & 0 \\ 0 & 1 & 3 & -1 \end{vmatrix}$

6. $\begin{vmatrix} 1 & 1 & 0 & 1 \\ 0 & -1 & 4 & -1 \\ -2 & 3 & 1 & -4 \\ 0 & 2 & 0 & 2 \end{vmatrix}$

7. Show that

$$\begin{vmatrix} a_1+b_1 & a_2+b_2 \\ c & d \end{vmatrix} = \begin{vmatrix} a_1 & a_2 \\ c & d \end{vmatrix} + \begin{vmatrix} b_1 & b_2 \\ c & d \end{vmatrix}$$

8. Prove that if a row or column of a square matrix consists entirely of zeros, the determinant of the matrix is zero. (*Hint*: Expand by cofactors.)

9. Prove that if matrix B is obtained by multiplying each element of a row of a square matrix A by a constant k, then $|B| = k|A|$.

10. Show that

$$\begin{vmatrix} ka_{11} & ka_{12} \\ a_{21} & a_{22} \end{vmatrix} = \begin{vmatrix} a_{11} & a_{12} \\ ka_{21} & ka_{22} \end{vmatrix} = k\begin{vmatrix} a_{11} & a_{12} \\ a_{21} & a_{22} \end{vmatrix}$$

11. Prove that if A is an $n \times n$ matrix and $B = kA$, where k is a constant, then $|B| = k^n|A|$.

12. Prove that if matrix B is obtained from a square matrix A by interchanging the rows and columns of A, then $|B| = |A|$.

11.6 Cramer's Rule

Determinants provide a convenient way of expressing formulas in many areas of mathematics, particularly in geometry. One of the better known uses of determinants is for solving systems of linear equations, a procedure known as *Cramer's Rule.*

In an earlier section, we solved systems of linear equations by the method of elimination. We now apply this method to the general system of two equations in two unknowns.

$$a_{11}x + a_{12}y = b_1 \qquad (1)$$

$$a_{21}x + a_{22}y = b_2 \qquad (2)$$

Let us multiply Equation (1) by a_{22}, Equation (2) by $-a_{12}$ and add. This eliminates y.

$$\begin{array}{l} a_{11}a_{22}x + a_{12}a_{22}y = b_1a_{22} \\ -a_{21}a_{12}x - a_{12}a_{22}y = -b_2a_{12} \\ \hline a_{11}a_{22}x - a_{21}a_{12}x = b_1a_{22} - b_2a_{12} \end{array}$$

Thus,

$$x(a_{11}a_{22} - a_{21}a_{12}) = b_1a_{22} - b_2a_{12}$$

or

$$x = \frac{b_1a_{22} - b_2a_{12}}{a_{11}a_{22} - a_{21}a_{12}}$$

Similarly, multiplying Equation (1) by a_{21}, Equation (2) by $-a_{11}$ and adding, we can eliminate x and solve for y.

$$y = \frac{b_2a_{11} - b_1a_{21}}{a_{11}a_{22} - a_{21}a_{12}}$$

The denominators in the expression for x and y are identical and can be written as the determinant of the matrix

$$|A| = \begin{vmatrix} a_{11} & a_{12} \\ a_{21} & a_{22} \end{vmatrix}$$

If we apply this same idea to the numerators, we have

$$x = \frac{\begin{vmatrix} b_1 & a_{12} \\ b_2 & a_{22} \end{vmatrix}}{|A|}, \quad y = \frac{\begin{vmatrix} a_{11} & b_1 \\ a_{21} & b_2 \end{vmatrix}}{|A|}, \quad |A| \neq 0$$

This formula is called **Cramer's Rule** and is a means of expressing the solution of a system of linear equations in determinant form. Let A_1 denote the matrix obtained by replacing the first column of A with the column of the right-hand sides of the equations. Furthermore, let A_2 denote the matrix obtained by replacing the second column of A again with the column of the right-hand sides.

We may summarize Cramer's Rule as follows:

Cramer's Rule for Two Unknowns

The solution to

$$a_{11}x + a_{12}y = b_1$$
$$a_{21}x + a_{22}y = b_2$$

is given by

$$x = \frac{|A_1|}{|A|}, \quad y = \frac{|A_2|}{|A|}, \quad |A| \neq 0$$

The following example outlines the steps for using Cramer's Rule.

EXAMPLE 1 *Cramer's Rule*

Solve by Cramer's Rule.

$$\begin{aligned} 3x - y &= 9 \\ x + 2y &= -4 \end{aligned}$$

SOLUTION

Step 1. Compute $\lvert A \rvert$, the determinant of the coefficient matrix A. If $A = 0$, Cramer's Rule cannot be used. Use Gaussian Elimination or Gauss-Jordan Elimination.	*Step 1.* $\lvert A \rvert = \begin{vmatrix} 3 & -1 \\ 1 & 2 \end{vmatrix} = 7$
Step 2. Compute $\lvert A_1 \rvert$, the determinant of the matrix obtained from A by replacing the column of coefficients of x, the first column unknown, with the column of right-hand sides of the equations. $x = \dfrac{\lvert A_1 \rvert}{\lvert A \rvert}$	*Step 2.* $x = \dfrac{\lvert A_1 \rvert}{\lvert A \rvert} = \dfrac{\begin{vmatrix} 9 & -1 \\ -4 & 2 \end{vmatrix}}{\lvert A \rvert} = \dfrac{18-4}{7} = \dfrac{14}{7} = 2$
Step 3. Compute $\lvert A_2 \rvert$, the determinant of the matrix obtained from A by replacing the column of coefficients of y, the second column unknown, with the column of right-hand sides of the equations. $y = \dfrac{\lvert A_2 \rvert}{\lvert A \rvert}$	*Step 3.* $y = \dfrac{\lvert A_2 \rvert}{\lvert A \rvert} = \dfrac{\begin{vmatrix} 3 & 9 \\ 1 & -4 \end{vmatrix}}{\lvert A \rvert} = \dfrac{-12-9}{7} = \dfrac{-21}{7} = -3$ Thus, $x = 2$, $y = -3$

■

Progress Check

Solve by Cramer's Rule.

$$\begin{aligned} 2x + 3y &= -4 \\ 3x + 4y &= -7 \end{aligned}$$

Answers

$x = -5$, $y = 2$

The steps outlined in Example 1 can be applied to solve *any* system of linear equations in which the number of equations is the same as the number of unknowns and in which $|A| \neq 0$. For example, assume A is 3×3. If A_3 is the matrix obtained by replacing the third column of A with the column of right-hand sides, then we have

Cramer's Rule for Three Unknowns

The solution to

$$\begin{aligned} a_{11}x + a_{12}y + a_{13}z &= b_1 \\ a_{21}x + a_{22}y + a_{23}z &= b_2 \\ a_{31}x + a_{32}y + a_{33}z &= b_3 \end{aligned}$$

is given by

$$x = \frac{|A_1|}{|A|}, \quad y = \frac{|A_2|}{|A|}, \quad z = \frac{|A_3|}{|A|}, \quad |A| \neq 0$$

EXAMPLE 2 *Cramer's Rule*

Solve by Cramer's Rule.

$$\begin{aligned} 3x + 2z &= -2 \\ 2x - y &= 0 \\ 2y + 6z &= -1 \end{aligned}$$

SOLUTION

We compute the determinant of the matrix of coefficients.

$$|A| = \begin{vmatrix} 3 & 0 & 2 \\ 2 & -1 & 0 \\ 0 & 2 & 6 \end{vmatrix} = -10$$

Then

$$x = \frac{|A_1|}{|A|} = \frac{\begin{vmatrix} -2 & 0 & 2 \\ 0 & -1 & 0 \\ -1 & 2 & 6 \end{vmatrix}}{|A|} = \frac{10}{-10} = -1$$

$$y = \frac{|A_2|}{|A|} = \frac{\begin{vmatrix} 3 & -2 & 2 \\ 2 & 0 & 0 \\ 0 & -1 & 6 \end{vmatrix}}{|A|} = \frac{20}{-10} = -2$$

$$z = \frac{|A_3|}{|A|} = \frac{\begin{vmatrix} 3 & 0 & -2 \\ 2 & -1 & 0 \\ 0 & 2 & -1 \end{vmatrix}}{|A|} = \frac{-5}{-10} = \frac{1}{2}$$

■

Solve by Cramer's Rule.

$$\begin{aligned} 3x \qquad\quad - z &= 1 \\ -6x + 2y \qquad &= -5 \\ -4y + 3z &= 5 \end{aligned}$$

Answers

$x = \frac{2}{3}, y = -\frac{1}{2}, z = 1$

a. Each equation of the linear system must be written in the form

$$ax + by + cz = k$$

before using Cramer's Rule.

b. If $|A| = 0$, Cramer's Rule cannot be used.

Exercise Set 11.6

In Exercises 1–8, solve the given linear system by using Cramer's Rule.

1. $$\begin{aligned} 2x + y + z &= -1 \\ 2x - y + 2z &= 2 \\ x + 2y + z &= -4 \end{aligned}$$

2. $$\begin{aligned} x - y + z &= -5 \\ 3x + y + 2z &= -5 \\ 2x - y - z &= -2 \end{aligned}$$

3. $$\begin{aligned} 2x + y - z &= 9 \\ x - 2y + 2z &= -3 \\ 3x + 3y + 4z &= 11 \end{aligned}$$

4. $$\begin{aligned} 2x + y - z &= -2 \\ -2x - 2y + 3z &= 2 \\ 3x + y - z &= -4 \end{aligned}$$

5. $$\begin{aligned} -x - y + 2z &= 7 \\ x + 2y - 2z &= -7 \\ 2x - y + z &= -4 \end{aligned}$$

6. $$\begin{aligned} 4x + y - z &= -1 \\ x - y + 2z &= 3 \\ -x + 2y - z &= 0 \end{aligned}$$

7. $$\begin{aligned} x + y - z + 2w &= 0 \\ 2x + y - w &= -2 \\ 3x + 2z &= -3 \\ -x + 2y + 3w &= 1 \end{aligned}$$

8. $$\begin{aligned} 2x + y - 3w &= -7 \\ 3x + 2z + w &= -1 \\ -x + 2y + 3w &= 0 \\ -2x - 3y + 2z - w &= 8 \end{aligned}$$

9. *Mathematics in Writing*: Give a step-by-step method for solving systems of equations by Cramer's Rule with your graphing calculator.

10. Redo Exercises 7 and 8 using the method you outlined in Exercise 9.

Chapter Summary

Key Terms, Concepts, and Symbols

Key Ideas for Review

Topic	Page	Key Idea
Matrices	600	A matrix is a rectangular array of numbers.
Addition and Subtraction	609 611	The sum and difference of two matrices A and B can be formed only if A and B are of the same dimension.
Multiplication	611	The product AB can be formed only if the number of columns of A is the same as the number of rows of B.
Systems of Linear Equations and Matrix Notation	614	A linear system can be written in the form $AX = B$, where A is the coefficient matrix, X is a column matrix of the unknowns and B is the column matrix of the right-hand sides. The elementary row operations are an abstraction of those operations that produce equivalent systems of equations.
Gaussian and Gauss-Jordan Elimination	605	Gaussian Elimination and Gauss-Jordan Elimination both involve the use of elementary row operations on the augmented matrix corresponding to a linear system. In the case of a system of three equations with three unknowns and a unique solution, the final matrices are of this form: $\left[\begin{array}{ccc\|c} * & * & * & * \\ 0 & * & * & * \\ 0 & 0 & * & * \end{array}\right]$ Gaussian Elimination; $\left[\begin{array}{ccc\|c} 1 & 0 & 0 & c_1 \\ 0 & 1 & 0 & c_2 \\ 0 & 0 & 1 & c_3 \end{array}\right]$ Gauss-Jordan Elimination. If Gaussian Elimination is used, back-substitution is then performed with the final matrix to obtain the solution. If Gauss-Jordan Elimination is used, the solution can be read from the final matrix.
Inverse of a Matrix	617	The $n \times n$ matrix B is said to be the inverse of the $n \times n$ matrix A if $AB = I_n$ and $BA = I_n$. We denote the inverse of A by A^{-1}. The inverse can be computed by using elementary row operations to transform the matrix $[A \mid I_n]$ to the form $[I_n \mid B]$, in which case $B = A^{-1}$.
Solving Linear Systems	621	If the linear system $AX = B$ has a unique solution, then $X = A^{-1}B$.

continues

Topic	Page	Key Idea
Determinants	628	Associated with every square matrix is a number called a determinant. The determinant of the 1×1 matrix $A = [a]$ is $\|A\| = a$. The rule for evaluating a determinant of order 2 is $\begin{vmatrix} a & b \\ c & d \end{vmatrix} = ad - bc$
Evaluation by Cofactors	629	For determinants of order greater than 2, the method of expansion by cofactors may be used to reduce the problem to that of evaluating determinants of order 2. When expanding by cofactors, choosing the row or column that contains the most zeros usually simplifies the arithmetic.
Properties	635	Some useful properties of determinants follow: 1. Interchange any two rows of A or interchange any two columns of A, and call the new matrix B. Then $\|B\| = -\|A\|$ 2. Multiply each element of any row of A or any column of A by a constant k, and call the new matrix B. Then $\|B\| = k\|A\|$ 3. Add k times one row to any other row or k times one column to any other column and call the new matrix B. Then $\|B\| = \|A\|$ 4. If A has a row or column with 0 elements or if A has two identical rows or two identical columns then $\|A\| = 0$ 5. Take the transpose of A, where we replace a_{ij} with a_{ji}, so that the rows become columns and the columns become rows. If we call the new matrix B, then $\|B\| = \|A\|$
Cramer's Rule	639	Cramer's Rule provides a means for solving a linear system by expressing the value of each unknown as a quotient of determinants.

Review Exercises

Exercises 1–4 refer to the matrix

$$A = \begin{bmatrix} -1 & 4 & 2 & 0 & 8 \\ 2 & 0 & -3 & -1 & 5 \\ 4 & -6 & 9 & 1 & -2 \end{bmatrix}$$

1. Determine the dimension of the matrix A.
2. Find a_{24}.
3. Find a_{31}.
4. Find a_{15}.

Exercises 5 and 6 refer to the linear system.

$$\begin{aligned} 3x - 7y &= 14 \\ x + 4y &= 6 \end{aligned}$$

5. Write the coefficient matrix of the linear system.
6. Write the augmented matrix of the linear system.

In Exercises 7 and 8, write a linear system corresponding to the augmented matrix.

7. $\left[\begin{array}{cc|c} 4 & -1 & 3 \\ 2 & 5 & 0 \end{array}\right]$

8. $\left[\begin{array}{ccc|c} -2 & 4 & 5 & 0 \\ 6 & -9 & 4 & 0 \\ 3 & 2 & -1 & 0 \end{array}\right]$

In Exercises 9–12, use back-substitution to solve the linear system corresponding to the given augmented matrix.

9. $\left[\begin{array}{cc|c} 1 & -2 & 7 \\ 0 & 1 & -4 \end{array}\right]$ 10. $\left[\begin{array}{cc|c} 1 & 2 & \frac{21}{2} \\ 0 & 1 & 5 \end{array}\right]$

11. $\left[\begin{array}{ccc|c} 1 & -4 & 2 & -18 \\ 0 & 1 & -2 & 5 \\ 0 & 0 & 1 & -1 \end{array}\right]$ 12. $\left[\begin{array}{ccc|c} 1 & -2 & 2 & -9 \\ 0 & 1 & 3 & -8 \\ 0 & 0 & 1 & -3 \end{array}\right]$

In Exercises 13–16, use matrix methods to solve the given linear system.

13. $\begin{aligned} x + y &= 2 \\ 2x - 4y &= -5 \end{aligned}$ 14. $\begin{aligned} 3x - y &= -17 \\ 2x + 3y &= -4 \end{aligned}$

15. $\begin{aligned} x + 3y + 2z &= 0 \\ -2x + 3z &= -12 \\ 2x - 6y - z &= 6 \end{aligned}$

16. $\begin{aligned} 2x - y - 2z &= 3 \\ -2x + 3y + z &= 3 \\ 2y - z &= 6 \end{aligned}$

In Exercises 17 and 18, solve for x.

17. $\begin{bmatrix} 5 & -1 \\ 3 & 2x \end{bmatrix} = \begin{bmatrix} 5 & -1 \\ 3 & -6 \end{bmatrix}$ 18. $\begin{bmatrix} 6 & x^2 \\ 4x & -2 \end{bmatrix} = \begin{bmatrix} 6 & 9 \\ -12 & -2 \end{bmatrix}$

Exercises 19–28 refer to the following matrices:

$A = \begin{bmatrix} 2 & -1 \\ 3 & 2 \end{bmatrix}$ $B = \begin{bmatrix} -1 & 5 \\ 4 & -3 \end{bmatrix}$

$C = \begin{bmatrix} -1 & 0 \\ 0 & 4 \\ 2 & -2 \end{bmatrix}$ $D = \begin{bmatrix} 1 & 3 & 4 \\ -1 & 0 & -6 \end{bmatrix}$

If possible, find the following:

19. $A + B$ 20. $B - A$

21. $A + C$ 22. $5D$

23. CD 24. DC

25. BC 26. CB

27. $A + 2B$ 28. $-AB$

In Exercises 29 and 30, find the inverse of the given matrix.

29. $\begin{bmatrix} -2 & 3 \\ 1 & 4 \end{bmatrix}$ 30. $\begin{bmatrix} 1 & 1 & -4 \\ -5 & -2 & 0 \\ 4 & 2 & -1 \end{bmatrix}$

In Exercises 31 and 32, solve the given system by finding the inverse of the coefficient matrix.

31. $\begin{aligned} 2x - y &= 1 \\ x + y &= 5 \end{aligned}$ 32. $\begin{aligned} x + 2y - 2z &= -4 \\ 3x - y \quad &= -2 \\ y + 4z &= -1 \end{aligned}$

In Exercises 33–38, evaluate the determinant of the given matrix.

33. $\begin{vmatrix} 3 & 1 \\ -4 & 2 \end{vmatrix}$ 34. $\begin{vmatrix} -1 & 2 \\ 0 & 6 \end{vmatrix}$

35. $\begin{vmatrix} 2 & -1 \\ 6 & -3 \end{vmatrix}$ 36. $\begin{vmatrix} 1 & 0 & -1 \\ 2 & 3 & -5 \\ 0 & 4 & 0 \end{vmatrix}$

37. $\begin{vmatrix} 1 & -1 & 2 \\ 0 & 5 & 4 \\ 2 & 3 & 8 \end{vmatrix}$ 38. $\begin{vmatrix} 1 & 2 & -1 \\ 0 & 3 & 4 \\ 0 & 0 & -1 \end{vmatrix}$

In Exercises 39–44, use Cramer's Rule to solve the given linear system.

39. $\begin{aligned} 2x - y &= -3 \\ -2x + 3y &= 11 \end{aligned}$ 40. $\begin{aligned} 3x - y &= 7 \\ 2x + 5y &= -18 \end{aligned}$

41. $\begin{aligned} x + 2y &= 2 \\ 2x - 7y &= 48 \end{aligned}$ 42. $\begin{aligned} 2x + 3y - z &= -3 \\ -3x \quad + 4z &= 16 \\ 2y + 5z &= 9 \end{aligned}$

43. $\begin{aligned} 3x \quad + z &= 0 \\ x + y + z &= 0 \\ -3y + 2z &= -4 \end{aligned}$ 44. $\begin{aligned} 2x + 3y + z &= -5 \\ 2y + 2z &= -3 \\ 4x + y - 2z &= -2 \end{aligned}$

Review Test

Exercises 1 and 2 refer to the matrix

$$A = \begin{bmatrix} -1 & 2 \\ -2 & 4 \\ 0 & 7 \end{bmatrix}$$

1. Find the dimension of the matrix A.
2. Find a_{31}.
3. Write the augmented matrix of the linear system

$$\begin{aligned} -7x \quad + 6z &= 3 \\ 2y - z &= 10 \\ x - y + z &= 5 \end{aligned}$$

4. Write a linear system corresponding to the augmented matrix

$$\left[\begin{array}{cc|c} -5 & 2 & 4 \\ 3 & -4 & 4 \end{array}\right]$$

5. Use back-substitution to solve the linear system corresponding to the augmented matrix

$$\left[\begin{array}{cc|c} 1 & 1 & 0 \\ 0 & 1 & \frac{1}{2} \end{array}\right]$$

6. Solve the linear system

$$\begin{aligned} -x + 2y &= 2 \\ \frac{1}{2}x + 2y &= -7 \end{aligned}$$

by applying Gaussian Elimination to the augmented matrix.

7. Solve the linear system

$$\begin{aligned} 2x - y + 3z &= 2 \\ x + 2y - z &= 1 \\ -x + y + 4z &= 2 \end{aligned}$$

by applying Gauss-Jordan Elimination to the augmented matrix.

8. Solve for x.

$$\begin{bmatrix} 2x-1 & 0 \\ 1 & -3 \end{bmatrix} = \begin{bmatrix} 5 & 0 \\ 1 & -3 \end{bmatrix}$$

Exercises 9–12 refer to the matrices

$$A = \begin{bmatrix} -4 & 0 & 3 \\ 6 & 2 & -3 \end{bmatrix} \qquad B = \begin{bmatrix} -1 \\ -3 \end{bmatrix}$$

$$C = \begin{bmatrix} 4 & 2 \\ -2 & 0 \\ 3 & -1 \end{bmatrix} \qquad D = \begin{bmatrix} 1 & -6 \\ 0 & 2 \\ 4 & -1 \end{bmatrix}$$

If possible, find the following:

9. $C - 2D$

10. AC

11. CB

12. BA

13. Find the inverse of the matrix

$$\begin{bmatrix} -1 & 0 & 4 \\ 2 & 1 & -1 \\ 1 & -3 & 2 \end{bmatrix}$$

14. Solve the given linear system by finding the inverse of the coefficient matrix.

$$\begin{aligned} 3x - 2y &= -8 \\ 2x + 3y &= -1 \end{aligned}$$

In Exercises 15 and 16, evaluate the determinant of the given matrix.

15. $\begin{vmatrix} -6 & -2 \\ 2 & 1 \end{vmatrix}$

16. $\begin{vmatrix} 0 & -1 & 2 \\ 2 & -2 & 3 \\ 1 & 4 & 5 \end{vmatrix}$

17. Use Cramer's Rule to solve the linear system

$$\begin{aligned} x + 2y &= -2 \\ -2x - 3y &= 1 \end{aligned}$$

Writing Exercises

1. Discuss how to solve a linear system in three unknowns if Cramer's Rule fails to hold.

2. Compare and contrast the additive properties of matrices with the additive properties of the real numbers.

3. Compare and contrast the multiplicative properties of square matrices with the multiplicative properties of the real numbers.

4. Compare and contrast Gauss-Jordan Elimination and Gaussian Elimination.

Chapter 11 Project

Manipulating images using computer technology is a major component of special effects in some of today's most popular films. The mathematics of matrices can help us see how images can be altered by increasing the contrast or adding two images together. One interesting use of the latter technique is a process called blue-screen chromakey, by which a character may appear to be in an environment which was actually photographed separately.

For this project, do the following exercises from this chapter: Section 11.1, 51, and Section 11.2, 37–39.

Now make up your own image matrix. Make it 20 pixels by 15 pixels, and let each pixel have 6 bits (this means each entry will be an integer between 0 and 63). Repeat the exercises using this matrix. Use your calculator to help you. If increasing the contrast results in an entry greater than 63, what should you do?

12 Topics in Algebra

There is a famous story from antiquity in which a grateful king offers a wise subject (a mathematician, no doubt!) any reward he can name. The subject requests to be given a chessboard with 64 squares, with one grain of rice on the first square, two grains on the second square, four grains on the third, and so on, doubling the number of grains upon each square. The king readily agrees, imagining that he will need only a few bushels to satisfy the request.

Imagine his surprise when the Royal Mathematician showed him how much grain would be needed for the last square of the chessboard! This problem is similar to Exercise 40 of Section 12.3. You are asked to find the answer at the end of the chapter.

If mathematical recreations interest you, take a look at some illusions and other fun items by visiting http://www.mathworld.wolfram.com and clicking "Recreational Mathematics." Check out the Parallelogram Illusion!

The topics in this chapter are related in that they all involve the set of natural numbers. As you might expect, despite our return to a simpler number system, the approach and results will be more advanced than in earlier chapters. For example, in discussing sequences, we will be dealing with functions whose domain is the set of natural numbers. Yet, sequences lead to considerations of series, and the underlying concepts of infinite series can be used as an introduction to calculus.

Another of the topics, mathematical induction, provides a means of proving certain theorems involving the natural numbers. As an example, we will use mathematical induction to prove that the sum of the first n consecutive positive integers is

$$\frac{n(n+1)}{2}$$

Yet another topic is the binomial theorem, which gives us a way to expand the expression $(a+b)^n$, where n is a natural number. One of the earliest results obtained in a calculus course requires the binomial theorem in its derivation.

Probability theory enables us to state the likelihood of the occurrence of a given event and has obvious applications to games of chance. The theory of permutations and combinations, which enables us to count the number of arrangements of a set of objects under various conditions, is a useful background to the study of probability theory.

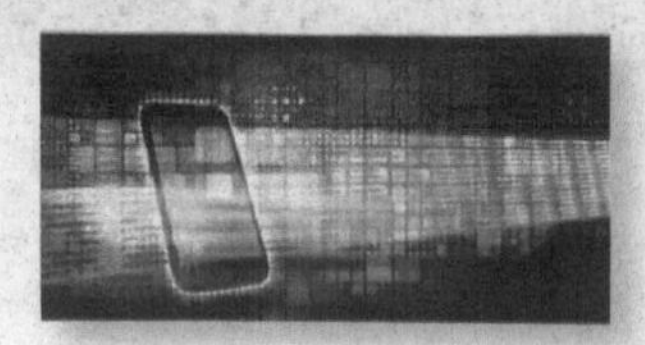

http://www.mathworld.wolfram.com

12.1 Sequences, Sigma Notation, and Series

12.1a *Infinite Sequences*

Can you see a pattern or relationship that describes this string of numbers

$$1, 4, 9, 16, 25, \ldots?$$

If we rewrite this string as

$$1^2, 2^2, 3^2, 4^2, 5^2, \ldots$$

we observe that these are the squares of successive natural numbers. Each number in the string is called a **term**. If we let a be the function defined by

$$a(n) = n^2$$

then $a(n)$ is the nth term of the list, where n is a natural number. Such a string of numbers is called an infinite sequence since the list is infinitely long.

> An **infinite sequence** is a function whose domain is the set of all natural numbers.

Sometimes we use the term "**sequence**" as being equivalent to *infinite sequence.*

The range of the function a is

$$a(1), a(2), a(3), \ldots, a(n), \ldots$$

which we write as

$$a_1, a_2, a_3, \ldots, a_n, \ldots$$

That is, we indicate a sequence by using subscript notation rather than function notation. We say that a_1 is the *first term* of the sequence, a_2 is the *second term,* and so on; and we write the ***n*th term** as $\boldsymbol{a_n}$ where $a_n = a(n)$.

EXAMPLE 1 ***Finding Specific Terms of a Sequence***

Write the first three terms and the tenth term of each of the sequences whose nth term is given.

a. $a_n = n^2 + 1$ b. $a_n = \dfrac{n}{n+1}$ c. $a_n = 2^n - 1$

SOLUTION

The first three terms are found by substituting $n = 1$, 2, and 3 in the formula for a_n. The tenth term is found by substituting $n = 10$.

a. $a_1 = 1^2 + 1 = 2$, $a_2 = 2^2 + 1 = 5$, $a_3 = 3^2 + 1 = 10$, $a_{10} = 10^2 + 1 = 101$

b. $a_1 = \dfrac{1}{1+1} = \dfrac{1}{2}$, $a_2 = \dfrac{2}{2+1} = \dfrac{2}{3}$, $a_3 = \dfrac{3}{3+1} = \dfrac{3}{4}$, $a_{10} = \dfrac{10}{10+1} = \dfrac{10}{11}$

c. $a_1 = 2^1 - 1 = 1$, $a_2 = 2^2 - 1 = 3$, $a_3 = 2^3 - 1 = 7$, $a_{10} = 2^{10} - 1 = 1023$ ■

Progress Check

Write the first three terms and the twelfth term of each of the sequences whose nth term is given.

a. $a_n = 3(1 - n)$ b. $a_n = n^2 + n + 1$ c. $a_n = 5$

Answers

a. $a_1 = 0$, $a_2 = -3$, $a_3 = -6$, $a_{12} = -33$

b. $a_1 = 3$, $a_2 = 7$, $a_3 = 13$, $a_{12} = 157$

c. $a_1 = a_2 = a_3 = a_{12} = 5$

An infinite sequence is often defined by a formula expressing the nth term by reference to preceding terms. Such a sequence is said to be defined *recursively,* or by a **recursion formula.**

EXAMPLE 2 *Finding Terms with a Recursion Formula*

Find the first four terms of the sequence defined by

$$a_n = a_{n-1} + 3 \quad \text{with} \quad a_1 = 2 \quad \text{and} \quad n \geq 2$$

SOLUTION

Any term of the sequence can be obtained if the preceding term is known. The recursion formula requires a starting point; and we are, indeed, given a_1. Then

$$a_1 = 2$$
$$a_2 = a_1 + 3 = 2 + 3 = 5$$
$$a_3 = a_2 + 3 = 5 + 3 = 8$$
$$a_4 = a_3 + 3 = 8 + 3 = 11$$

■

Progress Check

Find the first four terms of the infinite sequence

$$a_n = 2a_{n-1} - 1 \quad \text{with} \quad a_1 = -1 \quad \text{and} \quad n \geq 2$$

Answer

$-1, -3, -7, -15$

12.1b Summation Notation

In the following sections of this chapter, we will seek the sum of terms of a sequence such as

$$a_1 + a_2 + a_3 + \cdots + a_n$$

Since sums occur frequently in mathematics, a special notation has been developed that is defined in the following way:

Summation Notation

$$\sum_{k=1}^{n} a_k = a_1 + a_2 + a_3 + \cdots + a_n$$

This is often referred to as **sigma notation** since the Greek letter Σ indicates a sum of terms of the form a_k. The letter k is called the **index of summation** and always assumes successive integer values, starting with the value written under the Σ sign and ending with the value written above the Σ sign.

EXAMPLE 3 *Using Sigma Notation*

Evaluate.

a. $\sum_{k=1}^{3} 2^k(k+1)$ b. $\sum_{i=2}^{4} (i^2+2)$

SOLUTION

a. The terms are of the form

$$a_k = 2^k(k+1)$$

and the sigma notation indicates that we want the sum of the terms starting with a_1 and ending with a_3. Forming the terms and adding,

$$\begin{aligned}\sum_{k=1}^{3} 2^k(k+1) &= 2^1(1+1) + 2^2(2+1) + 2^3(3+1)\\ &= 4 + 12 + 32 = 48\end{aligned}$$

b. Any letter may be used for the index of summation. Here, the letter i is used.

$$\begin{aligned}\sum_{i=2}^{4} (i^2+2) &= (2^2+2) + (3^2+2) + (4^2+2)\\ &= 6 + 11 + 18 = 35\end{aligned}$$

Note that the index of summation can begin with an integer value other than 1. ■

FOCUS on Fibonacci Counting the Rabbits

The following is a problem that was first published in the year 1202:

A pair of newborn rabbits begins mating at age 1 month and thereafter produces one pair of offspring per month. If we start with a newly born pair of rabbits, how many rabbits will there be at the beginning of each month? This problem was posed by Leonardo Fibonacci of Pisa, and the resulting sequence is known as a *Fibonacci Sequence*.

Beginning of the Month	Pairs of Rabbits
0	P_1
1	P_1
2	$P_1 \to P_2$
3	$P_1 \to P_3 P_2$
4	$P_1 \to P_4$
	$P_2 \to P_5 P_3$
5	$P_1 \to P_6$
	$P_2 \to P_7$
	$P_3 \to P_8$
	$P_4 P_5$

The accompanying table helps in analyzing the problem. At the beginning of month zero, we have the pair of newborn rabbits P_1. At the beginning of month 1, we still have the pair P_1 since the rabbits do not mate until age 1 month. At the beginning of month 2, the pair P_1 has produced the offspring pair P_2. At the beginning of month 3, P_1 again produces offspring, pair P_3. Pair P_2, however, does not produce offspring during its first month. At the beginning of month 4, P_1 has offspring pair P_4, P_2 has offspring pair P_5, and pair P_3 does not produce offspring during its first month.

If we let a_n denote the number of pairs of rabbits at the beginning of month n, we see that

$$a_0 = 1,\ a_1 = 1,\ a_2 = 2,\ a_3 = 3,\ a_4 = 5,\ a_5 = 8,\ \ldots$$

The sequence has the property that each term is the sum of the two preceding terms; that is,

$$a_n = a_{n-1} + a_{n-2}$$

These Fibonacci numbers appear in nature quite frequently. For example, arrangements of seeds on sunflowers and leaves on some trees are related to Fibonacci numbers. Some researchers believe that cycle analysis, such as analysis of stock market prices, is also related in some way to Fibonacci numbers.

EXAMPLE 4 *Using Sigma Notation*

Write each sum using summation notation.

a. $\frac{1}{2} + \frac{1}{2 \cdot 2} + \frac{1}{2 \cdot 3} + \frac{1}{2 \cdot 4} + \frac{1}{2 \cdot 5}$

b. $\frac{2}{3} + \frac{3}{4} + \frac{4}{5} + \frac{5}{6}$

SOLUTION

a. The denominator of each term is of the form $2 \cdot k$, where k assumes integer values from 1 to 5. Then

$$\sum_{k=1}^{5} \frac{k}{2 \cdot k}$$

expresses the desired sum.

b. If the value of the numerator of a term is k, then the denominator is $k + 1$. Letting k range from 2 to 5,

$$\sum_{k=2}^{5} \frac{k}{k+1}$$

expresses the desired sum. ■

Progress Check

Write each sum using summation notation.

a. $x_1^2 + x_2^2 + x_3^2 + \cdots + x_{20}^2$ b. $2^3 + 3^4 + 4^5 + 5^6$

Answers

a. $\sum_{k=1}^{20} x_k^2$ b. $\sum_{k=2}^{5} k^{k+1}$

If a sequence is defined by $a_k = c$, where c is a constant, then

$$\begin{aligned}\sum_{k=1}^{n} a_k &= a_1 + a_2 + a_3 + \cdots + a_n \\ &= c + c + c + \cdots + c \\ &= nc\end{aligned}$$

This leads to the following rule:

For any constant c,

$$\sum_{k=1}^{n} c = nc$$

EXAMPLE 5 *Using Sigma Notation with a Constant*

Evaluate.

a. $\sum_{j=1}^{20} 5$ b. $\sum_{k=1}^{4} (k^2 - 2)$

SOLUTION

a. $\sum_{j=1}^{20} 5 = 20 \cdot 5 = 100$

b. $$\begin{aligned}\sum_{k=1}^{4} (k^2 - 2) &= (1^2 - 2) + (2^2 - 2) + (3^2 - 2) + (4^2 - 2) \\ &= -1 + 2 + 7 + 14 = 22\end{aligned}$$ ■

The following are properties of sums expressed in sigma notation:

Properties of Sums

For the sequences $a_1, a_2, \ldots,$ and $b_1, b_2, \ldots,$

1. $\sum_{k=1}^{n} (a_k + b_k) = \sum_{k=1}^{n} a_k + \sum_{k=1}^{n} b_k$

2. $\sum_{k=1}^{n} (a_k - b_k) = \sum_{k=1}^{n} a_k - \sum_{k=1}^{n} b_k$

3. $\sum_{k=1}^{n} ca_k = c\sum_{k=1}^{n} a_k$, where c is a constant.

EXAMPLE 6 ***Using Properties of Sigma Notation***

Use the properties of sums to evaluate

$$\sum_{k=1}^{4} (k^2 - 2)$$

SOLUTION

By Properties 2 and 3,

$$\sum_{k=1}^{4} (k^2 - 2) = \sum_{k=1}^{4} k^2 - \sum_{k=1}^{4} 2 = \sum_{k=1}^{4} k^2 - 8$$

$$= 1^2 + 2^2 + 3^2 + 4^2 - 8 = 30 - 8 = 22$$

Note that this is the same sum as was found in Example 5(b). ■

12.1c *Infinite Series*

Let us consider the sequence

$$a_1, a_2, a_3, \ldots, a_n, \ldots$$

We call the sum of terms

$$a_1 + a_2 + a_3 + \cdots + a_n + \cdots$$

the related **infinite series**, or simply, the related **series**. For this same sequence, we also define

$$S_1 = a_1$$
$$S_2 = a_1 + a_2$$
$$S_3 = a_1 + a_2 + a_3$$

and, in general,

$$S_n = \sum_{k=1}^{n} a_k = a_1 + a_2 + \cdots + a_n$$

The number S_n is called the ***n*th partial sum** of the sequence. The numbers

$$S_1, S_2, S_3, \ldots, S_n, \ldots$$

form a sequence called the **sequence of partial sums**. This type of sequence is studied in calculus courses, where methods are developed for analyzing infinite series.

EXAMPLE 7 ***Finding the nth Partial Sum***

Given the infinite sequence

$$a_n = n^2 - 1$$

find S_4.

SOLUTION

The first four terms of the sequence are

$$a_1 = 0, a_2 = 3, a_3 = 8, a_4 = 15$$

Then the sum S_4 is given by

$$S_4 = \sum_{k=1}^{4} a_k = 0 + 3 + 8 + 15 = 26$$

■

If a series alternates in sign, then a multiplicative factor of $(-1)^k$ or $(-1)^{k+1}$ can be used to produce the proper sign. For example, the series

$$-1^2 + 2^2 - 3^2 + 4^2$$

can be written in sigma notation as

$$\sum_{k=1}^{4} (-1)^k k^2$$

and the series

$$1^2 - 2^2 + 3^2 - 4^2$$

can be written as

$$\sum_{k=1}^{4} (-1)^{k+1} k^2$$

EXAMPLE 8 *Finding the General Term of a Sequence*

Consider the sequence

$$\sqrt{1}, -\sqrt{2}, \sqrt{3}, -\sqrt{4}, \ldots$$

Write an expression for the general term a_n and find the sum S_n using summation notation.

SOLUTION

If we multiply each term by $(-1)^{n+1}$, then the odd terms will be positive and the even terms negative. The general term a_n is then

$$a_n = (-1)^{n+1}\sqrt{n}$$

and the sequence is

$$\sqrt{1}, -\sqrt{2}, \sqrt{3}, -\sqrt{4}, \ldots, (-1)^{n+1}\sqrt{n}, \ldots$$

Finally, we have

$$S_n = \sum_{k=1}^{n} (-1)^{k+1}\sqrt{k}$$

■

FOCUS on Finding Areas Using Rectangles

Many textbooks introduce integral calculus using rectangles to approximate area. In the accompanying figure, we are interested in calculating the area under the graph of the function $f(x) = x^2$ that is bounded by the x-axis, the y-axis, and $x = 1$. The interval $[0, 1]$ is divided into n subintervals of equal width (here $n = 3$), and a rectangle is constructed in each interval as shown. We seek to use the sum of the areas of the rectangles as an approximation to the area under the curve.

Using 3 intervals

$x_1 = \frac{1}{3}$

$x_2 = \frac{2}{3}$

$x_3 = \frac{3}{3} = 1$

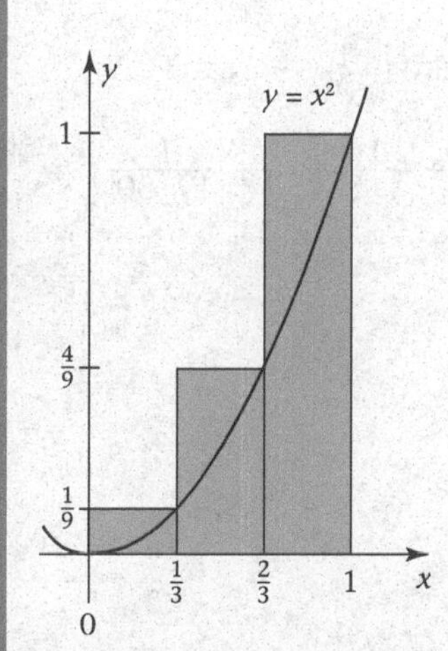

To calculate the area of a rectangle, we need to know the height and the width. Since the interval $[0, 1]$ has been divided into n parts of equal width, we see that

$$\text{Width of rectangle} = \frac{1-0}{n} = \frac{1}{n}$$

Now the height of the rectangle whose right endpoint is x_k (e.g., $x_2 = \frac{2}{3}$) is determined by the value of the function at that point, namely, $f(x_k) = x_k^2$ (e.g., $f\left(\frac{2}{3}\right) = \frac{4}{9}$). In general, for $k = 1, 2, \ldots, n$, the area of a "typical" rectangle whose right endpoint is $x_k = \frac{k}{n}$ is

$$(\text{Width})(\text{Height}) = \left(\frac{1}{n}\right)\left(\frac{k}{n}\right)^2$$

The sum of the areas of the rectangles is expressed in summation notation as

$$S_n = \sum_{k=1}^{n} \left(\frac{1}{n}\right)\left(\frac{k}{n}\right)^2 = \frac{1}{n^3}\sum_{k=1}^{n} k^2$$

Note that the greater the number of rectangles, the better the approximation. Verify that $S_3 = \frac{14}{27} \approx 0.5185$ and $S_{10} = \frac{77}{100} = 0.3850$. The actual area under the curve $f(x) = x^2$ in the interval $[0, 1]$ is $\frac{1}{3} \approx 0.333333\ldots$.

EXAMPLE 9 *Working with an Infinite Series*

Given the sequence

$$a_n = \frac{3}{10^n}$$

find the first five partial sums. Use these results to predict the sum of the infinite series

$$\sum_{k=1}^{\infty} \frac{3}{10^k}$$

SOLUTION

The terms of the sequence can be written as

$$0.3, 0.03, 0.003, 0.0003, 0.00003, \ldots$$

and the partial sums are then seen to be

$$S_1 = 0.3$$
$$S_2 = 0.3 + 0.03 = 0.33$$
$$S_3 = 0.3 + 0.03 + 0.003 = 0.333$$
$$S_4 = 0.3 + 0.03 + 0.003 + 0.0003 = 0.3333$$
$$S_5 = 0.3 + 0.03 + 0.003 + 0.0003 + 0.00003 = 0.33333$$

Since the decimal expansion of $\frac{1}{3}$ is $0.333333\ldots$, we predict that

$$\sum_{k=1}^{\infty} \frac{3}{10^k} = \frac{1}{3}$$

■

Exercise Set 12.1

In Exercises 1–12, find the first four terms and the twentieth term of the sequence whose nth term is given.

1. $a_n = 2n$
2. $a_n = 2n + 1$
3. $a_n = 4n - 3$
4. $a_n = 3n - 1$
5. $a_n = 5$
6. $a_n = 1 - \frac{1}{n}$
7. $a_n = \frac{n}{n+1}$
8. $a_n = \sqrt{n}$
9. $a_n = 2 + (0.1)^n$
10. $a_n = \frac{n^2 - 1}{n^2 + 1}$
11. $a_n = \frac{n^2}{2n+1}$
12. $a_n = \frac{2n+1}{n^2}$

In Exercises 13–18, a sequence is defined recursively. Find the indicated term of the sequence.

13. $a_n = 2a_{n-1} - 1$, $a_1 = 2$; find a_4
14. $a_n = 3 - 3a_{n-1}$, $a_1 = -1$; find a_3
15. $a_n = \frac{1}{a_{n-1} + 1}$, $a_3 = 2$; find a_6
16. $a_n = \frac{n}{a_{n-1}}$, $a_2 = 1$; find a_5
17. $a_n = (a_{n-1})^2$, $a_1 = 2$; find a_4
18. $a_n = (a_{n-1})^{n-1}$, $a_1 = 2$; find a_4

In Exercises 19–26, find the sum.

19. $\sum_{k=1}^{5} (3k - 1)$
20. $\sum_{k=1}^{5} (3 - 2k)$
21. $\sum_{k=1}^{6} (k^2 + 1)$
22. $\sum_{k=0}^{4} \frac{k}{k^2 + 1}$
23. $\sum_{k=3}^{5} \frac{k}{k-1}$
24. $\sum_{k=2}^{4} 4(2^k)$
25. $\sum_{j=1}^{4} 20$
26. $\sum_{i=1}^{10} 50$

In Exercises 27–36, use summation notation to express the sum. (The answer is not unique.)

27. $1 + 3 + 5 + 7 + 9$
28. $2 + 5 + 8 + 11 + 14$
29. $1 + 4 + 9 + 16 + 25$
30. $1 - 4 + 9 - 16 + 25$
31. $-1 + \frac{1}{\sqrt{2}} - \frac{1}{\sqrt{3}} + \frac{1}{\sqrt{4}}$
32. $\frac{1}{2 \cdot 4} + \frac{1}{2 \cdot 5} + \frac{1}{2 \cdot 6} + \frac{1}{2 \cdot 7}$
33. $\frac{1}{1^2 + 1} - \frac{2}{2^2 + 1} + \frac{3}{3^2 + 1} - \frac{4}{4^2 + 1}$
34. $2 - 4 + 8 - 16$
35. $1 + \frac{1}{x} + \frac{1}{x^2} + \frac{1}{x^3} + \cdots + \frac{1}{x^n}$
36. $\frac{1}{1 \cdot 2} + \frac{1}{2 \cdot 3} + \frac{1}{3 \cdot 4} + \frac{1}{4 \cdot 5} + \cdots + \frac{1}{49 \cdot 50}$

12.2 Arithmetic Sequences and Series

The sequence

$$2, 5, 8, 11, 14, 17, \ldots$$

is an example of a special type of sequence in which each successive term is obtained by adding a fixed number to the previous term.

Arithmetic Sequence

In an **arithmetic sequence**, there is a number d such that

$$a_n = a_{n-1} + d$$

for all $n > 1$. The number d is called the **common difference**.

An arithmetic sequence is also called an **arithmetic progression**. Returning to the sequence

$$2, 5, 8, 11, 14, 17, \ldots$$

the nth term can be defined recursively by

$$a_n = a_{n-1} + 3, \qquad a_1 = 2$$

This is an arithmetic progression with the first term equal to 2 and a common difference of 3.

EXAMPLE 1 *Finding the Terms of an Arithmetic Sequence*

Write the first four terms of an arithmetic sequence whose first term is −4 and whose common difference is −3.

SOLUTION

Beginning with −4, we add the common difference −3 to obtain

$$-4 + (-3) = -7 \qquad -7 + (-3) = -10 \qquad -10 + (-3) = -13$$

Alternatively, we note that the sequence is defined by

$$a_n = a_{n-1} - 3, \qquad a_1 = -4$$

which leads to the terms

$$a_1 = -4, \qquad a_2 = -7, \qquad a_3 = -10, \qquad a_4 = -13$$

■

Progress Check

Write the first four terms of an arithmetic sequence whose first term is 4 and whose common difference is $-\frac{1}{3}$.

Answer

$4, \frac{11}{3}, \frac{10}{3}, 3$

EXAMPLE 2 *An Arithmetic Sequence and Its Common Difference*

Show that the sequence

$$a_n = 2n - 1$$

is an arithmetic sequence, and find the common difference.

SOLUTION

We must show that the sequence satisfies

$$a_n - a_{n-1} = d$$

for some real number d. We have

$$a_n = 2n - 1$$
$$a_{n-1} = 2(n-1) - 1 = 2n - 3$$

so

$$a_n - a_{n-1} = 2n - 1 - (2n - 3) = 2$$

Therefore, we are dealing with an arithmetic sequence whose common difference is 2. ■

For a given arithmetic sequence, we now demonstrate one way to find a formula for the nth term a_n in terms of n and the first term a_1. Since

$$a_2 = a_1 + d$$

and

$$a_3 = a_2 + d$$

substituting for a_2

$$a_3 = (a_1 + d) + d = a_1 + 2d$$

Similarly, we can show that

$$a_4 = a_3 + d = (a_1 + 2d) + d = a_1 + 3d$$
$$a_5 = a_4 + d = (a_1 + 3d) + d = a_1 + 4d$$

We can obtain

The *n*th Term of an Arithmetic Sequence

Given common difference d and first term a_1, we have that the nth term

$$a_n = a_1 + (n-1)d \qquad n = 1, 2, 3, \ldots$$

Verify that the more general formula for a_n, when term a_k is given, becomes

$$a_n = a_k + (n-k)d \qquad k = 1, 2, 3, \ldots, n$$

EXAMPLE 3 *Finding a Specific Term of an Arithmetic Sequence*

Find the seventh term of the arithmetic progression whose first term is 2 and whose common difference is 4.

SOLUTION

We substitute $n = 7$, $a_1 = 2$, and $d = 4$ into the formula

$$a_n = a_1 + (n-1)d$$

obtaining

$$a_7 = 2 + (7-1)4 = 2 + 24 = 26$$ ■

Progress Check

Find the sixteenth term of the arithmetic progression whose first term is −5 and whose common difference is $\frac{1}{2}$.

Answer

$\frac{5}{2}$

EXAMPLE 4 *Finding a Specific Term of an Arithmetic Sequence*

Find the twenty-fifth term of the arithmetic sequence whose first and twentieth terms are −7 and 31, respectively.

SOLUTION

We can apply the given information to find d.

$$a_n = a_1 + (n-1)d$$
$$a_{20} = a_1 + (20-1)d$$
$$31 = -7 + 19d$$
$$d = 2$$

Now we use the formula for a_n to find a_{25}.

$$a_n = a_1 + (n-1)d$$
$$a_{25} = -7 + (25-1)2$$
$$a_{25} = 41$$

■

Progress Check

Find the sixtieth term of the arithmetic sequence whose first and tenth terms are 3 and $-\frac{3}{2}$, respectively.

Answer

$-\frac{53}{2}$

12.2a Arithmetic Series

The series associated with an arithmetic sequence is called an **arithmetic series**. Since an arithmetic sequence has a common difference d, we can write the nth partial sum S_n as

$$S_n = a_1 + (a_1 + d) + (a_1 + 2d) + \cdots + (a_n - 2d) + (a_n - d) + a_n \qquad (1)$$

where we write $a_2, a_3, \ldots$ in terms of a_1 and we write $a_{n-1}, a_{n-2}, \ldots$ in terms of a_n. Rewriting the right-hand side of Equation (1) in reverse order, we have

$$S_n = a_n + (a_n - d) + (a_n - 2d) + \cdots + (a_1 + 2d) + (a_1 + d) + a_1 \qquad (2)$$

We sum the corresponding sides of Equations (1) and (2), noting the cancellation of the terms with d.

$$2S_n = (a_1 + a_n) + (a_1 + a_n) + (a_1 + a_n) + \cdots + (a_1 + a_n)$$

Since there are n terms,

$$2S_n = n(a_1 + a_n)$$

Thus,

$$S_n = \frac{n}{2}(a_1 + a_n)$$

We now have two formulas.

The nth Partial Sum of an Arithmetic Series

For an arithmetic series with common difference d, first term a_1 and nth term a_n, we have that the nth partial sum:

$$S_n = \frac{n}{2}(a_1 + a_n)$$

$$S_n = \frac{n}{2}[2a_1 + (n - 1)d]$$

$$n = 1, 2, 3, \ldots$$

The choice of which formula to use depends on the available information. The following examples illustrate the use of the formulas.

EXAMPLE 5 *Summing an Arithmetic Series*

Find the sum of the first 30 terms of an arithmetic sequence whose first term is -20 and whose common difference is 3.

SOLUTION

We know that $n = 30$, $a_1 = -20$, and $d = 3$. Substituting into

$$S_n = \frac{n}{2}[2a_1 + (n - 1)d]$$

we obtain

$$S_{30} = \frac{30}{2}[2(-20) + (30 - 1)3]$$
$$= 15(-40 + 87)$$
$$= 705$$

■

Progress Check

Find the sum of the first 10 terms of the arithmetic sequence whose first term is 2 and whose common difference is $-\frac{1}{2}$.

Answer

$-\frac{5}{2}$

EXAMPLE 6 ***Working with an Arithmetic Series***

The first term of an arithmetic series is 2, the last term is 58, and the sum is 450. Find the number of terms and the common difference.

SOLUTION

We have $a_1 = 2$, $a_n = 58$, and $S_n = 450$. Substituting into

$$S_n = \frac{n}{2}(a_1 + a_n)$$

we have

$$450 = \frac{n}{2}(2 + 58)$$
$$900 = 60n$$
$$n = 15$$

Now we substitute into

$$a_n = a_1 + (n - 1)d$$
$$58 = 2 + (14)d$$
$$56 = 14d$$
$$d = 4$$

■

Progress Check

The first term of an arithmetic series is 6, the last term is 1, and the sum is $\frac{77}{2}$. Find the number of terms and the common difference.

Answers

$n = 11,\ d = -\frac{1}{2}$

Exercise Set 12.2

In Exercises 1–8, write the next two terms of each of the following arithmetic sequences:

1. 3, 6, 9, 12, ...
2. 2, −2, −6, −10, ...
3. $0, \frac{1}{4}, \frac{1}{2}, \frac{3}{4}, \ldots$
4. $y-4, y, y+4, y+8, \ldots$
5. 0, log 10, log 100, log 1000, ...
6. $4, \frac{11}{2}, 7, \frac{17}{2}, \ldots$
7. $\sqrt{5}-2, \sqrt{5}, \sqrt{5}+2, \sqrt{5}+4, \ldots$
8. 12, 8, 4, 0, ...

In Exercises 9–14, write the first four terms of the arithmetic sequence whose first term is a_1 and whose common difference is d.

9. $a_1 = 2, d = 4$
10. $a_1 = -2, d = -5$
11. $a_1 = 3, d = -\frac{1}{2}$
12. $a_1 = \frac{1}{2}, d = 2$
13. $a_1 = \frac{1}{3}, d = -\frac{1}{3}$
14. $a_1 = 6, d = \frac{5}{2}$

In Exercises 15–18, find the specified term of the arithmetic sequence whose first term is a_1 and whose common difference is d.

15. $a_1 = 4, d = 3$: eighth term
16. $a_1 = -3, d = \frac{1}{4}$: fourteenth term
17. $a_1 = 14, d = -2$: twelfth term
18. $a_1 = 6, d = -\frac{1}{3}$: ninth term

In Exercises 19–24, given two terms of an arithmetic sequence, find the specified term.

19. $a_1 = -2, a_{20} = -2$: twenty-fourth term
20. $a_1 = \frac{1}{2}, a_{12} = 6$: thirtieth term
21. $a_1 = 0, a_{61} = 20$: twentieth term
22. $a_1 = 23, a_{15} = -19$: sixth term
23. $a_1 = -\frac{1}{4}, a_{41} = 10$: twenty-second term
24. $a_1 = -3, a_{18} = 65$: thirtieth term

In Exercises 25–30, find the sum of the specified number of terms of the arithmetic sequence whose first term is a_1 and whose common difference is d.

25. $a_1 = 3, d = 2$: 20 terms
26. $a_1 = -4, d = \frac{1}{2}$: 24 terms
27. $a_1 = \frac{1}{2}, d = -2$: 12 terms
28. $a_1 = -3, d = -\frac{1}{3}$: 18 terms
29. $a_1 = 82, d = -2$: 40 terms
30. $a_1 = 6, d = 4$: 16 terms
31. How many terms of the arithmetic progression 2, 4, 6, 8, ... add up to 930?
32. How many terms of the arithmetic progression 44, 41, 38, 35, ... add up to 340?
33. The first term of an arithmetic series is 3, the last term is 90, and the sum is 1395. Find the number of terms and the common difference.
34. The first term of an arithmetic series is −3, the last term is $\frac{5}{2}$, and the sum is −3. Find the number of terms and the common difference.
35. The first term of an arithmetic series is $\frac{1}{2}$, the last term is $\frac{7}{4}$, and the sum is $\frac{27}{4}$. Find the number of terms and the common difference.
36. The first term of an arithmetic series is 20, the last term is −14, and the sum is 54. Find the number of terms and the common difference.
37. Find the sum of the first 16 terms of an arithmetic progression whose fourth and tenth terms are $-\frac{5}{4}$ and $\frac{1}{4}$, respectively.
38. Find the sum of the first 12 terms of an arithmetic progression whose third and sixth terms are 9 and 18, respectively.
39. Show that the sum of the first n natural numbers is
$$\frac{n(n+1)}{2}$$
40. Show that
$$1 + 3 + 5 + \cdots + (2n-1) = n^2$$

12.3 Geometric Sequences and Series

The sequence

$$3, 6, 12, 24, 48, \ldots$$

is an example of a special type of sequence in which each successive term is obtained by multiplying the previous term by a fixed number.

Geometric Sequence
In a **geometric sequence**, there is a number $r \neq 0$ and first term $a_1 \neq 0$ such that

$$a_n = ra_{n-1}$$

for all $n > 1$. The number r is called the **common ratio.**

A geometric sequence is also called a **geometric progression**. Returning to the sequence

$$3, 6, 12, 24, 48, \ldots$$

the nth term can be defined recursively by

$$a_n = 2a_{n-1}, \qquad a_1 = 3$$

This is a geometric progression with the first term equal to 3 and a common ratio of 2. In general, the common ratio r can be found by dividing any term a_k by the preceding term a_{k-1}.

In a geometric sequence, the common ratio r is given by

$$r = \frac{a_k}{a_{k-1}}$$

Let us look at successive terms of a geometric sequence whose first term is a_1 and whose common ratio is r. We have

$$a_2 = ra_1$$
$$a_3 = ra_2 = r(ra_1) = r^2a_1$$
$$a_4 = ra_3 = r(r^2a_1) = r^3a_1$$

The pattern suggests that the exponent of r is 1 less than the subscript of a on the left-hand side. We can obtain

The nth Term of a Geometric Sequence
Given a common ratio r and first term a_1, we observe that the nth term

$$a_n = a_1r^{n-1} \qquad n = 1, 2, 3, \ldots$$

Verify that the more general formula for a_n, when a_k is given, becomes

$$a_n = a_kr^{n-k} \qquad k = 1, 2, 3, \ldots, n$$

EXAMPLE 1 ***Finding a Specific Term of a Geometric Sequence***

Find the seventh term of the geometric sequence −4, −2, −1, ...

SOLUTION

Since

$$r = \frac{a_k}{a_{k-1}}$$

we see that

$$r = \frac{a_3}{a_2} = \frac{-1}{-2} = \frac{1}{2}$$

Substituting $a_1 = -4$, $r = \frac{1}{2}$, and $n = 7$, we have

$$a_n = a_1 r^{n-1}$$

$$a_7 = (-4)\left(\frac{1}{2}\right)^{7-1} = (-4)\left(\frac{1}{2}\right)^6$$

$$= (-4)\left(\frac{1}{64}\right) = -\frac{1}{16}$$

■

Progress Check

Find the sixth term of the geometric sequence 2, −6, 18, ...

Answer

−486

12.3a Geometric Mean

In a geometric sequence, the terms between the first and last terms are called **geometric means.** We will illustrate the method of calculating such means.

EXAMPLE 2 ***Inserting a Geometric Mean***

Insert three geometric means between 3 and 48.

SOLUTION

The geometric sequence must look like this.

$$3, a_2, a_3, a_4, 48, \ldots$$

Thus, $a_1 = 3$, $a_5 = 48$, and $n = 5$. Substituting into

$$a_n = a_1 r^{n-1}$$

$$48 = 3r^4$$

$$r^4 = 16$$

$$r = \pm 2$$

Thus, there are two geometric sequences with three geometric means between 3 and 48.

$$3, \ 6, 12, \ 24, 48, \ldots \text{ when } r = 2$$

$$3, -6, 12, -24, 48, \ldots \text{ when } r' = -2$$

■

Insert two geometric means between 5 and $\frac{8}{25}$.

Answer

2, $\frac{4}{5}$

12.3b Geometric Series

If $a_1, a_2, \ldots$ is a geometric sequence, then the nth partial sum

$$S_n = a_1 + a_2 + \cdots + a_n \tag{1}$$

is called a **geometric series.** Since each term of the series can be rewritten as $a_k = a_1 r^{k-1}$, we can rewrite Equation (1) as

$$S_n = a_1 + a_1 r + a_1 r^2 + \cdots + a_1 r^{n-2} + a_1 r^{n-1} \tag{2}$$

Multiplying each term in Equation (2) by r, we have

$$rS_n = a_1 r + a_1 r^2 + a_1 r^3 + \cdots + a_1 r^{n-1} + a_1 r^n \tag{3}$$

Subtracting Equation (3) from Equation (2) produces

$$S_n - rS_n = a_1 - a_1 r^n$$

Factoring both sides

$$S_n(1 - r) = a_1(1 - r^n)$$

If $r \neq 1$, we may divide both sides of the equation by $(1 - r)$ to obtain

$$S_n = \frac{a_1(1-r^n)}{1-r}$$

This may be summarized as

The *n*th Partial Sum of a Geometric Series

For a geometric series with common ratio $r \neq 1$ and first term a_1, we have that the nth partial sum

$$S_n = \frac{a_1(1-r^n)}{1-r}$$

EXAMPLE 3 *Summing a Geometric Series*

Find the sum of the first six terms of the geometric sequence whose first three terms are 12, 6, 3.

SOLUTION

The common ratio can be found by dividing any term by the preceding term.

$$r = \frac{a_k}{a_{k-1}} = \frac{a_2}{a_1} = \frac{6}{12} = \frac{1}{2}$$

Substituting $a_1 = 12$, $r = \frac{1}{2}$, and $n = 6$ into the formula for S_n, we have

$$S_6 = \frac{a_1(1-r^6)}{1-r} = \frac{12\left[1-\left(\frac{1}{2}\right)^6\right]}{1-\frac{1}{2}} = \frac{189}{8}$$

■

Progress Check

Find the sum of the first five terms of the geometric sequence whose first three terms are 2, $-\frac{4}{3}$, $\frac{8}{9}$.

Answer

$\frac{110}{81}$

EXAMPLE 4 *An Application of Geometric Series*

A father promises to give each child 2 cents on the first day, 4 cents on the second day, and to continue doubling the amount each day for a total of 8 days. How much will each child receive on the last day? How much will each child have received in total after 8 days?

SOLUTION

The daily payout to each child forms a geometric sequence 2, 4, 8, ... with $a_1 = 2$ and $r = 2$. The term a_8 is given by substituting into

$$a_n = a_1 r^{n-1}$$
$$a_8 = a_1 r^{8-1} = 2 \cdot 2^7 = 256$$

Thus, each child will receive \$2.56 on the last day. The total received by each child is given by

$$S_n = \frac{a_1(1-r^n)}{1-r}$$

$$S_8 = \frac{a_1(1-r^8)}{1-r} = \frac{2(1-2^8)}{1-2}$$

$$= \frac{2(1-256)}{-1} = 510$$

Each child will have received a total of \$5.10 after 8 days. ■

Progress Check

A ball is dropped from a height of 64 feet. On each bounce, it rebounds half the height it fell. (See Figure 12-1.) How high is the ball at the top of the fifth bounce? What is the total vertical distance the ball has traveled at the top of the fifth bounce?

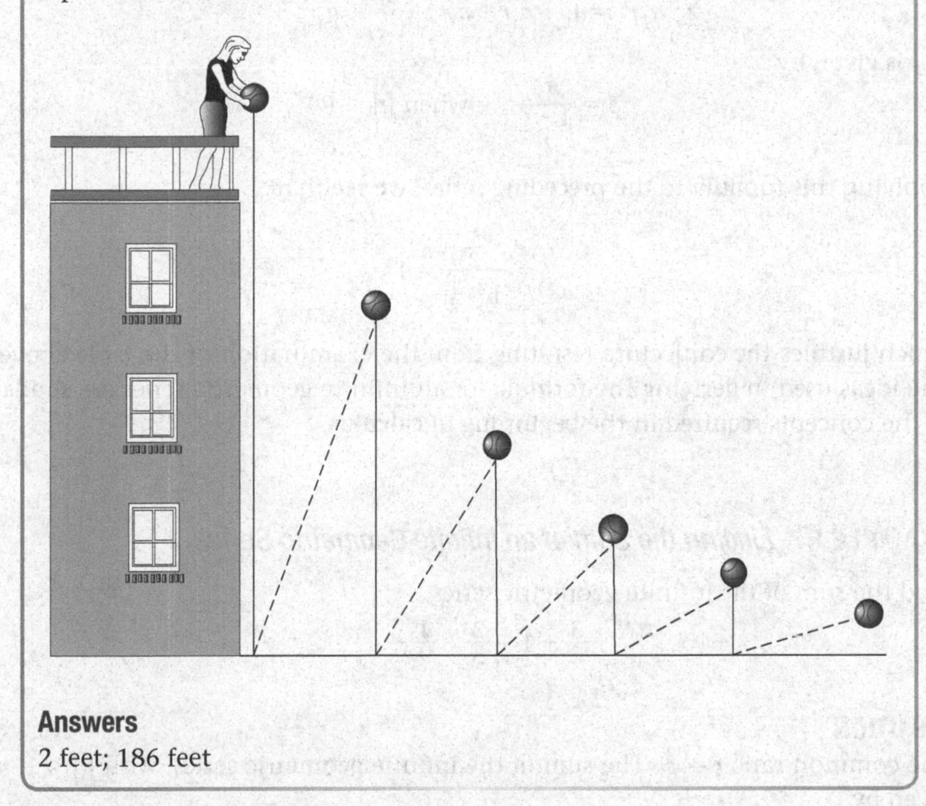

FIGURE 12-1
Bouncing Ball

Answers
2 feet; 186 feet

12.3c *Infinite Geometric Series*

We now want to focus on a geometric series for which $|r| < 1$, say

$$\frac{1}{2} + \frac{1}{4} + \frac{1}{8} + \cdots + \frac{1}{2^n} + \cdots$$

To see how the sum increases as n increases, we form a table of values of S_n to three decimal places.

n	1	2	3	4	5	6	7	8	9
S_n	0.500	0.750	0.875	0.938	0.969	0.984	0.992	0.996	0.998

We begin to suspect that S_n gets closer and closer to 1 as n increases. To see that this is really so, look at the formula

$$S_n = \frac{a_1(1 - r^n)}{1 - r}$$

when $|r| < 1$. When a number r that is less than 1 in absolute value is raised to higher and higher positive integer powers, the value of $|r^n|$ approaches zero. Thus, the term

r^n can be made as close to 0 as we like by choosing n sufficiently large. Since we are dealing with an infinite series, we say that "r^n approaches zero as n approaches infinity." We then replace r^n with 0 in the formula and denote the sum by S.

Sum of an Infinite Geometric Series
The sum S of the **infinite geometric series**

$$\sum_{k=0}^{\infty} a_1 r^k = a_1 + a_1 r + a_1 r^2 + \cdots + a_1 r^n + \cdots$$

is given by

$$S = \frac{a_1}{1-r} \qquad \text{when } |r| < 1$$

Applying this formula to the preceding series, we see that

$$S = \frac{\frac{1}{2}}{1-\frac{1}{2}} = 1$$

which justifies the conjecture resulting from the examination of the table above. The ideas used in deriving the formula for an infinite geometric series are similar to the concepts required in the beginning of calculus.

EXAMPLE 5 *Finding the Sum of an Infinite Geometric Series*

Find the sum of the infinite geometric series.

$$\frac{3}{2} + 1 + \frac{2}{3} + \frac{4}{9} + \cdots$$

SOLUTION
The common ratio $r = \frac{2}{3}$. The sum of the infinite geometric series, with $|r| < 1$, is given by

$$S = \frac{a_1}{1-r} = \frac{\frac{3}{2}}{1-\frac{2}{3}} = \frac{9}{2}$$ ■

Progress Check

Find the sum of the infinite geometric series

$$4 - 1 + \frac{1}{4} - \frac{1}{16} + \cdots$$

Answer
$\frac{16}{5}$

The notation

$$0.6525\overline{252}$$

indicates a **repeating decimal** with a pattern in which 52 is repeated indefinitely. Every repeating decimal can be written as a rational number. We will apply the formula for the sum of an infinite geometric series to find the rational number equal to a repeating decimal.

EXAMPLE 6 ***Repeating Decimals as an Infinite Geometric Series***

Find the rational number that is equal to $0.65252\overline{52}$.

SOLUTION

Note that

$$0.65252\overline{52} = 0.6 + 0.052 + 0.00052 + 0.0000052 + \cdots$$

We treat the sum

$$0.052 + 0.00052 + 0.0000052 + \cdots$$

as an infinite geometric series with $a = 0.052$ and $r = 0.01$. Then

$$S = \frac{a}{1-r} = \frac{0.052}{1-0.01} = \frac{0.052}{0.99} = \frac{52}{990}$$

and the repeating decimal is equal to

$$0.6 + \frac{52}{990} = \frac{6}{10} + \frac{52}{990} = \frac{646}{990} = \frac{323}{495}$$ ■

Progress Check

Write the repeating decimal $2.5454\overline{54}$ as a rational number.

Answer

$\frac{28}{11}$

Exercise Set 12.3

In Exercises 1–6, find the next term of the given geometric sequence.

1. $3, 6, 12, 24, \ldots$
2. $-4, 12, -36, 108, \ldots$
3. $-4, 3, -\frac{9}{4}, \frac{27}{16}, \ldots$
4. $2, -1, \frac{1}{2}, -\frac{1}{4}, \ldots$
5. $1.2, 0.24, 0.048, \ldots$
6. $\frac{1}{8}, \frac{1}{2}, 2, 8, \ldots$

In Exercises 7–12, write the first four terms of the geometric sequence whose first term is a_1 and whose common ratio is r.

7. $a_1 = 3, r = 3$
8. $a_1 = -4, r = 2$
9. $a_1 = 4, r = \frac{1}{2}$
10. $a_1 = 16, r = -\frac{3}{2}$
11. $a_1 = -3, r = 2$
12. $a_1 = 3, r = -\frac{2}{3}$

In Exercises 13–24, use the information given about a geometric sequence to find the item indicated.

13. If $a_1 = 3$ and $r = -2$, find a_8.
14. If $a_1 = 18$ and $r = -\frac{1}{2}$, find a_6.
15. If $a_1 = 16$ and $a_2 = 8$, find a_7.
16. If $a_1 = 15$ and $a_2 = -10$, find a_6.
17. If $a_1 = 3$ and $a_5 = \frac{1}{27}$, find a_7.
18. If $a_1 = 2$ and $a_6 = \frac{1}{16}$, find a_3.
19. If $a_1 = \frac{16}{81}$ and $a_6 = \frac{3}{2}$, find a_8.
20. If $a_4 = \frac{1}{4}$ and $a_7 = 1$, find r.
21. If $a_2 = 4$ and $a_8 = 256$, find r.
22. If $a_3 = 3$ and $a_6 = -81$, find a_8.
23. If $a_1 = \frac{1}{2}$, $r = 2$, and $a_n = 32$, find n.
24. If $a_1 = -2$, $r = 3$, and $a_n = -162$, find n.
25. Insert two geometric means between $\frac{1}{3}$ and 9.
26. Insert two geometric means between -3 and 192.
27. Insert two geometric means between 1 and $\frac{1}{64}$.
28. Insert three geometric means between $\frac{2}{3}$ and $\frac{32}{243}$.

In Exercises 29–32, find the partial sum for the geometric sequence whose first three terms are given.

29. If $a_1 = 3$, $a_2 = 1$, and $a_3 = \frac{1}{3}$, find S_7.
30. If $a_1 = \frac{1}{3}$, $a_2 = 1$, and $a_3 = 3$, find S_6.
31. If $a_1 = -3$, $a_2 = \frac{6}{5}$, and $a_3 = -\frac{12}{25}$, find S_5.
32. If $a_1 = 2$, $a_2 = \frac{4}{3}$, and $a_3 = \frac{8}{9}$, find S_6.

In Exercises 33–36, use the information given about a geometric sequence to find the requested partial sum.

33. If $a_1 = 4$ and $r = 2$, find S_8.
34. If $a_1 = -\frac{1}{2}$ and $r = -3$, find S_{10}.
35. If $a_1 = 2$ and $a_4 = -\frac{27}{4}$, find S_5.
36. If $a_1 = 64$ and $a_7 = 1$, find S_6.
37. A Christmas Club calls for savings of $5 by the end of January and twice as much in each successive month as in the previous month. How much money will have been saved by the end of November?
38. A city had 20,000 people in 2016. If the population increases 5% per year, how many people will the city have in 2026?
39. A city had 30,000 people in 2010. If the population increases 25% every 10 years, how many people will the city have in the year 2040?
40. For good behavior a child is offered a reward consisting of 1 cent on the first day, 2 cents on the second day, 4 cents on the third day, and so on. If the child behaves properly for two weeks, what is the total amount that the child will receive?

In Exercises 41–49, evaluate the sum of each infinite geometric series.

41. $1 + \frac{1}{2} + \frac{1}{4} + \frac{1}{8} + \ldots$
42. $\frac{4}{5} + \frac{1}{5} + \frac{1}{20} + \frac{1}{80} + \ldots$
43. $1 - \frac{1}{3} + \frac{1}{9} - \frac{1}{27} + \ldots$
44. $\frac{1}{2} - \frac{1}{4} + \frac{1}{8} - \frac{1}{16} + \ldots$
45. $2 + \frac{1}{2} + \frac{1}{8} + \frac{1}{32} + \ldots$
46. $1 + 0.1 + 0.01 + 0.001 + \ldots$
47. $0.5 + (0.5)^2 + (0.5)^3 + (0.5)^4 + \ldots$
48. $\frac{2}{5} + \frac{4}{25} + \frac{8}{125} + \frac{16}{625} + \ldots$
49. $\frac{1}{3} - \frac{2}{9} + \frac{4}{27} - \frac{8}{81} + \ldots$
50. Find the rational number equal to $3.666\overline{6}$.
51. Find the rational number equal to $0.3676\overline{767}$.
52. Find the rational number equal to $4.1414\overline{14}$.
53. Find the rational number equal to $0.325\overline{325}$.
54. Find the rational number equal to $0.999\overline{9}$.

12.4 Mathematical Induction

There are times when we wish to prove a result that is applicable to all of the natural numbers. If the set of natural numbers were finite, we could simply verify that result for each natural number and thereby conclude our proof. However, since the set of natural numbers is infinite, we need an alternative approach. **Mathematical induction** is a method of proof that permits us to accomplish this.

Consider the sums of the consecutive positive odd integers.

$$\begin{aligned} 1 &= 1 \\ 1+3 &= 4 \\ 1+3+5 &= 9 \\ 1+3+5+7 &= 16 \\ 1+3+5+7+9 &= 25 \end{aligned}$$

The first five terms of the sequence of these partial sums are

$$1, 4, 9, 16, 25$$

The sequence $a_n = n^2$ also has as its first five terms

$$1, 4, 9, 16, 25$$

Are these two sequences identical for all values of n? If we try $n = 6$, we obtain

$$1 + 3 + 5 + 7 + 9 + 11 = 36$$

Indeed, the sum of the first six positive odd integers is 36. This strengthens our suspicion that the result may hold in general. However, we cannot possibly verify a theorem for *all* positive integers by testing one integer at a time. At this point we turn to the principle of mathematical induction.

Principle of Mathematical Induction

Consider a statement involving a natural number n. We may prove that this statement is true for all natural numbers n if we do the following:

1. Prove that this statement is true when $n = 1$.
2. Prove that whenever this statement is true for the natural number $n = k$, it is also true for $n = k + 1$.

What does it mean when we say that a statement is true if it satisfies the principle of mathematical induction? Part 1 says that we must verify the validity of the statement for $n = 1$. Since this establishes the basis for the statement to be proved, it is called the *basis step*. Establishing Part 2, called the *induction step*, allows us to verify the validity of the statement for $n > 1$. In other words, if $n = k = 1$ in Part 2, the statement is also true for $n = k + 1 = 1 + 1 = 2$. However, if we now let $n = k = 2$ in Part 2, then the statement is also true for $n = k + 1 = 2 + 1 = 3$. The effect is similar to an endless string of dominoes, whereby the falling of the first domino causes each successive domino to fall. Thus, the principle has established the validity of the statement for *all* natural numbers.

Although the basis step is usually proved in a rather straightforward manner using substitution, it is the induction step that requires a certain degree of ingenuity. We outline the steps involved in applying the principle of mathematical induction in the following example.

EXAMPLE 1 *Applying the Principle of Mathematical Induction*

Prove that the sum of the first n consecutive positive integers is given by

$$\frac{n(n+1)}{2}$$

SOLUTION

Step 1. Verify that the statement is true for $n = 1$.	*Step 1.* The "sum" of the first integer is 1. Evaluating the formula for $n = 1$ yields $$\frac{1(1+1)}{2} = \frac{2}{2} = 1$$ which verifies the formula for $n = 1$.
Step 2. Assume the statement is true for $n = k$. Show that it is true for $n = k + 1$.	*Step 2.* For $n = k$, we have $$1 + 2 + 3 + \cdots + k = \frac{k(k+1)}{2}$$ Adding $k + 1$, the next consecutive integer, to both sides, we obtain $$1 + 2 + \cdots + k + (k+1) = \frac{k(k+1)}{2} + (k+1)$$ $$= (k+1)\left(\frac{k}{2}+1\right)$$ $$= (k+1)\left(\frac{k+2}{2}\right)$$ $$= \frac{1}{2}(k+1)(k+2)$$ Thus, the formula holds for $n = k + 1$. By the principle of mathematical induction, it is then true for all positive integer values of n. ■

EXAMPLE 2 *Applying the Principle of Mathematical Induction*

Prove that the sum of the first n consecutive odd positive integers is n^2.

SOLUTION

To verify the formula for $n = 1$, we need only observe that $1 = 1^2$.

The following table shows the correspondence between the natural numbers and the odd integers. We see that when $n = k$, the value of the nth consecutive odd integer is $2k - 1$. Since the formula is assumed to be true for $n = k$, we have

n	1	2	3	4	…	k
nth odd integer	1	3	5	7	…	$2k - 1$

$$1 + 3 + 5 + \cdots + (2k-1) = k^2$$

Adding $2k + 1$, the next consecutive odd integer, to both sides, we obtain

$$1 + 3 + \cdots + (2k-1) + (2k+1) = k^2 + (2k+1)$$

or

$$1 + 3 + \cdots + (2k+1) = (k+1)^2$$

Thus, the sum of the first $k + 1$ consecutive odd integers is $(k+1)^2$. By the principle of mathematical induction, the formula is true for all positive values of n. ■

Many of the theorems in this book can be proved using mathematical induction. Next is an example applied to a basic property of positive integer exponents.

EXAMPLE 3 *Applying the Principle of Mathematical Induction*

Prove that $(xy)^n = x^n y^n$ for all positive integer values of n.

SOLUTION

For $n = 1$, we have

$$(xy)^1 = xy = x^1 y^1$$

which verifies the validity of the statement for $n = 1$. Assuming that the statement holds for $n = k$, we have

$$(xy)^k = x^k y^k$$

To show that the statement holds for $n = k + 1$, we write

$(xy)^{k+1} = (xy)^k(xy)$	Definition of exponents
$= (x^k y^k)(xy)$	Statement holds for $n = k$
$= (x^k x)(y^k y)$	Associative and commutative laws
$= x^{k+1} y^{k+1}$	Definition of exponents

Thus, the statement holds for $n = k + 1$; and by the principle of mathematical induction, the statement holds for all integer values of n. ■

Exercise Set 12.4

In Exercises 1–10, prove that the statement is true for all positive integer values of n by using the principle of mathematical induction.

1. $2 + 4 + 6 + \cdots + 2n = n(n + 1)$
2. $1^2 + 3^2 + 5^2 + \cdots + (2n - 1)^2 = \dfrac{n(2n+1)(2n-1)}{3}$
3. $2 + 5 + 8 + \cdots + (3n - 1) = \dfrac{n(3n+1)}{2}$
4. $4 + 8 + 12 + \cdots + 4n = 2n(n + 1)$
5. $5 + 10 + 15 + \cdots + 5n = \dfrac{5n(n+1)}{2}$
6. $1^2 + 2^2 + 3^2 + \cdots + n^2 = \dfrac{n(n+1)(2n+1)}{6}$
7. $1 \cdot 2 + 2 \cdot 3 + 3 \cdot 4 + \cdots + n(n+1) = \dfrac{n(n+1)(n+2)}{3}$
8. $1^3 + 2^3 + 3^3 + \cdots + n^3 = \dfrac{n^2(n+1)^2}{4}$
9. $1 + 5 + 9 + \cdots + (4n - 3) = n(2n - 1)$
10. $\left(\dfrac{x}{y}\right)^n = \dfrac{x^n}{y^n}$
11. Prove that the nth term a_n of an arithmetic progression whose first term is a_1 and whose common difference is d is given by $a_n = a_1 + (n - 1)d$.
12. Prove that the nth term a_n of a geometric progression whose first term is a_1 and whose common ratio is r is given by $a_n = a_1 r^{n-1}$.
13. Prove that $2 + 2^2 + 2^3 + \cdots + 2^n = 2^{n+1} - 2$.
14. Prove that $a + ar + ar^2 + \cdots + ar^{n-1} = \dfrac{a(1-r^n)}{1-r}$.
15. Prove that $x^n - 1$ is divisible by $x - 1$, $x \neq 1$. [*Hint*: Recall that divisibility requires the existence of a polynomial $Q(x)$ such that $x^n - 1 = (x - 1)Q(x)$.]
16. Prove that $x^n - y^n$ is divisible by $x - y$, $x \neq y$. [*Hint*: Note that $x^{n+1} - y^{n+1} = (x^{n+1} - xy^n) + (xy^n - y^{n+1})$.]
17. If a is a real number such that $a > -1$, prove that

    $(1 + a)^n \geq 1 + na$ (Bernoulli's Inequality)

    for every positive integer n.

12.5 The Binomial Theorem

If we multiply each of the following equations by $(a + b)$, we obtain the equation just below it:

$$(a+b)^1 = a + b$$
$$(a+b)^2 = a^2 + 2ab + b^2$$
$$(a+b)^3 = a^3 + 3a^2b + 3ab^2 + b^3$$
$$(a+b)^4 = a^4 + 4a^3b + 6a^2b^2 + 4ab^3 + b^4$$
$$(a+b)^5 = a^5 + 5a^4b + 10a^3b^2 + 10a^2b^3 + 5ab^4 + b^5$$

The expression on the right-hand side of the equation is called the **binomial expansion of $(a + b)^n$**. We present in Table 12-1 a number of properties associated with these expansions that help us to compute the expansion of $(a+b)^n$ for any natural number n, without performing the actual multiplications.

TABLE 12-1 Properties of the Binomial Expansion $(a+b)^n$ $n = 1, 2, 3, \ldots$

1. The expansion has $n + 1$ terms.
2. The first term is a^n and the last term is b^n.
3. The sum of the exponents of a and b in each term is n.
4. In each successive term after the first, the exponent of a decreases by 1 and the exponent of b increases by 1.
5. For $k = 1, 2, 3, \ldots, n-1$ $\text{Coefficient of term } k+1 = \frac{(\text{Coefficient of term } k)(\text{Exponent of } a \text{ in term } k)}{k}$
6. The coefficients may also be obtained from the following array, which is known as **Pascal's Triangle.** (See Figure 12-2.) Each number, with the exception of those at the ends of the rows, is the sum of the two nearest numbers in the row above. (For example, $10 = 4 + 6$, and $3 = 2 + 1$, as shown in Figure 12-2.) The numbers at the ends of the rows are always 1.

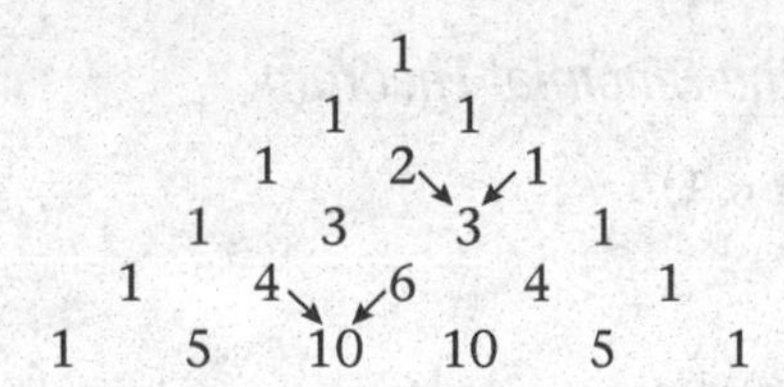

FIGURE 12-2
Pascal's Triangle

EXAMPLE 1 *Binomial Expansion*

Write the expansion of $(a + b)^6$.

SOLUTION

From Property 2, we know that the first term is a^6 and the last term is b^6. Thus,

$$(a+b)^6 = a^6 + \cdots + b^6$$

From Property 5, the next coefficient (term 2) is

$$\frac{1 \cdot 6}{1} = 6$$

By Property 4, the exponents of a and b in this term are 5 and 1, respectively. We have

$$(a+b)^6 = a^6 + 6a^5b + \cdots + b^6$$

Applying Property 5 again, we find that the next coefficient is

$$\frac{6 \cdot 5}{2} = 15$$

and by Property 4, the exponents of a and b in this term are 4 and 2, respectively. Thus,

$$(a+b)^6 = a^6 + 6a^5b + 15a^4b^2 + \cdots + b^6$$

Continuing in this manner, we see that

$$(a+b)^6 = a^6 + 6a^5b + 15a^4b^2 + 20a^3b^3 + 15a^2b^4 + 6ab^5 + b^6$$ ■

Progress Check

Write the first five terms in the expansion of $(a+b)^{10}$.

Answer

$a^{10} + 10a^9b + 45a^8b^2 + 120a^7b^3 + 210a^6b^4$

The expansion of $(a+b)^n$ that we have described is called the **binomial theorem**, or **binomial formula**, and can be written as

The Binomial Theorem

$$(a+b)^n = a^n + \frac{n}{1}a^{n-1}b + \frac{n(n-1)}{1 \cdot 2}a^{n-2}b^2 + \frac{n(n-1)(n-2)}{1 \cdot 2 \cdot 3}a^{n-3}b^3$$

$$+ \cdots + \frac{n(n-1)(n-2)\cdots(n-r+1)}{1 \cdot 2 \cdot 3 \cdots r}a^{n-r}b^r + \cdots + b^n$$

The binomial theorem can be proved by the method of mathematical induction discussed in the preceding section.

EXAMPLE 2 *Applying the Binomial Theorem*

Find the expansion of $(2x-1)^4$.

SOLUTION

Since $(2x-1)^4 = (2x+(-1))^4$, let $a = 2x$, $b = -1$, and apply the binomial theorem.

$$(2x-1)^4 = (2x)^4 + \frac{4}{1}(2x)^3(-1) + \frac{4 \cdot 3}{1 \cdot 2}(2x)^2(-1)^2 + \frac{4 \cdot 3 \cdot 2}{1 \cdot 2 \cdot 3}(2x)(-1)^3 + (-1)^4$$

$$= 16x^4 - 32x^3 + 24x^2 - 8x + 1$$ ■

Progress Check

Find the expansion of $(x^2-2)^4$.

Answer

$x^8 - 8x^6 + 24x^4 - 32x^2 + 16$

12.5a Factorial Notation

Note that the denominator of the coefficient in the binomial formula is always the product of the first n natural numbers. We use the symbol $n!$, which is read as **n factorial**, to indicate this type of product. For example,

$$1! = 1$$
$$2! = 2 \cdot 1 = 2$$
$$3! = 3 \cdot 2 \cdot 1 = 6$$
$$4! = 4 \cdot 3 \cdot 2 \cdot 1 = 24$$
$$5! = 5 \cdot 4 \cdot 3 \cdot 2 \cdot 1 = 120$$
$$6! = 6 \cdot 5 \cdot 4 \cdot 3 \cdot 2 \cdot 1 = 720$$

In general,

$$n! = n(n-1)(n-2) \cdot \cdots \cdot 4 \cdot 3 \cdot 2 \cdot 1 \qquad n = 1, 2, 3, \ldots$$

Since

$$(n-1)! = (n-1)(n-2)(n-3) \cdot \cdots \cdot 4 \cdot 3 \cdot 2 \cdot 1$$

we see that

$$n! = n(n-1)! \qquad n = 2, 3, 4, \ldots$$

For convenience, we define 0! as

$$0! = 1$$

EXAMPLE 3 *Working with Factorial Notation*

Evaluate each of the following:

a. $\frac{5!}{3!}$ b. $\frac{9!}{8!}$ c. $\frac{10!4!}{12!}$ d. $\frac{n!}{(n-2)!}$ e. $\frac{(2-2)!}{3!}$

SOLUTION

a. Since $5! = 5 \cdot 4 \cdot 3!$, we may write

$$\frac{5!}{3!} = \frac{5 \cdot 4 \cdot 3!}{3!} = 5 \cdot 4 = 20$$

b. $\frac{9!}{8!} = \frac{9 \cdot 8!}{8!} = 9$

c. $\frac{10!4!}{12!} = \frac{10!4!}{12 \cdot 11 \cdot 10!} = \frac{4!}{12 \cdot 11} = \frac{4 \cdot 3 \cdot 2 \cdot 1}{12 \cdot 11} = \frac{2}{11}$

d. $\frac{n!}{(n-2)!} = \frac{n(n-1)(n-2)!}{(n-2)!} = n(n-1) = n^2 - n$

e. $\frac{(2-2)!}{3!} = \frac{0!}{3 \cdot 2 \cdot 1} = \frac{1}{6}$ ■

Evaluate each of the following.

a. $\frac{12!}{10!}$ b. $\frac{6!}{4!2!}$ c. $\frac{10!8!}{9!7!}$ d. $\frac{n!(n-1)!}{(n+1)!(n-2)!}$ e. $\frac{8!}{6!(3-3)!}$

Answers

a. 132 b. 15 c. 80 d. $\frac{n-1}{n+1}$ e. 56

Using factorial notation, we have

The Binomial Theorem

$$(a+b)^n = a^n + \frac{n!}{1!(n-1)!}a^{n-1}b + \frac{n!}{2!(n-2)!}a^{n-2}b^2 + \frac{n!}{3!(n-3)!}a^{n-3}b^3 + \cdots + \frac{n!}{r!(n-r)!}a^{n-r}b^r + \cdots + b^n$$

The symbol

$$\binom{n}{r}$$

is called the **binomial coefficient**. It is defined as

Binomial Coefficient

$$\binom{n}{r} = \frac{n!}{r!(n-r)!} \qquad r = 0, 1, 2, \ldots, n$$

Note the special cases:

$$\binom{n}{0} = \frac{n!}{0!n!} = 1$$

and

$$\binom{n}{n} = \frac{n!}{n!0!} = 1$$

This symbol is useful in denoting the coefficients of the binomial expansion. Using this notation, the binomial theorem can be written as

The Binomial Theorem

$$(a+b)^n = \binom{n}{0}a^nb^0 + \binom{n}{1}a^{n-1}b^1 + \binom{n}{2}a^{n-2}b^2 + \cdots + \binom{n}{k}a^{n-k}b^k + \cdots + \binom{n}{n}a^0b^n$$
$$= \sum_{k=0}^{n}\binom{n}{k}a^{n-k}b^k$$

Sometimes we only want to find a certain term in the expansion of $(a+b)^n$. From the above presentation of the binomial theorem, we observe that the first term is

$$\binom{n}{0}a^nb^0$$

the second term is

$$\binom{n}{1}a^{n-1}b^1$$

and, in general, the $(k+1)$st term is

$$\binom{n}{k}a^{n-k}b^k, \qquad k = 0, 1, \ldots, n$$

EXAMPLE 4 ***Finding a Specified Term of a Binomial Expansion***

Find the fourth term in the expansion of $(x - 1)^5$.

SOLUTION

The fourth term of $(a + b)^5$ is

$$\binom{5}{3} a^2b^3$$

Since $a = x$ and $b = -1$, the fourth term is

$$\frac{5!}{3!2!}x^2 (-1)^3 = -10x^2$$

■

Progress Check

Find the third term in the expansion of

$$\left(\frac{x}{2} - 1\right)^8$$

Answer

$\frac{7}{16}x^6$

EXAMPLE 5 ***Finding a Specified Term of a Binomial Expansion***

Find the term in the expansion of $(x^2 - y^2)^6$ that involves y^8.

SOLUTION

Since $y^8 = (-y^2)^4$, we seek the term that involves b^4 in the expansion of $(a + b)^6$. Now b^4 occurs in the fifth term, which is

$$\binom{6}{4} a^2b^4$$

Since $a = x^2$ and $b = -y^2$, the desired term is

$$\frac{6!}{4!2!}(x^2)^2(-y^2)^4 = 15x^4y^8$$

■

Progress Check

Find the term in the expansion of $(x^3 - \sqrt{2})^5$ that involves x^6.

Answer

$-20\sqrt{2}x^6$

Exercise Set 12.5

In Exercises 1–12, expand and simplify.

1. $(3x+2y)^5$
2. $(2a-3b)^6$
3. $(4x-y)^4$
4. $\left(3+\frac{1}{2}x\right)^4$
5. $(2-xy)^5$
6. $(3a^2+b)^4$
7. $(a^2b+3)^4$
8. $(x-y)^7$
9. $(a-2b)^8$
10. $\left(\frac{x}{y}+y\right)^6$
11. $\left(\frac{1}{3}x+2\right)^3$
12. $\left(\frac{x}{y}+\frac{y}{x}\right)^5$

In Exercises 13–20, find the first four terms in the given expansion and simplify.

13. $(2+x)^{10}$
14. $(x-3)^{12}$
15. $(3-2a)^9$
16. $(a^2+b^2)^{11}$
17. $(2x-3y)^{14}$
18. $\left(a-\frac{1}{a^2}\right)^8$
19. $(2x-yz)^{13}$
20. $\left(x-\frac{1}{y}\right)^{15}$

In Exercises 21–32, evaluate the expression.

21. $5!$
22. $7!$
23. $\frac{12!}{11!}$
24. $\frac{13!}{12!}$
25. $\frac{11!}{8!}$
26. $\frac{7!}{9!}$
27. $\frac{10!}{6!}$
28. $\frac{9!}{6!}$
29. $\frac{6!}{3!}$
30. $\binom{8}{5}$
31. $\binom{10}{6}$
32. $\frac{(n+1)!}{(n-1)!}$

In Exercises 33–46, find the specified term in the expression.

33. The fourth term in $(2x-4)^7$.
34. The third term in $(4a+3b)^{11}$.
35. The fifth term in
$$\left(\frac{1}{2}x-y\right)^{12}$$
36. The sixth term in $(3x-2y)^{10}$.
37. The fifth term in
$$\left(\frac{1}{x}-2\right)^9$$
38. The next to last term in $(a+4b)^5$.
39. The middle term in $(x-3y)^6$.
40. The middle term in
$$\left(2a+\frac{1}{2}b\right)^6$$
41. The term involving x^4 in $(3x+4y)^7$.
42. The term involving x^6 in $(2x^2-1)^9$.
43. The term involving x^6 in $(2x^3-1)^9$.
44. The term involving x^8 in
$$\left(x^2+\frac{1}{y}\right)^8$$
45. The term involving x^{12} in
$$\left(x^3+\frac{1}{2}\right)^7$$
46. The term involving x^{-4} in
$$\left(y+\frac{1}{x^2}\right)^8$$
47. Evaluate $(1.3)^6$ to four decimal places by writing it as $(1+0.3)^6$ and using the binomial formula.
48. Using the method of Exercise 47, evaluate
    a. $(3.4)^4$
    b. $(48)^5$ (*Hint*: $48 = 50 - 2$.)

12.6 Counting: Permutations and Combinations

How many arrangements can be made using the letters a, b, c, and d taken three at a time? One way to solve this problem is to list all possible arrangements. The **tree diagram** shown in Figure 12-3 is one way to present this enumeration. The letters a, b, c, and d are listed at the top and represent the candidates for the first letter. The three branches emanating from these candidates lead to the possible choices for the second letter, and so on. For example, the portion of the tree shown in Figure 12-4 illustrates the arrangements *bda* and *bdc*. In this way, we determine that there are a total of 24 arrangements.

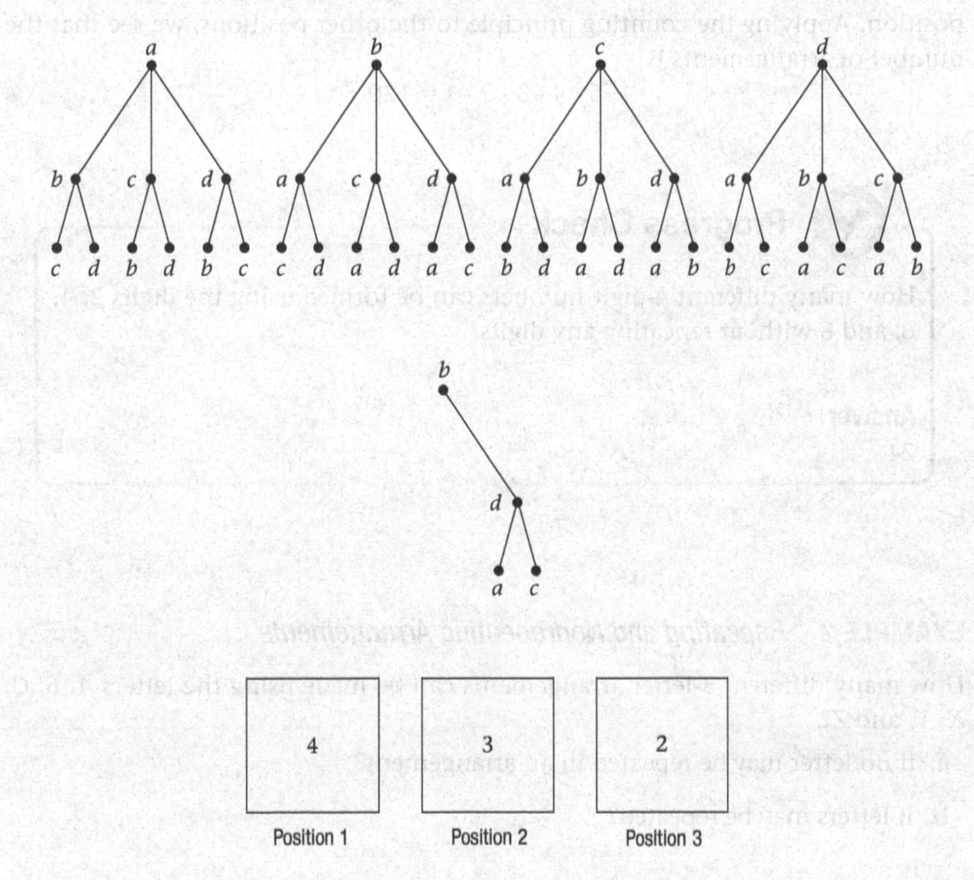

FIGURE 12-3 Arrangements of the Letters a, b, c, d Taken 3 at a Time

FIGURE 12-4 Arrangements *bda* and *bdc*

FIGURE 12-5 Counting Principle

There is a more efficient way to solve this problem. Each arrangement consists of a choice of letters to fill the three positions in Figure 12-5. Any one of the four letters a, b, c, or d can be assigned to the first position. Once a letter is assigned to the first position, any one of the three remaining letters can be assigned to the second position. Finally, either one of the remaining two letters can be assigned to the third position. Since each candidate for a position can be associated with any other candidate in another position, the *product*

$$4 \cdot 3 \cdot 2 = 24$$

yields the total number of arrangements. This example illustrates a very important principle.

> **Counting Principle**
> If one event can occur in m different ways and a second event can occur in n different ways, then both events can occur in mn different ways.

Note that the order of events is significant since each arrangement is counted as one of the "*mn* different ways."

EXAMPLE 1 ***Applying the Counting Principle***

In how many ways can 5 students be seated in a row of 5 seats?

SOLUTION

We have 5 positions to be filled. Any one of the 5 students can occupy the first position, after which any one of the remaining 4 students can occupy the next position. Applying the counting principle to the other positions, we see that the number of arrangements is

$$5 \cdot 4 \cdot 3 \cdot 2 \cdot 1 = 120$$ ■

Progress Check

How many different 4-digit numbers can be formed using the digits 2, 4, 6, and 8 without repeating any digits?

Answer

24

EXAMPLE 2 ***Repeating and Nonrepeating Arrangements***

How many different 3-letter arrangements can be made using the letters *A*, *B*, *C*, *X*, *Y*, and *Z*?

a. if no letter may be repeated in an arrangement?

b. if letters may be repeated?

SOLUTION

a. We need to fill 3 positions. Any one of the 6 letters may occupy the first position. Now, any one of the remaining 5 letters may occupy the second position since repetitions are not allowed. Finally, any one of the remaining 4 letters may occupy the third position. Thus, the total number of arrangements is $6 \cdot 5 \cdot 4 = 120$.

b. Any one of the 6 letters may fill any of the 3 positions since repetitions are allowed. The total number of arrangements is $6 \cdot 6 \cdot 6 = 216$. ■

The positions of president, secretary, and treasurer are to be filled from a class of 15 students. In how many ways can these positions be filled if no student may hold more than 1 position?

Answer

2730

12.6a *Permutations*

Each arrangement that can be made by using all or some of the elements of a set of objects *without repetition* is called a **permutation**. For example, the permutations of the letters a, b, and c taken 3 at a time include *bac* but exclude *aab*.

We will use the notation $\boldsymbol{P(n, r)}$ to indicate the number of permutations of n distinct objects taken r at a time. (Other notations in common use include ${}_nP_r$, P_r^n, ${}^nP_r$, and $P_{n,r}$.) If $r = n$, then using the counting principle, we see that

$$P(n, n) = n(n-1)(n-2) \cdots 2 \cdot 1$$

since any one of the n objects may fill the first position, any one of the remaining $(n-1)$ objects may fill the second position, and so on. Using factorial notation, we have,

$$P(n, n) = n!$$

Let us calculate $P(n, r)$, the number of permutations of n distinct objects taken r at a time when r is less than n. We may think of this as the number of ways of filling r positions with n candidates. Once again, we may fill the first position with any one of the n candidates, the second position with any one of the remaining $(n-1)$ candidates, and so on, obtaining

$$\begin{aligned} P(n, r) &= n(n-1)(n-2) \cdots (n-(r-1)) \\ &= n(n-1)(n-2) \cdots (n-r+1) \qquad (1) \end{aligned}$$

since $(n - r + 1)$ will be the rth factor. If we multiply the right-hand side of Equation (1) by

$$\frac{(n-r)!}{(n-r)!} = 1$$

we have

$$P(n, r) = \frac{n(n-1)(n-2)\cdots(n-r+1)(n-r)(n-r-1)\cdots 2\cdot 1}{(n-r)!}$$

or

Number of Permutations of *n* Distinct Objects Taken *r* at a Time

$$P(n, r) = \frac{n!}{(n-r)!}$$

EXAMPLE 3 *Using Permutation Notation*

Evaluate.

a. $P(5, 5)$ b. $P(5, 2)$ c. $\frac{P(6,2)}{3!}$

SOLUTION

a. $P(5, 5) = \frac{5!}{(5-5)!} = \frac{5!}{0!} = \frac{5 \cdot 4 \cdot 3 \cdot 2 \cdot 1}{1} = 120$

b. $P(5, 2) = \frac{5!}{(5-2)!} = \frac{5!}{3!} = \frac{5 \cdot 4 \cdot 3!}{3!} = 20$

c. $\frac{P(6,2)}{3!} = \frac{6!}{3!(6-2)!} = \frac{6!}{3!4!} = \frac{6 \cdot 5 \cdot 4!}{3 \cdot 2 \cdot 4!} = 5$ ■

Progress Check

Evaluate.

a. $P(4, 4)$ b. $P(6, 3)$ c. $\frac{2P(6,4)}{2!}$

Answers

a. 24 b. 120 c. 360

EXAMPLE 4 *Working with Permutations*

How many different arrangements can be made by taking 5 of the letters of the word *relation*?

SOLUTION

Since the word *relation* has 8 different letters, we are seeking the number of permutations of 8 objects taken 5 at a time, or $P(8, 5)$. Thus,

$$P(n, r) = \frac{n!}{(n-r)!}$$

$$P(8, 5) = \frac{8!}{(8-5)!} = \frac{8!}{3!} = 6720$$ ■

Progress Check

There is space on a bookshelf for displaying 4 books. If there are 6 different novels available, how many arrangements can be made?

Answer

360

EXAMPLE 5 *Working with Permutations*

How many arrangements can be made using all the letters of the word *quartz* if the vowels are always to remain adjacent to each other?

SOLUTION

If we treat the vowel pair *ua* (or *au*) as a unit, then there are five "letters" (q, ua, r, t, z) that can be arranged in $P(5, 5)$ ways. However, the vowels can themselves be arranged in $P(2, 2)$ ways. By the counting principle, the total number of arrangements is

$$P(5, 5) \cdot P(2, 2)$$

Since $P(5, 5) = 120$ and $P(2, 2) = 2$, the total number of arrangements is 240. ■

Progress Check

A bookshelf is to be used to display 5 new textbooks. There are 7 different mathematics textbooks and 4 different biology textbooks available. If we wish to put 3 mathematics books and 2 biology books on display, how many arrangements can be made if the books in each discipline must be kept together?

Answer

5040

Not all permutations of the word *state* are distinct since the letter t appears more than once. The next example illustrates the procedure for finding the number of distinct, or *distinguishable*, permutations.

EXAMPLE 6 *Permutations with Repeated Objects*

Find the number of distinguishable permutations of the letters in the word *remember*.

SOLUTION

The number of permutations of eight different letters is 8!. However in this example, because not all letters are different, permutations in which the two r's exchange places are *not* distinguishable. Similarly, permutations in which the three e's exchange places are not distinguishable, and neither are those in which the two m's exchange places. Since there are 2! permutations of the two r's, 3! permutations of the three e's, and 2! permutations of the two m's, the number of distinguishable permutations of the letters in the word *remember* is

$$\frac{8!}{2!3!2!} = 1680$$ ■

Progress Check

Find the number of distinguishable permutations of the letters in the word *Alaska*.

Answer

120

12.6b Combinations

Consider the arrangements of the letters a, b, and c taken two at a time.

$$ab \quad ba \quad ca$$
$$ac \quad bc \quad cb$$

Now we examine a different question: In how many ways can we select 2 of these letters when we *disregard* the order in which the letters are chosen? The result is then

$$ab \quad ac \quad bc$$

A set of r objects chosen from a set of n objects, regardless of order, is called a **combination**. We denote the number of combinations of r objects by $C(n, r)$. (Other notations in common use include ${}_nC_r$, C^n_r, ${}^nC_r$, and $C_{n,r}$.)

EXAMPLE 7 *Working with Combinations*

List the combinations of the letters a, b, c, and d taken 3 at a time.

SOLUTION

The combinations are seen to be

$$abc \quad abd \quad acd \quad bcd$$ ■

Progress Check

List the combinations of the letters a, b, c, and d taken 2 at a time.

Answer

ab, ac, ad, bc, bd, cd

Here is a rule that is helpful in determining whether a problem calls for the number of permutations or the number of combinations.

$P(n, r)$ or $C(n, r)$

Suppose we are interested in calculating the number of ways of choosing objects.

1. If the order of the chosen objects matters, we use permutations.
2. If the order of the chosen objects does not matter, we use combinations.

For example, suppose we want to determine the number of different 5-card hands that can be dealt from a deck of 52 cards. Since a hand consisting of 5 cards is the same hand regardless of the order of the cards, we must use combinations.

Let us find a formula for $C(n, r)$. There are three combinations of the letters a, b, and c, taken 2 at a time, namely

$$ab \quad ac \quad bc$$

so that $C(3, 2) = 3$. Now, each of these *combinations* can be arranged in 2! ways to yield the total list of *permutations*

$$ab \quad ba \quad ac \quad ca \quad bc \quad cb$$

Thus, $P(3, 2) = 6 = 2!\ C(3, 2)$. In general, each of the $C(n, r)$ combinations can be permuted in $r!$ ways, so by the counting principle, the total number of permutations is

$$P(n, r) = r!\ C(n, r)$$

Solving for $C(n, r)$, we have the following formula:

Number of Combinations of *n* Distinct Objects Taken *r* at a Time

$$C(n, r) = \frac{P(n,r)}{r!} = \frac{n!}{r!(n-r)!} = \binom{n}{r} \qquad r = 0, 1, 2, \ldots, n$$

From the above formula, we note that $C(n, r)$ is the $(r + 1)$st binomial coefficient in the expansion of $(a + b)^n$.

EXAMPLE 8 *Notation for Permutations and Combinations*

Evaluate.

a. $C(5, 2)$ b. $C(4, 4)$ c. $\dfrac{P(6,3)}{C(6,3)}$

SOLUTION

a. $C(5, 2) = \dfrac{5!}{2!(5-2)!} = \dfrac{5!}{2!3!} = \dfrac{5 \cdot 4 \cdot 3!}{2 \cdot 3!} = 10$

b. $C(4, 4) = \dfrac{4!}{4!(4-4)!} = \dfrac{4!}{4!0!} = 1$

c. $P(6, 3) = \dfrac{6!}{(6-3)!} = \dfrac{6!}{3!} = 6 \cdot 5 \cdot 4 = 120$

$C(6, 3) = \dfrac{6!}{3!(6-3)!} = \dfrac{6!}{3!3!} = \dfrac{6 \cdot 5 \cdot 4}{3 \cdot 2 \cdot 1} = 20$

$\dfrac{P(6,3)}{C(6,3)} = \dfrac{120}{20} = 20$ ■

Progress Check

Evaluate.

a. $C(6, 2)$ b. $C(10, 10)$ c. $\dfrac{P(3,2)}{3!C(5,4)}$

Answers

a. 15 b. 1 c. $\frac{1}{5}$

EXAMPLE 9 *An Application of Combinations*

In how many ways can a committee of 4 be selected from a group of 10 people?

SOLUTION

If A, B, C, and D constitute a committee, the arrangement B, A, C, D is not a different committee. Therefore, we are interested in computing $C(10, 4)$.

$$C(10, 4) = \frac{10!}{4!6!} = \frac{10 \cdot 9 \cdot 8 \cdot 7}{4 \cdot 3 \cdot 2 \cdot 1} = 210$$ ■

Progress Check

In how many ways can a 5-card hand be dealt from a deck of 52 cards?

Answer
2,598,960

FOCUS *on the 15 Puzzle*

The 15-puzzle was created in 1878 by Sam Lloyd. It is a square frame, consisting of 16 compartments. There are 15 numbered blocks, each occupying a compartment; and there is one vacant compartment. The object is to move the numbered blocks from a given arrangement to a specified arrangement by sliding numbered blocks into the empty compartment. The accompanying figure shows a starting arrangement and a desired ending arrangement.

The 15-puzzle was the rage of Europe and the United States in the late 1800s, and contests were held with substantial prizes being offered. The prizes were never collected because the specified arrangements could not be achieved.

1	2	3	4
5	6	7	8
9	10	11	12
13	14	15	

Starting arrangement

1	5	9	13
2	6	10	14
3	7	11	15
4	8	12	

Final arrangement

15	14	13	12
11	10	9	8
7	6	5	4
3	2	1	

Impossible arrangement

Since there are 16 locations on the board, there are 16! possible arrangements of the 15 numbered blocks and the vacant compartment (the sixteenth object). Thus, there are more than 20 trillion arrangements of these objects.

A permutation or arrangement of the set of numbers from 1 to n is said to have an *inversion* each time a larger number precedes a smaller one. Thus, the permutation 54132 of the numbers 1, 2, 3, 4, 5 has 8 inversions:

5 before 4	5 before 1	5 before 3	5 before 2
4 before 1	4 before 3	4 before 2	3 before 2

A permutation is said to be *even* if it has an even number of inversions, and *odd* if it has an odd number of inversions. Thus, the permutation 54132 is an even permutation. In 1879, the American mathematicians W. W. Johnson and W. E. Story showed that if the empty position remains fixed, then it is possible to go from any even (odd) arrangement to any other even (odd) arrangement and it is impossible to go from an even to an odd or from an odd to an even arrangement.

In the accompanying figures, the starting arrangement consists of the permutation

1 2 3 4 5 6 7 8 9 10 11 12 13 14 15

which has no inversions and is thus an even permutation. The desired final arrangement is the permutation

1 5 9 13 2 6 10 14 3 7 11 15 4 8 12

which has 36 inversions and is thus an even permutation. It is therefore possible to go from the starting arrangement to the final arrangement. The permutation

15 14 13 12 11 10 9 8 7 6 5 4 3 2 1

can never be achieved, since it has 105 inversions, and is thus an odd permutation.

EXAMPLE 10 *An Application of Combinations*

In how many ways can a committee of 3 girls and 2 boys be selected from a class of 8 girls and 7 boys?

SOLUTION

The girls can be selected in $C(8, 3)$ ways, and the boys can be selected in $C(7, 2)$ ways. By the counting principle, each choice of boys can be associated with each choice of girls.

$$C(8, 3) \cdot C(7, 2) = \frac{8!}{3!5!} \cdot \frac{7!}{2!5!} = (56)(21) = 1176$$ ■

Progress Check

In how many ways can 4 persons be chosen from 5 representatives of District A and 8 representatives of District B, if only 1 representative from District A is to be included?

Answer

280

EXAMPLE 11 *An Application of Combinations and Permutations*

A bookstore has 12 different French books and 9 different German books. In how many ways can a group of 6 books, consisting of 4 French and 2 German, be placed on a shelf?

SOLUTION

The French books can be selected in $C(12, 4)$ ways and the German books in $C(9, 2)$ ways. The 6 books can then be selected in $C(12, 4) \cdot C(9, 2)$ ways. Each *selection* of 6 books can then be *arranged* on the shelf in $P(6, 6)$ ways, so the total number of arrangements is

$$C(12, 4) \cdot C(9, 2) \cdot P(6, 6) = \frac{12!}{4!8!} \cdot \frac{9!}{2!7!} \cdot \frac{6!}{(6-6)!} = 495 \cdot 36 \cdot 720 = 12{,}830{,}400$$ ■

Progress Check

From 6 different consonants and 4 different vowels, how many 5-letter words can be made consisting of 3 consonants and 2 vowels? (Assume every arrangement is a "word.")

Answer

14,400

Exercise Set 12.6

1. How many different 5-digit numbers can be formed using the digits 1, 3, 4, 6, and 8?

2. How many different ways are there to arrange the letters in the word *study*?

3. An employee identification number consists of 2 letters of the alphabet followed by a sequence of 3 digits selected from the numbers 2, 3, 5, 6, 8, and 9. If repetitions are allowed, how many different identification numbers are possible?

4. In a psychological experiment, a subject has to arrange a cube, a square, a triangle, and a rhombus in a row. How many different arrangements are possible?

5. A coin is tossed 8 times, and the result of each toss is recorded. How many different sequences of heads and tails are possible?

6. A die is tossed 4 times, and the result of each toss is recorded. How many different sequences are possible?

7. A concert is to consist of 3 guitar pieces, 2 vocal numbers, and 2 jazz selections. In how many ways can the program be presented?

In Exercises 8–19, compute the given expression.

8. $P(6, 6)$
9. $P(6, 5)$
10. $P(4, 2)$
11. $P(8, 3)$
12. $P(5, 2)$
13. $P(10, 2)$
14. $P(8, 4)$
15. $\frac{P(9,3)}{3!}$
16. $\frac{4P(12,3)}{2!}$
17. $P(3, 1)$
18. $\frac{P(7,3)}{2!}$
19. $\frac{P(10,4)}{4!}$

20. Find the number of ways in which 5 men and 5 women can be seated in a row for the following:

    a. if any person may sit next to any other person

    b. if the men and women must be seated alternately

21. Find the number of permutations of the letters of the word *money*.

22. Find the number of distinguishable permutations of the letters of the word *goose*. (*Hint*: Permutations in which the letters *o* and *o* exchange places are not distinguishable.)

23. Find the number of distinguishable permutations of the letters of the word *needed*.

24. How many permutations of the letters *a*, *b*, *e*, *g*, *h*, *k*, and *m* are there when taken

    a. 2 at a time?
    b. 3 at a time?

25. How many 3-letter labels of new chemical products can be formed from the letters *a*, *b*, *c*, *d*, *f*, *g*, *l*, and *m* without repeating letters?

26. Find the number of distinguishable permutations that can be formed from the letters of the word *Mississippi*.

27. A family consisting of a mother, a father, and 3 children is having a picture taken. If all 5 people are arranged in a row, how many different photographs can be taken by rearranging family members?

28. List all the combinations of the numbers 4, 3, 5, 8, and 9 taken 3 at a time.

In Exercises 29–37, evaluate the given expression.

29. $C(9, 3)$
30. $C(7, 3)$
31. $C(10, 2)$
32. $C(7, 1)$
33. $C(7, 7)$
34. $C(5, 4)$
35. $C(n, n - 1)$
36. $C(n, n - 2)$
37. $C(n + 1, n - 1)$

38. In how many ways can a committee of 2 faculty members and 3 students be selected from 8 faculty members and 10 students?

39. In how many ways can a basketball team of 5 players be selected from among 15 candidates?

40. In how many ways can a 4-card hand be dealt from a deck of 52 cards?

41. Assuming letters in the English alphabet may be used, how many 3-letter moped plates on a local campus can be formed

    a. if no letters can be repeated?

    b. if letters can be repeated?

42. In a certain city each police car is staffed by 2 officers: one male and one female. A police captain, who needs to staff 8 cars, has 15 male officers and 12 female officers available. How many different teams can be formed?

43. How many different 10-card hands with 4 aces can be dealt from a deck of 52 cards?

44. A car manufacturer makes 3 different models, each of which is available in 5 different colors and with 2 different engines. How many cars must a dealer stock in the showroom to display the full line?

45. A penny, a nickel, a dime, a quarter, a half-dollar, and a silver dollar are to be arranged in a row. How many different arrangements can be formed if the penny and the dime must always be next to each other?

46. An automobile manufacturer who is planning an advertising campaign is considering 7 newspapers, 2 magazines, 3 radio stations, and 4 television stations. In how many ways can 5 advertisements be placed
    a. if all 5 are to be in newspapers?
    b. if 2 are to be in newspapers, 2 on radio, and 1 on television?

47. In a certain police station there are 12 prisoners and 10 police officers. How many possible lineups consisting of 4 prisoners and 3 officers can be formed?

48. The notation

$$\binom{n}{r}$$

is often used in place of $C(n, r)$. Show that

$$\binom{n}{r} = \binom{n}{n-r}$$

49. How many different 10-card hands with 6 red cards and 4 black cards can be dealt from a deck of 52 cards?

50. A bin contains 12 transistors, 7 of which are defective. In how many ways can 4 transistors be chosen so that
    a. all 4 are defective?
    b. 2 are good and 2 are defective?
    c. all 4 are good?
    d. 3 are defective and 1 is good?

12.7 Probability

12.7a Definition

There is a vast difference between the statements "It will probably rain today" and "It is equally probable that a tossed coin will come up heads or tails." The first statement conveys an expectation, but only in a vague sense. The latter statement quantifies the notion of probability.

Consider what happens when we toss a fair coin. The event has only 2 possible outcomes: heads or tails. Since heads represents 1 of 2 equally probable outcomes, we say that the probability of a head is $\frac{1}{2}$. Thus, we can define **probability** in the following way: If an event can occur in a total of t ways, all equally likely to occur, and S of these are considered successful, the probability of success is $\frac{s}{t}$. In short,

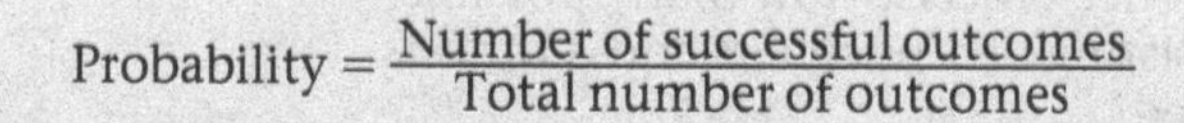

$$\text{Probability} = \frac{\text{Number of successful outcomes}}{\text{Total number of outcomes}}$$

EXAMPLE 1 ***Calculating the Probability of an Event***

A container holds 1 red ball, 2 white balls, and 2 blue balls. If 1 ball is drawn, what is the probability that it will be white?

SOLUTION

The selection of a ball represents a possible outcome, so there are a total of 5 possible outcomes. Since there are 2 ways of achieving a successful outcome (a white ball),

$$\text{Probability of selecting a white ball} = \frac{\text{Number of successful outcomes}}{\text{Total number of outcomes}} = \frac{2}{5}$$ ■

Progress Check

One card is drawn from an ordinary deck of 52 cards. What is the probability it is an ace?

Answer

$\frac{1}{13}$

EXAMPLE 2 ***Calculating the Probability of an Event***

A single die, whose faces are numbered 1, 2, 3, 4, 5, and 6, is tossed. What is the probability that the result is less than 5?

SOLUTION

There are 4 successful outcomes, occurring when the die shows a 1, 2, 3, or 4. Since there are 6 possible outcomes, we see that

$$\text{Probability} = \frac{\text{Number of successful outcomes}}{\text{Total number of outcomes}} = \frac{4}{6} = \frac{2}{3}$$ ■

A bag of coins contains 4 nickels, 5 dimes, and 10 quarters. If 1 coin is withdrawn, what is the probability that it will be worth less than 25 cents?

Answer

$\frac{9}{19}$

12.7b *Principles of Probability*

Consider a bag containing 3 red marbles and 5 brown marbles. The probability of drawing a red marble in a single draw is $\frac{3}{8}$, and the probability of drawing a brown marble in a single draw is $\frac{5}{8}$. What is the probability of drawing either a red marble or a brown marble? Since any of the 8 possible outcomes is considered a success, the probability is $\frac{8}{8} = 1$. What is the probability of drawing a black marble? Since there are no successful outcomes, the probability is $\frac{0}{8} = 0$. Generalizing these results, we can state the following **principles:**

1. If p is the probability that an event will happen, then $0 \leq p \leq 1$.
2. If p is the probability that an event will happen, then $1 - p$ is the probability that it will not happen.
3. If an event must happen (certainty), its probability is 1.
4. If an event cannot happen (impossibility), its probability is 0.

EXAMPLE 3 ***Probability of an Event Not Occurring***

While shuffling an ordinary deck of 52 cards, you drop 1 card. What is the probability that it is not a king?

SOLUTION

Since there are 4 kings in a deck, the probability of a king is $p = \frac{4}{52} = \frac{1}{13}$. Then the probability that it is *not* a king is $1 - p = \frac{12}{13}$. ■

Progress Check

Two people each throw a single die. If player A rolls a 4, what is the probability that player B will not also roll a 4?

Answer

$\frac{5}{6}$

12.7c Applications

The rules for computing permutations and combinations are useful in solving probability problems when items are drawn without being replaced.

EXAMPLE 4 *Combinations and Probability*

A bag contains 3 green, 5 white, and 7 yellow balls. If 3 balls are drawn at random, what is the probability that they will all be white?

SOLUTION

We can select 3 white balls from 5 white balls in $C(5, 3)$ ways. We can select 3 balls from the bag of 15 balls in $C(15, 3)$ ways. Then

$$\text{Probability of selecting three white balls} = \frac{C(5,3)}{C(15,3)} = \frac{\frac{5!}{3!2!}}{\frac{15!}{3!12!}} = \frac{10}{455} = \frac{2}{91} \quad \blacksquare$$

Progress Check

Three cards are drawn from an ordinary deck of 52 cards. What is the probability that they are all aces?

Answer

$\frac{1}{5525}$

Many problems in probability involve the tossing of a pair of fair dice. Since the faces of the dice are numbered 1, 2, 3, 4, 5, and 6, the sum of the numbers on the two dice can be any of the numbers 2 through 12. The sums, however, are not equally probable. In Table 12-2, we display the possible sums when tossing a pair of dice. In Table 12-3, we present the number of ways in which each sum can be obtained. The probability of tossing a sum of 3 with a pair of dice is therefore $\frac{2}{36}$, or $\frac{1}{18}$. The probability of tossing a sum of 7 is $\frac{6}{36}$, or $\frac{1}{6}$.

TABLE 12-2 Possible Sums When Tossing a Pair of Dice

	Die 2					
Die 1	1	2	3	4	5	6
1	2	3	4	5	6	7
2	3	4	5	6	7	8
3	4	5	6	7	8	9
4	5	6	7	8	9	10
5	6	7	8	9	10	11
6	7	8	9	10	11	12

TABLE 12-3 Frequency of Sums When Tossing a Pair of Dice

Sum of 2 dice	2	3	4	5	6	7	8	9	10	11	12
Number of ways	1	2	3	4	5	6	5	4	3	2	1

EXAMPLE 5 *Calculating Probability with Dice*

What is the probability of throwing a sum of 10 or higher with a pair of dice?

SOLUTION

The favorable outcomes are 10, 11, and 12. From Table 12-3, this can occur in a total of 6 ways. Then

$$\text{Probability} = \frac{\text{Number of successful outcomes}}{\text{Total number of outcomes}} = \frac{6}{36} = \frac{1}{6}$$ ■

Progress Check

What is the probability of throwing a sum no higher than a 5 with a pair of dice?

Answer

$\frac{5}{18}$

12.7d *Independent Events*

Suppose a card is drawn from a deck of 52 cards. After replacing that card in the deck, a second card is drawn. What is the probability that both cards will be aces? Note that these are **independent events** since the second outcome does not depend on the first outcome. Here is the principle that permits us to solve this type of problem:

> If p_1 and p_2 are the probabilities of the occurrence of two independent events, then p_1p_2 is the probability that both events will occur.

In our example, the probability p_1 that the first card drawn will be an ace is $\frac{4}{52} = \frac{1}{13}$. If the first card is replaced, the probability p_2 that the second card drawn will be an ace is also $\frac{1}{13}$. Then the probability of drawing aces successively is $p_1p_2 = \frac{1}{13} \cdot \frac{1}{13} = \frac{1}{169}$. We can extend this principle to more than two independent events by forming the corresponding product of the probabilities.

EXAMPLE 6 ***Calculating the Probability of Multiple Events***

What is the probability of throwing a sum of 7 twice in a row with a pair of dice?

SOLUTION

From Table 12-3, we see that a sum of 7 can occur in 6 ways, so the probability p_1 of the first 7 is $\frac{6}{36} = \frac{1}{6}$. Similarly, the probability p_2 of throwing a sum of 7 on the second roll is also $\frac{1}{6}$. Thus, the probability of throwing a sum of 7 on both rolls is $p_1p_2 = \frac{1}{6} \cdot \frac{1}{6} = \frac{1}{36}$. ■

Progress Check

What is the probability of throwing an 11 twice in succession with a pair of dice?

Answer

$\frac{1}{324}$

EXAMPLE 7 ***Calculating the Probability of Multiple Events***

What is the probability of throwing a 5 only on the first of 2 successive throws with a single die?

SOLUTION

The probability of throwing a 5 on the first toss is $p_1 = \frac{1}{6}$. A success on the second toss consists of *not* throwing a 5 and has a probability $p_2 = 1 - p_1 = \frac{5}{6}$. The probability of the desired result is

$$p_1p_2 = \frac{1}{6} \cdot \frac{5}{6} = \frac{5}{36}$$ ■

Progress Check

What is the probability of throwing a 7 or an 11 on the first of two successive throws with a pair of dice but not on the second throw?

Answer

$\frac{14}{81}$

EXAMPLE 8 ***Empirical Probability and Multiple Events***

A transistor manufacturer finds that 81 of every 1000 transistors made are defective. What is the probability that 2 transistors selected at random will both prove to be defective?

SOLUTION

This problem is an example of **empirical probability**, that is, probability obtained from experience or measurement rather than by theoretical means. The probability that a transistor is defective is $p_1 = \frac{81}{1000}$. The probability that a second transistor is defective is $p_2 = \frac{81}{1000}$. Thus, the probability that both transistors will be defective is

$$p_1 p_2 = \frac{81}{1000} \cdot \frac{81}{1000} = \frac{6561}{1,000,000}$$ ■

Progress Check

The probability of rain in a certain town on any given day is $\frac{1}{4}$. What is the probability of having a rainy Monday, a dry Tuesday, and a rainy Wednesday?

Answer

$\frac{3}{64}$

Exercise Set 12.7

1. If a single die is tossed, what is the probability that an odd number will appear?
2. If two dice are tossed, what is the probability of having at least one 5 showing on either die?
3. If a card is randomly selected from an ordinary deck of 52 cards, what is the probability that it will be the following?
   a. a red card
   b. a spade
   c. a king
4. Suppose that 2 coins are tossed. What is the probability of having the following tosses?
   a. both tails
   b. at least 1 heads
   c. neither tails
   d. 1 heads and 1 tails
5. If 2 dice are tossed, what is the probability that
   a. at least one of the dice will show a 4?
   b. the sum of the numbers on the dice will be 8?
   c. neither a 3 nor a 4 will appear on either die?
6. If a card is picked at random from a standard deck of 52 cards, what is the probability that it will be the following?
   a. a club
   b. a 4
   c. not an ace
   d. a 4 of spades
   e. either an ace or a king
   f. neither an ace nor a king
7. The quality control department of a calculator manufacturer determines that 1% of all calculators made are defective. What is the probability that a buyer of a calculator will get the following?
   a. a good calculator
   b. a defective calculator
8. A photography club consisting of 18 women and 12 men wishes to elect a steering committee composed of 3 members. If every member is equally likely to be elected, find the probability that
   a. all 3 members will be women
   b. none of the members will be a woman
   c. exactly 1 member will be a woman
   d. at least 1 member will be a woman
9. A box contains 97 good bulbs and 5 defective bulbs. If 3 bulbs are chosen at random without replacement, what is the probability of each of the following?
   a. all three bulbs will be defective
   b. exactly one of the bulbs will be defective
   c. none of the bulbs will be defective
10. Suppose that 2 cards are to be drawn in succession from a deck of 52 cards. What is the probability that both cards will be aces if the following occur?
    a. drawn cards are replaced
    b. drawn cards are not replaced
11. Suppose that 4 cards are selected, without replacement, from a deck of 52 cards. What is the probability that they are all hearts?
12. If the probability of getting an A in a course is 0.2, what is the probability of not getting an A?
13. The board of trustees of a university consists of 14 women and 12 men. Suppose that an executive committee of 6 persons is to be elected and that every trustee is equally likely to be elected. Find the probability that the committee will consist of 3 men and 3 women.
14. Suppose that the probability of a cloudy day in a certain town in England is 0.6.
    a. What is the probability of a clear day?
    b. What is the probability of 2 consecutive clear days?
15. A bag contains 6 blue marbles, 5 green marbles, and 7 yellow marbles. If 5 marbles are chosen without replacement, what is the probability that 2 will be blue, 2 will be green, and 1 will be yellow?
16. Suppose that 2 cards are drawn without replacement from a deck of 52 cards. What is the probability that 1 card is an ace and the other card is not a king?
17. If 2% of cameras made on a production line are defective, what is the probability that 4 cameras chosen at random will all be the following?
    a. good
    b. defective
18. A fraternity consists of 12 seniors, 10 juniors, and 14 sophomores. A steering committee of 7 members is randomly chosen. What is the probability that it consists of 3 seniors, 2 juniors, and 2 sophomores?

Chapter Summary

Key Terms, Concepts, and Symbols

Key Ideas for Review

Topic	Page	Key Idea
Sequences	650	A sequence is a function whose domain is restricted to the set of natural numbers. We generally write a sequence by using subscript notation a_n.
Recursive Definition	651	A sequence is defined recursively if each term is defined by reference to preceding terms.
Series	655	A series is the sum of the terms of a particular sequence.
Summation Notation	652	A sum of terms may be presented using sigma (Σ) or summation notation. The values written below and above the Σ indicate the starting and ending values of the index of summation, respectively.
Arithmetic Sequences	659	An arithmetic sequence has a common difference d between terms. We can define an arithmetic sequence by writing $$a_n = a_{n-1} + d \quad \text{or} \quad a_n = a_1 + (n-1)d$$
Sum S_n of the first n terms	662	$S_n = \frac{n}{2}(a_1 + a_n) \quad \text{or} \quad S_n = \frac{n}{2}[2a_1 + (n-1)d]$
Geometric Sequences	665	A geometric sequence has a common ratio r between terms. We can define a geometric sequence by writing $$a_n = ra_{n-1} \quad \text{or} \quad a_n = a_1 r^{n-1}$$
Sum S_n of the first n terms	667	$$S_n = \frac{a_1(1-r^n)}{1-r}$$
Infinite Geometric Series	669	If the common ratio r satisfies $-1 < r < 1$, then the infinite geometric series has the sum $$S = \frac{a_1}{1-r}$$
Mathematical Induction	673	Mathematical induction is useful in proving certain types of theorems involving natural numbers.

continues

Topic	Page	Key Idea
Factorial Notation	679	The notation $n!$ indicates the product of the natural numbers 1 through n. $$n! = n(n-1)(n-2)\cdots 2\cdot 1 \qquad \text{for } n \geq 1$$
Special Case	679	$0! = 1$
Binomial Theorem	677	The binomial theorem provides the terms of the expansion of $(a+b)^n$. $$(a+b)^n = a^n + \frac{n!}{1!(n-1)!}a^{n-1}b + \frac{n!}{2!(n-2)!}a^{n-2}b^2 + \frac{n!}{3!(n-3)!}a^{n-3}b^3 + \cdots + \frac{n!}{r!(n-r)!}a^{n-r}b^r + \cdots + b^n$$
Binomial Coefficient $\binom{n}{r}$	680	The notation $\binom{n}{r}$ is defined by $$\binom{n}{r} = \frac{n!}{r!(n-r)!}$$
Summation Notation	680	$(a+b)^n = \sum_{k=0}^{n} \binom{n}{k} a^{n-k}b^k$
Counting Principle	683	If one event can occur in m different ways and a second event can occur in n different ways, then both events can occur in mn ways.
Permutations	685	Permutations involve arrangements of the order of objects, so *abc* and *bac* are distinct permutations of the letters *a*, *b*, and *c*. The number $P(n, r)$ of permutations of n objects taken r at a time is $$P(n, r) = \frac{n!}{(n-r)!}$$
Combinations	688	Combinations involve selections of objects when the order is not significant. If we are selecting three letters from a box containing the letters *a*, *b*, *c*, and *d*, then *abc* and *bac* are the same combination. The number $C(n, r)$ of combinations of n objects taken r at a time is $$C(n, r) = \frac{n!}{r!(n-r)!} = \binom{n}{r}$$
Probability	694	*Probability is a way of expressing the likelihood of the occurrence of an event.* It is the ratio of the number of successful outcomes to the total number of outcomes, when all outcomes are equally likely to occur. Thus, if p is the probability that an event will happen, then $0 \leq p \leq 1$. Furthermore, $1 - p$ is the probability that this event will not happen.
Certainty	695	If an event must happen, its probability is 1.
Impossibility	695	If an event cannot happen, its probability is 0.
Independent Events	697	If p_1 and p_2 are the probabilities of the occurrence of two independent events, then p_1p_2 is the probability that both events will occur.

Review Exercises

In Exercises 1 and 2, write the first three terms and the tenth term of the sequence whose nth term is given.

1. $a_n = n^2 + n + 1$ 2. $a_n = \frac{n^3 - 1}{n + 1}$

In Exercises 3 and 4, find the fifth term of the recursive sequence.

3. $a_n = n - a_{n-1}$, $a_1 = 0$ 4. $a_n = na_{n-1}$, $a_1 = 1$

In Exercises 5–7, find the indicated sum.

5. $\sum_{k=1}^{4} (1 - 2k)$ 6. $\sum_{k=3}^{5} k(k+1)$

7. $\sum_{i=1}^{5} 10$

In Exercises 8–10, express the sum in sigma notation.

8. $\frac{1}{3} + \frac{2}{4} + \frac{3}{5} + \frac{4}{6}$ 9. $1 - x + x^2 - x^3 + x^4$

10. $\log x + \log 2x + \log 3x + \cdots + \log nx$

In Exercises 11 and 12, find the specified term of the arithmetic sequence whose first term is a_1 and whose common difference is d.

11. $a_1 = -2$, $d = 2$: twenty-first term

12. $a_1 = 6$, $d = -1$: sixteenth term

In Exercises 13 and 14, given two terms of an arithmetic sequence, find the specified term.

13. $a_1 = 4$, $a_{16} = 9$: thirteenth term

14. $a_1 = -4$, $a_{23} = -15$: twenty-sixth term

In Exercises 15 and 16, find the sum of the first 25 terms of the arithmetic sequence whose first term is a_1 and whose common difference is d.

15. $a_1 = -\frac{1}{3}$, $d = \frac{1}{3}$ 16. $a_1 = 6$, $d = -2$

In Exercises 17 and 18, determine the common ratio of the given geometric sequence.

17. $2, -6, 18, -54, \ldots$ 18. $-\frac{1}{2}, \frac{3}{4}, -\frac{9}{8}, \frac{27}{16}, \ldots$

In Exercises 19 and 20, write the first four terms of the geometric sequence whose first term is a_1 and whose common ratio is r.

19. $a_1 = 5$, $r = \frac{1}{5}$ 20. $a_1 = -2$, $r = -1$

21. Find the sixth term of the geometric sequence $-4, 6, -9, \ldots$

22. Find the eighth term of a geometric sequence for which $a_1 = -2$ and $a_5 = -32$.

23. Insert two geometric means between 3 and $\frac{1}{72}$.

24. Find the sum of the first six terms of the geometric sequence whose first three terms are $\frac{1}{3}, \frac{1}{6}, \frac{1}{12}$.

25. Find the sum of the first six terms of the geometric progression for which $a_1 = -2$ and $r = 3$.

In Exercises 26 and 27, find the sum of the infinite geometric series.

26. $5 + \frac{5}{2} + \frac{5}{4} + \cdots$ 27. $3 - 2 + \frac{4}{3} - \cdots$

28. Use the principle of mathematical induction to show that
$$3 + 6 + 9 + \cdots + 3n = \frac{3n(n+1)}{2}$$
is true for all positive integer values of n.

In Exercises 29–31, expand and simplify.

29. $(2x - y)^4$ 30. $\left(\frac{x}{2} - 2\right)^4$

31. $(x^2 + 1)^3$

In Exercises 32–37, evaluate the expression.

32. $6!$ 33. $\frac{13!}{11!12!}$

34. $\frac{(n-1)!(n+1)!}{n!n!}$ 35. $\binom{6}{4}$

36. $\binom{3}{0}$ 37. $\binom{10}{8}$

38. Four different novels have been selected for display on a shelf. How many different arrangements are possible?

39. Find the number of distinguishable permutations of the letters in the word *soothe*.

40. In how many ways can a tennis team of 6 players be selected from 10 candidates?

41. In how many ways can a consonant and a vowel be chosen from the letters in the word *fouled*?

42. If 2 dice are tossed, what is the probability that the sum of the numbers on the dice will be 7 or 11?

43. A box contains 3 red pens and 4 white pens. If 2 pens are selected at random without replacement, what is the probability that they will both be white?

44. Two cards are drawn in succession from a deck of 52 cards. What is the probability that the cards will be a king and an ace if the first card drawn is not replaced?

45. If 10% of the trees in a region are found to be diseased, what is the probability that 2 trees chosen at random are both free of disease?

46. Six husband–wife teams volunteer for an experiment in parapsychology. If 4 persons are selected at random to participate in the experiment, what is the probability that they will be two husband–wife teams?

Review Test

1. Write the first four terms of a sequence whose nth term is
$$a_n = \frac{n}{(n+1)^2}$$

2. Evaluate
$$\sum_{j=2}^{4} \frac{j}{j-1}$$

3. Write the first four terms of the arithmetic sequence whose first term is -1 and whose common difference is $\frac{3}{2}$.

4. Find the twenty-fifth term of the arithmetic sequence whose first term is -4 and whose common difference is $\frac{1}{2}$.

5. Find the fifteenth term of an arithmetic sequence whose first and tenth terms are -1 and 26, respectively.

6. Find the sum of the first ten terms of an arithmetic sequence whose first term is -4 and whose ninth term is 8.

7. Find the common ratio of the geometric sequence
$$12, 4, \frac{4}{3}, \frac{4}{9}, \ldots$$

8. Write the first four terms of the geometric sequence whose first term is $-\frac{2}{3}$ and whose common ratio is 2.

9. Find the tenth term of the sequence $2, -2, 2, \ldots$.

10. Insert two geometric means between -4 and 32.

11. Find the sum of the first seven terms of the geometric sequence whose first three terms are $-8, 4, -2$.

12. Find the sum of the infinite geometric series
$$-4 - \frac{4}{3} - \frac{4}{9} - \cdots$$

13. Use the principle of mathematical induction to show that $2 + 6 + 10 + \cdots + (4n - 2) = 2n^2$ is true for all positive integer values of n.

14. Find the first four terms in the expansion of
$$\left(a + \frac{1}{b}\right)^{10}$$

15. Evaluate
$$\frac{12!}{10!3!}$$

16. Evaluate $P(6, 4)$.

17. Evaluate $C(n + 1, n)$.

18. Three buses arrive simultaneously at a terminal that has 4 parking stalls. In how many ways can the buses be parked?

19. The 2010 census staff divided a region into 3 districts that were to be canvassed by 10, 8, and 15 staff members, respectively. If 40 staff members were available, in how many ways could they have been assigned to the 3 districts?

20. A particular sporting goods manufacturer distributes shoes to sponsored athletes. The company uses white, black, and green themes in its shoes, which are distributed randomly to athletes. If two rival athletes request shoes at the same time, what is the probability that they will be the same color?

21. Four marbles are removed at random without replacement from a box containing 4 purple, 3 blue, and 3 red marbles. What is the probability that these are 2 purple and 2 blue marbles?

Cumulative Review Exercises: Chapters 10–12

In Exercises 1–3, find all real solutions of the given system.

1. $x^2 - y^2 = 9$
   $x^2 + y^2 = 41$

2. $2x - 3y = 11$
   $3x + 5y = -12$

3. $x^2 + 3y^2 = 12$
   $x + 3y = 6$

4. A movie theater charges $6 admission for an adult and $3 for a child. If 600 tickets were sold and the total revenue received was $2700, how many tickets of each type were sold?

5. Find the partial-fraction decomposition of
$$\frac{x+5}{x^2-x-2}$$

6. Use back-substitution to solve the linear system corresponding to the augmented matrix
$$\left[\begin{array}{ccc|c} 1 & 3 & -1 & 0 \\ 0 & 1 & -2 & 5 \\ 0 & 0 & 1 & -3 \end{array}\right]$$

7. Solve by Gaussian Elimination:
$$\begin{aligned} 4x + y - z &= 0 \\ -x + 3y + 2z &= -3 \\ 3x - 2y - 2z &= -3 \end{aligned}$$

8. Given the system of equations
$$\begin{aligned} x - y &= 4 \\ 3x + 2ky &= 3 \end{aligned}$$
   a. for what values of k is there no solution?
   b. for what values of k are there infinitely many solutions?

9. Given the system of linear inequalities
$$\begin{aligned} x + y &\le 2 \\ 2x + y &\ge 0 \\ y &\ge 0 \end{aligned}$$
   draw the graph of the solution set.

10. Given the matrices
$$A = \begin{bmatrix} 1 & -1 & 2 \\ 0 & 2 & 3 \\ -2 & 1 & -2 \end{bmatrix} \quad B = \begin{bmatrix} 2 & -1 \\ 0 & 2 \\ -1 & 4 \end{bmatrix} \quad C = \begin{bmatrix} -1 \\ -3 \\ 0 \end{bmatrix}$$
   a. compute AB
   b. compute $|A|$
   c. solve $AX = C$

11. The sum of the digits of a two-digit number is 10. The square of the tens digit is one more than 16 times the ones digit. Find the number.

12. In a certain two-digit number, 3 times the tens digit is 2 more than 5 times the ones digit. If the digits are reversed, the new number is 18 less than the original number. Find the new number.

13. Evaluate.
   a. $\sum_{k=1}^{3} (2k-1)^2$ b. $\sum_{k=0}^{3} k(k-1)$

14. Find the first term a_1 of an arithmetic progression if $a_4 = 4$ and $a_7 = 3$.

15. If the first three terms of an arithmetic progression are $a_1 = 3 + 2k$, $a_2 = 2k - 1$, and $a_3 = -10k + 1$, find k.

16. If the first and third terms of a geometric progression are 5 and 2, respectively, find the second term.

17. Find the sum of the first four terms of a geometric progression if the first and second terms are 3 and 1, respectively.

18. If the sum of an infinite geometric progression is −6 and the first term is −5, find the common ratio r.

19. Find the third term in the binomial expansion of
$$\left(1 - \frac{1}{x}\right)^5$$

20. Use the method of induction to prove that $n < 2^n$ for all nonnegative integers.

Writing Exercises

1. Sometimes on the radio, the person announcing the weather says, "There is no probability of precipitation tomorrow." Comment on the announcer's understanding of probability.

2. What does it mean when the person announcing the weather says that there is a 50% chance of rain?

3. Explain the difference between a permutation and a combination. Give illustrations of each.

4. Describe an experimental procedure to determine if a coin is fair.

5. Write a report on the use of probability in the insurance industry.

Chapter 12 Project

Read the chapter opener about the grateful king. How many grains of rice were on that final square of the chessboard?

Let your graphing calculator help you. Use the SEQUENCE options to set this problem up as a sequence. Your input may look like the screenshot to below.

Now use the cumulative sum option to find the total number of grains on the board.

Appendix

Answers to Selected Odd-Numbered Exercises, Review Exercises, and Review Tests

Chapter 1

Exercise Set 1.1

1. $\{3, 4, 5, 6, 7\}$
3. $\{-9\}$
5. $\{1, 2\}$
7. $\{1, 3, 7\}$
9. F
11. F
13. T
15. T
17. T
19. F
21. F
23. commutative (addition)
25. distributive
27. associative (addition)
29. closure (multiplication)
31. commutative (multiplication)
33. commutative, associative (multiplication)
35. multiplicative inverse
41. symmetric
43. transitive or substitution
53. a. 5 b. -21 c. 21 d. -3 e. $\frac{10}{3}$ f. $\frac{-10}{3}$ g. $\frac{5}{4}$ h. $\frac{1}{2}$ i. $\frac{3}{4}$ j. $\frac{28}{33}$ k. $\frac{18}{175}$ l. $\frac{12}{35}$
55. a. 0.25 b. -0.6 c. $0.\overline{769230}$ d. $0.\overline{2857174}$
57. 35 miles
59. 4 feet and 6 feet
61. 13 ounces
63. \$256.10
65. \$10,652

Exercise Set 1.2

1. 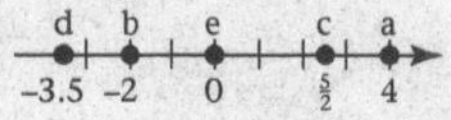

3. $A:1, B:2.5, C:-2, D:4, O:0, E:-3.5$

5. 4

7. −2

9. −5

11. 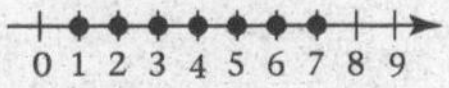

13. 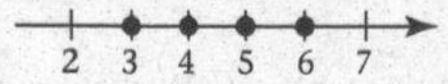

15. $10 > 9.99$

17. $a \geq 0$

19. $x > 0$

21. $\frac{1}{4} < a < \frac{1}{2}$

23. $b \geq 5$

25. multiplication by negative number reverses the inequality sign

27. multiplication by negative number reverses the inequality sign

29. multiplication by positive number preserves the inequality sign

31. 2

33. 1.5

35. −2

37. 1

39. 4

41. 2

43. $\frac{1}{5}$

45. 3

47. 2

49. $\frac{8}{5}$

53. $\{-2, -1, 0, 1, 2, 3, 4, 5, 6, 7, 8\}$

55. a. $1 < 7$ b. $-6 < 0$
c. $-2 < 10$ d. $5 > -25$
e. $1 > -5$ f. $-\frac{1}{2} < \frac{5}{2}$
g. $1 < 25$

57. $\frac{7}{2}$

59. −2

61. 0

63. a. $x \leq 3$ b. $x \leq \frac{2}{5}$

Exercise Set 1.3

1. 0

3. −8

5. $-\frac{1}{8}$

7. 13

9. a. \$2160 b. \$2080
c. \$2106.67

11. 9.37

13. −9

15. $\frac{3}{2}$

17. 0

19. b^7

21. $-20y^9$

23. $-3x^4$

25. a. 1 b. 100,000,000
c. 32 d. 7

29. 2; 3

31. $\frac{3}{5}$; 4

33. 3

35. 4

37. 11

39. 176.1971

41. $\frac{1}{2}bh$

43. cost of all purchases

45. $5x + 3$

47. $2s^2t^3 - 3s^2t^2 + 2s^2t + 3st^2 + st - s + 2t - 3$

49. $-2a^2bc + ab^2c - 2ab^3 + 3$

51. $-2x^4 + 4x^3 - x^2 + 4x - 4$

53. $6s^3 - s^2 - 11s + 6$

55. $-4y^5 - 2y^4 + 2y^3 - 5y^2 - 3y$

57. $4a^5 - 16a^4 + 14a^3 - 3a^2 - 14a + 15$

59. $3b^4 + 3ab^3 + 2b^3 - 7ab^2 + 2b^2 - 4ab - 6a$

61. $-6x^2 + 22x - 12$

63. $98x - 38y - 20z$

65. $x^2 + 2x - 3$

67. $4x^2 + 8x + 3$

69. $3x^2 - 5x + 2$

71. $x^2 + 2xy + y^2$

73. $9x^2 - 6x + 1$

75. $4x^2 - 1$

77. $x^4 + 2x^2y^2 + y^4$

79. a. 3^{11} b. 2^{2+n}

81. a. $\frac{4}{x^2} - 1$ b. $\frac{w^2x^2}{y^2} - \frac{2wxz}{y} + z^2$
c. $x^2 + 2xy + y^2 - z^2$

83. Eric

85. a. 392 b. 96
c. 16,777,216 d. 4,294,966,528

87. a. $v = 20, a = 14$ b. answers will vary

Exercise Set 1.4

1. $5(x-3)$
3. $-2(x+4y)$
5. $5b(c+5)$
7. $-y^2(3+4y^3)$
9. $3x^2(1+2y-3z)$
11. $(x+3)(x+1)$
13. $(y-3)(y-5)$
15. $(a-3b)(a-4b)$
17. $\left(y+\frac{1}{3}\right)\left(y-\frac{1}{3}\right)$
19. $(3+x)(3-x)$
21. $(x-7)(x+2)$
23. $\left(\frac{1}{4}+y\right)\left(\frac{1}{4}-y\right)$
25. $(x-3)^2$
27. $(x-10)(x-2)$
29. $(x+8)(x+3)$
31. $(x-2)(2x+1)$
33. $(a-3)(3a-2)$
35. $(2x+3)(3x+2)$
37. $(2m-3)(4m+3)$
39. $(2x-3)(5x+1)$
41. $(2a-3b)(3a+2b)$
43. $(2rs+t)(5rs+2t)$
45. $(4+3xy)(4-3xy)$
47. $(2n-5)(4n+1)$
49. $2(x-3)(x+2)$
51. $5(3x-2)(2x-1)$
53. $3m(2x+3)(3x+1)$
55. $2x^2(3+2x)(2-5x)$
57. $(x^2+y^2)(x+y)(x-y)$
59. $(b^2+4)(b^2-2)$
61. $(x+3y)(x^2-3xy+9y^2)$
63. $(3x-y)(9x^2+3xy+y^2)$
65. $(a+2)(a^2-2a+4)$
67. $\left(\frac{1}{2}m-2n\right)\left(\frac{1}{4}m^2+mn+4n^2\right)$
69. $(x+y-2)(x^2+2xy+y^2+2x+2y+4)$
71. $(2x^2-5y^2)(4x^4+10x^2y^2+25y^4)$
73. $4(y+2)(x-1)$
75. $-(x+2)^2(5x+31)$
81. a. $h(3x^2+3xh+h^2)$ b. $2^n(7)$
 c. $(4+9x^6)(2+3x^3)(2-3x^3)$
 d. $(z+x-y)(z-x+y)$
83. a. $C(R+1+r)(R+1-r)$
 b. $p(1-p)(a-b)^2$ c. $(X-L)(X-2L)$
 d. $(R_1+R_2)(R_1+R_2-2r)$ e. $-16(t-7)(t+3)$
85. $t\left(v-\frac{1}{2}at\right)$

Exercise Set 1.5

1. $\frac{1}{x-4},\ x\neq -4$
3. $x-4,\ x\neq 4$
5. $\frac{3x+1}{x+2},\ x\neq \frac{1}{2}$
7. $\frac{4}{9},\ x\neq 2$
9. $-2b(5+a),\ a\neq 5,\ b\neq -3$
11. $\frac{5y}{x-4},\ x\neq -2,\ y\neq 0$
13. $\frac{(2x+1)(x-2)}{(x-1)(x+1)}$
15. $\frac{(x+2)(2x+3)}{x+4}$
17. $\frac{(x+3)(x^2+1)}{x-2}$
19. $\frac{x+4}{(x-5)(x+1)}$
21. xy
23. $2a$
25. $(b-1)^2$
27. $(x+3)(x-2)$
29. $x(x+1)(x-1)$
31. $\frac{4}{a-2}$
33. $\frac{x+5}{3}$
35. $\frac{4(a+1)}{(a+2)(a-2)}$
37. $\frac{4y-15}{3xy}$
39. $\frac{5-2x}{2(x+3)}$
41. $\frac{23x+24}{6(x+3)(x-3)}$
43. $\frac{3x^2-4x-1}{(x-1)(x-2)(x+1)}$
45. $\frac{5x-3}{(x+2)(x-1)}$
47. $\frac{x^2+x+3}{(x+2)(x+3)(x+1)}$
49. $\frac{3x^2+10x+1}{(x+1)(x-1)(x+4)}$
51. $\frac{x+2}{x-3}$
53. $\frac{x(x+1)}{x-1}$
55. $4x(x+4)$
57. $\frac{a+2}{a+1}$
59. $a-b$
61. $\frac{x-2}{x}$
63. $\frac{(y+1)(y-2)}{(y-1)(y+2)}$
65. $\frac{x+1}{2x+1}$
67. a. $\frac{R_2R_3R_4+R_1R_3R_4+R_1R_2R_4+R_1R_2R_3}{R_1R_2R_3R_4}$
 b. $\frac{3c^2-2cd-6d^2}{(c-d)^3}$ c. $\frac{k(k+1)}{k+3}$
69. a. $\frac{a}{b+a}$ b. $\frac{a}{b+a}$
 c. left side cannot be simplified
 d. left side cannot be simplified
 e. $x^4-2x^{2y^2}+y^4$
 f. left side cannot be simplified

Exercise Set 1.6

1. x^6
3. b^4
5. $16x^4$
7. $-\frac{1}{128}$
9. y^{8n}
11. $-\frac{x^3}{y^3}$
13. x^{19}
15. $-32x^{10}$
17. x^{4n}
19. $\frac{1}{x^2}$
21. $30x^8$
23. 1
25. $\left(\frac{3}{2}\right)^n x^{2n}y^{3n}$
27. $(2x+1)^{10}$
29. $2^{2n}a^{4n}b^{6n}$
31. $\frac{4}{3}$
33. 3
35. 81
37. $-x^3$
39. y^6
41. 25
43. $\frac{1}{729}$
45. x^9
47. 32
49. $2x^2y$
51. $\frac{a^4b^6}{9}$
53. $\frac{-8y^{12}}{x^9}$
55. $\frac{a^9}{3b^4}$
57. $\frac{4a^{10}c^6}{b^8}$
59. $\frac{1}{a-2b^2}$
61. $\frac{(a-b)^2}{a+b}$
63. $\frac{b+a}{b-a}$
67. 0.074
69. 0.0113
71. 9.1×10^{-3}
73. 2.3×10^1
75. 8.0×10^{-4}
77. 0.000893
79. 145,000
81. 0.001253
83. 3.0521×10^{-3} cu in
85. 1.67×10^4 persons per square mile
87. a. $\frac{5}{12}$ b. $a(1+r+r^2)$ c. 3^{10m} d. 64 e. 4×10^{-13}

Exercise Set 1.7

1. 8
3. $\frac{1}{16}$
5. $2x^{13/12}$
7. $x^{5/36}$
9. x^2y^{12}
11. $\frac{x^9}{y^6}$
13. $\sqrt[5]{\frac{1}{16}}$
15. $\sqrt[4]{a^3}$
17. $\sqrt[3]{\frac{144x^6}{y^4}}$
19. $8^{3/4}$
21. $(-8)^{-2/5}$
23. $\left(\frac{4}{9}a^3\right)^{-\frac{1}{4}}$
25. $\frac{2}{3}$
27. not real
29. 5
31. $\frac{5}{4}$
33. 54.82
35. $\sqrt{13} \neq 5$
37. $4\sqrt{3}$
39. $3\sqrt[3]{2}$
41. $y^2\sqrt[3]{y}$
43. $2x^2\sqrt[4]{6}\sqrt{x}$
45. $x^2y\sqrt{xy}$
47. $2x^2y\sqrt[4]{y}$
49. $\frac{\sqrt{5}}{5}$
51. $\frac{\sqrt{3y}}{3y}$
53. $2x\sqrt{2x}$
55. $y^2\sqrt[3]{x^2y}$
57. $7\sqrt{3}$
59. $7\sqrt{x}$
61. $4\sqrt{3}$
63. $11\sqrt{5} - \sqrt[3]{5}$
65. $-5\sqrt{5}$
67. $3 + 4\sqrt{3}$
69. $3xy$
71. $5 - 2\sqrt{6}$
73. $3x - 4y - \sqrt{6xy}$
75. $\frac{3(3-\sqrt{2})}{7}$
77. $\frac{2}{13}(\sqrt{3}+4)$
79. $\frac{-3(3\sqrt{a}-1)}{9a-1}$
81. $\frac{-3(5-\sqrt{5y})}{5(5-y)}$
83. $3 + 2\sqrt{2}$
85. $3\sqrt{2} + 2\sqrt{3} + \sqrt{6} + 2$
87. $\frac{2}{2\sqrt{3}+\sqrt{10}}$
89. $\frac{-1}{\sqrt{x}+4}$
95. a. $\sqrt[8]{x^7}$ b. $\frac{(x-1)^2}{x}$ c. $\frac{\sqrt{1+x^2}}{2}$ d. $\frac{3}{5}$ e. $\frac{5}{(1+x^2)^{3/2}}$
97. $\frac{\sqrt{Lc_1c_2(c_1+c_2)}}{2\pi(c_1+c_2)}$

Exercise Set 1.8

1. 1
3. $-i$
5. $-i$
7. 1
9. i
11. $-\frac{3}{4} + 0i$
13. $0 + 5i$
15. $0 - 6i$
17. $3 - 7i$
19. $0.3 - 7\sqrt{2}\,i$
21. $-2 - 4i$
23. $x = \frac{2}{3}, y = -8$
25. $x = -1, y = -\frac{9}{2}$
27. $3 + i$
29. $5 + i$
31. $-5 - 4i$
33. $2 - 6i$
35. $-1 - \frac{1}{2}i$
37. 5
39. $2 + 14i$
41. $4 - 7i$
43. 5
45. 25
47. 20
49. $-\frac{13}{10} + \frac{11}{10}i$
51. $-\frac{7}{25} - \frac{24}{25}i$
53. $\frac{8}{5} - \frac{1}{5}i$
55. $\frac{5}{3} - \frac{2}{3}i$
57. $\frac{4}{5} + \frac{8}{5}i$
59. $\frac{4}{25} - \frac{3}{25}i$
61. $\frac{9}{10} + \frac{3}{10}i$
63. $0 + \frac{1}{5}i$
65. 0
67. 3
75. $y \geq 5$

Chapter 1 Review Exercises

1. $\{1, 2, 3, 4\}$
2. $\{-3, -2, -1\}$
3. $\{2\}$
4. T
5. F
6. F
7. F
8. additive inverse
9. distributive
10. commutative (addition)
11. multiplicative identity
12. 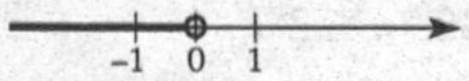 −1 0 1
13. 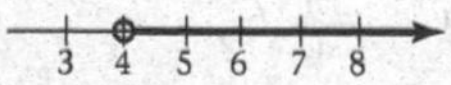 3 4 5 6 7 8
14. 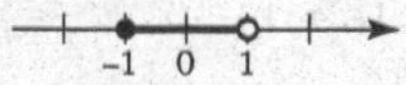 −1 0 1
15. -1
16. $\frac{3}{2}$
17. \$99
18. c
19. $-0.5, 7$
20. $-7, 5$
21. $a^2b^2 - 3a^2b + 4b$
22. $2x^3 + 3x^2 - 2x$
23. $12x^3 + 12x^2 + 3x$
24. $2(x + 1)(x - 1)$
25. $(x + 5y)(x - 5y)$
26. $(2a + 3b)(a + 3)$
27. $(4x - 1)(x + 5)$
28. $(x^4 + 1)(x^2 + 1)(x + 1)(x - 1)$
29. $(3r^2 + 2s^2)(9r^4 - 6r^2s^2 + 4s^4)$
30. $-\frac{6(y-1)}{xy^2(x-y)}$
31. $-\frac{3(2+x)}{2y}$
32. $\frac{3x(x+1)}{2x-1}$
33. $\frac{a-2b}{a-b}$
34. $2x^2(x + 2)(x - 2)$
35. $x(x - 1)^2$
36. $5y^2(x - 1)^2$
37. $4x^2(y - 1)(y + 1)^2$
38. $\frac{2(a^2-2)}{(a+2)(a-2)}$
39. $\frac{-2x-5}{(x+4)(x-4)}$
40. $\frac{x-7}{(x+2)(x-1)^2}$
41. $\frac{x^3-x^2+1}{x-1}$
42. $\frac{b^9}{8a^6}$
43. 2
44. $\frac{1}{x^4y^8}$
45. x^3
46. $4\sqrt{5}$
47. $\frac{\sqrt{3}}{3}$
48. $x^3y^2\sqrt{xy}$
49. $2x^2y\sqrt[4]{2y^2}$
50. $\frac{x-\sqrt{xy}}{x-y}$
51. 2.48×10^1
52. $\frac{1}{\sqrt{x}+\sqrt{y}}$
53. $3\sqrt{|xy|}$
54. $8 + 2\sqrt{15}$
55. a. 1.97 b. 15,689 c. 2.83 d. 9488.53 e. −1.93 f. −0.62 g. 1.94 h. 0.93 i. 0.20 j. 3.95
56. $x = -2, y = 4$
57. $-i$
58. $8 - i$
59. $3 + 4i$
60. $17 + 6i$
61. $-\frac{1}{13} - \frac{18}{13}i$
62. a. $\frac{bc+ac+ab}{abc}$ b. $\frac{cd(b+a)}{ab(d+c)}$
63. 1150 calories per day
64. a. 4
    b. (president, vice-president), (president, secretary), (president, treasurer), (vice-president, secretary, treasurer)
65. 6 children
66. \$20.00
67. $8x^2 - 2xy$
68. $(4 - 2s)(5 - 2s)s,\ 0 < s < 2$
69. a. $x^3 - y^3$ b. $x - y^4$ c. $x^5 - y^5$

70. $x^n - y^n = (x - y)(x^{n-1} + x^{n-2}y + x^{n-3}y^2 + \cdots + x^2y^{n-3} + xy^{n-2} + y^{n-1})$

71. $13 \times 17 = (20 - 7)(20 - 3)$
$= (20)(20) - (20)(3) - (20)(7) + 21$
$= 400 - 60 - 140 + 21$
$= 221$

72. $(c + b)(a - b)$; $(a - b)(c + b)$

73. F = 2, O = 9, R = 7, T = 8, Y = 6, E = 5, N = 0, S = 3, I = 1, X = 4

74. 496; 8128; 33,550,336

75. a. 3.3×10^{17} seconds b. 3.1536×10^7 seconds
c. 10.0464×10^{10} years

76. $[x + (x + x^{1/2})^{1/2}]^{1/2}$

77. yes

78. $\dfrac{1+\sqrt{5}}{2}$

79. a. $\dfrac{h+2}{h(\sqrt{x+h+1}+\sqrt{x-1})}$
b. $\dfrac{1}{\sqrt{3+x}+\sqrt{3}}$

80. $18 - 14i$ volts

Chapter 1 Review Test

1. $\{2, 4, 6, 8, 10, 12\}$
2. $\{3\}$
3. F
4. F
5. commutative (multiplication)
6. multiplicative inverse
7. 
8. 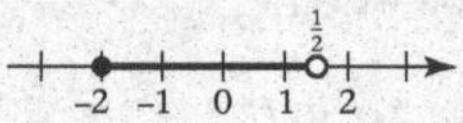
9. -1
10. 2
11. 25
12. $-\frac{7}{3}$
13. b
14. $-2.2, 5$
15. 14, 6
16. $4xy + 3x + 4y + 1$
17. $3a^3 + 5a^2 + 3a + 10$
18. $4a^2b(2ab^4 - 3a^3b + 4)$
19. $(2 + 3x)(2 - 3x)$
20. $\dfrac{6m^5}{n^2}$
21. $\dfrac{1-x}{x+1}$
22. $4x^2(x+1)(x-1)(x-2)$
23. $\dfrac{11x-15}{3(x+3)(x-3)}$
24. $\dfrac{2}{x+1}$
25. $\dfrac{1}{x^{17}}$
26. y^{n+1}
27. -1
28. $\dfrac{4a^4}{b^2}$
29. 0
30. $32 - 10\sqrt{7}$
31. $-\frac{11}{4}\sqrt{xy}$
32. $x \le 2$
33. $-1 + 0i$
34. $16 - 11i$
35. $\frac{8}{5} + \frac{9}{5}i$

Chapter 2

Exercise Set 2.1

1. T
3. T
5. -2
7. $-\frac{2}{3}$
9. 6
11. $\frac{-4}{3}$
13. $\frac{3}{2}$
15. $-\frac{10}{3}$
17. -2
19. 5
21. $-\frac{7}{2}$
23. 1
25. $\dfrac{8}{5-k}$
27. $\dfrac{6+k}{5}$
29. $\frac{10}{3}$
31. 1
33. 4
35. 4
37. 12
39. 2
41. $\frac{12}{7}$
43. no solution
45. I
47. C
49. T
51. F
53. T
55. a. $\frac{2}{9}$ b. $\frac{41}{333}$
c. $\frac{134}{99}$ d. 1
57. a. $w = \dfrac{cd}{c+d}$ b. $x = \dfrac{1}{a-c}$
c. $y = \dfrac{2c^2}{a-b-c}$

Exercise Set 2.2

1. $3+2n$
3. $6n-5=26$
5. 16, 28
7. 6, 7, 8
9. 68°
11. 4 meters and 8 meters
13. 10 nickels, 25 dimes
15. 300 children, 400 adults
17. 61 three-dollar tickets, 40 five-dollar tickets, 20 six-dollar tickets
19. $11,636.36 on 10-speeds, $4363.64 on 3-speeds
21. $7000
23. 20 hours
25. 50 miles per hour and 54 miles per hour
27. 40 kilometers per hour, 80 kilometers per hour
29. Ceylon: 2.4 ounces, Formosa: 5.6 ounces
31. 13.5 gallons
33. $\frac{1}{12}, \frac{1}{4}; -\frac{1}{4}, -\frac{1}{12}$
35. 2.4 hours
37. $4\frac{1}{2}$ days, 9 days
39. 8 hours
41. 140 miles per hour
43. $\frac{C}{2\pi}$
45. $\frac{5}{9}(F-32)$
47. $\frac{2A}{h}-b'$
49. $f_2=\frac{ff_1}{f_1-f}$
51. $L=\frac{a+Sr}{S+r}$
53. a. $2(r+s)$ b. $0.05|r-s|$ c. $2s-5$ d. $\frac{r}{s}$ e. r^2+s^2 f. $\frac{r+s}{2}$ g. $6r-4s$
55. 10, 12, 14
57. 4.29 quarts
59. 80 miles
61. Enrique
63. $W=\frac{k}{1-fk^2}$
65. $r=\frac{cR}{E-c}$

Exercise Set 2.3

1. 1, 2
3. 1, −2
5. −2, −4
7. 0, 4
9. $\frac{1}{2}, 2$
11. ± 2
13. $\frac{1}{3}, \frac{1}{2}$
15. ± 3
17. $\pm\sqrt{5}$
19. $\frac{-5\pm 2\sqrt{2}}{2}$
21. $\frac{5\pm 2\sqrt{2}}{3}$
23. $\pm\frac{8}{3}i$
25. 4, −2
27. $-\frac{1}{2}, 4$
29. $\frac{1}{3}, -3$
31. $\frac{-1\pm i}{2}$
33. $1, -\frac{3}{4}$
35. $\frac{-1\pm\sqrt{2}i}{3}$
37. $0, -\frac{3}{2}$
39. $\frac{2\pm\sqrt{11}i}{5}$
41. $\frac{2\pm\sqrt{21}i}{5}$
43. $\frac{2}{3}, -1$
45. $\frac{\pm 2\sqrt{3}}{3}$
47. $0, -\frac{3}{4}$
49. $\frac{-1\pm\sqrt{11}}{2}$
51. $-2, \frac{2}{3}$
53. $\frac{-5\pm\sqrt{7}i}{4}$
55. $\pm\frac{1}{2}$
57. $\frac{-1\pm\sqrt{11}i}{4}, 0$
59. $\pm\sqrt{c^2-a^2}$
61. $\frac{\pm\sqrt{3V\pi h}}{\pi h}$
63. $\frac{-v\pm\sqrt{v^2+2gs}}{g}$
65. two complex roots
67. one double, real root
69. two rational, real roots
71. two rational, real roots
73. two complex roots
75. two complex roots
77. two irrational, real roots
79. one double, real root
81. 4
83. 0, −8
85. 4
87. 3
89. 0, 4
91. 5
93. $u=-2, x=\pm\sqrt{-2}, x=\pm\sqrt{2}i$
95. $u=\frac{1}{2}; 2, -\frac{3}{2}$
97. $u=x^{1/5}; -32, -\frac{1}{32}$
99. $u=1+\frac{1}{x}; \frac{1}{3}, -\frac{2}{7}$
103. $-\frac{1}{2}$
105. −6
107. ± 9
109. 6
115. a. $n=10$ b. $n=8$ c. $\frac{n^2+n-2}{2}$
117. a. $n=11$ b. $1+3+5+7+9+11+13+15+17+19+21$

Exercise Set 2.4

1. A: 3 hours; B: 6 hours
3. roofer: 6 hours; assistant: 12 hours
5. $L = 12$ feet, $W = 4$ feet
7. $L = 8$ centimeters, $W = 6$ centimeters
9. 10 feet
11. 5 or $\frac{1}{5}$
13. 3, 7; −3, −7
15. 6, 8
17. 150 shares
19. 8 days
21. 16 centimeters from the end
23. 4.5 seconds
25. 14%
27. $v = 0.99c$
29. a. $t = 2$ seconds
 b. $t = \dfrac{-w_0 \pm \sqrt{w_0^2 - 2\alpha_0(\theta_0 - \theta)}}{\alpha_0}$
31. $p = 0.0078, 0.36$
33. $\sqrt{10}$
35. a. $L = \dfrac{5 \pm 5\sqrt{5}}{2}$ b. $W = \dfrac{-L \pm L\sqrt{5}}{2}$

Exercise Set 2.5

1. $[-5, 1)$
3. $(9, \infty)$
5. $[-12, -3]$
7. $(3, 7)$
9. $(-6, -4]$
11. $5 \le x \le 8$
13. $x > 3$
15. $x \le 5$
17. $x \ge 0$
19. $x < 4$
21. $x < -6$
23. $x \ge 5$ 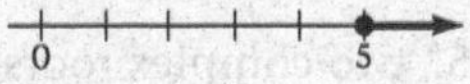
25. $a > -1$ 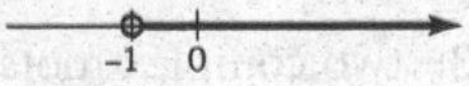
27. $y < -\frac{1}{2}$
29. $x \ge 0$
31. $r < 2$
33. $x \ge 1$ 
35. $x > \frac{5}{3}$ 
37. $(-\infty, 2]$
39. $[2, \infty)$
41. $\left(-\infty, \frac{3}{2}\right)$
43. $(-12, \infty)$
45. $\left(-\infty, \frac{9}{2}\right)$
47. $(-\infty, -6]$
49. $\left(-\infty, \frac{3}{2}\right)$
51. $(-\infty, 7]$
53. $\left(-\frac{1}{2}, \frac{5}{4}\right]$
55. $[-3, -2]$
57. $(-3, -1]$
59. $[-1, 2)$
61. 98
63. 5 or more
65. 2924
67. $L \le 20$ meters
69. $-4 \le x \le 1$ 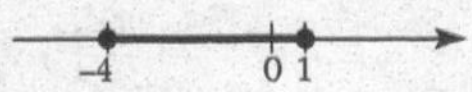
71. $x \le -\frac{2}{3}$ or $x \ge 2$ 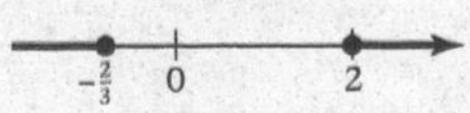
73. $r \le -4$ or $r \ge 0$ 
75. $x < -4$ or $x \ge 6$
77. $x \le 1$ or $x > \frac{3}{2}$ 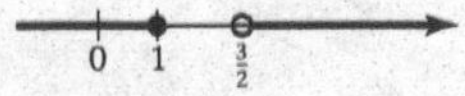
79. $x \le -\frac{5}{4}$ 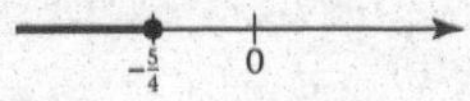
81. $-\frac{5}{2} \le x \le 2$ or $x \ge 4$ 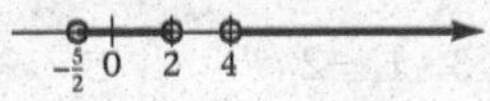
83. $x \le -2$ or $x \ge 5$ 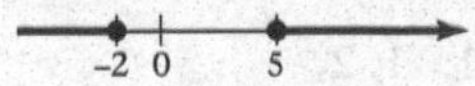
85. $-\frac{2}{3} \le x \le 2$ 
87. $x \le -\frac{2}{3}$ or $x > \frac{3}{2}$ 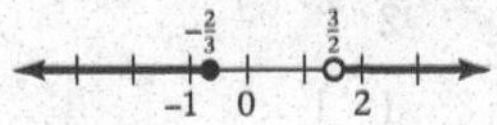
89. $-2 < x \le \frac{1}{2}$
91. $-2 \le x \le -\frac{1}{2}$ 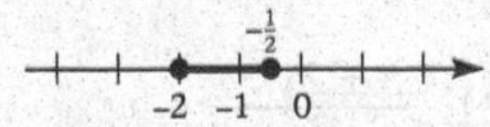
93. $x < -\frac{5}{2}$ or $-1 < x < \frac{2}{3}$
95. $-\frac{1}{2} \le x \le \frac{1}{2}$ or $x \ge 3$ 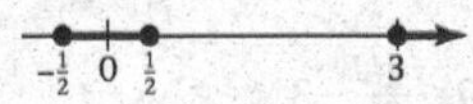
97. $x \le -\frac{1}{2}$ or $x \ge 3$
99. $x \le -1$ or $x \ge -\frac{1}{2}$
101. $\frac{1}{2} \le t \le 2$
103. 40 centimeters
105. 12 feet
107. 5 amperes
109. volume between 2 cubic inches and 3 cubic inches
111. 170
113. 14°F to 68°F
115. smallest = 6.6 inches, largest = 6.8 inches

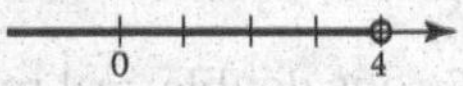

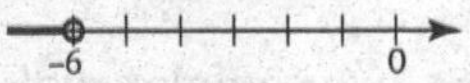

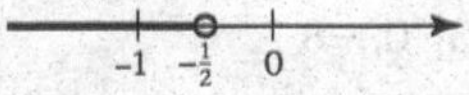

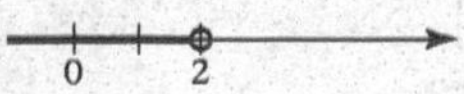

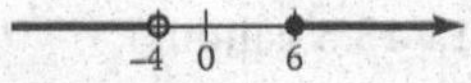

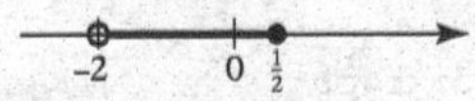

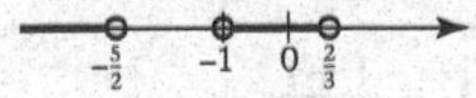

Exercise Set 2.6

1. $1, -5$
3. $3, 1$
5. $2, -\frac{4}{3}$
7. $\frac{3}{2}, -3$
9. $4, -2$
11. $x < -4$ or $x > 2$
13. $x < -\frac{1}{2}$ or $x > 1$
15. $-\frac{1}{3} < x < 1$
17. $(-\infty, -1] \cup [7, \infty)$
19. $\left(-\frac{7}{2}, \frac{9}{2}\right)$
21. no solution
23. $(-\infty, -7) \cup (17, \infty)$
25. $0, -1$
27. $2, 4$
31. $98 \le x \le 102$
33. $|x| > 6$
35. $|x - 3| \ge 2$
37. $x = 14, x = -6$
39. $x = \frac{22}{7}, x = \frac{14}{5}$
41. $\mu - k\sigma < x < \mu + k\sigma$

Chapter 2 Review Exercises

1. $\frac{8}{3}$
2. 0
3. $\frac{10}{3}$
4. $\frac{k}{4k+2}$
5. $\frac{10}{3}$ centimeters $\times \frac{8}{3}$ centimeters
6. 5 quarters, 14 dimes
7. 240 miles
8. 6 hours
9. F
10. F
11. $5, -4$
12. $\frac{1}{2}, \frac{4}{3}$
13. $1 \pm \sqrt{5}i$
14. $\frac{2 \pm \sqrt{2}i}{2}$
15. $-1, \frac{1}{3}$
16. $\pm\frac{3}{7}$
17. $\pm\frac{\sqrt{3\pi k}}{k}$
18. $-4, 3$
19. two rational, real roots
20. one real, double root
21. two complex roots
22. 4, no solution
23. 6
24. $\pm 1, \pm\sqrt{3}$
25. $-\frac{1}{2}, -1$
26. 60
27. $x \ge 1$
28. $-\frac{9}{2} \le x < \frac{5}{2}$
29. $(-\infty, 8)$
30. $\left(\frac{5}{2}, \infty\right)$
31. $[-9, \infty)$
32. $\frac{5}{3}, -3$
33. $x = -1, x = \frac{3}{2}$
34. $x > 3$ or $x < -4$
35. $\left(\frac{1}{5}, \frac{3}{5}\right)$
36. $\left(-\infty, -\frac{4}{3}\right] \cup \left[\frac{8}{3}, \infty\right)$
37. $x \le -\frac{3}{2}$ or $x \ge 2$
38. $[-5, 1]$
39. $(-\infty, -5) \cup \left[-\frac{1}{2}, \infty\right)$
40. $\left(-2, -\frac{3}{2}\right) \cup (3, \infty)$
41. 5 chairs
42. area of circle is greater.
43. 22 inches
44. $\overline{x} - \frac{1.96\sigma}{\sqrt{n}} < \mu < \overline{x} + \frac{1.96\sigma}{\sqrt{n}}$
45. a. 1 meter per second b. $\sqrt{2}$ meters per second
46. 8 inches $\times$ 6 inches $\times$ 2 inches

Chapter 2 Review Test

1. $\frac{3}{4}$
2. $\frac{8}{13}$
3. $\frac{18}{5}$ meters, $\frac{28}{5}$ meters, $\frac{29}{5}$ meters
4. \$6000 at 6.5%, \$6200 at 7.5%, \$12,300 at 9%
5. T
6. $-2, 7$
7. $\frac{1 \pm \sqrt{79}i}{10}$
8. $-\frac{3}{4}, \frac{1}{3}$
9. $\frac{5 \pm 3i}{2}$
10. $1, -\frac{3}{2}$
11. two rational, real roots
12. two complex roots
13. -4
14. $\pm\sqrt{2}i, \pm\frac{\sqrt{3}}{3}$
15. 8 meters $\times$ 12 meters
16. $-2 \le x < 1$
17. $\left(-\infty, -\frac{15}{2}\right]$
18. $[-4, 4]$
19. $\frac{5}{2}, -2$
20. $-2 \le x \le 3$
21. $\left(-\infty, -\frac{4}{3}\right) \cup (2, \infty)$
22. $x \le \frac{1}{3}$ or $x \ge 1$
23. $\left(-\infty, \frac{1}{2}\right] \cup [1, \infty)$
24. $\left[-2, \frac{2}{3}\right] \cup [1, \infty)$
25. $(-\infty, -1) \cup (2, \infty)$

Chapter 3

Exercise Set 3.1

1.

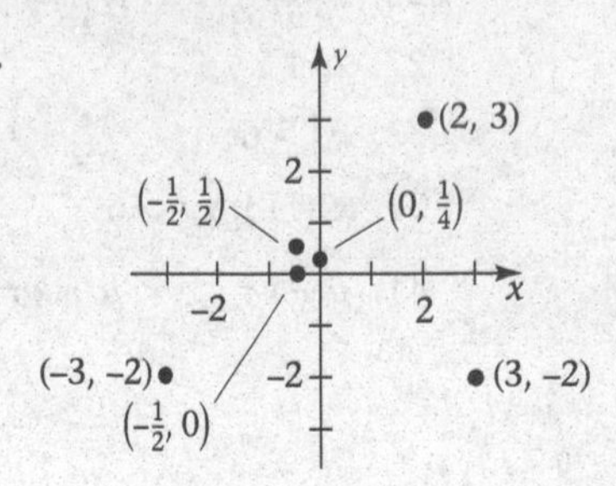

3. $3\sqrt{2}; \left(\frac{7}{2}, \frac{5}{2}\right)$

5. $4\sqrt{2}; (-3, -3)$

7. $\frac{\sqrt{1345}}{6}; \left(-\frac{2}{3}, -\frac{5}{4}\right)$

9. $\overline{BC} = \sqrt{37}$

11. $\overline{RS} = \frac{\sqrt{2}}{2}$

13. not a right triangle

15. is a right triangle

17. A, B, C are collinear

19. A, B, C are collinear

21. $2\sqrt{10} + 7 + 5\sqrt{2} + \sqrt{37}$

25. 0, 5

27. x-intercept: −2
y-intercept: 4

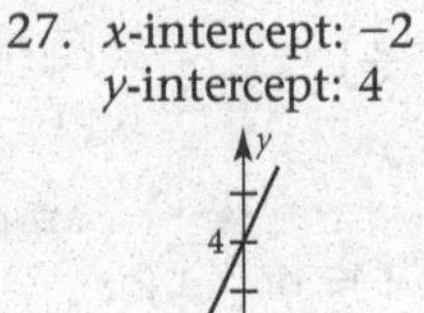

29. x-intercept: 0
y-intercept: 0

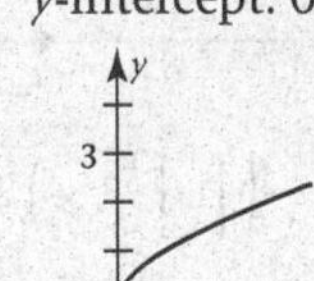

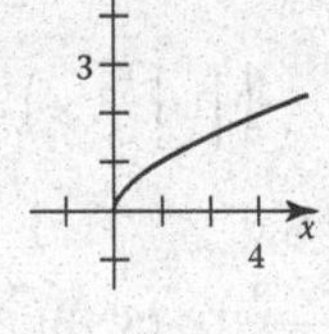

31. x-intercept: −3
y-intercept: 3

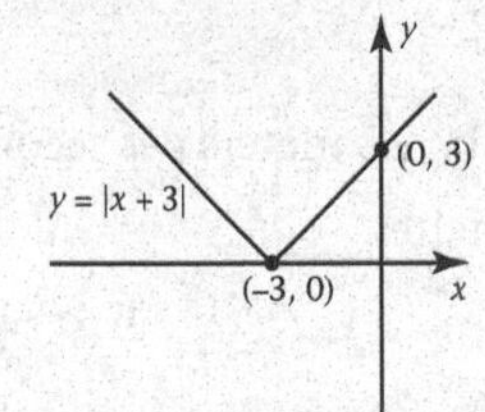

33. x-intercept: $\pm\sqrt{3}$
y-intercept: 3

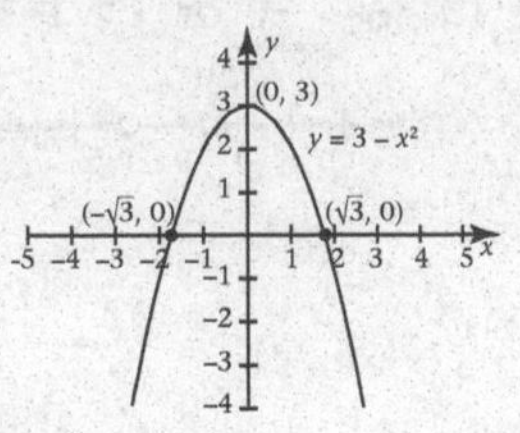

35. x-intercept: −1
y-intercept: 1

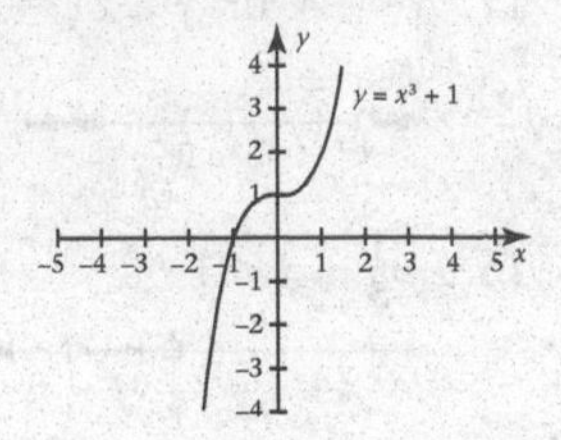

37. x-intercept: −1
y-intercept: ±1

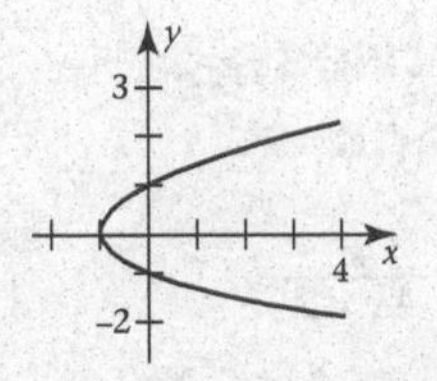

39. XSCL = 5, YSCL = 5
x-intercept: 7
y-intercept: 7

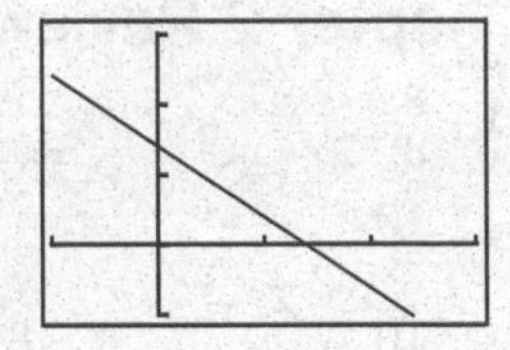

41. XSCL = 1, YSCL = 1
x-intercept: −1, 0, 1
y-intercept: 0

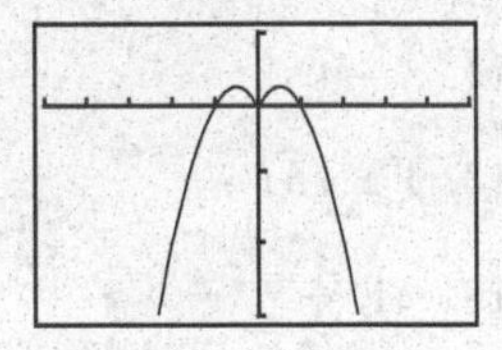

43. XSCL = 1, YSCL = 1
x-intercept: −0.6, 0, 1.6
y-intercept: 0

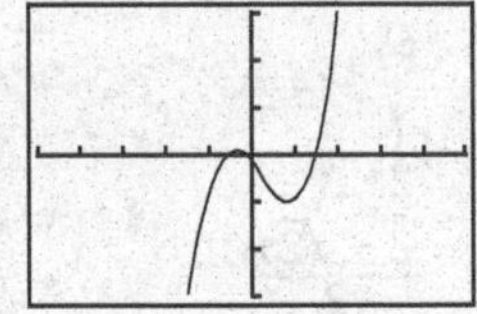

45. none

47. x-axis

49. x-axis

51. x-axis

53. none

55. y-axis

57. all

59. origin

63. (1, 2)

65. $x = -2, y = 4$

Exercise Set 3.2

1. domain: all reals
range: all reals

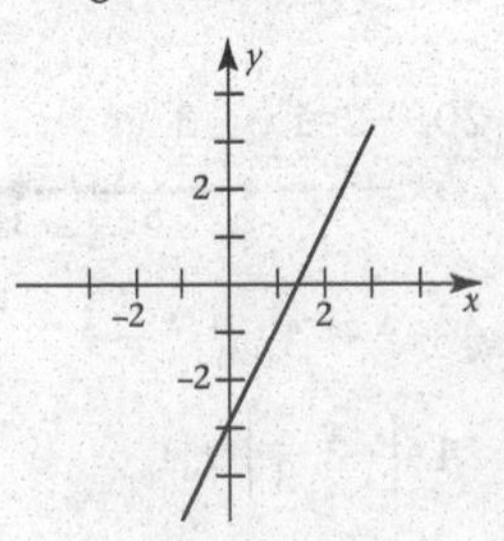

3. domain: all reals
range: all reals

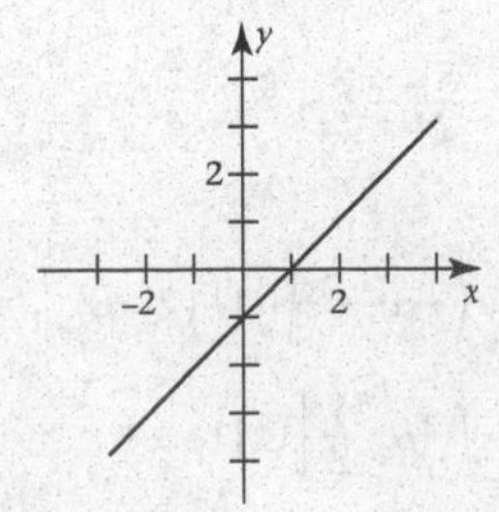

5. domain: $x \ge 1$
range: $y \ge 0$

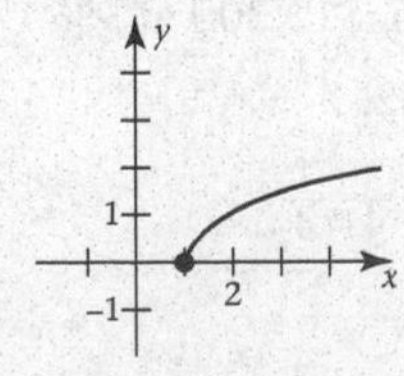

7. $\left\{x \middle| x \geq \frac{3}{2}\right\}$
WINDOW: [0, 9.5] × [−2, 5]

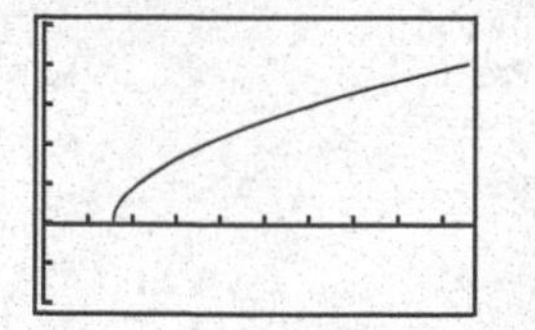

9. $\{x|x > 2\}$
WINDOW: [0, 5] × [−1, 4]

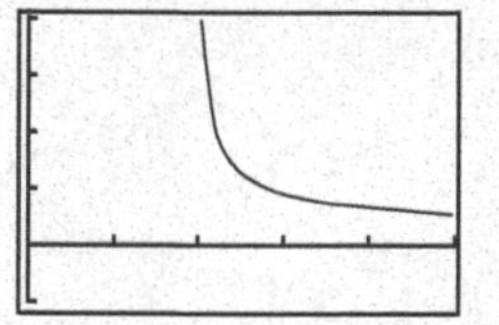

11. $\{x|1 \leq x < 2 \text{ or } x \leq 2\}$
WINDOW: [0, 5] × [−5, 5]

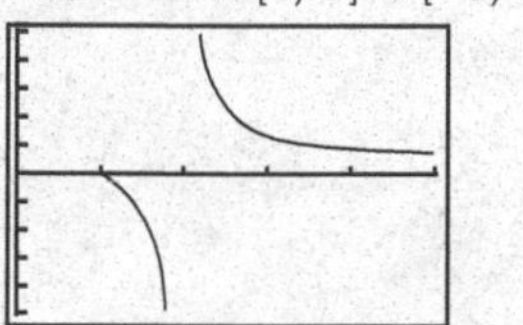

13. $\frac{7}{2}$
15. $\frac{3}{2}$
17. 5
19. $2a^2 + 5$
21. $6x^2 + 15$
23. $4a + 2h$
25. $\frac{1+2x}{x^2}$
27. $x^2 - 2x$
29. $2a + h + 2$
31. $\frac{3x-1}{x^2+1}$
33. $\frac{2(4x^2+1)}{6x-1}$
35. −0.210
37. $\frac{2(a-1)}{4a^2+4a-3}$
39. $\frac{a-1}{a(a+4)}$
41. $I(x) = 0.28x$
43. $d(c) = \frac{C}{\pi}$
45. $V = 4x^3 - 44x^2 + 120x$
47. a. $S(x) = 120000 + 1000x - 50x^2$
b. $125,000
49. a. H = 0.06*w*
b. O = 0.94*w*
c. 15 grams of H and 235 grams of O
51. a. −53.76
b. −63.09, −57.63, −50.25, −46.31, −45.57

Exercise Set 3.3

1.

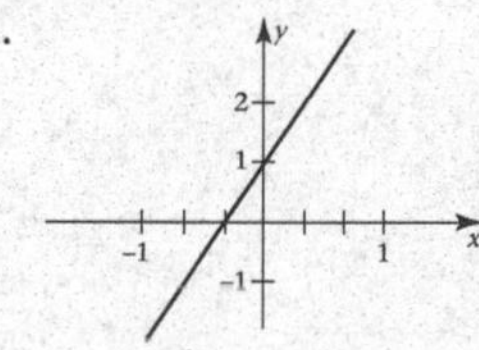

increasing: $(-\infty, \infty)$

3.

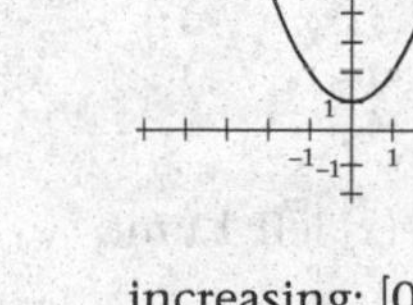

increasing: $[0, \infty)$
decreasing: $(-\infty, 0]$

5.

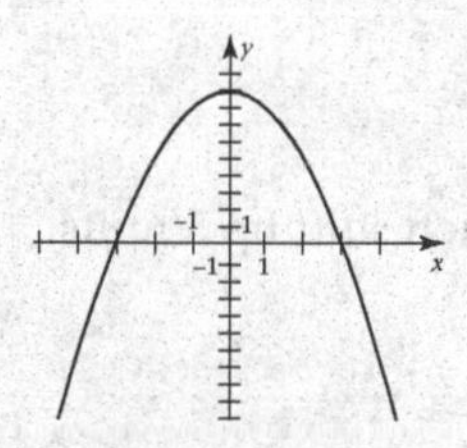

increasing: $(-\infty, 0]$
decreasing: $[0, \infty)$

7.

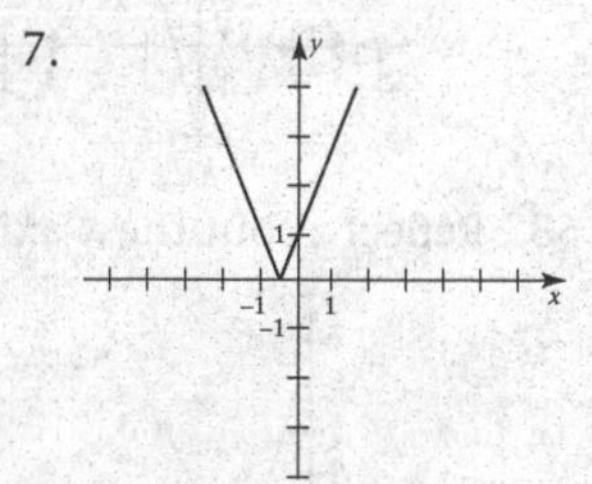

increasing: $\left[-\frac{1}{2}, \infty\right)$
decreasing: $\left(-\infty, -\frac{1}{2}\right]$

9.

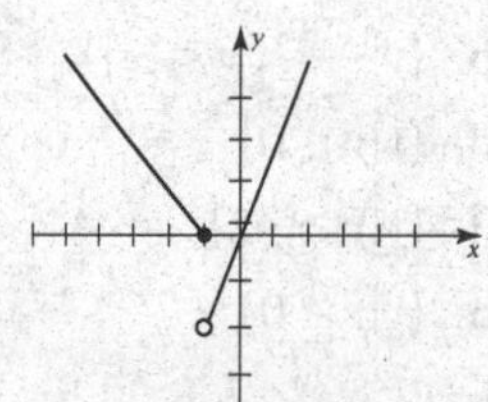

increasing: $(-1, \infty)$
decreasing: $(-\infty, -1]$

11.

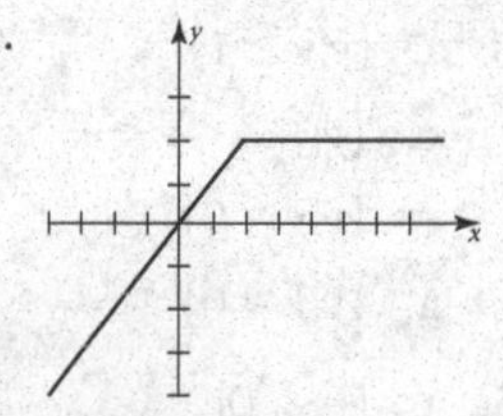

increasing: $(-\infty, 2]$
constant: $[2, \infty)$

13.

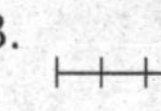

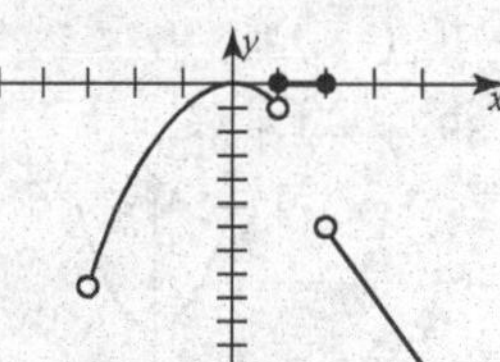

increasing: $(-3, 0]$
decreasing: $[0, 1), (2, \infty)$
constant: $[1, 2]$

15.

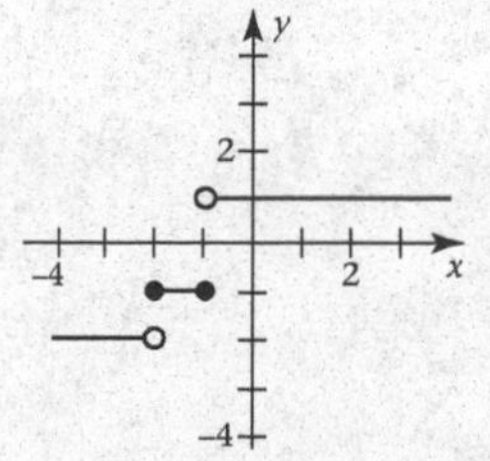

constant: $(-\infty, -2), [-2, -1], (-1, \infty)$

17.

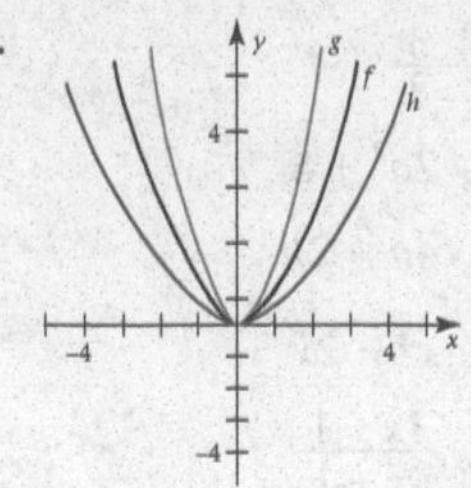

19.

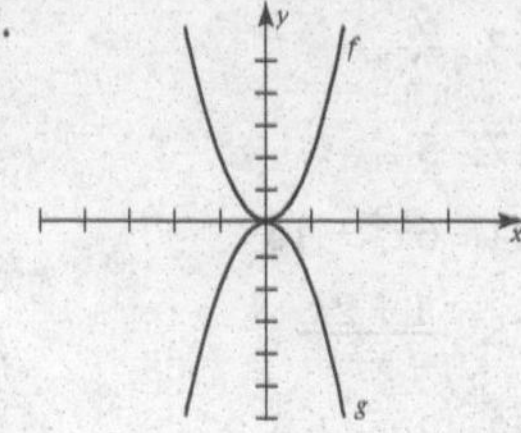

21.

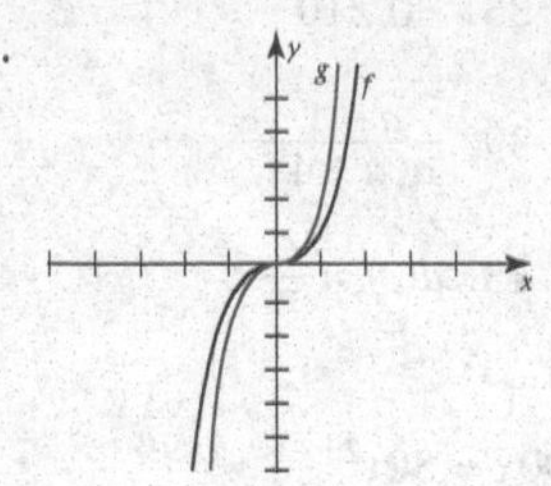

23.

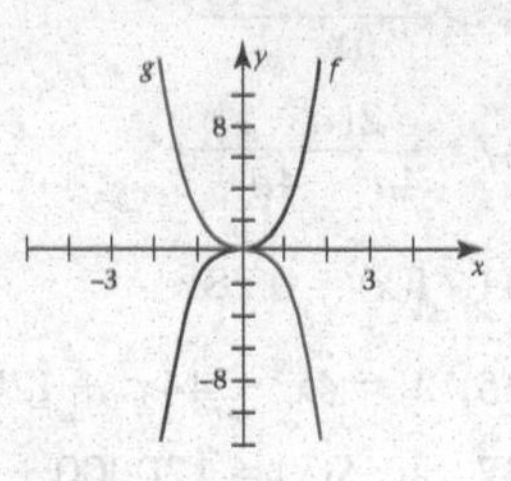

25. $C(u) = \begin{cases} 6.50 & \text{if} \quad 0 \le u \le 100 \\ 0.50 + 0.06u & \text{if} \quad 100 < u \le 200 \\ 2.50 + 0.05u & \text{if} \quad u > 200 \end{cases}$

27. $R(x) = \begin{cases} 30{,}000 & \text{if} \quad 0 \le x \le 100 \\ 400x - x^2 & \text{if} \quad x > 100 \end{cases}$

29. a. $C(m) = 24 + 0.15(1000) = 39$
    b. Domain $\{m|m \ge 0\}$
    c. 39

33. a. $\{y|y \ge 1\}$ b. $\{y|-\infty < y < \infty\}$
    c. $\{y|-\infty < y < \infty\}$ d. $\{y|y \le 0\}$
    e. $\{y|y \ge 0\}$ f. $\{y|y \ge -2\}$
    g. $\{y|y \ge 0\}$ h. $\{y|y \ge 0\}$

35. a. $(-\infty, 0]$ b. never
    c. $(-\infty, \infty)$ d. $[0, \infty)$
    e. $(-\infty, 0]$ f. $(-\infty, 0]$
    g. $(-\infty, -3]$, $[0, 3]$ h. $(-\infty, 1]$

37. $y = -3$

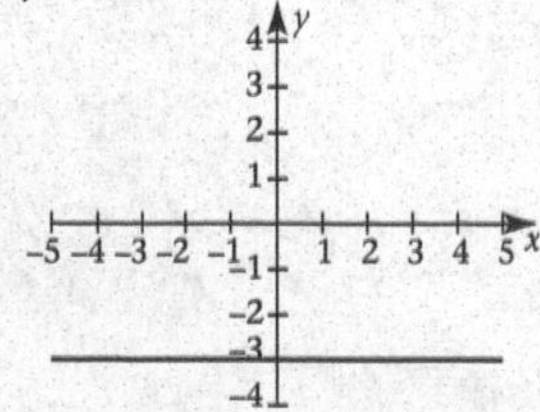

39. $y = 1 + x^2$

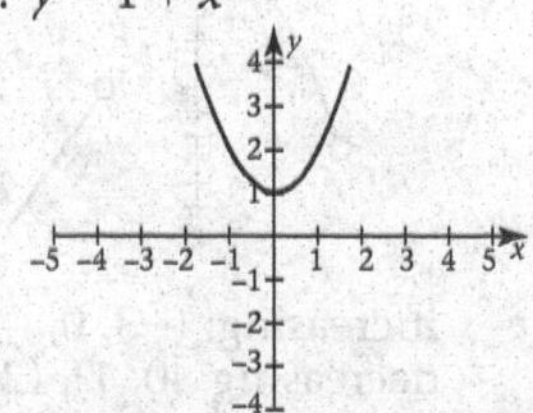

41. Shift $f(x)$ up 2 units.

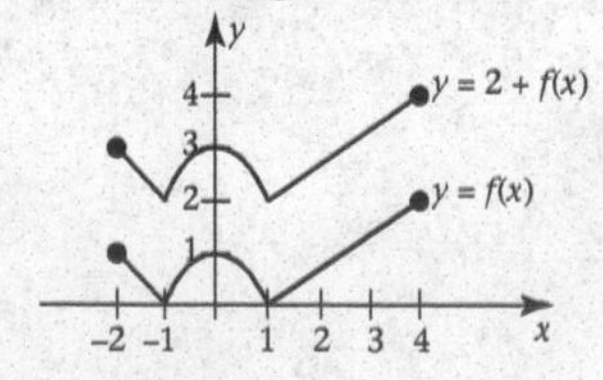

43. Reflect $f(x)$ about the x-axis.

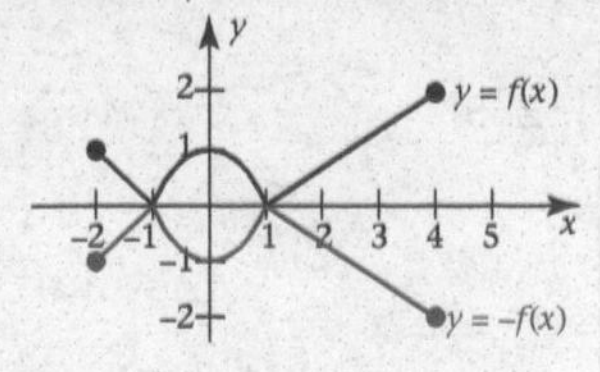

45. Shift $f(x)$ down 3 units.

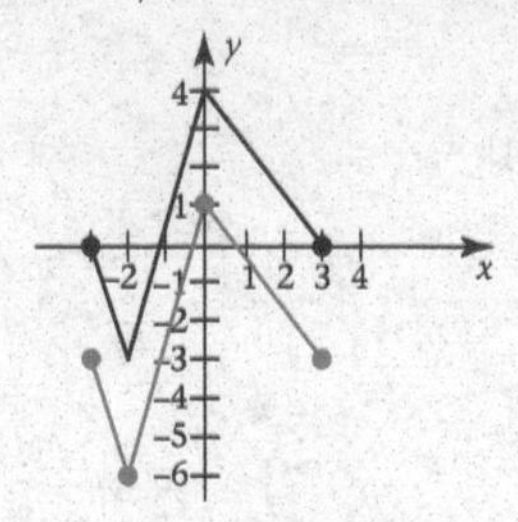

47. Reflect $f(x)$ about the x-axis.

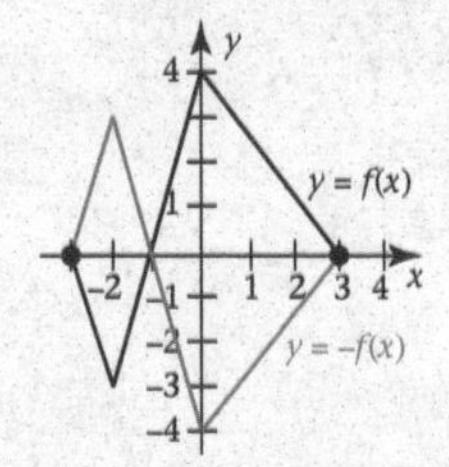

49. Reflect $f(x)$ about the y-axis.

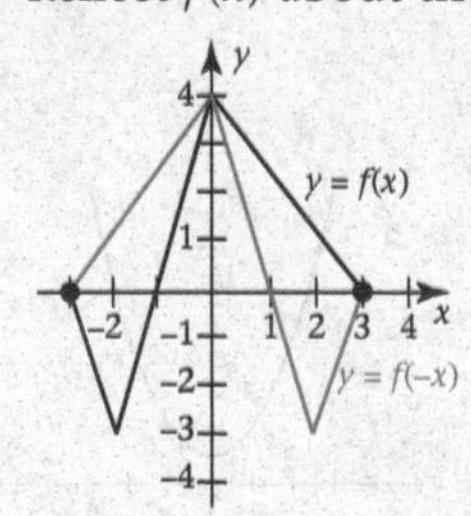

51. Shift $f(x)$ left 1 unit.

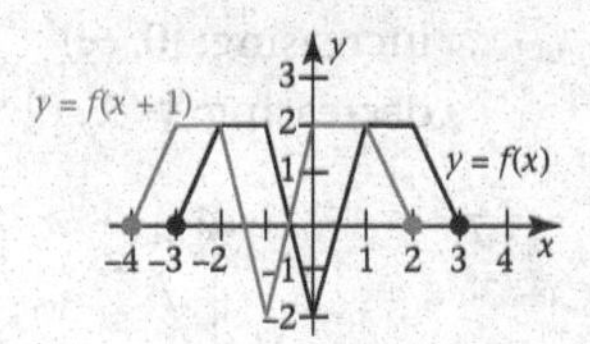

53. Reflect about the x-axis, then shift up 5 units.

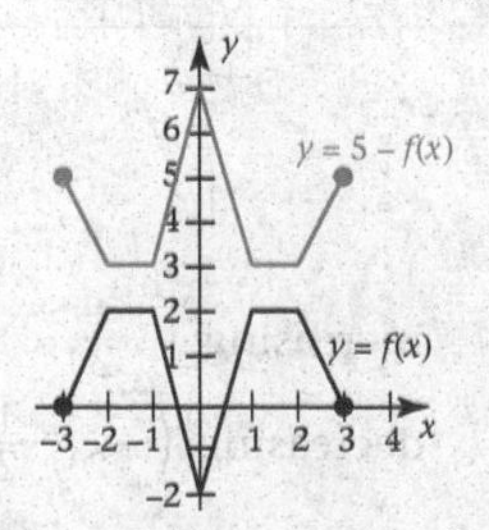

55. Reflect $f(x)$ about the y-axis.

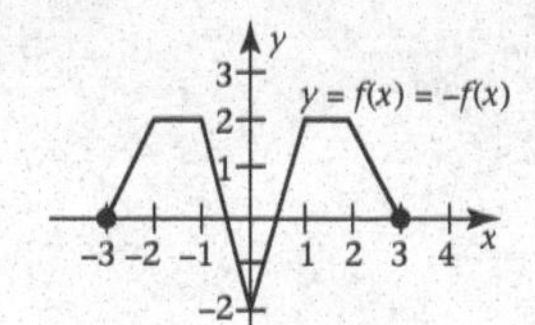

57. $S = -16t^2 + 80t + 96$

a.

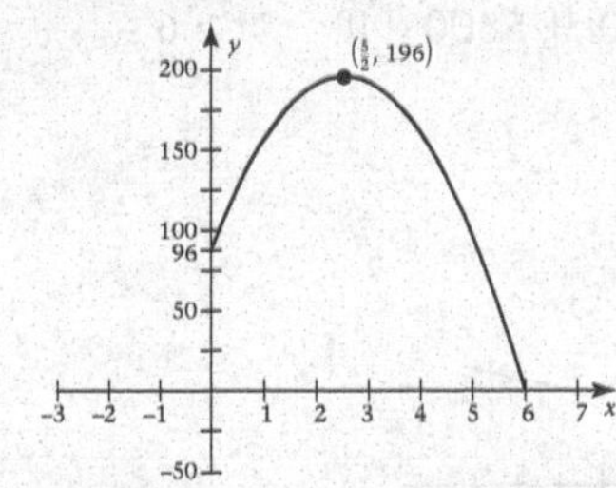

b. 196 ft

c. $\frac{5}{2}$ sec

d. 6 sec

Exercise Set 3.4

1. 2; increasing
3. $-\frac{3}{2}$, decreasing
5. -1; decreasing
9. $2x + 5$
11. $y = 3x$
13. $y = 2x$
15. $y = \frac{2}{3}x$
17. $y = 2x$
19. $y = 3x + 2$
21. $y = 2$
23. $y = \frac{1}{3}x - 5$
25. $m = -\frac{3}{4}$, $b = \frac{5}{4}$
27. $m = 0$, $b = 4$
29. $m = -\frac{3}{4}$, $b = -\frac{1}{2}$
31. a. $y = 3$ b. $x = -6$
33. a. $y = 0$ b. $x = -7$
35. a. $y = -9$ b. $x = 9$
37. a. -3 b. $\frac{1}{3}$
39. a. $\frac{4}{3}$ b. $-\frac{3}{4}$
41. a. $y = -3x + 6$ b. $y = \frac{1}{3}x + \frac{8}{3}$
43. a. $y = -\frac{3}{5}x + \frac{1}{5}$ b. $y = \frac{5}{3}x + 7$
45. a. $F = \frac{9}{5}C + 32$ b. 68°F
47. $1,000,000
49. 5
51. $f(x) = 8x + 13$
61. a.

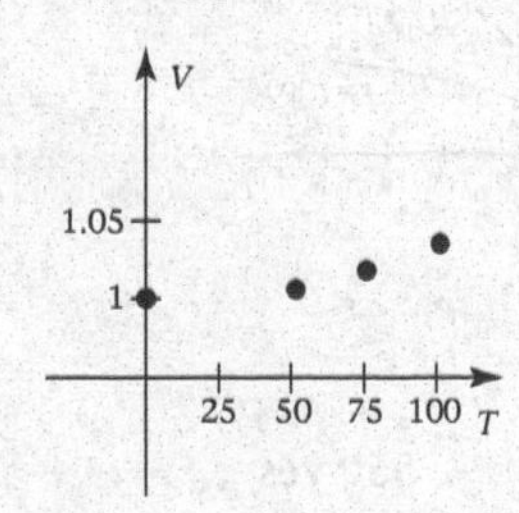

b. not linear

c. no line to graph

63. a.

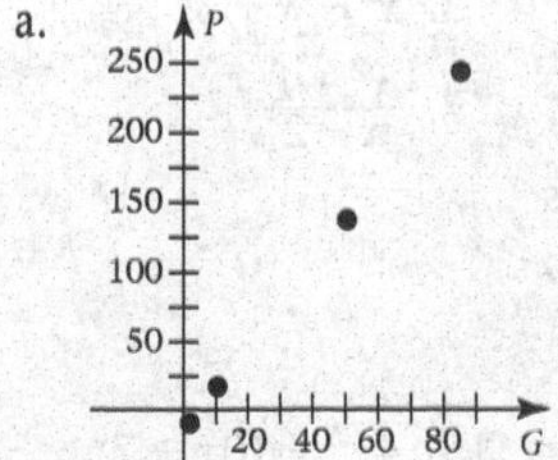

b. $P = 3G - 10$

c. XMIN = 0, XMAX = 100, XSCL = 10
YMIN = −50, YMAX = 250, YSCL = 50

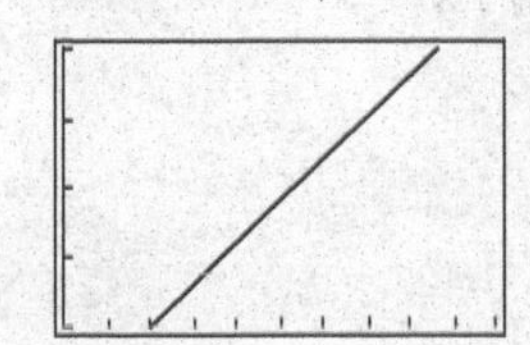

65. 356.35 miles per hour
67. no
69. a. 1 b. 0
c. 1 d. 0.57
e. 0.5 f. 0.7
g. It is at rest during this time.
71. a. $p = \frac{4}{9}d + 15$ b. 4815 pounds per in^2
c. 189 feet d.

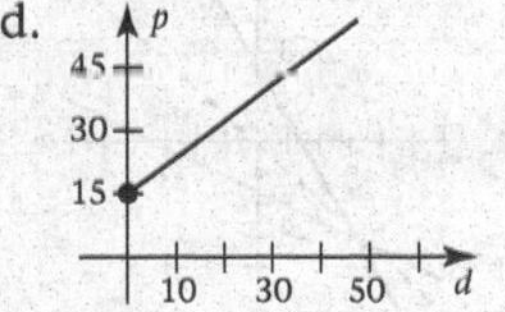

73. a. $V = \begin{cases} -2{,}000t + 20{,}500 & \text{if } 0 < t \le 6 \\ 450t + 5800 & \text{if } t > 6 \end{cases}$

b.

c. $10,300

Exercise Set 3.5

1. $x^2 + x - 1$

3. $x^2 - x + 3$

5. $x^3 - 2x^2 + x - 2$

7. $\dfrac{x^2+1}{x-2}$

9. dom_f: all real numbers
   dom_g: all real numbers

11. $4x^2 + 2x + 1$

13. 21

15. $4x^2 + 10x + 7$

17. $8x^2 - 6x + 1$

19. $x + 6,\ x \ge -2$

21. 29

23. domain: $\{x \mid x \in \Re\}$

25. $(f \circ g)(x) = x + 1$; $(g \circ f)(x) = x + 1$

27. $(f \circ g)(x) = \dfrac{x-1}{x}$

    $(g \circ f)(x) = -\dfrac{x+1}{x}$

29. $f(x) = x + 3$; $g(x) = x^2$

31. $f(x) = x^8$; $g(x) = 3x + 2$

33. $f(x) = x^{1/3}$; $g(x) = x^3 - 2x^2$

35. $f(x) = |x|$; $g(x) = x^2 - 4$

37. $f(x) = \sqrt{x}$; $g(x) = 4 - x$

45. $f^{-1}(x) = \dfrac{x-3}{2}$

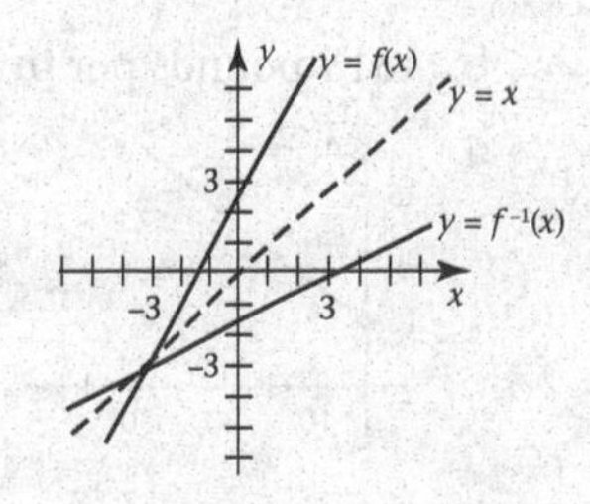

47. $f^{-1}(x) = \dfrac{-x+3}{2}$

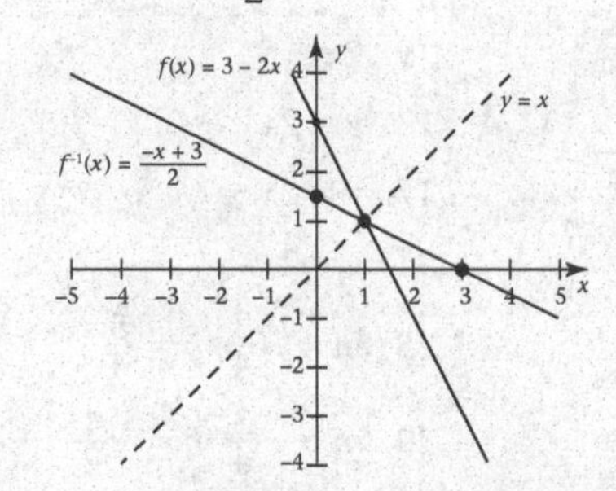

49. $f^{-1}(x) = 3x + 15$

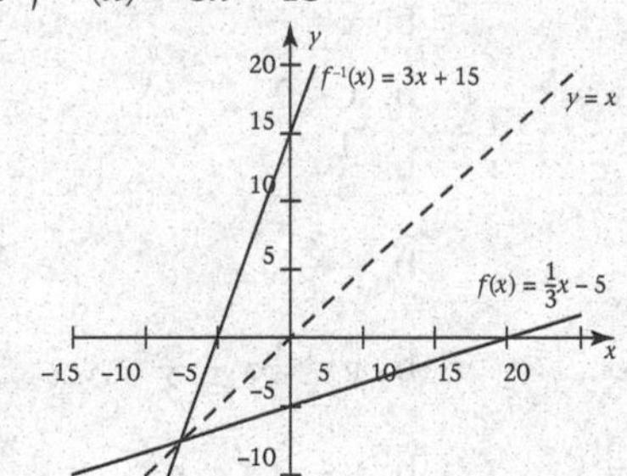

51. $f^{-1}(x) = \sqrt[3]{x-1}$

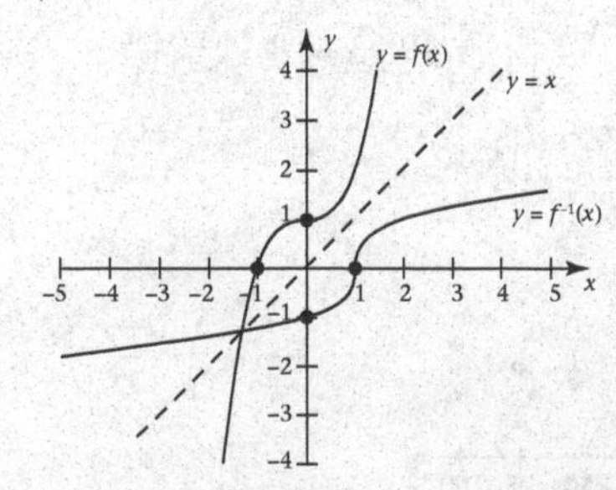

53. (b), (d), (e), (h), (i)

55. yes

57. no

59. yes

61. no

65. a. $f^{-1}(x) = x - 3$
    b. $f^{-1}(x) = -x - 3$

Exercise Set 3.6

1. a. 4 b. $y = 4x$

c.

x	8	12	20	30
y	32	48	80	120

3. a. $-\frac{1}{32}$ b. $-\frac{3}{8}$

5. a. $\frac{1}{10}$ b. $\frac{5}{2}$

7. a. -3 b. $-\frac{1}{4}$

9. a. 512 b. $\frac{512}{125}$

11. a. $M = \frac{r^2}{s^2}$ b. $\frac{36}{25}$

13. a. $T = \frac{16pv^3}{u^2}$ b. $\frac{2}{3}$

15. a. 400 feet b. 7 seconds

17. $\frac{40}{3}$ ohms

19. a. $\frac{800}{9}$ candlepower b. 8 feet

21. 6

23. 120 candlepower per square foot

25. a. $I = \frac{k}{d^2}$

b. 7.18 watts per square meter

27. 1456.88

29. a. 23.4375 kg per cm^3

b. 111.54°F

31. a. $L = \frac{kwl^2}{d}$

b. $\frac{32}{3}$ pounds per square foot

c. 800 pounds

d. The load is quadrupled.

e. The load is quadrupled.

Chapter 3 Review Exercises

1. $\sqrt{61}$

2. $\sqrt{65}$

3.

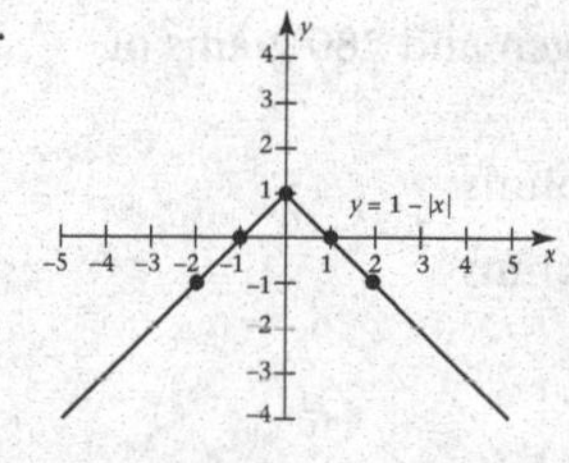

4.

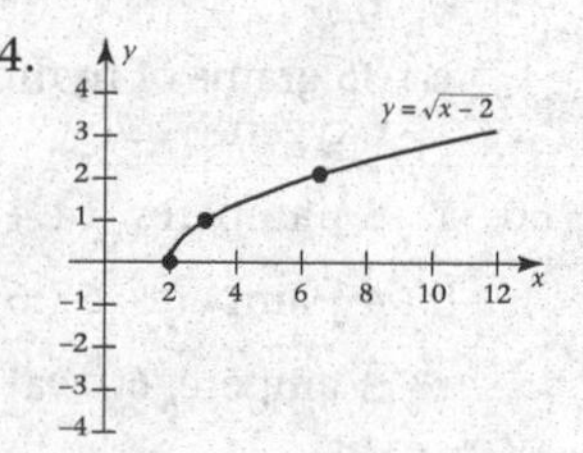

5. x-axis

6. all

7. yes

8. yes

9. $x \geq \frac{5}{3}$

10. $x \neq -1$

11. 226

12. ± 3

13. $y = 5x - 4$

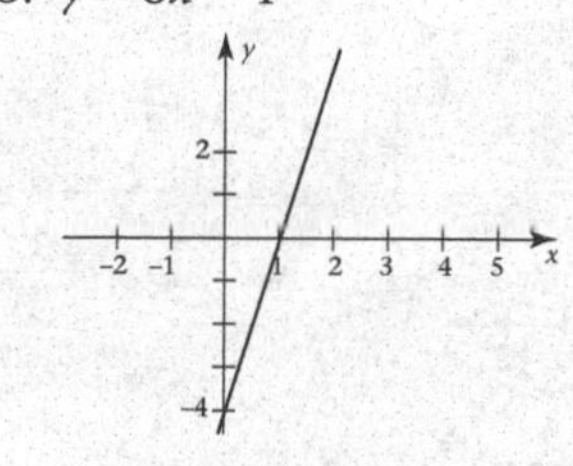

14. $y = 3x^3 + 2$

15. $y = x - x^2$

16. $y = |x - x^2|$

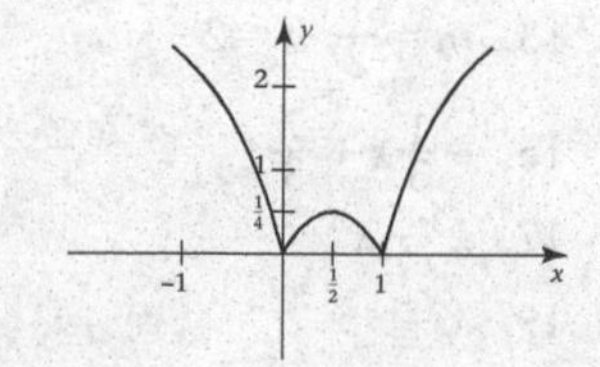

17. $y = x - 3$

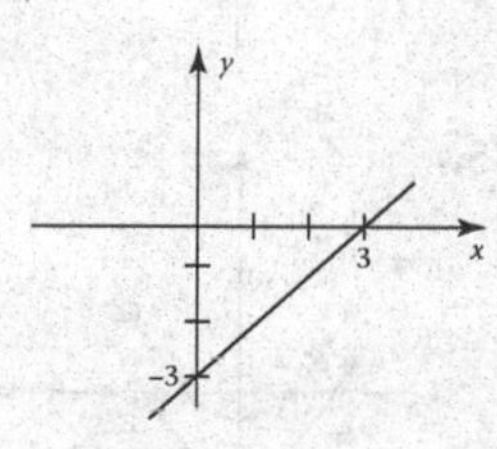

18. $y = |x - 3|$

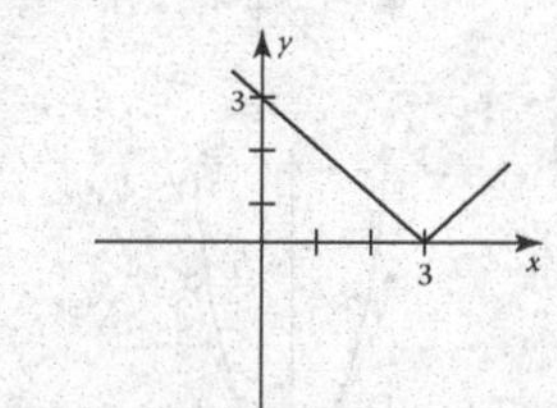

19. $x \neq 3$

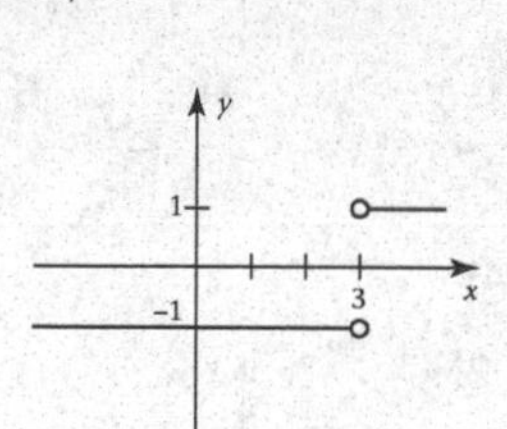

20. $y = 2\sqrt{x} + 7$

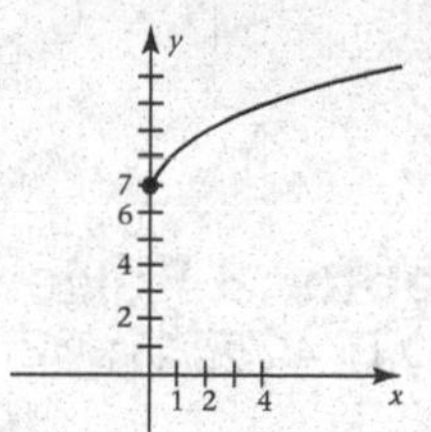

21. $y = 2\sqrt{(x + 7)}$

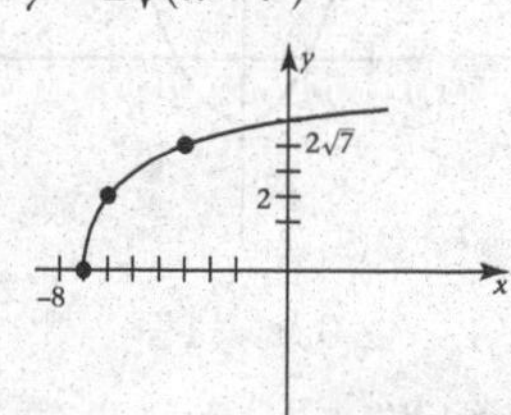

22. $y = \frac{5}{x^2} + 1$

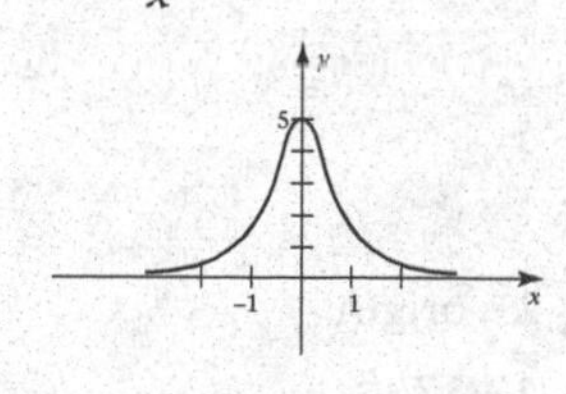

23. 12

24. $y^2 - 3y + 2$

25. $3 + h$

26. 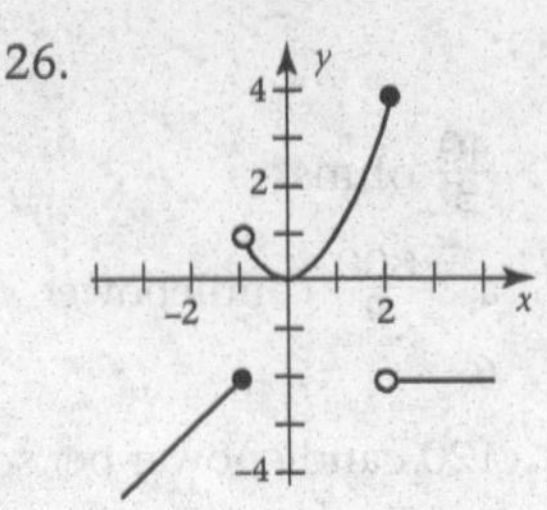

36. 3

37. $y = 3x + 6$

38. $x = -4$

39. $y = 3$

40. $y = 4x + 10$

41. $y = 2x + 5$

42. $x^2 + x$

43. 0

44. $\dfrac{1}{x-1}$

45. $x \neq \pm 1$

46. $x^2 + 2x$

47. 4

48. $|x| - 2$

49. $x - 4\sqrt{x} + 4$

27. increasing: $x \leq -1, 0 \leq x \leq 2$
    decreasing: $-1 < x \leq 0$
    constant: $x > 2$

50. 0

51. not defined

52. x

53. 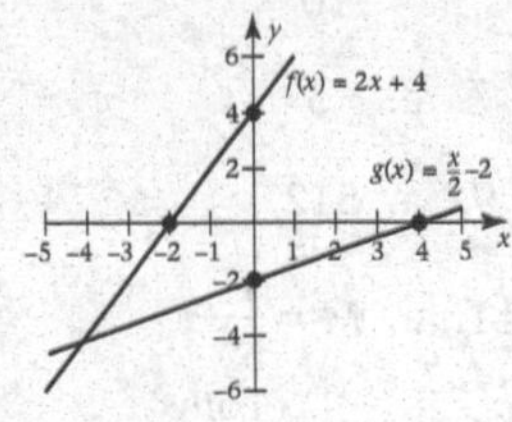

28. -5

29. -2

30. 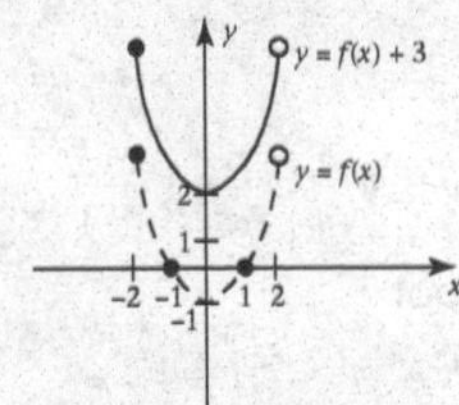

31. 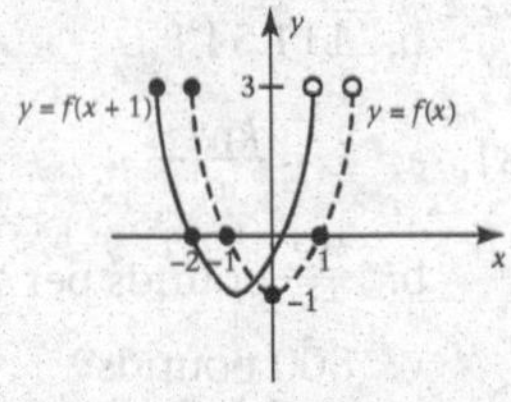

54. 160

55. 1

56. $-\dfrac{1}{4}$

32. 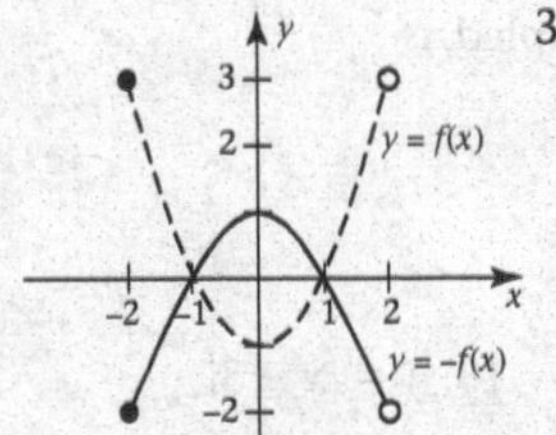

33. 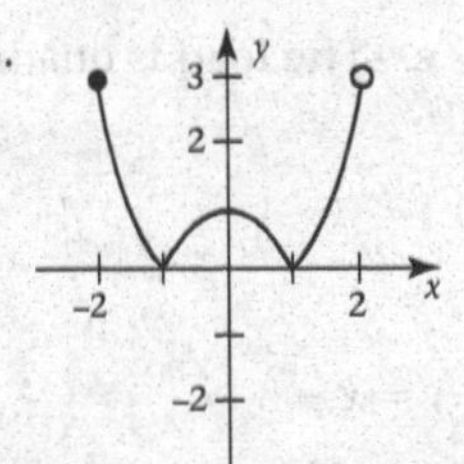

57. a. $525 + 0.145x$
    b. \$670

58. a. $G = 0.004S - 1.8$
    b. $250G + 450 = S$
    c. 1450

59. a. $\dfrac{1}{9}w$
    b. $\dfrac{8}{9}w$
    c. 35 grams of hydrogen and 280 grams of oxygen

34. 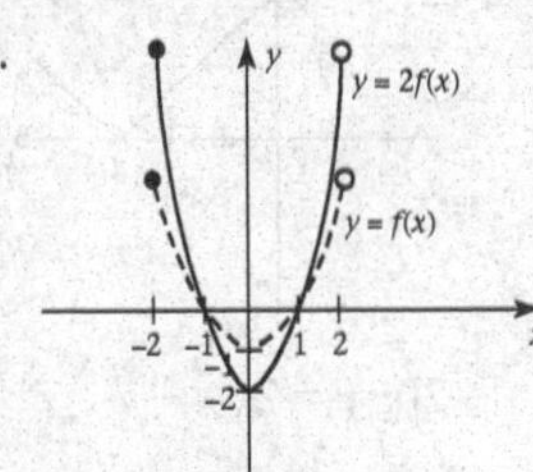

35. 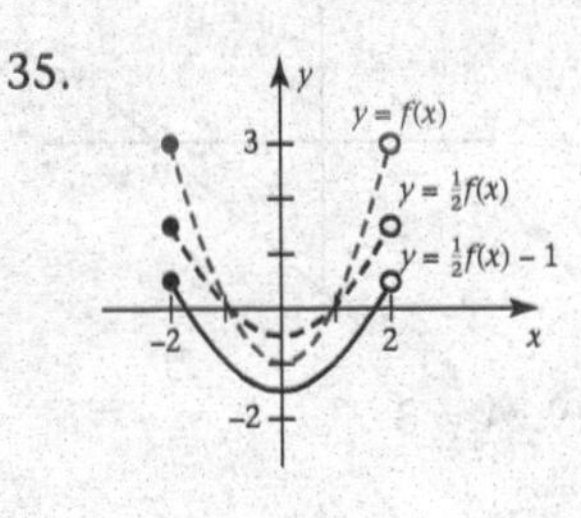

60. a. $8\frac{1}{3}$ amperes, 14.4 ohms
    b. $4\frac{1}{6}$ amperes, 57.6 ohms
    c. $\frac{1}{2}$ ampere, 60 watts

Chapter 3 Review Test

1. $\sqrt{41} + \sqrt{26} + 3$

2. 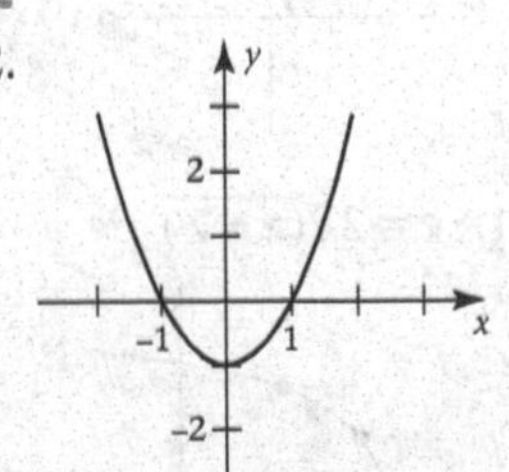

10. 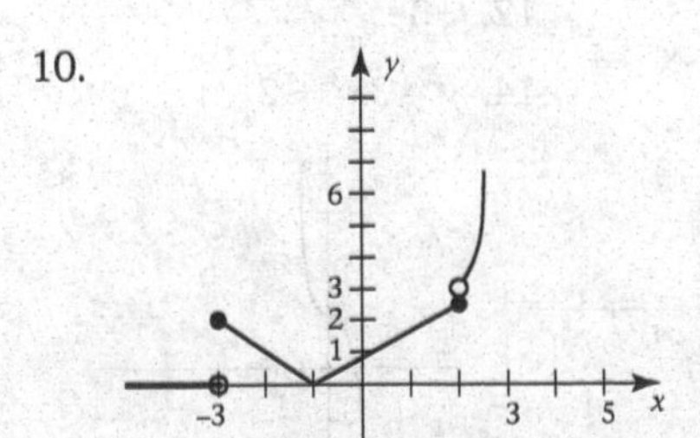

3. origin

4. $x \geq 0, x \neq 1$

11. $\dfrac{3}{2}x + \dfrac{19}{2}$

12. $x = -3$

5. 17

6. $8t^2 + 3$

13. $m = \dfrac{1}{2}$; $b = 2$

14. $y = -1$

7. increasing: $x \geq 0$
   decreasing: $-2 \leq x \leq 0$
   constant: $x < -2$

15. $-\dfrac{1}{3}x + \dfrac{7}{3}$

16. -3

17. $x^3 - x^2$

18. $\dfrac{1}{4}$

8. 0

9. 2

19. x

20. -1024

21. 65,536

Cumulative Review Exercises

Chapters 1–3

1. polynomial
2. not a polynomial
3. not a polynomial
4. polynomial
5. polynomial
6. $a^2b\sqrt{b}$
7. $\dfrac{21+3\sqrt{x}}{49-x}$
8. $\dfrac{a^3}{b^{7/2}}$
9. $\dfrac{3-x-x^3}{x(x^2+1)}$
10. $-\dfrac{x+3}{x+2}$
11. $u=-5,\ v=1$
12. $5-12i$
13. $26+2i$
14. $88+234i$
15. $x^{11/6}-x^{5/2}$
16. $3\sqrt{3}$
17. $\dfrac{-2(x+1)}{x}$
18. $\dfrac{1+x^2}{x-3x^2}$
19. $x(3x-2)(x+1)$
20. $(x-1)^{2/3}(2x-1)$
21. $\dfrac{2\sqrt{x}+2\sqrt{2}}{x-2}$
22. $x \le \frac{5}{2}$
23. $h<-\frac{1}{2}$ or $h>\frac{3}{2}$
24. $t\le-\frac{3}{2}$ or $t\ge 4$
25. $-\sqrt{5}\le x\le -1$ or $1\le x\le\sqrt{5}$
26. $x>\frac{1}{2}$
27. $\{x|x\ne\pm1\}$
28. $\sqrt{61}$
29. a. $-4a^2+4a$ b. ±4 c. $-2t-h$
30. a. 0 b. $\frac{1}{2}$ c.

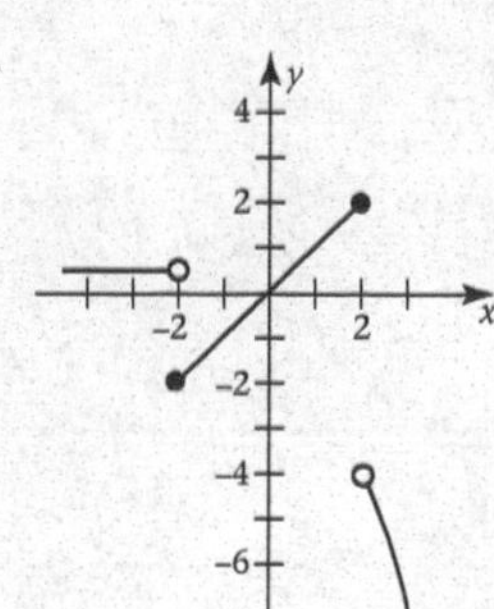

31. $0.05x+0.07y+0.035z$
32. $-1, \frac{3}{2}$
33. $\dfrac{1\pm i\sqrt{31}}{4}$
34. 16
35. $-\frac{5}{9}$
36. $-\frac{4}{3}, \frac{1}{2}$
37. 7, −1
38. −1
39. $-2\le x\le\frac{3}{2}$
40. $x\le-17$ or $x\ge 23$
41. $-2\le x\le 5$
42. $\{x|x>2\}$
43. $y=-1$
44. $y=-4x+4$
45. a. −1 b. $y=-x+1$ c. $x=-1$ d. $y=x+3$
46. $2\sqrt{10}$
47. x-intercepts: $-1, \frac{2}{3}$; y-intercepts: 2
48. $y=-x+a$
49. \$300, 40 months
50. Route 1
51. a. Domain: $\{x|0\le x\le 1000\}$
    b.

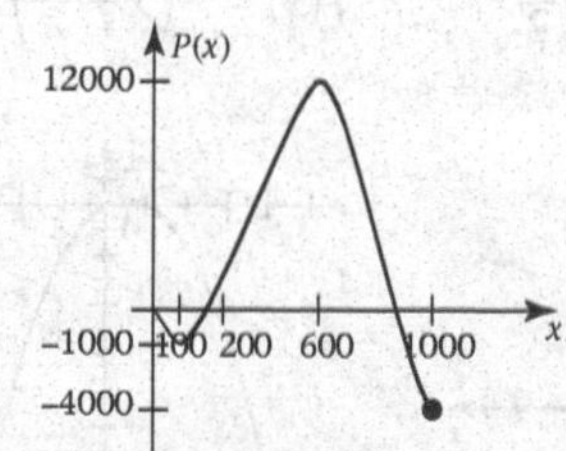

    c. i. [−4000, 12000]
       ii. 600 books
       iii. \$12,000
    d. 200 books
52. a. 0.6 b. 45°F

Chapter 4

Exercise Set 4.1

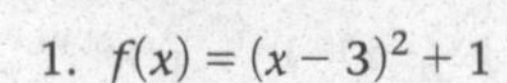

1. $f(x) = (x - 3)^2 + 1$

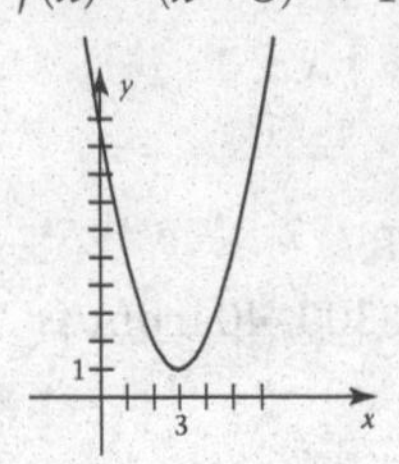

3. $f(x) = -2(x - 1)^2 - 3$

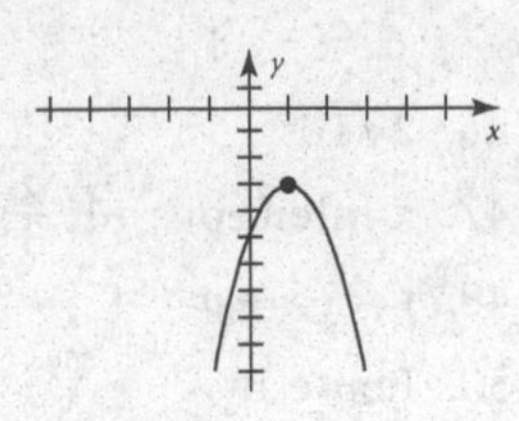

5. $f(x) = 2\left(x + \frac{3}{2}\right)^2 + \frac{1}{2}$

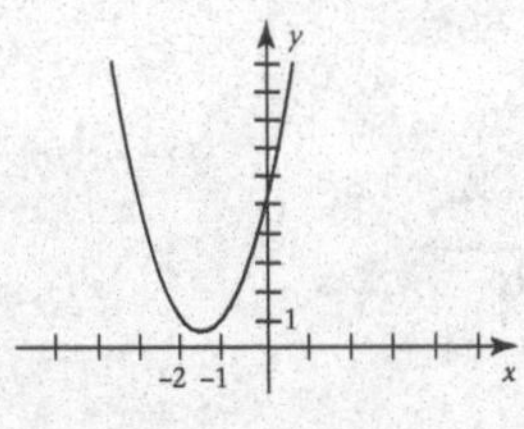

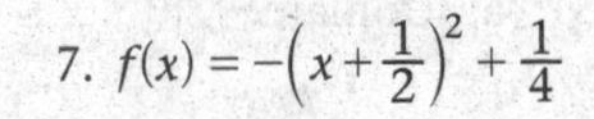

7. $f(x) = -\left(x + \frac{1}{2}\right)^2 + \frac{1}{4}$

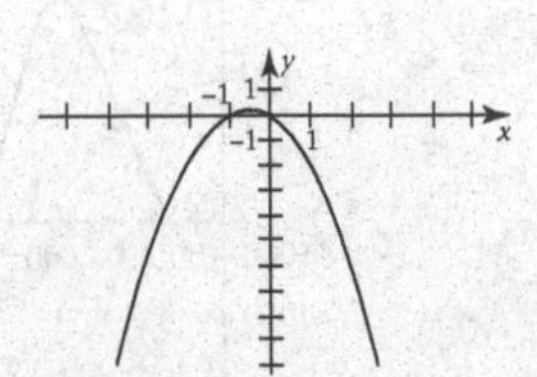

9. $f(x) = -2(x - 0)^2 + 5$

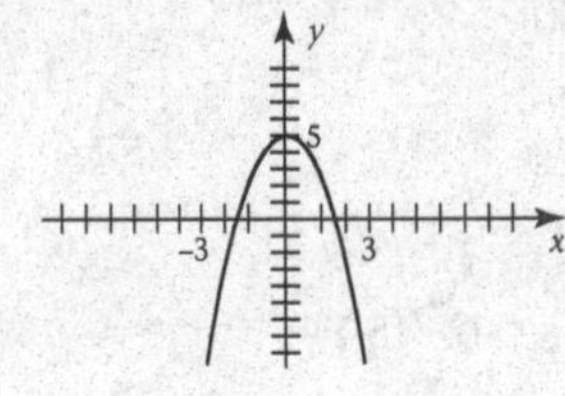

11. vertex: $(1, -2)$

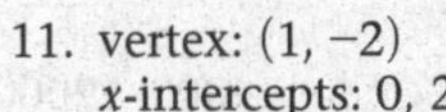

x-intercepts: 0, 2
y-intercept: 0

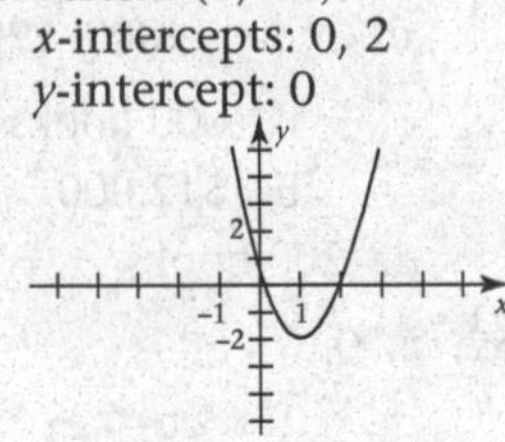

13. vertex: $\left(\frac{1}{2}, 0\right)$
x-intercept: $\frac{1}{2}$
y-intercept: -1

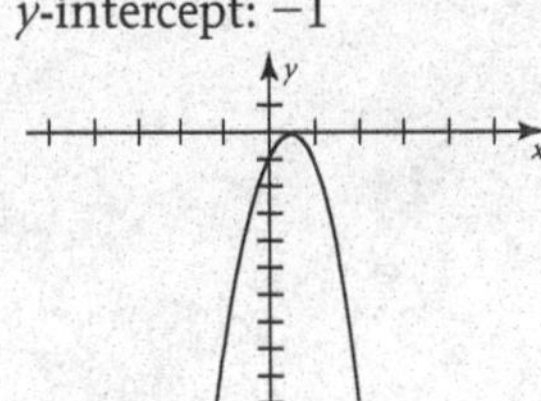

15. vertex: $(-2, 2)$

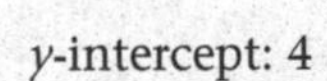

y-intercept: 4

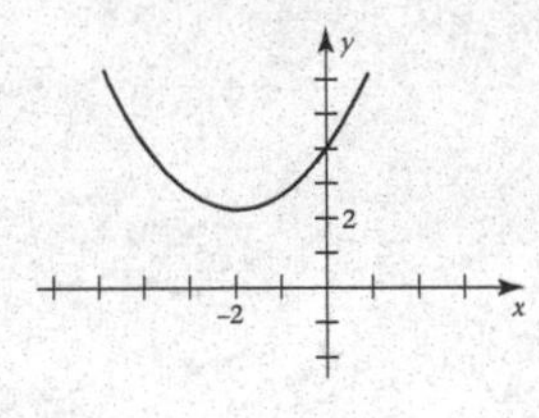

17. vertex: $\left(3, \frac{1}{2}\right)$
x-intercept: 4, 2
y-intercept: -4

In Exercises 19–26, the following RANGE values are being used:

XMIN = −10
XMAX = 10
XSCL = 1
YMIN = −10
YMAX = 10
YSCL = 1

The linear factors determine the x-intercepts of the graph.

If $(x - r)$ is a factor then $x = r$ is an x-intercept.

19. $Y = (X + 1)(X - 2)$

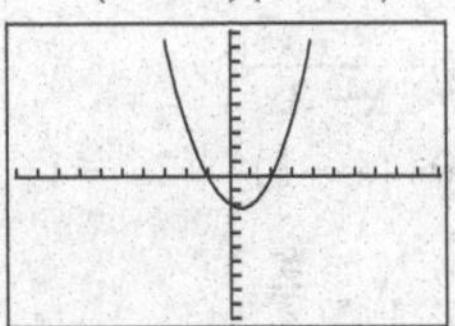

21. $Y = X(X + 5)$

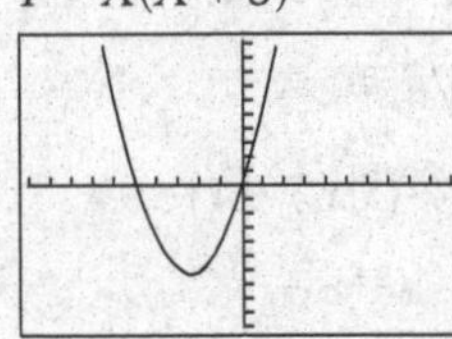

23. $Y = (X - 3)(X - 7)$

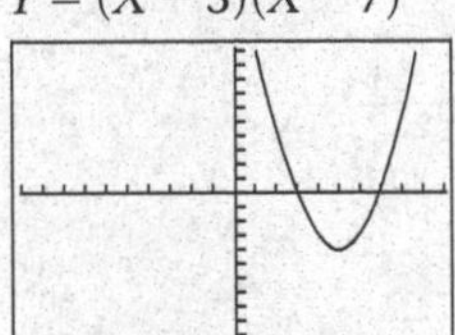

25. $Y = 0.3(X - 3)(X - 7)$

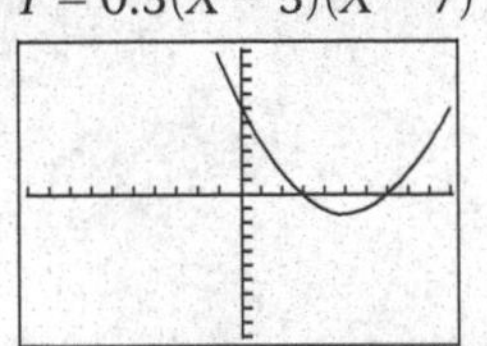

27. a. $r_1 = -4$
$r_2 = 5$
b. $f(x) = (x + 4)(x - 5) = x^2 - x - 20$

29. a. $r_1 = -3$
$r_2 = -8$
b. $f(x) = -(x + 3)(x + 8) = -x^2 - 11x - 24$

31. a. minimum b. $\frac{1}{3}$
c. $\frac{11}{3}$

33. a. maximum b. 0
c. -5

35. a. minimum b. $-\frac{5}{2}$
c. $-\frac{25}{4}$

37. a. minimum
b. $\frac{1}{8}$
c. $-\frac{49}{32}$

39. 10, 10

41. 25, 25

43. $\left(\frac{1}{2}, \frac{\sqrt{2}}{2}\right)$

45. 250 feet × 500 feet

47. 10 feet × 10 feet

49. 2.5 seconds, 100 feet

51. \$0.75

55. 100

Exercise Set 4.2

1. $P(-2)$ and $P(-1)$ are of opposite sign; therefore, $P(x)$ has at least one root in the interval $[-2, -1]$.

3. $P(1)$ and $P(2)$ are of opposite sign; therefore, $P(x)$ has at least one root in the interval $[1, 2]$.

5. $P(1)$ and $P(2)$ are of opposite sign; therefore, $P(x)$ has at least one root in the interval $[1, 2]$.

7. $P(-2)$ and $P(-1)$ are of opposite sign; therefore, $P(x)$ has at least one root in the interval $[-2, -1]$.

In Exercises 9–16, the WINDOW settings for graph (a), (b), and (c) are as follows:

a. XMIN = −10
XMAX = 10
XSCL = 1
YMIN = −100
YMAX = 100
YSCL = 0

b. XMIN = −10
XMAX = 10
XSCL = 1
YMIN = −10,000
YMAX = 10,000
YSCL = 0

c. XMIN = −10
XMAX = 10
XSCL = 1
YMIN = −100,000
YMAX = 100,000
YSCL = 0

9. leading term: x^7
large values of $|x|$, $x > 0$: U
large values of $|x|$, $x < 0$: D

a.
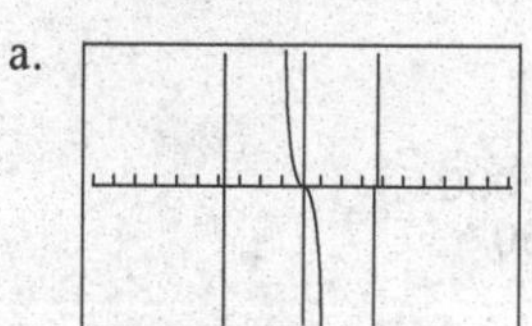

b.
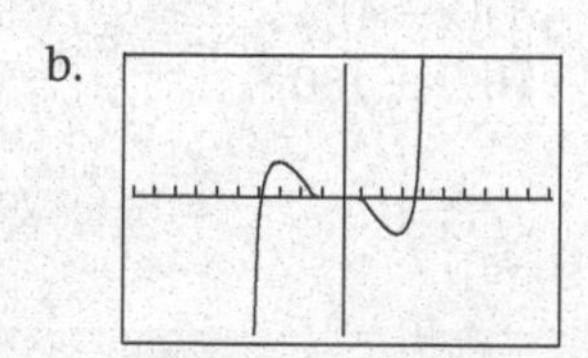

c.
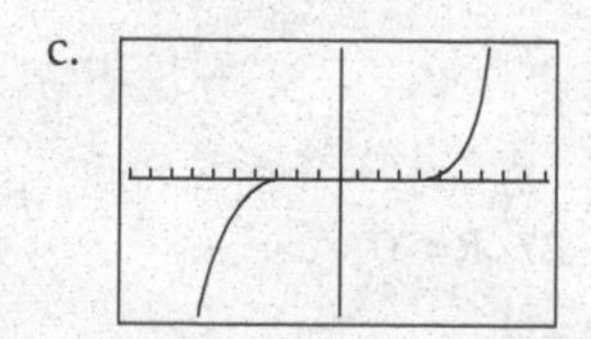

11. leading term: $-8x^3$
large values of $|x|$, $x > 0$: D
large values of $|x|$, $x < 0$: U

a.
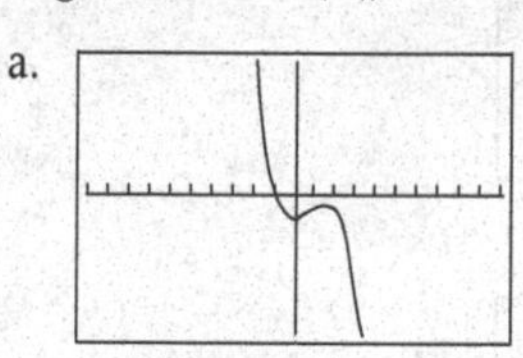

b.
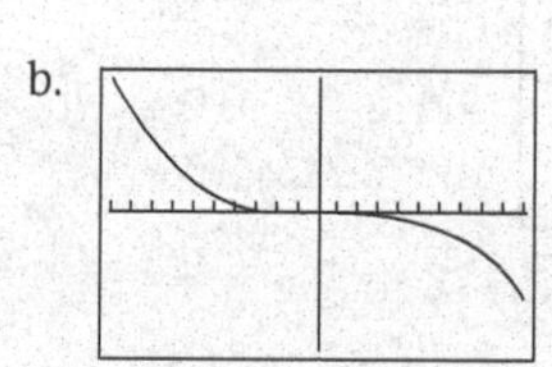

c.
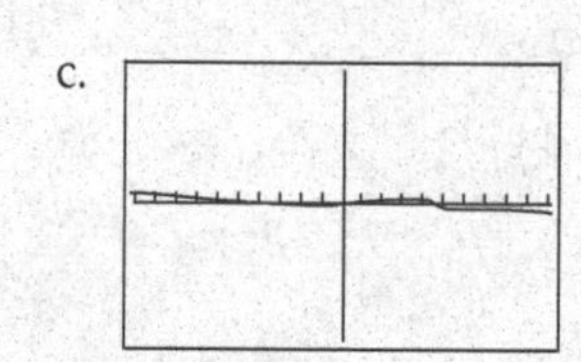

13. leading term: $-5x^{10}$
large values of $|x|$, $x > 0$: D
large values of $|x|$, $x < 0$: D

a.
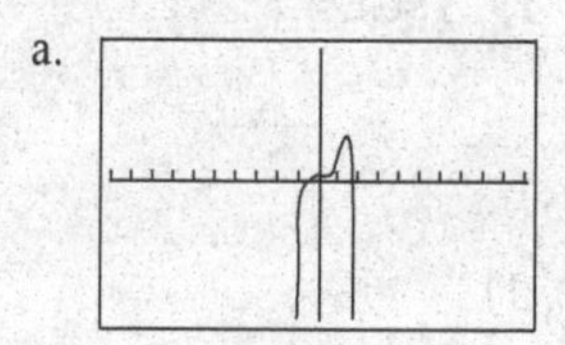

b.
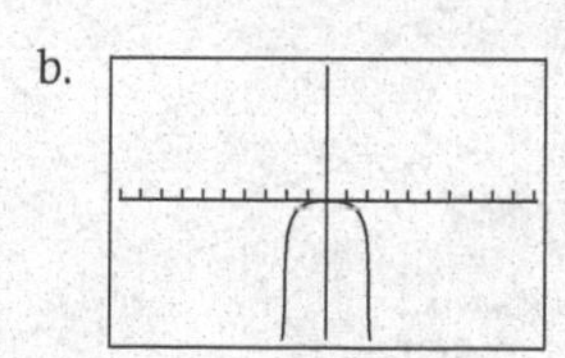

c.
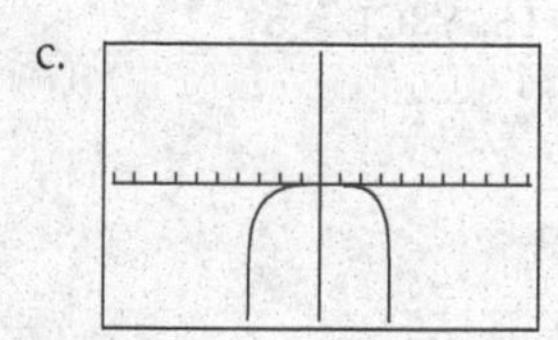

15. leading term: $4x^8$
large values of $|x|$, $x > 0$: U
large values of $|x|$, $x < 0$: U
a.
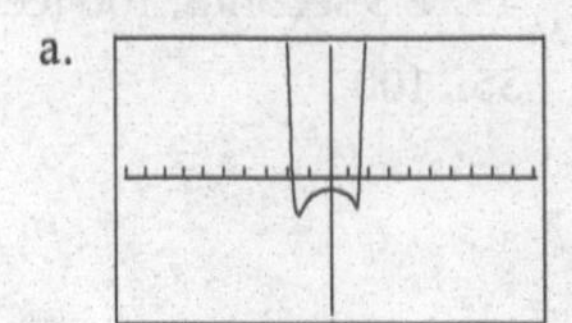
b.
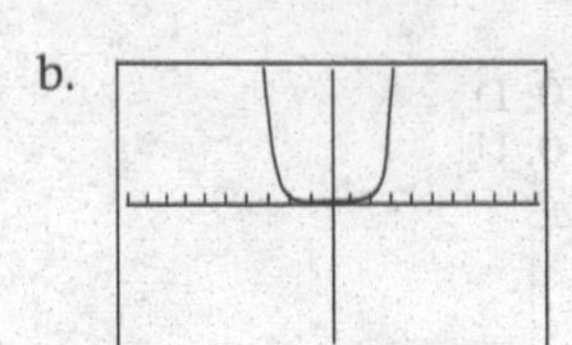
c.
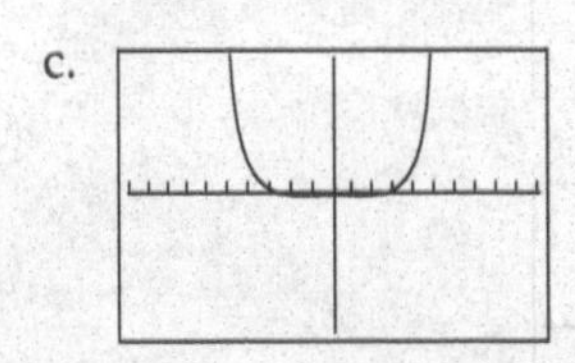

17. x-intercepts: $-2, \frac{1}{2}, 3$
$P(x) > 0: \left(-2, \frac{1}{2}\right), (3, \infty)$
$P(x) < 0: (-\infty, -2), \left(\frac{1}{2}, 3\right)$
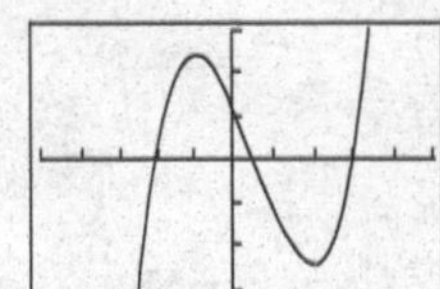
XMIN = −5, XMAX = 5, XSCL = 1
YMIN = −15, YMAX = 15, YSCL = 5

19. x-intercepts: $-\frac{5}{2}, 0, 1$
$P(x) > 0: \left(-\frac{5}{2}, 0\right), (1, \infty)$
$P(x) < 0: \left(-\infty, -\frac{5}{2}\right), (0, 1)$
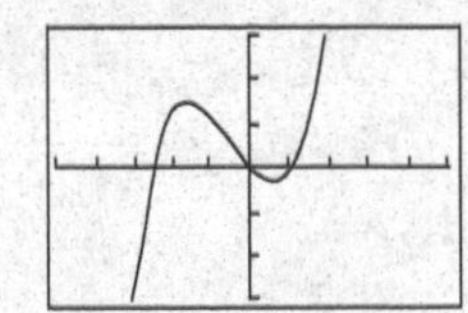
XMIN = −5, XMAX = 5, XSCL = 1
YMIN = −15, YMAX = 15, YSCL = 5

21. x-intercepts: $-2, 0, 3$
$P(x) > 0: (-\infty, -2), (3, \infty)$
$P(x) < 0: (-2, 0), (0, 3)$
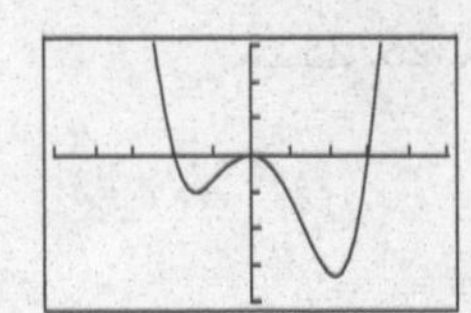
XMIN = −5, XMAX = 5, XSCL = 1
YMIN = −20, YMAX = 15, YSCL = 5

23. $x^3 - 2x^2 - 16x + 32$

25. $x^3 + 6x^2 + 11x + 6$

27. $x^3 - 6x^2 + 6x + 8$

29.
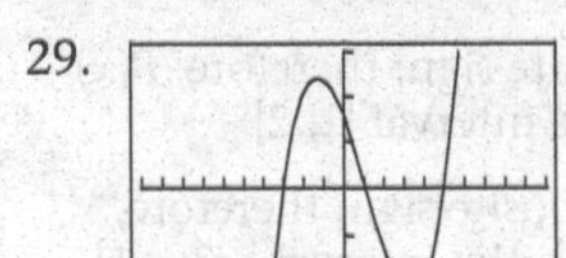
XSCL = 1, YSCL = 10
x-intercepts: $-3, 1, 5$
y-intercepts: 15

31.
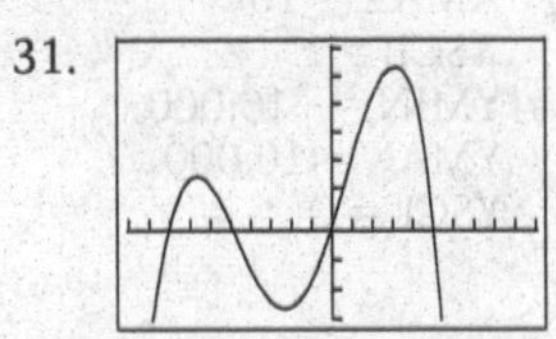
XSCL = 1, YSCL = 100
x-intercepts: $-8, -5, 0, 5$
y-intercepts: 0

33.
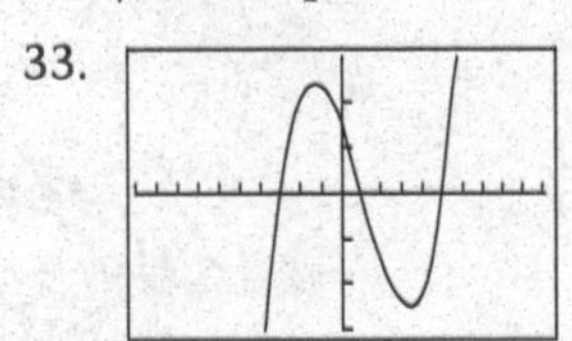
XSCL = 10, YSCL = 10,000
x-intercepts: $-30, 10, 50$
y-intercepts: 15,000

35. a. $y = (x + 9)(x + 5)(x - 1)(x - 4)$
$= x^4 + 9x^3 - 21x^2 - 169x + 180$

Exercise Set 4.3

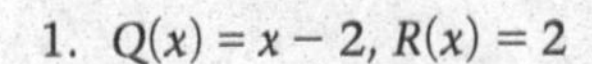

1. $Q(x) = x - 2$, $R(x) = 2$
3. $Q(x) = 2x - 4$, $R(x) = 8x - 4$
5. $Q(x) = 3x^3 - 9x^2 + 25x - 75$, $R(x) = 226$
7. $Q(x) = 2x - 3$, $R(x) = -4x + 6$
9. $Q(x) = x^2 - x + 1$, $R(x) = 0$
11. $Q(x) = x^2 - 3x$, $R = 5$
13. $Q(x) = x^3 + 3x^2 + 9x + 27$, $R = 0$
15. $Q(x) = 3x^2 - 4x + 4$, $R = 4$
17. $Q(x) = x^4 - 2x^3 + 4x^2 - 8x + 16$, $R = 0$
19. $Q(x) = 6x^3 + 18x^2 + 53x + 159$, $R = 481$

Exercise Set 4.4

1. −7
3. −34
5. 0
7. −1
9. 0
11. −62
13. yes
15. no
17. yes
19. yes
21. yes
23. yes
25. no
27. −1, 2
29. 1, $\frac{1}{2}$
31. $\frac{1}{2}, \frac{1}{2}, \frac{1}{2}$
33. $r = 3, -1$
35. $\frac{5}{2}$

Exercise Set 4.5

1. 5
3. 25
5. 20
7. $-\frac{13}{10} + \frac{11}{10}i$
9. $-\frac{7}{25} - \frac{24}{25}i$
11. $\frac{8}{5} - \frac{1}{5}i$
13. $\frac{5}{3} - \frac{2}{3}i$
15. $\frac{4}{5} + \frac{8}{5}i$
17. $\frac{4}{25} - \frac{3}{25}i$
19. $\frac{9}{10} + \frac{3}{10}i$
21. $0 + \frac{1}{5}i$
25. $x^3 - 2x^2 - 16x + 32$
27. $x^3 + 6x^2 + 11x + 6$
29. $x^3 - 6x^2 + 6x + 8$
31. $\frac{x^3}{3} + \frac{x^2}{3} - \frac{7x}{12} + \frac{1}{6}$
33. $x^3 - 4x^2 - 2x + 8$
35. 3, −1, 2
37. −2, 4, −4
39. −2, −1, $-\frac{1}{2}$
41. 5, 5, 5, −5, −5
43. $x^3 + 6x^2 + 12x + 8$
45. $4x^4 + 4x^3 - 3x^2 - 2x + 1$
47. 2, −1
49. $\frac{3}{2} + \frac{\sqrt{3}}{2}i, \frac{3}{2} - \frac{\sqrt{3}}{2}i$
51. −1, −2, 4

In Exercises 53–60, set the RANGE as follows:

XMIN = −10
XMAX = 10
XSCL = 1
YMIN = −100
YMAX = 100
YSCL = 10

53. $y = (x + 2)(x - 3)$

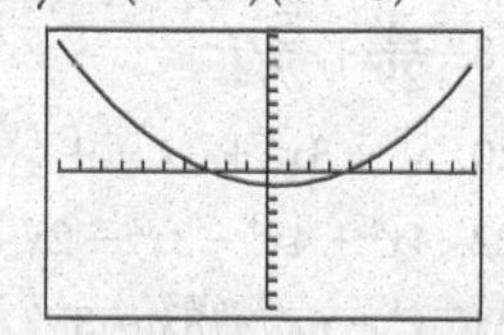

55. $y = (x + 2)(x - 3)^3$

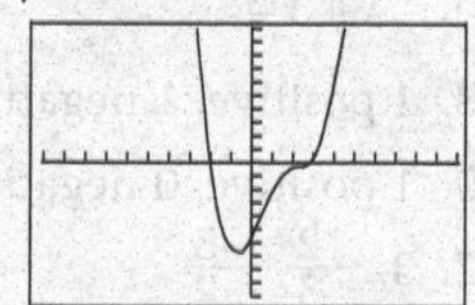

57. $y = (x + 2)^3(x - 3)$

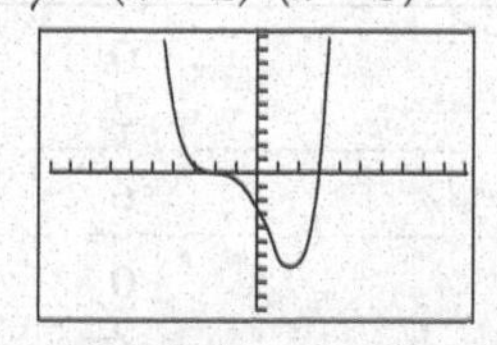

59. $y = (x + 2)^3(x - 3)^2$

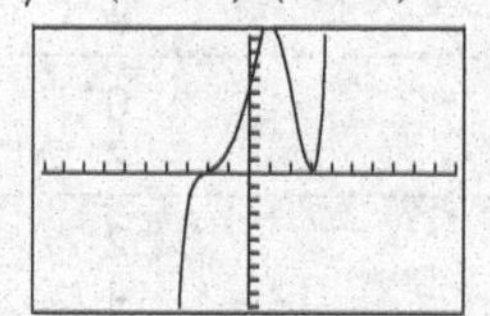

61.

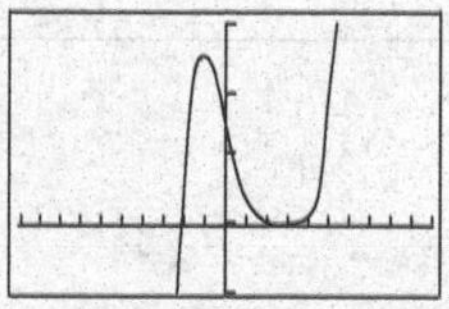

XSCL = 1
YMIN = −100, YMAX = 300, YSCL = 100

63.

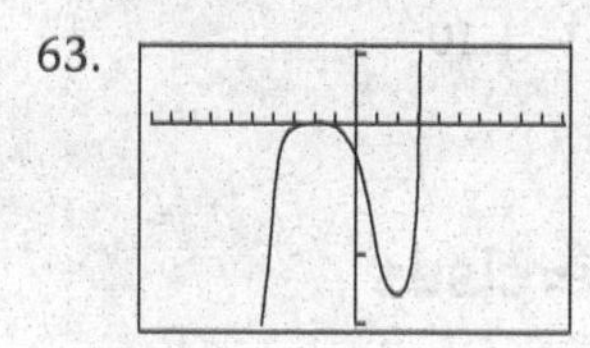

XSCL = 1
YMIN = −300, YMAX = 100, YSCL = 100

65.

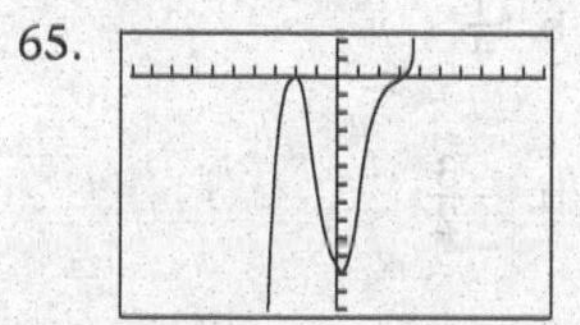

XSCL = 1
YMIN = −700, YMAX = 100, YSCL = 100

67.

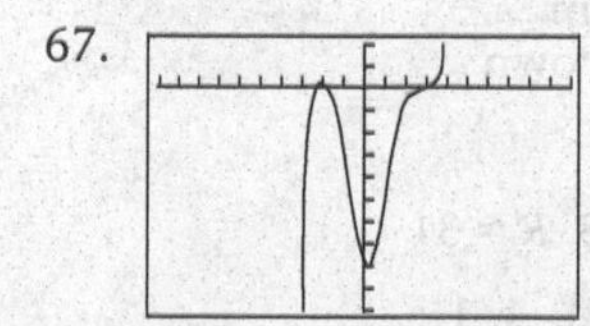

XSCL = 1
YMIN = −5000, YMAX = 1000, YSCL = 500

69. $x^2 + (1 - 3i)x - (2 + 6i)$

71. $x^2 - 3x + (3 + i)$

73. $x^3 + (1 + 2i)x^2 + (-8 + 8i)x + (-12 + 8i)$

75. $(x^2 - 6x + 10)(x - 1)$

77. $(x^2 + 2x + 5)(x^2 + 2x + 4)$

79. $(x + 2)(x^2 + 6x + 10)(x - 3)(x - 2)$

81. $x - (a + bi)$

85. a. 10, 30

b. Maximum value occurs at $x = 10$

c. It is where the volume curve turns again

87. 11

Exercise Set 4.6

	Positive Roots	Negative Roots	Complex Roots
1.	3	1	0
	1	1	2
3.	0	0	6
5.	3	2	0
	1	2	2
	3	0	2
	1	0	4
7.	1	2	0
	1	0	2
9.	2	0	2
	0	0	4
11.	1	1	6

13. $1, -2, 3$

15. $2, -1, -\frac{1}{2}, \frac{2}{3}$

17. $1, -1, -1, \frac{1}{5}$

19. $1, -\frac{3}{4}$

21. $3, 3, \frac{1}{2}$

23. $-1, \frac{3}{4}, \pm i$

25. $\frac{3}{5}, \pm 2, \pm\sqrt{2}i$

27. $0, \frac{1}{2}, \frac{2}{3}, -1$

29. $\frac{1}{2}, -4, 2 \pm \sqrt{2}$

31. $3, 3, \pm i$

33. $2, -2, \pm\sqrt{2}$

35. $-\frac{1}{2}, -\frac{1}{2}, \pm 1$

37. $k = 3, r = -2; k = -\frac{3}{2}, r = 1; k = -\frac{5}{3}, r = 2$

39. $k = 7, r = 1; k = -7, r = -1$

Exercise Set 4.7

1. -1.09

3. 1.73

5. 1.43

7. -1.87

9. 1.40

11. 1.10

13. -1.08

15. 1.73

17. 1.43

19. -1.87

21. 1.40

23. 1.10

Chapter 4 Review Exercises

1. vertex: $(-2, 4)$; $x = 0, -4$; $y = 0$

2. vertex: $\left(\frac{5}{2}, \frac{3}{4}\right)$; no x-intercepts; $y = 7$

3. a. minimum b. $\frac{1}{4}$

c. $\frac{7}{8}$

4. a. maximum b. $-\frac{3}{2}$

c. $\frac{5}{4}$

5. large values of $|x|$, $x > 0$: down
large values of $|x|$, $x < 0$: up

6. large values of $|x|$, $x > 0$: up
large values of $|x|$, $x < 0$: down

7. $Q(x) = 2x^2 + 2x + 8$, $R = 4$

8. $Q(x) = x^3 - 5x^2 + 10x - 18$, $R = 31$

9. $46, -8$

10. $4, 1$

13. $\frac{6}{25} - \frac{17}{25}i$

14. $-\frac{1}{5} + \frac{2}{5}i$

15. $-\frac{5}{2} + \frac{5}{2}i$

16. $\frac{1}{10} - \frac{3}{10}i$

17. $0 + \frac{1}{4}i$

18. $\frac{2}{29} + \frac{5}{29}i$

19. $x^3 + 6x^2 + 11x + 6$

20. $x^3 - 3x^2 + 3x - 9$

21. $x^4 + x^3 - 5x^2 - 3x + 6$

22. $4x^4 + 4x^3 - 3x^2 - 2x + 1$

23. $x^4 + 2x^2 + 1$

24. $x^4 - 6x^2 - 8x - 3$

25. $-\frac{1}{2}, 3$

26. $-1 \pm \sqrt{2}$

27. $4, 2 + i, 2 - i$

28. 1 positive, 1 negative

29. 5 positive, 0 negative

30. 1 positive, 0 negative

31. 2 positive, 2 negative

32. $3, -\frac{2}{3}, -\frac{3}{2}$

33. $1, -2, \frac{2}{3}, \frac{3}{2}$

34. none

35. $-1, \frac{-9 \pm \sqrt{321}}{12}$

36. $2, \frac{3}{2}, -1 \pm \sqrt{2}$

Chapter 4 Review Test

1. vertex: $\left(\frac{1}{3}, \frac{2}{3}\right)$; x-intercept: none; y-intercept: 1; minimum value: $\frac{2}{3}$
2. extends infinitely downward
3. extends infinitely upward
4. $Q(x) = 2x^2 - 5$, $R(x) = 11$
5. $Q(x) = 3x^3 - 7x^2 + 14x - 28$, $R(x) = 54$
6. -25
7. -165
9. $x^3 - 2x^2 - 5x + 6$
10. $x^4 - 6x^3 + 6x^2 + 6x - 7$
11. $2, \pm i$
12. $-1, -1, \frac{3 \pm \sqrt{17}}{2}$
13. $x^5 + 3x^4 - 6x^3 - 10x^2 + 21x - 9$
14. $16x^5 - 8x^4 + 9x^3 - 9x^2 - 7x - 1$
15. $x^2 - (1 + 2i)x + (-1 + i)$
16. $\frac{1}{2}, \frac{1}{2}$
17. $-1, \pm i$
18. $(x^2 - 4x + 5)(x - 2)$
19. 2
20. 1
21. none
22. $1, 1, -1, -1, \frac{1}{2}$
23. $\frac{2}{3}, -3, \pm i$

Chapter 5

Exercise Set 5.1

1. domain: $\{x | x \neq 1\}$; intercepts: $(0, 0)$
3. domain: $\{x | x \neq 0, 2\}$; intercepts: none
5. domain: $\{x | x \text{ is real}\}$; intercepts: $(\pm\sqrt{3}, 0)$, $(0\ -1)$
7. $x = 4$, $y = 0$

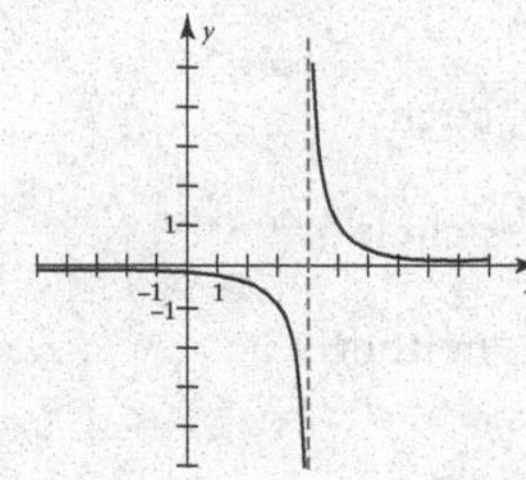

9. $x = -2$, $y = 0$

11. $x = -1$, $y = 0$

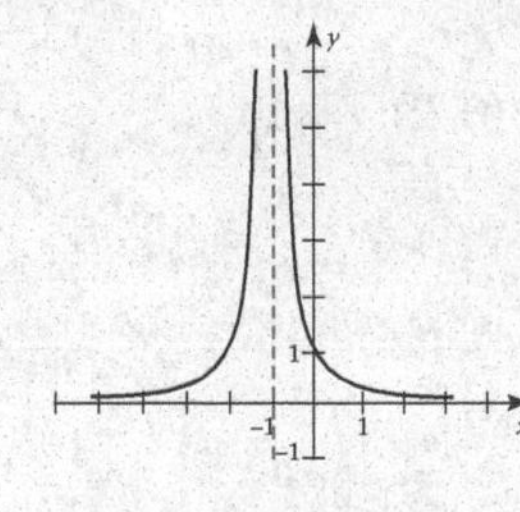

13. $x = 2$, $y = 1$

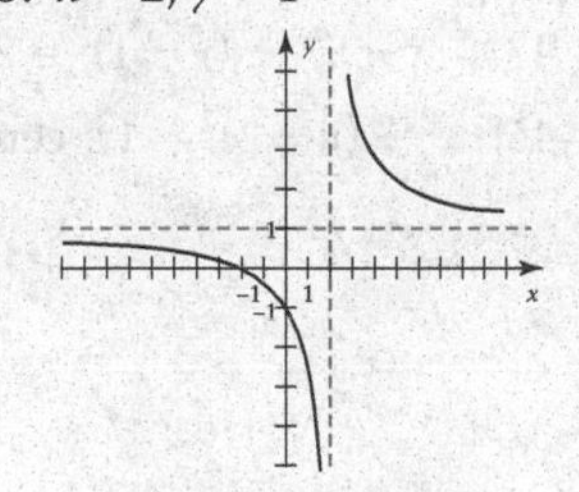

15. $x = 2$, $x = -2$, $y = 2$

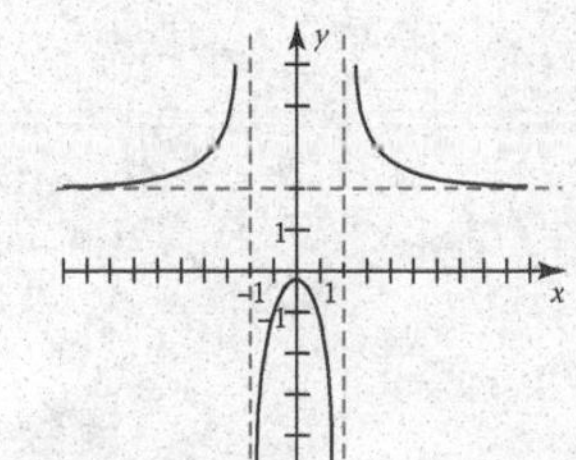

17. $x = \frac{-3}{2}$, $x = 2$, $y = \frac{1}{2}$

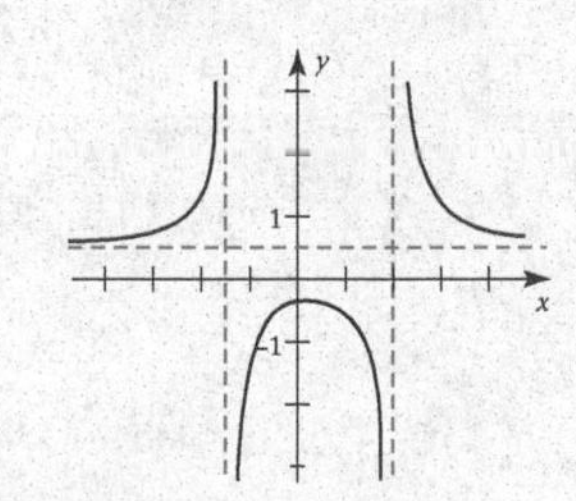

19. $x = 1$

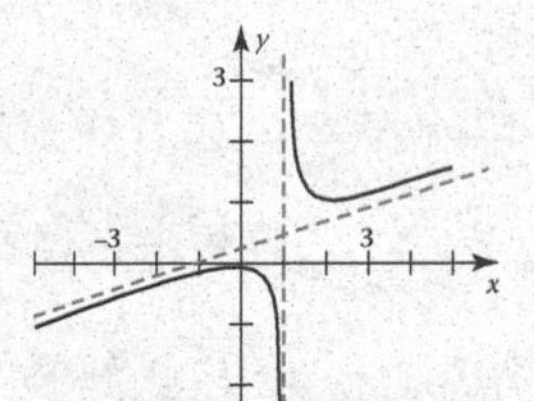

21. $x = 5$, $x = -5$

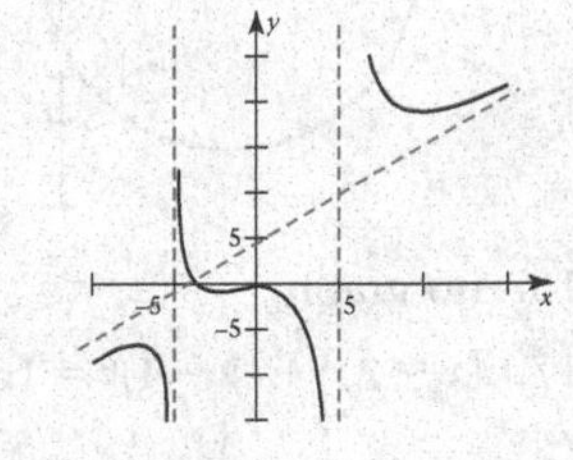

23. $x \neq -2$

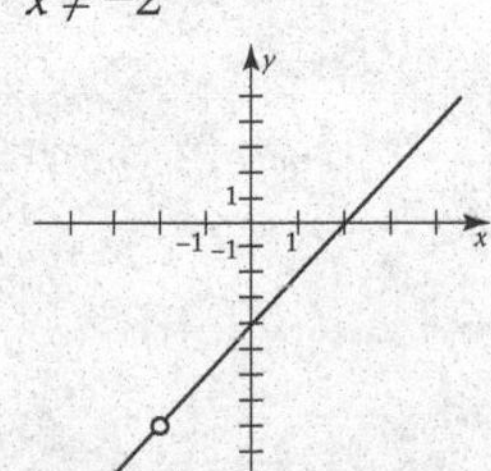

25. $x \neq 2$

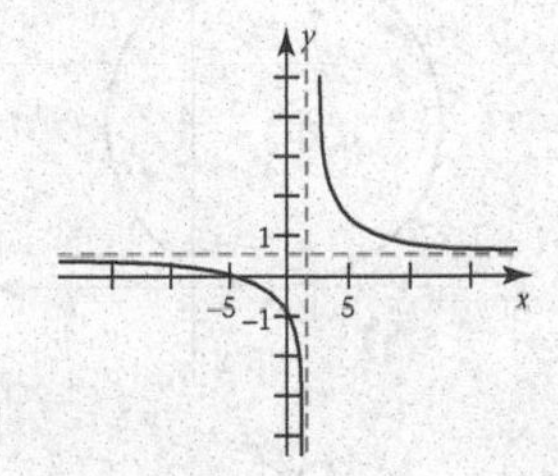

27. $x \neq 0$, $x \neq -1$

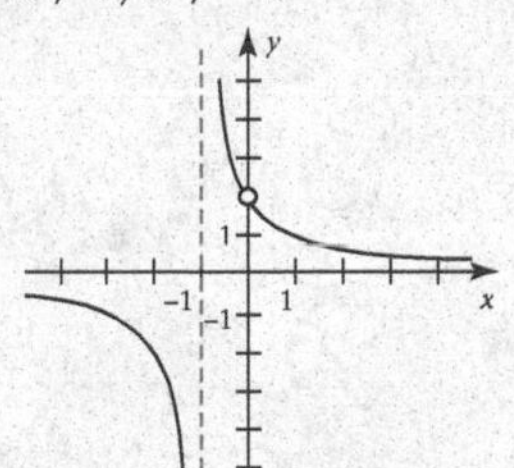

Exercise Set 5.2

1. $(x-2)^2+(y-3)^2=4$
3. $(x+2)^2+(y+3)^3=5$
5. $x^2+y^2=9$
7. $(x+1)^2+(y-4)^2=8$
9. $(h, k)=(2, 3);\ r=4$

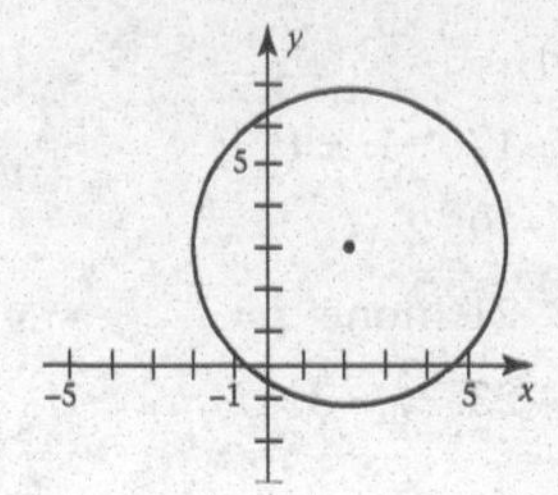

11. $(h, k)=(2, -2);\ r=2$

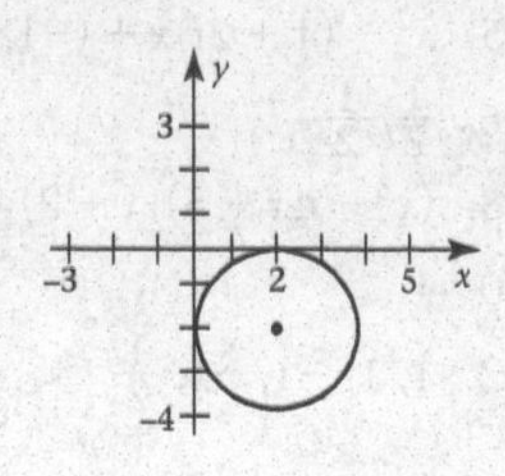

13. $(h, k)=\left(-4, \frac{-3}{2}\right);\ r=3\sqrt{2}$

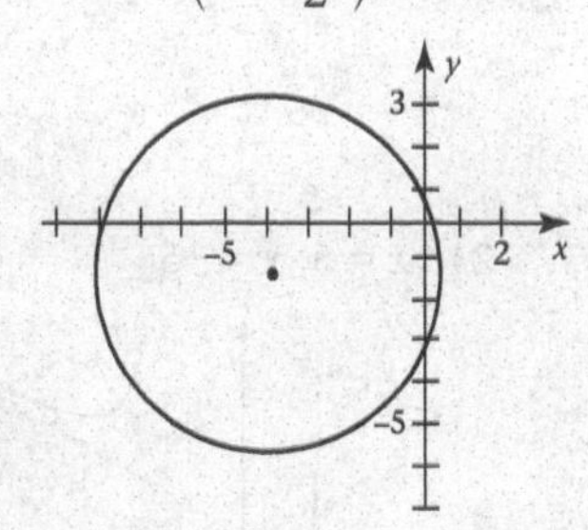

15. no graph
17. $(x-2)^2+(y-4)^2=16;\ (h, k)=(-2, 4);\ r=4$

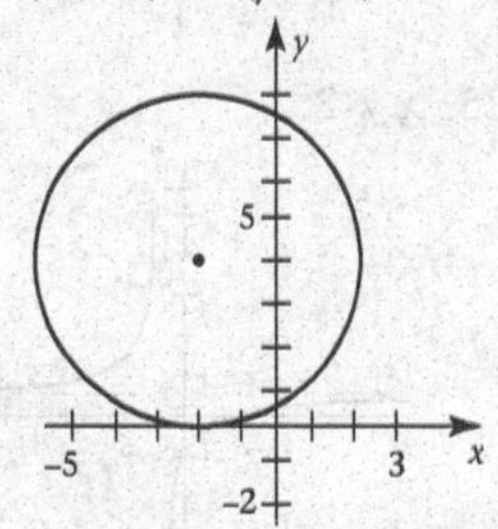

19. $\left(x-\frac{3}{2}\right)^2+\left(y-\frac{5}{2}\right)^2=\frac{11}{2};\ (h, k)=\left(\frac{3}{2}, \frac{5}{2}\right);\ r=\frac{\sqrt{22}}{2}$

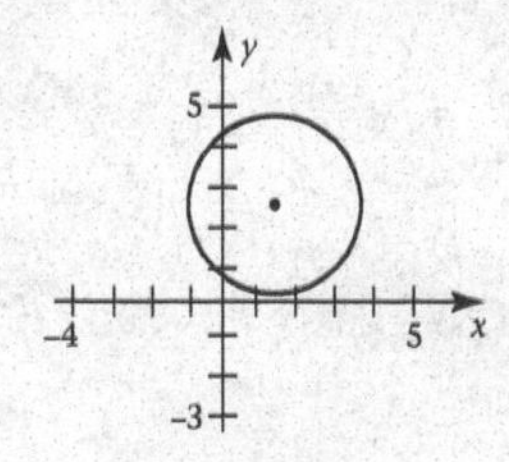

21. $(x-1)^2+y^2=\frac{7}{2};\ (h, k)=(1, 0);\ r=\frac{\sqrt{14}}{2}$

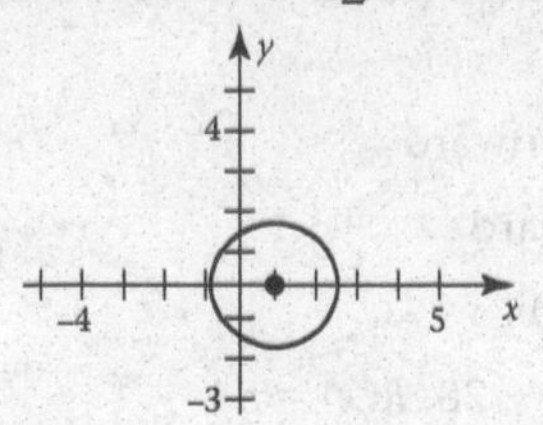

23. $(x-2)^2+(y+3)^2=8;\ (h, k)=(2, -3);\ r=2\sqrt{2}$

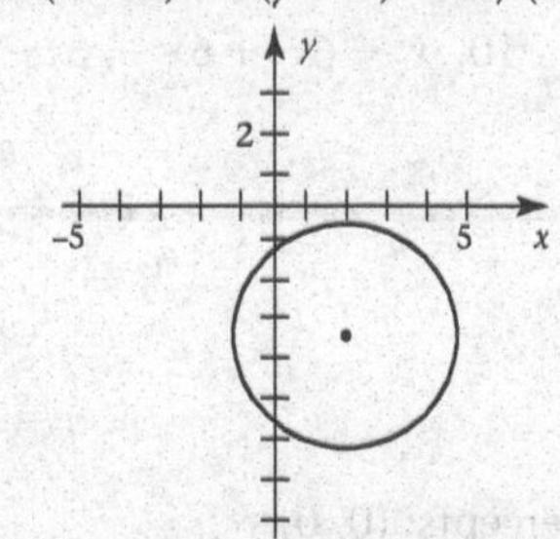

25. $(x-3)^2+(y+4)^2=0$; point at $(3, -4)$
27. $\left(x+\frac{3}{2}\right)^2+\left(y-\frac{5}{2}\right)^2=\frac{3}{2}$, circle
29. $(x-3)^2+y^2=11$, circle
31. $\left(x-\frac{3}{2}\right)^2+(y-1)^2=\frac{17}{4}$, circle
33. $(x+2)^2+\left(y-\frac{2}{3}\right)^2=\frac{100}{9}$, circle
35. $\left(x+\frac{3}{2}\right)^2+\left(y-\frac{5}{2}\right)^2=-1$, neither
37. 9π
41. $(x+5)^2+(y-2)^2=8$
43. $(x-5)^2+(y-1)^2=20$
45. $x^2+(y-2)^2=1^2$; center: $(0, 2)$; $r+1$

Exercise Set 5.3

1. focus: (0, 1); directrix: $y = -1$

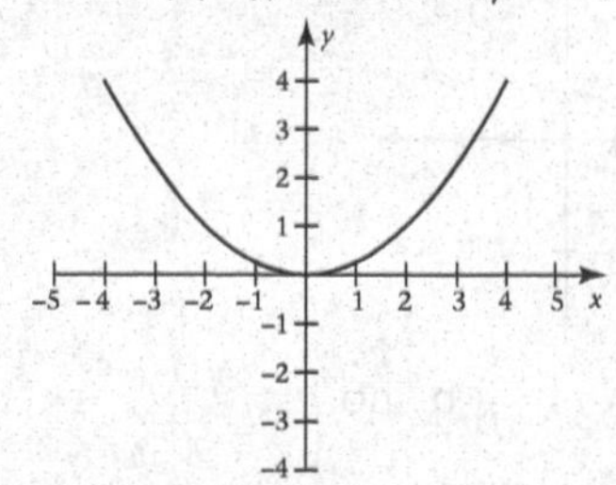

3. focus: $\left(\frac{1}{2}, 0\right)$; directrix: $x = -\frac{1}{2}$

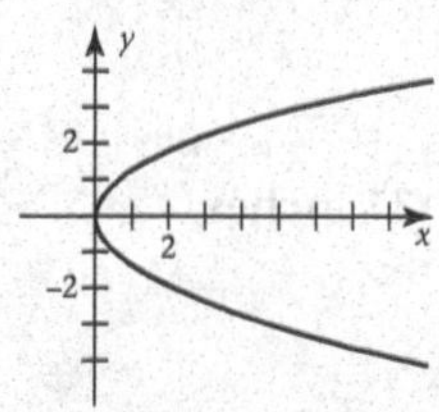

5. focus: $\left(0, -\frac{5}{4}\right)$; directrix: $y = \frac{5}{4}$

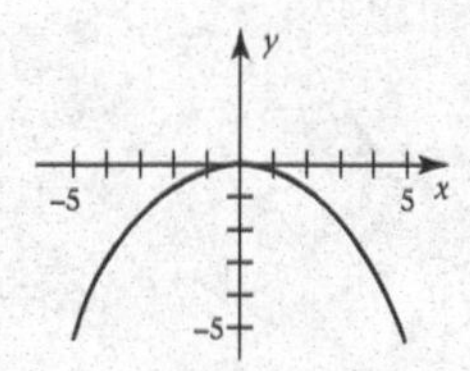

7. focus: (3, 0); directrix: $x = -3$

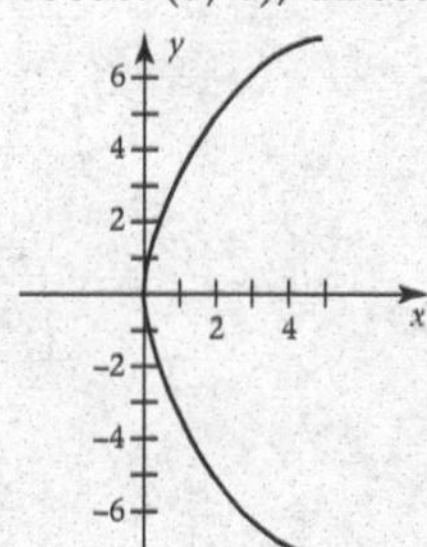

9. $y^2 = 4x$
11. $y^2 = 6x$
13. $x^2 = -8y$
15. $y^2 = -5x$
17. $y^2 = -4x$
19. $y^2 = x$
21. downward
23. to the left
25. $(x-1)^2 = 3(y-2)$; vertex: (1, 2); axis: $x = 1$; direction: up

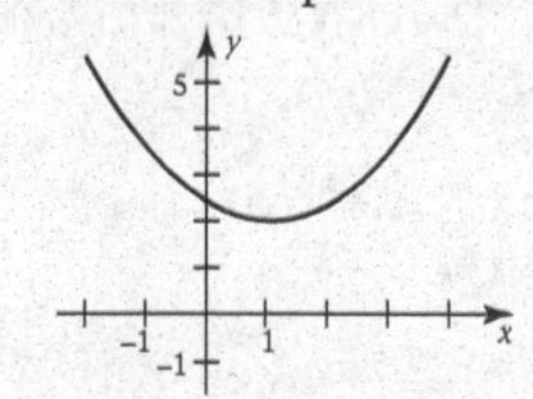

27. $(y-4)^2 = -2(x-2)$; vertex: (2, 4); axis: $y = 4$; direction: left

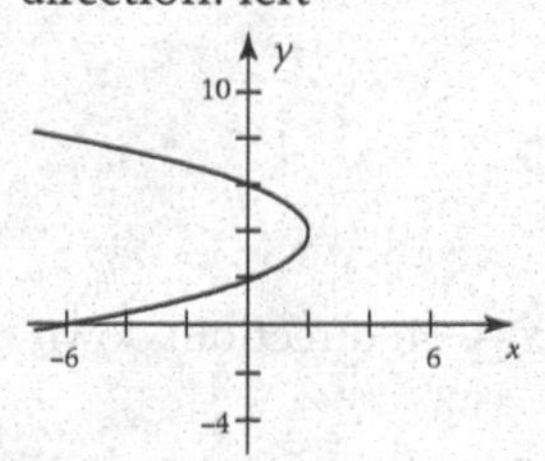

29. $\left(x-\frac{1}{2}\right)^2 = -3\left(y+\frac{1}{4}\right)$; vertex: $\left(\frac{1}{2}, -\frac{1}{4}\right)$; axis: $x = \frac{1}{2}$; direction: down

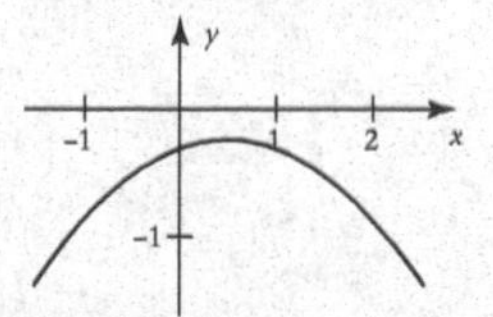

31. $(y-5)^2 = 3\left(x+\frac{1}{3}\right)$; vertex: $\left(-\frac{1}{3}, 5\right)$; axis: $y = 5$; direction: right

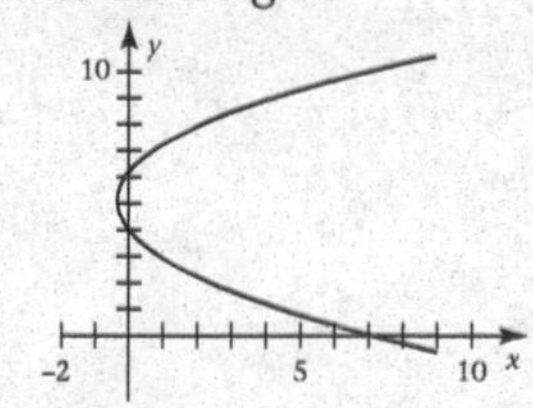

33. $\left(x-\frac{3}{2}\right)^2 = 3\left(y+\frac{5}{12}\right)$; vertex: $\left(\frac{3}{2}, -\frac{5}{12}\right)$; axis: $x = \frac{3}{2}$; direction: up

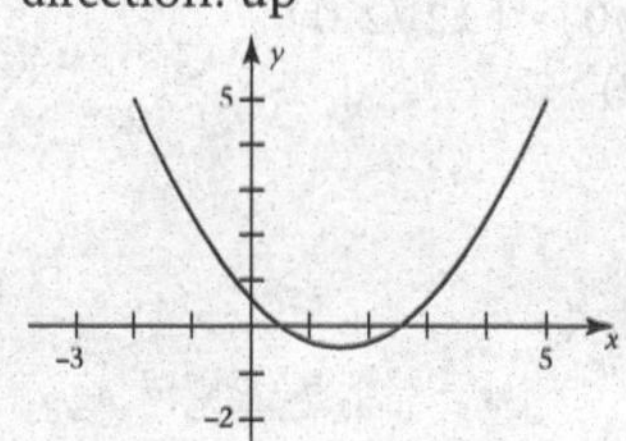

35. $(y+3)^2 = -\frac{1}{2}(x-4)$; vertex: (4, −3); axis: $y = -3$; direction: left

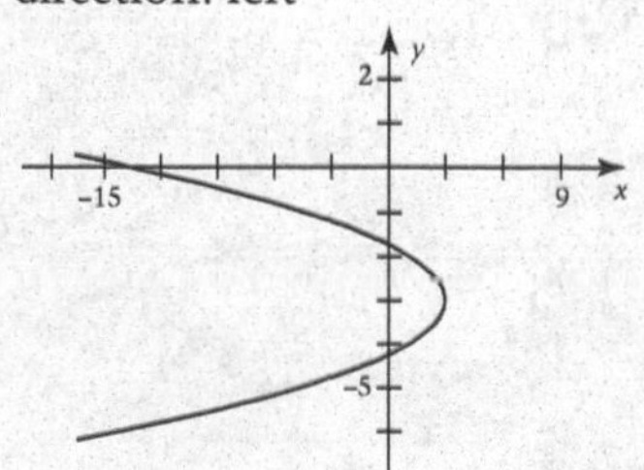

37. $(x+1)^2 = -2(y+1)$; vertex: (−1, −1); axis: $x = -1$; direction: down

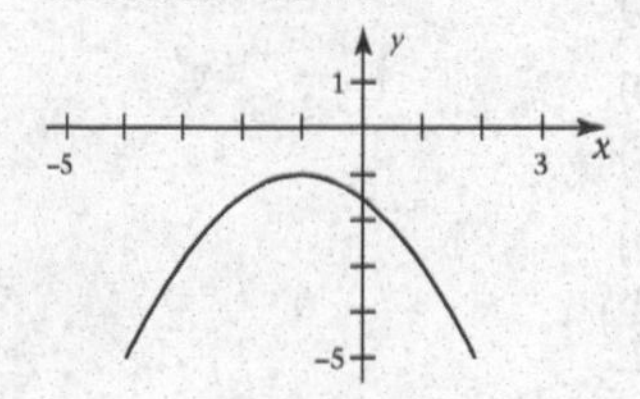

39. vertex: $(2, -1)$; axis: $x = 2$; direction: up

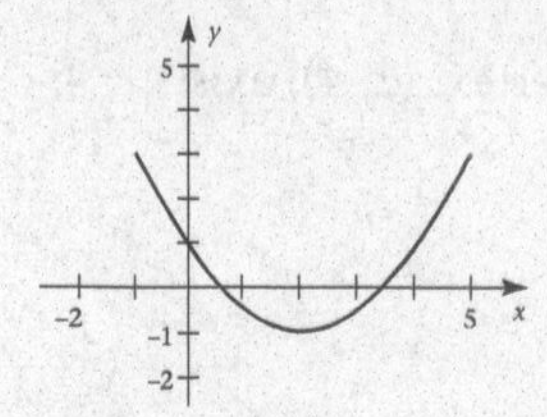

41. vertex: $(-4, -2)$; axis: $x = -4$; direction: down

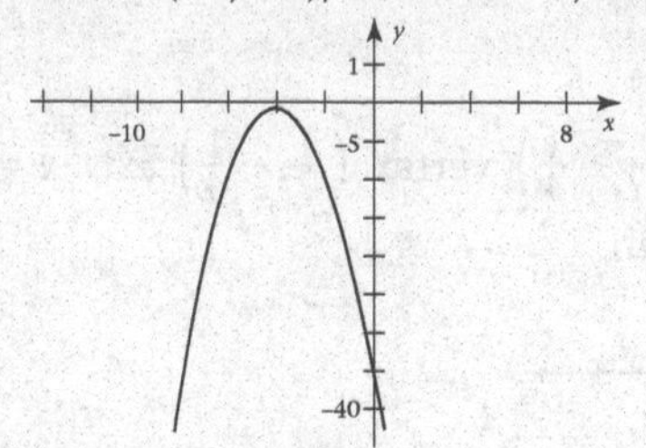

43. vertex: $(-1, 0)$; axis: $y = 0$; direction: left

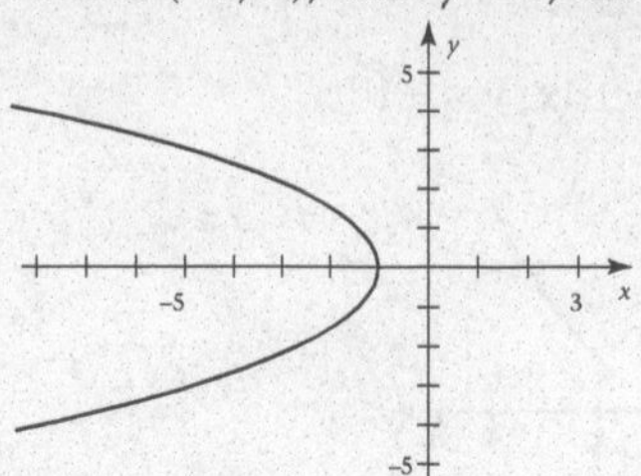

45. a.

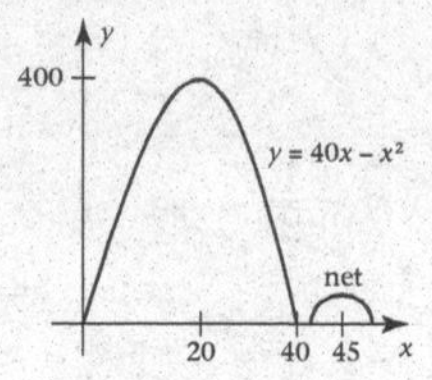

b. no

47. a. 49 inches b. 31 inches

Exercise Set 5.4

1. x-intercepts: $(\pm 5, 0)$
   y-intercepts: $(0, \pm 2)$

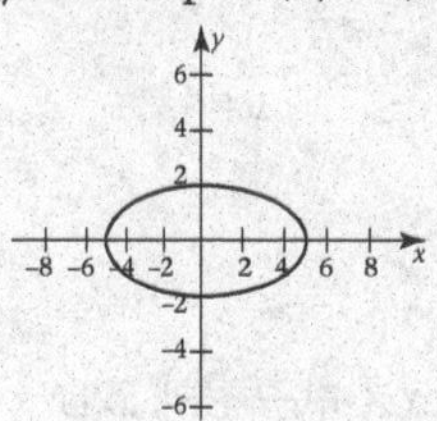

3. x-intercepts: $(\pm\sqrt{8}, 0) = (\pm 2\sqrt{2}, 0)$
   y-intercepts: $(0, \pm 2)$

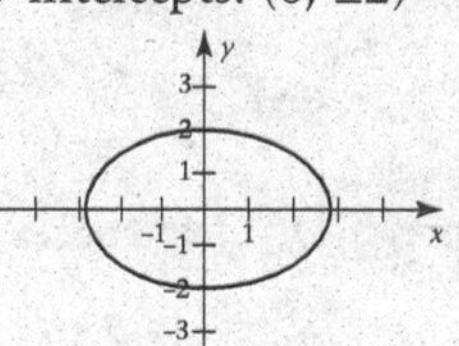

5. x-intercepts: $(\pm 4, 0)$
   y-intercepts: $(0, \pm 5)$

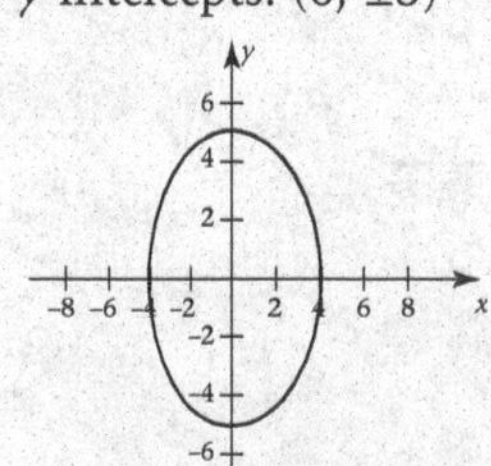

7. x-intercepts: $(\pm 3, 0)$
   y-intercepts: $(0, \pm 2)$

9. x-intercepts: $(\pm 2, 0)$
   y-intercepts: $(0, \pm 1)$

11. x-intercepts: $(\pm 1, 0)$
    y-intercepts: $\left(0, \pm \frac{1}{2}\right)$

13. x-intercepts: $(\pm\sqrt{3}, 0)$
    y-intercepts: $(0, \pm 2)$

15. x-intercepts: $\left(\pm \frac{1}{2}, 0\right)$
    y-intercepts: $\left(0, \pm \frac{3\sqrt{2}}{4}\right)$

17. y-intercepts: $(0, \pm 4)$

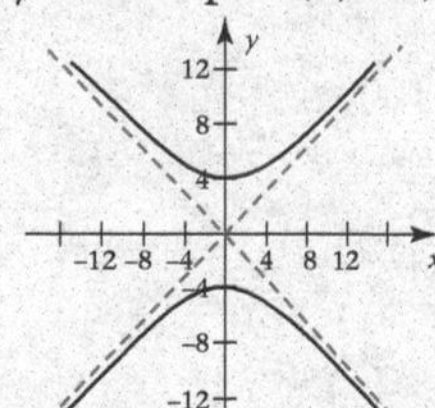

19. x-intercepts: $(\pm 6, 0)$
    viewing rectangle: 3 times the EQUAL viewing rectangle
    XSCL = 3
    YSCL = 3
    GRAPH: $Y = \sqrt{((X^2 \div 36) - 1)}$
    and
    $Y = -\sqrt{((X^2 \div 36) - 1)}$

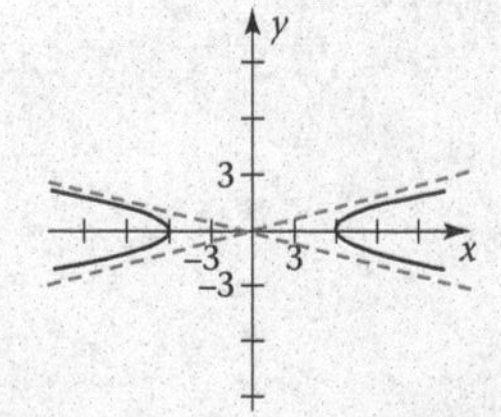

21. x-intercepts: $(\pm\sqrt{6}, 0)$

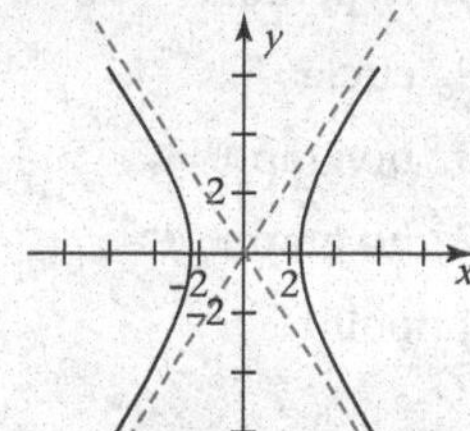

23. $\frac{x^2}{4} - \frac{y^2}{64} = 1$; $(\pm 2, 0)$

33. x-intercepts: $(\pm 5, 0)$

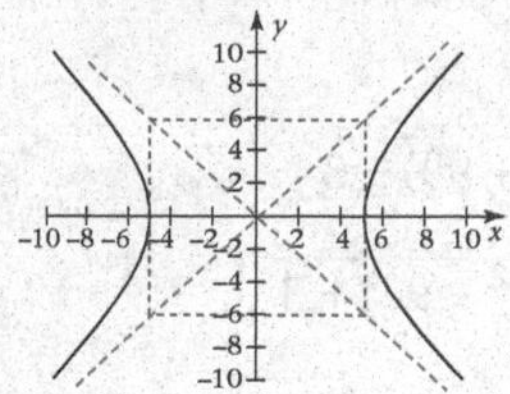

37. $\frac{\sqrt{299}}{2}$ inches

25. $\frac{y^2}{\frac{1}{4}} - \frac{x^2}{\frac{1}{4}} = 1$; $\left(0, \pm\frac{1}{2}\right)$

27. $\frac{x^2}{5} - \frac{y^2}{4} = 1$; $(\pm\sqrt{5}, 0)$

29. x-intercepts: $(\pm 3, 0)$

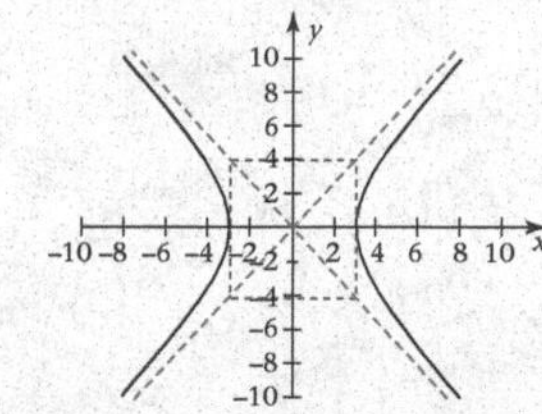

31. y-intercepts: $(0, \pm 3)$

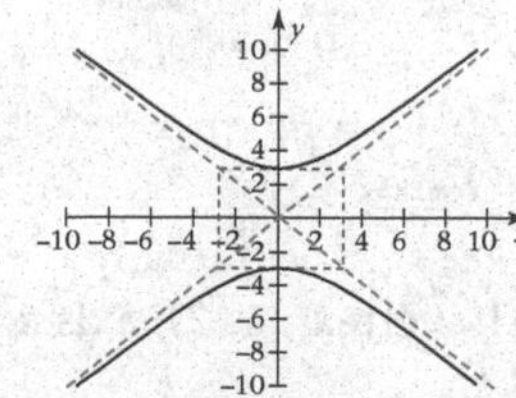

39. $\frac{80\sqrt{6}}{7}$ feet

41. $\frac{x^2}{9} + \frac{y^2}{4} = 1$

43. $\frac{x^2}{9} + \frac{y^2}{25} = 1$

45. $\frac{x^2}{9} - y^2 = 1$

Exercise Set 5.5

1. $(1, -4)$

3. $(5, -1)$

5. $(-3, 4)$

7. $(1, 7)$

9. viewing rectangle: 3 times the ZDECIMAL viewing rectangle
   XSCL = 3
   YSCL = 3
   GRAPH: $Y = \frac{\sqrt{(36X^2 + 216X + 99)}}{10}$
   and
   $Y = -\frac{\sqrt{(36X^2 + 216X + 99)}}{10}$

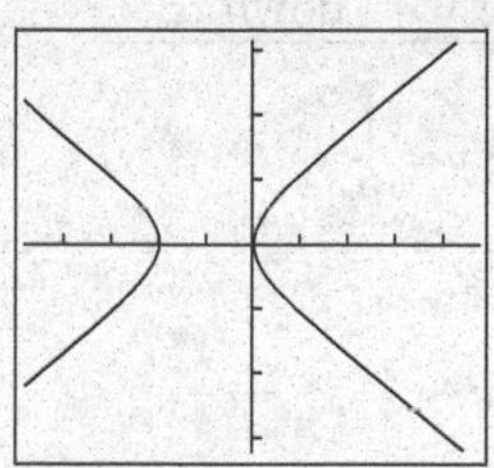

11. viewing rectangle: 3 times the ZDECIMAL viewing rectangle
   XSCL = 3
   YSCL = 3
   GRAPH $Y = X^2 + 4X + 5$

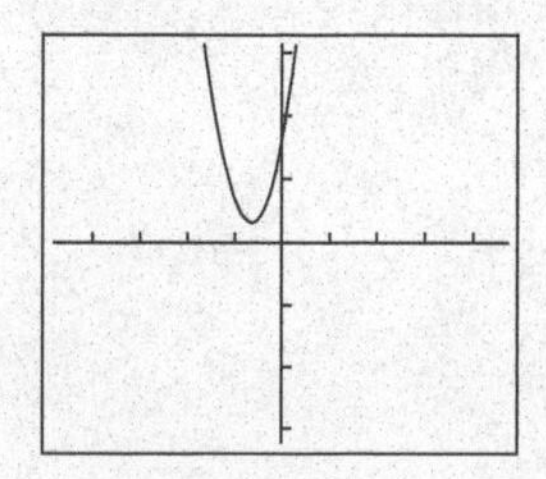

13. XMIN = −20
   XMAX = 0
   XSCL = 1
   YMIN = −10
   YMAX = 10
   YSCL = 1
   GRAPH: $Y = \sqrt{(-2X - 15)}$
   and
   $Y = -\sqrt{(-2X - 15)}$

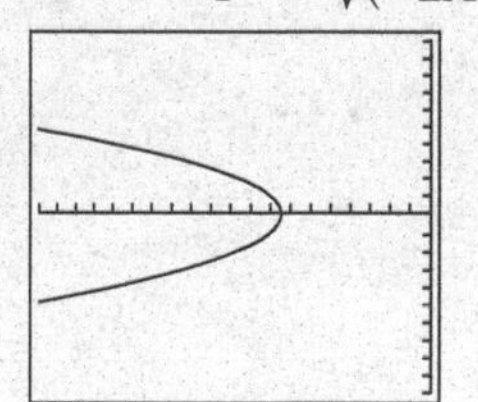

15. Viewing rectangle: 2 times the ZDECIMAL viewing rectangle
   XSCL = 2
   YSCL = 2
   GRAPH: $Y = \sqrt{(4 - ((X+5)^2 \div 4))} + 1$
   and
   $Y = -\sqrt{(4 - ((X+5)^2 \div 4))} + 1$

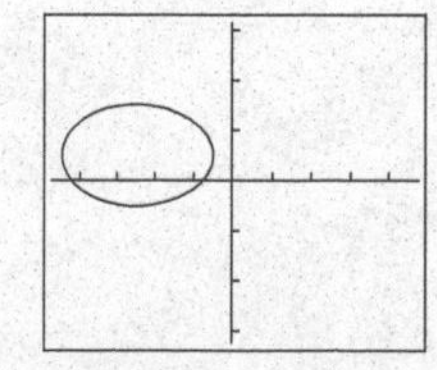

17. Viewing rectangle: 2 times the ZDECIMAL viewing rectangle
XSCL = 2
YSCL = 2
GRAPH: $Y = \sqrt{(1-(X^2 \div 9))}+3$
and
$Y = -\sqrt{(1-(X^2 \div 9))}+3$

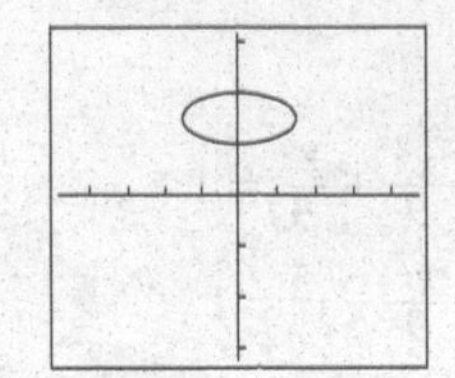

19. parabola
21. ellipse
23. hyperbola
25. hyperbola
27. parabola
29. circle
31. hyperbola
33. no graph
35. no graph
37. no graph
39. hyperbola
41. point

Chapter 5 Review Exercises

1. 

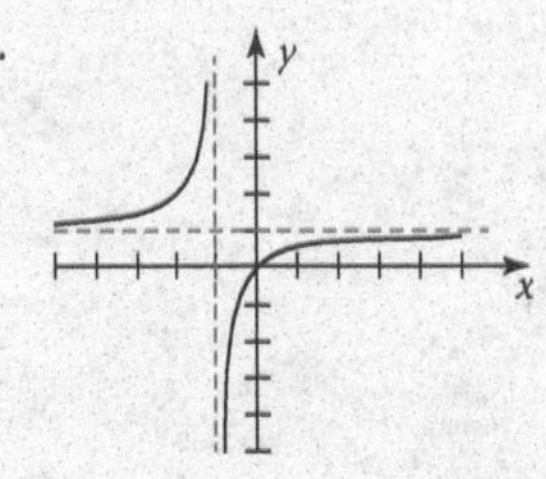

2. 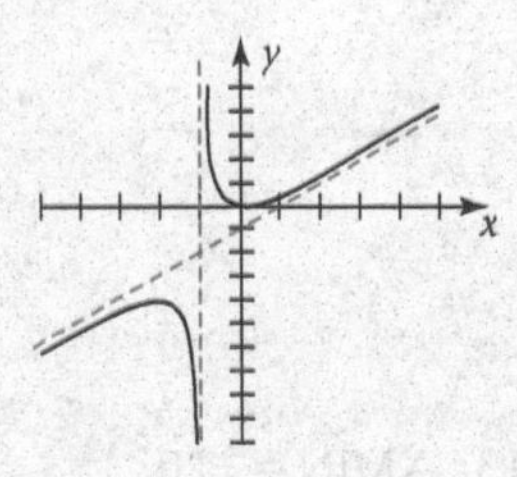

3. 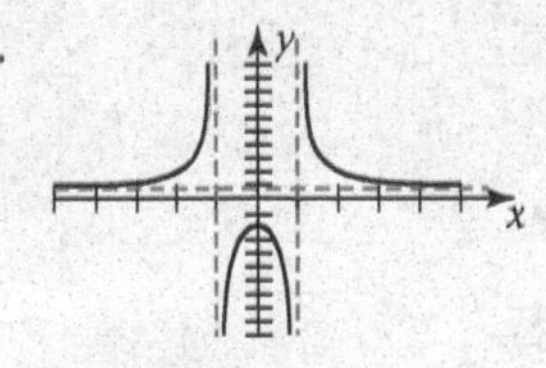

4. $(x+5)^2 + (y-2)^2 = 16$
5. $(x+3)^2 + (y-3)^2 = 4$
6. center: $(2, -3)$, radius: 3
7. center: $\left(-\frac{1}{2}, 4\right)$, radius: $\frac{1}{3}$
8. center: $(-2, 3)$, radius: $\sqrt{3}$
9. center: $(1, -1)$, radius: $\frac{\sqrt{2}}{2}$
10. center: $(0, 3)$, radius: $\sqrt{6}$
11. center: $(1, 1)$, radius: $\sqrt{10}$
12. vertex: $\left(\frac{3}{2}, -5\right)$; axis $y = -5$

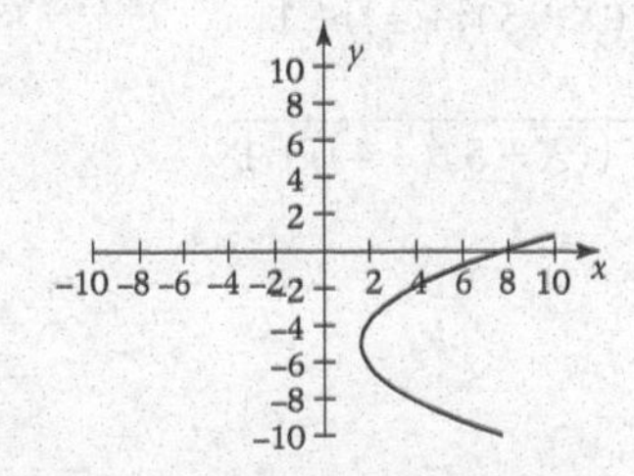

13. vertex: $(1, 2)$; axis $x = 1$

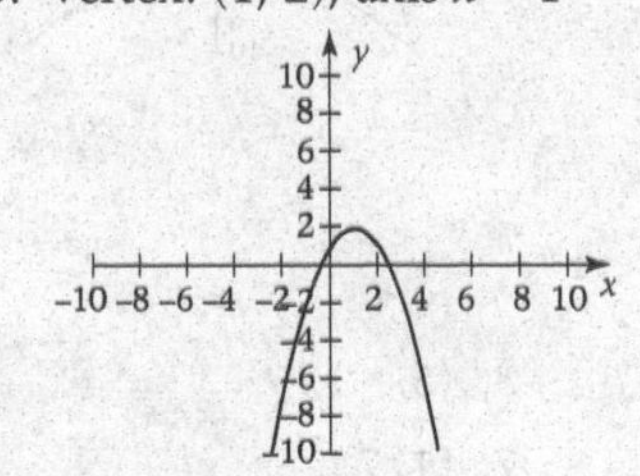

	VERTEX	AXIS	DIRECTION
14.	$(-3, 0)$	$y = 0$	left
15.	$(2, -2)$	$y = -2$	left
16.	$(3, -2)$	$x = 3$	up
17.	$\left(-2, -\frac{1}{2}\right)$	$x = -2$	down
18.	$(0, 1)$	$y = 1$	right
19.	$(-3, 0)$	$x = -3$	down

20. focus: $\left(0, -\frac{1}{6}\right)$; directrix: $y = \frac{1}{6}$

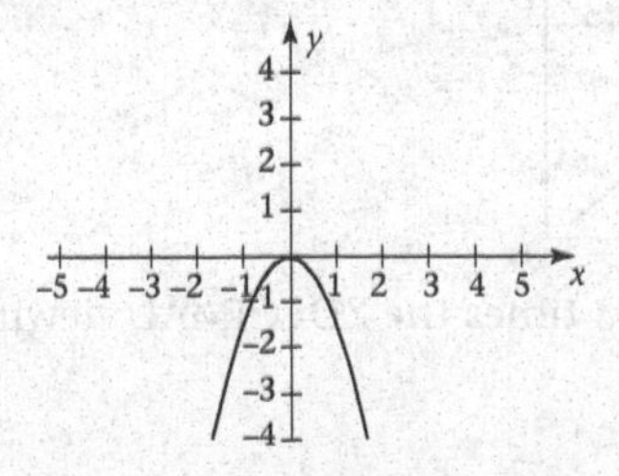

21. focus: $\left(-\frac{1}{6}, 0\right)$; directrix: $x = \frac{1}{6}$

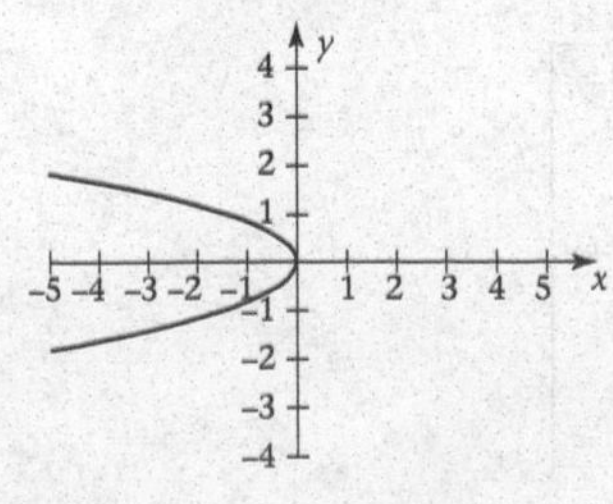

22. $x^2 = -7y$
23. $x^2 = \frac{2}{5}y$

24. $\frac{x^2}{4}-\frac{y^2}{9}=1$; $(\pm 2, 0)$

25. $\frac{x^2}{1}+\frac{y^2}{9}=1$; $(\pm 1, 0)$, $(0, \pm 3)$

26. $\frac{x^2}{7}+\frac{y^2}{5}=1$; $(\pm\sqrt{7},0),(0,\pm\sqrt{5})$

27. $\frac{x^2}{16}-\frac{y^2}{9}=1$; $(\pm 4, 0)$

28. $\frac{x^2}{3}+\frac{y^2}{\frac{9}{4}}=1$; $(\pm\sqrt{3},0),\left(0,\pm\frac{3}{2}\right)$

29. $\frac{y^2}{\frac{20}{3}}-\frac{x^2}{4}=1$; $\left(0,\frac{\pm 2\sqrt{15}}{3}\right)$

30.

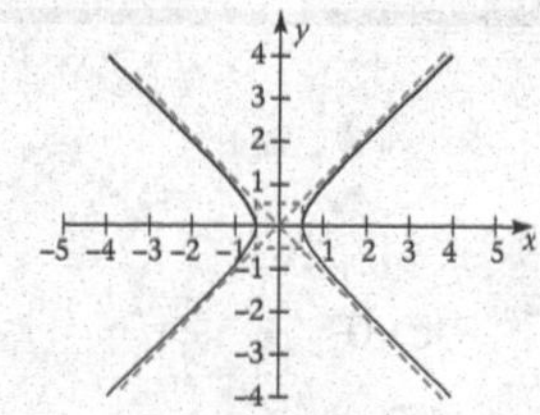

31.

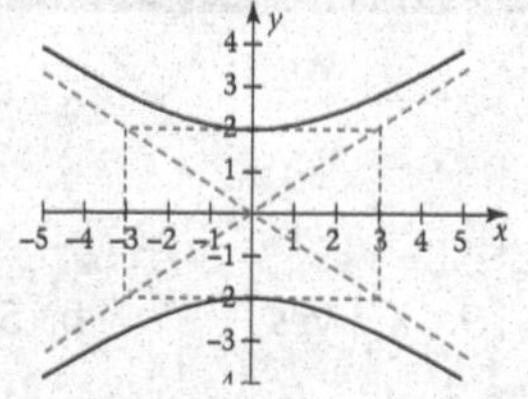

32. $\frac{x^2}{4}-\frac{y^2}{\frac{4}{5}}=1$

33. ellipse

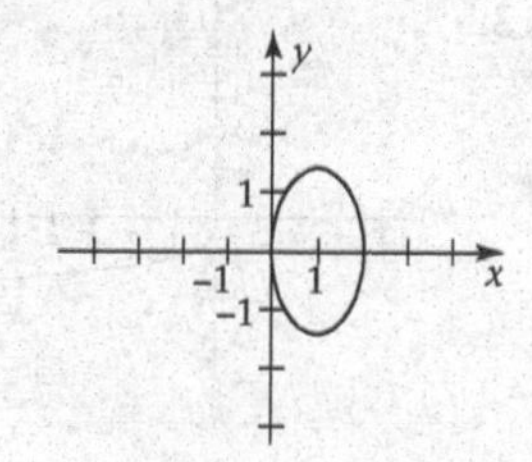

34. parabola

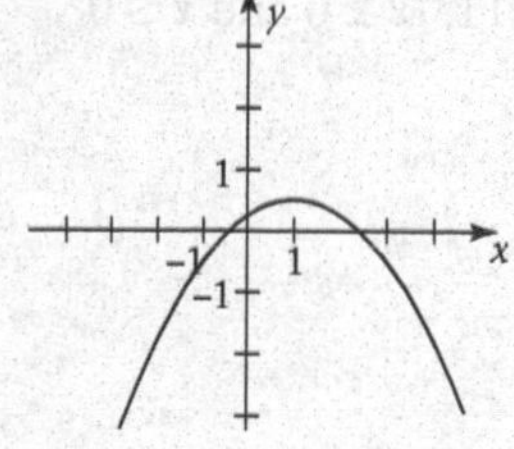

35. no graph

36. hyperbola

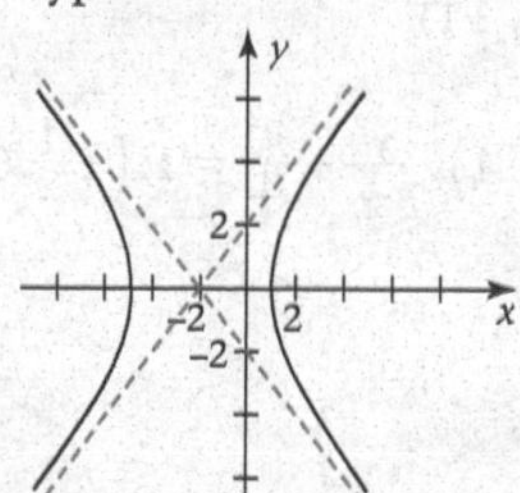

37. two lines

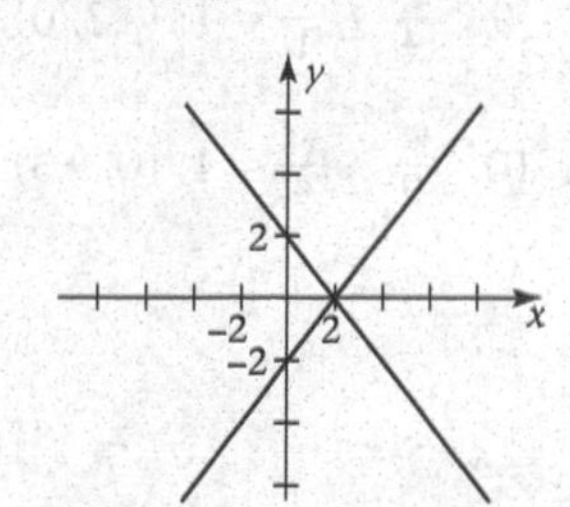

38. hyperbola

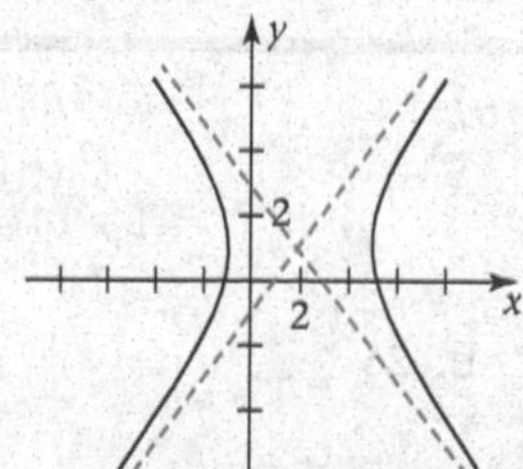

39. parabola

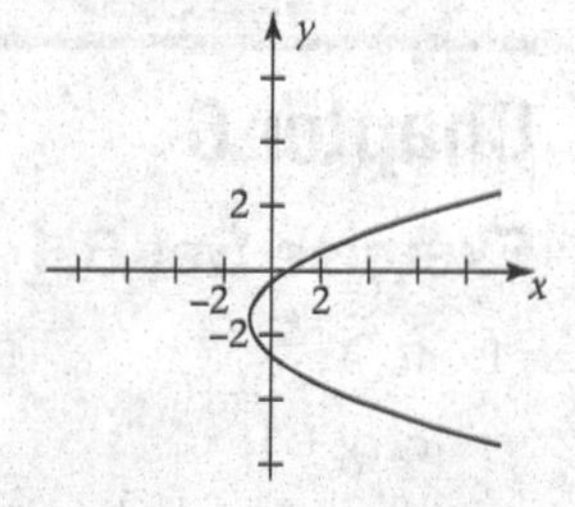

40. hyperbola

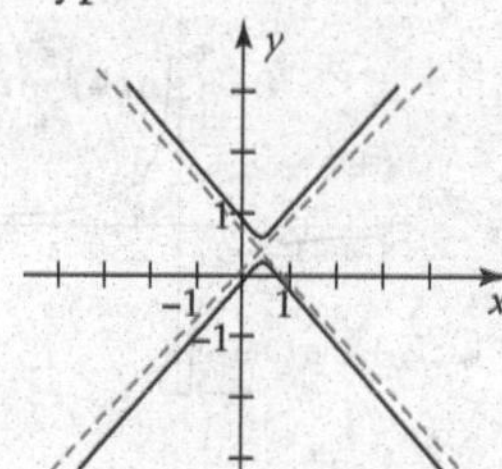

41. ellipse

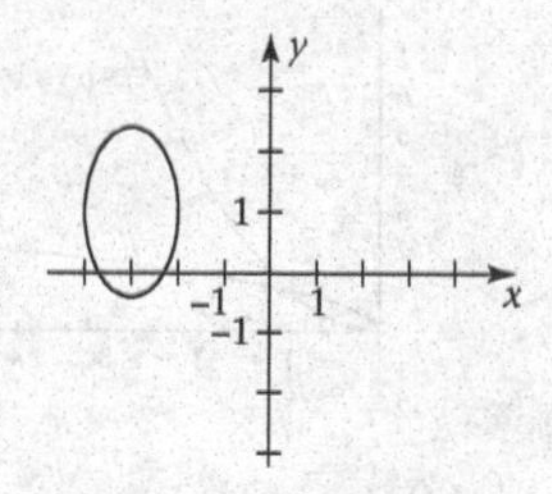

42. no graph

Chapter 5 Review Test

1.

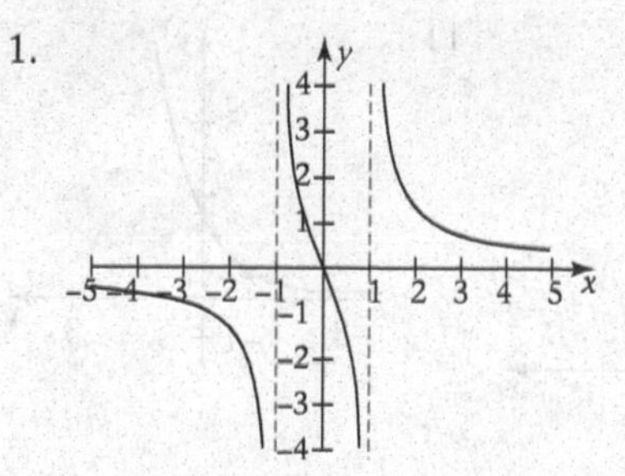

2. $(x-2)^2+(y+3)^2=36$

3. center: $(1, -2)$, radius: 2

4. center: $(2, 0)$, radius: $\sqrt{5}$

5. vertex: $(-3, 1)$; axis: $x=-3$

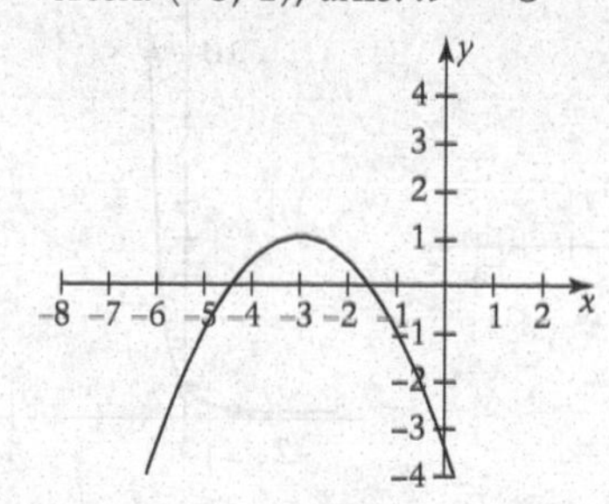

6. vertex: $(1, 2)$; axis: $y = 2$

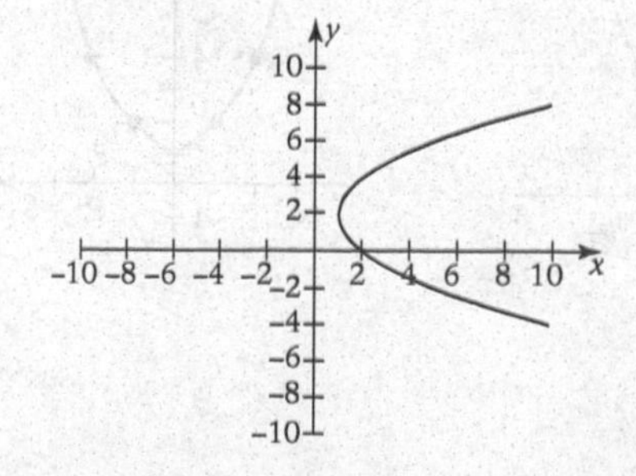

7. vertex: $(3, 2)$; axis: $x=3$; direction: down

8. vertex: $(-2, -4)$; axis: $y = -4$; direction: right

9. $\frac{x^2}{4} + \frac{y^2}{1} = 1$; $(\pm 2, 0)$, $(0, \pm 1)$

10. $\frac{y^2}{9} - \frac{x^2}{4} = 1$; $(0, \pm 3)$

11. $\frac{x^2}{\frac{1}{4}} - \frac{y^2}{\frac{1}{4}} = 1$; $\left(\pm\frac{1}{2}, 0\right)$

12.

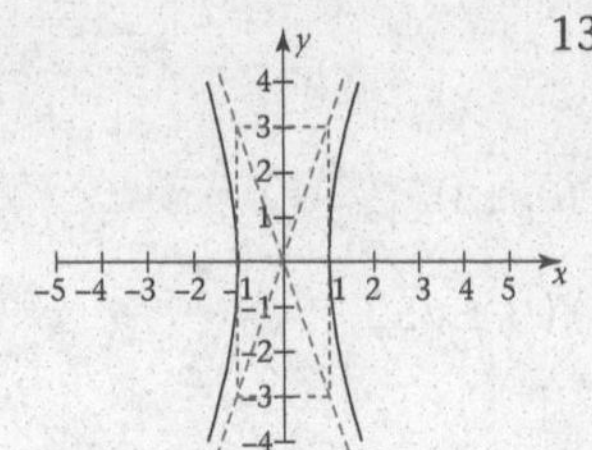

13.

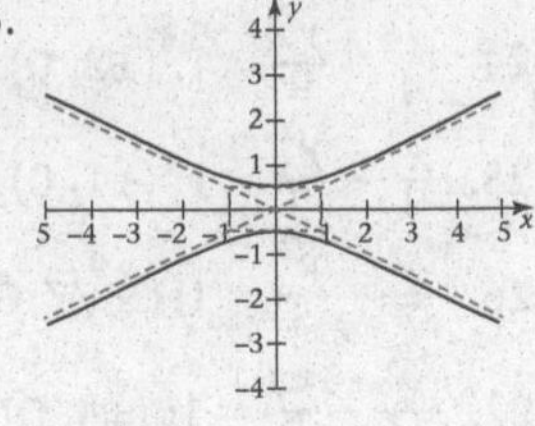

14. hyperbola

15. parabola

16. circle

17. ellipse

Chapter 6

Exercise Set 6.1

1. a. 4 b. 25 c. $y + 1$

3. yes, $f^{-1}(x) = x^2$, $x \geq 0$

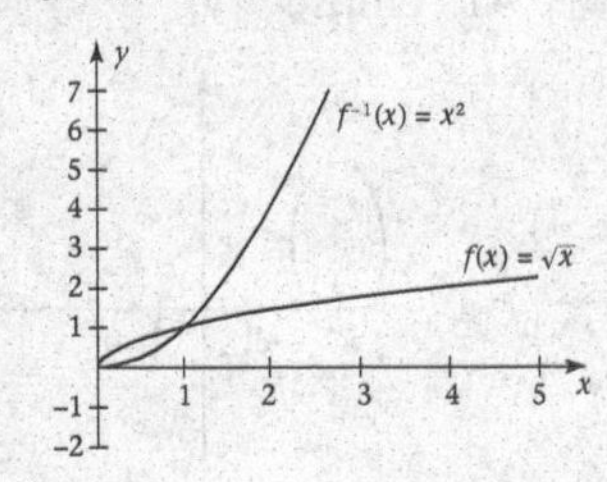

5. yes, $f^{-1}(x) = \frac{1}{x}$

7. 0

9. a. yes b. 5 c. 0

d.

x	17	20	−4	3	10
$G^{-1}(x)$	−10	−5	0	5	10

11. $x \geq 0$ and $x \leq 0$

13.

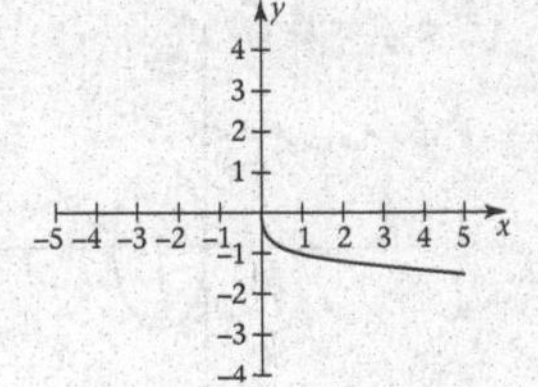

Exercise Set 6.2

1.

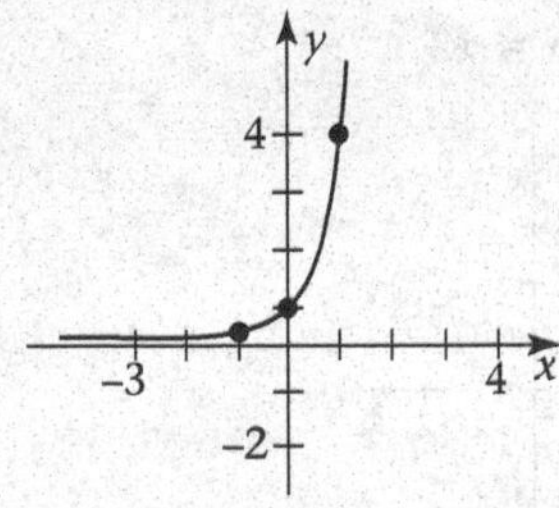

3.

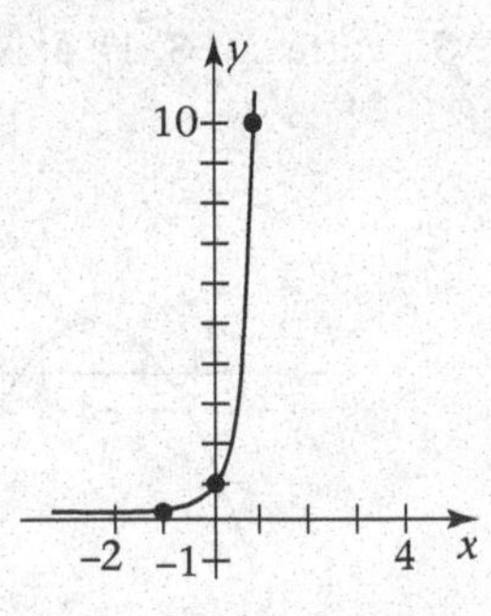

5.

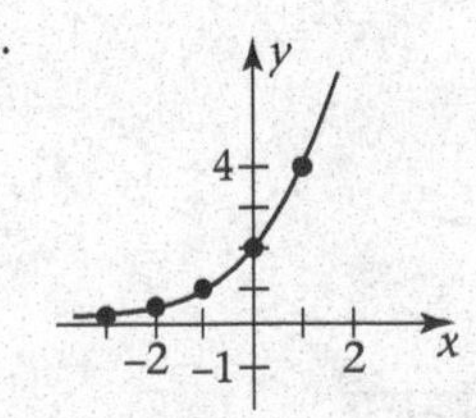

7.

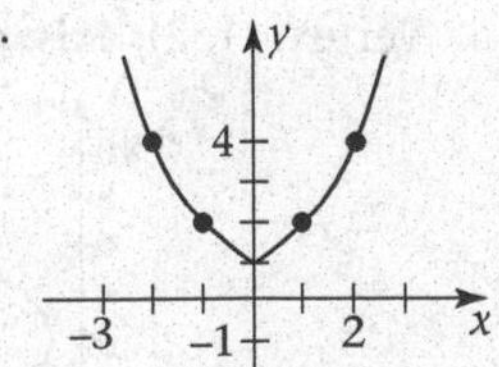

9.

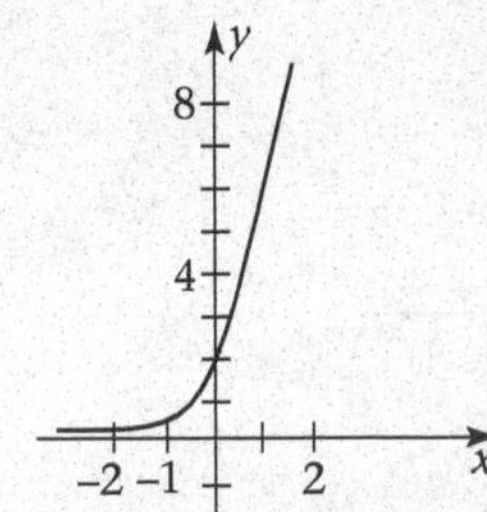

11.

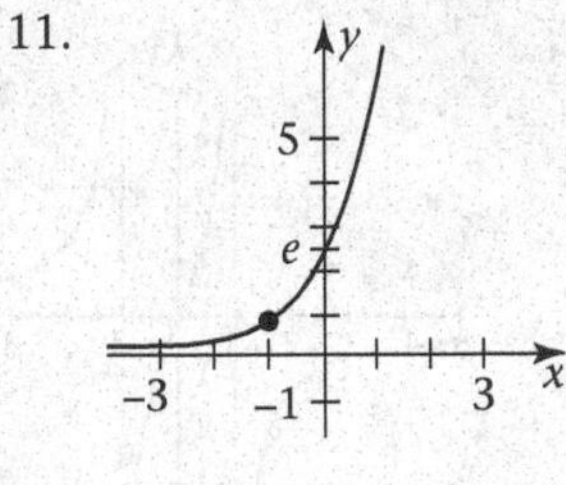

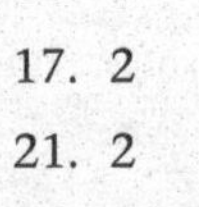

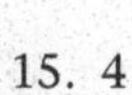

13. 3

15. 4

17. 2

19. 4

21. 2

23. 1

25. 7.389

27. 10.068

29. .833

31. 3^{π}

33. a. 717.81 b. 722.52
c. 724.97 d. 726.65
e. 727.47 f. 727.49
g. 727.5
h. (a) \$500 compounded annually at 7.5% for 5 years
(b) \$500 compounded semi-annually at 7.5% for 5 years
(c) \$500 compounded quarterly at 7.5% for 5 years
(d) \$500 compounded monthly at 7.5% for 5 years
(e) \$500 compounded daily at 7.5% for 5 years
(f) \$500 compounded hourly at 7.5% for 5 years
(g) \$500 compounded continuously at 7.5% for 5 years

35. a.
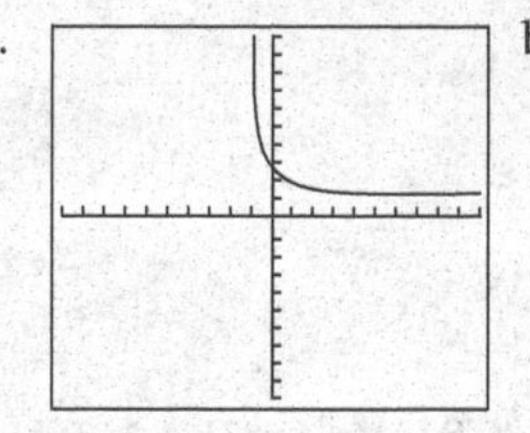
XSCL = 1, YSCL = 1

b.
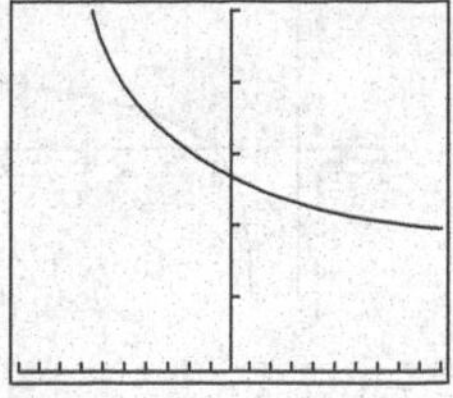
XSCL = 0.1, YSCL = 1

c.
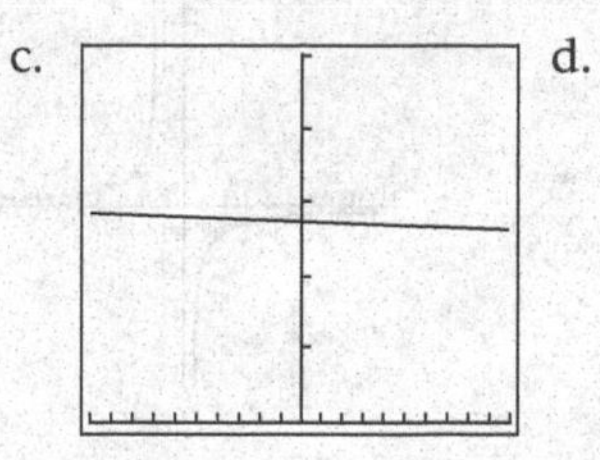
XSCL = 0.01, YSCL = 1

d.
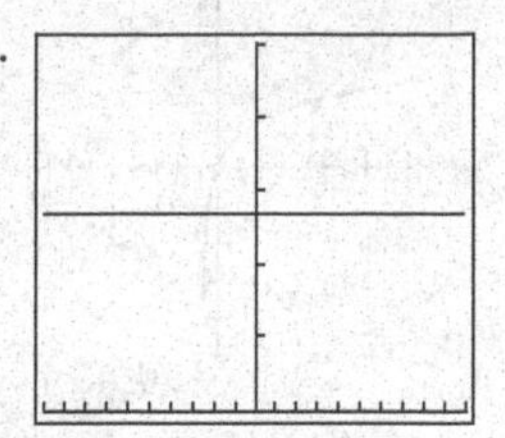
XSCL = 0.000001, YSCL = 1

37. symmetric with respect to the y-axis

39. 1073.83

41. $q_0 = 2500$, $Q7 = 320{,}000$

43. a. 200 b. 29685.6
c. $Q(1) = 256$, $Q(4) = 543$, $Q(8) = 1477$, $Q(10) = 2436$

45. 10.93 billion 47. 670.32 grams

49. 17125.53 51. 30524.91

53. 116.18 55. 68132.95

57. 1323.13 59. a. \$9.35

Exercise Set 6.3

1. $2^2 = 4$ 3. $9^{-2} = \frac{1}{81}$

5. $e^3 \approx 20.09$ 7. $10^3 = 1000$

9. $e^0 = 1$ 11. $3^{-3} = \frac{1}{27}$

13. $\log_5 25 = 2$ 15. $\log_{10} 10{,}000 = 4$

17. $\log_2 \frac{1}{8} = -3$ 19. $\log_2 1 = 0$

21. $\log_{36} 6 = \frac{1}{2}$ 23. $\log_{16} 64 = \frac{3}{2}$

25. $\log_{27} \frac{1}{3} = -\frac{1}{3}$ 27. 25

29. $\frac{1}{5}$ 31. e^2

33. $e^{-1/2}$ 35. −2

37. 512 39. 124

41. 2 43. 3

45. 6 47. 2

49. 3 51. $\frac{1}{2}$

53. 2 55. 1

57. 0 59. −2

61. 4 63. 2

65. 0.2041 67. 1.2920

69. 501.7167

71.
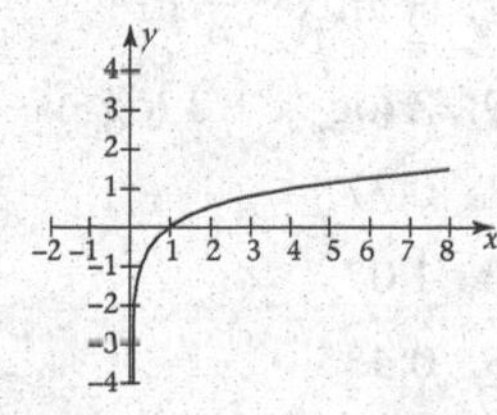

73.
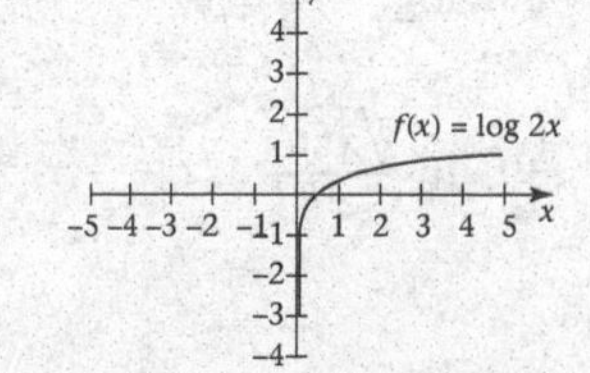

75.
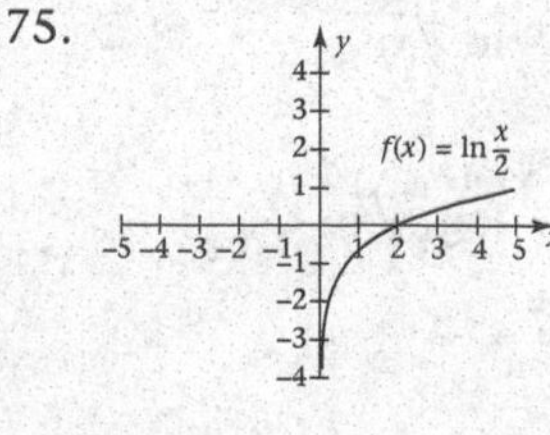
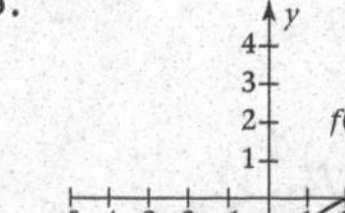

77.

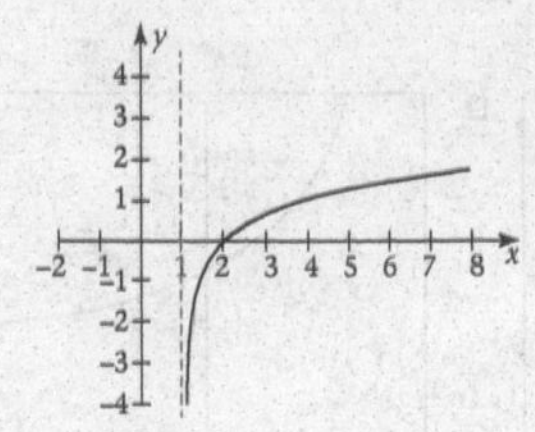

79. $\{x \mid x < 1\}$

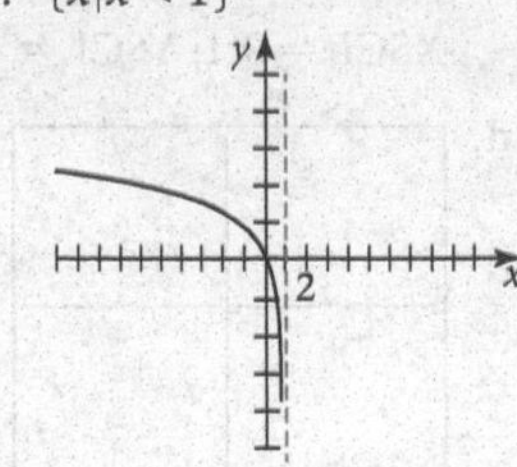

81. $\{x \mid x < 0, x > 1\}$

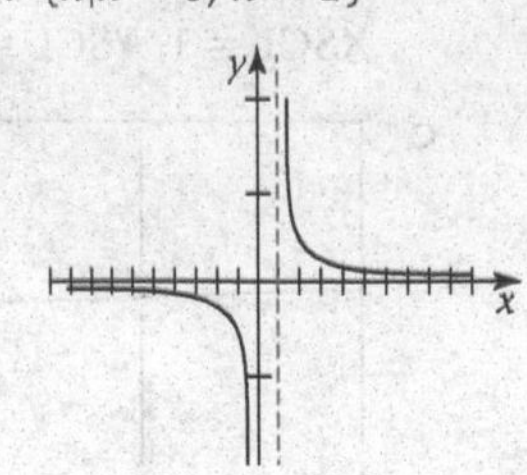

83. $\{x \mid x \text{ is real}\}$

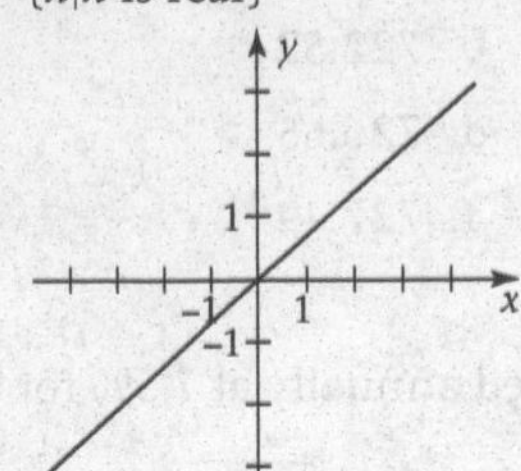

85. Since the domain has no elements, there is no graph.

87. 18.4 years

89. 5.5%

91.

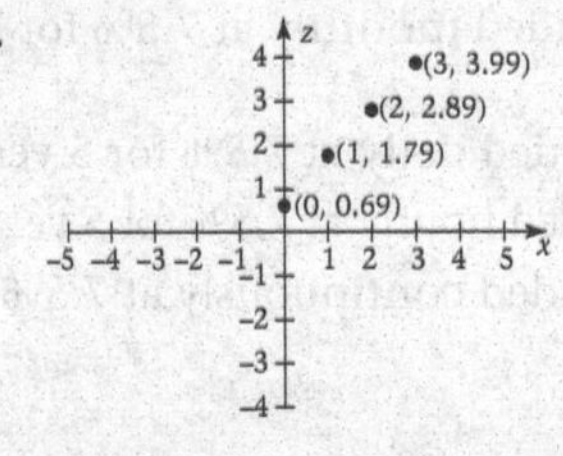

93. $y = 2(3^x)$

95. a. 100
b. 6.3
c. 15.8

Exercise Set 6.4

1. $4 \log_{10} 2 + 3 \log_{10} 3 + 1$

3. 4

5. $\log_a 2 + \log_a x + \log_a y$

7. $\log_a x - \log_a y - \log_a z$

9. $5 \ln x$

11. $2 \log_a x + 3 \log_a y$

13. $\frac{1}{2} \log_a x + \frac{1}{2} \log_a y$

15. $2 \ln x + 3 \ln y + 4 \ln z$

17. $\frac{1}{2} \ln x + \frac{1}{3} \ln y$

19. $2 \log_a x + 3 \log_a y - 4 \log_a z$

21. 0.77

23. 0.94

25. 1.07

27. 0.87

29. 0.435

31. $\log x^2 \sqrt{y}$

33. $\ln \sqrt[3]{xy}$

35. $\log_a \dfrac{x^{\frac{1}{3}} y^2}{z^{\frac{3}{2}}}$

37. $\log_a \sqrt{xy}$

39. $\ln \dfrac{\sqrt[3]{x^2 y^4}}{z^3}$

41. $\log_a \dfrac{\sqrt{x-1}}{(x+1)^2}$

43. $\log_a \dfrac{x^3 (x+1)^{\frac{1}{6}}}{(x-1)^2}$

45. 1.2304

47. 4.5046

49. 2.3892

51. a. 1.255
b. −.352
c. 0.699

53. 1.4307

55. 1.4037

57.

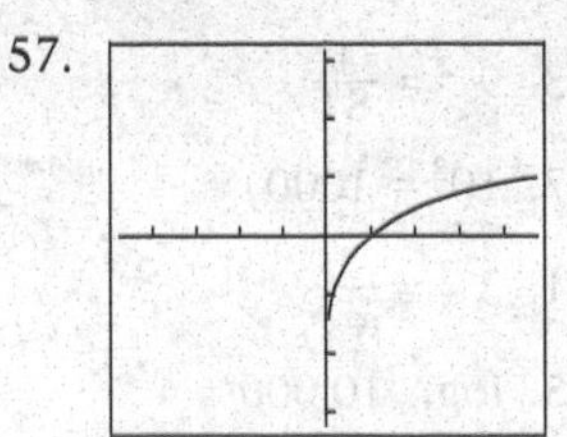

XSCL = 1
YSCL = 1
GRAPH $Y = \ln X \div \ln 5$

Exercise Set 6.5

1. $\dfrac{\log 18}{\log 5}$

3. $\dfrac{\log 7}{\log 2}+1$

5. $\dfrac{\log 46}{2\log 3}$

7. $\dfrac{\log 564}{2\log 5}+\dfrac{5}{2}$

9. $\dfrac{\log 3-\log 2}{\log 3-2\log 2}$

11. $\dfrac{-\log 15}{\log 2}$

13. $\dfrac{\log 12}{\log 4}+\dfrac{1}{2}$

15. $\ln 18$

17. $\dfrac{-3+\ln 30}{2}$

19. 500

21. $\frac{1}{2}$

23. 5

25. 3

27. 8

29. $-1+\sqrt{17}$

31. $\ln\left(y+\sqrt{y^2+1}\right)$

33. 37 years

35. 12.6 hours

37. 8.84 years

39. 27.5 days

41. 1.39 days

43. $y=e^{x^2+x}-1$

45. $y=\sqrt{xe^3+1}$

Chapter 6 Review Exercises

1. $f^{-1}(x)=3(x+2)$

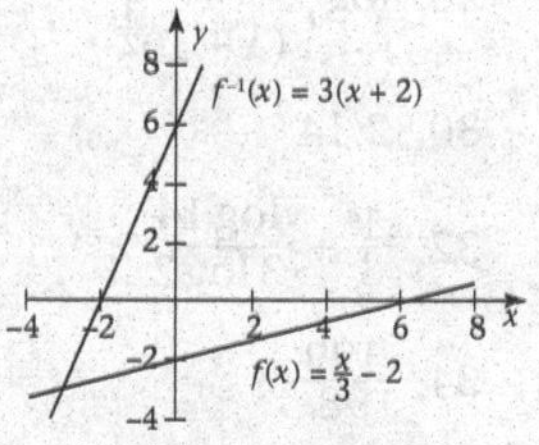

2. $f(x)$ is not one-to-one

3.

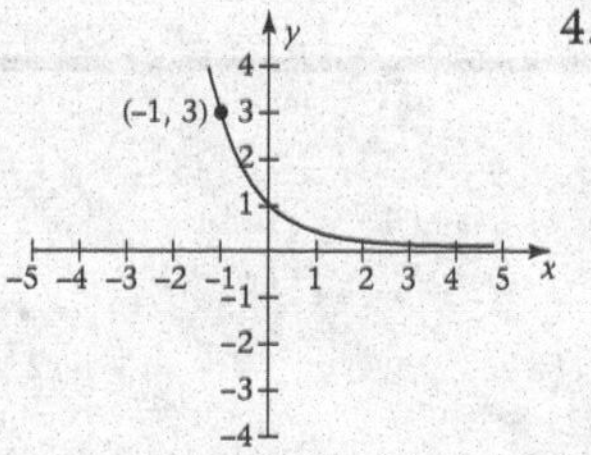

4. 3

5. 2

6. \$12,750.78

7. $\log_9 27=\frac{3}{2}$

8. $64^{\frac{1}{2}}=8$

9. $2^{-3}=\frac{1}{8}$

10. $\log_6 1=0$

11. 2

12. −2

13. e^{-4}

14. 26

15. 5

16. $-\frac{1}{3}$

17. −1

18. 3

19.

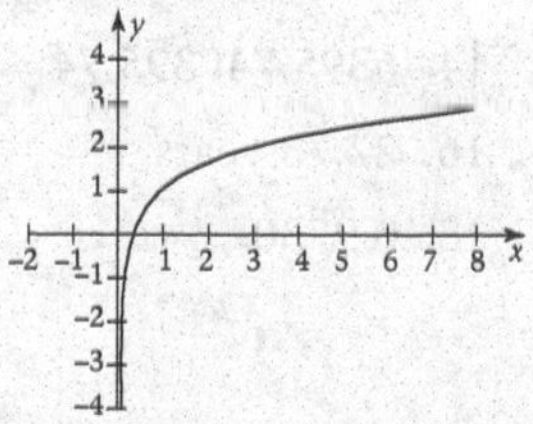

20. $\frac{1}{2}\log_a(x-1)-\log_a 2-\log_a x$

21. $\log_a x+2\log_a(2-x)-\frac{1}{2}\log_a(y+1)$

22. $4\ln(x+1)+2\ln(y-1)$

23. $\frac{2}{5}\log y+\frac{1}{5}\log z-\frac{1}{5}\log(z+3)$

24. 0.0067

25. 0.3466

26. −2.4709

27. 12,188.60

28. 12382.581

29. 2.3219

30. 1.2619

31. 32

32. 6.5809

33. 0.2691

34. Joe, by \$527.73

35. 3.12 hours

36. 2,566,558

37. 1,142.113

38. Bank A

39. a. 1.15

b.

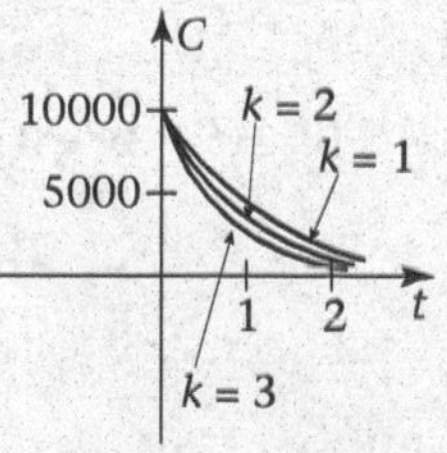

40. 40 decibels

41. 30 decibels

Chapter 6 Review Test

1. $f^{-1}(x) = -\frac{1}{4}x + \frac{1}{2}$

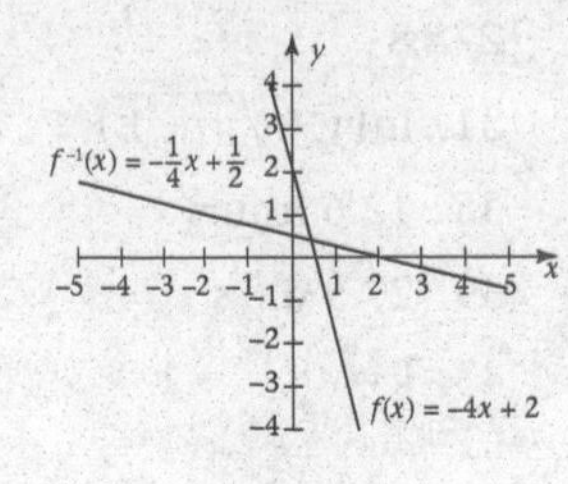

2. $f^{-1}(x) = -\frac{1}{x}$

$f(x) = f^{-1}(x) = \frac{1}{x}$

3.

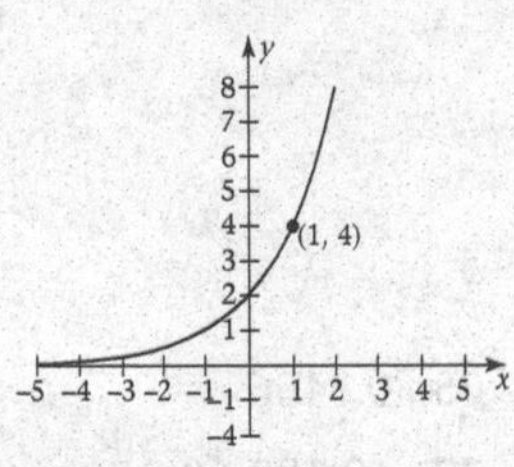

4. $-\frac{2}{3}$

5. $3^{-2} = \frac{1}{9}$

6. $\log_{16} 64 = \frac{3}{2}$

7. 3

8. −1

9. $\frac{5}{2}$

10. $\frac{1}{2}$

11. $3 \log_a x - 2 \log_a y - \log_a z$

12. $2 \log x + \frac{1}{2} \log (2y - 1) - 3 \log y$

13. 0.7

14. 0.45

15. $\log \frac{x^2}{(y+1)^3}$

16. $\log_a \left(\frac{x+3}{x-3}\right)^{\frac{2}{3}}$

17. 34.66 hours

18. $530.76

19. 200

20. 4

21. 1.15

22. 0.55

23. 0.39

24. −0.15

25. $\log_a \frac{x^{\frac{1}{3}}}{y^{\frac{1}{2}}}$

26. $\log (x^2 - x)^{\frac{4}{3}}$

27. $\ln \frac{3xy^2}{z}$

28. $\log_a \frac{(x+2)^2}{(x+1)^{\frac{3}{2}}}$

29. 1.67

30. 2.14

31. 11.55 hours

32. $\frac{1}{3} + \frac{\log 14}{3 \log 2}$

33. $50\sqrt{2}$

34. $\frac{199}{98}$

Cumulative Review Exercises

Chapters 4–6

1. a. 7 b. $2x^2 + 3x - 3$, 6

2. $\frac{1}{5}$

3. $(\sqrt{3}, 0), (-\sqrt{3}, 0)$

4. a. $\{x | x \neq \pm 3\}$

   b. vertical asymptote: $x = 3$
   horizontal asymptote: $y = 2$

   c.

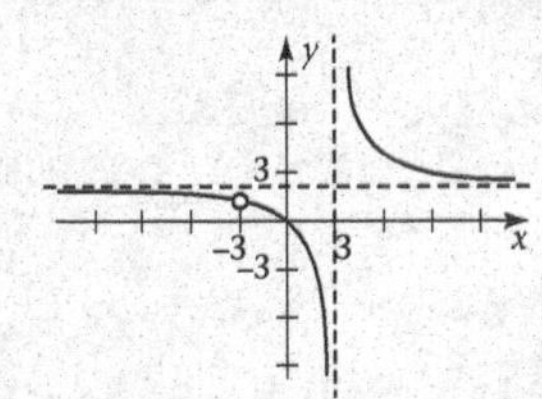

5. $\frac{9}{5}$

6. 3

7. 9

8. 7.9 years

9.

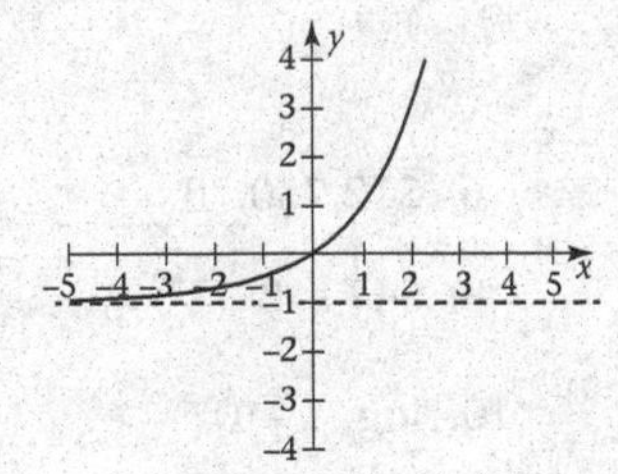

10. a. $s + 2t$ b. $u - 2s$

    c. $t - s$ d. $\frac{1}{2}(t + u - s)$

11. $960.79

12. $2 \ln x + \frac{1}{2} \ln(x + 1) - \ln(x - 1)$

13. 9.24%

14. 6395.74, 395.74

15. $16211.68

16. 27.73 years

17. Bond B

18. 0.9730

Chapter 7

Exercise Set 7.1

1. IV
3. I
5. II
7. I
9. III
11. II
13. II
15. III
17. I
19. $\frac{\pi}{6}$
21. $-\frac{5\pi}{6}$
23. $\frac{5\pi}{12}$
25. $-\frac{5\pi}{2}$
27. $\frac{3\pi}{4}$
29. $\frac{2\pi}{3}$
31. 0.251π
33. 0.685π
35. 45°
37. 270°
39. −90°
41. 240°
43. 450°
45. −300°
47. 98.55°
49. T
51. F
53. F
55. 50°
57. 20°
59. 85°
61. $\frac{2\pi}{5}$
63. 72°
65. $\frac{\pi}{4}$
67. $\frac{4}{7}$, 32.74°
69. 1.9 m
71. 6.81 feet; 775.33 rotations
73. 10 ribs
75. $\frac{6\pi}{5}$
77. 7.25°

Exercise Set 7.2

	$\sin\theta$	$\cos\theta$	$\tan\theta$	$\csc\theta$	$\sec\theta$	$\cot\theta$
1.	$\frac{3}{5}$	$\frac{4}{5}$	$\frac{3}{4}$	$\frac{5}{3}$	$\frac{5}{4}$	$\frac{4}{3}$
3.	$\frac{4}{5}$	$\frac{3}{5}$	$\frac{4}{3}$	$\frac{5}{4}$	$\frac{5}{3}$	$\frac{3}{4}$
5.	$\frac{2\sqrt{5}}{5}$	$\frac{\sqrt{5}}{5}$	2	$\frac{\sqrt{5}}{2}$	$\sqrt{5}$	$\frac{1}{2}$
7.	$\frac{h}{5}$	$\frac{\sqrt{25-h^2}}{5}$	$\frac{h\sqrt{25-h^2}}{25-h^2}$	$\frac{5}{h}$	$\frac{5\sqrt{25-h^2}}{25-h^2}$	$\frac{\sqrt{25-h^2}}{h}$

9. $5\sin\theta$
11. $6.5\cot\theta$
13. $3.7\csc\theta$
15. 62.23
17. 30.30
19. 33.91

Exercise Set 7.3

1. III
3. II
5. II
7. II
9. 0
11. $\frac{\pi}{7}$
13. $\frac{3\pi}{2}$
15. $\frac{5\pi}{6}$
17. π
19. $\frac{7\pi}{5}$
21.

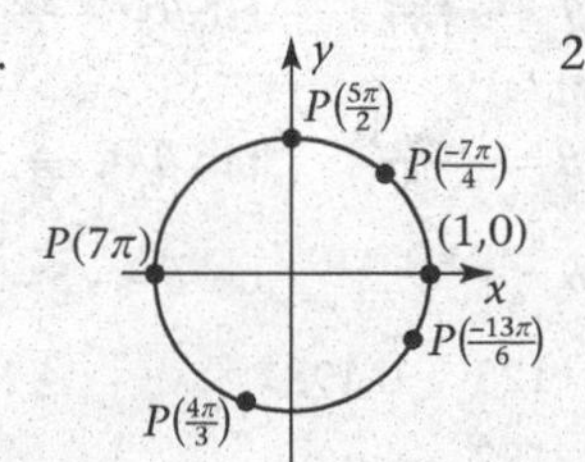

23.

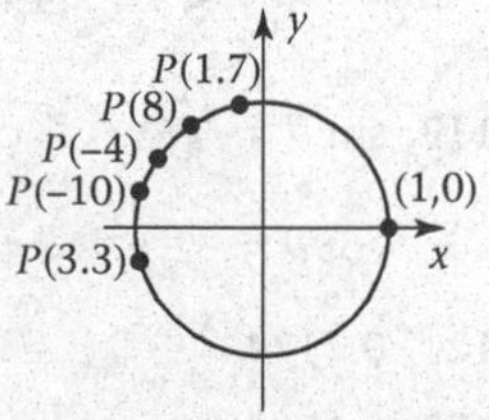

25. (−1, 0)
27. (0, −1)
29. (0, −1)
31. (1, 0)
33. (0, 1)
35. (0, 1)
37. $\sin t = \frac{4}{5}$, $\cos t = -\frac{3}{5}$
39. $\sin t = -\frac{1}{2}$, $\cos t = \frac{\sqrt{3}}{2}$
41. $\sin t = \frac{\sqrt{2}}{2}$, $\cos t = -\frac{\sqrt{2}}{2}$
43. $\sin t = -\frac{1}{4}$, $\cos t = \frac{\sqrt{15}}{4}$

	$\tan t$	$\cot t$	$\sec t$	$\csc t$
45.	$-\frac{4}{3}$	$-\frac{3}{4}$	$-\frac{5}{3}$	$\frac{5}{4}$
47.	$-\frac{\sqrt{3}}{3}$	$-\sqrt{3}$	$\frac{2\sqrt{3}}{3}$	−2
49.	−1	−1	$-\sqrt{2}$	$\sqrt{2}$
51.	$-\frac{\sqrt{15}}{15}$	$-\sqrt{15}$	$\frac{4\sqrt{15}}{15}$	−4

Exercise Set 7.4

1. 70°
3. 30°
5. $\frac{\pi}{5}$
7. 25°
9. 47°
11. $\frac{\pi}{7}$
13. $\sin t = -\frac{\sqrt{3}}{2}$, $\cos t = \frac{1}{2}$
15. $\sin t = 0$, $\cos t = -1$
17. $\sin t = -\frac{\sqrt{2}}{2}$, $\cos t = \frac{\sqrt{2}}{2}$
19. $\sin t = \frac{\sqrt{3}}{2}$, $\cos t = -\frac{1}{2}$
21. $\sin t = -\frac{\sqrt{3}}{2}$, $\cos t = \frac{1}{2}$
23. $\sin t = -\frac{\sqrt{2}}{2}$, $\cos t = -\frac{\sqrt{2}}{2}$
25. $\sin 135^\circ = \frac{\sqrt{2}}{2}$, $\cos 135^\circ = -\frac{\sqrt{2}}{2}$, $\tan 135^\circ = -1$, $\cot 135^\circ = -1$, $\sec 135^\circ = -\sqrt{2}$, $\csc 135^\circ = \sqrt{2}$
27. $\sin(-30^\circ) = -\frac{1}{2}$, $\cos(-30^\circ) = \frac{\sqrt{3}}{2}$, $\tan(-30^\circ) = -\frac{\sqrt{3}}{3}$, $\cot(-30^\circ) = -\sqrt{3}$, $\sec(-30^\circ) = \frac{2\sqrt{3}}{3}$, $\csc(-30^\circ) = -2$
29. $\sin\frac{\pi}{3} = \frac{\sqrt{3}}{2}$, $\cos\frac{\pi}{3} = \frac{1}{2}$, $\tan\frac{\pi}{3} = \sqrt{3}$, $\cot\frac{\pi}{3} = \frac{\sqrt{3}}{3}$, $\sec\frac{\pi}{3} = 2$, $\csc\frac{\pi}{3} = \frac{2\sqrt{3}}{3}$
31. $\sin\frac{\pi}{4} = \frac{\sqrt{2}}{2}$, $\cos\frac{\pi}{4} = \frac{\sqrt{2}}{2}$, $\tan\frac{\pi}{4} = 1$, $\cot\frac{\pi}{4} = 1$, $\sec\frac{\pi}{4} = \sqrt{2}$, $\csc\frac{\pi}{4} = \sqrt{2}$
33. $\sin\frac{5\pi}{6} = \frac{1}{2}$, $\cos\frac{5\pi}{6} = -\frac{\sqrt{3}}{2}$, $\tan\frac{5\pi}{6} = -\frac{\sqrt{3}}{3}$, $\cot\frac{5\pi}{6} = -\sqrt{3}$, $\sec\frac{5\pi}{6} = -\frac{2\sqrt{3}}{3}$, $\csc\frac{5\pi}{6} = 2$
35. $\sin\frac{3\pi}{2} = -1$, $\cos\frac{3\pi}{2} = 0$, $\tan\frac{3\pi}{2}$ is undefined, $\cot\frac{3\pi}{2} = 0$, $\sec\frac{3\pi}{2}$ is undefined, $\csc\frac{3\pi}{2} = -1$
37. $\sin\frac{3\pi}{4} = \frac{\sqrt{2}}{2}$, $\cos\frac{3\pi}{4} = -\frac{\sqrt{2}}{2}$, $\tan\frac{3\pi}{4} = -1$, $\cot\frac{3\pi}{4} = -1$, $\sec\frac{3\pi}{4} = -\sqrt{2}$, $\csc\frac{3\pi}{4} = \sqrt{2}$
39. $\sin\left(-\frac{5\pi}{4}\right) = \frac{\sqrt{2}}{2}$, $\cos\left(-\frac{5\pi}{4}\right) = \frac{-\sqrt{2}}{2}$, $\tan\left(-\frac{5\pi}{4}\right) = -1$, $\cot\left(-\frac{5\pi}{4}\right) = -1$, $\sec\left(-\frac{5\pi}{4}\right) = -\sqrt{2}$, $\csc\left(-\frac{5\pi}{4}\right) = \sqrt{2}$
41. $(-1, 0)$
43. $\left(\frac{\sqrt{2}}{2}, -\frac{\sqrt{2}}{2}\right)$
45. $\left(-\frac{\sqrt{2}}{2}, -\frac{\sqrt{2}}{2}\right)$
47. $\left(-\frac{1}{2}, -\frac{\sqrt{3}}{2}\right)$
49. $\left(-\frac{1}{2}, -\frac{\sqrt{3}}{2}\right)$
51. $\left(-\frac{\sqrt{3}}{2}, -\frac{1}{2}\right)$
53. $\left(-\frac{\sqrt{3}}{2}, -\frac{1}{2}\right)$
55. $\left(\frac{1}{2}, \frac{\sqrt{3}}{2}\right)$
57. $\pi, -\pi$
59. $\frac{3\pi}{4}, -\frac{5\pi}{4}$
61. $\frac{5\pi}{6}, -\frac{7\pi}{6}$
63. $\frac{5\pi}{3}, -\frac{\pi}{3}$
65. a. $\left(-\frac{3}{5}, -\frac{4}{5}\right)$ b. $\left(\frac{4}{5}, -\frac{3}{5}\right)$ c. $\left(\frac{3}{5}, -\frac{4}{5}\right)$ d. $\left(-\frac{3}{5}, \frac{4}{5}\right)$
69. II
71. III
73. III
75. III
77. II
79. III
81. III
83. $-\frac{3}{2}$
85. -1
87. $-\frac{\sqrt{2}}{2}$
89. $-\frac{\sqrt{3}}{2}$
91. $-\sqrt{3}$
93. $-\frac{\sqrt{3}}{2}$
95. a. $\left(-\frac{12}{13}, -\frac{5}{13}\right)$ b. $\left(\frac{5}{13}, -\frac{12}{13}\right)$ c. $\left(\frac{12}{13}, -\frac{5}{13}\right)$ d. $\left(-\frac{12}{13}, \frac{5}{13}\right)$
97. a. a b. $-b$
101. $-\frac{3}{4}$
103. $-\frac{12}{13}$
105. $-\frac{3}{5}$
107. $\frac{4}{5}$
109. $\sin\theta = \frac{12}{13}$, $\cos\theta = -\frac{5}{13}$, $\tan\theta = -\frac{12}{5}$, $\cot\theta = -\frac{5}{12}$, $\sec\theta = -\frac{13}{5}$, $\csc\theta = \frac{13}{12}$
111. $\sin\theta = -\frac{\sqrt{2}}{2}$, $\cos\theta = -\frac{\sqrt{2}}{2}$, $\tan\theta = 1$, $\cot\theta = 1$, $\sec\theta = -\sqrt{2}$, $\csc\theta = -\sqrt{2}$
113. $\sin\theta = \frac{3}{5}$, $\cos\theta = -\frac{4}{5}$, $\tan\theta = -\frac{3}{4}$, $\cot\theta = -\frac{4}{3}$, $\sec\theta = -\frac{5}{4}$, $\csc\theta = \frac{5}{3}$
115. $\sin\theta = -\frac{5}{13}$, $\cos\theta = \frac{12}{13}$, $\tan\theta = -\frac{5}{12}$, $\cot\theta = -\frac{12}{5}$, $\sec\theta = \frac{13}{12}$, $\csc\theta = -\frac{13}{5}$
117. $\sin\theta = -\frac{5}{13}$, $\cos\theta = -\frac{12}{13}$, $\tan\theta = \frac{5}{12}$, $\cot\theta = \frac{12}{5}$, $\sec\theta = -\frac{13}{12}$, $\csc\theta = -\frac{13}{5}$
119. $\sin\theta = \frac{\sqrt{5}}{5}$, $\cos\theta = -\frac{2\sqrt{5}}{5}$, $\tan\theta = -\frac{1}{2}$, $\cot\theta = -2$, $\sec\theta = -\frac{\sqrt{5}}{2}$, $\csc\theta = \sqrt{5}$
121. 0.7174
123. −0.1987
125. 0.1003
127. $-\sin t$
129. $\sin t$
131. $\frac{\cos t}{\tan t}$
133. $\sin^2 t$
135. $\tan^2 t$

Exercise Set 7.5

1. $\frac{\pi}{2}, \frac{3\pi}{2}$
3. $\frac{\pi}{4}, \frac{5\pi}{4}$
5. $\frac{\pi}{4}, \frac{3\pi}{4}$
7. $\frac{5\pi}{6}, \frac{7\pi}{6}$
9. $\frac{\pi}{3}, \frac{4\pi}{3}$
11. $\frac{\pi}{3}, \frac{2\pi}{3}$
13. $\frac{7\pi}{6}, \frac{11\pi}{6}$
15. no solution
17. $\frac{\pi}{6}, \frac{5\pi}{6}$
19. 0
21. $\frac{7\pi}{6}, \frac{11\pi}{6}$
23. $\frac{\pi}{4}, \frac{5\pi}{4}$
25. $\frac{3\pi}{4}, \frac{7\pi}{4}$
27. $\frac{\pi}{4}, \frac{7\pi}{4}$
29. $\frac{5\pi}{6}, \frac{11\pi}{6}$
31. $\frac{5\pi}{6}$
33. $\frac{2\pi}{3}$
35. $\frac{5\pi}{4}$
37. $\frac{3\pi}{4}$
39.

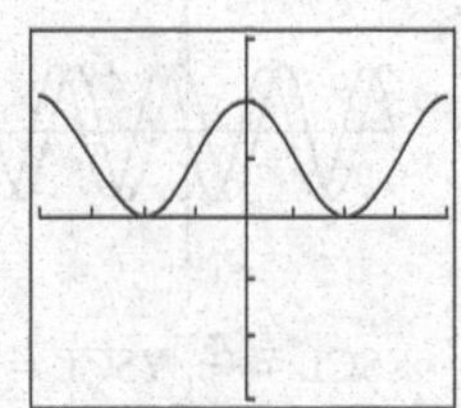

XMIN = -2π, XMAX = 2π, XSCL = $\frac{\pi}{2}$
YMIN = -3, YMAX = 3, YSCL = 1

41.

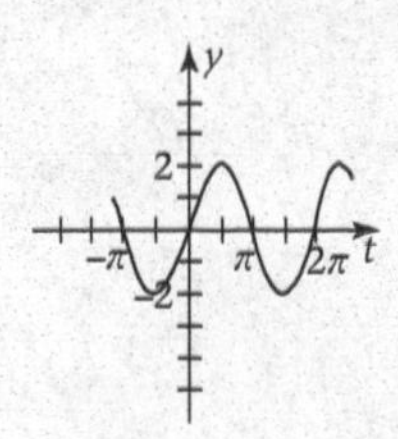

XMIN = -2π, XMAX = 2π, XSCL = $\frac{\pi}{2}$
YMIN = -3, YMAX = 3, YSCL = 1

43.

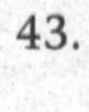

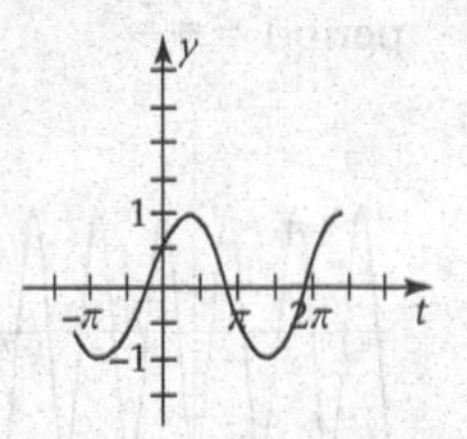

XMIN = -2π, XMAX = 2π, XSCL = $\frac{\pi}{2}$
YMIN = -3, YMAX = 3, YSCL = 1

45.

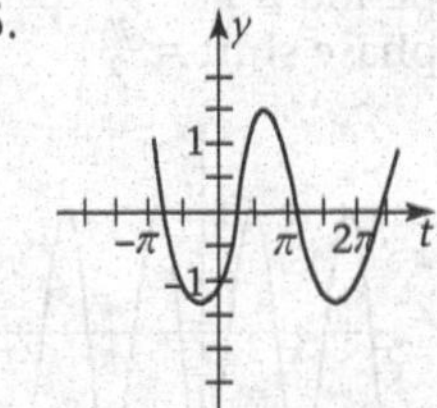

XMIN = -2π, XMAX = 2π, XSCL = $\frac{\pi}{2}$
YMIN = -3, YMAX = 3, YSCL = 1

47.

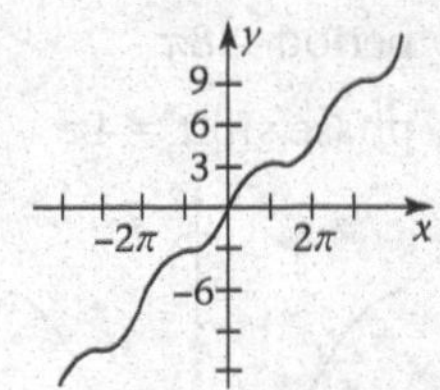

XMIN = -15, XMAX = 15, XSCL = 5
YMIN = -15, YMAX = 15, YSCL = 5

Exercise Set 7.6

1. amplitude = 3
period = 2π

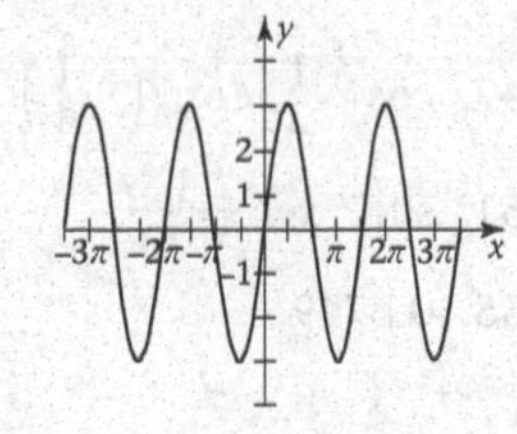

3. amplitude = 1
period = $\frac{\pi}{2}$

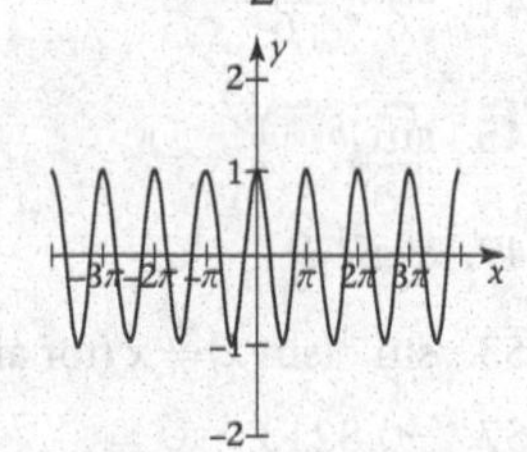

5. amplitude = 2
period = $\frac{\pi}{2}$

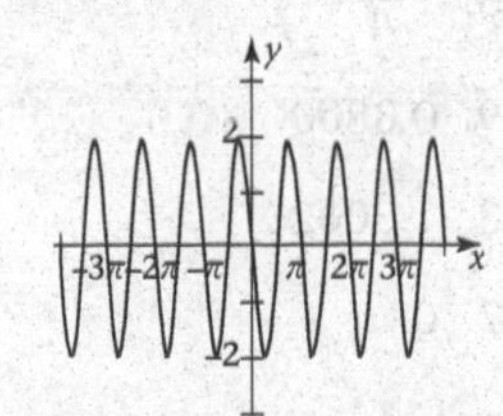

7. amplitude = 2
period = 6π

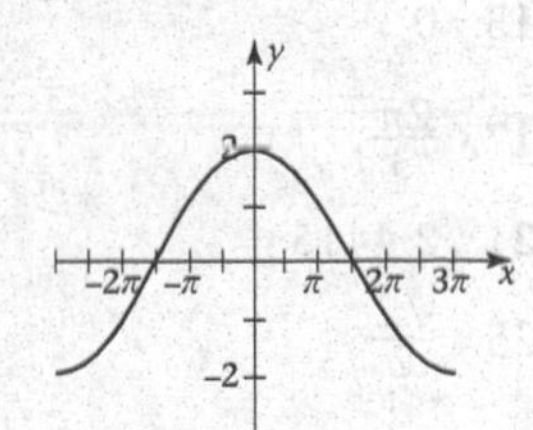

9. amplitude $= \frac{1}{4}$
period $= 8\pi$

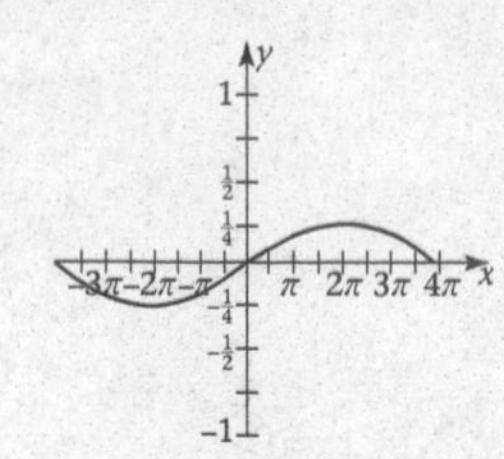

11. amplitude $= 3$
period $= \frac{2\pi}{3}$

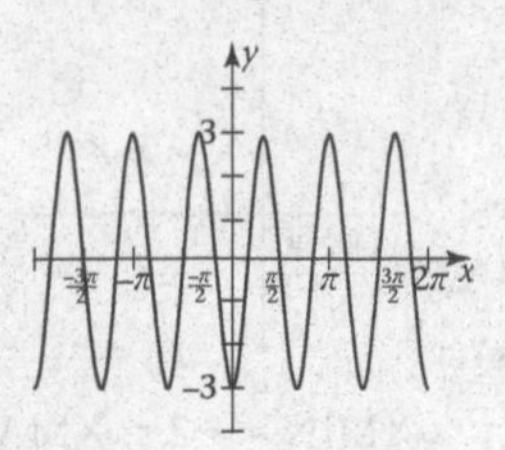

13. amplitude $= 2$
period $= 2\pi$
phase shift $= \pi$

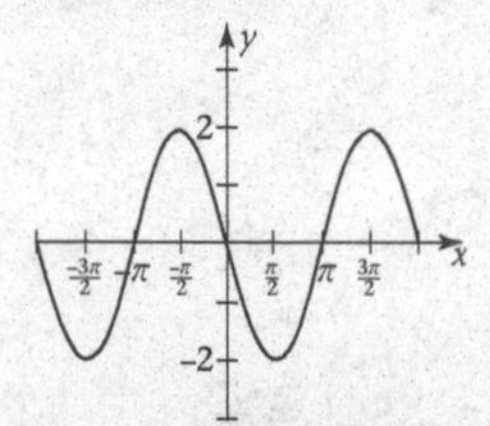

15. amplitude $= 3$
period $= \pi$
phase shift $= \frac{\pi}{2}$

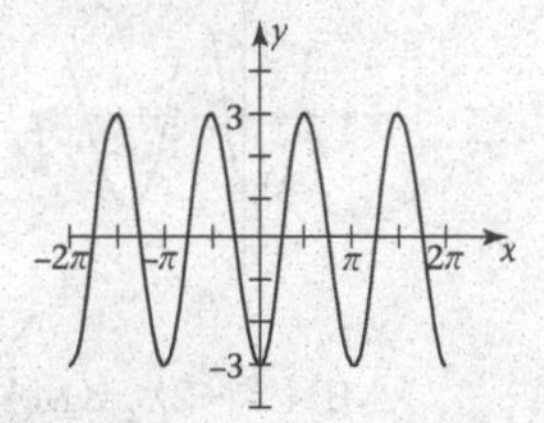

17. amplitude $= \frac{1}{3}$
period $= \frac{2\pi}{3}$
phase shift $= -\frac{\pi}{4}$

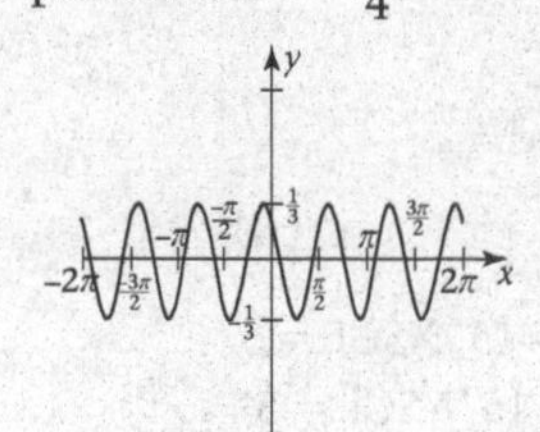

19. amplitude $= 2$
period $= 8\pi$
phase shift $= 4\pi$

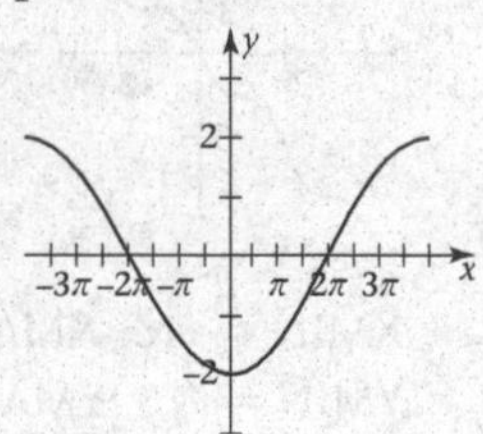

21. $f(x) = 3 \sin x$

23. $f(x) = 5 \sin 4x$

25. $y = 2 \sin 3x$

27. $y = \tan\left(x - \frac{\pi}{3}\right)$

29.

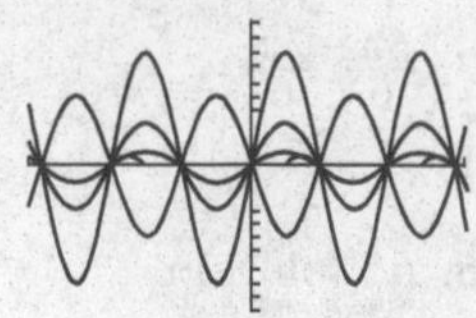

XSCL = 1, YSCL = 1

31. i.

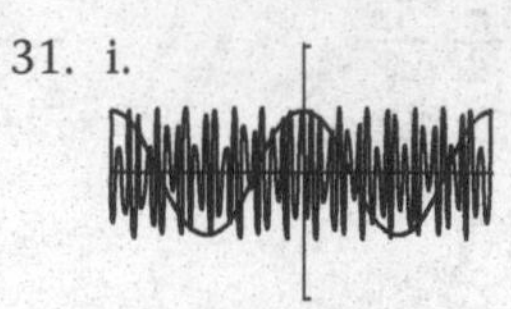

XSCL $= \frac{\pi}{2}$, YSCL = 1

ii.

XSCL $= \frac{\pi}{2}$, YSCL = 1

33.

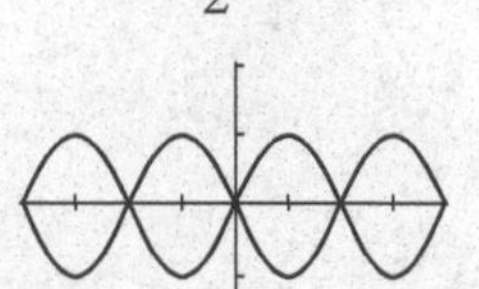

XSCL $= \frac{\pi}{2}$, YSCL = 1

35.

XSCL $= \frac{\pi}{2}$, YSCL = 1

37. $y = x - \frac{x^3}{3!} + \frac{x^5}{5!} - \frac{x^7}{7!} + \frac{x^9}{9!} - \frac{x^{11}}{11!} + \frac{x^{13}}{13!} - \frac{x^{15}}{15!} + \frac{x^{17}}{17!} - \frac{x^{19}}{19!} + \frac{x^{21}}{21!}$

Exercise Set 7.7

1. $-\frac{\pi}{6}$
3. $\frac{\pi}{3}$
5. $-\frac{\pi}{4}$
7. $\frac{5\pi}{6}$
9. $-\frac{\pi}{2}$
11. $\frac{\pi}{2}$
13. 0
15. $-\frac{\pi}{4}$
17. $\frac{2\pi}{3}$
19. 0.3800
21. 2.4415
23. 1.3002
25. $\frac{\sqrt{2}}{2}$
27. 0
29. $\frac{\pi}{4}$
31. $\frac{2\pi}{3}$
33. $-\frac{5}{12}$
35. $\frac{3}{5}$
37. $\frac{\pi}{3}$
39. $\frac{\pi}{6}$
41. $\frac{7\pi}{6}, \frac{11\pi}{6}$
43. $\frac{4\pi}{3}$
45. $\sin^{-1} \frac{-\sqrt{7}}{7}$
47. $\cos^{-1} \frac{1}{3}, \cos^{-1}\left(-\frac{1}{4}\right)$
49. $\sin^{-1} \frac{2}{3}$
51. $\sin^{-1} x \neq \frac{1}{\sin x}$
53. $\sin^{-1}(\sin x) \neq x$ (for all x)
55. 0.6749
57. −0.8213

Exercise Set 7.8

1. 36°52′12″
3. 53°7′
5. 60°57′
7. 7767 feet
9. 969.71 meters
11. 39.09°, 50.91°
13. 18.7 cm
15. 53 ft
17. 24.2 miles

Chapter 7 Review Exercises

1. $-\frac{\pi}{3}$
2. 270°
3. −75°
4. $\frac{\pi}{4}$
5. Yes
6. No
7. Yes
8. IV
9. II
10. II
11. I
12. 50°
13. 5°
14. 45°
15. 1.4 radians
16. $\frac{15}{4}$ centimeters
17. $\frac{\pi}{2}$
18. $\frac{\pi}{2}$
19. 0
20. $\frac{5\pi}{3}$
21. $\frac{5}{13}$
22. $\frac{4}{3}$
23. $\frac{\sqrt{65}}{7}$
24. 2
25. −1
26. $-\frac{1}{2}$
27. $\frac{\sqrt{2}}{2}$
28. $\left(-\frac{\sqrt{3}}{2}, -\frac{1}{2}\right)$
29. $\left(-\frac{1}{2}, -\frac{\sqrt{3}}{2}\right)$
30. $\left(-\frac{\sqrt{3}}{2}, \frac{1}{2}\right)$
31. $\left(\frac{\sqrt{2}}{2}, \frac{\sqrt{2}}{2}\right)$
32. $\left(\frac{\sqrt{3}}{2}, -\frac{1}{2}\right)$
33. $\frac{\sqrt{3}}{2}$
34. $-\sqrt{2}$
35. $-\frac{\sqrt{3}}{3}$
36. −2
37. $\left(-\frac{4}{5}, \frac{3}{5}\right)$
38. $\left(\frac{3}{5}, \frac{4}{5}\right)$
39. $\left(\frac{4}{5}, \frac{3}{5}\right)$
40. $\left(-\frac{3}{5}, -\frac{4}{5}\right)$
41. $\left(-\frac{4}{5}, -\frac{3}{5}\right)$
42. $\frac{5\pi}{4}$
43. $\frac{11\pi}{6}$
44. $\frac{\pi}{3}$
45. $\frac{2\pi}{3}$
46. IV
47. III
48. $-\frac{3}{4}$
49. $-\frac{5}{3}$
50. $-\frac{12}{5}$
51. $\frac{13}{12}$
52. tan t
53. (tan t)(sec t)
54. −1.8343
55. 0.4247
56. 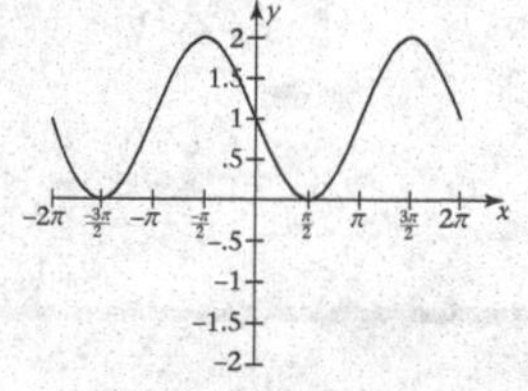
57. 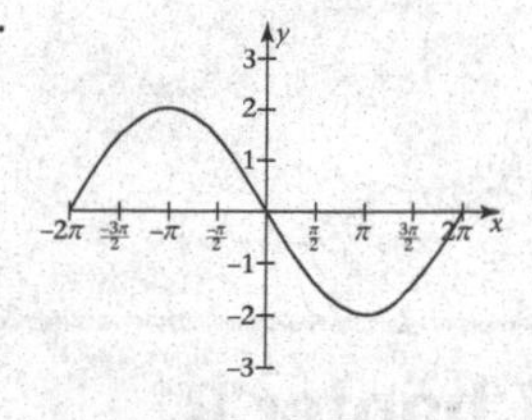
58. amplitude = 1
period = π
phase shift = $\frac{\pi}{2}$
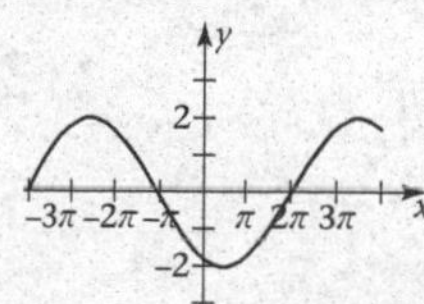
59. amplitude = 4
period = 2π
phase shift = $\frac{\pi}{2}$
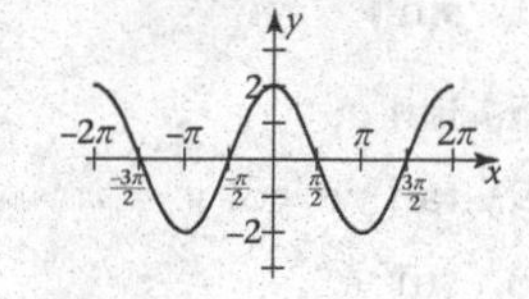
60. amplitude = 2
period = 6π
phase shift = $-\pi$
61. $-\frac{\pi}{6}$
62. 0
63. 5
64. $\cos^{-1}\left(\frac{-2\sqrt{5}}{5}\right)$
65. 39.81°
66. 14.56
67. 25.38
68. 16.55
69. 5.44 meters
70. 68.2°
71. 36°
72. $y = 2 + \cos 3x$
73. $y = \cos\left(x + \frac{3\pi}{4}\right)$

Chapter 7 Review Test

1. 300°
2. $-\frac{10\pi}{9}$
3. $\frac{5\pi}{12}$
4. 335°
5. 45°
6. 20°
7. $\frac{\pi}{4}$
8. 0.8
9. $\frac{\pi}{3}$
10. 0
11. $\frac{7}{5}$
12. 3
13. −1
14. −1
15. 2
16. $\left(-\frac{\sqrt{3}}{2}, \frac{1}{2}\right)$

17. $\left(\frac{1}{2}, -\frac{\sqrt{3}}{2}\right)$
18. $\left(\frac{5}{13}, -\frac{12}{13}\right)$
19. $\left(\frac{12}{13}, \frac{5}{13}\right)$
20. $\left(-\frac{5}{13}, -\frac{12}{13}\right)$
21. $\frac{1}{2}$
22. $-\frac{2\sqrt{3}}{3}$
23. $\frac{5\pi}{4}$
24. $\frac{7\pi}{4}$
25. $-\frac{5}{13}$
26. $-\frac{5}{4}$
27. $\frac{\cos x - \sin x}{\cos x}$
28. -0.5973
29. -0.2509
30. 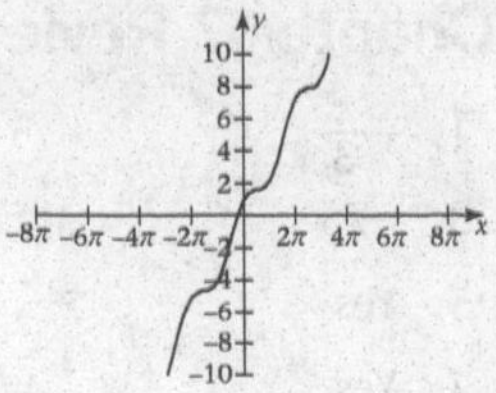
31. amplitude: 2; period: 2π; phase shift: π
32. amplitude: 2; period: 4π; phase shift: π
33. $-\frac{\pi}{3}$
34. $\frac{1}{2}$
35. $\arctan\frac{2}{3}$, $\arctan\frac{3}{2}$
36. 56.44°
37. 23.58
38. 53.13°
39. 137.64 meters

Chapter 8

Exercise Set 8.1

1. $\sin y$
3. $\frac{1+\sin v}{\cos v}$
5. $\tan \alpha$
7. $2 - \tan^2 x$
9. $\tan y + \cot y$
11. 1
13. $\sin^2 \theta$
15. $\sec y$
17. $\sin^2 w$
19. 2
21. $\frac{\sin^4 v + 1}{\sec^2 v}$
23. $1 - \cos \alpha$
25. $\frac{1-\sin t}{\cos t}$
27. 1
29. $\sin y \cos y$
31. $-\cot t$
33. $\csc x$
35. $\frac{\sec x}{\csc x}$
37. $\frac{1-\sin t}{1+\sin t}$
39. $\tan^2 w$
41. $\frac{\sin y - \cos y}{\sin y + \cos y}$
43. $1 - \cos y$
45. $2\sec^2 x$
47. $-1 \neq 1$
49. $2 \neq 1$

Exercise Set 8.2

1. $\frac{\sqrt{2}}{2} \neq -\frac{\sqrt{2}}{2}$
3. $-1 = 1$
5. $-3.73 \neq 2.73$
7. $\frac{\sqrt{6}-\sqrt{2}}{4}$
9. $\frac{\sqrt{2}+\sqrt{6}}{4}$
11. $-\frac{\sqrt{3}}{2}$
13. $\sqrt{3}$
15. $\frac{-\sqrt{2}+\sqrt{6}}{4}$
17. $\frac{-\sqrt{6}+\sqrt{2}}{4}$
19. $-\frac{1}{2}$
21. $2-\sqrt{3}$
23. $\cos 43°$
25. $\cot\frac{\pi}{3}$
27. $\sin\frac{\pi}{6}$
29. $-\frac{4}{5}$
31. -7
33. -2.3
35. $-\frac{16}{65}$
37. $\frac{2}{29}$

Exercise Set 8.3

1. $\frac{7}{25}$
3. $-\frac{\sqrt{3}}{2}$
5. $-\frac{240}{161}$
7. $-\frac{24}{25}$
9. $-\frac{161}{289}$
11. 0.1022
13. $\frac{\sqrt{2-\sqrt{3}}}{2}$
15. $\sqrt{2}-1$
17. $2\sqrt{2+\sqrt{3}}$
19. $-\frac{2\sqrt{5}}{5}$
21. $\frac{\sqrt{6}}{3}$
23. -2
25. $-\frac{\sqrt{6}}{3}$
47. $\frac{24}{25}$
49. $\frac{120}{119}$

Exercise Set 8.4

1. $\sin 6\alpha + \sin 4\alpha$
3. $\frac{\cos 5x - \cos x}{2}$
5. $-(\cos 7\theta + \cos 3\theta)$
7. $\frac{\cos 2\alpha + \cos 2\beta}{2}$
9. $\frac{-2-\sqrt{2}}{4}$
11. $\frac{\sqrt{3}}{4}$
13. $2 \sin 3x \cos 2x$
15. $2 \cos 4\theta \cos 2\theta$
17. $2 \sin \alpha \cos \beta$
19. $2 \cos(5x) \sin(2x)$
21. $\frac{\sqrt{6}}{2}$
23. $-\sqrt{2}$
35. $\frac{\sin(a+b)x + \sin(a-b)x}{2}$

Exercise Set 8.5

1. $\frac{\pi}{6}, \frac{5\pi}{6}$; 30°, 150°
3. π; 180°
5. $\frac{\pi}{6}, \frac{5\pi}{6}, \frac{7\pi}{6}, \frac{11\pi}{6}$; 30°, 150°, 210°, 330°
7. $\frac{\pi}{6}, \frac{5\pi}{6}, \frac{7\pi}{6}, \frac{11\pi}{6}$; 30°, 150°, 210°, 330°
9. $0, \frac{\pi}{6}, \frac{5\pi}{6}\pi$; 0°, 30°, 150°, 180°
11. $0, \frac{\pi}{3}, \frac{5\pi}{3}$; 0°, 60°, 300°
13. $\frac{\pi}{10}, \frac{\pi}{2}, \frac{9\pi}{10}, \frac{13\pi}{10}, \frac{17\pi}{10}$; 18°, 90°, 162°, 234°, 306°
15. $\frac{\pi}{3}, \frac{5\pi}{3}$; 60°, 300°
17. $\frac{\pi}{12}, \frac{5\pi}{12}, \frac{13\pi}{12}, \frac{17\pi}{12}$; 15°, 75°, 195°, 255°
19. 0; 0°
21. $\frac{\pi}{6} \pm \pi n, \frac{5\pi}{6} \pm n$
23. $\frac{\pi}{3} + \pi n, \frac{2\pi}{3} + \pi n$
25. $\frac{\pi}{6} + \pi n; \frac{5\pi}{6} + \pi n$
27. $\frac{\pi n}{4}$
29. $\frac{\pi}{12} + \frac{\pi n}{2}, \frac{5\pi}{12} + \frac{\pi n}{2}$
31. $\frac{\pi}{2} + \pi n$
33. $\frac{\pi}{2} + 2\pi n, \frac{7\pi}{6} + 2\pi n, \frac{11\pi}{6} + 2\pi n$
35. $\pi n, \frac{\pi}{4} + \pi n$
37. $\frac{\pi}{6} + 2\pi n, \frac{5\pi}{6} + 2\pi n$
39. 0.83 radians, 2.31 radians, 3.71 radians, 5.71 radians
41. 2.18 radians, 2.91 radians, 5.32 radians, 6.05 radians
43. $x \approx 0.739085$
45. $x \approx 1.4284922$; $x \approx 3.8011932$; $x \approx 4.9447768$

Chapter 8 Review Exercises

4. $\frac{\sqrt{2}+\sqrt{6}}{4}$
5. $-\frac{\sqrt{2}}{2}$
6. $-2-\sqrt{3}$
7. $\frac{\sqrt{2}+\sqrt{6}}{4}$
8. sec 75°
9. sin 67°
10. $\cos \frac{3\pi}{8}$
11. $\cot \frac{3\pi}{14}$
12. $\frac{5}{13}$
13. $\frac{10(4\sqrt{3}+3)}{39}$
14. $\frac{3}{4}$
15. 70
16. $\frac{16}{65}$
17. $-\frac{7}{25}$
18. $-\frac{24}{25}$
19. $\frac{24}{25}$
20. $-\frac{\sqrt{3}}{2}$
21. $\frac{120}{169}$
22. $-\frac{\sqrt{10}}{10}$
23. $-\frac{1}{3}$
24. $-\frac{\sqrt{30}}{6}$
25. $\frac{\sqrt{2+\sqrt{3}}}{2}$
26. $\frac{\sqrt{2-\sqrt{2}}}{2}$
27. $\frac{\sqrt{\sqrt{2}+1}}{\sqrt{\sqrt{2}-1}}$
31. $\frac{\cos \alpha - \cos 2\alpha}{2}$
32. $-2 \sin 2x \sin x$
33. $\frac{1}{4}$
34. 0
35. $\frac{\pi}{4}, \frac{3\pi}{4}, \frac{5\pi}{4}, \frac{7\pi}{4}$
36. $0, \frac{\pi}{2}, \pi, \frac{3\pi}{2}$
37. $0, \frac{\pi}{3}, \pi, \frac{5\pi}{3}$
38. 90° + 180°n
39. 45° + 60°n
40. 30° + 90°n, 60° + 90°n
41. $x \approx 3.1830868$

Chapter 8 Review Test

2. $\frac{1}{2}$
3. $-2+\sqrt{3}$
4. cos 43°
5. $\frac{3}{5}$
6. $-\frac{81}{76}$
7. $-\frac{119}{169}$
8. $\frac{7}{25}$
9. $-\frac{\sqrt{2-\sqrt{3}}}{2}$
10. $\frac{\sqrt{2-\sqrt{3}}}{\sqrt{2+\sqrt{3}}}$
12. $2 \sin \frac{5x}{2} \cos \frac{x}{2}$
13. 0
14. $\frac{\pi}{3}, \frac{2\pi}{3}, \frac{4\pi}{3}, \frac{5\pi}{3}$
15. 45° + 90°n

Chapter 9

Exercise Set 9.1

1. 29.06
3. 7.24
5. 15.71
7. none
9. 10.91
11. 8.34, 1.56
13. 98.88 meters
15. 681.78 meters
17. 40.49 miles
19. 18.73 meters
21. 114.95 centimeters

Exercise Set 9.2

1. 41.2°
3. 14.38
5. 17.84
7. 62.48°
9. 90°
11. 82.1°
13. 29.1 miles, S 29°E
15. 32.83 miles
17. 68.48

Exercise Set 9.3

1. $\sqrt{13}$
3. $\sqrt{2}$
5. $2\sqrt{10}$
7. $3\sqrt{2}\left(\cos\frac{7\pi}{4}+i\sin\frac{7\pi}{4}\right)$
9. $2\left(\cos\frac{11\pi}{6}+i\sin\frac{11\pi}{6}\right)$
11. $\sqrt{2}\left(\cos\frac{3\pi}{2}+i\sin\frac{3\pi}{2}\right)$
13. $4(\cos\pi+i\sin\pi)$
15. -4
17. $-1+i$
19. $-5i$
21. $6(\cos 360^\circ+i\sin 360^\circ)$
23. $2\left(\cos\frac{9\pi}{20}+i\sin\frac{9\pi}{20}\right)$
25. $\sqrt{2}\left(\cos\frac{7\pi}{4}+i\sin\frac{7\pi}{4}\right)$, $2\left(\cos\frac{\pi}{2}+i\sin\frac{\pi}{2}\right)$, $2+2i$
27. $4\left(\cos\frac{2\pi}{3}+i\sin\frac{2\pi}{3}\right)$, $3\sqrt{2}\left(\cos\frac{\pi}{4}+i\sin\frac{\pi}{4}\right)$, $(-6-6\sqrt{3})+(6\sqrt{3}-6)i$
29. $5(\cos 0+i\sin 0)$, $2\sqrt{2}\left(\cos\frac{5\pi}{4}+i\sin\frac{5\pi}{4}\right)$, $-10-10i$
31. $\frac{5\sqrt{2}}{4}\left(\cos\frac{3\pi}{4}+i\sin\frac{3\pi}{4}\right)$, $-\frac{5}{4}+\frac{5}{4}i$
33. $\frac{3\sqrt{2}}{4}\left(\cos\frac{19\pi}{12}+i\sin\frac{19\pi}{12}\right)$, $\frac{3}{8}(-1+\sqrt{3})+\frac{3}{8}(-1-\sqrt{3})i$
35. $\frac{2}{\sqrt{2}}\left(\cos\frac{3\pi}{4}+i\sin\frac{3\pi}{4}\right)$, $-1+i$
37. $0+512i$
39. $16-16i$
41. $-8+8i$
43. $\sqrt{2}+\sqrt{2}i, -\sqrt{2}+\sqrt{2}i, -\sqrt{2}-\sqrt{2}i, \sqrt{2}-\sqrt{2}i$
45. $\sqrt{2}\left(\cos\frac{5\pi}{6}+i\sin\frac{5\pi}{6}\right)$, $\sqrt{2}\left(\cos\frac{11\pi}{6}+i\sin\frac{11\pi}{6}\right)$
47. $-2, 1\pm\sqrt{3}i$
49. $\pm 2i, \pm 2$

Exercise Set 9.4

1.

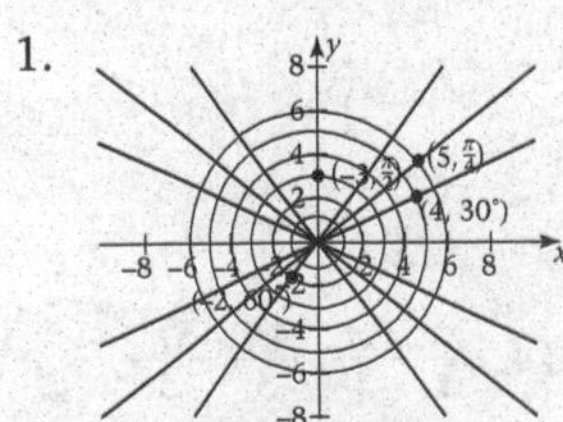

3. a. (6, 495°), (6, −225°) b. (−2, 480°), (2, 300°)
   c. $\left(4, \frac{17\pi}{6}\right), \left(-4, \frac{11\pi}{6}\right)$ d. $\left(-4, \frac{\pi}{4}\right), \left(4, -\frac{3\pi}{4}\right)$
5. a. (2, 210°) b. (4, 120°)
   c. $\left(3, \frac{5\pi}{3}\right)$ d. $\left(1, -\frac{\pi}{6}\right)$
7. a. yes b. no
   c. yes d. no
9. a. $\left(\frac{5\sqrt{3}}{2}, -\frac{5}{2}\right)$ b. $(0, -2)$
   c. $(2\sqrt{3}, 2)$ d. $\left(\frac{3}{2}, \frac{3\sqrt{3}}{2}\right)$
11. a. $\left(2\sqrt{2}, \frac{3\pi}{4}\right)$ b. $\left(2, \frac{5\pi}{3}\right)$
    c. $\left(2, \frac{\pi}{6}\right)$ d. $\left(4\sqrt{2}, \frac{5\pi}{4}\right)$

13. 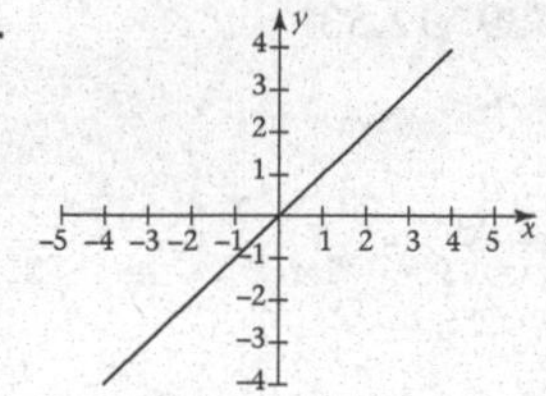

15. 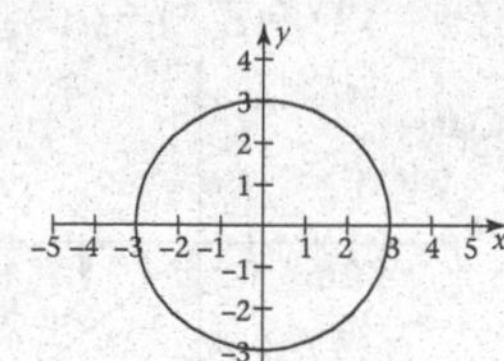

17. 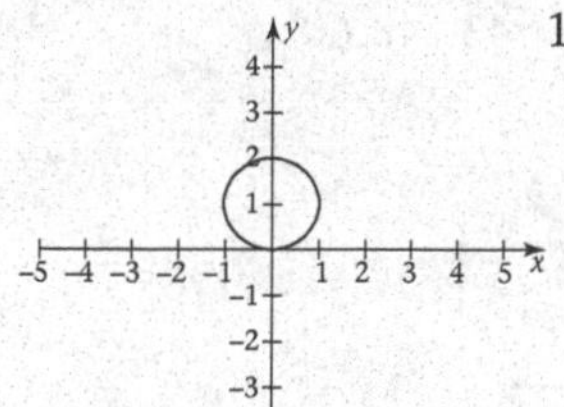

19. 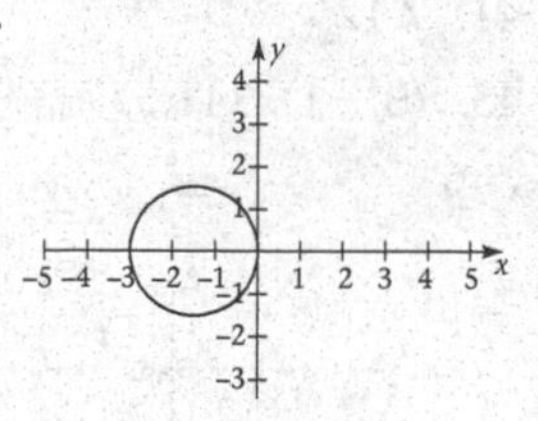

21. 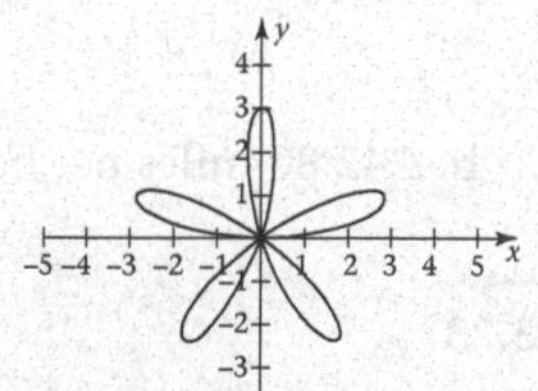

23. 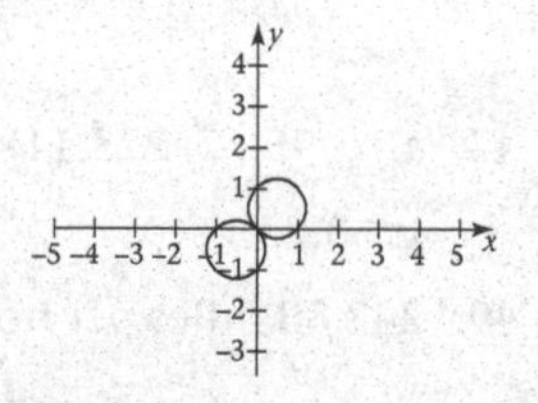

25. 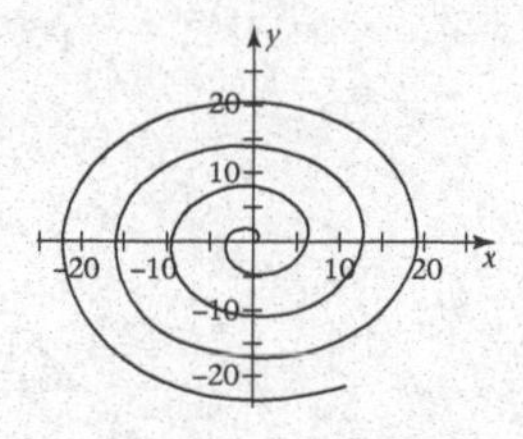

27. $x^2 + y^2 = 16$

29. $x^2 + y^2 + 2y = 0$

31. $x = 2$

33. $r = 5$

35. $r = -5 \sec \theta$

37. $\tan \theta = 3$

39. a. 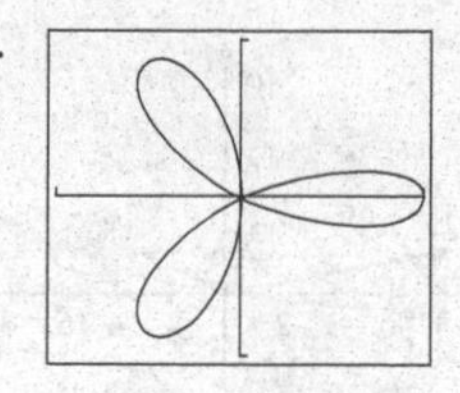

XMIN = −1, XMAX = 1, XSCL = 1
YMIN = −1, YMAX = 1, YSCL = 1

b. 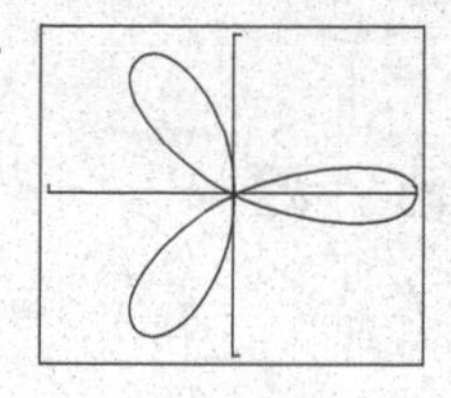

XMIN = −1, XMAX = 1, XSCL = 1
YMIN = −1, YMAX = 1, YSCL = 1
The graphs do not differ.

41. a. 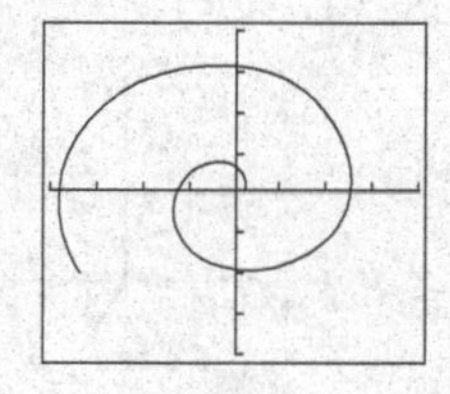

XMIN = −20, XMAX = 20, XSCL = 5
YMIN = −20, YMAX = 20, YSCL = 5

b. 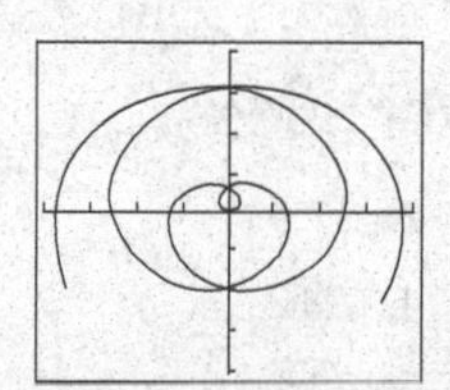

XMIN = −20, XMAX = 20, XSCL = 5
YMIN = −20, YMAX = 20, YSCL = 5
Half of the graph is missing when negative values of θ are not used.

Exercise Set 9.5

1. 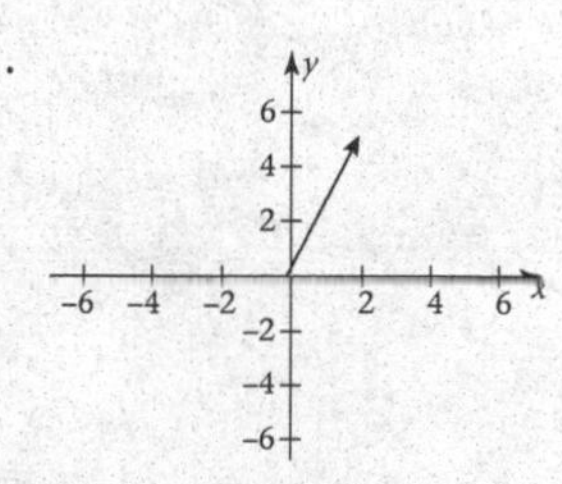

3. 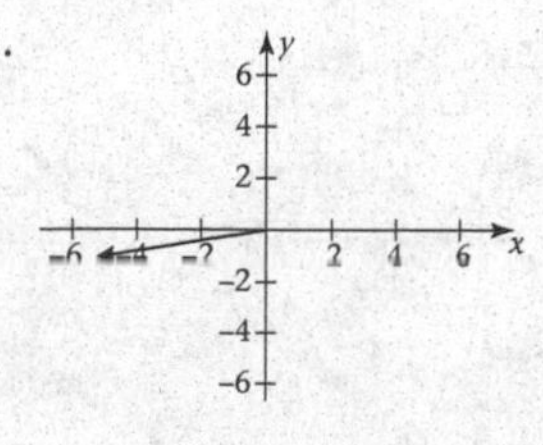

5. 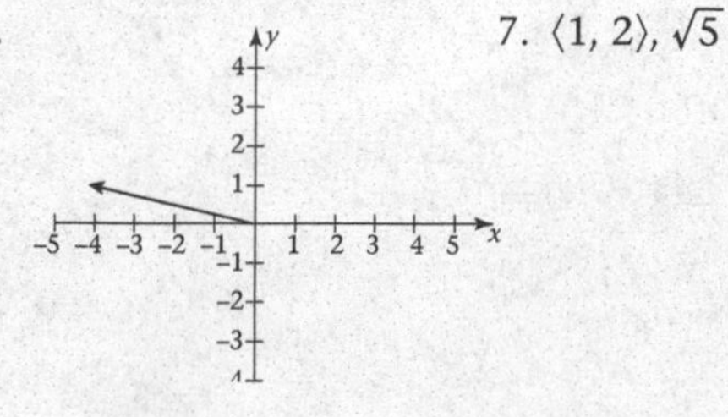

7. $\langle 1, 2 \rangle, \sqrt{5}$

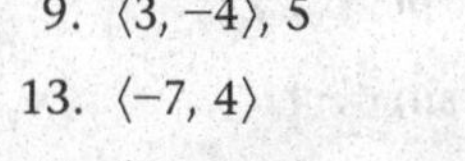

9. $\langle 3, -4 \rangle, 5$

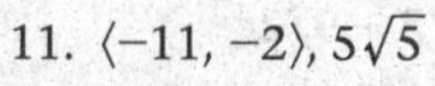

11. $\langle -11, -2 \rangle, 5\sqrt{5}$

13. $\langle -7, 4 \rangle$

15. $\langle -8, 6 \rangle$

17. $\langle 17, -18 \rangle$

19. $\langle -7. 6 \rangle$

21. 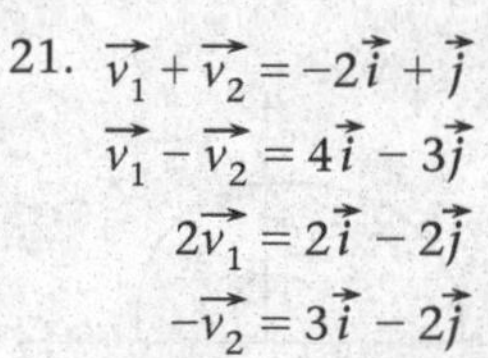

$\vec{v_1} + \vec{v_2} = -2\vec{i} + \vec{j}$

$\vec{v_1} - \vec{v_2} = 4\vec{i} - 3\vec{j}$

$2\vec{v_1} = 2\vec{i} - 2\vec{j}$

$-\vec{v_2} = 3\vec{i} - 2\vec{j}$

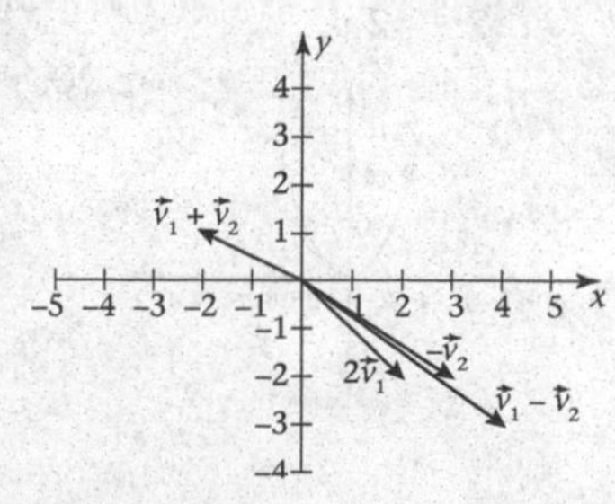

23. $\vec{v_1} + \vec{v_2} = -7\vec{i} + \vec{j}$
$\vec{v_1} - \vec{v_2} = -\vec{i} + 3\vec{j}$
$2\vec{v_1} = -8\vec{j} + 2\vec{j}$
$-\vec{v_2} = 3\vec{i} + 2\vec{j}$

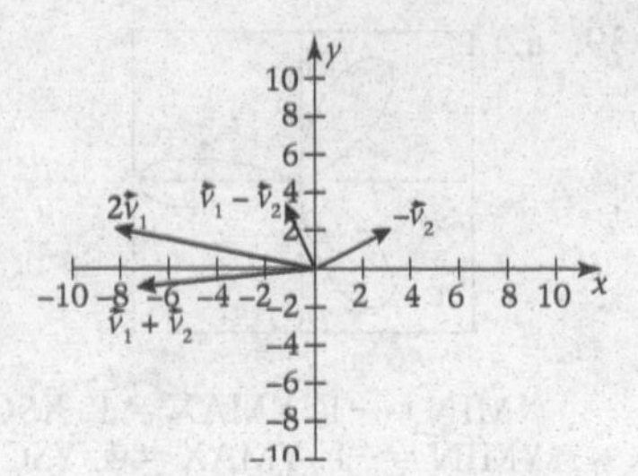

25. $\vec{U_1} = \langle 1, 0 \rangle, \vec{U_2} = \langle -1, 0 \rangle$

27. $\vec{U_1} = \left\langle \frac{1}{\sqrt{5}}, \frac{2}{\sqrt{5}} \right\rangle, \vec{U_2} = \left\langle -\frac{1}{\sqrt{5}}, -\frac{2}{\sqrt{5}} \right\rangle$

29. $\vec{U_1} = \left\langle -\frac{3}{\sqrt{13}}, \frac{2}{\sqrt{13}} \right\rangle, \vec{U_2} = \left\langle \frac{3}{\sqrt{13}}, -\frac{2}{\sqrt{13}} \right\rangle$

31. $\vec{U_1} = \left\langle -\frac{3}{\sqrt{10}}, -\frac{1}{\sqrt{10}} \right\rangle, \vec{U_2} = \left\langle \frac{3}{\sqrt{10}}, \frac{1}{\sqrt{10}} \right\rangle$

33. $\left\langle \frac{3}{2}, \frac{3\sqrt{3}}{2} \right\rangle$

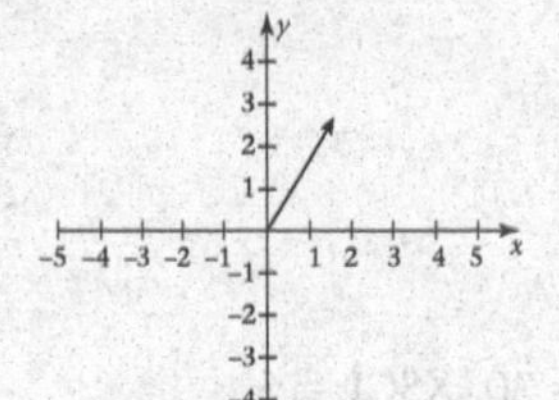

35. $\langle 1, -\sqrt{3} \rangle$

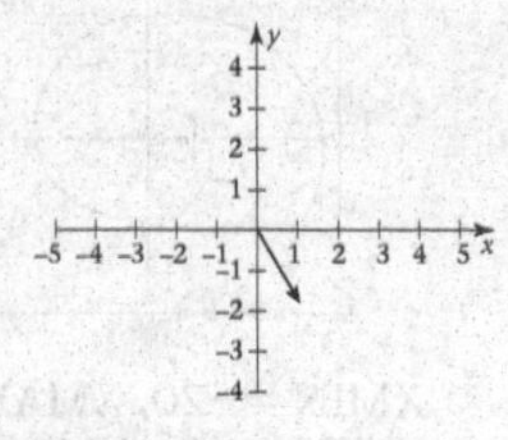

37. $\langle 4\sqrt{2}, -4\sqrt{2} \rangle$

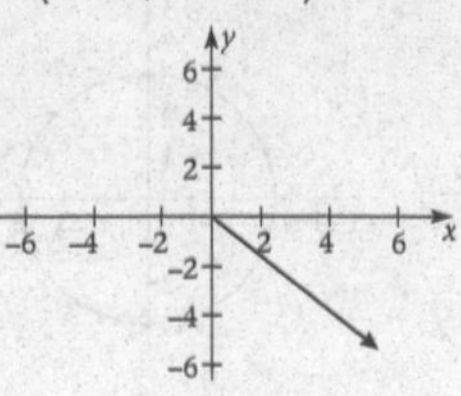

39. 12.53°

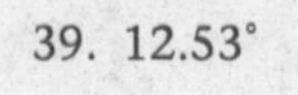

41. 7.12°

43. 43.60°

45. $\langle 3, -1 \rangle$, 341.57°

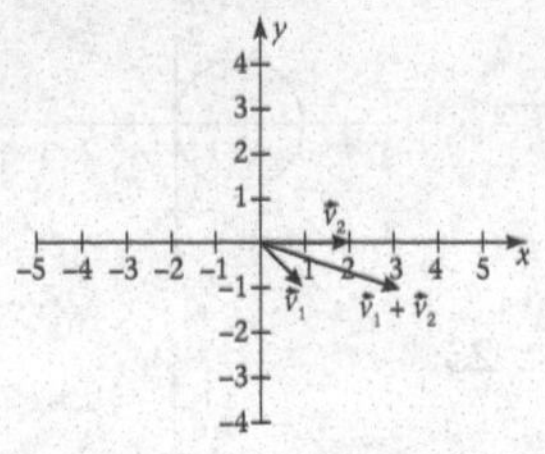

47. a. $\langle -95.77, 223.11 \rangle$ b. 242.80 miles per hour
c. 3.23°

49. 242.80 miles per hour 3.23°

Chapter 9 Review Exercises

1. 51.75°
2. 37.91°
3. 14.07
4. 7.12
5. $\sqrt{5}$
6. $\sqrt{13}$
7. $\sqrt{41}$
8. $3\sqrt{2}\left(\cos\frac{3\pi}{4} + i\sin\frac{3\pi}{4}\right)$
9. $2i$
10. $1 - i$
11. $2(\cos \pi + i \sin \pi)$
12. $24(\cos 37° + i \sin 37°)$
13. $\frac{5}{3}(\cos 21° + i \sin 21°)$
14. $2(\cos 90° + i \sin 90°)$
15. $-972 + 972i$
16. $0 - 8i$
17. $3\left(\cos\frac{\pi}{2} + i\sin\frac{\pi}{2}\right), 3\left(\cos\frac{3\pi}{2} + i\sin\frac{3\pi}{2}\right)$
18. $1, -\frac{1}{2} \pm \frac{\sqrt{3}}{2}i$
19.

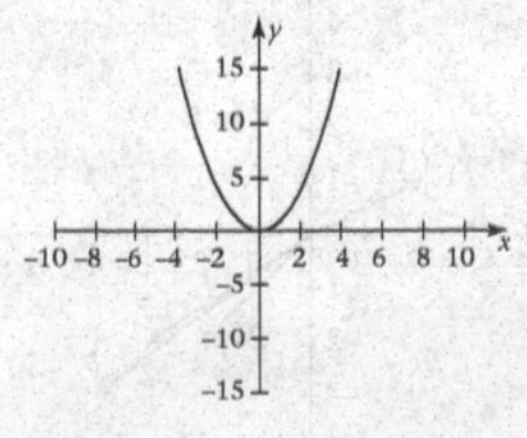

20.

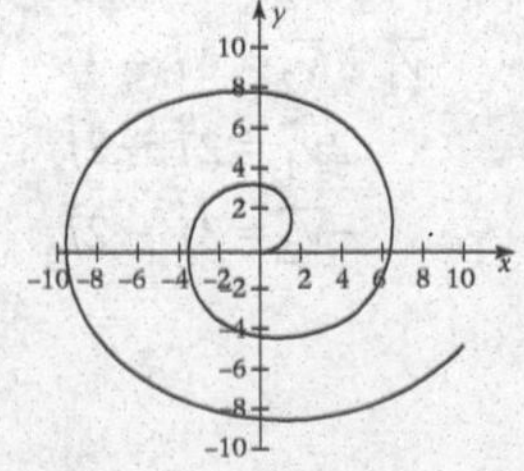

21. $y = 3$
22. $r^2 = \frac{9}{\cos^2\theta + 9\sin^2\theta}$

23. a.

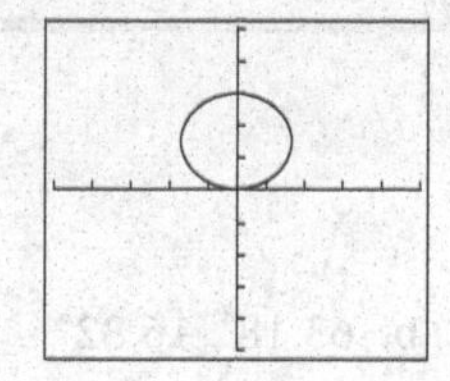

XMIN = −5, XMAX = 5, XSCL = 1
YMIN = −5, YMAX = 5, YSCL = 1

b.

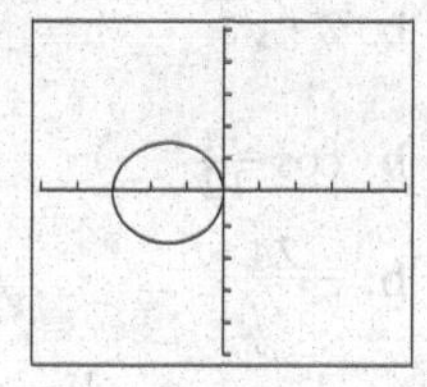

XMIN = −5, XMAX = 5, XSCL = 1
YMIN = −5, YMAX = 5, YSCL = 1

c.

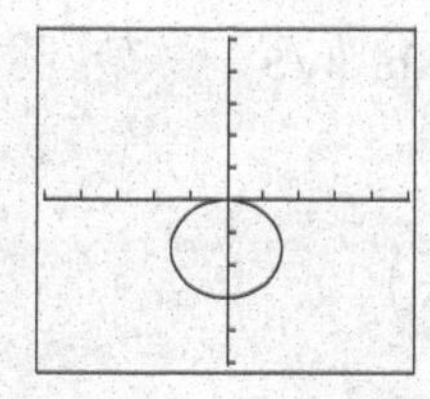

XMIN = −5, XMAX = 5, XSCL = 1
YMIN = −5, YMAX = 5, YSCL = 1

d.

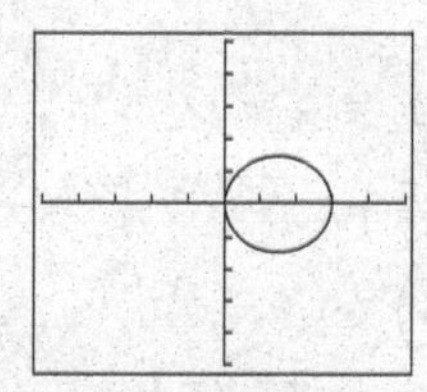

XMIN = −5, XMAX = 5, XSCL = 1
YMIN = −5, YMAX = 5, YSCL = 1

24. a.

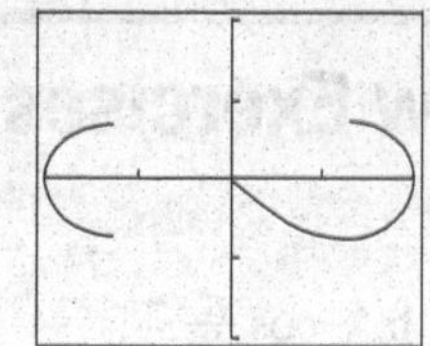

XMIN = −2, XMAX = 2, XSCL = 1
YMIN = −2, YMAX = 2, YSCL = 1

b.

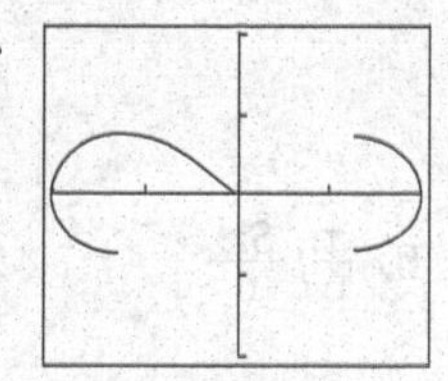

XMIN = −2, XMAX = 2, XSCL = 1
YMIN = −2, YMAX = 2, YSCL = 1

25. $-13\vec{i} + 2\vec{j}$

26. $\sqrt{13}$, 236.31°

27. 150.26°

Chapter 9 Review Test

1. 22.33°
2. 14.53
3. 48°63′
4. 5.76
5. 5
6. $2\sqrt{2}$
7. $5(\cos 86° + i \sin 86°)$
8. $\frac{1}{2}(\cos 77° + i \sin 77°)$
9. $5\left(\cos\frac{13\pi}{40} + i\sin\frac{13\pi}{40}\right)$
10. $2\left(\cos\frac{137\pi}{70} + i\sin\frac{137\pi}{70}\right)$
11. $-\frac{1}{1250} + \frac{\sqrt{3}}{1250}i$
12. $-64 + 0i$
13. $\cos\frac{5\pi}{3} + i\sin\frac{5\pi}{3}$
14. $3(\cos 240° + i \sin 240°)$
15.

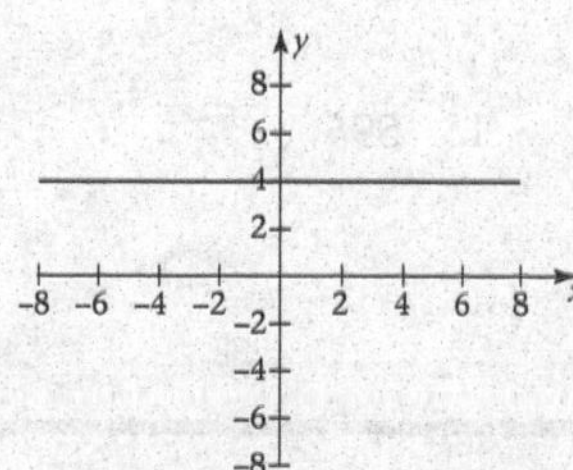

16. $y^2 + 4x - 4 = 0$
17. $r^2 = \frac{2}{2 + \sin 2\theta}$
18. 125.84°

Cumulative Review Exercises

Chapters 7–9

1. a. $\sin\frac{\pi}{5}$ b. $-\cos\frac{\pi}{3}$
   c. $\tan\frac{3\pi}{7}$
2. a. 2 b. $-\sqrt{3}$
   c. $-\frac{1}{2}$
3. $f(x) = 3\sin\left(2x+\frac{\pi}{2}\right)$
4. a. $\frac{3\pi}{4}, \frac{5\pi}{4}$ b. $\frac{\pi}{6}, \frac{5\pi}{6}$
   c. $\frac{2\pi}{3}, \frac{5\pi}{3}$
5. a. $\frac{4}{3}$ b. $\frac{13}{12}$
   c. π
6. a. $\frac{4\pi}{3}$ b. $\frac{5\pi}{3}$
   c. none
8.

	(a)	(b)
amplitude	1	3
period	4π	π
phase shift	$-\pi$	$\frac{\pi}{4}$

9. a. $x = \sin^{-1}\left(\frac{\sqrt{6}}{3}\right)$ or $x = \sin^{-1}\left(-\frac{\sqrt{6}}{3}\right)$
   b. no solution or $x = \cos^{-1}(-1)$
10. 1.29, 0.58
11. a. 85° b. $\frac{\pi}{3}$
    c. 60°
12. a. $\frac{\pi}{36}$ b. −30°
    c. $\left(\frac{360}{\pi}\right)^{\circ}$
13. a. $\frac{\sqrt{34}}{5}$ b. $\frac{5\sqrt{34}}{34}$
    c. $\frac{5}{3}$
14. $\frac{240}{\pi}$ feet 15. 59°
16. 56.3°
17. a. $\gamma = 28.96°$ b. 63.18°, 16.82°
19. a. $\frac{\sqrt{3}}{2}\cos\theta+\frac{1}{2}\sin\theta$ b. $-\frac{1-\tan\theta}{1+\tan\theta}$
20. a. $\frac{\sqrt{2+\sqrt{2}}}{2}$ b. $2+\sqrt{3}$
21. a. tan 28° b. $\cos\frac{5\pi}{12}$
22. a. $\frac{\sqrt{10}}{10}$ b. $-\frac{24}{7}$
23. 2 cos 30° cos 10° 24. $\frac{1}{2}\sin 5t-\frac{1}{2}\sin t$
25. $\frac{\pi}{6}, \frac{5\pi}{6}, \frac{3\pi}{2}$
26. a. $\sqrt{13}$ b. $3\sqrt{5}$
27. a. $3\left(\cos\frac{3\pi}{2}+i\sin\frac{3\pi}{2}\right)$
    b. $4\sqrt{2}\left(\cos\frac{7\pi}{4}+i\sin\frac{7\pi}{4}\right)$
28. $\frac{5\sqrt{3}}{2}+\frac{5}{2}i$
29. $z^2+4=0$
    $z^2=-4$
    $-4 = 4(\cos\pi + i\sin\pi)$
    $u_0=\sqrt{4}\left(\cos\frac{\pi}{2}+i\sin\frac{\pi}{2}\right)$
    $=2\left(\cos\frac{\pi}{2}+i\sin\frac{\pi}{2}\right)$
    $u_1=\sqrt{4}\left(\cos\frac{\pi+2\pi}{2}+i\sin\frac{\pi+2\pi}{2}\right)$
    $=2\left(\cos\frac{3\pi}{2}+i\sin\frac{3\pi}{2}\right)$
30. [graph]

31. hyperbola: $x^2 - y^2 = 1$
32. 46.91° 33. ⟨−0.64, −0.77⟩

Chapter 10

Exercise Set 10.1

1. $x=2, y=-1$ 3. no solution
5. $x=1, y=-4$
7. $x=3, y=2; x=3, y=-2$
9. no solution 11. $x=2, y=-1$
13. $x=3, y=2; x=\frac{1}{5}, y=-\frac{18}{5}$
15. $x=1, y=1; x=\frac{9}{16}, y=-\frac{3}{4}$
17. $x=1, y=2; x=\frac{13}{5}, y=-\frac{6}{5}$
19. $x=\frac{-1+\sqrt{5}}{2}, y=\frac{1+\sqrt{5}}{2}; x=\frac{-1-\sqrt{5}}{2}, y=\frac{1-\sqrt{5}}{2}$
21. $x=3, y=-1$ 23. no solution
25. $x=3, y=2; x=3, y=-2$ 27. no solution

29. $x = 3, y = 2; x = -3, y = 2; x = 3, y = -2; x = -3, y = -2$

31. I

33. I

35. I

37. C; all points on the line $3x - y = 18$

39. C; $x = 1, y = -1; x = \frac{5}{2}, y = \frac{1}{2}$

41. I

Exercise Set 10.2

1. 25 nickels, 15 dimes
3. color: \$11.00; black and white: \$4.00
5. \$40,000 in bond A, \$20,000 in bond B
7. 10 rolls of 12 inches, 4 rolls of 15 inches
9. 20 lbs of \$7.60/lb coffee and 4 lbs of \$10.00/lb coffee
11. speed of bicycle: $\frac{105}{8}$ miles per hour; wind speed: $\frac{15}{8}$ miles per hour
13. 34
15. 34 lbs of raisins and 16 lbs of nuts
17. \$60,000 in type A, \$120,000 in type B
19. 5 units Epiline I, 4 units Epiline II
21. a. $R = 125x$

    b.

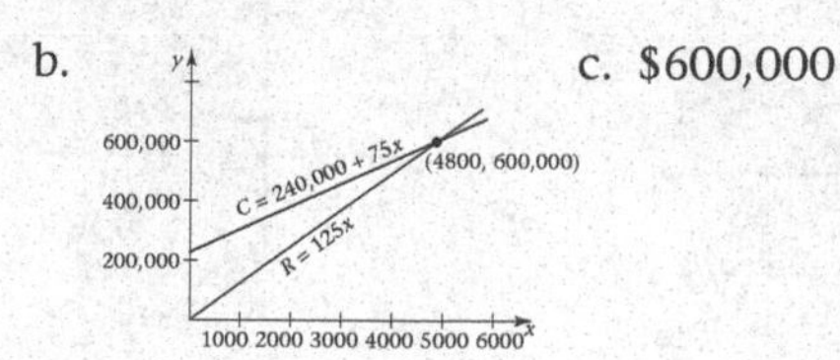

    c. \$600,000

23. a. $p = \$2000$

    b. 12,000 scooters
25. 4, 5
27. 22 nickels, 12 quarters
29. 1.2 lbs of nuts and 0.8 lbs of raisins
31. 125 type-A, 75 type-B
33. 6m × 8m
35. $y = 2x$
37. $y = \frac{2}{3}x$
39. $y = 2x$

Exercise Set 10.3

1. $x = 2, y = -1, z = -2$
3. $x = 1, y = \frac{2}{3}, z = -\frac{2}{3}$
5. inconsistent
7. $x = 1, y = 2, z = 2$
9. $x = 1, y = 1, z = 0$
11. $x = 1, y = \frac{27}{2}, z = -\frac{5}{2}$
13. inconsistent
15. inconsistent
17. $x = -15, y = 5, z = 20$
19. 2 units of A, 3 units of B, 3 units of C
21. three 40-inch sets, eight 46-inch sets, five 52-inch sets

Exercise Set 10.4

1. $\frac{3}{x+2} - \frac{1}{x-3}$
3. $\frac{3}{3x-1} - \frac{1}{2x-1}$
5. $-\frac{2}{x} + \frac{2}{x-1} + \frac{1}{x+1}$
7. $\frac{2}{x} - \frac{1}{x^2} - \frac{2}{x+2}$
9. $\frac{\frac{1}{2}}{x-1} + \frac{\frac{1}{2}}{x+1} - \frac{2}{(x+1)^2}$
11. $\frac{\frac{1}{4}}{x} - \frac{\frac{1}{4}x+2}{x^2+4}$
13. $\frac{2x-1}{x^2+3} + \frac{3-5x}{(x^2+3)^2}$
15. $\frac{\frac{1}{3}}{x+1} + \frac{\frac{1}{3}-\frac{1}{3}x}{x^2-3x-1}$
17. $-\frac{1}{x+1} + \frac{2x-2}{x^2+2} + \frac{x-1}{(x^2+2)^2}$
19. $x - \frac{2}{x+1} + \frac{2x}{x^2+1}$

Exercise Set 10.5

1.

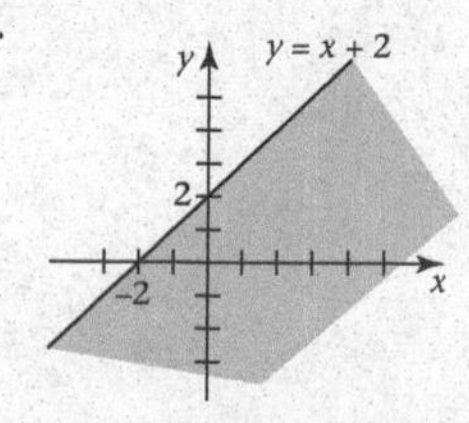

3.

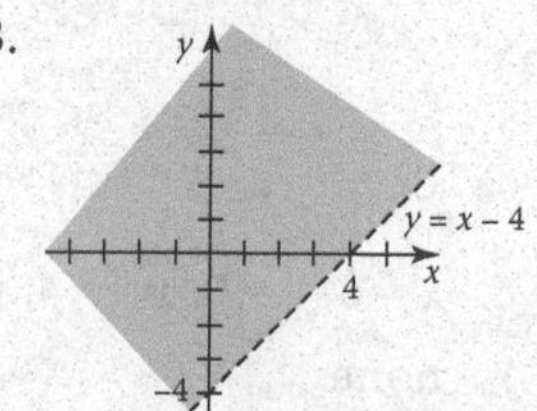

5.

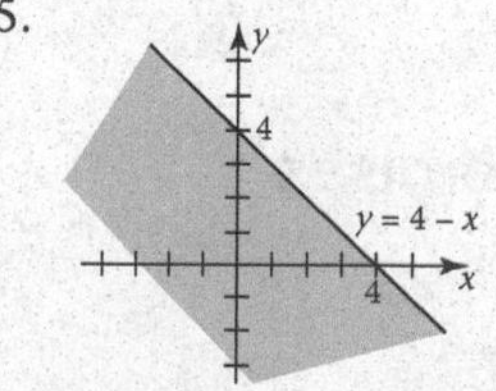

7.

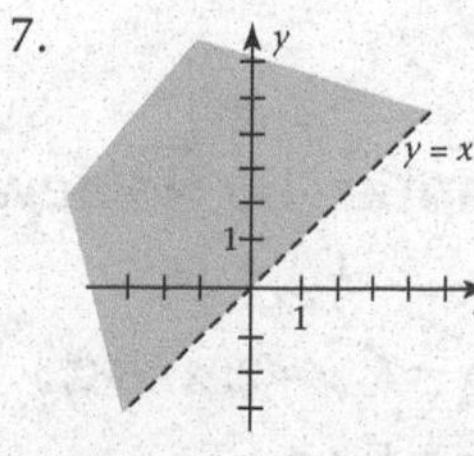

9.

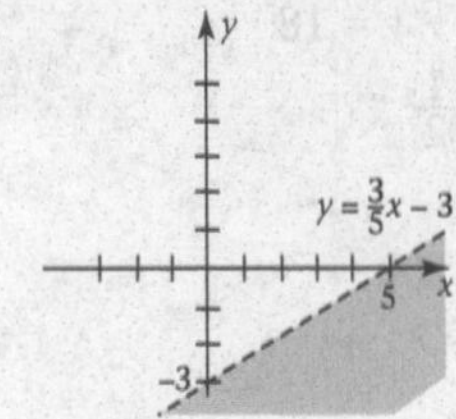

11.

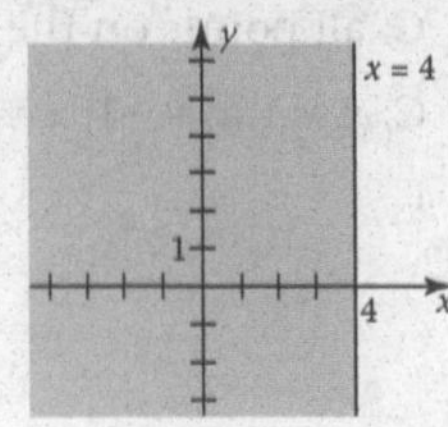

13.

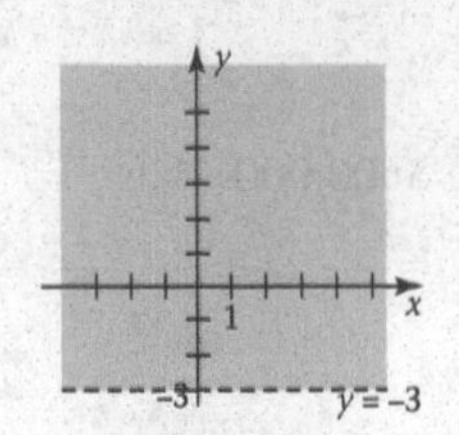

15.

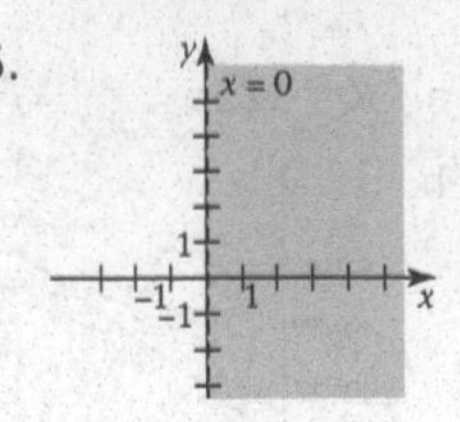

17.

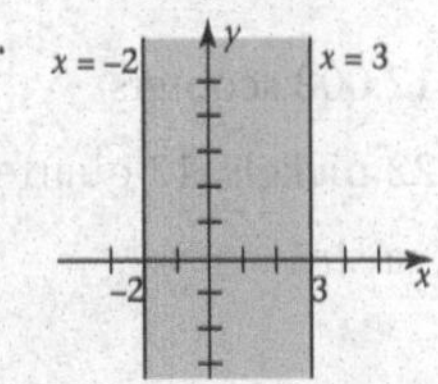

19. $2x + 5y \leq 15;\ x \geq 0;\ y \geq 0$

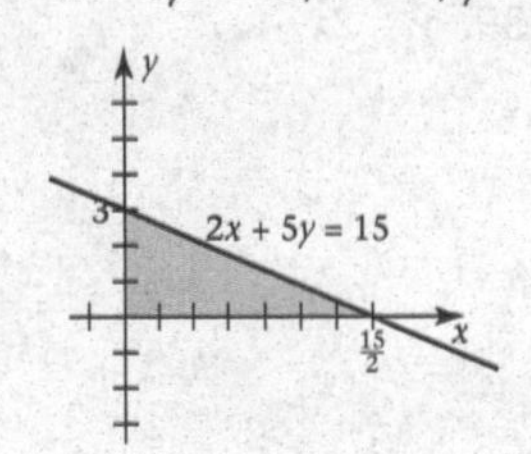

21.

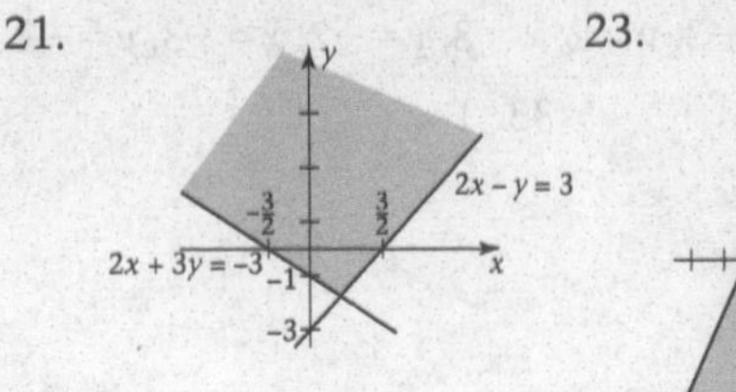

23.

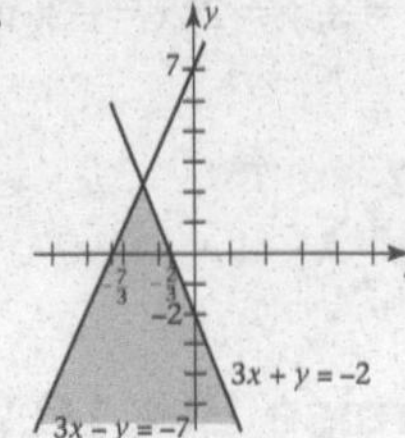

25.

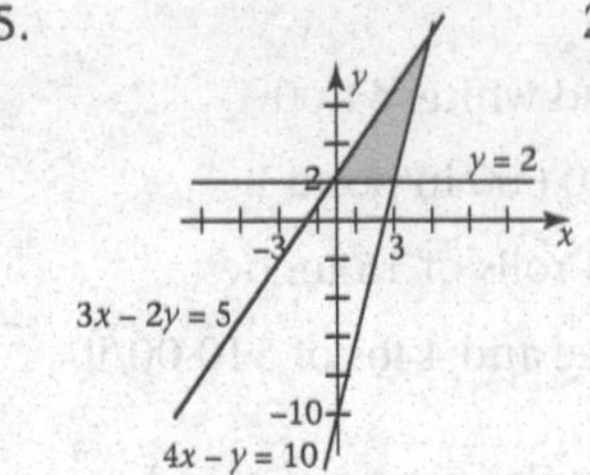

27.

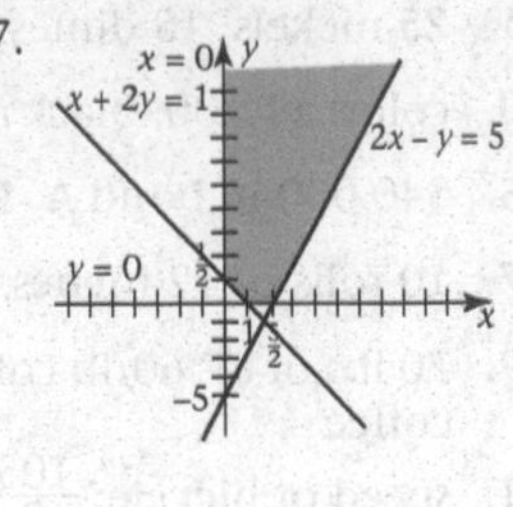

29. no solution

31.

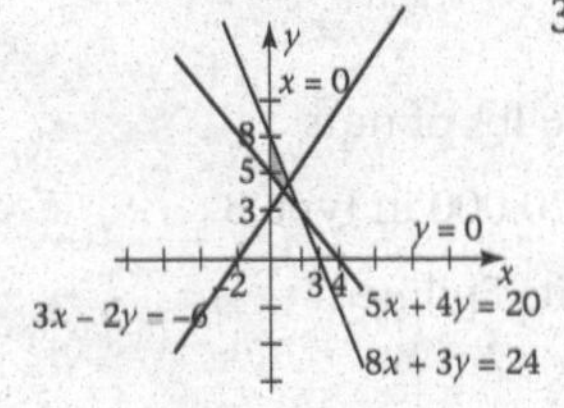

33.

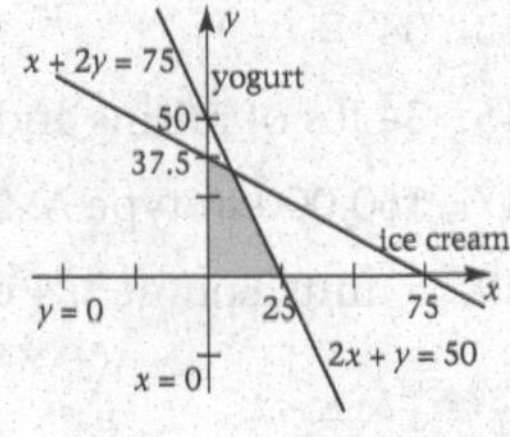

35.

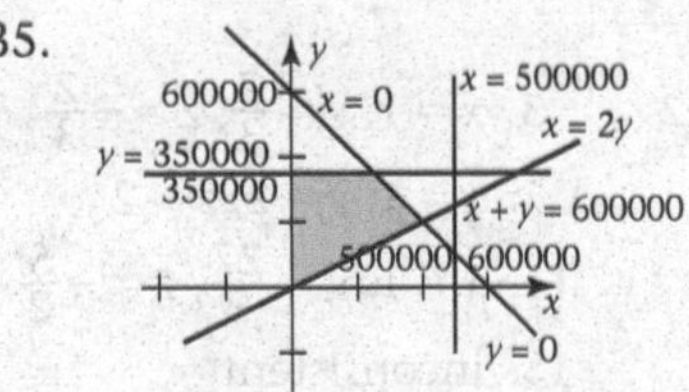

37. $y > 2x + 4$

39. $y < -x + 5$

Exercise Set 10.6

	MINIMUM	MAXIMUM
1.	$-2; (5, 14)$	$5; (5, 0)$
3.	$-3; (2, 2)$	$3; (6, 0)$
5.	$\frac{6}{7}; \left(\frac{6}{7}, \frac{6}{7}\right)$	$\frac{19}{2}; \left(4, -\frac{3}{2}\right)$
7.	$\frac{1}{2}; \left(3, \frac{11}{2}\right)$	$14; (8, 2)$

9. minimum: preferred; $63\frac{1}{3}$ square feet
maximum: regular: 30 square feet

11. large: 120 small: 260

13. Java: 250 pounds Columbian: 250 pounds

15. crop A: 60 acres crop B: 40 acres

17. pack A: 6 pounds pack B: 12 pounds

Chapter 10 Review Exercises

1. $x = -\frac{1}{2}, y = 1$
2. $x = 5, y = 2; x = 10, y = -3$
3. $x = 5, y = -1$
4. $x = -4, y = \frac{3}{2}$
5. $x = \frac{1}{4}, y = -\frac{1}{2}$
6. $x = 5, y = 0; x = -4, y = 3$
7. none
8. $x = 4, y = 4; x = \frac{36}{25}, y = -\frac{12}{5}$
9. $x = -3, y = 5$
10. $x = 2, y = -2$
11. $x = 4, y = -1$
12. $x = -2, y = 3$

13. $x = 1, y = -1; x = 5, y = 3$
14. $x = 0, y = 3$
15. 45
16. 72
17. steak: \$10.00 per pound; hamburger: \$3.00 per pound
18. 600 kilometers per hour
19. 3, 11
20. 575, \$57,500
21. $x = -3, y = 1, z = 4$
22. $x = -2, y = \frac{1}{2}, z = 3$
23. $x = 1, y = -1, z = 2$
24. $x = 3, y = \frac{1}{4}, z = -\frac{1}{3}$
25. $x = -3, y = 4$
26. $x = -\frac{5}{3}, y = \frac{5}{6}$
27. $x = -2, y = -1, z = -3$
28. $x = \frac{1}{2}, y = -1, z = 1$
29. $\frac{3}{2x-1} - \frac{2}{x+2}$
30. $\frac{3x}{x^2+1} + \frac{2x-1}{(x^2+1)^2}$
31. $2x + 1 + \frac{4}{x-1} + \frac{1}{(x-1)^2}$

32.

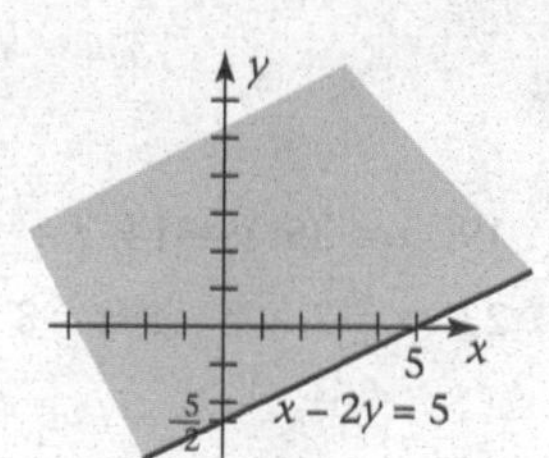

33.

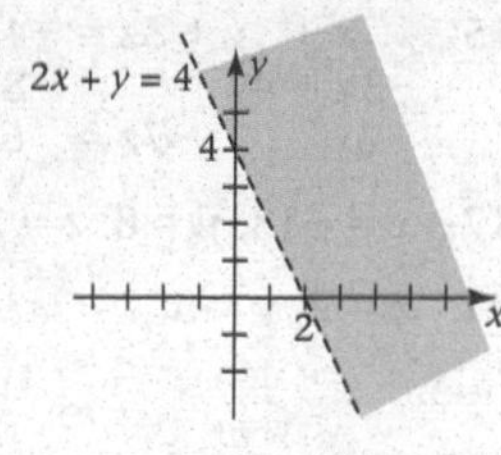

34.

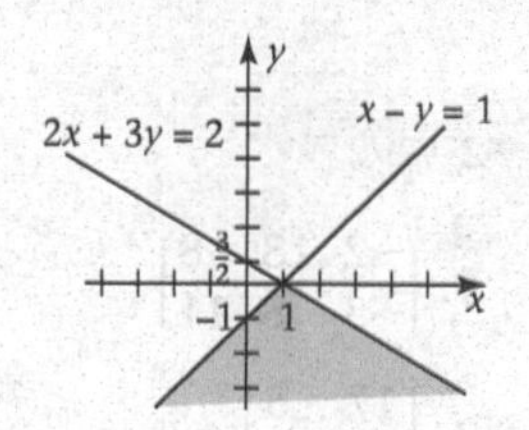

35.

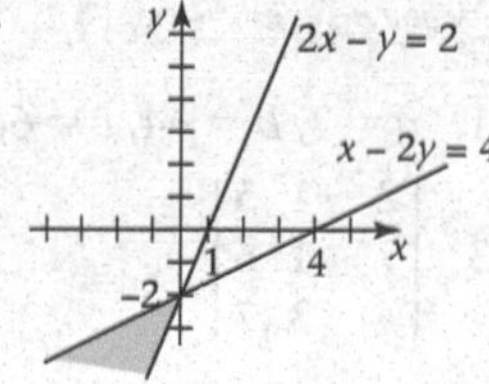

36.

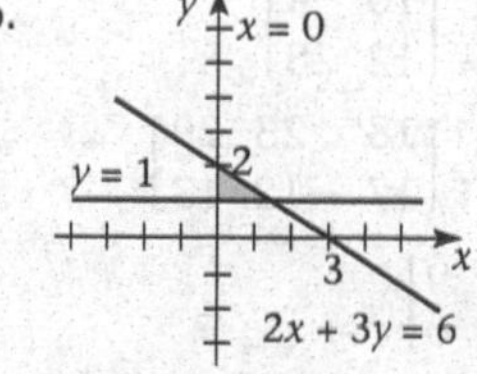

37.

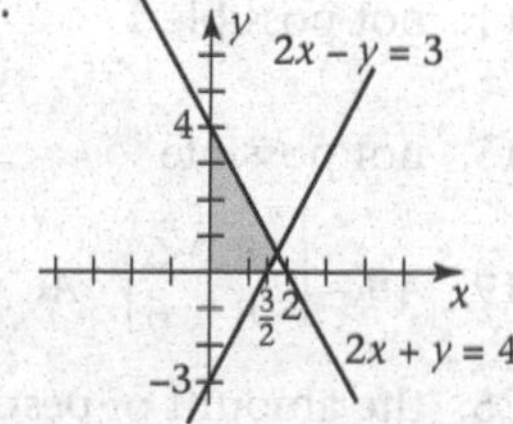

38. $x = 4, y = 6, z = 26$
39. $x = \frac{11}{2}, y = 2, z = \frac{27}{2}$

Chapter 10 Review Test

1. $x = -5, y = 2$
2. $x = -1, y = 6$
3. $x = 2, y = \pm\sqrt{10}; x = 3, y = \pm\sqrt{15}$
4. $x = 1, y = -3$
5. $x = 3, y = \pm 4; x = -3, y = \pm 4$
6. 38
7. shirts: \$60; ties: \$50
8. 3000 meals
9. $x = -2, y = 4, z = 6$
10. $x = -\frac{1}{3}, y = -1$
11. $x = \frac{2}{3}, y = 2, z = -2$
12. $\frac{-3}{x+3} - \frac{2}{x-2}$
13.

14.

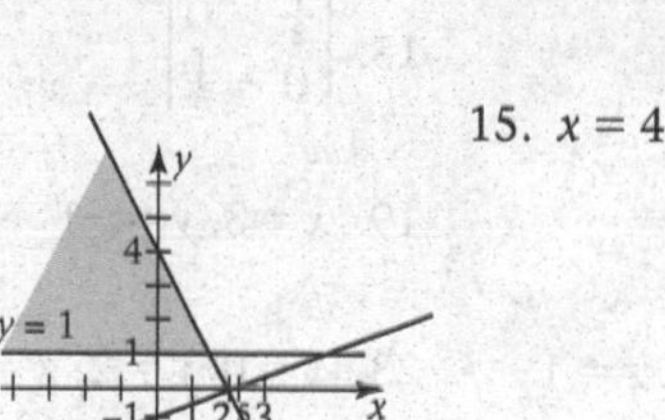

15. $x = 4, y = 8, z = 36$

Chapter 11

Exercise Set 11.1

1. 2×2
3. 4×3
5. 3×3
7. a. -4 b. 7 c. 6 d. -3
9. $\begin{bmatrix} 3 & -2 \\ 5 & 1 \end{bmatrix}, \left[\begin{array}{cc|c} 3 & -2 & 12 \\ 5 & 1 & -8 \end{array}\right]$
11. $\begin{bmatrix} \frac{1}{2} & 1 & 1 \\ 2 & -1 & -4 \\ 4 & 2 & -3 \end{bmatrix}, \left[\begin{array}{ccc|c} \frac{1}{2} & 1 & 1 & 4 \\ 2 & -1 & -4 & 6 \\ 4 & 2 & -3 & 8 \end{array}\right]$
13. $\frac{3}{2}x + 6y = -1$
$4x + 5y = 3$

15. $\begin{aligned} x + y + 3z &= -4 \\ -3x + 4y \quad &= 8 \\ 2x \quad + 7z &= 6 \end{aligned}$

17. $x = -13, y = 8, z = 2$

19. $x = 35, y = 14, z = -4$

21. $x = 2, y = 3$

23. $x = 2, y = -1, z = 3$

25. $x = 3, y = 2, z = -1$

27. $x = -5, y = 2, z = 3$

29. $x = -\frac{5}{7}, y = -\frac{2}{7}, z = -\frac{3}{7}, w = \frac{2}{7}$

31. $x = 2, y = 3$

33. $x = 2, y = -1, z = 3$

Exercise Set 11.2

1. $a = 3, b = -4, c = 6, d = -2$

3. $\begin{bmatrix} 2 & -1 & 5 \\ 7 & 1 & 6 \\ 4 & 3 & 7 \end{bmatrix}$

5. $\begin{bmatrix} -2 & 18 & 8 \\ -3 & 8 & 11 \end{bmatrix}$

7. not possible

9. $\begin{bmatrix} 17 & 5 \\ 10 & 12 \end{bmatrix}$

11. not possible

13. $\begin{bmatrix} 10 & 4 \\ 12 & 28 \end{bmatrix}$

15. not possible

17. $\begin{bmatrix} 18 & 23 & 29 \\ 17 & -12 & 13 \end{bmatrix}$

19. $AB = \begin{bmatrix} 8 & -6 \\ -8 & 6 \end{bmatrix}$ $AC = \begin{bmatrix} 8 & -6 \\ -8 & 6 \end{bmatrix}$

25. the amount of pesticide 2 eaten by herbivore 3

27. $A = \begin{bmatrix} 3 & 4 \\ 3 & -1 \end{bmatrix}, \quad X = \begin{bmatrix} x \\ y \end{bmatrix} \quad B = \begin{bmatrix} -3 \\ 5 \end{bmatrix}$

29. $A = \begin{bmatrix} 3 & -1 & 4 \\ 2 & 2 & \frac{3}{4} \\ 1 & -\frac{1}{4} & 1 \end{bmatrix}, \quad X = \begin{bmatrix} x \\ y \\ z \end{bmatrix} \quad B = \begin{bmatrix} 5 \\ -1 \\ \frac{1}{2} \end{bmatrix}$

31. $\begin{aligned} x_1 - 5x_2 &= 0 \\ 4x_1 + 3x_2 &= 2 \end{aligned}$

33. $\begin{aligned} 4x_1 + 5x_2 - 2x_3 &= 2 \\ 3x_2 - x_3 &= -5 \\ 2x_3 &= 4 \end{aligned}$

35. $A = \begin{bmatrix} a_{11} & a_{12} & a_{13} \\ a_{21} & a_{22} & a_{23} \\ a_{31} & a_{32} & a_{33} \end{bmatrix}$

37. $A = \begin{bmatrix} 0 & 4 & 6 \\ 5 & 0 & 1 \\ 7 & 2 & 3 \end{bmatrix}$

$2A = \begin{bmatrix} 0 & 8 & 12 \\ 10 & 0 & 2 \\ 14 & 4 & 6 \end{bmatrix}$

Exercise Set 11.3

1. no

3. yes

5. $\begin{bmatrix} \frac{2}{3} & \frac{5}{6} \\ \frac{1}{3} & \frac{1}{6} \end{bmatrix}$

7. $\begin{bmatrix} 1 & -1 \\ 2 & -1 \end{bmatrix}$

9. $\begin{bmatrix} 8 & 7 & -1 \\ -4 & -4 & 1 \\ -5 & -5 & 1 \end{bmatrix}$

11. $\begin{bmatrix} \frac{4}{7} & -\frac{3}{7} \\ \frac{1}{7} & \frac{1}{7} \end{bmatrix}$

13. none

15. $\begin{bmatrix} \frac{1}{2} & 0 \\ 0 & -\frac{1}{3} \end{bmatrix}$

17. $\begin{bmatrix} -1 & 1 & -1 \\ 2 & -1 & 2 \\ -2 & 1 & -1 \end{bmatrix}$

19. $x = 3, y = -1$

21. $x = 2, y = -3, z = 1$

23. $x = 0, y = 2, z = -3$

25. $x = 2, y = 3$

27. $x = 2, y = -1, z = 3$

29. $x = 3, y = 2, z = -1$

31. $x = -5, y = 2, z = 3$

33. $x = -\frac{5}{7}, y = -\frac{2}{7}, z = -\frac{3}{7}, w = \frac{2}{7}$

35. $x = \begin{bmatrix} 25 \\ 3 \\ 1 \end{bmatrix}, \begin{bmatrix} -2 \\ 5 \\ 10 \end{bmatrix}$

37. We cannot perform any elementary row operations upon rows 1 and 3 that will produce a nonzero pivot element for a_{22}.

Exercise Set 11.4

1. 22

3. −8

5. 0

7. a. −6
   b. −1
   c. 1
   d. 7

9. a. −6
   b. 1
   c. 1
   d. 7

11. 52

13. −3

15. −70

17. −6

19. −70

Exercise Set 11.5

1. 0

3. −12

5. 0

Exercise Set 11.6

1. $x = 1, y = -2, z = -1$
3. $x = 3, y = 2, z = -1$
5. $x = -3, y = 0, z = 2$
7. $x = -\frac{5}{7}, y = -\frac{2}{7}, z = -\frac{3}{7}, w = \frac{2}{7}$

Chapter 11 Review Exercises

1. 3×5
2. -1
3. 4
4. 8
5. $\begin{bmatrix} 3 & -7 \\ 1 & 4 \end{bmatrix}$
6. $\left[\begin{array}{cc|c} 3 & -7 & 14 \\ 1 & 4 & 6 \end{array}\right]$
7. $\begin{aligned} 4x - y &= 3 \\ 2x + 5y &= 0 \end{aligned}$
8. $\begin{aligned} -2x + 4y + 5z &= 0 \\ 6x - 9y + 4z &= 0 \\ 3x + 2y - z &= 0 \end{aligned}$
9. $x = -1, y = -4$
10. $x = \frac{1}{2}, y = 5$
11. $x = -4, y = 3, z = -1$
12. $x = -1, y = 1, z = -3$
13. $x = \frac{1}{2}, y = \frac{3}{2}$
14. $x = -5, y = 2$
15. $x = 3, y = \frac{1}{3}, z = -2$
16. $x = 3 + \frac{5t}{4}, y = 3 + \frac{t}{2}$, t = any number
17. -3
18. -3
19. $\begin{bmatrix} 1 & 4 \\ 7 & -1 \end{bmatrix}$
20. $\begin{bmatrix} -3 & 6 \\ 1 & -5 \end{bmatrix}$
21. not possible
22. $\begin{bmatrix} 5 & 15 & 20 \\ -5 & 0 & -30 \end{bmatrix}$
23. $\begin{bmatrix} -1 & -3 & -4 \\ -4 & 0 & -24 \\ 4 & 6 & 20 \end{bmatrix}$
24. $\begin{bmatrix} 7 & 4 \\ -11 & 12 \end{bmatrix}$
25. not possible
26. $\begin{bmatrix} 1 & -5 \\ 16 & -12 \\ -10 & 16 \end{bmatrix}$
27. $\begin{bmatrix} 0 & 9 \\ 11 & -4 \end{bmatrix}$
28. $\begin{bmatrix} 6 & -13 \\ -5 & -9 \end{bmatrix}$
29. $\begin{bmatrix} \frac{-4}{11} & \frac{3}{11} \\ \frac{1}{11} & \frac{2}{11} \end{bmatrix}$
30. $\begin{bmatrix} \frac{2}{5} & \frac{-7}{5} & \frac{-8}{5} \\ -1 & 3 & 4 \\ \frac{-2}{5} & \frac{2}{5} & \frac{3}{5} \end{bmatrix}$
31. $x = 2, y = 3$
32. $x = -1, y = -1, z = \frac{1}{2}$
33. 10
34. -6
35. 0
36. 12
37. 0
38. -3
39. $x = \frac{1}{2}, y = 4$
40. $x = 1, y = -4$
41. $x = 10, y = -4$
42. $x = -4, y = 2, z = 1$
43. $x = \frac{1}{3}, y = \frac{2}{3}, z = -1$
44. $x = \frac{1}{4}, y = -2, z = \frac{1}{2}$

Chapter 11 Review Test

1. 3×2
2. 0
3. $\left[\begin{array}{ccc|c} -7 & 0 & 6 & 3 \\ 0 & 2 & -1 & 10 \\ 1 & -1 & 1 & 5 \end{array}\right]$
4. $\begin{aligned} -5x + 2y &= 4 \\ 3x - 4y &= 4 \end{aligned}$
5. $x = -\frac{1}{2}, y = \frac{1}{2}$
6. $x = -6, y = -2$
7. $x = \frac{1}{2}, y = \frac{1}{2}, z = \frac{1}{2}$
8. 3
9. $\begin{bmatrix} 2 & 14 \\ -2 & -4 \\ -5 & 1 \end{bmatrix}$
10. $\begin{bmatrix} -7 & -11 \\ 11 & 15 \end{bmatrix}$
11. $\begin{bmatrix} -10 \\ 2 \\ 0 \end{bmatrix}$
12. not possible
13. $\begin{bmatrix} \frac{1}{27} & \frac{4}{9} & \frac{4}{27} \\ \frac{5}{27} & \frac{2}{9} & \frac{-7}{27} \\ \frac{7}{27} & \frac{1}{9} & \frac{1}{27} \end{bmatrix}$
14. $x = -2, y = 1$
15. -2
16. 27
17. $x = 4, y = -3$

Chapter 12

Exercise Set 12.1

1. 2, 4, 6, 8; 40
3. 1, 5, 9, 13; 77
5. 5, 5, 5, 5; 5
7. $\frac{1}{2}, \frac{2}{3}, \frac{3}{4}, \frac{4}{5}; \frac{20}{21}$
9. 2.1, 2.01, 2.001, 2.0001; $2 + (0.1)^{20}$
11. $\frac{1}{3}, \frac{4}{5}, \frac{9}{7}, \frac{16}{9}; \frac{400}{41}$
13. 9
15. $\frac{4}{7}$
17. 256
19. 40
21. 97
23. $\frac{49}{12}$
25. 80
27. $\sum_{k=1}^{5} (2k - 1)$
29. $\sum_{k=1}^{5} k^2$

31. $\sum_{k=1}^{4} \frac{(-1)^k}{\sqrt{k}}$

33. $\sum_{k=1}^{4} \frac{(-1)^{k+1}k}{k^2+1}$

35. $\sum_{k=0}^{n} \frac{1}{x^k}$

Exercise Set 12.2

1. 15, 18
3. $1, \frac{5}{4}$
5. log(10,000), log(100,000)
7. $\sqrt{5}+6, \sqrt{5}+8$
9. 2, 6, 10, 14
11. $3, \frac{5}{2}, 2, \frac{3}{2}$
13. $\frac{1}{3}, 0, -\frac{1}{3}, -\frac{2}{3}$
15. 25
17. −8
19. −2
21. $\frac{19}{3}$
23. $\frac{821}{160}$
25. 440
27. −126
29. 1720
31. 30
33. $n = 30, d = 3$
35. $n = 6, d = \frac{1}{4}$
37. −2

Exercise Set 12.3

1. 48
3. $-\frac{81}{64}$
5. 0.0096
7. 3, 9, 27, 81
9. $4, 2, 1, \frac{1}{2}$
11. −3, −6, −12, −24
13. −384
15. $\frac{1}{4}$
17. $\frac{1}{243}$
19. $\frac{27}{8}$
21. ±2
23. 7
25. 1, 3
27. $\frac{1}{4}, \frac{1}{16}$
29. $\frac{1093}{243}$
31. $-\frac{1353}{625}$
33. 1020
35. $\frac{55}{8}$
37. $10,235
39. 58,594 people
41. 2
43. $\frac{3}{4}$
45. $\frac{8}{3}$
47. 1
49. $\frac{1}{5}$
51. $\frac{182}{495}$
53. $\frac{325}{999}$

Exercise Set 12.4

1. $n = (k+1)(k+2)$
Thus the formula holds for $n = k + 1$.
Thus the statement holds for all n.

3. $n = \frac{(k+1)[3(k+1)+1]}{2}$
Thus the formula holds for $n = k + 1$.
Thus the statement holds for all n.

5. $n = \frac{5(k+1)(k+2)}{2}$
Thus the formula holds for $n = k + 1$.
Thus the statement holds for all n.

7. $n = \frac{(k+1)(k+2)(k+3)}{3}$
Thus the formula holds for $n = k + 1$.
Thus the statement holds for all n.

9. $n = (k+1)[2(k+1) - 1]$
Thus the formula holds for $n = k + 1$.
Thus the statement holds for all n.

11. $n = a_1 + kd$
Thus the formula holds for $n = k + 1$.
Thus the statement holds for all n.

13. $n = 2^{k+2} - 2$
Thus the formula holds for $n = k + 1$.
Thus the statement holds for all n.

Exercise Set 12.5

1. $243x^5 + 810x^4y + 1080x^3y^2 + 720x^2y^3 + 240xy^4 + 32y^5$
3. $256x^4 - 256x^3y + 96x^2y^2 - 16xy^3 + y^4$
5. $32 - 80xy + 80x^2y^2 - 40x^3y^3 + 10x^4y^4 - x^5y^5$
7. $a^8b^4 + 12a^6b^3 + 54a^4b^2 + 108a^2b + 81$
9. $a^8 - 16a^7b + 112a^6b^2 - 448a^5b^3 + 1120a^4b^4 - 1792a^3b^5 + 1792a^2b^6 - 1024ab^7 + 256b^8$
11. $\frac{1}{27}x^3 + \frac{2}{3}x^2 + 4x + 8$
13. $1024 + 5120x + 11{,}520x^2 + 15{,}360x^3 + \ldots$
15. $19{,}683 - 118{,}098a + 314{,}928a^2 - 489{,}888a^3 + \ldots$
17. $16{,}384x^{14} - 344{,}064x^{13}y + 3{,}354{,}624x^{12}y^2 - 20{,}127{,}744x^{11}y^3 + \ldots$
19. $8192x^{13} - 53{,}248x^{12}yz + 159{,}744x^{11}y^2z^2 - 292{,}864x^{10}y^3z^3 + \ldots$

21. 120
23. 12
25. 990
27. 5040
29. 120
31. 210
33. $-35{,}840x^4$
35. $\frac{495}{296}x^8y^4$
37. $\frac{2016}{x^5}$
39. $-540x^3y^3$
41. $181440x^4y^3$
43. $-144x^6$
45. $\frac{35x^{12}}{8}$
47. 4.8268

Exercise Set 12.6

1. 120
3. 146,016
5. 256
7. 5040
9. 720
11. 336
13. 90
15. 84
17. 3
19. 210
21. 120
23. 60
25. 336
27. 120
29. 84
31. 45
33. 1
35. n
37. $\frac{n^2+n}{2}$
39. 3003
41. a. 15,600 b. 17,576
43. 12,271,512
45. 240
47. 59,400
49. 3.4419×10^9

Exercise Set 12.7

1. $\frac{1}{2}$
3. a. $\frac{1}{2}$ b. $\frac{1}{4}$ c. $\frac{1}{13}$
5. a. $\frac{11}{36}$ b. $\frac{5}{36}$ c. $\frac{4}{9}$
7. a. $\frac{99}{100}$ b. $\frac{1}{100}$
9. a. $\frac{1}{17{,}170}$ b. $\frac{1164}{8585}$ c. $\frac{7372}{8585}$
11. $\frac{11}{4165}$
13. $\frac{8}{23}$
15. $\frac{25}{204}$
17. a. 0.922 b. 1.6×10^{-7}

Chapter 12 Review Exercises

1. 3, 7, 13; 111
2. 0, $\frac{7}{3}$, $\frac{13}{2}$; $\frac{999}{11}$
3. 2
4. 120
5. −16
6. 62
7. 50
8. $\sum_{k=1}^{4} \frac{k}{k+2}$
9. $\sum_{k=0}^{4} (-1)^k x^k$
10. $\sum_{k=1}^{n} \log kx$
11. 38
12. −9
13. 8
14. $-\frac{33}{2}$
15. $\frac{275}{3}$
16. −450
17. −3
18. $-\frac{3}{2}$
19. 5, 1, $\frac{1}{5}$, $\frac{1}{25}$
20. −2, 2, −2, 2
21. $\frac{243}{8}$
22. ±256
23. 3, $\frac{1}{2}$, $\frac{1}{12}$, $\frac{1}{72}$
24. $\frac{21}{32}$
25. −728
26. 10
27. $\frac{9}{5}$
29. $16x^4 - 32x^3y + 24x^2y^2 - 8xy^3 + y^4$
30. $\frac{x^4}{16} - x^3 + 6x^2 - 16x + 16$
31. $x^6 + 3x^4 + 3x^2 + 1$
32. 720
33. 78
34. $\frac{n+1}{n}$
35. 15
36. 1
37. 45
38. 24
39. 360
40. 210
41. 9
42. $\frac{2}{9}$
43. $\frac{2}{7}$
44. $\frac{8}{663}$
45. 0.81
46. $\frac{1}{33}$

Chapter 12 Review Test

1. $\frac{1}{4}$, $\frac{2}{9}$, $\frac{3}{16}$, $\frac{4}{25}$
2. $\frac{29}{6}$
3. −1, $\frac{1}{2}$, 2, $\frac{7}{2}$
4. 8
5. 41
6. $\frac{55}{2}$
7. $\frac{1}{3}$
8. $-\frac{2}{3}$, $-\frac{4}{3}$, $-\frac{8}{3}$, $-\frac{16}{3}$
9. −2
10. 8, −16
11. $-\frac{43}{8}$
12. −6
14. $a^{10} + \frac{10a^9}{b} + \frac{45a^8}{b^2} + \frac{120a^7}{b^3} + \cdots$
15. 22
16. 360
17. $n + 1$
18. 24
19. 8.46×10^{20}
20. $\frac{1}{3}$
21. $\frac{3}{35}$

Cumulative Review Exercises

Chapters 10–12

1. $x = 5, y = \pm 4$; $x = -5, y = \pm 4$
2. $x = 1, y = -3$
3. $x = 0, y = 2$; $x = 3, y = 1$
4. 300 of each
5. $\frac{\frac{7}{3}}{x-2} - \frac{\frac{4}{3}}{x+1}$
6. $x = 0, y = -1, z = -3$
7. $x = -1, y = 2, z = -2$
8. a. $k = -\frac{3}{2}$ b. none
9.

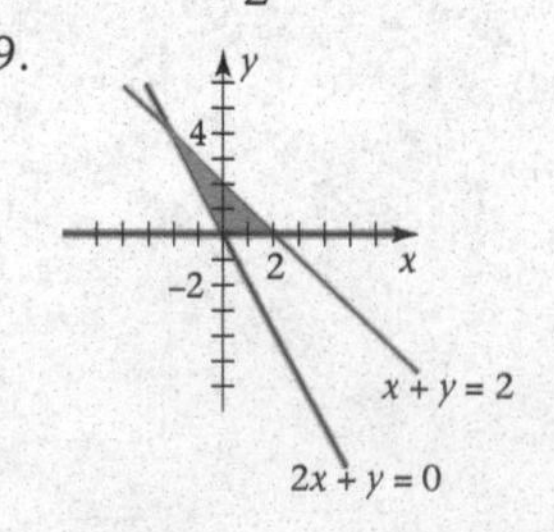

10. a. $\begin{bmatrix} 0 & 5 \\ -3 & 16 \\ -2 & -4 \end{bmatrix}$ b. 7 c. $\begin{bmatrix} 1 \\ 0 \\ -1 \end{bmatrix}$
11. 73
12. 24
13. a. 35 b. 8
14. 5
15. $\frac{1}{2}$
16. $\pm\sqrt{10}$
17. $\frac{40}{9}$
18. $\frac{1}{6}$
19. $\frac{10}{x^2}$

Index

A

B

C

S

T

U

V

W

X

Z